Nanocarbon-Based Composites and Their Thermal, Electrical, and Mechanical Properties

Nanocarbon-Based Composites and Their Thermal, Electrical, and Mechanical Properties

Collection Editor

Gil Goncalves

MDPI

Basel • Beijing • Wuhan • Barcelona • Belgrade • Novi Sad • Cluj • Manchester

Collection Editor
Gil Goncalves
Department of Mechanical
Engineering
University of Aveiro
Aveiro
Portugal

Editorial Office
MDPI AG
Grosspeteranlage 5
4052 Basel, Switzerland

This is a reprint of the Topical Collection, published open access by the journal *C* (ISSN 2311-5629), freely accessible at: https://www.mdpi.com/journal/carbon/topical_collections/nanocarbon_based_composites_and_their_thermal_electrical_and_mechanical_properties.

For citation purposes, cite each article independently as indicated on the article page online and as indicated below:

Lastname, A.A.; Lastname, B.B. Article Title. *Journal Name* **Year**, *Volume Number*, Page Range.

ISBN 978-3-7258-3685-7 (Hbk)
ISBN 978-3-7258-3686-4 (PDF)
https://doi.org/10.3390/books978-3-7258-3686-4

Contents

About the Editor

Gil Goncalves

Gil Goncalves graduated in 2003 in Industrial Chemistry from the University of Aveiro (Portugal) and obtained his Master's degree in Materials Science (EMMS, Joint European Master's Programme) in 2008. In 2012, he received his Ph.D. in Mechanical Engineering from the University of Aveiro with a thesis focused on nanocomposite materials for biomedical applications. After obtaining a Marie Curie research fellowship in 2016, he started working at the Institute of Material Science of Barcelona, High Council of Spanish Research (ICMAB-CSIC), on the development of ultra-sensitive nanotherapeutic anticancer agents for neutron capture therapy. He has been a short-term visitor on various occasions at the Laboratory of Applied Nuclear Energy (LENA, University of Pavia) and the Department of Chemical and Pharmaceutical Sciences at the University of Trieste. Currently, he is working at the Centre for Mechanical Technology and Automation (TEMA, University of Aveiro) as a researcher on the development of new carbon-based nanocomposite materials with applications in many different fields, such as heterogeneous catalysis (catalysis), water remediation (environment), scaffolds for biomedical applications (biomaterials), substrates for selective biomolecule detection (sensors), and nanoplatforms for cancer bioimaging and therapy (therapeutic agents). Dr. Gil has (co-) authored numerous scientific papers (h-index 29 and >3750 citations (Scopus)) and communications at national and international conferences, and he is also a regular reviewer for high-impact scientific journals in the field of multifunctional carbon nanomaterials. He is a member of the Editorial Board of *Scientific Reports* (Nature Publishing Group) in the field of chemical physics.

Preface

This collection explores the development of advanced nanocomposite materials reinforced with carbon nanostructures to achieve superior mechanical, thermal, and electrical properties. It highlights innovations in multifunctional nanocomposites across various matrices through the integration of graphene-based materials, carbon nanotubes, nanodiamonds, and other carbon nanostructures.

Renowned for their exceptional mechanical strength, high electrical and thermal conductivity, and versatile surface chemistry, carbon nanomaterials offer unique opportunities to enhance dispersibility and compatibility with various matrices. Their ability to be functionalized at the atomic level enables precise interface modulation, significantly improving composite performance.

This collection provides an in-depth analysis of the remarkable property enhancements resulting from the synergistic interactions between carbon nanostructures and diverse matrices. Key topics include manufacturing techniques, computational modelling, predictive simulations, and structural characterization of carbon-based nanocomposites.

This collection constitutes an invaluable resource for the scientific community, students, industry professionals and the general public interested in the recent technologies and breakthroughs in high-performance nanocarbon-based composites.

Gil Goncalves
Collection Editor

C Journal of Carbon Research

MDPI

Editorial

Nanocarbon-Based Composites and Their Thermal, Electrical, and Mechanical Properties

Gil Gonçalves [1,2]

[1] Centre for Mechanical Technology and Automation (TEMA), University of Aveiro, 3810-193 Aveiro, Portugal; ggoncalves@ua.pt
[2] Intelligent Systems Associate Laboratory (LASI), 4800-058 Guimarães, Portugal

Carbon materials have played a pivotal role in humanity's progress since ancient times. Their significance can be traced throughout history, shaping the evolution of civilization. One of the earliest uses of carbon materials was artistic expression, in which charcoal served as a pigment for cave paintings, contributing to the social and symbolic development of early societies [1]. In the 18th century, coal became a driving force behind scientific and technological advancements, ultimately leading to the First Industrial Revolution [2]. The invention of the steam engine, powered by coal, revolutionized manufacturing and transportation, and fueled economic growth and industrialization. More recently, carbon fibers have been widely explored for commercial applications, particularly in the form of polymeric composites [3]. These advanced composite materials exhibit a synergistic effect between their components, which allows them to outperform in industries such as the transportation and prosthetics industries, among others.

The discovery of carbon nanomaterials has expanded the potential of carbon-based materials for high-tech applications. This journey began in 1985 with the identification of cage-like fullerenes [4]. Another significant breakthrough occurred in 1991 when Sumio Iijima discovered multi-walled carbon nanotubes (CNTs) consisting of multiple carbon sheets rolled into tubes with closed ends [5]. This discovery ignited a global search for new allotropic forms of carbon. Today, a vast array of carbon nanostructures, including nano-onions, carbon nanohorns, graphene and its derivatives, and carbon dots, amongst others, are applied across diverse fields, including in healthcare, environmental science, energy storage, sensing, and electronics.

The importance of these discoveries was recognized by the scientific community with the awarding of the 1996 Nobel Prize in Chemistry to Professor Robert F. Curl of Rice University, Houston, USA [6]. Later, in 2010, the Nobel Prize in Physics was awarded to Andre Geim and Konstantin Novoselov of the University of Manchester (UK) for their groundbreaking discovery of graphene, the first two-dimensional material with atomic thickness to be isolated under room conditions [7,8].

Carbon nanomaterials, with their diverse dimensions and structures, exhibit excellent mechanical strength, electrical and thermal conductivities, and show a high versatility for chemical surface functionalization [9,10]. This adaptability is crucial for modulating interfaces at the atomic level and enhancing dispersibility and compatibility with various matrices. X-ray photoelectron spectroscopy (XPS) is a powerful tool for analyzing the atomic composition of carbon-based materials. David J. Morgan recently reviewed XPS spectral analysis of graphitic materials, presenting a simple and logical approach to increase confidence in functionalization insights derived from carbon core-level spectra [11].

Graphene and its derivatives have attracted significant interest in the development of advanced nanocomposites with enhanced performances [12]. Navaratnarajah K. et al.

Received: 20 February 2025
Revised: 25 February 2025
Accepted: 6 March 2025
Published: 8 March 2025

Citation: Gonçalves, G. Nanocarbon-Based Composites and Their Thermal, Electrical, and Mechanical Properties. *C* **2025**, *11*, 21. https://doi.org/10.3390/c11010021

recently reviewed major synthetic methods for graphene-based materials, providing a systematic analysis of their physicochemical properties and applications, including batteries, supercapacitors, electrochemical sensors, transparent electrodes, and environmental remediation [13].

Graphene-based materials exhibit high thermal conductivity, making them ideal for developing multifunctional nanocomposites with enhanced thermal stability and for use as solid lubricants for heat dissipation [14]. Alexander A. Balandin investigated the use of graphene as a reinforcing agent in epoxy resin to improve thermal diffusivity and conductivity. This work demonstrated that epoxy nanocomposites containing randomly oriented few-layer and single-layer graphene achieved a remarkable thermal conductivity enhancement of 15% to 25% [15]. Jacob S. Lewis studied the equilibrium temperatures of engineered device substitutes fixed to passive heat sink solutions with varying concentrations of few-layer graphene (FLG) as thermal interface materials (TIMs). His results showed that increasing FLG TIM concentration from 0 to 7.3 vol.% reduced the passively cooled heat sink's thermal resistance from 4.23 °C/W to 2.93 °C/W, leading to a decrease in operating temperature from approximately 108 °C to 85 °C at a heat dissipation rate of 20 W [16].

Graphene-based materials have also been explored as electrode materials for electrochemical capacitors or supercapacitors (SCs) for applications in energy storage devices, electrical vehicles, and flexible electronics [17,18]. Alexander Tkach et al. developed a self-supported paper of reduced graphene oxide (rGO) with polycarbonate (PC) as a free-standing single electrode. The flexible graphene-based symmetric supercapacitors showed a higher areal capacitance value then PET/Cu/C substrate [19]. rGO attracted considerable attention for the development of high-performance sensing materials. A recent study demonstrated the development of rGO-based electrodes using laser irradiation for thin flexible electrochemical sensors. The surface impedance of the rGO electrode was sensitive to the humidity of the working environment. However, it still presented some operational limitations, including difficulty in recovering after exposure to environmental humidity [20]. It was also observed that graphene can be utilized as a processing aid in molten polystyrene. A recent study compared the performance of different graphene materials and clays with different lateral structures to reduce viscosity. Graphene in a loose agglomerated configuration showed a higher reduction in polymer viscosity. This decrease in viscosity was attributed to the superlubricity effects generated by the atomic structure of graphene nanoparticles. Another example reported by Grigoriev et al. is the synthesis of a novel rGO-doped platinum nanoparticle electrocatalyst. Their study found that prior modification of rGO with nitrogen-containing heteroatoms (pyridine and quaternary) promoted the nucleation of Pt nanoparticles while preventing their agglomeration. This enhancement resulted in a high catalytic efficiency of the Pt nanoparticles, surpassing that of the bare rGO-supported catalyst and comparable to that of Pt/C [21].

CNTs are one-dimensional nanomaterials classified based on the number of graphene layers into single-walled (SWCNTs) [22], double-walled (DWCNTs) [23], or multi-walled (MWCNTs) [24]. SWCNTs have diameters ranging from 0.4 to 3.0 nm and can reach lengths of several micrometers. Composed of a single sp^2-hybridized carbon lattice, SWCNTs exist in three distinct chiral forms: armchair, chiral, and zigzag. Indeed, their structure directly influences their electrical, thermal, and mechanical properties [25]. Recently, Chensong Dong reported a new, simple method for determining the effective elastic modulus of wavy SWCNTs by finite element analysis [26]. MWCNTs typically have an inner diameter ranging from 0.4 nm to a few nanometers, with outer diameters reaching up to 30 nm. Their length can vary from 1 μm to several centimeters [27]. Rosario A. Gerhardt recently reviewed MWCNTs and thin films' fabrication methodologies, as well as their applications as electrode materials for electrochemical double-layer supercapacitors [28].

CNTs exhibit remarkable properties, including a high mechanical tensile modulus, strength (~1 TPa), and excellent electrical (>10^6 S/m) and thermal conductivity (>3000 W/m K) [29]. However, their performance and reactivity are highly dependent on factors such as structure, surface area, surface charge, size distribution, surface chemistry, agglomeration state, and purity. Andrew R. Barron et al. investigated the effects of both thermal and voltage-induced annealing on the electrical conductivity of CNT fibers. Their study found that fibers annealed at 500 °C or subjected to voltage annealing under similar conditions exhibited a three-order-of-magnitude reduction in electrical resistance. This improvement was primarily attributed to the selective removal of non-conductive or less-conductive amorphous carbon and the self-repair mechanisms of CNTs, which enhanced their conductivity [30]. To expand their range of applications, CNTs can undergo chemical functionalization on their surface (exohedral) or within their internal cavity (endohedral) to create various novel hybrid materials for different applications [31]. Kenji Hata et al. investigated the correlation between CNT/Cu electrical performance by preparing three different types of CNT-Cu composites using Cu electrodeposition: SWCNT/Cu wires, SWCNT/Cu pillars, and MWCNT/Cu wires [32]. Their study revealed that SWCNT/Cu wires and pillars outperformed MWCNT/Cu wires, exhibiting approximately three times higher room-temperature, measured using a four-point probe (reaching 30–40% of pure Cu conductivity). Additionally, the SWCNT/Cu composites demonstrated up to four times lower temperature coefficients of resistance, indicating more stable conductivity across temperature variations compared to MWCNT/Cu. These findings also suggest that few-walled, small-diameter nanotubes contribute to superior temperature-stable CNT/Cu conductivities.

Many other examples of carbon-based composite materials with enhanced performance have been reported over the years. Several carbon-reinforcing agents have been studied, including carbon fibers [33], carbon black [34], diamond [35,36], and nano graphite [37]. Additionally, more unconventional allotropes, such as tetragonal hybrid sp^3/sp^2 carbon structures [38], have also been explored. These studies highlight the exceptional mechanical, thermal, and electrical properties that emerge from synergistic interactions between carbon nanostructures and various matrices. These properties have significant scientific and technological relevance across multiple fields.

This Special Issue, Nanocarbon-Based Composites and Their Thermal, Electrical, and Mechanical Properties, published in *C—Journal of Carbon Research*, compiles 20 original research articles and review papers. These contributions provide valuable insights into the development of carbon-based composite materials with improved performance in diverse applications. It is anticipated that articles on this Special Issue will offer readers a strong foundational understanding of this rapidly advancing field.

Funding: This research was funded by Fundação para a Ciência e a Tecnologia (FCT) trough the projects 10.54499/UIDB/00481/2020 (https://doi.org/10.54499/UIDB/00481/2020, accessed on 20 February 2025) and 10.54499/UIDP/00481/2020 (https://doi.org/10.54499/UIDP/00481/2020, accessed on 20 February 2025) and the project CarboNCT 2022.03596.PTDC (https://doi.org/10.54499/2022.03596.PTDC, accessed on 20 February 2025).

Conflicts of Interest: The author declares no conflict of interest.

References

1. Callaway, E. Is This Cave Painting Humanity's Oldest Story? *Nature* **2019**. [CrossRef] [PubMed]
2. Fernihough, A.; O'Rourke, K.H. Coal and the European Industrial Revolution. *Econ. J.* **2021**, *131*, 1135–1149. [CrossRef]
3. Zhang, J.; Lin, G.; Vaidya, U.; Wang, H. Past, Present and Future Prospective of Global Carbon Fibre Composite Developments and Applications. *Compos. B Eng.* **2023**, *250*, 110463. [CrossRef]

4. Kroto, H.W.; Heath, J.R.; O'Brien, S.C.; Curl, R.F.; Smalley, R.E. C60: Buckminsterfullerene. *Nature* **1985**, *318*, 162–163. [CrossRef]
5. Iijima, S. Helical Microtubules of Graphitic Carbon. *Nature* **1991**, *354*, 56–58. [CrossRef]
6. Bali, P. Fullerenes Finally Score as Nobel Committee Honours Chemists. *Nature* **1996**, *383*, 561. [CrossRef]
7. Gerstner, E. Nobel Prize 2010: Andre Geim & Konstantin Novoselov. *Nat. Phys.* **2010**, *6*, 836. [CrossRef]
8. It's Still All about Graphene. *Nat. Mater.* **2011**, *10*, 1. [CrossRef]
9. Sun, Y.P.; Fu, K.; Lin, Y.; Huang, W. Functionalized Carbon Nanotubes: Properties and Applications. *Acc. Chem. Res.* **2002**, *35*, 1096–1104. [CrossRef]
10. Xiao, Y.; Pang, Y.X.; Yan, Y.; Qian, P.; Zhao, H.; Manickam, S.; Wu, T.; Pang, C.H. Synthesis and Functionalization of Graphene Materials for Biomedical Applications: Recent Advances, Challenges, and Perspectives. *Adv. Sci.* **2023**, *10*, 2205292. [CrossRef]
11. Morgan, D.J. Comments on the XPS Analysis of Carbon Materials. *C* **2021**, *7*, 51. [CrossRef]
12. Punetha, V.D.; Rana, S.; Yoo, H.J.; Chaurasia, A.; McLeskey, J.T.; Ramasamy, M.S.; Sahoo, N.G.; Cho, J.W. Functionalization of Carbon Nanomaterials for Advanced Polymer Nanocomposites: A Comparison Study between CNT and Graphene. *Prog. Polym. Sci.* **2017**, *67*, 1–47. [CrossRef]
13. Santhiran, A.; Iyngaran, P.; Abiman, P.; Kuganathan, N. Graphene Synthesis and Its Recent Advances in Applications—A Review. *C* **2021**, *7*, 76. [CrossRef]
14. Govindaraj, P.; Sokolova, A.; Salim, N.; Juodkazis, S.; Fuss, F.K.; Fox, B.; Hameed, N. Distribution States of Graphene in Polymer Nanocomposites: A Review. *Compos. B Eng.* **2021**, *226*, 109353. [CrossRef]
15. Lewis, J.S.; Perrier, T.; Mohammadzadeh, A.; Kargar, F.; Balandin, A.A. Power Cycling and Reliability Testing of Epoxy-Based Graphene Thermal Interface Materials. *C* **2020**, *6*, 26. [CrossRef]
16. Lewis, J.S. Reduction of Device Operating Temperatures with Graphene-Filled Thermal Interface Materials. *C* **2021**, *7*, 53. [CrossRef]
17. Wu, Z.S.; Parvez, K.; Feng, X.; Müllen, K. Graphene-Based in-Plane Micro-Supercapacitors with High Power and Energy Densities. *Nat. Commun.* **2013**, *4*, 2487. [CrossRef]
18. Huang, Y.; Liang, J.; Chen, Y. An Overview of the Applications of Graphene-Based Materials in Supercapacitors. *Small* **2012**, *8*, 1805–1834. [CrossRef]
19. Okhay, O.; Bastos, A.C.; Andreeva, K.; Tuukkanen, S.; Tkach, A. Reduced Graphene Oxide—Polycarbonate Electrodes on Different Supports for Symmetric Supercapacitors. *C* **2022**, *8*, 12. [CrossRef]
20. Rowley, A.; Stehle, Y.; Kilby, L.; Bashant, C. Graphene Oxide Membranes: Controlled Laser Reduction for Sensing Applications. *C* **2023**, *9*, 74. [CrossRef]
21. Kozlova, M.; Butrim, S.; Solovyev, M.; Pushkarev, A.; Pushkareva, I.; Kalinichenko, V.; Akelkina, S.; Grigoriev, S. Structural and Electrochemical Characteristics of Platinum Nanoparticles Supported on Various Carbon Carriers. *C* **2022**, *8*, 14. [CrossRef]
22. Ouyang, M.; Huang, J.L.; Lieber, C.M. Fundamental Electronic Properties and Applications of Single-Walled Carbon Nanotubes. *Acc. Chem. Res.* **2002**, *35*, 1018–1025. [CrossRef]
23. Moore, K.E.; Tune, D.D.; Flavel, B.S. Double-Walled Carbon Nanotube Processing. *Adv. Mater.* **2015**, *27*, 3105–3137. [CrossRef] [PubMed]
24. Andrews, R.; Jacques, D.; Qian, D.; Rantell, T. Multiwall Carbon Nanotubes: Synthesis and Application. *Acc. Chem. Res.* **2002**, *35*, 1008–1017. [CrossRef] [PubMed]
25. Odom, T.W.; Huang, J.L.; Kim, P.; Lieber, C.M. Atomic Structure and Electronic Properties of Single-Walled Carbon Nanotubes. *Nature* **1998**, *391*, 62–64. [CrossRef]
26. Dong, C. Effective Elastic Modulus of Wavy Single-Wall Carbon Nanotubes. *C* **2023**, *9*, 54. [CrossRef]
27. Lehman, J.H.; Terrones, M.; Mansfield, E.; Hurst, K.E.; Meunier, V. Evaluating the characteristics of multiwall carbon nanotubes. *Carbon* **2011**, *49*, 2581–2602. [CrossRef]
28. Jiang, K.; Gerhardt, R.A. Fabrication and Supercapacitor Applications of Multiwall Carbon Nanotube Thin Films. *C* **2021**, *7*, 70. [CrossRef]
29. De Volder, M.F.L.; Tawfick, S.H.; Baughman, R.H.; Hart, A.J. Carbon Nanotubes: Present and Future Commercial Applications. *Science* **2013**, *339*, 535–539. [CrossRef]
30. Gangoli, V.S.; Barnett, C.J.; McGettrick, J.D.; Orbaek White, A.; Barron, A.R. Increased Electrical Conductivity of Carbon Nanotube Fibers by Thermal and Voltage Annealing. *C* **2022**, *8*, 1. [CrossRef]
31. Liu, D.; Shi, L.; Dai, Q.; Lin, X.; Mehmood, R.; Gu, Z.; Dai, L. Functionalization of Carbon Nanotubes for Multifunctional Applications. *Trends Chem.* **2024**, *6*, 186–210. [CrossRef]
32. Sundaram, R.; Sekiguchi, A.; Chen, G.; Futaba, D.; Yamada, T.; Kokubo, K.; Hata, K. Influence of Carbon Nanotube Attributes on Carbon Nanotube/Cu Composite Electrical Performances. *C* **2021**, *7*, 78. [CrossRef]
33. Ofoegbu, S.U.; Ferreira, M.G.S.; Nogueira, H.I.S.; Zheludkevich, M. Comparison of the Electrochemical Response of Carbon-Fiber-Reinforced Plastic (CFRP), Glassy Carbon, and Highly Ordered Pyrolytic Graphite (HOPG) in Near-Neutral Aqueous Chloride Media. *C* **2023**, *9*, 7. [CrossRef]

34. Vidakis, N.; Petousis, M.; Velidakis, E.; Mountakis, N.; Fischer-Griffiths, P.E.; Grammatikos, S.; Tzounis, L. Fused Filament Fabrication Three-Dimensional Printing Multi-Functional of Polylactic Acid/Carbon Black Nanocomposites. *C* **2021**, *7*, 52. [CrossRef]
35. Volosova, M.A.; Okunkova, A.A. Study of the Influence of Silicon-Containing Diamond-like Carbon Coatings on the Wear Resistance of SiAlON Tool Ceramics. *C* **2023**, *9*, 50. [CrossRef]
36. Lebedev, O.V.; Kurkin, T.S.; Golubev, E.K.; Vasiliev, A.L.; Gatin, A.K.; Goncharuk, G.P.; Ozerin, A.N. Detonation Synthesis Nanodiamond Soot as a Promising Filler for Polymer Composites. *C* **2022**, *8*, 69. [CrossRef]
37. Kausar, A. Avant-Garde Polymer and Nano-Graphite-Derived Nanocomposites—Versatility and Implications. *C* **2023**, *9*, 13. [CrossRef]
38. Matar, S.F.; Solozhenko, V.L. Novel Superhard Tetragonal Hybrid Sp3/Sp2 Carbon Allotropes Cx (x = 5, 6, 7): Crystal Chemistry and Ab Initio Studies. *C* **2024**, *10*, 64. [CrossRef]

C Journal of Carbon Research

MDPI

Article

Novel Superhard Tetragonal Hybrid sp^3/sp^2 Carbon Allotropes C_x (x = 5, 6, 7): Crystal Chemistry and Ab Initio Studies

Samir F. Matar [1] and Vladimir L. Solozhenko [2,*]

1 Lebanese German University (LGU), Jounieh P.O. Box 206, Lebanon
2 LSPM–CNRS, Université Sorbonne Paris Nord, 93430 Villetaneuse, France
* Correspondence: vladimir.solozhenko@univ-paris13.fr

Abstract: Novel superhard tetragonal carbon allotropes C_5, C_6, and C_7, characterized by the presence of sp^3- and sp^2-like carbon sites, have been predicted from crystal chemistry and extensively studied by quantum density functional theory (DFT) calculations. All new allotropes were found to be cohesive, with crystal densities and cohesive energies decreasing along the C_5-C_6-C_7 series due to the greater openness of the structures resulting from the presence of C=C ethene and C=C=C propadiene subunits, and they were mechanically stable, with positive sets of elastic constants. The Vickers hardness evaluated by different models qualifies all allotropes as superhard, with H_v values ranging from 90 GPa for C_5 to 79 GPa for C_7. Phonon band structures confirm that the new allotropes are also dynamically stable. The electronic band structures reveal their metallic-like behavior due to the presence of sp^2-hybridized carbon.

Keywords: carbon allotropes; DFT; topology; crystal structure; equation of state; elastic constants; hardness; phonons; electronic band structures

Citation: Matar, S.F.; Solozhenko, V.L. Novel Superhard Tetragonal Hybrid sp^3/sp^2 Carbon Allotropes C_x (x = 5, 6, 7): Crystal Chemistry and Ab Initio Studies. *C* **2024**, *10*, 64. https://doi.org/10.3390/c10030064

Academic Editor: Giuseppe Cirillo

Received: 26 April 2024
Revised: 23 June 2024
Accepted: 9 July 2024
Published: 16 July 2024

1. Introduction

In recent years, significant efforts in the search for novel phases with advanced properties have benefited from structure prediction programs such as USPEX [1] and CALYPSO [2] and, more recently, from the crystallographic machine learning platform *CrystalMELA* [3]. The topic of carbon research, especially regarding periodic structures, shows a sustained interest in the search for new allotropes, especially those characterized by ultrahigh hardness close to that of diamond. In view of the large number of carbon structures claimed as novel, a database, SACADA [4], has been created to group all known carbon allotropes, thus helping researchers in their endeavors on the one hand, and avoiding claims of known phases as novel allotropes on the other.

Another avenue of investigation that could be complementary to modern materials science software lies in the crystal chemistry rationale called "structural engineering". Recently, the body-centered tetragonal *tet*-C_4 (Figure 1a) in space group *I*-4*m*2 was proposed by us as one of the simplest three-dimensional (3D) carbon networks to serve as a template for the design of other novel allotropes and chemical compounds [5]. Using the TopCryst crystallography package [6], *tet*-C_4 was identified with **dia** (i.e., diamond-like) topology. Several chemical compounds also have a **dia** topology, as categorized in the Reticular Chemistry Structure Resource (RCSR) database [7].

Recently, Wei et al. proposed a superhard *tet*-C_5 allotrope with mixed carbon hybridization (sp^2 and sp^3) and metallic character based on structure prediction programs and first-principles calculations within DFT [8]. However, such 'pentacarbon' should be unstable, because it has some negative elastic constants. Later, in 2020, cubic 'pentadiamond' was reported as a novel allotrope with mechanical properties close to diamond, but the paper was retracted due to serious errors in the optimized structure of 'pentadiamond' under mechanical deformation admitted by the authors themselves [9]. Such interest in

C_5 allotropes with mixed carbon hybridization arises from the possible change in the electronic structure of the insulating diamond, leading to metallic-like behavior with potential applications in electronics [10].

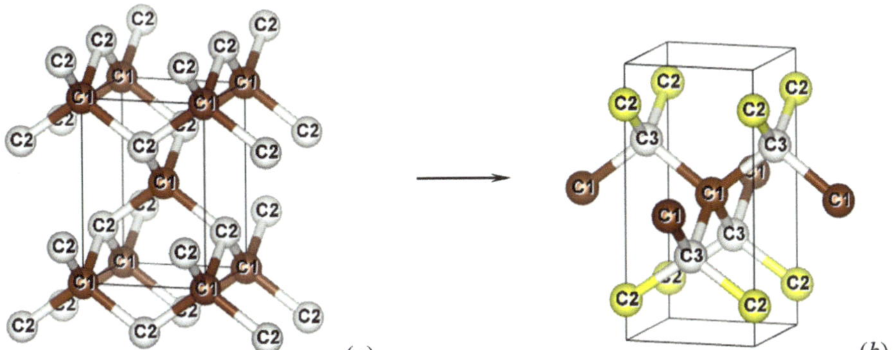

(a) (b)

Figure 1. From *tet*-C_4 (**a**) to *tet*-C_5 (**b**) by changing the atomic positions. The carbon atoms numbers correspond to the numbering in Table 1.

Table 1. C_4-to-C_5 crystal structure transformation of tetragonal carbon allotropes.

Space Group Topology	*tet*-C_4 [5] *I*-4*m*2 (No. 119) dia	Simple Tetragonal C_4		*tet*-C_5 *P*-4*m*2 (No. 115) 3,4^2T1-CA
a, Å	2.527	2.527		2.48
c, Å	3.574	3.574		4.99
Atomic positions	C1 (2a) 0, 0, 0 C2 (2d) $\frac{1}{2}$, 0, $\frac{1}{4}$	C1a ($\frac{1}{2}$, $\frac{1}{2}$, $\frac{1}{2}$) C1b (0, 0, 0) C2a ($\frac{1}{2}$, 0, $\frac{1}{4}$)] C2b (0, $\frac{1}{2}$, $\frac{3}{4}$)]	→ → →	C1 (1c) $\frac{1}{2}$, $\frac{1}{2}$, $\frac{1}{2}$ C2 (2e) 0, 0, ±z C3 (2g) 0, $\frac{1}{2}$, ±z′

In the present work, starting from *tet*-C_4 [5], we construct novel tetragonal carbon allotrope C_5 by crystal chemistry modifications, followed by full geometry relaxations using calculations based on the quantum mechanics density functional theory (DFT) [11,12]. The proposed *tet*-C_5 was found to be cohesive and stable both mechanically and dynamically and to exhibit an original undocumented topology. In a subsequent step, new tetragonal C_6 and C_7 allotropes were proposed and analyzed simultaneously with *tet*-C_5. From a chemical point of view, C_5 and C_6 emphasize the double bond >C=C< *ene*, as in the ethene molecule, C_2H_4. Remarkably linear C3 >C=C=C<, resembling the allene or *propadiene* molecule, C_3H_4, characterizes C_7. Note that the central carbon in C3 is sp^1 hybridized while the end carbons are sp^2.

After this contextual introduction and Section 2, which presents the computational framework, the crystal chemistry rationale leading to the novel tetragonal C_5, C_6, and C_7 allotropes is discussed. Section 3 is devoted to the crystal engineering protocol used to develop the various allotropes. Section 4 qualitatively illustrates the charge density projections on and between the atoms. The mechanical, dynamic, and thermal properties are detailed in Sections 5, 7 and 8, respectively, and the electronic band structures are discussed in Section 9.

2. Computational Framework

The structural investigations using structure prediction programs [1–3] or requiring crystal engineering, as here, need quantitative support from quantum mechanical calculations. Such investigations are mostly carried out within the framework of DFT [11,12], which is widely accepted by the scientific community. Briefly, the DFT was developed in two works: for the theoretical framework, by Hohenberg and Kohn in 1964 [11]; this was followed by Kohn and Sham in 1965 [12], who established the Kohn–Sham equations for solving the wave equation practically within computational codes built around the DFT. The Vienna Ab initio Simulation Package (VASP) code [13,14] and the Projector Augmented Wave (PAW) method [14,15] for the atomic potentials were used for the identification of the ground-state structures corresponding to the energy minima and the subsequent prediction of their mechanical and dynamical properties. DFT exchange-correlation (XC) effects were considered using the generalized gradient approximation (GGA) [16]. Relaxation of the atoms to the ground-state structures was performed with the conjugate gradient algorithm according to Press et al. [17]. The Blöchl tetrahedron method [18] with corrections according to the Methfessel and Paxton scheme [19] was used for geometry optimization and energy calculations, respectively. Brillouin zone (BZ) integrals were approximated by a special **k**-point sampling according to Monkhorst and Pack [20]. Structural parameters were optimized until atomic forces were below 0.02 eV/Å and all stress components were <0.003 eV/Å^3. The calculations were converged at an energy cutoff of 400 eV for the plane-wave basis set in terms of the **k**-point integration in the reciprocal space from $kx(6) \times ky(6) \times kz(6)$ up to $kx(12) \times ky(12) \times kz(12)$ to obtain a final convergence and relaxation to zero strains for the original stoichiometries presented in this work. In the post-processing of the ground-state electronic structures, the charge density projections were operated on the lattice sites.

The mechanical stability criteria were obtained from the calculations of the elastic constants. The treatment of the results was performed using the ELATE online tool [21], dedicated to the analysis of the elastic tensors. The program provides the bulk (*B*), shear (*G*), and Young's (*E*) moduli along different averaging methods; here, the Voigt method [22] was used. Two empirical models, the Mazhnik–Oganov [23] model and the Chen–Niu model [24], have been used to estimate Vickers hardness (H_V) from elastic constants.

Vickers hardness was also evaluated in the framework of the thermodynamic model [25,26], which is based on thermodynamic properties and crystal structure, and using the Lyakhov–Oganov approach [27], which takes into account the topology of the crystal structure, the strength of covalent bonds, the degree of ionicity, and directionality. The fracture toughness (K_{Ic}) was estimated using the Mazhnik–Oganov model [23].

The dynamic stabilities of the new allotropes were confirmed by the positive values of the phonon frequencies. The corresponding phonon band structures were obtained from a high resolution of the tetragonal Brillouin zones according to Togo et al. [28]. The electronic band structures were obtained using the all-electron DFT-based ASW method [29] and the GGA XC functional [16]. The VESTA (Visualization for Electronic and STructural Analysis) program [30] was used to visualize the crystal structures and charge densities.

3. Crystal Structure Engineering

The template *tet*-C_4 (space group *I*-4*m*2, No. 119) is characterized by two 2-fold Wyckoff atomic positions for carbon, as shown in Table 1 [5]. The tetragonal body-centered structure in Figure 1a shows the corresponding carbon atoms labeled C1 (brown spheres) at the origin and body center and C2 (white spheres) forming tetrahedra around C1. Considering the structure in the simple tetragonal configuration, the second column of Table 1 explains the four carbon sites. The transformation to C_5 retains the tetragonal symmetry, albeit lowered from body-centered ("I" centering) to primitive ("P" centering) and resolved in space group *P*-4*m*2, No. 115. This consists of keeping the body-centered carbon (C1 at $\frac{1}{2}$, $\frac{1}{2}$, $\frac{1}{2}$) so that it goes in the C(1c) Wyckoff position—follow the transformation arrows—as well as the two C2s (C2a and C2b), which become C3(2g), where the z coordinate becomes $\pm z'$ instead of $\pm\frac{1}{4}$. The additional carbon C2 is provided by parameterizing the z position of C1b, resulting in a two-fold C2 (2e) at 0, 0, $\pm$z. After full geometry relaxation, the resulting structure is shown in Figure 1b, with colored spheres corresponding to the three carbon sites given with the coordinates in Table 1, i.e., with well-determined z = ±0.854 and z′ = ±0.314. The topology of *tet*-C_5 is found to be **3,4^2T1-CA**, documented in the SACADA database, while the template *tet*-C_4 has a **dia** topology. Now, the cohesive energy per atom, E_{coh}/atom = −2.06 eV, is lower than that of *tet*-C_4, which is characterized by a large cohesive energy, E_{coh}/atom = −2.49 eV (the same as in diamond).

The polyhedral projection of the *tet*-C_5 structure (Figure 2a, center) shows four tetrahedra. More details are observed in the right panel of Figure 2a with a double cell along the *c*-direction, highlighting the C2-C2 bonds between the tetrahedra in these two cells. The C2-C2 segments form double bonds like C=C *ene* units, which will be further discussed in Section 4 relevant to the charge density projections.

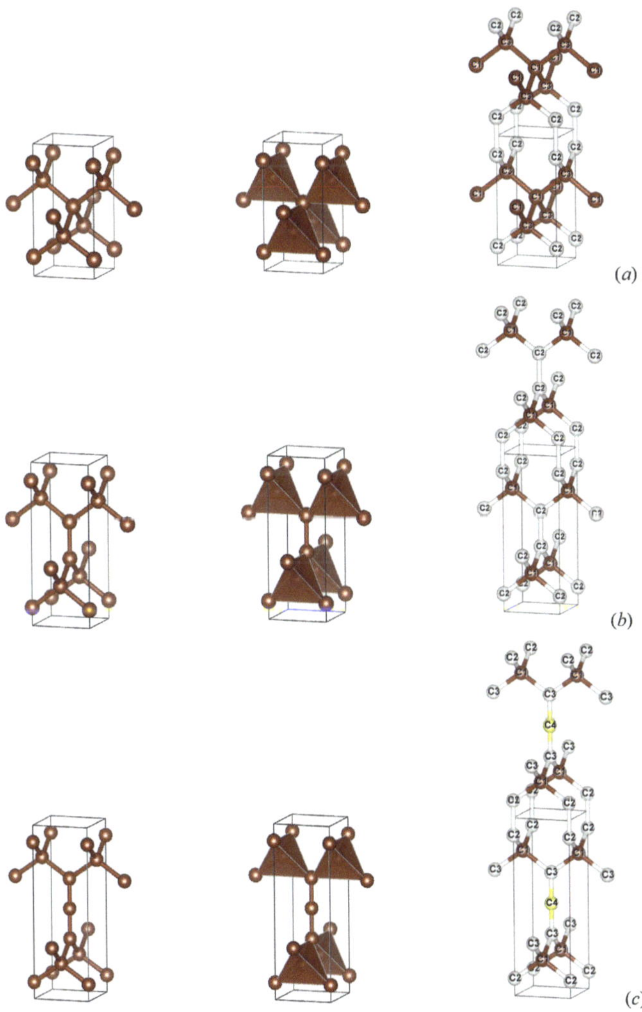

Figure 2. Crystal structures of new tetragonal carbon allotropes in three representations: ball-and-stick (**left**), polyhedral (**middle**), and double cells with numbering of atoms as in Table 2: C_5 (**a**); C_6 (**b**); C_7 (**c**).

Using *tet*-C_5 as a template, a new tetragonal C_6 allotrope was constructed with the modification of the central carbon position from $\frac{1}{2}, \frac{1}{2}, \frac{1}{2}$ to $\frac{1}{2}, \frac{1}{2}$, z with z ~ 0.4, thus transforming it into a two-fold position (2f) in the primitive tetragonal space group *P*-4*m*2, No. 115. However, after full geometry optimization and space group analysis, the (2e) and (2f) positions were merged by symmetry into a single four-fold position with body-center space group *I*-4*m*2, No. 119, as shown in Table 2. The topological analysis of the new allotrope revealed a **tfa** topology similar to that of the recently reported ultrahard C_6 allotrope called 'neoglitter' [31]. The cohesive energy of the new *tet*-C_6 is E_{coh}/atom = −1.73, which allows us to describe it as cohesive but less stable than *tet*-C_5. The structure of *tet*-C_6 is shown in Figure 2b in three representations. Clearly, the tetrahedra are now connected by C2-C2 segments along two adjacent cells, as in C_5, but also within the cell, thus exhibiting higher symmetry compared to *tet*-C_5.

Table 2. Crystal structure parameters of new tetragonal carbon allotropes.

Space Group Topology	C_5 ***P*-4*m*2 (No. 115)** **3,4^2T1-CA**	C_6 ***I*-4*m*2 (No. 119)** **tfa**	C_7 ***P*-4*m*2 (No. 115)** **tfa**
a, Å	2.48	2.456	2.563
c, Å	4.998	6.415	7.471
V_{cell}, Å^3	30.75	38.7	49.07
Density, g/cm^3	3.25	3.12	2.84
Shortest bond length, Å	1.46/1.50/1.55	1.44/1.51	1.30/1.45/1.53
Atomic positions	C1 (1c) $\frac{1}{2}, \frac{1}{2}, \frac{1}{2}$ C2 (2e) 0, 0, 0.854 C3 (2g) 0, $\frac{1}{2}$, 0.314	C1 (2c) 0, $\frac{1}{2}, \frac{1}{4}$ C2 (4e) 0, $\frac{1}{2}$, 0.888	C1 (1c) $\frac{1}{2}, \frac{1}{2}, \frac{1}{2}$ C2 (2e) 0, 0, 0.903 C3 (2f) $\frac{1}{2}, \frac{1}{2}$, 0.326 C4 (2g) 0, $\frac{1}{2}$, 0.210
E_{total}, eV E_{coh}/atom, eV	−43.29 −2.06	−50.06 −1.73	−58.07 −1.69

N.B. E(C) = −6.6 eV. E_{coh}/atom (diamond) = −2.49 eV.

The addition of a central carbon atom to *tet*-C_6 results in C_7 stoichiometry. The fully geometry-optimized lattice parameters are given in Table 2. There are now four different Wyckoff positions in the primitive tetragonal space group *P*-4*m*2, No. 115. The topological analysis led to the assignment of **tfa** topology, as in the case of *tet*-C_5. The structure of *tet*-C_7 is shown in Figure 2c, with the aligned C=C=C connecting the tetrahedra within the cell. As in C_5 and C_6, two adjacent cells are connected via C2-C2. However, *tet*-C_7 has the exception of having both C=C *ene* and C=C=C propadiene subunits. The cohesive energy E_{coh}/atom = −1.69 eV is lower than that of *tet*-C_6.

The density changes drastically along the series: ρ(C_5) = 3.25 > ρ(C_6) = 3.12 > ρ(C_7) = 2.84 g/cm^3. This is due to the increase in volume caused by the introduction of a larger number of sp^2-hybridized carbon atoms into the unit cell. The shortest interatomic distance (1.30 Å) is observed in the case of C_7 for the propadiene-like C=C=C subunit, whereas in C_6, d(C=C) = 1.44 Å. The other distances shown in Table 2 are in the range of values observed for tetrahedral carbon allotropes. The relevant effects on the phonon frequencies are shown in Section 7.

Simulated X-ray diffraction patterns of three new tetragonal carbon allotropes are shown in Figure 3. It is evident that they all exhibit distinct topologies that differ significantly from diamond.

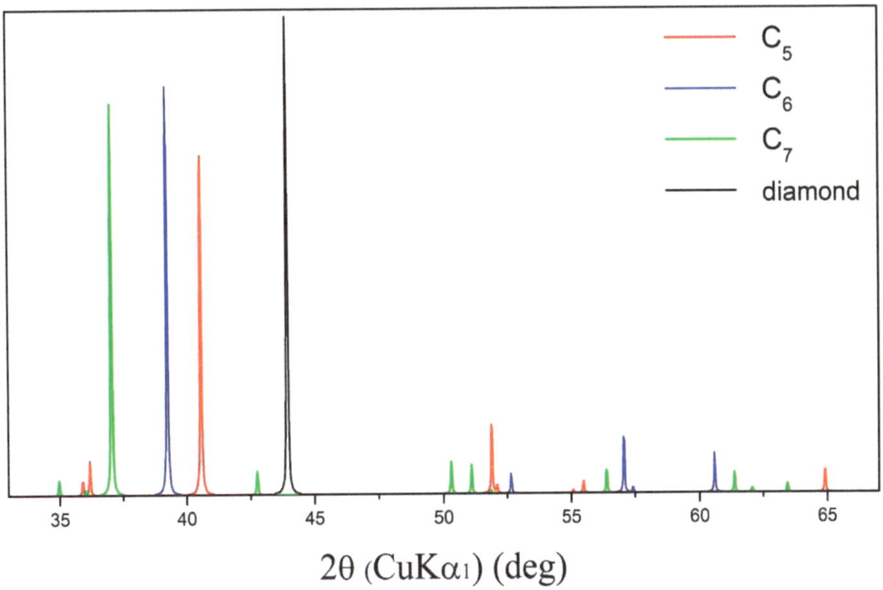

Figure 3. Simulated X-ray diffraction patterns of new tetragonal carbon allotropes.

4. Projections of the Charge Densities

To illustrate the peculiarities introduced by the C=C and C=C=C subunits into the lattice, the analysis was extended to a qualitative illustration of the charge densities. Despite the similar chemical nature of the constituents, i.e., carbon, the different hybridizations (sp^3 and sp^2) lead to different behaviors of the carbon sites in terms of charge transfer. Figure 4 illustrates the charge density distribution.

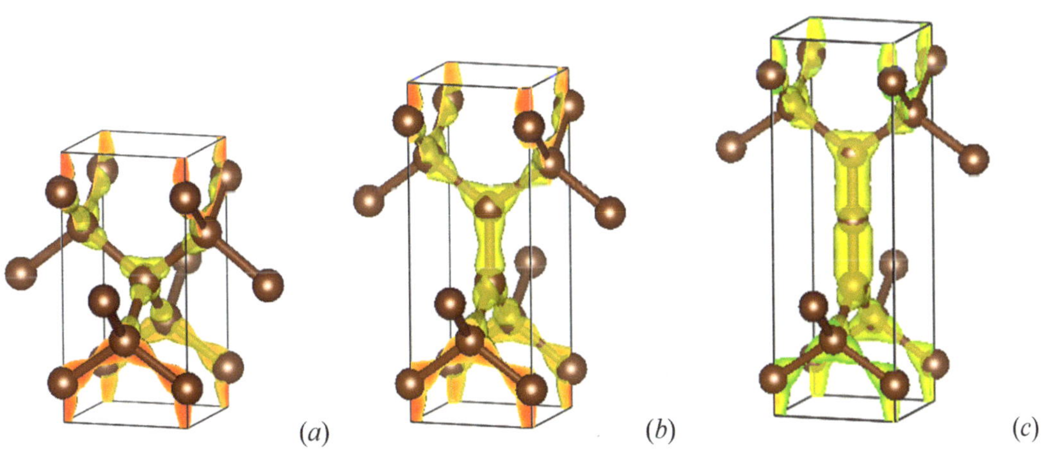

Figure 4. Charge density projections of tetragonal C_5 (**a**); C_6 (**b**); and C_7 (**c**).

The most relevant feature in C_5 (Figure 4a) is the red charge cross sections, which indicate the continuous charge density of the red traces through the eight corners toward the adjacent cells along the tetragonal *c* direction (vertical direction). This feature is also observed within the C_6 cell, which shows features similar to C_5 regarding the eight corners, as well as the central >C=C< with characteristic sp^2 hybridization. Going to C_7, the central part of the unit cell contains the >C=C=C< propadiene subunit, while the >C=C< *ene* subunits are located between the two adjacent cells (Figure 4c). However, the charge density becomes weaker as one moves to the next cell, as shown by the green areas, which indicate less charge between adjacent cells; the charge is now concentrated around the propadiene subunit. This may explain why C_7 is the least cohesive of the series studied.

5. Mechanical Properties

An analysis of the mechanical behavior of new tetragonal carbon allotropes has been carried out by calculating the elastic properties by inducing finite distortions of the lattice. The allotropes are fully described by the calculated sets of elastic constants C_{ij} given in Table 3. All C_{ij} values are positive, indicating mechanically stable phases. The major (i.e., C_{11} and C_{33}) C_{ij} values of C_5 are systematically larger than those of C_6 and C_7. The analysis of the elastic tensors was performed using ELATE software [21], which provides the bulk (*B*), shear (*G*), and Young's (*E*) moduli, as well as the Poisson's ratio (ν) along different averaging methods; in the present work, the Voigt approach [22] was chosen. Table 4 shows the calculated elastic moduli, with values that follow the trends observed for C_{ij}.

Table 3. Elastic constants (C_{ij}) of new tetragonal carbon allotropes (all values are in GPa).

	C_{11}	C_{12}	C_{13}	C_{33}	C_{44}	C_{66}
C_5	943	9	136	1194	198	337
C_6	784	19	135	1313	144	270
C_7	715	70	131	1286	119	268

The Vickers hardness values evaluated using the four contemporary hardness models are summarized in Table 4. A clear tendency of a decrease in hardness in the C_5–C_6–C_7 series due to the increasing openness of the crystal structures is observed for all four models. Since it has been shown earlier that the thermodynamic model is the most reliable in the case of the superhard compounds of light elements [32], and it shows perfect agreement with the available experimental data [33], it is obvious that the hardness values calculated within the empirical models are strongly underestimated. The Oganov–Lyakhov model also gives underestimated (by 10–20%) values.

Table 4. Mechanical properties of new tetragonal carbon allotropes compared with diamond: Vickers hardness (H_V), bulk modulus (*B*), shear modulus (*G*), Young's modulus (*E*), Poisson's ratio (ν), fracture toughness (K_{Ic}).

	H_V				*B*		G_V	E_V	ν_V	K_{Ic} ‡‡
	T *	LO †	MO ‡	CN §	B_0 *	B_V				
	GPa									$MPa{\cdot}m^{\frac{1}{2}}$
Diamond	98	90	100	93	445 **		530 **	1138 ††	0.074 ††	6.7 ‡‡
C_5 #115	90	81	49	45	410	405	333	784	0.177	7.1
C_6 #119	86	71	35	35	391	384	284	684	0.203	6.1
C_7 #115	79	62	30	31	360	376	260	634	0.219	5.8

* Thermodynamic model [26]. † Lyakhov–Oganov model [27]. ‡ Mazhnik–Oganov model [23]. § Chen–Niu model [24]. ** Ref [34]. †† Calculated from B_V and G_V using isotropic approximation. ‡‡ Ref. [35].

The fracture toughness of the new carbon allotropes decreases from 7.1 $MPa{\cdot}m^{\frac{1}{2}}$ for *tet*-C_5, which is higher than that of diamond (K_{Ic} = 6.7 $MPa{\cdot}m^{\frac{1}{2}}$ [35]), down to 5.8 $MPa{\cdot}m^{\frac{1}{2}}$ for *tet*-C_7.

6. Equations of State

The comparative energy trends of new tetragonal carbon allotropes have been determined from their equations of state, based on a series of calculations of total energy as a function of volume. The resulting $E(V)$ curves fitted to the third-order Birch equations of state [36] are shown in Figure 5. As can be seen, all three tetragonal carbon allotropes are metastable over the entire range of experimentally accessible pressures. However, the formation of these allotropes is possible at high pressures and high temperatures as a result of alternative metastable behavior, most likely under highly non-equilibrium conditions.

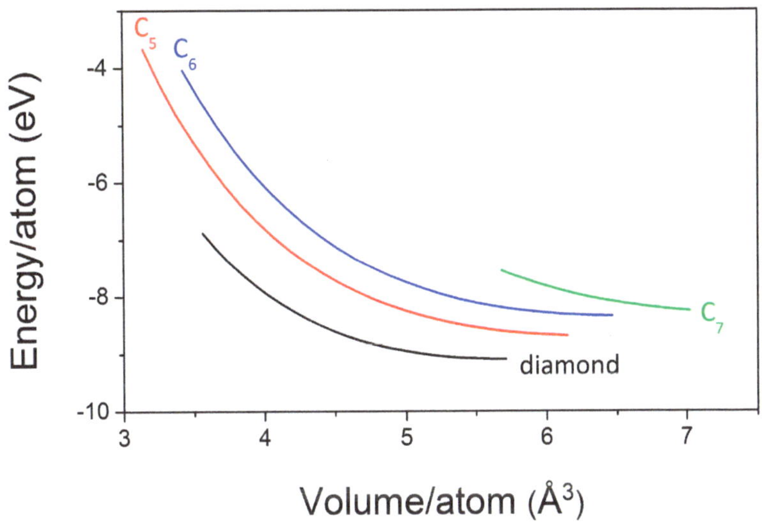

Figure 5. Calculated total energy per atom as a function of volume for new tetragonal carbon allotropes and diamond.

7. Phonon Band Structures

To verify the dynamic stability of the novel carbon allotropes, their phonon properties were studied. The phonon band structures obtained from a high resolution of the tetragonal Brillouin zone according to Togo et al. [28] are shown in Figure 6. The bands (red lines) develop along the main directions of the tetragonal Brillouin zone (horizontal x-axis), separated by vertical lines for better visualization, and the vertical direction (y-axis) shows the frequencies ω, given in terahertz (THz).

The band structures include 3N bands: three acoustic modes starting from zero energy ($\omega = 0$) at the Γ point (the center of the Brillouin zone) up to a few terahertz and 3N-3 optical modes at higher energies. The acoustic modes correspond to the rigid translation modes of the crystal (two transverse and one longitudinal). All panels in Figure 6 show positive phonon frequencies, indicating the dynamic stability of all three new carbon allotropes.

In Figure 6a, relevant to C_5, the general aspect is that of dispersed bands with the highest magnitude just below 40 THz. This changes for C_6 (Figure 6b), where the highest frequency is closer to 40 THz, corresponding to 1650 cm^{-1}. This value is exactly in the range of C=C vibrations with a strong signal, i.e., 1625–1680 cm^{-1} in the molecular state, as listed in the web-available "RAMAN Band Correlation Table". Due to the twice-larger amounts of C=C bonds with sp^2 hybridization, there are also more flat lines towards the top of the panel and less dispersion of the bands. These features are confirmed upon inspection of the Figure 6c relevant to C_7, where the highest bands are flat, with the highest line found around 55 THz corresponding to the short C–C distance of 1.30 Å in the C=C=C subunit, also identified in C_3H_4 propadiene (~1900 cm^{-1}).

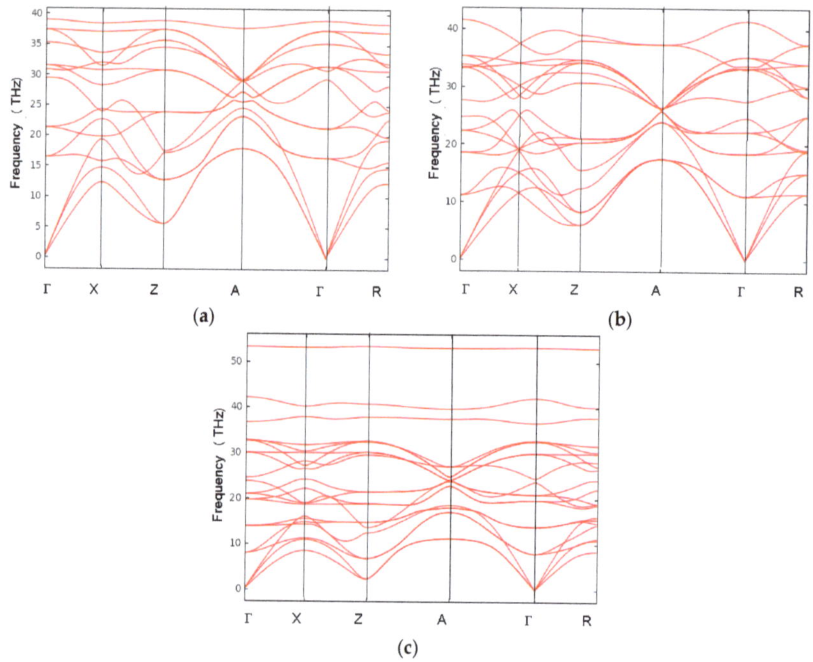

Figure 6. Phonon band structures of new carbon allotropes along the major directions of the simple tetragonal Brillouin zone: C_5 (**a**); C_6 (**b**); C_7 (**c**).

8. Thermodynamic Properties

The thermodynamic properties of novel carbon allotropes were calculated from the phonon frequencies using the statistical thermodynamic approach [37] on a high-precision sampling mesh in the Brillouin zone. The temperature dependencies of the heat capacity at constant volume (C_v) and entropy (S) are shown in Figure 7 in comparison with available experimental C_p data for diamond [38,39]. For all three allotropes, the heat capacity and entropy are higher than those of diamond and increase in the C_5–C_6–C_7 series, which is to be expected due to the increasing openness of the crystal structures compared to diamond, on the one hand, and the respective conducting versus insulator electronic properties, on the other hand.

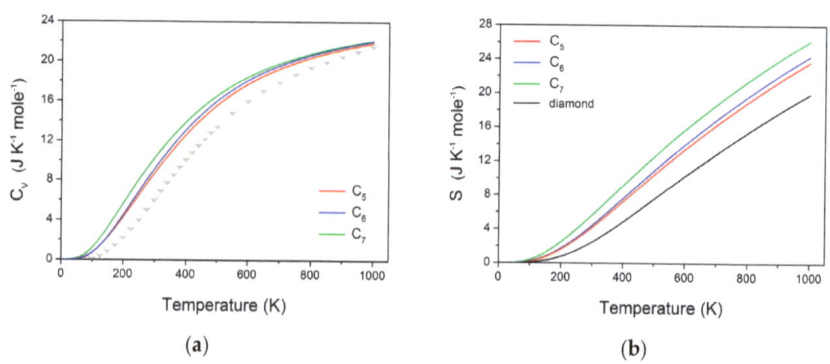

Figure 7. Heat capacities at constant volume (C_V) (**a**) and entropies (S) (**b**) of new tetragonal carbon allotropes compared to diamond. Experimental C_V data of diamond [38,39] are shown as gray symbols.

9. Electronic Band Structures

The electronic band structures of three novel carbon allotropes were calculated using the all-electron DFT-based augmented spherical method (ASW) [29]. The results are shown in Figure 8. The bands (blue lines) develop along the main directions of the tetragonal Brillouin zone shown in Figure 8d. The zero energy along the vertical axis is considered with respect to the Fermi-level E_F, since all three allotropes, C_5, C_6, and C_7, are metallic-like with no energy gap separating the valence band VB from the conduction band CB. Such behavior can be explained by the delocalized nature of the π electrons resulting from the presence of C(sp^2) atoms of the C=C and C=C=C subunits.

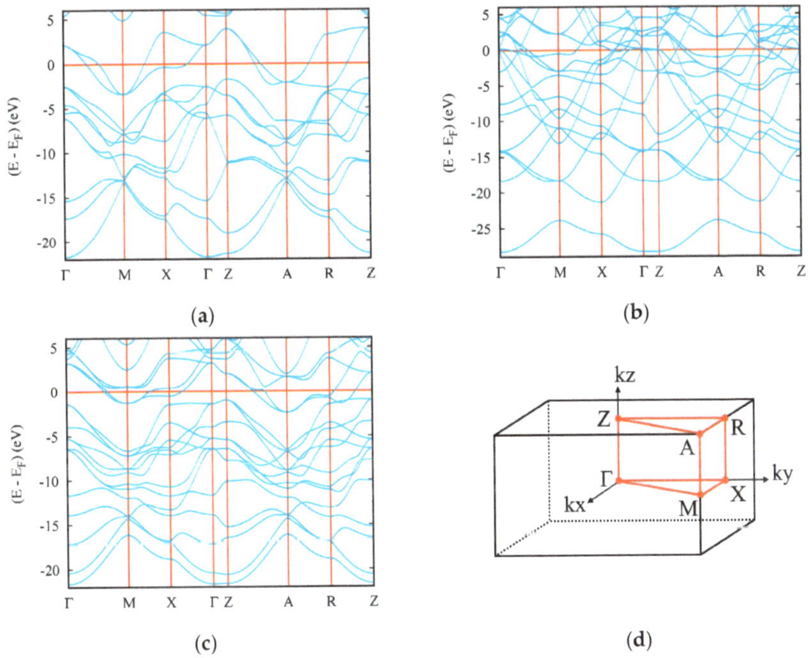

Figure 8. Electronic band structures of new tetragonal carbon allotropes: C_5 (**a**); C_6 (**b**); C_7 (**c**); the tetragonal Brillouin zone (**d**).

10. Conclusions

In the present work, the novel superhard tetragonal carbon allotropes C_5, C_6, and C_7, characterized with mixed sp^2 and sp^3 carbon hybridizations, were found to be cohesive and stable both mechanically (elastic constants and their combinations) and dynamically (phonon band structures). Whereas C_5 and C_6 contain ethene-like >C=C< subunits, C_7 shows both ethene-like and propadiene-like >C=C=C< subunits, with signatures observed from the highest phonon frequencies correlating with Raman vibrations of ethene and propadiene molecules. The density decreases along the C_5–C_6–C_7 series due to the increasing openness of the structures and consequently leads to a decrease in the mechanical properties, in particular the Vickers hardness, which decreases from 90 GPa for C_5 to 79 GPa for C_7. The energy-volume equations of state show metastability of all three allotropes over the whole range of experimentally accessible pressures, but this does not exclude their possible formation under high pressure—high temperature conditions. Thermodynamically, the temperature dependencies of the heat capacity and entropy of all three allotropes were found to be higher than those of diamond due to the insulating electronic properties of the latter compared to the conducting C_5, C_6, and C_7 as identified from their electronic band structures. Delocalized π electrons of $C(sp^2)$ atoms present in the three new allotropes are considered responsible for their metallic-like conductivity.

Author Contributions: Conceptualization, S.F.M.; methodology, S.F.M. and V.L.S.; investigation, S.F.M. and V.L.S.; formal analysis, S.F.M. and V.L.S.; data curation, S.F.M. and V.L.S.; visualization, S.F.M. and V.L.S.; validation, S.F.M. and V.L.S.; resources, S.F.M.; writing—original draft preparation, S.F.M.; writing—review and editing, V.L.S. All authors have read and agreed to the published version of the manuscript.

Funding: This research received no external funding.

Data Availability Statement: The data presented in this study are available on reasonable request.

Conflicts of Interest: The authors declare no conflicts of interest.

References

1. Glass, C.W.; Oganov, A.R.; Hansen, N. USPEX—Evolutionary crystal structure prediction. *Comput. Phys. Commun.* **2006**, *175*, 713–720. [CrossRef]
2. Wang, Y.; Lv, J.; Zhu, L.; Ma, Y. CALYPSO: A method for crystal structure prediction. *Comput. Phys. Commun.* **2012**, *183*, 2063–2070. [CrossRef]
3. Correiro, N.; Rizzi, R.; Settembre, R.G.; Del Buono, N.; Diacono, D. *CrystalMELA*: A new crystallographic machine learning platform for crystal system determination. *J. Appl. Cryst.* **2023**, *56*, 409–419. [CrossRef] [PubMed]
4. Hoffmann, R.; Kabanov, A.A.; Golov, A.A.; Proserpio, D.M. Homo Citans and carbon allotropes: For an ethics of citation. *Angew. Chem. Int. Ed.* **2016**, *55*, 10962–10976. [CrossRef]
5. Matar, S.F.; Solozhenko, V.L. The simplest dense carbon allotrope: Ultra-hard body centered tetragonal C_4. *J. Solid State Chem.* **2022**, *314*, 123424. [CrossRef]
6. Shevchenko, A.P.; Shabalin, A.A.; Karpukhin, I.Y.; Blatov, V.A. Topological representations of crystal structures: Generation, analysis and implementation in the *TopCryst* system. *Sci. Technol. Adv. Mater.* **2022**, *2*, 250–265. [CrossRef]
7. O'Keeffe, M.; Peskov, M.A.; Ramsden, S.J.; Yaghi, O.M. The reticular chemistry structure resource (RCSR) database of, and symbols for, crystal nets. *Acc. Chem. Res.* **2008**, *41*, 1782–1789. [CrossRef]
8. Wei, Q.; Zhang, Q.; Yan, H.; Zhang, M.; Wei, B. A new tetragonal superhard metallic carbon allotrope. *J. Alloys Compd.* **2018**, *769*, 347–352. [CrossRef]
9. Fujii, Y.; Maruyama, M.; Cuong, N.T.; Okada, S. Retraction: Pentadiamond: A hard carbon allotrope of a pentagonal network of sp^2 and sp^3 C atoms [Phys. Rev. Lett. 125, 016001 (2020)]. *Phys. Rev. Lett.* **2020**, *125*, 079901. [CrossRef] [PubMed]
10. Yang, H.; Ma, Y.; Dai, Y. Progress of structural and electronic properties of diamond: A mini review. *Funct. Diam.* **2021**, *1*, 150–159. [CrossRef]
11. Hohenberg, P.; Kohn, W. Inhomogeneous electron gas. *Phys. Rev. B* **1964**, *136*, 864–871. [CrossRef]
12. Kohn, W.; Sham, L.J. Self-consistent equations including exchange and correlation effects. *Phys. Rev. A* **1965**, *140*, 1133–1138. [CrossRef]
13. Kresse, G.; Furthmüller, J. Efficient iterative schemes for ab initio total-energy calculations using a plane-wave basis set. *Phys. Rev. B* **1996**, *54*, 11169. [CrossRef] [PubMed]
14. Kresse, G.; Joubert, J. From ultrasoft pseudopotentials to the projector augmented wave. *Phys. Rev. B* **1999**, *59*, 1758–1775. [CrossRef]
15. Blöchl, P.E. Projector augmented wave method. *Phys. Rev. B* **1994**, *50*, 17953–17979. [CrossRef]
16. Perdew, J.; Burke, K.; Ernzerhof, M. The Generalized Gradient Approximation made simple. *Phys. Rev. Lett.* **1996**, *77*, 3865–3868. [CrossRef] [PubMed]
17. Press, W.H.; Flannery, B.P.; Teukolsky, S.A.; Vetterling, W.T. *Numerical Recipes*, 2nd ed.; Cambridge University Press: New York, NY, USA, 1986.
18. Blöchl, P.; Jepsen, O.; Anderson, O. Improved tetrahedron method for Brillouin-zone integrations. *Phys. Rev. B* **1994**, *49*, 16223–16233. [CrossRef] [PubMed]
19. Methfessel, M.; Paxton, A.T. High-precision sampling for Brillouin-zone integration in metals. *Phys. Rev. B* **1989**, *40*, 3616–3621. [CrossRef] [PubMed]
20. Monkhorst, H.J.; Pack, J.D. Special k-points for Brillouin Zone integration. *Phys. Rev. B* **1976**, *13*, 5188–5192. [CrossRef]
21. Gaillac, R.; Pullumbi, P.; Coudert, F.X. ELATE: An open-source online application for analysis and visualization of elastic tensors. *J. Phys. Condens. Matter* **2016**, *28*, 275201. [CrossRef]
22. Voigt, W. Über die Beziehung zwischen den beiden Elasticitätsconstanten isotroper Körper. *Annal. Phys.* **1889**, *274*, 573–587. [CrossRef]
23. Mazhnik, E.; Oganov, A.R. A model of hardness and fracture toughness of solids. *J. Appl. Phys.* **2019**, *126*, 125109. [CrossRef]
24. Chen, X.Q.; Niu, H.; Li, D.; Li, Y. Modeling hardness of polycrystalline materials and bulk metallic glasses. *Intermetallics* **2011**, *19*, 1275–1281. [CrossRef]

25. Mukhanov, V.A.; Kurakevych, O.O.; Solozhenko, V.L. The interrelation between hardness and compressibility of substances and their structure and thermodynamic properties. *J. Superhard Mater.* **2008**, *30*, 368–378. [CrossRef]
26. Mukhanov, V.A.; Kurakevych, O.O.; Solozhenko, V.L. Hardness of materials at high temperature and high pressure. *Philos. Mag.* **2009**, *89*, 2117–2127. [CrossRef]
27. Lyakhov, A.O.; Oganov, A.R. Evolutionary search for superhard materials: Methodology and applications to forms of carbon and TiO_2. *Phys. Rev. B* **2011**, *84*, 092103. [CrossRef]
28. Togo, A.; Tanaka, I. First principles phonon calculations in materials science. *Scr. Mater.* **2015**, *108*, 1–5. [CrossRef]
29. Eyert, V. Basic notions and applications of the augmented spherical wave method. *Int. J. Quantum Chem.* **2000**, *77*, 1007–1031. [CrossRef]
30. Momma, K.; Izumi, F. VESTA3 for three-dimensional visualization of crystal, volumetric and morphology data. *J. Appl. Crystallogr.* **2011**, *44*, 1272–1276. [CrossRef]
31. Matar, S.F.; Solozhenko, V.L. Novel ultrahard sp^2/sp^3 hybrid carbon allotrope from crystal chemistry and first principles: Body-centered tetragonal C_6 ('neoglitter'). *Diam. Relat. Mater.* **2023**, *133*, 109747. [CrossRef]
32. Solozhenko, V.L.; Matar, S.F. Prediction of novel ultrahard phases in the B-C-N system from first principles: Progress and problems. *Materials* **2023**, *16*, 886. [CrossRef]
33. Munro, R.; Freiman, S.; Baker, T. *Fracture Toughness Data for Brittle Materials;* NIST Interagency/Internal Report No. 6153; NIST: Gaithersburg, MD, USA, 1998.
34. Brazhkin, V.V.; Solozhenko, V.L. Myths about new ultrahard phases: Why materials that are significantly superior to diamond in elastic moduli and hardness are impossible. *J. Appl. Phys.* **2019**, *125*, 130901. [CrossRef]
35. Matar, S.F.; Solozhenko, V.L. Crystal chemistry and *ab initio* prediction of ultrahard rhombohedral B_2N_2 and BC_2N. *Solid State Sci.* **2021**, *118*, 106667. [CrossRef]
36. Birch, F. Finite strain isotherm and velocities for single-crystal and polycrystalline NaCl at high pressures and 300 K. *J. Geophys. Res.* **1978**, *83*, 1257–1268. [CrossRef]
37. Dove, M.T. *Introduction to Lattice Dynamics;* Cambridge University Press: New York, NY, USA, 1993.
38. DeSorbo, W. Specific heat of diamond at low temperatures. *J. Chem. Phys.* **1953**, *21*, 876–880. [CrossRef]
39. Victor, A.C. Heat capacity of diamond at high temperatures. *J. Chem. Phys.* **1962**, *36*, 1903–1911. [CrossRef]

C Journal of Carbon Research

MDPI

Article

Low-Temperature Deposition of Diamond Films by MPCVD with Graphite Paste Additive

Stephen Yang-En Guu, Fu-Cheng Lin, Yu-Sen Chien, Alen Jhang and Yon-Hua Tzeng *

Institute of Microelectronics, Department of Electrical Engineering, National Cheng Kung University, One University Road, Tainan City 70101, Taiwan; jespgl@gmail.com (S.Y.-E.G.); c0958654796@gmail.com (F.-C.L.); simonmoney199892@gmail.com (Y.-S.C.); bibi1001ofjhang@gmail.com (A.J.)
* Correspondence: tzengyo@mail.ncku.edu.tw

Abstract: Modern integrated circuits (ICs) take advantage of three-dimensional (3D) nanostructures in devices and interconnects to achieve high-speed and ultra-low-power performance. The choice of electrical insulation materials with excellent dielectric strength, electrical resistivity, strong mechanical strength, and high thermal conductivity becomes critical. Diamond possesses these properties and is recently recognized as a promising dielectric material for the fabrication of advanced ICs, which are sensitive to detrimental high-temperature processes. Therefore, a high-rate low-temperature deposition technique for large-grain, high-quality diamond films of the thickness of a few tens to a few hundred nanometers is desirable. The diamond growth rate by microwave plasma chemical vapor deposition (MPCVD) decreases rapidly with lowering substrate temperature. In addition, the thermal conductivity of non-diamond carbon is much lower than that of diamond. Furthermore, a small-grain diamond film suffers from poor thermal conductivity due to frequent phonon scattering at grain boundaries. This paper reports a novel MPCVD process aiming at high growth rate, large grain size, and high sp^3/sp^2 ratio for diamond films deposited on silicon. Graphite paste containing nanoscale graphite and oxy-hydrocarbon binder and solvent vaporizes and mixes with gas feeds of hydrogen, methane, and carbon dioxide to form plasma. Rapid diamond growth of diamond seeds at 450 °C by the plasma results in large-grained diamond films on silicon at a high deposition rate of 200 nm/h.

Keywords: low temperature; MPCVD; diamond; integrated circuit; graphite paste

Citation: Guu, S.Y.-E.; Lin, F.-C.; Chien, Y.-S.; Jhang, A.; Tzeng, Y.-H. Low-Temperature Deposition of Diamond Films by MPCVD with Graphite Paste Additive. *C* **2024**, *10*, 39. https://doi.org/10.3390/c10020039

Academic Editor: Gil Goncalves

Received: 5 February 2024
Revised: 14 March 2024
Accepted: 7 April 2024
Published: 16 April 2024

1. Introduction

Microwave plasma chemical vapor deposition (MPCVD) of diamond exhibits strong effects of substrate temperature on the growth rate [1,2], secondary nucleation rate, and the ratio of sp^3/sp^2 carbon contents in a diamond film [3]. For most micrometers-thick diamond films, MPCVD is carried out at a substrate temperature higher than 700 °C and typically around 900 °C [4]. For integrated-circuit applications, diamond films must be deposited at a low temperature, which does not cause degradation of already fabricated structures. For this application, diamond films need to be deposited at around 450 °C. At a lower substrate temperature, both the growth rate and the ratio of sp^3/sp^2 carbon contents of a diamond film decrease while the secondary nucleation rate increases. The increase in the secondary nucleation rate leads to smaller diamond grains. At a substrate temperature below 700 °C, a diamond film grown by MPCVD typically has sub-micrometer-sized diamond grains. Both small diamond grain sizes and low sp^3/sp^2 ratios in a diamond film are detrimental to phonon transport, leading to a thermal conductivity much reduced from 2000 W/m-K for single-crystalline diamond [5,6]. For example, for a diamond film with diamond grain sizes of 2–5 nm known as ultra-nano-crystalline diamond (UNCD), the thermal conductivity is only about 1% of that for single crystalline diamond. On the other hand, the largest grain size of a polycrystalline diamond film is limited by the film thickness when the lateral size of a diamond grain increases proportionally with the vertical height, and only one diamond grain grows from one diamond seed. Therefore, a diamond film of 100 nanometers thick consists of diamond grains of at most 100 nm in size. Small diamond grain size thus is a severe challenge to achieving high thermal conductivity. Diamond grains grow larger and larger in lateral sizes from diamond seeds on a substrate. A diamond grain is, therefore, smaller in size on the nucleation interface than on the growth surface. The bottom portion of a diamond film with smaller diamond grains and more grain boundaries exhibits lower thermal conductivity than the top portion of a diamond film [7]. Furthermore, voids between diamond grains are filled with low sp^3/sp^2 ratio carbon [8], which exhibits much lower thermal conductivity than a single crystalline diamond. Therefore, it is desirable to promote fast growth of diamond grains, especially in the lateral direction, and suppress the secondary nucleation of polycrystalline diamond grains on a growing diamond grain from the very beginning of the diamond deposition process. Alternatively, larger diamond seeds may be planted in the substrate to allow the interfacial side of a diamond film to be made up with larger diamond grains.

Many attempts have been made to achieve the deposition of diamond at low temperatures. MPCVD in varied gas compositions has been reported. For example, H_2-CH_4 [9], H_2-CH_4-CO_2 [10], H_2-CH_4-O_2 [11–14], CH_4-CO_2 [15], and H_2-CH_4-Ar [16] were used. Microwave plasma produces C, H, and O-containing radicals, which play essential roles in the deposition of diamond and etching of non-diamond carbon phases. CH_3 and C_2 are known to be two main species for diamond growth [17,18]. The termination of dangling bonds on the growing diamond surface by atomic hydrogen stabilizes the surface and prevents surface reconstruction to form undesirable sp^2 carbon-carbon bonds [19]. However, when atomic hydrogen occupies the diamond surface, carbon-containing radicals cannot attach onto the diamond surface. Low substrate temperature is not favorable for the thermal desorption of atomic hydrogen on the diamond surface. There are non-thermal ways of removing the atomic hydrogen from the diamond surface; for example, abstraction of surface atomic hydrogen by another atomic hydrogen coming from plasma. Oxygen-containing radical abstract surface atomic hydrogen and preferentially etch non-diamond carbon phases. Low-energy ions and energetic photons can also stimulate desorption of surface atomic hydrogen. The addition of oxygen-containing gases, such as CO_2, O_2, and CO, has been applied to enhance the ratio of sp^3/sp^2 carbon phases in the deposited diamond film. Steigler et al. carried out a study on the effect of nitrogen and oxygen on low-temperature diamond growth and concluded that in the H_2/CH_4 system, nitrogen promotes the growth rate of diamond. They also reported that when there is appreciable nitrogen in the plasma, oxygen addition can suppress the secondary diamond nucleation caused by nitrogen and lower the defect density in the diamond decreased [20]. Martyanov et al. reported that a small amount of nitrogen stimulates secondary nucleation [21]. Bachmann et al. proposed a phase diagram to help predict whether a gas mixture results in diamond, non-diamond carbon, or no carbon growth based on the percentage of C, H, and O in the gas feeds [2]. Fine-tuning of the C, H, and O contents in the gas feed of MPCVD is thus helpful for optimizing the quality and high growth rate for MPCVD diamond films.

Varga et al. performed diamond growth with a polyvinyl alcohol $(C_2H_4O)_x$ additive, which is a type of polymer. They reported that the main role of the polymer is to serve as a carbon source for the promotion of diamond nucleation and growth [22]. Chen et al. performed diamond growth with a poly (phenyl carbyne) $(C_6H_5C)_x$ additive, which is a polymer with a random structure network in which each carbon is connected to three different carbon atoms and one phenyl substituent. They reported that the phenyl substituent can be removed from the polymer when temperature is increased to approximately 200 °C, and totally removed at 450 °C. After the removal of the phenyl substituent, a dangling bond forms, allowing a sp^3 carbon-carbon bond to be linked to the dangling bond. They also stated that with the help of atomic hydrogen, a sp^2 carbon-carbon bond may be converted to a sp^3 bond [23]. Bianconi et al. stated that the optimal stoichiometry of a polymer precursor for diamond growth is in the form of (CH)x due to the network backbone being only functionalized by hydrogen. They performed diamond growth with poly(hydrocarbon) additive, which has the empirical formula of $(CH)_x$ [24]. Adamantane ($C_{10}H_{16}$) has a tetrahedral structure similar to diamond and is a stable crystalline compound. It is a diamondoid ($C_{(4x+6)}H_{(4x+12)}$), where x = 0, 1, 2, 3. . .), and has point group symmetry. Tirawi et al. compared silicon substrates evaporation-coated with adamantane and bare silicon substrates. They found that after MPCVD at a low temperature of 530 °C, the diamond density for the silicon coated with adamantane was five times higher than that of the bare silicon substrate. They stated that adamantane may decompose in plasma, and due to hydrogen abstraction, may react with the methane species to grow diamond [25].

In this paper, we report the application of nano-graphite in organic solvent and binder in the form of a graphite paste to promote low-temperature diamond growth around 450 °C in the initial 20 min of MPCVD diamond growth from 3 nm diamond seeds in a gas mixture of CH_4, CO_2, and hydrogen. Time-dependent optical emission spectra of the MPCVD plasma were recorded. The diamond grain size, film thickness, and Raman spectra were measured. Diamond films of tens to a few 100 nanometers thick compatible with IC applications were deposited at 450 °C, exhibiting desirable characteristics of diamond films for high thermal conductivity. Further optimization of the deposition process for the fabrication of high-thermal-conductivity diamond films of a few hundred nanometers thick by making use of the novel graphite paste-assisted MPCVD technique is being undertaken and will be reported elsewhere.

2. Experimental Details

2.1. Diamond Seeding

P-type (001) silicon wafers were utilized for diamond nucleation through seeding and growth. Seeding was performed using an ultrasonic electrostatic seeding method. Before diamond growth, silicon wafers were seeded with a commercial nanodiamond solution from the Nano Carbon Research Institute, Japan (nominal 3.7 nm crystalline diamond particles, ζ potential of +45~53 mV) in an ultrasonic bath for 50 min. After the 50 min seeding, the wafers were additionally shaken in the ultrasonic bath using methanol and DI water for 10 s each to remove agglomerated and unbonded particles. The seeded wafers were then dried with N_2.

2.2. Diamond Growth

Diamond film deposition was carried out in a 2.45 GHz Astex AX5010 Seki Technotron Corp. MPCVD reactor. Before the deposition, a seeded silicon wafer was adhered to a molybdenum plate by graphite paste for better temperature uniformity. The molybdenum was placed on a water-cooled substrate holder. The graphite paste was from Pin Shuo Tan Corp., Taipei, Taiwan. The contents of the graphite paste included 30% methyl ethyl ketone (solvent), 30% propylene glycol methyl ether acetate (solvent), 30% modified epoxy resin (binder), and 10% synthetic graphite. The process gases used were H_2, CH_4, and CO_2. The growth parameters included 950 W microwave power, 30 torr gas pressure, 160 sccm total gas flow rate, and 450 °C substrate temperature. At the beginning of diamond deposition, pure hydrogen was used to ignite plasma at about 1 torr gas pressure. After igniting the plasma, the gas pressure was adjusted to be at 30 torr by a needle throttle valve. CH_4 and CO_2 were introduced into the chamber once the gas pressure was controlled at 30 torr. It typically took about 2 min after the plasma was ignited. The deposition time was measured from the introduction of the CH_4 and CO_2 gas sources.

The added graphite paste vaporizes and supplies additional C, O, and H-containing radicals when the substrate temperature rises in the microwave plasma. This transient increase of vapor from the graphite paste greatly affects the initial phase of diamond growth from diamond seeds. Experiments were conducted to investigate the effects of graphite paste on MPCVD diamond deposition. We controlled the added amount of graphite paste in individual experiments. Besides the C, O, and H additives in the initial phase, the following two gas compositions were applied after diamond deposition: gas composition

#1: 5% CH_4 and 1% CO_2 in H_2, and gas composition #2: 5% CH_4 in H_2. Every period of diamond deposition was 20 min. One or more periods were carried out under the same or varied experimental conditions. Some experiments included repeated growth periods with graphite paste being added at the beginning of each period.

2.3. Measurement Techniques and Characterization

Optical emission spectroscopy (OES, Ocean Optics USB4000) was used to detect light emission from excited radicals in plasma. SEM (Hitachi SU8000) was used for recording images of diamond films. The grain size and size distribution of deposited diamond films were observed from SEM plane-view images. The thicknesses of the diamond films were observed from SEM cross-sectional images. Raman spectroscopy (excited by a 458 nm laser) was used for the analysis of the carbon contents in diamond films. A dual-color optical pyrometer (Williamson Pro Series, Concord, MA USA) was used for substrate temperature measurements.

3. Results and Discussion

3.1. Optical Emission Spectroscopy Analysis

Table 1 lists peak wavelengths of OES signal intensity measured from microwave plasma for diamond deposition. Figure 1 shows the OES of background light. The OES had peaks of 440 nm, 550 nm, and 625 nm. Figure 2 shows the OES of hydrogen plasma. Figure 3 shows the OES of microwave plasma for diamond deposition in gas composition #2 without the addition of graphite paste. Using gas composition #1 (5% CH_4 and 1% CO_2 in H_2), 0.018 g graphite paste was applied to adhere silicon substrates onto a molybdenum plate. The molybdenum plate was then placed on a water-cooled substrate holder. After the hydrogen plasma was ignited and before CH_4 and CO_2 were fed to the plasma, a strong CH peak at 392 nm started to appear and reached the maximum intensity (at 1 min 30 s after hydrogen plasma was ignited) in the hydrogen plasma. After reaching the maximum intensity, the CH peak started to decrease gradually until it completely disappeared in about 10 min after the amount of graphite paste additive was diluted by the hydrogen gas flow. The CH peak disappeared after a longer time when more graphite paste was added. Peaks at 363 nm, 420 nm, and 518 nm also increased in intensity at the beginning of diamond deposition along with the CH peak at 392 nm. The peaks at 420 nm and 518 nm are attributed to CH and C_2, respectively. This shows that graphite paste dissociates in plasma and releases radicals and molecules, which are beneficial to the deposition of diamond film. Once all the graphite paste was dissociated and diluted by the hydrogen gas flow, the CH peaks at 392 nm disappeared. The C_2 peak remained longer into the next 20 min period for further deposition of diamond in CH_4 and H_2 gas flow with or without CO_2 additive.

Table 1. OES Signal.

Peak (nm)	Signal Source
392,420	CH
518,550	C_2 (550 nm peak overlaps with background light)
587	H_2
656	Hα
485	Hβ
440,550,625	Background light

Figure 4a shows the OES when the CH peak reached the maximum intensity without CH_4 and CO_2. This was measured at 1 min 30 s after the plasma was ignited. Figure 4b shows the OES of plasma in 5% CH_4 and 1% CO_2 balanced by hydrogen after the plasma was ignited for 20 min. Excluding the C_2 peak, other peaks attributed to graphite paste had completely disappeared.

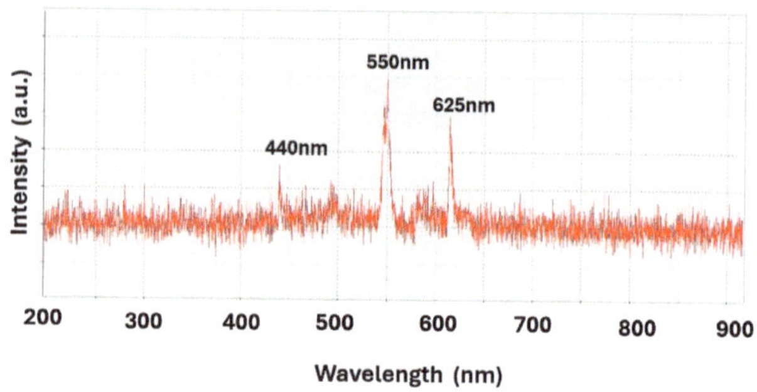

Figure 1. OES of background light.

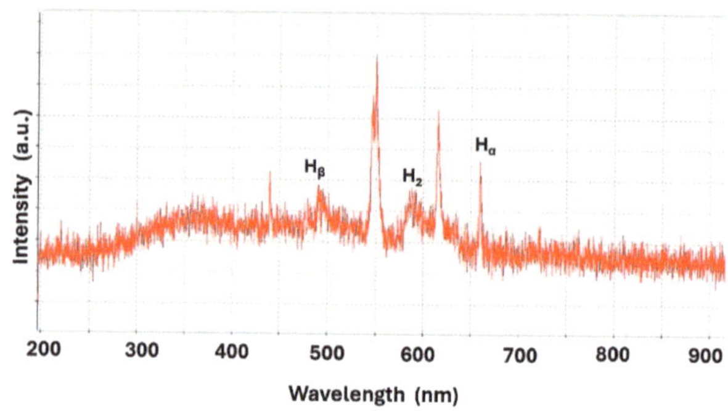

Figure 2. OES of hydrogen plasma measured at 1 min 30 s after the hydrogen plasma was ignited.

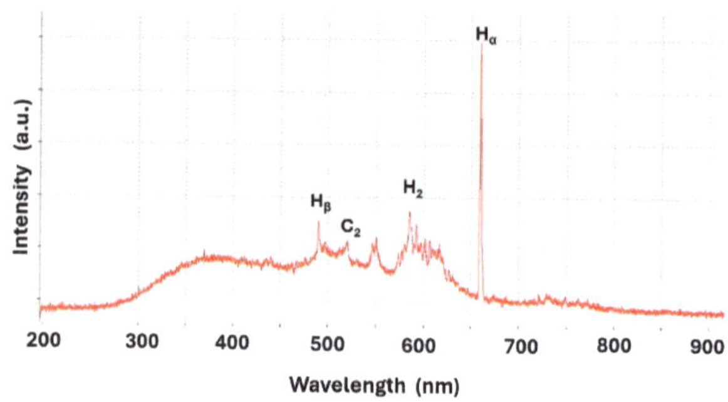

Figure 3. OES of plasma in CH_4 and H_2 (gas composition #2) without graphite paste additive (measured 20 min after the plasma was ignited).

The peaks at 440 nm, 550 nm, and 625 nm were from the background lights of the room, which existed before plasma was ignited. The peak at 550 nm should have been another peak attributed to C_2, but the position overlapped with the background signals; thus, C_2 was determined by the peak at 518 nm. The C_2 peak at 518 nm showed a nearly constant intensity during the growth. OES peak positions of CH and CN overlapped at both 392 nm and 420 nm. For the experiments with graphite paste addition, there was a small peak at 363 nm. OES peak positions of CN and N_2 overlapped at 363 nm. This indicated the possible existence of CN and N_2 originating from the contents of the graphite paste.

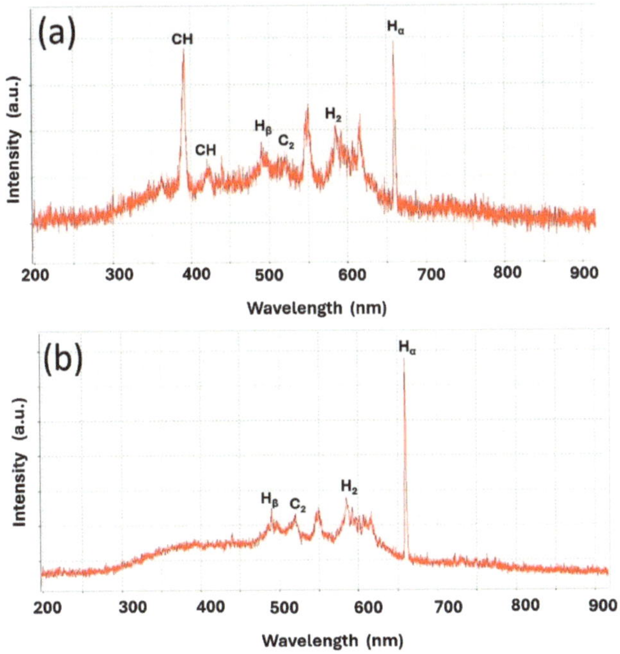

Figure 4. OES of plasma in gas composition #1 with 0.018 g graphite paste additive. (**a**) OES measured before CH_4 and CO_2 injection. (**b**) OES measured after CH_4 and CO_2 injection.

Using gas composition #2 (5% CH_4 in H_2 without CO_2 additive), 0.018 g of graphite paste was applied to adhere samples onto the molybdenum plate before placing the molybdenum plate on a water-cooled substrate holder for diamond growth. The diamond growth results using gas composition #2 were similar to those obtained using gas composition #1. The graphite paste appeared to dominate the diamond growth process, leading the difference between gas composition #1 and composition #2 to exhibit minor differences in the diamond growth results. After the hydrogen plasma was ignited and before CH_4 was fed into the plasma, OES showed that the CH peak at 392 nm started to appear and reached the maximum intensity (at 1 min 10 s after hydrogen plasma was ignited) in the hydrogen plasma. After reaching the maximum intensity, the CH peak at 392 nm gradually decreased until it completely disappeared. The CH peak at 420 nm and C_2 peak at 518 nm also increased in intensity along with the CH peak at 392 nm.

Once all the graphite paste was dissociated and diluted by gas feeds, the CH peaks at 392 nm and 420 nm disappeared, which also occurred approximately 10 min after the plasma was ignited. The C_2 peak remained during diamond growth.

Figure 5a shows the OES of plasma with CH_4 additive when the CH peak reached the maximum intensity. It occurred at 1 min and 10 s after the plasma was ignited. Figure 5b shows the OES of plasma in 5% CH_4 balanced by hydrogen at the time of 20 min after the ignition of the plasma. Besides the C_2 peak, the other peaks originated from graphite paste had completely disappeared. The C_2 peak at 518 nm showed a constant intensity during the growth. There was no apparent difference between the OES of the plasma in gas compositions #1 and #2, probably because the composition difference was only 1% CO_2 in the gas mixture.

In plasma with 5% CH_4 balanced by H_2 without the addition of graphite paste, OES did not show strong CH peaks at 392 nm and 420 nm nor the strong C_2 peak at 518 nm. This confirms that strong CH peaks and C_2 peaks can be attributed to the plasma dissociation of graphite paste. When 0.018 g graphite paste additive was applied, the effects of graphite paste took place during the first 10 min of diamond nucleation and growth. When more than 0.018 g graphite paste is added, the effects of graphite paste will last longer than 10 min.

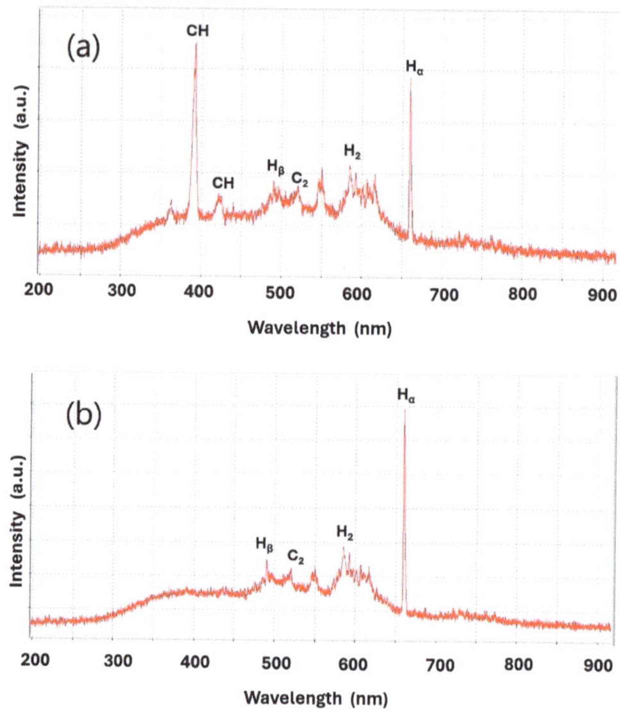

Figure 5. OES of plasma in gas composition #2 with 0.018 g graphite paste additive. (**a**) OES measurement taken before CH_4 injection. (**b**) OES measurement taken after CH_4 injection.

3.2. *SEM Analysis*

SEM was employed to measure diamond grain size and film thickness. The graphite paste dissociated within about 20 min of the MPCVD process. To carry out diamond growth for a longer time under the influence of graphite paste, the substrate was taken out of the MPCVD chamber every 20 min to reapply the same amount of graphite paste. For the growth by MPCVD in gas compositions #1 and #2 with graphite paste additive, 0.018 g of graphite paste was reapplied between two consecutive 20 min growth periods. As a comparison, samples with a small amount of 0.008 g graphite paste applied only once before continuous 1 h growth were also obtained in gas compositions #1 and #2. These were further compared with diamond deposition without graphite paste additives.

3.2.1. Diamond Grain Size

Table 2 shows a comparison of diamond grain sizes using gas composition #1 and gas composition #2. The average grain size for diamond films grown in gas composition #1 was 52.07 nm with standard deviation of 19.85 nm. The average grain size for diamond films grown in gas composition #2 was 53.25 nm with standard deviation of 19.52 nm. The similar results are attributed to the fact that although the gas compositions were slightly different, the amount of graphite paste additive, which dominated the diamond growth process, was the same for both cases.

Table 2. Comparison of diamond grain sizes.

Growth in Gas Compositions		Grain Sizes (nm)
#1 or #2		
#1	Without graphite paste, 20 min growth	16.6
#1	0.018 g graphite paste, 20 min growth	48.75
#2	Without graphite paste, 20 min growth	23.75
#2	0.018 g graphite paste, 20 min growth	52.5
#1	0.018 g graphite paste, 20 min growth, twice	65
#2	0.018 g graphite paste, 20 min growth, twice	67.5
#1	0.018 g graphite paste, 20 min growth, three times	75
#2	0.018 g graphite paste, 20 min growth, three times	80
#1	0.008 g graphite paste, 1 h continuous growth	55
#2	0.008 g graphite paste, 1 h continuous growth	42.5
#1	Average	52.07
#1	Standard deviation	19.85
#2	Average	53.25
#2	Standard deviation	19.52

SEM plane-view images of diamond films grown in gas composition #1 are shown in Figure 6, and SEM plane-view images of diamond films grown in gas composition #2 are shown in Figure 7. Diamond films grown in gas compositions #1 and #2 without graphite paste additive exhibited an average grain size of 15–25 nm. The CO_2 additive resulted in a higher diamond nucleation density. When graphite paste was added, the diamond grain sizes were about 50 nm after the initial 20 min MPCVD process in both gas compositions #1 and #2. A strong OES peak from CH radical due to graphite paste additive promoted the growth rate of diamond grains from the 3 nm diamond seeds.

After terminating the initial 20 min MPCVD process for reapplying graphite paste followed by additional 20 min MPCVD process, the diamond grain sizes increased to about 65 nm. This process for reapplication of graphite paste was repeated again to grow diamond grains of about 80 nm in size. The growth rate with 0.018 g of graphite paste added for every 20 min resulted in the growth of diamond grain sizes by about 10–20 nm every 20 min. This trend was similar for both gas composition #1 and #2. This indicates the essential role of graphite paste additive in the growth of diamond at low temperatures.

Diamond grain sizes synthesized in 1 h continuous MPCVD were about the same as those produced by diamond MPCVD for only 20 min but with two times the amount of graphite paste additive. The importance of graphite pastes in the growth of diamond seeds to form diamond films at the substrate temperature around 450 °C was clearly demonstrated.

3.2.2. Diamond Film Thickness

Table 3 shows a comparison of the thickness of diamond films grown under different conditions. The average thickness of diamond films grown in gas composition #1 was 97.88 nm with standard deviation of 58.82 nm. The average thickness of diamond films grown in gas composition #2 was 95.53 nm with standard deviation of 52.47 nm.

SEM cross-sectional-view images of diamond films grown in gas composition #1 are shown in Figure 8, and SEM cross-sectional-view images of diamond films grown in gas composition #2 are shown in Figure 9. The diamond films that grew continuously for 1 h with 0.008 g graphite paste additive were thinner than diamond films grown for 20 min with 0.018 g graphite paste. For both gas composition #1 and #2, the diamond films after 20 min growth with 0.018 g graphite paste had a thickness of about 60 nm. After two repetitive growth periods of 20 min each, the diamond film grown in gas composition #1 was 85 nm thick. As a comparison, the diamond film grown in gas composition #2 was 103.5 nm. After three repetitive growth periods of 20 min each, the diamond film grown in gas composition #1 was 197 nm. As a comparison, diamond films grown in gas composition #2 were 178 nm. In repetitive 20 min growth periods with graphite paste being applied in each growth period, diamond films of nearly 200 nm were deposited within 1 h at a low temperature around 450 °C. In contrast, the diamond film grown for 20 min without graphite paste was too thin for the measurement of the thickness by SEM.

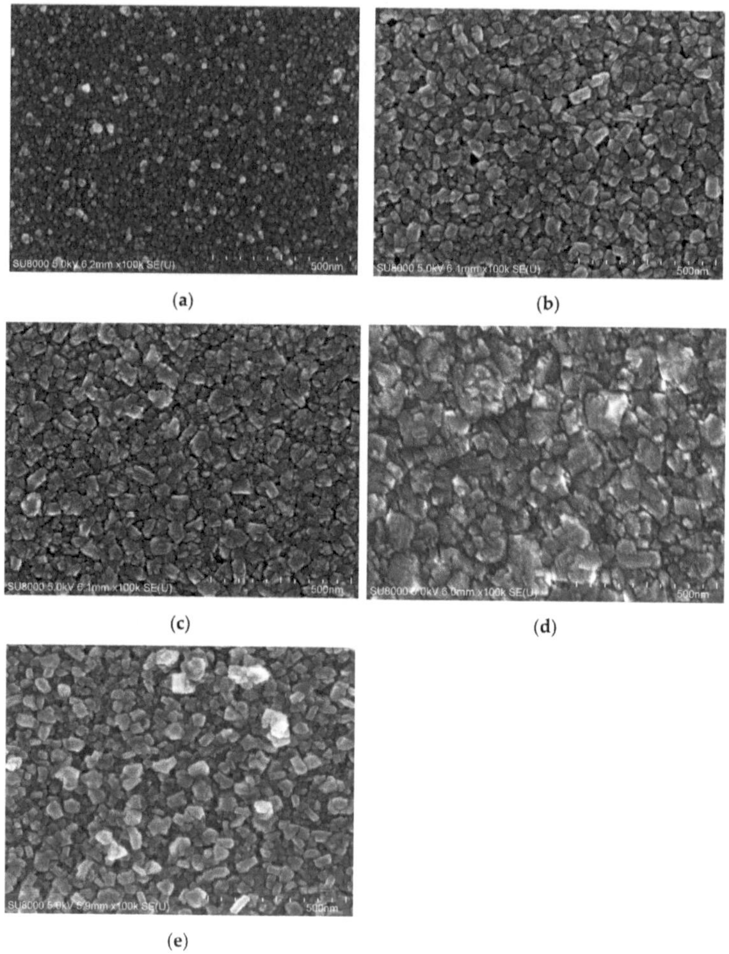

Figure 6. SEM plane-view images of diamond films grown in gas composition #1: (**a**) after 20 min growth without graphite paste additive; (**b**) after 20 min growth with 0.018 g graphite paste additive; (**c**) after two 20 min growth periods with 0.018 g graphite paste added in each growth period; (**d**) after three 20 min growth periods with 0.018 g graphite paste added in each growth period; (**e**) after 1 h continuous growth with 0.008 g graphite paste additive. The scale bars are all 500 nm.

3.3. Raman Spectroscopy Analysis

Raman spectroscopy excited by a 458 nm light source was employed to characterize diamond films. Relevant Raman peak positions are listed in Table 4. The peaks at 1445 cm^{-1} and 1556 cm^{-1} were background Raman scattering signals attributed to the quartz optical components excited by the 458 nm laser, as shown in Figure 10. Raman spectra for diamond films grown in gas composition #1 are shown in Figures 11 and 12, while Raman spectra for diamond films grown in gas composition #2 are shown in Figures 13 and 14.

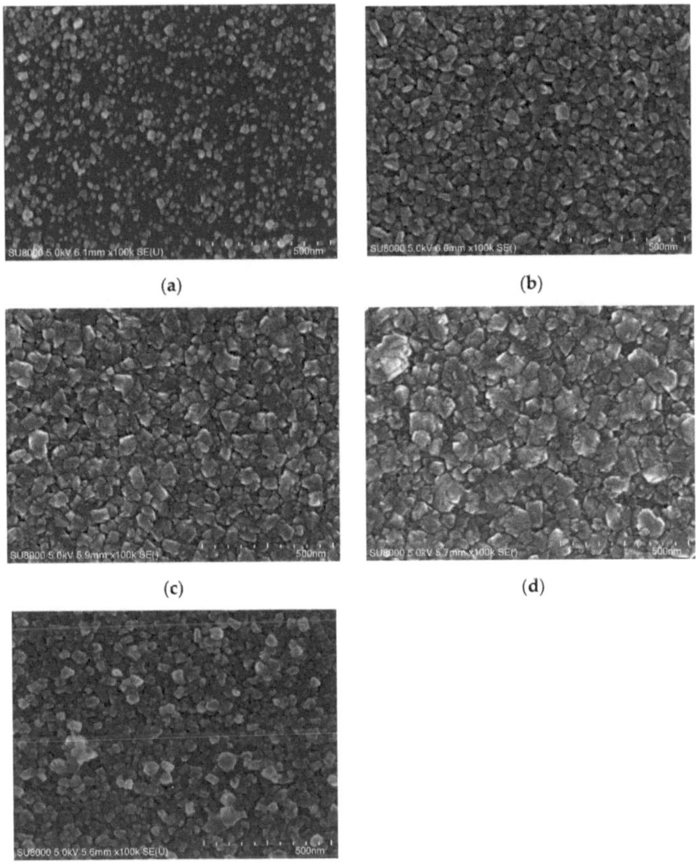

Figure 7. SEM plane-view images of diamond films grown in gas composition #2: (**a**) after 20 min growth without graphite paste additive; (**b**) after 20 min growth with 0.018 g graphite paste additive; (**c**) after two 20 min growth periods with 0.018 g graphite paste added in each growth period; (**d**) after three 20 min growth periods with 0.018 g graphite paste added in each growth period; (**e**) after 1 h continuous growth with 0.008 g graphite paste additive. The scale bars are all 500 nm.

Table 3. Thickness of diamond films.

Growth in Gas Compositions	Film Thickness (nm)
#1 and #2	
#1 0.018 g graphite paste, 20 min growth	61.8
#2 0.018 g graphite paste, 20 min growth	55.6
#1 0.018 g graphite paste, 20 min growth, twice	85.4
#2 0.018 g graphite paste, 20 min growth, twice	103.5
#1 0.018 g graphite paste, 20 min growth, three times	197
#2 0.018 g graphite paste, 20 min growth, three times	178
#1 0.008 g graphite paste, 1 h continuous growth	47.3
#2 0.008 g graphite paste, 1 h continuous growth	45
#1 Average	97.88
#1 Standard deviation	58.82
#2 Average	95.53
#2 Standard deviation	52.47

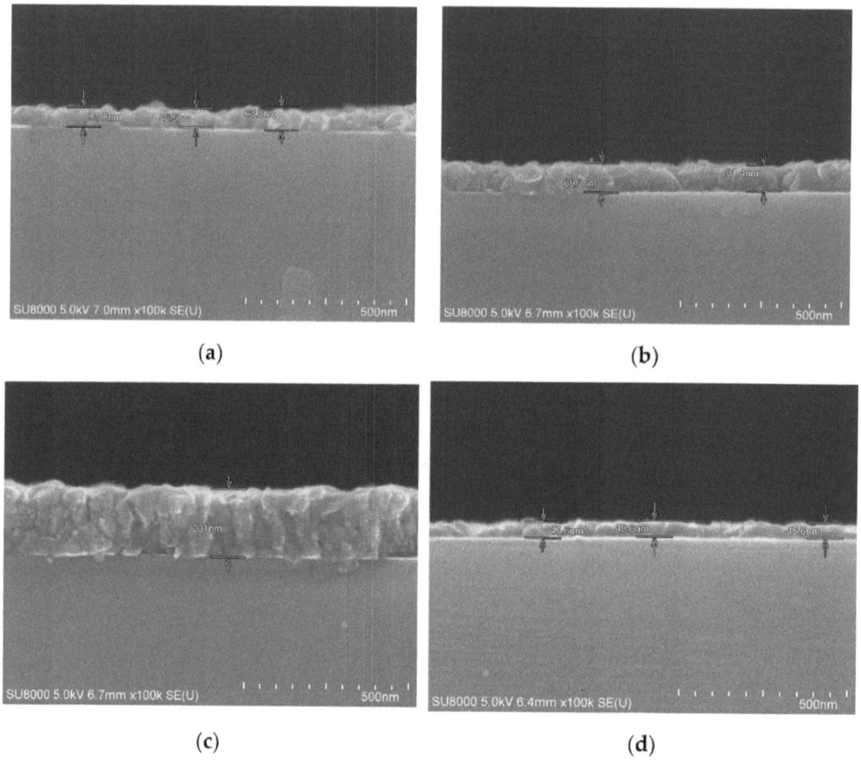

Figure 8. SEM cross-sectional-view images of diamond films grown in gas composition #1: (**a**) after 20 min growth with 0.018 g graphite paste additive; (**b**) after two 20 min growth periods with 0.018 g graphite paste added in each growth period; (**c**) after three 20 min growth periods with 0.018 g graphite paste added in each growth period; (**d**) after 1 h continuous growth with 0.008 g graphite paste additive.

For both gas compositions, Raman spectra of diamond films grown for 20 min without graphite paste did not display the diamond peak at 1332 cm^{-1}, as shown in Figures 11 and 13a. However, those for the diamond films grown with graphite paste additive displayed the diamond peak at 1332 cm^{-1} after 20 min diamond growth. The addition of graphite paste to the MPCVD process contributes to the formation of sp^3-bonded diamond.

Diamond films grown for 20 min in gas composition #1 with a small amount of 0.003 g graphite paste additive did not display the D band and the G band. When a larger amount of 0.018 g graphite paste was added, the D band and G band appeared, as shown in Figure 11.

Diamond films grown for 20 min in gas composition #2 with 0.0045 g graphite paste additive displayed the D band and G band after 20 min diamond growth, as shown in Figure 13b. In comparison with diamond films grown in gas composition #1, diamond films grown in gas composition #2 with 0.018 g graphite paste also displayed the D band and the G band Raman signal, as shown in Figure 13.

From the comparison between Figures 12d and 14d, diamond films grown for 1 h continuously with 0.008 g graphite paste added at the beginning displayed the G band signal, but the D band signal disappeared. This revealed that when too little (0.008 g) graphite paste was added, the D band Raman signal disappeared after a 1 h growth process.

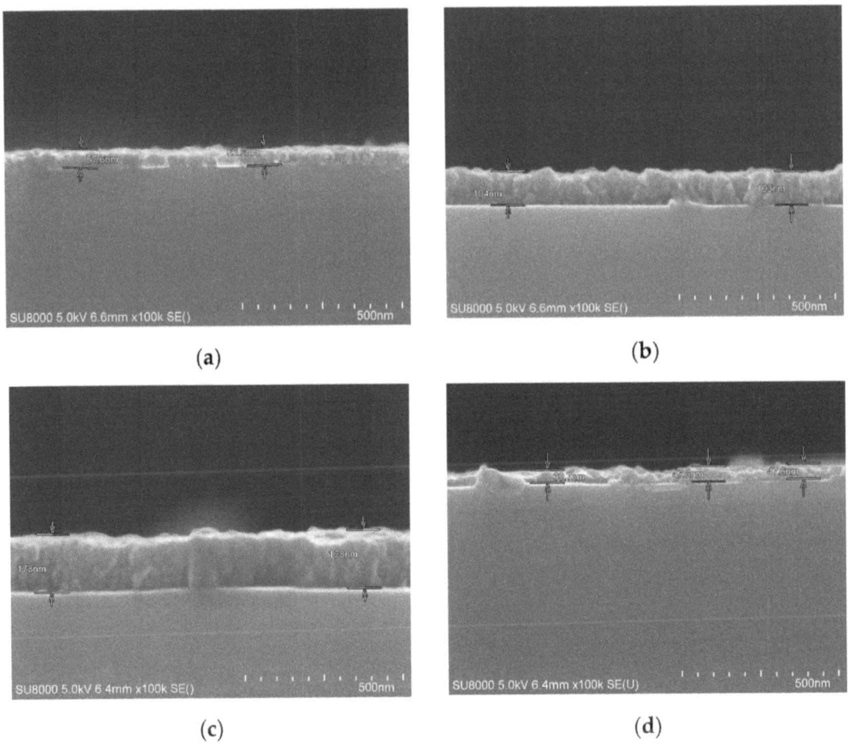

Figure 9. SEM cross-sectional-view images of diamond films grown in gas composition #2: (**a**) after 20 min growth with 0.018 g graphite paste additive; (**b**) after two 20 min growth periods with 0.018 g graphite paste added in each growth period; (**c**) after three 20 min growth periods with 0.018 g graphite paste added in each growth period; (**d**) after 1 h continuous growth with 0.008 g graphite paste additive.

Table 4. Raman scattering peaks.

Wavenumber (cm^{-1})	Raman Peak
1332	Sp3
1350	D Band
1445	Background signal attributed to laser excitation of quartz optical components
1480	Transpolyacetylene (TPA)
1500	Amorphous carbon
1556	Background signal attributed to laser excitation of quartz optical components
1600	G Band
1710	C=O

Figures 12 and 14 show that Raman scattering of diamond films grown by the repetitive termination of the growth process for additional graphite paste additive displayed the sp3 diamond Raman peak. A larger amount of graphite paste contributed to the formation of sp3-bonded diamond along with more intense G bands and D bands. However, Raman spectra for diamond films grown by two time and three-time repetitive growth processes with graphite paste additives showed that the G band signal strength at

1600 cm^{-1} decreased, and the Raman scattering from TPA and amorphous carbon at 1500 cm^{-1} became more apparent. Diamond films grown in both gas composition #1 and #2 displayed Raman scattering from TPA at diamond grain boundaries.

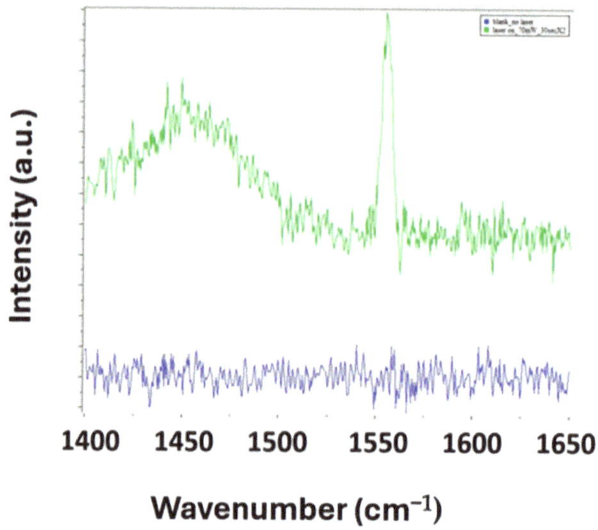

Figure 10. Background Raman peaks measured without a test sample due to laser scattering by quartz optical components in the Raman system.

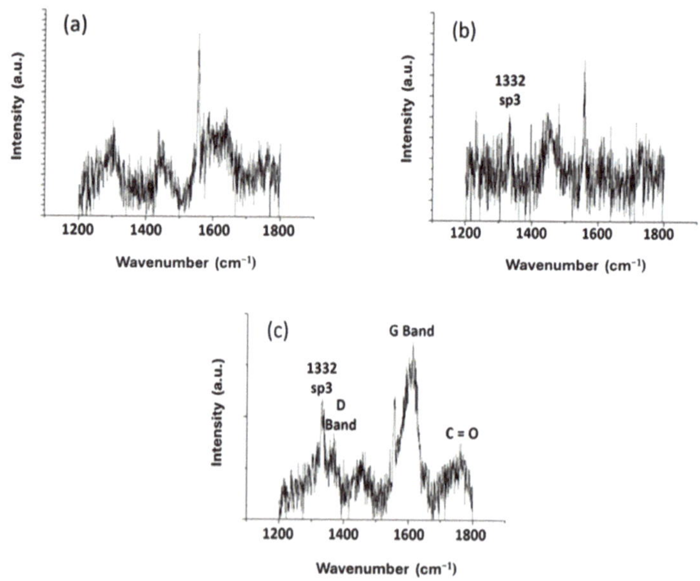

Figure 11. Raman spectra of diamond films grown in gas composition #1: (**a**) 20 min diamond growth without graphite paste additive; (**b**) 20 min diamond growth with 0.003 g graphite paste additive; (**c**) 20 min growth with 0.018 g graphite paste additive.

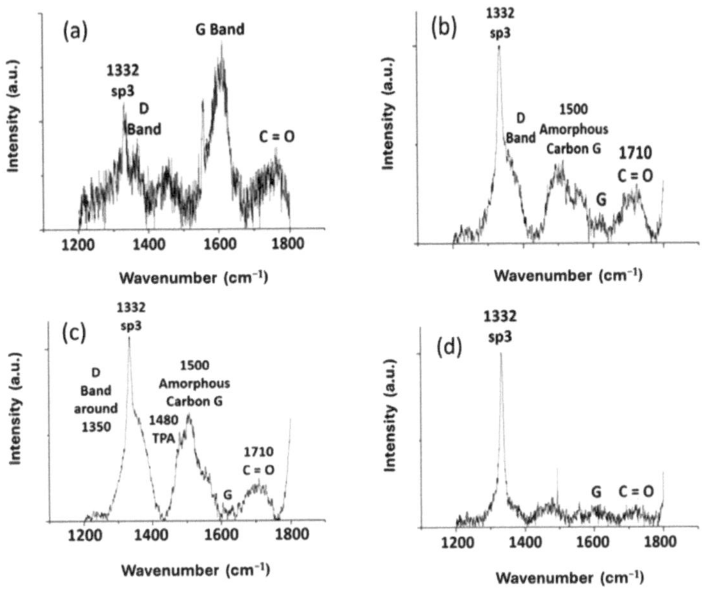

Figure 12. Raman spectra of diamond films grown in gas composition #1: (**a**) 20 min growth with 0.018 g graphite paste additive; (**b**) two repetitive 20 min growth periods with 0.018 g graphite paste additive for each period; (**c**) three repetitive 20 min growth periods with 0.018 g graphite paste additive for each period; (**d**) 1 h continuous growth with 0.008 g graphite paste additive only at the beginning.

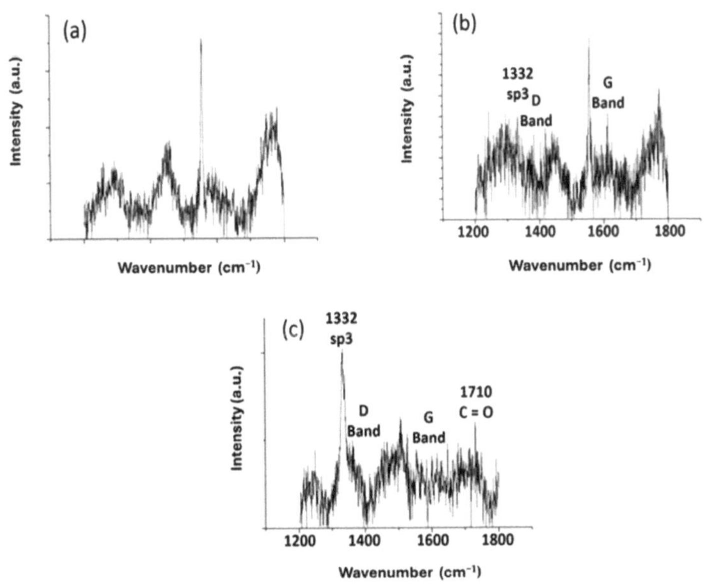

Figure 13. Raman spectra of diamond films grown in gas composition #2: (**a**) 20 min growth without graphite paste; (**b**) 20 min growth with 0.0045 g graphite paste; (**c**) 20 min growth with 0.018 g graphite paste.

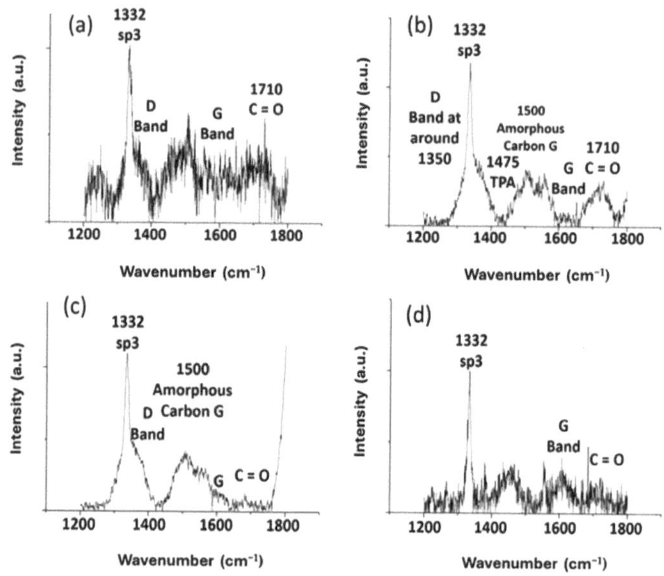

Figure 14. Raman spectra of diamond films grown in gas composition #2: (**a**) 20 min growth with 0.018 g graphite paste; (**b**) two repetitive 20 min growth with 0.018 g graphite paste additive for each period; (**c**) three repetitive 20 min growth with 0.018 g graphite paste additive for each period; (**d**) 1 h continuous growth with 0.008 g graphite paste additive at the beginning.

Raman spectra of repetitive 20 min growth of diamond films for a total of 1 h with resupply of graphite paste at 0.018 g for each period were compared with those of diamond films grown continuously for 1 h with 0.008 g graphite paste added only once at the beginning. Raman spectra for diamond grown by the repetitive process with resupply of 0.008 g graphite paste each time displayed Raman peaks of G-band and D band at 1500 cm^{-1}, while the diamond films grown by continuous process with graphite paste supplied only once at the beginning displayed only the G band without the D band.

Except for the small difference between diamond films grown for 20 min in gas composition #1 with 0.003 g graphite paste additive and those grown in gas composition #2 with 0.0045 g graphite paste additive in growth condition 1 and 0.0045 g growth with growth condition 2 for 20 min, no significant difference was observed between diamond films grown in two different gas compositions. Significant differences in diamond film characteristics were observed when the amount of graphite paste additive varied. Graphite paste affected diamond growth at low temperature more than the 1% CO_2 additive in both gas compositions.

The application of an oxy-hydrocarbon vapor such as acetone with CO_2 and O_2 additives for MPCVD growth of diamond was reported at temperatures between 600 °C and 1400 °C higher than those in this work [26]. After fine-tuning the percentage of C, H, and O in the gas mixture, Bachmann et al. [2] summarized the growth zone in the C, H, and O phase diagram. Effects of substrate temperature on diamond growth in the low-temperature range of 450 °C were not clearly understood. Nevertheless, desorption or abstraction of atomic hydrogen termination on the diamond growing surface was seen as desirable in order to create active carbon sites for the attachment of carbon-containing radicals. Oxygen was expected to play a role in activating the diamond growing surface to promote the diamond growth rate. An appropriate amount of oxygen was seen as beneficial to diamond growth at low temperature, which is consistent with results presented in this work.

Nanoscale graphite was reported to be beneficial for diamond growth. For example, Li et al. applied highly oriented pyrolytic graphite (HOPG), and graphite fibers as precursors to promote diamond deposition on graphite-seeded silicon wafers. They reported that the number density of diamond nuclei was enhanced by the presence of graphite, and that diamond grew preferentially on the prism planes of graphite,; i.e., on the edges of graphite. There was an epitaxial relationship between graphite and diamond. There was etching of graphite when diamond was nucleated [27]. Xie et al. performed simulation work and suggested that graphite preferentially turned into hexagonal diamond instead of cubic diamond [28]. Lambrecht et al. performed calculations and suggested that diamond can nucleate on graphite with subsequent hydrogenation along the edges of graphite [29]. In this work, the graphite paste included nanoscale graphite as a conductivity additive. It was thought that some of the nanographite was released into the plasma along with the vaporized solvent and binder of the graphite paste. Hydrogen and oxygen would react with nanoscale graphite to form carbon-containing radicals, which help promote the diamond growth. The latter research is consistent with the experimental results presented in this work demonstrating the dominating effects of graphite paste in the growth of diamond.

We observed that graphite paste contributed to the growth of sp3 bonded carbon. Too much graphite paste additive also contributed to the formation of Raman scattering of the D band and G band signal. The effects of graphite paste added to the MPCVD of diamond films at a low temperature of 450 °C have been revealed to significantly enhance the diamond growth rate from the beginning of the deposition process. This promotion converts the undesirable small-diamond-grain diamond films, which suffer from frequent phonon scattering at the abundant grain boundaries, into large-diamond-grain diamond films beginning from the silicon-diamond interface. Optimization of the thermal conductivity, dielectric strength, leakage current, and mechanical strength of low-temperature diamond films deposited by this novel process is being carried out. Preliminary results indicate high thermal conductivity of the low-temperature diamond films of several tens to a few hundred nanometers thick are feasible. Detailed results will be published elsewhere in near future.

4. Conclusions

A novel MPCVD process, leading to enhanced growth rate, grain size, and sp^3/sp^2 ratio for nano-diamond thin films on silicon, was demonstrated. Graphite paste containing nano-scale graphite, an oxy-hydrocarbon binder, and solvent vaporized and mixed with gas feeds of hydrogen, methane, and carbon dioxide to form plasma. Rapid grain growth from 3 nm diamond seeds embedded on silicon resulted in the growth of diamond films with large diamond grains on the nucleation side at a rate exceeding 200 nm/h at around 450 °C. Quantitative study on the effects of the graphite paste additive on low-temperature diamond deposition was carried out and reported based on SEM, Raman scattering of diamond films, and OES analysis of the plasma chemistry. As a comparison, without the graphite paste additive, the diamond growth rate and diamond grain size were impractically small. This novel process helps make diamond a more promising dielectric material for future IC fabrication in the semiconductor industry. Optimization of the thermal conductivity, dielectric strength, leakage current, and mechanical strength of low-temperature diamond films synthesized by this novel process is being carried out and will be reported elsewhere.

Author Contributions: Conceptualization, Y.-H.T.; methodology, Y.-H.T. and S.Y.-E.G.; validation, S.Y.-E.G., A.J., F.-C.L. and Y.-S.C.; formal analysis, Y.-H.T., S.Y.-E.G., F.-C.L. and A.J.; investigation, S.Y.-E.G., F.-C.L., A.J. and Y.-S.C.; resources, Y.-H.T.; data curation, S.Y.-E.G., Y.-S.C., F.-C.L. and A.J.; writing—original draft preparation, Y.-H.T. and S.Y.-E.G.; writing—review and editing, S.Y.-E.G., F.-C.L. and Y.-H.T.; visualization, S.Y.-E.G.; supervision, Y.-H.T.; project administration, Y.-H.T.; funding acquisition, Y.-H.T. All authors have read and agreed to the published version of the manuscript.

Funding: This research was funded by Taiwan Semiconductor Manufacturing Company, Taiwan under the contract of NCKU-TSMC JDP 2022-2024.

Data Availability Statement: Data are contained within the article.

Acknowledgments: Special thanks to Hui-Chen Kung of National Cheng Kung University, Tainan Taiwan for letting us use the 458 nm Raman system in her laboratory, the Precious Equipment Center of National Cheng Kung University, Tainan, Taiwan for our uses of the SEM, and Hsiaokang Chang, Hsin-Yen Huang, Shao-Kuan Lee, Cheng-Chin Lee, and Kuang-Wei Yang of the Interconnect Thin Film Department, Interconnect Pathfinding Division, Taiwan Semiconductor Manufacturing Company, Hsinchu, Taiwan for very inspiring and helpful discussion.

Conflicts of Interest: The authors declare no conflicts of interest.

References

1. Reshi, B.A.; Misra, A.; Varma, R. Investigation of diamond deposition on the diamond, silicon and quartz substrates by microwave plasma chemical vapor deposition and Monte Carlo simulations. *Mater. Res. Express* **2019**, *6*, 046407. [CrossRef]
2. Bachmann, P.K.; Leers, D.; Lydtin, H. Towards a general concept of diamond chemical vapour deposition. *Diam. Relat. Mater.* **1991**, *1*, 1–12. [CrossRef]
3. Tu, R.; Xu, T.; Li, D.; Zhang, S.; Yang, M.; Li, Q.; Zhang, L.; Shimada, T.; Goto, T.; Shi, J. Morphology and mechanical behavior of diamond films fabricated by IH-MPCVD. *RSC Adv.* **2018**, *8*, 16061–16068. [CrossRef] [PubMed]
4. Mallik, A.K.; Shih, W.-C.; Pobedinskas, P.; Haenen, K. Early periods of low-temperature linear antenna CVD nucleation and growth study of nanocrystalline diamond films. *Coatings* **2024**, *14*, 184. [CrossRef]
5. Graebner, J.E.; Jin, S.; Kammlott, G.W.; Herb, J.A.; Gardinier, C.F. Unusually high thermal conductivity in diamond films. *Appl. Phys. Lett.* **1992**, *60*, 1576. [CrossRef]
6. Liu, W.L.; Shamsa, M.; Calizo, I.; Balandin, A.A.; Ralchenko, V.; Popovich, A.; Saveliev, A. Thermal conduction in nanocrystal-line diamond films: Effects of the grain boundary scattering and nitrogen doping. *Appl. Phys. Lett.* **2006**, *89*, 171915. [CrossRef]
7. Dai, B.; Zhao, J.; Ralchenko, V.; Khomich, A.; Popovich, A.; Liu, K.; Shu, G.; Gao, G.; Mingqi, S.; Yang, L.; et al. Thermal conductivity of free-standing CVD diamond films by growing on both nuclear and growth sides. *Diam. Relat. Mater.* **2017**, *76*, 9–13. [CrossRef]
8. Keblinski, P.; Phillpot, S.R.; Wolf, D.; Gleiter, H. On the nature of grain boundaries in nanocrystalline diamond. *Nanostructured Mater.* **1999**, *12*, 339–344. [CrossRef]
9. Dong, L.; Ma, B.; Dong, G. Diamond deposition at low temperature by using CH_4/H_2 gas mixture. *Diamond Relat. Mater.* **2002**, *11*, 1697–1702. [CrossRef]
10. Stiegler, J.; Lang, T.; Nygård-Ferguson, M.; Von Kaenel, Y.; Blank, E. Low temperature limits of diamond film growth by microwave plasma-assisted CVD. *Diam. Relat. Mater.* **1996**, *5*, 226–230. [CrossRef]
11. Liou, Y.; Inspektor, A.; Weimer, R.; Knight, D.; Messier, R.J. The effect of oxygen in diamond deposition by microwave plasma enhanced chemical vapor deposition. *Mater. Res.* **1990**, *5*, 2305–2312. [CrossRef]
12. Liou, Y.; Weimer, R.; Knight, D.; Messier, R. Effect of oxygen in diamond deposition at low substrate temperatures. *Appl. Phys. Lett.* **1990**, *56*, 437–439. [CrossRef]
13. Liou, Y.; Inspektor, A.; Weimer, R.; Messier, R. Low-temperature diamond deposition by microwave plasma-enhanced chemical vapor deposition. *Appl. Phys. Lett.* **1989**, *55*, 631–633. [CrossRef]
14. Malakoutian, M.; Zheng, X.; Woo, K.; Soman, R.; Kasperovich, A.; Pomeroy, J.; Kuball, M.; Chowdhury, S. Low thermal budget growth of near-isotropic diamond grains for heat spreading in semiconductor devices. *Adv. Funct. Mater.* **2022**, *32*, 2208997. [CrossRef]
15. Petherbridge, J.; May, P.; Pearce, S.; Rosser, K.; Ashfold, M.J. Low temperature diamond growth using CO_2/CH_4 plasmas: Molecular beam mass spectrometry and computer simulation investigations. *Appl. Phys.* **2001**, *89*, 1484–1492. [CrossRef]
16. Goyal, V.; Sumant, A.V.; Teweldebrhan, D.; Balandin, A.A. Direct low-temperature integration of nanocrystalline diamond with GaN substrates for improved thermal management of high-power electronics. *Adv. Funct. Mater.* **2012**, *22*, 1525–1530. [CrossRef]
17. Su, C.H.; Chang, C.Y. Effect of CH/C_2 species density on surface morphology of diamond film grown by microwave plasma jet chemical vapor deposition. *Mater. Trans.* **2008**, *49*, 1380–1384. [CrossRef]
18. Redfern, P.; Horner, D.A.; Curtiss, L.A.; Gruen, D.M. Theoretical studies of growth of diamond (110) from dicarbon. *J. Phys. Chem.* **1996**, *100*, 11654–11663. [CrossRef]
19. Angus, J.H.; Hayman, C.C. Low-pressure, metastable growth of diamond and "diamondlike" phases. *Science* **1988**, *241*, 913–921. [CrossRef]
20. Stiegler, J.; Bergmaier, A.; Michler, J.; Laufer, S.; Dollinger, G.; Blank, E. The effect of nitrogen on low temperature growth of diamond films. *Thin Solid Film.* **1999**, *352*, 29–40. [CrossRef]
21. Martyanov, A.K.; Tiazhelov, I.; Savin, S.; Voronov, V.; Konov, V.; Sedov, V. Synthesis of polycrystalline diamond films in microwave plasma at ultrahigh concentrations of methane. *Coatings* **2023**, *13*, 751. [CrossRef]

22. Varga, M.; Potocky, S.; Tesarek, P.; Babchenko, O.; Davydova, M.; Kromka, A. Diamond growth on copper rods from polymer composite nanofibers. *Appl. Surf. Sci.* **2014**, *312*, 220–225. [CrossRef]
23. Chen, J.; Wang, X.; Zheng, Z.; Yan, F. Nucleation and growth of diamond on silicon substrate coated with polymer. *Thin Solid Film.* **1999**, *346*, 120–124.
24. Bianconi, P.; Joray, S.; Aldrich, B.; Sumranjit, J.; Duffy, D.; Long, D.; Lazorcik, J.; Raboin, L.; Kearns, J.; Smulligan, S.; et al. Diamond and diamond-like carbon from a preceramic polymer. *Am. Chem. Soc.* **2004**, *126*, 3191–3202. [CrossRef] [PubMed]
25. Tiwari, R.N.; Chang, L. Chemical precursor for the synthesis of diamond films at low temperature. *Appl. Phys. Express* **2010**, *3*, 045501. [CrossRef]
26. Chein, T.H.; Tzeng, Y. CVD diamond grown by microwave plasma in mixtures of acetone/oxygen and acetone/carbon dioxide. *Diam. Relat. Mater.* **1999**, *8*, 1393–1401. [CrossRef]
27. Li, Z.; Wang, L.; Suzuki, T.; Argoitia, A.; Pirouz, P.; Angus, J.C. Orientation relationship between chemical vapor deposited diamond and graphite substrates J. *Appl. Phys.* **1993**, *73*, 711–715. [CrossRef]
28. Xie, Y.P.; Zhang, X.J.; Liu, Z.P. Graphite to diamond: Origin for kinetics selectivity J. *Am. Chem. Soc.* **2017**, *139*, 2545–2548. [CrossRef]
29. Lambrech, W.; Lee, C.; Segal, B.; Angus, J.; Li, Z.; Sunkara, M. Diamond nucleation by hydrogenation of the edges of graphitic precursors. *Nature* **1993**, *364*, 607–610. [CrossRef]

C Journal of Carbon Research

MDPI

Article

Comprehensive Study on Carbon-Coated Silver for Improved Tribo-Electrical and Wetting Performance

Bruno Alderete *, Frank Mücklich and Sebastian Suarez

Department of Materials Science and Engineering, Saarland University, Campus D3.3, 66123 Saarbrücken, Germany; muecke@matsci.uni-sb.de (F.M.); s.suarez@mx.uni-saarland.de (S.S.)
* Correspondence: bruno.alderete@uni-saarland.de; Tel.: +49-681-3027-0544

Abstract: The rise in electrification has considerably increased the demand for high-efficiency and durable electrical contact materials. Carbon nanoparticles (CNP) are a promising coating material due to their intrinsic transport properties (thus minimizing the impact on conductivity), their proven solid lubricity (potentially improving tribological performance), and their hydrophobic wetting behavior (potentially providing atmospheric protection). In this study, carbon nanotube and nanohorn coatings are produced via electrophoretic deposition on silver-plated surfaces, followed by tribo-electrical and wetting characterization. The proposed coatings do not negatively affect the conductivity of the substrate, showing resistance values on par with the uncoated reference. Tribo-electrical characterization revealed that the coatings reduce adhesive wear during fretting tests while maintaining stable and constant electrical contact resistance. Furthermore, CNP-coated surfaces show a hydrophobic wetting behavior toward water, with graphite and carbon nanotube (CNT) coatings approaching super-hydrophobicity. Prolonged exposure to water droplets during sessile drop tests caused a reduction in contact angle (CA) measurement; however, CNT coatings' CA reduction after five minutes was only approximately 5°. Accordingly, CNP (specifically CNT) coatings show auspicious results for their application as wear and atmospheric protective barriers in electrical contacts.

Keywords: carbon nanotubes; carbon nanohorns; electrical contact resistance; graphite; tarnished silver

Citation: Alderete, B.; Mücklich, F.; Suarez, S. Comprehensive Study on Carbon-Coated Silver for Improved Tribo-Electrical and Wetting Performance. *C* **2024**, *10*, 16. https://doi.org/10.3390/c10010016

Academic Editor: Gil Goncalves

Received: 13 December 2023
Revised: 18 January 2024
Accepted: 1 February 2024
Published: 4 February 2024

1. Introduction

Silver is very reputed as an electrical contact material not only due to its extraordinary conductivity but also due to its chemical stability. A key component in silver's ubiquitous use in high-reliability connectors is that it does not oxidize. The lack of oxidation is advantageous since it precludes the formation of non-conductive or semi-conducting oxide films, as is the case with other commonly used contact materials (i.e., copper, tin, aluminum, etc.). Nonetheless, prolonged atmospheric exposure causes the formation of a semi-conductive silver sulfide (tarnishing) film—reaction shown in (1) [1–12]. Many methods have been developed to hinder or prevent this reaction, primarily in the form of protective coatings; however, alkanethiols have garnered the most widespread industrial adaptation [13–16]. The use of alkanethiols (e.g., hexadecanethiol and octadecanethiol, HDT and ODT, respectively) delays the formation of silver sulfide by saturating the silver surface with sulfhydryl groups. HDT and ODT consist of long chains of unsaturated oils where sulfhydryl groups bond with carbon terminals, thus allowing HDT and ODT to bond with transition metals. Sulfhydryl passivation forms Ag-S covalent bonds, forming a dense self-assembled monolayer, thus preventing atmospheric sulfur from reacting with the silver contact material by saturating the silver surface with the sulfur from the sulfhydryl groups [16]. However, the efficacy of these self-assembled monolayers is reliant on the orientation of the free-standing HDT or ODT tail. After the sulfhydryl group bonds with silver, the tail is free to rotate, allowing it to orient itself in an upstanding orderly position or lying on the silver surface in a disordered fashion. These monolayers are more efficient at preventing silver from tarnishing when the tails are oriented in an upstanding position due to increased thiol bonds with the substrate [16].

$$2Ag + S \rightarrow Ag_2S \quad (1)$$

Carbon nanoparticles (CNP) present a viable alternative to other passivation techniques which cannot only improve on the shortcomings of these techniques but could also present added advantages. Carbon nanostructures have excellent transport properties, a characteristic of utmost importance when dealing with electrical contact material [17–22]. Furthermore, nanocarbons present the advantage of outstanding lubricating capabilities, with extensive use as solid lubricants and lubricant oil additives [23–32].

Previous work carried out by Loyd et al. delineated the lubricating capabilities of graphite, graphene, and CNT specifically for their implementation in low-voltage and low-current applications [33]. Moreover, previous studies on CNP coatings over copper substrates have shown favorable results in terms of tribological and electrical performance, as well as for atmospheric protection due to the hydrophobic tendency of the CNP [34–37]. Therefore, it is the objective of this study to evaluate the impact, drawbacks, and advantages of CNP coatings on silver substrates. The CNP chosen for this analysis are carbon nanotubes (CNT) and carbon nanohorns (CNH), nanostructures that have shown the most promising results on copper substrates with regards to their effect on the system's conductivity, as well as wear and atmospheric protection. However, since silver is used in demanding, high-reliability applications, and considering the significant financial investment that silver contacts incur, it is crucial to comprehensively understand the impact that CNP coatings could have on this outstanding conductor, as well as the benefits that could be gained from CNP-coated silver surfaces. In order to accurately appraise the feasibility of the proposed coatings, benchmarking samples were also evaluated. Accordingly, the CNT and CNH coatings were compared with a reference sample, a tarnished sample (i.e., a silver surface with a thin silver sulfide tarnishing film), and a sample coated with graphite flakes (GF). The proposed coatings were not only compared to a clean reference sample (optimal conductivity) but also to a tarnished sample (worsened conductivity) to gain complete insight into the performance of the CNT and CNH coatings within this conductivity spectrum. Furthermore, GF-coated samples serve as an industrially accepted benchmark (due to graphite's solid lubricity and conductivity), which will better reflect on the feasibility and applicability of the proposed coatings.

Prior to characterization, all samples were chemically cleaned. The cleaned samples were subsequently tarnished or coated. The tarnishing process is a chemical process that involves subjecting the silver surfaces to a saturated sulfur atmosphere, thus rapidly generating a silver sulfide film on the sample's surface. The coatings were produced via electrophoretic deposition (EPD). This method was chosen due to its simplicity, ease in controlling coating thicknesses, its ability to coat geometrically complex surfaces, as well as its scalability. All surfaces were tribo-electrically characterized via load-dependent electrical contact resistance (ECR) and static ECR development during fretting wear. Furthermore, roughness, coating thickness, and assessment of the fretting marks were carried out via confocal laser scanning microscopy (CLSM). The latter were additionally characterized via scanning electron microscopy (SEM) and energy dispersive X-ray spectroscopy (EDS) for extensive chemical analysis. Moreover, wetting behavior was characterized via sessile drop test using two different media: deionized water and a potassium sulfide-deionized water solution. These characterization techniques allow for an in-depth analysis of the impact on the system's conductivity, as well as the potential wear and atmospheric protection provided by the CNP coatings in relation to the reference, tarnished, and GF-coated samples.

2. Materials and Methods

2.1. Sample Preparation

Three CNP coatings were the subject of this study (namely, GF, CNT, and CNH). The GF (Alfa Aesar GmbH, Kandel, Germany) have a median size between 7 and 10 μm a carbon purity of 99.8%. The CNT used were chemical vapor deposition-grown multi-walled CNT (Graphene Supermarket, New York, NY, USA). The as-received CNT have a length ranging from 10 to 15 μm, an outer diameter distribution between 50 and 85 nm, and a carbon purity over 94%. Furthermore, the CNH used are single-walled, dahlia-type nanostructures (Carbonium SRL, Brescia, Italy). These particles were produced without a catalyst by rapid condensation of carbon atoms, resulting in high carbon purity. Horn diameters range from 3 to 5 nm, with horn lengths from 30 to 50 nm and cluster diameters from 60 to 120 nm.

The substrates and reference material were silver-plated copper surfaces. A focused ion beam (FIB) cross-section was conducted (FEI Helios NanoLab600 Dual Beam Setup Hillsboro, OR, USA) to determine the plating thickness. The cross-section revealed an average silver plating thickness of 4 μm, as well as a nickel interlayer between the silver and copper base material (see Figure S1). Due to the thinness of the silver plating, the samples were neither ground nor polished to avoid the removal of the plating material Vickers hardness of the substrate was measured using a microhardness tester (Dura Scan 50, Struers Inc., Cleveland, OH, USA). A total of 35 indentations were carried out using a load of approximately 2 N and a holding time of 15 s. The imprints were optically observed using a 40× objective, resulting in an average hardness of 1.33 ± 0.04 GPa.

Accelerated corrosion (sulfidation) of the silver surfaces was carried out using the methodology proposed by Huo et al. [38]. This method consists of subjecting the silver surfaces to a saturated atmosphere of sulfur vapor. This methodology ensures a well-controlled and repeatable sulfur-rich atmosphere, which promotes and accelerates the formation of the silver sulfide tarnishing film. Therefore, an oversupply of sulfur powder (8 g) was placed in a ceramic container within a beaker. The vessel was covered with a watch glass and heated to 150 °C, thus evaporating the sulfur and saturating the atmosphere within the vessel. Thereupon the silver samples were placed inside the sulfur-rich atmosphere for 10 min. After the accelerated tarnishing process, the samples present a highly heterogeneous silver sulfide layer (see Figure S2). To ensure reproducibility, all tarnished samples used in this study were sulfurized simultaneously. Therefore, the degree to which the samples were tarnished did not significantly differ, thus guaranteeing similar behavior regardless of the sample that was used.

EPD was carried out at a constant voltage of 300 V (potentiostatic process) with an inter-electrode distance of 15 mm. Prior to the deposition process, the CNP must be dispersed in isopropyl alcohol by homogenization and ultrasound [35,36]. Triethylamine ($C_6H_{15}N$) was added to the colloid prior to dispersion. The use of this additive enhances the stability of the colloidal dispersion as well as providing the CNP with a superficial charge through the attachment of functional groups (i.e., positive charge, anodic deposition). The colloid, dispersion, and deposition parameters for all four sample types are detailed in Table S1 [35]. The four resulting coatings are a 10 min GF coating, a 5 min CNT coating, a 10 min CNT coating, and a 10 min CNH coating (henceforth GF10, CNT5, CNT10, and CNH10, respectively). To ensure consistent results, a fresh dispersion was prepared prior to each coating process. Consequently, the CNP's concentration and dispersion quality of the colloid can also be considered constant. Although the use of other solvents (e.g., ethylene glycol) results in a more stable dispersion [39], isopropyl alcohol was chosen due to its lower viscosity. A solvent with lower viscosity will minimize viscous drag in the colloid, which will consequently minimize the CNP's resistance toward electrophoresis, thus improving the deposition rate. Moreover, the lower vapor pressure of isopropyl alcohol guarantees fast and efficient drying of the coated samples after removal from the colloid.

Prior to EPD and accelerated tarnishing, the silver platelets were chemically cleaned to remove any passivating elements that were added (e.g., octadecanethiol) as a final step in the manufacturing process. The cleaning process consists of reducing the silver sulfide layer formed during passivation [40,41]. Removing any potential passivation elements on the samples' surface ensures that maximum electrophoresis is achieved during EPD and that the accelerated tarnishing process is effective without requiring excessive exposure times. After chemical cleaning, the samples were thoroughly rinsed with deionized water, followed by ultrasonication in isopropyl alcohol for 5 min, thus removing potential residue from the cleaning process.

2.2. Characterization Techniques

The substrate, tarnished, and coated samples were micrographed via CLSM (OLS4100, Olympus, Tokyo, Japan). CLSM micrographs were acquired using 20× and 50× magnification and a laser wavelength of 405 nm. These micrographs were used to analyze the roughness of the samples (shown in Table 1) and coating thicknesses (shown in Table 2), as well as to qualitatively evaluate fretting marks on both electrodes. Furthermore, fretting marks were also micrographed via SEM (Helios™ G4 PFIB CXe DualBeam™ Super FIB/SEM equipped with an energy dispersive X-ray spectroscopy detector EDAX Octane Elite Super, Thermo-Fisher Scientific, Eindhoven, The Netherlands) using ion conversion and electron (ICE) detector with an acceleration voltage of 5 kV. The same instrument was used to carry out chemical composition maps of the wear marks via EDS at an acceleration voltage of 15 kV. With this technique, a qualitative analysis of material transfer can be conducted. Furthermore, it can be used to determine if the wear resulting from the fretting tests was severe enough to reach the copper base material.

Table 1. Root mean squared roughness of samples, measured using CLSM.

Surface	Roughness/μm
Reference	0.26 ± 0.03
Tarnished	0.40 ± 0.04
GF10	1.40 ± 0.12
CNT5	0.91 ± 0.16
CNT10	0.62 ± 0.02
CNH10	0.51 ± 0.01

As the results from Table 2 show, the resulting coating thicknesses are relatively thin. Previous studies evaluating CNP-coated copper substrates using the same deposition technique produced considerably thicker coatings [35,37,42,43]. It was hypothesized that residue left behind by the chemical cleaning process might affect the conductivity of the silver samples, thus reducing the deposition rate of CNP onto their surfaces. To rule this possibility out, an as-received silver sample was coated—without prior cleaning—using the same parameters. The resulting coating thickness measured coincided with the samples that were chemically cleaned. The high roughness present in the silver samples could also play a role in the thinness of the coatings (260 nm against 10–20 nm in mirror-polished copper) since the electric field generated during EPD is concentrated at the asperities. The field's concentration can thus disrupt and diminish the deposition rate of the CNP. However, this is unlikely to be the cause since it has been previously studied and reported that EPD has the capability to coat geometrically complex surfaces [44–46].

Table 2. Coating thicknesses measured via CLSM.

Coating	Thickness/μm
GF10	0.34 ± 0.10
CNT5	0.14 ± 0.04
CNT10	0.41 ± 0.05
CNH10	0.32 ± 0.08

Wetting behavior was characterized by measuring the contact angle (CA) via sessile drop test using a drop shape analyzer (Krüss—DSA100B, Hamburg, Germany). A 3 μL droplet of deionized water and potassium sulfide (K_2S) was placed onto the surfaces via a pipette. The volume of 3 μL suppresses gravitation distortion of the droplet's shape and, at the same time, covers a sufficiently large area on the sample's surface to even out topographical or chemical inhomogeneities [47,48]. The potassium sulfide solution consists of dissolving the compound in deionized water at a concentration of 3 wt.% at 23 °C (according to factory standard LK2070). Once each droplet was settled onto the sample's surface, the CA was recorded by fitting its contour using an elliptical shape within the ADVANCE Software (version 1.14.3, Krüss, Germany). At least two droplets were measured per sample, measuring CA after 10, 30, 60, and 300 s. Measuring at different time intervals provides insight into the dynamic, time-resolved wetting performance of the surfaces in question and its reliance on the liquid media used during sessile drop test. All CA measurements were carried out at temperatures and humidity ranging from 19–20 °C and 24–26% RH, respectively.

Tribo-electrical characterization of the reference, tarnished, and CNP-coated samples was conducted using a custom testing rig [49]. Characterization was divided into two experimental methodologies: i.e., load-dependent ECR and ECR evolution during fretting test. Load-dependent ECR consisted of one loading and unloading cycle sourcing 100 mA_{DC} and measuring voltage drop between the electrodes (loading cycle: 0.5, 1, 2, 3, 4, 5, 6, 7, 8, 9, 10 N). This current level was chosen so that the measurement process does not cause microstructural modification of the contacting surfaces, thus affecting subsequent measurements (i.e., dry circuit conditions) [50]. In this loading range, a maximum Hertzian contact pressure of 230 and 623 MPa were achieved for 0.5 and 10 N, respectively. These values were determined using Hertzian contact theory for sphere-on-plane contact [51,52]. Fretting tests were carried out at a load of 1 N (maximum Hertzian contact pressure of 289 MPa), with an amplitude of 35 μm and an oscillation frequency of 8 Hz. A total of 5000 fretting cycles were conducted with periodic galvanostatic (100 mA_{DC}) ECR measurements carried out every 100 cycles (static ECR). All tribo-electrical tests were carried out using a Keithley 2400 source and a Keithley 2182a nanovoltmeter with a range of 0.1 V (Cleveland, OH, USA). In all cases, the voltage drop was recorded ten times and averaged, with each test being repeated at least three times to ensure reproducibility. Tribo-electrical characterization was conducted under ambient conditions, i.e., 21 ± 1 °C and 35 ± 5% RH. Since humidity plays an important role in the conductivity of CNT [44,53–56], all electrical measurements for each coating type were conducted on the same day. Therefore, errors in the measurements due to changes in atmospheric conditions were minimized.

The counter electrodes used for tribo-electrical characterization are silver-nickel core ($AgNi_{0.15}$) and hard-gold-coated ($AuCo_{0.2}$) rivets (Adam Bornbaum GmbH, Neuhausen, Germany). The hard-gold coating has an average thickness of 6.47 ± 0.18 μm. The rivets have a hemispherical head, with a mean radius of curvature at its tip of 4 mm, a root mean squared roughness of 0.26 μm (measured via CLSM), and a hardness of 1.38 ± 0.01 GPa (measured via Vickers microhardness). A new rivet was used for each measurement.

3. Results and Discussion

3.1. Load-Dependent ECR

Electrical characterization is an important component when considering coatings for electrical contacts since these layers play a pivotal role in the conductivity and overall efficiency of the connector. A key aspect that can be tailored during connector design is the contact load. Therefore, load-dependent ECR curves of the materials used provide crucial information for connector and connector housing design. The ECR curves for the reference, tarnished, and coated samples are shown in Figure 1. Immediately, it is clear that the tarnished sample stands out from the reference and coated samples. Throughout the loading range, the tarnished sample consistently shows ECR values above 20 mΩ, whereas the reference and coated samples show ECR values between 11 and 20 mΩ. It is interesting to highlight the variability in the ECR of the tarnished sample in the loading semi-cycle, whereas the ECR values are constant in the unloading semi-cycle. Between 0.5 N and 2 N, there is a sharp reduction in ECR. This is due to the puncturing of the silver sulfide tarnishing film by the counter electrode's asperities [41,57]. However, as loads gradually increase (above 2 N), so does the ECR. The counter electrode's geometry plays a role in this observation. As the load increases, the apparent contact diameter increases on account of the hemispherical shape of the rivets used. As the apparent contact area increases, so do the contacting sites between the counter electrode and the less-conductive (semi-conductive) tarnishing film, increasing the quasi-metallic contact [57,58]. The newly formed quasi-metallic contacts between the rivet's lower asperities and the tarnished sample increase the predominance of the film resistance, thus playing a more crucial role in the overall electrical resistance of the system. For sufficiently high loads (from 4 N), however, the resistance of the film takes a less dominant role due to the puncturing of the film and well-established metallic contact between the counter electrode and the silver surface beneath the silver sulfide film. This results in the constant ECR measured from 4 N (in the loading semi-cycle) until the end of the measurement cycle. This behavior in the tarnished sample is solely caused by the puncturing of the tarnishing film since the sharp reduction in ECR is characteristic of this sample, with the clean silver sample not presenting this behavior. In other words, the asperities in contact with the silver sulfide film undergo additional constriction of the current, which is eased as the film is punctured by the individual asperities (between 0.5 and 2 N). The additional constriction of the current is a consequence of the higher resistivity of the tarnishing film compared with the resistivity of the bulk material. Between 2 N and 4 N, new quasi-metallic contacts are established between lower asperities and the tarnishing film, once again increasing constriction resistance's dominance in the system. For loads above 4 N, most of the asperities have punctured the film, resulting in lower constriction resistance due to the current flowing from the counter electrode's asperities toward the silver bulk material. Moreover, the current level used to measure ECR is sufficiently high to overcome the loss of ohmicity produced by not only the tarnishing film but also other contaminant films and the CNP coatings (current-dependent ECR curves shown in Figure S3).

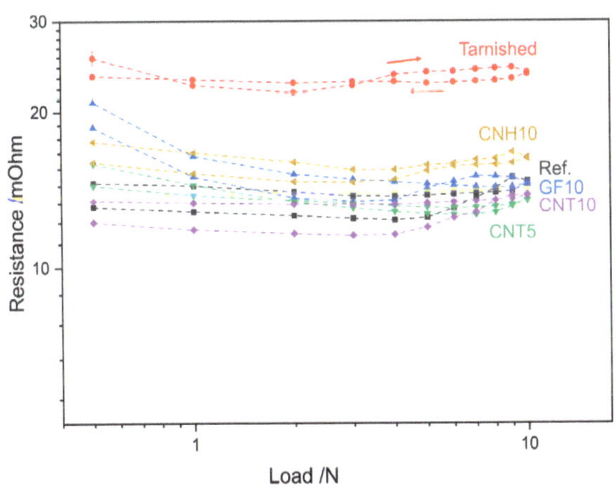

Figure 1. Load-dependent ECR of reference, tarnished, and coated samples. The arrows indicate the direction in which the loading and unloading semi-cycles progress.

The CNH coating shows the highest resistance among the coated samples, with the only exception being the GF coating at low loads. However, at 10 N, the ECR of CNH10 is only 2 mΩ higher than the reference sample at the same load. Therefore, the CNH coating does not significantly impact the electrical performance of the silver substrate. Nonetheless, the CNH coating is considerably thin (approximately 0.32 µm). Moreover, the CNH coating shows the most similarity in terms of ECR values between loading and unloading semi-cycle. This is due to the high elasticity of the coating, thus showing low electrical hysteresis. In this regard, the use of CNH coatings for electrical contacts could promote reproducible contact areas irrespective of the normal load applied while also mitigating topographic factors [48]. The high elasticity observed in CNH coatings is due to two factors: (1) the nanoparticles' morphology and (2) the microporous network obtained by EPD. The former allows the CNT-like horns to deform inward from tip to base when undergoing compressive stress, thus enabling this nanostructure to return to its original shape after removing the compressive load (elastic restitution) [59]. This characteristic of the CNH justifies the low electrical hysteresis observed. The latter is a characteristic of the coatings obtained. The CNH coatings do not present large voids (as is the case with other CNP coatings) but rather a complex microporous network (see coating cross-section in [35]). This allows the voids to be rapidly closed during compaction with minimal CNH readjustment required, thus explaining the relatively constant ECR values measured during loading and unloading.

GF produces the thickest CNP coating; nonetheless, the resistance values are not severely impacted by the coating for loads above 1 N. Although the ECR values for the GF-coated sample are higher than the reference, the ECR coincide at 9 and 10 N. The CNT-coated samples, on the other hand, show exceptional electrical performance. The thinner CNT coating shows marginally higher ECR values than the reference sample; however, past 5 N CNT5 shows lower ECR. The thicker CNT coating outperformed the reference sample for the entire loading range, highlighting the promising potential of this coating type. The improved performance can be explained by two aspects: (1) the conductivity and (2) the elasticity of the CNT. These two aspects combined increase the real contact area between the counter electrode and the sample. The increased conductivity of the nanotubes, in conjunction with their elasticity, enables more pathways for electron motion between the two electrodes. In other words, although the apparent contact area may be the same, the highly elastic CNT coating increases the real contact area between the electrodes, thus reducing the ECR. Since CNT5 is thinner than CNT10, this phenomenon does not occur initially, thus requiring higher loads to establish an adequate electrical contact between the nanotubes and the counter electrode. Furthermore, the CNT tend to adhere to the counter electrode after loading, which also promotes lower electrical hysteresis and overall ECR (CLSM micrographs of counter electrodes post-ECR are shown in Figure S4).

3.2. ECR Evolution during Fretting Test

Tribo-electrical characterization aims to determine the effectiveness of the proposed coatings at reducing wear with negligible influence on ECR. Consequently, tracking the evolution of ECR as fretting cycles progress is of utmost importance. The results are shown in Figure 2. The plot depicts that irrespective of the sample type, all samples showed higher ECR initially and a slight reduction as fretting cycles progressed. However, this was to be expected since prior to fretting, the ECR is a combination of the film resistance (contaminant film, tarnishing film, or the coating itself), constriction resistance, and bulk material resistance. After the fretting test begins, contaminant films are constantly being broken down. Therefore, the resistance after the initial fretting cycles is primarily due to constriction resistance and the bulk material's resistance. Particularly in the case of the coated samples, the CNP are displaced due to the fretting motion, which reduces the amount of CNP present at the contacting site, thus marginally reducing ECR [42,43]. It is noteworthy to highlight, however, that the ECR values after the first few hundred fretting cycles are highly stable for all samples. In other words, there are no abrupt fluctuations in ECR due to fretting. This could be related to the fact that silver and gold do not form oxide films (as opposed to other metals, e.g., copper). It could also be related to the generation of a stable conductive carbonaceous tribofilm, which stabilizes the interface. Figure 2 exhibits that two distinct groupings took place. Namely, the tarnished and GF-coated samples present higher ECR during fretting than the reference, CNT, and CNH-coated samples. As was the case with load-dependent ECR (Figure 1), both CNT coatings outperformed the reference sample. CNT5 initially has a slightly higher ECR than the reference sample. However, after the first one hundred fretting cycles, it quickly falls below the values observed in the reference sample. CNT10, on the other hand, has a lower ECR throughout the measurement. The CNH coating, nonetheless, also outperformed the reference sample during fretting tests with an electrical performance similar to that of CNT5, albeit this could be caused by the heterogeneous nature of the EPD coatings and should not be over-interpreted.

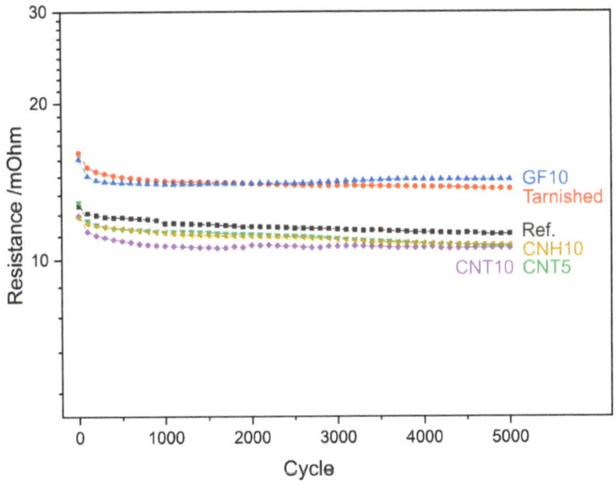

Figure 2. Evolution of ECR as fretting cycles progress of reference, tarnished, and coated samples. Note: error in all ECR measurements below ±0.2 mΩ.

Post-fretting micrographs of the marks left behind on the samples and counter electrodes are shown in Figure 3. In the reference sample, a considerable amount of material was transferred from the counter electrode toward the sample (Figure 3a), whereas marginal amounts of silver were transferred toward the counter electrode (Figure 3b). Silver debris is found primarily in the perimeter of the fretting marks and, to a minor extent, within the gross slip regions in the center of the fretting mark [60]. Gold is only observed adhered onto the sample's surface without gold debris surrounding the fretting marks. Bidirectional material transfer was expected during wear tests since the silver sample and the hard-gold counter electrode have similar hardness values (less than 10% difference—1.33 and 1.38 GPa, respectively) [60]. The tarnished sample fretting mark (Figure 3c,d) strongly contrasts the wear track observed in the reference sample. Although both fretting marks are approximately the same size, the mark on the tarnished sample does not have adhered gold. Furthermore, the counter electrode that contacted the tarnished sample shows minimal silver debris particles. The lack of material transfer can be explained by the lubricating capacity of the silver sulfide tarnishing film, albeit the lubricating capacity of the tarnishing film does not significantly reduce worn area [41]. Nonetheless, the presence of the tarnishing film strongly reduces gross slip, with both the sample and the counter electrode showing minimal wear. Furthermore, black particles observed in the counter electrode are silver sulfide particles that adhered during tribological testing.

The micrographs of the GF-coated sample (Figure 3e,f) depict that considerable gross slip took place in the central region of the fretting mark (in both the sample and the counter electrode). This is likely due to the ease of GF displacement and the detachment of the GF coating. Nonetheless, as shown in Figure 3f, large amounts of detached GF particles adhered to the counter electrode (lower-right region of the fretting mark in Figure 3e,f). Consequently, minimal amounts of GF remain within the fretting mark, therefore reducing this coating's capacity to reduce wear. CNT and CNH-coated samples' fretting marks (Figure 3g–i) highly contrast with GF-coated samples. These coatings do not show significant displacement or detachment. Consequently, after 5000 fretting cycles, there are regions within the fretting marks that still have CNP which provide protection from wear (regions highlighted by cyan ovals in CLSM micrographs). Consequently, the CNPs within the fretting mark protect from wear, as proven by the enclosed region in the counter electrodes (i.e., regions where partial slip or no damage took place). Moreover, traces of CNT and CNH adhered to the counter electrode, further promoting wear protection. However, the amounts of transferred CNT and CNH are negligible compared to the quantity of GF that was transferred.

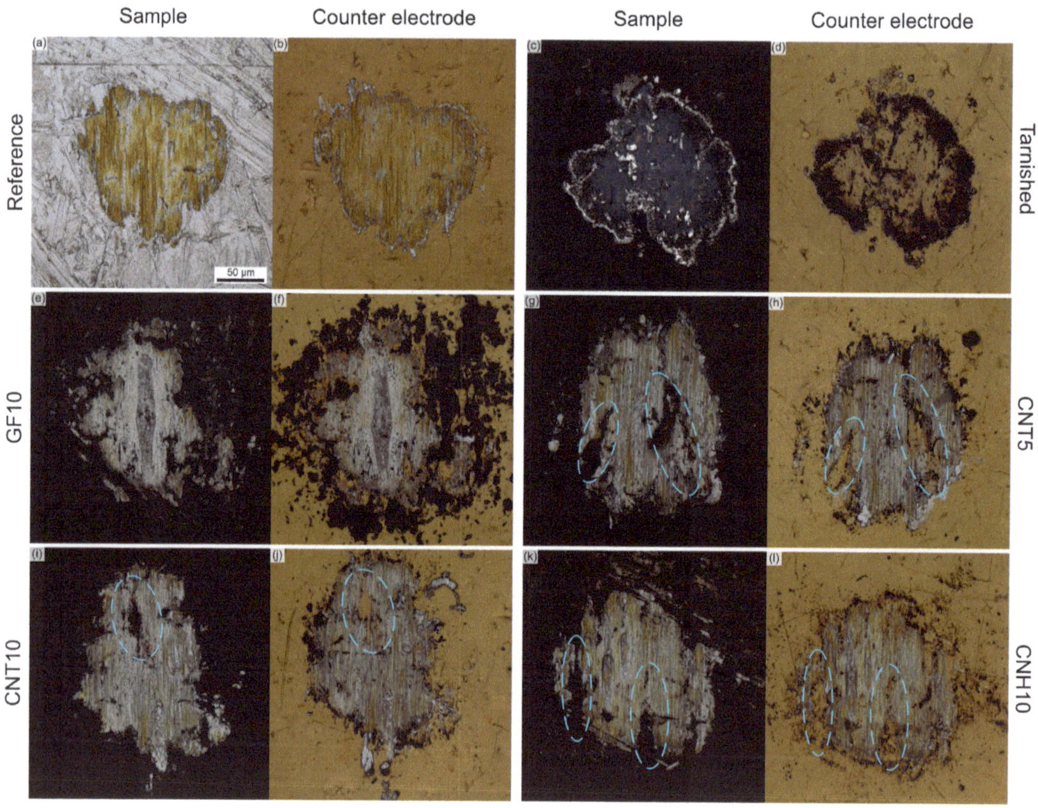

Figure 3. CLSM micrograph after 5000 fretting cycles of reference (**a**) sample and (**b**) counter electrode, tarnished (**c**) sample and (**d**) counter electrode, GF10 (**e**) sample and (**f**) counter electrode, CNT5 (**g**) sample and (**h**) counter electrode, CNT10 (**i**) sample and (**j**) counter electrode, and CNH10 (**k**) sample and (**l**) counter electrode. Cyan dashed line highlights regions in the sample where CNP remain within the fretting mark and their corresponding region in the counter electrode where only partial slip took place rather than gross slip.

A key distinction between the reference sample and the CNP-coated samples is the permutation of the dominant transfer electrode. In other words, the former shows that gold was predominantly transferred from the counter electrode toward the silver sample, whereas the latter shows that silver was predominantly transferred from the sample toward the counter electrode. Silver was not transferred in the form of debris (there are minimal amounts of silver debris present in the CNP-coated samples), but rather, silver adhered to the gold counter electrode. The material transfer observed during fretting is caused by adhesive wear, in other words, cold welding. After initial wear, contaminant films are removed from both surfaces (silver and gold). As fretting cycles progress, the elimination of the contaminant films enables the adhesion of the two metals [61]. However, the presence of carbon in the coated samples reduces the microscopic temperature of the asperities in the silver sample due to the outstanding thermal conductivity of the CNP. Furthermore, the carbon coating effectively separates the two contacting surfaces, reducing the likelihood of metal-metal contact but rather generating a contacting site constituted of metal-carbon-metal. Therefore, it is reasonable to infer that the permutation of the deposition electrode is caused by the increased temperature in the counter electrode and a consequent reduction on the silver surfaces. This hypothesis is further supported by the lack of material transfer observed in the tarnished sample (Figure 3c,d). Since the silver sulfide tarnishing film lubricates the contact site, gold adhesion did not take place in the silver sample, whereas small silver debris particles can be seen in the counter electrode's micrographs surrounding the fretting mark.

To further analyze the mechanism behind material transfer, EDS mappings were carried out. The EDS maps are shown in Figure 4, with the corresponding secondary electron micrograph shown in Figure S5 and the sulfur EDS map of the tarnished sample shown in Figure S6. The EDS maps confirm that gold was transferred from the counter electrode towards the CNP-coated silver samples. Although EDS is a qualitative technique, it is accurate to state that the presence of the CNP coatings reduces the amount of gold transferred onto the samples due to the lack of gold signal detected within the fretting marks in spots where carbon and/or silver signals are stronger. Furthermore, the EDS maps of the tarnished sample (Figure 4b) also confirm that the lubricating capacity of the silver sulfide layer prevents the adhesion of gold. Moreover, the EDS maps prove that after 5000 fretting cycles, the base material was not reached in any of the samples analyzed. Although copper was detected within the fretting mark in the CNT5 sample (Figure 4d), the areas enclosed by the white dashed lines showcase that the plating material's thickness throughout the samples is heterogeneous. Consequently, there are regions (unaffected by fretting wear) where the copper base material was detected via EDS. Therefore, the copper detected within the fretting mark in CNT5 cannot be attributed to a lack of wear protection from the coating but rather heterogeneity in the plating process.

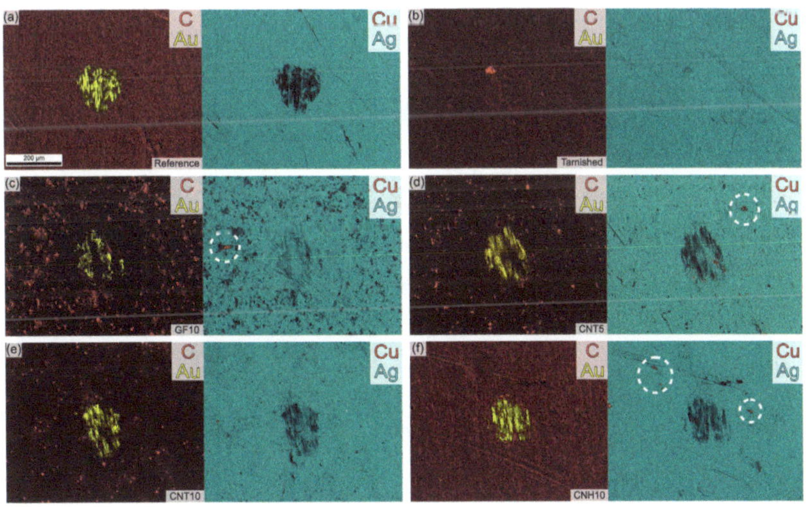

Figure 4. EDS elemental maps of (**a**) reference, (**b**) tarnished, (**c**) GF10, (**d**) CNT5, (**e**) CNT10, and (**f**) CNH10 samples. The left image shows an overlay of carbon (red) and gold (yellow), whereas the right image shows an overlay of silver (turquoise) and copper (orange). The regions enclosed by the dashed lines are regions unaffected by fretting wear where the copper base material was detected.

3.3. Wetting Behavior

Water contact angle measurements were carried out to evaluate the wetting behavior of the reference, tarnished, and CNP-coated surfaces (Figure 5). All surfaces show a hydrophobic wetting behavior within five minutes of placing the water droplet. Nonetheless, the reference sample shows a marginally hydrophobic behavior, with stable CA values of approximately 100° within the first minute. After five minutes, however, the CA decreases to approximately 92°, thus showing borderline hydrophobicity. The tarnished sample initially shows a higher CA; however, the value quickly decreases as time passes, reaching 90° after five minutes. GF and CNT show a strong hydrophobic behavior, approaching superhydrophobic wetting behavior after prolonged droplet exposure. CNH also shows high CA values; however, its performance does not reach as high CA as GF and CNT. The slightly stronger affinity toward water from CNH can be caused by the particle's morphology. Due to the dahlia-type nanostructures, the nanohorns have a strong strain-induced curvature, which delocalizes the high mobility π orbitals, thus increasing this nanostructure's reactivity and consequently affecting its wetting behavior [62]. Nonetheless, CNH coatings produced via EPD generate a coating with a complex microporous network [35]. Therefore, despite the reactivity of the CNH, coating morphology prevents the penetration of the water droplet, consequently producing a surface-droplet interface that resembles a Cassie-Baxter wetting state. Based on a previous study, thicker CNH coatings do not significantly impact its wetting behavior [43]. Therefore, even for thin CNH coatings, the microporous network is quickly established and is effective at preventing droplet infiltration.

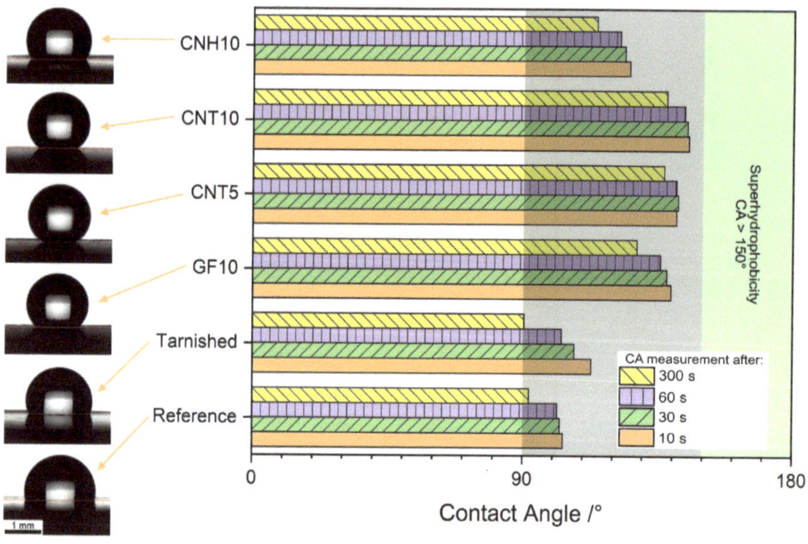

Figure 5. Water contact angle measurements 10 s (orange), 30 s (green), 60 s (violet), and 300 s (yellow) after placing the droplets onto the surfaces. To the left, the macrographs of a droplet on each surface and the baseline (red) are shown. The shaded regions highlight the surfaces with hydrophobic (gray) and superhydrophobic (green) wetting behavior.

GF shows near-superhydrophobic wetting behavior; however, after five minutes, the CA decreases by approximately 10°. This reduction indicates that the water droplet is slowly infiltrating the coating, primarily due to pores on the surface of the coating. Notwithstanding, the complexity of the porous network within the coating retains the hydrophobic wetting behavior of the GF-coated surface, resembling the performance observed for the CNH-coated surface. On the other hand, both CNT coatings show, to a certain extent, stable

CA values within the timeframe evaluated. After five minutes of setting the droplet onto the CNT surfaces, the CA decreases by approximately 5° as opposed to over 10° in the GF- and CNH-coated surfaces. This performance highlights the hydrophobic nature of the nanotubes, with thicker coatings showing a tendency toward marginally higher CA. Therefore, prolonging the deposition process (and thus obtaining a thicker coating) could promote super-hydrophobicity in the CNT-coated surfaces. Moreover, a smoother substrate could further increase the CA, improving hydrophobicity [63]. Additionally, Korczeniewski et al. previously reported that larger CNT agglomerate reduces CA, however, prolonging dispersion times (homogenization and ultrasound) negatively impacts the structure of the nanotubes [39,64]. The damage incurred by the CNT (in the form of dangling bonds, among other structural defects) could prove counterproductive by increasing the nanostructure's reactivity, thus reducing its CA.

Based on the wetting behavior observed with water droplets, it was of interest to evaluate the effectiveness of the coatings with a harsher medium (i.e., potassium sulfide). As was done with deionized water, 3 μL of the potassium sulfide solution was pipetted onto the six surfaces (CA results shown in Figure 6). However, the CNP-coated surfaces did not prevent the infiltration of the solution. The droplet rapidly penetrated the coatings and dispersed over the surfaces, producing CA below 40°. The GF showed an improved performance compared to the reference and tarnished samples. After 10 s, GF showed a CA of approximately 60°. However, the CA rapidly decreased to approximately 50° after 60 S and below 40° after 5 min. Nonetheless, this is a higher CA of 10–20° after five minutes of exposure to the highly aggressive potassium sulfide solution.

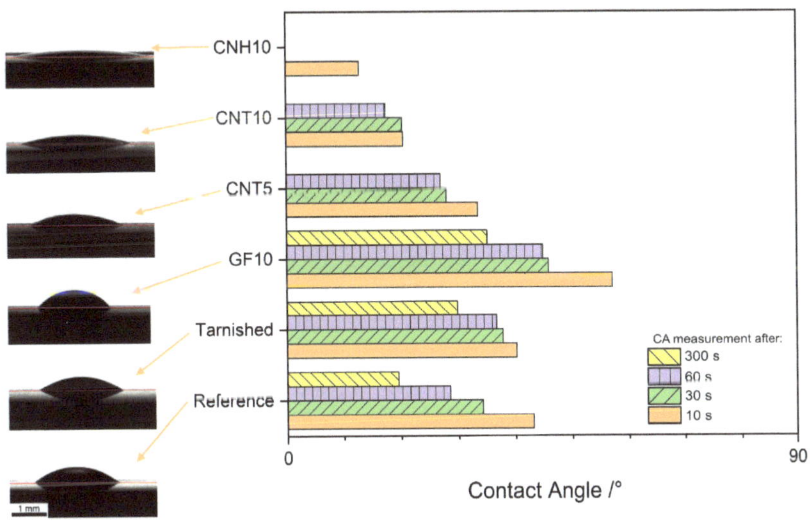

Figure 6. Contact angle measurements 10 s (orange), 30 s (green), 60 s (violet), and 300 s (yellow) after placing a 3 μL droplet of K_2S 3 wt.% solution onto the surfaces. To the left, the macrographs of a droplet on each surface and the baseline (red) are shown. Due to the hydrophilic wetting behavior, it was not possible to measure CA after 300 s in CNT5 and CNT10. Only CA after 10 s could be measured in CNH10.

Although the wetting results obtained when measuring CA with potassium sulfide do not suggest a favorable outcome, it should be noted that the solution is an extreme case used to analyze the potential protection offered by the carbon coatings. In other words, this is an accelerated method used to simulate prolonged exposure to sulfur-rich atmospheres. As previously studied, nanocarbon coatings show exceptional wetting behavior when measuring CA with other media than water (i.e., salt water) [43]. Therefore, the results from the potassium sulfide tests should be used to fully dismiss the potential use cases of carbon coatings, particularly GF.

4. Conclusions

In this work, silver-plated electrical connectors were coated with CNT and CNH. The tribo-electrical and wetting performance of these coatings was compared to reference, tarnished (corroded), and GF-coated samples. Load-dependent electrical contact resistance proves that the presence of thin CNP films does not hinder the efficiency of the connectors. CNT-coated samples showed exceptional electrical performance, with similar ECR values to that of the reference sample throughout the entirety of the loading ranges analyzed. Although CNH and GF presented slightly higher ECR values than the reference, these two coatings did not increase the resistance as significantly as the corrosion film (silver sulfide tarnishing film). Likewise, the tarnished sample showed poor electrical performance during fretting wear tests. The tarnishing film, however, protects the silver surface from severe wear, minimizing adhesive wear and gross slip. The CNP-coated surfaces reduced material transfer during fretting while maintaining consistently low ECR. However, coating displacement hinders wear protection (primarily affecting GF coatings). Certain regions of the CNT and CNH-coated samples (and counter electrode) were protected from severe fretting wear due to the presence of these nanostructures within the contacting site after 5000 fretting cycles. Moreover, the CNP-coated surfaces exhibit a hydrophobic (near-superhydrophobic in the case of GF and CNT) wetting behavior in sessile drop tests measured using deionized water, outperforming the tarnished and reference samples. Prolonged water droplet exposures tend to decrease the CA measured; however, the reduction in CA for the CNT-coated samples is significantly lower than in the other surfaces studied. Moreover, the CA measurements with potassium sulfide indicated that although CNT and CNH do not offer significant protection, GF coatings did show higher CA than the reference and tarnished samples, even after five minutes of exposure. Nonetheless, it should be reiterated that the medium used (i.e., potassium sulfide) is extremely aggressive and was chosen based on the fact that this solution would simulate prolonged exposure to sulfur-rich atmospheres.

This study features the exceptional characteristics of nanocarbon coatings and their potential benefits for the electrical contact industry. However, further studies are required to improve on the aspects presented in this work. Namely, improving wear and atmospheric protection without impacting electrical performance. The former may be improved by evaluating different EPD additives that improve coating adhesion (e.g., magnesium nitrate), thus reducing coating displacement during wear tests. The latter, on the other hand, may benefit from chemical alterations of the CNP (e.g., CNP decoration), consequently tailoring the reactivity of the nanostructures for specific atmospheric conditions.

Supplementary Materials: The following supporting information can be downloaded at: https://www.mdpi.com/article/10.3390/c10010016/s1, Figure S1: FIB cross section of silver-plated copper electrical contact. The platinum layer was deposited using the SEM with the purpose of protecting the sample and to minimize the curtaining effect during ion milling; Figure S2: Silver sample (a) before and (b) after being subjected to sulfur-rich atmosphere. Highlighted regions show SEM micrographs of the surface before and after the accelerated tarnishing process; Table S1: Colloid, dispersion, and EPD parameters; Figure S3: Current-dependent ECR of references, tarnished, and coated samples; Figure S4: CLSM micrograph of counter electrodes after ECR measurements of (a) reference, (b) tarnished, (c) GF10, (d) CNT5, (e) CNT10, and (f) CNH10 sample; Figure S5: Secondary electron micrograph of fretting marks after 5000 cycles of (a) reference sample, (b) tarnished sample, (c) GF10, (d) CNT5, (e) CNT10, and (f) CNH10. Micrographs (a) and (b) were acquired at 15 kV due to the presence of surface contaminants, whereas the rest were acquired at 5 kV acceleration voltage; Figure S6: EDS map of the fretting mark after 5000 cycles in the tarnished sample showing the signal for sulfur.

Author Contributions: Conceptualization, B.A. and S.S.; Data curation, B.A.; Formal analysis, B.A.; Funding acquisition, F.M. and S.S.; Investigation, B.A.; Methodology, B.A. and S.S.; Project administration, F.M. and S.S.; Resources, F.M.; Supervision, S.S.; Validation, B.A.; Visualization, B.A.; Writing—original draft, B.A.; Writing—review and editing, S.S. All authors have read and agreed to the published version of the manuscript.

Funding: This research received no external funding.

Data Availability Statement: Data available upon reasonable request from the corresponding author.

Acknowledgments: B. Alderete wishes to acknowledge the support from the German Academic Exchange Service (DAAD) and the Roberto Rocca Education Program (RREP). The authors gratefully acknowledge funding in the ZuMat project, supported by the State of Saarland from the European Regional Development Fund (Europäischen Fonds für Regionale Entwicklung, EFRE). Funding for the PFIB/SEM instrument by the German Research Foundation is greatly acknowledged (INST 256/510-1 FUGG).

Conflicts of Interest: The authors declare no conflict of interest.

References

1. Outka, D.A.; Madix, R.J.; Fisher, G.B.; DiMaggio, C. Oxidation of Sulfur Dioxide on Silver (110): Vibrational Study of the Structure of Intermediate Complexes Formed. *J. Phys. Chem.* **1986**, *90*, 4051–4057. [CrossRef]
2. Bauer, R. Sulfide Corrosion of Silver Contacts during Satellite Storage. *J. Spacecr. Rocket.* **1988**, *25*, 439–440. [CrossRef]
3. Lilienfeld, S.; White, C.E. A Study of the Reaction between Hydrogen Sulfide and Silver. *J. Am. Chem. Soc.* **1930**, *52*, 885–892. [CrossRef]
4. Russ, G. Electrical Characteristics of Contacts Contaminated with Silver Sulfide Film. *IEEE Trans. Parts Mater. Packag.* **1970**, *6*, 129–137. [CrossRef]
5. Hindin, B.S. Silver Sulfide Corrosion Control Using Corrosion Prevention Compounds. In Proceedings of the CORROSION 2006, San Diego, CA, USA, 12–16 March 2006; p. NACE-06264.
6. Denaburg, C.R. Corrosion in Aerospace Electrical/Electronic Components in the Early Years of the Apollo Program, Kennedy Space Center, Florida. In Proceedings of the CORROSION 2000, Orlando, FL, USA, 26–31 March 2000; p. NACE-00718.
7. Schmitt, W.; Franz, S.; Heber, J.; Lutz, O.; Behrens, V. Formation of Silver Sulfide Layers and Their Influence on the Electrical Characteristics of Contacts in the Field of Information Technology. In Proceedings of the ICEC, Innsbruck, Austria, 19–22 August 2008; pp. 489–494.
8. Wagner, C. Investigations on Silver Sulfide. *J. Chem. Phys.* **1953**, *21*, 1819–1827. [CrossRef]
9. Hebb, M.H. Electrical Conductivity of Silver Sulfide. *J. Chem. Phys.* **1952**, *20*, 185–190. [CrossRef]
10. Sahraoui, K.; Benramdane, N.; Khadraoui, M.; Miloua, R.; Mathieu, C. Characterization of Silver Sulphide Thin Films Prepared by Spray Pyrolysis Using a New Precursor Silver Chloride. *Sens. Transducers* **2014**, *27*, 319–325.
11. Zhang, S.; Osterman, M.; Shrivastava, A.; Kang, R.; Pecht, M.G. The Influence of H2S Exposure on Immersion-Silver-Finished PCBs under Mixed-Flow Gas Testing. *IEEE Trans. Device Mater. Reliab.* **2010**, *10*, 71–81. [CrossRef]
12. Abbott, W.H. The Development and Performance Characteristics of Mixed Flowing Gas Test Environment. *IEEE Trans. Compon. Hybrids Manuf. Technol.* **1988**, *11*, 22–35. [CrossRef]
13. Ying, T. An Advanced Anti-Tarnish Process for Silver Coins and Silverware—Monomolecular Octadecanethiol Protective Film. *Tribol. Trans.* **2021**, *64*, 341–349. [CrossRef]
14. Liang, C.; Yang, C.; Huang, N. Tarnish Protection of Silver by Octadecanethiol Self-Assembled Monolayers Prepared in Aqueous Micellar Solution. *Surf. Coat. Technol.* **2009**, *203*, 1034–1044. [CrossRef]
15. Liang, C.H.; Yang, C.J.; Huang, N.B.; Wu, B. Comparison of Four Antitarnishing Self-Assembled Monolayers on Silver Coin. *Surf. Eng.* **2011**, *27*, 199–204. [CrossRef]
16. Phillips, A.C.; Cowley, A. Octadecanthiol for Tarnish-Resistant Silver Coatings. In Proceedings of the Modern Technologies in Space- and Ground-Based Telescopes and Instrumentation II, Amsterdam, The Netherlands, 1–6 July 2012; Navarro, R., Cunningham, C.R., Prieto, E., Eds.; Volume 8450, p. 84503W.
17. Planes, E.; Flandin, L.; Alberola, N. Polymer Composites Bipolar Plates for PEMFCs. *Energy Procedia* **2012**, *20*, 311–323. [CrossRef]
18. Saito, R.; Dresselhaus, G.; Dresselhaus, M.S. *Physical Properties of Carbon Nanotubes*; Imperial College Press: London, UK, 1998; ISBN 978-1-86094-093-4.
19. Dresselhaus, M.S.; Dresselhaus, G.; Saito, R. Physics of Carbon Nanotubes. *Carbon* **1995**, *33*, 883–891. [CrossRef]
20. Popov, V. Carbon Nanotubes: Properties and Application. *Mater. Sci. Eng. R Rep.* **2004**, *43*, 61–102. [CrossRef]
21. Ebbesen, T.W. Carbon Nanotubes. *Annu. Rev. Mater. Sci.* **1994**, *24*, 235–264. [CrossRef]
22. Nasir, S.; Hussein, M.; Zainal, Z.; Yusof, N. Carbon-Based Nanomaterials/Allotropes: A Glimpse of Their Synthesis, Properties and Some Applications. *Materials* **2018**, *11*, 295. [CrossRef] [PubMed]
23. Lee, C.-G.; Hwang, Y.-J.; Choi, Y.-M.; Lee, J.-K.; Choi, C.; Oh, J.-M. A Study on the Tribological Characteristics of Graphite Nano Lubricants. *Int. J. Precis. Eng. Manuf.* **2009**, *10*, 85–90. [CrossRef]

24. Zhang, Z.; Simionesie, D.; Schaschke, C. Graphite and Hybrid Nanomaterials as Lubricant Additives. *Lubricants* **2014**, *2*, 44–65. [CrossRef]
25. Cornelio, J.A.C.; Cuervo, P.A.; Hoyos-Palacio, L.M.; Lara-Romero, J.; Toro, A. Tribological Properties of Carbon Nanotubes as Lubricant Additive in Oil and Water for a Wheel–Rail System. *J. Mater. Res. Technol.* **2016**, *5*, 68–76. [CrossRef]
26. Dassenoy, F.; Joly-Pottuz, L.; Martin, J.M.; Mieno, T. Carbon Nanotubes as Advanced Lubricant Additives. In *Carbon Nanotubes*; Popov, V.N., Lambin, P., Eds.; Kluwer Academic Publishers: Dordrecht, The Netherlands, 2006; pp. 237–238.
27. Kałużny, J.; Merkisz-Guranowska, A.; Giersig, M.; Kempa, K. Lubricating Performance of Carbon Nanotubes in Internal Combustion Engines—Engine Test Results for CNT Enriched Oil. *Int. J. Automot. Technol.* **2017**, *18*, 1047–1059. [CrossRef]
28. Maharaj, D.; Bhushan, B.; Iijima, S. Effect of Carbon Nanohorns on Nanofriction and Wear Reduction in Dry and Liquid Environments. *J. Colloid. Interface Sci.* **2013**, *400*, 147–160. [CrossRef]
29. Sethi, S.; Dhinojwala, A. Superhydrophobic Conductive Carbon Nanotube Coatings for Steel. *Langmuir* **2009**, *25*, 4311–4313. [CrossRef]
30. Rahman, M.M.; Islam, M.; Roy, R.; Younis, H.; AlNahyan, M.; Younes, H. Carbon Nanomaterial-Based Lubricants: Review of Recent Developments. *Lubricants* **2022**, *10*, 281. [CrossRef]
31. De Nicola, F.; Castrucci, P.; Scarselli, M.; Nanni, F.; Cacciotti, I.; Crescenzi, M. De Super-Hydrophobic Multi-Walled Carbon Nanotube Coatings for Stainless Steel. *Nanotechnology* **2015**, *26*, 145701. [CrossRef] [PubMed]
32. Chen, X.; Li, J. Superlubricity of Carbon Nanostructures. *Carbon* **2020**, *158*, 1–23. [CrossRef]
33. Loyd, A.; Hemond, J.; Martens, R. A Preliminary Investigation of Graphite, Graphene and Carbon Nanotubes (CNT's) as Solid State Lubricants. In Proceedings of the 2011 IEEE 57th Holm Conference on Electrical Contacts (Holm), Minneapolis, MN, USA, 11–14 September 2011; IEEE: Piscataway, NJ, USA, 2011; pp. 1–9.
34. Reinert, L.; Varenberg, M.; Mücklich, F.; Suárez, S. Dry Friction and Wear of Self-Lubricating Carbon-Nanotube-Containing Surfaces. *Wear* **2018**, *406–407*, 33–42. [CrossRef]
35. Alderete, B.; Mücklich, F.; Suarez, S. Characterization and Electrical Analysis of Carbon-Based Solid Lubricant Coatings. *Carbon Trends* **2022**, *7*, 100156. [CrossRef]
36. Alderete, B.; MacLucas, T.; Espin, D.; Brühl, S.P.; Mücklich, F.; Suarez, S. Near Superhydrophobic Carbon Nanotube Coatings Obtained via Electrophoretic Deposition on Low-Alloy Steels. *Adv. Eng. Mater.* **2021**, *23*, 2001448. [CrossRef]
37. Alderete, B.; Lößlein, S.M.; Bucio Tejeda, D.; Mücklich, F.; Suarez, S. Feasibility of Carbon Nanoparticle Coatings as Protective Barriers for Copper—Wetting Assessment. *Langmuir* **2022**, *38*, 15209–15219. [CrossRef]
38. Huo, Y.; Fu, S.-W.; Chen, Y.-L.; Lee, C.C. A Reaction Study of Sulfur Vapor with Silver and Silver–Indium Solid Solution as a Tarnishing Test Method. *J. Mater. Sci. Mater. Electron.* **2016**, *27*, 10382–10392. [CrossRef]
39. Reinert, L.; Zeiger, M.; Suárez, S.; Presser, V.; Mücklich, F. Dispersion Analysis of Carbon Nanotubes, Carbon Onions, and Nanodiamonds for Their Application as Reinforcement Phase in Nickel Metal Matrix Composites. *RSC Adv.* **2015**, *5*, 95149–95159. [CrossRef]
40. Philadelphia Museum of Art Finishing Techniques in Metalwork—Hands-Free Silver "Polishing". Available online: https://www.philamuseum.org/booklets/7_44_85_1.html?page=2 (accessed on 18 January 2021).
41. Alderete, B.; Mücklich, F.; Suarez, S. Tarnishing (Ag2S) Layer on Silver-Plated Electrical Contacts: Its Influence on Electrical Contact Resistance. *IEEE Trans. Compon. Packag. Manuf. Technol.* **2023**, *13*, 45–58. [CrossRef]
42. Alderete, B.; Mücklich, F.; Suarez, S. Wear Reduction via CNT Coatings in Electrical Contacts Subjected to Fretting. *Tribol. Lett.* **2023**, *71*, 54. [CrossRef]
43. Alderete, B.; Suarez, S.; Tejeda, D.B.; Mücklich, F. Fretting and Electrical Contact Resistance Characteristics of Carbon Nanoparticle-Coated Cu Electrical Contacts. In Proceedings of the 2022 IEEE 67th Holm Conference on Electrical Contacts (HLM), Tampa, FL, USA, 23–26 October 2022; IEEE: Piscataway, NJ, USA, 2022; pp. 1–8.
44. Thomas, B.J.C.; Boccaccini, A.R.; Shaffer, M.S.P. Multi-Walled Carbon Nanotube Coatings Using Electrophoretic Deposition (EPD). *J. Am. Ceram. Soc.* **2005**, *88*, 980–982. [CrossRef]
45. Boccaccini, A.R.; Cho, J.; Roether, J.A.; Thomas, B.J.C.; Jane Minay, E.; Shaffer, M.S.P. Electrophoretic Deposition of Carbon Nanotubes. *Carbon* **2006**, *44*, 3149–3160. [CrossRef]
46. Alderete, B.; Nayak, U.P.; Mücklich, F.; Suarez, S. Influence of Topography on Electrical Contact Resistance of Copper-Based Materials. *Surf. Topogr.* **2023**, *11*, 025027. [CrossRef]
47. Vafaei, S.; Podowski, M.Z. Theoretical Analysis on the Effect of Liquid Droplet Geometry on Contact Angle. *Nucl. Eng. Des.* **2005**, *235*, 1293–1301. [CrossRef]
48. Krainer, S.; Hirn, U. Contact Angle Measurement on Porous Substrates: Effect of Liquid Absorption and Drop Size. *Colloids Surf. A Physicochem. Eng. Asp.* **2021**, *619*, 126503. [CrossRef]
49. Alderete, B.; Puyol, R.; Slawik, S.; Espin, E.; Mücklich, F.; Suarez, S. Multipurpose Setup Used to Characterize Tribo-Electrical Properties of Electrical Contact Materials. *MethodsX* **2021**, *8*, 101498. [CrossRef] [PubMed]
50. Bock, E.M. Low-Level Contact Resistance Characterization. *AMP J. Technol.* **1993**, *3*, 64–68.
51. Johnson, K.L. *Contact Mechanics*; Cambridge University Press: Cambridge, UK, 1987; ISBN 0521347963.
52. Callister, W.D., Jr. *Materials Science and Engineering: An Introduction*, 7th ed.; John Wiley & Sons, Ltd.: Hoboken, NJ, USA, 2006; Volume 94, ISBN 9780470054888.

53. Shooshtari, M.; Salehi, A.; Vollebregt, S. Effect of Humidity on Gas Sensing Performance of Carbon Nanotube Gas Sensors Operated at Room Temperature. *IEEE Sens. J.* **2021**, *21*, 5763–5770. [CrossRef]
54. Tsai, J.T.H.; Lu, C.-C.; Li, J.G. Fabrication of Humidity Sensors by Multi-Walled Carbon Nanotubes. *J. Exp. Nanosci.* **2010**, *5*, 302–309. [CrossRef]
55. Lee, Y.; Yoon, J.; Kim, Y.; Kim, D.M.; Kim, D.H.; Choi, S.-J. Humidity Effects According to the Type of Carbon Nanotubes. *IEEE Access* **2021**, *9*, 6810–6816. [CrossRef]
56. Ling, Y.; Gu, G.; Liu, R.; Lu, X.; Kayastha, V.; Jones, C.S.; Shih, W.-S.; Janzen, D.C. Investigation of the Humidity-Dependent Conductance of Single-Walled Carbon Nanotube Networks. *J. Appl. Phys.* **2013**, *113*, 024312. [CrossRef]
57. Holm, R. *Electric Contacts*, 4th ed.; Springer: Berlin/Heidelberg, Germany, 1967; ISBN 978-3-642-05708-3.
58. Endres, H. *Praxishandbuch Steckverbinder*, 2nd ed.; Vogel COmmunications Group: Würzburg, Germany, 2021; ISBN 978-3-8343-3501-2.
59. Karousis, N.; Suarez-Martinez, I.; Ewels, C.P.; Tagmatarchis, N. Structure, Properties, Functionalization, and Applications of Carbon Nanohorns. *Chem. Rev.* **2016**, *116*, 4850–4883. [CrossRef] [PubMed]
60. Slade, P.G. *Electrical Contacts*, 2nd ed.; Slade, P.G., Ed.; CRC Press: Boca Raton, FL, USA, 2014; ISBN 9781315216829.
61. Merstallinger, A.; Sales, M.; Semerad, E.; Dunn, B.D. Assessment of Cold Welding between Separable Contact Surfaces Due to Impact and Fretting under Vacuum. *ESA Sci. Tech. Memo.* **2009**, *279*, 57.
62. Haddon, R.C. Chemistry of the Fullerenes: The Manifestation of Strain in a Class of Continuous Aromatic Molecules. *Science* **1993**, *261*, 1545–1550. [CrossRef]
63. Korczeniewski, E.; Zięba, M.; Zięba, W.; Kolanowska, A.; Bolibok, P.; Kowalczyk, P.; Wiertel-Pochopień, A.; Zawała, J.; Boncel, S.; Terzyk, A.P. Electrophoretic Deposition of Layer-by-Layer Unsheathed Carbon Nanotubes—A Step Towards Steerable Surface Roughness and Wettability. *Materials* **2020**, *13*, 595. [CrossRef]
64. Hilding, J.; Grulke, E.A.; George Zhang, Z.; Lockwood, F. Dispersion of Carbon Nanotubes in Liquids. *J. Dispers. Sci. Technol.* **2003**, *24*, 1–41. [CrossRef]

C Journal of Carbon Research

MDPI

Article

Graphene and Nanoclay as Processing Aid Agents: A Study on Rheological Behavior in Polystyrene

Julie Genoyer [1,2], Emna Helal [1,2,*], Giovanna Gutierrez [2], Nima Moghimian [2], Eric David [1] and Nicole R. Demarquette [1]

[1] Ecole de Technologie Supérieure, Department of Mechanical Engineering, Montreal, QC H3C 1K3, Canada; julie.genoyer.1@ens.etsmtl.ca (J.G.); eric.david@etsmtl.ca (E.D.); nicoler.demarquette@etsmtl.ca (N.R.D.)
[2] NanoXplore Inc., Montreal, QC H4R 2P2, Canada; giovanna.gutierrez@nanoxplore.ca (G.G.); nima.moghimian@nanoxplore.ca (N.M.)
* Correspondence: emna.helal@nanoxplore.ca

Abstract: The effectiveness of layered particles as processing aid agents in molten polystyrene was studied. Three graphene grades and two clays of different lateral size were selected for this purpose. The morphologies of the composites were observed using scanning electron microscopy. Steady shear measurements were carried out and the Carreau–Yasuda model with yield stress was applied to the experimental results. A decrease in viscosity was observed at 2 wt.% of particle content for almost all composites. The most efficient particle for reducing viscosity was found to be graphene in a loose agglomerated configuration. Graphene and clay particles with similar dispersion states had a similar effect on the viscosity, inducing a decrease by 29% and 22%, respectively, suggesting comparable efficiency as processing aid agents. The observed decrease in viscosity is attributed to the phenomenon of superlubricity, which is a lubricating mechanism that is closely linked to the atomic structure of the particles.

Keywords: graphene; clay; thermoplastic polymer; processability; processing aid; rheology

Citation: Genoyer, J.; Helal, E.; Gutierrez, G.; Moghimian, N.; David, E.; Demarquette, N.R. Graphene and Nanoclay as Processing Aid Agents: A Study on Rheological Behavior in Polystyrene. *C* **2023**, *9*, 96. https://doi.org/10.3390/c9040096

Academic Editor: Gil Gonçalves

Received: 31 May 2023
Revised: 15 September 2023
Accepted: 18 September 2023
Published: 7 October 2023

1. Introduction

Processing thermoplastics requires the polymer to be molten at high temperatures, which needs a high amount of energy because of the poor thermal conductivity of those materials. To ease the process, some processing aid agents can be added to improve flowability, mold filing, and mold release, as well as to reduce shark skin and postpone gross melt fracture to significantly higher shear rates [1–3]. In other words, adding a processing aid agent, or lubricant, can help to increase output or decrease energy consumption during processing.

Several types of processing agents exist, such as fluoropolymers, stearates, and boron nitride [1]. Hexagonal boron nitride in particular was shown to reduce storage and loss moduli and melt fracture and head pressure in extrusion [2,3]. For example, the addition of only 0.05 wt.% of h-BN has been shown to decrease the transient shear viscosity and extensional stress growth of HDPE [4]. On the other hand, other studies showed an increase in complex viscosity upon the addition of 5 to 30 wt.% of h-BN in PEI and PEEK, indicating that low contents of h-BN are necessary to act as a processing aid [5,6]. As with many 2D particles, it is also a solid lubricant that can be encountered in tribology studies as it helps in reducing friction and wear between two parts [7,8].

Other layered nanoparticles of great interest in tribology are carbon-based fillers such as graphene, which is also a 2D layered nanoparticle. Their low-friction behavior was discovered in 1928 [9,10]. First attributed to low shear resistance between adjacent atomic layers, it is now believed to be due to what is called the superlubricity phenomenon. Indeed, the interlayer shear strength of graphite was estimated to be between 0.2 MPa and 7 GPa, values which are not reached during common rheological tests [11]. On the other hand, superlubricity, occurring when friction between two surfaces vanishes or nearly vanishes, depends on the angle between two particles and does not require high shear. More specifically, this variation in friction has been shown to depend on the commensurability of the contact surfaces [12,13]. It means that depending on the atomic structure of the two surfaces, there are angles between two particles, known as commensurate angles, for which the atomic structure is aligned, leading to friction, and there are also incommensurate angles where ultra-low friction, and thus superlubricity, can be achieved [14]. Because of this interesting capacity, graphite and graphene are nowadays used as solid lubricants as well as additives in oil and water for lubricating fluids. To our knowledge, graphene is not commonly used in molten polymers as a processing aid agent or lubricant, but it could be an interesting candidate given its tribological properties [15]. In addition, it would complement the unique combination of functionalities of graphene, such as high mechanical strength, electrical conductivity [16–18], UV resistance [19], piezoelectric response enhancement [20], as well as the valorization of recycled plastics and mixed plastic waste streams [21,22]. Furthermore, recent toxicity studies performed on few-layer graphene industrial grades produced by the mechanochemical exfoliation of graphite have much lower health and safety concerns than other nanocarbons such carbon nanotube and carbon black [23].

As of today, few papers deal with the possibility of using solid particles aside from boron nitride as processing aid agents in molten polymers [24–27]. Although Yang Bai et al. showed that the addition of 16 wt.% of graphite led to a significant decrease from 2.10^3 to 7.10^2 Pa.s of complex viscosity whereas the addition of 30 wt.% of BN did not decrease significantly the complex viscosity, this effect was not explicitly discussed in the paper [24]. Recently, graphite, multi-layered graphene oxide, and carbon nanotubes were shown to decrease the viscosity of molten polymers, thus improving flowability [25–27]. Ferreira et al. [25] concluded that it was the superlubricity ability which promoted slipping between loose agglomerates which in turn led to lower viscosity. In their work, graphite oxide and multi-layered graphene oxide had the most efficient lubricant effect, whereas graphite did not significantly decrease the viscosity of the polyolefin matrices. It was attributed to the fact that the oxidation of graphite caused fragmentation in numerous pieces, which created incommensurate contacts that were less numerous in graphite.

This superlubricity phenomenon is not exclusive to graphene and has been shown to exist in other 2D particles such as Molybdenum disulfide or even mica [28,29]. In particular, clays have been shown to display the superlubricity phenomenon as well, even if it has been rarely discussed. In another study, it was briefly pointed out that adding different clays into PMMA led to a decrease in viscosity in molten state. However, the mechanism was not further investigated in that paper as it was not the main focus. The paper instead focused on the effect of nanoclay as a compatibilizer in PMMA/PS blends [30]. Still, clay could be an interesting candidate as a processing aid agent.

In this work, the effect of the addition of graphene of different lateral sizes on the viscosity of polystyrene (PS) was studied. First, the morphology of graphene composites was observed using scanning electron microscopy (SEM). The shear viscosity was then obtained using a rotational rheometer, and Carreau–Yasuda's model with yield stress was applied to quantify the effect of the filler particles on the viscosity. Throughout the paper, this model will be called modified Carreau–Yasuda. Finally, the addition of montmorillonite and laponite, two smectic clays, was studied as well and compared to graphene.

2. Materials and Methods

2.1. Materials

In this study, a polystyrene (PS), grade Styron^{TM} 685D from AM Styrenics, The Woodlands, TX, USA, was used as the matrix. Five different layered particles were added to it. First, GrapheneBlack^{TM} 3X and GrapheneBlack^{TM} 0X graphene grades from NanoXplore, Montreal, QC, Canada, were used and will be referred to as 3X and 0X, respectively. As a means of comparison, another grade of graphene (xGnP H-25) previously acquired from

XG Sciences, Lansing, MI, USA, currently owned by NanoXplore, was used as well. This graphene will be designed as graphene H throughout the paper. Finally, two layered clays were used: a laponite (L) and a montmorillonite (MMT), from BYK, Wesel, Germany. All the information regarding the particles used is reported in Table 1.

Table 1. Properties of graphenes and clays.

Type of Particle	Supplier	Grade	Abbreviation	Primary Particle Lateral Size (μm) *
Graphene	NanoXplore Inc.	GrapheneBlack™ 3X GrapheneBlack™ 0X	3X 0X	1–2 0.5–1
	XG Sciences Inc.	xGnP H-25	H	25
Montmorillonite Laponite	BYK	Cloisite Na^+ Laponite RD	MMT L	<25 0.030

* values reported by supplier.

2.2. Processing

A masterbatch of PS with 10 wt.% of each filler was prepared by melt-mixing using a Haake Rheomix OS PTW16 twin-screw extruder (Thermo Fisher Scientific Inc., Waltham, MA, USA) at 220 °C in all zones. The screw speed was adjusted to 100 rpm for all compositions. Prior to the masterbatch extrusion, the nanoclay powders were dried to avoid the moisture effect during processing. Then, dilutions into PS composites with concentrations varying from 0 to 2 wt.% were prepared using an internal mixer at 220 °C and 100 rpm. Neat PS underwent the same process in order to have the same thermomechanical history as the composites.

The composites were compression molded into discs of 25 mm diameter and 1.2 mm thickness at 10 MPa and 200 °C for 10 min.

The composites were named Psparticle-concentration. For example, PS0X-2 is PS containing 2 wt.% of GrapheneBlack™ 0X.

2.3. Characterizations

The morphology was observed by scanning electron microscopy (SEM) using a Hitachi TM3000 SEM at an accelerating voltage of 15 kV. Discs were fractured at room temperature and coated with gold prior to observations.

Rheological data were obtained using a stress-controlled rheometer MCR 501 from Anton Paar; the tests were carried out at 200 °C and under a nitrogen atmosphere. Steady shear tests were obtained using a cone-plate geometry with an angle of 0.985° and a gap of 0.048 mm. The shear rate varied from 0.001 to 100 s^{-1} using a logarithmic ramp rate. For all formulations, three samples were tested to obtain a mean value of viscosity over the shear rate.

X-ray diffraction (XRD) was carried out using a PANalytical diffractometer, model X'Pert Pro, with a cobalt radiation source of wavelength of 0.179 nm. Samples were tested at ambient temperature with 2θ ranging from 5 to 70 °C. From the obtained diffractometers, Braag's law (Equation (1)) can be used to calculate the interlayer spacing between graphene sheets.

$$d = \frac{n\lambda}{2sin\theta} \quad (1)$$

where λ is the X-ray wavelength in nanometers (nm), θ is the scattering angle, and n the diffraction order.

3. Results and Discussion

3.1. Morphology of Graphene Composites

In order to study the effect of graphene nanoplatelets' average lateral size on the viscosity of a molten polymer matrix, three graphenes of different lateral sizes were used.

To evaluate the dispersion state of the different particles in the prepared composites, XRD and SEM were performed on rheology disks in order to draw appropriate conclusions.

XRD was carried out on samples containing 2 wt.% of each graphene. The results are reported in Table 2. A sharp peak at 30.9 ° corresponding to the (002) Bragg reflection of graphene platelets was observed on the diffractograms. Using Braag's law (1), the interlayer spacing was calculated to be 0.335 nm for each one of the samples, which is the same as graphite commonly found in the literature [31–33]. As the primary particles of graphene 0X and 3X were originally in stacks of 6 to 10 layers (supplier information), it indicates no exfoliation during extrusion. In order to visualize and conclude on the dispersion state of graphene, SEM imaging was used.

Table 2. XRD results.

	2θ (°C)	Interlayer Spacing (nm)
PS0X−2	30.99	0.335
PS3X−2	30.99	0.335
PSH−2	30.94	0.335

Figure 1 shows the morphology of polystyrene (PS) containing 2 wt.% of graphene 0X, graphene 3X, and graphene H. It was observed that graphene 0X tended to form loose agglomerates and create a rich area in the sample, as shown in Figure 1b. The whole image is a graphene-rich area with tightly packed and randomly oriented flakes. Aside from these areas, there were also well-distributed graphene regions (see Figure 1a). In the case of graphene 3X (Figure 1c) and graphene H (Figure 1d), the distribution was visually more homogeneous even if the size of platelets was larger on average.

Graphene is an allotrope of carbon where sp2 carbon atoms are arranged in a honeycomb lattice structure. Graphene structure is very stable and can be considered as the minimal unit of carbon materials [34]. π-π interaction, a type of Van der Waals force, is known to hold graphene sheets together at a distance of 0.335 nm [35,36]. Given that polystyrene has a mer unit containing an aromatic ring, a mechanism of interaction with graphene through π-π and CH-π interactions is possible [37]. However, the density in aromatic rings in PS is less than in graphene given that they are more spaced out by the carbon main chain of the macromolecule. This may explain why graphene keeps agglomerating instead of dispersing in PS.

In each sample, the average lateral size of the particles or the flakes, composed of few individual particles, was measured over fifty units. The values are reported in Table 3. In the case of graphene 3X, the flakes distributed in PS are larger, with a D50 of 12.6 μm, compared to 8.3 μm for 0X. However, the 3X was overall more homogeneously distributed: there were no rich phases nor poor phases in PS3X-2. Graphene H resulted in a dispersion and distribution similar to the one of graphene 3X even if graphene H had the largest initial particle size (25 μm as given by supplier in Table 1). In PS, graphene H is fairly well distributed, with a D50 of 9.6 μm, which is lower than the one of graphene 3X. However, the D90 of PSH-2 is higher than the one of PS3X-2. By taking into account their initial particle size reported in Table 1 and SEM images reported in Figure 1, it seems that the smallest graphene particles, i.e., graphene 0X, tend to form a less homogeneous distribution and more graphene-rich areas covered with loose agglomerates.

Table 3. Size of agglomerations of graphene 0X, 3X, and H in PS observed using SEM.

Sample	D50 (μm)	D90 (μm)
PS0X-2	8.3 *	15.6 *
	>22 **	>55 **
PS3X-2	12.6	40.1
PSH-2	9.6	48.8

* in dispersed area; ** in highly agglomerated area.

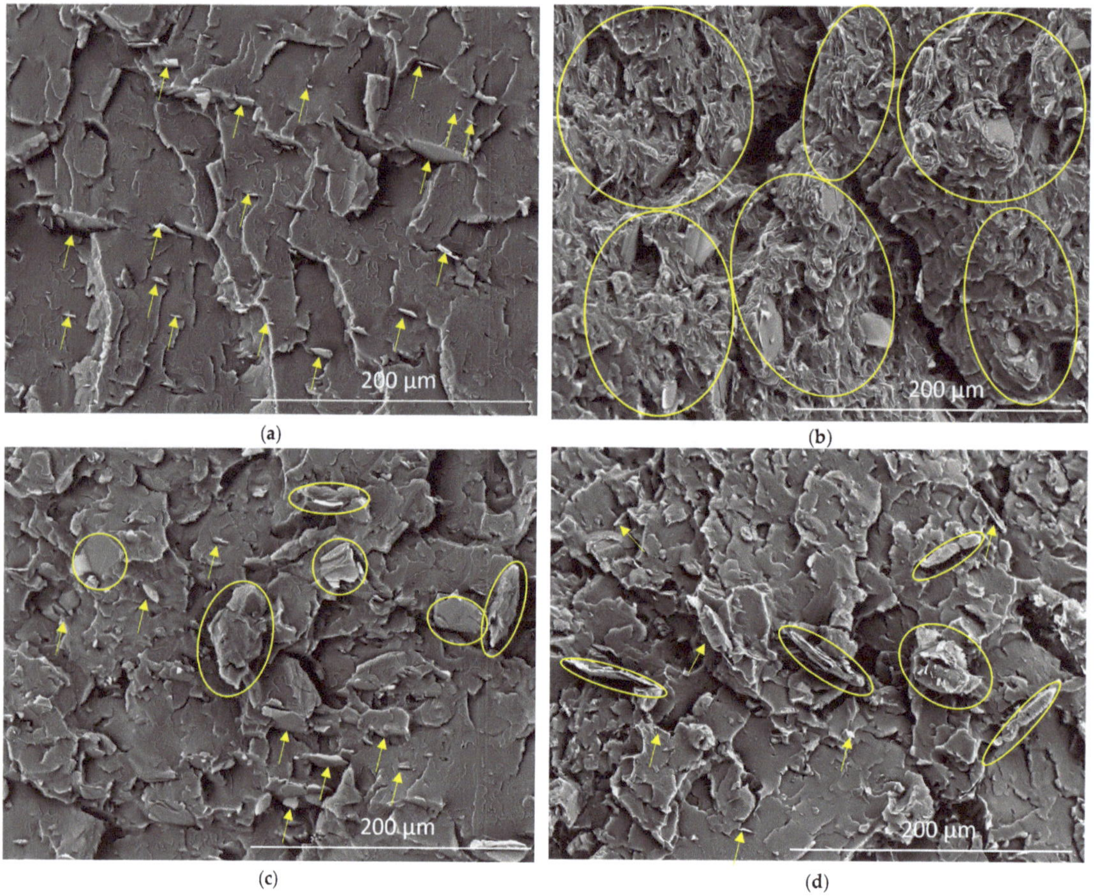

Figure 1. Scanning electron microscopy images of PS with 2 wt.% of (**a**) graphene 0X (dispersed phase), (**b**) graphene 0X (agglomerated phase), (**c**) graphene 3X, (**d**) graphene H. Platelets are indicated by arrows and circles.

In order to associate the lubricant effect and the morphology, the viscosity of each composite was measured using a rotational rheometer and is discussed below.

3.2. *Effect of Graphene on Shear Viscosity*

Figure 2 compares the viscosity as a function of shear rate obtained in steady shear for all composites containing 0 to 5 wt.% of graphene 0X, 3X, and H. It can be seen that the shape of the curve depends on the graphene content. For low contents, the viscosity is similar to the one of neat PS. But more significant changes in the shape of the curve are observed with increasing graphene loadings.

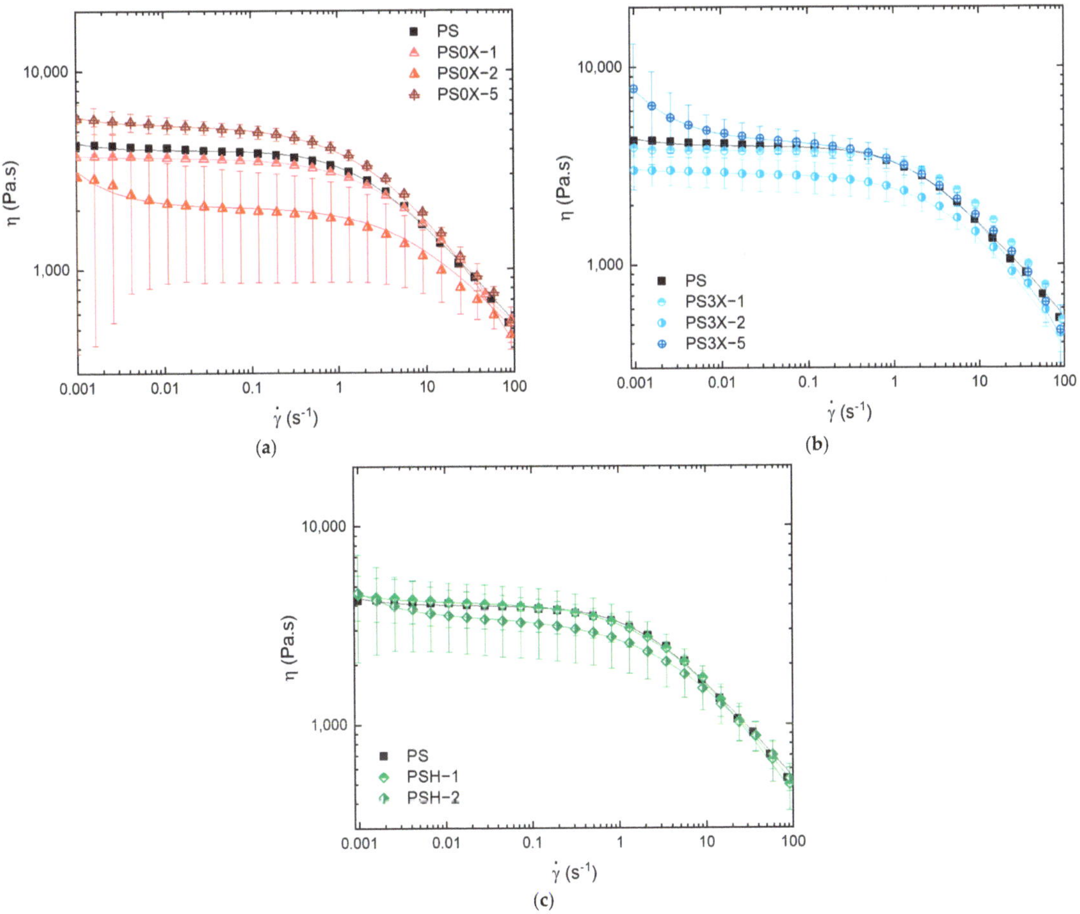

Figure 2. Steady shear viscosity as a function of shear rate of (**a**) PS0X, (**b**) PS3X, and (**c**) PSH composites at 200 °C. Lines represent the fitting of modified Carreau–Yasuda model.

To better quantify the effect of graphene on the viscosity curve, Carreau–Yasuda's model can be used. This model is used widely for shear-thinning polymers and can be fitted to the shear viscosity with the following expressions:

$$\eta = \eta_\infty + (\eta_0 - \eta_\infty)\left(1 + (\lambda\dot{\gamma})^a\right)^{\frac{n-1}{a}} \quad (2)$$

$$\eta = \eta_0\left(1 + (\lambda\dot{\gamma})^a\right)^{\frac{n-1}{a}} \quad (3)$$

where η_∞ is the viscosity at infinite shear, η_0 the zero-shear viscosity, λ the relaxation time, n the power law index, and a the Yasuda parameter. The fit is made so that all parameters should be positive or equal to zero. More specifically, n is fixed to 0 and $0 \leq \lambda \leq 10$.

Equation (2) is the original Carreau–Yasuda's model developed in 1979 widely used to describe the shear-thinning behavior of most thermoplastic polymers [38]. In most cases, η_∞ can be set to zero as the second plateau value of the viscosity, as a very high shear rate is difficult to reach experimentally. Simplifying the expression of Carreau–Yasuda leads to Equation (2), which can be fitted to the pure matrix or composites of low contents. It adequately describes the plateau region at low shear rates followed by the shear-thinning region at higher shear rates. The most interesting parameters for this study are the zero shear viscosity, η_0, and the relaxation time, λ. The zero-shear viscosity corresponds to the plateau value at low shear rates, which is used as a viscosity reference value for polymers. In this paper, the evaluation of the lubricant effect was performed by studying the reduction in zero-shear viscosity. The relaxation time is also of interest, as the presence of particles will affect the relaxation behavior of the molten polymer matrix.

However, Equation (3) does not take into account the increase at low shear rates that can be observed at a higher graphene content. To take this effect into account, Vergnes proposed adding an apparent yield stress, σ_0 [39]. Even though this previous work was conducted on the model for oscillatory shear, mathematically, the same can be done for the model in the case of steady shear.

The model becomes:

$$\eta = \frac{\sigma_0}{\dot{\gamma}} + \eta_0\left(1 + (\lambda\dot{\gamma})^a\right)^{\frac{n-1}{a}} \quad (4)$$

This apparent yield stress is an indicator of the presence of a stiff nanoparticle network which dominates the response at low shear rates [39–41].

The zero-shear viscosities, obtained by fitting the modified Carreau–Yasuda model to the shear viscosities of the graphene composites, are reported in Figure 3. At 1 wt.%, the zero-shear viscosities of the composites are similar to those of neat PS, indicating that graphene has no significant effect at this content.

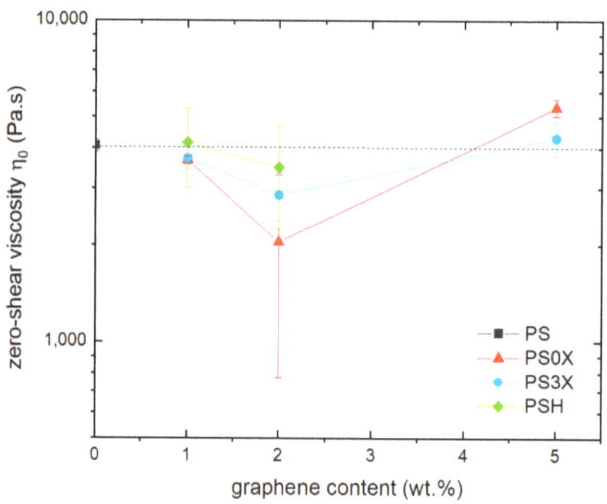

Figure 3. Zero-shear viscosity, η_0, obtained by fitting the modified Carreau–Yasuda model to the shear viscosities of composites containing graphene 0X, 3X, and H.

At 5 wt.%, the viscosities of 3X and 0X composites are higher than those of neat PS. In this case, the graphene platelets are numerous enough to increase the viscosity as commonly observed in many other composites, which was predicted by Einstein to scale with particle volume fraction [42].

However, at 2 wt.%, the viscosities of composites containing graphene 0X and 3X are consistently lower than those of neat PS, while the decrease due to graphene H seems to be less significant. It can be concluded that there is a processing aid effect occurring at an optimal graphene concentration and that the efficiency of the particles, i.e., their ability to decrease viscosity, depends on the graphene used. As shown by SEM imaging, the distribution of graphene 0X is less homogeneous compared to the other graphene grade and forms significantly, yet it is the most efficient at decreasing viscosity. Also, graphene 3X was shown to agglomerate less than 0X but more than H. As graphene 3X has a lubricant effect in between 0X and H, it leads to the idea that the state of dispersion combined with the actual agglomerates' size may be the key factor to control the lubricant effect.

Ferreira et al. [25,43] suggested that the superlubricity property of graphene and graphite plays a role in the reduction in viscosity. As discussed in the introduction, graphene, like many other 2D particles, is known to exhibit this superlubricity effect, which makes it a suitable lubricant in diverse applications. Superlubricity is when friction vanishes or nearly vanishes, and it is currently believed to be due to the atomic structure of graphene. As illustrated in Figure 4, when two flakes are in commensurate contact, i.e., at angles of 0°, 60°, 120°, 180°, 240°, and 300°, the flakes align in a structured manner, causing friction. When the flakes are in incommensurate contact (all other angles), the flakes' structures do not align and there is slipping between the flakes. This is due to the 60 ° recurrence in the structure of the graphene layer. In a random distribution of graphene in a molten polymer, there is a much higher chance of being in incommensurate contact rather than commensurate contact. When particles slip suddenly, free volume increases, resulting in higher chain mobility and thus a lower viscosity.

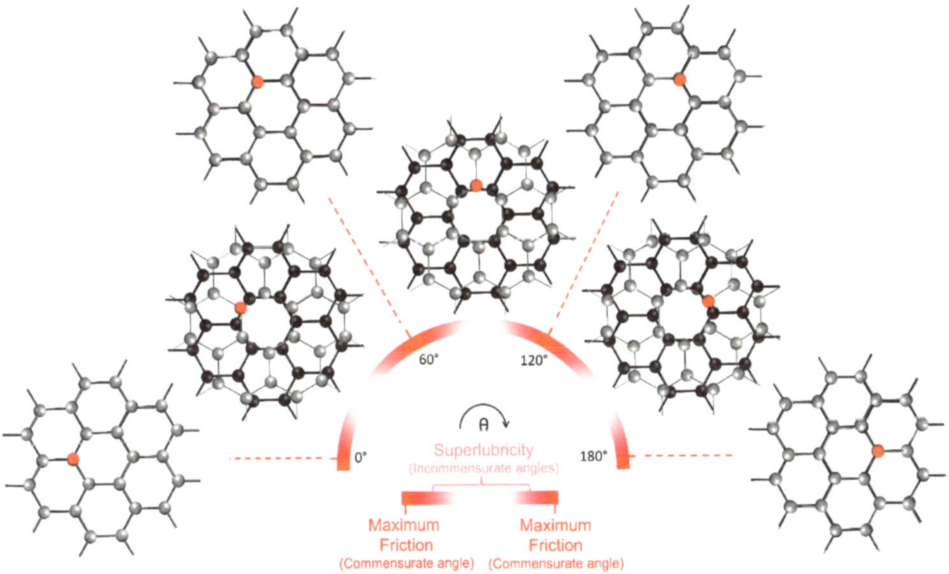

Figure 4. Superlubricity phenomenon in graphene as a function of flake-to-flake angle, θ. Flake 1 in black, flake 2 in gray. One carbon of flake 2 is in red as a reference point.

It is worth mentioning that the π-π interactions discussed earlier seem to play an additional role at particular spatial arrangements, such as in the local friction phenomena described by the superlubricity theory. The presumptive additional noncovalent interaction could corroborate the superlubricity mechanism and be part of the explanation of why there is friction at specific angles.

In this study, the agglomeration of graphene seems to promote the lubricant effect, which is what was also observed in Ferreira et al.'s work [25]. It is believed that incommensurate contacts are promoted in between agglomerates rather than in between well-distributed particles. It can also be concluded that the concentration of graphene should be high enough to promote agglomeration but not so high as to create a particle network which would lead to an increase in viscosity. In terms of the size of graphene, in this case, the smallest initial lateral particle led to the bigger reduction in viscosity because it was more prone to agglomeration and not creating a network. Also, at the same weight of particles, the efficiency of slipping may be more significant for smaller particles, as the number of particles, and thus the quantity of slipping, will be greater.

Aside from the zero-shear viscosity, two other parameters from the modified Carreau–Yasuda model are of interest when investigating the effect of particles in a molten polymer matrix. The first one is the yield stress, σ_0. This parameter allows the increase of viscosity to be taken into account at low shear rates only. It is commonly associated with the formation of a stiff particle network opposing the flow at low shear rates. At higher shear rates, the particle network can break, gradually releasing the particles that in time orient in the flow direction.

The yield stresses, obtained by fitting the modified Carreau–Yasuda model to the composite shear viscosities, are reported in Figure 5. It can be seen that even a minor increase in viscosity at low shear rates, as reported in Figure 2 for 2 wt.% of graphene 0X and H, induces a significant increase in yield stress. At 5 wt.%, graphene 3X induces the highest increase in yield stress, indicating that a stiff network of particles is created. It is expected that 5 wt.% of graphene H would lead to a significant increase in yield stress as well given its good distribution. It can be noticed that at 2 wt.% graphene 0X induces a decrease of viscosity as well as an increase in yield stress, indicating a competition between the lubricant effect and network stiffness in the viscosity curve.

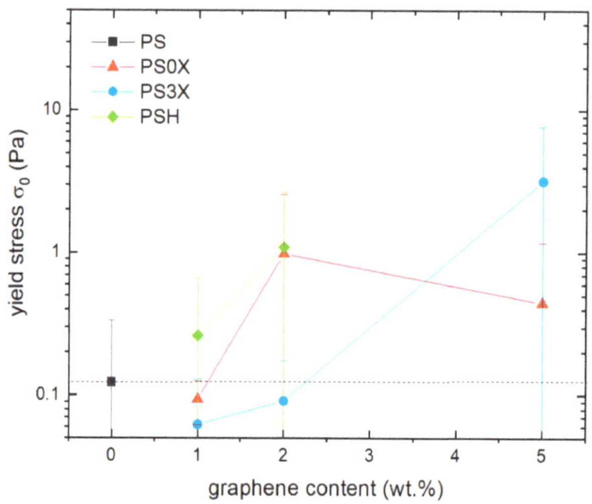

Figure 5. Yield stress, σ_0, obtained by fitting the modified Carreau–Yasuda model to the shear viscosities of composites containing graphene 0X, 3X, and H.

Another interesting parameter to discuss is the relaxation time, λ. This parameter represents the transition between the Newtonian plateau and the power law behavior in shear viscosity. In other words, it is the transition between the entangled state and the disentanglement of polymer chains due to the flow [44]. In neat polymer matrices, entanglement and disentanglement states are known to depend on the molecular weight and polydispersity of the polymer chains. The relaxation times extracted from the fitting of the modified Carreau–Yasuda model on composites are reported in Figure 6. It can clearly be seen that adding 2 wt.% of graphene of any type decreases the relaxation time. It means that the addition of graphene leads to a faster relaxation of the polymer chains. It means that, while it is not obvious in Figure 2, the transition between the plateau region and power law behavior shifts from 16 s^{-1} to 50 s^{-1}. Contrary to the reduction in viscosity, this decrease seems to be independent of the type of graphene added.

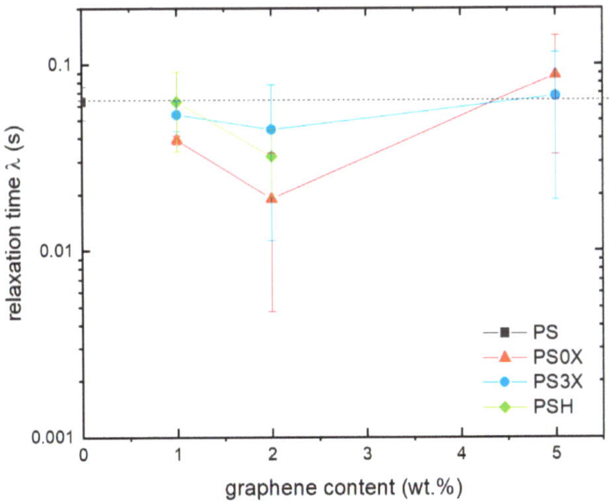

Figure 6. Relaxation time, λ, obtained by fitting the modified Carreau–Yasuda model to the shear viscosities of composites containing graphene 0X, 3X, and H.

Of course, the addition of graphene cannot change the chain length, but it could help to disentangle molecular chains or increase free volume, as suggested in works where the incorporation of cross-linked PS nanoparticles to linear PS reduced viscosity [42,45]. In our case, the hypothesis that superlubricity is responsible for the viscosity decrease could be explained by the effect of increasing free volume: the sudden slipping of the graphene particles has the potential to create more free space, resulting in higher chain mobility and thus a smaller relaxation time.

Finally, the standard deviations of the zero-shear viscosity and yield stress are significantly larger in composites compared to neat PS. It is especially true for PS0X-2, where it was shown that the morphology is not homogeneous. It can also be noticed that for lower concentrations, the standard deviation is smaller, such as the PS standard deviation. The presence of loose agglomerates randomly distributed may be the key for obtaining the lubricant effect, but it leads to less stable measurements as well.

This superlubricity phenomenon is not exclusive to graphene and has been shown to exist in other 2D particles such as Molybdenum disulfide or even mica [28].

3.3. *Effect of Smectic Clays on Shear Viscosity*

Clays such as micas or smectites are sheet-like materials with a tetrahedral outer layer and they differ by their inner layers and interlayers. The outer layer is composed of a tetrahedral arrangement of oxygen atoms inside of which a Si, Al, or Fe atom is placed [46]. In particular, micas have been shown to display the superlubricity phenomenon because of the structure of their outer tetrahedral layer composed of SiO_4 [28]. As all clays' outer layers have the same structure composed of tetrahedra, the superlubricity effect due to the structure of the Si layer of mica should also be present in smectite clays such as montmorillonite or laponite.

Moreover, the structure of the base of the tetrahedra constitutes the outer layer that will be in contact with another clay particle. This structure is shown and compared to the one of graphene in Figure 7. Even if it is composed by oxygen atoms rather than carbon atoms, the structure is similar to graphene as it has the same 60° orientation recurrence. As a means of comparison for graphene and clays, Figure 7 presents two commensurate contacts: in AA stacking, so a perfect alignment (Figure 7a,d), and AB stacking, so a shifted alignment (Figure 7c,f). Those stackings correspond to commensurate contact angles of 0, 60, 180, 240, and 300°. Figure 7 also shows an example of incommensurate contact in Figure 7b,e.

Because the structure is similar, it is believed that clays could also act as lubricants in molten polymer if the superlubricity phenomenon is really the main lubrication mechanism.

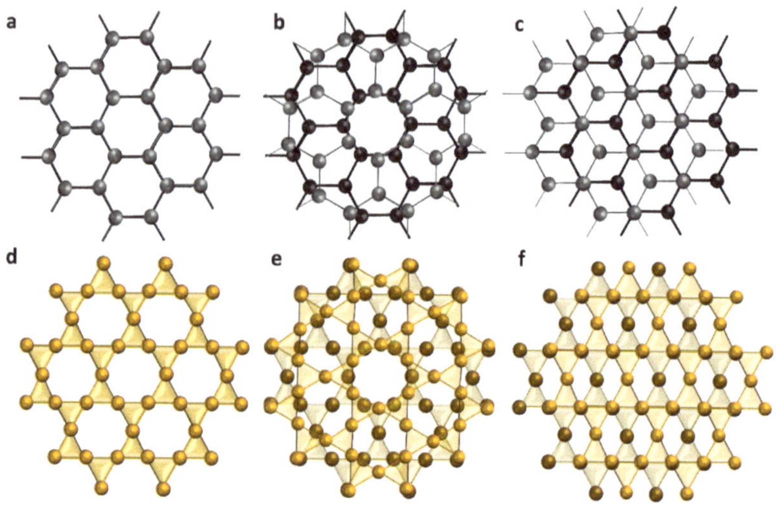

Figure 7. Two outer layers of (**a**) graphene and (**d**) clays in commensurate contact, AA stacking. Two outer layers of (**b**) graphene and (**e**) clay in incommensurate contact. Two outer layers of (**c**) graphene and (**f**) clay in commensurate contact, AB stacking.

As a matter of fact, a previous work of Genoyer et al. [30] showed a decrease in the magnitude of the complex viscosity when montmorillonite (MMT) and laponite (L) were added in PMMA. The complex viscosities obtained by oscillatory shear in this study are shown in Figure 8, showing that some of the PMMA composites displayed this lubricant phenomenon.

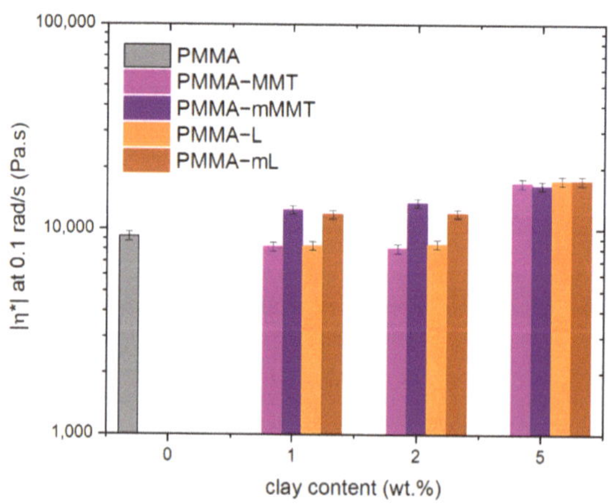

Figure 8. Magnitude of complex viscosities at 0.1 rad/s and 200 °C of PMMA containing 0 to 5 wt.% of clay particles: montmorillonite (MMT), modified montmorillonite (mMMT), laponite (L), and modified laponite (mL) [30].

In particular, 1 and 2 wt.% of MMT and L led to a significant decrease in viscosity by 10 to 12%, respectively. However, mMMT (modified montmorillonite) and mL (modified laponite) at the same concentrations led to an increase in viscosity. The only difference between neat clays (MMT and L) and modified clays (mMMT and mL) is the chemical modification. Clays are modified using di(hydrogenated tallow)dimethylammonium chloride, which intercalates in between the silicate sheets and helps to exfoliate and thus to improve the particles' dispersion. Based on the analysis, it can be inferred that a higher degree of dispersion results in a decreased lubricating effect. Such an observation provides further evidence for the conclusion drawn for graphene and supports the idea that clays can also induce a lubricant effect.

As this previous study was carried out with a different polymer matrix and the rheological measurement was in oscillatory mode instead of the constant shear used in the present study, it was necessary to verify the clay's ability in the exact same condition as graphene. To do so, MMT and L, which were the most efficient at reducing viscosity in the previous study, were added to the same PS matrix as the studied graphene grades and underwent the same processing history.

Figure 9 shows the steady shear viscosities of PS composites containing 1 and 2 wt.% of MMT and L. At first glance, it can be seen that the increase at low shear rates is more significant for clays, indicating that they produce a particle network with lower contents than graphene does.

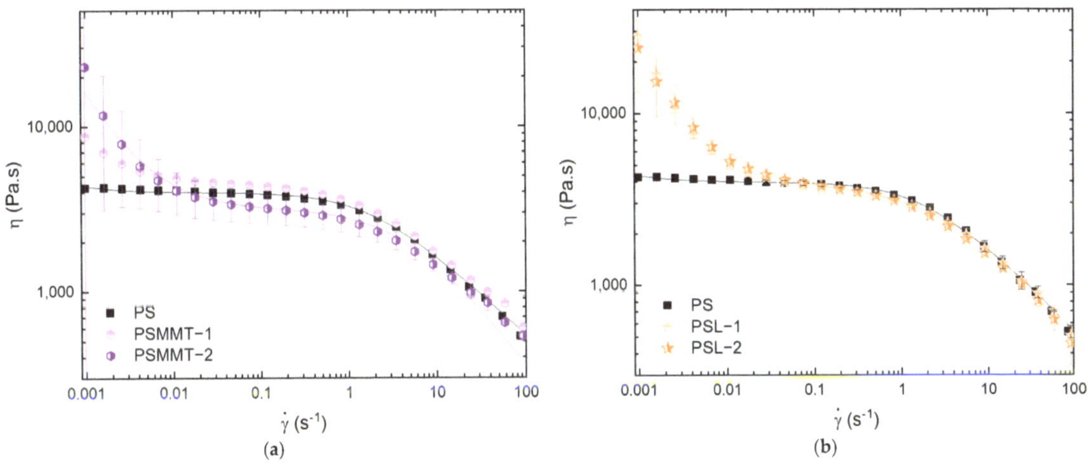

Figure 9. Steady shear viscosity as a function of shear rate of (**a**) PSMMT and (**b**) PSL composites. Lines represent the fitting of modified Carreau–Yasuda model to shear viscosities.

To better assess the clays' effect, the Carreau–Yasuda with yield stress model, Equation (3), was fitted to the shear viscosities. The zero-shear viscosities, relaxation times, and yield stresses obtained are reported in Figure 10. It can be seen that the viscosities are all close to those of neat PS; however, a slight decrease is achieved for 2 wt.% of MMT and 1 wt.% of L. The relaxation times of all composites are lower than those of neat PS. Finally, the yield stresses (Figure 10b) of those composites clearly indicate that a network of particles is created starting at 1 wt.% of L and 2 wt.% of MMT. Those results seem to point out a very good distribution of L which percolates at a lower content. Also, at 2 wt.% of MMT, the zero-shear viscosity decreases while the yield stress value increases, indicating a competition between the lubricant effect and network formation.

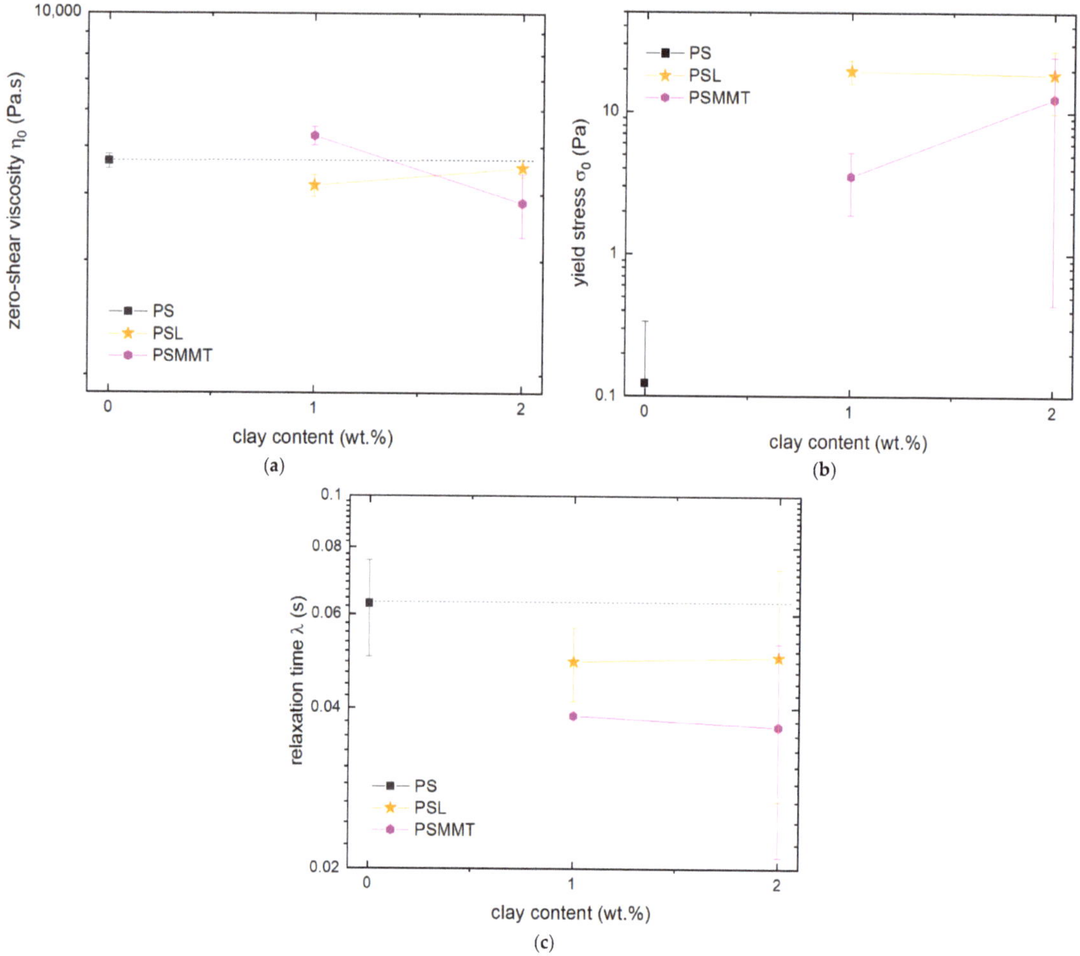

Figure 10. (**a**) Zero-shear viscosities, (**b**) yield stresses, and (**c**) relaxation times obtained by fitting modified Carreau–Yasuda model to shear viscosities of PS composites containing MMT and L.

SEM was used to observe the dispersion state of MMT and L in PS, and the pictures are shown in Figure 11. The lateral size of the particles was measured, and it is reported in Table 4. L particles did aggregate, and the lateral size measured by SEM was larger than the one given by the supplier. Figure 11 illustrates some agglomerates of L and MMT found within the samples. The overall size of L agglomerates is smaller than for MMT as the D50 of PSL-2 is significantly lower than that of PSMMT-2. Contrary to the observations made in 0X composites, for both MMT and L composites, no rich zones of particles were encountered in the matrix.

Table 4. Size of agglomerates of MMT and L in PS observed using SEM.

Sample	D50 (μm)	D90 (μm)
PSMMT−2	11.8	35.2
PSL−2	4.0	37.2

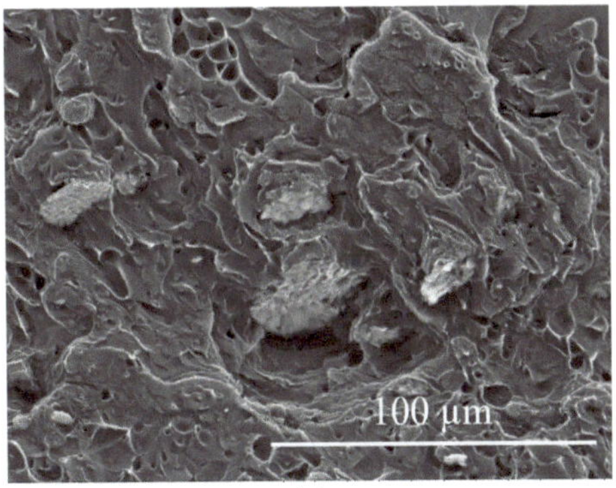

(a)

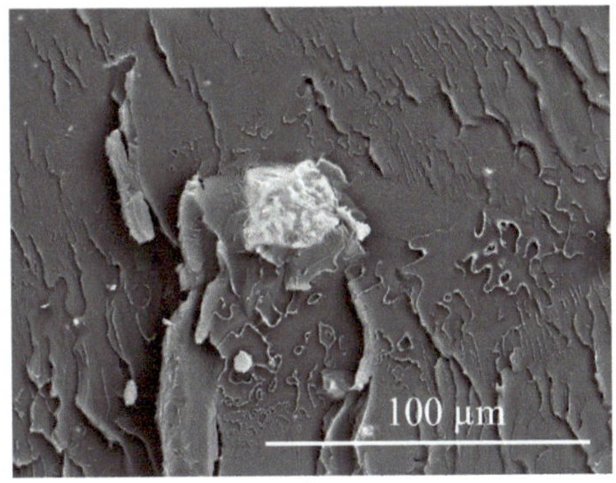

(b)

Figure 11. Scanning electron microscopy images of PS with 2 wt.% of (**a**) MMT and (**b**) L.

In the case of clays such as montmorillonite and laponite, the layers are held together by dipole–dipole interactions of the hydrogen bonding, creating stronger interactions than other Van der Waals bonding. The clays' outer layer in contact with the polymer is composed of a SiO_4 tetrahedra structure with oxygen atoms on the outer line (see Figure 7), which enable the formation of hydrogen bonds. However, PS cannot form hydrogen bonds, so the interactions between PS and clays are limited to Van der Waals, which also explains clays' tendency to agglomerate instead of dispersing in PS. To help dispersion in polymers, clay organic modification can be conducted by cation exchange with organic molecules [30]. However, chemical modification of the molecules was not considered in this work since a certain degree of particle agglomeration is a key factor for witnessing a lubricant effect.

3.4. Comparison of Graphene and Clays

The reduction in zero-shear viscosities as a function of D50 for all composites at a content of 2 wt.% of particles is reported in Figure 12. As discussed previously, graphene 0X leads to the largest reduction in viscosity by 49%. Meanwhile, graphene 3X and MMT have a similar effect, leading to 29% and 22% reduction, respectively, and the least efficient lubricants are graphene H and L, which show no significant reduction in zero-shear viscosity.

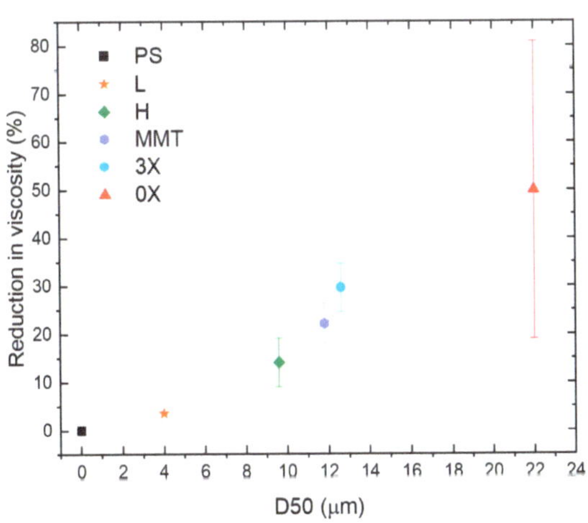

Figure 12. Reduction in zero-shear viscosity of composites containing 2 wt.% of particles as a function of the D50 of agglomerates measured by SEM. The point at 0 µm corresponds to neat PS.

By plotting those reductions in viscosity as a function of the D50 of the agglomerated particles, it becomes clear that the bigger the D50, the more reduction in viscosity. Indeed graphene 3X and MMT lead to similar reductions in the viscosity of the PS matrix and similar D50 values of 12.6 μm and 11.8 μm. Graphene H and L have very different initial primary particle lateral sizes of 25 μm and 0.03 μm, but as their D50s are the two lowest D50s of the group with 9.6 and 4.0 μm, respectively, their efficiency as processing aids is also the lowest.

To conclude, the composites with clays and graphene with a similar dispersion state have similar viscosity, thus confirming that the mechanism of lubrication is not linked to chemistry but rather to the structure and dispersion state of particles. Because of that, the superlubricity ability of such materials is a strong proposition. Indeed, as superlubricity is due to slipping between layered particles, an agglomerated state of those particles will promote the effect, because in an agglomerated state, particles are all close to one another without being strongly linked. Furthermore, random orientation, and thus a random contact angle, will enhance the superlubricity, as only a few specific angles induce friction, as described in Figure 4.

It can also be concluded that in this work, graphene was more suitable than clay as a lubricant because it was more susceptible to presenting randomly distributed agglomerations

4. Conclusions

Introducing low contents of layered particles in polymer matrices reduces viscosity, enabling applications related to eased thermoplastics processing, energy savings, and improved process efficiency. In this study, we investigated the impact of adding graphene and clays with varying lateral sizes on the viscosity of a melted polymer matrix.

First, it was shown that the optimum particle concentration to reduce the viscosity of the molten polymer matrix was 2 wt.% for both graphene and clay. At higher contents, the creation of a strong particle network dominates the viscoelastic response, leading to a competition between the lubricant effect, i.e., a reduction in zero-shear viscosity, and the creation of a network, i.e., the increase at low shear rates represented by the yield stress.

Second, at the optimum concentration, it was found that the bigger the agglomerates of particles, the more significant the reduction in viscosity. Graphene 0X was the most efficient particle at reducing the viscosity on the PS system by 49% because this graphene tended to produce rich filler zones with many agglomerates in close contact during the melt-mixing process. Overall, it was shown that the percentage of viscosity reduction increases as a function of the agglomerates' D50.

Finally, it was concluded that the lubricant effect mechanism of both graphene and clay particles is similar and likely to be a so-called superlubricity phenomenon. The phenomenon of superlubricity, where layered particles slide smoothly against each other, creates extra space between the particles, allowing the polymer chains to relax more quickly and resulting in lower zero-shear viscosity.

Author Contributions: Conceptualization, J.G., E.H., G.G. and N.M.; methodology, J.G., N.R.D. and E.D.; validation, J.G., E.H., E.D. and N.R.D.; formal analysis, J.G.; investigation, J.G.; resources, E.H., G.G., E.D., N.M. and N.R.D.; data curation, J.G.; writing—original draft preparation, J.G.; writing—review and editing, E.H., G.G., N.R.D., E.D. and N.M.; visualization, J.G.; supervision, E.H., G.G., N.R.D., E.D. and N.M.; project administration, E.H., G.G., E.D. and N.R.D.; funding acquisition, E.D. and N.R.D. All authors have read and agreed to the published version of the manuscript.

Funding: This research was funded by NanoXplore Inc.; Natural Sciences and Engineering Research Council, grant number CRDPJ 538482-18; PRIMA Quebec, grand number R18-46-001; Fonds de Recherche du Québec-Nature et Technologies, grant number 318813.

Institutional Review Board Statement: Not applicable.

Data Availability Statement: The author confirms that the data supporting the reported results can be found within the article. Raw data that support the findings of this work can be requested from the corresponding authors upon reasonable request.

Acknowledgments: The author would like to thank Catherine Dufour for the preliminary results obtained during her internship.

Conflicts of Interest: The authors declare no conflict of interest.

References

1. Achilleos, E.C.; Georgiou, G.; Hatzikiriakos, S.G. Role of processing aids in the extrusion of molten polymers. *J. Vinyl Addit. Technol.* **2002**, *8*, 7–24. [CrossRef]
2. Kazatchkov, I.B.; Yip, F.; Hatzikiriakos, S.G. The effect of boron nitride on the rheology and processing of polyolefins. *Rheol. Acta* **2000**, *39*, 583–594. [CrossRef]
3. Yip, F.; Diraddo, R.; Hatzikiriakos, S.G. Effect of combining boron nitride with fluoroelastomer on the melt fracture of HDPE in extrusion blow molding. *J. Vinyl Addit. Technol.* **2000**, *6*, 196–204. [CrossRef]
4. Adesina, A.A.; Nasser, M.S.; Hussein, I.A. Comparative Analysis of the Effect of Organoclay, Boron Nitride, and Fluoropolymer on the Rheology and Instabilities in the Extrusion of High Density Polyethylene. *Int. J. Polym. Sci.* **2015**, *2015*, 190537. [CrossRef]
5. Gul, S.; Arican, S.; Cansever, M.; Beylergil, B.; Yildiz, M.; Okan, B.S. Design of Highly Thermally Conductive Hexagonal Boron Nitride- Reinforced PEEK Composites with Tailored Heat Conduction Through-Plane and Rheological Behaviors by a Scalable Extrusion. *Appl. Polym. Mater.* **2023**, *5*, 329–341. [CrossRef]
6. Emre Bozkurt, Y.; Yıldız, A.; Turkarslan, O.; Sasal, F.N.; Cebeci, H. Thermally conductive h-BN reinforced PEI composites: The role of processing conditions on dispersion states. *Mater. Today Commun.* **2021**, *29*, 102854. [CrossRef]
7. Çelik, O.N.; Ay, N.; Göncü, Y. Effect of nano hexagonal boron nitride lubricant additives on the friction and wear properties of AISI 4140 steel. *Part. Sci. Technol.* **2013**, *31*, 501–506. [CrossRef]
8. Sagbas, B. Tribological performance of peek with green lubricant enhanced by nano hexagonal boron nitride powder. *Ind. Lubr. Tribol.* **2020**, *72*, 203–210. [CrossRef]
9. Bragg, W. An Introduction to Crystal Analysis. *Nature* **1928**, *122*, 915–916.
10. Liu, L.; Zhou, M.; Jin, L.; Li, L.; Mo, Y.; Su, G.; Tian, Y. Recent advances in friction and lubrication of graphene and other 2D materials: Mechanisms and applications. *Friction* **2019**, *7*, 199–216. [CrossRef]
11. Liu, Z.; Zhang, S.M.; Yang, J.R.; YangLiu, J.Z.; Yang, Y.L.; Zheng, Q.S. Interlayer shear strength of single crystalline graphite. *Acta Mech. Sin.* **2012**, *28*, 978–982. [CrossRef]
12. Dienwiebel, M.; Pradeep, N.; Verhoeven, G.S.; Zandbergen, H.W.; Frenken, J.W.M. Model experiments of superlubricity of graphite. *Surf. Sci.* **2005**, *576*, 197–211. [CrossRef]
13. Dienwiebel, M.; Verhoeven, G.S.; Pradeep, N.; Frenken, J.W.M. Superlubricity of Graphite. *Phys. Rev. Lett.* **2004**, *92*, 126101. [CrossRef]
14. Feng, X.; Kwon, S.; Park, J.Y.; Salmeron, M. Superlubric Sliding of Graphene Nanoflakes on Graphene. *ACS Nano* **2013**, *7*, 1718–1724. [CrossRef]
15. Karatas, E.; Gul, O.; Karsli, N.G.; Yilmaz, T. Synergetic effect of graphene nanoplatelet, carbon fiber and coupling agent addition on the tribological, mechanical and thermal properties of polyamide 6,6 composites. *Compos. Part B Eng.* **2019**, *163*, 730–739. [CrossRef]
16. Kurusu, R.S.; Helal, E.; Moghimian, N.; David, E.; Demarquette, N. The Role of Selectively Located Commercial Graphene Nanoplatelets in the Electrical Properties, Morphology, and Stability of EVA/LLDPE Blends. *Macromol. Mater. Eng.* **2018**, *303*, 1–9. [CrossRef]
17. Helal, E.; Kurusu, R.S.; Moghimian, N.; Gutierrez, G.; David, E.; Demarquette, N.R. Correlation between morphology, rheological behavior, and electrical behavior of conductive cocontinuous LLDPE/EVA blends containing commercial graphene nanoplatelets. *J. Rheol.* **2019**, *63*, 961–976. [CrossRef]
18. Salari, M.; Sansone, N.D.; Razzaz, Z.; Taromsari, S.M.; Leroux, M.; Park, C.B.; Lee, P.C. Insights into synergy-induced multifunctional property enhancement mechanisms in hybrid graphene nanoplatelet reinforced polymer composites. *Chem. Eng. J.* **2023**, *463*, 142406. [CrossRef]
19. Karimi, S.; Helal, E.; Gutierrez, G.; Moghimian, N.; Madinehei, M.; David, E.; Samara, M.; Demarquette, N.R. A review on graphene's light stabilizing effects for reduced photodegradation of polymers. *Crystals* **2021**, *11*, 3. [CrossRef]
20. Silibin, M.; Karpinsky, D.; Bystrov, V.; Zhaludkevich, D.; Bazarova, M.; Mirzadeh Vaghefi, P.; Marques, P.A.A.P.; Singh, B.; Bdikin, I. Preparation, Stability and Local Piezoelectrical Properties of P(VDF-TrFE)/Graphene Oxide Composite Fibers. *C* **2019**, *5*, 48. [CrossRef]
21. Diallo, A.K.; Helal, E.; Gutierrez, G.; Madinehei, M.; David, E.; Demarquette, N.; Moghimian, N. Graphene: A multifunctional additive for sustainability. *Sustain. Mater. Technol.* **2022**, *33*, e00487. [CrossRef]
22. Sultana, S.M.N.; Helal, E.; Gutiérrez, G.; David, E.; Moghimian, N.; Demarquette, N.R. Effect of Few-Layer Graphene on the Properties of Mixed Polyolefin Waste Stream. *Crystals* **2023**, *13*, 358. [CrossRef]
23. Moghimian, N.; Nazarpour, S. The future of carbon: An update on graphene's dermal, inhalation, and gene toxicity. *Crystals* **2020**, *10*, 718. [CrossRef]

24. Bai, Y.; Zhou, S.; Lei, X.; Zou, H.; Liang, M. Enhanced thermal conductivity of polycarbonate-based composites by constructing a dense filler packing structure consisting of hybrid boron nitride and flake graphite. *J. Appl. Polym. Sci.* **2022**, *139*, e52895. [CrossRef]
25. Ferreira, E.H.C.; Andrade, R.J.E.; Fechine, G.J.M. The 'Superlubricity State' of Carbonaceous Fillers on Polyethylene-Based Composites in a Molten State. *Macromolecules* **2019**, *52*, 9620–9631. [CrossRef]
26. Muñoz, P.A.R.; de Oliveira, C.F.P.; Amurin, L.G.; Rodriguez, C.L.C.; Nagaoka, D.A.; Tavares, M.I.B.; Domingues, S.H.; Andrade, R.J.E.; Fechine, G.J.M. Novel improvement in processing of polymer nanocomposite based on 2D materials as fillers. *Express Polym. Lett.* **2018**, *12*, 930–945. [CrossRef]
27. Vega, J.F.; Martinez-Salazar, J.; Trujillo, M.; Arnal, M.L.; Muller, A.J.; Bredeau, S.; Dubois, P. Rheology, processing, tensile properties, and crystallization of polyethylene/carbon nanotube nanocomposites. *Macromolecules* **2009**, *42*, 4719–4727. [CrossRef]
28. Hirano, M.; Shinjo, K. Superlubricity and frictional anisotropy. *Wear* **1993**, *168*, 121–125. [CrossRef]
29. Claerbout, V.E.P.; Polcar, T.; Nicolini, P. Superlubricity achieved for commensurate sliding: MoS 2 frictional anisotropy in silico. *Comput. Mater. Sci.* **2019**, *163*, 17–23. [CrossRef]
30. Genoyer, J.; Demarquette, N.R.; Soulestin, J. Effect of clay particles size and location on coalescence in PMMA/PS blends. *J. Rheol.* **2019**, *63*, 883–893. [CrossRef]
31. Seehra, M.S.; Geddam, U.K.; Schwegler-Berry, D.; Stefaniak, A.B. Detection and quantification of 2H and 3R phases in commercial graphene-based materials. *Carbon* **2015**, *95*, 818–823. [CrossRef]
32. Kim, T.H.; Jeon, E.K.; Ko, Y.; Jang, B.Y.; Kim, B.S.; Song, H.K. Enlarging the d-spacing of graphite and polarizing its surface charge for driving lithium ions fast. *J. Mater. Chem. A* **2014**, *2*, 7600–7605. [CrossRef]
33. Low, I.; Albetran, H.M.; Degiorgio, M. Structural Characterization of Commercial Graphite and Graphene Materials. *J. Nanotechnol. Nanomater.* **2020**, *1*, 23–30.
34. Jian, M.Q.; Xie, H.H.; Xia, K.L.; Zhang, Y.Y. Challenge and Opportunities of Carbon Nanotubes. In *Industrial Applications of Carbon Nanotubes*; Peng, H., Li, Q., Chen, T., Eds.; Elsevier Inc.: Amsterdam, The Netherlands, 2017; pp. 433–476.
35. Pérez, E.M.; Martín, N. π-π Interactions in carbon nanostructures. *Chem. Soc. Rev.* **2015**, *44*, 6425–6433. [CrossRef] [PubMed]
36. Georgakilas, V.; Tiwari, K.; Kemp, K.C.; Perman, J.A.; Bourlinos, A.B.; Kim, K.S.; Zboril, R. Noncovalent Functionalization of Graphene and Graphene Oxide for Energy Materials, Biosensing, Catalytic, and Biomedical Applications. *Chem. Rev.* **2016**, *116*, 5464–5519. [CrossRef] [PubMed]
37. Apátiga, J.L.; Del Castillo, R.M.; Del Castillo, L.F.; Calles, A.G.; Espejel-Morales, R.; Favela, J.F.; Compan, V. Non-covalent interactions on polymer-graphene nanocomposites and their effects on the electrical conductivity. *Polymers* **2021**, *13*, 1714. [CrossRef]
38. Macosko, C.W. *Rheology: Principles, Measurements, and Applications*; Wiley-VCH: New York, NY, USA, 1994.
39. Vergnes, B. The use of apparent yield stress to characterize exfoliation in Polymer Nanocomposites. *Int. Polym. Process.* **2011**, *26*, 229–232. [CrossRef]
40. Lertwimolnun, W.; Vergnes, B. Influence of compatibilizer and processing conditions on the dispersion of nanoclay in a polypropylene matrix. *Polymer* **2005**, *46*, 3462–3471. [CrossRef]
41. Genoyer, J.; Lentzakis, H.; Demarquette, N.R. Effect of the addition of cellulose filaments on the relaxation behavior of thermoplastics. *J. Rheol.* **2021**, *65*, 779–789. [CrossRef]
42. Mackay, M.E.; Dao, T.T.; Tuteja, A.; Ho, D.L.; Van Horn, B.; Kim, H.; Hawker, C.J. Nanoscale effects leading to non-einstein-like decrease in viscosity. *Nat. Mater.* **2003**, *2*, 762–766. [CrossRef]
43. Ferreira, E.H.C.; de Lima, L.P.; Fechine, G.J.M. The 'Superlubricity State' of Carbonaceous Fillers on Polymer Composites. *Macromol. Chem. Phys.* **2020**, *52*, 9620–9631.
44. Dealy, J.M.; Wang, J. *Melt Rheology, and Its Applications in the Plastics Industry*; Springer: Berlin/Heidelberg, Germany, 2013.
45. Tuteja, A.; Mackay, M.E.; Hawker, C.J.; Van Horn, B. Effect of ideal, organic nanoparticles on the flow properties of linear polymers: Non-einstein-like behavior. *Macromolecules* **2005**, *38*, 8000–8011. [CrossRef]
46. Brigatti, M.F.; Galán, E.; Theng, B.K.G. Structure and Mineralogy of Clay Minerals. In *Developments in Clay Science*; Bergaya, F., Lagaly, G., Eds.; Elsevier Inc.: Amsterdam, The Netherlands, 2013; Volume 5, pp. 21–81.

C Journal of Carbon Research

MDPI

Article

Electrical Resistance Evolution of Graphite and Talc Geological Heterostructures under Progressive Metamorphism

Augusto Gonçalves Nobre [1,*], Fabio Ramos Dias de Andrade [2], Andres Fabian Salazar-Naranjo [2], Josue Neroti Rigue [1], Ricardo Barreto da Silva [1], Silvio Roberto Farias Vlach [2] and Romulo Augusto Ando [3]

1 Center for Natural and Exact Sciences, Federal University of Santa Maria, Santa Maria 97105-900, RS, Brazil
2 Institute of Geosciences, University of São Paulo, São Paulo 05508-080, SP, Brazil
3 Institute of Chemistry, University of São Paulo, São Paulo 05508-900, SP, Brazil
* Correspondence: augusto.nobre@ufsm.br

Abstract: The electrical properties of isolated graphene established precedents for studies of electrical superconducting materials at room temperature. After the discovery of stabilized graphene and graphite nanoplatelets in a geological context, the interest in characterizing the properties of these minerals arose. This work evaluates the electrical resistance evolution of mineral graphite and talc heterostructures under progressive metamorphism simulated in the laboratory. The experiments were conducted on an end-loaded piston-cylinder apparatus. This equipment allows for the application of equal pressure in all sample directions (lithostatic pressure) and heating, simulating geological phenomena. The behavior of two sets of mineral samples were compared: graphite and talc in billets and powder. Samples in billets were submitted to treatments at 400 °C and 4 kbar; 400 °C and 6 kbar; and 700 °C and 9 kbar. The powder samples were subjected to 700 °C and 9 kbar, with two ways of disposing the mineral powders (mixed and in adjacent contact) beyond 900 °C and 9 kbar (in adjacent contact). The results show that the samples in billets had lower electrical resistance when compared to the powder samples. The lowest electrical resistance was observed in the sample treated at 400 °C and 6 kbar, conditions that are consistent with metamorphic mineral assemblage observed in the field. Powdered samples showed better cleavage efficiency during the experiment, resulting in thinner flakes and even graphene, as pointed out by Raman spectroscopy. However, these flakes were not communicating, which resulted in high electrical resistance, due to the need for an electrical current to pass through the talc, resulting in a Joule effect. The maximum electrical resistance obtained in the experiment was obtained in the sample submitted to 900 °C, in which talc decomposed into other mineral phases that were even more electrically insulating. This work demonstrates that electrical resistance prospecting can be an efficient tool to identify potential target rocks with preserved mineral nanometric heterostructures that can form an important raw material for the nanotechnology industry.

Keywords: applied mineralogy; experimental petrology; mineral technology; natural graphene; mineral feedstock

Citation: Nobre, A.G.; de Andrade, F.R.D.; Salazar-Naranjo, A.F.; Rigue, J.N.; da Silva, R.B.; Vlach, S.R.F.; Ando, R.A. Electrical Resistance Evolution of Graphite and Talc Geological Heterostructures under Progressive Metamorphism. C **2023**, 9, 75. https://doi.org/10.3390/c9030075

Academic Editor: Gil Gonçalves

Received: 21 June 2023
Revised: 22 July 2023
Accepted: 26 July 2023
Published: 30 July 2023

1. Introduction

The graphite mechanical microcleavage to obtain graphene was responsible for the emergence of two-dimensional (2D) materials science [1]. Among the outstanding properties of 2D materials are the electrical superconductivity at room temperature and high transparency, initially observed in graphene [2–4], which earned its authors the 2010 Nobel Prize in Physics [5]. Graphene's electrical and optical properties are key to the development of high-speed electronics [6–8], single molecule detection [9–11], high-capacity sensors [12,13], field-effect devices [14–16], and transparent electrodes devices [17–19].

Despite its strategic properties, graphene is still not widely available in industry [20,21]. The two main problems with the popularization of graphene devices are instability and the high cost of producing the material [22]. Graphene is unstable and tends to recombine with itself or other flakes in the sample. To avoid this phenomenon, graphene needs to be anchored in a substrate or dispersed in a solvent that reduces its surface energy [23,24]. Graphene production processes are still expensive, especially when dependent on synthetic raw materials [25]. In addition, most large-scale processes have low profitability and logistical difficulties (with storage and transport) [26].

Bøggild et al. [27] pointed out that the price of an ideal monolayer graphene flake in 2006 was 1 EUR/μm^2. In 2016, this value dropped to 1 EUR/cm^2, which represents a factor 10^8 reduction. The evolution of both top-down (from bulk precursors) and bottom-up (from chemical reagents) techniques led graphene production processes to become progressively more profitable, allowing the production of larger volumes of material at a lower operating cost. Even so, it is still of high value to make graphene an everyday material.

A new way of obtaining graphene and other 2D substances via top-down techniques is the extraction from mineral deposits [28]. There are tectono-structural contexts that allow graphite-bearing rocks to be sheared, forming graphite nanoplatelets and graphene [29]. These contexts are recorded in metamorphic rocks. Metamorphism is the process in which any type of rock undergoes structural and compositional transformations (mineralogically) in a solid state, becoming denser and more compact. This process occurs as a result of the increase in temperature, pressure, presence and composition of fluids, in addition to the time the rock was kept under these deformation conditions [30].

The thin crystals of graphite and graphene found in rocks have crystalline defects and compositional impurities but are stabilized in rock, remaining stable for over a hundred million years [31]. These nanomaterials require simple and low-cost processing and have the potential to be used for various functionalities that do not depend on perfect crystallinity and analytical purity [32–35].

In a geological context, graphene stabilization also needs to occur in a substrate. It is necessary that the rock has another mineral of placoid structure with van der Waals forces to anchor the exfoliated graphite [36]. Clay minerals [37,38], micas [39,40], molybdenite [41,42] and talc [43,44] are examples of minerals with these characteristics.

Talc is proving to be a prominent candidate as a mineral substrate for geological graphenes [45]. The epitaxial relationship between bulk minerals favors the maintenance of thin flakes, in addition to the relative abundance of talc and graphite in metamorphic rocks deformed by shear zones [45–48]. Therefore, talc was chosen as the graphite substrate in this study's experiments.

The possibility of prospecting, identifying, and extracting graphene or its by-products from rocks opens possibilities to transform it into a mineral commodity. This would be an important step towards popularizing its use in everyday products. These are mineral deposits of smaller dimensions, with less environmental impact for their extraction, with simpler processing than other mineral commodities that require large mining plants. This would enable the production of large volumes of raw material for the nanotechnology industry [49,50].

Graphene geology had its first mention in 2011 [51] and is an emerging interdisciplinary area between Material Sciences and Geosciences. Research in this area has been highlighted by demonstrating that the petrological genesis of graphite directly affects the properties of graphene and graphene oxide produced from it [52]. At the same time, this research line demonstrated the existence of rocks bearing stabilized graphene [33,47] and showed that the presence of graphenes and graphite nanoplatelets added quality to lithic artifacts [53].

As the variation in electrical resistance is one of the main mechanisms for recognizing the presence of graphene in rock during fieldworks, this property was chosen to characterize the samples of the proposed studies. Field reconnaissance is the first tool used in mineral resource prospecting surveys. In graphitic rocks, significant reductions in electrical resistivity are indicative that the mineral was cleaved in a privileged way, possibly being a graphene-bearing rock.

This experimental study proposes the formation simulation of fine talc and graphite heterostructures in different metamorphic contexts to evaluate which one presents the lowest electrical resistivity. In this way, the developed studies aimed to evaluate the electrical resistance behavior of bulk and nanometric talc and graphite geological heterostructures. The precursor minerals (graphite and talc) were treated under different conditions of pressure and temperature (simulating the metamorphic process) that favor minerals' mutual diffusion in the solid state. The minerals were placed in different ways in the sample holder so that the arrangement of precursor minerals as controllers of electrical resistance could be analyzed.

The cleavage efficiency of graphite was analyzed by Raman spectroscopy. Raman spectroscopy gained notoriety for the study of graphitic materials as it allows the relative measurement of the covalent bonds and van der Waals forces abundance in their structure. This analytical technique can represent whether an originally multilayer graphite passed to a 2D system after treatment or deformation.

The diffusion zones formed under experimental conditions were subjected to electrical resistance measurements. These results allowed us to infer which metamorphism conditions and which distribution of minerals in the rock are more favorable for the natural availability of heterostructures with low electrical resistance.

2. Materials and Methods

2.1. Precursor Minerals and Piston-Cylinder Apparatus

The study included the preparation and analysis of six samples. The thermobarometric treatment conditions chosen were 400 °C and 4 kbar; 400 °C and 6 kbar; 700 °C and 9 kbar; 900 °C and 9 kbar, which correspond to conditions of progressive metamorphism existing inside the Earth's continental crust.

The samples treated under 700 °C and 9 kbar were arranged in three different ways in the sample holder: (i) minerals in billets (named B); (ii) minerals in adjacent powders (named P); and (iii) minerals in homogeneous powder mixture (named MP). These configurations allowed for the evaluation of the contact relationship between precursor minerals for electrical resistance analysis purposes.

In this way, the nomenclature of the samples and their experimentation conditions were as follows:

- **B-400-4**: Graphite and talc in billets, treated at 400 °C and 4 kbar;
- **B-400-6**: Graphite and talc in billets, treated at 400 °C and 6 kbar;
- **B-700-9**: Graphite and talc in billets, treated at 700 °C and 9 kbar;
- **MP-700-9**: Graphite and talc in a homogeneous mixture of powders, treated at 700 °C and 9 kbar;
- **P-700-9**: Graphite and talc in adjacent contact powders, treated at 700 °C and 9 kbar;
- **P-900-9**: Graphite and talc in adjacent contact powders, treated at 900 °C and 9 kbar.

The end-loaded piston-cylinder apparatus is a set of equipment adapted from commercial devices that allows for the simulation of temperature and pressure conditions found in the Earth's crust and upper mantle [54]. Traditionally, it is used for melting and crystallization studies of geological materials [55]. In this study, the equipment simulated processes for obtaining and stabilizing natural graphite nanoplatelets and graphenes in rocks [45].

The studies were carried out with an end-loaded Bristol-type piston-cylinder apparatus, and the temperature was set up and monitored using a B-type thermocouple and a 2404 Eurotherm PID controller.

The precursor minerals were placed in contact to yield diffusion and mutual exfoliation inside the sample holder. The sample holder was a tube 6 mm long (L) and 2.6 mm in diameter (w), made of a $Au_{75}Pd_{25}$ alloy, which was welded at its ends to avoid material exchange with the outside. This metal capsule was inserted into a 3/4" concentric furnace assembly consisting of crushable MgO, which acts as a sample holder fixer, graphite heater, pyrex (B-silicate) glass (spacer), and NaCl sleeves to maintain the almost friction-free sample, besides steel plugs. An alumina two-bore insulator allowed thermocouple insertion and temperature measures on the capsule top-end, isolated from the thermocouple probe with a thin Al disk [56]. The 3/4″ furnace assembly used in the studies can be seen in Figure 1.

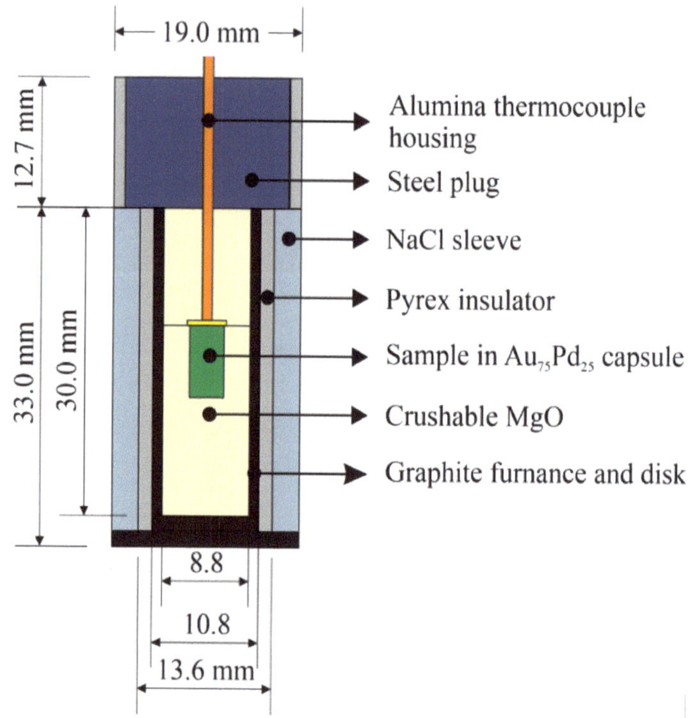

Figure 1. The 3/4″ furnace assembly used in samples processing at the piston-cylinder apparatus.

All samples were subjected to thermobarometric treatment for 24 h so that it was possible to observe diffusion zones. The experiments were carried out up to a pressure of 9 kbar, which ensures the phase stability of graphite [57]. Experiments carried out in temperatures up to 700 °C aimed to keep the talc stable in the system, as it transitions to other phases at higher temperatures, given the pressure conditions used [58]. The experiment conducted at 900 °C analyzed whether the progressive phase change from talc (Mg_3 Si_4 $O_{10}(OH)_2$) to enstatite ($MgSiO_3$), silica (SiO_2), and water (H_2O) was relevant to the electrical resistance of graphite nanoplatelets and graphene formed in the diffusion zone.

After completing the 24 h period, the power supply to the equipment was cut and the furnace set was cooled down to room temperature at a rate of approximately −30 °C/s. This cooling rate was fast enough to avoid phase re-equilibrium, which preserved the structures formed under the chosen thermobarometric conditions.

2.2. Diffusion Zones Characterization

The AuPd capsules after the experiments were sawn by microtome to expose the minerals and their diffusion zone and were fixed in epoxy resin to facilitate handling during characterization.

A diffusion zone between the mineral precursors contacts was observed with reflected-light optical microscopy (RLOM, Zeiss Axio Imager, equipped with the Zeiss AxioCam system) and scanning electron microscopy (SEM, LEO 440I) with compositional mode of backscattered electrons (BSE). The SEM analytical conditions were 20 kV voltage, working distance of 25 mm, and a sample current between 600 and 1200 pA.

Contact zones between graphite and talc were studied by Raman spectroscopy using a WITec Raman microscope alpha 300 R with a 633 nm excitation laser, with 30 accumulations in an integration time of 3 s.

Raman spectroscopy is a strategic characterization technique used to evaluate the systematic cleavage of van der Waals interactions in the graphite structure [59,60]. This technique measures the relative intensity between discrete graphite–graphene spectral bands, namely D (approximately 1350 cm^{-1}, corresponding to defects in covalent bonds), G (approximately 1580 cm^{-1}, associated with the sp^2 carbon atoms' plane vibration), and 2D (approximately 2700 cm^{-1}, linked to the structural arrangement of the two-dimensional plane) [61,62].

2.3. Electrical Resistance Tests

The electrical resistances of the samples were obtained using a two-probe method (in a modified version of the four-wire Kelvin method) [63,64]. Due to the small dimensions of the sample plane surfaces, the same pair of probes, separated by a distance (d) of 1.30 mm, was used to impinge the probe current and to measure the voltage on the device under test (DUT). The constant current values were adjusted with a precision digital ammeter connected in series with the DC source and the DUT, so the voltage drop was measured directly on the DUT using a precision digital voltmeter and a separated pair of wires. The high resistance of the digital voltmeter ensures the probe current flows predominantly through the DUT and that the measured voltage is the voltage drop (V)2 across the DUT, allowing an accurate resistance measurement. Figure 2 shows a scheme of the experimental setup.

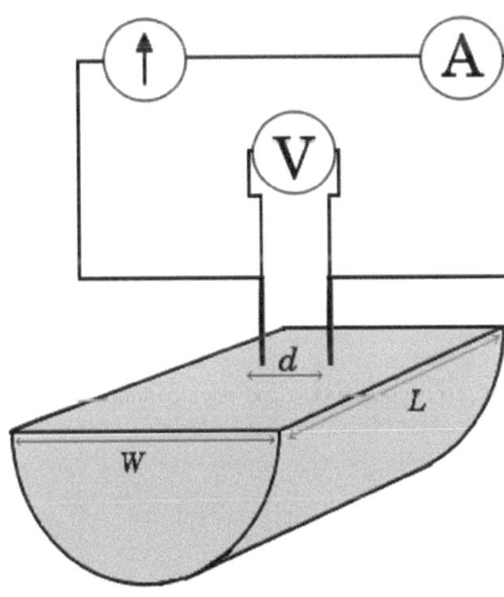

Figure 2. Experimental setup used for the electrical resistance measurement. L = 6 mm; w = 2.6 mm; d ≅ 1.30 mm; V: voltmeter; A: ammeter.

3. Results

3.1. Microscopy of Treated Samples

Figure 3 presents the RLOM result. In these images, it is possible to verify the distribution of talc and graphite, in addition to the formation of diffusion zones between the precursor minerals.

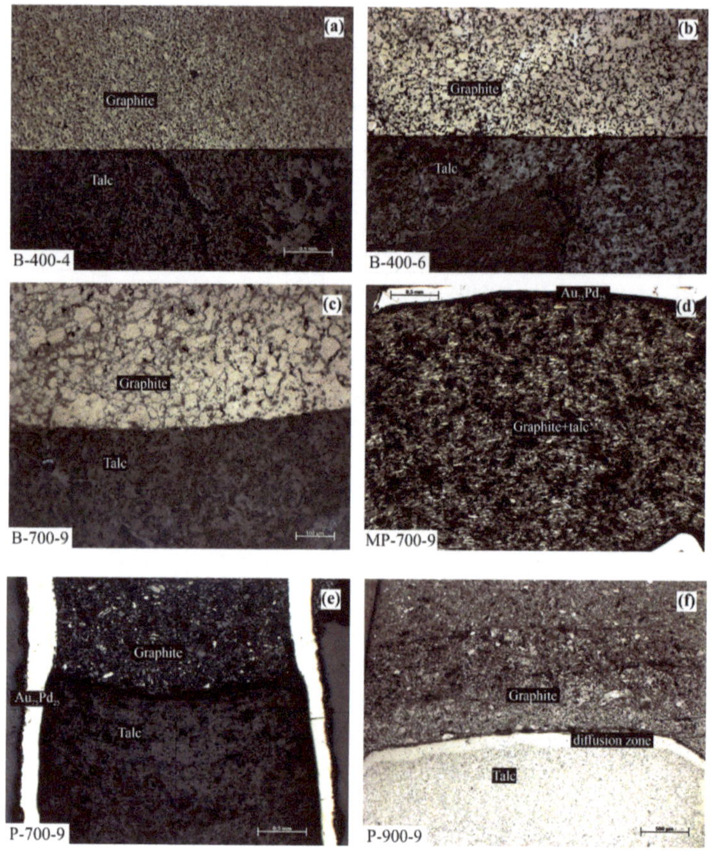

Figure 3. RLOM photomicrographs. (**a**) B-400-4; (**b**) B-400-6; (**c**) B-700-9; (**d**) MP-700-9; (**e**) P-700-9; (**f**) P-900-9.

Details of the interface zone between graphite and talc can be seen in the BSE-SEM images in Figure 4.

It was possible to verify that in all six samples, there was the formation of a diffusion zone, proving the success of experiments in simulating metamorphic rock structures from which the electrical resistance was measured.

The microscopy images showed that the samples treated under lower-pressure and temperature conditions (B-400-4 and B400-6) practically did not deform the flat surface of contact between the precursor minerals. This is indicative of an efficient mineral diffusion and intercalation process with low precursor crystal deformation, while in the other samples (MP-700-9, B-700-9, P-700-9, and P-900-9), despite having formed zones of efficient diffusion, the sample deformation was significant. The least efficient diffusion occurred in the MP-700-9 sample, as the minerals were already uniformly mixed before the thermobarometric treatment. The main phenomenon that occurred that was observed in this sample was the minerals' dynamic recrystallization. Among the samples treated equally at 700 °C and 9 kbar, sample B-700-9 had the formation of a low-strain diffusion zone with greater efficiency, while P-700-9 showed strong deformation and formation of microfractures in the diffusion zone.

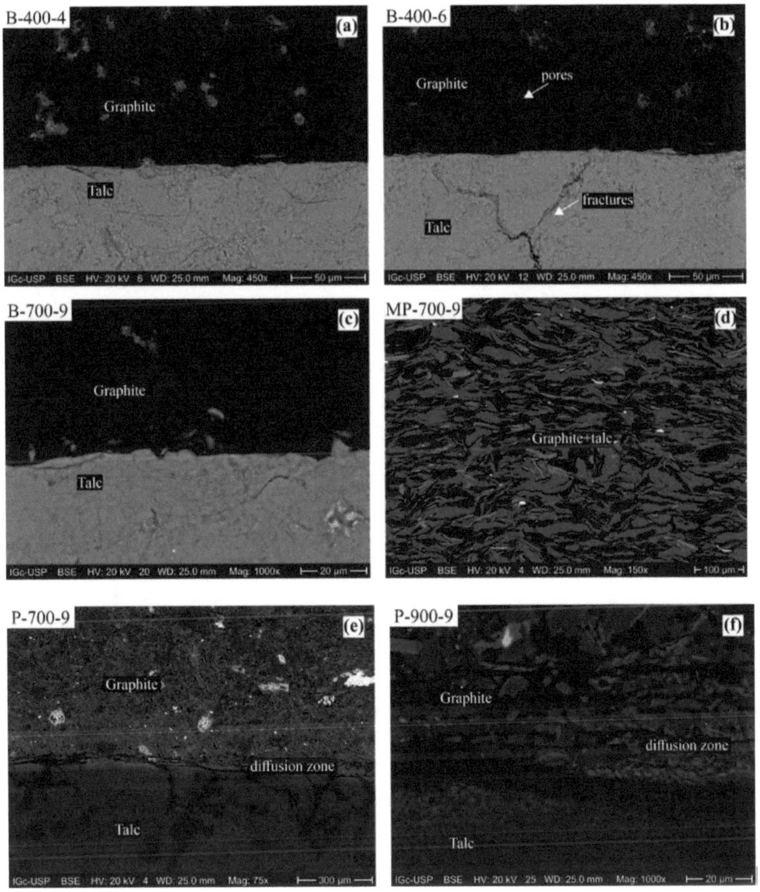

Figure 4. BSE-SEM photomicrographs. (**a**) B-400-4; (**b**) B-400-6; (**c**) B-700-9; (**d**) MP-700-9; (**e**) P-700-9; (**f**) P-900-9.

3.2. Raman Spectroscopy of Graphite and Graphene

Raman spectroscopy aimed to analyze the evolution of the graphite–graphene system from the pure graphite zone to the interior of the talc–graphite diffusion zone to show an effective disruption of the van der Waals forces in the graphitic structure. Graphene and graphite nanoplatelet formations are evidence of efficient exfoliation and, consequently, a reduction in electrical resistance in flakes in heterostructures. Figure 5 displays the Raman spectra of the six studied samples.

It is possible to observe that in all samples, there is a relative increase in the intensity of the 2D band in relation to the graphite G band in the diffusion zone. This is evidence of the efficient cleavage of graphite in all simulations of metamorphic conditions. It was also possible to observe an increase in the intensity of the D band in sample B-400-6 where the diffusion zone begins, revealing that before cleaving, the graphite deforms to form heterostructures. This happens because the graphitic structure is stressed before breaking the van der Waals interactions in the formation of heterostructures.

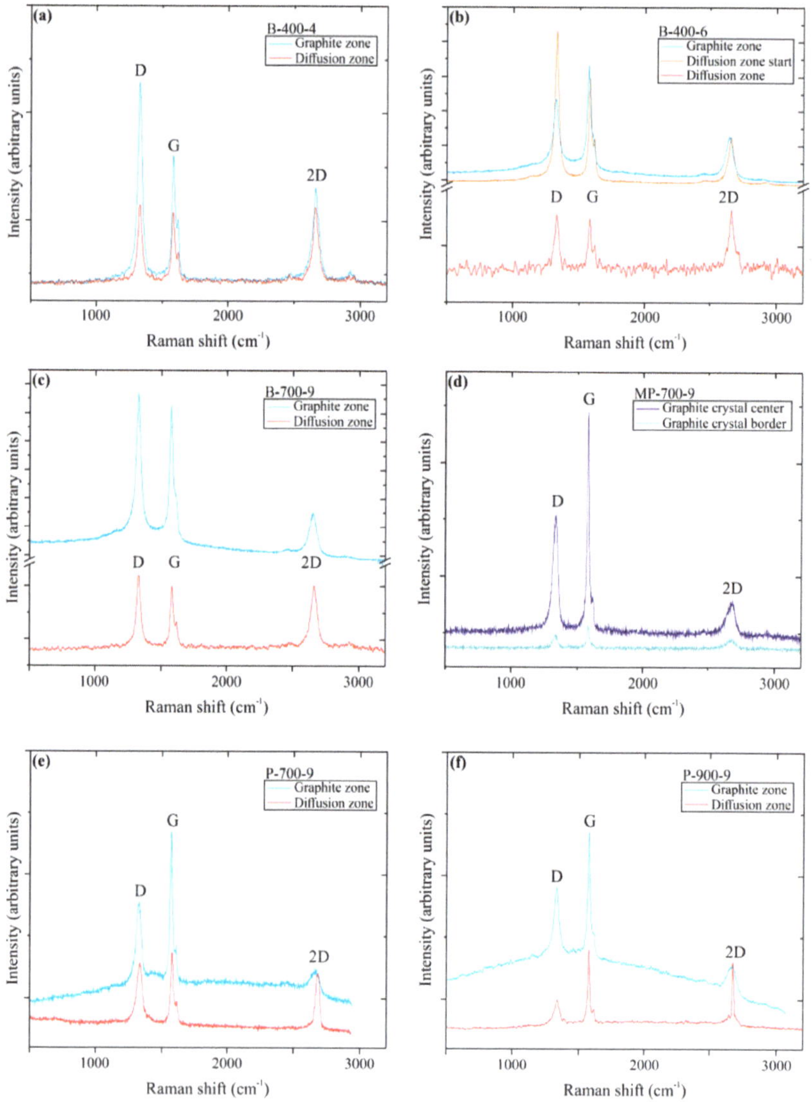

Figure 5. Results of graphite–graphene system Raman spectroscopy. (**a**) B-400-4; (**b**) B-400-6; (**c**) B-700-9; (**d**) MP-700-9; (**e**) P-700-9; (**f**) P-900-9.

3.3. Electrical Resistance Tests

The V vs. I curves are shown in the Figure 6. The symbols represent the experimental data, while the lines are the linear fits obtained using commercial data analysis software. The probe current values were varied from zero up to 5 mA and 2 mA for the low (Figure 6a) and high resistance (Figure 6b) samples, respectively. In these ranges of current values, all samples presented ohmic behavior, so the electrical resistance could be obtained from the Ohm law (V = RI) as the linear coefficients of the linear fits. The results are shown in Table 1.

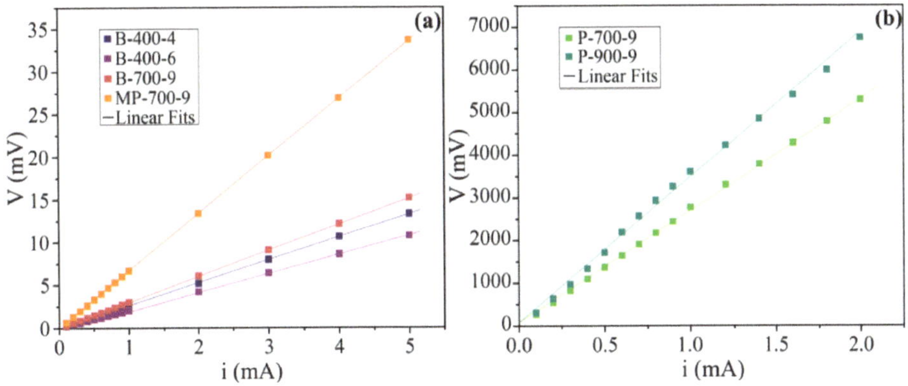

Figure 6. Measurements of electrical resistance as a function of V and I. (**a**) The V vs. I curves for the low-resistance samples. (**b**) The V vs. I curves for the high-resistance samples.

Table 1. Resistance values obtained from the linear fit of the V vs. I curves.

Sample	Resistance (Ω)
B-400-4	2.67
B-400-6	2.17
B-700-9	3.05
MP-700-9	6.51
P-700-9	2655.00
P-900-9	3583.00

4. Discussion

Six samples were processed using a piston-cylinder apparatus to simulate the metamorphic behavior of natural talc and graphite for the formation of heterostructures. The choice of 400 °C and 4 kbar refers to conditions where these structures have already been reported in the literature [65,66]. Based on these conditions, the aim was to assess the progression of metamorphism for the cleavage and generation of graphite heterostructures in another mineral substrate.

All experiments were successful in forming a diffusion zone after 24 h of processing. The micrographs reveal that the samples treated at 700 °C and 900 °C deformed severely, losing the flat faces between the precursor minerals. The experiments at 400 °C, on the other hand, formed diffusion zones but preserved the original structure of the precursor minerals in the sample holder.

The graphite zone showed uniform behavior in all samples following Raman spectroscopy. The G band in all graphite zones was always more intense than the 2D band. Similarly, Raman spectra acquired from the diffusion zone of all experimental products revealed an approximation of the intensity of the G and 2D bands. This represents a significant reduction in the van der Waals forces of the mineral structure compared to its covalent bonds, which is indicative of efficient cleavage.

Considering the potential of graphene and its associated materials as an electrical superconductor at room temperature, the electrical resistivity of the heterostructures formed in the diffusion zone was measured to verify which metamorphism conditions allowed the formation of less resistive flakes.

Figure 7 shows the electrical resistance as a function of the samples. In Figure 7a, it is possible to observe the electrical resistance of the samples in billets, showing that the conditions of lower resistance are those observed in B-400-6. Figure 7b reveals the electrical behavior of the powdered samples, highlighting the greater resistance obtained in the experiments on the P-900-9 sample. As there was the beginning of the talc phase transition and the formation of other more resistive minerals, the diffusion zone did not allow the fine flakes formed to communicate and reduce the electrical conductivity in this sample. Figure 7c shows the comparison of samples treated at 700 °C and 9 kbar under different disposition of precursor minerals. Figure 7c shows the comparison of samples treated at 700 °C and 9 kbar under different disposition of precursor minerals. It is possible to observe that the samples in adjacent powders are not efficient in communicating fine heterostructures of low electrical resistance, whereas in the samples in billets, diffusion occurs in the most prominent way and maintains the communicating flakes. The sample in homogeneous mixed powders presents communicant flakes; however, the exfoliation of the graphitic structure is less prominent, which justifies the intermediate resistance. Figure 7d summarizes all electrical characterization results as a function of the pressure and temperature conditions of each sample.

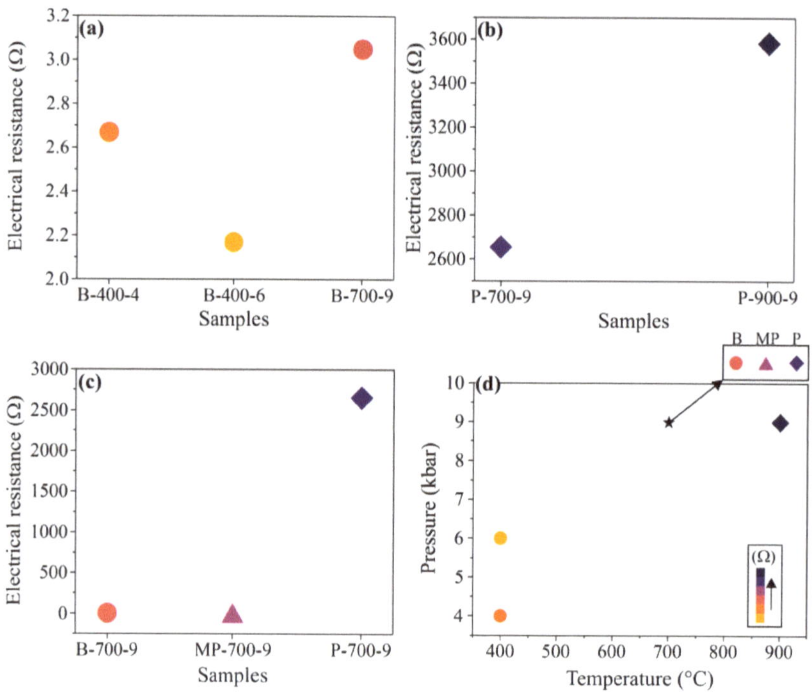

Figure 7. Compilation of electrical resistance data as a function of thermobarometric treatment. (**a**) Samples in billets. (**b**) Adjacent powder samples. (**c**) Comparison of samples arranged differently in the sample holder and uniformly treated at 700 °C and 9 kbar. (**d**) Compilation of electrical resistance as a function of pressure and temperature of each evaluated sample.

The highest values of electrical resistance were obtained in samples P-700-9 and P-900-9. As the P-900-9 sample surpassed the talc stability temperature at 9 kbar, there was the beginning of the formation of other mineral phases (such as enstatite and quartz) that do not have a planar structure and do not act as an efficient substrate for exfoliated graphites. Despite the treatment conditions of the P-700-9 sample maintaining the stability of the precursor minerals, the high pressure is more efficient in deforming the precursor crystals than in allowing the interdigitation of talc and graphite, fundamental for the formation of heterostructures with lower electrical resistance.

5. Conclusions

The results indicate that graphene formation is feasible in metamorphism and shearing of graphite and talc-bearing rocks. If talc is the mineral substrate for natural graphenes, the most suitable metamorphism temperature occurs below 700 °C, guaranteeing the thermal stability of talc.

To ensure the barometric stability of graphite, it is important that the pressure does not exceed 9 kbar. At the same time, it was possible to observe that pressures above 6 kbar are severe enough to deform the minerals in a non-planar way, which compromises the formation of communicating heterostructures that are, consequently, less electrically resistive.

The most efficient way to form heterostructures of low electrical resistance is from precursor minerals in billets. This arrangement favors the formation of communicating heterostructures that are more electrically conductive.

It was possible to verify that graphene and other 2D structures can be prospected in natural rocks when there are favorable mineralogical and tectonic contexts for their formation and stabilization. The electrical resistance survey can be used as a preliminary analysis tool, in fieldwork or even in a laboratory environment, to assess whether a given metamorphic graphite-bearing rock may have stabilized graphite nanoplatelets and graphene in its structure. Based on systematic surveys and discoveries of areas where this phenomenon is complete, it will be possible to obtain graphene in abundance as mineral resources.

Author Contributions: Conceptualization, A.G.N.; Methodology, A.G.N. and F.R.D.d.A.; Validation, A.G.N. and F.R.D.d.A.; Formal analysis, A.G.N., F.R.D.d.A., A.F.S.-N., J.N.R., R.B.d.S., S.R.F.V. and R.A.A.; Investigation, A.G.N., F.R.D.d.A., A.F.S.-N., J.N.R., R.B.d.S., S.R.F.V. and R.A.A.; Resources, A.G.N., F.R.D.d.A., A.F.S.-N., J.N.R., R.B.d.S., S.R.F.V. and R.A.A.; Data curation, A.G.N., F.R.D.d.A., A.F.S.-N., J.N.R., R.B.d.S., S.R.F.V. and R.A.A.; Writing—original draft, A.G.N. and F.R.D.d.A.; Writing—review & editing, A.G.N., F.R.D.d.A. and A.F.S.-N. Supervision, A.G.N. and F.R.D.d.A.; Project administration, A.G.N. Funding acquisition, S.R.F.V. All authors have read and agreed to the published version of the manuscript.

Funding: This research received no external funding.

Data Availability Statement: All the raw data used in this work can be obtained by contacting the corresponding author via email request.

Acknowledgments: The authors would like to thank the Laboratory of Petrology and Experimental Geochemistry of the Multiusuary Geoanalytical Center (CMGeo), in addition to the Laboratory of Petrographic Microscopy Laboratory and Laboratory of Scanning Electronic Microscopy (LabMev) of the IGc-USP for carrying out, respectively, the test on the piston-cylinder apparatus and microscopic characterization of the samples; the Laboratory of Molecular Spectroscopy (LEM) at IQ-USP for the Raman spectroscopy characterization; and the Laboratory of Magnetism and Magnetic Materials at Federal University of Santa Maria (UFSM) for electrical tests and Professor. The authors also thank Romario Trentin from UFSM who made the initial electrical tests possible.

Conflicts of Interest: The authors declare no conflict of interest. The funders had no role in the design of the study; in the collection, analyses, or interpretation of data; in the writing of the manuscript; or in the decision to publish the results.

References

1. Geim, A.K.; Novoselov, K.S. The rise of graphene. *Nat. Mater.* **2007**, *6*, 183–191. [CrossRef]
2. Novoselov, K.S.; Geim, A.K.; Morozov, S.V.; Jiang, D.; Zhang, Y.; Dubonos, S.V.; Grigorieva, I.V.; Firsov, A.A. Electric Field Effect in Atomically Thin Carbon Films. *Science* **2004**, *306*, 666–669. [CrossRef] [PubMed]
3. Schwierz, F. Graphene Transistors: Status, Prospects, and Problems. *Proc. IEEE* **2013**, *101*, 1567–1584. [CrossRef]
4. Tiwari, S.K.; Sahoo, S.; Wang, N.; Huczko, A. Graphene research and their outputs: Status and prospect. *J. Sci. Adv. Mater. Devices* **2020**, *5*, 10–29. [CrossRef]
5. Geim, A.K. Graphene: Status and Prospects. *Science* **2009**, *324*, 1530–1534. [CrossRef]
6. Liao, L.; Lin, Y.C.; Bao, M.; Cheng, R.; Bai, J.; Liu, Y.; Qu, Y.; Wang, K.L.; Huang, Y.; Duan, X. High-speed graphene transistors with a self-aligned nanowire gate. *Nature* **2010**, *467*, 305–308. [CrossRef]
7. Sreeprasad, T.S.; Berry, V. How Do the Electrical Properties of Graphene Change with its Functionalization? *Small* **2012**, *9*, 341–350. [CrossRef]

8. Cusati, T.; Fiori, G.; Gahoi, A.; Passi, V.; Lemme, M.C.; Fortunelli, A.; Iannaccone, G. Electrical properties of graphene-metal contacts. *Sci. Rep.* **2017**, *7*, 5109. [CrossRef]
9. Guan, J.; Jia, C.; Li, Y.; Liu, Z.; Wang, J.; Yang, Z.; Gu, C.; Su, D.; Houk, K.N.; Zhang, D.; et al. Direct single-molecule dynamic detection of chemical reactions. *Sci. Adv.* **2018**, *4*, eaar2177. [CrossRef]
10. Aguillon, F.; Marinica, D.C.; Borisov, A.G. Molecule Detection with Graphene Dimer Nanoantennas. *J. Phys. Chem. C* **2020**, *124*, 28210–28219. [CrossRef]
11. Yang, C.; Yang, C.; Guo, Y.; Feng, J.; Guo, X. Graphene–molecule–graphene single-molecule junctions to detect electronic reactions at the molecular scale. *Nat. Protoc.* **2023**, *18*, 1958–1978. [CrossRef] [PubMed]
12. Nag, A.; Mitra, A.; Mukhopadhyay, S.C. Graphene and its sensor-based applications: A review. *Sens. Actuators A Phys.* **2018**, *270*, 177–194. [CrossRef]
13. Liu, J.; Bao, S.; Wang, X. Applications of Graphene-Based Materials in Sensors: A Review. *Micromachines* **2022**, *13*, 184. [CrossRef] [PubMed]
14. Fu, W.; Jiang, L.; van Geest, E.P.; Lima, L.M.C.; Schneider, G.F. Sensing at the Surface of Graphene Field-Effect Transistors. *Adv. Mater.* **2016**, *29*, 1603610. [CrossRef] [PubMed]
15. Giubileo, F.; Di Bartolomeo, A. The role of contact resistance in graphene field-effect devices. *Prog. Surf. Sci.* **2017**, *92*, 143–175. [CrossRef]
16. Lu, N.; Wang, L.; Li, L.; Liu, M. A review for compact model of graphene field-effect transistors. *Chin. Phys. B* **2017**, *26*, 036804. [CrossRef]
17. Ma, Y.; Zhi, L. Graphene-Based Transparent Conductive Films: Material Systems, Preparation and Applications. *Small Methods* **2018**, *3*, 1800199. [CrossRef]
18. Suriani, A.B.; Muqoyyanah; Mohamed, A.; Alfarisa, S.; Mamat, M.H.; Ahmad, M.K.; Birowosuto, M.D.; Soga, T. Synthesis, transfer and application of graphene as a transparent conductive film: A review. *Bull. Mater. Sci.* **2020**, *43*, 310. [CrossRef]
19. Miao, J.; Fan, T. Flexible and stretchable transparent conductive graphene-based electrodes for emerging wearable electronics. *Carbon* **2023**, *202*, 495–527. [CrossRef]
20. Allen, M.J.; Tung, V.C.; Kaner, R.B. Honeycomb Carbon: A Review of Graphene. *Chem. Rev.* **2010**, *110*, 132–145. [CrossRef]
21. Novoselov, K.S. Graphene: Materials in the Flatland (Nobel Lecture). *Angew. Chem. Int. Ed.* **2011**, *50*, 6986–7002. [CrossRef] [PubMed]
22. Kin, Y.J.; Kin, Y.; Novoselov, K.; Hong, B.H. Engineering electrical properties of graphene: Chemical approaches. *2D Mater.* **2015**, *2*, 042001. [CrossRef]
23. Kiesel, M.L.; Platt, C.; Hanke, W.; Abanin, D.A.; Thomale, R. Competing many-body instabilities and unconventional superconductivity in graphene. *Phys. Rev. B* **2012**, *86*, 020507. [CrossRef]
24. Han, T.H.; Kim, H.; Kwon, S.J.; Lee, T.W. Graphene-based flexible electronic devices. *Mater. Sci. Eng. R* **2017**, *118*, 1–43. [CrossRef]
25. Kong, W.; Kum, H.; Bae, S.H.; Shim, J.; Kim, H.; Kong, L.; Meng, Y.; Wang, K.; Kim, C.; Kim, J. Path towards graphene commercialization from lab to market. *Nat. Nanotechnol.* **2019**, *14*, 927–938. [CrossRef] [PubMed]
26. Zurutuza, A.; Marinelli, C. Challenges and opportunities in graphene commercialization. *Nat. Nanotechnol.* **2014**, *9*, 730–734. [CrossRef] [PubMed]
27. Bøggild, P.; Mackenzie, D.M.A.; Whelan, P.R.; Petersen, D.H.; Buron, J.D.; Zurutuza, A.; Gallop, J.; Hao, L.; Jepsen, P.U. Mapping the electrical properties of large-area graphene. *2D Mater.* **2017**, *4*, 042003. [CrossRef]
28. Nobre, A.G.; Martínez, J.A.E.; Florêncio, O. Mineral Nanotechnology in Circular Economy. In *Smart Innovation, Systems and Technologies*; Iano, Y., Saotome, O., Kemper, G., Seixas, A.C.M., Oliveira, G.G., Eds.; Springer: Cham, Switzerland, 2021; Volume 233, pp. 220–226. [CrossRef]
29. Nobre, A.G.; da Silva, L.P.N.; Andrade, F.R.D. Graphene Geology and the Fourth Industrial Revolution. In *Smart Innovation, Systems and Technologies*; Iano, Y., Saotome, O., Vásquez, G.L.K., Pezzuto, C.C., Arthur, R., Oliveira, G.G., Eds.; Springer: Cham, Switzerland, 2022; Volume 207, pp. 342–348. [CrossRef]
30. Yardley, B.; Warren, C. *An Introduction to Metamorphic Petrology*; Cambridge University Press: Cambridge, UK, 2021. [CrossRef]
31. Nobre, A.G.; Martínez, J.A.E.; Terence, M.C.; Florêncio, O. The action of shear zones in the natural availability of graphite nanoplatelets: The example of the metadolomites of the Itaiacoca Group and the mica schist of the Dom Silverio Group. *Braz. J. Anim. Environ. Res.* **2020**, *3*, 3108–3118. [CrossRef]
32. Alaferdov, A.V.; Gholamipour-Shirazi, A.; Canesqui, M.A.; Danilov, Y.A.; Mashkalev, S.A. Size-controlled synthesis of graphite nanoflakes and multi-layer graphene by liquid phase exfoliation of natural graphite. *Carbon* **2014**, *69*, 525–535. [CrossRef]
33. Kottegoda, I.R.M.; Gao, X.; Nayanajith, L.D.C.; Manorathne, C.H.; Wang, J.; Wang, J.Z.; Liu, H.K.; Gofer, Y. Comparison of Few-layer Graphene Prepared from Natural Graphite through Fast Synthesis Approach. *J. Mater. Sci. Technol.* **2015**, *31*, 907–912. [CrossRef]
34. Suragtkhuu, S.; Tserendavag, O.; Vandandoo, U.; Bati, A.S.R.; Bat-Erdene, M.; Shapter, J.G.; Batmunkh, M.; Davaasambuu, S. Efficiency and stability enhancement of perovskite solar cells using reduced graphene oxide derived from earth-abundant natural graphite. *RSC Adv.* **2020**, *10*, 9133–9139. [CrossRef]
35. Mu, Y.; Han, M.; Li, J.; Liang, J.; Yu, J. Growing vertical graphene sheets on natural graphite for fast charging lithium-ion batteries. *Carbon* **2021**, *173*, 477–484. [CrossRef]

36. Frisenda, R.; Niu, Y.; Gant, P.; Muñoz, M.; Castellanos-Gomez, A. Naturally occurring van der Waals materials. *npj 2D Mater. Appl.* **2020**, *4*, 38. [CrossRef]
37. Zhou, Y.; LaChance, A.M.; Smith, A.T.; Cheng, H.; Liu, Q.; Sun, L. Strategic Design of Clay-Based Multifunctional Materials: From Natural Minerals to Nanostructured Membranes. *Adv. Funct. Mater.* **2019**, *29*, 1807611. [CrossRef]
38. Saha, K.; Deka, J.; Gogoi, R.K.; Datta, K.K.R.; Raidongia, K. Applications of Lamellar Membranes Reconstructed from Clay Mineral-Based Nanosheets: A Review. *ACS Appl. Nano Mater.* **2022**, *5*, 15972–15999. [CrossRef]
39. Liu, S.; Zou, D.; Yu, X.; Wang, Z.; Yang, Z. Transfer-Free PZT Thin Films for Flexible Nanogenerators Derived from a Single-Step Modified Sol–Gel Process on 2D Mica. *ACS Appl. Mater. Interfaces* **2020**, *12*, 54991–54999. [CrossRef]
40. Piao, H.; Choi, G.; Rejinold, N.S.; Vinu, A.; Choy, J.H. g-C3N4 Monolayer/2D Mica Nanohybrids with Highly Effective UV–HEV-Screening Function. *Adv. Mater. Interfaces* **2023**, *10*, 2202486. [CrossRef]
41. Ulian, G.; Moro, D.; Valdrè, G. Electronic and optical properties of graphene/molybdenite bilayer composite. *Compos. Struct.* **2021**, *255*, 112978. [CrossRef]
42. Li, L.; Wang, H.; Liang, T.; Cao, J.M.; Yan, C.; Wu, X.L. Natural ore molybdenite as a high-capacity and cheap anode material for advanced lithium-ion capacitors. *Nanotechnology* **2022**, *33*, 255401. [CrossRef]
43. Vasić, B.; Czibula, C.; Kratzer, M.; Neves, B.R.A.; Matković, A.; Teichert, C. Two-dimensional talc as a van der Waals material for solid lubrication at the nanoscale. *Nanotechnology* **2021**, *32*, 265701. [CrossRef]
44. Zhao, J.; Gao, T.; Dang, J.; Cao, W.; Wang, Z.; Li, S.; Shi, Y. Using Green, Economical, Efficient Two-Dimensional (2D) Talc Nanosheets as Lubricant Additives under Harsh Conditions. *Nanomaterials* **2022**, *12*, 1666. [CrossRef] [PubMed]
45. Nobre, A.G.; Salazar-Naranjo, A.F.; Andrade, F.R.D.; Vlach, S.R.F.; Ando, R.A. Simulation of geological graphene genesis by the piston-cylinder apparatus. *Matéria* **2022**, *27*, e20220122. [CrossRef]
46. Alencar, A.B.; Barboza, A.P.M.; Archanjo, B.S.; Chacham, H.; Neves, B.R.A. Experimental and theoretical investigations of monolayer and few-layer talc. *2D Mater.* **2015**, *2*, 015004. [CrossRef]
47. Mania, E.; Alencar, A.B.; Cadore, A.R.; Carvalho, B.R.; Watanabe, K.; Taniguchi, T.; Neves, B.R.A.; Chacham, H.; Campos, L.C. Spontaneous doping on high quality talc-graphene-hBN van der Waals heterostructures. *2D Mater.* **2017**, *4*, 031008. [CrossRef]
48. Barcelos, I.D.; Cadore, A.R.; Alencar, A.B.; Maia, F.C.B.; Mania, E.; Oliveira, R.F.; Bufon, C.C.B.; Malachias, A.; Freitas, R.O.; Moreira, R.L.; et al. Infrared Fingerprints of Natural 2D Talc and Plasmon–Phonon Coupling in Graphene–Talc Heterostructures. *ACS Photonics* **2018**, *5*, 1912–1918. [CrossRef]
49. Yakobson, B.I.; Ding, F. Observational Geology of Graphene, at the Nanoscale. *ACS Nano* **2011**, *5*, 1569–1574. [CrossRef]
50. Wong, C.H.A.; Sofer, Z.; Pumera, M. Geographical and Geological Origin of Natural Graphite Heavily Influence the Electrical and Electrochemical Properties of Chemically Modified Graphenes. *Chem.-Eur. J.* **2015**, *21*, 8435–8440. [CrossRef]
51. Prinsloo, L.C.; Bordes, L.; Mauran, G.; Lombard, M.; Wadley, L. Graphite and multilayer graphene detected on ~70,000-year-old stone tools: Geological origin or constituent of hafting resin? *J. Raman Spectrosc.* **2022**, *54*, 182–190. [CrossRef]
52. Boccard, N. Commodities & Sustainability. *SSRN* **2021**, *1*, 1–37. [CrossRef]
53. Hushko, S.; Botelho, J.M.; Maksymova, I.; Slusarenko, K.; Kulishov, V. Sustainable development of global mineral resources market in Industry 4.0 context. *IOP Conf. Ser. Earth Environ. Sci.* **2021**, *628*, 012025. [CrossRef]
54. Boyd, F.R.; England, J.L. Apparatus for phase-equilibrium measurements at pressures up to 50 kilobars and temperatures up to 1750 °C. *J. Geophys. Res.* **1960**, *65*, 741–748. [CrossRef]
55. Holloway, J.R.; Wood, B.J. *Simulating the Earth: Experimental Geochemistry*, 1st ed.; Springer Science & Business Media: London, UK, 2012.
56. Vlach, S.R.F.; Salazar-Naranjo, A.F.; Torres-Corredor, J.S.; Carvalho, P.R.; Mallmann, G. Calibration of high-temperature furnace assemblies for experiments between 200 and 600 MPa with end-loaded piston-cylinder apparatuses. *Braz. J. Geol.* **2019**, *49*, 1–7. [CrossRef]
57. Pawley, A.R.; Wood, B.J. The high-pressure stability of talc and 10 Å phase: Potential storage sites for H_2O in subduction zones. *Am. Mineral.* **1995**, *9*, 998–1003. [CrossRef]
58. Oganov, A.R.; Hemley, R.J.; Hazen, R.M.; Jones, A.P. Structure, Bonding, and Mineralogy of Carbon at Extreme Conditions. *Rev. Mineral. Geochem.* **2013**, *75*, 47–77. [CrossRef]
59. Tuinstra, F.; Koenig, J.L. Raman Spectrum of Graphite. *J. Chem. Phys.* **1970**, *53*, 1126–1130. [CrossRef]
60. Ferrari, A.C.; Meyer, J.C.; Scardaci, V.; Casiraghi, C.; Lazzeri, M.; Mauri, F.; Piscanec, S.; Jiang, D.; Novoselov, K.S.; Roth, S.; et al. Raman spectrum of graphene and graphene layers. *Phys. Rev. Lett.* **2006**, *97*, 187401. [CrossRef] [PubMed]
61. Ferrari, A.C. Raman spectroscopy of graphene and graphite: Disorder, electron–phonon coupling, doping and nonadiabatic effects. *Solid State Commun.* **2007**, *143*, 47–57. [CrossRef]
62. Singh, V.; Joung, D.; Zhai, L.; Das, S.; Khondaker, S.; Seal, S. Graphene based materials: Past, present and future. *Prog. Mater. Sci.* **2011**, *56*, 1178–1271. [CrossRef]
63. Martin, J.P. Measuring resistance/impedance with the four-wire Kelvin method. *Watts Curr. Tech. Bull.* **2014**, *1*, 1–11.
64. Heaney, M.B. Electrical Conductivity and Resistivity. In *Electrical Measurement, Signal Processing, and Displays*; CRC Press: Boca Raton, FL, USA, 2003.

65. Szabó, G.A.J.; Andrade, F.R.D.; Guimarães, G.B.; Carvalho, F.M.S.; Moya, F.A. As Jazidas de Talco no Contexto da Históri Metamórfica dos Metadolomitos do Grupo Itaiacoca, PR. *Geol. USP Série Científica* **2006**, *5*, 13–31. [CrossRef]
66. Saunite, D.M.; Bello, R.M.S.; Andrade, F.R.D.; Szabó, G.A.J. Metadolomitos talcificados do Grupo Itaiacoca, Paraná: Regime d fluidos e implicações genéticas. *Geol. USP Série Científica* **2011**, *11*, 171–187. [CrossRef]

C Journal of Carbon Research

MDPI

Article

Graphene Oxide Membranes: Controlled Laser Reduction for Sensing Applications

Aiden Rowley, Yijing Stehle *, Luke Kilby and Caleb Bashant

Department Mechanical Engineering, Union College, Schenectady, NY 12308, USA; rowleya@union.edu (A.R.); kilbyl@union.edu (L.K.)
* Correspondence: stehley@union.edu

Abstract: Reduced graphene oxide (rGO) has attracted attention as an active electrode material for flexible electrochemical devices due to its high electric conductivity and large surface area. Compared to other reduction processes, laser reduction is a precise, low-cost, and chemical-free process that is directly applied to graphene oxide (GO) membranes. This study aims to develop rGO through laser irradiation for application as electrodes in thin flexible electrochemical sensors. Laser irradiation parameters will be optimized to achieve reduction of a low oxygen to carbon (O/C) ratio and surface impedance. The influence of humidity on the impedance of rGO electrodes will be studied. The observed instability of the rGO electrode is related to incomplete reduction and oxygenated defects involved in reduction. Partially removed oxygenated functional groups not only influence the impedance of the electrode but make it sensitive to the humidity of the working environment. The result provides references for GO's laser reduction optimization, demonstrates the potential of applying rGO as an electrode in sensing applications, but also reveals the limitation of applying the laser reduced rGO electrode in a non-constant humidity environment.

Keywords: reduced graphene oxide (rGO); graphene oxide (GO); laser reduction; electrode; impedance; humidity sensing

Citation: Rowley, A.; Stehle, Y.; Kilby, L.; Bashant, C. Graphene Oxide Membranes: Controlled Laser Reduction for Sensing Applications. C **2023**, 9, 74. https://doi.org/10.3390/c9030074

Academic Editor: Gil Gonçalves

Received: 1 June 2023
Revised: 4 July 2023
Accepted: 24 July 2023
Published: 30 July 2023

1. Introduction

GO is a monolayer carbon graphene sheet with abundant oxygenated functional groups (OFGs) such as hydroxyl, epoxy, ketone, and carboxyl groups. The OFGs present along the membrane impart electrically insulating and hydrophilic properties of GO. The reduction process of GO involves decomposition and subsequent removal of OFGs in the form of H_2O, CO, and CO_2 from the structure. This causes an increase in conductivity and reduces the hydrophilicity of the GO [1,2]. There are a variety of GO reduction methods ranging from thermal, irradiation, chemical, electrochemical, bacterial degradation, etc. [3–5]. The quality of rGO is limited compared to pristine graphene (unoxidized form) due to incomplete reduction, defects, and chemical residues involved in the reduction process. The application of rGO as a potential electrode material for flexible electrochemical devices and sensors has received attention due to the high electrical conductivity, high specific surface area, and low cost when compared with pristine graphene [6,7]. Compared to chemical and thermal reduction methods, laser-induced reduction of GO (LrGO) has many advantages, such as repeatability, speed, low cost, non-hazardousness, and an automated fabrication process [8,9]. The presence of partially removed OFGs also makes LrGO sensitive to water molecules. Additionally, there are a lack of systematic studies and publications observing the relationship between the laser parameters and the quality of the LrGO, making it difficult to characterize appropriate laser operational settings. Using a high-speed pulse CO_2 laser beam with a wavelength of 10.6 µm, Trusovas et al. [10] determined that a laser power of 50 mW and a scanning speed of 30 mm/s yield optimal reduction results. Alternatively, Dhrubajyoti et al. [11] found that a laser power of 30 W and scanning speed of 10 mm/s fully reduced a 15 μm GO membrane. Provided that OFGs are a key factor in tailoring electrical resistance and hydrophilicity of GO, the presence of remaining OFGs on the LrGO tends to absorb water vapor molecules in the air. This absorption process impacts the electrochemical performance of the LrGO-based electrochemical supercapacitors and sensors, with working environments of variable humidity. In addition, high-speed pulsed CO_2 lasers tend to generate either foam-like or ablative surface structuring during GO membrane reduction. Combined with the deoxygenation effect, these porous carbon structures serve as an effective electrode candidate for disposable and flexible devices such as supercapacitors and sensors.

In this study, a 200 μm width x 5 cm length LrGO electrode was prepared via a GO membrane (~12 μm thick), with a CO_2 pulse laser beam operating under ambient conditions. Laser parameters including power, scan rate, and focal distance will be optimized to achieve a minimum O/C ratio. The surface morphology and composition of the rGO following laser treatment will be explored via scanning electron microscope (SEM) imaging to further understand the limitations of laser-induced reduction and its effects. Furthermore, the impact of humidity variation on the surface impedance of the rGO electrode will be evaluated in addition to potential sensor applications. Future study will be discussed regarding the application of advanced gold nanoparticles for improved sensitivity.

2. Materials and Methods

Free-standing GO membranes [12] with varying thicknesses between 10 and 16 μm were adhered on flexible substrates with GO solution and left to dry overnight. The GO membrane was reduced using a conventional CO_2 laser engraver (Epilog Fusion Pro, wavelength: 1065 nm) under ambient conditions (21 °C, 30% RH). The surface morphology of the as-prepared pure GO and laser-reduced GO was characterized by Zeiss EVO SEM. The elemental analysis was investigated using Energy Dispersive Spectroscopy (EDS) equipped with SEM. The AC impedance of the LrGO electrode was collected through a four-point probe station using Electrochemistry Impedance Spectra (EIS) (CHI608E) in the frequency range of 1 Hz–0.1 MHz. In the four-point probe station, current was forced through the load via one set of source leads, while the voltage across the load was measured through a second set of leads. To investigate the impact of the humidity of the working environmental on the electric property of the as-prepared rGO electrode, the electrical impedance response to environmental humidity (RH) was measured. This was conducted using saturated salts to produce relatively stable humid environments in an enclosure at a constant temperature of 21 °C [13,14]. To ensure the reliability of the humidity, RH was monitored with a temperature and humidity sensor (Arduino DHT22). The LrGO electrodes were placed into a two-point configuration test mount with two leads attached directly to the electrochemical analyzer, and the impedance was recorded for varying frequencies between 0.01 Hz and 100 kHz at a constant humidity.

3. Results and Discussion

As shown in Figure 1a, the GO membrane surface exhibited a large amount of wrinkle. The "brick-and-mortar" stacking of GO sheets during membrane formation led to 'peak and valley' undulations, which appear as wrinkles on the surface [15]. The surface height variation of the GO membrane due to the wrinkles could result in laser to sample surface distance variation during laser-induced reduction. Laser reduction has its advantage as the only chemical-free, dry, and fast deoxygenation reaction, but also has the disadvantages of incomplete reduction and non-uniformity. In general, laser-induced reduction of GO with a laser wavelength above 800 nm is a photothermal reaction, where local GO flakes are heated above their thermal reduction threshold with the laser beam [16,17]. As shown in Figure 1b, c, two parallel lines separated by a 220 μm distance are directly applied to the GO membrane with a constant laser frequency of 20% and scan speed of 100%. The photothermal and photochemical reduction induced by the laser beam causes the laminated GO sheets to curl up. For laser power below 1 W, only the top few layers of the GO sheet along the laser beam path curl up, as shown in Figure 1b. The depth of the laser reduction increases with higher laser power. Through optimizing the laser power, frequency, and the planar distance, the resulting LrGO structure (Figure 1c) shows an optimal "bloom", or "raising pastry". The reduction is further confirmed via elemental analysis and resistance measurements. While, visually, the surface of the LrGO appears to be optimal, surface morphology change is not by itself enough to prove that laser irradiation is the most effective reduction method. In our study, the focused laser beam with a smaller diameter generated a large amount of energy on the GO membrane surface to enable instantaneous reduction, which appeared as a "blooming" around the position of the laser dot rather than as uniform exfoliation.

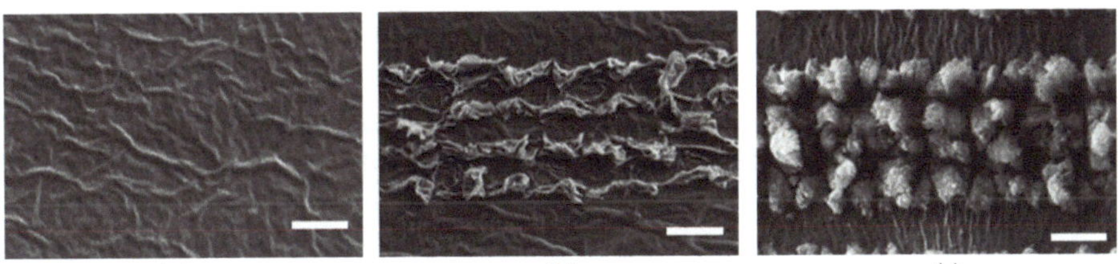

(a) (b) (c)

Figure 1. SEM image of a pristine GO membrane (**a**) under reduced LrGO electrode (**b**) and optimized LrGO electrode (**c**). White scale bars indicate 100 μm.

Further carbon and oxygen elemental maps can be applied to confirm the removal of the oxygenated functional groups. Figure 2 depicts an SEM image and the corresponding energy dispersive x-ray analysis (EDX) maps for carbon (green) and oxygen (blue) at the interface of GO and rGO obtained via line beam irradiation. Comparing the LrGO (bloom) region in Figure 2a with the GO area below it, an oxygen atomic ratio decrease is observed while the carbon atomic ratio does not have a visible difference. It is also observed that 100% reduction is not achieved; there are still plenty of unreduced OFGs present within the GO membrane. This oxygen atomic decrease could greatly influence the effective conductivity of the reduced rGO. The composition is differentiated based on the EDX mapping results. The O/C ratio undergoes significant reduction from the GO region (0.4~0.5) to the rGO region (0.1~0.2), indicating successful laser-induced removal of the majority of the OFGs.

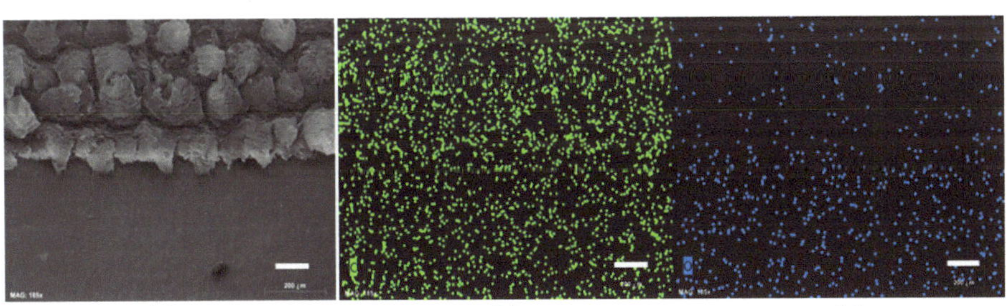

(a) (b) (c)

Figure 2. SEM image (**a**), carbon (**b**), and oxygen (**c**) EDX mapping of an LrGO electrode. White scale bars indicate 200 μm.

The carbon and oxygen element map of GO demonstrates a uniform atom distribution. Upon further observation in Figure 3, a lower carbon concentration and oxygen concentration variation at the location of the laser dot is shown, which indicates non-uniform reduction of the GO due to the nature of the pulse laser. In addition, there is potential carbon removal from overreduction during the laser irradiation process. The hydrophilicity of GO originated from its hydrophilic OFGs, and the removal of OFGs through reduction processes will decrease its hydrophilicity, as shown in Figure 3d. It is observed that the laser reduction process decreases the hydrophilicity by increasing the average contact angle from 46° to 67° for GO and rGO, respectively. The hydrophilic GO membrane is dispersible in water as GO sheets after sonication, but the LrGO is not dispersible. It is worth mentioning that the hydrophilicity partially contributed to the "bloom" structure of the rGO electrode. There is a large volume increase during the reduction that will increase the porosity of the as-fabricated LrGO electrode. However, the LrGO "bloom" does not hold the same mechanical strength as the GO membrane and can be squeezed through by a gentle push or collected through a brush, which limits the application of the as-fabricated LrGO electrode-based sensor. If the sensing materials' composite with the GO membrane before laser reduction, the sensing materials will undergo "blooming" with the graphene sheet, which could form many nanosized micro-sensors with sensing layers pre-coordinated with the LrGO nanosheets.

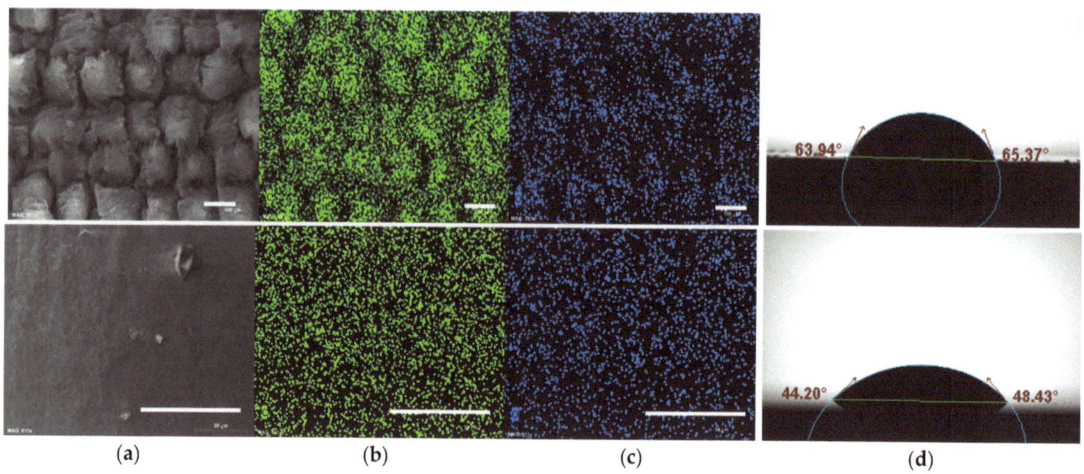

Figure 3. SEM image (**a**), carbon (**b**), and oxygen (**c**) EDX map, and water contact angle (**d**) of an LrGO electrode (top) and GO membrane (bottom). White scale bars indicate 100 μm (top) and 50 μm (bottom).

With either laser power increase or frequency decrease, the laser beam cuts through the GO membrane directly along its path, rather than inducing and forming a "wrapping" effect through deoxidation. Figure 4 illustrates a typical front view (a) and cross-section view (b) of a laser cut GO membrane, with a corresponding laser power of 3.6 W. Figure 4b demonstrates a membrane thickness increase by a factor of 400 from the GO region to the fully reduced LrGO region when exposed to laser irradiation. Additionally, it is noted that the effective width of laser reduction is limited to within 300 μm. Currently, there is not an effective strategy to form perfect uniform laser reduction of GO membrane on a large area and from top to bottom, which is critical to develop high quality LrGO-based sensors. Therefore, three-dimensional and in situ characterizations of LrGO electrode are needed to better understand the "blooming" mechanism of the GO membrane during the reduction.

As the OFGs on the GO sheet will cut the charge transfer route of the graphene sheet of the GO, the optimized laser reduction should result in the lowest resistance. Here, the Electrochemical Impedance Spectrum (EIS) is applied to evaluate a reduction-induced resistance decrease. The Nyquist plot of an as-prepared LrGO electrode is shown as part of two semicircles. The direct result from EIS was complex impedance ($Z = Zr + i \cdot Zi$), which includes both the real part (Zr) and the imaginary part (Zi), as shown in Figure 5a. An equivalent circuit inserted in Figure 5a was applied to analyze the effect of laser reduction of the electrode. The equivalent circuit consists of a series resistance, Rs, from the probe connections; interface resistance, Rct, which represents the charge transfer resistivity of the rGO; and capacitance, C, formed by a rGO/GO/rGO capacitive effect from non-uniform reduction. Figure 5b summarizes the fitted series resistance of the LrGO electrode of the same width and length but with a different laser power and frequency. The longer times and higher temperatures lead to lower resistivity due to the removal of more polymer residue. It was observed that the optimized laser parameter will be 20% frequency, 100% speed, and 2 W power for a 12 μm thick GO membrane.

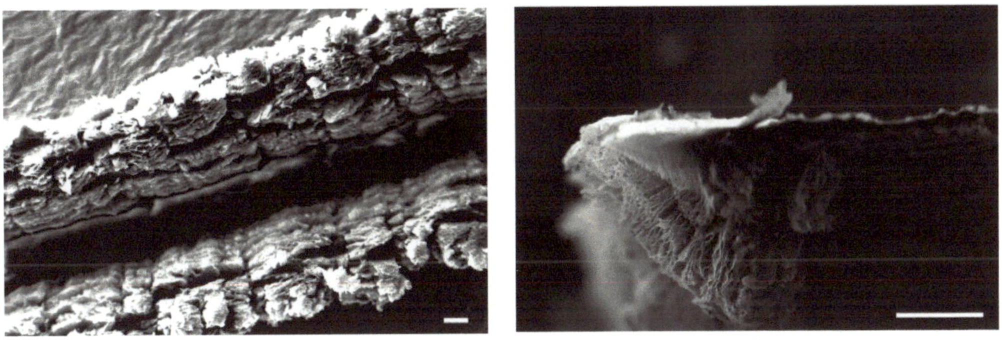

(a) (b)

Figure 4. Typical front view (**a**) and cross-section (**b**) SEM image of an LrGO membrane cut. White scale bars indicate 100 μm.

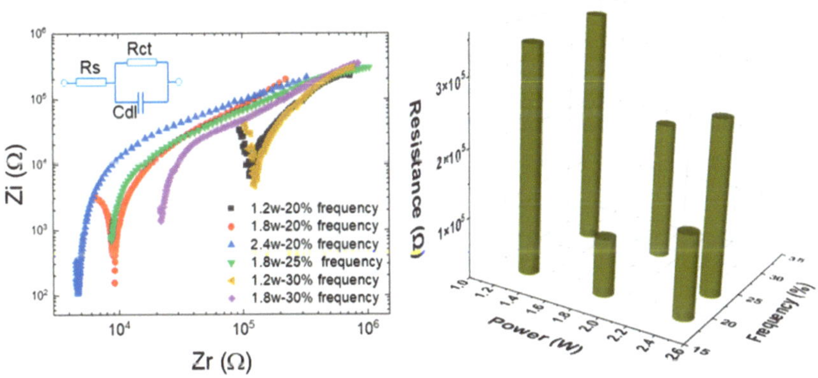

(a) (b)

Figure 5. Typical Nyquist plot (**a**) of LrGO electrode reduced with 20% frequency and 1.2 W intensity, a simplified Randles equivalent circuit for impedance fitting; (**b**) relationship between laser power, frequency, and fitted resistance.

The as-prepared LrGO electrode demonstrates promising OFGs removal without overreduction, called "overburn", to the graphene surface of GO. Further power increase either cuts the membrane through or causes burning of the membrane. However, the OFG residue of the LrGO electrode could both lead to water vapor sensitivity and difficulty of recovery following sensing since it will be hard to desorb water after it is bonded with residue OFGs. To evaluate the impact of the humidity of the working environment on the surface impedance of the LrGO electrode, a relative humidity (RH) variation from 30% under ambient condition to 90% under working condition is utilized for a sensing test. Figure 6a shows the impedance of the LrGO electrode as a function of time for a 25 s exposure to a 90% RH working environment, followed by 25 s of recovery. Exposure was under ambient condition (30% RH). It was observed that the total impedance across the surface of the LrGO electrode decreased 17% over five cycles. While the LrGO electrode responds to humidity change, the impedance drops after working in a high humidity environment, limiting its potential application in a non-dry condition. Similar instability of LrGO is observed in other humidity ranges and temperatures. Such instability from the OFG residue can be solved through either further reduction optimization or by applying sensing layers such as gold nanoparticles above the electrode.

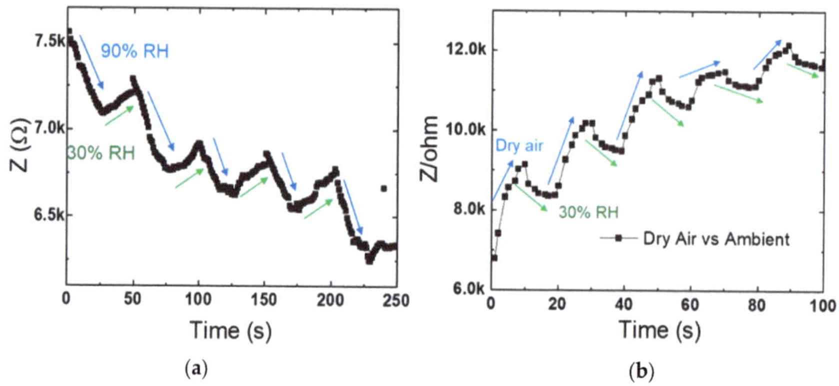

Figure 6. The time–impedance curve of an LrGO electrode exposed to liquid water (**a**) and dry air (**b**) under different humidity environments at room temperature.

The application of LrGO as the electrode in the sensor attracts much attention due to its good electrical conductivity, large surface area, and high stability [17,18]. The humidity sensor made with an LrGO electrode exhibits high sensitivity, stability, reliability, and fast response and recovery in a wide relative humidity (RH) range (11~97%) [19–21]. However, the "bloom" morphology of LrGO represents a non-uniform reduction of the GO membrane. As laser reduction is dependent on the distance from laser tip to the sample surface, the tip of the "bloom" tends to have fewer OFGs than the bottom of the "bloom". Generally, rGO is visualized as hole doped (p-type) semiconductor and charge transport through proton hopping mechanism [22]. As the amount of water adsorbed on the LrGO electrode increases, RH increases and hence the concentration of H^+ ions increase, which will facilitate the proton hopping between hydroxyl groups on the LrGO surface and result in an impedance decrease. In the humidity range of our study, the hydronium ion, H_3O^+, is the major charge carrier, and the charge transport occurs by releasing a proton from each H_3O^+ ion and it bonding to neighboring water molecules, which accept it while releasing another proton by the Grotthuss mechanism [23]. At equilibrium state, the amount of water molecules that physically adsorb and desorb from the LrGO surface is equivalent. As the humidity increases, more water molecules are adsorbed to the LrGO electrode than desorbed until they reach the new equilibrium. The residue OFGs on the LrGO has a tendency to absorb the water molecule, which means it takes a longer time for the LrGO electrode to reach a new equilibrium impedance when switching to a new RH.

Generally, the absorption of water molecules to the carbon surface can be explained with chemisorption and physisorption [24,25]. Chemisorption involves the formation of a chemical bond and is thus difficult to break, while physisorption forms weak van der Waals bonds, and thus physiosorbed water molecules can be easily removed. As shown in Figure 6a, the impedance drop caused by water molecules adsorbed by the LrGO electrode from a high humidity environment is larger than the impedance increases from water molecular desorption after leaving the high humidity range during the same time period. Similarly, in Figure 6b, the impedance increases of the LrGO electrode caused by water molecular desorption when moved to a chamber with dry air humidity is larger than the impedance decreases from slowly absorbing the water molecular back after returning to the ambient condition. The combination of Figure 6a,b indicates that the slope does not represent the adsorption and desorption rate difference of the LrGO electrode. Instead, it can be considerate as it is easier to interrupt an adsorption/desorption equilibrium than for it to recover after it reaches equilibrium during the humidity change. Such a slope could be a sensing pattern of the LrGO electrode with "bloom "morphology, but more experimental study is needed to better understand its sensing mechanism.

4. Conclusions

In this study, laser-reduced graphene oxide electrodes were fabricated with a CO_2 pulse laser under ambient conditions. The optimized laser irradiation parameter of 2~2.4 W power, 20% frequency, and 100% speed for reducing the GO membrane of 12~16 μm thickness were identified via SEM, EDX, and impedance analyses. The complex impedance of the as-prepared LrGO is fitted through a simplified Randles circuit for valid resistance data. The surface impedance of the as-prepared LrGO electrode is sensitive to the humidity of the working environment and has difficultly recovering after being influenced by the working environment humidity change. The RH%-dependent impedance of the LrGO electrodes impacts their potential application in a non-dry working environment and demonstrates the potential of the LrGO electrode for application as a non-contact sensor. This study promotes the understanding of laser reduction condition optimization and the sensing behavior limitations of LrGO electrodes. Further study on the sensing behavior of LrGO electrode-based sensors with the application of added sensing material will be explored. Under the influence of this complex working environment, a better understanding of the limitations of LrGO and the potential solutions is necessary in future research.

Author Contributions: A.R. methodology, investigation, writing—original draft preparation; Y.S. Conceptualization, methodology, visualization, writing—review and editing, validation; L.K. methodology; C.B. methodology. All authors have read and agreed to the published version of the manuscript.

Funding: This work was supported by The Bender Scientific Fund of The Community Foundation for the Greater Capital Region.

Data Availability Statement: The data presented in this study are available on request from the corresponding author.

Acknowledgments: Undergraduate student research funding support from Union College to A.R., L.K., C.B. is gratefully acknowledged.

Conflicts of Interest: The authors declare no conflict of interest.

References

1. Pei, S.; Cheng, H.-M. The Reduction of Graphene Oxide. *Carbon* **2012**, *50*, 3210–3228. [CrossRef]
2. He, H.; Riedl, T.; Lerf, A.; Klinowski, J. Solid-State NMR Studies of the Structure of Graphene Oxide. *J. Phys. Chem.* **1996**, *100*, 19735–19739. [CrossRef]
3. Chua, C.K.; Pumera, M. Chemical Reduction of Graphene Oxide: A Synthetic Chemistry Viewpoint. *Chem. Soc. Rev.* **2014**, *43*, 291–312. [CrossRef] [PubMed]
4. Lee, S.W.; Mattevi, C.; Chhowalla, M.; Sankaran, R.M. Plasma-Assisted Reduction of Graphene Oxide at Low Temperature and Atmospheric Pressure for Flexible Conductor Applications. *J. Phys. Chem. Lett.* **2012**, *3*, 772–777. [CrossRef]
5. Cai, J.; Lv, C.; Aoyagi, E.; Ogawa, S.; Watanabe, A. Laser Direct Writing of a High-Performance All-Graphene Humidity Sensor Working in a Novel Sensing Mode for Portable Electronic. *ACS Appl. Mater. Interfaces* **2018**, *10*, 23987–23996. [CrossRef]
6. El-Kady, M.F.; Kaner, R.B. Scalable Fabrication of High-Power Graphene Micro-Supercapacitors for Flexible and On-Chip Energy Storage. *Nat. Commun.* **2013**, *4*, 1475. [CrossRef]
7. In, J.B.; Hsia, B.; Yoo, J.-H.; Hyun, S.; Carraro, C.; Maboudian, R.; Grigoropoulos, C.P. Facile fabrication of flexible all solid-state micro-supercapacitor by direct laser writing of porous carbon in polyimide. *Carbon* **2015**, *83*, 144–151. [CrossRef]
8. Gao, W.; Singh, N.; Song, L.; Liu, Z.; Reddy, A.L.M.; Ci, L.; Vajtai, R.; Zhang, Q.; Wei, B.; Ajayan, P.M. Direct Laser Writing of Mirco-Supercapacitors on Hydrated Graphene Oxide Films. *Nat. Nanotechnol.* **2011**, *6*, 496–500. [CrossRef]
9. Yang, D.; Bock, C. Laser Reduced Graphene for Supercapacitor Applications. *J. Power Sources* **2017**, *337*, 73–81. [CrossRef]
10. Trusovas, R.; Ratautas, K.; Račiukaitis, G.; Barkauskas, J.; Stankevičienė, I.; Niaura, G.; Mažeikienė, R. Reduction of Graphite Oxide to Graphene with Laser Irradiation. *Carbon* **2013**, *52*, 574–582. [CrossRef]

11. Bhattacharjya, D.; Kim, C.-H.; Kim, J.-H.; You, I.-K.; In, J.B.; Lee, S.-M. Fast and controllable reduction of graphene oxide by low-cost CO_2 laser for supercapacitor application. *Appl. Surf. Sci.* **2018**, *462*, 353–361. [CrossRef]
12. Stehle, Y.Y.; Robertson, E.; Cortez, R.; Bucinell, R.B.; Vlassiouk, I.V.; Olsson, K.; Kilby, L. Using Al^{3+} to Taylor Graphene Oxide Nanochannels: Impact on Membrane Stability and Permeability. *Membranes* **2022**, *12*, 871. [CrossRef] [PubMed]
13. O'Brien, F.E.M. The Control of Humidity by Saturated Salt Solutions. *J. Sci. Instrum.* **1948**, *25*, 73. [CrossRef]
14. Greenspan, L. Humidity Fixed Points of Binary Saturated Aqueous Solutions. *J. Res. Natl. Bur. Stand A Phys. Chem.* **1977**, *81A* 89–96. [CrossRef]
15. Shen, X.; Lin, X.; Yousefi, N.; Jia, J.; Kim, J. Wrinkling in Graphene Sheets and Graphene Oxide Papers. *Carbon* **2014**, *66*, 84–92 [CrossRef]
16. Tran, T.X.; Choi, H.; Che, C.H.; Sul, J.H.; Kim, I.G.; Lee, S.-M.; Kim, J.-H.; In, J.B. Laser-induced reduction of graphene oxide by intensity-modulated line beam for supercapacitor applications. *ACS Appl. Mater. Interfaces* **2018**, *10*, 39777–39784. [CrossRef] [PubMed]
17. Orekhov, N.D.; Bondareva, J.v.; Potapov, D.O.; Dyakonov, P.v.; Dubinin, O.N.; Tarkhov, M.A.; Diudbin, G.D.; Maslakov, K.I. Logunov, M.A.; Kvashnin, D.G.; et al. Mechanism of Graphene Oxide Laser Reduction at Ambient Conditions: Experimental and ReaxFF Study. *Carbon* **2022**, *191*, 546–554. [CrossRef]
18. Zhu, J.; Huang, X.; Song, W. Physical and Chemical Sensors on the Basis of Laser-Induced Graphene: Mechanisms, Applications, and Perspectives. *ACS Nano* **2021**, *15*, 18708–18741. [CrossRef]
19. Ait Lahcen, A.; Rauf, S.; Beduk, T.; Durmus, C.; Aljedaibi, A.; Timur, S.; Alshareef, H.N.; Amine, A.; Wolfbeis, O.S.; Salama, K.N. Electrochemical sensors and biosensors using laser-derived graphene: A comprehensive review. *Biosens. Bioelectron.* **2020**, *168*, 112565. [CrossRef]
20. Zhu, C.; Tao, L.-Q.; Wang, Y.; Zheng, K.; Yu, J.; Xiandong, L.; Chen, X.; Huang, Y. Graphene oxide humidity sensor with laser-induced graphene porous electrodes. *Sens. Actuators B-Chem.* **2020**, *325*, 128790. [CrossRef]
21. Romero, F.J.; Rivadeneyra, A.; Salinas-Castillo, A.; Ohata, A.; Morales, D.P.; Becherer, M.; Rodriguez, N. Design, fabrication and characterization of capacitive humidity sensors based on emerging flexible technologies. *Sens. Actuators B-Chem.* **2019**, *287*, 459–467. [CrossRef]
22. Yuan, W.; Liu, A.; Huang, L.; Li, C.; Shi, G. High-Performance NO_2 Sensors Based on Chemically Modified Graphene. *Adv. Mater.* **2012**, *25*, 766–771. [CrossRef] [PubMed]
23. Miyake, T.; Rolandi, M. Grotthuss mechanisms: From proton transport in proton wires to bioprotonic devices. *J. Phys. Condens. Matter* **2015**, *28*, 023001. [CrossRef] [PubMed]
24. Tulliani, J.-M.; Inserra, B.; Ziegler, D. Carbon-Based Materials for Humidity Sensing: A Short Review. *Micromachines* **2019**, *10*, 232. [CrossRef] [PubMed]
25. Han, M.; Ding, X.; Duan, H.; Luo, S.; Chen, G. Ultrasensitive Humidity Sensors with Synergy between Superhydrophilic Porous Carbon Electrodes and Phosphorus-Doped Dielectric Electrolyte. *ACS Appl. Mater. Interfaces* **2023**, *15*, 9740–9750. [CrossRef] [PubMed]

C Journal of Carbon Research

MDPI

Article

Effective Elastic Modulus of Wavy Single-Wall Carbon Nanotubes

Chensong Dong

School of Civil and Mechanical Engineering, Curtin University, GPO Box 1987, Perth 6845, Western Australia, Australia; c.dong@curtin.edu.au

Abstract: A simple method for determining the effective elastic modulus of wavy single-wall carbon nanotubes (SWCNTs) is presented in this paper. The effective modulus of curved SWCNTs is derived using Castigliano's theorem. The effect of curvature on the effective modulus is studied. This method is verified by finite element analysis (FEA). The distributions of effective moduli are studied by Monte Carlo simulation. The effective modulus of a general wavy SWCNT is derived by considering the SWCNT as a number of curved SWCNT sections.

Keywords: carbon nanotube (CNT); modulus; waviness; Monte Carlo simulation

1. Introduction

Since the discovery of multi-wall carbon nanotubes (MWCNTs) in 1991 by Iijima [1], and subsequent synthesis of single-wall carbon nanotubes (SWCNTs) [2,3], numerous experimental and theoretical studies have been carried out to investigate the electronic, chemical, and mechanical properties of CNTs. SWNT–polymer composites are theoretically predicted to have both exceptional mechanical and special functional properties that carbon fibre–polymer composites cannot offer [4].

Previous studies on the mechanical properties of nanotubes utilized computational methods, such as molecular dynamics and ab initio models. In general, these computational studies have found nominal values for the axial Young's modulus on the order of 1 TPa [1,2]. Deng et al. [5] found the effective modulus is in the range 530–700 GPa from Raman characterisation of highly aligned electrospun SWNT/PVA nanocomposite fibres and SWNT/PVA nanocomposite films.

It has been found that CNTs are highly wavy when dispersed in a polymer. Waviness refers to the degree of bending or curvature present in the structure of individual nanotubes. Carbon nanotubes can exhibit different degrees of waviness depending on their synthesis method, growth conditions, and structural defects. The waviness of CNTs is influenced by several factors, including the diameter, length, chirality (structural arrangement of carbon atoms), and the presence of impurities or structural defects.

In general, CNTs can have varying degrees of waviness. Some nanotubes are relatively straight and have minimal curvature, while others exhibit more pronounced bending or twisting. The waviness of CNTs can be influenced by external forces, such as mechanical stress or interactions with neighbouring nanotubes or other materials.

In general, waviness in CNTs can have a detrimental effect on their properties and make them less useful for certain applications [6]. Waviness in CNTs can alter their electronic structure, leading to changes in their electronic transport properties, such as their conductivity and density of states. This can result in a reduction of the intrinsic carrier mobility and an increase in the resistance [7,8]. Waviness can also affect the mechanical properties of CNTs, leading to decreased stiffness and strength. This can result in a reduced ability of CNTs to withstand loads and deformations.

When being used to reinforce a polymer, the waviness of CNTs can limit the modulus enhancement of the composite, resulting in improvements that are less than predicted by traditional theories. Thus, waviness needs to be considered when studying the mechanical

Citation: Dong, C. Effective Elastic Modulus of Wavy Single-Wall Carbon Nanotubes. C **2023**, 9, 54. https://doi.org/10.3390/c9020054

Academic Editors: Gil Gonçalves and Stefano Bellucci

Received: 1 March 2023
Revised: 14 May 2023
Accepted: 25 May 2023
Published: 26 May 2023

properties of CNTs. Fisher et al. [9] and Bradshaw et al. [10] extended the Mori–Tanaka method to composites consisting of more than one type of inclusion type via the method introduced by Tendon and Weng [11] so that the distribution of CNT waviness can be considered. They also demonstrated the effect of CNT waviness distribution by introducing a discrete probability distribution. Maghsoudlou et al. [12] demonstrated the necessity of consideration of three important parameters, including SWCNTs' interphase, curvature, and agglomerations. Ginga et al. [13] demonstrated that the waviness of the CNT played a critical role in the significant increase of the compliance values (very low effective modulus) by orders of magnitude. Shi et al. [6] found that waviness and agglomeration might reduce the stiffening effect of nanotubes significantly. Shao et al. [14] studied the effects of the waviness of the CNTs and the interfacial debonding between them and the matrix on the effective moduli of CNT-reinforced composites. A simple analytical model was presented to investigate the influence of the waviness on the effective moduli. Zhu and Narh [15] utilized numerical simulation to study the impact of nanotube curvature on the tensile modulus of polymer composites reinforced with carbon nanotubes. Results from their simulations indicated that the tensile modulus of a nanotube-reinforced composite undergoes a significant decrease when nanotubes deviate from their axial direction. Moreover, the presence of curved nanotubes in the polymer matrix has a substantial impact on reducing the composite modulus.

The curvature of SWCNTs also has a significant effect on buckling. Shima [16] presented a review on the buckling of CNTs. Li et al. [17] studied the nonlinear in-plane instability of shallow circular arches made of functionally graded carbon-nanotube-reinforced composite that were limited by rotational symmetry and put under constant radial stress in a temperature environment.

In this paper, a simple method based on Castigliano's theorem for determining the effective elastic modulus of SWCNTs is presented. This method is verified by finite element analysis (FEA). Using this method, the effect of curvature on the effective modulus is studied. The statistical distributions of the effective moduli are studied by Monte Carlo simulation. A general wavy SWCNT is considered as a number of curved SWCNT sections being connected in series. The effective modulus for each section can be derived using the developed approach, and the effective modulus of the general wavy SWCNT can be derived accordingly.

2. Approach

2.1. Analytical Model

A simple analytical model for predicting the effective modulus of SWCNTs is developed based on Castigliano's theorem. For simplicity, a curved SWCNT is considered, as schematically shown in Figure 1. The SWCNT is subjected to point forces, P, at both ends. The diameter of SWCNTs varies. Chen et al. [18] presented a method to both precisely and continuously control the average diameter of single-walled carbon nanotubes in a forest ranging from 1.3 to 3.0 nm with 1 A° resolution. Fagan et al. [19] demonstrated the effective separation of single-wall carbon nanotube (SWCNT) species with diameters 1.3–1.6 nm through multistage aqueous two-phase extraction (ATPE), including isolation at the near-monochiral species level up to at least the diameter range of SWCNTs synthesized by electric arc synthesis. Charlier and Lambin [20] obtained SWCNTs of the diameter 1.3 nm. Park et al. [21] synthesised SWCNTs with diameter distributions peaked at 0.9 and 1.3 nm. Shi et al. [22] found that the SWCNTs had nearly the same diameter of 1.3 nm and belonged to armchair (n, n)-type carbon nanotubes (n = 8, 9, 10 and 11). Jinno et al. [23] also found that the diameter of SWCNTs was approximately 1.3 nm. In this study, the outer diameter and wall thickness of the SWCNT are taken to be 1.3 nm and 0.142 nm, respectively [18–26]. The curve is characterised by an arc. Given the chord length, L, and the arc height, h, the radius of the arc, R, is given by

$$R = \frac{L^2}{8h} + \frac{h}{2} = \frac{L}{8\lambda} + \frac{L\lambda}{2} \tag{1}$$

where $\lambda = \frac{h}{L}$.

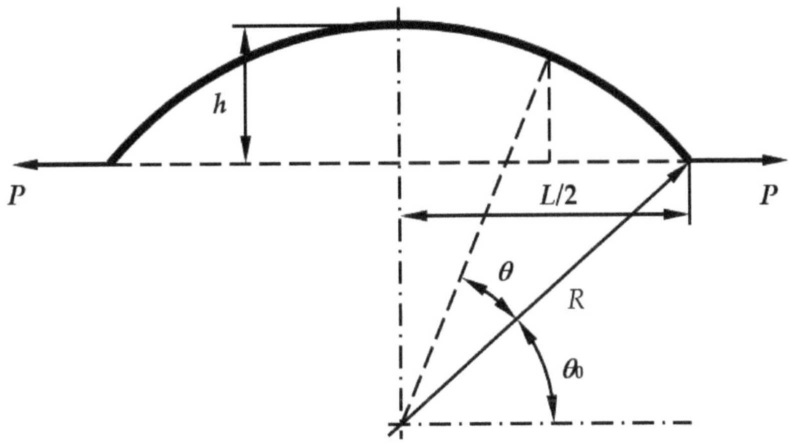

Figure 1. A curved SWCNT.

The angle θ_0 is given by

$$\theta_0 = \cos^{-1}\left(\frac{L}{2R}\right) = \cos^{-1}\left(\frac{4Lh}{L^2 + 4h^2}\right) = \cos^{-1}\left(\frac{4\lambda}{1 + 4\lambda^2}\right) \quad (2)$$

Applying Castigliano's theorem, the displacement at one point is found by differentiating the strain energy, U, with respect to the load, P, acted on that point, i.e.,

$$\delta = \frac{dU}{dP} \quad (3)$$

The strain energy is due to bending, tension and shear, i.e.,

$$U = U_{bending} + U_{tension} + U_{shear} \quad (4)$$

Because of the symmetry, the strain energy due to bending is given by

$$U_{bending} = \frac{P^2R^3}{E_{CNT}I}\int_0^{\pi/2-\theta_0} [\sin(\theta + \theta_0) - \sin\theta_0]^2 d\theta \quad (5)$$

where E_{CNT} and I are the nominal elastic modulus and second moment of inertia, respectively.

The strain energy due to tension is given by

$$U_{tension} = \frac{P^2R}{E_{CNT}A}\int_0^{\pi/2-\theta_0} \sin^2(\theta + \theta_0) d\theta \quad (6)$$

where A is the cross-sectional area.

The strain energy due to shear is given by

$$U_{shear} = \frac{P^2R}{G_{CNT}A\gamma}\int_0^{\pi/2-\theta_0} \cos^2(\theta + \theta_0) d\theta \quad (7)$$

where G_{CNT} is the nominal shear modulus, and γ is the Timoshenko shear coefficient (approximately 0.9).

The total displacement is given by

$$\begin{aligned}\delta &= \delta_{bending} + \delta_{tension} + \delta_{shear} \\ &= \frac{2PR^3}{E_{CNT}I}\left[\frac{1}{2}\left(\frac{\pi}{2} - \theta_0\right) - \frac{3}{4}\sin 2\theta_0 + \left(\frac{\pi}{2} - \theta_0\right)\sin^2\theta_0\right] \\ &+ \frac{2PR}{E_{CNT}A}\left(\frac{\pi}{4} - \frac{\theta_0}{2} + \frac{\sin 2\theta_0}{4}\right) + \frac{2PR}{G_{CNT}A\gamma}\left(\frac{\pi}{4} - \frac{\theta_0}{2} - \frac{\sin 2\theta_0}{4}\right)\end{aligned} \tag{8}$$

The effective modulus is then given by

$$E_{CNTeff} = \frac{PL}{\delta A} = \frac{E_{CNT}L}{2R\left\{\begin{array}{l}\frac{R^2A}{I}\left[\frac{1}{2}\left(\frac{\pi}{2} - \theta_0\right) - \frac{3}{4}\sin 2\theta_0 + \left(\frac{\pi}{2} - \theta_0\right)\sin^2\theta_0\right] \\ +\left(\frac{\pi}{4} - \frac{\theta_0}{2} + \frac{\sin 2\theta_0}{4}\right) + \frac{2(1+\nu_f)}{\gamma}\left(\frac{\pi}{4} - \frac{\theta_0}{2} - \frac{\sin 2\theta_0}{4}\right)\end{array}\right\}} \tag{9}$$

where ν_f is the Poisson's ratio, which can be taken to be around 0.3.

2.2. Finite Element Analysis

The nanocomposite reinforced by a curved SWCNT is modelled by finite element analysis (FEA) for comparison purpose. Epoxy is chosen to be the matrix, and its modulus is 3.1 GPa. For simplicity, the nanocomposite is modelled as a rectangular prism, and the SWCNT is modelled as a curved hollow cylinder. Prefect bond is assumed between the SWCNT and the matrix. As an example, when L = 50 nm and h = 1 nm, the FEA model is shown in Figure 2. A 3D solid analysis is chosen, and a quarter of the composite is modelled by applying symmetry boundary conditions to the two planes of symmetry. The SWCNT volume fraction is determined from the volumes of the SWCNT and the matrix. An axial displacement 0.1 nm is applied to the end face. The reaction force due to this displacement is obtained from the FEA. The effective modulus of the composite is given by

$$E_{eff} = \frac{FL}{\delta A} \tag{10}$$

where F is the reaction force, L is the half-length of the rectangular prism (because symmetry is applied), δ is the axial displacement, and A is the cross-sectional area of the end face.

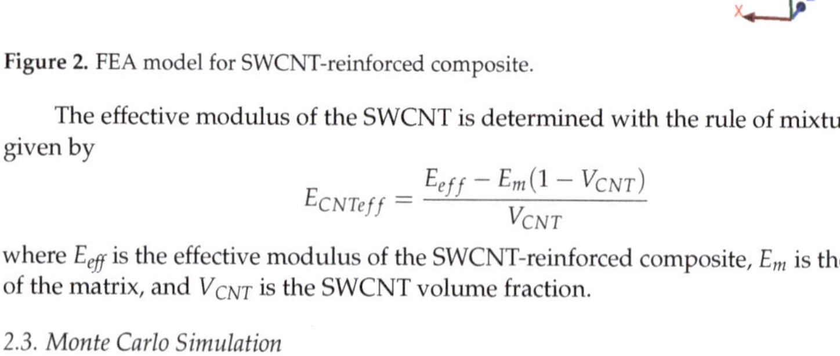

Figure 2. FEA model for SWCNT-reinforced composite.

The effective modulus of the SWCNT is determined with the rule of mixtures (RoM), given by

$$E_{CNTeff} = \frac{E_{eff} - E_m(1 - V_{CNT})}{V_{CNT}} \tag{11}$$

where E_{eff} is the effective modulus of the SWCNT-reinforced composite, E_m is the modulus of the matrix, and V_{CNT} is the SWCNT volume fraction.

2.3. Monte Carlo Simulation

The effective moduli of SWCNTs are statistically studied via Monte Carlo simulation using Equation (9). h/L is assumed to be uniformly distributed, and the SWCNT length is assumed to be normally distributed. Different distributions are considered, as shown in

Table 1. In total, eight simulations are run to study the effect of the range of h/L ratios, the average of the lengths of SWCNTs, and the standard deviation of the lengths of SWCNTs. In each simulation, for h/L or L, random sampling is performed by generating 1000 random numbers based on the given distribution. The resulting effective moduli of SWCNTs are analysed statistically.

Table 1. Monte Carlo simulations.

Set	Simulation	h/L	SWCNT Length (nm)
1	1	$U(0, 0.0025)$	$N(500, 100)$
	2	$U(0, 0.005)$	$N(500, 100)$
	3	$U(0, 0.01)$	$N(500, 100)$
	4	$U(0, 0.02)$	$N(500, 100)$
	5	$U(0, 0.01)$	$N(1000, 100)$
	6	$U(0, 0.01)$	$N(300, 100)$
	7	$U(0, 0.01)$	$N(1000, 50)$
	8	$U(0, 0.01)$	$N(1000, 200)$
2	1	$N(0.005, 0.001)$	$N(500, 100)$
	2	$N(0.01, 0.001)$	$N(500, 100)$
	3	$N(0.02, 0.001)$	$N(500, 100)$
	4	$N(0.01, 0.002)$	$N(500, 100)$
	5	$N(0.01, 0.003)$	$N(500, 100)$
	6	$N(0.01, 0.001)$	$N(300, 100)$
	7	$N(0.01, 0.001)$	$N(1000, 100)$
	8	$N(0.01, 0.001)$	$N(1000, 50)$
	9	$N(0.01, 0.001)$	$N(1000, 200)$
3	1	$U(0, 0.0025)$	$U(300, 700)$
	2	$U(0, 0.005)$	$U(300, 700)$
	3	$U(0, 0.01)$	$U(300, 700)$
	4	$U(0, 0.02)$	$U(300, 700)$
	5	$U(0, 0.01)$	$U(100, 500)$
	6	$U(0, 0.01)$	$U(800, 1200)$

2.4. Waviness

A general wavy SWCNT can be decomposed into several curved SWCNT sections being connected in series. An example is given in Figure 3, which is decomposed into three curved SWCNT sections. For a general wavy SWCNT, its effective modulus is given by

$$\frac{1}{E_{CNTeff}} = \sum_{i=1}^{N} \frac{1}{E_{CNTeffi}} \tag{12}$$

where $E_{CNTeffi}$ is the effective modulus of the i-th SWCNT section, which can be determined from L_i and h_i/L_i.

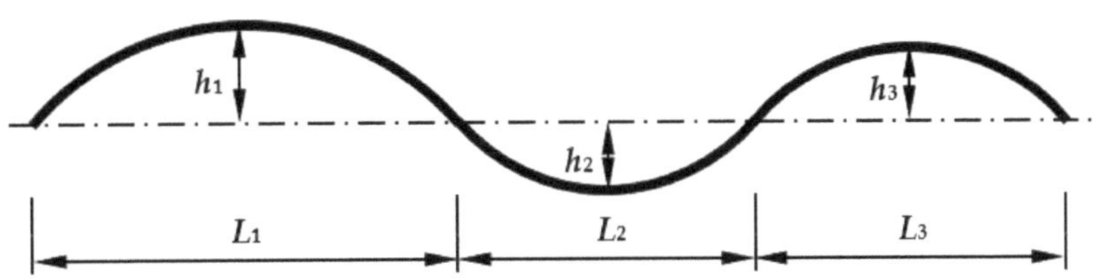

Figure 3. A general waved SWCNT.

3. Results

3.1. Effect of Curvature

The effect of curvature on the effective modulus is graphically shown in Figure 4, from which it is seen that the effective modulus decreases with both increasing L and increasing h/L. Since a fixed SWCNT diameter is used in the simulation, longer SWCNTs have larger aspect ratios and thus are easier to be bent. Therefore, a small increase in h/L results in a sharp decrease in the effective modulus.

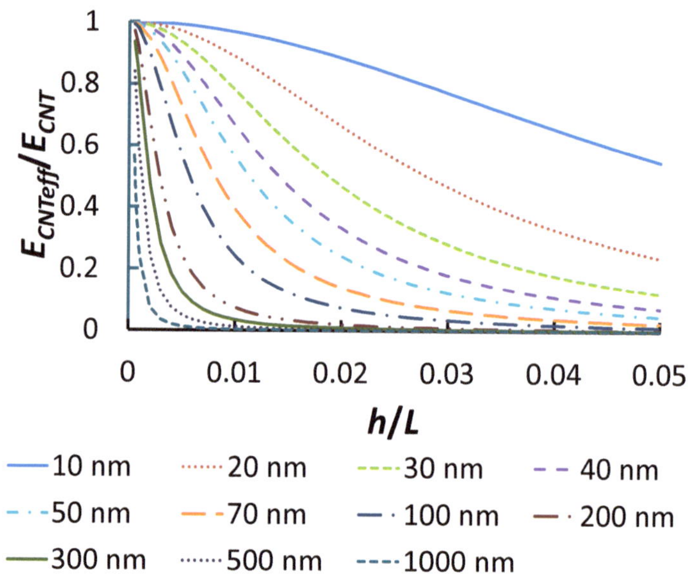

Figure 4. Effective modulus vs. h/L for different SWCNT lengths.

3.2. Results from FEA

When L = 50 nm and h = 1 nm, the displacement in the x direction from FEA is shown in Figure 5, in which the contours stand for the displacement in nm. The load is shown being transferred to the SWCNT near the end of the SWCNT. Figure 6 shows the stress distribution in the composite, in which the contours stand for the stress in MPa.

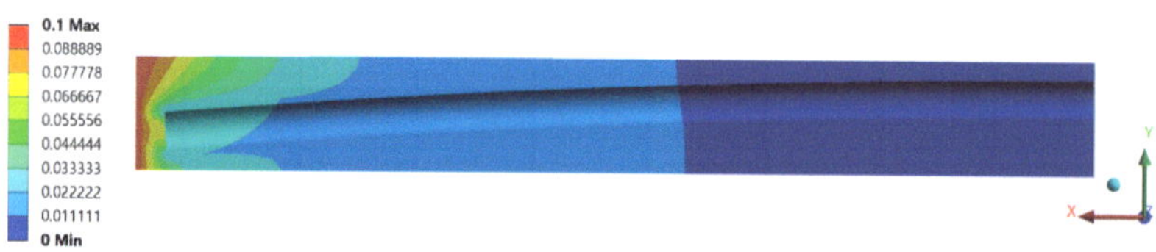

Figure 5. Displacements in SWCNT-reinforced composite.

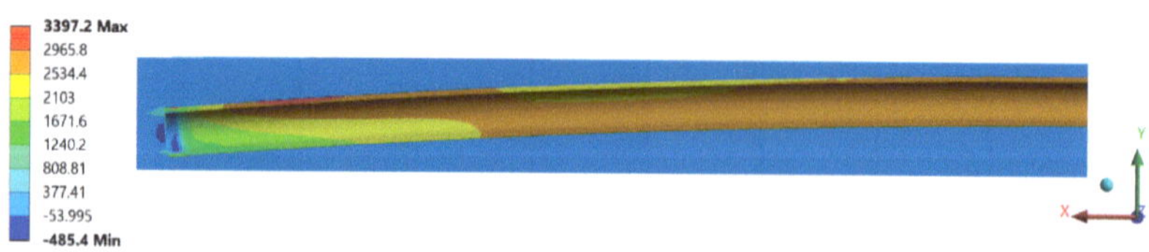

Figure 6. Stresses in SWCNT-reinforced composite.

For three different SWCNT lengths, the effective moduli from the analytical model and FEA are shown in Figure 7 and Table 2. It is seen that the analytical results and FEA results are in good agreement, with the relative difference being less than 10%. These results prove that the effective modulus of a curved SWCNT is significantly less than the nominal value.

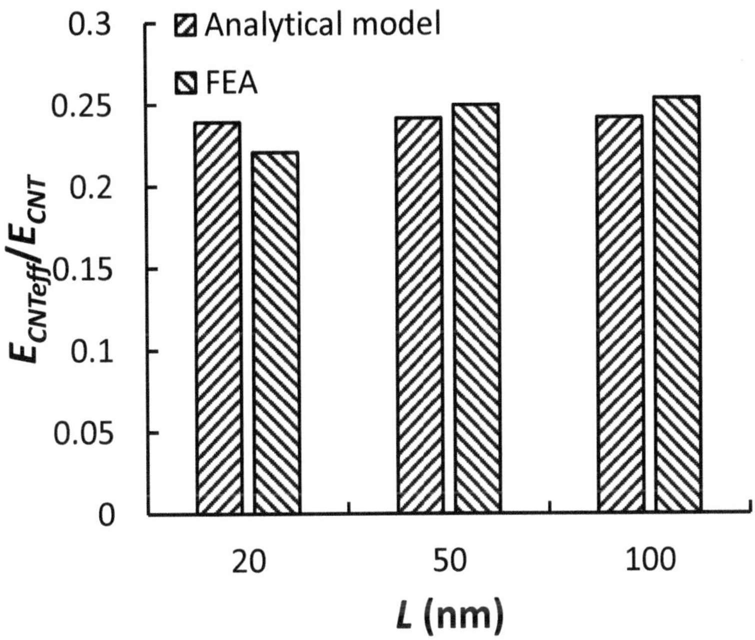

Figure 7. Effective moduli from analytical model and FEA.

Table 2. Effective moduli from analytical model and FEA.

	Effective Modulus, E_{CNTeff}/E_{CNT}		
L (nm)	**Analytical Model**	**FEA**	**Relative Difference**
20	0.2390	0.2205	8.37%
50	0.2414	0.2495	−3.24%
100	0.2417	0.2534	−4.63%

3.3. *Effect of Statistical Distributions*

When h/L follows $U(0, 0.01)$ and L follows $N(500, 100)$, the histogram of the effective moduli is shown in Figure 8, together with the cumulative percentage.

The data for the effective moduli are fitted into Weibull distributions. The shape parameters, scale parameters, means and medians from Monte Carlo simulations are given in Appendix A. When the SWCNT length follows $N(500, 100)$, and h/L follows uniform distributions of different ranges, the fitted Weibull distributions are shown in Figure 9. It is noted that when h/L follows $U(0, 0.0025)$, the maximum probability occurs at E_{CNTeff}/E_{CNT} = 0.3478. As the range of h/L increases, the maximum probability tends to occur when the effective modulus approaches zero. Figure 9 also shows the mean and median of the effective moduli vs. the range of h/L. When the range of h/L increases, both the mean and median of the effective moduli decrease.

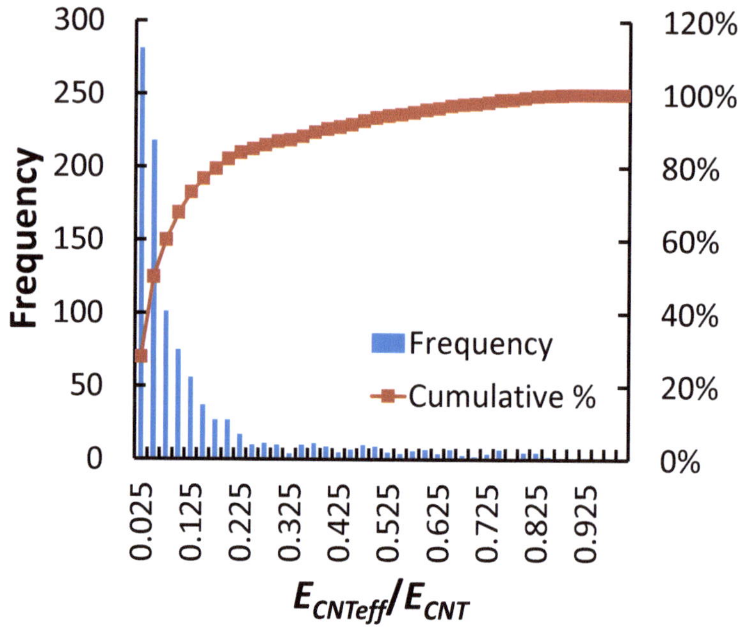

Figure 8. Histogram and cumulative percentage of effective moduli of SWCNTs when h/L follows $U(0, 0.01)$ and L follows $N(500, 100)$.

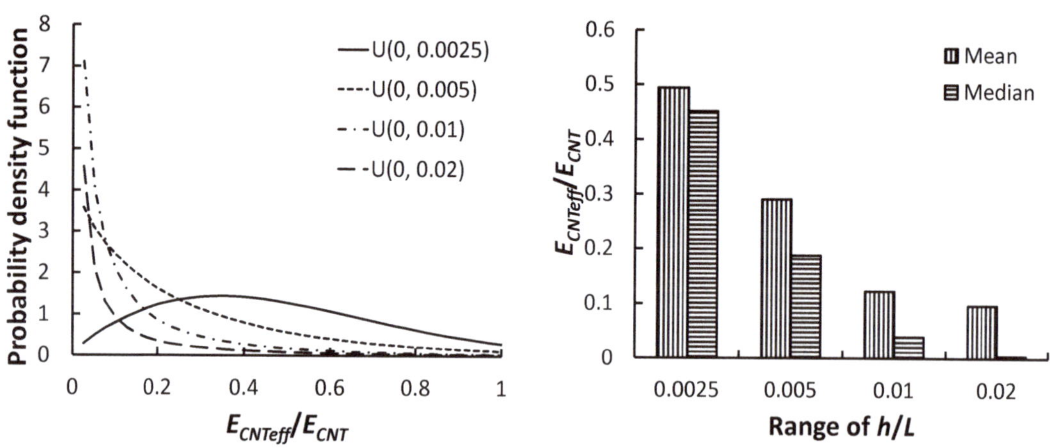

Figure 9. Effect of h/L range on Weibull distribution for effective SWCNT moduli.

The fitted Weibull distributions are shown in Figure 10 for when h/L follows $U(0, 0.01)$, and the SWCNT length follows normal distributions of different means and the same standard deviation, 100 nm. Figure 10 also shows the mean and median of the effective moduli vs. the mean of SWCNT lengths. It is noted that when the mean of the SWCNT lengths increases, both the mean and median of the effective moduli decrease.

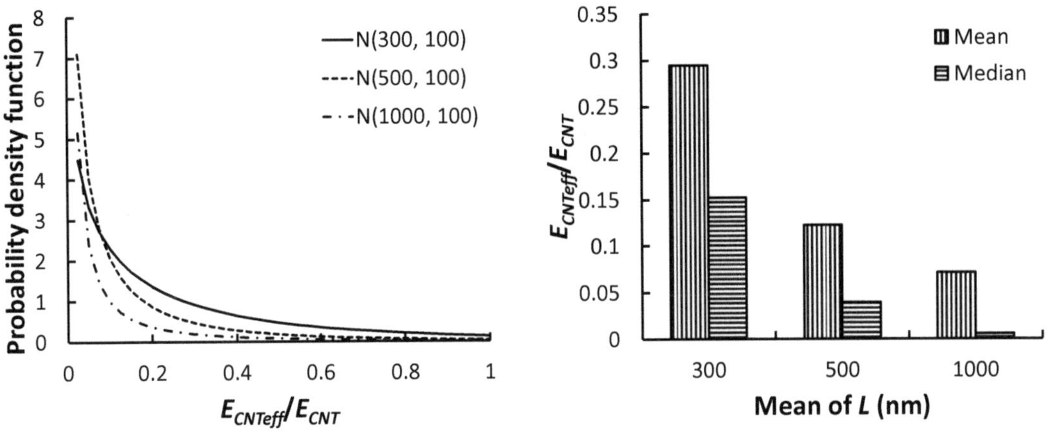

Figure 10. Effect of average SWCNT length on Weibull distribution for effective SWCNT moduli.

The fitted Weibull distributions are shown in Figure 11 for when h/L follows $U(0, 0.01)$, and the SWCNT length follows normal distributions of different standard deviations and the same mean, 1000 nm. Figure 11 also shows the mean and median of the effective moduli vs. the standard deviation of SWCNT lengths. It is noted that the standard deviation of the SWCNT lengths has little effect on either the mean or the median of the effective moduli. Overall, all Weibull distributions skew to the right.

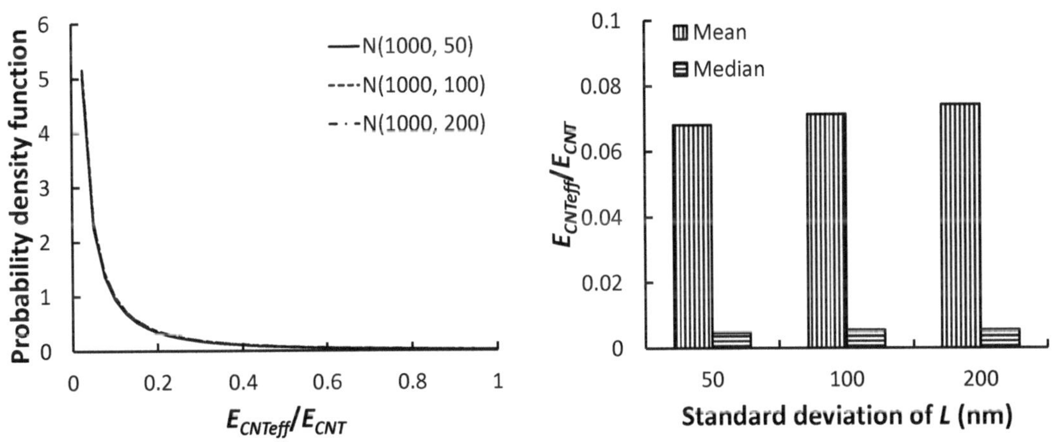

Figure 11. Effect of SWCNT length standard deviation on Weibull distribution for effective SWCNT moduli.

The fitted Weibull distributions are shown in Figure 12 for when the SWCNT length follows $N(500, 100)$ and h/L follows normal distributions of different means. Figure 12 also shows the mean and median of the effective moduli vs. the range of h/L. When the mean of h/L increases, both the mean and median of the effective moduli decrease.

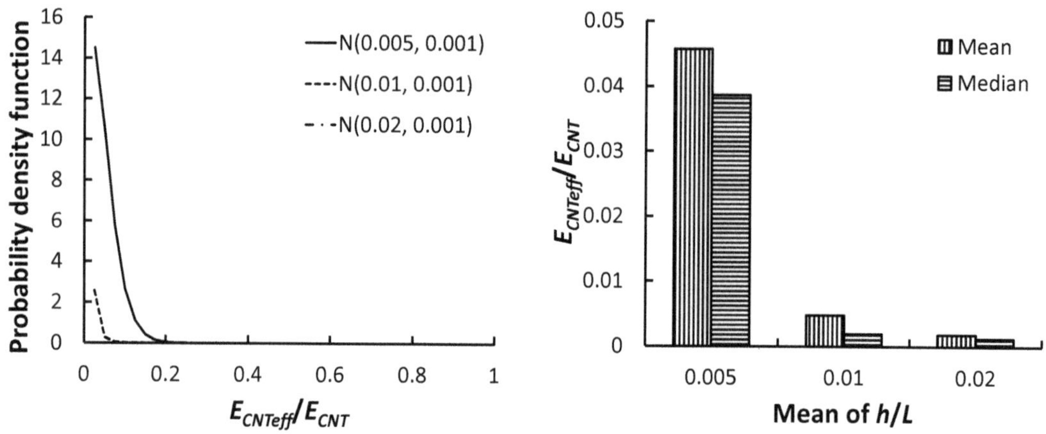

Figure 12. Effect of h/L mean on Weibull distribution for effective SWCNT moduli.

The fitted Weibull distributions are shown in Figure 13 for when the SWCNT length follows $N(500, 100)$ and h/L follows normal distributions of different standard deviations. Figure 13 also shows the mean and median of the effective moduli vs. the range of h/L. It is noted that the standard deviation of the SWCNT lengths has little effect on the median of the effective moduli. For the mean of the effective moduli, it is shown that the mean of the effective moduli increases slightly when the standard deviation of h/L increases from 0.001 to 0.002. A more significant increase is seen when the standard deviation of h/L further increases to 0.003.

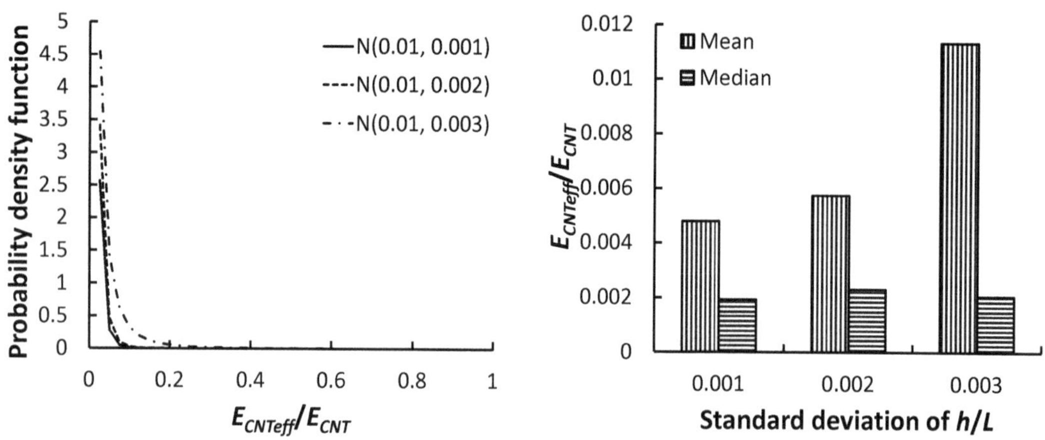

Figure 13. Effect of h/L standard deviation on Weibull distribution for effective SWCNT moduli.

3.4. Effect of Waviness

To study the effect of waviness, SWCNTs of different lengths are assumed to consist of different numbers of equal curved sections of different h/L ratios. The effective moduli are given in Figure 14. When h/L is 0.005, the effective modulus decreases with the increase in the number of curved sections when the SWCNT length is 50 nm or 100 nm. When the SWCNT length is 200 nm, the maximum effective modulus occurs when the number of curved sections is 2. When the SWCNT length is 500 nm, the maximum effective modulus occurs when the number of curved sections is 5. When h/L is 0.01, the effective modulus decreases with the increase in the number of curved sections when the SWCNT length is

50 nm. When the SWCNT length is 100 nm, the maximum effective modulus occurs when the number of curved sections is 2. When the SWCNT length is 200 nm, the maximum effective modulus occurs when the number of curved sections is 3. When the SWCNT length is 500 nm, the effective modulus increases with the increase in the number of curved sections. When h/L is 0.02, the maximum effective modulus occurs when the number of curved sections is 2 when the SWCNT length is 50 nm. When the SWCNT length is 100 nm, the maximum effective modulus occurs when the number of curved sections is 3. When the SWCNT length is 200 nm, the maximum effective modulus occurs when the number of curved sections is 7. When the SWCNT length is 500 nm, the effective modulus increases with increasing number of curved sections. When h/L is 0.05, the maximum effective modulus occurs when the number of curved sections is 5 when the SWCNT length is 50 nm. For the other SWCNT lengths, the effective modulus increases with the increase in the number of curved sections.

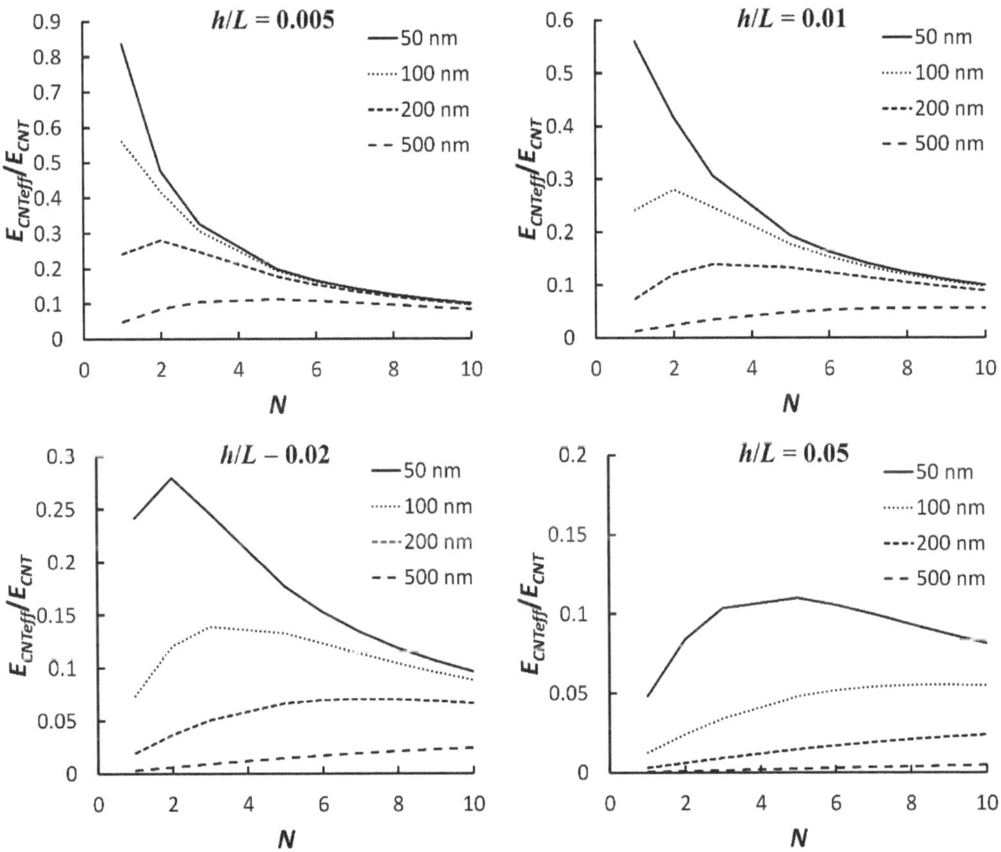

Figure 14. Effective modulus vs. number of curved SWCNT sections.

For evaluating the practical scenario, randomly generated section lengths are used to calculate the effective modulus of SWCNTs, and the results show the effective modulus when waviness is present is typically between 150 and 250 MPa, which is similar to the modulus of carbon fibre.

4. Conclusions

A simple method for determining the effective elastic modulus of curved and wavy SWCNTs is presented in this paper. The effective modulus of curved SWCNTs is derived using Castigliano's theorem. It is shown that the effective modulus decreases with increasing curvature and the length of SWCNT. For longer SWCNTs, a small increase in the curvature results in a sharp decrease in the effective modulus. The SWCNT reinforced epoxy composites are modelled with the aid of the FEA, and it is seen that the effective moduli from the FEA are in good agreement with those calculated using the presented method. Monte Carlo simulations are run with h/L ratio, and SWCNT length follow various distributions It is shown that the resulting effective moduli of SWCNTs follow Weibull distribution. The mean of the effective SWCNT moduli depends on the means of h/L and SWCNT lengths, not the actual distribution type. The effective modulus of a general wavy SWCNT can be conveniently derived by considering the SWCNT as a number of curved SWCNT sections The effective modulus when waviness is present is similar to the modulus of carbon fibre It should be noted that only 2D curvature is considered in this study. The SWCNTs with 3D twist should be addressed in future studies.

Funding: This research received no external funding.

Data Availability Statement: All data generated or analysed during this study are included in this published article.

Acknowledgments: Not applicable.

Conflicts of Interest: The authors declare no conflict of interest.

Appendix A

Table A1. Shape parameters, scale parameters, means and medians from Monte Carlo simulations.

Set	Simulation	h/L	SWCNT Length (nm)	β	η	E_{CNTeff}/E_{CNT} Mean	Median
1	1	U(0, 0.0025)	N(500, 100)	1.77307	0.55544	0.49432	0.45171
	2	U(0, 0.005)	N(500, 100)	0.92055	0.27978	0.29086	0.18789
	3	U(0, 0.01)	N(500, 100)	0.56502	0.07486	0.12238	0.03913
	4	U(0, 0.02)	N(500, 100)	0.31703	0.01327	0.09695	0.00418
	5	U(0, 0.01)	N(1000, 100)	0.35652	0.01508	0.07119	0.00539
	6	U(0, 0.01)	N(300, 100)	0.75143	0.24804	0.29487	0.15230
	7	U(0, 0.01)	N(1000, 50)	0.34914	0.01340	0.06799	0.00469
	8	U(0, 0.01)	N(1000, 200)	0.35343	0.01524	0.07411	0.00540
2	1	N(0.005, 0.001)	N(500, 100)	1.40284	0.05015	0.04570	0.03862
	2	N(0.01, 0.001)	N(500, 100)	0.63766	0.00343	0.00478	0.00193
	3	N(0.02, 0.001)	N(500, 100)	1.00000	0.00180	0.00180	0.00125
	4	N(0.01, 0.002)	N(500, 100)	0.63626	0.00410	0.00574	0.00230
	5	N(0.01, 0.003)	N(500, 100)	0.45245	0.00463	0.01134	0.00206
	6	N(0.01, 0.001)	N(300, 100)	0.93528	0.03499	0.03609	0.02365
	7	N(0.01, 0.001)	N(1000, 100)	1.00000	0.00033	0.00033	0.00023
	8	N(0.01, 0.001)	N(1000, 50)	1.00000	0.00033	0.00033	0.00023
	9	N(0.01, 0.001)	N(1000, 200)	1.00000	0.00033	0.00033	0.00023
3	1	U(0, 0.0025)	U(300, 700)	1.7050	0.5424	0.48384	0.43747
	2	U(0, 0.005)	U(300, 700)	0.8992	0.2718	0.28611	0.18081
	3	U(0, 0.01)	U(300, 700)	0.5054	0.0827	0.16220	0.04006
	4	U(0, 0.02)	U(300, 700)	0.3146	0.0124	0.09352	0.00387
	5	U(0, 0.01)	U(100, 500)	0.7780	0.2521	0.29160	0.15742
	6	U(0, 0.01)	U(800, 1200)	0.3289	0.0131	0.08298	0.00431

References

1. Iijima, S. Helical microtubules of graphitic carbon. *Nature* **1991**, *354*, 56–58. [CrossRef]
2. Iijima, S. Single-shell carbon nanotubes of 1-nm diameter. *Nature* **1993**, *363*, 603–605. [CrossRef]
3. Bethune, D.S.; Kiang, C.H.; Devries, M.S.; Gorman, G.; Savoy, R.; Vazquez, J.; Beyers, R. Cobalt-catalyzed growth of carbon nanotubes with single-atomic-layerwalls. *Nature* **1993**, *363*, 605–607. [CrossRef]
4. Thostenson, E.T.; Chou, T.-W. On the elastic properties of carbon nanotube-based composites: Modelling and characterization. *J. Phys. D Appl. Phys.* **2003**, *36*, 573. [CrossRef]
5. Deng, L.; Eichhorn, S.J.; Kao, C.-C.; Young, R.J. The Effective Young's Modulus of Carbon Nanotubes in Composites. *ACS Appl. Mater. Interfaces* **2011**, *3*, 433–440. [CrossRef]
6. Shi, D.-L.; Feng, X.-Q.; Huang, Y.Y.; Hwang, K.-C.; Gao, H. The Effect of Nanotube Waviness and Agglomeration on the Elastic Property of Carbon Nanotube-Reinforced Composites. *J. Eng. Mater. Technol.* **2004**, *126*, 250–257. [CrossRef]
7. Li, C.; Thostenson, E.T.; Chou, T.-W. Effect of nanotube waviness on the electrical conductivity of carbon nanotube-based composites. *Compos. Sci. Technol.* **2008**, *68*, 1445–1452. [CrossRef]
8. Bartels, J.; Jürgens, J.-P.; Kuhn, E.; Ploshikhin, V. Effects of curvature and alignment of carbon nanotubes on the electrical conductivity of carbon nanotube-reinforced polymers investigated by mesoscopic simulations. *J. Compos. Mater.* **2018**, *53*, 1033–1047. [CrossRef]
9. Fisher, F.T.; Bradshaw, R.D.; Brinson, L.C. Fiber waviness in nanotube-reinforced polymer composites—I: Modulus predictions using effective nanotube properties. *Compos. Sci. Technol.* **2003**, *63*, 1689–1703. [CrossRef]
10. Bradshaw, R.D.; Fisher, F.T.; Brinson, L.C. Fiber waviness in nanotube-reinforced polymer composites—II: Modeling via numerical approximation of the dilute strain concentration tensor. *Compos. Sci. Technol.* **2003**, *63*, 1705–1722. [CrossRef]
11. Tandon, G.P.; Weng, G.J. Average stress in the matrix and effective moduli of randomly oriented composites. *Compos. Sci. Technol.* **1986**, *27*, 111–132. [CrossRef]
12. Maghsoudlou, M.A.; Barbaz Isfahani, R.; Saber-Samandari, S.; Sadighi, M. Effect of interphase, curvature and agglomeration of SWCNTs on mechanical properties of polymer-based nanocomposites: Experimental and numerical investigations. *Compos. Part B Eng.* **2019**, *175*, 107119. [CrossRef]
13. Ginga, N.J.; Chen, W.; Sitaraman, S.K. Waviness reduces effective modulus of carbon nanotube forests by several orders of magnitude. *Carbon* **2014**, *66*, 57–66. [CrossRef]
14. Shao, L.H.; Luo, R.Y.; Bai, S.L.; Wang, J. Prediction of effective moduli of carbon nanotube–reinforced composites with waviness and debonding. *Compos. Struct.* **2009**, *87*, 274–281. [CrossRef]
15. Zhu, L.; Narh, K.A. Numerical simulation of the effect of nanotube orientation on tensile modulus of carbon-nanotube-reinforced polymer composites. *Polym. Int.* **2004**, *53*, 1461–1466. [CrossRef]
16. Shima, H. Buckling of Carbon Nanotubes: A State of the Art Review. *Materials* **2011**, *5*, 47–84. [CrossRef]
17. Li, C.; Zhu, C.; Lim, C.W.; Li, S. Nonlinear in-plane thermal buckling of rotationally restrained functionally graded carbon nanotube reinforced composite shallow arches under uniform radial loading. *Appl. Math. Mech.* **2022**, *43*, 1821–1840. [CrossRef]
18. Chen, G.; Seki, Y.; Kimura, H.; Sakurai, S.; Yumura, M.; Hata, K.; Futaba, D.N. Diameter control of single-walled carbon nanotube forests from 1.3–3.0 nm by arc plasma deposition. *Sci. Rep.* **2014**, *4*, 3804. [CrossRef]
19. Fagan, J.A.; Hároz, E.H.; Ihly, R.; Gui, H.; Blackburn, J.L.; Simpson, J.R.; Lam, S.; Hight Walker, A.R.; Doorn, S.K.; Zheng, M. Isolation of >1 nm Diameter Single-Wall Carbon Nanotube Species Using Aqueous Two-Phase Extraction. *ACS Nano* **2015**, *9*, 5377–5390. [CrossRef]
20. Charlier, J.C.; Lambin, P. Electronic structure of carbon nanotubes with chiral symmetry. *Phys. Rev. B* **1998**, *57*, R15037–R15039. [CrossRef]
21. Park, K.H.; Chhowalla, M.; Iqbal, Z.; Sesti, F. Single-walled carbon nanotubes are a new class of ion channel blockers. *J. Biol. Chem.* **2003**, *278*, 50212–50216. [CrossRef] [PubMed]
22. Shi, Z.; Lian, Y.; Zhou, X.; Gu, Z.; Zhang, Y.; Iijima, S.; Zhou, L.; Yue, K.T.; Zhang, S. Mass-production of single-wall carbon nanotubes by arc discharge method11This work was supported by the National Natural Science Foundation of China, No. 29671030. *Carbon* **1999**, *37*, 1449–1453. [CrossRef]
23. Jinno, M.; Ando, Y.; Bandow, S.; Fan, J.; Yudasaka, M.; Iijima, S. Raman scattering study for heat-treated carbon nanotubes: The origin of ≈1855cm−1 Raman band. *Chem. Phys. Lett.* **2006**, *418*, 109–114. [CrossRef]
24. Jorio, A.; Saito, R. Raman spectroscopy for carbon nanotube applications. *J. Appl. Phys.* **2021**, *129*, 021102. [CrossRef]
25. Chen, J.; Xu, X.; Zhang, L.; Huang, S. Controlling the Diameter of Single-Walled Carbon Nanotubes by Improving the Dispersion of the Uniform Catalyst Nanoparticles on Substrate. *Nano-Micro Lett.* **2015**, *7*, 353–359. [CrossRef]
26. Navas, H.; Picher, M.; Andrieux-Ledier, A.; Fossard, F.; Michel, T.; Kozawa, A.; Maruyama, T.; Anglaret, E.; Loiseau, A.; Jourdain, V. Unveiling the Evolutions of Nanotube Diameter Distribution during the Growth of Single-Walled Carbon Nanotubes. *ACS Nano* **2017**, *11*, 3081–3088. [CrossRef]

C Journal of Carbon Research

Article

Study of the Influence of Silicon-Containing Diamond-like Carbon Coatings on the Wear Resistance of SiAlON Tool Ceramics

Marina A. Volosova * and Anna A. Okunkova

Department of High-Efficiency Processing Technologies, Moscow State University of Technology STANKIN, Vadkovskiy per. 3A, 127055 Moscow, Russia
* Correspondence: m.volosova@stankin.ru; Tel.: +7-499-972-94-29

Abstract: DLC coatings have low adhesive bond strength with the substrate and a high level of residual stresses. This paper is devoted to researching a complex of characteristics of a DLC-Si coating deposited on samples of SiAlON ceramics with intermediate coatings (CrAlSi)N pre-formed to improve the adhesive bond strength employing vacuum-plasma spraying. DLC-Si coatings were formed by chemical vapor deposition in a gas mixture of acetylene, argon, and tetramethylsilane supplied through a multichannel gas purge system controlling the tetramethylsilane volume by 1, 4, 7, and 10%. The SiAlON samples with deposited (CrAlSi)N/DLC-Si coatings with different silicon content in the DLC layer were subjected to XPS and EDX analyses. Tribological tests were carried out under conditions of high-temperature heating at 800C. The nanohardness and elasticity modulus of the rational (CrAlSi)N/DLC-Si coating with Si-content of 4.1% wt. were 26 ± 1.5 GPa and 238 ± 6 GPa, correspondingly. The rational composition of (CrAlSi)N/DLC-Si coating was deposited on cutters made of SiAlON ceramics and tested in high-speed machining of aircraft nickel-chromium alloy compared to uncoated and DLC-coated samples. The average operating time (wear resistance) of (CrAlSi)N/DLC-Si(4.1% wt.)-coated end mills before reaching the accepted failure criterion was 15.5 min when it was 10.5 min for the original cutters.

Keywords: ceramic cutters; cutting ceramics; coefficient of friction; (CrAlSi)N sublayer; DLC-Si coatings; high-temperature wear; SiAlON; sp3- and sp2-hybridizations; wear resistance

Citation: Volosova, M.A.; Okunkova, A.A. Study of the Influence of Silicon-Containing Diamond-like Carbon Coatings on the Wear Resistance of SiAlON Tool Ceramics. C **2023**, 9, 50. https://doi.org/10.3390/c9020050

Academic Editor: Gil Gonçalves

Received: 22 February 2023
Revised: 3 May 2023
Accepted: 9 May 2023
Published: 16 May 2023

1. Introduction

Nowadays, diamond-like carbon (DLC) coatings are widely used in mechanical engineering as wear-resistant layers, the primary function of which is to increase the hardness significantly and reduce the coefficient of the friction of the product's working surfaces during machining under various contact loads and intense frictional interaction of counter bodies with various materials [1–5]. DLC coatings consist of carbon atoms with sp^3- and sp^2-hybridizations combined into an amorphous structure. The high content of carbon atoms with diamond sp^3-bonds in the presence of graphite-like sp^2-bonds provides a unique set of parameters for coatings of this type [6–9].

DLC coatings have apparent positive features but many disadvantages at the same time. For example, the majority of the diamond-like carbon coatings including both hydrogen-free (a-C) and hydrogenated (a-C:H) have low adhesive bond strength with the substrate and a high level of residual stresses, which are aggravated by increasing DLC coatings' thickness [10–14]. In addition, a significant limitation of the coatings' application areas is related to their relatively low thermal stability, which results in the coatings rapidly losing their properties under conditions of exposure to high thermal loads on the contact surfaces [15,16].

Various technological approaches have become widespread among the leading research groups to minimize these shortcomings. The most effective is alloying with various elements during the DLC coating formation. For example, introducing metal and metalloid atoms into a diamond-like carbon matrix can significantly reduce internal stresses and improve coatings' mechanical and tribological properties since DLC coatings have a high hardness and low friction coefficient when alloying elements are characterized by high ductility. In addition, introducing additional elements can increase the operation efficiency of DLC coatings under thermal loads due to the thermally stable oxides' formation upon heating [17–19]. Another technological approach that has proved its advantages in practice is the formation of intermediate coatings before DLC coating's deposition to increase the total thickness of the coating while ensuring the high strength of the adhesive bond with the substrate [20–23]. In addition, the intermediate coating can protect the product's contact surfaces from premature wear for an additional time after the abrasion of the outer DLC layer.

There is no universal technological solution with relatively intensive development of improving DLC coatings. Moreover, one must proceed from the specific operating conditions of the product (substrate) on which the coating is applied and pay special attention to the type and substrate properties to develop a DLC coating and its deposition technique. Evaluation of the effectiveness of the chosen approach to the DLC coating formation should be carried out under thermomechanical loads corresponding to the nature and level of operational ones [24,25].

DLC coatings' deposition on metal and metal-ceramic alloys is a reasonably well-studied area, but the deposition of these coatings on ceramics remains outside the main focus of research groups. Technical ceramics are structurally inhomogeneous composite materials with high hardness and heat resistance, products which are operated under the simultaneous action of large mechanical and thermal loads in many cases [26–30]. The influence of various thin-film coatings based on nitrides of refractory metals on the wear resistance of ceramics has been studied by various authors [31–34]. The results of studies of the potential possibilities of using DLC coatings to improve the wear resistance of ceramic products are in areas of scientific and practical interest. Preliminary studies performed by the authors of this work have shown that the formation of intermediate coatings, in particular, multicomponent nitride such as (CrAlSi)N before DLC coatings' deposition, as well as doping of the outer DLC coating of silicon, is an approach that has a particular potential and needs more in-depth study [35–39].

One of the most modern SiAlON ceramics, which exhibit improved physical and mechanical properties among industrially produced ceramic materials and is used to manufacture various types of metalworking ceramic tools, was chosen as substrates for DLC coatings' deposition [39–44]. At the same time, end mills designed for high-speed machining aviation-type nickel-chromium alloys, the so-called superalloys, are of particular practical interest among tools made from SiAlON.

Thus, among their various advantages, DLC-coatings exhibit a low level of adhesion to the ceramic substrate and a high level of stress. These shortcomings only increase with increasing coating thickness and ultimately affect the durability of the cutting tool, which is especially important in the high-temperature machining of nickel-containing alloys.

This work is aimed at solving this problem and proposes an original integrated approach to improve the tool-life parameters, namely to use the method of alloying the DLC-coating deposited by a chemical method and the formation of an intermediate coating to improve the adhesive bond.

Namely, the study is devoted to the research of the effect of DLC coatings with different silicon contents formed by chemical vapor deposition on the wear resistance of SiAlON tool ceramics where DLC-Si was deposited on a preformed (CrAlSi)N intermediate coating. The volume content of tetramethylsilane in the gas mixture was varied at the level of 1, 4, 7, and 10% vol. to provide different content of the alloying element in DLC-Si. The (CrAlSi)N/DLC-Si coatings were subjected to XPS and EDX analyses, nanohardness and elastic modulus measurements, as well as tribological tests at high-temperature friction sliding. The rational composition of the (CrAlSi)N/DLC-Si coating (from the point of view of high-temperature tribological tests' results, in other words, the coating option that has shown the best results in tests) was deposited on end-mill samples made of SiAlON tool ceramics, which were tested during high-speed machining of aircraft NiCr20TiAl nickel-chromium alloy. During testing, data were obtained on wear resistance and prospects for further application of these coatings.

2. Materials and Methods

2.1. Characteristics of SiAlON-Ceramic Substrates and End Mills Made of It

When conducting research, SiAlON-based ceramics for tool purposes was used as a substrate material for (CrAlSi)N/DLC-Si coating. It has the following physical and mechanical properties: hardness of 15.0 GPa; density of 3.45 g/cm^3; bend strength of 870 MPa; Young's modulus of 280 GPa; fracture toughness K_{1c} of 6.9 MPa·m$^{1/2}$. The technology of high-speed spark plasma sintering of the initial ceramic powders α-SiAlON and β-SiAlON was used to obtain samples based on SiAlON under the conditions of the appearance of spark plasma between the particles of the sintered workpiece and the passage of pulsed currents through it [35].

Figure 1a shows the SEM image of the microstructure of samples of sintered SiAlON ceramics obtained on a Tescan VEGA 3 LMH scanning electron microscope (Brno, Czech Republic). Samples in the form of washers of 19 mm in diameter and 7 mm in height were made from sintered ceramic blanks, which were subsequently coated with (CrAlSi)N/DLC-Si compositions for various types of analysis and tribological tests. In addition, samples of end mills with a diameter of 10 mm and a length of 60 mm were made from sintered ceramic blanks. An SEM image of its cutting part is shown in Figure 1b. A photograph of their general view is presented in Figure 1c. The technology of diamond sharpening of end mills is described in detail in [35]. (CrAlSi)N/DLC-Si coatings were deposited on the end mills, which showed the best performance in high-temperature tribological tests.

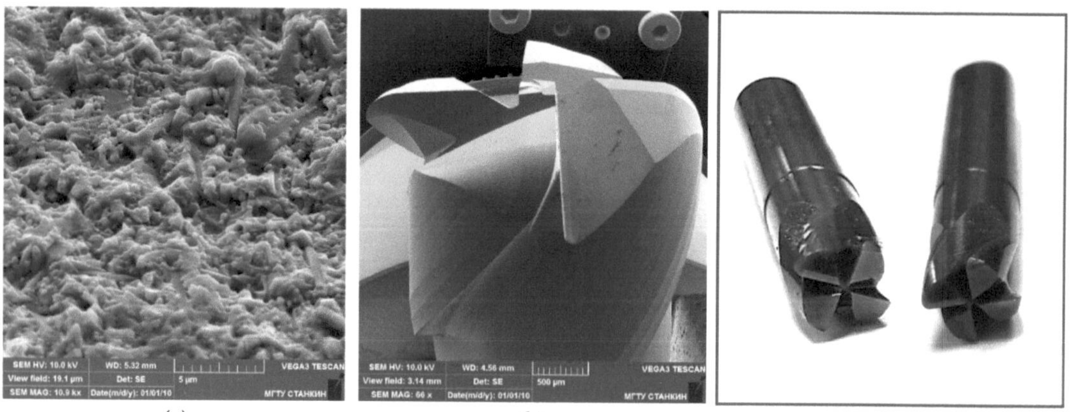

(a) (b) (c)

Figure 1. SEM images of the microstructure of SiAlON ceramics (**a**), the cutting part of the end mill made from it (**b**) and a general view of the ceramic end mills (**c**).

The X-ray diffraction pattern of a sintered sample made of SiAlON ceramics used in the research is shown in Figure 2. X-ray phase analysis was carried out on a Bruker D8 Advance diffractometer (Bruker AXS GmbH, Karlsruhe, Germany) with monochromatic CuKα radiation when shooting in the angle range of 20–90° 2Θ with a step of 0.05° and exposure to each point for 5 s. The analysis of the obtained results using the ICDD PDF-2 database made it possible to reveal the main structural components of the sintered SiAlON ceramics, which is represented by the Si_5AlON_7 and Si_3N_4 phases.

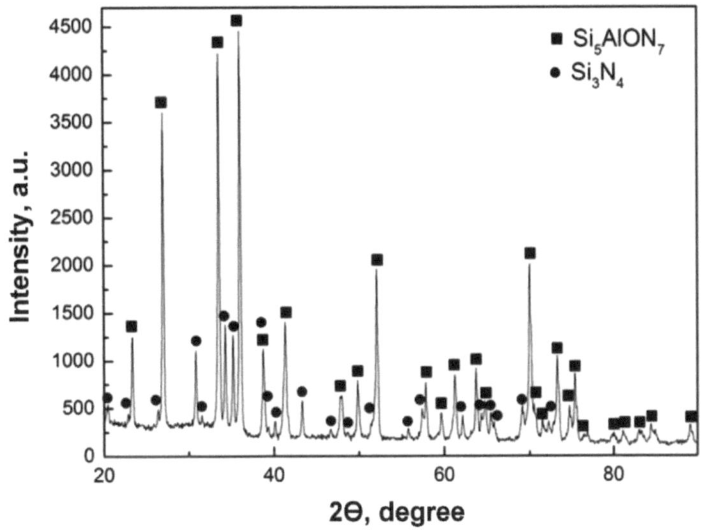

Figure 2. X-ray diffraction pattern of a sample made of SiAlON ceramics used in research.

2.2. (CrAlSi)N/DLC-Si Coating Technique for SiAlON Ceramic Samples

Coating of SiAlON-ceramic samples was carried out on an experimental and industrial technological unit developed at Moscow State University of Technology STANKIN (Moscow, Russia), equipped with a vacuum-arc discharge plasma generation device for the deposition of a nitride transition nitride coating (CrAlSi)N by evaporation of AlSi and Cr cathodes, a multichannel system gas inlet and glow discharge generation system, which ensures the formation of a coating based on DLC layers by gas-phase deposition in glow discharge plasma through a chemical reaction and decomposition of the components of a gas mixture of acetylene, argon, and tetramethylsilane. The original unit is based on a set of solutions created by the research group of the university and allows the deposition of a wide range of transition coatings based on refractory metal nitrides [45–49]. The implemented-in-the-unit DLC deposition technique is similar to the approach developed and implemented by PLATIT (Selzach, Switzerland).

The general view and internal design of the processing unit used for the deposition of CrAlSi)N/DLC-Si coatings are shown in Figure 3. The volume content of tetramethylsilane in the gas mixture was varied in the volume of 1, 4, 7, and 10% to provide different content of the alloying element (silicon) in DLC-Si composition. The total thickness of the deposited coatings was controlled by the deposition time and amounted to ~3.9 μm including ~1.9 μm for (CrAlSi)N and ~2.0 μm for DLC. A block diagram of the technological process and the sequence of operations with the modes of (CrAlSi)N/DLC-Si coating deposition on samples made of SiAlON ceramics is shown in Figure A1 (Appendix A).

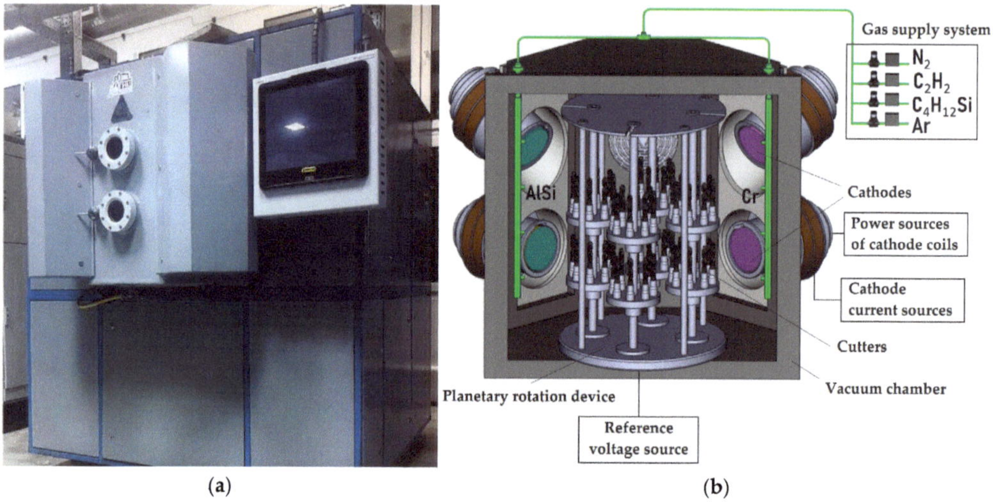

Figure 3. General view (**a**) and internal design of the processing unit (**b**) used for (CrAlSi)N/DLC-Si coatings deposition on SiAlON-ceramic samples.

2.3. Methods for Studying the Structure, Elemental Composition, and Tribological Testing of Samples of SiAlON Ceramics with (CrAlSi)N/DLC-Si Coatings

X-ray Photoelectron Spectroscopy (XPS) analysis was used to study carbon atoms' chemical and electronic state on the surface of ceramic samples with (CrAlSi)N/DLC-Si coatings. The DLC and (CrAlSi)N/DLC-Si coated samples and uncoated ones were heated to 800 °C in a high-temperature vacuum furnace Thermionic T1 (JSC Thermionika, Podolsk, Moscow region, Russia) and analyzed. Preheating was carried out to predict the transformation of the initial structure of the (CrAlSi)N/DLC-Si coatings during cutting, which is accompanied by the development of high temperatures.

XPS analysis was performed on a Thermo Scientific K-ALPHA X-ray photoelectron spectrometer (Thermo Fisher Scientific Inc., Bremen, Germany). High-resolution imaging was carried out using monochromatic and polychromatic Mg Kα and Al Kα radiation at a power of 200–300 W, a voltage of 14 kV, and a chamber pressure of 7×10^{-8} Pa. Studies were carried out to identify various peaks of the DLC layer in the composition of the prominent C1s peak: sp2 and sp3 peaks characterizing carbon hybridizations, and C=O and C-O peaks characterizing various compounds of components with oxygen and other elements. In addition, the peaks of phases containing silicon compounds were analyzed. A set of characteristics was evaluated during XPS analysis such as start binding energy (SBE), binding energy (BE), end binding energy (EBE), and atomic percent (Atomic %) for each component of the DLC layer.

The study of the elemental composition of the surface layer of ceramic samples with (CrAlSi)N/DLC-Si coatings was carried out using the Oxford Instruments INCA Energy energy-dispersive X-ray spectroscopy (EDX) system with which a VEGA 3 LMH scanning electron microscope (Tescan, Brno, Czech Republic) is equipped. The atoms were excited using a beam of electrons in the EDX process and emitted x-rays specific to each chemical element. When studying the energy spectrum of this radiation using a specialized program, data were obtained on the samples' semi-quantitative elemental composition.

The study of the structure of the formed coatings was carried out by transmission electron microscopy using JEM-2100F equipment (JEOL, Tokyo, Japan). The surface microrelief of ceramic samples with (CrAlSi)N/DLC-Si coatings was studied on a DektakXT stylus profilometer (Bruker, Karlsruhe, Germany).

The nanohardness of (CrAlSi)N/DLC-Si coatings formed on Si-AlON-ceramic samples was estimated on the Martens scale by indentation with a diamond indenter on a CSEM instruments' nanohardness tester (CSM Instruments SA, Peseux, Switzerland) according to the method of Oliver and Farr, [50–53]. The cycle "load–unload" duration was 50 s, and the applied load was 4.0 mN. The nanohardness and modulus of elasticity of the coatings were calculated based on the obtained experimental data.

Wear resistance of coated ceramic specimens under conditions of high-temperature friction-sliding in contact with a counterbody (ceramic ball of 6 mm in diameter) at a load of 1 N, a sliding speed of 10 cm/s, a displacement radius of 2 mm, and a heating temperature of 800 °C was evaluated on a TNT tribometer S-AX0000 (Anton Paar TriTec SA, Wien, Austria). The change in the coefficient of friction over time was determined with the rotation of the test sample relative to a fixed counterbody. The reduced volumetric wear of the coated samples was determined using the average value of the cross-sectional area of the wear track, which was estimated from the results of measuring five transverse profiles of the wear track on the DektakXT profilometer.

2.4. Wear Test Technique for End Mills of SiAlON Ceramics with (CrAlSi)N/DLC-Si Coatings

Laboratory testing of uncoated and (CrAlSi)N/DLC-Si coated SiAlON-ceramic end mills was carried out by high-speed cutting of aviation nickel-chromium alloy – NiCr20TiAl (EN 10269), which had the shape of a cylinder with a diameter of 50 mm. The chemical composition of the NiCr20TiAl alloy is shown in Table 1. Physical and mechanical properties of the nickel-chromium alloy to be machined are as follows: HRC hardness of 32–33; ultimate strength (tensile strength) of 1130–1160 MPa; density of 5.66 g/cm^3.

Table 1. Chemical composition (%) of NiCr20TiAl nickel-chromium alloy.

Ni	Cr	Ti	Al	Fe	Si	Co	Mn	Cu	C	Other Elements
73	20	2.5	1.0	1.0	0.6	0.5	0.4	0.2	0.07	0.73

The tests were carried out on a CTX 1250TC multi-axis turning and milling machining center from DMG (Bielefeld, Germany). The positioning of SiAlON-ceramic end mills relative to the workpiece made of NiCr20TiAl alloy during wear resistance tests is shown in Figure 4a. The experiments were carried out under the following milling modes: cutting speed of 376.8 m/min (rotational speed 12,000 rpm), feed per minute of 1500 mm/min, and feed per tooth of 0.03 mm/tooth. The "dry" machining scheme without cutting fluids was used [54].

(a) (b)

Figure 4. Positioning of end mills made of SiAlON ceramics relative to the workpiece made of NiCr20TiAl alloy for wear resistance testing (**a**) and the wear center location on the back surface of the cutting tooth and measuring h_f in quantifying wear (**b**).

The critical size of the wear area h_f along the flank surface of the cutter tooth equal to 0.4 mm was taken as the criterion of the end mills' failure (Figure 4b). The wear resistance of the tool was defined as the cutting time until the cutter reaches the specified wear. When the mentioned criterion is exceeded, the cases of chipping of the cutting edge increase many times over. A SteREO Discovery V12 metallographic optical microscope from Zeiss (Oberkochen, Germany) was used to quantify wear. The wear pad was monitored on each of the tested ceramic cutter teeth. The arithmetic mean values were calculated based on the obtained data, which were used to draw the "wear-cutting time" curves.

3. Results and Discussion

3.1. Structure and Morphology of (CrAlSi)N/DLC-Si Coatings Deposited on SiAlON Ceramics

The results of studies performed by means of transmission electron microscopy of the cross-section of (CrAlSi)N/DLC-Si coatings deposited on SiAlON ceramics make it possible to reveal the structural features of the DLC-Si outer layer (Figure 5). The DLC-Si layer formed by chemical vapor deposition in a mixture of acetylene, argon, and tetramethylsilane is characterized by a homogeneous amorphous structure, on which the gradient DLC layer and the main DLC layer are well pronounced. It should be noted that no microstructural differences were observed in the coatings formed at different contents of tetramethylsilane $Si(CH_3)_4$.

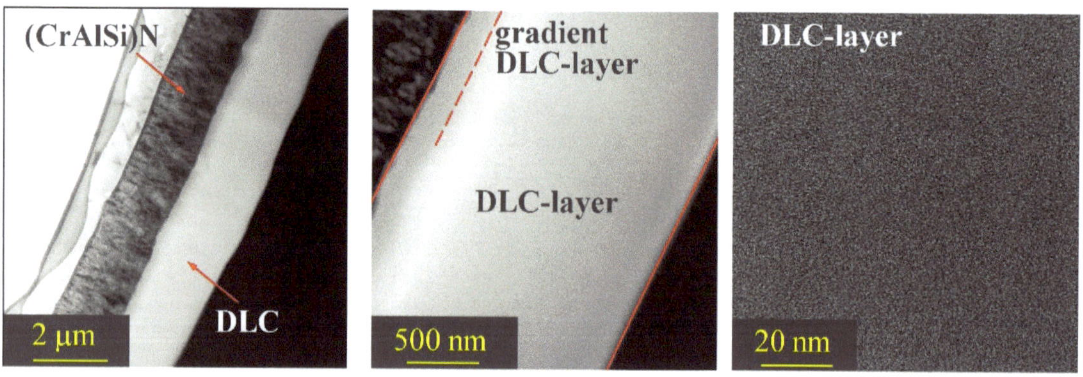

(a) (b) (c)

Figure 5. TEM images of the cross-section of the (CrAlSi)N/DLC-Si coating (**a**,**b**) and the DLC-Si outer layer (**c**).

The distinctive features of the (CrAlSi)N/DLC-Si coatings are revealed by SEM analysis and 3D surface profilometry (Figure 6). The surface of (CrAlSi)N/DLC-Si coatings has a characteristic morphology of densely packed rounded granule-like particles with a diameter of 300–900 nm. A similar morphological pattern was observed by the authors of [55] in studying DLC coatings.

3.2. Structure and Elemental Composition of the Surface Layer of (CrAlSi)N/DLC-Si Coatings Deposited on SiAlON Ceramics

Tables 2–5 summarize the results of the XPS analysis of (CrAlSi)N/DLC-Si coatings formed on SiAlON-ceramic at various contents of 1, 4, 7, and 10% vol. of tetramethylsilane $Si(CH_3)_4$ in a gas mixture with acetylene and argon. Considering that an excess amount of oxygen on the surface of the samples complicates the assessment of sp2 and sp3 hybridizations of carbon, a thin surface layer of the coating was etched with argon ions with an energy of 2 keV to a depth of about 0.005 µm immediately before analysis. The error in determining the binding energy of electronic levels in the XPS analysis was 0.1 eV.

The XPS analysis also revealed an O1s peak during scanning in the energy range from 525 to 545 eV, which characterizes the presence of various forms of oxygen with increased binding energies, 531.6–533.5 eV. The studied samples contain oxygen in the form of surface-adsorbed groups. The proportion of the oxygen component for the samples in the initial state is about 7–10%. It should be noted that another component in the main oxygen peak O1s was identified as silicon dioxide SiO_2 for samples on which the DLC-Si layer was deposited in a gas mixture containing 7 and 10% vol. of $Si(CH_3)_4$. Its share was about 1%.

In addition, a Si2p3 silicon photoelectron peak was detected, indicating the presence of Si, Si-N, Si-C type compounds with energies connections of 100.3–100.5 when scanning (CrAlSi)N/DLC-Si coatings in the energy range from 95 to 110 eV in the XPS analysis. For the samples at the initial state, its proportion increased from 1.1 to 7.06% as $Si(CH_3)_4$ increased from 1 to 10% vol. in the gas mixture during the DLC-Si layer deposition.

The last column of Tables 2–5 shows the calculated data of the XPS analysis of the samples after heating in a vacuum furnace to a temperature of 800 °C and holding for 30 min. The data obtained indicate a significant transformation in the DLC-Si layer structure upon thermal treatment. An analysis of these changes is of particular interest since the surface layer of tools made of SiAlON ceramics is subjected to intense thermal impact during operation at high cutting speeds. The main changes identified are as follows:

- Firstly, pronounced changes in the C1s carbon peak were established as follows: The share of sp3/sp2 hybridization among the total composition of the components of the coatings under study decreases by 9–13%. Simultaneously, the proportion of diamond hybridization (sp3) significantly decreases by 17–23%, while the proportion of graphite hybridization (sp2) increases by 6–10%. The proportion of various carbon-containing impurities in the DLC-Si layer increases by 4–7%. The specified tendency for the transition of sp3- to sp2-hybridization under thermal exposure is due to the level of thermal stability and features of the interatomic bonds of carbon atoms and is consistent with the experimental data of other researchers [59–62].
- Secondly, significant changes in the O1s oxygen peak were revealed. Its fraction in the composition of the DLC-Si layer's components for samples subjected to high-temperature treatment increases by about 2 times compared to the initial samples. At the same time, the proportion of silicon dioxide SiO_2 in the oxygen peak O1s for the samples after thermal treatment increases significantly as $Si(CH_3)_4$ increases in the gas mixture during the DLC-Si layer deposition. If $Si(CH_3)_4$ content is 1% vol., the proportion of SiO_2 is about 1%. Then, the share of silicon dioxide in the DLC-Si layer's composition is 3.3, 5.6, and 6% at 4, 7, and 10% vol. of $Si(CH_3)_4$, respectively. Simultaneously, there is a significant decrease in the proportion of the Si2p3 silicon peak (compounds of the Si, Si-N, Si-C type) in the surface layer of the samples after heating. It indicates the oxidative reactions' occurrence with the thermostable SiO_2 phases' formation that reduces the compressive stress in the films and improves the film fracture toughness [63]. The authors of [64] reported an increase in the wear resistance of DLC:Si films two times in comparison with pure DLC films. The authors of [65] noted that analysis of the XPS peaks of DLC:Si films containing 4–29 at.% of Si deposited by reactive magnetron sputtering revealed the presence of Si–C bonds and a significant amount of Si–O–C and $Si–O_x$ bonds. However, the rest of the mentioned study [65] is devoted to researching the optical properties of those coatings. The intensification of the formation of these phases occurs, starting from temperatures of 800 °C as follows from the data of works [66,67].

Since the performed XPS analysis provides information on the presence of various compounds in a thin near-surface layer and does not provide information on the elemental composition and content of the alloying element (silicon) along the depth of the surface layer, the EDX analysis of the samples was additionally performed. The EDX maps of the distribution of chemical elements in a tool SiAlON-ceramic sample coated with (CrAlSi)N/DLC-Si (as an example, a characteristic pattern for the deposition of a DLC-Si layer at 4% vol. in a gas mixture) is presented in Figure 7. It allows us to qualitatively assess the various elements' concentration in the surface layer from contrast images. In particular the content of silicon is traced in the bulk of the SiAlON-ceramic, in the preliminarily formed nitride layer (CrAlSi)N, and directly on the surface of the formed DLC-Si layer.

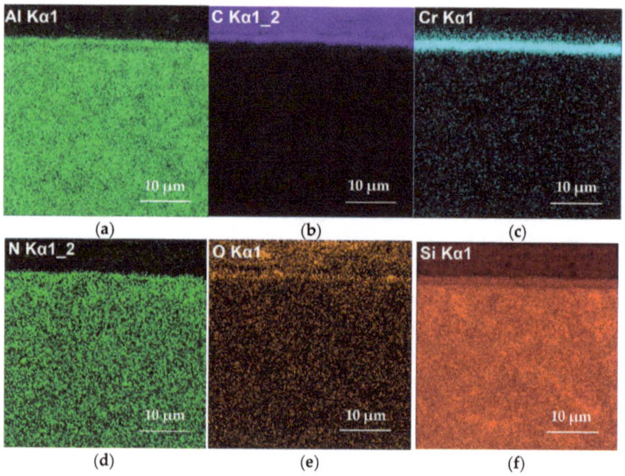

Figure 7. EDX-maps of the chemical elements' distribution in the surface layer of a sample made of SiAlON ceramics with (CrAlSi)N/DLC-Si coatings: (**a**) aluminum; (**b**) carbon; (**c**) chromium; (**d**) nitrogen; (**e**) oxygen; (**f**) silicon.

Figure 8 shows the distribution of elements over the cross-section of a (CrAlSi)N/DLC-Si coating formed on a sample of SiAlON tool ceramics. It is seen that the silicon concentration increases upon transition from the main DLC layer to the gradient DLC layer (the layer structure is visible in Figure 5b), then it slightly decreases along the depth of the intermediate (CrAlSi)N layer and sharply increases along the depth of the SiAlON ceramics.

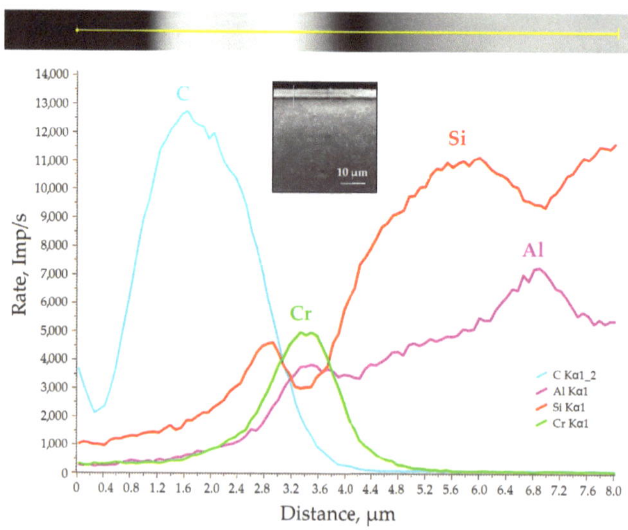

Figure 8. Elements' distribution over the cross-section of the (CrAlSi)N/DLC-Si coating deposited on a sample of tool SiAlON-ceramic. (Data obtained by EDX analysis.)

Figure 9 shows the relationship between the content of tetramethylsilane in the gas mixture used during the primary (outer) DLC-Si layer deposition and the silicon content in it, revealed by the results of XPS analysis. It can be seen that the dependence is linear, and the content of the mass fraction of Si (% wt.) increases monotonically with an increase in the $Si(CH_3)_4$ content in the gas mixture (% vol.). Thus, the presence of 1, 4, 7, and 10% vol. $Si(CH_3)_4$ in the gas mixture ensures the Si content in the DLC-Si layer is about 1.5, 4.1, 6.2, and 7.7% wt.

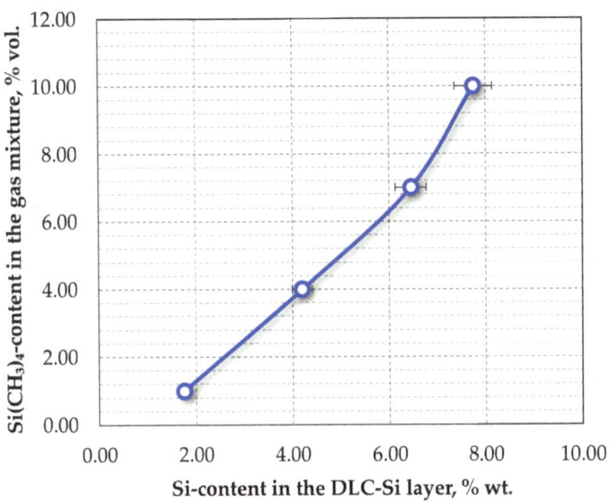

Figure 9. Dependence of the Si mass content in the main DLC-Si layer on the volume content of $Si(CH_3)_4$ in the gas mixture. (Data obtained by EDX analysis.)

3.3. Physical, Mechanical, and Tribological Characteristics of the Surface Layer of (CrAlSi)N/DLC-Si Coatings Deposited on SiAlON Ceramics

The dominant role in forming DLC coatings' various properties such as hardness and elastic modulus coefficient of friction, and volumetric wear, is played by the structural configurations of carbon and alloying/doping elements' compounds present in the deposited coating. Table 6 shows the results of the assessment of nanohardness and modulus of elasticity for polished samples of SiAlON ceramics with (CrAlSi)N/DLC-Si coatings with different Si content in the outer DLC layer, provided by varying the content of $Si(CH_3)_4$ in a gas mixture after deposition. The results were obtained at a penetration depth of the Berkovich indenter of about 400 nm. It can be seen that as the proportion of carbon sp3 bonds decreases and the proportion of silicon compounds increases, the nanohardness (H) and elastic modulus (E) of the DLC-Si layer decrease compared to the "pure" DLC film deposited in the absence of a silicon-containing gas. For example, while the values of H and E for the DLC layer are 28 and 246 GPa, these values decrease to 19 and 198 GPa, respectively, for the DLC-Si layer at Si, 7.7% wt.

Table 6. Characteristics of (CrAlSi)N/DLC-Si coatings deposited on SiAlON-ceramic substrates with different Si contents in the outer layer.

Sample	Si-Content in DLC Layer, % wt.	Nanohardness H, GPa	Modulus of Elasticity E, GPa
Original	-	28 ± 1.5	246 ± 6
No. 1	1.5	27 ± 1	241 ± 8
No. 2	4.1	26 ± 1.5	238 ± 6
No. 3	6.2	22 ± 1	210 ± 6
No. 4	7.7	19 ± 1	198 ± 4

It should be noted that the obtained nanohardness (H) and elastic modulus (E) of the DLC-Si layer correlates with the data obtained in [63] for hardness and elastic modulus of DLC films doped with Si deposited on the glass substrate and even surpass them even for the classic DLC-coating: 24 ± 2 and 159 ± 4 GPa, correspondingly.

However, it is impossible to predict the behavior of (CrAlSi)N/DLC-Si coatings under the simultaneous action of thermal and mechanical loads typical for the operation of a cutting tool only based on the established values of nanohardness and elastic modulus. Thus, the evaluating results of the effect of the Si content in the DLC-Si layer on the coefficient of friction and volumetric wear of the surface layer of SiAlON samples under high-temperature exposure are of interest.

Figure 10 shows the results of tribological tests for samples made of SiAlON ceramics with (CrAlSi)N/DLC-Si coatings with different Si-content (1.5, 4.1, 6.2, and 7.7% wt.). Additional data obtained for uncoated samples made of SiAlON ceramics and samples with single-layer coatings such as (CrAlSi)N and "pure" DLC are shown in Figure 11 for comparison.

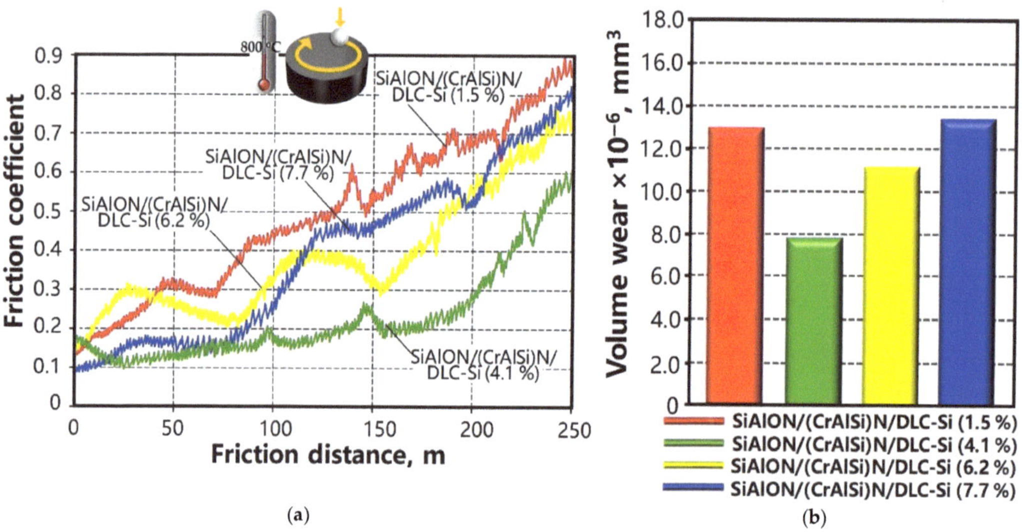

Figure 10. Change in the coefficient of friction depending on the testing time of samples made of SiAlON ceramics with (CrAlSi)N/DLC-Si coatings (**a**) and volumetric wear of samples after passing a distance of 250 m at high-temperature friction-sliding (**b**).

The presented results allow us to conclude (Figure 10) that the deposited (CrAlSi)N/DLC-Si coatings with different silicon content reduce the coefficient of friction of SiAlON ceramics over a 150 m test distance. After that, it increases significantly for samples with a Si content of 1.5, 6.2, and 7.7% wt. Upon reaching 250 m, it already has values characteristic of the original SiAlON ceramics and "pure" DLC coatings (Figure 11). The coefficient of friction of silicon-free DLC coatings begins to increase sharply already after passing a distance of 50 m and quickly reaches values corresponding to the ceramic substrate. It can be explained by the well-known low thermal stability of "pure" DLC coatings. Volumetric wear of the (CrAlSi)N/DLC-Si coated samples with the Si content of 1.5 and 7.7% wt was slightly less than for those made of SiAlON ceramics. Only a sample with a silicon content of about 4.1% wt had a lower coefficient of friction throughout the entire friction distance than SiAlON ceramics and "pure" DLC coatings. The same coating was distinguished by the most negligible volumetric wear, determined in the cross-section of the worn friction path. The authors of [63] also noted the primary influence of the content of SiO_x (O/Si ratio in DLC:SiO_x films was 0.286 ± 0.008) on the wear resistance of coatings for sliding/friction parts such as bearings or for precision gages.

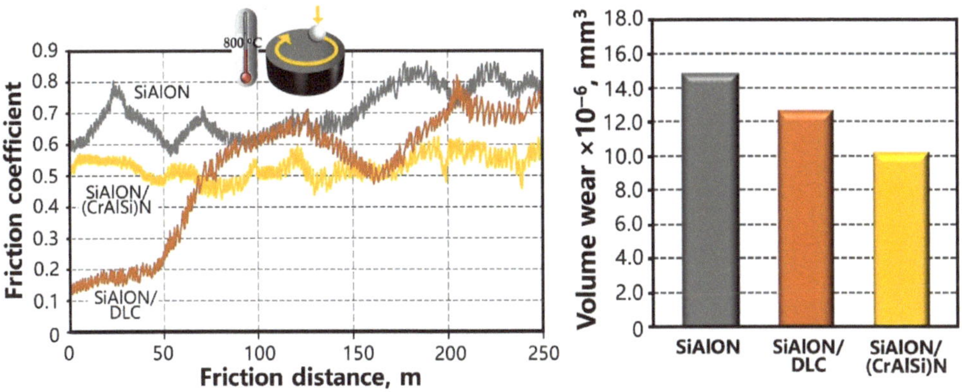

(a) (b)

Figure 11. Change in the coefficient of friction depending on the testing time of uncoated samples made of SiAlON ceramics and with the deposited single-layer (CrAlSi)N and DLC coatings (**a**) and volumetric wear of samples after passing a distance of 250 m at high-temperature friction-sliding (**b**).

The results obtained are related to the fact that samples with (CrAlSi)N/DLC-Si coatings with a silicon content of about 4.1% wt. have the most minor decrease in the proportion of diamond sp3 hybridization in the structure upon heating (Table 3), and doping silicon in carbon coating makes it possible to increase its thermal stability due to the formation of SiO_2 compounds under thermal action, which was identified by the XPS analysis described above. It can be assumed that for a sample with a lower silicon content of 1.5% wt., the amount of SiO_2 formed during heating is still insufficient for a noticeable increase in the thermal stability of the surface layer (Table 2). The insufficient effect of the use of (CrAlSi)N/DLC-Si coatings with a higher silicon content (6.2 and 7.7% wt.) can be attributed to the fact that despite an increase in the amount of SiO_2 formed during thermal exposure (Tables 4 and 5), this compound possessing thermal stability exhibits also increased brittleness reported by various authors [68,69]. Knowing the wear-resistant properties of SiO_x content in DLC coatings, the authors [70] even propose the deposition of SiO_2 quartz film as an independent layer on Al_2O_3 ceramics in combination with DLC coating on top of it. Perhaps that is why the rather intensive wear is observed under the mechanical action of the counterbody on the surface layer of these samples. In addition, the lowest values of H and E were recorded for these samples (Table 6), which will significantly decrease during the thermal action on the surface layer.

When analyzing the behavior of (CrAlSi)N/DLC-Si coatings during high-temperature friction-sliding, the possible role of the intermediate (CrAlSi)N nitride coating, which in itself reduces the friction coefficient and the volumetric wear of the surface layer in the friction track, should be noted (Figure 11). Multi-component coatings of this type have proven well during operation under conditions of temperature influence, and their high thermal stability among nitride coatings was reported in [71,72]. It can be assumed that the outer DLC-Si layer of the two-layer (CrAlSi)N/DLC-Si coating composition performs antifriction functions and protects the surface layer from intense wear at the initial stage. After it loses its wear-resistant properties, the intermediate layer (CrAlSi)N perceives effective thermal and mechanical loads, providing less intensive volumetric wear of SiAlON ceramics.

It was shown that the (CrAlSi)N/DLC-Si coating with a Si content of about 4.1% wt exhibits the best wear resistance considering the results of tribological tests. This coating deposited on SiAlON-ceramic end mills has been selected for laboratory testing in milling of NiCr20TiAl nickel-chromium alloy.

3.4. Wear Resistance of SiAlON-Ceramic End Mills with (CrAlSi)N/DLC-Si Coatings

Two groups of SiAlON-ceramic end mills (each group included four cutters) were subjected to wear resistance tests during cutting: the samples without coatings and with (CrAlSi)N/DLC-Si coatings (Si of ~4.1% wt.). Figure 12 shows cutting tool flank wear versus cutting time in milling of NiCr20TiAl alloy at a cutting speed of 376.8 m/min, a feed per minute of 1500 mm/min, and a feed per tooth of 0.03 mm/tooth. It can be seen that the deposition of (CrAlSi)N/DLC-Si coatings significantly (by 1.47) increases the wear resistance of cutters made of SiAlON ceramics (up to a failure criterion of 400 μm). Figure 13 shows optical images of the development of wear on the flank surface of the teeth of uncoated and coated (CrAlSi)N/DLC-Si end mills with an increase in cutting time, clearly demonstrating the decrease in the intensity of wear in the presence of the coating.

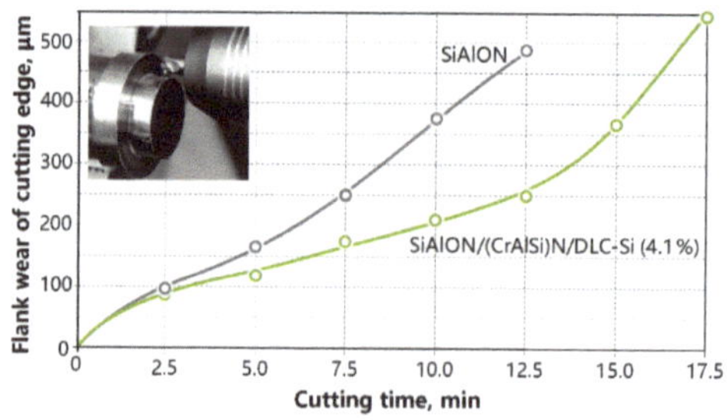

Figure 12. Dependences of flank wear of various SiAlON-ceramic end mills on cutting time in milling of NiCr20TiAl alloy.

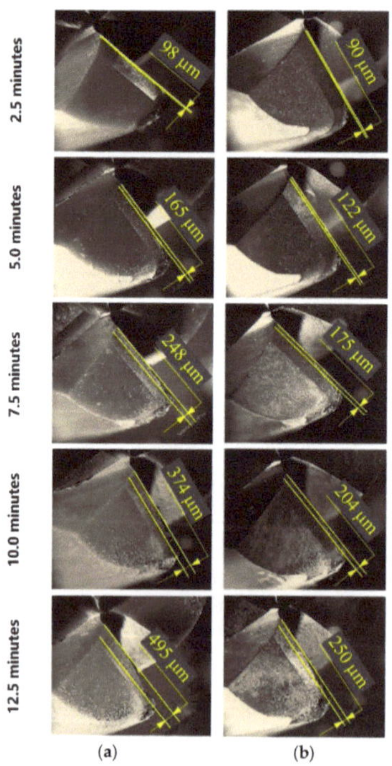

Figure 13. Optical images of the development of wear areas on the tooth flanks of uncoated (**a**) and (CrAlSi)N/DLC-Si-coated (**b**) SiAlON-ceramic end mills with an increase in cutting time.

Thus, the average operating time (wear resistance) of coated end mills before reaching the accepted failure criterion was 15.5 min and 10.5 min for the original cutters. Considering the extremely high cost of ceramic end mills, the achieved result of increasing wear resistance through the formation of (CrAlSi)N/DLC-Si coatings can provide a significant economic benefit.

It should be noted that a single-layer DLC coating is thermally unstable when heated at the mentioned temperature, as can be seen from high-temperature friction-sliding tests. However, the Si-doped DLC-coating combined with a three-component nitride sublayer demonstrated higher thermal stability. At the same time, the role of the substrate cannot be ignored. In this study, the substrate is a high-temperature-resistant material. It is obvious that if the studies were performed not for oxide-nitride ceramics but for less thermo-resistant material (for example, metal-based alloy), then the coating would demonstrate a different temperature tolerance.

4. Conclusions

This study has demonstrated that silicon doping of DLC coatings formed by gas-phase deposition in combination with the pre-formed interlayer coating deposition is an integrated approach that can expand the possible areas of application of diamond-like carbon coatings and increase the efficiency of their use for products operating under high thermal loads (primarily for cutting tools).

The EDX analysis showed that the presence of $Si(CH_3)_4$in the gas mixture in a volume of 1, 4, 7, and 10% ensures the formation of DLC coatings with a Si content of about 1.5, 4.1, 6.2, and 7.7% wt on SiAlON-ceramic samples, respectively.

XPS analysis showed that the main component of the structural state of the DLC-Si coatings formed by gas-phase deposition is diamond sp3 (share of 48–56%) and graphite sp2 (15–18%) hybridization of carbon. Under increased thermal exposure (heating up to 800 °C in a vacuum furnace), a significant transformation in the DLC-Si structure was revealed as follows: the fraction of sp3 significantly decreased (up to 23%), while the fractions of sp2 and carbon-containing impurities increased. In addition, a twofold increase in the fraction of the oxygen component in the DLC-Si coatings is noted upon heating. Simultaneously, oxidative reactions occur in the surface layer of the samples during heating with the formation of thermostable SiO_2 phases. Their fraction depends on the content of $Si(CH_3)_4$ in the gas mixture in the range of 1–6%.

Studies of the (CrAlSi)N/DLC-Si coatings' behavior in high-temperature friction-sliding have shown that the samples whose DLC coatings with ~4.1% wt. of silicon demonstrated the smallest coefficient of friction over the entire friction distance and the minimum volumetric wear of the surface layer at the friction path.

Tests of SiAlON-ceramic end mills in milling of NiCr20TiAl nickel-chromium alloy showed that the deposition of coatings of the rational (CrAlSi)N/DLC-Si composition (Si of ~4.1% wt.) made it possible to increase the operating time of the cutters by a factor of 1.47 compared to the uncoated ones.

The original experimental data can be used to develop DLC coatings for metalworking needs further.

Author Contributions: Conceptualization, M.A.V.; methodology, M.A.V.; software, M.A.V. and A.A.O.; validation, M.A.V.; formal analysis M.A.V.; investigation, M.A.V.; resources, M.A.V.; data curation, A.A.O.; writing—original draft preparation, M.A.V.; writing—review and editing, M.A.V.; visualization, M.A.V. and A.A.O.; supervision, M.A.V.; project administration, M.A.V.; funding acquisition, M.A.V. All authors have read and agreed to the published version of the manuscript.

Funding: This work was supported financially by the Ministry of Science and Higher Education of the Russian Federation (project No. FSFS-2023-0003).

Data Availability Statement: Data are available in a publicly accessible repository.

Acknowledgments: This study was carried out on the equipment of the Center of collective use of MSUT "STANKIN" supported by the Ministry of Higher Education of the Russian Federation (project No. 075-15-2021-695 from 26 July 2021, unique identifier RF 2296.61321 × 0013).

Conflicts of Interest: The authors declare no conflict of interest.

Appendix A

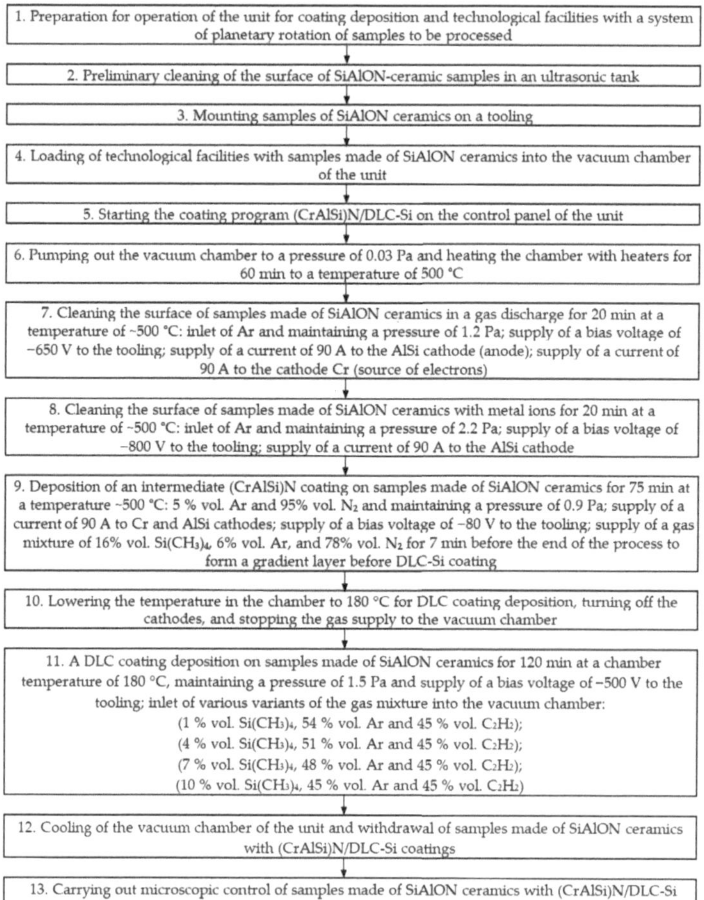

Figure A1. Flowchart of the technological process and the sequence of operations for (CrAlSi)N/DLC-Si coatings' deposition to samples made of SiAlON ceramics.

References

1. Tyagi, A.; Walia, R.S.; Murtaza, Q.; Pandey, S.M.; Tyagi, P.K.; Bajaj, B. A critical review of diamond like carbon coating for wear resistance applications. *Int. J. Refract. Met. Hard Mater.* **2019**, *78*, 107–122. [CrossRef]
2. Martins, P.S.; Magalhães, P.A.A.; Carneiro, J.R.G.; Talibouya Ba, E.C.; Vieira, V.F. Study of Diamond-Like Carbon coating application on carbide substrate for cutting tools used in the drilling process of an Al–Si alloy at high cutting speeds. *Wear* **2022**, *498–499*, 204326. [CrossRef]
3. Plotnikov, V.A.; Dem'yanov, B.F.; Yeliseeyev, A.P.; Makarov, S.V.; Zyryanova, A.I. Structural state of diamond-like amorphous carbon films, obtained by laser evaporation of carbon target. *Diam. Relat. Mater.* **2019**, *91*, 225–229. [CrossRef]
4. Fominski, V.Y.; Grigoriev, S.N.; Celis, J.P.; Romanov, R.I.; Oshurko, V.B. Structure and mechanical properties of W-Se-C/diamond-like carbon and W-Se/diamond-like carbon bi-layer coatings prepared by pulsed laser deposition. *Thin Solid Films* **2012**, *520*, 6476–6483. [CrossRef]

5. Rajak, D.K.; Kumar, A.; Behera, A.; Menezes, P.L. Diamond-Like Carbon (DLC) Coatings: Classification, Properties, and Applications. *Appl. Sci.* **2021**, *11*, 4445. [CrossRef]
6. Ohtake, N.; Hiratsuka, M.; Kanda, K.; Akasaka, H.; Tsujioka, M.; Hirakuri, K.; Hirata, A.; Ohana, T.; Inaba, H.; Kano, M.; et al. Properties and Classification of Diamond-Like Carbon Films. *Materials* **2021**, *14*, 315. [CrossRef]
7. Li, C.; Huang, L.; Yuan, J. Effect of sp^3 Content on Adhesion and Tribological Properties of Non-Hydrogenated DLC Films. *Materials* **2020**, *13*, 1911. [CrossRef]
8. Grigoriev, S.N.; Volosova, M.A.; Fedorov, S.V.; Mosyanov, M. Influence of DLC coatings deposited by PECVD technology on the wear resistance of carbide end mills and surface roughness of $AlCuMg_2$ and $41Cr_4$ workpieces. *Coatings* **2020**, *10*, 1038. [CrossRef]
9. Hatada, R.; Flege, S.; Ashraf, M.N.; Timmermann, A.; Schmid, C.; Ensinger, W. The Influence of Preparation Conditions on the Structural Properties and Hardness of Diamond-Like Carbon Films, Prepared by Plasma Source Ion Implantation. *Coatings* **2020**, *10*, 360. [CrossRef]
10. Gómez, I.; Claver, A.; Santiago, J.A.; Fernandez, I.; Palacio, J.F.; Diaz, C.; Mändl, S.; Garcia, J.A. Improved Adhesion of the DLC Coating Using HiPIMS with Positive Pulses and Plasma Immersion Pretreatment. *Coatings* **2021**, *11*, 1070. [CrossRef]
11. Liang, J.H.; Milne, Z.; Rouhani, M.; Lin, Y.-P.; Bernal, R.A.; Sato, T.; Carpick, R.W.; Jeng, Y.R. Stress-dependent adhesion and sliding-induced nanoscale wear of diamond-like carbon studied using in situ TEM nanoindentation. *Carbon* **2022**, *193*, 230–241. [CrossRef]
12. Wu, Y.; Li, H.; Ji, L.; Ye, Y.; Chen, J.; Zhou, H. Vacuum tribological properties of a-C:H film in relation to internal stress and applied load. *Tribol. Int.* **2014**, *71*, 82–87. [CrossRef]
13. Liu, X.; Yang, J.; Hao, J.; Zheng, J.; Gong, Q.; Liu, W. A near-frictionless and extremely elastic hydrogenated amorphous carbon film with self-assembled dual nanostructure. *Adv. Mater.* **2012**, *24*, 4614–4617. [CrossRef]
14. Zhang, T.; Pu, J.; Xia, Q.; Son, M.; Kim, K. Microstructure and nano-wear property of si-doped diamond-like carbon films deposited by a hybrid sputtering system. *Mater. Today Proc.* **2016**, *3*, S190–S196. [CrossRef]
15. Wang, L.; Nie, X.; Hu, X. Effect of Thermal Annealing on Tribological and Corrosion Properties of DLC Coatings. *J. Mater. Eng Perform* **2013**, *22*, 3093–3100. [CrossRef]
16. He, M.; Lee, S.; Yeo, C.-D. Investigating atomic structure of thin carbon film under mechanical stress and frictional heat generation. *Surf. Coat. Technol.* **2015**, *261*, 79–85. [CrossRef]
17. Meng, W.J.; Meletis, E.I.; Rehn, L.E.; Baldo, P.M. Inductively coupled plasma assisted deposition and mechanical properties of metal-free and Ti-containing hydrocarbon coatings. *J. Appl. Phys.* **2000**, *87*, 2840–2848. [CrossRef]
18. Zou, C.W.; Wang, H.J.; Feng, L.; Xue, S.W. Effects of Cr concentrations on the microstructure, hardness, and tempera-ture-dependent tribological properties of Cr-DLC coatings. *Appl. Surf. Sci.* **2013**, *286*, 137–141. [CrossRef]
19. Jeon, Y.; Park, Y.S.; Kim, H.J.; Hong, B.; Choi, W.S. Tribological properties of ultrathin DLC films with and without metal interlayers. *J. Korean Phys. Soc.* **2007**, *51*, 1124–1128. [CrossRef]
20. Bociaga, D.; Sobczyk-Guzenda, A.; Komorowski, P.; Balcerzak, J.; Jastrzebski, K.; Przybyszewska, K.; Kaczmarek, A. Surface Characteristics and Biological Evaluation of Si-DLC Coatings Fabricated Using Magnetron Sputtering Method on Ti6Al7Nb Substrate. *Nanomaterials* **2019**, *9*, 812. [CrossRef]
21. Wang, K.; Zhou, H.; Zhang, K.; Liu, X.; Feng, X.; Zhang, Y.; Chen, G.; Zheng, Y. Effects of Ti interlayer on adhesion property of DLC films: A first principle study. *Diam. Relat. Mater.* **2021**, *111*, 108188. [CrossRef]
22. Lubwama, M.; Corcoran, B.; McDonnell, K.; Dowling, D.; Kirabira, J.; Sebbit, A.; Sayers, K. Flexibility and frictional behaviour of DLC and Si-DLC films deposited on nitrile rubber. *Surf. Coat. Technol.* **2014**, *239*, 84–94. [CrossRef]
23. Almeida, E.A.; Milan, J.C.G.; da Costa, C.E.; Binder, C.; de Mello, J.D.B.; Costa, H.L. Combined Use of Surface Texturing, Plasma Nitriding and DLC Coating on Tool Steel. *Coatings* **2021**, *11*, 201. [CrossRef]
24. Aijaz, A.; Ferreira, F.; Oliveira, J.; Kubart, T. Mechanical properties of hydrogen free diamond-like carbon thin films deposited by high power impulse magnetron sputtering with Ne. *Coatings* **2018**, *8*, 385. [CrossRef]
25. Grigoriev, S.; Volosova, M.; Fyodorov, S.; Lyakhovetskiy, M.; Seleznev, A. DLC-coating Application to Improve the Durability of Ceramic Tools. *J. Mater. Eng. Perform.* **2019**, *28*, 4415–4426. [CrossRef]
26. Volosova, M.A.; Grigor'ev, S.N.; Kuzin, V.V. Effect of Titanium Nitride Coating on Stress Structural Inhomogeneity in Oxide-Carbide Ceramic. Part 4. Action of Heat Flow. *Refract. Ind. Ceram.* **2015**, *56*, 91–96. [CrossRef]
27. Gong, F.; Zhao, J.; Pang, J. Evolution of cutting forces and tool failure mechanisms in intermittent turning of hardened steel with ceramic tool. *Int. J. Adv. Manuf. Technol.* **2017**, *89*, 1603–1613. [CrossRef]
28. Bian, R.; Ding, W.; Liu, S.; He, N. Research on High Performance Milling of Engineering Ceramics from the Perspective of Cutting Variables Setting. *Materials* **2019**, *12*, 122. [CrossRef]
29. Vereschaka, A.A.; Grigoriev, S.N.; Volosova, M.A.; Batako, A.; Vereschaka, A.S.; Sitnikov, N.N.; Seleznev, A.E. Nano-scale multi-layered coatings for improved efficiency of ceramic cutting tools. *Int. J. Adv. Manuf. Technol.* **2017**, *90*, 27–43. [CrossRef]
30. Palmero, P. Structural Ceramic Nanocomposites: A Review of Properties and Powders' Synthesis Methods. *Nanomaterials* **2015**, *5*, 656–696. [CrossRef]
31. Vereschaka, A.S.; Grigoriev, S.N.; Sotova, E.S.; Vereschaka, A.A. Improving the efficiency of the cutting tools made of mixed ceramics by applying modifying nano-scale multilayered coatings. *Adv. Mat. Res.* **2013**, *712*, 391–394.
32. Bensouilah, H.; Aouici, H.; Meddour, I.; Yallese, M.A.; Mabrouki, T.; Girardin, F. Performance of coated and uncoated mixed ceramic tools in hard turning process. *Measurement* **2016**, *82*, 1–18. [CrossRef]

33. Volosova, M.; Grigoriev, S.; Metel, A.; Shein, A. The role of thin-film vacuum-plasma coatings and their influence on the efficiency of ceramic cutting inserts. *Coatings* **2018**, *8*, 287. [CrossRef]
34. Grigoriev, S.N.; Vereschaka, A.A.; Vereschaka, A.S.; Kutin, A.A. Cutting tools made of layered composite ceramics with nano-scale multilayered coatings. *Procedia CIRP* **2012**, *1*, 301–306. [CrossRef]
35. Grigoriev, S.N.; Volosova, M.A.; Fedorov, S.V.; Okunkova, A.A.; Pivkin, P.M.; Peretyagin, P.Y.; Ershov, A. Development of DLC-Coated Solid SiAlON/TiN Ceramic End Mills for Nickel Alloy Machining: Problems and Prospects. *Coatings* **2021**, *11*, 532 [CrossRef]
36. Sobol, O.V.; Andreev, A.A.; Grigoriev, S.N.; Gorban, V.F.; Volosova, M.A.; Aleshin, S.V.; Stolbovoi, V.A. Effect of high-voltage pulses on the structure and properties of titanium nitride vacuum-arc coatings. *Met. Sci. Heat Treat.* **2012**, *54*, 195–203. [CrossRef]
37. Lanigan, J.; Zhao, H.; Morina, A.; Neville, A. Tribochemistry of silicon and oxygen doped, hydrogenated diamond-like carbon in fully-formulated oil against low additive oil. *Tribol. Int.* **2015**, *82*, 431–442. [CrossRef]
38. Nakazawa, H.; Kamata, R.; Miura, S.; Okuno, S. Effects of frequency of pulsed substrate bias on structure and properties of silicon-doped diamond-like carbon films by plasma deposition. *Thin Solid Film.* **2015**, *574*, 93–98. [CrossRef]
39. Grigoriev, S.N.; Volosova, M.A.; Vereschaka, A.A.; Sitnikov, N.N.; Milovich, F.; Bublikov, J.I.; Fyodorov, S.V.; Seleznev, A.E. Properties of (Cr,Al,Si)N-(DLC-Si) composite coatings deposited on a cutting ceramic substrate. *Ceram. Int.* **2020**, *46*, 18241–18255. [CrossRef]
40. Uhlmann, E.; Hübert, C. Tool grinding of end mill cutting tools made from high performance ceramics and cemented carbides. *CIRP Ann.* **2011**, *60*, 359–362. [CrossRef]
41. Smirnov, K.L. Sintering of SiAlON ceramics under high-speed thermal treatment. *Powder Metall. Met. Ceram.* **2012**, *51*, 76–82. [CrossRef]
42. Chi, I.S.; Bux, S.K.; Bridgewater, M.M.; Star, K.E.; Firdosy, S.; Ravi, V.; Fleurial, J.-P. Mechanically robust SiAlON ceramics with engineered porosity via two-step sintering for applications in extreme environments. *MRS Adv.* **2016**, *1*, 1169–1175. [CrossRef]
43. Zheng, G.; Zhao, J.; Zhou, Y.; Li, A.; Cui, X.; Tian, X. Performance of graded nano-composite ceramic tools in ultra-high-speed milling of Inconel 718. *Int. J. Adv. Manuf. Technol.* **2013**, *67*, 2799–2810. [CrossRef]
44. Molaiekiya, F.; Stolf, P.; Paiva, J.M.; Bose, B.; Goldsmith, J.; Gey, C.; Engin, S.; Fox-Rabinovich, G.; Veldhuis, S.C. Influence of process parameters on the cutting performance of SiAlON ceramic tools during high-speed dry face milling of hardened Inconel 718. *Int. J. Adv. Manuf. Technol.* **2019**, *105*, 1083–1098. [CrossRef]
45. Metel, A.; Bolbukov, V.; Volosova, M.; Grigoriev, S.; Melnik, Y. Equipment for deposition of thin metallic films bombarded by fast argon atoms. *Instrum. Exp. Technol.* **2014**, *57*, 345–351. [CrossRef]
46. Vereschaka, A.A.; Volosova, M.A.; Grigoriev, S.N.; Vereschaka, A.S. Development of wear-resistant complex for high-speed steel tool when using process of combined cathodic vacuum arc deposition. *Procedia CIRP* **2013**, *9*, 8–12. [CrossRef]
47. Metel, A.; Grigoriev, S.; Melnik, Y.; Panin, V.; Prudnikov, V. Cutting Tools Nitriding in Plasma Produced by a Fast Neutral Molecule Beam. *Jpn. J. Appl. Phys.* **2011**, *50*, 08JG04. [CrossRef]
48. Vereschaka, A.; Tabakov, V.; Grigoriev, S.; Sitnikov, N.; Milovich, F.; Andreev, N.; Bublikov, J. Investigation of wear mechanisms for the rake face of a cutting tool with a multilayer composite nanostructured Cr–CrN-(Ti,Cr,Al,Si)N coating in high-speed steel turning. *Wear* **2019**, *438*, 203069. [CrossRef]
49. Vereschaka, A.; Tabakov, V.; Grigoriev, S.; Sitnikov, N.; Milovich, F.; Andreev, N.; Sotova, C.; Kutina, N. Investigation of the influence of the thickness of nanolayers in wear-resistant layers of Ti-TiN-(Ti,Cr,Al)N coating on destruction in the cutting and wear of carbide cutting tools. *Surf. Coat. Technol.* **2020**, *385*, 125402. [CrossRef]
50. Evaristo, M.; Fernandes, F.; Jeynes, C.; Cavaleiro, A. The Influence of H Content on the Properties of a-C(W):H Coatings. *Coatings* **2023**, *13*, 92. [CrossRef]
51. Vereschaka, A.; Grigoriev, S.; Tabakov, V.; Migranov, M.; Sitnikov, N.; Milovich, F.; Andreev, N. Influence of the Nanostructure of Ti-TiN-(Ti,Al,Cr)N Multilayer Composite Coating on Tribological Properties and Cutting Tool Life. *Tribol. Int.* **2020**, *150*, 106388. [CrossRef]
52. Haq, A.J.; Munroe, P.R.; Hoffman, M.; Martin, P.J.; Bendavid, A. Berkovich indentation of diamondlike carbon coatings on silicon substrates. *J. Mater. Res.* **2008**, *23*, 1862–1869. [CrossRef]
53. Grigoriev, S.; Vereschaka, A.; Zelenkov, V.; Sitnikov, N.; Bublikov, J.; Milovich, F.; Andreev, N.; Mustafaev, E. Specific features of the structure and properties of arc-PVD coatings depending on the spatial arrangement of the sample in the chamber. *Vacuum* **2022**, *200*, 111047. [CrossRef]
54. Vereschaka, A.A.; Vereschaka, A.S.; Grigoriev, S.N.; Kirillov, A.K.; Khaustova, O.U. Development and research of environmentally friendly dry technological machining system with compensation of physical function of cutting fluids. *Procedia CIRP* **2013**, *7*, 311–316. [CrossRef]
55. Kyzioł, K.; Jabłoński, P.; Niemiec, W.; Prażuch, J.; Kottfer, D.; Łętocha, A.; Kaczmarek, Ł. Deposition, morphology and functional properties of layers based on DLC:Si and DLC:N on polyurethane. *Appl. Phys. A* **2020**, *126*, 751. [CrossRef]
56. Ahmed, S.F.; Mitra, M.K.; Chattopadhyay, K.K. Low-macroscopic field emission from siliconincorporated diamond-like carbon film synthesized by dc PECVD. *Appl. Surf. Sci.* **2007**, *253*, 5480–5484. [CrossRef]
57. Puttichaem, C.; Souza, G.P.; Ruthe, K.C.; Chainok, K. Characterization of Ultra-Thin Diamond-Like Carbon Films by SEM/EDX. *Coatings* **2021**, *11*, 729. [CrossRef]

58. Zeng, Q.; Ning, Z. High-temperature tribological properties of diamond-like carbon films: A review. *Rev. Adv. Mater. Sci.* **2021**, *60*, 276–292. [CrossRef]
59. He, M.; Yeo, C. Evaluation of Thermal Degradation of DLC Film Using a Novel Raman Spectroscopy Technique. *Coatings* **2018**, *8*, 143. [CrossRef]
60. Huang, B.; Liu, L.-T.; Han, S.; Du, H.-M.; Zhou, Q.; Zhang, E.-G. Effect of deposition temperature on the microstructure and tribological properties of Si-DLC coatings prepared by PECVD. *Diam. Relat. Mater.* **2022**, *129*, 109345. [CrossRef]
61. Moolsradoo, N.; Abe, S.; Watanabe, S. Thermal Stability and Tribological Performance of DLC-Si–O Films. *Adv. Mater. Sci. Eng.* **2011**, *2011*, 483437. [CrossRef]
62. Zeng, Q.; Eryilmaz, O.; Erdemir, A. Superlubricity of the DLC films-related friction system at elevated temperature. *RSC Adv.* **2015**, *5*, 93147–93154. [CrossRef]
63. Zajíčková, L.; Buršíková, V.; Peřina, V.; Macková, A.; Janča, J. Correlation between SiOx content and properties of DLC:SiOx films prepared by PECVD. *Surf. Coat. Technol.* **2003**, *174–175*, 281–285. [CrossRef]
64. Grischke, M.; Hieke, A.; Morgenweck, F.; Dimigen, H. Variation of the Wettability of DLC-coatings by NetworkModification Using Silicon and Oxygen. *Diam. Relat. Mater.* **1998**, *7*, 454–458. [CrossRef]
65. Meškinis, Š.; Vasiliauskas, A.; Andrulevičius, M.; Peckus, D.; Tamulevičius, S.; Viskontas, K. Diamond Like Carbon Films Containing Si: Structure and Nonlinear Optical Properties. *Materials* **2020**, *13*, 1003. [CrossRef]
66. Maser, K. Varying internal parameters in the thermal silicon oxidation. *J. Solid State Electrochem.* **2019**, *23*, 2589–2593. [CrossRef]
67. Gerlach, G.; Maser, K. A Self-Consistent Model for Thermal Oxidation of Silicon at Low Oxide Thickness. *Adv. Condens. Matter Phys.* **2016**, *2016*, 7545632. [CrossRef]
68. Hatty, V.; Kahn, H.; Heuer, A.H. Fracture Toughness, Fracture Strength, and Stress Corrosion Cracking of Silicon Dioxide Thin Films. *J. Microelectromech. Syst.* **2008**, *17*, 943–947. [CrossRef]
69. Zarchi, M.; Ahangarani, S. Optical and morphological properties of silicon dioxide thin films. *J. Mater. Sci: Mater. Electron.* **2016**, *27*, 1165–1170. [CrossRef]
70. Jeong, J.S.; Cho, C.H.; Kim, J.O.; Yeo, D.H.; Choi, W.Y. Preparation and characterization of DLC/SiO_2/Al_2O_3 nanofiltration membrane. *Bull. Mater. Sci.* **2013**, *36*, 1133–1138. [CrossRef]
71. Grigoriev, S.; Vereschaka, A.; Milovich, F.; Tabakov, V.; Sitnikov, N.; Andreev, N.; Sviridova, T.; Bublikov, J. Investigation of multicomponent nanolayer coatings based on nitrides of Cr, Mo, Zr, Nb, and Al. *Surf. Coat. Technol.* **2020**, *401*, 126258. [CrossRef]
72. Vetter, J.; Eriksson, A.O.; Reiter, A.; Derflinger, V.; Kalss, W. Quo Vadis: AlCr-Based Coatings in Industrial Applications. *Coatings* **2021**, *11*, 344. [CrossRef]

C Journal of Carbon Research

MDPI

Review

Graphene Nanocomposites in Space Sector—Fundamentals and Advancements

Ayesha Kausar [1,2,3,*], Ishaq Ahmad [1,2,3], M. H. Eisa [4] and Malik Maaza [2]

1 NPU-NCP Joint International Research Center on Advanced Nanomaterials and Defects Engineering, Northwestern Polytechnical University Xi'an, Xi'an 710060, China
2 UNESCO-UNISA Africa Chair in Nanosciences/Nanotechnology, iThemba LABS, Somerset West 7129, South Africa
3 NPU-NCP Joint International Research Center on Advanced Nanomaterials and Defects Engineering, National Centre for Physics, Islamabad 44000, Pakistan
4 Department of Physics, College of Science, Imam Mohammad Ibn Saud Islamic University (IMSIU), Riyadh 13318, Saudi Arabia
* Correspondence: dr.ayeshakausar@yahoo.com

Abstract: Graphene is one of the most significant carbon nanomaterials, with a one-atom-thick two-dimensional nanostructure. Like other nanocarbons, graphene has been used as a polymer reinforcement. This review explores the impact of graphene and graphene-based nanocomposites on aerospace applications. The fabrication and indispensable features of graphene-derived nanocomposites have been considered. Numerous polymers and nanocomposites have been employed for aerospace systems such as reinforced thermosetting/thermoplastic polymers and epoxy/graphene nanocomposites. Moreover, graphene-modified carbon-fiber-based composites have been discussed for the space sector. Aerospace nanocomposites with graphene have been investigated for superior processability, structural features, morphology, heat stability, mechanical properties, flame resistance, electrical/thermal conductivity, radiation protection, and adhesion applications. Subsequently, epoxy and graphene-derived nanocomposites have been explored for heat/mechanically stable aerospace engineering structures, radiation-shielding materials, adhesives, coatings, etc.

Keywords: graphene; epoxy; nanocomposite; fabrication; aerospace; mechanical; thermal

Citation: Kausar, A.; Ahmad, I.; Eisa, M.H.; Maaza, M. Graphene Nanocomposites in Space Sector—Fundamentals and Advancements. *C* **2023**, *9*, 29. https://doi.org/10.3390/c9010029

Academic Editor: Gil Gonçalves

Received: 17 January 2023
Revised: 15 February 2023
Accepted: 28 February 2023
Published: 3 March 2023

1. Introduction

Polymer composites/nanocomposites have found potential for aerospace-related technological and engineering areas [1]. In the aeronautical field, fiber-based composite structures were initially applied in space structures [2]. Later research focused on the significance of polymers and nanofiller/additive-based nanocomposites in this sector [3,4]. Accordingly, using nanofillers (instead of macrofillers) remarkably enhanced the durability, fatigue resistance, strength, and toughness properties of aeronautical materials [5]. Moreover, using lightweight nanocomposites in aerospace vehicles has advantages of reducing fuel consumption and improved performance compared to heavy metal space structures [6,7]. Most importantly, carbon nanoparticles such as graphene, carbon nanotubes, and nanodiamond nanofillers have been used to form aerospace nanocomposites [8]. Among carbon nanoparticles, graphene is recognized as an important carbon nanostructure [9]. It has carbon atoms arranged in a honeycomb lattice structure. A graphene nanosheet is one-atom-thick with exceptional structural and physical properties. Polymer/graphene nanocomposites offer fine processability, resilience, frivolity, mechanical stability, thermal conductivity, and electrical conductivity properties [10]. Owing to the remarkable features of graphene and polymer/graphene nanocomposites, their potential applications have been discovered for aerospace [11].

In essence, this advanced review focused on the design, features, and potential of graphene and derived nanomaterials for aerospace. Incidentally, various combinations of polymer/graphene and epoxy/graphene nanocomposites, related fabrication approaches, and possible utilizations were considered. Some previous literature reports were observed on the design and performance of polymer/graphene nanocomposites; however, the reported literature is not updated enough to portray the true current state of polymer/graphene materials and graphene-modified carbon fiber composites for aerospace [12]. Likewise, the previous literature does not depict the main progress in this field during recent years. In this regard, future developments in functional polymer/graphene nanocomposite are not possible for aerospace-related researchers without gaining prior knowledge of the recent relevant literature. Significant literature reports on aerospace-related polymer/graphene nanocomposites have been found between 2015 and 2023, which are now focused on in this overview [13]. To the best of our knowledge, such a specific review on aerospace-related polymer/graphene nanocomposites and graphene-modified carbon fiber composites has not been seen in the literature before with similarly well-arranged interpretations and outlines of the recent literature. Hence, this review portrays scientific developments and advancements in the field of polymer/graphene nanocomposites for aerospace. In a few words, this article offers a groundbreaking review on polymer/graphene nanomaterials in aerospace. Despite the remarkable properties and vast potential of polymer/graphene nanomaterials, devoted future research efforts are desirable to form high-performance nanocomposites to overcome the associated challenges.

2. Graphene

Graphene is a two-dimensional nanosheet made up of sp^2-hybridized carbon atoms [14]. Single-layer graphene was produced and recognized in 2004 by Andre Geim and Konstantin Novoselov [15], although it was theoretically explored by P. R. Wallace in 1947 and experimentally by Hanns-Peter Boehm and his coworkers in 1962. Graphene has been prepared through various approaches such as graphite exfoliation, graphite mechanical cleavage, chemical vapor deposition, laser techniques, and numerous other organic synthesis routes [16]. Graphene has revealed exceptional structural and physical features. First of all, graphene (a one-atom-thick nanosheet) is the thinnest known material [17]. Moreover, graphene has a high electron mobilization of ~200,000 $cm^2V^{-1}s^{-1}$. It also shows a high thermal conductivity of ~3000–5000 W/mK [18]. The significantly high Young's modulus of graphene has been observed as 1 TPa, i.e., 200 times stronger than steel [19]. Owing to van der Waals forces, graphene nanosheets possess a wrinkling tendency, and so can easily crumple [20]. To resolve the nanosheet wrinkling and dispersion issues, graphene has been modified to generate functional groups on the surface [21]. Consequently, graphene has been oxidized to form graphene oxide as a modified form with hydroxyl, epoxide, carbonyl, and carboxylic acid functionalities on the surface. Figure 1 portrays the general structures of graphene nanosheets, whereas Figure 2 displays the structures of graphene oxide. The majority of functional groups in graphene oxide have been found as epoxy and hydroxyl functionalities. Graphene has superior characteristics such as mechanical consistency, thermal stability, chemical stability, and electrical and thermal conductivity [22]. The properties of graphene have been further improved through incorporation in nanocomposites. Due to the unique structure and properties, graphene and graphene-derived nanomaterials have been applied in electronics [23–25], sensors [26], energy devices [27], membranes [28], etc. In addition, the impact of graphene has also been analyzed for aerospace purposes [29].

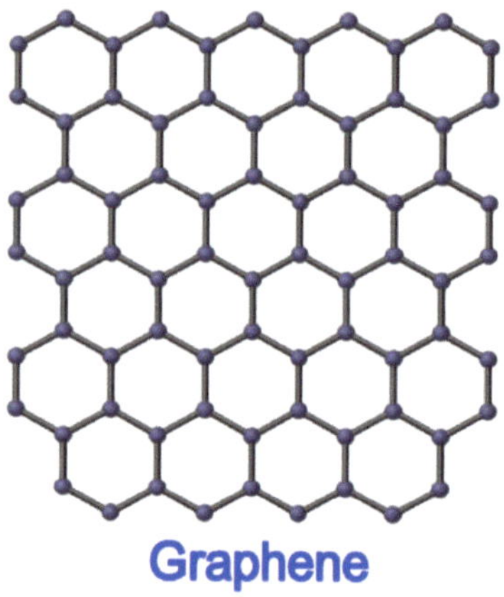

Figure 1. Graphene and graphene oxide.

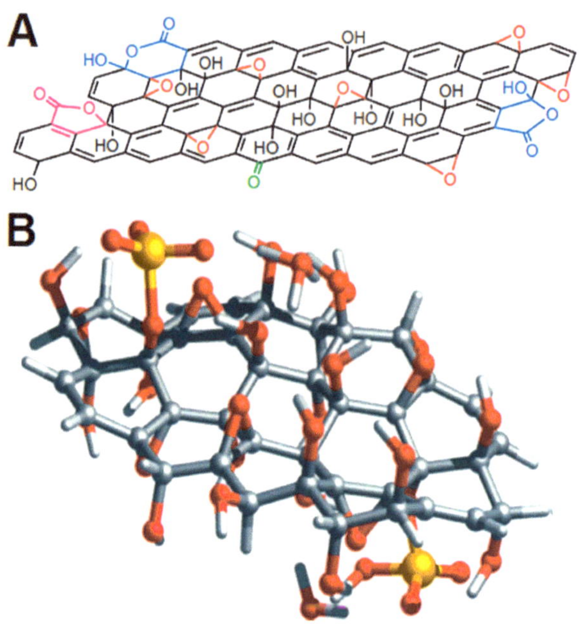

Figure 2. (**A**) Structural model of graphene oxide, taking into account five- and six-membered lactol rings (blue), ester of a tertiary alcohol (purple), hydroxyl (black), epoxy (red), and ketone (green) functionalities. The relative ratios are likely to be 115 (hydroxyl and epoxy): 3 (lactol O–C–O): 63 (graphitic sp2 carbon): 10 (lactol/ester/acid carbonyl): 9 (ketone carbonyl). Reprinted with permission from [30]. 2009, Nature. The model here only shows the chemical connectivity, and not the steric orientation, of these functionalities and (**B**) Structure of graphene oxide with epoxy and hydroxyl as dominant functional groups. Reprinted with permission from [31]. 2013, Wiley.

3. Composite/Nanocomposite Materials Regarding Aerospace

For technical and engineering industries, polymeric composites/nanocomposites have revealed several design and property advantages [32]. Before the application of composites, pristine polymers were used in aerospace structural parts [33,34]. Later, the development of composites and nanocomposites was focused on for aerospace purposes [35]. Compared with unfilled polymers, composite/nanocomposite materials showed better thermal and mechanical properties. In this regard, a number of thermoplastics, thermosets, and rubbery matrices have been applied in the space sector [36,37]. Composites/nanocomposites have the advantage of being lightweight in aerospace, relative to metal-based structures [38]. Using low-density materials also decreased the fuel ingestion of the spacecrafts [39,40]. Moreover, these materials have high strength, modulus, toughness, thermal stability, friction resistance, fatigue performance, and shear resistance characteristics [41–43]. Consequently, the structural durability and working life of aerospace structures have been improved [44–46]. Multifunctional composites/nanocomposites have been found to be valuable in increasing the capability of space structures in bearing shocks and jerks [47,48] and lightning strikes [49], in radiation shielding [50], and as high-temperature stable engine components [51]. In aerospace nanocomposites, carbon nanoparticles (carbon nanotubes, carbon nanofibers, etc.) and inorganic nanoparticles (metal, metal oxide, etc.) have been used as reinforcements [52,53]. In particular, widely used polymer/carbon nanotube nanocomposites have revealed reasonably high heat stability, flame defiance, thermal conductivity, strength, and mechanical stability for aerospace structures [54–56]. Nevertheless, research on aerospace-related composites/nanocomposites is continuously growing, in search of new high-performance structures [57].

4. Graphene in Polymeric Nanocomposites for Space Relevance

Graphene and related nanofillers have effectively increased the electrical conductivity, thermal conductivity, and mechanical features of nanocomposites [58–60]. For aerospace applications, nanocomposites have been investigated for physical property improvements and morphological profiles [61]. Various thermosetting and thermoplastic polymers have been used to form graphene-derived nanocomposites for aerospace structures [62]. Among thermosetting polymers, epoxy resins have been extensively used in aerospace-related applications [63]. Graphene and related nanofillers have been reinforced in epoxy matrices to enhance the essential conducting, thermal, and mechanical properties [64–66]. According to studies, up to 20 wt.% graphene nanofiller contents may increase the high thermal conductivity, toughness, and fatigue resistance properties of epoxy nanocomposites [67–69]. Moreover, graphene oxide nanofiller also improves the tensile strength, fatigue resistance, toughness, heat stability, and tribological properties of epoxy matrices [70,71]. Most importantly, epoxy nanocomposites with graphene and related nanofillers have been used in structural components, adhesives, coatings, etc., for aerospace vehicles [72].

Thermoplastic polymers such as polyethylene, polypropylene, poly(methyl methacrylate), poly(vinyl alcohol), and others have been applied in space applications [73]. Suner and coworkers [74] fabricated polyethylene and graphene-oxide-derived nanocomposites for enhanced mechanical stability. The addition of 0.5 wt.% nanofiller significantly improved the mechanical characteristics of the nanocomposites due to matrix–nanofiller compatibility [75]. Song et al. [76] reported on the polypropylene/graphene nanocomposites fabricated using the solution casting and melt method. The conditions and processes involved in the preparation of polypropylene/graphene nanocomposites are displayed in Figure 3. Graphene oxide was obtained from graphite flakes using the Hummers method [77]. Then, graphene oxide was processed through a solution as well as melt routes to form the nanocomposites. The mechanical properties of the polypropylene/graphene nanocomposites are displayed in Table 1. The inclusion of up to 1 wt.% nanofiller content was found to enhance the yield strength and tensile strength of the nanocomposites due to the mechanical interlocking of the polymer chains with the graphene nanostructure. Moreover, the enhanced mechanical properties were attributed to better graphene dispersion and load transfer between the matrix and the nanofiller [78]. Moreover, the increasing graphene loading was found to improve the crystallinity of the nanocomposite due to orderly nanofiller dispersion and interaction with the polymer matrix [79]. Nanocomposites have been found to be valuable for aerospace applications.

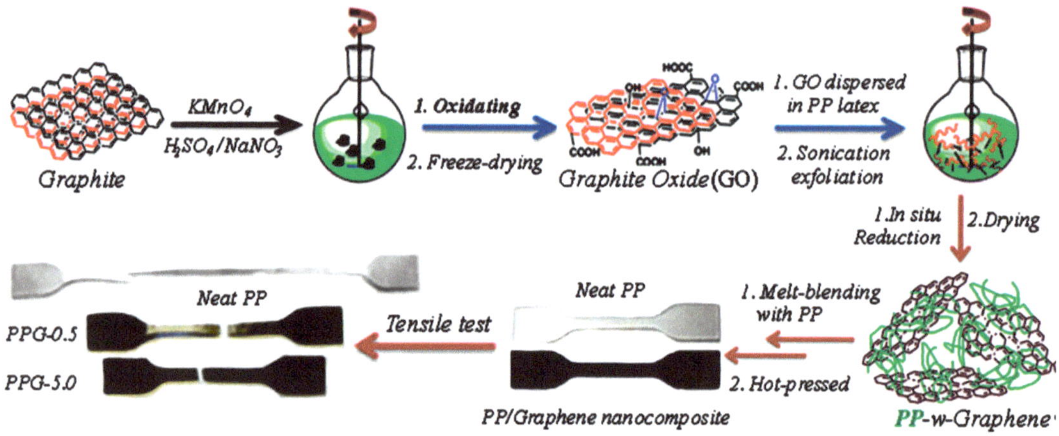

Figure 3. Schematic of polypropylene/graphene nanocomposite fabrication. PP = polypropylene; G = graphene; PPG = polypropylene/graphene. Reprinted with permission from [76]. 2011, Elsevier.

Table 1. Mechanical properties of polypropylene/graphene oxide nanocomposites with various loadings. PP = polypropylene; YS = yield strength; TS = tensile strength; %YS = percent increment of yield strength. Reprinted with permission from [76]. 2011, Elsevier.

Sample	YS (MPa)	TS (MPa)	YS (%)
Neat PP	22 ± 0.8	24 ± 0.8	-
PP + 0.1 wt.% nanofiller	30 ± 1.1	33 ± 1.4	36
PPG + 0.5 wt.% nanofiller	36 ± 1.2	36 ± 1.5	50
PP + 1.0 wt.% nanofiller	38 ± 1.7	37 ± 1.6	75

Huang and coresearchers [80] produced flame-retardant poly(methyl methacrylate) nanocomposites for aerospace. An in situ method was used to develop the nanocomposites [81]. Graphene and layered double hydroxide were added as intumescent flame retardants in the poly(methyl methacrylate) matrix to enhance the flame retardancy of the matrix. The poly(methyl methacrylate) matrix with the 1 wt.% graphene and 5 wt.% layered hydroxide considerably enhanced the nonflammability properties through improving the limiting oxygen index. Xu et al. [82] developed poly(vinyl alcohol)/graphene oxide nanocomposites via the vacuum filtration procedure. As compared to the Young's modulus of the neat poly(vinyl alcohol) matrix (2.1 MPa), the addition of 3 wt.% graphene oxide increased the property up to 4.8 GPa (by 128%). Moreover, including 3 wt.% nanofiller boosted the tensile strength by 110 MPa (70%), as compared to the neat polymer (65 MPa). The increase in the mechanical properties was credited to the interfacial interactions in the matrix–nanofiller, leading to relevance in aerospace. Hence, different polymers reinforced with graphene nanofillers have enhanced mechanical, thermal, and conducting profiles and are found to be useful as aerospace materials.

In the aerospace industry, high-performance matrices such as poly(ether ether ketone) (PEEK) [83,84], poly(ether ketone ketone) (PEKK) [85,86], polyetherimide [87], and polysulfone [88] have been found to be desirable due to high mechanical, thermal, and electrical properties. Puértolas et al. [89] fabricated PEEK/graphene nanocomposites using solvent-free melt-blending and injection-molding techniques. The 1–10 wt.% graphene nanofiller was included in PEEK matrix. The inclusion of graphene caused a 60% improvement in the hardness of the material. The coefficient of friction and the wear factor were decreased by 38% and 83%, respectively. Thus, graphene was found to be an important nanofiller to enhance the surface hardness and tribological properties of the PEEK matrix, desirable for aerospace uses. Alvaredo et al. [90] introduced graphene nanoplatelets in PEEK using the melt-blending technique. The 1–10 wt.% nanofiller was included in the PEEK matrix. The inclusion of graphene nanoplatelets enhanced the complex viscosity and rheological behavior of the nanocomposites and they were found to be effective for the space sector. Wang et al. [91] produced PEKK/graphene nanomaterials with enhanced electrical conductivity and mechanical performance, suitable for high-tech industrial relevance. Sun et al. [92] fabricated high-performance polyetherimide/graphene oxide nanocomposites. The inclusion of graphene oxide nanofiller improved the tensile strength and temperature-dependent tensile behaviors of the polyetherimide/graphene oxide nanocomposites. These materials have fine suitability for the space sector. Ionita et al. [93] developed polysulfone/graphene oxide nanocomposites using the phase inversion technique. The inclusion of 0.25–2 wt.% nanofiller was studied. The homogeneous nanofiller dispersion considerably improved the thermal stability of the nanocomposites. Janire Peña-Bahamonde et al. [94] also prepared polysulfone/graphene oxide nanocomposites through the solvent-free extrusion–injection technique. The addition of 3 wt.% nanofiller upsurged the dispersion, rheological properties, and toughness of the nanocomposites. These materials can be feasible candidates for aerospace applications.

5. Epoxy/Graphene Nanocomposites for Aerospace Applications

Epoxy composites reinforced with carbon or glass fibers have been widely used in the aerospace sector [95–97]. Carbon nanoparticles are a remarkable type of carbonaceous nanomaterial, also known as carbon nanoparticles. To attain improved structural, mechanical, and thermal properties in the epoxy composites, carbon nanoparticles such as carbon nanotubes, graphene, and graphene-derived nanofillers have been reinforced in epoxy resins [98]. In some attempts, fibers have been modified with graphene and used in aerospace materials [99,100]. Modified fillers have been applied to improve the fatigue resistance, fracture toughness, and other mechanical properties of epoxy matrices. Shiu and coworkers [101] fabricated epoxy/graphene nanocomposites with high mechanical robustness properties. Compared with the neat epoxy resins, the glass transition temperature of the epoxy/graphene was also found to be higher [102]. Shadlou et al. [103] also explored epoxy/graphene nanocomposites for enhanced tensile strength and Young's modulus for application in aerospace. A better mechanical performance was observed due to the fine dispersion and compatibility of the nanofiller in the epoxy resin. Bustero and coworkers [104] fabricated epoxy and graphene-nanoplatelet-derived nanocomposites for high-end aerospace applications. The nanocomposite was prepared through the facile hand lay-up technique (Figure 4) [105,106]. Initially, an epoxy resin layer was applied in the mold, which was then coated with the graphene nanoplatelet layer. After that, again, an epoxy resin layer was applied to form the layered nanocomposite, which was cured at 120 °C. The thermal stability of the epoxy and graphene nanoplatelet nanocomposites was analyzed using thermogravimetric analysis (Figure 5). The inclusion of graphene nanoplatelets enhanced the thermal stability of the epoxy resins. The highest improvement in the heat stability was observed with the inclusion of 30 wt.% nanofiller in the epoxy resin. The improvement in the thermal stability of the aerospace matrix was attributed to the reinforcement effect of the graphene nanofiller [107].

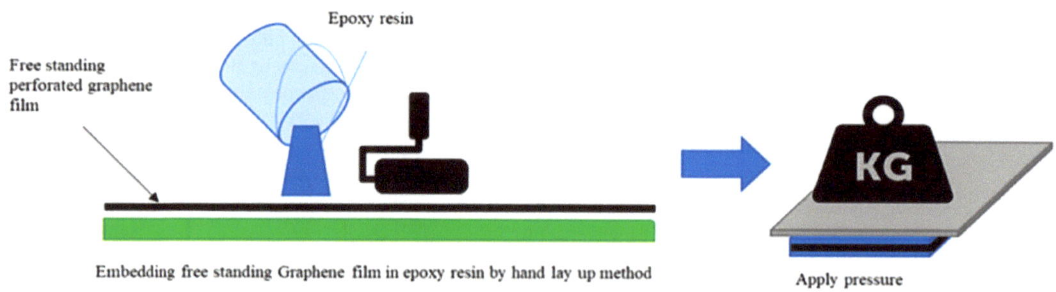

Figure 4. Fabrication process of epoxy/graphene nanocomposite. Reprinted with permission from [104]. 2020, Springer.

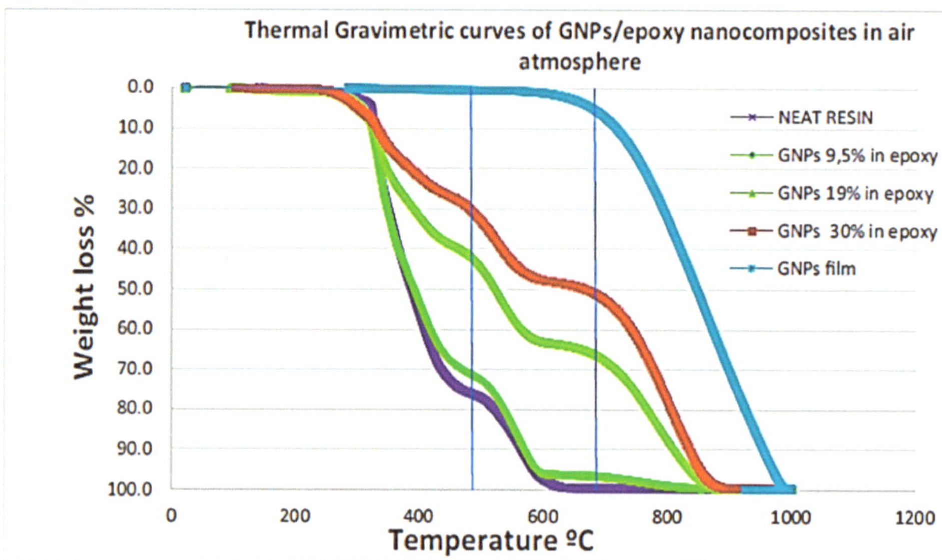

Figure 5. Thermogravimetric curves of graphene nanoplatelets/epoxy (GNPs/epoxy) nanocomposites in air. Reprinted with permission from [104]. 2020, Springer.

García-Martínez et al. [108] developed neat epoxy and epoxy/graphene nanocomposite coatings using the solution route. The coatings were applied on Al 2024-T3 samples and tested for surface properties. To study the hydrophobicity of the neat epoxy and epoxy/graphene nanocomposite coating, the water contact angle was analyzed (Table 2). The epoxy/graphene nanocomposite coating revealed a higher contact angle than the neat epoxy coating due to the enhanced barrier properties. Consequently, the graphene nanofiller enhanced the adhesion properties of the coatings without affecting the curing behavior.

Table 2. Water contact angle values for graphene-filled and unfilled epoxy resin. Reprinted with permission from [108]. 2021, MDPI.

Sample	Contact Angle (Degree)
Unfilled epoxy resin	60.4 ± 1
Epoxy resin/graphene	75.3 ± 1

Graphene oxide has been used as a reinforcement in epoxy matrices for aerospace [109]. The epoxy/graphene oxide nanocomposites possess light weight, high thermal stability, and high mechanical properties relative to neat epoxy resin [110]. In addition, graphene oxide has also been modified to reinforce epoxy resins [111]. Yousefi et al. [112] fabricated epoxy/reduced graphene oxide nanocomposites for the space sector. The epoxy/reduced graphene oxide nanocomposites showed high dielectric properties and a low percolation threshold value of 0.12 vol%. The electromagnetic interference (EMI)-shielding efficiency was found as 38 dB. EMI-shielding epoxy/reduced graphene oxide nanocomposites have found important use in the aerospace sector. Yousefi et al. [113] prepared a waterborne epoxy and reduced-graphene-oxide-based nanocomposite. The set-up for the formation of the reduced graphene oxide and uniformly dispersed nanocomposite is given in Figure 6. In the aligned epoxy and reduced-graphene-oxide-based nanocomposite, covalent bonding occurs between the epoxy and nanofiller. Moreover, π-π stacking interactions exist between bisphenol-A epoxy resin and reduced graphene oxide to enhance the matrix–nanofiller compatibility. The transmission electron microscopy images of the monolayer graphene oxide nanosheet and the epoxy/reduced graphene oxide nanocomposite are given in Figure 7.

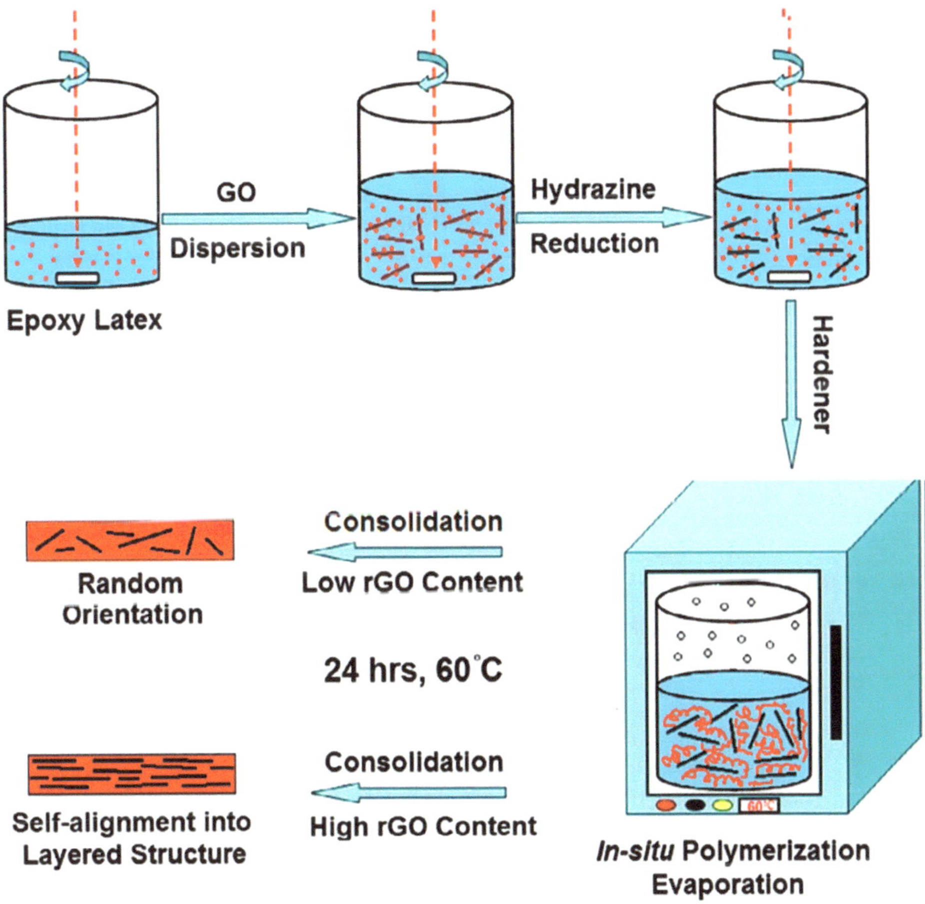

Figure 6. Schematic of nanocomposite preparation. Reprinted with permission from [113]. 2013, Elsevier.

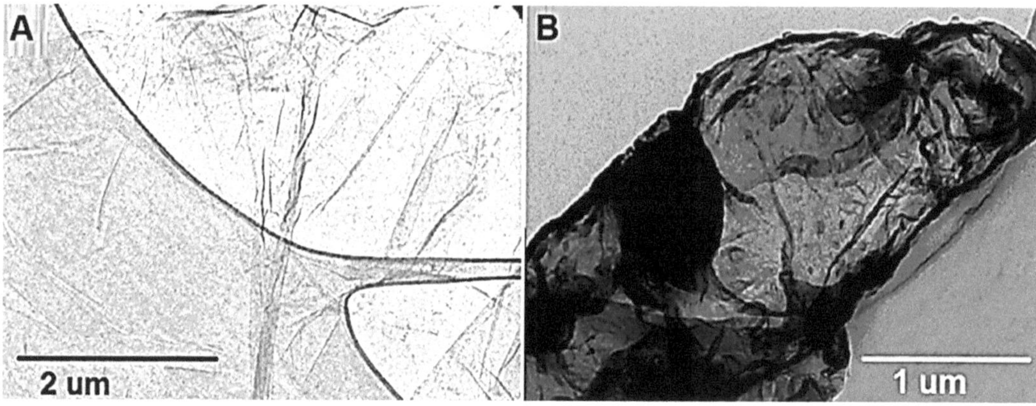

Figure 7. Transmission electron microscopy images of (**A**) monolayer graphene oxide nanosheets; and (**B**) epoxy and reduced graphene oxide nanocomposite. Reprinted with permission from [113]. 2013, Elsevier.

The reduced graphene oxide was produced from graphene oxide and used as a reinforcement in the epoxy resin. In reduced graphene oxide, the elimination of oxygen containing compounds may lead to a higher specific surface area compared with graphene oxide. Moreover, graphene oxide has strong hydrophilic behavior, while reduced graphene oxide has hydrophobic behavior. Consequently, graphene oxide has better dispersibility and colloidal properties compared with reduced graphene oxide. Accordingly, a single graphene oxide nanosheet was found as a homogeneous transparent layer in the micrograph. On the other hand, the reduced graphene oxide nanosheet seemed to be covered with the epoxy matrix. According to the mechanical property analysis, the inclusion of 1.5 wt.% reduced graphene oxide in the epoxy matrix enhanced the strength of the nanocomposite by 500%. The results pointed to fine interfacial adhesion between the epoxy matrix and reduced graphene oxide nanosheets.

Li et al. [114] prepared modified nanofillers of graphene oxide, i.e., 3-aminopropyltrimethoxysilane-functionalized graphene oxide (APTS-GO) and 3-glycidoxypropyltrimethoxysilane-functionalized graphene oxide (GPTS-GO). The APTS-GO and GPTS-GO were used as reinforcements in the epoxy matrix. Figure 8 presents the in situ solution routes to the formation of APTS-GO- and GPTS-GO-based epoxy nanocomposites. The silane coupling agents were capable of grafting to the graphene oxide surface and also developing covalent bonding with the epoxide groups. This led to the formation of well-compatible interface to enhance the mechanical properties of the nanocomposites [115]. Figure 9 shows the fracture surfaces of the APTS-GO- and GPTS-GO-based epoxy nanocomposites. The micrographs revealed the formation of twisting patterns for the APTS-GO and GPTS-GO nanocomposites. As compared with the neat epoxy, changes in morphology were observed for the nanocomposites which suggested energy absorption through the crack propagation mechanism [116]. Consequently, the addition of 0.2 wt.% functional nanofiller significantly improved the fracture toughness and fracture energy to 1.46 MPam$^{1/2}$ and 0.62 kJ/m^2, respectively. Epoxy/graphene nanocomposites have been used as effective adhesives for aluminum or metal substrates in aerospace engineering [117].

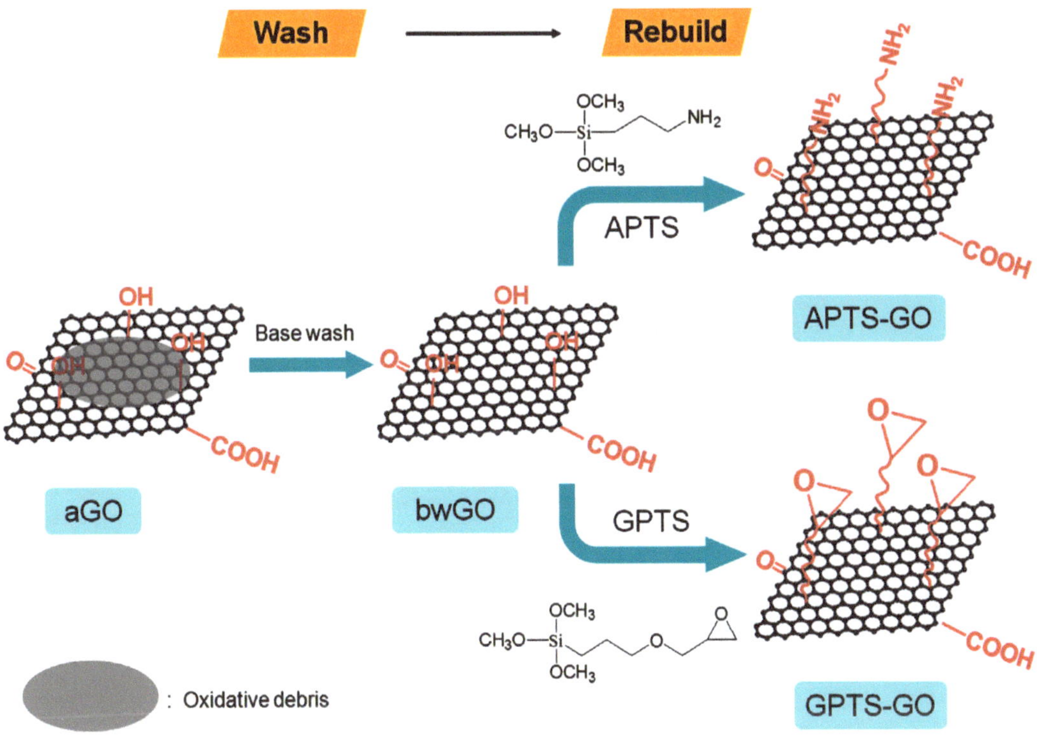

Figure 8. Schematic illustration of the "wash-and-rebuild" process. GO = graphene oxide; APTS = 3-aminopropyltrimethoxysilane; GPTS = 3-glycidoxypropyltrimethoxysilane; APTS-GO = 3-aminopropyltrimethoxysilane-functionalized GO; GPTS-GO = 3-glycidoxypropyltrimethoxysilane-functionalized GO. Reprinted with permission from [114]. 2013, Elsevier.

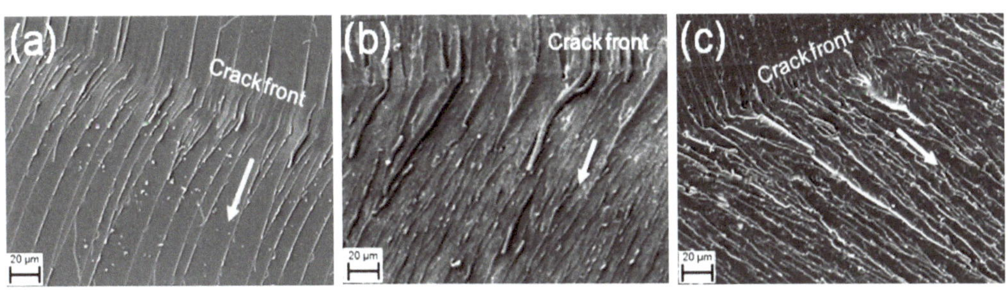

Figure 9. Fracture surfaces of specimens after fracture toughness testing (**a**) neat epoxy; (**b**) 0.2 wt.% APTS-GO/epoxy nanocomposites; and (**c**) 0.2 wt.% GPTS-GO/epoxy nanocomposites. The white arrows indicate the direction for crack propagation. APTS-GO/epoxy = 3-aminopropyltrimethoxysilane-functionalized graphene oxide/epoxide; GPTS-GO/epoxy = 3-glycidoxypropyltrimethoxysilane-functionalized graphene oxide/epoxy. Reprinted with permission from [114]. 2013, Elsevier.

Abdullah and coworkers [118] fabricated epoxy and graphene-oxide-derived nanocomposites through the direct mixing and curing technique. The 1.5 to 6 vol.% nanofiller loading improved the tensile strength and Young's modulus of the nanocomposites. The improvement in mechanical properties was credited to the interface formation between the epoxy and graphene oxide, which enhanced the mechanical properties [119]. Prolongo and coworkers [120] fabricated an epoxy/graphene nanoplatelet nanocomposite and an amine-functional graphene nanoplatelet nanocomposite. The nanomaterials were used as aerospace adhesives. The amine-functional graphene nanoplatelet had an ability to react with the epoxide groups of the epoxy resin to develop covalent linking. The covalently linked epoxy/graphene nanoplatelet nanocomposite was effectively used as an aerospace adhesive to recover the failure surfaces of joints (Figure 10). The nanocomposite showed fine adhesion to the aluminum surface, compared with the neat epoxy resin as well as the nonfunctional graphene nanoplatelet nanomaterial. Accordingly, the epoxy and graphene-nanofiller-derived nanocomposites revealed multifunctional applications in the aerospace sector [121].

Figure 10. Failure surfaces of epoxy adhesives on aluminum joints: (**a′**) epoxy; (**b′**) 6 wt.% GNP-NH_2; (**c′**) and 6 wt.% GNP. GNP = graphene nanoplatelets; GNP-NH_2 = amine-functionalized graphene. Reprinted with permission from [120]. 2018, Elsevier.

As a consequence, various industrial aspects of using graphene nanomaterials in military space systems, space defense technology, and aerospace engineering have been discovered [122]. Versatile graphene-based fibrous systems have been developed for military space applications [123]. Graphene nanoplatelets have resulted in high electrical, thermal, and mechanical properties which enlarge the application of nanocomposites to space-related military relevance [124]. In this regard, carbon, jute, or glass fibers modified with graphene revealed a high EMI-shielding efficiency > 50 dB. Such materials have been found to be valuable for several military areas including vehicles and personal protective equipment. In space defense applications, high-performance, lightweight, graphene-derived nanomaterials revealed low fuel consumption and offered cost and weight savings [125]. Due to their unique properties, polymer/graphene nanocomposites have been successfully applied for the ballistic protection and lightweight personnel armor [126]. Because of the considerably high strength and Young's modulus of fiber/graphene-derived nanocomposites, relative to conventionally used metallic components, aerospace structures have shown improved longevity and aircrafts have shown improved fuel efficiency. Thus, graphene and its derivatives have found suitability in the materials used for aerospace engineering.

6. Polymers with Graphene-Modified Carbon Fiber for Aerospace

Carbon-fiber-reinforced polymer (CFRP) composites consisting of carbon fibers as reinforcements in the polymer matrices have been used in airplanes, automobiles, trains, ships, and wind turbines [127]. The increasing needs of cost efficiency and ecological strategies have led to the replacement of metallic structures with composite structures in these industries [128]. The mechanical properties of carbon fibers have been considered as key factors for the superior performance of CFRP [129]. Moreover, the high specific surface area, chemical inertness, and high interfacial strength of carbon fibers have been found essential to enhance the characteristics of CFRP [130–132]. The strong interfacial bonding facilitated the stress transfer from the matrix to the carbon fiber, leading to less fiber debonding and pull-out which improve the properties of CFRP [133]. Therefore, a suitably engineered matrix–fiber interface has been found essential to ensure the efficient load transfer the between matrix and the reinforcements to enhance the mechanical behavior of CFRP. For improved properties, carbon fiber modification has been used to improve the compatibility between the fiber and the matrix. The surface modification of carbon fiber has been performed through sizing [134,135], coating [136,137], chemical grafting [138,139], plasma treatments [140,141], and electrophoretic deposition [142,143].

An important surface modification method is the incorporation of secondary nanoscale fillers (e.g., graphene and carbon nanotubes) at the interface between the matrix and fiber [144]. In this regard, the roughening of the carbon fiber surface and direct growth of carbon nanotube has been considered [145]. However, using catalysts and thermally activated mechanochemical modification on the carbon fiber surface may reduce the strength properties [146]. Using graphene for the surface modification of carbon fiber has been found advantageous, since no catalyst is required for the growth of graphene [147]. Graphene and graphene derivatives such as graphene oxide, graphene nanoflakes, etc., have been used to modify carbon fibers [148]. Cui et al. [149] produced carbon fibers coated with graphene and TiAl-alloy-based composites. The powder metallurgy, melt spinning, and vacuum-melting techniques were used to form the aerospace-grade composites. The inclusion of graphene-modified carbon fiber in the TiAl alloy matrix enhanced the strength from 1801 to 2312 MPa. Compared with the neat matrix, the fracture strain of the composites was enhanced by 26.27% and density was decreased for aerospace uses. Zhang et al. [150] formed graphene-oxide-modified carbon fibers for epoxy composites. The carbon fibers were modified through sizing using 5 wt.% graphene oxide. The modified carbon fiber developed a matrix–filler interface to enhance the interfacial properties of epoxy/carbon fiber composites. Significant enhancements in the interfacial shear strength, interlaminar shear strength, and tensile properties of the graphene-oxide-modified carbon-fiber-based composites were attained, compared with the nonmodified CFRP. The enhanced mechanical performance was found useful for the aerospace industry. Karakassides et al. [151] used the one-step microwave plasma technique for the direct growth of graphene nanoflakes on the surface of carbon fiber. The inclusion of graphene-nanoflake-modified carbon fiber improved the tensile strength by 28% and interfacial shear strength by 101.5%, relative to nonmodified CFRP. The cost-effective and lightweight CFRP composites were applied in aerospace structures. Qiu et al. [152] used magnetic graphene-oxide-modified carbon fibers to improve the interfacial properties of CFRP. The Fe_3O_4 functional graphene oxide prevented nanoparticle aggregation and promoted the controlled sizing of the carbon fiber. The magnetic graphene-oxide-modified carbon fibers enhanced the interlaminar shear strength and interfacial shear strength by 56% and 467%, compared to the nonmodified CFRP. The materials were potentially applicable for aerospace structures. Thus, graphene-modified carbon fibers have the potential to improve the interfacial shear strength of epoxy composites by 100%, as demanded for aerospace applications [153]. The effect was observed due to the improved interlocking between the graphene and epoxy matrix. Moreover, graphene-modified carbon fibers have been used to improve the mechanical properties of thermoplastic matrices such as polypropylene and nylon [154]. Modified carbon fibers have enhanced matrix–nanofiller bonding, leading to significant improvements in the tensile and interfacial shear strength of the aerospace composites [155]. Moreover, thermoplastic polymer/graphene-modified carbon fiber composites have enhanced electromagnetic shielding effectiveness for aerospace applications [156].

7. Graphene Nanomaterials and Present Aerospace Industry

The aerospace industry is one of the foremost adopters of advanced composite materials [157]. Graphene is indeed a wonderful material for the aerospace sector. Graphene combines lightness, flexibility, and strength, making it an ideal material for aerospace. Many commercial-level applications of graphene nanomaterials have been discovered. Firstly, graphene, due to its mechanical strength and lightness, has been considered as an excellent candidate for solar sails which may lead to the exploration of star systems [158–160]. In future, sails weighing only a few pounds will be produced while preserving their structural integrity. Secondly, for space lifts, extremely resistant cables have been developed using graphene nanomaterials [161]. In future, such cables will be used to carry ounces of fuel, astronauts, satellites, pieces of space stations, etc., up to thousands of kilometers high. Thirdly, graphene has been used in microgravity to form cooling systems for satellites [162]. In this regard, Italian aerospace company Leonardo, the National Research Council, and the universities of Cambridge and Brussels have performed successful experiments (2017) [163]. Fourthly, the Graphene Flagship has emerged as the largest European research initiative [164,165]. The Graphene Flagship and the European Space Agency performed experiments on the use of graphene in zero-gravity conditions in 2017. The use of graphene was verified in the thermal devices in satellites through a parabolic flight in gravity-free conditions. Then, the European Space Agency (ESA) and Novespace (France) developed graphene-based coatings as heat exchangers and cooling systems in satellites [166].

The applications of graphene are definitely surprising and limitless in the aeronautical field [162]. Graphene has been used to create light and impact-resistant aircrafts and helicopters [167]. Owing to high electrical conductivity, graphene-based deicing systems have been integrated in the wings [168]. Due to the shielding capacity, graphene nanomaterials have been used to build wind farms to prevent disturbance towards radars. Engineers from the University of Central Lancashire (UK) in collaboration with the space industry developed a drone fully coated with graphene. The drone was presented at the Futures Day event at Farnborough Air Show 2018. In addition, graphene nanomaterials are receiving positive feedback in the aeronautical field. Aerospace experts are continuously working to create superlight and resistant space structures.

8. Forthcoming and Conclusions

Polymeric nanocomposites have found potential for different types of aerospace uses [169–171]. Graphene is an imperative nanocarbon with excellent characteristics, suitable for aerospace structures. In this regard, epoxy has been a frequently applied material. Epoxy resin is a high-performance thermosetting polymer with good mechanical properties. It has been widely used in structural composites for space industries. Like nanofillers, the type of epoxy resin also affects the final nanocomposite properties. Consequently, the structure of the epoxy resin and the crosslinking agent used for curing (in addition to the nanofiller type/contents) influence the material characteristics. The melt method may result in poor graphene nanoparticle dispersion, thus affecting the material properties. In particular, epoxy nanocomposites have been designed with superior flame resistance, thermal stability, strength, conductivity, nonflammability, radiation shielding, and other high-performance features [172,173]. Moreover, the choice of fabrication method used for epoxy/graphene also affects the nanocomposite properties. For epoxy/graphene-based nanocomposite coatings and adhesives, the solution, direct mixing, in situ, and melt methods have been used. The solution and in situ routes have resulted in homogeneous nanofiller dispersion, thus leading to better matrix–nanofiller interactions and enhanced physical properties. For the large-scale processing of epoxy/graphene-based structural composites, the lay-up technique has been preferred. This method also involves the dispersion of graphene in an appropriate solvent prior to layering. Furthermore, polymer/graphene nanocomposites must have the capability of bearing lightning strikes, friction, and jerks/shocks. Nanocomposites must have high durability and performance for aerospace structures [174,175]. Incidentally, functional graphene has been used as a filler in polymers to improve the properties desired in space such as high thermal stability, mechanical stability at elevated temperatures, corrosion protection, radiation shielding, etc. Furthermore, graphene-modified carbon fiber composites have been fabricated with significantly enhanced mechanical and radiation-shielding properties for aerospace applications. For space-related graphene nanocomposites and graphene carbon fiber composites, various efficient and reliable characterization techniques have been used. For thermal stability analysis, thermogravimetric analysis and differential scanning calorimetry have been used. The morphology analysis of the fracture surfaces of specimens has been performed with the help of high-resolution scanning/transmission electron microscopic techniques. A number of other characterization techniques for mechanical testing, damage/shock analysis, conductivity measurement, EMI shielding analysis, contact angle measurements, etc., have been applied to study the behavior of the nanomaterials.

The main challenges in designing polymer/graphene nanocomposites involve the uniform dispersion of graphene in the matrix and the scalable production of these materials. Both these challenges can be overcome by the choice of an appropriate processing technique. In this regard, facile and low-cost processing techniques have been preferred. Despite the challenges, polymer/graphene nanomaterials have tunable functional surfaces, a facile processing method, and superior physical properties compared to polymer/carbon nanotube nanocomposites for aerospace structures. Accordingly, homogeneous graphene dispersion has been found to be essential to enhance the foremost aerospace properties. In graphene-modified carbon-fiber-based composites, new fiber sizing and modification techniques need to be developed. The dispersion of graphene on the carbon fiber surface has been found to be essential to enhance the matrix–fiber bonding and the ensuing mechanical features. Subsequently, the resulting polymer/graphene nanocomposites and composites have potential in high-performance aerospace structures including next-generation airplanes, jets, missiles, and space shuttles. Thus, future research must focus on using modified epoxy and functional graphene nanofiller for commercial aerospace applications.

This up-to-date review discusses numerous polymer and graphene-derived nanocomposites within the aerospace field. In addition to thermoplastics, epoxy matrices have been reinforced with graphene nanofiller, resulting in high-performance nanocomposites for the aerospace industry. The main emphasis of this review is on the design and probability of polymer/graphene nanocomposites, especially epoxy/graphene nanocomposites, for aerospace applications. Moreover, a viewpoint on graphene-modified carbon fiber-based composites is presented. Homogeneous graphene dispersion has been found important to enhance space-related material properties. Graphene has also been used to modify the fibers for epoxy reinforcement. In this regard, numerous processing strategies have been used, as discussed in this article. Accordingly, the choice of processing method has been found to be important for enhancing the desired material properties. The resulting high-performance polymer/graphene nanocomposites have been found to be valuable for several space-related industries such as military space systems, space defense technology, and the engineering of aerospace structural parts.

Author Contributions: Conceptualization, A.K.; data curation, A.K.; writing of original draft preparation, A.K.; review and editing, A.K., I.A., M.H.E. and M.M. All authors have read and agreed to the published version of the manuscript.

Funding: This research received no external funding.

Data Availability Statement: Not applicable.

Conflicts of Interest: The authors declare no conflict of interest.

References

1. Lebedev, O.V.; Kurkin, T.S.; Golubev, E.K.; Vasiliev, A.L.; Gatin, A.K.; Goncharuk, G.P.; Ozerin, A.N. Detonation Synthesis Nanodiamond Soot as a Promising Filler for Polymer Composites. *C* **2022**, *8*, 69. [CrossRef]
2. Padmanaban, D.B.; Mohan, L.; Giri, P.; Bera, P.; Anandan, C.; Barshilia, H.C. Effect of Molybdenum Content on Mechanical and Tribological Properties of Diamond-Like Carbon Coatings over Titanium β-21S Alloy. *C* **2020**, *7*, 1. [CrossRef]
3. Gbordzoe, S.; Adusei, P.K.; Chauhan, D.; Alvarez, N.T.; Haase, M.R.; Mansari, K.; Kanakaraj, S.N.; Hsieh, Y.-Y.; Shanov, V. A Simple Two-Step Process for Producing Strong and Aligned Carbon Nanotube-Polymer Composites. *C* **2019**, *5*, 35. [CrossRef]
4. Susi, B.T.; Tu, J.F. Digital Synthesis of Realistically Clustered Carbon Nanotubes. *C* **2022**, *8*, 34. [CrossRef]
5. Driscoll, A.J.; Johnson, P.A. Numerical modeling of analyte diffusion and adsorption behavior in microparticle and nanoparticle based biosensors. *Chem. Eng. Sci.* **2018**, *184*, 141–148. [CrossRef]
6. Rambabu, P.; Prasad, N.E.; Kutumbarao, V.; Wanhill, R. Aluminium alloys for aerospace applications. In *Aerospace Materials and material Technologies*; Springer: Singapore, 2017; pp. 29–52.
7. Gorbatikh, L.; Wardle, B.L.; Lomov, S.V. Hierarchical lightweight composite materials for structural applications. *Mrs Bull.* **2016**, *41*, 678–682. [CrossRef]
8. Bilisik, K.; Akter, M. Graphene nanocomposites: A review on processes, properties, and applications. *J. Ind. Text.* **2022**, *51*, 3718S–3766S. [CrossRef]
9. Razaq, A.; Bibi, F.; Zheng, X.; Papadakis, R.; Jafri, S.H.M.; Li, H. Review on graphene-, graphene oxide-, reduced graphene oxide-based flexible composites: From fabrication to applications. *Materials* **2022**, *15*, 1012. [CrossRef]
10. Al Faruque, M.A.; Syduzzaman, M.; Sarkar, J.; Bilisik, K.; Naebe, M. A review on the production methods and applications of graphene-based materials. *Nanomaterials* **2021**, *11*, 2414. [CrossRef]
11. Kausar, A. Aerospace applications of polymer/carbonaceous nanofiller nanocomposites: Mechanical, thermal, nonflammability, and physical aspects. In *Polymeric Nanocomposites with Carbonaceous Nanofillers for Aerospace Applications*; Elsevier: Amsterdam, The Netherlands, 2022; p. 85.
12. Liu, Y.; Du, H.; Liu, L.; Leng, J. Shape memory polymers and their composites in aerospace applications: A review. *Smart Mater. Struct.* **2014**, *23*, 023001. [CrossRef]
13. Kausar, A. Aeronautical composites and materials. In *Polymeric Nanocomposites with Carbonaceous Nanofillers for Aerospace Applications*; Elsevier: Amsterdam, The Netherlands, 2022; p. 1.
14. Gao, Y.; Zhang, Y.; Chen, P.; Li, Y.; Liu, M.; Gao, T.; Ma, D.; Chen, Y.; Cheng, Z.; Qiu, X. Toward single-layer uniform hexagonal boron nitride–graphene patchworks with zigzag linking edges. *Nano Lett.* **2013**, *13*, 3439–3443. [CrossRef] [PubMed]
15. Novoselov, K.S.; Geim, A.K.; Morozov, S.V.; Jiang, D.-e.; Zhang, Y.; Dubonos, S.V.; Grigorieva, I.V.; Firsov, A.A. Electric field effect in atomically thin carbon films. *Science* **2004**, *306*, 666–669. [CrossRef] [PubMed]
16. Wei, C.; Negishi, R.; Ogawa, Y.; Akabori, M.; Taniyasu, Y.; Kobayashi, Y. Turbostratic multilayer graphene synthesis on CVD graphene template toward improving electrical performance. *Jpn. J. Appl. Phys.* **2019**, *58*, SIIB04. [CrossRef]

17. Narayanam, P.K.; Botcha, V.D.; Ghosh, M.; Major, S.S. Growth and Photocatalytic Behaviour of Transparent Reduced GO-ZnO Nanocomposite Sheets. *Nanotechnology* **2019**, *30*, 485601. [CrossRef]
18. Shen, X.J.; Zeng, X.L.; Dang, C.Y. Graphene Composites. *Handb. Graphene* **2019**, *1*, 1–25.
19. Zandiatashbar, A.; Lee, G.-H.; An, S.J.; Lee, S.; Mathew, N.; Terrones, M.; Hayashi, T.; Picu, C.R.; Hone, J.; Koratkar, N. Effect of defects on the intrinsic strength and stiffness of graphene. *Nat. Commun.* **2014**, *5*, 3186. [CrossRef]
20. Zhou, Q.; Xia, G.; Du, M.; Lu, Y.; Xu, H. Scotch-tape-like exfoliation effect of graphene quantum dots for efficient preparation of graphene nanosheets in water. *Appl. Surf. Sci.* **2019**, *483*, 52–59. [CrossRef]
21. Pei, S.; Cheng, H.-M. The reduction of graphene oxide. *Carbon* **2012**, *50*, 3210–3228. [CrossRef]
22. Lee, H.; Lee, K.S. Interlayer Distance Controlled Graphene, Supercapacitor and Method of Producing the Same. U.S. Patent 10,214,422, 26 February 2019.
23. Han, J.T.; Jang, J.I.; Cho, J.Y.; Hwang, J.Y.; Woo, J.S.; Jeong, H.J.; Jeong, S.Y.; Seo, S.H.; Lee, G.-W. Synthesis of nanobelt-like 1-dimensional silver/nanocarbon hybrid materials for flexible and wearable electronics. *Sci. Rep.* **2017**, *7*, 4931. [CrossRef]
24. Schwierz, F. Graphene transistors: Status, prospects, and problems. *Proc. IEEE* **2013**, *101*, 1567–1584. [CrossRef]
25. Marconcini, P.; Macucci, M. Transport Simulation of Graphene Devices with a Generic Potential in the Presence of an Orthogonal Magnetic Field. *Nanomaterials* **2022**, *12*, 1087. [CrossRef] [PubMed]
26. Panwar, N.; Soehartono, A.M.; Chan, K.K.; Zeng, S.; Xu, G.; Qu, J.; Coquet, P.; Yong, K.-T.; Chen, X. Nanocarbons for biology and medicine: Sensing, imaging, and drug delivery. *Chem. Rev.* **2019**, *119*, 9559–9656. [CrossRef] [PubMed]
27. Tang, C.; Titirici, M.-M.; Zhang, Q. A review of nanocarbons in energy electrocatalysis: Multifunctional substrates and highly active sites. *J. Energy Chem.* **2017**, *26*, 1077–1093. [CrossRef]
28. Bao, Q.; Zhang, H.; Yang, J.x.; Wang, S.; Tang, D.Y.; Jose, R.; Ramakrishna, S.; Lim, C.T.; Loh, K.P. Graphene–polymer nanofiber membrane for ultrafast photonics. *Adv. Funct. Mater.* **2010**, *20*, 782–791. [CrossRef]
29. Wazalwar, R.; Sahu, M. Novel applications of graphene in the aerospace industry. In *Novel Applications of Carbon Based Nano-Materials*; CRC Press: Boca Raton, FL, USA, 2022; pp. 180–198.
30. Gao, W.; Alemany, L.B.; Ci, L.; Ajayan, P.M. New insights into the structure and reduction of graphite oxide. *Nat. Chem.* **2009**, *1*, 403–408. [CrossRef] [PubMed]
31. Eigler, S.; Dotzer, C.; Hof, F.; Bauer, W.; Hirsch, A. Sulfur species in graphene oxide. *Chem. Eur. J.* **2013**, *19*, 9490–9496. [CrossRef] [PubMed]
32. Bailey, E.J.; Winey, K.I. Dynamics of polymer segments, polymer chains, and nanoparticles in polymer nanocomposite melts: A review. *Prog. Polym. Sci.* **2020**, *105*, 101242. [CrossRef]
33. Zhao, Q.; Qi, H.J.; Xie, T. Recent progress in shape memory polymer: New behavior, enabling materials, and mechanistic understanding. *Prog. Polym. Sci.* **2015**, *49*, 79–120. [CrossRef]
34. Pilate, F.; Toncheva, A.; Dubois, P.; Raquez, J.-M. Shape-memory polymers for multiple applications in the materials world. *Eur. Polym. J.* **2016**, *80*, 268–294. [CrossRef]
35. Ghori, S.W.; Siakeng, R.; Rasheed, M.; Saba, N.; Jawaid, M. The role of advanced polymer materials in aerospace. In *Sustainable Composites for Aerospace Applications*; Jawaid, M., Thariq, M., Eds.; Woodhead Publishing, Elsevier: Amsterdam, The Netherlands, 2018; pp. 19–34.
36. Ramli, N.; Norkhairunnisa, M.; Ando, Y.; Abdan, K.; Leman, Z. Advanced Polymer Composite for Aerospace Engineering Applications. In *Advanced Composites in Aerospace Engineering Applications*; Springer: Berlin/Heidelberg, Germany, 2022; pp. 1–21.
37. Zeyrek, B.; Aydogan, B.; Dilekcan, E. Review of Thermoplastic Composites in Aerospace Industry. *Int. J. Eng. Tech. Inf.* **2022**, *3*, 1–6.
38. Norkhairunnisa, M.; Chai Hua, T.; Sapuan, S.; Ilyas, R. Evolution of Aerospace Composite Materials. In *Advanced Composites in Aerospace Engineering Applications*; Springer: Berlin/Heidelberg, Germany, 2022; pp. 367–385.
39. Pervaiz, M.; Panthapulakkal, S.; Sain, M.; Tjong, J. Emerging trends in automotive lightweighting through novel composite materials. *Mater. Sci. Appl.* **2016**, *7*, 26. [CrossRef]
40. Zhang, J.; Lin, G.; Vaidya, U.; Wang, H. Past, present and future prospective of global carbon fibre composite developments and applications. *Compos. Part B Eng.* **2022**, *250*, 110463. [CrossRef]
41. McIlhagger, A.; Archer, E.; McIlhagger, R. Manufacturing processes for composite materials and components for aerospace applications. In *Polymer Composites in the Aerospace Industry*; Elsevier: Amsterdam, The Netherlands, 2020; pp. 59–81.
42. Long, V.T.; Tung, H.V. Thermomechanical Nonlinear Buckling of Pressurized Shear Deformable FGM Cylindrical Shells Including Porosities and Elastically Restrained Edges. *J. Aerosp. Eng.* **2021**, *34*, 04021011. [CrossRef]
43. Weber, T.A.; Ruff-Stahl, H.-J.K. Advances in composite manufacturing of helicopter parts. *Int. J. Aviat. Aeronaut. Aerosp.* **2017**, *4*, 5. [CrossRef]
44. Tiwari, K.A.; Raisutis, R.; Samaitis, V. Hybrid signal processing technique to improve the defect estimation in ultrasonic non-destructive testing of composite structures. *Sensors* **2017**, *17*, 2858. [CrossRef]
45. Duodu, E.; Gu, J.; Shang, Z.; Ding, W.; Tang, S. Damage induced by high-velocity impact on composite structures using finite element simulation. *Iran. J. Sci. Technol. Trans. Mech. Eng.* **2017**, *41*, 97–107. [CrossRef]
46. Rana, S.; Fangueiro, R. *Advanced Composite Materials for Aerospace Engineering: Processing, Properties and Applications*; Woodhead Publishing: Sawston, UK, 2016.

47. Wang, Z.; Zulifqar, A.; Hu, H. Auxetic composites in aerospace engineering. In *Advanced Composite Materials for Aerospace Engineering*; Elsevier: Amsterdam, The Netherlands, 2016; pp. 213–240.
48. Carey, J.; Melenka, G.; Hunt, A.; Cheung, B.; Ivey, M.; Ayranci, C. Braided composites in aerospace engineering. In *Advanced Composite Materials for Aerospace Engineering*; Elsevier: Amsterdam, The Netherlands, 2016; pp. 175–212.
49. Kamiyama, S.; Hirano, Y.; Okada, T.; Ogasawara, T. Lightning strike damage behavior of carbon fiber reinforced epoxy, bismaleimide, and polyetheretherketone composites. *Compos. Sci. Technol.* **2018**, *161*, 107–114. [CrossRef]
50. Adam, T.J.; Liao, G.; Petersen, J.; Geier, S.; Finke, B.; Wierach, P.; Kwade, A.; Wiedemann, M. Multifunctional composites for future energy storage in aerospace structures. *Energies* **2018**, *11*, 335. [CrossRef]
51. Padture, N.P. Advanced structural ceramics in aerospace propulsion. *Nat. Mater.* **2016**, *15*, 804. [CrossRef]
52. Ghoshal, S. Polymer/carbon nanotubes (CNT) nanocomposites processing using additive manufacturing (three-dimensional printing) technique: An overview. *Fibers* **2017**, *5*, 40. [CrossRef]
53. Safaei, B.; Moradi-Dastjerdi, R.; Behdinan, K.; Chu, F. Critical buckling temperature and force in porous sandwich plates with CNT-reinforced nanocomposite layers. *Aerosp. Sci. Technol.* **2019**, *91*, 175–185. [CrossRef]
54. Rafique, I.; Kausar, A.; Anwar, Z.; Muhammad, B. Exploration of epoxy resins, hardening systems, and epoxy/carbon nanotube composite designed for high performance materials: A review. *Polym.-Plast. Technol. Eng.* **2016**, *55*, 312–333. [CrossRef]
55. Ji, C.; Yan, C.; Wang, Y.; Xiong, S.; Zhou, F.; Li, Y.; Sun, R.; Wong, C.-P. Thermal conductivity enhancement of CNT/MoS2/graphene— epoxy nanocomposites based on structural synergistic effects and interpenetrating network. *Compos. Part B Eng.* **2019**, *163*, 363–370. [CrossRef]
56. Wang, F.; Zhang, K.; Liang, W.; Wang, Z.; Tay, T.E.; Lu, S.; Yang, B. Epoxy/CNT@ X nanocomposite: Improved quasi-static, dynamic fracture toughness, and conductive functionalities by non-ionic surfactant treatment. *Polym. Test.* **2020**, *81*, 106256. [CrossRef]
57. Leng, J.; Lau, A.K.-T. *Multifunctional Polymer Nanocomposites*; CRC Press: Boca Raton, FL, USA, 2010.
58. Cao, M.; Xiong, D.B.; Yang, L.; Li, S.; Xie, Y.; Guo, Q.; Li, Z.; Adams, H.; Gu, J.; Fan, T. Ultrahigh electrical conductivity of graphene embedded in metals. *Adv. Funct. Mater.* **2019**, *29*, 1806792. [CrossRef]
59. Kausar, A. Shape memory polymer/graphene nanocomposites: State-of-the-art. *e-Polym.* **2022**, *22*, 165–181. [CrossRef]
60. Garg, K.K.; Pandey, S.; Kumar, A.; Rana, A.; Sahoo, N.G.; Singh, R.K. Graphene nanosheets derived from waste plastic for cost-effective thermoelectric applications. *Results Mater.* **2022**, *13*, 100260. [CrossRef]
61. Evgin, T.; Turgut, A.; Hamaoui, G.; Spitalsky, Z.; Horny, N.; Micusik, M.; Chirtoc, M.; Sarikanat, M.; Omastova, M. Size effects of graphene nanoplatelets on the properties of high-density polyethylene nanocomposites: Morphological, thermal, electrical, and mechanical characterization. *Beilstein J. Nanotechnol.* **2020**, *11*, 167–179. [CrossRef]
62. Sing, S.L.; Yeong, W.Y. Emerging Materials for Additive Manufacturing. *Materials* **2022**, *16*, 127. [CrossRef]
63. Ahmadi, Z. Epoxy in nanotechnology: A short review. *Prog. Org. Coat.* **2019**, *132*, 445–448. [CrossRef]
64. Ahmad, H.; Laini, V.V.; Qian, T.Z.; Jelani, R.M.; Rosli, F.A.; Kamaruzaman, S. Efficient removal of lead from aqueous using hybrid graphite nanoflakes/mesoporous silica nanoparticles, amine functionalized mesoporous silica and graphite nanoflakes as adsorbents. *Malays. J. Anal. Sci.* **2020**, *24*, 236–246.
65. Nie, Q.; Wei, X.; Qin, X.; Huang, Y.; Chen, G.; Yang, L.; Wang, B.; Tang, W. Microstructure and properties of graphite nanoflakes/Cu matrix composites fabricated by pressureless sintering and subsequent thermo-mechanical treatment. *Diam. Relat. Mater.* **2020**, *108*, 107948. [CrossRef]
66. Dimic-Misic, K.; Buffiere, J.; Imani, M.; Nieminen, K.; Sixta, H.; Gane, P. Improved stabilisation of graphite nanoflake dispersions using hydrothermally-produced nanocellulose. *Colloids Surf. A Physicochem. Eng. Asp.* **2021**, *610*, 125668. [CrossRef]
67. Eayal Awwad, K.; Yousif, B.; Fallahnezhad, K.; Saleh, K.; Zeng, X. Influence of Graphene Nanoplatelets on Mechanical Properties and Adhesive Wear Performance of Epoxy-Based Composites. *Friction* **2020**, *9*, 856–875. [CrossRef]
68. Vincent, V.A.; Kailasanathan, C.; Shanmuganathan, V.; Kumar, J.S.P.; Prakash, V.A. Strength characterization of caryota urens fibre and aluminium 2024-T3 foil multi-stacking sequenced SiC-toughened epoxy structural composite. *Biomass Convers. Biorefin.* **2020**, *12*, 4009–4019. [CrossRef]
69. Jayabalakrishnan, D.; Saravanan, K.; Ravi, S.; Prabhu, P.; Maridurai, T.; Prakash, V.A. Fabrication and characterization of acrylonitrile butadiene rubber and stitched E-glass fibre tailored Nano-silica epoxy resin composite. *Silicon* **2020**, *13*, 2509–2517. [CrossRef]
70. Qiu, J.; Wang, S. Enhancing polymer performance through graphene sheets. *J. Appl. Polym. Sci.* **2011**, *119*, 3670–3674. [CrossRef]
71. Kausar, A. Nanodiamond Reinforced Polymer Nanocomposite: Prospective Corrosion Protection Materials. In *Phenomena and Theories in Corrosion Science, Methods of Prevention*; Andras, G., Ed.; NOVA Science Publishers Inc.: Hauppauge, NY, USA, 2019; p. 181.
72. Xue, G.; Zhang, B.; Sun, M.; Zhang, X.; Li, J.; Wang, L.; Song, C. Morphology, thermal and mechanical properties of epoxy adhesives containing well-dispersed graphene oxide. *Int. J. Adhes. Adhes.* **2019**, *88*, 11–18. [CrossRef]
73. Gopanna, A.; Rajan, K.P.; Thomas, S.P.; Chavali, M. Polyethylene and polypropylene matrix composites for biomedical applications. In *Materials for Biomedical Engineering*; Elsevier: Amsterdam, The Netherlands, 2019; pp. 175–216.
74. Suner, S.; Joffe, R.; Tipper, J.; Emami, N. Ultra high molecular weight polyethylene/graphene oxide nanocomposites: Thermal, mechanical and wettability characterisation. *Compos. Part B Eng.* **2015**, *78*, 185–191. [CrossRef]

75. Bhawal, P.; Ganguly, S.; Das, T.K.; Mondal, S.; Choudhury, S.; Das, N. Superior electromagnetic interference shielding effectiveness and electro-mechanical properties of EMA-IRGO nanocomposites through the in-situ reduction of GO from melt blended EMA-GO composites. *Compos. Part B Eng.* **2018**, *134*, 46–60. [CrossRef]
76. Song, P.; Cao, Z.; Cai, Y.; Zhao, L.; Fang, Z.; Fu, S. Fabrication of exfoliated graphene-based polypropylene nanocomposites with enhanced mechanical and thermal properties. *Polymer* **2011**, *52*, 4001–4010. [CrossRef]
77. Feicht, P.; Biskupek, J.; Gorelik, T.E.; Renner, J.; Halbig, C.E.; Maranska, M.; Puchtler, F.; Kaiser, U.; Eigler, S. Brodie's or Hummers' method: Oxidation conditions determine the structure of graphene oxide. *Chem. Eur. J.* **2019**, *25*, 8955–8959. [CrossRef] [PubMed]
78. Saha, M.; Tambe, P.; Pal, S. Thermodynamic approach to enhance the dispersion of graphene in epoxy matrix and its effect on mechanical and thermal properties of epoxy nanocomposites. *Compos. Interfaces* **2016**, *23*, 255–272. [CrossRef]
79. Yang, J.; Tang, L.-S.; Bao, R.-Y.; Bai, L.; Liu, Z.-Y.; Yang, W.; Xie, B.-H.; Yang, M.-B. Largely enhanced thermal conductivity of poly (ethylene glycol)/boron nitride composite phase change materials for solar-thermal-electric energy conversion and storage with very low content of graphene nanoplatelets. *Chem. Eng. J.* **2017**, *315*, 481–490. [CrossRef]
80. Huang, G.; Chen, S.; Song, P.; Lu, P.; Wu, C.; Liang, H. Combination effects of graphene and layered double hydroxides on intumescent flame-retardant poly (methyl methacrylate) nanocomposites. *Appl. Clay Sci.* **2014**, *88*, 78–85. [CrossRef]
81. Singho, N.D.; Lah, N.A.C.; Johan, M.R.; Ahmad, R. FTIR studies on silver-poly (methylmethacrylate) nanocomposites via in-situ polymerization technique. *Int. J. Electrochem. Sci.* **2012**, *7*, 5596–5603.
82. Xu, Y.; Hong, W.; Bai, H.; Li, C.; Shi, G. Strong and ductile poly (vinyl alcohol)/graphene oxide composite films with a layered structure. *Carbon* **2009**, *47*, 3538–3543. [CrossRef]
83. Chen, B.; Berretta, S.; Evans, K.; Smith, K.; Ghita, O. A primary study into graphene/polyether ether ketone (PEEK) nanocomposite for laser sintering. *Appl. Surf. Sci.* **2018**, *428*, 1018–1028. [CrossRef]
84. Arif, M.; Alhashmi, H.; Varadarajan, K.; Koo, J.H.; Hart, A.; Kumar, S. Multifunctional performance of carbon nanotubes and graphene nanoplatelets reinforced PEEK composites enabled via FFF additive manufacturing. *Compos. Part B Eng.* **2020**, *184*, 107625. [CrossRef]
85. Xu, F.; Wu, X.; Wang, H.; Liu, H.; Ye, Z. Experimental research on moulding of graphene/PEKK composite powder by spark plasma sintering technology. *Appl. Phys. A* **2022**, *128*, 122. [CrossRef]
86. Cortes, L.Q.; Racagel, S.; Lonjon, A.; Dantras, E.; Lacabanne, C. Electrically conductive carbon fiber/PEKK/silver nanowires multifunctional composites. *Compos. Sci. Technol.* **2016**, *137*, 159–166. [CrossRef]
87. Shan, C.; Wang, L.; Han, D.; Li, F.; Zhang, Q.; Zhang, X.; Niu, L. Polyethyleneimine-functionalized graphene and its layer-by-layer assembly with Prussian blue. *Thin Solid Film.* **2013**, *534*, 572–576. [CrossRef]
88. Sali, S.; Mackey, H.R.; Abdala, A.A. Effect of graphene oxide synthesis method on properties and performance of polysulfone-graphene oxide mixed matrix membranes. *Nanomaterials* **2019**, *9*, 769. [CrossRef] [PubMed]
89. Puértolas, J.; Castro, M.; Morris, J.; Ríos, R.; Ansón-Casaos, A. Tribological and mechanical properties of graphene nanoplatelet/PEEK composites. *Carbon* **2019**, *141*, 107–122. [CrossRef]
90. Alvaredo, Á.; Martín, M.I.; Castell, P.; Guzmán de Villoria, R.; Fernández-Blázquez, J.P. Non-isothermal crystallization behavior of PEEK/graphene nanoplatelets composites from melt and glass states. *Polymers* **2019**, *11*, 124. [CrossRef]
91. Wang, Q.; Jia, D.; Pei, X.; Wu, X.; Xu, F.; Wang, H.; Cao, M.; Chen, H. Investigation of electromagnetic pulse compaction on conducting graphene/PEKK composite powder. *Materials* **2021**, *14*, 636. [CrossRef]
92. Sun, Z.; Li, Y.-Q.; Huang, P.; Cao, H.-J.; Zeng, W.; Li, J.; Li, F.; Sun, B.-G.; Shi, H.-Q.; Zhou, Z.-l. Temperature-dependent mechanical properties of polyetherimide composites reinforced by graphene oxide-coated short carbon fibers. *Compos. Struct.* **2021**, *270*, 114075. [CrossRef]
93. Ionita, M.; Pandele, A.M.; Crica, L.; Pilan, L. Improving the thermal and mechanical properties of polysulfone by incorporation of graphene oxide. *Compos. Part B Eng.* **2014**, *59*, 133–139. [CrossRef]
94. Peña-Bahamonde, J.; San-Miguel, V.; Baselga, J.; Fernández-Blázquez, J.P.; Gedler, G.; Ozisik, R.; Cabanelas, J.C. Effect of polysulfone brush functionalization on thermo-mechanical properties of melt extruded graphene/polysulfone nanocomposites. *Carbon* **2019**, *151*, 84–93. [CrossRef]
95. Gokulkumar, S.; Thyla, P.; Sathish, S.; Karthi, N.; Prabhu, L.; Vignesh, P.; Vishnu, R.; Srikanth, K.; Prasath, S.V. Acoustical and mechanical optimization of Camellia Sinensis/Ananas Comosus/GFRP/Epoxy composites by TOPSIS & Weighted Aggregates Sum Product Assessment method. *Mater. Today Proc.* **2021**, *45*, 7980–7985.
96. Vinod, A.; Gowda, Y.; Vijay, R.; Gupta, M.K.; Jamil, M.; Kushvaha, V.; Siengchin, S. Novel Muntingia Calabura bark fiber reinforced green-epoxy composite: A sustainable and green material for cleaner production. *J. Clean. Prod.* **2021**, *294*, 126337. [CrossRef]
97. Sadowski, Ł.; Hoła, J.; Czarnecki, L.; Mathia, T.G. New paradigm in the metrology of concrete surface morphology: Methods, parameters and applications. *Measurement* **2021**, *169*, 108497. [CrossRef]
98. Alwash, D.; Kalfat, R.; Du, H.; Al-Mahaidi, R. Development of a new nano modified cement based adhesive for FRP strengthened RC members. *Constr. Build. Mater.* **2021**, *277*, 122318. [CrossRef]
99. Kostagiannakopoulou, C.; Loutas, T.; Sotiriadis, G.; Kostopoulos, V. Effects of graphene geometrical characteristics to the interlaminar fracture toughness of CFRP laminates. *Eng. Fract. Mech.* **2021**, *245*, 107584. [CrossRef]

100. Ghazanlou, S.I.; Eghbali, B.; Petrov, R. Microstructural Evolution and Strengthening Mechanisms in Al7075/Graphene Nanoplates/Carbon Nano-tubes Composite Processed through Accumulative Roll Bonding. *Mater. Sci. Eng. A* **2021**, *807*, 140877 [CrossRef]
101. Shiu, S.-C.; Tsai, J.-L. Characterizing thermal and mechanical properties of graphene/epoxy nanocomposites. *Compos. Part B Eng.* **2014**, *56*, 691–697. [CrossRef]
102. Kumar, A.; Sharma, K.; Dixit, A.R. A review of the mechanical and thermal properties of graphene and its hybrid polymer nanocomposites for structural applications. *J. Mater. Sci.* **2019**, *54*, 5992–6026. [CrossRef]
103. Shadlou, S.; Ahmadi-Moghadam, B.; Taheri, F. The effect of strain-rate on the tensile and compressive behavior of graphene reinforced epoxy/nanocomposites. *Mater. Des.* **2014**, *59*, 439–447. [CrossRef]
104. Bustero, I.; Gaztelumendi, I.; Obieta, I.; Mendizabal, M.A.; Zurutuza, A.; Ortega, A.; Alonso, B. Free-standing graphene films embedded in epoxy resin with enhanced thermal properties. *Adv. Compos. Hybrid Mater.* **2020**, *3*, 31–40. [CrossRef]
105. Yogeshwaran, S.; Natrayan, L.; Rajaraman, S.; Parthasarathi, S.; Nestro, S. Experimental investigation on mechanical properties of epoxy/graphene/fish scale and fermented spinach hybrid bio composite by hand lay-up technique. *Mater. Today Proc.* **2021**, *37*, 1578–1583. [CrossRef]
106. Jaafar, C.A.; Zainol, I.; Rizal, M.M.; Sultan, U. Preparation and characterisation of epoxy/silica/kenaf composite using hand lay-up method. In Proceedings of the 27th Scientific Conference of the Microscopy Society Malaysia (27th SCMSM 2018), Melaka, Malaysia, 3–4 December 2018; pp. 2–6.
107. Peng, J.; Huang, C.; Cao, C.; Saiz, E.; Du, Y.; Dou, S.; Tomsia, A.P.; Wagner, H.D.; Jiang, L.; Cheng, Q. Inverse nacre-like epoxy-graphene layered nanocomposites with integration of high toughness and self-monitoring. *Matter* **2020**, *2*, 220–232. [CrossRef]
108. García-Martínez, V.; Gude, M.R.; Calvo, S.; Ureña, A. Enhancing an Aerospace Grade Benzoxazine Resin by Means of Graphene Nanoplatelets Addition. *Polymers* **2021**, *13*, 2544. [CrossRef] [PubMed]
109. He, X.; Feng, L.; Zhang, Z.; Hou, X.; Ye, X.; Song, Q.; Yang, Y.; Suo, G.; Zhang, L.; Fu, Q.-G. High-Performance Multifunctional Carbon–Silicon Carbide Composites with Strengthened Reduced Graphene Oxide. *ACS Nano* **2021**, *15*, 2880–2892. [CrossRef] [PubMed]
110. Guan, F.; Ren, H.; Yu, L.; Cui, Q.; Zhao, W.; Liu, J. Nitrated Graphene Oxide Derived from Graphite Oxide: A Promising Energetic Two-Dimensional Material. *Nanomaterials* **2021**, *11*, 58. [CrossRef] [PubMed]
111. Yuan, W.; Hu, Q.; Zhang, J.; Huang, F.; Liu, J. Hydrophobic Modification of Graphene Oxide and Its Effect on the Corrosion Resistance of Silicone-Modified Epoxy Resin. *Metals* **2021**, *11*, 89. [CrossRef]
112. Yousefi, N.; Sun, X.; Lin, X.; Shen, X.; Jia, J.; Zhang, B.; Tang, B.; Chan, M.; Kim, J.K. Highly aligned graphene/polymer nanocomposites with excellent dielectric properties for high-performance electromagnetic interference shielding. *Adv. Mater.* **2014**, *26*, 5480–5487. [CrossRef]
113. Yousefi, N.; Lin, X.; Zheng, Q.; Shen, X.; Pothnis, J.R.; Jia, J.; Zussman, E.; Kim, J.-K. Simultaneous in situ reduction, self-alignment and covalent bonding in graphene oxide/epoxy composites. *Carbon* **2013**, *59*, 406–417. [CrossRef]
114. Liu, W.; Wei, J.; Chen, Y.; Huo, P.; Wei, Y. Electrospinning of poly (L-lactide) nanofibers encapsulated with water-soluble fullerenes for bioimaging application. *ACS Appl. Mater. Interfaces* **2013**, *5*, 680–685. [CrossRef]
115. Fioravante, A.; de Oliveira, P.C.; Siqueira, E.J.; Vitorino, L.S.; Brito, P.P.; Oréfice, R.L.; Isaac, A. From brittle-to-ductile fracture of polymer composites: The incorporation of energy dissipation mechanisms by carbon nanotubes-based multilayered interface. *J. Appl. Polym. Sci.* **2020**, *137*, 49348. [CrossRef]
116. Sethy, D.; Makireddi, S.; Varghese, F.; Balasubramaniam, K. Piezoresistive behaviour of graphene nanoplatelet (GNP)/PMMA spray coated sensors on a polymer matrix composite beam. *Express Polym. Lett.* **2019**, *13*, 1018–1025. [CrossRef]
117. Ahmadi, Z. Nanostructured epoxy adhesives: A review. *Prog. Org. Coat.* **2019**, *135*, 449–453. [CrossRef]
118. Abdullah, S.I.; Ansari, M. Mechanical properties of graphene oxide (GO)/epoxy composites. *Hbrc J.* **2015**, *11*, 151–156. [CrossRef]
119. Wu, H.; Cheng, L.; Liu, C.; Lan, X.; Zhao, H. Engineering the interface in graphene oxide/epoxy composites using bio-based epoxy-graphene oxide nanomaterial to achieve superior anticorrosion performance. *J. Colloid Interface Sci.* **2020**, *587*, 755–766. [CrossRef]
120. Salom, C.; Prolongo, M.; Toribio, A.; Martínez-Martínez, A.; de Cárcer, I.A.; Prolongo, S. Mechanical properties and adhesive behavior of epoxy-graphene nanocomposites. *Int. J. Adhes. Adhes.* **2018**, *84*, 119–125. [CrossRef]
121. Xiao, W.; Li, B.; Yan, J.; Wang, L.; Huang, X.; Gao, J. Three dimensional graphene composites: Preparation, morphology and their multi-functional applications. *Compos. Part A Appl. Sci. Manuf.* **2022**, *165*, 107335. [CrossRef]
122. Theravalappil, R.; Rahaman, M. Patents on graphene-based polymer composites and their applications. In *Polymer Nanocomposites Containing Graphene*; Elsevier: Amsterdam, The Netherlands, 2022; pp. 615–638.
123. Liu, J.; Hui, D.; Lau, D. Two-dimensional nanomaterial-based polymer composites: Fundamentals and applications. *Nanotechnol. Rev.* **2022**, *11*, 770–792. [CrossRef]
124. Abreu, I.; Ferreira, D.P.; Fangueiro, R. Versatile graphene-based fibrous systems for military applications. *NATO's Sci. Technol. Organ.* **2019**, *1*, 8–14.
125. Hawal, T.T.; Patil, M.S.; Swamy, S.; Kulkarni, R.M. A Review on Synthesis, Functionalization, Processing and Applications of Graphene Based High Performance Polymer Nanocomposites. *Curr. Nanosci.* **2022**, *18*, 167–181. [CrossRef]

126. Lynch, J. High Performance, Lightweight Graphene Enhanced Polymer Matrix Composites for Defense Applications. *AVT* **2020**, *2*, 18–28.
127. Zhang, J.; Chevali, V.S.; Wang, H.; Wang, C.-H. Current status of carbon fibre and carbon fibre composites recycling. *Compos. Part B Eng.* **2020**, *193*, 108053. [CrossRef]
128. Rajak, D.K.; Pagar, D.D.; Menezes, P.L.; Linul, E. Fiber-reinforced polymer composites: Manufacturing, properties, and applications. *Polymers* **2019**, *11*, 1667. [CrossRef]
129. Jayan, J.S.; Appukuttan, S.; Wilson, R.; Joseph, K.; George, G.; Oksman, K. An introduction to fiber reinforced composite materials. In *Fiber Reinforced Composites*; Elsevier: Amsterdam, The Netherlands, 2021; pp. 1–24.
130. González, C.; LLorca, J. Mechanical behavior of unidirectional fiber-reinforced polymers under transverse compression: Microscopic mechanisms and modeling. *Compos. Sci. Technol.* **2007**, *67*, 2795–2806. [CrossRef]
131. Sethi, S.; Ray, B.C. Environmental effects on fibre reinforced polymeric composites: Evolving reasons and remarks on interfacial strength and stability. *Adv. Colloid Interface Sci.* **2015**, *217*, 43–67. [CrossRef] [PubMed]
132. ElKhoury, L.; Berg, J.C. The tailoring of interfacial adhesion and its effects on material properties for carbon-fiber reinforced polymeric composites. *J. Adhes. Sci. Technol.* **2023**, 1–22. [CrossRef]
133. Zhao, J.; Liebscher, M.; Schneider, K.; Junger, D.; Mechtcherine, V. Effect of surface profiling on the mechanical properties and bond behaviour of mineral-impregnated, carbon-fibre (MCF) reinforcement based on geopolymer. *Constr. Build. Mater.* **2023**, *367*, 130199. [CrossRef]
134. Yuan, H.; Zhang, S.; Lu, C. Surface modification of carbon fibers by a polyether sulfone emulsion sizing for increased interfacial adhesion with polyether sulfone. *Appl. Surf. Sci.* **2014**, *317*, 737–744. [CrossRef]
135. Sokolowski, K.; Pszczola, I.; Blazewicz, S.; Bernasik, A.; Kornaus, K.; Fraczek-Szczypta, A. Effect of Various Methods of Pre-Treatment of carbon Fibers on the mechanical Properties of PIP-Based ceramics/Carbon Composites. *Res. Sq.* **2023**, *1*, 1–31.
136. Sharma, M.; Gao, S.; Mäder, E.; Sharma, H.; Wei, L.Y.; Bijwe, J. Carbon fiber surfaces and composite interphases. *Compos. Sci. Technol.* **2014**, *102*, 35–50. [CrossRef]
137. Yang, L.; Chen, Y.; Xu, Z.; Xia, H.; Natuski, T.; Xi, Y.; Ni, Q. Effect of surface modification of carbon fiber based on magnetron sputtering technology on tensile properties. *Carbon* **2023**, *204*, 377–386. [CrossRef]
138. Zhou, H.; Zhang, B.; Liu, W.; Xue, Y.; Zhao, C.; Zhang, X.; Cui, X.; Tang, Z. Grafting glycidyl methacrylate onto high-modulus carbon fiber by acid activation synergized with ionization radiation. *Appl. Surf. Sci.* **2023**, *610*, 155593. [CrossRef]
139. Li, N.; Cheng, S.; Wang, B.; Zong, L.; Bao, Q.; Wu, G.; Hu, F.; Wang, J.; Liu, C.; Jian, X. Chemical grafting of graphene onto carbon fiber to produce composites with improved interfacial properties via sizing process: A step closer to industrial production. *Compos. Sci. Technol.* **2023**, *231*, 109822. [CrossRef]
140. Dilsiz, N. Plasma surface modification of carbon fibers: A review. *J. Adhes. Sci. Technol.* **2000**, *14*, 975–987. [CrossRef]
141. Qiao, Y.; Shin, Y.; Pallaka, M.R.; Nickerson, E.K.; Merkel, D.R.; Seffens, R.J.; Ortiz, A.; Ramos, J.L.; Simmons, K.L. Plasma surface modification coupled with thermal and step-over distance effects on significant fracture improvement of adhesively-bonded metal-CFRTP dissimilar materials. *Compos. Sci. Technol.* **2023**, *232*, 109833. [CrossRef]
142. Xia, S.; Guo, Q.; Yu, Y.; Li, Y.; Wang, S.; Dong, D.; Liu, Z.; Zhou, H.; Zhou, X.; Liu, Z. Surface modification of carbon fiber cloth with graphene oxide through an electrophoresis method for lithium metal anode. *Carbon* **2023**, *203*, 743–752. [CrossRef]
143. Huang, L.; Dong, Y.; Tian, Y.; Zhang, X. Electrophoretic deposition of Ti3C2Tx MXene nanosheets onto high modulus carbon fiber enhances their composites interfacial properties. *Mater. Today Commun.* **2023**, *34*, 105429. [CrossRef]
144. Tkalya, E.E.; Ghislandi, M.; de With, G.; Koning, C.E. The use of surfactants for dispersing carbon nanotubes and graphene to make conductive nanocomposites. *Curr. Opin. Colloid Interface Sci.* **2012**, *17*, 225–232. [CrossRef]
145. Qin, J.; Wang, C.; Liu, R.; Su, S.; Yao, Z.; Zheng, L.; Gao, Q.; Wang, Y.; Wang, Q.; Wei, H. Uniform growth of carbon nanotubes on carbon fiber cloth after surface oxidation treatment to enhance interfacial strength of composites. *Compos. Sci. Technol.* **2020**, *195*, 108198. [CrossRef]
146. Steiner III, S.A.; Li, R.; Wardle, B.L. Circumventing the mechanochemical origins of strength loss in the synthesis of hierarchical carbon fibers. *ACS Appl. Mater. Interfaces* **2013**, *5*, 4892–4903. [CrossRef]
147. Karakasidis, A. *Hierarchical Carbon-Fibre Reinforced Polymer (CFRP) Composites Utilising Directly Grown Graphene Nanoflakes*; Ulster University: Belfast, UK, 2022.
148. Adstedt, K.; Buxton, M.L.; Henderson, L.C.; Hayne, D.J.; Nepal, D.; Gogotsi, Y.; Tsukruk, V.V. 2D graphene oxide and MXene nanosheets at carbon fiber surfaces. *Carbon* **2023**, *203*, 161–171. [CrossRef]
149. Cui, S.; Cui, C.; Xie, J.; Liu, S.; Shi, J. Carbon fibers coated with graphene reinforced TiAl alloy composite with high strength and toughness. *Sci. Rep.* **2018**, *8*, 2364. [CrossRef]
150. Zhang, X.; Fan, X.; Yan, C.; Li, H.; Zhu, Y.; Li, X.; Yu, L. Interfacial microstructure and properties of carbon fiber composites modified with graphene oxide. *ACS Appl. Mater. Interfaces* **2012**, *4*, 1543–1552. [CrossRef] [PubMed]
151. Karakassides, A.; Ganguly, A.; Tsirka, K.; Paipetis, A.S.; Papakonstantinou, P. Radially grown graphene nanoflakes on carbon fibers as reinforcing interface for polymer composites. *ACS Appl. Nano Mater.* **2020**, *3*, 2402–2413. [CrossRef]
152. Qiu, B.; Zhang, X.; Xia, S.; Sun, T.; Ling, Y.; Zhou, S.; Guang, H.; Chen, Y.; Xu, Z.; Liang, M. Magnetic graphene oxide/carbon fiber composites with improved interfacial properties and electromagnetic interference shielding performance. *Compos. Part A Appl. Sci. Manuf.* **2022**, *155*, 106811. [CrossRef]

153. Maurya, A.; Sinha, S.; Kumar, P.; Singh, V. A review: Impact of surface treatment of nanofillers for improvement in thermo mechanical properties of the epoxy based nanocomposites. *Mater. Today Proc.* 2023, *in press*. [CrossRef]
154. Raagulan, K.; Ghim, J.; Braveenth, R.; Chai, K.Y.; Kim, B.M. Improving the EMI shielding of graphene oxide (GNO)-coated glass-fiber–GNO–MA-grafted polypropylene (PP) composites and nylon 1D–2D nanocomposite foams. *RSC Adv.* **2022**, *12*, 15316–15328. [CrossRef] [PubMed]
155. Ghaemi, F.; Yunus, R.; Salleh, M.; Rashid, S.; Ahmadian, A.; Lim, H. Effects of the surface modification of carbon fiber by growing different types of carbon nanomaterials on the mechanical and thermal properties of polypropylene. *RSC Adv.* **2015**, *5*, 28822–28831. [CrossRef]
156. Lin, C.-L.; Li, J.-W.; Chen, Y.-F.; Chen, J.-X.; Cheng, C.-C.; Chiu, C.-W. Graphene Nanoplatelet/Multiwalled Carbon Nanotube/Polypyrrole Hybrid Fillers in Polyurethane Nanohybrids with 3D Conductive Networks for EMI Shielding. *ACS Omega* **2022**, *7*, 45697–45707. [CrossRef]
157. Stefan, A.; Pelin, G.; Nae, C.C.; Pelin, C.E.; Cristea, G.C.; Botan, M.; Bocioaga, M. Suitability assessment of the component materials of advanced composite system for aircraft structural application. In Proceedings of the AIAA SCITECH 2023 Forum, National Harbor, MD, USA, 23–27 January 2023; p. 2122.
158. Gaudenzi, R.; Stefani, D.; Cartamil-Bueno, S.J. Light-induced propulsion of graphene-on-grid sails in microgravity. *Acta Astronaut.* **2020**, *174*, 204–210. [CrossRef]
159. Johnson, L.; Barnes, N.; Ceriotti, M.; Chen, T.Y.; Davoyan, A.; Friedman, L.; Garber, D.; Kezerashvili, R.; Kobayashi, K.; Matloff, G. Solar Sail Propulsion by 2050: An Enabling Capability for Heliophysics Missions. *arXiv* **2023**, arXiv:2301.01297.
160. Viale, A.; McInnes, C.R. Attitude control actuator scaling laws for orbiting solar reflectors. *Adv. Space Res.* **2023**, *71*, 604–623. [CrossRef]
161. Ahmed, K.; Avi, S.D.; Tanvir, M.S.; Rahman, M.M.; Shufian, A.; Sagor, M.M.I.; Farrok, O. Transportation in Between the Earth and Space by Using Carbon Nanotubes as the Elevator Cable. In Proceedings of the 2019 5th International Conference on Advances in Electrical Engineering (ICAEE), Dhaka, Bangladesh, 26–28 September 2019; pp. 426–430.
162. Scalia, T.; Bonventre, L.; Terranova, M.L. From Protosolar Space to Space Exploration: The Role of Graphene in Space Technology and Economy. *Nanomaterials* **2023**, *13*, 680. [CrossRef] [PubMed]
163. Stutzmann, M.; Csoklich, C. *The Physics of Renewable Energy*; Springer Nature: Berlin/Heidelberg, Germany, 2023.
164. Döscher, H.; Reiss, T. Graphene roadmap briefs (No. 1): Innovation interfaces of the Graphene Flagship. *2D Mater.* **2021**, *8*, 022004. [CrossRef]
165. Roche, S.; Åkerman, J.; Beschoten, B.; Charlier, J.-C.; Chshiev, M.; Dash, S.P.; Dlubak, B.; Fabian, J.; Fert, A.; Guimarães, M. Graphene spintronics: The European Flagship perspective. *2D Mater.* **2015**, *2*, 030202. [CrossRef]
166. Hong, H.; Part, F.; Nowack, B. Prospective Dynamic and Probabilistic Material Flow Analysis of Graphene-Based Materials in Europe from 2004 to 2030. *Environ. Sci. Technol.* **2022**, *56*, 13798–13809. [CrossRef]
167. Alam, M.A.; Ya, H.; Sapuan, S.; Mamat, O.; Parveez, B.; Yusuf, M.; Masood, F.; Ilyas, R. Recent Advancements in Advanced Composites for Aerospace Applications: A Review. In *Advanced Composites in Aerospace Engineering Applications*; Springer: Cham, Switzerland, 2022; pp. 319–339.
168. Wang, Z.; Shen, Y.; Tao, J.; Liu, S.; Jiang, J.; Xu, Y.; Liu, W.; Li, H. An integrated superhydrophobic anti/de-icing film heater with low energy consumption: Interpenetration behavior of components based on wet-film spraying method. *Appl. Therm. Eng.* **2023**, *223*, 120028. [CrossRef]
169. Rout, D.; Nayak, R.K.; Praharaj, S. Aerospace and vehicle industry. In *Handbook of Polymer Nanocomposites for Industrial Applications*; Elsevier: Amsterdam, The Netherlands, 2021; pp. 399–417.
170. Kurahatti, R.; Surendranathan, A.; Kori, S.; Singh, N.; Kumar, A.; Srivastava, S. Defence applications of polymer nanocomposites. *Def. Sci. J.* **2010**, *60*, 551–563. [CrossRef]
171. Asyraf, M.; Ilyas, R.; Sapuan, S.; Harussani, M.; Hariz, H.; Aiman, J.; Baitaba, D.M.; Sanjay, M.; Ishak, M.; Norkhairunnisa, M. Advanced Composite in Aerospace Applications: Opportunities, Challenges, and Future Perspective. In *Advanced Composites in Aerospace Engineering Applications*; Springer: Cham, Switzerland, 2022; pp. 471–498.
172. BUENDIA, L.; TORRES, I.; CASTELLANOS, A. Novel Testing Technique to Evaluate the Durability of Textile Composites at Extreme Environments. In Proceedings of the American Society for Composites—Thirty-Fifth Technical Conference, Tucson, AZ, USA, 19–21 September 2022.
173. Kausar, A. Aeronautical Impact of Epoxy/Carbon Nanotube Nanocomposite. In *Nanotechnology in Aerospace and Structural Mechanics*; IGI Global: Hershey, PA, USA, 2019; pp. 80–102.
174. Patel, V.; Mahajan, Y. Polymer nanocomposites: Emerging growth driver for the global automotive industry. In *Handbook of Polymernanocomposites. Processing, Performance and Application*; Springer: Berlin/Heidelberg, Germany, 2014; pp. 511–538.
175. Farahani, R.D.; Dubé, M.; Therriault, D. Three-dimensional printing of multifunctional nanocomposites: Manufacturing techniques and applications. *Adv. Mater.* **2016**, *28*, 5794–5821. [CrossRef]

C Journal of Carbon Research

MDPI

Article

Joule Heating of Carbon-Based Materials Obtained by Carbonization of *Para*-Aramid Fabrics

Daniel Karthik *, Jiri Militky, Yuanfeng Wang and Mohanapriya Venkataraman

Department of Material Engineering, Technical University of Liberec, Studentska 1402/2, 46117 Liberec, Czech Republic
* Correspondence: daniel.karthik@tul.cz

Abstract: The Joule heating behavior of carbon-based materials obtained by the process of carbonization of industrial para-aramid fabric wastes are investigated in the present work. Carbonization involves a process of thermally decomposing organic material, thereby altering its physical and chemical properties to obtain carbon-rich materials that are electrically conductive and display Joule heating behavior. The principle of Joule heating is based on the intrinsic electrical resistance of the material across an applied voltage. Here, para-aramid woven fabric wastes are converted into activated carbon materials through straightforward, controlled, single-step thermal treatments by three different kinds of atmosphere, i.e., in the CO_2 evolved from charcoal, a mixture of gases from ammonium bicarbonate salt (NH_4HCO_3), and Nitrogen gas (N_2), respectively, inside a high-temperature furnace. The carbonization temperatures were varied from 800 to 1100 °C. The carbonization process variables were optimized to obtain carbon-rich materials with lower electrical resistivity. The results of electrical resistivity measurements show that for all three methods, the electrical resistivity decreases with increasing carbonization temperatures. An experimental setup consisting of an infrared (IR) camera, positioned over the surface of the fabric specimen to record the surface temperature of the material connected to a DC power supply, was employed. The kinetics of Joule heating and subsequent cooling were also analyzed at a fixed voltage of 5 V by recording the changes in surface temperature with respect to time. The heating–cooling cycle is described by a simple kinetic model of first order.

Keywords: Joule heating; carbonization; activated carbon; electrical resistivity; para-aramid wastes

Citation: Karthik, D.; Militky, J.; Wang, Y.; Venkataraman, M. Joule Heating of Carbon-Based Materials Obtained by Carbonization of *Para*-Aramid Fabrics. C **2023**, *9*, 23. https://doi.org/10.3390/c9010023

Academic Editor: Gil Gonçalves

Received: 12 January 2023
Revised: 10 February 2023
Accepted: 13 February 2023
Published: 15 February 2023

1. Introduction

In recent years, the growing demand for more multi-functional, techno-economic and sustainable materials for various applications has attracted enormous research interest. Electrothermal and Ohmic heating performance of textiles in particular have compelled much attention in various heating applications. Carbonaceous materials in various forms (fibers, fabrics, or particles) have drawn considerable attention as effective heating elements due to their intrinsic thermal and electrical properties [1,2]. These materials have versatile applications in areas such as smart sensors, grounding materials to dissipate static charge, structural health monitoring, de-icing in aerospace and wind-turbine structures, and electromagnetic interference (EMI) shielding materials, among others [3–6].

Joule heating (also referred to as Ohmic heating or resistive heating), based on the law of conservation of energy, is a phenomenon that occurs when a voltage difference is applied across a conductive material such that electrical energy is consumed in overcoming the resistance in the material between the electrons and the atoms, causing this energy to be generated in the form of heat. This heating is referred to as Joule's effect. The amount of heat produced by the material is proportional to the square of the current, the resistance of the circuit, and the duration for which the current flows through the circuit [7], shown mathematically as follows:

$$Q = I^2Rt \quad (1)$$

where Q is the heat produced by the material in joules; I is the electrical current flowing through the conductor, in amperes; R is the electrical resistance, in ohms; t is the elapsed time, in seconds. Joule heating has been extensively used in various applications where thermal energy produced by the resistance in the conductor is put to use in the form of conduction, convection, or radiation.

It is quite rhetorical to query how the processes of pyrolysis and carbonization of various polymeric materials have been constantly carried out over time, in order to obtain carbon-rich products conducive to various areas of application. Thermal decomposition through these two phases involves aromatic evolution and polymerization of the organic material, through bond cleavage reactions and the rearrangement of compounds in an inert atmosphere over a wide temperature range [8–12]. Non-carbon volatiles in the form of nitrogen, hydrogen, methane, carbon dioxide, carbon monoxide, ammonia, water, etc., are eliminated during the process of carbonization, which contributes to almost 50 percent or more by weight of the precursor material. The maximum temperature of carbonization depends on the required type of carbon and its specific application. Fundamental modifications can be observed in the chemical composition as well as the physical properties [13–17].

The type of the precursor and the selected process parameters greatly influence the final product of carbonization. Having said that, para-aramids (1, 4 p-phenylene terephthalamide), more commonly recognized as Kevlar, have shown to be suitable precursors to obtain carbon fibers by various methods of carbonization and activation (physical and chemical) [10,18,19]. The carbonaceous materials obtained displayed significant electrical and thermal properties. In the present experiment, we utilized feedstock material, gathered from a local industry, as Kevlar fabric wastes and carbonized them to obtain carbon fabric produce.

In the present study, a comparative analysis of the Ohmic heating behavior and electrical properties of carbon fabrics obtained by the carbonization of para-aramid fabrics by three different procedures was conducted. Here, we utilize charcoal and ammonium bicarbonate salt (in separate methods), which are inexpensive and commonly available, in comparison to N_2 gas, in order to potentially create an oxygen-free atmosphere inside the furnace to support activation during the carbonization process. The electrical resistivity and Joule heating behavior of the activated carbon are analyzed. A kinetics model is further used to analyze the heating and cooling cycles of the Joule-heated materials with respect to voltage and time. The para-aramid fabrics used for this study were industrial feedstock material, hence contributing towards sustainability and the potential utilization of textile wastes.

2. Experimental

2.1. Materials

Kevlar 29 para-aramid woven fabric was obtained from Veba Textile Mills Ltd. (Broumov), Czech Republic, in the form of discarded short fabric wastes with a yarn count of 1710 dtex, density of 1.43 g/cm^3, and 5.2 wt % moisture content (parameters specified by the manufacturing company).

2.2. Methodology

Carbonization of Kevlar was carried out using three different methods based on the atmosphere created in the furnace during the process of pyrolysis, as described below.

Method 1: Charcoal Method

The carbonization of Kevlar fibrous wastes was carried out under the layer of charcoal, creating a CO_2 atmosphere, in a high-temperature furnace. The Kevlar fabric was directly carbonized without any intermediate stabilization step. The final temperatures of 800 °C, 900 °C, 1000 °C, and 1100 °C, with heating rate of 300 °C/h and without any holding time, were adopted during the process of carbonization. This method is described more precisely in our previous work [20].

Method 2: Ammonium bicarbonate salt Method

Carbonization of Kevlar was carried out in a smelting furnace in a combined gas atmosphere generated by using ammonium bicarbonate salt. Ammonium bicarbonate is an inorganic compound with formula NH_4HCO_3, and it decomposes when heated above 36 °C, releasing ammonia, water, and carbon dioxide gases as follows: $NH_4HCO_3 \rightarrow NH_3 + H_2O + CO_2$, so the atmosphere created inside the furnace is mainly a mixture of gasses: CO_2 and NH_3, replacing air.

The Kevlar fabric samples in rolled form, were placed inside a cylindrical crucible, positioned in the center of the furnace. A total of 10 g of ammonium bicarbonate salt was added into the crucible containing the Kevlar sample, in order to create a mixed gas atmosphere. The Kevlar fabric samples were carbonized to four final temperatures from 800 to 1100°C, with an interval of 100 °C.

Method 3: N_2 Method

Carbonization of Kevlar was carried out in a high-temperature furnace, in the presence of an inert N_2 flow atmosphere; a heating rate of 10 °C/min and holding time of 20 min at the final temperature was adopted. The Kevlar fabric samples were carbonized to four final temperatures from 800 to 1100 °C, with an interval of 100 °C.

The carbon samples obtained from the three methods mentioned above are listed in Table 1, with details of their yields in percentages. For all three methods, the final product at every selected final carbonization temperature, although weak, retained its fabric form and possessed structural integrity, making it suitable to handle for further investigations.

Table 1. Details of carbonized samples.

Sample Code	Atmosphere Type	Carbonization Method	Final Carbonization Temperature (°C)	Fabric Yield (%)	Carbon Yield (%)
C8	CO_2	1	800	58.24 ± 1.3	69.1 ± 0.9
C9	CO_2	1	900	47.26 ± 1.6	76.4 ± 1.2
C10	CO_2	1	1000	35.69 ± 2.1	83.4 ± 0.8
C11	CO_2	1	1100	31.44 ± 1.8	89.1 ± 0.4
A8	Gas mixture	2	800	30.21 ± 1.1	83.7 ± 0.6
A9	Gas mixture	2	900	26.31 ± 1.7	87.1 ± 0.6
A10	Gas mixture	2	1000	23.49 ± 2.5	88.5 ± 0.8
A11	Gas mixture	2	1100	19.89 ± 2.8	90.2 ± 0.8
N8	Nitrogen	3	800	46.23 ± 1.1	67.2 ± 0.9
N9	Nitrogen	3	900	41.79 ± 1.2	73.4 ± 0.8
N10	Nitrogen	3	1000	34.47 ± 1.9	84.1 ± 0.6
N11	Nitrogen	3	1100	30.60 ± 1.5	88.2 ± 0.8

Energy dispersive X-ray (EDX) analysis was carried out using an LZ 5 EDX detector, Oxford Instruments, High Wycombe, UK, in order to calculate the relative percentages of different elements in the activated material after the carbonization process at different temperatures. The elemental analysis of the starting material, Kevlar, in terms of weight percentages on dry basis, was 63.85% C, 15.09% N, 0.8% S, and 18.84% O [20]. Thermal degradation advances with increasing temperatures by the removal of volatile compounds. The increase in carbon content (as a percentage) is of primary importance when enhancing the electrical conductivity and Joule heating of the final product (shown in Table 1). The remaining composition of the carbon fabrics obtained after carbonization is made up of trivial amounts of impurities such as Na, K, Cl, and Ca.

2.3. Ohmic Heating Measurement

The Joule heating behavior of the carbon fabric samples were characterized using a test setup as illustrated in Figure 1. The test sample was clamped in position along its width by a pair of 6 cm long electrodes at both ends, which were 8 cm apart. These electrodes were connected to a DC power supply (TIPA PS3010), TIPA, spol. sr.o, Opava, Czech Republic. An IR camera (FLIR-E6390), Teledyne FLIR LLC, Oregon, USA. was placed above the test sample at a height of 40 mm, in order to monitor the surface temperature and heat distribution on the sample surface. The current 1 A was kept constant, and the voltage was gradually increased from 0 V to 10 V, at intervals of 1 V, with a holding time of 10 s, and the average temperature over the middle section of the material between the two sets of electrodes was considered.

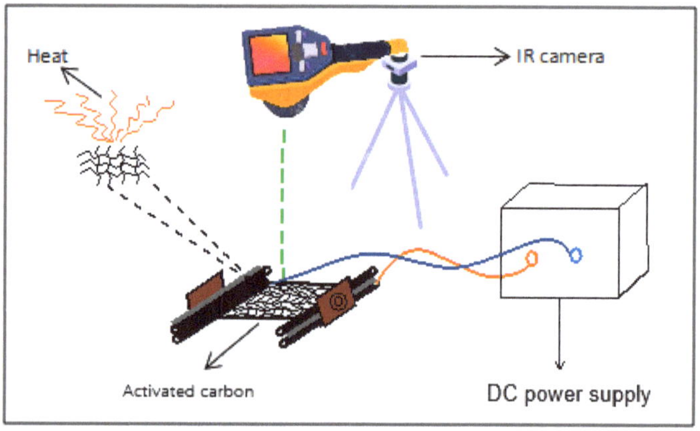

Figure 1. Joule heating measurement setup.

Furthermore, to investigate the Joule heating performance of the carbon fabrics obtained by different methods of carbonization, a constant voltage of 5 V was applied to evaluate the heating/cooling kinetics based on the time-temperature characteristics of the materials.

2.4. Electrical Resistivity

Correspondingly, the test sample was clamped the same way as mentioned in the previous section to measure the Joule heating and was connected to a digital multimeter, which displayed the electrical resistance (in ohms) of the material. A current flow of 1 A was set by regulating the voltage and the resistance value was recorded after 10 s. Furthermore, the electrical resistivity (in ohm.cm) was calculated mathematically as follows [7]:

$$\rho = \frac{A \times R}{L} \tag{2}$$

where ρ is the electrical resistivity (in ohm.cm); A is the cross-sectional area (along the width of the fabric sample in our case); R is the resistance measured (by the multimeter); and L is the length of the fabric sample.

3. Results and Discussions

3.1. Electrical Resistivity

Figure 2 shows the electrical resistivity of the fabrics carbonized to four final temperatures ranging from 800–1100 °C, using methods 1, 2, and 3. The electrical resistivity of all samples was measured five times per sample. It is evident that for all three methods, the electrical resistivity decreased progressively with increasing carbonization temperatures.

This is attributed to the higher carbon yield due to carbonization at higher temperatures, due to the removal of organic compounds in the form of volatiles [21,22]. These carbon-rich fabrics acquire a dense micro current network and create a better passage for current flow. Migration and hopping of electrons through the fibrous network, being a potential means of electron transport, is thought to be responsible for the higher electrical conductivity or lower resistivity [20,23]. For all three methods, there was no significant difference in the resistivity between fabrics carbonized at 800 °C to 1000 °C. The electrical resistivity of fabrics processed by methods 1 and 2 were nearly the same at 1000 °C and 1100 °C carbonization temperatures. Overall, in all three methods the carbonized fabrics displayed lower resistivity, or rather greater conductivity at higher carbonization temperatures, which indicates higher current flow through these materials, thus contributing to increased Joule heating.

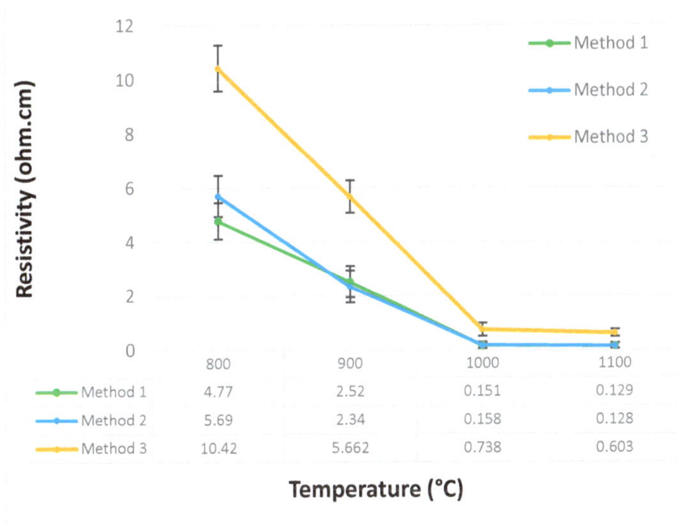

Figure 2. Electrical resistivity of activated carbon fabrics from methods 1, 2, and 3.

3.2. Joule Heating Characteristics

Temperature–Voltage (T-V) relation

The surface temperature of activated carbon fabrics was investigated at applied voltages ranging from 0 to 10 V, at intervals of 1 V for constant current I = 1 A; the surface temperature was recorded 10 s after setting the specific voltage. Figure 3a–c show the T–V curves of the fabrics carbonized by methods 1, 2, and 3, respectively. For each sample, the average value of five test measurements was considered. For fabrics obtained by all three methods, observations made from the T–V curves demonstrate that, initially, the surface temperatures stayed nearly the same, without a significant rise in temperature when a low voltage (V < 3 V) was supplied. Subsequently, with increasing voltage, the surface temperature of the fabrics gradually increased. Similarly, for all three methods, the fabrics carbonized at 1000 °C and 1100 °C showed a considerably steep increase in surface temperature with increasing voltage. The surface temperatures for fabrics carbonized at 1000 °C and 1100 °C were nearly alike over the entire range of applied voltage (0–10 V). At the maximum applied voltage (10 V), methods 1, 2, and 3 showed surface temperatures of 244 °C, 264 °C, and 210 °C, for fabrics carbonized at 1000 °C.

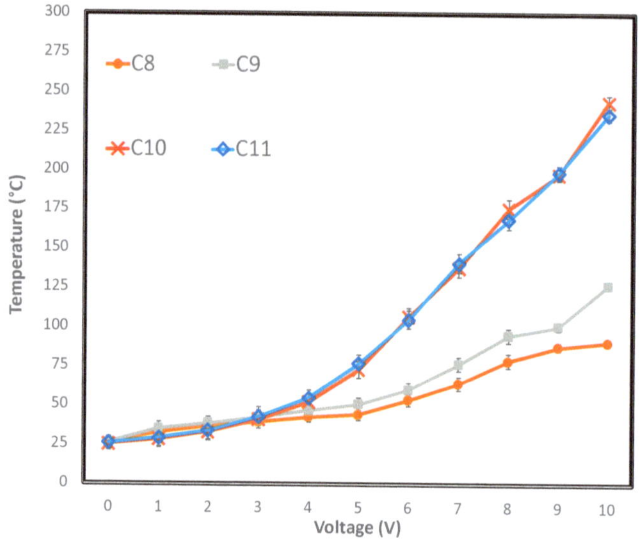

(a)

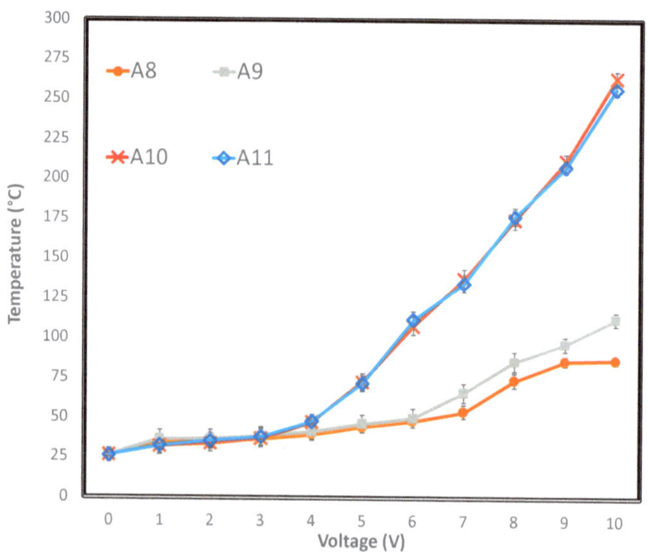

(b)

Figure 3. *Cont.*

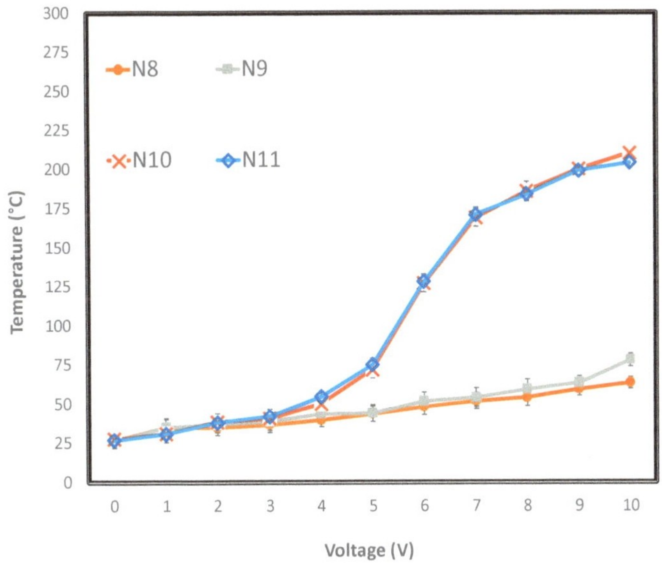

(c)

Figure 3. Temperature–voltage relation of activated carbon fabrics from (**a**) method 1; (**b**) method 2; and (**c**) method 3.

Heating/Cooling Kinetics: Temperature–Time (T–I) relation

The purpose of this segment was to analyze the kinetics of continuous heating and cooling of the carbon fabrics, obtained by time-dependent temperature (T–I) relation curves (shown in Figure 4). For each sample, the average value of five test measurements was considered. It is evident that for all the samples, the surface temperature increased steeply once the voltage (set at 5 V) was applied starting at 0 s, and then, after 30 s, the rate of temperature growth gradually increased and reached the maximum temperature at 120 s. Consecutively, after switching off the electrical power supply, the temperature decreased rapidly to room temperature. The stable increase in temperature by Joule heating for fabrics obtained by all three methods greatly depended on the final carbonization temperature. Furthermore, the leveling-off temperature was higher for fabrics carbonized at 1000 °C and 1100 °C, for all three methods at the same applied voltage (5 V). It is clear that under the same applied voltage, the magnitudes of the maximum temperatures were nearly the same for fabrics carbonized at 1000 °C and 1100 °C for all three methods. For this reason, the fabrics carbonized at 1000 °C, using the three methods (C10, A10, and N10), were taken into consideration for comparison (Figure 4d). The maximum temperature of C10 was slightly higher than that of A10 and N10. Furthermore, the initial increase in surface temperature with respect to time was significantly higher for C10 than for A10 and N10. This can be attributed to the more uniform distribution of graphite layers with the fibrous structures for higher current flow. The delay in the increase of surface temperature before reaching a maximum, especially for A10 and N10, is possibly be due to the initial heat loss due to convention or radiation, which means it takes longer for current to be distributed uniformly as possible, throughout the surface of the fabric, and thus takes longer to stabilize and reach the maximum temperature.

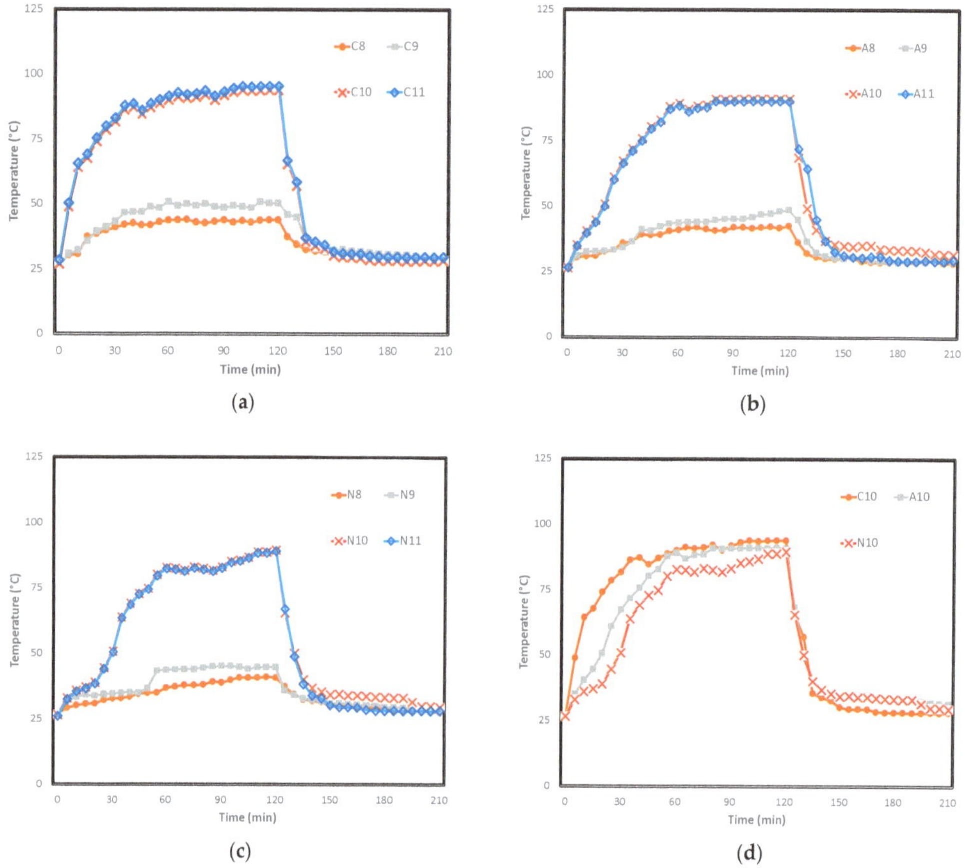

Figure 4. Temperature–time relation of activated carbon fabrics from (**a**) method 1; (**b**) method 2; (**c**) method 3; and (**d**) comparison of the three methods for fabrics carbonized at 1000 °C.

Here, three heating/cooling kinetic parameters, namely, heating time constant, efficiency of heat transfer, and cooling time constant were obtained to further investigate the Joule heating properties of the carbon fabrics. The T–I curves for fabrics obtained by all three methods shown in Figure 4 can be divided into three sections: (i) temperature growth (heating; 0–60 s), (ii) the region of equilibrium (maximum temperature; 60–120 s), and (iii) temperature decay (cooling; 120–210 s) [24,25].

The heating/cooling kinetics can be simply expressed by a rate equation of the first order, where rate of temperature change *dT/dt* is proportional to actual temperature *T*. For the heating phase, the rate equation is of the form [24,25]:

$$\frac{dT}{dt} = -k_h(T_m - T) \tag{3}$$

where T_m is maximum temperatue in equlibrium and k_h is the rate constant of heating, which is evidently replaced by the time constant $\tau_g = 1/\mathrm{k_h}$. Integration of this differntial equation in the range from ambient temperature T_0, at time $t = 0$, to temperature T_t, at time t, leads to the following equation [24,25]:

$$\frac{T_t - T_0}{T_m - T_0} = 1 - \exp(-t/\tau_g) \tag{4}$$

The initial heating segment is expressed by Equation (4) and the parameters T_m, τ_g are obtained by nonlinear regression.

For the cooling phase, starting from time t_0, it can be shown that [24,25]:

$$\frac{T_t - T_0}{T_m - T_0} = \exp(-(t - t_0)/\tau_d) \tag{5}$$

where cooling time constant is $\tau_d = 1/k_c$.

The resulting values for all the activated carbon fabrics are listed in Table 2. The values of τ_g for N10 and N11 are greater than those of A10, A11, C10, and C11 because of their higher resistivity.

Table 2. Characteristic parameters (τ_g, *hr+c*, and τ_d) for Joule heating performance of all activated carbon fabric samples under applied voltages.

Carbonization Method	Carbonization Temperature	τ_g (S)	h_{r+c} (W/°C)	τ_d (S)
1	800	19.60	0.09	89.83
	900	24.11	0.08	107.52
	1000	19.84	0.07	57.82
	1100	19.63	0.07	60.66
2	800	30.36	0.04	100.77
	900	41.01	0.04	126.59
	1000	25.99	0.10	92.24
	1100	28.92	0.10	89.12
3	800	44.82	0.03	126.82
	900	38.86	0.03	105.76
	1000	42.17	0.09	88.94
	1100	41.68	0.10	70.21

The τ_d values obtained from nonlinear regression of experimental time-dependent temperature decay data, using Model (5), are summarized in Table 2. The τ_g and τ_d values of C10 and C11 were noticeably lower than those of A10, A11, N10, and N11. This means that C10 and C11 exhibited comparatively more rapid temperature responses to applied voltages and were easier to control.

In the maximum temperature zone (region of equilibrium), where the temperature as a function of time becomes constant, the heat gain by electric power is counteracted by the heat emitted to the surroundings through radiation and/or convection, due to conservation of energy. As such, the heat transferred through radiation and convection, h_{r+c}, can be expressed as [24,25]:

$$h_{r+c} = \frac{I_c V_0}{T_m - T_0} \tag{6}$$

where I_c is the steady-state current and V_0 is the applied voltage. It can be seen in Table 2 that the h_{r+c} increases with A10, A11, N10, and N11 as compared to C10 and C11, indicating that comparatively more electrical energy is required to maintain the maximum temperatures for these materials. In the third segment, the material at maximum temperature is left to cool down to the ambient temperature according to Equation (5).

4. Conclusions

In this work, simple and straightforward methods to utilize textile wastes in a sustainable manner with high carbon yield, electrical conductivity and joule heating behavior has been achieved. Carbonization of industrial para-aramid fabric wastes was carried out using three unadorned and straightforward methods to produce activated carbon fabric which was then investigated for the realization of electrical resistivity and Joule heating. Inexpensive and commonly available compounds such as charcoal and ammonium bicarbonate salt (methods 1 and 2) were utilized to potentially create an oxygen-free atmosphere inside a furnace to assist in activation during the carbonization process, in comparison to N2 gas, which was used in method 3. All three methods were shown to be effective at producing activated carbon fabrics from para-aramid. The electrical resistivity of the activated carbon fabrics processed by all three methods was shown to decrease significantly with increasing carbonization temperature. Methods 1 and 2 showed lower electrical resistivity in comparison to method 3. Both electrical resistivity and Joule heating characteristics were greatly influenced by the final carbonization temperatures for all three methods. For fabrics obtained by all three methods, it can be seen that in the T–V curves the surface temperature remained low with no significant changes when a low voltage ($V < 3$ V) was supplied. However, the surface temperature of the fabrics increased with higher voltage ($V > 3$ V). At maximum voltage (10 V), C10 and A10 showed the highest surface temperatures of 244 °C and 264 °C due to lower electrical resistivity, or greater conductivity. To further investigate the thermal properties of the carbonized materials, we expressed three heating/cooling kinetic parameters. The increase in temperature by Joule heating for fabrics obtained by all three methods greatly depended on the final carbonization temperature. The leveling-off temperature was higher for fabrics carbonized at 1000 °C and 1100 °C, for all three methods at the same applied voltage (5 V). The maximum temperature of C10 was slightly higher than that of A10 and N10. Moreover, the initial increase in surface temperature with respect to time was significantly higher for C10 than for A10 and N10. Cumulatively, we can say that methods 1 and 2, which utilized charcoal and ammonium bicarbonate salt, were comparatively more desirable in terms of both electrical and Joule heating characteristics, while also being cost-effective and sustainable. We concluded that this material retains structural integrity after carbonization at specified temperatures (800 to 1100°C), excellent electrical conductivity and Joule heating ability, and can thus be proposed as a material for protective panels, enabling EMI shielding and heat generation via external powering towards thermally and/or electrically regulated sensors.

Author Contributions: Conceptualization, D.K. and J.M.; data curation, Y.W.; formal analysis, D.K. and Y.W.; funding acquisition, M.V.; investigation, D.K.; methodology, D.K.; project administration, J.M. and M.V.; resources, Y.W. and M.V.; supervision, J.M.; visualization, D.K.; writing—original draft, D.K.; writing—review and editing, J.M. All authors have read and agreed to the published version of the manuscript.

Funding: This work was supported by the Czech Science Foundation (GACR)-project "Advanced structures for thermal insulation under extreme conditions" (Reg. No. 21-32510M).

Institutional Review Board Statement: Not applicable.

Informed Consent Statement: Not applicable.

Data Availability Statement: Additional details of the experiments and data may be asked via email to the corresponding author.

Conflicts of Interest: The authors declare no conflict of interest.

References

1. Kim, M.; Sung, D.H.; Kong, K.; Kim, N.; Kim, B.-J.; Park, H.W.; Park, Y.-B.; Jung, M.; Lee, S.H.; Kim, S.G. Characterization of resistive heating and thermoelectric behavior of discontinuous carbon fiber-epoxy composites. *Compos. Part B Eng.* **2016**, *90*, 37–44. [CrossRef]
2. Zhang, Q.; Yu, Y.; Yang, K.; Zhang, B.; Zhao, K.; Xiong, G.; Zhang, X. Mechanically robust and electrically conductive graphene-paper/glass-fibers/epoxy composites for stimuli-responsive sensors and Joule heating deicers. *Carbon* **2017**, *124*, 296–307. [CrossRef]
3. Isaji, S.; Bin, Y.; Matsuo, M. Electrical conductivity and self-temperature-control heating properties of carbon nanotubes filled polyethylene films. *Polymer* **2009**, *50*, 1046–1053. [CrossRef]
4. Lopes, H.; Ribeiro, J.E. Structural Health Monitoring in Composite Automotive Elements. In *New Advances in Vehicular Technology and Automotive Engineering*; Carmo, J., Ed.; InTech: London, UK, 2012. [CrossRef]
5. Zanjani, J.S.M.; Okan, B.S.; Pappas, P.-N.; Galiotis, C.; Menceloglu, Y.Z.; Yildiz, M. Tailoring viscoelastic response, self-heating and deicing properties of carbon-fiber reinforced epoxy composites by graphene modification. *Compos. Part A Appl. Sci. Manuf.* **2018**, *106*, 1–10. [CrossRef]
6. Athanasopoulos, N.; Kostopoulos, V. Resistive heating of multidirectional and unidirectional dry carbon fibre preforms. *Compos. Sci. Technol.* **2012**, *72*, 1273–1282. [CrossRef]
7. Redondo, O.; Prolongo, S.; Campo, M.; Sbarufatti, C.; Giglio, M. Anti-icing and de-icing coatings based Joule's heating of graphene nanoplatelets. *Compos. Sci. Technol.* **2018**, *164*, 65–73. [CrossRef]
8. Jiménez, V.; Sánchez, P.; Romero, A. Materials for activated carbon fiber synthesis. In *Activated Carbon Fiber and Textiles*; Woodhead Publishing: Sawston, UK, 2017; pp. 21–38. [CrossRef]
9. Ko, K.S.; Park, C.W.; Yoon, S.-H.; Oh, S.M. Preparation of Kevlar-derived carbon fibers and their anodic performances in Li secondary batteries. *Carbon* **2001**, *39*, 1619–1625. [CrossRef]
10. Mosquera, M.E.; Jamond, M.; Martinez-Alonso, A.; Tascon, J.M. Thermal Transformations of Kevlar Aramid Fibers During Pyrolysis: Infrared and Thermal Analysis Studies. *Chem. Mater.* **2002**, *6*, 1918–1924. Available online: https://pubs.acs.org/doi/pdf/10.1021/cm00047a006 (accessed on 5 October 2021). [CrossRef]
11. Yang, M.; Zhu, X.L.; Liang, G. Pyrolysis Process of Kevlar Fibers with Thermogravimetric Analysis coupled and Fourier Transform Infrared Spectroscopy. *Guang Pu* **2016**, *36*, 1374–1377. [PubMed]
12. Ramgobin, A.; Fontaine, G.; Bourbigot, S. Thermal Degradation and Fire Behavior of High Performance Polymers. *Polym. Rev.* **2019**, *59*, 55–123. [CrossRef]
13. Chen, J.; Harrison, I. Modification of polyacrylonitrile (PAN) carbon fiber precursor via post-spinning plasticization and stretching in dimethyl formamide (DMF). *Carbon* **2002**, *40*, 25–45. [CrossRef]
14. EFitzer, E.; Frohs, W.; Heine, M. Optimization of stabilization and carbonization treatment of PAN fibres and structural characterization of the resulting carbon fibres. *Carbon* **1986**, *24*, 387–395. [CrossRef]
15. Edie, D.D. The effect of processing on the structure and properties of carbon fibers. *Carbon* **1998**, *36*, 345–362. [CrossRef]
16. Huang, X. Fabrication and Properties of Carbon Fibers. *Materials* **2009**, *2*, 2369–2403. [CrossRef]
17. Chen, J.Y. *Activated Carbon Fiber and Textiles*; Woodhead Publishing: Sawston, UK, 2016; p. 342.
18. Choma, J.; Osuchowski, L.; Marszewski, M.; Dziura, A.; Jaroniec, M. Developing microporosity in Kevlar®-derived carbon fibers by CO_2 activation for CO_2 adsorption. *J. CO2 Util.* **2016**, *16*, 17–22. [CrossRef]
19. Brown, J.R.; Power, A.J. Thermal degradation of aramids: Part I—Pyrolysis/gas chromatography/mass spectrometry of poly(1,3-phenylene isophthalamide) and poly(1,4-phenylene terephthalamide). *Polym. Degrad. Stab.* **1982**, *4*, 379–392. [CrossRef]
20. Karthik, D.; Baheti, V.; Militky, J.; Naeem, M.S.; Tunakova, V.; Ali, A. Activated Carbon Derived from Carbonization of Kevlar Waste Materials: A Novel Single Stage Method. *Materials* **2021**, *14*, 6433. [CrossRef] [PubMed]
21. Suárez-García, F.; Martinez-Alonso, A.; Tascón, J.M. Activated carbon fibers from Nomex by chemical activation with phosphoric acid. *Carbon* **2004**, *42*, 1419–1426. [CrossRef]
22. Conte, G.; Stelitano, S.; Policicchio, A.; Minuto, F.D.; Lazzaroli, V.; Galiano, F.; Agostino, R.G. Assessment of activated carbon fibers from commercial Kevlar® as nanostructured material for gas storage: Effect of activation procedure and adsorption of CO_2 and CH_4. *J. Anal. Appl. Pyrolysis* **2020**, *152*, 104974. [CrossRef]
23. Naeem, S.; Baheti, V.; Tunakova, V.; Militky, J.; Karthik, D.; Tomkova, B. Development of porous and electrically conductive activated carbon web for effective EMI shielding applications. *Carbon* **2017**, *111*, 439–447. [CrossRef]
24. El-Tantawy, F.; Kamada, K.; Ohnabe, H. In situ network structure, electrical and thermal properties of conductive epoxy resin–carbon black composites for electrical heater applications. *Mater. Lett.* **2002**, *56*, 112–126. [CrossRef]
25. Kong, K.; Deka, B.K.; Kim, M.; Oh, A.; Kim, H.; Park, Y.-B.; Park, H.W. Interlaminar resistive heating behavior of woven carbon fiber composite laminates modified with ZnO nanorods. *Compos. Sci. Technol.* **2014**, *100*, 83–91. [CrossRef]

C Journal of Carbon Research

Article

Galling-Free Fine Blanking of Titanium Plates Using Carbon-Supersaturated High-Speed Steel Punch

Tatsuhiko Aizawa [1,*] and Kenji Fuchiwaki [2]

1 Surface Engineering Design Laboratory, Shibaura Institute of Technology, Tokyo 144-0045, Japan
2 Hatano Precision, Co., Ltd., Hadano City 257-0015, Japan
* Correspondence: taizawa@sic.shibaura-it.ac.jp; Tel.: +81-3-6424-8615

Abstract: A carbon-supersaturated (CS-) high-speed steel punch was prepared using low-temperature plasma carburizing for fine blanking of pure titanium plates. The bare high-speed steel punch was also prepared as a reference to describe the adhesion and abrasive galling in the fine blanking of the titanium plates, even in a single shot. The CS-punch was free from severe chemical galling, even after repeatedly fine-blanking the pure titanium plates. A microstructure analysis, element mapping and a chemical composition analysis demonstrated that titanium debris fragments slightly deposited at the CS-punch edges in the presence of agglomerated free-carbon film due to the CS-punch. This galling-free fine-blanking behavior came from the in situ formation of free carbon tribofilms. This in situ lubrication resulted in the completely burnished surfaces of pure titanium blanks.

Keywords: fine blanking; galling-free; high-speed steel punch; titanium; carbon supersaturation; free carbon tribofilm; in situ solid lubrication; high qualification of blanks

Citation: Aizawa, T.; Fuchiwaki, K. Galling-Free Fine Blanking of Titanium Plates Using Carbon-Supersaturated High-Speed Steel Punch. C **2023**, 9, 15. https://doi.org/10.3390/c9010015

Academic Editor: Gil Gonçalves

Received: 28 December 2022
Revised: 19 January 2023
Accepted: 22 January 2023
Published: 25 January 2023

1. Introduction

Fine blanking has been widely utilized in various manufacturing fields, especially in automotive industries [1]. The punched-out product has a fully burnished surface without geometric distortion since the work plate is fixed under the compressive stress state [2]. However, this strict constraint of work plates to the die set, results in the shortage of punch life [3], e.g., the chipping of the punch edge corners, the abrasive wear of the punch edge and the plastic distortion of the punch edge profile. These defects have a direct influence on product quality. The severity of wear at the punch edges is enhanced when fine blanking high-strength ductile metals and alloys. In particular, adhesion wear, or galling, often occurs and significantly shortens the punch life; the fresh fragments of work materials adhere to the punch edges and side surfaces, and their oxide debris particles deposit onto the punch surface [4]. As stated in [4,5], various coatings, including Si-bearing diamond-like carbon (DLC-Si), experience high friction and the sticking of titanium fragments. Most bulk ceramic tools, such as ZrO_2, Al_2O_3, SiC and Si_3N_4, have no capacity to improve the deep drawability against the titanium sheets [6].

Solid lubrication provides a way to reduce friction and wear in general [7]. Considering that galling occurs at the hot spot under the highly stressed interface between the tool and titanium work materials, the solid lubricants, such as h-BN and graphitic powders, must be deposited to form their tribofilms on those hot spots. In nature, this solid lubrication has a risk of not forming a reliable enough tribofilm to prevent punch and die surfaces from galling [8]. The in situ solid lubrication of carbon-supersaturated (CS-) steel dies has been proposed to reduce the friction and adhesive wear of dies during dry and cold forging, even at a high reduction level of thickness [9–11]. Free-carbon solutes diffuse from CS-steel dies, agglomerate themselves and form a free-carbon tribofilm on the hot spots of the contact interface between the CS-tool surface and titanium work material [11,12]. Through this in situ solid lubrication, the low friction and low work hardening of titanium materials have been sustained through the forging process [13,14] and the fine-blanking process [15].

In the present paper, in situ slid lubrication via free-carbon tribofilm formation is investigated in the fine-blanking of thick titanium plates by using CS-high-speed steel punches. Both bare and CS-high-speed punches are prepared for these fine-blanking experiments in order to describe the effect of in situ solid lubrication on the adhesive wear behavior of punches. In particular, Scanning Electron Microscopy (SEM)–Electron-Dispersive X-ray Spectroscopy (EDX) and a chemical component analysis are performed to analyze the contact interface between the CS-punch and titanium work material. The difference in the in situ-formed free-carbon tribofilms caused by the upsetting and fine-blanking processes is discussed to consider the effect of applied stresses during metal forming on in situ solid lubrication.

2. Materials and Methods

A low-temperature plasma carburizing system was utilized for the carbon supersaturation of punches. A fine-blanking system was used to punch out pure titanium plates under normal operating conditions. SEM–EDX and a chemical composition analysis were employed to characterize the tribofilm that was in situ deposited onto the punch surfaces.

2.1. Carbon Supersaturation Process

The plasma carburizing system (YS-Electric Industry, Co., Ltd., Kofu, Japan) is schematically illustrated in Figure 1a. The punch was located laterally in the hollow cathode. With reference to the previous studies in [16,17], this hollow cathode device was employed to increase the carbon-ion and CH-radical densities to more than 3×10^{17} ions/m^3 and to increase the carbon-supersaturated layer thickness. This was estimated to be 40 mm in the case of plasma carburizing at 673 K for 14.4 ks by 70 Pa.

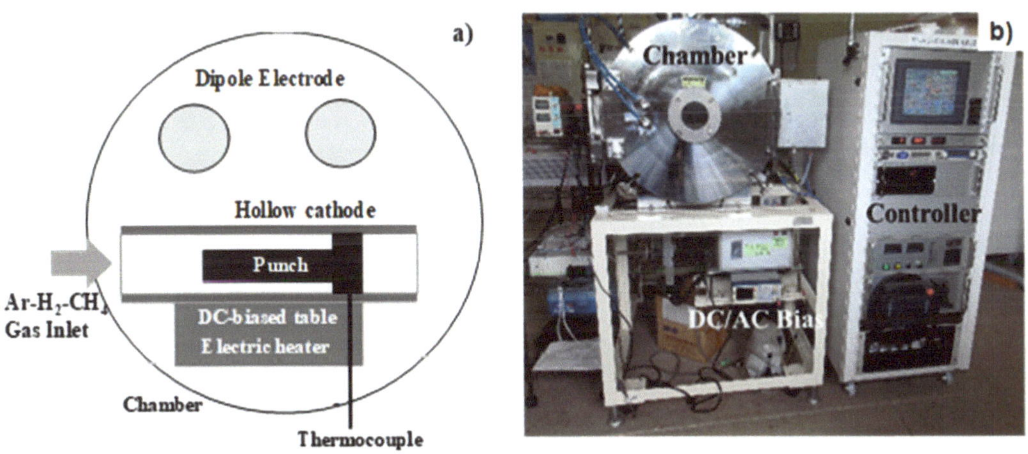

Figure 1. Carbon supersaturation processing system using low-temperature plasma carburizing: (**a**) schematic view of the carbon supersaturation into steel punches and (**b**) overview of the plasma carburizing system.

Dipole electrodes were utilized to ignite the radio-frequency (RF) plasma, while the direct-current (DC) plasma was also induced by applying the bias voltage to the bottom of the hollow cathode. In the following experiments, the RF voltage and the DC bias were constant at +220 V and −400 V, respectively. After evacuation down to the base pressure of 0.01 Pa, argon gas was introduced into a chamber, as shown in Figure 1b, at room temperature (RT) to clean the punch and die surfaces. After increasing the process temperature to 673 K under the argon atmosphere, hydrogen gas and argon gas were also introduced at flow rates of 20 mL/min and 160 mL/min, respectively. The total pressure was constant at 70 Pa.

After presputtering by the DC plasmas for 1.8 ks, methane gas was introduced as a carbon source into the argon and hydrogen gas mixture at a flow rate of 20 mL/min. At the specified duration of 14.4 ks, the specimen was cooled down in the chamber under a nitrogen atmosphere before evacuation down to atmospheric pressure.

The processing temperature was in situ monitored using a thermocouple, which was embedded into the base plate below the hollow cathode device, as shown in Figure 1a. The total pressure and each flow rate of the argon, hydrogen and methane gasses were also measured for process control. The deviation of temperature and pressure during operation was $\pm$ 0.1 K and $\pm$ 0.05 Pa, respectively.

Afterward [10,13,14], no iron or chromium carbides were synthesized in the plasma-carburized layer. The peak shift of a-ion to the lower 2q angles in the XRD analysis revealed that the a-lattices in the carburized steels expanded by themselves through the supersaturation of the carbon solutes into steel punches.

2.2. Fine-Blanking System

A pure titanium plate with a thickness of 2.0 mm was blanked with a narrow clearance of 4% as illustrated in Figure 2a. A mechanical stamper (FB 160-FDE; Mori Steel Works Co., Ltd.; Saga, Japan), specially accommodated for the fine-blanking process, was used for this experiment, as shown in Figure 2b.

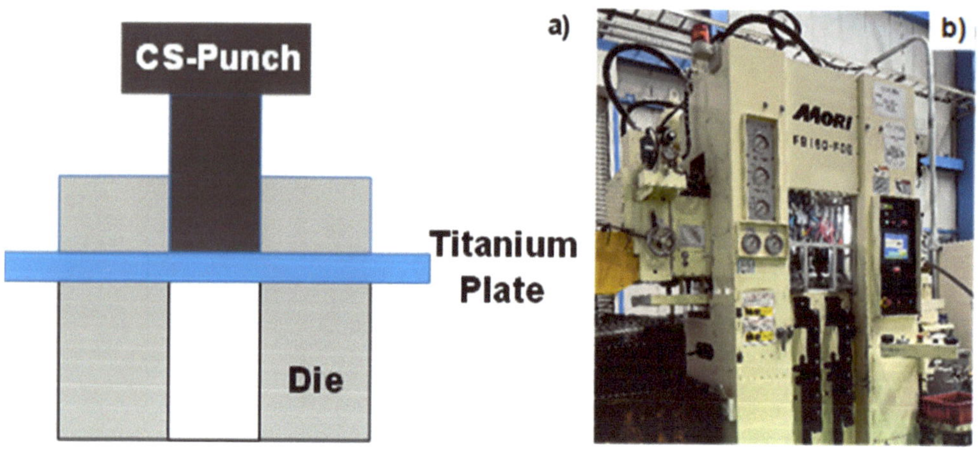

Figure 2. Fine-blanking system: (**a**) schematic view of the fine-blanking process and (**b**) overview of the fine-blanking system.

The maximum loading capacity was 1600 kN. The loading sequence for the fine blanking was CNC-programmed. The punches with and without carbon supersaturation were used to describe the effect of the carbon supersaturation on the galling behavior. Each punch was fixed into a punch holder, which was further set up in the upper die set. In the following experiments, FBH9-HMC, with a viscosity of 101 m^2/s, was utilized as a lubricating oil. A high-speed steel type YXR7 punch is depicted in Figure 3. This high-speed steel had the following chemical composition: carbon by 0.8 mass%, silicon by 0.8 mass%, manganese by 0.3 mass%, chromium by 4.7 mass%, tungsten by 1.3 mass%, molybdenum by 5.5 mass%, vanadium by 1.3 mass% and iron in balance.

Figure 3. High-speed steel type YXR7 punch for fine blanking the square hole.

2.3. Materials

Pure titanium wires and plates were utilized as work materials for the upsetting and fine-blanking experiments, respectively. Their chemical compositions consist of hydrogen by 0.0012 mass%, oxygen by 0.097 mass%, nitrogen by 0.007 mass%, iron by 0.042 mass%, carbon by 0.007 mass% and titanium for balance.

2.4. Characterization

SEM–EDX (JOEL, Co., Ltd.; Kanagawa, Japan) was used for a microstructure analysis and element mapping of the contact interface between the forging and fine-blanking tools and the titanium work materials. The chemical components of the tribofilm were analyzed pointwise at the selected positions on the punch side surfaces.

3. Results

Bare and CS-YXR7 punches were utilized for the fine blanking of pure titanium plates. The adhesive wear of titanium fragments deposited onto the punches was compared between the two punches. In particular, the contact interface between the CS-YXR7 punch and the titanium work material was analyzed to describe the formation of free-carbon tribofilms at the contact interface of the CS-YXR7 punch. The punched-out blanks were compared between the two punches to investigate the effect of in situ solid lubrication on the quality of the punched-out blanks.

3.1. Fine Blanking of Pure Titanium Plates Using Bare YXR7 Punch

The bare YXR7 punch before fine blanking, as shown in Figure 3, was compared to the punch after the fine blanking of pure titanium plates, with a thickness of 2 mm, in a single shot, as shown in Figure 4. No wear was seen on either of the punch heads after the fine blanking in a single shot. However, their side surfaces, with a length of around 2 mm, were almost covered by the fragments of the titanium work material. In particular, as seen in Figure 4b, the titanium fragments adhered to and eroded into the YXR7 punch, causing detrimental damage to the punch.

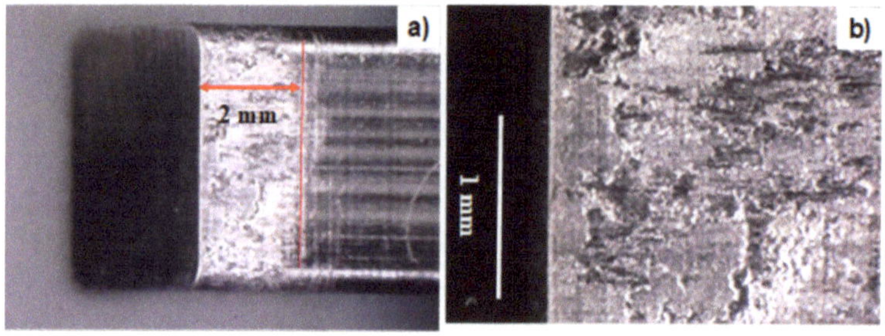

Figure 4. The bare YXR7 punch just after fine blanking a pure titanium plate with a thickness of 2 mm in a single shot: (**a**) overview of the bare YXR7 punch and (**b**) XYR7 punch side surface with severe adherence of titanium fragments.

It is difficult to ground and polish the debris particles that adhere to the punch surfaces during fine blanking. This galling behavior proves that the freshly sheared titanium work material adheres easily onto punch side surfaces under high normal and shear stress conditions and narrow clearance during fine blanking, even under lubricating conditions. When fine blanking the carbon steels, the lubricating oil worked well to protect the YXR7 punches from abrasive wear, even under the same fine-blanking conditions. This severe adhesive wear is intrinsic to the fine blanking of pure titanium plates.

3.2. Fine Blanking of Pure Titanium Plates Using CS-YXR7 Punch

A CS-YXR7 punch was used for the fine blanking of AISI316 plates with a thickness of 3 mm in three shots and then for the fine blanking of pure titanium plates with a thickness of 2 mm in two shots. Figure 5a shows an overview of the CS-YXR7 punch after a series of fine-blanking shots. In contrast to the severe adhesion of titanium fragments to the bare YXR7 punch as shown in Figure 4a,b, the CS-XYR7 punch is completely free from the galling of the sheared titanium work material. As shown in Figure 5a, a thin adhesive film is formed on each side surface only within the region from the punch head to 2 mm away from the punch edge. This implies that a tribofilm formed on the punch side surfaces to prevent them from severe galling by titanium fragments. Figure 5b shows the punch edges and corners after fine blanking. No significant wear was seen on these punch edges and corners.

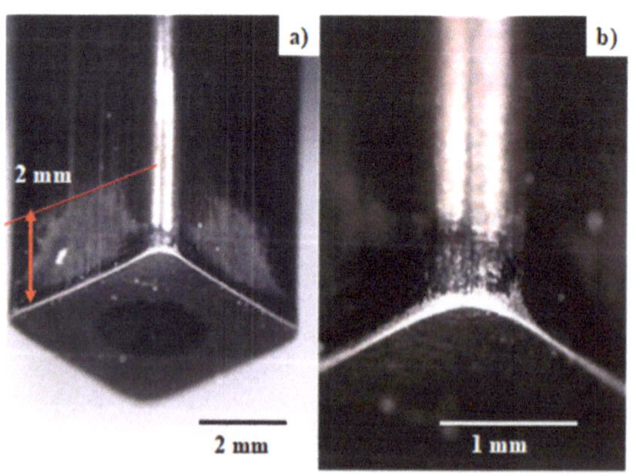

Figure 5. The CS-YXR7 punch after fine blanking AISI304 plates with a thickness of 3 mm in three shots and pure titanium plates with a thickness of 2 mm in two shots: (**a**) overview of CS-YXR7 punch after fine blanking and (**b**) the punch edges and punch corner.

3.3. SEM-EDX Analysis of the Contact Interface

The contact interface between the CS-YXR7 punch and the titanium work material was analyzed by using SEM-EDX to describe the microstructure of the thin tribofilm that formed on the punch side surfaces. At first, SEM and Yttrium Aluminum Garnet (YAG) scintillator images were utilized to describe the tribofilm that formed from the punch edge to a position 1 mm away from this edge. During the fine-blanking process, the punch edge often suffers from hard contact with work materials under high-stress conditions. This SEM-YAG observation around the edge was expected to describe the tribofilm that forms under severe contact conditions. However, the SEM-YAG observation of the middle of the contact interface describes the stationary formed tribofilm under normal and shear stresses.

As depicted in Figure 6, a tribofilm was formed even from the punch edge in the shearing direction on the punch side surface. A few grooves were seen on this surface; these were cut into the tribofilm when the punch was ejected from the punch holder. This suggests that this tribofilm was thin and soft enough to be cut into.

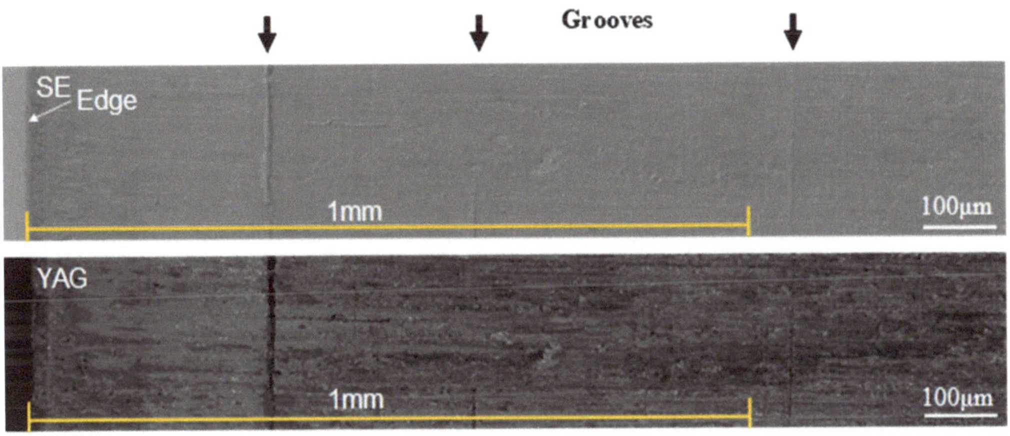

Figure 6. SEM and YAG images of the punch side surface from the punch edge to a position 1 mm away from this.

SEM and YAG analyses were performed around the punch edge and at a particular position away from this edge at X = 1 mm. Figure 7 compares the microstructures analyzed at both positions. Adhesive films, indicated by "a" in Figure 7, formed as a layer with a length of 20 mm and a width of 50 mm in the vicinity of the punch edge. This layer formed long stripes, formed along the axial direction or in the X-axis.

SEM-EDX was also utilized for element mapping. Figure 8 compares the iron, the titanium and the carbon maps of the contact interface. Fine titanium layers were detected in the vicinity of the punch edge. Away from the punch edge, titanium was detected as stripes along the X-axis. Figures 7 and 8 reveal that the black zones, as indicated by "a" in the YAG images, correspond to the titanium maps and that everything apart from the a-zones correspond to the original CS-YXR7 punch side surface. This demonstrates that the adhesion of the titanium debris fragments during the fine blanking was significantly reduced at the contact interface between the CS-YXR7 punch and the titanium work material. The carbon map is also depicted in Figure 8; the carbon concentration was not clearly detected by this wide scanning.

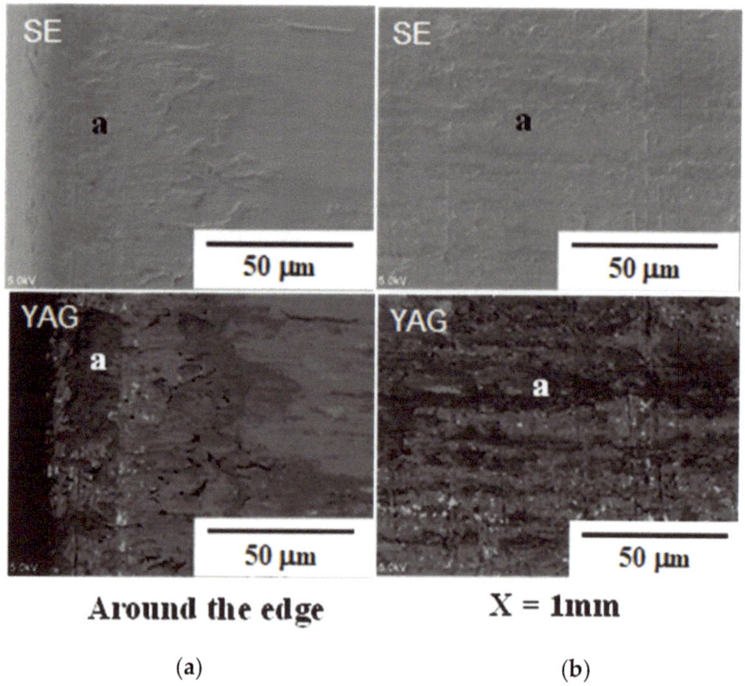

Figure 7. SEM and YAG images of the punch side surface with a higher magnification around the punch edge and at x = 1 mm from this edge: (**a**) SEM and YAG images around the punch edge and (**b**) SEM and YAG images at x = 1 mm from the punch edge.

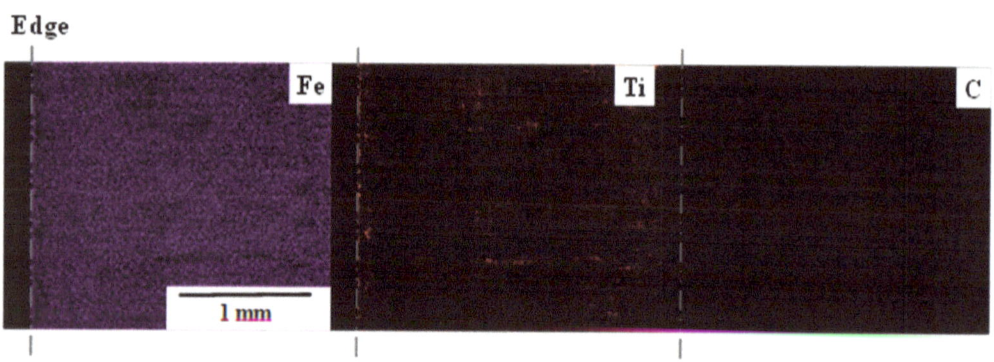

Figure 8. Element mapping of the punch side surface around the punch edge.

3.4. Chemical Component Analysis

The chemical components were analyzed in the vicinity of the punch edge and at a position 1 mm away from the punch edge at the contact interface between the CS-YXR7 punch and the titanium work material. EDX was also utilized for this pointwise analysis. Table 1 compares the analyzed chemical compositions at both positions with reference to the original chemical compositions of the YXR7 substrate.

Table 1. Comparison of the chemical compositions analyzed in the vicinity of the edge and at a position 1 mm away from the punch edge at the contact interface between the CS-YXR7 punch and the titanium work material, with reference to the chemical compositions of the bare YXR7.

Chemical Element	In the Vicinity of Punch Edge (mass%)	1 mm Away from the Edge (mass%)	Bare YXR7 Substrate (mass%)
Carbon	1.49	3.38	0.8
Oxygen	0.58	0.60	—
Silicon	0.20	0.59	0.8
Tungsten	2.62	6.65	1.3
Phosphorous	0.02	0.06	—
Molybdenum	1.63	3.48	5.5
Sulfur	0.13	0.23	—
Titanium	34.67	4.31	—
Vanadium	0.76	1.42	1.3
Chromium	3.40	3.94	4.7
Manganese	0.07	0.24	0.3
Iron	53.99	75.03	85.3
Nickel	0.44	0.06	—

Phosphorus, sulfur and nickel were found to be impurities in the analysis. Since the fine blanking was performed at RT in the cold, the oxygen content remained sufficiently low and had nothing to do with the tribofilm formation. A high titanium content was detected in the vicinity of the punch edge; titanium debris fragments adhered to the punch edge. At a position 1 mm away from this punch edge, the detected contents of molybdenum, chromium, vanadium and iron were similar to their original concentrations in the YXR7 substrate, together with a significant reduction in the titanium content. This reveals that the thickness and width of the adhesive titanium layer significantly reduced along the X-direction from the punch edge in accordance with the reduction in the titanium content, as shown in Figures 7 and 8.

Notably, the carbon content increased in the X-direction; a thin carbon film formed on the punch surface. As shown in Figures 6 and 7, this carbon film formed on the punch side surface but not in the a-zones. These films also formed as stripes along the punching direction, together with the thin titanium deposits.

3.5. Estimate of the Tribofilm Thickness

We estimated the free-carbon tribofilm thickness. The original surface roughness of the CS-YXR7-punch was reduced by the formation of a tribofilm on the sheared surface of the punch. Hence, this tribofilm thickness was estimated by measuring the maximum surface roughness on the outside and inside of the sheared punch surface. InfiniteFocusG5 (EuroTechno; Tokyo, Japan) was utilized to measure the roughness profile along the length of 160 mm at positions 1 mm and 5 mm away from the punch edge. The maximum roughness, as an average of the four punch side surfaces, was 0.366 mm at 5 mm and 0.218 mm at 1 mm. The tribofilm thickness was estimated to be around 0.15 mm or 150 nm.

3.6. Quality Evaluation of the Punched-Out Blanks

The galling-free behavior suggests that the quality of the punched-out blank is improved by the fine-blanking process. Figure 9 depicts the top and side surfaces of the punched-out titanium blank. No fractured regions are detected on any of the four side surfaces of the blank. This proves that the CS-YXR7 punch is highly qualified and suitable for the fine blanking of titanium plates to produce products.

Figure 9. Observation of the sheared square blank using the CS-YXR7 punch: (**a**) top view of punched-out titanium blank and (**b**) side surface of punched-out titanium blank.

A three-dimensional profilometer (Keyence; Tokyo, Japan) was utilized to measure the surface profile of the sheared surface of the titanium blanks, as shown in Figure 9b. A three-dimensional profile was developed from the measured optical microscopic images. A bird's eye view of the developed profile is shown in Figure 10. No fractured areas were detected, even at the trace level, by this microscopic analysis of the sheared surface of the titanium blank.

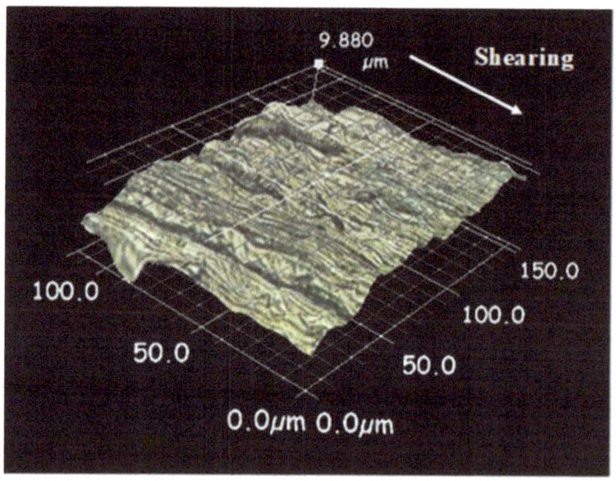

Figure 10. Three-dimensional surface profile of the sheared square blank using the CS-YXR7 punch.

4. Discussion

The fine blanking of titanium and titanium alloys, oxygen-free copper alloys and austenitic stainless steels is considered difficult [18]. Severe adhesive wear caused by the work material debris fragments deposited onto the punch and die is common in the fine-blanking processes of these materials, even under lubricating conditions. As shown in Figure 6, the high-speed steel type YXR7 punch suffered from severe galling with detrimental damage when fine blanking a pure titanium plate with a thickness of 2 mm, even when using lubricating oils. If a solid lubricating tribofilm is not present on the hot spots at the contact interfaces between steel punches and these work materials, severe galling can never be prevented during the fine-blanking process. This is because high shearing and normal stresses applied on the punch side surfaces force titanium debris fragments to adhere to them. The carbon supersaturation of a tool-steel punch and die is a unique solution for the in situ formation of a tribofilm at the contact interface and preventing the punch and die from severe galling.

In conventional solid lubrication, graphite or h-BN powders are deposited or sprayed as a sufficiently thin layer onto the fine-blanking punch surfaces with the appropriate interval of shots. The punch experiences high shearing stress transient in every shot; most of the deposited or sprayed powders are at risk of delamination due to the punch side surfaces inducing partial galling via titanium debris particles. As shown in Figures 5–7, a thin film with a thickness of 0.15 mm only formed on the sheared area of the punch side surfaces in this study. This tribofilm worked as a solid lubricating layer throughout the fine-blanking shots without intermission. Since the formed tribofilm was preserved through the fine-blanking process, no additional depositions or sprays were necessary.

As studied in [9–14], the free-carbon tribofilm was in situ-formed by the agglomeration of diffusing carbon solutes from the CS-punch and CS-die. A high stress gradient during the dry forging processes induced this carbon solute diffusion and agglomeration. In the fine blanking with a narrow clearance, the CS-punch side surfaces experienced high shearing and normal stress gradients, so the supersaturated carbon solutes in the CS-YXR7 punch diffused to the punch surfaces and agglomerated on the surfaces to form a free-carbon tribofilm.

The free-carbon tribofilm formation analyzed in Figures 7 and 8 and Table 1 is commonly observed in both fine-blanking and forging processes using CS-punches. As stated in [12–14], free-carbon layers formed in the direction of the metal flow at the contact interface between the CS-die and the titanium and b-phase titanium alloy work materials. Only carbon agglomerates were detected at the hot spots of the contact interface when using Raman spectroscopy. These findings are also confirmed by the SEM-YAG analyses in Figure 7, the element mapping in Figure 8 and the chemical composition analysis in Table 1. As shown in Figure 4, no iron element, which is a constituent of the YXR7 punch, was detected on the punch surface galled by titanium debris particles. However, the iron element was uniformly distributed on the punch side surfaces in Figure 8; the titanium element was slightly detected in the vicinity of the punch edge. This reveals that the adhesion of the titanium work material fragments was significantly reduced on the CS-YXR7 punch surfaces. Carbon was uniformly distributed on the sheared surface of the punch in Figure 7. After the chemical composition analysis, as shown in Table 1, a higher carbon content was detected on these punch surfaces. Remembering that YXR7 is a low-carbon high-speed steel, this high carbon concentration, detected at the interface between the CS-YXR7 punch and the titanium work material, reveals that free-carbon solute diffused from the CS-YXR7 agglomerates on the highly stressed hot spots at the contact interface and worked as a solid lubricant to protect the YXR7 punch substrate from adhesion wear.

As keenly studied in [13,14], this free-carbon tribofilm works to significantly reduce the friction at the contact interface and to lower the work hardening of titanium and b-phase titanium alloy materials during the forging process with a reduction in thickness higher than 70%. In the case of the fine-blanking process with a CS-punch and die, the friction coefficient also decreases by preventing the adhesion of work material fragments and by preserving the narrow clearance. As shown in Figures 9 and 10, the surface roughness of the ground and polished CS-punch was transcribed onto the blank surfaces without fractured zones. This fully burnished blank surface proves that the low frictional state was preserved during the shearing process of the fine blanking.

DEFORM-3D was utilized to estimate the normal and shear stress distributions in the punch [19]. The punch edge experienced stresses higher than 2750 MPa when fine blanking the AISI304 stainless steels with a thickness of 3 mm. With a narrow clearance during the fine blanking of the titanium plates, as shown in Figure 2, the shearing stress around the YXR7-punch achieved a level equivalent to that of the shear yield stress of titanium.

As discussed in [12], the normal and shear stresses applied to the contact interface drive the supersaturated carbon in the CS-punch to diffuse and agglomerate onto the contact interface. During forging, the normal compressive stress is mainly applied onto the contact interface, the area of which increases with a more severe reduction in thickness. However, normal compressive stress and severe shear stress are also applied to the contact interface, the area of which is constant during fine blanking. This difference in the applied stress level and the geometric configuration must have an influence on the in situ solid lubrication via the free-carbon tribofilm formation. In the free-carbon mapping, free carbon was distinctly detected at the contact interface as a stripe pattern in the case of forging. However, the free carbon was distributed slightly but uniformly at the contact interface in the case of fine blanking. This difference in carbon mapping is due to the normal stress effect on the isolation process of the free-carbon solutes from the CS-YXR7 substrate, as well as the larger contact interface area. That is, the normal stress applied to the contact interface was lower than that applied during forging processes. This lower stress gradient at the interface resulted in a lower content of free-carbon agglomerates on a larger contact area, as demonstrated by the EDX analysis results in Figure 8c, where the free carbon tribofilm was not distinctly detected by EDX in wide scanning.

The fully burnished surface and dimensional accuracy characterize fine-blanked products as suitable for mechanical parts and medical tools. In conventional fine-blanking processes, these most attractive aspects are diminished when fine-blanking high-strength and ductile work materials, such as titanium and titanium alloys, oxygen-free copper alloys and stainless steels. This CS-punch and die technology is expected to provide a new solution to galling-less fine blanking in order to make the net shaping of titanium and titanium alloy gears for robotic arms, to punch out copper electrodes for fuel-cell automotive parts and to yield austenitic stainless-steel pedals for braking units.

5. Conclusions

Carbon supersaturation (CS) in tooling via plasma carburizing at 673 K for 14.4 ks provides a solution to galling-less fine blanking against titanium and titanium alloy work materials, which have been identified as being difficult-to-work materials. A bare high-speed steel punch suffered from severe galling on its side surfaces only after fine blanking in a single shot. The titanium work material fragments adhered to the punch surfaces during severe normal and shearing stresses. Titanium debris fragments were slightly detected only around the CS-punch edges, even after repeatedly fine blanking the titanium plates. A carbon content higher than 3 mass% was detected on the sheared surface of the CS-punch, proving that a free-carbon tribofilm was in situ-formed on the CS-punch side surfaces to prevent the high-speed punch substrate from adhesive wear and to significantly reduce friction and wear damage.

This in situ solid lubrication was reflected in the quality of the punched-out blanks. They were completely free from geometric distortion, which is often induced by galling behavior when fine blanking titanium plates. They had fully burnished surfaces with a well-defined thickness and parallelism, suitable for automotive parts and medical tools. In particular, this geometric accuracy is suitable for lightweight, high-strength gears with a high chemical stability, as well as bio-compatible medical tools.

Author Contributions: Conceptualization, T.A. and K.F.; methodology, T.A.; software, K.F.; validation, T.A. and K.F.; formal analysis, K.F.; investigation, T.A. and K.F.; resources, T.A. and K.F.; data curation, K.F.; writing—original draft preparation, T.A.; writing—review and editing, T.A. and K.F.; visualization, K.F.; supervision, T.A.; project administration, T.A.; funding acquisition, K.F. All authors have read and agreed to the published version of the manuscript.

Funding: This research received no external funding.

Data Availability Statement: Not applicable.

Acknowledgments: The authors would like to express their gratitude to S. Kobayashi (Hatano Precision, Co., Ltd.) for his help in the fine-blanking experiments and to Y. Aihara, S. Hirabayashi and Y. Suzuki (Komatsu-Seiki Kosakusho, Co., Ltd.) for their help in the analyses.

Conflicts of Interest: The authors declare no conflict of interest.

References

1. Kuhlman, G.W. Forging of Titanium Alloys. In *Metalworking: Bulk Forming*; ASM International: Novelty, OH, USA, 2005; Volume 14.
2. Chandrasekaran, M. Forging of metals and alloys for biomedical applications. In *Metals for Biomedical Devices*, 2nd ed.; Woodhead Publishing: Sawston, UK, 2019; pp. 293–310.
3. Nielsen, C.V.; Martins, P.A.F. *Metal Forming: Formability, Simulation and Tool Design*; Elsevier: Amsterdam, The Netherlands, 2021.
4. Kihara, T. Visualization of deforming process of titanium and titanium alloy using high speed camera. In Proceedings of the Japanese Society of Technology of Plasticity, Kyoto, Japan, 9 June 2019; pp. 41–42.
5. Dohda, K.; Aizawa, T. Tribo-characterization of silicon doped and nano-structured DLC coatings by metal forming simulators. *Manuf. Lett.* **2014**, *2*, 82–85. [CrossRef]
6. Kataoka, S.; Murakawa, M.; Aizawa, T.; Ike, H. Tribology of dry deep-drawing of various metal sheets wit use of ceramic tools. *Surf. Coat. Technol.* **2004**, *174*, 582–590. [CrossRef]
7. Hutchings, I.; Shipway, P. *Tribology: Friction and Wear of Engineering Materials*, 2nd ed.; Elsevier: Oxford, UK, 2017.
8. Dohda, K.; Yamamoto, M.; Hu, C.; Dubar, L.; Ehman, K.F. Galling phenomena in metal forming. *Friction* **2020**, *9*, 686–696. [CrossRef]
9. Aizawa, T.; Yoshino, T.; Suzuki, Y.; Shiratori, T. Anti-galling cold dry forging of pure titanium by plasma carburized AISI420J2 dies. *Appl. Sci.* **2021**, *11*, 595. [CrossRef]
10. Aizawa, T.; Yoshino, T.; Suzuki, Y.; Shiratori, T. Free-forging of pure titanium with high reduction of thickness by plasma carburized SKD11 dies. *J. Mater.* **2021**, *14*, 2536. [CrossRef] [PubMed]
11. Aizawa, T.; Ishiguro, S.; Shiratori, T.; Yoshino, T. Near-net forging of titanium and titanium alloys by the plasma carburized SKD11 dies. *Key Eng. Mater.* **2022**, *926*, 1143–1150. [CrossRef]
12. Aizawa, T.; Shiratori, T. In situ solid lubrication in cold dry forging of titanium by isolated free carbon from carbon- supersaturated dies. *Friction*, 2022; *in press*.
13. Ishiguro, S.; Aizawa, T.; Funazuka, T.; Shiratori, T. Green forging of titanium and titanium alloys by using the carbon supersaturated SKD11 dies. *J. Appl. Mech.* **2022**, *3*, 724–739. [CrossRef]
14. Aizawa, T.; Funazuka, T.; Shiratori, T. Near-net forging of titanium and titanium alloys with low friction and low work hardening by using carbon-supersaturated SKD11 dies. *Lubricants* **2022**, *10*, 203. [CrossRef]
15. Aizawa, T.; Fuchiwaki, K.; Dohda, K. Galling-free fine blanking of titanium plates by carbon-supersaturated tool steel punch. In Proceedings of the 26th ESAFORM, Krakof, Poland, 18–21 April 2023. *in press*.
16. Aizawa, T. Low Temperature Plasma Nitriding of Austenitic Stainless Steels. In *Stainless Steels and Alloys*; IntechOpen: London, UK, 2019; pp. 31–50.
17. Geobel, D.M.; Becatti, C.; Mikellides, I.G.; Ortega, A.L. Plasma hollow cathodes. *J. App. Phys.* **2021**, *130*, 050902. [CrossRef]
18. Adamus, J.; Lacki, P.; Wieckowaski, W.; Tytanowych, D.B. Numerical simulation of the fine blanking process of sheet titanium. *Arch. Metall. Mater.* **2011**, *56*, 431–437. [CrossRef]
19. Fuchiwaki, K. Edge profile control of the fine blanking punches. In Proceedings of the 348th Symposium on the Metal Forming, Tokyo, Japan, 1 December 2022; pp. 21–26.

C Journal of Carbon Research

MDPI

Review

Avant-Garde Polymer and Nano-Graphite-Derived Nanocomposites—Versatility and Implications

Ayesha Kausar [1,2,3]

[1] NPU-NCP Joint International Research Center on Advanced Nanomaterials and Defects Engineering, National Centre for Physics, Islamabad 44000, Pakistan; dr.ayeshakausar@yahoo.com
[2] iThemba LABS, UNESCO-UNISA Africa Chair in Nanosciences/Nanotechnology, Somerset West 7129, South Africa
[3] NPU-NCP Joint International Research Center on Advanced Nanomaterials and Defects Engineering, Northwestern Polytechnical University, Xi'an 710072, China

Abstract: Graphite (stacked graphene layers) has been modified in several ways to enhance its potential properties/utilities. One approach is to convert graphite into a unique 'nano-graphite' form. Nano-graphite consists of few-layered graphene, multi-layered graphene, graphite nanoplatelets, and other graphene aggregates. Graphite can be converted to nano-graphite using physical and chemical methods. Nano-graphite, similar to graphite, has been reinforced in conducting polymers/thermoplastics/rubbery matrices to develop high-performance nanocomposites. Nano-graphite and polymer/nano-graphite nanomaterials have characteristics that are advantageous over those of pristine graphitic materials. This review basically highlights the essential features, design versatilities, and applications of polymer/nano-graphite nanocomposites in solar cells, electromagnetic shielding, and electronic devices.

Keywords: polymer; nano-graphite; nanocomposite; solar cell; electronics

Citation: Kausar, A. Avant-Garde Polymer and Nano-Graphite-Derived Nanocomposites—Versatility and Implications. *C* **2023**, *9*, 13. https://doi.org/10.3390/c9010013

Academic Editor: Gil Gonçalves

Received: 22 December 2022
Revised: 10 January 2023
Accepted: 16 January 2023
Published: 19 January 2023

1. Introduction

Natural graphite has stimulated dynamic research interest towards hybrid polymeric materials [1]. Various polymeric matrices have been characteristically reinforced by graphite filler: polyaniline/graphite, epoxy/graphite, polystyrene graphite, PMMA/graphite, and many others [2]. The inclusion of graphite in polymers has remarkably improved their electrical, thermal, mechanical, and other essential physical properties [3–5]. Pristine graphite- and polymer-based materials have been applied in electronics and energy devices, packaging, anticorrosion coatings, membranes, etc. [6]. However, the stacked layered structure of graphite has limited dispersion in polymers and ensuing final high performance. In this regard, various research attempts have focused on the altered forms of compact graphite structures.

Nano-graphite is an exceptional modified form of graphite, having a mixture of layered graphene, nanosheets, and nanoplatelet components [7]. Nano-graphite has been obtained from pristine graphite by using simple techniques, ranging from physical grinding and sonication to chemical conversion techniques. Similar to graphite, it has been reinforced in polymeric matrices in order to form polymer/nano-graphite nanocomposites [8,9]. However, the properties of the resulting nanocomposites have been found to be better than those of pristine-derived polymeric composites. In particular, nano-graphite has been suggested to possess enhanced dispersion in various polymeric matrices, matrix–nanofiller interactions, and end properties. The recent literature has shown the upgraded utility of nano-graphite-derived nanocomposites in technical application areas, including electronics, EMI, and solar cells [10]. However, no review article has been published in the literature to date regarding high-performance polymer/nano-graphite nanocomposites. Hence, this state-of-the-art review definitely contributes important knowledge towards the future design possibilities and methodological potential of novel polymer/nano-graphite nanocomposites.

2. Nano-Graphite

Graphite is a naturally abundant carbon allotrope [11]. It is made up of stacked layers of sp^2-hybridized carbon atoms, also known as graphene [12]. The graphene layers are usually held together through weak van der Waals forces. The parallel stacking of carbon layers may form a three-dimensional lattice structure.

Nano-graphite is a simply modified form of graphite. The structure of nano-graphite is actually a mixture of nanocarbon nanomaterials with graphitic nanoplatelets or nanosheets and layered graphene [13]. In nano-graphite, the few-layered graphene may consist of two to ten layers or nanosheets. Moreover, graphene nanoplatelets and few-layered graphite oxides also exist. The nanoplatelets may have a thickness of 30–50 graphene layers. Figure 1 displays a simple illustration of the nano-graphite structure. Nano-graphite has fascinating physicochemical characteristics. The electrical, thermal, and mechanical properties of nano-graphite have been found in between the graphite and graphene layers [14]. The preparation of nano-graphite from graphite has been performed through numerous straightforward approaches, such as mechanical exfoliation (cavitation [15], shear mixing [16], grinding [17], high-pressure homogenization [18], and microfluidization [19]) and sonication, as well as various thermal and chemical reduction methods. The mechanical exfoliation methods are preferred for the conversion of graphite to nano-graphite due to their low-cost, no use of toxic chemicals, and environmental friendliness [20]. Blomquist and co-workers [21] applied a metal-free aqueous method to obtain nano-graphite. The detonation technique, involving the use of oxygen-deficient explosives, has also been adopted to form nano-graphite [22]. Ioni et al. [23] applied high-power sonication as a single-step method to attain nano-graphite dispersion in various solvents. In another study, transmission electron microscopy, scanning electron microscopy, atomic force microscopy, X-ray diffraction, and Raman spectroscopy were used to examine the nano-graphite nanostructure. The nano-graphite had 30–50 graphene layers in its structure [24]. The lateral sizes of the nano-graphite platelets were found to be in the range of 300–500 nm. The thicknesses of the nano-graphite platelets were ~20–40 nm.

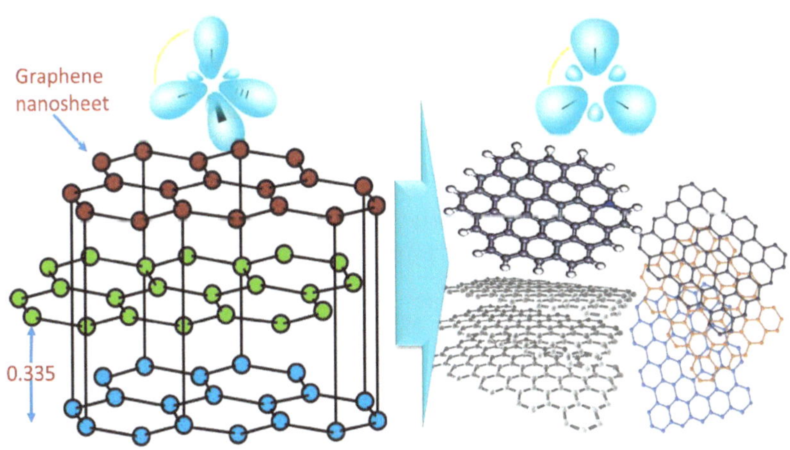

Figure 1. Nano-graphite.

Due to its unique nanostructure, the charge transference properties of nano-graphite have been found to be remarkably high, so it is useful for various related technical applications [25]. The applications of nano-graphite have been observed in thermal/electrical conductors, microwave absorbents, lubricants, coatings, absorbents, electromagnetic shields, gaskets, etc. In polymers, graphite has been applied as an electrically conducting and strengthening filler.

3. Polymer and Nano-Graphite-Derived Nanocomposites: Variations and Features

In bulk form, graphite is a layered material with a high electrical conductivity of ~10^6 S/m [26]. Graphite has been widely used as a filler in polymeric composites to improve their electrical, thermal, mechanical, and other physical properties. Polymer/graphite nanocomposites have been fabricated using facile approaches, such as in situ polymerization, solution blending, and melt processing [27]. The graphite dispersion in polymers depends on the choice of the processing technique [28]. It has been observed that graphite particles may develop an interconnecting network to support the percolation threshold of conductivity in the matrix. Nylon 6, polyamides, polystyrene, and other thermoplastics have been effectively filled with graphite using solution mixing in situ and melt routes. The fine graphite content and dispersion have led to the application of polymer/graphite composites in wide a range of applications, from automobile parts to electronics [29].

The performance of polymer/nano-graphite nanocomposites depends on the manufacturing techniques used and the related processing parameters [30]. Polymer/nano-graphite nanocomposites have been mostly processed using solution, melt, in situ, and other facile manufacturing techniques. The choice of technique determines the nanofiller dispersion, matrix–nanofiller interactions, and interface formation, affecting the enhancements in the final material properties [31]. Most importantly, the preparation method and the parameters determine the extent of nanofiller scattering in the polymers. Solution blending has the advantages of fine nanofiller dispersion and the homogeneous morphology of the nanocomposites; however, its major disadvantage is the use of toxic organic solvents [32]. The melt processing system can be considered ecofriendly; however, it may have a nanofiller aggregation problem [33]. The in situ technique has also been successfully used to form polymer/nano-graphite nanocomposites, and it can use eco-friendly solvents [34]. Therefore, the processing of polymer/nano-graphite nanocomposites by using an appropriate manufacturing strategy may form homogeneous microstructures with enhanced electrical, mechanical, thermal, and other physical properties [35].

3.1. Conducting Polymer/Nano-Graphite

Similar to graphite, the reinforcement behavior and features of nano-graphite have been investigated in nanocomposites. Consequently, nano-graphite has been filled in various polymers using suitable processing techniques, such as solution mixing, melt blending, and in situ polymerization [36]. Conducting polymers and graphite-based nanocomposites have been reported [37]. Most importantly, graphite has been frequently used to enhance the electrical conductivity of conjugated polymers [38]. Singhal et al. [39] designed a poly(3,4-ethylenedioxythiophene): poly(styrenesulfonate)- and nano-graphite-based nanocomposite using the solution method. To consistently disperse the nano-graphite, the swift heavy ion method was used with a Ni ion beam of 80 MeV. The swift heavy ion method actually breaks the nano-graphite accumulation due to van der Waals interactions. Figure 2 depicts the solution processing of the poly(3,4-ethylenedioxythiophene): poly(styrenesulfonate)/nano-graphite nanocomposite. After solution processing, the nanocomposite films were cast on a glass substrate through solution evaporation. Figure 3 displays the transmission electron microscopy images of the pristine nano-graphite and the nanocomposite. A cross-section image of the poly(3,4-ethylenedioxythiophene): poly(styrenesulfonate)/nano-graphite nanocomposite after irradiation showed that the nano-graphite stack had 7–8 graphene layers with a 2.6 nm thickness. Moreover, the nano-graphite had an inter-layer distance of 0.335 nm. Shinde and co-workers [40] prepared nano-graphite via planetary ball milling. Then, the polyaniline/nano-graphite nanocomposites were shaped in the form of pellets via cold pressing. The nanocomposites were characterized for their electrical conductivity and hardness. Mo et al. [41] sonicated expanded graphite in a 95% alcohol solution to obtain nano-graphite. Polypyrrole/nano-graphite nanocomposites were developed through in situ polymerization. The electrical conductivity of the polypyrrole/nano-graphite nanocomposites was enhanced with the nanofiller loading. The maximum electrical conductivity of 80.0 Scm^{-1} was achieved with a 3 wt.% nanofiller. Accordingly, the electrical conductivity of the conducting polymers was further boosted with graphite filler loading.

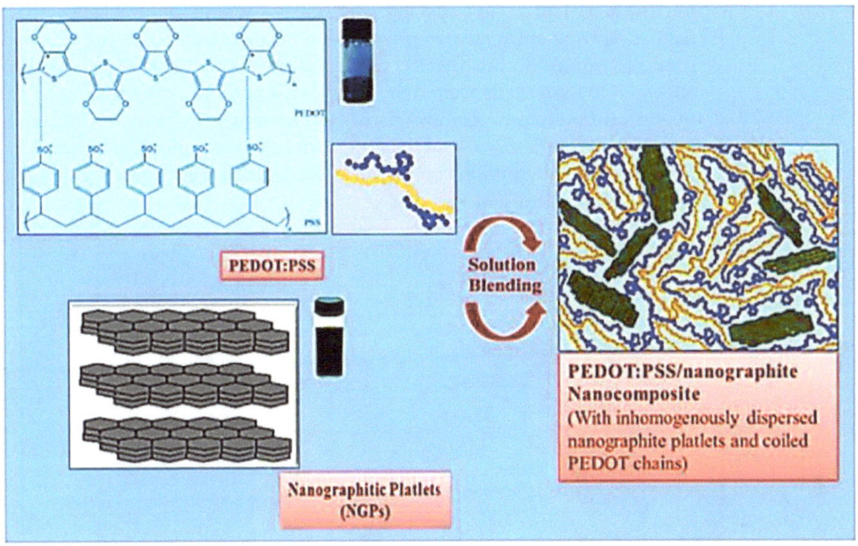

Figure 2. Schematic of poly(3,4-ethylenedioxythiophene):poly(styrenesulfonate)/nano-graphite nanocomposite formation using solution blending technique [39]. Reproduced with permission from ACS.

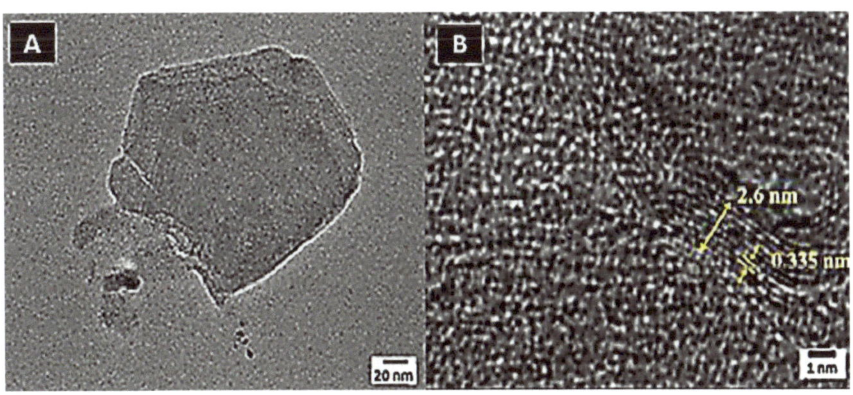

Figure 3. Transmission electron microscopy images of irradiated poly(3,4-ethylenedioxythiophene): poly(styrenesulfonate)/nano-graphite nanocomposite: (**A**) nano-graphite and (**B**) cross-section of nanocomposite showing 2.6 nm-thick nano-graphite stack with characteristic 0.34 nm inter-laminar spacing [39]. Reproduced with permission from ACS.

3.2. *Polystyrene/Nano-Graphite*

Polystyrene has been widely used as an important matrix for pristine graphite, similarly used with nano-graphite [42–44]. Xuemei et al. [45] adopted the detonation method to form nano-graphite. A nanoparticle size of 16 nm and a specific surface area of 583.6 $m^2 \cdot g^{-1}$ were attained by using the Brunauer–Emmett–Teller (BET) equation. The polystyrene/nano-graphite nanocomposite was prepared via Pickering emulsion polymerization, with azobisisobutylonitrile as the initiator in an aqueous solution. Figure 4 displays the optical micrographs of the polystyrene/nano-graphite Pickering emulsion microspheres. The micrographs clearly indicate that the polystyrene microspheres were homogeneously surrounded with a layer of detonation nano-graphite particles. The resulting polystyrene microspheres had a fine dispersity, spherical shape, and uniform size of up to 100 μm. The nanocomposite-encapsulated polystyrene microspheres were polydispersive due to the sonication emulsification method used. A large number of nano-graphite particles densely anchored to form shells around the polystyrene droplets. The use of nano-graphite instead of graphite led to the recompences of better interactions between the nanofiller particles and the polymeric microspheres. The better matrix–nanofiller association and interface formation directly influenced the nano-graphite dispersion and final physical property improvement of the nanocomposite.

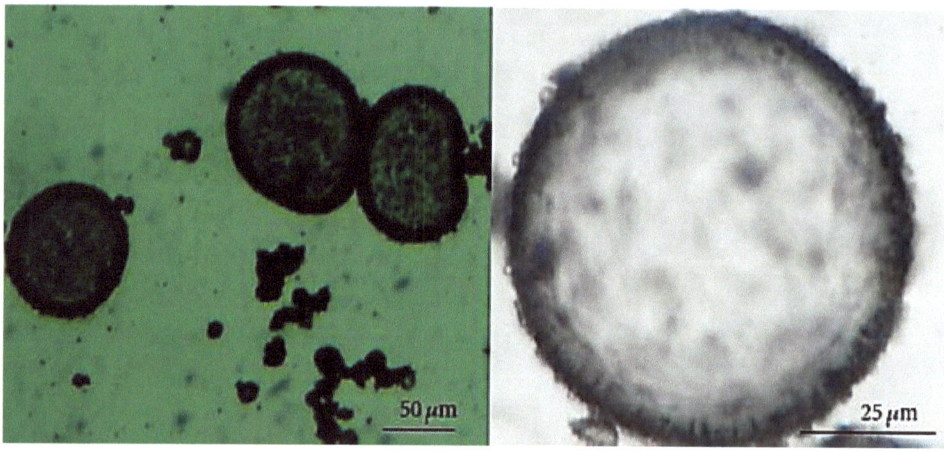

Figure 4. Optical picture of droplet of styrene-in-water Pickering emulsion stabilized by detonation nano-graphite particles [45]. Reproduced with permission from Hindawi.

3.3. *Poly(Methyl Methacrylate)/Nano-Graphite*

Poly(methyl methacrylate) has been applied as a vital thermoplastic matrix to form nanocomposites with neat graphite [46–48]. Likewise, nano-graphite nanofiller has also been reinforced in poly(methyl methacrylate) matrices. Singhi et al. [49] used commercially obtained nano-graphite of ~400 nm. A poly(methyl methacrylate)/nano-graphite nanocomposite was formed using the in situ polymerization of methylmethacrylate monomer and nano-graphite along with sonication. The nano-graphite was loaded in the range of 0.25–1.5 wt.%. The thermal stability of the neat poly(methyl methacrylate) and poly(methyl methacrylate)/nano-graphite nanocomposites was determined by conducting a thermogravimetric analysis (TGA) and differential scanning calorimetry (DSC). The results of the thermal studies are given in Table 1. According to TGA, the initial decomposition temperature of the poly(methyl methacrylate)/nano-graphite nanocomposites increased with the nanofiller loading up to 154 °C relative to that of the neat polymer (114 °C). The DSC thermograms also revealed improvements in the glass transition temperature of up to 114 °C compared with that of the neat polymer (110 °C). The increase in the thermal constancy and glass transition of the nanocomposite can be attributed to the polymer chain stiffness and stability due to the addition of the nano-graphite nanofiller. Soni et al. [50] produced a poly(methyl methacrylate)/nano-graphite nanocomposite by using the solution casting technique. The nano-graphite nanofiller was loaded in 1–5 wt.% contents. The nanocomposite was studied for its optical properties. The optical properties of the poly(methyl methacrylate)/nano-graphite nanocomposite were not affected by the nanofiller loading;

however, the optical properties of the films decreased with an increase in the temperature. The reason for this seems to be the decrease in the transmittance and energy bandgap at higher temperatures. Poly(methyl methacrylate)/nano-graphite nanocomposites have been fabricated through in situ and solution methods. The effect of nano-graphite loading has mostly been considered for the enhancement of optical and thermal properties.

Table 1. Results of thermal characterization of poly(methyl methacrylate)/nano-graphite nanocomposite [49]. Reproduced with permission from Wiley.

Nano-Graphite (wt.%)	Glass Transition Temperature (°C)	Initial Thermal Decomposition Temperature (°C)
Neat	110.6	114.5
0.25	107.8	97.72
0.50	112.1	153.9
0.75	116.6	132.5
1.0	112.0	126.2
1.5	–	88.67

3.4. *Poly(Vinyl Chloride)/Nano-Graphite*

Poly(vinyl chloride)- and graphite-based materials have gained considerable research interest [51–53]. Poly(vinyl chloride) is another thermoplastic matrix nanocomposite with nano-graphite. Zhang et al. [54] adopted commercial nano-graphite to form a nanocomposite with a titania and poly(vinyl chloride) matrix. The photocatalytic degradation mechanism of the poly(vinyl chloride) nanocomposite with nano-graphite and titania is demonstrated in Figure 5. The nano-graphite and titania were consistently distributed in the polymeric matrix and, thus, improved the conductivity and photocatalytic activity of the nanomaterial. Figure 6 shows that the photocatalytic degradation weight loss of the nanocomposite was significantly more enhanced than that of the neat poly(vinyl chloride) and pristine nano-graphite. The inclusion of the combination of nanocarbon and titania not only upsurged the photocatalytic efficiency due to electron/photon transfer but also promoted structural stability due to interfacial interactions. Although few poly(vinyl chloride)/nano-graphite systems have been explored to date, enhancements in dispersion, thermal, and photocatalytic behaviors have been observed with nanofiller loading.

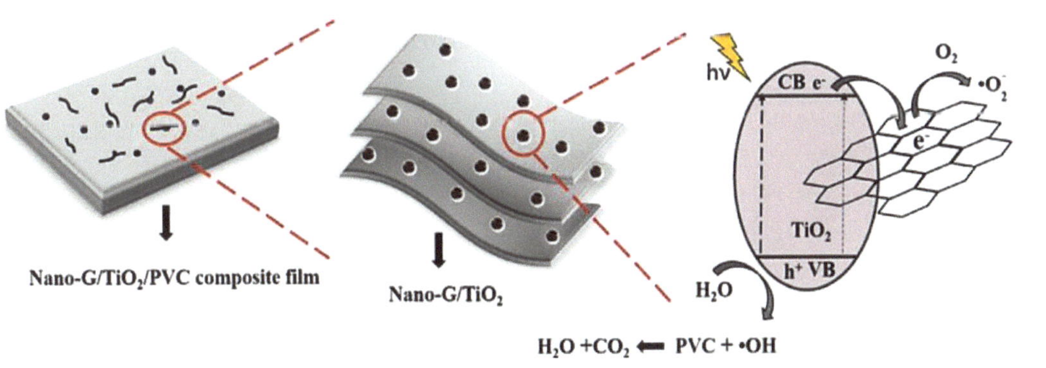

Figure 5. The photocatalytic degradation mechanism of the nanocomposite. Nano-G = nano-graphite; PVC = poly(vinyl chloride); Nano-G/TiO_2/PVC = nano-graphite/titania/poly(vinyl chloride) [54]. Reproduced with permission from Elsevier.

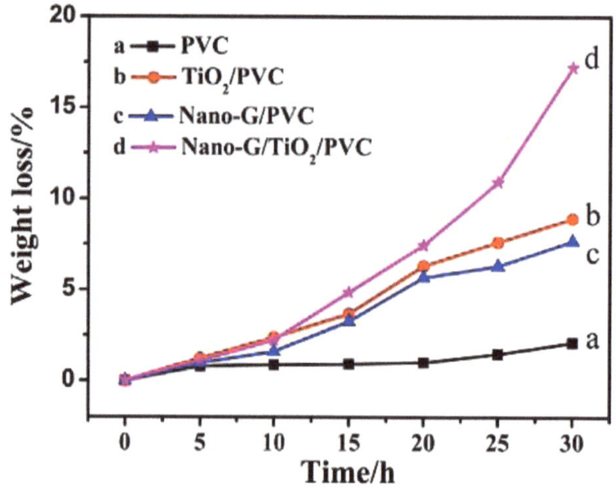

Figure 6. Photocatalytic degradation weight loss curves: (a) poly(vinyl chloride) (PVC); (b) titania/poly(vinyl chloride) (TiO_2/PVC); (c) nano-graphite/poly(vinyl chloride) (Nano-G/PVC); (d) nano-graphite/titania/poly(vinyl chloride) (Nano-G/TiO_2/PVC) [54]. Reproduced with permission from Elsevier.

3.5. Poly(Vinylidene Fluoride)/Nano-Graphite

Into the bargain, the literature reports on poly(vinylidene fluoride)- and graphite-based materials were examined [55–57]. Poly(vinylidene fluoride) has also been adopted to form a nanocomposite with a nano-graphite nanofiller. Li et al. [58] developed nanocomposites of poly(vinylidene fluoride) and nano-graphite via solution casting and compression molding. The nanofiller contents at the percolation threshold were efficient in improving the dielectric permittivity of the nanocomposite. Abdelaziz et al. [59] reported the development of a poly(vinylidene fluoride)-, hydroxypropyl cellulose-, and nano-graphite-based nanocomposite via the solution method. The addition of nano-graphite generated the β-phase polymorph in the poly(vinylidene fluoride) and matrix–nanofiller interface. Due to the β-phase and interface effects, the dielectric and thermal properties of the nanocomposite were positively affected. Again, in the case of the poly(vinylidene fluoride)/nano-graphite nanocomposite, the literature is limited. Further efforts are required in order to uncover the features and technical potential of poly(vinylidene fluoride)/nano-graphite nanocomposites.

3.6. Poly(Lactic Acid)/Nano-Graphite

The literature also reports poly(lactic acid)- and graphite-derived composites [60–62]. Correspondingly, research has led to the inclusion of nano-graphite in poly(lactic acid) nanocomposites. Gardella et al. [63] produced nano-graphite through the sonication of graphite in an ionic liquid, 1-butyl-3-methylimidazoliumhexa-fluorophosphate (80 °C). The ionic liquid was also used as a medium to disperse the nano-graphite in order to form a nanocomposite with poly(lactic acid). The poly(lactic acid)/nano-graphite nanocomposite was processed via melt blending. The ionic liquid solvent was stable and did not decompose during the nanocomposite processing. According to a scanning electron microscopy analysis, nano-graphite aggregates of 300 nm were observed in the matrix. Guo et al. [64] used a commercial nano-graphite with a diameter of 1.5–2.0 μm. Poly(lactic acid)/nano-graphite nanocomposites were designed via Fused Deposition Modeling. Figure 7 shows the thermal conductivity of the poly(lactic acid)/nano-graphite nanocomposites with varying nanofiller contents. The thermal conductivity increased with an increase in the nano-graphite content. The reason for this was the formation of transportation conducting pathways due to the nanofiller linking in the polymeric matrices. Owing to the high thermal/electrical conductivity, the poly(lactic acid)/nano-graphite nanocomposites were printed using Fused Deposition Modeling. Figure 8 shows the printed models on paper and flexible materials. The printed models had a fine flexibility and adhesion to the substrates due to fine matrix–nanofiller interactions and interface formation.

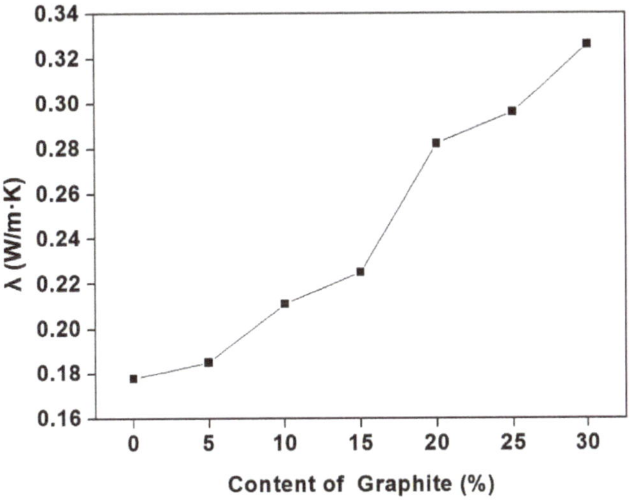

Figure 7. Thermal conductivity of poly(lactic acid)/nano-graphite nanocomposites with different nano-graphite loadings [64]. Reproduced with permission from MDPI.

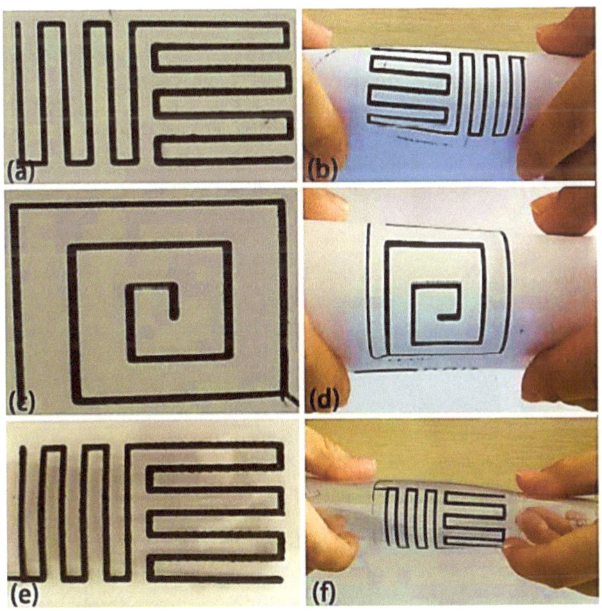

Figure 8. Models printed via Fused Deposition Modeling on different baseplates: (**a**,**c**) different shapes printed on paper; (**b**,**d**) the bend parts; (**e**) models printed on flexible materials; and (**f**) the bend parts [64]. Reproduced with permission from MDPI.

3.7. Polyurethane/Nano-Graphite

Polyurethane has also been applied as an important nanocomposite matrix. Mishra et al. [65] developed a polyurethane/nano-graphite nanocomposite for energy devices. The electrical conductivity of the polyurethane/nano-graphite nanocomposite with varying nanofiller contents is shown in Figure 9.

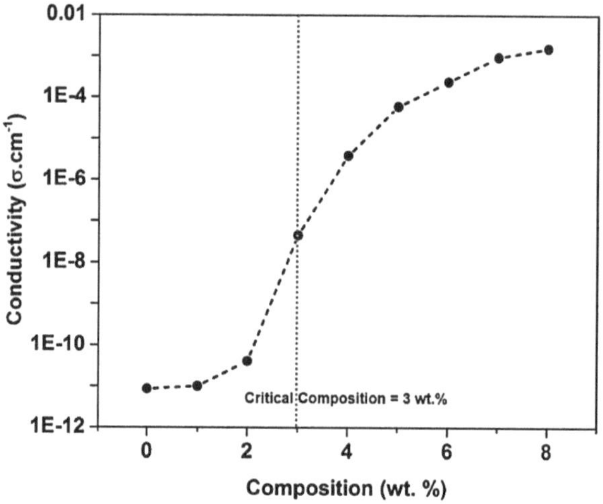

Figure 9. Variation in electrical conductivity of polyurethane/nano-graphite nanocomposite with varying nanofiller contents [65]. Reproduced with permission from Springer.

The percolation threshold of the nano-graphite in the polyurethane matrix was analyzed. The electrical conductivity of the polyurethane/nano-graphite nanocomposite was enhanced after the addition of 3 wt.% nanofiller, and it was stabilized after 5 wt.%. The 3 wt.% nano-graphite content was used as the percolation threshold value for energy device applications. Although there have been limited attempts to develop polyurethane/nano-graphite nanocomposites, the effect of the nanofiller on the conductivity properties points to the important technical significance of these nanomaterials.

3.8. Rubber/Nano-Graphite

Rubber- and graphite-based composites have gained vital importance due to their thermal/mechanical properties and engineering applications [66,67]. El-Nashar et al. [68] prepared nano-graphite using a poly(ethylene glycol)-based method. In acrylonitrilebutadiene rubber, the nano-graphite nanoparticles enhanced the mechanical properties (Young's modulus and hardness) and thermal stability of the matrix. According to dielectric studies, the permittivity and dielectric loss were found to increase with a rapid increase in the percolation threshold upon the addition of the nano-graphite. The acrylonitrilebutadiene rubber/nano-graphite nanocomposite was found to be useful for antistatic applications. Tiwari et al. [69] loaded nano-graphite in a chlorobutyl elastomer matrix. A nano-graphite loading of 6 phr was used as the percolation threshold in order to enhance the electrical conductivity of the nanocomposite. Scanning electron microscopy revealed some agglomerative nano-graphite nanoparticles, as well as the formation of an unremitting network in the matrix. The storage modulus and dielectric properties of the rubber/nano-graphite nanocomposite were dependent on the matrix–nanofiller interactions and were also affected by the rubber–nano-graphite network formation. Thus, rubber/nano-graphite nanocomposites have been investigated due to their advantageous dispersion, conductivity, dielectric, thermal, and mechanical features for relevant practical applications.

Hence, various polymers, such as poly(3,4-ethylenedioxythiophene): poly(styrenesulfonate), polyaniline, polypyrrole, polystyrene, poly(methyl methacrylate), poly(vinyl chloride), poly(vinylidene fluoride), poly(lactic acid), polyurethane, acrylonitrilebutadiene rubber, chlorobutyl elastomer, and silicone rubber, have been employed as matrices for nano-graphite nanofillers. The dispersion of the nanofiller in the conducting polymer matrices formed via in situ methods have been found to be superior. Moreover, the nanofiller dispersion in thermoplastic matrices via solution methods has also been found to be homogeneous. All the conducting polymers, thermoplastics, and rubbery matrices behave differently with nano-graphite in terms of physical properties, depending on the backbone composition. Hence, advantageous properties can be obtained using various polymers and nano-graphite nanofillers, and no single polymer/nano-graphite nanocomposite category can be referred to as the best. However, nanofiller dispersion has been found to be dependent on the amount of nanofiller loaded, as well as on the processing technique used for the nanocomposite formation. The fine nanofiller dispersion attained via the facile processing technique may lead to enhanced properties for all the polymer/nano-graphite categories.

The structure of the nano-graphite affects the final properties and performance of the nanocomposites. Therefore, nano-graphite with a fine quality, the least impurities, and less structural disorders plays an essential role in the improvement of nanofiller dispersion, the provision of superior physical properties, and the production of high-performance nanocomposites. Consequently, the addition of nano-graphite influences the microstructure, surface morphology, and phase composition of nanocomposites. Subsequently, nano-graphite affects the compatibility of the polymer–nanofiller in a nanomaterial. In particular, the thermal stability, glass transition temperature, thermal conductivity, electrical conductivity, and mechanical behavior have been tuned with the addition of nano-graphite nanofiller. The better nano-graphite interactions with the polymer chains lead to mechanical interlocking, which enhances the mechanical properties. Moreover, the polymer chain rigidity and stability improve, leading to an upgraded heat stability and glass transition temperature. The inclusion of nanofiller in polymers through physical/covalent means also affects electron and thermal transportation through the nanocomposite systems.

4. Application Arenas of High-Performance Polymer/Nano-Graphite Nanocomposites

4.1. In Dye-Sensitized Solar Cells

A dye-sensitized solar cell (DSSC) is usually composed of counter electrodes as crucial components [70]. Platinum (Pt) has been commercially used as a counter electrode in DSSCs in addition to indium tin oxide or fluorine tin oxide electrodes [71]. Pt-electrode-based DSSCs have the disadvantages of a high cost and weight. Pt electrodes have been beneficially replaced with carbon-based materials [72]. For instance, graphite, carbon black, carbon nanotubes, etc., have been used as DSSC electrodes, showing high conductivity [73,74]. To constructively use carbon materials in electrodes, conducting polymers, such as poly(3,4-ethylenedioxythiophene):poly(styrene sulfonate) [75], polyaniline [76], and polypyrrole [77], have been used. Polyaniline-based nanocomposites are frequently fabricated for DSSC electrodes [78]. Polyaniline/graphite nanocomposites have been used in DSSC electrodes [79]. Correspondingly, Huang et al. [80] applied a polyaniline/nano-graphite nanocomposite in a DSSC counter electrode. The polyaniline/nano-graphite nanocomposite was prepared by using electro-polymerization. The power-conversion efficiency of the polyaniline/nano-graphite-nanocomposite-based DSSC was high at ~7.07%, i.e., comparable to the platinum electrode DSSC (7.19%). Yue et al. [81] prepared various counter electrodes for DSSCs, such as polypyrrole (PPy), platinum (Pt), nano-graphite (NG), and nano-graphite/polypyrrole (NG/PPy) counter electrodes. Figure 10 shows the comparative electrochemical impedance spectra (EIS) of the counter electrodes studied. The electrochemical performance of the NG/PPy electrode revealed a low charge-transfer resistance at the electrolyte–electrode interface, as well as excellent electrocatalyst performance. From the EIS spectra, the electrocatalytic activity of NG/PPy was determined from a semicircle diameter to be 0.1–100 kHz. The photovoltaic performance of the DSSC is given in Figure 11 and Table 2. Among all the electrodes, the DSSC with the NG/PPy electrode had a higher short-circuit current density and a power conversion efficiency of 14.83 mA·cm^{-2} and 7.40%, respectively. In solar cells, more novel polymer/nano-graphite combinations have been found to be desirable to attain a high power conversion efficiency of >8%.

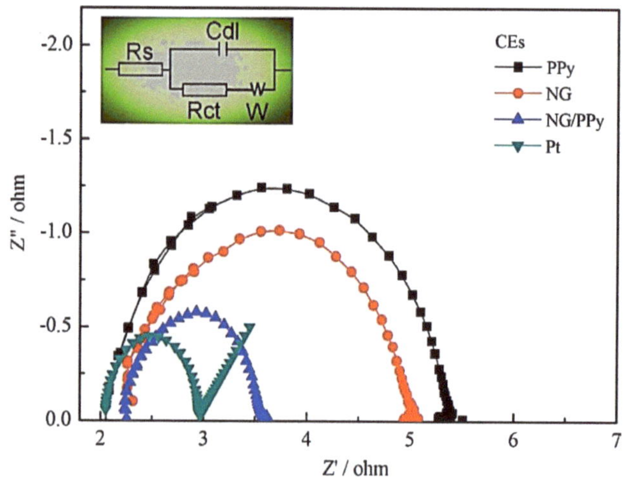

Figure 10. Nyquist plots of counter electrodes (CEs), i.e., graphite/polypyrrole (NG/PPy), polypyrrole (PPy), platinum (Pt), and nano-graphite (NG). Rs = Ohmic serial resistance; Rct = charge-transfer resistance of single electrode; Cdl = double-layer capacitance; W = diffusion impedance [81]. Reproduced with permission from Elsevier.

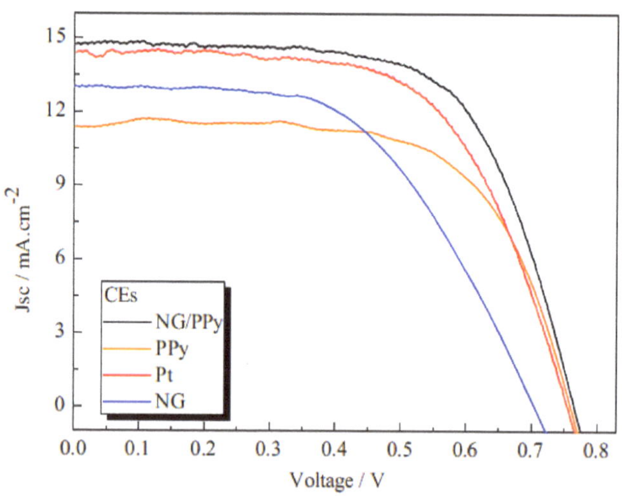

Figure 11. Photocurrent–voltage characteristics of dye-sensitized solar cell with nano-graphite/polypyrrole (NG/PPy)-, polypyrrole (PPy)-, platinum (Pt)-, and nano-graphite (NG)-based counter electrodes (CEs) under illumination of AM 1.5 G [81]. Reproduced with permission from Elsevier.

Table 2. Photocurrent-voltage parameters of dye-sensitized solar cell with nano-graphite/polypyrrole (NG/PPy), polypyrrole (PPy), Platinum (Pt), and nano-graphite (NG) under the illumination of AM 1.5 [81]. Voc = open-circuit voltage; Jsc = short-circuit current densit; FF = fill factor; PCE = power conversion efficiency. Reproduced with permission from Elsevier.

Sample	Voc (v)	Jsc ($mA \cdot cm^{-2}$)	FF	PCE (%)
PPy	0.760	11.74	0.632	5.64
NG	0.705	13.04	0.582	5.35
Pt	0.755	14.54	0.636	6.98
NG/PPy	0.765	14.83	0.652	7.40

4.2. In Electronics

Electronic devices, such as stretchable sensors, actuators, and flexible electrical devices, have been focused on for the use of innovative nanomaterials [82,83]. Polymer- and graphite-derived materials have been applied in next-generation electronic devices [84]. A polytetrafluoroethylene/nano-graphite grid was prepared and compared with a commercial lead alloy (Figure 12). The polytetrafluoroethylene/nano-graphite grid had a smooth surface (Figure 12B,C), and it was more efficiently used as a negative current collector in batteries compared to the commercial lead-based grids (Figure 12A). Room-temperature-vulcanized (RTV) silicone rubber/nano-graphite-based stretchable strain sensors and actuators have been industrialized [85]. Previously, carbon nanotubes have been used in RTV silicone rubber devices to augment their stiffness to some extent [86].

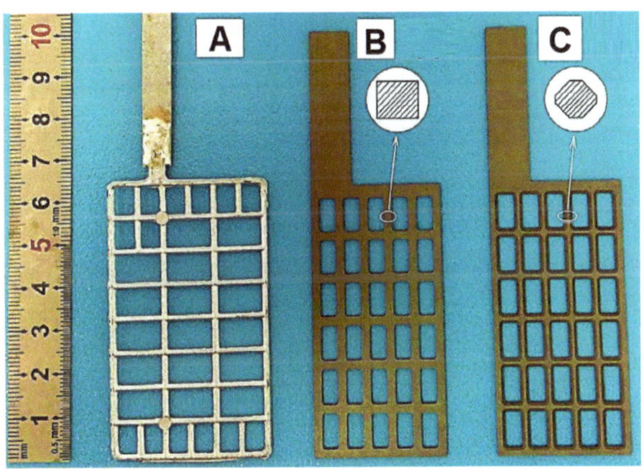

Figure 12. Photographs of (**A**) lead alloy grid and (**B**,**C**) nano-graphite grids with different alignments [84]. Reproduced with permission from Wiley.

Nevertheless, the use of new nano-graphite materials has suitably adjusted the flexibility and stretchability of RTV silicone rubber devices [87]. The noteworthy piezo-resistive strain-sensing and actuation performances of RTV silicone rubber/nano-graphite devices have been experiential. Nano-graphite with a few layers of three-dimensional nanoplatelets with a diameter of 12–15 nm was filled in RTV silicone rubber devices. The addition of 5 phr of nano-graphite boosted the compressive modulus and tensile modulus of the RTV devices by 135% and 125%, respectively [88]. The actuation displacement of the RTV silicone rubber/nano-graphite devices was also improved by 32% via a growing voltage supply from 2 to 8 kV. The strain sensor had a high stretchability of >100% and a durability of up to 5000 cycles [89]. In this field, further design investigations into polymer/nano-graphite nanomaterials have been considered significant to the discovery of important uses in next-generation flexible sensors, actuators, and related electronic devices.

4.3. EMI Shielding

The electromagnetic interference pollution from electronic/telecommunication instruments is continuously affecting human, biological, and environmental systems [90]. Electromagnetic interference (EMI)-shielding materials have been developed to reduce adverse radiation affects [91]. EMI shielding reduces incident radiation via the reflection and absorption phenomenon through a material. Consequently, EMI-shielding materials provide protection against incident radiation. Electrical conductivity has been considered to be an important property of EMI-shielding materials [92]. Initially, metals were applied as EMI shields [93]. However, metallic materials have the disadvantages of a high cost, a heavy weight, corrosion, and processability problems [94]. In this regard, electrically conducting polymeric composites have been used to overawe the limits of metal shields [95]. Carbon fillers have high aspect ratios, permittivity, permeability, and intrinsic conductivity for a high EMI-shielding performance [96]. Accordingly, conducting fillers, such as graphite, carbon black, and carbon nanotubes, have been filled in polymeric matrices to potentially improve EMI shielding [97,98]. Graphite-based materials have been preferably applied in EMI shields [99]. Jeddi and co-researchers [100] developed silicon rubber/nano-graphite nanocomposites and studied their EMI-shielding effectiveness. The nanomaterials had a high electrical conductivity and low percolation threshold values, contributing to radiation protection. According to X-ray band frequency studies, the absorption loss was found to be a leading mechanism for the intensity attenuation of incident electromagnetic waves. In another attempt, Jeddi et al. [101] designed silicon rubber/nano-graphite/carbon black nanocomposites. The nanomaterials had a fine conducting network that promoted electrical properties and, thus, EMI shielding. The dielectric permittivity of the silicon rubber with the nano-graphite, carbon black, and nano-graphite/carbon black is shown in Figure 13. The values of the dielectric permittivity increased with an increase in filler contents. The combination of nano-graphite/carbon black in the silicon rubber was effective in enhancing the dielectric permittivity relative to that of the silicon rubber/nano-graphite nanocomposite. The effect was observed due to the attainment of the percolation threshold and the polarization of the matrix–nanofiller interface. The absorption loss was found to be a major mechanism for the attenuation of incident EMI radiation in these nanocomposites. However, to date, very limited nano-graphite-based design options have been discovered for EMI shields. The development of novel polymer/nano-graphite nanocomposites may result in practically important high-performance radiation-shielding materials.

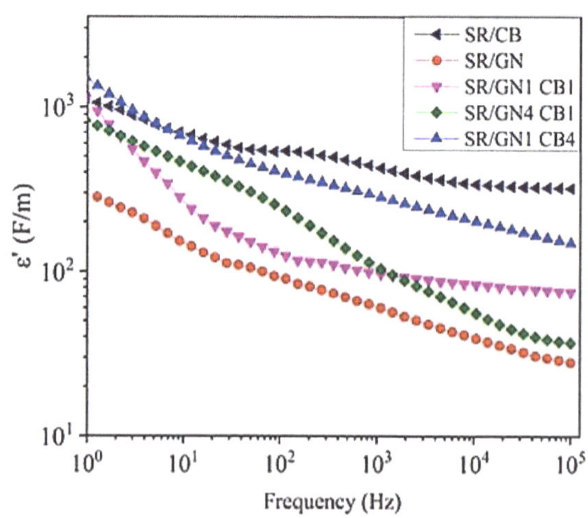

Figure 13. Real permittivity (ε′) properties of silicon rubber (SR) nanocomposites with sample nano-graphite (GN) and carbon black (CB) nanofillers [101]. Reproduced with permission from Elsevier.

5. Future Perspectives, Challenges, and Conclusions

Table 3 demonstrates the processing strategies, noteworthy features, and technical applications of polymer/nano-graphite nanocomposites. Nano-graphite nanofiller has been reinforced in polymer matrices to improve structural, morphological, electrical conductivity, thermal conductivity, dielectric characteristics, thermal stability, mechanical properties, photovoltaics, sensing/actuating, radiation shielding, and other physical properties. Nano-graphite nanoparticles have been prepared using various facile physical and chemical methods.

Table 3. Specifications and impact of polymer/nano-graphite nanocomposites.

Polymer	Nanofiller/Preparation Method	Nanocomposite Fabrication	Property/Application	Ref
Poly(3,4-ethylenedioxythiophene):poly(styrenesulfonate)	Nano-graphite; 7–8 graphene layers; 2.6 nm thickness	Solution method	Dispersion; swift heavy ions	[39]
Polyaniline	Nano-graphite; planetary ball milling	Cold pressing	Electrical conductivity; hardness	[40]
Polypyrrole	Nano-graphite; sonication in 95% alcohol solution	In situ polymerization	Electrical conductivity 80.0 Scm^{-1}	[41]
Polystyrene	Nano-graphite; detonation method; size 16 nm; specific surface area 583.6 $m^2 \cdot g^{-1}$	Pickering emulsion	Morphology; polystyrene microspheres encapsulated with nano-graphite shield	[45]
Poly(methyl methacrylate)	Commercial nano-graphite; ~400 nm	In situ polymerization	Thermogravimetric analysis; initial decomposition temperature ~154 °C	[49]
Poly(methyl methacrylate)	Nano-graphite	Solution casting technique	Optical properties; thermal stability	[50]
Poly(vinyl chloride)	Commercial nano-graphite; Titania	Solution casting technique	Conductivity; photocatalytic activity	[54]
Poly(vinylidene fluoride)	Nano-graphite	Solution casting; compression molding	Percolation threshold; dielectric permittivity	[58]
Poly(vinylidene fluoride)	Nano-graphite	Solution method	β-phase; interface effects	[59]
Poly(lactic acid)	Nano-graphite; sonication of graphite in ionic liquid, 1-butyl-3-methylimidazoliumhexa-fluorophosphate	Melt blending	Aggregates of 300 nm	
Poly(lactic acid)	Commercial. nano-graphite; diameter 1.5–2.0 μm	Fused Deposition Modeling	Thermal conductivity; conducting pathways	[64]
Polyurethane	Nano-graphite	Solution method	Percolation threshold; electrical conductivity	[65]
Acrylonitrilebutadiene rubber	Nano-graphite in poly(ethylene glycol)	Melt method	Young's modulus; hardness; thermal stability; permittivity; dielectric loss; antistatic application	[68]

Table 3. *Cont.*

Polymer	Nanofiller/Preparation Method	Nanocomposite Fabrication	Property/Application	Ref
Chlorobutyl elastomer	Nano-graphite	Melt method	Percolation threshold; storage modulus; dielectric properties	[69]
Polyaniline	Nano-graphite	Electro-polymerization	DSSC counter electrode; power-conversion efficiency ~7.07%	[80]
Polypyrrole	Nano-graphite	Electro-polymerization	Electrochemical impedance spectroscopy; power-conversion efficiency ~7.40%	[81]
Room temperature vulcanized silicone rubber	Nano-graphite; 12–15 nm diameter	Melt; solution; printing	Sensors/actuators; piezo-resistive strain sensing; compressive modulus; tensile modulus stretchability >100%; durability of up to 5000 cycles	[86–89]
Silicon rubber	Nano-graphite; carbon black	Melt route	Electromagnetic interference shielding	[100,101]

The future of polymer/nano-graphite nanocomposites relies on their potential applications in various technical fields. The potential application of polymer/nano-graphite nanocomposites has been observed for DSSCs. High-performance solar cells can be achieved due to the geometric and electronic effects of the nano-graphite in the conducting polymers. A high power conversion efficiency can be achieved for these nanocomposites. In future, a higher efficiency can be attained by using functional nano-graphite nanostructures. Nano-graphite-based nanocomposites also possess potential for use in electronics, sensors, and actuator devices. However, future efforts are needed to design high-performance polymer/nano-graphite nanocomposites for commercial electronic device applications. Similarly, the potential application of polymer/nano-graphite nanocomposites has been observed for EMI nanomaterials. Future efforts may lead to a high EMI shielding efficiency of >50 dB for commercial applications. Consequently, due to the restricted research in various fields, several future technical grounds still need to be explored for polymer/nano-graphite nanocomposites. Forthcoming research on these innovative nanomaterials may lead to applications in batteries, aerospace/automobiles, construction materials, and biomedical areas.

Challenges regarding the use of polymer/nano-graphite nanocomposites have been found related to their properties, performance, and structure–property relationship. The nanofiller dispersion has been pointed out as the main challenge in the application of polymer/nano-graphite nanocomposites. The dispersion issues can be overcome by using the appropriate processing technique and controlled processing parameters. Moreover, the use of optimum nanofiller contents and functional nano-graphite may result in better matrix–nanofiller interactions, interface formation, and a homogeneous dispersion in the polymers. In polymer/nano-graphite nanocomposites, the augmentation of the physical property in fact relies on the nano-graphite dispersion in the polymers. Additionally, it is important to understand the mechanisms of physical and chemical interactions and interface formation in polymer/nano-graphite nanocomposites in order to develop high-performance nanomaterials.

Succinctly, this appraisal provides a novel insight into polymer/nano-graphite nanocomposites. Nano-graphite has been found to be practical as a unique nanofiller for polymeric nanocomposites. Numerous facile strategies have been used to design nano-graphite nanoparticles and polymer/nano-graphite nanocomposites. Polymer/nano-graphite nanocomposites have uncovered important structural, morphological, electrical, thermal, mechanical, and other essential properties. The applications of polymer/nano-graphite nanocomposites have been extended to DSSCs, electronics, and EMI-shielding devices. Future research on functional nano-graphite and polymer/nano-graphite nanocomposites may lead to enormous progressions in the related technical fields.

Funding: This research received no external funding.

Institutional Review Board Statement: Not applicable.

Informed Consent Statement: Not applicable.

Data Availability Statement: Not applicable.

Conflicts of Interest: The authors declare no conflict of interest.

References

1. Morgan, D.J. Comments on the XPS analysis of carbon materials. *C* **2021**, *7*, 51. [CrossRef]
2. de Souza, F.M.; Choi, J.; Bhoyate, S.; Kahol, P.K.; Gupta, R.K. Expendable graphite as an efficient flame-retardant for novel partial bio-based rigid polyurethane foams. *C* **2020**, *6*, 27. [CrossRef]
3. Cazzanelli, M.; Basso, L.; Cestari, C.; Bazzanella, N.; Moser, E.; Orlandi, M.; Piccoli, A.; Miotello, A. Fluorescent Nanodiamonds Synthesized in One-Step by Pulsed Laser Ablation of Graphite in Liquid-Nitrogen. *C* **2021**, *7*, 49. [CrossRef]
4. Odusanya, A.; Rahaman, I.; Sarkar, P.K.; Zkria, A.; Ghosh, K.; Haque, A. Laser-Assisted Growth of Carbon-Based Materials by Chemical Vapor Deposition. *C* **2022**, *8*, 24. [CrossRef]
5. Kausar, A. Poly (methyl methacrylate) nanocomposite reinforced with graphene, graphene oxide, and graphite: A review. *Polym.-Plast. Technol. Mater.* **2019**, *58*, 821–842. [CrossRef]
6. Kausar, A.; Rafique, I.; Muhammad, B. Aerospace application of polymer nanocomposite with carbon nanotube, graphite, graphene oxide, and nanoclay. *Polym.-Plast. Technol. Mater.* **2017**, *56*, 1438–1456. [CrossRef]
7. Blomquist, N.; Koppolu, R.; Dahlström, C.; Toivakka, M.; Olin, H. Influence of substrate in roll-to-roll coated nanographite electrodes for metal-free supercapacitors. *Sci. Rep.* **2020**, *10*, 5282. [CrossRef] [PubMed]
8. Kumar, V.; Forsberg, S.; Engström, A.-C.; Nurmi, M.; Andres, B.; Dahlström, C.; Toivakka, M. Conductive nanographite–nanocellulose coatings on paper. *Flexib. Prin. Electron.* **2017**, *2*, 035002. [CrossRef]
9. Wan, C.; Yao, C.; Li, J.; Duan, H.; Zhan, S.; Jia, D.; Li, Y.; Yang, T. Friction and wear behavior of polyimide matrix composites filled with nanographite. *J. Appl. Polym. Sci.* **2022**, *139*, 52058. [CrossRef]
10. Kumar, V.; Alam, M.N.; Manikkavel, A.; Choi, J.; Lee, D.J. Investigation of silicone rubber composites reinforced with carbon nanotube, nanographite, their hybrid, and applications for flexible devices. *J. Vin. Add. Technol.* **2021**, *27*, 254–263. [CrossRef]
11. Luo, P.; Zheng, C.; He, J.; Tu, X.; Sun, W.; Pan, H.; Zhou, Y.; Rui, X.; Zhang, B.; Huang, K. Structural Engineering in Graphite-Based Metal-Ion Batteries. *Adv. Funct. Mater.* **2022**, *32*, 2107277. [CrossRef]
12. Maier, P.; Xavier Jr, N.F.; Truscott, C.L.; Hansen, T.; Fouquet, P.; Sacchi, M.; Tamtögl, A. How does tuning the van der Waals bonding strength affect adsorbate structure? *Phys. Chem. Chem. Phys.* **2022**, *24*, 29371–29380. [CrossRef] [PubMed]
13. Bianco, A., Cheng, H.-M., Enoki, T., Gogotsi, Y., Hurt, R.H., Koratkar, N.; Kyotani, T.; Monthioux, M.; Park, C.R.; Tascon, J.M. All in the graphene family—A recommended nomenclature for two-dimensional carbon materials. *Carbon* **2013**, *65*, 1–6. [CrossRef]
14. Blomquist, N.; Engström, A.-C.; Hummelgård, M.; Andres, B.; Forsberg, S.; Olin, H. Large-scale production of nanographite by tube-shear exfoliation in water. *PLoS ONE* **2016**, *11*, e0154686. [CrossRef] [PubMed]
15. Yi, M.; Li, J.; Shen, Z.; Zhang, X.; Ma, S. Morphology and structure of mono-and few-layer graphene produced by jet cavitation. *Appl. Phys. Lett.* **2011**, *99*, 123112. [CrossRef]
16. Paton, K.R.; Varrla, E.; Backes, C.; Smith, R.J.; Khan, U.; O'Neill, A.; Boland, C.; Lotya, M.; Istrate, O.M.; King, P. Scalable production of large quantities of defect-free few-layer graphene by shear exfoliation in liquids. *Nat. Mater.* **2014**, *13*, 624–630. [CrossRef] [PubMed]
17. Knieke, C.; Berger, A.; Voigt, M.; Taylor, R.N.K.; Röhrl, J.; Peukert, W. Scalable production of graphene sheets by mechanical delamination. *Carbon* **2010**, *48*, 3196–3204. [CrossRef]
18. Nacken, T.; Damm, C.; Walter, J.; Rüger, A.; Peukert, W. Delamination of graphite in a high pressure homogenizer. *RSC Adv.* **2015**, *5*, 57328–57338. [CrossRef]
19. Karagiannidis, P.G.; Hodge, S.A.; Lombardi, L.; Tomarchio, F.; Decorde, N.; Milana, S.; Goykhman, I.; Su, Y.; Mesite, S.V.; Johnstone, D.N. Microfluidization of graphite and formulation of graphene-based conductive inks. *ACS Nano* **2017**, *11*, 2742–2755. [CrossRef]

20. Koppolu, R.; Blomquist, N.; Dahlström, C.; Toivakka, M. High-Throughput Processing of Nanographite–Nanocellulose-Based Electrodes for Flexible Energy Devices. *Ind. Eng. Chem. Res.* **2020**, *59*, 11232–11240. [CrossRef]
21. Blomquist, N.; Wells, T.; Andres, B.; Bäckström, J.; Forsberg, S.; Olin, H. Metal-free supercapacitor with aqueous electrolyte and low-cost carbon materials. *Sci. Rep.* **2017**, *7*, 39836. [CrossRef] [PubMed]
22. Shenderova, O. *Detonation Nanodiamonds: Science and Applications*; CRC Press: Boca Raton, FL, USA, 2014.
23. Ioni, Y.V.; Tkachev, S.; Bulychev, N.; Gubin, S. Preparation of finely dispersed nanographite. *Inorg. Mater.* **2011**, *47*, 597–602. [CrossRef]
24. Chen, Q.; Wang, X.; Wang, Z.; Liu, Y.; You, T. Preparation of water-soluble nanographite and its application in water-based cutting fluid. *Nanoscale Res. Lett.* **2013**, *8*, 52. [CrossRef] [PubMed]
25. Prasad, B.; Sato, H.; Enoki, T.; Hishiyama, Y.; Kaburagi, Y.; Rao, A.; Sumanasekera, G.U.; Eklund, P. Intercalated nanographite: Structure and electronic properties. *Phys. Rev. B* **2001**, *64*, 235407. [CrossRef]
26. Frost, B.R.; Fyfe, W.S.; Tazaki, K.; Chan, T. Grain-boundary graphite in rocks and implications for high electrical conductivity in the lower crust. *Nature* **1989**, *340*, 134–136. [CrossRef]
27. Mokhena, T.C.; Mochane, M.J.; Sefadi, J.S.; Motloung, S.V.; Andala, D.M. *Thermal conductivity of graphite-based polymer composites. Impact of Thermal Conductivity on Energy Technologies*; IntechOpen: London, UK, 2018; Volume 181.
28. Salavagione, H.J.; Martínez, G.; Ellis, G. Recent advances in the covalent modification of graphene with polymers. *Macromol. Rapid Commun.* **2011**, *32*, 1771–1789. [CrossRef]
29. Sadasivuni, K.K.; Ponnamma, D.; Thomas, S.; Grohens, Y. Evolution from graphite to graphene elastomer composites. *Prog. Polym. Sci.* **2014**, *39*, 749–780. [CrossRef]
30. Gorshenev, V. Influence of Technological Conditions in the Formation of Electrically Conductive Thermoplastic Polymer-Graphite Composites. *Inorg. Mater. Appl. Res.* **2022**, *13*, 515–522. [CrossRef]
31. Yan, H.; Tao, X.; Yang, Z.; Li, K.; Yang, H.; Li, A.; Cheng, R. Effects of the oxidation degree of graphene oxide on the adsorption of methylene blue. *J. Hazard. Mater.* **2014**, *268*, 191–198. [CrossRef]
32. Liao, Y.; Cao, L.; Wang, Q.; Li, S.; Lin, Z.; Li, W.; Zhang, P.; Yu, C. Enhanced tribological properties of PEEK-based composite coatings reinforced by PTFE and graphite. *J. Appl. Polym. Sci.* **2022**, *139*, 51878. [CrossRef]
33. Jena, K.K.; AlFantazi, A.; Mayyas, A.T. Efficient and cost-effective hybrid composite materials based on thermoplastic polymer and recycled graphite. *Chem. Eng. J.* **2022**, *430*, 132667. [CrossRef]
34. Jiang, Z.; Liu, X.; Xu, Q.; Zhou, C.; Shang, Y.; Zhang, H. Thermal conductive segregated multi-scale network constructed by ball-milling and in-situ polymerization in PEEK/MWCNT/graphite composite. *Compos. Communicat.* **2022**, *29*, 101035. [CrossRef]
35. Seyedjamali, H.; Pirisedigh, A. Well-dispersed polyimide/TiO_2 nanocomposites: In situ sol–gel fabrication and morphological study. *Coll. Polym. Sci.* **2012**, *290*, 653–659. [CrossRef]
36. Psarras, G.C. Nanographite–Polymer Composites. In *Carbon Nanomaterials*; CRC Press: Boca Raton, FL, USA, 2018; pp. 647–673.
37. Radzuan, N.A.M.; Sulong, A.B.; Sahari, J. A review of electrical conductivity models for conductive polymer composite. *Int. J. Hydrog. Energy* **2017**, *42*, 9262–9273. [CrossRef]
38. Krause, P.; Nowoświat, A. Experimental studies involving the impact of solar radiation on the properties of expanded graphite polystyrene. *Energies* **2019**, *13*, 75. [CrossRef]
39. Singhal, P.; Rattan, S. Swift heavy ion irradiation as a tool for homogeneous dispersion of nanographite platelets within the polymer matrices: Toward tailoring the properties of PEDOT: PSS/nanographite nanocomposites. *J. Phys. Chem. B* **2016**, *120*, 3403–3413. [CrossRef]
40. Shinde, H.; Kakani, S.M.; Shinde, S.; More, A.; Kher, J.; Patil, S.R. Mechanically Stable Nano-graphite Polyaniline Composites: Synthesis and Characterization. *Int. J. Innov. Res. Sci. Eng. Technol.* **2014**, *3*, 14164–14171.
41. Mo, Z.L.; Xie, T.-T.; Zhao, Y.-X.; Sun, W.-H.; Guo, R.-B.; He, J.-X. Synthesis of PPy/nano-graphite sheets/TbCl3 composites with high conductivity. *Mater. Manuf. Process.* **2012**, *27*, 1343–1347. [CrossRef]
42. Zhu, Z.-M.; Rao, W.-H.; Kang, A.-H.; Liao, W.; Wang, Y.-Z. Highly effective flame retarded polystyrene by synergistic effects between expandable graphite and aluminum hypophosphite. *Polym. Degrad. Stab.* **2018**, *154*, 1–9. [CrossRef]
43. Han, X.; Cheng, Q.; Bao, F.; Gao, J.; Yang, Y.; Chen, T.; Yan, C.; Ma, R. Synthesis of low-density heat-resisting polystyrene/graphite composite microspheres used as water carrying fracturing proppants. *Polym.-Plast. Technol. Mater.* **2014**, *53*, 1647–1653. [CrossRef]
44. Rymansaib, Z.; Iravani, P.; Emslie, E.; Medvidović-Kosanović, M.; Sak-Bosnar, M.; Verdejo, R.; Marken, F. All-polystyrene 3D-printed electrochemical device with embedded carbon nanofiber-graphite-polystyrene composite conductor. *Electroanalysis* **2016**, *28*, 1517–1523. [CrossRef]
45. Xuemei, H.; Hao, Y. Fabrication of polystyrene/detonation nanographite composite microspheres with the core/shell structure via pickering emulsion polymerization. *J. Nanomater.* **2013**, *2013*, 8. [CrossRef]
46. Soni, G.; Soni, P.; Jangra, P.; Vijay, Y. Optical, mechanical and thermal properties of PMMA/graphite nanocomposite thin films. *Mater. Res. Exp.* **2019**, *6*, 075315. [CrossRef]
47. Ali, E.A.G.E.; Kim, T.; Adhha, M.A. Effects of graphite milling time and composition to tensile properties of poly-methyl methacrylate (PMMA)/graphite composite. *J. Sustain. Sci. Manag.* **2019**, *14*, 4–11.
48. Wang, J.; Shi, S.; Yang, J.; Zhang, W. Multiscale analysis on free vibration of functionally graded graphene reinforced PMMA composite plates. *Appl. Math. Model.* **2021**, *98*, 38–58. [CrossRef]

49. Singhi, M.; Fahim, M. Dielectric properties of nanographite-filled PMMA composites prepared by in situ polymerization. *Polym. Compos.* **2012**, *33*, 675–682. [CrossRef]
50. Soni, G.; Jangir, R.K. Effect of temperature nano graphite doped polymethylmethacrylate (PMMA) composite flexible thin films prepared by solution casting: Synthesis, optical and electrical properties. *Optik* **2021**, *226*, 165915. [CrossRef]
51. Raza, J.; Hamid, A.; Khan, M.; Hussain, F.; Tiehu, L.; Fazil, P.; Zada, A.; Wahab, Z.; Ali, A. Spectroscopic characterization of biosynthesized lead oxide (PbO) nanoparticles and their applications in PVC/graphite-PbO nanocomposites. *Z. Für Phys. Chem.* **2022**, *236*, 619–636. [CrossRef]
52. Focke, W.W.; Muiambo, H.; Mhike, W.; Kruger, H.J.; Ofosu, O. Flexible PVC flame retarded with expandable graphite. *Polym. Degrad. Stab.* **2014**, *100*, 63–69. [CrossRef]
53. Miao, F.; Liu, Y.; Gao, M.; Yu, X.; Xiao, P.; Wang, M.; Wang, S.; Wang, X. Degradation of polyvinyl chloride microplastics via an electro-Fenton-like system with a TiO_2/graphite cathode. *J. Hazard. Mater.* **2020**, *399*, 123023. [CrossRef]
54. Zhang, Y.; Sun, T.; Zhang, D.; Shi, Z.; Zhang, X.; Li, C.; Wang, L.; Song, J.; Lin, Q. Enhanced photodegradability of PVC plastics film by codoping nano-graphite and TiO_2. *Polym. Degrad. Stab.* **2020**, *181*, 109332. [CrossRef]
55. Takahashi, K.; Higa, K.; Mair, S.; Chintapalli, M.; Balsara, N.; Srinivasan, V. Mechanical degradation of graphite/PVDF composite electrodes: A model-experimental study. *J. Electrochem. Soc.* **2015**, *163*, A385. [CrossRef]
56. Iqbal, N.; Lee, S. Mechanical failure analysis of graphite anode particles with PVDF binders in Li-ion batteries. *J. Electrochem. Soc.* **2018**, *165*, A1961. [CrossRef]
57. Li, X.; Wang, X.; Weng, L.; Yu, Y.; Zhang, X.; Liu, L.; Wang, C. Dielectrical properties of graphite nanosheets/PVDF composites regulated by coupling agent. *Mater. Today Commun.* **2019**, *21*, 100705. [CrossRef]
58. Li, Y.; Tjong, S.C.; Li, R. Dielectric properties of binary polyvinylidene fluoride/barium titanate nanocomposites and their nanographite doped hybrids. *eXPRESS Polym. Lett.* **2011**, *5*, 526–534. [CrossRef]
59. Abdelaziz, M.; Abdelrazek, E. The effect of nanographite addition on the physical properties of poly (vinylidene fluoride)/hydroxypropyl cellulose blend. *J. Mater. Sci. Mater. Electron.* **2014**, *25*, 5481–5490. [CrossRef]
60. Wang, G.; Zhao, G.; Wang, S.; Zhang, L.; Park, C.B. Injection-molded microcellular PLA/graphite nanocomposites with dramatically enhanced mechanical and electrical properties for ultra-efficient EMI shielding applications. *J. Mater. Chem. C* **2018**, *6*, 6847–6859. [CrossRef]
61. Przekop, R.E.; Kujawa, M.; Pawlak, W.; Dobrosielska, M.; Sztorch, B.; Wieleba, W. Graphite modified polylactide (PLA) for 3D printed (FDM/FFF) sliding elements. *Polymers* **2020**, *12*, 1250. [CrossRef]
62. Kjelgård, K.G.; Wisland, D.T.; Lande, T.S. 3D printed wideband microwave absorbers using composite graphite/PLA filament. In Proceedings of the 2018 48th European Microwave Conference, Madrid, Spain, 23–27 September 2018; IEEE: Piscataway, NJ, USA, 2018; pp. 859–862.
63. Gardella, L.; Furfaro, D.; Galimberti, M.; Monticelli, O. On the development of a facile approach based on the use of ionic liquids: Preparation of PLLA (sc-PLA)/high surface area nano-graphite systems. *Green Chem.* **2015**, *17*, 4082–4088. [CrossRef]
64. Guo, R.; Ren, Z.; Bi, H.; Xu, M.; Cai, L. Electrical and thermal conductivity of polylactic acid (PLA)-based biocomposites by incorporation of nano-graphite fabricated with fused deposition modeling. *Polymers* **2019**, *11*, 549. [CrossRef]
65. Mishra, P.; Bhat, B.R.; Bhattacharya, B.; Mehra, R. Synthesis and Characterization of High-Dielectric-Constant Nanographite–Polyurethane Composite. *JOM* **2018**, *70*, 1302–1306. [CrossRef]
66. ElFaham, M.M.; Alnozahy, A.M.; Ashmawy, A. Comparative study of LIBS and mechanically evaluated hardness of graphite/rubber composites. *Mater. Chem. Phys.* **2018**, *207*, 30–35. [CrossRef]
67. Aguilar-Bolados, H.; Brasero, J.; López-Manchado, M.A.; Yazdani-Pedram, M. High performance natural rubber/thermally reduced graphite oxide nanocomposites by latex technology. *Compos. Part B Eng.* **2014**, *67*, 449–454. [CrossRef]
68. El-Nashar, D.E.; Rozik, N.N.; Abd El-Messieh, S.L. Mechanical and electrical properties of acrylonitrile-butadiene rubber filled with treated nanographite. *Polimery* **2014**, *59*, 834–844. [CrossRef]
69. Tiwari, S.K.; Choudhary, R.N.P.; Mahapatra, S.P. Dynamic mechanical and dielectric relaxation studies of chlorobutyl elastomer nanocomposites: Effect of nanographite loading and temperature. *High Perform. Polym.* **2015**, *27*, 274–287. [CrossRef]
70. Nazeeruddin, M.K.; Baranoff, E.; Grätzel, M. Dye-sensitized solar cells: A brief overview. *Solar Energy* **2011**, *85*, 1172–1178. [CrossRef]
71. Yun, S.; Hagfeldt, A.; Ma, T. Pt-free counter electrode for dye-sensitized solar cells with high efficiency. *Adv. Mater.* **2014**, *26*, 6210–6237. [CrossRef]
72. Zambrzycki, M.; Piech, R.; Raga, S.R.; Lira-Cantu, M.; Fraczek-Szczypta, A. Hierarchical carbon nanofibers/carbon nanotubes/NiCo nanocomposites as novel highly effective counter electrode for dye-sensitized solar cells: A structure-electrocatalytic activity relationship study. *Carbon* **2022**, *203*, 97–110. [CrossRef]
73. Cruz-Gutiérrez, C.A.; Félix-Navarro, R.M.; Calva-Yañez, J.C.; Silva-Carrillo, C.; Lin-Ho, S.W.; Reynoso-Soto, E.A. Carbon nanotube-carbon black hybrid counter electrodes for dye-sensitized solar cells and the effect on charge transfer kinetics. *J. Solid State Electrochem.* **2021**, *25*, 1479–1489. [CrossRef]
74. Brennan, L.J.; Byrne, M.T.; Bari, M.; Gun'ko, Y.K. Carbon nanomaterials for dye-sensitized solar cell applications: A bright future. *Adv. Ener. Mater.* **2011**, *1*, 472–485. [CrossRef]

75. Lee, C.-P.; Lin, C.-A.; Wei, T.-C.; Tsai, M.-L.; Meng, Y.; Li, C.-T.; Ho, K.-C.; Wu, C.-I.; Lau, S.-P.; He, J.-H. Economical low-light photovoltaics by using the Pt-free dye-sensitized solar cell with graphene dot/PEDOT: PSS counter electrodes. *Nano Energy* **2015** *18*, 109–117. [CrossRef]
76. Kuppu, S.V.; Senthilkumaran, M.; Sethuraman, V.; Balaji, M.; Saravanan, C.; Ahmed, N.; Mohandoss, S.; Lee, Y.R.; Anandharaj, J. Stalin, T. The surfactants mediated electropolymerized poly (aniline)(PANI)-reduced graphene oxide (rGO) composite counter electrode for dye-sensitized solar cell. *J. Phys. Chem. Solids* **2022**, *173*, 111121. [CrossRef]
77. Rafique, S.; Rashid, I.; Sharif, R. Cost effective dye sensitized solar cell based on novel Cu polypyrrole multiwall carbon nanotubes nanocomposites counter electrode. *Sci. Rep.* **2021**, *11*, 14830. [CrossRef]
78. Shih, Y.-C.; Lin, H.-L.; Lin, K.-F. Electropolymerized polyaniline/graphene nanoplatelet/multi-walled carbon nanotube composites as counter electrodes for high performance dye-sensitized solar cells. *J. Electroanal. Chem.* **2017**, *794*, 112–119. [CrossRef]
79. Reza, M.; Utami, A.N.; Amalina, A.N.; Benu, D.P.; Fatya, A.I.; Agusta, M.K.; Yuliarto, B.; Kaneti, Y.V.; Ide, Y.; Yamauchi, Y. Significant role of thorny surface morphology of polyaniline on adsorption of triiodide ions towards counter electrode in dye-sensitized solar cells. *New J. Chem.* **2021**, *45*, 5958–5970. [CrossRef]
80. Huang, K.-C.; Huang, J.-H.; Wu, C.-H.; Liu, C.-Y.; Chen, H.-W.; Chu, C.-W.; Lin, C.-L.; Ho, K.-C. Nanographite/polyaniline composite films as the counter electrodes for dye-sensitized solar cells. *J. Mater. Chem.* **2011**, *21*, 10384–10389. [CrossRef]
81. Yue, G.; Zhang, X.A.; Wang, L.; Tan, F.; Wu, J.; Jiang, Q.; Lin, J.; Huang, M.; Lan, Z. Highly efficient and stable dye-sensitized solar cells based on nanographite/polypyrrole counter electrode. *Electrochim. Acta* **2014**, *129*, 229–236. [CrossRef]
82. Chen, D.; Pei, Q. Electronic muscles and skins: A review of soft sensors and actuators. *Chem. Rev.* **2017**, *117*, 11239–11268. [CrossRef]
83. Jang, Y.; Kim, S.M.; Spinks, G.M.; Kim, S.J. Carbon nanotube yarn for fiber-shaped electrical sensors, actuators, and energy storage for smart systems. *Adv. Mater.* **2020**, *32*, 1902670. [CrossRef]
84. Zhang, S.; Zhang, H.; Cheng, J.; Zhang, W.; Cao, G.; Zhao, H.; Yang, Y. Novel polymer-graphite composite grid as a negative current collector for lead-acid batteries. *J. Power Source* **2016**, *334*, 31–38. [CrossRef]
85. Kumar, V.; Kumar, A.; Han, S.S.; Park, S.-S. RTV silicone rubber composites reinforced with carbon nanotubes, titanium-di-oxide and their hybrid: Mechanical and piezoelectric actuation performance. *Nano Mater. Sci.* **2021**, *3*, 233–240. [CrossRef]
86. Kumar, V.; Lee, G.; Choi, J.; Lee, D.-J. Studies on composites based on HTV and RTV silicone rubber and carbon nanotubes for sensors and actuators. *Polymer* **2020**, *190*, 122221. [CrossRef]
87. Costa, P.; Nunes-Pereira, J.; Oliveira, J.; Silva, J.; Moreira, J.A.; Carabineiro, S.; Buijnsters, J.; Lanceros-Mendez, S. High-performance graphene-based carbon nanofiller/polymer composites for piezoresistive sensor applications. *Compos. Sci. Technol.* **2017**, *153*, 241–252. [CrossRef]
88. Wei, X.-F.; Liu, J.-X.; Bao, W.; Qin, Y.; Li, F.; Liang, Y.; Xu, F.; Zhang, G.-J. High-entropy carbide ceramics with refined microstructure and enhanced thermal conductivity by the addition of graphite. *J. Eur. Ceram. Soc.* **2021**, *41*, 4747–4754. [CrossRef]
89. Irani, F.S.; Shafaghi, A.H.; Tasdelen, M.C.; Delipinar, T.; Kaya, C.E.; Yapici, G.G.; Yapici, M.K. Graphene as a Piezoresistive Material in Strain Sensing Applications. *Micromachines* **2022**, *13*, 119. [CrossRef]
90. Breen, M.S.; Long, T.C.; Schultz, B.D.; Williams, R.W.; Richmond-Bryant, J.; Breen, M.; Langstaff, J.E.; Devlin, R.B.; Schneider, A.; Burke, J.M. Air pollution exposure model for individuals (EMI) in health studies: Evaluation for ambient PM2. 5 in Central North Carolina. *Environ. Sci. Technol.* **2015**, *49*, 14184–14194. [CrossRef]
91. Chen, Y.; Yang, Y.; Xiong, Y.; Zhang, L.; Xu, W.; Duan, G.; Mei, C.; Jiang, S.; Rui, Z.; Zhang, K. Porous aerogel and sponge composites: Assisted by novel nanomaterials for electromagnetic interference shielding. *Nano Today* **2021**, *38*, 101204. [CrossRef]
92. Cao, M.-S.; Wang, X.-X.; Cao, W.-Q.; Yuan, J. Ultrathin graphene: Electrical properties and highly efficient electromagnetic interference shielding. *J. Mater. Chem. C* **2015**, *3*, 6589–6599. [CrossRef]
93. Wang, H.; Li, S.; Liu, M.; Li, J.; Zhou, X. Review on shielding mechanism and structural design of electromagnetic interference shielding composites. *Macromolecul. Mater. Eng.* **2021**, *306*, 2100032. [CrossRef]
94. Chung, D. Materials for electromagnetic interference shielding. *Mater. Chem. Phys.* **2020**, *255*, 123587. [CrossRef]
95. Yao, Y.; Jin, S.; Zou, H.; Li, L.; Ma, X.; Lv, G.; Gao, F.; Lv, X.; Shu, Q. Polymer-based lightweight materials for electromagnetic interference shielding: A review. *J. Mater. Sci.* **2021**, *56*, 6549–6580. [CrossRef]
96. Safdar, F.; Ashraf, M.; Javid, A.; Iqbal, K. Polymeric textile-based electromagnetic interference shielding materials, their synthesis, mechanism and applications—A review. *J. Ind. Text.* **2022**, *51*, 7293S–7358S. [CrossRef]
97. Thomassin, J.-M.; Jerome, C.; Pardoen, T.; Bailly, C.; Huynen, I.; Detrembleur, C. Polymer/carbon based composites as electromagnetic interference (EMI) shielding materials. *Mater. Sci. Eng. R Rep.* **2013**, *74*, 211–232. [CrossRef]
98. Zhang, W.; Wei, L.; Ma, Z.; Fan, Q.; Ma, J. Advances in waterborne polymer/carbon material composites for electromagnetic interference shielding. *Carbon* **2021**, *177*, 412–426. [CrossRef]
99. Shakir, M.F.; Khan, A.N.; Khan, R.; Javed, S.; Tariq, A.; Azeem, M.; Riaz, A.; Shafqat, A.; Cheema, H.M.; Akram, M.A. EMI shielding properties of polymer blends with inclusion of graphene nano platelets. *Res. Phys.* **2019**, *14*, 102365. [CrossRef]

100. Jeddi, J.; Katbab, A.A. AC electrical conductivity, shielding effectiveness and viscoelastic characteristics of nanocomposites based on RTV silicon rubber and nano graphite sheets/carbon black hybrid system. *AIP Conf. Proc.* **2017**, *1914*, 030013.
101. Jeddi, J.; Katbab, A.A.; Mehranvari, M. Investigation of microstructure, electrical behavior, and EMI shielding effectiveness of silicone rubber/carbon black/nanographite hybrid composites. *Polym. Compos.* **2019**, *40*, 4056–4066. [CrossRef]

C Journal of Carbon Research

MDPI

Article

Comparison of the Electrochemical Response of Carbon-Fiber-Reinforced Plastic (CFRP), Glassy Carbon, and Highly Ordered Pyrolytic Graphite (HOPG) in Near-Neutral Aqueous Chloride Media

Stanley Udochukwu Ofoegbu [1,2,*], Mário Guerreiro Silva Ferreira [2], Helena I. S. Nogueira [3] and Mikhail Zheludkevich [4,5]

[1] Centre for Mechanical Technology and Automation (TEMA), Department of Mechanical Engineering, University of Aveiro, Campus Universitário de Santiago, 3810-193 Aveiro, Portugal
[2] Department of Materials and Ceramic Engineering, CICECO-Aveiro Institute of Materials, University of Aveiro, Campus Universitário de Santiago, 3810-193 Aveiro, Portugal
[3] Department of Chemistry, CICECO-Aveiro Institute of Materials, University of Aveiro, Campus Universitário de Santiago, 3810-193 Aveiro, Portugal
[4] Institute of Surface Science, Helmholtz-Zentrum Hereon, Max-Planck-Strasse 1, 21502 Geesthacht, Germany
[5] Institute for Materials Science, Faculty of Engineering, Kiel University, 24103 Kiel, Germany
[*] Correspondence: ofoegbu.stanley@ua.pt; Tel.: +351-915-248-642

Abstract: Carbon-fiber-reinforced polymers (CFRP), being conductive, are capable of supporting cathodic oxygen reduction reactions (ORR) and thus promote galvanic corrosion when coupled to many metallic materials. Hence, understanding cathodic processes at carbon surfaces is critical to developing new strategies for the corrosion protection of multi-material assemblies. In the present work, the electrochemical responses of CFRP, glassy carbon, and HOPG (Highly Ordered Pyrolytic Graphite) have been evaluated in a quiescent 50 mM NaCl solution, and their respective activities towards ORR have been ranked. Employing the averages of the specific charges (CFRP, 129.52 mC cm^{-2}; glassy carbon, 89.95 mC cm^{-2}; HOPG, 60.77 mC cm^{-2}) passed during 1 h polarization of each of the 3 carbon surfaces at -1000 mV_{SCE} in the test media as a ranking criterion, the propensities of the 3 carbon surfaces (CFRP, GC, and HOPG) to support cathodic activities that can lead to anodic metal dissolution on galvanic coupling to metallic materials are ranked thusly; CFRP > GC > HOPG. This ranking is consistent with the trend of capacitance values obtained in this work: CFRP (19.5 to 34.5 μF cm^{-2}), glassy carbon (13.6 to 85.5 μF cm^{-2}), and HOPG (1.4 to 1.8 μF cm^{-2}). A comparison of electrochemical data at potentials relevant to galvanic coupling to metals indicated that at these cathodic potential(s) the CFRP surface is the most electrochemically active of the studied carbon surfaces. On the basis of the values and trends of the electrochemical parameters evaluated, it is postulated that the observed differences in the electrochemical responses of these 3 carbon-rich surfaces to ORR are significantly due to differences in the proportions of edge sites present on each carbon surface. These results could provide valuable insights on plausible strategies for designing carbon surfaces and carbon fiber composites with reduced activity toward ORR for corrosion protection applications or enhanced activity towards ORR for energy applications.

Keywords: galvanic coupling; CFRP; carbon surfaces; cathodic polarization; oxygen reduction reactions (ORR); multi-material combinations

Citation: Ofoegbu, S.U.; Ferreira, M.G.S.; Nogueira, H.I.S.; Zheludkevich, M. Comparison of the Electrochemical Response of Carbon-Fiber-Reinforced Plastic (CFRP), Glassy Carbon, and Highly Ordered Pyrolytic Graphite (HOPG) in Near-Neutral Aqueous Chloride Media. C **2023**, 9, 7. https://doi.org/10.3390/c9010007

Academic Editor: Simonpietro Agnello

Received: 2 October 2022
Revised: 29 December 2022
Accepted: 4 January 2023
Published: 7 January 2023

1. Introduction

Carbon, and particularly carbon surfaces, are in high demand for a wide range of technological applications. Carbon materials are used in a variety of applications as anode or cathode materials due to their low cost and chemical stability in a variety of environments, good thermal and electrical conductivity, and wide potential window (2.5 to

3.5 V [1,2]). Carbon electrodes and surfaces are employed in a wide variety of applications including structural reinforcements [3–5], electroanalysis (i.e., sensor applications) [6–9], electrosynthesis [10–12], energy applications [13,14], environmental remediation and water treatment [13,15], and many more. The use of carbon in the form of carbon fibers in electrochemical systems can be considered a miniaturization step that yields a variety of advantages: the possibility of concurrent electroanalysis in small volumes and at high speeds with enhanced signal quality due to the "microelectrode effect" [16,17] and the subsequent emergence of micro-electrode arrays (MEAs) [18–20]. The incorporation of nanostructured carbon such as nanotubes into carbon-reinforced composites [21,22], the growth of carbon nanotubes on carbon fiber [23,24], and other carbon substrate surfaces [25,26] have been reported to yield remarkable improvements in electrochemical activity [24], and subsequently led to the emergence of carbon-based nanoelectrode arrays with a very wide range of applications [26,27].

In a lot of these applications, the chemical and electrochemical reactivity of carbon surfaces is an important performance-determining indicator. Hence, insights into the electrochemical response of a variety of carbon surfaces are important in order to maximize material surface capabilities in the design of highly efficient application-specific engineered carbon surfaces. In this work, the electrochemical response of three carbon-rich surfaces (glassy carbon (GC), highly ordered pyrolytic graphite, and carbon fiber-reinforced polymer composite surface (CFRP)) towards oxygen reduction reactions (ORR) in quiescent aqueous media in configurations consistent with conditions in many corrosion systems has been studied, and the respective responses have been compared. Carbon-fiber-reinforced polymer composites are often components of multi-material structures increasingly employed in the aeronautical and automobile industries in a bid to realize the weight-to-strength ratios necessary to achieve reduced emission targets. In such multi-material combinations, the galvanic coupling of conductive carbon surfaces to metallic materials is likely [28]. In this configuration, galvanic corrosion (anodic dissolution) of the coupled metal is promoted by cathodic reactions (primarily ORR) that take place on conductive carbon surfaces that act as a cathode.

Glassy carbon has been described as a graphene-rich form of elemental carbon derived from carbon-rich polymer pyrolysis and comprising three-dimensionally arranged curved graphene fragments together with fractions of disordered carbon and voids [29,30]. The presence of voids in glassy carbon is thought to account for its low density ($\approx$1.5 g cm^{-3}) in comparison to other graphites ($\approx$2.3 g cm^{-3}) [29,31]. Glassy carbon is regarded as a chemically stable form of entirely sp^2-bonded carbon with locally ordered domains and is considered to be the intermediate material between graphite and diamond [32]. Glassy carbon is reported to contain graphite-like nanostructures that, depending on its mode of production, can present as a complex solid containing mixtures of micro-graphitized and fullerenized zones where sp^2 carbon atoms predominate [33–36].

Pyrolytic graphites manifest a high degree of anisotropy in both their electrical and heat conductivity and are impervious to gases [37,38]. Pyrolytic graphites are characterized by large strains in the c directions and a preferred orientation that increases with increasing deposition temperature (e.g., <102:1 to as high as 104:1 as the temperature is increased from 1700 °C to 2500 °C), nearly random layer order (turbostratic), average crystallite dimensions (Lc) ranging from below 100 Å to up to 265 Å depending on deposition temperature, and a relatively strain-free basal plane [38]. Highly oriented pyrolytic graphite (HOPG) is a synthetic graphite prepared by the thermal and/or stress annealing of pyrolytic graphite [39]. The thermal and/or stress annealing of pyrolytic graphite at high temperatures ($\geq$ 2500 °C) results in a material (HOPG) with more uniform interlayer spacing ($\approx$ 3.35 Å) across its thickness and a drastically reduced stacking disorder factor/mosaic spread (the angle between the tiles of graphite) that tends towards zero) [40]. Treated pyrolytic graphite with a mosaic spread of less than 1 is classified as HOPG [41]. HOPG is often used as a well-defined model carbon material for studying the chemical and electrochemical behavior of carbon surfaces [42,43].

Oxygen reduction on carbon surfaces is of great relevance in a variety of applications. In some applications, like fuel cells and other energy applications, a high rate of ORR on carbon surfaces is desirable. However, in other applications, like the corrosion of metals galvanically coupled to metallic materials (multi-material combinations), it is undesirable and needs to be prevented. Whether desirable or undesirable, a better understanding of the mechanism(s) and kinetics of ORR on carbonaceous surfaces under application-relevant conditions is critical for manipulating the ORR phenomenon on carbon surfaces to advantage. Consequently, application-specific designs of carbon surfaces require information-driven surface engineering or modification(s). Due to their potential benefits, there has been sustained research interest in metal-free carbon-based electrocatalysts for oxygen reduction reactions [44–49] as replacements for costly Pt-based ORR catalysts. The advantages of carbon over other electrocatalysts include its stability in neutral solution, low cost, abundance, good electrical conductivity, and light weight [50], making it a good candidate material for the oxygen reduction reactions required in fuel cells.

Carbon-fiber-reinforced polymers (CFRPs) are composite materials made of carbon fibers in a matrix of polymers, usually, epoxy, polyester, and vinyl ester. The carbon fibers can be continuous or discontinuous and can be arranged in a variety of orientations [51,52]. Due to the high strength-to-weight ratios achievable, CFRPs are replacing metallic materials in a lot of applications, particularly in the aeronautical and automobile industies. Carbon-fiber-reinforced polymer (CFRP) is much exploited in the aeronautical [53–56] and automobile [57–62] industries in weight-optimized multi-material combinations that offer high strength-to-weight ratios. In such CFRP-containing multi-material combinations, galvanic coupling of the metal or alloy(s) component(s) with the conductive CFRP composite is often inevitable. Such galvanic coupling of CFRP to the metallic component(s) of multi-material assemblies results in accelerated anodic metal dissolution due to the ability of CFRP (actually, its carbon surfaces) to support cathodic reactions such as oxygen reduction.

Irrespective of the history of any carbon material, carbon surfaces are known to present two types of surface sites with different reactivities: edge-plane sites and basal plane sites [63,64]. Due to the known differences in the electrochemical behavior of edge and basal sites [63–65], their technological applications vary. In all applications, the carbon surface presents either predominantly edge sites, basal sites, or a combination of both. It is on the more electrochemically active edge-plane sites [63,66] that most electrochemical activity (including ORR) predominantly occurs on carbon and carbon-rich surfaces.

Chu and Kinoshita [63] have demonstrated that the electrochemical behavior of carbon surfaces can be enhanced by surface modifications that introduce more edge sites or defects on the carbon surface(s). In contrast, it has been demonstrated that the electrochemical behavior of carbon-rich surfaces can be suppressed by surface modifications that employ surfactants that tend to interact and block these active sites [67–72] or by precipitating inhibitors that deposit on carbon surfaces and thus interfere with the transport of electroactive species to the active sites [73–75].

ORR involves electron transfer steps. It is reported [76] that a carbon matrix with a low degree of graphitization is not conducive to electron transfer and conduction. Some reports [77–79] indicate that the electrochemical oxidation of carbon (glassy carbon) activates the electrode surface, leading to faster electron transfer kinetics, which is desirable for electrochemical sensor applications. Interestingly, Matsumoto et al. [80] prepared contaminant-free graphene oxide (eGO) electrodes via controlled oxidation (by electrolysis in pure water under high voltage) and electrochemically reduced the graphene oxide (eGO) electrodes to obtain reduced eGO (r-eGO). The reduced graphene oxide (r-eGO) thus obtained is reported to display high double-layer charging capacitance, high electrocatalytic activity for oxygen reduction reactions (ORR), and n-type semiconductor electrode behavior. They attributed [80] these excellent properties of reduced graphene oxide (r-eGO) to carbon defects and/or OH groups produced on the reduction of epoxide groups formed at the basal plane. Carbon surfaces are sought after as metal-free electrocatalysts of oxygen reduction reactions (ORR) in alkaline media [81–84], and reports suggest that surface modification to achieve the presence of nitrogen-containing surface groups on carbon materials enhances catalytic activity towards ORR [85–89].

Under sufficient applied polarization and/or in the presence of strong oxidizers [90–93], the graphite layers on carbon surfaces can be oxidized. This oxidation results in oxidized graphitic sheets with the significant presence of hydroxyl (-OH), ketone (C-O), carbonyl (C=O), carboxylic (O=C-O), and epoxide groups on their graphite surface [93–99] and highly oxidized polycyclic molecules (tannic acids or oxidized debris) that apparently result from side reactions during graphite oxidation [93,100]. Through $\pi-\pi$ interactions and hydrogen bonding [93], this debris can remain adsorbed to the oxidized graphite or graphene surface and thus influence the physical and electrochemical responses (electron transfer kinetics) of these carbon surfaces. Wu et al. [101] carried out the nitric acid oxidation of carbon fibers and reported the formation of acidic functional groups on PAN-based high-strength carbon fibers and the adherence of partially oxidized graphitic fragments that are insoluble in water. However, they reported that these partially oxidized graphitic fragments are sufficiently solvated and dissolved in aqueous sodium hydroxide and hence are removable by contact with NaOH (high pH).

Literature data suggest that the formation of these surface species is the initial step in the degradation of carbon surfaces. Yi et al. [102] investigated the electrochemical corrosion of a glassy carbon electrode in alkaline, neutral, and acidic media under conditions similar to those which occur during oxygen reduction reactions (ORR) and oxygen evolution reactions (OER) and discovered that glassy carbon degradation begins with the formation of surface oxides via acid-catalyzed process(es) in acidic media. This process(es) leads to ring opening in the graphitic structure and ultimately oxidation in the bulk material. With respect to degradation in alkaline media, they [102] posited that this occurs through OH radicals' preferential attack on alkyl side chains, which leads to the oxidation of the edges of carbon layers; hence, they become hydrophilic and dissolve.

Literature on the cathodic modification of carbon surfaces appears to be scarce. However, insights from previous works on carbon surface polarization in aqueous media indicate that, under cathodic polarization, the local environment around carbon surface(s) is amenable to being different from the bulk environment and becoming alkaline. Furthermore, refs. [28,51,67] guides us to an appreciation of feasible changes to carbon surfaces subjected to cathodic polarization. A report [80] indicates that, at the limits of the cathodic potentials employed in this work, the electrochemical reduction of any graphitic oxides formed by prior anodic polarization or oxidation might be feasible. The alkaline oxidation of carbon surface(s) is deemed to operate through mechanism(s) that are different from its oxidation in acidic media. For oxidation in acidic media, Yi et al. [103] advanced an acid-catalyzed electrophilic substitution reaction mechanism. Other authors proposed a variety of mechanisms for carbon fiber (carbon surface) oxidation in an alkaline environment, based primarily on OH radicals promoting oxidation [103–105], and oxygen (as an oxidizing agent) enabling the oxidation of the carbon fiber anode in the case of water electrolysis [104,106–108].

The major cathodic process on electrodes immersed in aqueous media under ambient conditions relevant to galvanic coupling with metals is the oxygen reduction reaction (ORR). Based on the electrode material and the solution composition (and pH), the oxygen reduction reaction proceeds either as a direct four-electron process or a two-electron process [109,110]. The mechanism(s) of oxygen reduction processes on carbon surfaces have been severally reported [79,89,109,111–118], but there is no consensus among researchers on any of the proposed mechanisms, the rate-determining step (RDS), or the adsorbed intermediate(s) involved in the electrochemical reduction of oxygen on carbon surfaces [115]. Nevertheless, there appears to be a consensus that the process involves the adsorption of oxygen and/or superoxide. The review article by Šljukić et al. [115] contains details on the most prominent mechanisms proposed for oxygen reduction on carbon surfaces. Oxygen reduction in aqueous media is reported to be influenced by the test media and the nature of the electrode material [119,120]. Morcos and Yeager [121] reported a

Tafel slope of −120 mV/decade at 25 °C in alkaline media, with a stoichiometric number of 2 for the overall two-electron reduction of O_2 to HO_2^- on pyrolytic graphite surfaces. For the reduction of oxygen to HO_2^- and OH^- on glassy carbon, Zhang et al. [122] reported a Tafel slope of −60 mV/decade and a stoichiometric number of 1. Drawing on these works [121,122], Yeager [110] concluded that there are apparent mechanistic differences between O_2 reduction on pyrolytic graphite surfaces and glassy carbon. Major factors reportedly affecting the reactivity of carbon surfaces are mainly the microstructure of the carbon material [42,123–125], the cleanliness of the electrode surface [77,126,127], and the types of functional groups present on the surface [128–134]. This possibility of differences in the mechanism of cathodic processes on different carbon surfaces is the main driving force behind the present research effort.

In a recent publication [51], carbon-fiber-reinforced polymer (CFRP), which is a component in some multi-material assemblies, was characterized in near-neutral aqueous chloride media in an effort to evaluate and understand its ability to sustain cathodic activities (oxygen reduction reactions) under cathodic polarization consistent with galvanic coupling to metals in multi-material combinations. Oxygen-diffusion-limited cathodic current densities in the range of 30 to 40 μA cm^{-2} have been reported [73]. Discounting self-corrosion and assuming equal surface areas, such cathodic current densities (40 μA cm^{-2}) had been estimated to be capable of supporting corrosion rates of about 0.599, 0.436, and 0.132 mm yr^{-1}, respectively, or mass loss rates of 11.708, 3.223, and 2.85 gm^{-2} d^{-1}, respectively, on zinc, aluminium, and iron, respectively, galvanically coupled to CFRP [73]. Efforts aimed at mitigating such cathodic activities at potentials relevant to galvanic coupling to metals by the use of a variety of inhibitors have been reported [66,67], with marginal success obtained with the surfactant: sodium dodecyl sulphate (SDS), Na_3PO_4, lanthanum acetate, and inhibitor combinations comprising sodium nitrate with benzotriazole and cerium acetate with benzotriazole. The reported marginal suppression of cathodic activity on CFRP in the presence of these inhibitors was attributed to two mechanisms. Both mechanisms result in some reduction in the electrochemically active area, either via the interaction of inhibitor molecules with the CFRP surface, most probably by adsorption, and/or oxygen diffusion to the CFRP surface due to high-local-pH-triggered precipitation of inhibitor compounds on the CFRP surface [73].

Since the electrochemical activity of CFRP is attributable to its carbon content (and/or carbon fraction) in the composite in the form of carbon fibers, it can be argued that electrochemical data acquired from various carbon surfaces can be valuable for the design of high-performance multi-material combinations containing CFRP and optimized for both specific strength and the reduction in the electrochemical activity of the CFRP towards oxygen reduction reactions (ORR). To test this hypothesis, the electrochemical response of CFRP with 65% carbon fiber content was compared with electrochemical data acquired from glassy carbon and highly ordered pyrolytic graphite (HOPG), and the results are presented herein. Since the CFRP surface is a composite surface comprised of carbon fibers (of calculable surface coverage) and the polymer matrix and since the carbon fibers are deemed to be the conductive constituent (and plausibly the electrochemically active surface), its electrochemical response is compared with those of other carbon surfaces such as glassy carbon (GC) and highly ordered pyrolytic graphite (HOPG) with a known surface area. By comparison and analysis of the current profile and total quantity of charges passed (from chronoamperometric tests) across the CFRP–, GC–, and HOPG–solution interfaces under cathodic polarization during a fixed time, these carbon surfaces are ranked with respect to their plausible abilities to support cathodic processes and hence the concomitant anodic dissolution of coupled metal(s) in galvanic and/or multi-material combinations.

This work is focused on ORR, which is the predominant reaction at the cathodic potential ranges that CFRP is polarized to, on galvanic coupling to most metals in multi-material assemblies that are now common in the aerospace and transport industries. To mitigate the galvanic corrosion of coupled metallic components in such hybrid structures, it is desirable to engineer coupling carbon surfaces to have a lower response to ORR.

The present work demonstrates and quantifies the implications and magnitude of the technological challenge arising from the well-known differences in the electrochemical activity of basal and edge carbon sites of carbon materials on corrosion mitigation in multi-material combinations. It provides quantitative data that clearly indicates that, in spite of its lower carbon content (about 65%), CFRP is more electrochemically active with regards to ORR compared to carbon surfaces with a greater surface area (glassy carbon and HOPG). The data generated in this study can serve as baselines for evaluating research efforts at modifying carbon-reinforced composite surfaces to achieve reduced response to ORR.

2. Materials and Methods

2.1. Materials Preparation

Pultruded CFRP rods of diameters 1 and 8 mm with 65% fiber content made of Tenax Total GmbH HT 24 K carbon fiber and epoxy vinyl matrix material were obtained from Modulor Material. These were cut to the required lengths and mounted in a polymeric resin (Epokwick) after attaching an insulated wire at one end for electrical contact. Before each test, a fresh surface was exposed by wet-grinding progressively with silicon carbide paper of grit sizes 240, 360, 400, 500, 600, and 1200, washing in between with copious amounts of distilled water. A glassy carbon electrode was prepared from glassy carbon samples with a cross-section of 10 × 10 mm that were mounted in polymeric resin (Epokwick) as for CFRP above, and the surfaces were prepared for testing using the same polishing procedure employed for CFRP. For tests with highly ordered pyrolytic graphite (HOPG), an HOPG sample of dimensions 10 × 10 × 2 mm was mounted on a conductive substrate and fresh surface-cleaved prior to each new test, with a test surface area of ≈ 0.28 cm^2 demarcated using an O-ring of internal diameter ≈ 5.5 mm and an in-house fabricated test rig.

2.2. Test Procedures

Electrochemical tests were carried out using an Autolab PGSTAT302N potentiostat, employing the 3-electrode method with a saturated calomel reference electrode, platinum wire as the counter electrode, and CFRP as the working electrode.

2.2.1. Open-Circuit Potential Measurements

Open-circuit potentials of the CFRP, glassy carbon (GC), and highly ordered pyrolytic graphite (HOPG) in the 50 mM NaCl test solutions were monitored at 60 s intervals immediately after immersion until 10 h at least 3 times respectively.

2.2.2. Potentiodynamic Tests

Triplicate potentiodynamic tests were carried out on CFRP, glassy carbon (GC), and highly ordered pyrolytic graphite (HOPG) respectively, in quiescent 50 mM NaCl test solutions with a potential step size of 2.44 mV and a scan rate of 1 mV s^{-1} after 1 h immersion at OCP (without any conditioning), from 20 mV anodic of OCP towards cathodic potentials up to −2500 mV_{SCE}.

2.2.3. Cyclic Voltammetry

Cyclic voltammetric tests were carried out on each of CFRP, glassy carbon (GC), and highly ordered pyrolytic graphite (HOPG) respectively, using a step size of 2.44 mV and a scan rate of 50 mV s^{-1} from −1000 to +1000 mV_{SCE} (from cathodic to more anodic potentials).

2.2.4. Chronoamperometric Measurements

Hextuple chronoamperometric tests were carried out on each of CFRP, glassy carbon (GC), and highly ordered pyrolytic graphite (HOPG) at a polarization of −1000 mV_{SCE} in quiescent 50 mM NaCl for 1 h with current sampled every 0.5 s.

2.2.5. Electrochemical Impedance Spectroscopy

At OCP, electrochemical impedance spectroscopy was performed on CFRP, glassy carbon (GC), and highly ordered pyrolytic graphite (HOPG) using a single sine wave with a amplitude of 10 mV (RMS) in the frequency range of 100,000 Hz to 10 mHz. To check the effect of time on the impedance spectra, EIS tests were performed sequentially after 0.083 (5 min), 0.5, 1, 2, 3, 4, 5, and 6 h of immersion.

2.2.6. Confocal Raman Spectroscopy

Raman spectra were acquired from the respective carbon surfaces before and after polarization using a WITec Alpha 300 RA+ confocal Raman system (WITec, Ulm, Germany) using a 532 nm laser and a laser power of 1 mW.

2.2.7. Scanning Electron Microscopy and Energy-Dispersive X-Ray Spectroscopy (EDXS)

The surfaces of the respective carbon surfaces before and after 1 h cathodic polarization in 50 mM NaCl at -1000 mV_{SCE} were studied using a Hitachi TM4000Plus tabletop scanning electron microscope with an energy-dispersive X-ray spectroscopy (EDXS) capability provided by a Bruker Quantax 75 EDS system. The elemental mapping of similar surface areas of the respective samples' surfaces enabled the acquisition of comparative surface compositions of the carbon and oxygen on the respective carbon surfaces. Since the CFRP presents a composite surface, in addition to surface mapping, point elemental analysis on the carbon fibers on the CFRP surface was carried out to enable a more realistic comparison of the carbon surface composition before and after polarization.

3. Results

3.1. Comparison of the Open-Circuit-Potential Profiles of Carbon Surfaces in 50 mM NaCl

A comparison of the repeated measurements of the OCP profiles of the three different carbon surfaces (Figure 1) shows that the OCP profiles of CFRP and glassy carbon are very similar, but there is a significant difference in the OCP profiles of HOPG, which is predominantly anodic compared to the other carbon surfaces in the test media. This is principally attributed to poor electrode kinetics of the HOPG basal plane surface. However, for metallic surfaces, such an anodic shift in the evolution of the OCP profiles of HOPG would have been suggestive of a passivated surface and leads us to suspect the possibility of some passivation of the freshly cleaved HOPG surface. Some reports [102,103] appear to suggest that carbon surfaces can be passivated, particularly in acidic media, in a manner similar to that observed with some metallic materials, apparently due to the formation of a covering oxide layer (oxidized debris). Granted that, the nature of metallic surfaces and carbon surfaces differ, however, similar changes can occur on carbon surfaces on polarization. On polarization in aqueous media, the surface of carbon can be changed by (a) changes in the functional groups attached to carbon atoms [97,135], (b) electrochemically induced exfoliation of graphite/graphene layers (most probable in surfaces with a predominance of edge sites) [136,137], (c) erosion of graphite layers and creation of ledges, etc. (most probable in surfaces with a predominance of basal sites) [138–140], and (d) deposition of surface films on the carbon surface [141,142].

This position is further supported by the recent report of Patel et al. [143,144]. The earlier held traditional consensus was that (a) the basal surface of HOPG is characterized by very poor electrode kinetics [145] or even nil electroactivity [42,64,146–152] compared to edge plane graphite for a wide range of redox couples and (b) the intersection of step edges with the basal surface is essentially responsible for all of the sites for electron transfer (ET) for a range of redox couples [33,42,64,147,153–156]. Contesting these points, Patel et al. [153,154] argued strongly and with some merit that freshly cleaved pristine HOPG is much more electrochemically active than previously thought (in the step-edge model of HOPG activity) but also that the HOPG basal surface readily passivates in a variety of ways.

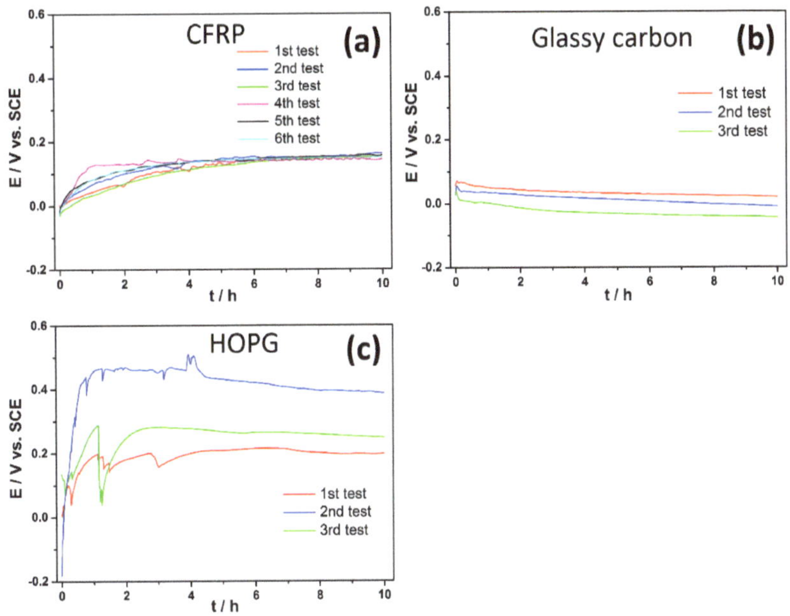

Figure 1. Open−circuit−potential profiles of (**a**) CFRP, (**b**) HOPG, and (**c**) glassy carbon in a 50 mM NaCl solution.

3.2. Comparison of Cathodic Potentiodynamic Polarization Curves of Carbon Surfaces in 50 mM NaCl

From the potentiodynamic polarization curves for the three carbon surfaces (Figure 2), it can be observed that, whereas there is a marked resemblance in the cathodic polarization curves of CFRP and glassy carbon, that of HOPG is markedly different. The major difference in the cathodic polarization curves of HOPG in the test media compared to CFRP and GC (Figure 2d) is the absence of a clearly defined oxygen-diffusion-limited plateau region. For HOPG, oxygen reduction appeared to proceed without transport constraints on its surface until hydrogen evolution became the dominant cathodic reaction (at potentials ≤ -1500 mV$_{SCE}$ for all three carbon surfaces studied). The absence of a defined oxygen-diffusion-limited plateau region in the polarization curve for HOPG is indicative of very sluggish oxygen reduction kinetics on HOPG surfaces, which ensures that oxygen reduction does not come under mass transport (i.e., diffusion) control on HOPG surfaces. This phenomenon can be attributed to the preponderance of the less electrochemically active basal sites in the plane of the highly ordered pyrolytic graphite sample tested and apparently lends support to the "step-edge model" of HOPG activity.

In addition, for HOPG, the potentiodynamic curves (Figure 2d) indicate that the oxygen reduction reaction kinetics on its cleaved surface (basal plane) are slower compared to glassy carbon and CFRP. Morcos and Yeager [121] reported inhibited oxygen reduction and peroxide oxidation on the cleaved surfaces (basal planes) of the oriented graphite relative to edge surfaces in their studies on the kinetics of the oxygen–peroxide couple on pyrolytic graphite. Interestingly, unlike for glassy carbon and CFRP, the polarization curve for HOPG shows two different slopes in the kinetics-controlled region of the curve (from about 0 mV$_{SCE}$ to about -1250 mV$_{SCE}$). Specifically, at potentials more negative (i.e., more cathodic) than -500 mV$_{SCE}$, the slope of the cathodic curve for HOPG (Figure 2d) is observed to change significantly before reaching the diffusion-controlled region around -1250 mV$_{SCE}$. This change in slope is indicative of a change in the predominant reduction mechanism or reaction for HOPG [157]. We attribute the first slope (moving from OCP and terminating around -500 mV$_{SCE}$) to be related to the predominant reduction of O_2 to H_2O_2, and the second slope to the predominant reduction of H_2O_2 to H_2O. These attributions find support from Qiang et al. [158], who in their studies of the electrochemical generation of hydrogen peroxide from dissolved oxygen in acidic solutions using graphite electrodes reported that the optimal potential for the electro-generation of H_2O_2 is a cathodic potential of -500 mV_{SCE}, that the decomposition of H_2O_2 is slow in the absence of metal ions, and that significant H_2O_2 self-decomposition occurs only at high pH (pH > 9) and at high temperatures (>23 °C). In addition, with consideration of the lower electrochemical activity of the HOPG basal plane, a recent report [51] indicates that the polarization of the HOPG-cleaved surface to -500 mV_{SCE} is unlikely to raise the local pH beyond pH 9 so that H_2O_2 persists in solution for its subsequent electroreduction to water.

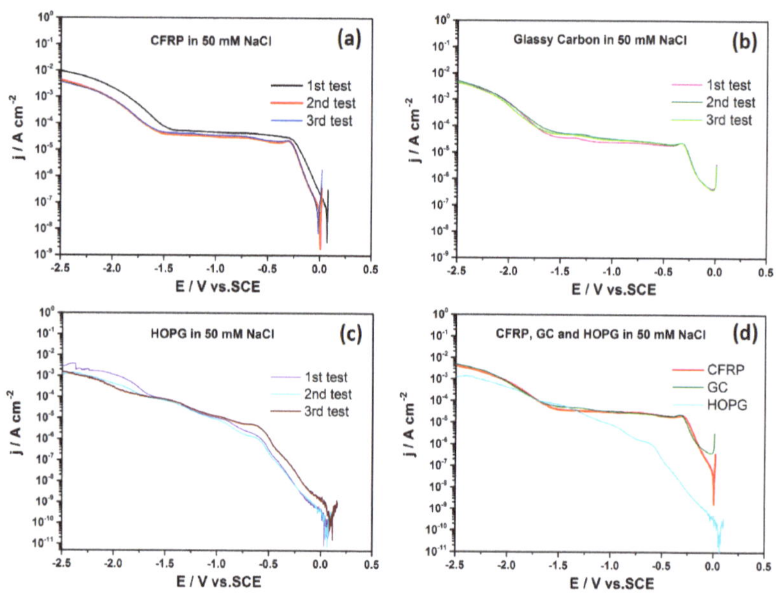

Figure 2. Comparison of cathodic potentiodynamic polarization curves of CFRP (**a**), and glassy carbon (**b**) HOPG (**c**), CFRP + GC+HOPG (**d**), in 50 mM NaCl acquired at scan rates of 1 mV s^{-1}.

3.3. Comparison of Cyclic Voltammograms in 50 mM NaCl

The cyclic voltammograms of the three carbon surfaces (Figure 3) in solutions at different pH further confirms the low electrochemical activity of HOPG compared to other carbon surfaces and its insensitivity to pH in the pH range studied. For CFRP and glassy carbon, their cyclic voltammograms were marked by a prominent cathodic peak around -350 mV_{SCE} at pH $\approx$ 7, 10, and 12. This peak around -350 mV_{SCE} is attributed to a two-electron reduction of oxygen to HO_2^- and OH^- (Eo ≈ -309 mV_{SCE}) [159]. However, at pH $\approx$ 12, a second peak is observed in the cyclic voltammogram of glassy carbon around -700 mV_{SCE}, which we infer to be most probably linked to the reduction of H_2O_2 produced at less negative potentials (less cathodic potentials) [159]. Insights from the analysis of the cathodic polarization curves for HOPG (Figure 2d) indicate that H_2O_2 production is likely to be more predominant in the potential range from 0 mV_{SCE} to -500 mV_{SCE}. The absence of this second peak on CFRP at the same pH (pH $\approx$ 12) is inferred to be indicative of some differences in the mechanism(s) of oxygen reduction on glassy carbon and CFRP surfaces. Earlier works [159–161] reported two peaks in the polarization curves of glassy carbon and polished pyrolytic carbon in oxygen-saturated alkaline solutions and attributed them to oxygen reduction on two different surface sites.

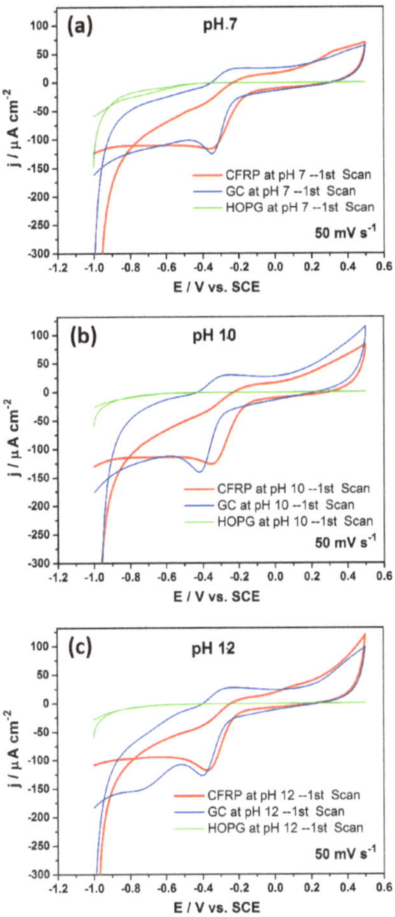

Figure 3. Comparison of cyclic voltammograms of CFRP, HOPG, and glassy carbon in 50 mM NaCl adjusted to pH 7 (**a**), 10 (**b**), and 12 (**c**) respectively, and at scan rates of 50 mV s^{-1}.

3.4. Comparison of Electrochemical Impedance Spectra in 50 mM NaCl

The impedance spectra of the three respective carbon surfaces present a single time constant, as can be seen from the Nyquist plots (Figure 4) and the Bode plots (Figure 5).

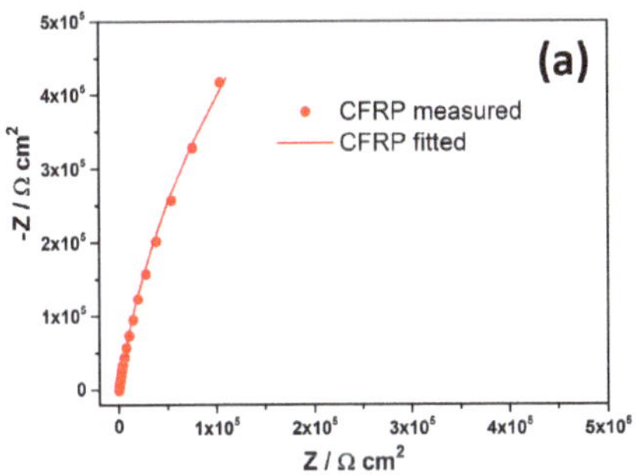

Figure 4. *Cont.*

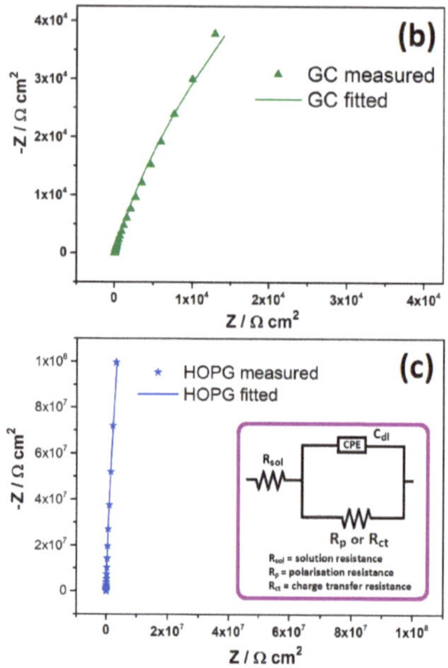

Figure 4. Nyquist plots of CFRP (**a**), glassy carbon (**b**), HOPG (**c**) in 50 mM NaCl adjusted to pH 7 after 1-h immersion at OCP (inset in c is the equivalent circuit used to fit the spectra).

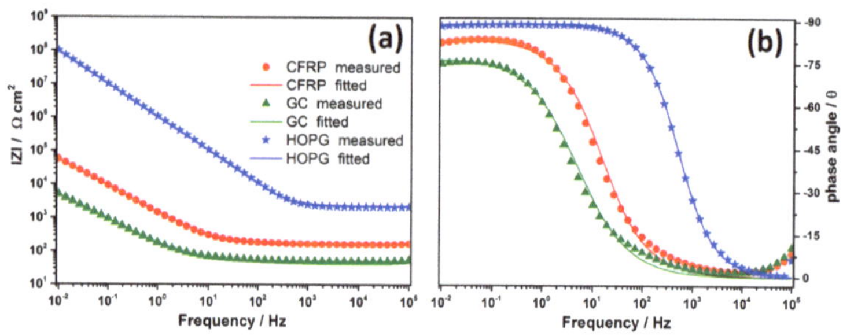

Figure 5. Bode plots of CFRP, HOPG, and glassy carbon in 50 mM NaCl adjusted to pH 7 after 1-h immersion at OCP presented as (**a**) IZI and (**b**) phase angle plots (inset in Figure 4c is the equivalent circuit used to fit the spectra).

The impedance spectra were fitted using the one-time constant equivalent circuit (inset in Figure 4c). In this equivalent circuit, a constant-phase element (CPE) is used instead of a capacitance to simulate the interface capacitance (C_{dl}) generated on the carbon surfaces in order to account for surface inhomogeneities that can lead to a distribution of relaxation time constants and hence deviations from ideal capacitor behavior [162–164]. The CPE is expressed and related to the capacitance according to Equation (1);

$$\frac{1}{Z} = Y = Y_o\,(j\,\omega)^n \tag{1}$$

where Z is the impedance of the CPE, and Y_o and n are the characteristic parameters of the CPE, j is a mathematical operator (imaginary number equal to square root (−1)) associated with the phase information (imaginary component of the circuit), ω is the angular frequency, and the exponent n is valued ($0 < n < 1$); exponent $n = 1$ for the ideal capacitor response.

Capacitances were calculated from impedance data using the relation (Equation (2)) proposed by Brug et al. [165].

$$C = Y_o^{\frac{1}{n}} \left[\frac{1}{R_s} \right]^{\frac{n-1}{n}} \quad (2)$$

where Y_o is the fitted value of CPE element, R_s is the resistance in series with a CPE element (in this case, the solution resistance), and n the CPE exponent.

From the time evolution of low-frequency impedance (at 10^{-2} Hz) (Figure 6a), it can be observed that HOPG presents the highest global impedance of all three carbon surfaces studied, even after accounting for the differences in the "solution resistance". A high global impedance compared to other carbon surfaces is indicative of a lower electrochemical response of HOPG to oxygen reduction, which is sustained during the 6 h test duration By fitting the acquired impedance spectra using the equivalent circuit (Figure 4c inset), it was possible to de-convolute the global impedances into their respective resistive components (charge transfer resistance, Rct) and their capacitive component (the double layer capacitances) and to follow their evolution as a function of time for each carbon surface (Figure 6b,c). From Figure 6b, it is obvious that the polarization resistance is markedly and consistently higher for HOPG than for CFRP or glassy carbon, ranking in the order; HOPG >> CFRP > GC, and indicative of a higher resistance to charge transport across the HOPG–solution interface. Contrary to this hypothesis and in agreement with capacitance values reported by earlier workers for HOPG (1.7 to 2.8 μF cm^{-2} for ZYA-HOPG [143], 0.7 μF cm^{-2} for mechanically cleaved HOPG [46], 1.9 μF cm^{-2} for low-defect HOPG [166], and 3.4 to 7.1 μF cm^{-2} for the ZYH grade of HOPG, and the high-value range (>3 μF cm^{-2}) deemed to be indicative of higher defect concentrations [33,156,166,167]), markedly low double-layer capacitances of 1.39 to 1.77 μF cm^{-2} were obtained in the present work for HOPG compared to 19.5 to 34.4 μF cm^{-2} for CFRP and 13.6 to 85.5 μF cm^{-2} for glassy carbon. Considering the HOPG surface to be predominated by basal-plane graphite sites, the capacitance values obtained for HOPG compared to GC and CFRP is consistent with McCreery and co-workers' report [166] that edge-plane graphite exhibits capacitance values that can be more than 30-fold higher than that of basal-plane graphite. It is worth noting that recent studies postulate that capacitance is a weak indicator of the surface quality of carbon surfaces [168,169].

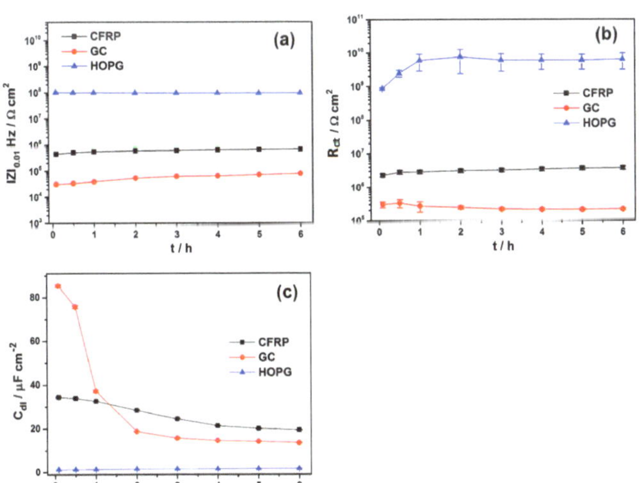

Figure 6. Comparison of the time evolution of measured and fitted parameters from EIS data from CFRP, HOPG, and glassy carbon in 50 mM NaCl adjusted to pH 7, (**a**) low-frequency impedance at 0.01 Hz ($|Z|_{0.01\,Hz}$), (**b**) charge transfer resistance (Rct), and (**c**) interface capacitance (C_{dl}) (error bars indicate fitting errors).

In general, the capacitance values decreased slowly with immersion time in all three carbon surfaces, with the exception of glassy carbon, for which the decrease was fast during the first two hours. Randin and Yeager [170] reported a similar decrease in capacitance with time. The observation (from Figure 6b) that, during the first 1 h, in which the decrease in capacitance was most prominent for glassy carbon and CFRP surfaces (regarded to have more edge sites), coincided with a progressively increasing trend in the parallel resistance (Rct in the inset of Figure 4c) leads us to conclude that this is most likely due to some surface modification(s) on the carbon surfaces that result in an increase in surface area, which might be most prominent on the glassy carbon surface. The increased surface area is bound to increase the value of A (area of plates) in Equation (3), which is directly proportional to the capacitance and is bound to result in a reduction in capacitance as observed. This scenario could explain both the observed decrease in capacitance with time until the surface stabilizes, as well as the capacitance and charge transfer resistance(s) stabilizing to lower and more stable values at extended immersion times ($t \to \infty$).

$$C = \frac{\varepsilon_o \, \varepsilon_r \, A}{d} \tag{3}$$

where C is capacitance, εo is the permittivity of free space (8.854×10^{-12} F m^{-1}), εr is the dielectric constant (of the electrolyte), and A is the distance between the two plates (electrode surfaces).

3.5. *Comparison of Chronoamperomteric Profiles in 50 mM NaCl*

Hextuple chronoamperometric measurements were carried out on CFRP, glassy carbon (GC), and highly ordered pyrolytic graphite (HOPG) in quiescent 50 mM NaCl by polarizing the samples to -1000 mV$_{SCE}$ for 1 h while measuring the current in 0.5 s intervals.

Averaged values of the six chronoamperometric profile measurements are presented in Figure 7, which shows that CFRP was most electrochemically active towards ORR under test conditions. The individual repeated chronoamperometric profiles for each carbon surface are presented in Supplementary Material Figure S1. By the integration of the currents measured over the test duration (1 h), the quantity of charge passed on each test for each carbon surface was determined, normalized using the exposed sample surface to obtain the specific charge passed, and presented as Supplementary Materials Figure S2 and Table S1. Using GraphPad version 6, this data was statistically analyzed using one-way analysis of variance (ANOVA) and Bonferroni's multiple comparison test with significance set at $p < 0.05$. The results from the statistical analysis (inset in Figure 7 and Supplementary Material Table S2) indicate that the averages (CFRP, 129.52 mC cm^{-2}; glassy carbon, 89.95 mC cm^{-2}; HOPG, 60.77 mC cm^{-2}) of the charges passed through each carbon surface are statistically significant. Since during galvanic corrosion with coupled metals (and hence, cathodic polarization), charges need to be passed between the cathode (example CFRP) and the anode (metallic material) to bring about enhanced anodic metal dissolution, a measure of the quantity of charges passed across each carbon surface under same impressed cathodic polarization can be taken as an indicator of the magnitudes of cathodic activities supportable by each carbon surface. Employing this as a ranking criterion, the propensities of the three carbon surfaces (CFRP, GC, and HOPG) to support cathodic activities that can lead to anodic metal dissolution on galvanic coupling to metallic materials can be visualized (as inset in Figure 7).

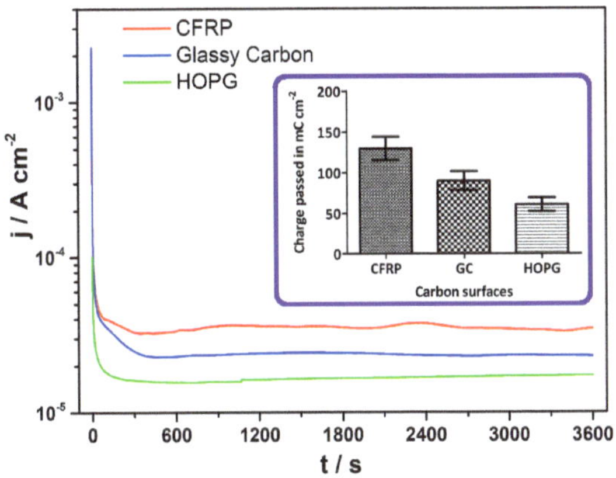

Figure 7. Averaged chronoamperometric profile from hextuple measurements of each carbon surface (Inset is the output from the statistical comparison of charges passed during 1-h polarization of CFRP, HOPG, and glassy carbon in 50 mM NaCl at -1000 mV$_{SCE}$ from hextuple chronoamperometric tests, respectively). (Error bars indicate standard error of means (SEM)).

3.6. Confocal Raman Studies of Carbon Surfaces

From Figure 8, it can be observed that, generally, the polarization of the respective carbon surfaces had little effect on the Raman spectra of the carbon surfaces, as no new peaks were observed after polarization compared to the corresponding unpolarized samples. However, it can be observed that, for glassy carbon (Figure 8b), all the peaks observed in the Raman spectra of the unpolarized sample became much more prominent (manifesting higher Raman intensities) after polarization. In contrast, for HOPG, a marginal decrease in the Raman intensity of all the peaks present in the Raman spectra of an unpolarized HOPG surface (freshly cleaved surface) was observed after polarization.

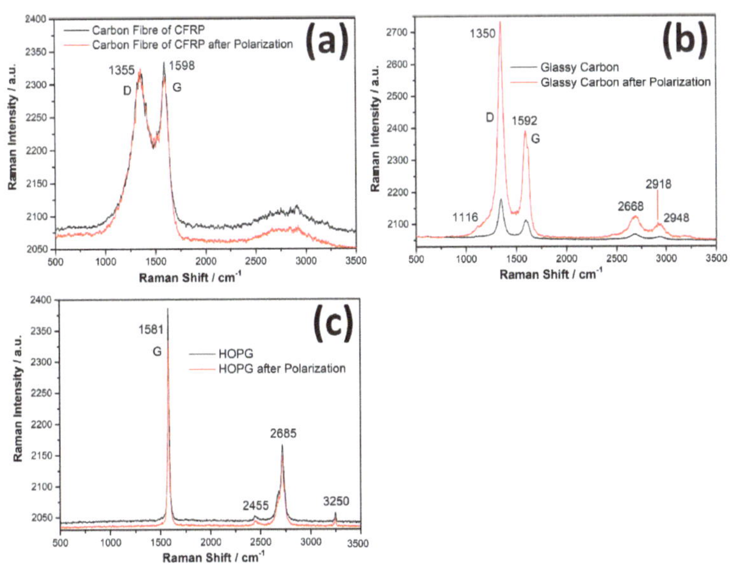

Figure 8. Confocal Raman spectra of (**a**) carbon fiber surfaces in CFRP, (**b**) glassy carbon surfaces, and (**c**) HOPG surfaces before and after 1 h polarization in 50 mM NaCl at -1000 mV$_{SCE}$, respectively.

3.7. Energy-Dispersive Spectroscopy of Carbon Surfaces

From the EDS data of the carbon surfaces in Table 1, it can be observed that the surface composition of each of the three carbon surfaces changed after polarization compared with values measured in unpolarized samples. The oxygen content of the carbon surfaces (atom. %) increased after polarization. A very small increase in surface oxygen concentration was observed on the HOPG surface after polarization. Surface oxygen concentration of the carbon surfaces increased in the following order; CFRP > GC > HOPG. The observed trend of increased oxygen concentration on the post-polarized carbon surfaces is consistent with the ranking of their responses to ORR. Consequently, it is concluded that their respective responses to ORR can be significantly attributed to differences in their propensities to undergo surface modification under the test conditions. The differences in their respective abilities to undergo surface modification can in turn be linked to initial differences in the structural and chemical compositions of the respective carbon surfaces.

Table 1. Table of elemental compositions of carbon surfaces determined by energy-dispersive X-ray spectroscopy (EDS) before and after polarization in a 50 mM NaCl solution.

Carbon Surface	Status	Elements	Atomic No.	Counts	Mass [%]	Mass Norm. [%]	Atom [%]	abs. Error [%] (1σ)	Rel. Error [%] (1σ)
CFRP	polished	C *	6	50,588	95.25	95.25	96.39	4.12	4.33
	polished	O *	8	822	4.75	4.75	3.61	0.31	6.62
	polished	C **	6	41,255	95.73	95.73	96.76	4.14	4.33
	polished	O **	8	593	4.27	4.27	3.24	0.31	7.15
CFRP	**polarized**	**C ***	**6**	**98,313**	**93.28**	**93.28**	**94.89**	**4.05**	**4.35**
	polarized	**O ***	**8**	**2392**	**6.64**	**6.64**	**5.07**	**0.38**	**5.72**
	polarized	**C ****	**6**	**83,433**	**93.04**	**93.04**	**94.68**	**4.04**	**4.35**
	polarized	**O ****	**8**	**2135**	**6.96**	**6.96**	**5.32**	**0.40**	**5.77**
Glassy Carbon	polished	C #	6	61,582	97.61	97.61	98.20	4.19	4.29
	polished	O #	8	463	2.37	2.37	1.79	0.18	7.61
	polished	C ##	6	76,278	97.98	97.98	98.47	4.20	4.28
	polished	O ##	8	482	2.02	2.02	1.53	0.15	7.64
Glassy Carbon	**polarized**	**C**	**6**	**55,184**	**96.66**	**96.66**	**97.54**	**4.18**	**4.33**
		O	**8**	**566**	**3.09**	**3.09**	**2.34**	**0.22**	**7.23**
		Na	**11**	**266**	**0.19**	**0.19**	**0.10**	**0.03**	**15.56**
		Cl	**17**	**201**	**0.07**	**0.07**	**0.02**	**0.01**	**17.09**
HOPG	Freshly cleaved	C	6	266,167	99.65	99.65	99.74	4.23	4.25
		O	8	276	0.35	0.35	0.26	0.04	11.97
HOPG	**polarized**	**C**	**6**	**125,020**	**99.78**	**99.78**	**99.84**	**4.25**	**4.25**
		O	**8**	**76**	**0.21**	**0.21**	**0.16**	**0.06**	**26.78**

[* denotes elemental data values from EDS mapping of CFRP surface comprised of carbon fibers and the epoxy matrix. ** denotes elemental data values from point analysis on carbon fiber on the CFRP surface. # and ## denote the values from the first and second measurements of glassy carbon, respectively, after polishing. Data from polarized samples are in bold].

4. Discussion

In the chronoamperometric measurements in quiescent 50 mM NaCl solutions in Figure 7 (and supplementary materials Figures S1 and S2), the charge passed per square centimeter of the exposed surface during 1 h polarization of CFRP, glassy carbon, and HOPG was in the range of ≈115–147, 80–105, and 51.4–73.6 mC cm^{-2}, respectively, while the averaged values were ≈129.5, 90, and 60.8 mC cm^{-2}, respectively. The fact that there was no overlap in the values measured for the different carbon surfaces in six different measurements on each surface confirms clear differences in the reactivity of the different carbon surfaces under impressed cathodic polarization and under quiescent conditions.

Interestingly, it is observed that the charges passed during 1 h for CFRP are higher than those for glassy carbon in the chronoamperometric tests (Figure 7 and Supplementary Material Figure S1), while their cathodic current densities are similar in the polarization curves (Figure 2). These can be clarified by repeated chronoamperometric testing at different agitation rates in order to control oxygen transport conditions. However, such dynamic tests (under forced oxygen mass transport conditions) are inconsistent with our current focus on the galvanic-induced corrosion of metallic members of multi-material structures containing carbon-based material components under quiescent electrolyte conditions and hence, are beyond the scope of the present study.

The comparison of the quantity of charges passed across CFRP, GC, and HOPG during 1 h under quiescent conditions indicates that CFRP, in spite of its lower fraction of carbon surface, was the most electrochemically active at the impressed cathodic polarization, followed by glassy carbon (GC) and then HOPG. This order of the electrochemical activity of the "carbon surfaces" (CFRP > GC > HOPG) is attributed to differences in the distribution of the more electrochemically active edge plane sites on the different carbon surfaces. This ranking, in which HOPG manifests the least electrochemical activity, is consistent with the report of Cline et al. [171], who reported anomalously slow electron transfer kinetics on HOPG surfaces.

The very low (essentially inert) response of the HOPG surface to ORR can be attributed to the low content of the edge plane sites, which on HOPG surfaces emanate from surface defect sites. According to previous works [64,172,173], a given HOPG surface consists, on average, of 1–10% edge-plane sites. while the remaining surface (90–99%) consists of much less electrochemically active basal-plane sites. The equally low capacitance values observed are consistent with earlier reports (Table 2).

Table 2. Reported capacitance values for different carbon surfaces and sites.

Material	Edge Plane Capacitance ($\mu F\ cm^{-2}$)	Basal Plane Capacitance ($\mu F\ cm^{-2}$)	Material Capacitance ($\mu F\ cm^{-2}$)	References
Graphene	1.0×10^5 (nano-effects could be present)	4		[65]
Graphite	2.8	1.8		[174]
HOPG		0.81–1.0		[145]
Graphite	1–3	50–100		[170,175,176]
Glassy carbon			≥13	[170]
Carbon felt	47.8	3.2		[177]
Glassy carbon			10–25	[177]
Glassy carbon			35	[69]
Hydrogenated glassy carbon			20	[69]
HOPG			6	[69]
Diamond			5	[69]
CFRP			**19.5**–34.5	This work
Glassy carbon			**13.6**–85.5	This work
HOPG			**1.4**–1.8	This work

[In bold are more stable capacitance values obtained after 6 hours' immersion in this work].

According to McDermott et al. [145], glassy carbon has faster kinetics than the HOPG basal plane due to its higher edge plane density. The results from this work indicate that CFRP is more electrochemically active towards ORR than glassy carbon and much more electrochemically active than HOPG (basal plane sites). In addition, the rankings of electrochemical responses and parameters for glassy carbon and HOPG obtained in this work are also consistent with the ranking and (capacitance) values reported by Xu et al. [69].

Typical Raman spectra of carbon materials are often composed in most instances of at least three bands denoted as D (observed around 1350 cm^{-1}), G (observed around 1385 cm^{-1}), and 2D, often also denoted as G′ (observed around 2700 cm^{-1}), respectively [178,179]. The D-band is attributed to the presence of defects (vacancies and dislocations in the graphene layer and at its edges) [178–181]. The G band has been attributed to the in-plane vibration of sp^2-hybridized carbon atoms [182–185] and the 2D or G′ band to the number of graphene layers [186,187] in the carbonaceous material. Besides these, additional bands have been reported (D′ around 1620 cm^{-1}, peak denoted as D* or G* around 2440 cm^{-1}, and other peaks denoted as D + D′ or D + G and 2D′ or G + D) by the deconvolution of the Raman spectra, and additional information on the carbonaceous material has been extracted by the comparison of the ratios of the intensities of these peaks [179,181,188,189]. Employing a laser source with a wavelength similar to that used in this work (532 nm), Kaniyoor and Ramaprabhu [179] reported that the Raman spectra of graphite oxide (GO) present intense D and G bands and a flat 2D region and show that the D band is more intense than the G band (with I_D/I_G = 1.1). In this work (Figure 8b), the I_D/I_G ratios of glassy carbon increased from 1.03 in unpolarized samples to 1.14 in the polarized sample, which suggests the presence of graphite oxide after cathodic polarization in the test media. Both graphene oxide (GO) and reduced graphene oxide (rGO) are reported to be catalytic towards ORR by causing it to proceed via a two-electron mechanism at low overpotentials [190]. However, reduced graphene oxide (rGO) is much more catalytic toward ORR than graphene oxide (GO) [190,191]. According to Kauppila et al. [192], graphene oxide can be electrochemically reduced in an aqueous solution by applying cathodic polarization in the range of −711 mV vs. Ag/AgCl (≡−756 mV vs. SCE) at pH 2 to −1416 mV vs. Ag/AgCl (≡−1661 mV vs. SCE) at pH 12. The reduction of graphene oxide in this cathodic potential range has been corroborated by other workers [190,193–195]. Based on the fact that (a) the cathodic potential used in this work falls within the range reported for the graphene oxide (GO) electrochemical reduction to reduced graphene oxide (rGO) in an aqueous solution, (b) the presence of graphite/graphene oxide post anodization on glassy carbon can be inferred from the Raman spectra (Figure 8b), and (c) increased oxygen contents in all three carbon surfaces were detected by EDS analysis after cathodic polarization (Table 1), we postulate the likely presence of graphene oxide and/or reduced graphene oxide in situ, in varying proportions on the carbon surfaces during cathodic polarization. This can partly account for the observed differences in their respective responses to ORR under the test conditions.

The high electrochemical activity of CFRP demonstrated in this work can be linked to the production process of the carbon fibers and their uni-directional arrangement (in the samples used in this work), which favor the presence of more edge sites on the exposed (transverse) ends of the carbon fibers in the CFRP samples [64]. In light of this, it is obvious that the electrochemical activity of CFRP samples with multi-directional carbon fibers is likely to differ from the values reported herein. The intermediate electrochemical activity of glassy carbon is attributed to the fact that the glassy carbon surface is composed of a less steep distribution of the more electrochemically active edge sites and the less electrochemically active basal sites. The lower measured electrochemical activity of the HOPG is attributed to the predominance of the less electrochemically active basal plane sites [196] on the surface of the highly ordered carbon surface tested. It is noteworthy that, in spite of the differences in the electrochemical activity of the different carbon surfaces, the cathodic current densities (attributable to oxygen reduction processes) varied insignificantly for CFRP and glassy carbon in the polarization plots (Figure 2d). Based on of the values and trends of electrochemical parameters (primarily capacitance) obtained in this work for CFRP (19.5 to 34.5 μF cm^{-2}), glassy carbon (13.6 to 85.5 μF cm^{-2}), and HOPG (1.4 to 1.8 μF cm^{-2}), as well as values reported by Yuan et al. [65] for edge plane sites

(1.0 × 10^5 μF cm^{-2}) and basal plane sites (4 μF cm^{-2}), and by other workers (presented in Table 1) on carbon materials, we posit that the observed differences in electrochemical responses of these three carbon-rich surfaces to ORR is significant due to differences in the proportions of edge sites present on each carbon surface, as illustrated in Figure 9.

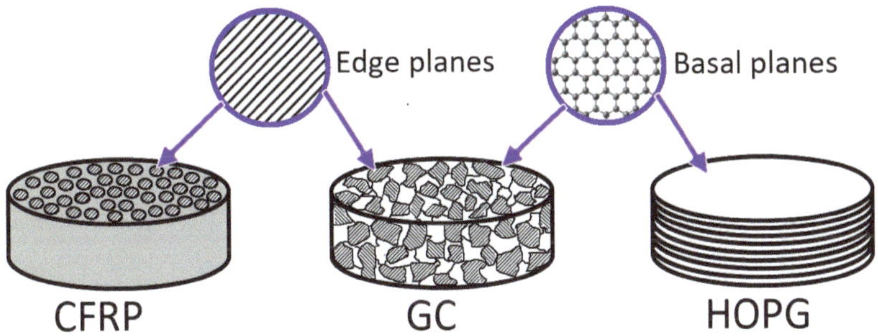

Figure 9. Schematic representation of the hypothesized differences in electrochemically active surfaces of the carbon surfaces studied.

5. Conclusions

A comparison of the electrochemical data of the studied carbon surfaces at potentials relevant to galvanic coupling to metals indicated that, at this cathodic potential, the CFRP surface is the most electrochemically active of the studied carbon surfaces, with the ranking of the surfaces following the order; CFRP > GC > HOPG. This ranking is attributed to differences in the ratios of the more electrochemically active edge sites compared to the less active basal sites on each carbon surface. Based on these results, it is concluded that the use of data acquired from non-composite ("homogenous") carbon surfaces to estimate the electrochemical activity of carbon-fiber-reinforced polymers (CFRP) in the design of corrosion-resistant multi-material assemblies is likely to lead to an under-estimation of the severity of the cathodic activities on the CFRP surface that support anodic dissolution of metals galvanically coupled to CFRP in such hybrid assemblies. When collecting such data from CFRP surfaces is difficult, data collected from the glassy carbon surface are advocated as a viable alternative based on the strength of trends observed in the potentiodynamic polarization and electrochemical impedance spectroscopy tests in this work. Engineering carbon-based surfaces to appear as dispersed micrometer-sized arrays of predominantly carbon edge plane sites can be a feasible technological approach for applications requiring fast oxidation rates for ORR. In contrast, for corrosion mitigation, in which the suppression of ORR is a goal, surface engineering to achieve a predominance of carbon basal sites can be a suitable solution. It is envisaged that these results will motivate future efforts at engineering CFRP surfaces that have a reduced response to ORR similar to that observed on HOPG, as this would result in more corrosion-resistant hybrid structures. In addition, the data from this work can be valuable as a baseline for ranking the performance of carbon surfaces engineered for reduced response to ORR.

Supplementary Materials: The following supporting information can be downloaded at: https://www.mdpi.com/article/10.3390/c9010007/s1, Figure S1: Hextuple chronoamperometic profiles of the 3 carbon surfaces during 1 h polarization at −1000 mV_{SCE} in 50 mM NaCl; Figure S2: Comparison of charges passed during 1 h polarization of CFRP, HOPG, and glassy carbon in 50 mM NaCl at −1000 mV_{SCE} in hextuple chronoamperometric tests; Table S1: Table of integrated charges passed in hextuple chronoamperomteric measurements; Table S2: Table of outputs from ANOVA tests for significance.

Author Contributions: Conceptualization, S.U.O. and M.Z.; methodology, S.U.O.; validation, S.U.O., M.G.S.F., H.I.S.N., and M.Z.; formal analysis, S.U.O. and H.I.S.N.; investigation, S.U.O. and H.I.S.N.;

resources, M.Z. and H.I.S.N.; data curation, S.U.O.; writing—original draft preparation, S.U.O. writing—review and editing, M.G.S.F. and M.Z.; visualization, S.U.O.; supervision, M.G.S.F. and M.Z.; project administration, M.Z.; funding acquisition, M.Z. All authors have read and agreed to the published version of the manuscript.

Funding: S.U. Ofoegbu acknowledges Fundação para a Ciência e a Tecnologia (FCT) Portugal for the doctoral grant (SFRH/BD/75167/2010). Funding from FCT project: "Corrosion and Corrosion Protection in Multi-material Systems", (PTDC/CTM/108446/2008), and European FP7 project: "Active PROtection of multi-material assemblies for AIRcrafts" ["PROAIR" (PIAPP-GA-2013-612415)] are acknowledged. This work was developed within the scope of the project CICECO-Aveiro Institute of Materials, POCI-01-0145-FEDER-007679 (FCT Ref. UID/CTM/50011/2013), financed by national funds through the FCT/MEC and, when appropriate, co-financed by FEDER under the PT2020 Partnership Agreement.

Institutional Review Board Statement: Not applicable.

Informed Consent Statement: Not applicable.

Data Availability Statement: The raw data generated in the course of this study and supporting reported results are publicly archived in the Mendeley data repository and can be accessed using the link: https://doi.org/10.17632/x8c3kpc49g.2.

Acknowledgments: Rui Sampaio and Alexandre C. Bastos both of the Department of Materials and Ceramic Engineering, CICECO-Aveiro Institute of Materials, Aveiro (Aveiro, Portugal), are gratefully acknowledged for graciously providing the glassy carbon samples used, Kiryl Yasakau of the Department of Materials and Ceramic Engineering, CICECO-Aveiro Institute of Materials, Aveiro (Aveiro, Portugal), is gratefully acknowledged for graciously loaning us the HOPG sample.

Conflicts of Interest: The authors declare no conflict of interest. The funders had no role in the design of the study; in the collection, analyses, or interpretation of data; in the writing of the manuscript; or in the decision to publish the results.

References

1. Honda, K.; Rao, T.N.; Tryk, D.A.; Fujishima, A.; Watanabe, M.; Yasui, K.; Masuda, H. Impedance characteristics of the nanoporous honeycomb diamond electrodes for electrical double-layer capacitor applications. *J. Electrochem. Soc.* **2001**, *148*, A668. [CrossRef]
2. Yoshimura, M.; Honda, K.; Kondo, T.; Uchikado, R.; Einaga, Y.; Rao, T.N.; Tryk, D.A.; Fujishima, A. Factors controlling the electrochemical potential window for diamond electrodes in non-aqueous electrolytes. *Diam. Relat. Mater.* **2002**, *11*, 67–74. [CrossRef]
3. Torokhov, V.G.; Chukov, D.I.; Tcherdyntsev, V.V.; Sherif, G.; Zadorozhnyy, M.Y.; Stepashkin, A.A.; Larin, I.I.; Medvedeva, E.V. Mechanical and Thermophysical Properties of Carbon Fiber-Reinforced Polyethersulfone. *Polymers* **2022**, *14*, 2956. [CrossRef] [PubMed]
4. Barile, C.; Casavola, C.; Pappalettera, G.; Kannan, V.P.; Renna, G. Investigation of Interlaminar Shear Properties of CFRP Composites at Elevated Temperatures Using the Lempel-Ziv Complexity of Acoustic Emission Signals. *Materials* **2022**, *15*, 4252. [CrossRef] [PubMed]
5. Guo, F.; Li, W.; Jiang, P.; Chen, F.; Liu, Y. Deep Learning Approach for Damage Classification Based on Acoustic Emission Data in Composite Materials. *Materials* **2022**, *15*, 4270. [CrossRef]
6. Brett, C.M.; Angnes, L.; Liess, H.D. Carbon film resistors as electrodes: Voltammetric properties and application in electroanalysis. *Electroanal. Int. J. Devoted Fundam. Pract. Asp. Electroanal.* **2001**, *13*, 765–769. [CrossRef]
7. Rana, A.; Baig, N.; Saleh, T.A. Electrochemically pretreated carbon electrodes and their electroanalytical applications—A review. *J. Electroanal. Chem.* **2019**, *833*, 313–332. [CrossRef]
8. Magesh, V.; Sundramoorthy, A.K.; Ganapathy, D. Recent advances on synthesis and potential applications of carbon quantum dots. *Front. Mater.* **2022**, 383. [CrossRef]
9. Murugan, P.; Nagarajan, R.D.; Sundramoorthy, A.K.; Ganapathy, D.; Atchudan, R.; Nallaswamy, D.; Khosla, A. Electrochemical detection of H2O2 using an activated glassy carbon electrode. *ECS Sens. Plus* **2022**, *1*, 034401. [CrossRef]
10. Groult, H.; Devilliers, D.; Vogler, M.; Hinnen, C.; Marcus, P.; Nicolas, F. Electrochemical behaviour and surface analysis of crude and modified carbon electrodes for fluorine production. *Electrochim. Acta* **1993**, *38*, 2413–2421. [CrossRef]
11. Aryal, N.; Ammam, F.; Patil, S.A.; Pant, D. An overview of cathode Materials for microbial electrosynthesis of chemicals from carbon dioxide. *Green Chem.* **2017**, *19*, 5748–5760. [CrossRef]
12. Kornienko, G.V.; Kolyagin, G.A.; Kornienko, V.L.; Parfenov, V.A. Graphitized carbon Materials for electrosynthesis of H_2O_2 from O_2 in gas-diffusion electrodes. *Russ. J. Electrochem.* **2016**, *52*, 983–987. [CrossRef]

13. Le, T.X.H.; Bechelany, M.; Cretin, M. Carbon felt based-electrodes for energy and environmental applications: A review. *Carbon* **2017**, *122*, 564–591. [CrossRef]
14. Kamedulski, P.; Lukaszewicz, J.P.; Witczak, L.; Szroeder, P.; Ziolkowski, P. The importance of structural factors for the electrochemical performance of graphene/carbon nanotube/melamine powders towards the catalytic activity of oxygen reduction reaction. *Materials* **2021**, *14*, 2448. [CrossRef]
15. Liu, Y.; Ren, Y.; You, S. Electrified carbon nanotube membrane technology for water treatment. In *Electrochemical Membrane Technology for Water and Wastewater Treatment*; Elsevier: Amsterdam, The Netherlands, 2022; pp. 111–140. [CrossRef]
16. Hosford, P.S.; Wells, J.A.; Christie, I.N.; Lythgoe, M.F.; Millar, J.; Gourine, A.V. Electrochemical carbon fiber-based technique for simultaneous recordings of brain tissue PO_2, pH, and extracellular field potentials. *Biosens. Bioelectron.* **2019**, *3*, 100034. [CrossRef]
17. Huffman, M.L.; Venton, B.J. Carbon-fiber microelectrodes for in vivo applications. *Analyst* **2009**, *134*, 18–24. [CrossRef]
18. Liu, R.; Feng, Z.Y.; Li, D.; Jin, B.; Lan, Y.; Meng, L.Y. Recent trends in carbon-based microelectrodes as electrochemical sensors for neurotransmitter detection: A review. *TrAC Trends Anal. Chem.* **2022**, *148*, 116541. [CrossRef]
19. Chen, R.; Li, Y.; Huo, K.; Chu, P.K. Microelectrode arrays based on carbon nanomaterials: Emerging electrochemical sensors for biological and environmental applications. *RSC Adv.* **2013**, *3*, 18698–18715. [CrossRef]
20. Beidaghi, M.; Chen, W.; Wang, C. Electrochemically activated carbon micro-electrode arrays for electrochemical micro-capacitors. *J. Power Sources* **2011**, *196*, 2403–2409. [CrossRef]
21. Kamae, T.; Drzal, L.T. Carbon Fiber/Epoxy Composite Property Enhancement through Incorporation of Carbon Nanotubes at the Fiber-Matrix Interphase–Part II: Mechanical and Electrical Properties of Carbon Nanotube Coated Carbon Fiber Composites. *Compos. Part A: Appl. Sci. Manuf.* **2022**, *160*, 107023. [CrossRef]
22. Du, X.; Zhou, H.; Sun, W.; Liu, H.Y.; Zhou, G.; Zhou, H.; Mai, Y.W. Graphene/epoxy interleaves for delamination toughening and monitoring of crack damage in carbon fibre/epoxy composite laminates. *Compos. Sci. Technol.* **2017**, *140*, 123–133. [CrossRef]
23. Xiang, L.; Yu, P.; Hao, J.; Zhang, M.; Zhu, L.; Dai, L.; Mao, L. Vertically aligned carbon nanotube-sheathed carbon fibers as pristine microelectrodes for selective monitoring of ascorbate in vivo. *Anal. Chem.* **2014**, *86*, 3909–3914. [CrossRef] [PubMed]
24. Morais, E.A.D.; Alvial, G.; Longuinhos, R.; Figueiredo, J.M.A.; Lacerda, R.G.; Ferlauto, A.S.; Ladeira, L.O. Enhanced electrochemical activity using vertically aligned carbon nanotube electrodes grown on carbon fiber. *Mater. Res.* **2011**, *14*, 403–407. [CrossRef]
25. Park, S.; Dong-Won, P.; Yang, C.S.; Kim, K.R.; Kwak, J.H.; So, H.M.; Ahn, C.W.; Kim, B.S.; Chang, H.; Lee, J.O. Vertically aligned carbon nanotube electrodes directly grown on a glassy carbon electrode. *ACS Nano* **2011**, *5*, 7061–7068. [CrossRef] [PubMed]
26. Akinoglu, E.M.; Kätelhön, E.; Pampel, J.; Ban, Z.; Antonietti, M.; Compton, R.G.; Giersig, M. Nanoscopic carbon electrodes: Structure, electrical properties and application for electrochemistry. *Carbon* **2018**, *130*, 768–774. [CrossRef]
27. Tu, Y.; Lin, Y.; Yantasee, W.; Ren, Z. Carbon nanotubes based nanoelectrode arrays: Fabrication, evaluation, and application in voltammetric analysis. *Electroanalysis* **2005**, *17*, 79–84. [CrossRef]
28. Ofoegbu, S.U.; Ferreira, M.G.; Zheludkevich, M.L. Galvanically stimulated degradation of carbon-fiber reinforced polymer composites: A critical review. *Materials* **2019**, *12*, 651. [CrossRef]
29. Jenkins, G.M.; Kawamura, K.; Ban, L.L. Formation and structure of polymeric carbons. *Proc. R. Soc. Lond. A. Math. Phys. Sci.* **1972**, *327*, 501–517. [CrossRef]
30. Sharma, S.; Shyam Kumar, C.N.; Korvink, J.G.; Kübel, C. Evolution of glassy carbon microstructure: In situ transmission electron microscopy of the pyrolysis process. *Sci. Rep.* **2018**, *8*, 1–12. [CrossRef]
31. Yamada, S.; Sato, H. Some physical properties of glassy carbon. *Nature* **1962**, *193*, 261–262. [CrossRef]
32. Diaf, H.; Pereira, A.; Melinon, P.; Blanchard, N.; Bourquard, F.; Garrelie, F.; Donnet, C.; Vondráčk, M. Revisiting thin film of glassy carbon. *Phys. Rev. Mater.* **2020**, *4*, 066002. [CrossRef]
33. McCreery, R.L. Advanced carbon electrode Materials for molecular electrochemistry. *Chem. Rev.* **2008**, *108*, 2646–2687. [CrossRef] [PubMed]
34. Jenkins, G.M.; Kawamura, K. Structure of glassy carbon. *Nature* **1971**, *231*, 175–176. [CrossRef] [PubMed]
35. Harris, P.J.F. Fullerene-related structure of commercial glassy carbons. *Philos. Mag.* **2004**, *84*, 3159–3167. [CrossRef]
36. Uskoković, V. A historical review of glassy carbon: Synthesis, structure, properties and applications. *Carbon Trends* **2021**, *5*, 100116. [CrossRef]
37. Robertson, S.D. Graphite formation from low temperature pyrolysis of methane over some transition metal surfaces. *Nature* **1969**, *221*, 1044–1046. [CrossRef]
38. Guentert, O.J. X-ray Study of Pyrolytic Graphites. *J. Chem. Phys.* **1962**, *37*, 884–891. [CrossRef]
39. Moore, A.W. Pyrolytic Carbon and Graphite. In *Encyclopedia of Materials: Science and Technology*; Jürgen Buschow, K.H., Cahn, R.W., Flemings, M.C., Ilschner, B., Kramer, E.J., Mahajan, S., Veyssière, P., Eds.; Elsevier: Amsterdam, The Netherlands, 2001; pp. 7933–7937. [CrossRef]
40. Moore, A.W.; Ubbelohde, A.R.J.P.; Young, D.A. Stress recrystallization of pyrolytic graphite. *Proc. R. Soc. Lond. Ser. A. Math. Phys. Sci.* **1964**, *280*, 153–169. [CrossRef]
41. Devi, M.; Rawat, S.; Sharma, S. A comprehensive review of the pyrolysis process: From carbon nanomaterial synthesis to waste treatment. *Oxf. Open Mater. Sci.* **2021**, *1*, itab014. [CrossRef]
42. Bowling, R.J.; Packard, R.T.; McCreery, R.L. Activation of highly ordered pyrolytic graphite for heterogeneous electron transfer: Relationship between electrochemical performance and carbon microstructure. *J. Am. Chem. Soc.* **1989**, *111*, 1217–1223. [CrossRef]

43. Iamprasertkun, P.; Ejigu, A.; Dryfe, R.A. Understanding the electrochemistry of "water-in-salt" electrolytes: Basal plane highly ordered pyrolytic graphite as a model system. *Chem. Sci.* **2020**, *11*, 6978–6989. [CrossRef]
44. Edvardsen, L.; Grandcolas, M.; Lædre, S.; Yang, J.; Lange, T.; Bjørge, R.; Gawel, K. Conductive epoxy/carbon nanofiber coatings for scale control. *Surf. Coat. Technol.* **2021**, *425*, 127694. [CrossRef]
45. Zhou, Y.; Yan, L.; Hou, J. Nanosheets with High-Performance Electrochemical Oxygen Reduction Reaction Revived from Green Walnut Peel. *Molecules* **2022**, *27*, 328. [CrossRef] [PubMed]
46. Sawant, S.Y.; Han, T.H.; Cho, M.H. Metal-free carbon-based materials: Promising electrocatalysts for oxygen reduction reaction in microbial fuel cells. *Int. J. Mol. Sci.* **2017**, *18*, 25. [CrossRef] [PubMed]
47. de Falco, G.; Florent, M.; Jagiello, J.; Cheng, Y.; Daemen, L.L.; Ramirez-Cuesta, A.J.; Bandosz, T.J. Alternative view of oxygen reduction on porous carbon electrocatalysts: The substance of complex oxygen-surface interactions. *Iscience* **2021**, *24*, 102216. [CrossRef]
48. Liu, Y.; Hu, M.; Xu, W.; Wu, X.; Jiang, J. Catalytically active carbon from cattail fibers for electrochemical reduction reaction. *Front. Chem.* **2019**, 786. [CrossRef]
49. Chen, P.; Wang, L.K.; Wang, G.; Gao, M.R.; Ge, J.; Yuan, W.J.; Shen, Y.H.; Xie, A.J.; Yu, S.H. Nitrogen-doped nanoporous carbon nanosheets derived from plant biomass: An efficient catalyst for oxygen reduction reaction. *Energy Environ. Sci.* **2014**, *7*, 4095–4103. [CrossRef]
50. Nie, Y.; Li, L.; Wei, Z. Recent advancements in Pt and Pt-free catalysts for oxygen reduction reaction. *Chem. Soc. Rev.* **2015**, *44*, 2168–2201. [CrossRef]
51. Ofoegbu, S.U.; Quevedo, M.C.; Bastos, A.C.; Ferreira, M.G.S.; Zheludkevich, M.L. Electrochemical characterization and degradation of carbon fibre reinforced polymer in quiescent near neutral chloride media. *Npj Mater. Degrad.* **2022**, *6*, 1–14. [CrossRef]
52. Rahmani, H.; Najafi, S.H.M.; Ashori, A. Mechanical performance of epoxy/carbon fiber laminated composites. *J. Reinf. Plast. Compos.* **2014**, *33*, 733–740. [CrossRef]
53. Soutis, C. Carbon fiber reinforced plastics in aircraft construction. *Mater. Sci. Eng. A* **2005**, *412*, 171–176. [CrossRef]
54. Williams, G.; Trask, R.; Bond, I. A self-healing carbon fibre reinforced polymer for aerospace applications. *Compos. Part A Appl. Sci. Manuf.* **2007**, *38*, 1525–1532. [CrossRef]
55. Botelho, E.C.; Silva, R.A.; Pardini, L.C.; Rezende, M.C. A review on the development and properties of continuous fiber/epoxy/aluminum hybrid composites for aircraft structures. *Mater. Res.* **2006**, *9*, 247–256. [CrossRef]
56. Marsh, G. Airbus takes on Boeing with reinforced plastic A350 XWB. *Reinf. Plast.* **2007**, *51*, 26–29. [CrossRef]
57. Rezaei, F.; Yunus, R.; Ibrahim, N.A.; Mahdi, E.S. Development of short-carbon-fiber-reinforced polypropylene composite for car bonnet. *Polym. Plast. Technol. Eng.* **2008**, *47*, 351–357. [CrossRef]
58. Jacob, G.C.; Fellers, J.F.; Simunovic, S.; Starbuck, J.M. Energy absorption in polymer composites for automotive crashworthiness. *J. Compos. Mater.* **2002**, *36*, 813–850. [CrossRef]
59. Fuchs, E.R.; Field, F.R.; Roth, R.; Kirchain, R.E. Strategic Materials selection in the automobile body: Economic opportunities for polymer composite design. *Compos. Sci. Technol.* **2008**, *68*, 1989–2002. [CrossRef]
60. Alves, C.; Silva, A.J.; Reis, L.G.; Freitas, M.; Rodrigues, L.B.; Alves, D.E. Ecodesign of automotive components making use of natural jute fiber composites. *J. Clean. Prod.* **2010**, *18*, 313–327. [CrossRef]
61. Al-Qureshi, H.A. Automobile leaf springs from composite materials. *J. Mater. Process. Technol.* **2001**, *118*, 58–61. [CrossRef]
62. Talib, A.A.; Ali, A.; Badie, M.A.; Lah, N.A.C.; Golestaneh, A.F. Developing a hybrid, carbon/glass fiber-reinforced, epoxy composite automotive drive shaft. *Mater. Des.* **2010**, *31*, 514–521. [CrossRef]
63. Chu, X.; Kinoshita, K. Surface modification of carbons for enhanced electrochemical activity. *Mater. Sci. Eng. B* **1997**, *49*, 53–60. [CrossRef]
64. Banks, C.E.; Davies, T.J.; Wildgoose, G.G.; Compton, R.G. Electrocatalysis at graphite and carbon nanotube modified electrodes: Edge-plane sites and tube ends are the reactive sites. *Chem. Commun.* **2005**, *36*, 829–841. [CrossRef] [PubMed]
65. Yuan, W.; Zhou, Y.; Li, Y.; Li, C.; Peng, H.; Zhang, J.; Liu, Z.; Dai, L.; Shi, G. The edge-and basal-plane-specific electrochemistry of a single-layer graphene sheet. *Sci. Rep.* **2013**, *3*, 1–7. [CrossRef] [PubMed]
66. Shen, A.; Zou, Y.; Wang, Q.; Dryfe, R.A.; Huang, X.; Dou, S.; Wang, S. Oxygen reduction reaction in a droplet on graphite: Direct evidence that the edge is more active than the basal plane. *Angew. Chem.* **2014**, *126*, 10980–10984. [CrossRef]
67. Ofoegbu, S.U.; Yasakau, K.; Kallip, S.; Nogueira, H.I.; Ferreira, M.G.S.; Zheludkevich, M.L. Modification of carbon fibre reinforced polymer (CFRP) surface with sodium dodecyl sulphate for mitigation of cathodic activity. *Appl. Surf. Sci.* **2019**, *478*, 924–936. [CrossRef]
68. Brownson, D.A.; Metters, J.P.; Kampouris, D.K.; Banks, C.E. Graphene electrochemistry: Surfactants inherent to graphene can dramatically effect electrochemical processes. *Electroanalysis* **2011**, *23*, 894–899. [CrossRef]
69. Xu, J.; Chen, Q.; Swain, G.M. Anthraquinonedisulfonate electrochemistry: A comparison of glassy carbon, hydrogenated glassy carbon, highly oriented pyrolytic graphite, and diamond electrodes. *Anal. Chem.* **1998**, *70*, 3146–3154. [CrossRef]
70. Arnold, C.L.; Eyckens, D.J.; Servinis, L.; Nave, M.D.; Yin, H.; Marceau, R.K.; Pinson, J.; Demir, B.; Walsh, T.R.; Henderson, L.C. Simultaneously increasing the hydrophobicity and interfacial adhesion of carbon fibres: A simple pathway to install passive functionality into composites. *J. Mater. Chem. A* **2019**, *7*, 13483–13494. [CrossRef]
71. Delamar, M.; Hitmi, R.; Pinson, J.; Saveant, J.M. Covalent modification of carbon surfaces by grafting of functionalized aryl radicals produced from electrochemical reduction of diazonium salts. *J. Am. Chem. Soc.* **1992**, *114*, 5883–5884. [CrossRef]

72. Allongue, P.; Delamar, M.; Desbat, B.; Fagebaume, O.; Hitmi, R.; Pinson, J.; Savéant, J.M. Covalent modification of carbon surfaces by aryl radicals generated from the electrochemical reduction of diazonium salts. *J. Am. Chem. Soc.* **1997**, *119*, 201–207. [CrossRef]
73. Ofoegbu, S.U. Corrosion and Corrosion Inhibition in Multi-material Combinations. Ph.D. Thesis, University of Aveiro, Aveiro, Portugal, 2018; pp. 117–178. Available online: http://hdl.handle.net/10773/24097 (accessed on 12 March 2021).
74. Rinat, J.; Korin, E.; Soifer, L.; Bettelheim, A. Electrocrystallization of calcium carbonate on carbon-based electrodes. *J. Electroanal. Chem.* **2005**, *575*, 195–202. [CrossRef]
75. Shimizu, K.; Sepunaru, L.; Compton, R.G. Innovative catalyst design for the oxygen reduction reaction for fuel cells. *Chem. Sci.* **2016**, *7*, 3364–3369. [CrossRef] [PubMed]
76. Choi, C.H.; Baldizzone, C.; Polymeros, G.; Pizzutilo, E.; Kasian, O.; Schuppert, A.K.; Ranjbar Sahraie, N.; Sougrati, M.T.; Mayrhofer, K.J.; Jaouen, F. Minimizing operando demetallation of Fe-NC electrocatalysts in acidic medium. *ACS Catal.* **2016**, *6*, 3136–3146. [CrossRef]
77. Engstrom, R.C.; Strasser, V.A. Characterization of electrochemically pretreated glassy carbon electrodes. *Anal. Chem.* **1984**, *56*, 136–141. [CrossRef]
78. Anjo, D.M.; Kahr, M.; Khodabakhsh, M.M.; Nowinski, S.; Wanger, M. Electrochemical activation of carbon electrodes in base: Minimization of dopamine adsorption and electrode capacitance. *Anal. Chem.* **1989**, *61*, 2603–2608. [CrossRef]
79. Chen, P.; McCreery, R.L. Control of electron transfer kinetics at glassy carbon electrodes by specific surface modification. *Anal. Chem.* **1996**, *68*, 3958–3965. [CrossRef]
80. Matsumoto, Y.; Tateishi, H.; Koinuma, M.; Kamei, Y.; Ogata, C.; Gezuhara, K.; Hatakeyama, K.; Hayami, S.; Taniguchi, T.; Funatsu, A. Electrolytic graphene oxide and its electrochemical properties. *J. Electroanal. Chem.* **2013**, *704*, 233–241. [CrossRef]
81. Paul, R.; Dai, Q.; Hu, C.; Dai, L. Ten years of carbon-based metal-free electrocatalysts. *Carbon Energy* **2019**, *1*, 19–31. [CrossRef]
82. Jung, H.; Choung, S.; Han, J.W. Design principles of noble metal-free electrocatalysts for hydrogen production in alkaline media: Combining theory and experiment. *Nanoscale Adv.* **2021**, *3*, 6797–6826. [CrossRef] [PubMed]
83. Xu, H.; Yang, J.; Ge, R.; Zhang, J.; Li, Y.; Zhu, M.; Dai, L.; Li, S.; Li, W. Carbon-based bifunctional electrocatalysts for oxygen reduction and oxygen evolution reactions: Optimization strategies and mechanistic analysis. *J. Energy Chem.* **2022**, *71*, 234–265. [CrossRef]
84. Wu, K.H.; Wang, D.; Lu, X.; Zhang, X.; Xie, Z.; Liu, Y.; Su, B.J.; Chen, J.M.; Su, D.S.; Qi, W.; et al. Highly selective hydrogen peroxide electrosynthesis on carbon: In situ interface engineering with surfactants. *Chem* **2020**, *6*, 1443–1458. [CrossRef]
85. Alexeyeva, N.; Shulga, E.; Kisand, V.; Kink, I.; Tammeveski, K. Electroreduction of oxygen on nitrogen-doped carbon nanotube modified glassy carbon electrodes in acid and alkaline solutions. *J. Electroanal. Chem.* **2010**, *648*, 169–175. [CrossRef]
86. Gong, K.; Du, F.; Xia, Z.; Durstock, M.; Dai, L. Nitrogen-doped carbon nanotube arrays with high electrocatalytic activity for oxygen reduction. *Science* **2009**, *323*, 760–764. [CrossRef] [PubMed]
87. He, W.; Jiang, C.; Wang, J.; Lu, L. High-Rate Oxygen Electroreduction over Graphitic-N Species Exposed on 3D Hierarchically Porous Nitrogen-Doped Carbons. *Angew. Chem.* **2014**, *126*, 9657–9661. [CrossRef]
88. Singh, S.K.; Takeyasu, K.; Nakamura, J. Active sites and mechanism of oxygen reduction reaction electrocatalysis on nitrogen-doped carbon materials. *Adv. Mater.* **2019**, *31*, 1804297. [CrossRef]
89. Guo, D.; Shibuya, R.; Akiba, C.; Saji, S.; Kondo, T.; Nakamura, J. Active sites of nitrogen-doped carbon Materials for oxygen reduction reaction clarified using model catalysts. *Science* **2016**, *351*, 361–365. [CrossRef] [PubMed]
90. Li, X.; Yang, X.; Jia, L.; Ma, X.; Zhu, L. Carbonaceous debris that resided in graphene oxide/reduced graphene oxide profoundly affect their electrochemical behaviors. *Electrochem. Commun.* **2012**, *23*, 94–97. [CrossRef]
91. Liyev, E.; Filiz, V.; Khan, M.M.; Lee, Y.J.; Abetz, C.; Abetz, V. Structural characterization of graphene oxide: Surface functional groups and fractionated oxidative debris. *Nanomaterials* **2019**, *9*, 1180. [CrossRef]
92. Bonanni, A.; Ambrosi, A.; Chua, C.K.; Pumera, M. Oxidation debris in graphene oxide is responsible for its inherent electroactivity. *ACS Nano* **2014**, *8*, 4197–4204. [CrossRef]
93. Rourke, J.P.; Pandey, P.A.; Moore, J.J.; Bates, M.; Kinloch, I.A.; Young, R.J.; Wilson, N.R. The real graphene oxide revealed: Stripping the oxidative debris from the graphene-like sheets. *Angew. Chem. Int. Ed.* **2011**, *50*, 3173–3177. [CrossRef]
94. He, H.; Riedl, T.; Lerf, A.; Klinowski, J. Solid-state NMR studies of the structure of graphite oxide. *J. Phys. Chem.* **1996**, *100*, 19954–19958. [CrossRef]
95. Lerf, A.; He, H.; Riedl, T.; Forster, M.; Klinowski, J. ^{13}C and ^{1}H MAS NMR studies of graphite oxide and its chemically modified derivatives. *Solid State Ion.* **1997**, *101*, 857–862. [CrossRef]
96. Lerf, A.; He, H.; Forster, M.; Klinowski, J. Structure of graphite oxide revisited. *J. Phys. Chem. B* **1998**, *102*, 4477–4482. [CrossRef]
97. Roubaud, E.; Lacroix, R.; Da Silva, S.; Esvan, J.; Etcheverry, L.; Bergel, A.; Basséguy, R.; Erable, B. Industrially scalable surface treatments to enhance the current density output from graphite bioanodes fueled by real domestic wastewater. *Iscience* **2021**, *24*, 102162. [CrossRef]
98. Szabó, T.; Berkesi, O.; Forgó, P.; Josepovits, K.; Sanakis, Y.; Petridis, D.; Dékány, I. Evolution of surface functional groups in a series of progressively oxidized graphite oxides. *Chem. Mater.* **2006**, *18*, 2740–2749. [CrossRef]
99. Buono, C.; Davies, P.R.; Davies, R.J.; Jones, T.; Kulhavý, J.; Lewis, R.; Morgan, D.J.; Robinson, N.; Willock, D.J. Spectroscopic and atomic force studies of the functionalisation of carbon surfaces: New insights into the role of the surface topography and specific chemical states. *Faraday Discuss.* **2014**, *173*, 257–272. [CrossRef] [PubMed]

100. Thomas, H.R.; Day, S.P.; Woodruff, W.E.; Vallés, C.; Young, R.J.; Kinloch, I.A.; Morley, G.W.; Hanna, J.V.; Wilson, N.R.; Rourke, J.P. Deoxygenation of graphene oxide: Reduction or cleaning? *Chem. Mater.* **2013**, *25*, 3580–3588. [CrossRef]
101. Wu, Z.; Pittman, C.U., Jr.; Gardner, S.D. Nitric acid oxidation of carbon fibers and the effects of subsequent treatment in refluxing aqueous NaOH. *Carbon* **1995**, *33*, 597–605. [CrossRef]
102. Yi, Y.; Tornow, J.; Willinger, E.; Willinger, M.G.; Ranjan, C.; Schlögl, R. Electrochemical degradation of multiwall carbon nanotubes at high anodic potential for oxygen evolution in acidic media. *ChemElectroChem* **2015**, *2*, 1929–1937. [CrossRef]
103. Yi, Y.; Weinberg, G.; Prenzel, M.; Greiner, M.; Heumann, S.; Becker, S.; Schlögl, R. Electrochemical corrosion of a glassy carbon electrode. *Catal. Today* **2017**, *295*, 32–40. [CrossRef]
104. Yang, Y.; Lin, Z.G. In situ FTIR characterization of the electrooxidation of glassy carbon electrodes. *J. Appl. Electrochem.* **1995**, *25*, 259–266. [CrossRef]
105. Besenhard, J.O.; Jakob, J.; Krebber, U.; Moeller, P.; Sauter, R.F.; Kurtze, A.; Kanani, N.; Meyer, H.; Hoerber, J.K.H.; Jannakoudakis, A.D. Anodische Oberflächen-und Volumenoxidation graphitischer Materialien in neutralen und alkalischen wäßrigen Lösungen/Anodic Surface and Bulk Oxidation of Graphitic Materials in Neutral and Basic Aqueous Solutions. *Z. Für Nat. B* **1989**, *44*, 729–735. [CrossRef]
106. Hollemann, A.F.; Wiberg, E.N.; Wiberg, N. *Lehrbuch der Anorganischen Chemie*, 91st ed.; Walter de Gruyter: Berlin, NY, USA, 1985.
107. Yue, Z.R.; Jiang, W.; Wang, L.; Gardner, S.D.; Pittman, C.U., Jr. Surface characterization of electrochemically oxidized carbon fibers. *Carbon* **1999**, *37*, 1785–1796. [CrossRef]
108. Nose, M.; Kinumoto, T.; Choo, H.S.; Miyazaki, K.; Abe, T.; Ogumi, Z. Electrochemical oxidation of highly oriented pyrolytic graphite in sulphuric acid solution under potential pulse condition. *Fuel Cells* **2009**, *9*, 284–290. [CrossRef]
109. Yeager, E. Electrocatalysts for O_2 reduction. *Electrochim. Acta* **1984**, *29*, 1527–1537. [CrossRef]
110. Yeager, E. Dioxygen electrocatalysis: Mechanisms in relation to catalyst structure. *J. Mol. Catal.* **1986**, *38*, 5–25. [CrossRef]
111. Strelko, V.V.; Kartel, N.T.; Dukhno, I.N.; Kuts, V.S.; Clarkson, R.B.; Odintsov, B.M. Mechanism of reductive oxygen adsorption on active carbons with various surface chemistry. *Surf. Sci.* **2004**, *548*, 281–290. [CrossRef]
112. Maldonado, S.; Stevenson, K.J. Influence of nitrogen doping on oxygen reduction electrocatalysis at carbon nanofiber electrodes. *J. Phys. Chem. B* **2005**, *109*, 4707–4716. [CrossRef]
113. Sidik, R.A.; Anderson, A.B.; Subramanian, N.P.; Kumaraguru, S.P.; Popov, B.N. O_2 reduction on graphite and nitrogen-doped graphite: Experiment and theory. *J. Phys. Chem. B* **2006**, *110*, 1787–1793. [CrossRef]
114. Zhao, Y.; Yang, L.; Chen, S.; Wang, X.; Ma, Y.; Wu, Q.; Jiang, Y.; Qian, W.; Hu, Z. Can boron and nitrogen co-doping improve oxygen reduction reaction activity of carbon nanotubes? *J. Am. Chem. Soc.* **2013**, *135*, 1201–1204. [CrossRef]
115. Šljukić, B.; Banks, C.E.; Compton, R.G. An overview of the electrochemical reduction of oxygen at carbon-based modified electrodes. *J. Iran. Chem. Soc.* **2005**, *2*, 1–25. [CrossRef]
116. Wiggins-Camacho, J.D.; Stevenson, K.J. Mechanistic discussion of the oxygen reduction reaction at nitrogen-doped carbon nanotubes. *J. Phys. Chem. C* **2011**, *115*, 20002–20010. [CrossRef]
117. Nugent, J.M.; Santhanam, K.S.V.; Rubio, A.; Ajayan, P.M. Fast electron transfer kinetics on multiwalled carbon nanotube microbundle electrodes. *Nano Lett.* **2001**, *1*, 87–91. [CrossRef]
118. Vaik, K.; Sarapuu, A.; Tammeveski, K.; Mirkhalaf, F.; Schiffrin, D.J. Oxygen reduction on phenanthrenequinone-modified glassy carbon electrodes in 0.1 M KOH. *J. Electroanal. Chem.* **2004**, *564*, 159–166. [CrossRef]
119. Sawyer, D.T.; Chiericato, G.; Angelis, C.T.; Nanni, E.J.; Tsuchiya, T. Effects of media and electrode Materials on the electrochemical reduction of dioxygen. *Anal. Chem.* **1982**, *54*, 1720–1724. [CrossRef]
120. Benck, J.D.; Pinaud, B.A.; Gorlin, Y.; Jaramillo, T.F. Substrate Selection for Fundamental Studies of Electrocatalysts and Photoelectrodes: Inert Potential Windows in Acidic, Neutral, and Basic Electrolyte. *PLoS ONE* **2014**, *9*, e107942. [CrossRef]
121. Morcos, I.; Yeager, E. Kinetic studies of the oxygen—Peroxide couple on pyrolytic graphite. *Electrochim. Acta* **1970**, *15*, 953–975. [CrossRef]
122. Zhang, Z.W.; Tyrk, D.A.; Yeager, E. Quinone-like surfaces structures and O_2 reduction on glassy carbon and graphite surfaces in alkaline solutions. In *Proceedings of the Workshop on the Electrochemistry of Carbon, Cleveland, OH, USA, 17–19 August 1983*; Sarangapani, S., Akridge, J.R., Schumm, B., Eds.; The Electrochemical Society: Pennington, NJ, USA, 1984; pp. 158–178.
123. McCreery, R.L. *Electroanalytical Chemsitry*; Bard, A.J., Ed.; Dekker: New York, NY, USA, 1991; Volume 17.
124. Bowling, R.J.; McCreery, R.L.; Pharr, C.M.; Engstrom, R.C. Observation of kinetic heterogeneity on highly ordered pyrolytic graphite using electrogenerated chemiluminescence. *Anal. Chem.* **1989**, *61*, 2763–2766. [CrossRef]
125. Pontikos, N.M.; McCreery, R.L. Microstructural and morphological changes induced in glassy carbon electrodes by laser irradiation. *J. Electroanal. Chem.* **1992**, *324*, 229–242. [CrossRef]
126. Hu, I.F.; Kuwana, T. Oxidative mechanism of ascorbic acid at glassy carbon electrodes. *Anal. Chem.* **1986**, *58*, 3235–3239. [CrossRef]
127. Fagan, D.T.; Hu, I.F.; Kuwana, T. Vacuum heat-treatment for activation of glassy carbon electrodes. *Anal. Chem.* **1985**, *57*, 2759–2763. [CrossRef]
128. Evans, J.F.; Kuwana, T. Radiofrequency oxygen plasma treatment of pyrolytic graphite electrode surfaces. *Anal. Chem.* **1977**, *49*, 1632–1635. [CrossRef]
129. Tse, D.C.S.; Kuwana, T. Electrocatalysis of dihydronicotinamide adenosine diphosphate with quinones and modified quinone electrodes. *Anal. Chem.* **1978**, *50*, 1315–1318. [CrossRef]

130. Deakin, M.R.; Stutts, K.J.; Wightman, R.M. The effect of pH on some outer-sphere electrode reactions at carbon electrodes. *J. Electroanal. Chem. Interfacial Electrochem.* **1985**, *182*, 113–122. [CrossRef]
131. Cabaniss, G.E.; Diamantis, A.A.; Murphy, W.R., Jr.; Linton, R.W.; Meyer, T.J. Electrocatalysis of proton-coupled electron-transfer reactions at glassy carbon electrodes. *J. Am. Chem. Soc.* **1985**, *107*, 1845–1853. [CrossRef]
132. Barbero, C.; Silber, J.J.; Sereno, L. Studies of surface-modified glassy carbon electrodes obtained by electrochemical treatment: Its effect on $Ru(bpy)^{2+}{}_3$ adsorption and the electron transfer rates of the Fe^{2+}/Fe^{3+} couple. *J. Electroanal. Chem. Interfacial Electrochem.* **1988**, *248*, 321–340. [CrossRef]
133. Armstrong, F.A.; Bond, A.M.; Hill, H.A.; Oliver, B.N.; Psalti, I.S. Electrochemistry of cytochrome c, plastocyanin, and ferredoxin at edge-and basal-plane graphite electrodes interpreted via a model based on electron transfer at electroactive sites of microscopic dimensions in size. *J. Am. Chem. Soc.* **1989**, *111*, 9185–9189. [CrossRef]
134. Boehm, H.P. Surface oxides on carbon and their analysis: A critical assessment. *Carbon* **2002**, *40*, 145–149. [CrossRef]
135. Byon, H.R.; Gallant, B.M.; Lee, S.W.; Shao-Horn, Y. Role of oxygen functional groups in carbon nanotube/graphene freestanding electrodes for high performance lithium batteries. *Adv. Funct. Mater.* **2013**, *23*, 1037–1045. [CrossRef]
136. Lowe, S.E.; Shi, G.; Zhang, Y.; Qin, J.; Jiang, L.; Jiang, S.; Al-Mamun, M.; Liu, P.; Zhong, Y.L.; Zhao, H. The role of electrolyte acid concentration in the electrochemical exfoliation of graphite: Mechanism and synthesis of electrochemical graphene oxide. *Nano Mater. Sci.* **2019**, *1*, 215–223. [CrossRef]
137. Aghamohammadi, H.; Eslami-Farsani, R.; Torabian, M.; Amousa, N. Recent advances in one-pot functionalization of graphene using electrochemical exfoliation of graphite: A review study. *Synth. Met.* **2020**, *269*, 116549. [CrossRef]
138. Hathcock, K.W.; Brumfield, J.C.; Goss, C.A.; Irene, E.A.; Murray, R.W. Incipient electrochemical oxidation of highly oriented pyrolytic graphite: Correlation between surface blistering and electrolyte anion intercalation. *Anal. Chem.* **1995**, *67*, 2201–2206. [CrossRef]
139. Sinitsyna, O.V.; Meshkov, G.B.; Grigorieva, A.V.; Antonov, A.A.; Grigorieva, I.G.; Yaminsky, I.V. Blister formation during graphite surface oxidation by Hummers' method. *Beilstein J. Nanotechnol.* **2018**, *9*, 407–414. [CrossRef] [PubMed]
140. Matsumoto, M.; Manako, T.; Imai, H. Electrochemical STM investigation of oxidative corrosion of the surface of highly oriented pyrolytic graphite. *J. Electrochem. Soc.* **2009**, *156*, B1208. [CrossRef]
141. Tang, J.; Chen, S.; Yuan, Y.; Cai, X.; Zhou, S. In situ formation of graphene layers on graphite surfaces for efficient anodes of microbial fuel cells. *Biosens. Bioelectron.* **2015**, *71*, 387–395. [CrossRef]
142. Liu, J.; Liu, Y.; Feng, C.; Wang, Z.; Jia, T.; Gong, L.; Xu, L. Enhanced performance of microbial fuel cell using carbon microspheres modified graphite anode. *Energy Sci. Eng.* **2017**, *5*, 217–225. [CrossRef]
143. Patel, A.N.; Collignon, M.G.; O'Connell, M.A.; Hung, W.O.; McKelvey, K.; Macpherson, J.V.; Unwin, P.R. A new view of electrochemistry at highly oriented pyrolytic graphite. *J. Am. Chem. Soc.* **2012**, *134*, 20117–20130. [CrossRef]
144. Lai, S.C.; Patel, A.N.; McKelvey, K.; Unwin, P.R. Definitive Evidence for Fast Electron Transfer at Pristine Basal Plane Graphite from High-Resolution Electrochemical Imaging. *Angew. Chem. Int. Ed.* **2012**, *51*, 5405–5408. [CrossRef]
145. McDermott, M.T.; Kneten, K.; McCreery, R.L. Anthraquinonedisulfonate adsorption, electron-transfer kinetics, and capacitance on ordered graphite electrodes: The important role of surface defects. *J. Phys. Chem.* **1992**, *96*, 3124–3130. [CrossRef]
146. Banks, C.E.; Moore, R.R.; Davies, T.J.; Compton, R.G. Investigation of modified basal plane pyrolytic graphite electrodes: Definitive evidence for the electrocatalytic properties of the ends of carbon nanotubes. *Chem. Commun.* **2004**, *10*, 1804–1805. [CrossRef]
147. Davies, T.J.; Moore, R.R.; Banks, C.E.; Compton, R.G. The cyclic voltammetric response of electrochemically heterogeneous surfaces. *J. Electroanal. Chem.* **2004**, *574*, 123–152. [CrossRef]
148. Davies, T.J.; Hyde, M.E.; Compton, R.G. Nanotrench arrays reveal insight into graphite electrochemistry. *Angew. Chem. Int. Ed.* **2005**, *44*, 5121–5126. [CrossRef] [PubMed]
149. Ji, X.; Banks, C.E.; Crossley, A.; Compton, R.G. Oxygenated Edge Plane Sites Slow the Electron Transfer of the Ferro-/Ferricyanide Redox Couple at Graphite Electrodes. *ChemPhysChem* **2006**, *7*, 1337–1344. [CrossRef]
150. Ji, X.; Buzzeo, M.C.; Banks, C.E.; Compton, R.G. Electrochemical response of cobalt (II) in the presence of ammonia. *Electroanalysis* **2006**, *18*, 44–52. [CrossRef]
151. Moore, R.R.; Banks, C.E.; Compton, R.G. Electrocatalytic detection of thiols using an edge plane pyrolytic graphite electrode. *Analyst* **2004**, *129*, 755–758. [CrossRef] [PubMed]
152. Hyde, M.E.; Davies, T.J.; Compton, R.G. Fabrication of random assemblies of metal nanobands: A general method. *Angew. Chem.* **2005**, *117*, 6649–6654. [CrossRef]
153. Dumitrescu, I.; Unwin, P.R.; Macpherson, J.V. Electrochemistry at carbon nanotubes: Perspective and issues. *Chem. Commun.* **2009**, *41*, 6886–6901. [CrossRef]
154. Pumera, M. Graphene-based nanomaterials and their electrochemistry. *Chem. Soc. Rev.* **2010**, *39*, 4146–4157. [CrossRef]
155. Lee, C.Y.; Guo, S.X.; Bond, A.M.; Oldham, K.B. Effect of heterogeneity on the dc and ac voltammetry of the $[Fe(CN)_6]^{3-/4-}$ solution-phase process at a highly ordered pyrolytic graphite electrode. *J. Electroanal. Chem.* **2008**, *615*, 1–11. [CrossRef]
156. Lee, C.Y.; Bond, A.M. Evaluation of levels of defect sites present in highly ordered pyrolytic graphite electrodes using capacitive and faradaic current components derived simultaneously from large-amplitude Fourier transformed ac voltammetric experiments. *Anal. Chem.* **2008**, *81*, 584–594. [CrossRef]

157. Shinagawa, T.; Garcia-Esparza, A.T.; Takanabe, K. Insight on Tafel slopes from a microkinetic analysis of aqueous electrocatalysis for energy conversion. *Sci. Rep.* **2015**, *5*, 13801. [CrossRef]
158. Qiang, Z.; Chang, J.H.; Huang, C.P. Electrochemical generation of hydrogen peroxide from dissolved oxygen in acidic solutions. *Water Res.* **2002**, *36*, 85–94. [CrossRef] [PubMed]
159. Taylor, R.J.; Humffray, A.A. Electrochemical studies on glassy carbon electrodes: II. Oxygen reduction in solutions of high pH (pH > 10). *J. Electroanal. Chem. Interfacial Electrochem.* **1975**, *64*, 63–84. [CrossRef]
160. Paliteiro, C.; Hamnett, A.; Goodenough, J.B. The electroreduction of oxygen on pyrolytic graphite. *J. Electroanal. Chem. Interfacial Electrochem.* **1987**, *233*, 147–159. [CrossRef]
161. Baez, V.B.; Pletcher, D. Preparation and characterization of carbon/titanium dioxide surfaces—The reduction of oxygen. *J. Electroanal. Chem.* **1995**, *382*, 59–64. [CrossRef]
162. Hirschorn, B.; Orazem, M.E.; Tribollet, B.; Vivier, V.; Frateur, I.; Musiani, M. Constant-phase-element behavior caused by resistivity distributions in films: II. Applications. *J. Electrochem. Soc.* **2010**, *157*, C458. [CrossRef]
163. Alexander, C.L.; Tribollet, B.; Vivier, V.; Orazem, M.E. Contribution of surface distributions to constant-phase-element (CPE) behavior: 3. Adsorbed intermediates. *Electrochim. Acta* **2017**, *251*, 99–108. [CrossRef]
164. Orazem, M.E.; Frateur, I.; Tribollet, B.; Vivier, V.; Marcelin, S.; Pébère, N.; Bunge, A.L.; White, E.A.; Riemer, D.P.; Musiani, M. Dielectric properties of Materials showing constant-phase-element (CPE) impedance response. *J. Electrochem. Soc.* **2013**, *160*, C215. [CrossRef]
165. Brug, G.J.; Van Den Eeden, A.L.G.; Sluyters-Rehbach, M.; Sluyters, J.H. The analysis of electrode impedances complicated by the presence of a constant phase element. *J. Electroanal. Chem. Interfacial Electrochem.* **1984**, *176*, 275–295. [CrossRef]
166. Rice, R.J.; McCreery, R.L. Quantitative relationship between electron transfer rate and surface microstructure of laser-modified graphite electrodes. *Anal. Chem.* **1989**, *61*, 1637–1641. [CrossRef]
167. Güell, A.G.; Tan, S.Y.; Unwin, P.R.; Zhang, G. Electrochemistry at Highly Oriented Pyrolytic Graphite (HOPG): Toward a New Perspective. In *Electrochemistry of Carbon Electrodes*; Wiley: Hoboken, NJ, USA, 2015; pp. 31–82. [CrossRef]
168. Luque, N.B.; Schmickler, W. The electric double layer on graphite. *Electrochim. Acta* **2012**, *71*, 82–85. [CrossRef]
169. Zhou, Y.; Holme, T.; Berry, J.; Ohno, T.R.; Ginley, D.; O'Hayre, R. Dopant-induced electronic structure modification of HOPG surfaces: Implications for high activity fuel cell catalysts. *J. Phys. Chem. C* **2010**, *114*, 506–515. [CrossRef]
170. Randin, J.P.; Yeager, E. Differential capacitance study of stress-annealed pyrolytic graphite electrodes. *J. Electrochem. Soc.* **1971**, *118*, 711. [CrossRef]
171. Cline, K.K.; McDermott, M.T.; McCreery, R.L. Anomalously slow electron transfer at ordered graphite electrodes: Influence of electronic factors and reactive sites. *J. Phys. Chem.* **1994**, *98*, 5314–5319. [CrossRef]
172. McDermott, M.T.; McCreery, R.L. Scanning tunneling microscopy of ordered graphite and glassy carbon surfaces: Electronic control of quinone adsorption. *Langmuir* **1994**, *10*, 4307–4314. [CrossRef]
173. Chang, H.; Bard, A.J. Observation and characterization by scanning tunneling microscopy of structures generated by cleaving highly oriented pyrolytic graphite. *Langmuir* **1991**, *7*, 1143–1153. [CrossRef]
174. Velický, M.; Toth, P.S.; Woods, C.R.; Novoselov, K.S.; Dryfe, R.A. Electrochemistry of the basal plane versus edge plane of graphite revisited. *J. Phys. Chem. C* **2019**, *123*, 11677–11685. [CrossRef]
175. Randin, J.P.; Yeager, E. Differential capacitance study on the basal plane of stress-annealed pyrolytic graphite. *J. Electroanal. Chem. Interfacial Electrochem.* **1972**, *36*, 257–276. [CrossRef]
176. Randin, J.P.; Yeager, E. Differential capacitance study on the edge orientation of pyrolytic graphite and glassy carbon electrodes. *J. Electroanal. Chem. Interfacial Electrochem.* **1975**, *58*, 313–322. [CrossRef]
177. Rabbow, T.J.; Whitehead, A.H. Deconvolution of electrochemical double layer capacitance between fractions of active and total surface area of graphite felts. *Carbon* **2017**, *111*, 782–788. [CrossRef]
178. Wang, Y.; Alsmeyer, D.C.; McCreery, R.L. Raman spectroscopy of carbon materials: Structural basis of observed spectra. *Chem. Mater.* **1990**, *2*, 557–563. [CrossRef]
179. Kaniyoor, A.; Ramaprabhu, S. A Raman spectroscopic investigation of graphite oxide derived graphene. *Aip Adv.* **2012**, *2*, 032183. [CrossRef]
180. Kumar, R.; Mehta, B.R.; Bhatnagar, M.; Mahapatra, S.; Salkalachen, S.; Jhawar, P. Graphene as a transparent conducting and surface field layer in planar Si solar cells. *Nanoscale Res. Lett.* **2014**, *9*, 1–9. [CrossRef] [PubMed]
181. Xu, L.; Cheng, L. Graphite oxide under high pressure: A Raman spectroscopic study. *J. Nanomater.* **2013**, *2013*, 47. [CrossRef]
182. Tuinstra, F.; Koenig, J.L. Raman spectrum of graphite. *J. Chem. Phys.* **1970**, *53*, 1126–1130. [CrossRef]
183. Jorio, A.; Souza Filho, A.G.; Dresselhaus, G.; Dresselhaus, M.S.; Swan, A.K.; Ünlü, M.S.; Goldberg, B.B.; Pimenta, M.A.; Hafner, J.H.; Lieber, C.M.; et al. G-band resonant Raman study of 62 isolated single-wall carbon nanotubes. *Phys. Rev. B* **2002**, *65*, 155412. [CrossRef]
184. Casari, C.S.; Bassi, A.L.; Baserga, A.; Ravagnan, L.; Piseri, P.; Lenardi, C.; Tommasini, M.; Milani, A.; Fazzi, D.; Bottani, C.E.; et al. Low-frequency modes in the Raman spectrum of $sp - sp^2$ nanostructured carbon. *Phys. Rev. B* **2008**, *77*, 195444. [CrossRef]
185. Smith, M.W.; Dallmeyer, I.; Johnson, T.J.; Brauer, C.S.; McEwen, J.S.; Espinal, J.F.; Garcia-Perez, M. Structural analysis of char by Raman spectroscopy: Improving band assignments through computational calculations from first principles. *Carbon* **2016**, *100*, 678–692. [CrossRef]

186. Yoon, D.; Moon, H.; Cheong, H.; Choi, J.S.; Choi, J.A.; Park, B.H. Variations in the Raman spectrum as a function of the number of graphene layers. *J. Korean Phys. Soc.* **2009**, *55*, 1299–1303. [CrossRef]
187. Sharbidre, R.S.; Lee, C.J.; Hong, S.G.; Ryu, J.K.; Kim, T.N. Comparison of existing methods to identify the number of graphene layers. *Korean J. Mater. Res.* **2016**, *26*, 704–708. [CrossRef]
188. Muzyka, R.; Drewniak, S.; Pustelny, T.; Chrubasik, M.; Gryglewicz, G. Characterization of graphite oxide and reduced graphene oxide obtained from different graphite precursors and oxidized by different methods using Raman spectroscopy. *Materials* **2018**, *11*, 1050. [CrossRef] [PubMed]
189. Muzyka, R.; Drewniak, S.; Pustelny, T.; Sajdak, M.; Drewniak, Ł. Characterization of graphite oxide and reduced graphene oxide obtained from different graphite precursors and oxidized by different methods using Raman spectroscopy statistical analysis. *Materials* **2021**, *14*, 769. [CrossRef]
190. Toh, S.Y.; Loh, K.S.; Kamarudin, S.K.; Daud, W.R.W. The impact of electrochemical reduction potentials on the electrocatalytic activity of graphene oxide toward the oxygen reduction reaction in an alkaline medium. *Electrochim. Acta* **2016**, *199*, 194–203. [CrossRef]
191. Liu, D.; Long, Y.T. Superior catalytic activity of electrochemically reduced graphene oxide supported iron phthalocyanines toward oxygen reduction reaction. *ACS Appl. Mater. Interfaces* **2015**, *7*, 24063–24068. [CrossRef] [PubMed]
192. Kauppila, J.; Kunnas, P.; Damlin, P.; Viinikanoja, A.; Kvarnström, C. Electrochemical reduction of graphene oxide films in aqueous and organic solutions. *Electrochim. Acta* **2013**, *89*, 84–89. [CrossRef]
193. Shao, Y.; Wang, J.; Engelhard, M.; Wang, C.; Lin, Y. Facile and controllable electrochemical reduction of graphene oxide and its applications. *J. Mater. Chem.* **2010**, *20*, 743–748. [CrossRef]
194. Toh, S.Y.; Loh, K.S.; Kamarudin, S.K.; Daud, W.R.W. Graphene production via electrochemical reduction of graphene oxide: Synthesis and characterisation. *Chem. Eng. J.* **2014**, *251*, 422–434. [CrossRef]
195. Ramesha, G.K.; Sampath, S. Electrochemical reduction of oriented graphene oxide films: An in situ Raman spectroelectrochemical study. *J. Phys. Chem. C* **2009**, *113*, 7985–7989. [CrossRef]
196. Brownson, D.A.; Kampouris, D.K.; Banks, C.E. Graphene electrochemistry: Fundamental concepts through to prominent applications. *Chem. Soc. Rev.* **2012**, *41*, 6944–6976. [CrossRef]

C Journal of Carbon Research

MDPI

Article

Pull-Out of Pristine and Functionalized Carbon Nanotubes from Cement: A Molecular Modelling Study

Isabel Lado-Touriño

Engineering Department, School of Architecture, Engineering and Design, Universidad Europea de Madrid, 28760 Villaviciosa de Odón, Spain; misabel.lado@universidadeuropea.es

Abstract: Carbon nanotubes (CNTs) are widely used as reinforcements in cement-based composites. The improvement in the mechanical properties of the resulting materials depends on the characteristics of the interface formed between CNTs and the cement matrix. The experimental characterization of the interfacial properties of these composites is still limited and hard to achieve with currently available technologies. In this work, molecular dynamics and molecular mechanics pull-out simulations of pristine and functionalized CNTs, taken from a tobermorite crystal, were carried out to study interfacial shear strength (ISS) from an atomic perspective. ISS was calculated from the potential energy of the systems. The effects of the CNT diameter and the degree of functionalization on the pull-out process were analyzed according to the ISS and non-bonded energy results. The influence of H-bonding and electrostatic interactions between the CNT and the matrix were also studied. The results show that ISS decreases with increasing CNT radius for pristine CNTs and depends upon the number of H-bonds for functionalized CNTs. ISS values are positively correlated to $E_{non\text{-}bonded\ energy}$, which is related to the number of carboxyl groups on the CNT surface. A high degree of functionalization increases both the number of H-bonds and the number of Ca^{2+}-O interactions between the CNT and the tobermorite surface. This results in a stronger interfacial interaction and, therefore, an elevated ISS value.

Keywords: carbon nanotubes; cement; molecular mechanics; molecular dynamics; pull-out; interfacial shear strength

Citation: Lado-Touriño, I. Pull-Out of Pristine and Functionalized Carbon Nanotubes from Cement: A Molecular Modelling Study. *C* **2022**, *8*, 80. https://doi.org/10.3390/c8040080

Academic Editor: Gil Gonçalves

Received: 21 November 2022
Accepted: 13 December 2022
Published: 16 December 2022

1. Introduction

Cementitious composites are widely used materials with high compressive strength and simple production methods. However, they also have many disadvantages such as low tensile strength, high permeability to water and other harmful substances, and high cracking tendency. Reinforcements are used to overcome these drawbacks. Due to their remarkable mechanical properties, carbon nanotubes (CNTs) are promising candidates as reinforcing materials. During the last few years, there have been an increasing number of research studies on composites made of cement and CNTs [1–5]. Mechanical testing has shown that CNTs have elastic moduli greater than 1 TPa and tensile strengths as high as 63 GPa [6,7]. When appropriate dispersion of CNTs within the cement matrix is achieved, the resulting composites show denser structures [8], reduced water permeability [9], improved mechanical properties [10] and good durability [11]. The improved mechanical properties are attributed to the crack-bridging effect of CNTs. The addition of CNTs to a cement matrix results in a delay in cracks propagation. A recent study [12] shows that the use of functionalized CNTs improves compressive strength of cement pastes without the need for ultrasonication. Thus, good dispersion of the reinforcement into the matrix and improved interfacial interactions, enhanced by the functional groups, are the key factors in obtaining materials with better properties. CNTs have also been used as additives to enhance the bonding strength between concrete- and carbon fiber-reinforced polymers [13], which in turn increased the bond strength, ultimate slip, and interfacial fracture energy of the composites.

It is well known that the interface between CNTs and the cement matrix plays a critical role in determining the effectiveness of mechanical reinforcement, as the interface is usually the weakest part in a composite and determines its fracture behavior [14]. Apart from diverse experimental results, recent coarse-grained and all-atoms molecular dynamics studies [15,16] suggest that the addition of CNT to a cement matrix changes the way of fracturing and significantly improves the fracture energy and other mechanical properties.

Interfacial bond strength can be tailored in many ways, one of which is functionalizing the CNT surface [17,18]. The functionalization processes can be classified into two broad groups depending on the nature of the interactions established between the CNTs and the molecules linked to them [19]: covalent functionalization, which implies the formation of strong chemical bonds between the CNTs and the functional groups, and non-covalent functionalization, which is based on different adsorption forces (hydrogen bonds, van der Waals forces, electrostatic forces or π-stacking interactions) and does not modify the properties of CNTs [20]. Non-covalent functionalization is weaker than covalent functionalization, which may result in low efficiency of the load transfer in composites [21]. Acid treatment of the CNTs allows for covalent modification and is commonly used to introduce carboxyl groups to the sidewalls of CNTs [22], thereby facilitating dispersion as well as improving the bonding between CNTs and the cement matrix [23,24].

A direct measure of the interfacial bond strength can be estimated from the interfacial shear strength (ISS), which can be obtained from micromechanical and macromechanical tests, such as fragmentation, pull-out, microdroplet, push-out and push-in tests [25–30]. A consensus is yet to emerge on the best experimental technique to measure the ISS, and most experimental studies have focused on CNTs-reinforced polymer composites.

Due to the difficulties in obtaining experimental results to study the CNT–cement interface, the interfacial properties of CNTs-reinforced cementitious composites have mainly been studied by molecular mechanics (MM) and molecular dynamics (MD) simulations. It is important to note that this kind of simulations can take advantage of providing atomic information of the interface between CNTs and matrices, which is unattainable with experimental methods. Many authors have calculated ISS using a pull-out model [31]. This is because the CNT–matrix interface strength is essentially related to the pulling-out behavior, which determines how strongly the CNT is bonded to the matrix, and how effective the reinforcement is. As in the case of experimental results, ISS of CNTs-reinforced polymer composites has been extensively studied by molecular modelling [32–36], and only several results on carbonaceous nanomaterials–cement matrices have been reported in the literature, most of which related to graphene [37–40]. As the interface between CNTs and matrices play a critical role in enhancing the mechanical properties of the composites, the influence of the functionalization of the CNT surface on the pull-out process has also been studied by many authors [40–45]. Both simulation and experimental results show that cement composites, containing functionalized CNTs, result in improved mechanical properties due to a better adhesion at the interface. However, technical difficulties associated with the manipulation of the nanotubes in cement matrices, hinder the measurement of ISS of these composites.

After carrying out a review of the existing literature, we noticed that there are many experimental results on ISS of CNT-containing composites, but most of them have focused on polymeric matrices, as ISS of CNT–cement composites are difficult to measure experimentally. On the other hand, simulation studies of cementitious composites are more numerous, especially those that contain graphene and graphene-derived reinforcements. There are some MD calculations of mechanical properties of CNT–cement composites, but almost none focuses on ISS of functionalized CNTs. Therefore, in this work, the ISS of CNT/tobermorite nanocomposites was systematically studied by using a series of CNT pull-out simulations based on MD and MM approaches. The results are explained as a function of non-bonded energies, which in turn, are influenced by the chemical interactions established between the CNTs and the matrix at the composite interface. Pristine and functionalized armchair CNTs, with a radius in the range of 2.70 to 4.74 Å and a length of

19.68 Å, were inserted into a tobermorite matrix. The effect of CNT diameter and degree of functionalization with carboxyl groups on ISS was assessed and related to non-bonded interactions between the CNT and the matrix. To the best of our knowledge, molecular simulations of a CNT pull-out from a cement matrix, unlike in the case of polymer matrices or graphene, are scarce. With this work, we are aiming to get more insight into the atomic interactions existing at CNT–cement interfaces, which are primarily responsible for the enhancements and failures of the mechanical properties of these composites.

2. Materials and Methods

2.1. Model Systems

A simulation cell, with a periodic boundary condition in the y–z plane (29.54 × 67.46 Å^2), was used for the pull-out calculations. The size of this cell was large enough (500 Å) in the axial direction of the CNT (*x* direction), and included a vacuum layer to eliminate the interaction between periodic images in neighboring cells and leave space for pulling out the CNT (Figure 1).

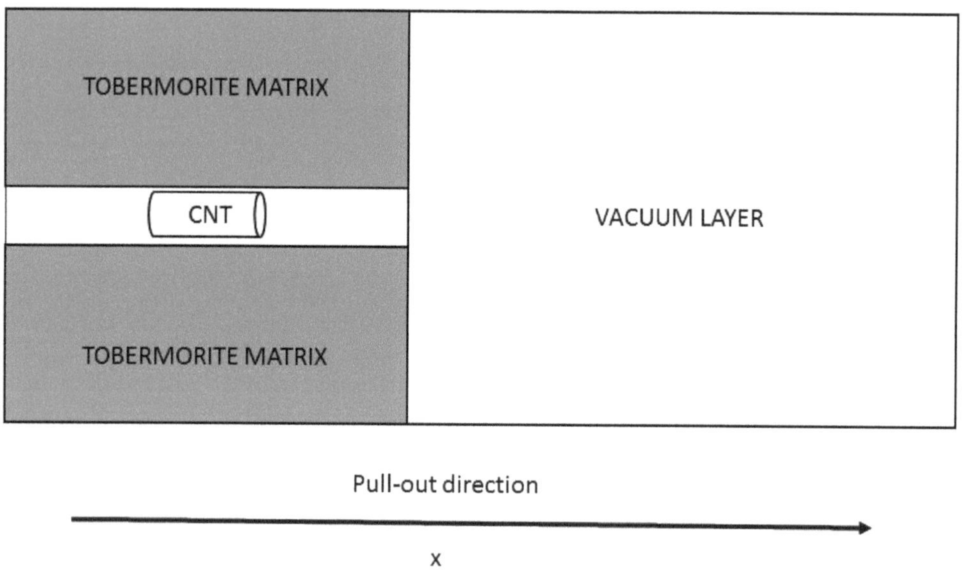

Figure 1. Schematic cell for the pull-out simulations.

To represent the cement matrix, a tobermorite 11 Å structure was used (left side of Figure 2). This structure contained double silicon oxide chains interspersed with calcium oxide sheets and hydration water molecules; 11 Å indicates the basal spacing of the calcium oxide layers. Although it is well known that real cementitious systems present poor crystalline order with variable stoichiometry [46], tobermorite has been regarded as a good model of cement structures by many different authors [47–49].

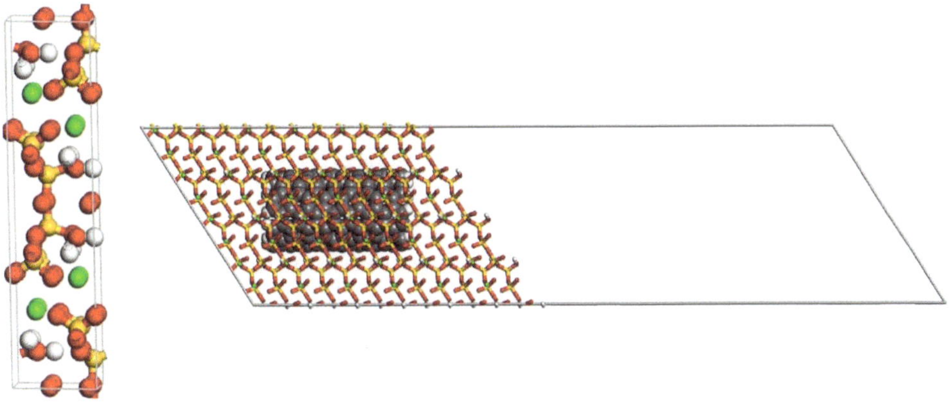

Figure 2. Tobermorite crystal (left) and top view of a tobermorite/CNT composite (right).

Since the interface between the tobermorite and the CNT is the focus of this research, we think this model is reasonable and can work well for our purposes.

The composite material was built by inserting a CNT into the tobermorite matrix (right side of Figure 2). CNTs with radius ranging from 2.71 to 4.74 Å (CNT (4, 4) = 2.70 Å, CNT (5, 5) = 3.39 Å, CNT (6, 6) = 4.07 Å and CNT (7, 7) = 4.74 Å) and a length of 19.68 Å, functionalized with different numbers of carboxyl groups (0, 2, 4, 6 and 10), were used to study the influence of diameter and degree of functionalization on ISS. The geometrical characteristics of the CNTs are presented in Table 1, and a few representative examples of the CNTs used for the simulations are shown in Figure 3.

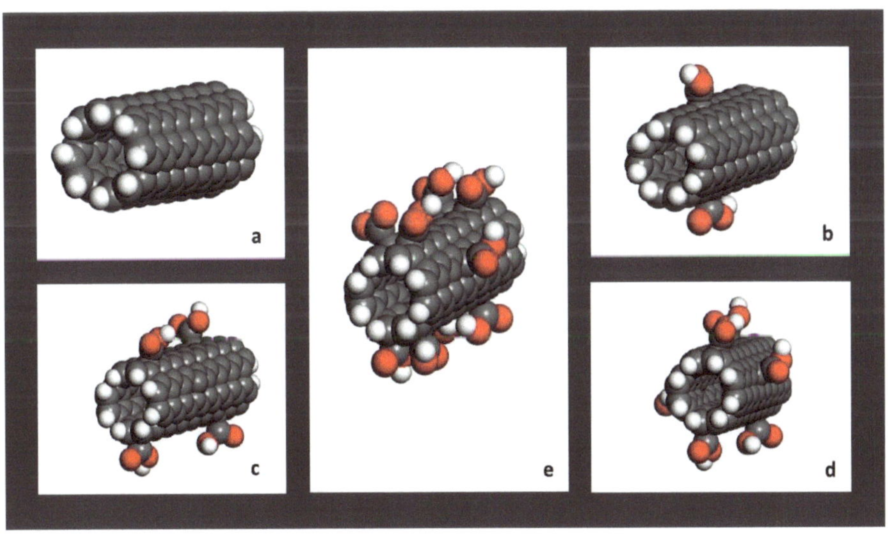

Figure 3. Pristine (a) and functionalized CNTs: 2 COOH (b), 4 COOH (c), 6 COOH (d) and 10 COOH (e).

Table 1. Geometrical characteristics of the CNTs used in the simulations.

System	Radius (Å)	Length (Å)	Number of COOH Groups
CNT (4, 4)	2.71		0
CNT (5, 5)	3.39		2
CNT (6, 6)	4.07	9.84	4
CNT (7, 7)	4.74		10

2.2. Calculation Method

Initially, a 100 ps calculation in the NPT ensemble at 298 K and atmospheric pressure was performed to optimize the parameters of the simulation lattice and relax the system. The temperature was controlled by a Nose–Hoover thermostat [50], and a Berendsen barostat [51] was used to maintain the pressure at 1 atm. A time step of 1 fs was used during the entire relaxation. Then, the resulting relaxed structures were used as inputs to simulate the pull-out process. This was done after increasing the size of the cell in the axial direction of the CNT (x direction) and including a vacuum layer, while keeping the lattice parameters in the other two directions unchanged. These new models were further optimized by molecular mechanics (MM) before the pull-out process. The models used to simulate the pull-out process were created by displacing the CNT along the x axis in increments of 10 Å. Snapshots of the models for one of the CNT/tobermorite systems during the CNT pull-out process are shown in Figure 4. After each pull-out step, the energy of the resulting molecular structure was calculated by MM. Materials Studio 9.0 software [52] was used to perform the simulations.

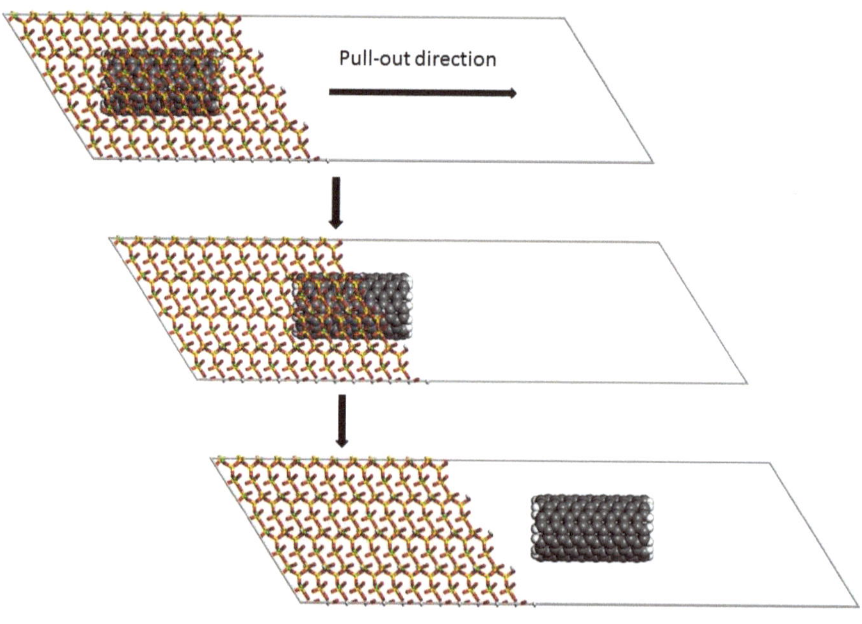

Figure 4. Snapshots of the pull-out process in the CNT/tobermorite composites.

The COMPASSII force field was used to calculate the interactions between the tobermorite and the CNT [53]. COMPASSII has been successfully applied in the simulation of systems containing CNTs and different materials derived from cement [38,54–57]. Coulomb interactions were calculated by the Ewald summation method, and an atom-based cut-off method was applied for the calculation of van der Waal interactions [58]. The cutoff distance for both kind of interactions was set to 12.5 Å.

The interfacial shear strength (ISS = τ) was calculated based on the total pull-out energy, and is given by [59]:

$$\tau = \frac{E_{pull\text{-}out}}{\pi\, r\, L^2} \tag{1}$$

τ, r, and L are the interfacial shear strength, CNT radius and CNT length. $E_{pull\text{-}out}$ is the energy difference between the fully embedded configuration of the CNT and the complete pull-out configuration (see Figure 5). It should be noted that this formula gives the average ISS.

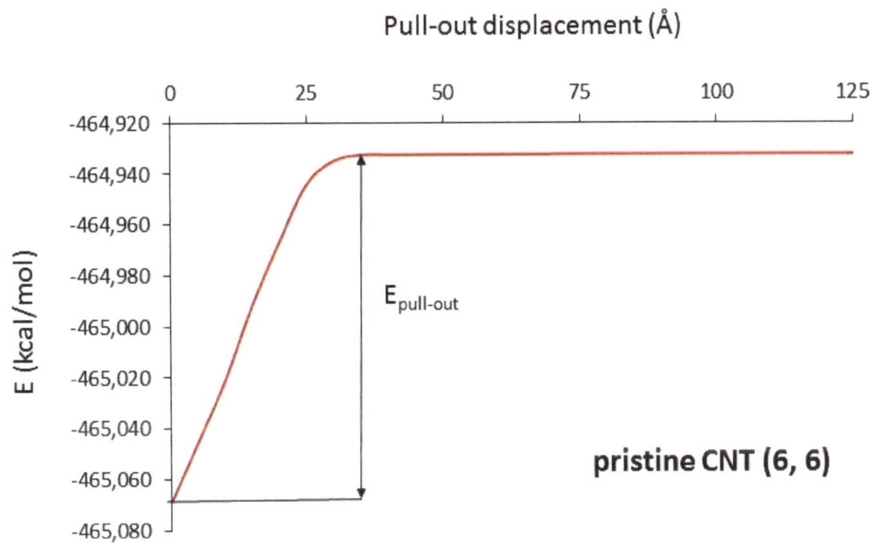

Figure 5. Variation of total potential energy with pull-out displacement.

3. Results and Discussion

3.1. Interfacial Shear Strength

Figure 5 shows the variation in total potential energy of the composite, with pull-out displacement for one of the models. As the CNT was pull out from the tobermorite matrix, the total potential energy increased toward a constant value, which indicated that there was no interaction between the CNT and the matrix beyond a certain pull-out distance. Thus, in all cases studied in this work, the energy of the fully embedded configuration was lower than the energy of the complete pull-out configuration. As indicated in Figure 5, $E_{pull\text{-}out}$ is the energy difference between the fully embedded configuration of the CNT and the complete pull-out configuration. A similar trend was also observed in previous studies [33,36–38]. Increasing the pull-out displacement, the total potential energy decreased since the interfacial interaction between the CNT and the matrix progressively reduced to zero.

The ISS was calculated from Equation (1), and the results are presented in Figure 6 for all systems. The figures above each column correspond to the number of H-bonds existing at the interface between the CNT and the matrix, and the mean distance (Å) between both structures. The geometrical criteria for the existence of an H-bond are the following: the distance between the H of the donor group (D) and the O of the acceptor group (A) is less than 2.5 Å and the DHA angle is greater than 90°.

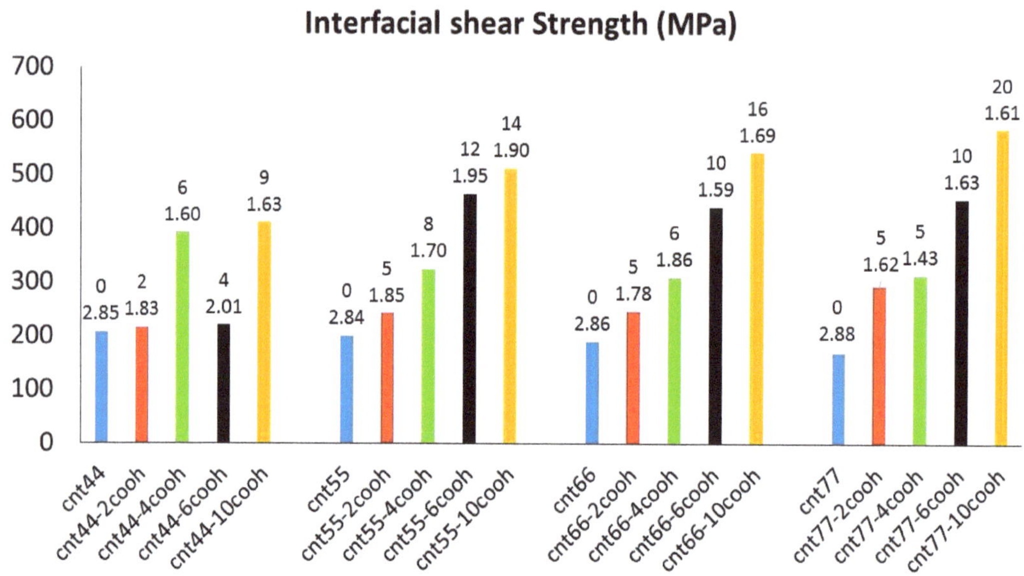

Figure 6. ISS as a function of CNT diameter and number of COOH groups.

It can be seen that ISS values decrease with increasing CNT radius for pristine CNTs. Li et al. [34] found similar results for the pull-out of pristine CNTs from a polyethylene matrix. However, functionalized CNTs do not follow the same trend, and ISS values for these systems seem to be correlated to the number of H-bonds established between the carboxyl groups of the CNT and the tobermorite structure. In general, the higher the number of carboxyl groups, the higher the ISS value. This is because more H-bonds can be formed at the interface, a phenomenon which reinforces the interaction between the CNT and the tobermorite matrix. Some exceptions can be observed. For example, the ISS value for the CNT (4, 4)-4COOH system (392.56 MPa), which presents six H-bonds, is higher than that calculated for the CNT (5, 5)-4COOH system, with eight H-bonds (325.27 MPa). This may be because the mean distance for H-bonds in the CNT (4, 4) system is somewhat smaller (1.60 Å) than that found for the CNT (5, 5) system (1.70 Å), which weakens the interaction.

Some authors analyzed the influence of groups, capable of forming H-bonds with the matrix, on the mechanical properties of graphene-reinforced cementitious composites [37,40,48,60,61]. In all cases, the surface functionalization led to composites with enhanced strength and good binding affinity between the graphene and the cement matrix compared to that of the non-functionalized graphene. In a previous work [62], we also found that the calculated binding energies and experimental results of the mechanical properties of functionalized CNTs are better than those of pristine CNTs. Additionally, other authors found that the functionalized C atoms move out of the nanostructure surface (the hybridization of the functionalized C atom changes from sp^2 to sp^3). The result is that the flat carbon structures turn into wrinkled structures, which mechanically interlock with the matrix, contributing to the enhancement of the mechanical strength [63].

In this work, the calculated ISS values range from 170 to 586 MPa, depending upon the CNT diameter and the number of carboxyl groups bonded to the CNT surface. Other authors found values ranging from 2 to 940 MPa using MD studies of CNT/polymer matrices [64] and references therein]. It would be useful to compare our results with experimental data of single-CNT pull-outs from cement matrices tests. To the best of our knowledge, none are currently available. This may be due to the technical difficulties in designing the experiments and characterizing the interface at the nanoscale. Fan et al. [37] modeled the pull-out of graphene and graphene oxide sheets from a tobermorite crystal, finding ISS values as high as 647 MPa. Their results are comparable to ours. The ISS values calculated by Alkhateb et al. [38] for pulling-out graphene functionalized with hydroxyl, amine and carboxyl groups varied between 1.2 and 13.5 GPa.

3.2. Non-Bonded Energy

As our models do not include chemical bonds between the CNT and the tobermorite, ISS values can be mainly correlated to changes in non-bonded energy (Van der Waals and electrostatic energies) as the CNT is pulled out from the tobermorite matrix. Figure 7 shows $\Delta E_{\text{non-bonded energy}} = E_{\text{non-bonded energy pulled-out}} - E_{\text{non-bonded energy fully embedded}}$ (non-bonded energy difference between the complete pull-out configuration and the fully embedded configuration of the CNT) for all systems. The higher the value, the stronger the interaction between the CNT and the matrix. Again, the figures above each column correspond to the number of H-bonds existing at the interface between the CNT and the matrix, and the mean distance (Å) between both structures. $\Delta E_{\text{non-bonded energy}}$ follows the same trend as that of ISS. High $\Delta E_{\text{non-bonded energy}}$ values correlate well with the number of H-bonds, indicating a strong interfacial interaction and, therefore, an elevated ISS. The positive correlation between $\Delta E_{\text{non-bonded energy}}$ and ISS is clearly seen in Figure 8. Using MD simulations, Lushnikova et al. [65] inserted a functionalized CNT into a tobermorite structure and found improved bulk modulus, shear modulus, plane stress, and Young's modulus. They attributed the results to the interactions established between the O atoms of the tobermorite and the CNT surface, that is to say, to high non-bonded energy values.

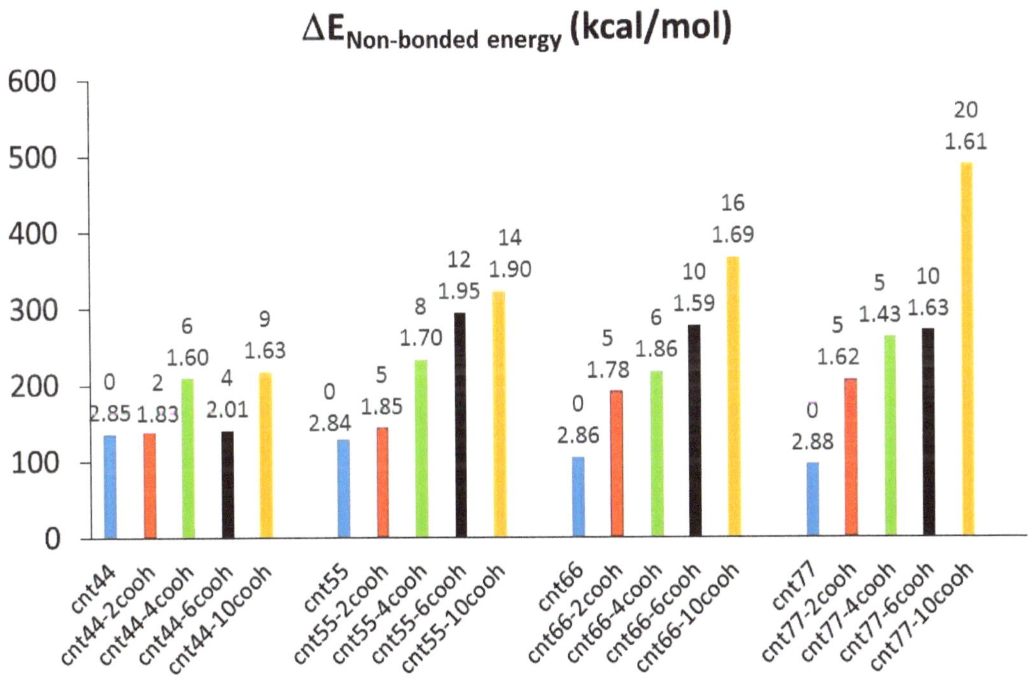

Figure 7. $\Delta E_{\text{non-bonded energy}}$ as a function of CNT diameter and number of COOH groups.

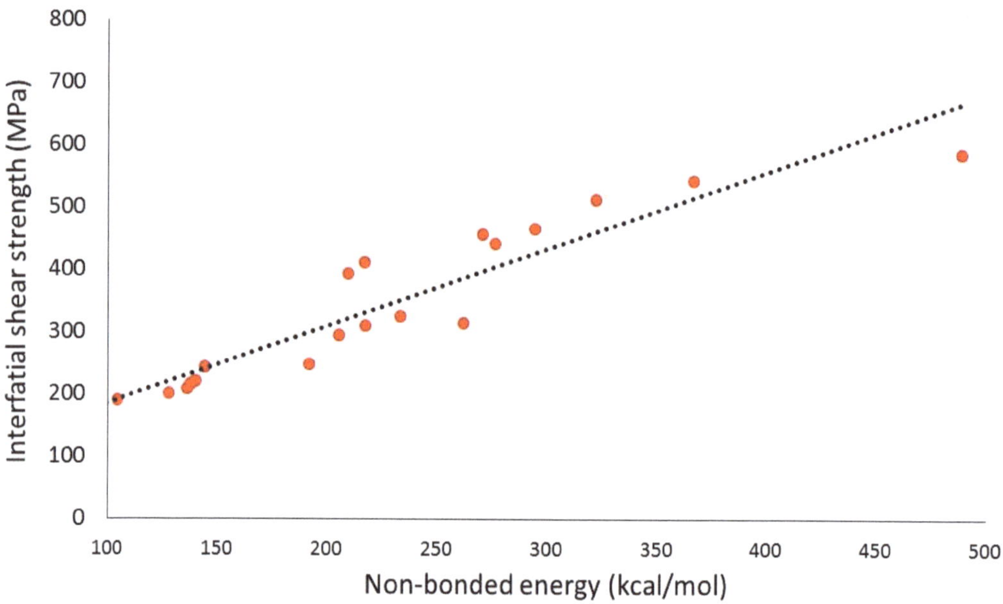

Figure 8. ISS versus $\Delta E_{\text{non-bonded energy}}$.

3.3. *Ca^{2+}-O Interactions*

Ca^{2+} ions in the tobermorite structure can interact with the functional groups of the CNT, acting as bridges that connect the O atoms of the carboxyl groups with the O atoms of the tobermorite structure. Thus, as in the case of H-bonds, this kind of interaction, mediated by Ca^{2+} ions, improves the adhesion between the CNT and the tobermorite surface. Hou et al. [48] studied the strengthening mechanisms of calcium silicate hydrate (C-S-H) and graphene oxide (GO) using reactive force field molecular dynamics and found that Ca^{2+} ions could bridge O atoms in silicate chains and hydroxyl groups in GO, which resulted in the construction of a reinforcing silicate–carbon skeleton in the interface region. Kai et al. [40] found that the Young's modulus and strength of C-S-H were enhanced by 52.6% and 23.3%, respectively, by the incorporation of hydroxyl groups into the GO. This came because of Ca^{2+} coordination and the formation of hydrogen bonds with the hydroxyl groups of GO. The abovementioned value of 647 MPa, found by Fan et al. [35], is attributed to the strong interactions established between the O atoms of a functionalized GO sheet and the Ca^{2+} ions in the tobermorite structure.

The mean coordination number of Ca^{2+} ions at the interfacial region, the number of Ca^{2+}-carboxylic O close contacts for each CNT, and the mean distance between Ca^{2+} ions and O atoms (either in the carboxyl groups or in the silicate chains) are shown in Table 2 for all systems. As can be seen in Table 2, coordination numbers vary from 5 to 6, with the latter being the preferred one in most structures. It is well known that Ca^{2+} ions in crystal structures generally bind to O atoms in ligands, with coordination numbers mostly ranging from 6 to 8 [66]. Bond lengths are similar to those found in compounds with 6-fold oxygen coordination [67]. In all cases, one or two of the O ligands coordinating a Ca^{2+} ion belong to a carboxyl group of the CNT surface. Furthermore, the number of Ca^{2+} ions that are coordinated by a carboxylic O increases with the number of carboxyl groups, as expected (right column of Table 2). This electrostatic interaction contributes, together with the H-bonds, to the strengthening of the composite interface. The structure of a 6-fold coordinated Ca^{2+} at the interface between the CNT and the tobermorite matrix is shown in Figure 9 for one of the models. For the sake of clarity, some of the atoms of the structure have been removed.

Table 2. Coordination number of Ca^{2+} ions, mean Ca^{2+}-O distance and number of Ca^{2+}-carboxylic O close contacts.

System	Mean Coordination Number	d (Å)	Number of Ca^{2+}-Carboxylic O Close Contacts
cnt44-2cooh	5.0	2.45	2
cnt44-4cooh	5.7	2.53	3
cnt44-6cooh	5.3	2.46	3
cnt44-10cooh	5.3	2.46	4
cnt55-2cooh	5.5	2.34	2
cnt55-4cooh	5.5	2.37	4
cnt55-6cooh	6.0	2.5	5
cnt55-10cooh	6.0	2.47	10
cnt66-2cooh	6.0	2.56	2
cnt66-4cooh	6.0	2.52	4
cnt66-6cooh	6.0	2.48	6
cnt66-10cooh	6.3	2.57	7
cnt77-2cooh	5.5	2.50	2
cnt77-4cooh	5.6	2.45	5
cnt77-6cooh	5.7	2.48	7
cnt77-10cooh	5.9	2.45	8

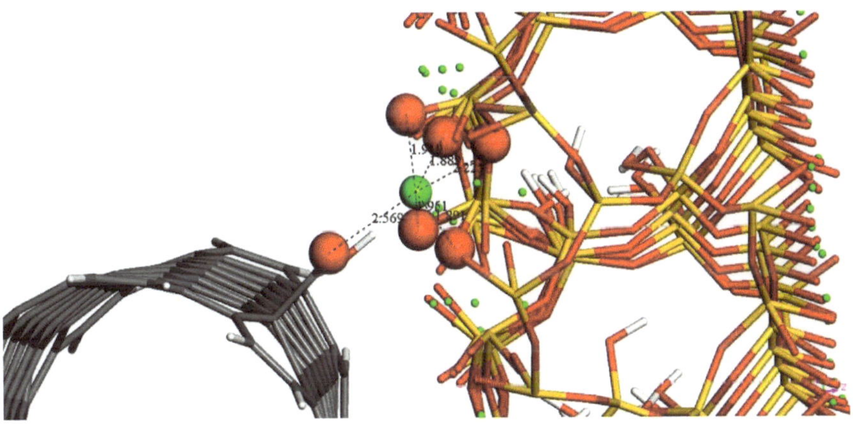

Figure 9. Close look at the interface: 6-fold Ca^{2+} coordination and Ca^{2+}-O bond lengths. Red: oxygen; white: hydrogen; yellow: silicon; grey: carbon; green: calcium.

4. Conclusions

A series of pull-out simulations of pristine and functionalized CNTs, taken from a tobermorite matrix, were carried out to study the interfacial characteristics of the resulting composites. Only non-bonded interactions were considered. The results allow for the following conclusions:

- ISS decreased with increasing CNT radius for pristine CNTs and depended upon the number of H-bonds for functionalized CNTs.
- In the absence of chemical bonds between the CNT and the matrix, ISS values were positively correlated to $\Delta E_{\text{non-bonded energy}}$.
- High, positive $\Delta E_{\text{non-bonded energy}}$ values were mainly derived from the destruction of H-bonds and electrostatic Ca^{2+}-O interactions, as the CNT was pulled out from the

matrix. Specifically, the complete pull-out configuration was less stable than the fully embedded configuration.

- The higher the number of carboxyl groups, the higher the ISS value, as the functionalization of CNTs with these polar groups resulted in both more H-bonds and more Ca^{2+}-O interactions.

Funding: This research received no external funding.

Institutional Review Board Statement: Not applicable.

Informed Consent Statement: Not applicable.

Data Availability Statement: Not applicable.

Conflicts of Interest: The author declares no conflict of interest.

References

1. Reales, O.A.M.; Toledo Filho, R.D. A review on the chemical, mechanical and microstructural characterization of carbon nanotubes-cement based composites. *Constr. Build. Mater.* **2017**, *154*, 697–710. [CrossRef]
2. Shi, T.; Li, Z.; Guo, J.; Gong, H.; Gu, C. Research progress on CNTs/CNFs-modified cement-based composites—A review. *Constr. Build. Mater.* **2019**, *202*, 290–307. [CrossRef]
3. Rocha, V.V.; Ludvig, P.; Trindade, A.C.C.; de Andrade Silva, F. The influence of carbon nanotubes on the fracture energy, flexural and tensile behavior of cement based composites. *Constr. Build. Mater.* **2019**, *209*, 1–8. [CrossRef]
4. Makul, N. Advanced smart concrete—A review of current progress, benefits and challenges. *J. Clean Prod.* **2020**, *274*, 122899. [CrossRef]
5. Metaxa, Z.S.; Tolkou, A.K.; Efstathiou, S.; Rahdar, A.; Favvas, E.P.; Mitropoulos, A.C.; Kyzas, G.Z. Nanomaterials in Cementitious Composites: An Update. *Molecules* **2021**, *26*, 1430. [CrossRef] [PubMed]
6. Li, G.Y.; Wang, P.M.; Zhao, X. Mechanical behavior and microstructure of cement composites incorporating surfacetreated multi-walled carbon nanotubes. *Carbon* **2005**, *43*, 1239–1245. [CrossRef]
7. Li, G.Y.; Wang, P.M.; Zhao, X. Pressure-sensitive properties and microstructure of carbon nanotube reinforced cement composites. *Cem. Concr. Compos.* **2007**, *29*, 377–382. [CrossRef]
8. Gao, F.; Tian, W.; Wang, Z.; Wang, F. Effect of diameter of multi-walled carbon nanotubes on mechanical properties and microstructure of the cement-based materials. *Constr. Build. Mater.* **2020**, *260*, 120452. [CrossRef]
9. Gao, Y.; Jing, H.; Zhou, Z.; Shi, X.; Li, L.; Fu, G. Roles of carbon nanotubes in reinforcing the interfacial transition zone and impermeability of concrete under different water-to-cement ratios. *Constr. Build. Mater.* **2021**, *272*, 121664. [CrossRef]
10. Adhikary, S.K.; Rudžionis, Z.; Rajapriya, R. The Effect of carbon nanotubes on the flowability, mechanical, microstructural and durability properties of cementitious composite: An overview. *Sustainability* **2020**, *12*, 8362. [CrossRef]
11. Du, Y.; Gao, P.; Yang, J.; Shi, F.; Shabaz, M. Experimental Analysis of Mechanical Properties and Durability of Cement-Based Composite with Carbon Nanotube. *Adv. Mater. Sci. Eng.* **2021**, *2021*, 8777613. [CrossRef]
12. Silvestro, L.; Ruviaro, A.; Lima, G.; de Matos, P.; de Azevedo, A.R.G.; Monteiro, S.N.; Gleize, P. Influence of Ultrasonication of Functionalized Carbon Nanotubes on the Rheology, Hydration, and Compressive Strength of Portland Cement Pastes. *Materials* **2021**, *14*, 5248. [CrossRef] [PubMed]
13. Jongvivatsakul, P.; Thongchom, C.; Mathuros, A.; Prasertsri, T.; Adamu, M.; Orasutthikul, S.; Lenwari, A.; Charainpanitkul, T. Enhancing bonding behavior between carbon fiber-reinforced polymer plates and concrete using carbon nanotube reinforced epoxy composites. *Case Stud. Constr. Mater.* **2022**, *17*, e01407. [CrossRef]
14. Yu, Z.; Lau, D. Evaluation on mechanical enhancement and fire resistance of carbon nanotube (CNT) reinforced concrete. *Coupled Syst. Mech.* **2017**, *6*, 335–349.
15. Qin, R.; Zhou, A.; Yu, Z.; Wang, Q.; Lau, D. Role of carbon nanotube in reinforcing cementitious materials: An experimental and coarse-grained molecular dynamics study. *Cem. Concr. Res.* **2021**, *147*, 106517. [CrossRef]
16. Sindu, B.S.; Sasmal, S. Molecular dynamics simulations for evaluation of surfactant compatibility and mechanical characteristics of carbon nanotubes incorporated cementitious composite. *Constr. Build. Mater.* **2020**, *253*, 119190.
17. Balasubramaniam, B.; Mondal, K.; Ramasamy, K.; Palani, G.S.; Iyer, N.R. Hydration Phenomena of Functionalized Carbon Nanotubes (CNT)/Cement Composites. *Fibers* **2017**, *5*, 39. [CrossRef]
18. Sarvandani, M.M.; Mahdikhani, M.; Aghabarati, H.; Fatmehsari, M.H. Effect of functionalized multi-walled carbon nanotubes on mechanical properties and durability of cement mortars. *J. Build. Eng.* **2021**, *41*, 102407. [CrossRef]
19. Balasubramanian, K.; Burghard, M. Chemically functionalized carbon nanotubes. *Small* **2005**, *1*, 180–192. [CrossRef]
20. Mallakpour, S.; Soltaniana, S. Surface functionalization of carbon nanotubes: Fabrication and applications. *RSC Adv.* **2016**, *6*, 109916. [CrossRef]
21. Liu, P. Modifications of carbon nanotubes with polymer. *Eur. Polym. J.* **2005**, *41*, 2693–2703. [CrossRef]

22. Peng, H.; Alemany, L.B.; Margrave, J.L.; Khabashesku, V.N. Sidewall Carboxylic Acid Functionalization of Single-Walled Carbon Nanotubes. *J. Am. Chem. Soc.* **2003**, *125*, 15174–15182. [CrossRef] [PubMed]
23. Kang, S.T.; Seo, J.Y.; Park, S.H. The Characteristics of CNT/Cement Composites with Acid-Treated MWCNTs. *Adv. Mater. Sci. Eng.* **2015**, *2015*, 308725. [CrossRef]
24. Ruan, Y.; Han, B.; Yu, X.; Zhang, W.; Wang, D. Carbon nanotubes reinforced reactive powder concrete. *Compos. Part A Appl. Sci. Manuf.* **2018**, *112*, 371–382. [CrossRef]
25. Yilmaz, Y.L. Analyzing single fiber fragmentation test data by using stress transfer model. *J. Compos. Mater.* **2002**, *36*, 537–551. [CrossRef]
26. Zhao, Y.R.; Xing, Y.M.; Lei, Z.K. Interfacial stress transfer behavior in a specially-shaped fiber/matrix pullout test. *Acta Mech. Sin.* **2010**, *26*, 113–119. [CrossRef]
27. Zu, M.; Li, Q.; Zhu, Y. The effective interfacial shear strength of carbon nanotube fibers in an epoxy matrix characterized by a microdroplet test. *Carbon* **2012**, *50*, 1271–1279. [CrossRef]
28. Battisti, A.; Esqué-de los Ojos, D.; Ghisleni, R. Single fiber push-out characterization of interfacial properties of hierarchical CNT-carbon fiber composites prepared by electrophoretic deposition. *Compos. Sci. Technol.* **2014**, *95*, 121–127. [CrossRef]
29. Rodríguez, M.; Molina-Aldareguía, J.M.; González, C. A methodology to measure the interface shear strength by means of the fiber push-in test. *Compos. Sci. Technol.* **2012**, *72*, 1924–1932. [CrossRef]
30. Ji-Hong, J.; Chan, Y.; Chan-Gi, P. Bonding Characteristics of Macrosynthetic Fiber in Latex-Modified Fiber-Reinforced Cement Composites as a Function of Carbon Nanotube Content. *Int. J. Polym. Sci.* **2016**, *2016*, 7324975.
31. Li, Y.; Wang, Q.; Wang, S. A review on enhancement of mechanical and tribological properties of polymer composites reinforced by carbon nanotubes and graphene sheet: Molecular dynamics simulations. *Compos. B Eng.* **2019**, *160*, 348–361. [CrossRef]
32. Liao, K.; Li, S. Interfacial characteristics of a carbon nanotube–polystyrene composite system. *Appl. Phys. Lett.* **2001**, *79*, 4225. [CrossRef]
33. Chowdhury, S.C.; Okabe, T. Computer simulation of carbon nanotube pull-out from polymer by the molecular dynamics method. *Compos. Part A Appl. Sci. Manuf.* **2007**, *38*, 747–754. [CrossRef]
34. Li, Y.; Liu, Y.; Peng, X.; Yan, C.; Liu, S.; Hu, N. Pull-out simulations on interfacial properties of carbon nanotube-reinforced polymer nanocomposites. *Comput. Mater. Sci.* **2011**, *50*, 1854–1860. [CrossRef]
35. Chandra, Y.; Scarpa, F.; Adhikari, S.; Zhang, J.; Flores, E.S.; Peng, H.X. Pullout strength of graphene and carbon nanotube/epoxy composites. *Compos. B Eng.* **2016**, *102*, 1–8. [CrossRef]
36. Chawla, R.; Sharma, S. Molecular dynamics simulation of carbon nanotube pull-out from polyethylene matrix. *Compos. Sci. Technol.* **2017**, *144*, 169–177. [CrossRef]
37. Fan, D.; Lue, L.; Yang, S. Molecular dynamics study of interfacial stress transfer in graphene-oxide cementitious composites. *Comput. Mater. Sci.* **2017**, *139*, 56–64. [CrossRef]
38. Alkhateb, H.; Al-Ostaz, A.; Cheng, A.H.D.; Li, X. Materials genome for graphene-cement nanocomposites. *J. Nanomechanics Micromechanics* **2013**, *3*, 67–77. [CrossRef]
39. Eftekhari, M.; Mohammadi, S. Molecular dynamics simulation of the nonlinear behavior of theCNT-reinforced calcium silicate hydrate (C–S–H) composite. *Int. J. Impact. Eng.* **2016**, *82*, 78–87.
40. Kai, M.; Zhang, L.; Liew, K. Graphene and graphene oxide in calcium silicate hydrates: Chemical reactions, mechanical behaviors and interfacial sliding. *Carbon* **2019**, *146*, 181–193. [CrossRef]
41. Lv, C.; Xue, Q.; Xia, D.; Ma, M.; Xie, J.; Chen, H. Effect of chemisorption on the interfacial bonding characteristics of graphene–polymer composites. *J. Phys. Chem. C* **2010**, *114*, 6588–6594. [CrossRef]
42. Shiu, S.C.; Tsai, J.L. Characterizing thermal and mechanical properties of graphene/ epoxy nanocomposites. *Compos. B Eng.* **2014**, *56*, 691–697. [CrossRef]
43. Li, Y.; Wang, S.; Wang, Q. Enhancement of tribological properties of polymer composites reinforced by functionalized graphene. *Compos. B Eng.* **2017**, *120*, 83–91. [CrossRef]
44. Nikkhah, S.J.; Moghbeli, M.R.; Hashemianzadeh, S.M. Investigation of the interface polyethylene and functionalized graphene: A computer simulation study. *Curr. Appl. Phys.* **2015**, *15*, 1188–1199. [CrossRef]
45. Zhang, J.; Jiang, D. Molecular dynamics simulation of mechanical performance of graphene/graphene oxide paper based polymer composites. *Carbon* **2014**, *67*, 784–791. [CrossRef]
46. Bauchy, M.; Qomi, M.A.; Ulm, F.J.; Pellenq, R.M. Order and Disorder in Calcium-silicate-hydrate. *J. Chem. Phys.* **2014**, *140*, 214503. [CrossRef]
47. Richardson, I.G. Tobermorite/jennite- and tobermorite/calcium hydroxide-based models for the structure of C–S–H: Applicability to hardened pastes of tricalcium silicate, -dicalcium silicate, Portland cement, and of Portland cement with blast-furnace slag, metakaolin, or silica fume. *Cem. Concr. Res.* **2004**, *34*, 1733–1777.
48. Hou, D.; Lu, Z.; Li, X.; Ma, H.; Li, Z. Reactive molecular dynamics and experimental study of graphene-cement composites: Structure, dynamics and reinforcement mechanisms. *Carbon* **2017**, *115*, 188–208. [CrossRef]
49. Allen, A.J.; Thomas, J.J.; Jennings, H.M. Composition and density of nanoscale calcium–silicate–hydrate in cement. *Nat. Mater.* **2007**, *6*, 311–316. [CrossRef]
50. Hoover, W.G. Canonical dynamics: Equilibrium phase-space distributions. *Phys. Rev. A* **1985**, *31*, 1695–1697. [CrossRef]

51. Berendsen, H.J.C.; Postma, J.P.M.; Van Gunsteren, W.F.; DiNola, A.; Haak, J.R. Molecular dynamics with coupling to an external bath. *J. Chem. Phys.* **1984**, *81*, 3684–3690. [CrossRef]
52. Materials Studio Materials Modeling & Simulation Application | Dassault Systèmes BIOVIA. Available online: https://www.3ds.com/products-services/biovia/products/molecular-modeling-simulation/biovia-materials-studio/ (accessed on 21 November 2022).
53. Sun, H.; Ren, P.; Fried, J.R. The COMPASS force field: Parameterization and validation for phosphazenes. *Comput. Theor. Polym. Sci.* **1998**, *8*, 229–246. [CrossRef]
54. Bhuvaneshwari, B.; Palani, G.S.; Karunya, R.; Iyer, N.R. Nano mechanical properties on the mineralogical array of calcium silicate hydrates and calcium hydroxide through molecular dynamics-CSIR-SERC. *Curr. Sci.* **2015**, *108*, 1058–1065.
55. Du, J.; Bu, Y.; Shen, Z. Interfacial properties and nanostructural characteristics of epoxy resin in cement matrix. *Constr. Build. Mater.* **2018**, *164*, 103–112. [CrossRef]
56. Al-Ostaz, A.; Wu, W.; Cheng, A.-D.; Song, C. A molecular dynamics and microporomechanics study on the mechanical properties of major constituents of hydrated cement. *Compos. Part B Eng.* **2010**, *41*, 543–549. [CrossRef]
57. Hajilar, S.; Shafei, B. Nano-scale investigation of elastic properties of hydrated cement paste constituents using molecular dynamics simulations. *Comput. Mater. Sci.* **2015**, *101*, 216–226. [CrossRef]
58. Tosi, M. Cohesion of ionic solids in the Born model. In *Solid State Physics*; Elsevier: Amsterdam, The Netherlands, 1964; Volume 16, pp. 1–120.
59. Gou, J.; Minaie, B.; Wang, B.; Liang, Z.; Zhang, C. Computational and experimental study of interfacial bonding of single-walled nanotube reinforced composites. *Comput. Mater. Sci.* **2004**, *31*, 225–236. [CrossRef]
60. Al-Muhit, B.; Sanchez, F. Nano-engineering of the mechanical properties of tobermorite 14 Å with graphene via molecular dynamics simulations. *Constr. Build. Mater.* **2020**, *233*, 117237. [CrossRef]
61. Sanchez, F.; Zhang, L. Molecular dynamics modeling of the interface between surface functionalized graphitic structures and calcium–silicate–hydrate: Interaction energies, structure, and dynamics. *J. Colloid. Interface Sci.* **2008**, *323*, 349–358. [CrossRef]
62. Merodio-Perea, R.G.; Páez-Pavón, A.; Lado-Touriño, I. Reinforcing cement with pristine and functionalized carbon nanotubes: Experimental and simulation studies. *Int. J. Smart Nano Mater.* **2020**, *11*, 370–386. [CrossRef]
63. Wang, M.C.; Lai, Z.B.; Galpaya, D.; Yan, C.; Hu, N.; Zhou, L.M. Atomistic simulation of surface functionalization on the interfacial properties of graphene-polymer nanocomposites. *J. Appl. Phys.* **2014**, *115*, 123520. [CrossRef]
64. Mani, A.; Sharma, S. Interfacial shear strength of carbon nanotube reinforced polymer composites: A review. *Mater. Today Proc.* **2022**, *50*, 1774–1780. [CrossRef]
65. Lushnikova, A.; Zaoui, A. Improving mechanical properties of C–S–H from inserted carbon nanotubes. *J. Phys. Chem. Solids* **2017**, *105*, 72–80. [CrossRef]
66. Katz, A.K.; Glusker, J.P.; Beebe, S.A.; Bock, C.W. Calcium Ion Coordination: A CoMParison with That of Beryllium, Magnesium, and Zinc. *J. Am. Chem. Soc.* **1996**, *118*, 5752–5763. [CrossRef]
67. Sowrey, F.E.; Skipper, L.J.; Pickup, D.M.; Drake, K.O.; Lin, Z.; Smith, M.E.; Newport, R.J. Systematic empirical analysis of calcium–oxygen coordination environment by calcium K-edge XANES. *Phys. Chem. Chem. Phys.* **2004**, *6*, 188–192. [CrossRef]

C Journal of Carbon Research

Review

Technology Features of Diamond Coating Deposition on a Carbide Tool

Evgeny Ashkinazi [1], Sergey Fedorov [2,*], Alexander Khomich [3], Vladimir Rogalin [4], Andrey Bolshakov [1], Dmitry Sovyk [1], Sergey Grigoriev [2] and Vitaly Konov [1]

1 Prokhorov General Physics Institute of the Russian Academy of Sciences, 38 Vavilov st., 119991 Moscow, Russia
2 Moscow State University of Technology "STANKIN", 1 Vadkovsky per., 127055 Moscow , Russia
3 Fryazinsky Branch of the V.A. Kotelnikov Institute of Radio Engineering and Electronics of the Russian Academy of Sciences, Moscow region, 1 Vvedensky Square, 141190 Fryazino, Russia
4 Institute of Electric Phisics and Electric Power Engineering of the Russian Academy of Sciences, Dvortsovaya quay, 191186 St. Petersburg, Russia
* Correspondence: sv.fedorov@icloud.com; Tel.: +7-(916)-290-2607

Abstract: The production of carbide tools with polycrystalline diamond coatings, which are used for processing modern carbon composite materials, includes a number of technological techniques that ensure reliable adhesion of the coating to the substrate. This review examines these features of substrate-surface pretreatment to improve adhesion, which includes chemical etching, mechanical hardening, modification by ion beams, plasma treatment and application of buffer layers between the substrate and the coating. This review also discusses the advantages and disadvantages of the most common methods for obtaining polycrystalline diamond coatings using hot filament and deposition of coatings from microwave plasma.

Keywords: carbon fiber reinforced polymer materials (CFRP); polycrystalline diamond coatings; tungsten carbide; adhesion; buffer layer; diamond tools; hardness; wear resistance; CVD

Citation: Ashkinazi, E.; Fedorov, S.; Khomich, A.; Rogalin, V.; Bolshakov, A.; Sovyk, D.; Grigoriev, S.; Konov, V. Technology Features of Diamond Coating Deposition on a Carbide Tool. *C* **2022**, *8*, 77. https://doi.org/10.3390/c8040077

Academic Editor: Gil Gonçalves

Received: 17 October 2022
Accepted: 6 December 2022
Published: 9 December 2022

1. Introduction

With the advent of new materials, the question of their machinability inevitably arises. Carbon Fiber Reinforced Polymer materials (CFRP) are characterized by highly specific stiffness and strength, low weight, high corrosion resistance, and a low coefficient of thermal expansion (CTE). They are widely used for various engineering applications, especially in the aviation and aerospace industries [1–4].

Traditionally, the machinability of hard-to-process materials increases due to wear-resistant coatings [5]. In the case of CFRP, the diamond-coated tool has proven itself in the best way. Progress in the creation of diamond tools with increased wear resistance is associated with the need to overcome the technological barrier for the existing tool, initiated by the rapid development of new materials with unique properties designed to work in critical operating conditions. Overcoming the frame's super-strength combination with the bundle's superelasticity—subject to one of the main requirements—the rejection of using coolants during processing is possible only with record-low friction coefficients and maximum thermal conductivity characteristic of a diamond coating.

Diamond has unique properties (high hardness and thermal conductivity, excellent abrasion resistance, low coefficient of friction [6,7]) and is widely used for processing materials. The development of methods for applying durable polycrystalline diamond coatings (DC) directly to the working surface of the base materials makes it possible to manufacture tools using the best properties of diamond. The most common method of applying DC to hard alloys based on tungsten carbide (WC-Co) is deposition from the gas phase by chemical vapor deposition (CVD) [8]. Nevertheless, obtaining a solid layer of diamond coating on the surface of a hard alloy remains a difficult technological task.

To date, such a tool is the most promising substitute for a cutting tool based on polycrystals of diamond (PCD) obtained by diamond powder sintering. The main difference is the method of diamond synthesis. CVD diamond coating with a thickness of 10–20 μm has a denser polycrystalline structure with a tensile strength in the range of 7000–12,000 kg/mm^2 and a friction coefficient of 0.05–0.15. Often, the hardness of the coating is 15% higher than PCD. The advantages of diamond coatings also include the possibility of their synthesis on an instrument of complex shape. PCD tools, due to technological limitations, are made in the form of plates only.

The typical cutting speeds for a diamond–coated tool are in the range of 200–2000 m/min, while the operating speed range of the PCD tool is 780–900 m/min.

Figure 1 shows a schematic diagram of obtaining DC on WC-Co alloy [9]. However, numerous experiments have shown that it is practically impossible to get a high-quality diamond coating with good adhesion by direct deposition. Cobalt catalytically interacts with carbon atoms at a diamond deposition temperature of 600–900 °C and stimulates the formation of a softening graphite intermediate layer at the coating–substrate interface [7,10], which noticeably worsens the adhesion of the DC to the substrate.

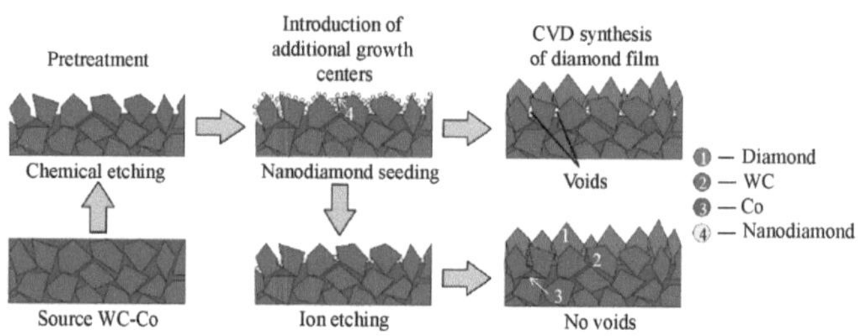

Figure 1. Schematic diagram of the production of diamond films on WC-Co alloy [6].

As a rule, the service life of the DC is determined by the peeling of the coating, not by its wear [11]. Several types of substrate surface pretreatment have been developed to solve the adhesion problem, including the preliminary chemical, mechanical, laser, or plasma treatments, as well as the creation of intermediate layers between the DC and the substrate [12–15]. After chemical etching, a middle buffer layer must apply to the substrate to inhibit the diffusion of Co to the inner surface and dampen the difference in the substrate CTE and the diamond, causing the appearance of significant mechanical stresses in the coating layer that can lead to peeling.

Table 1 shows the CTE values of some materials suitable for creating buffer layers. Unfortunately, there are practically no materials close to diamond in terms of CTE, except invar (an alloy of nickel and iron), which is not suitable for this task since Fe, similar to Co, chemically interacts actively with carbon. In the works [16–20], the choice of tungsten as a buffer layer material is justified by its smaller CTE compared to other metals.

Table 1. Coefficient of thermal expansion values of materials suitable for making the buffer.

Substrate	CTE 10^{-6} sm^{-1} (0 °C)
WC	3.8
WC-Co	4.6–5
W	4.3
Co	12
Ta	6.5
Cr	5.8
Diamond	1.2
Si	5.1
Graphite	1.5
TaC	4.6

First of all, the surface of the tungsten-containing hard alloy is maximally cleaned of cobalt by chemical etching. Unfortunately, at the same time, the strength properties of the cobalt-depleted layer inevitably deteriorate, and it is desirable to strengthen it, for example, mechanically. Then a buffer layer is applied to the hard alloy, and its surface is seeded with nano-sized diamond particles, which are necessary as a seed for growing DC.

This paper presents an overview of the prominent publications on the technological processes of manufacturing tools based on tungsten carbide with diamond coating and preparing a carbide substrate for its deposition.

2. Chemical Methods of Etching Cobalt on the Surface of WC-Co Substrate

The chemical methods of etching the surface of WC-Co substrate are used to remove cobalt from its near-surface layer to increase the density of diamond nucleation and improve the adhesion of DC to WC-Co alloy [21]. The etching procedure is as follows: first, the surface layer is treated with a Murakami reagent ($K_3[Fe(CN)_6]$:KOH:H_2O in a ratio of 1:1:10), and then in a solution of peroxide mono sulfuric acid H_2SO_5 (Caro's acid) [22–29].

The Murakami reagent selectively etches tungsten carbide grains and removes cobalt from the surface layer while creating the necessary surface roughness of the hard alloy. Caro acid eroded cobalt to soluble Co^{2+} compounds, reducing the concentration of Co on the surface [30] and oxidizing a layer of super stoichiometric carbon, interfering with adhesion in amorphous form (sp_2) to carbon dioxide gas. An increase in surface roughness also increases the effective area, which contributes to an increase in DC adhesion [15]. At the same time, careful monitoring of the thickness of the etched zone is necessary since the residual pores formed as a result of etching reduce the adequate thickness of the layer, providing adhesion [31] and reduces the viscosity of the near-surface area, which causes embrittlement of a hard alloy [30,32].

Selective etching Co with acids involves a wide range of different pretreatment methods. The Murakami and Caro's reagent treatment time depends on the WC grains size and the Co% concentration [15,33]. Moreover, additional factors must be considered to obtain good adhesion of DC with WC-Co [26,34].

The adhesion of DC to the carbide base weakens due to the presence of a certain number of micropores at the grain boundaries, the graphite phase, and high residual stresses at the DC–WC interface [35]. The authors [26,35] recommend using WC-Co with a low Co content and paying attention to the sufficient roughness of the substrate because Co catalyzes the synthesis of graphitized sp_2-bound carbon rather than the formation of diamond sp_3-carbon and leads to the growth of non-diamond (graphite) phases acting as softening layers [34–36], which reduces the deposition rate of diamond [35].

There are three main ways to neutralize the effect of cobalt on the adhesion of DC:

- The coating of an intermediate layer can prevent the diffusion of Co to the surface of the substrate;
- The creation of stable intermetallic cobalt compounds. This method is implemented using gas or liquid phase reactions;
- The removal of cobalt from the surface of a hard alloy [37,38] by laser ablation, selective dissolution of cobalt with aggressive chemicals, and by combined heat treatment (heating WC-Co at 1000 °C in an atmosphere of 0.25% CH_4/H_2) with subsequent selective removal of cobalt in Caro acid.

Another way to remove Co from the WC-Co surface is to treat it with microwave CO_2–O_2 plasma at 900 °C. As a result of plasma action, $CoWO_4$ and WO_3 compounds are formed. These phases are removed in an alkaline solution [25,39].

The advanced chemical treatment used before the DC coating to WC-Co is a two-stage method based on the etching of WC-Co substrates in the Murakami reagent followed by the dissolution of Co by selective etching in Caro acid [27,39–41].

In ref [42], the results of various chemical treatment regimes were compared to identify the most effective process of Co etching. Three modes with other concentrations of reagents and different etching times were selected for Murakami etching. The concentration of $K_3[Fe(CN)_6]$ varied in the range of 10, 20, and 30 wt.%. The concentration of KOH varied in the range of 10, 20, and 30 wt.%, and the etching times were 10, 15, and 20 min. A total of 11 experiments were conducted. For the second stage using HCl, three parameters were also changed: HCl concentration (40, 53, 65%), H_2O_2 concentration (10, 21, 32%), and etching time (10, 15, 20 min). As a result, it has been shown that—to develop more roughness—it is necessary to increase the etching time to 20 min in a Murakami mixture of 30 wt.% $K_3[Fe(CN)_6]$ and 10 wt.% KOH. The most effective for complete (0%) removal of cobalt from the surface was etching for 1 h in a modified mixture of Caro (65 vol.% HNO_3 + 32 vol.% H_2O_2) in an ultrasonic bath.

In ref [43], the effect of the substrate surface pretreatment on the cutting efficiency of the processed material during the micro treatment of a carbon CFRP with a WC-Co (10%) tool coated with micro crystalline diamond was experimentally investigated. Etching in the Murakami solution for 120 min was the most effective way to remove Co and improve the quality of microcrystalline DC.

In ref [44], the effect of etching duration on the adhesion of DC on WC-Co drills (6 wt.%) was studied. The drills were first subjected to ultrasonic treatment in a mixture of 10 mL of 98% H_2SO_4 + 30 mL of 38% H_2O_2 for 10, 20, 30, 60, and 90 s. The process was carried out at a temperature of 120 °C to improve the etching efficiency. The removal of Co was accompanied by the appearance of a light pink liquid, characteristic of Co salts, so all samples were thoroughly washed with distilled water in an ultrasonic bath until a clear liquid was obtained. The drills were treated in an ultrasound with a solution of ultrafine detonation diamond nanoparticles of 8–10 nm in size to increase the density of diamond nucleation. Raman scattering spectra showed that a high-quality microcrystalline DC was successfully grown after etching for 30 s. An increase in the duration of preliminary processing led to the appearance of a porous structure, island films of DC, and deterioration of the operating parameters of the drills.

In ref [45], the process of preliminary chemical treatment of WC-Co alloy before DC deposition was optimized. The size of WC grains, which significantly affects the thickness of the etched Co layer, was determined depending on the time of the process in two ways: by determining the mass of dissolved Co and by the depth of profiling of the etched surface. Although the removal of Co is beneficial, it causes a negative effect of WC-Co surface layer softening, during which a porous tungsten carbide skeleton forms [15].

The authors of the work [46] used the 6% Co alloy, initially purified for 10 min in an ultrasonic bath with ethanol. The treatment of the hard alloy is carried out with Murakami solution in an ultrasonic bath, followed by washing with deionized water. Cobalt was removed from the surface of the hard alloy with Caro's acid for 7, 11, or 15 min. Next, nanodiamond nuclei were seeded on the surface of the hard alloy in a diamond suspension in an ultrasonic bath. The authors believe that the etching of a hard alloy in a Caro's solution has little effect on its roughness. The best results demonstrated the treatment mode in Murakami solution for 7 min + Caro's acid for 11 min for the WC-6 mass % Co. It is necessary to optimize the acid etching time to improve the quality of the DC.

Thus, in the works [21–46], the modes of preliminary chemical treatment of a hard alloy were established, ensuring the best adhesion of the tool with the applied DC.

3. Chemical Treatment of WC-Co Substrates with a High Cobalt Content

Most studies of the adhesion of DC to hard alloys were performed on substrates with a low concentration of Co (up to 6 wt.%). However, for hard-to-process materials, a tool with a cobalt concentration of more than 6 wt.% is required. In ref [47], the pretreatment processes of WC-9% Co substrates were investigated. This process is based on two-stage chemical etching followed by seeding with diamond nanoparticles and studying the methods of nucleation and growth of DC. For such substrates, the authors proposed to reduce the etching time of a hard alloy in a Murakami solution to 10 min with simultaneous exposure to ultrasound, which is sufficient for the manifestation of WC granularity and time of Co removed at the second stage of etching in Caro solution. The optimal time of the second etching stage is 5 min.

In ref [48], the etching processes of a tungsten carbide-based hard alloy with a Co content of 6, 10, and 12 wt.% were experimentally studied. The transformations of Co during treatment in Murakami solution and DC deposition by the hot thread method were investigated, including nucleation, growth, and heating of processed materials without a carbon source. A two-stage pretreatment method was used: ultrasonic etching in a Murakami solution with different dilution levels and duration of stay in an ultrasonic bath and etching of the sample using Caro's solution with varying levels of dilution. An additional processing stage is proposed to increase the concentration of nucleation sites. This stage consisted of wiping the sample with a paste of diamond chips (3 microns) in glycerin to remove the loose surface layer.

As a result of the research, a technological scheme was proposed that includes seven main stages and describes the behavior of Co in the manufacture of a tool with DC, taking into account the stage of pretreatment, nucleation, and growth of the diamond coating. The mechanisms of Co diffusion in the hard alloy surface layer when heated in a hydrogen atmosphere and the evolution of Co compounds in an untreated solid-alloy tool with the same parameters of nucleation and heating processes considered. Relatively high substrate temperatures (800 and 950 °C) will stimulate (due to diffusion from the volume to the surface) an increase in the cobalt concentration on pretreated surfaces and negatively affect the rate of nucleation and the quality of DC. For untreated samples at a very high temperature (950 °C), the Co content on the substrate surface decreases, probably due to the sublimation of Co. During deposition, a dynamic balance is formed between the diffusion of Co from the volume to the surface and the subsequent substitution of Co, while the contribution of diffusion prevails. However, the diffusion of Co from the interface to the DC gradually decreases and even disappears because of the deposited diamond crystals' barrier effect with an increase in the growth duration. Thus, Co diffusion is a factor that cannot prevent the formation of embryos, but the growth parameters and DC quality near the film–substrate interface are comparatively worse. The depth of removal of Co from the surface layer for WC-Co alloys with a high Co content should be at least 8–9 mkm. It is also necessary to use not only etching but also other operations of preliminary processing of WC-Co substrates, as was proposed, for example, in ref [48].

4. Physical-Mechanical Methods of W-Co Alloys Surface Pretreatment

Before the application of DC, several WC-Co treatment methods were used to increase adhesion, including chemical processes, mechanical treatment, plasma methods, and laser ablation. Thus, in ref [49], the high efficiency of impact hydrotreatment for improving the adhesion of DC on the WC-Co surface is shown. Mechanical treatment methods often remove Co from the WC-Co surface [50]. In ref [51], a complex pretreatment method was tested, including sandblasting, two-stage pretreatment, shot blasting with solid particles (SiO_2 or Al_2O_3) and ultrasonic cavitation, and nanocrystalline DC deposited on the surface of a hard alloy treated in this way. Before DC deposition, a Cr/CrN/Cr buffer layer is applied to the hard alloy [52]. Subsequent mechanical treatment with diamond powders increased the roughness of the hard alloy. It increased the effective surface area of the coating deposition, which also increased the nucleation density necessary to accelerate the deposition of CVD diamond and increased the adhesion of DC. However, the thick buffer layer reduces the efficiency of the instruments with DC cutting edges.

In refs [25,50,53,54], the effect of laser and plasma treatment and the adhesion of the coating to WC-Co were investigated. Laser treatment of the surface provided adhesive strength comparable to the preliminary microblasting. Pulsed laser irradiation creates an optimal surface texture for good adhesion of the DC [55].

Figure 2 shows images of diamond coatings on the surface of a hard alloy.

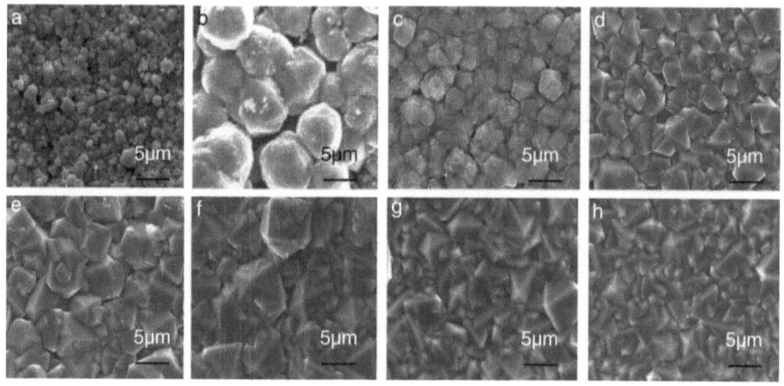

Figure 2. SEM images of diamond coating deposited on WC-Co substrates pretreated by: (**a**) abrading; (**b**) Cr interlayer deposition; (**c**) Nb interlayer deposition; (**d**) Ta interlayer deposition; (**e**) chemical etching; (**f**) chemical etching and Cr interlayer deposition; (**g**) chemical etching and Nb interlayer deposition; and (**h**) chemical etching and Ta interlayer deposition [28].

Ref [56] presents the results of a preliminary combination of a WC-Co treatment with a high Co content, which combined chemical etching and sandblasting, and created a carbide-forming metal layer (Figure 3). The adhesion of DC to a hard alloy was higher than when using only chemical etching or applying a sublayer. The carbide-forming metal forms related carbides during DC growth. Despite the high temperature required to accelerate the diamond nucleation process, carbides prevent the diffusion of Co and suppress the forming graphite phase. The layers of pure Cr, Nb, and Ta are usually used in this case.

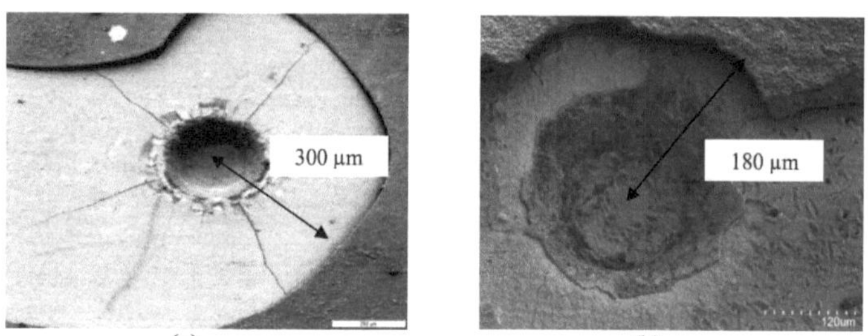

Figure 3. Effect of sandblasting with diamond powder on Cr-coated WC-Co substrate: (**a**) delamination area without sandblasting and (**b**) the elamination area with sandblasting [56].

In ref [27], it has been shown that chemical etching and the CrN/Cr buffer layer contribute to an increase in the adhesion of DC. It has also been found that after the chemical removal of Co from the surface of a hard alloy, the deposited chromium not only prevents the diffusion of Co but also improves the porous structure resulting from chemical etching, which increases the adhesion of DC [28]. This result is of considerable interest, as it confirms the unique adhesive properties of chromium, which is widely used as a sublayer when applying a variety of coatings.

In ref [57], the chemical etching pretreatment, laser etching, and acid treatment of WC-Co, followed by laser etching, were investigated. Laser irradiation leads to a periodic relief of the irradiated surface, which increases its roughness and reduces mechanical stress between the substrate and the deposited DC. The properties of the DC and the magnitude of the residual compression stresses in the coatings were studied by RAMAN spectroscopy. Chemical etching was used to remove Co from the surface, but the surface roughness created by etching did not sufficiently reduce mechanical stresses in the DC. In constant, laser etching reduced the number of mechanical stresses, but the concentration of Co remained high. The combination of chemical etching and laser treatment methods ensured the production of a homogeneous high-quality DC, high adhesion of the coating to the substrate, and a low residual stress level.

Plasma modifies only a few atomic layers of the substrate without changing the bulk properties of the hard alloy [52,58,59]. In ref [60], pooled plasma and chemical etching were used to prepare WC-Co before the deposition of (TiAl)N layers by vacuum spraying. Plasma etching increases the surface roughness and slightly reduces the Co content. It concluded that the processing modes studied in this work, including plasma etching, are suitable for applying a sublayer (TiAl)N to a WC-Co tool, increasing the cutting time by up to 59%.

In ref [61], sandblasting + acid etching was used for the pretreatment of WC-Co before the deposition of DC. The surface roughness varied depending on the parameters of sandblasting. The porous Co-less layer after sandblasting and acid etching was thinner than after the two-stage chemical treatment.

The effect of surface sandblasting on nucleation during chemical gas-phase deposition of the coating has also been studied. Mechanical tests have shown that sandblasting increases the adhesion of the coating compared to two-stage chemical treatment. X-ray diffraction analysis and RAMAN spectroscopy have recorded a decrease in residual stresses in the DC; tests on a cutting machine have shown that tools with DC that have been sandblasted and etched in acid have a smaller wear area of the flank and coating. Thus, in ref [61], it was concluded that sandblasting is a stable, safe, and inexpensive technology of the pretreatment for the tool preceding the CVD process.

5. Suppression of Co Diffusion and Increasing of DC Adhesion by Buffer Layers Formation

The etching of the surface of a hard alloy is not consistent with the tendency to increase the Co content in WC-Co to improve the properties of the carbide tool. The use of etching alone is ineffective if the Co content in the hard alloy exceeds 6 wt.%. In addition, at a high deposition temperature, cobalt from the volume can diffuse to the etched surface [62]. Therefore, to suppress the carbon-cobalt reaction during the deposition of the diamond coating by the CVD method, buffer layers are used between the carbide substrate and the DC [21,63–69]. A properly selected intermediate (buffer) layer serves as a diffusion barrier for Co, which can improve the adhesion of DC to WC-10% Co substrates.

The buffer layer structure affected the adhesion of DC to a hard alloy [70]. In ref [71], the buffer layers TiN, TiC, (TiSi)N, Si_3N_4, a-SiC, and a-SiC_xN_y were investigated, and it was found that DC has insufficient adhesion to TiN, and a pulsed arc deposited an a-C (laser-arc). The DC has the best adhesion to the SiC and Si_3N_4 layers due to the best correspondence of the CTE gradients. In ref [21], a comparative study of CrN, ZrN, NbN, and TaN (with a thickness of 1 mkm as barrier layers for Co diffusion during DC deposition at 880 °C on WC-12 wt.% substrates) shows that NbN and TaN interact with carbon during DC deposition, forming carbides, and are the most effective barriers. A small diffusion depth of Co and weak carburization of the buffer layer were observed when using the ZrN layer, and the CrN layer was wholly transformed due to carburization.

Co-diffusion blocker's amorphous layers demonstrate high values of yield, tensile strength, and corrosion resistance [72,73]. In ref [71], amorphous layers a-SiC, a-Si_3N_4, and a-SiC_xN_y were synthesized by CVD, on which diamond coatings were deposited. It established that amorphous buffer layers of SiC improve adhesion, friction properties, and the cutting efficiency of DC by WC-6 wt.% Co [74,75].

Improved adhesion to the substrate is also achieved after pretreatment, which leads to the formation of stable Co compounds, such as borides or silicides, which reduce the pressure of Co vapors and mitigate the effects of the volatile Co phase [76]. Various metals and ceramics were also used as buffer layers: tungsten [16–20,67,77], aluminum [78,79], chromium [80], NbC [81], Si_3N_4 [82], TiN, Ti(CN), and CrN [21,81,83]. In ref [69], good DC adhesion with a thickness of 8 and 20 microns was demonstrated using a CrN buffer layer.

The buffer layers blocking Co diffusion methods are not limited only by physical (PVD) or chemical vapor deposition (CVD) technologies [53]. It is known that some intermediate layers hinder the nucleation of the diamond and inhibit the growth of DC. In this case, the additional WC-Co treatment increases the density of diamond nuclei [84] and complicates the entire technological process.

In refs [85,86], CrN was proposed as the optimal buffer layer material for DC on WC-10% Co substrates, which contradicts the results of ref [21]. The use of CrN effectively prevents the diffusion of Co from the substrate, preventing the catalytic activity of Co during the deposition of DC, which ensures good chemical and mechanical adhesion to DC due to the formation of a thin layer of CrC [85–87]. It is shown in ref [88] that the adhesion of DC and WC-10% Co with a Cr-N interlayer sharply depends on the diamond layer deposition temperature. In addition, the formation of a chemical bond at the interface and residual stresses in the films have opposite temperature dependences, which makes it possible to optimize the deposition temperature of the diamond-based coating for better adhesion. Though less strong adhesion was observed at the diamond/Cr-N interface.

The effect of a diamond coating deposition temperature on adhesion to WC-10% Co substrates with a Cr-N buffer layer was studied in ref [77]. The DC was deposited at various temperatures (550, 650, and 750 °C) in a CVD reactor with a hot thread. It was found that the carbon diffusion through the diamond/CrN film significantly depends on the deposition temperature and leads to partial carburization. Nitrogen in CrN is replaced by carbon, and a CrC layer is formed. Its formation is influenced by both the rate of growth (coalescence) of alkali grains and the rate of diffusion of carbon into the interlayer, which have different temperature dependences. The rate of DC growth linearly depends on the deposition temperature [88], and the rate of carbon diffusion has an exponential temperature dependence $D = D_0 exp(E_a/kT)$. As the substrate temperature increases, the DC growth rate increases, which reduces the duration of the carburization process by replacing Cr-N with Cr-C, while the rate of carbon diffusion into the Cr-N layer also increases exponentially with temperature. This leads to the fact that carbon diffuses to a greater depth, forming a thicker Cr-C layer before the diamond grains form a continuous film. Over time, the concentration of active carbon trends to zero due to the coalescence of diamond grains and the formation of a solid diamond film, which limits the supply of reagents for further carburization. In addition, the density of the Cr-C layer formed at the interface (6.68 g/cm^3) is higher than that of Cr-N (5.9 g/cm^3), which also leads to a decrease in the diffusion rate of carbon and nitrogen. The formation of a Cr-C layer provides a strong chemical bond between the DC and the Cr-N layer, which contributes to its adhesive strength. At the same time, an increase in the Cr-C layer thickness is undesirable since the difference between the CTE of the diamond and Cr_3C_2 is higher than in the diamond–Cr-N pair. According to the results of Rockwell hardness measurement, the best adhesion was obtained for films applied at 650 °C. At deposition temperatures of 550 °C and above 750 °C, the adhesion of the DC was worse. In ref. [88], this is explained by the fact that, at a temperature of 550 °C, too thin of a layer of Cr-C grows. In contrast, at 750 °C, adhesion deteriorates due to residual thermal stresses. A Cr-C layer of optimal thickness is formed at a temperature of 650 °C. It guarantees the absence of cracks on the DC surface or at the surface–buffer layer interface. This does not extend to the Cr-C/Cr-N interface, which prevents the peeling of the coating.

In ref [89], the effect of magnetron sputtering CrN and TiN layers on the adhesion of DC film to WC-Co substrate was investigated. Opposite data about the effectiveness of buffer CrN layers was obtained. DC synthesis was carried out in a CVD reactor with a glow discharge. DC adhesion was determined by Rockwell hardness measurement results. It was found that TiN does not react with DC, but CrN (according to small-angle X-ray diffraction) almost entirely turns into chromium carbide Cr_3C_2. However, adhesion tests had shown that the efficiency of the buffer layers studied is significantly lower than when the substrate was pretreated with Murakami and Caro solutions. It concluded that using only the TiN layer to increase the adhesion of the DC is impractical due to the lack of chemical bonds between the DC and TiN layers. The use of the CrN barrier layer also proved ineffective due to the conversion of chromium nitride into chromium carbide and the embrittlement of DC.

In ref [90], comparative studies were carried out on the adhesion of (111)-textured DC deposited by the hot thread method on buffer layers TiN, CrN, TiC, and SiN deposited on mirror-polished substrates from WC-Co. The adhesion of the coatings was determined by the Vickers microhardness measurement method. The best results were obtained by a sample with a buffer layer of CrN 800 nm thick, pretreated by chemical etching for 10 min. It is shown in ref. [91] that Cr_2O_3 is also an effective blocker of Co diffusion, preventing the formation of graphite at the DC/WC-Co boundary. Cr_2O_3 was deposited at 300 °C by spraying Cr targets (99.95%) in oxygen and argon plasma. A thin layer of Cr was deposited on it for 5 min to increase adhesion after the deposition of Cr_2O_3. This Cr layer allowed a carbide layer to form at the border with DC. At the same time, the need for careful control of the methane dosage in the gas mixture was noted to ensure the optimal concentration of carbon on the surface during the diamond particle nucleation, which is relevant when carbide-forming materials are used as an intermediate layer.

In ref [92], the properties of DC deposited on WC-Co with or without TiN, CrN, TiC, and SiN buffer layers were analyzed. It could be expected that the most significant mechanical stresses in DC would occur when using CrN, TiN, TiC, and, especially, CrN buffer layers, whose KTE differs as much as possible from diamond CTE due to the difference in the CTE of the WC-Co and diamond buffer layers. The maximum residual stresses were observed in the DC with the SiN buffer layer. Microhardness measurements have shown that the SiN buffer layer, which does not chemically interact with the diamond and the substrate, does not provide sufficient adhesion of the DC with the substrate. Another reason for the negative effect of SiN on adhesion may be its amorphous structure, which does not ensure the alignment of crystal lattices according to CTE, and its high hardness. The interaction of TiN and TiC buffer layers with DC increases the adhesion of the coating to the substrate. Still, the maximum positive effect on the adhesion of DC is provided by buffer layers that chemically interact with the diamond. During etching, pits form on the surface of the substrate, and filling these pits on the grain boundaries with a buffer layer leads to mechanical adhesion of the substrate to the buffer layer. This effect was observed on samples with all the buffer layers studied. In contrast, chemical reactions with diamond occurred only in the CrN buffer layer, which provided better adhesion of the DC with the CrN layer.

In ref [48], an increase in the adhesion of DC to WC-Co provided by magnetron sputtering of homogeneous buffer layers of TiB_2 with a thickness of about 0.1–1 μm, effectively preventing the diffusion of Co from the substrate into the coating discussed. At the same time, the density of the embryos and the growth rate of the diamond layer increased. An increase in adhesion was observed at the optimal thickness of the TiB_2 layer of 200 nm.

In ref. [81], the deposition of DC at 780 °C on CrN_x and NbC layers with a thickness of 1 μm was studied. The choice of these layers was due to their high hardness and proximity of CTE to its value of diamond and WC-Co [93,94]. The minimum level of mechanical stresses in the coatings during its cooling at the stage of the growth process's completion was provided. Studies of the barrier properties of the CrN_x buffer layer have shown that, during heat treatment in a hydrogen atmosphere, a significant amount of Co diffuses through this layer, having a porous structure. Thus, using a CrN_x layer to obtain a DC with good adhesion requires the application of an additional barrier layer that prevents the diffusion of Co.

The properties of the NbC buffer layer significantly differ from the CrN_x layer. The NbC layer is thermally stable when heated in a hydrogen atmosphere, although traces of Co diffusion through NbC was observed after exposure for 60 h. Sufficient adhesion of DC to NbC was achieved only after increasing the roughness of WC-Co before Co poisoning and precipitation of NbC. The appropriate WC-Co pretreatment is necessary to obtain sufficient adhesion before the deposition of NbC. A significant improvement in adhesion was observed after the Murakami reagent or sandblasting + Murakami reaction. In ref [81], it concluded that a double buffer layer of Cr_xC_y and CrN_x can be optimal for the successful application of DC. The Cr_xC_y layer is a diffusion barrier for Co, and the CrN_x layer, after transformation into a porous structure, provides good adhesion of the DC.

6. Two-Layer and Gradient Buffer Layers

In ref [95], dense and uniform gradient coatings from Mo/Mo_2C layers were synthesized on WC-Co substrates by their two-stage ion-plasma treatment in argon with the addition (at the second stage) of methane. The deposition of coatings was carried out on an ion-plasma installation with two cathodes. The formation of Mo_2C occurs due to the reaction of Mo with C in the atmosphere of CH_4. Diamond-coating deposition was carried out in a high-frequency (2.45 GHz) plasma CVD reactor. The morphology, phase composition, adhesion of buffer layers, and their influence on the subsequent deposition of DC were studied in this work. A diffusion layer of Mo/Mo_2C with a thickness of ~4.0 μm was applied to the WC-Co nano substrate, and a DC layer with a thickness of ~2.7 μm was deposited on it by the CVD method. Analysis of the distribution of chemical elements by depth showed that the concentration of Mo gradually decreased, and Co was entirely blocked in the diffusion layer and did not affect the formation of DC. The gradient diffusion layer had good adhesion with both the intermediate layer and the nanocrystalline DC since the parameters of the crystal lattices of Mo and diamond are pretty close. Thus, the results of ref [95] demonstrated that intermediate gradient layers of Mo/Mo_2C could be considered buffer layers for applying DC to WC-Co.

Figure 4 shows the cross-section morphology and the distribution of elements over the intermediate layer Mo/Mo_2C.

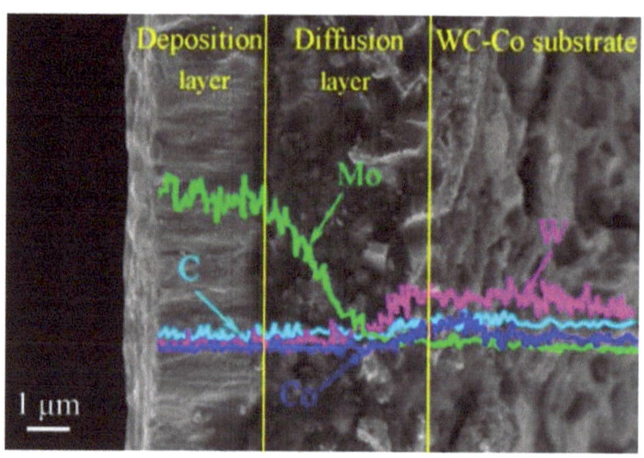

Figure 4. Cross-sectional morphology and the corresponding elemental depth distributions in the diamond coating deposited on Mo/Mo2 C sublayer [95].

A two-layer HfC/HfC-SiC diffusion barrier deposited on WC-Co by reactive plasma alloying was also used [96]. The inner nanostructured HfC layer was ~1.0 μm thick, and the outer layer of HfC-SiC had a thickness of ~0.7 μm. After reactive plasma alloying, the microhardness of the surface increased from 1802 to 2857 $HV_{0.1}$. The Co diffusion from WC-Co was effectively suppressed due to the formation of a CoHf and a nanostructured HfC layer. In ref [96], it was concluded that using a diffusion-modified two-layer HfC/HfC-SiC system is one of the most effective ways to increase the adhesion of DC to a WC-Co substrate.

7. Buffer Layers Based on Silicon Compounds

The authors [97,98] believe that the cubic carbide of silicon β-SiC can be used as a barrier layer to prevent the diffusion of cobalt. In ref [98], the effect of the SiC buffer layer on morphology, phase composition, and adhesive strength was investigated. Cobalt reacted with SiC to form the silicides Co_2Si and CoSi, which suggests that SiC can suppress the harmful effects of Co without compromising the characteristics of DC. The β-SiC layer reduces residual stresses in diamond coatings, reducing the KTE between WC-Co and the diamond coating [99]. It was shown in ref [100] that hard alloy cutters with DC deposited on the β-SiC buffer layer proved to be more durable than similar cutters hardened with sintered diamond. Using microwave plasma on W, Mo, and WC-Co substrates, diamond/β-SiC nanocomposite structures with increased adhesion were synthesized [101]. The deposition of intermediate layers is provided by a controlled change in the composition of the gas mixture [70,101]. The diamond/SiC interlayers deposited by the CVD method are also used as buffer layers [102]. Composite layers significantly hindered the diffusion of Co from the carbide substrate and allowed the deposition of microcrystalline DC on the buffer layer. The best operational characteristics are achieved with a SiC/diamond phase ratio of about 45% [102].

In ref [103], an improvement in the adhesion of DC to WC-Co substrate was also achieved by using a β-SiC buffer layer. At the same time, it claimed that the β-SiC layer contributed to the origin and growth of DC because, firstly, β-SiC has a diamond-like cubic lattice and, secondly, when activated by atomic hydrogen, sp_3 carbon formed from the gas phase due to the substitution of Si atoms in the β-SiC lattice with C atoms. In ref [103], the effect of the deposition temperature and the mode of magnetron sputtering of the β-SiC buffer layer on the processes of DC nucleation and growth was investigated. The magnetron sputtering reduces the temperature forming a buffer layer of silicon carbide. At the same time, it establishes that an increase in the deposition temperature contributes to the elimination of cauliflower-type clusters on the surface of the DC and improves the quality of the β-SiC barrier layer. The β-SiC layer blocks Co diffusion and stimulates the nucleation process during DC deposition. The content of sp_3-C bonds in DC increases with the concentration of the β-SiC phase in the buffer layer. DC nucleation and growth on the β-SiC layer can conditionally divide into six stages: initialization, activation, substitution, formation and evolution of embryos, and DC growth. The most critical process determining the appearance of sp_3-carbon is the reaction of substitution by C atoms of Si atoms in the β-Si lattice during the activation of CH_x + groups and atomic hydrogen.

In refs [75,104], the influence of amorphous ceramic buffer layers based on silicon on the tribological properties of DC on WC-Co was studied. Amorphous buffer films, a-SiO_2 and a-SiC, were synthesized by pyrolysis by CVD on a burning filament of mixtures of tetraethoxysilane, $C_8H_{20}O_4Si$ and dimethyl-diethoxy silane $C_6H_{16}O_2Si$, and then DC was deposited on these layers. For comparison, the DC is worse without an intermediate layer. The grain size and orientation of the crystallites were almost the same in both coatings. The upper layers of the DC on amorphous ceramic layers had less roughness than the coatings applied without an intermediate layer. Films of a two-layer (amorphous ceramic + diamond layer) coating exhibit a lower coefficient of friction than films without a middle layer. As a result, it shows that ceramic buffer layers could act as an amorphous binder to fill the space between diamond grains and brittle WC particles, reducing the surface roughness and improving the DC's tribological characteristics. The a-SiC coating had the lowest surface roughness and the lowest coefficient of friction [75].

8. Buffer Layers Based on Tungsten and Tantalum

In refs [16–20], barrier layers of tungsten are applied at the formation stage of diamond nuclei to eliminate the increased diffusional activity of the binder and the clustering of Co. When choosing the method of applying barrier layers of tungsten, it was taken into account that as the layer thickness increases, the internal stresses leading to embrittlement of the barrier layer accumulate. The ion-plasma coating deposition method, when the substrate temperature reaches ~700 °C, is replaced by the cold magnetron deposition when the substrate temperature does not exceed 50–80 °C to prevent cracking of the barrier layer. In this case, the tungsten coating, sprayed at a speed of 1 nm/s, repeats the profile of the complex substrate relief, and no temperature stresses occur during cooling. At the same time, the adhesion strength is provided by a natural 3D relief formed by a combination of WC grain boundaries. The technology of applying a tungsten film consists of stages using a thin (10–30 nm) tungsten sublayer to a substrate activated by an ion beam at operating pressures of a magnetron discharge in argon (less than 0.25 Pa), with subsequent formation of a tungsten film at a given thickness at pressures of 0.5–0.7 Pa.

The features of the wear mechanism of nanoscale and microcrystalline DC deposited on tungsten carbide substrates were studied by RAMAN spectroscopy [20].

Deposition conditions have a significant influence on the grain size of DC. Figure 5 shows the photographs of the two-layer coating microstructure of micro and nanodiamonds deposited on a tungsten buffer layer at different methane concentrations in a gas mixture.

Figure 5. SEM images of diamond coatings (DC) deposited at a concentration of methane in the gas mixture: (**a**) 4% (microcrystalline DC) and (**b**) 15% (nanocrystalline DC) [20].

Figure 6 shows the RAMAN spectra of single-layer nanocrystalline, and microcrystalline DC measured in the entire region (spectra 1 and 3) and in the area of maximum wear (spectra 2 and 4) after tribological tests with a counter body ball of Si_3N_4 [16–20]. Spectrum 5 in Figure 6 was measured on a natural diamond crystal implanted with nickel ions [105].

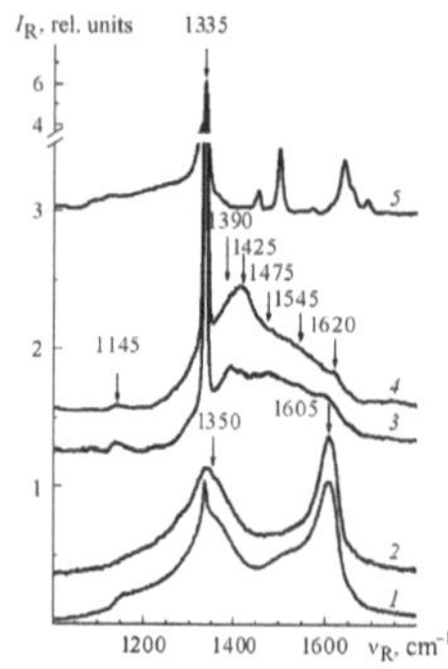

Figure 6. Raman spectra of single-layer nano- and microcrystalline diamond coatings were measured in the entire region (1, 3) and the region of maximum wear (2, 4) after tribological tests with a Si3 N4 counter-ball. Spectrum 5 is measured in a natural diamond crystal implanted with nickel ions [105]. The spectra are shifted vertically relative to each other for clarity [20].

In the RAMAN spectra of microcrystalline diamond (Figure 6, spectrum 3), the diamond peak with a maximum at $\Delta\nu R = 1335$ cm^{-1} is dominant. In addition to the low-intensity bands, with maxima at 1155, 1350, and 1475 cm^{-1}, it is observed in the spectrum of nanocrystalline DC that relatively narrow bands with maxima around 1390, 1425 and 1620 cm^{-1} appear in the microcrystalline coating (spectra 3 and 4). The relative intensity of these bands varies along the DC surface. It was probably determined by the morphological features of the layer consisting of diamond crystallites with a size of ~0.5–5 μm. A band of about 1620 cm^{-1} in the spectra may indicate the inclusions of microcrystalline graphite. After tribological tests, the width of the diamond peak increases, and the band 1145 cm^{-1} weakens slightly and shifts to the region of lower frequencies. The band with a maximum of ~1425 cm^{-1} (spectrum 4) is caused by defects in the diamond lattice in the twinning regions.

In refs [21,106], the layers of Ta and its compounds with Cr and Nb were studied as a buffer coating for DC. The nucleation density of diamond particles on the Ta sublayer was the highest, apparently due to the carbide formation. The intermediate layer of Ta effectively suppressed the diffusion of Co. However, the adhesion of Ta to WC-Co was poor due to thermal stresses occurring at the interface with the substrate.

Nevertheless, it is believed that the chemical compound of tantalum with carbon can be used as a buffer layer due to its high hardness and good resistance to thermal shock [107]. It is confirmed in ref [108] that the intermediate layers of Ta_xC and mixtures of Ta_2C and TaC are used. In ref [109], the effect on adhesion of the deposition temperature of DC in microwave plasma (temperature up to 950 °C) on buffer nanocrystalline layers of Ta_xC deposited on WC-Co was studied. It shows that the obtained DC has good adhesion, high hardness, and maximum wear resistance when the coating temperature of tantalum carbide is 800 °C.

A more complex variant of the Ta-based buffer layer with a second layer of ZrN and Mo was proposed in ref [110]. The best results were obtained for a system consisting of 9 periods of a two-layer TaN/ZrN method with a thickness of each layer of 30 nm, on which a two-layer system of TaN (0.5 μm) and Mo (0.5 μm) was applied. When applied, each layer of deposited TaN with a thickness of 100 nm was subjected to ion etching using a reactive mixture of Ar-6 vol.% N_2 at a total pressure of 0.02 Pa for 1 h. The sputtering–etching processes repeated until a layer of TaN with a thickness of ~0.5 μm was obtained. The upper layer of Mo (~0.5 μm) was then deposited, providing the highest nucleation of DC. Thus, it shows that the DC's high wear resistance is ensured by preventing the propagation of cracks in a high-hardness, multi-layer nanometer buffer coating.

Some works for tungsten carbide pretreatment before DC are summarized in Table 2.

Table 2. Summary of the works on tungsten carbides pretreatment before DC.

Reference	Pretreatment type	Process
[21–46]	Selective dissolution of cobalt with aggressive chemicals	Murakami reagent ($K_3[Fe(CN)_6]$:KOH:H_2O in a ratio of 1:1:10) and then in a solution of peroxide mono sulfuric acid H_2SO_5 (Caro's acid)
[37,38]		Combined heat treatment (heating WC-Co at 1000 °C in an atmosphere of 0.25% CH_4/H_2) with subsequent selective removal of cobalt in Caro's acid
[25,39]		Treating with microwave CO_2–O_2 plasma at 900 °C. As a result of plasma action, $CoWO_4$ and WO_3 compounds formed and removed in an alkaline solution
[42]		etching for 1 h in a modified mixture (65 vol.% HNO_3 + 32 vol.% H_2O_2) in an ultrasonic bath
[43]		Ultrasonic treatment in a mixture of 10 mL 98% H_2SO_4 + 30 mL 38% H_2O_2
[25,37,38,52–55,58–60]	Physical-mechanical	The laser and plasma irradiation
[49]		The impact hydrotreatment
[50,51]		The shot blasting with solid particles (SiO_2 or Al_2O_3) and ultrasonic cavitation
[61]		The sandblasting + acid etching
[50]		The mechanical treatment with diamond powders
[56]		The chemical etching, sandblasting, and creation of a carbide-forming metal layer
[57]		The chemical etching pretreatment, laser etching, and acid treatment pooled plasma and chemical etching of WC-Co, followed by laser etching
[27,52,80,85–87]	The buffer layers formation	The Cr/CrN/Cr buffer layer, Cr, and CrC
[71]		The buffer layers TiN, TiC, (TiSi)N, Si_3N_4, a-SiC, and a-SiC_xN_y
[21]		The CrN, ZrN, NbN, and TaN with a thickness of 1 μm
[16–20,67,77]		W
[78,79]		Al
[82]		Si_3N_4
[81]		NbC
[21,81,83]		TiN, Ti(CN)
[21,106]		Ta
[108]		The mixtures of Ta_2C and TaC
[110]		TaN/ZrN
[89]		The magnetron sputtering CrN and TiN layers
[48]		The buffer layers of TiB_2 with a thickness of about 0.1–1 μm
[97,98,102,103]		SiC buffer layer
[96]		HfC/HfC-SiC
[95]		The gradient coatings from Mo/Mo_2C
[75,104]		The amorphous buffer films, a-SiO_2 and a-SiC

9. CVD Diamond Coatings Deposition by the Hot Thread Method

In most publications devoted to the application of DC to WC-Co tools, the coating of carbide tools is carried out by the hot thread method. First of all, this is due to economic considerations since hot-thread installations allow the diamond coating to be applied to a sufficiently large area and reduces the cost of products [6,10,77,90,104,111–120]. However, the inherent disadvantages of the hot thread method do not allow to use of it cutting tools with acceptable quality for several unique applications.

So, the high-temperature alloy thread is heated to a temperature of 2000 °C during deposition and is a source of DC pollution. A decrease in the filament temperature inevitably leads to a reduction in the coating's deposition rate and, as a rule, to an increase in the proportion of non-diamond carbon in the DC. In addition, the service life of the thread is short, which reduces the productivity of the growth installation.

It should also be noted that the hot-thread installations have significant limitations on the possibility of multi-layer and gradient DC synthesis operational control [10]. The use of double-layer DC significantly increases the performance characteristics of the cutting tool. The nanocrystalline top layer provides a slight roughness, low coefficient of friction values, and high bending strength. The lower layer is usually deposited in the microcrystalline diamond growth mode, and such layers have better adhesion to WC-Co and high hardness and elasticity. In addition, two-layer DC, compared to single-layer, have a higher thermal conductivity and increased resistance to cracking. Currently, two-layer DC is deposited on WC-Co by the hot thread method. At the same time, the grain size can be changed by adjusting the bias voltage or by simultaneously changing the methane concentration and the gas mixture pressure. However, judging by the results [10], when the deposition parameters change, the formation of a layer of sp_2-carbon usually occurs on the boundary between micro- and nanocrystalline DC, which can negate all the advantages of two-layer DC deposited by the hot thread method.

10. Diamond Coating by CVD with Activation of the Gas Phase by Plasma in a Microwave Discharge

Recently, there has been a clear trend in the broader application of DC deposition technology to carbide tools in microwave plasma. The hot thread method is noticeably inferior to this technology in terms of parameters, such as gas consumption, growth rate, high purity of the gas mixture, and the possibility of alloying DC in the deposition process. However, a serious problem in ensuring same thickness and grain size of the diamond film during the deposition by the microwave plasma method remains difficult. The successful deposition of DC with a thickness of 20 to 40 μm from mixtures of methane (0.5–1.0 vol.%) and hydrogen was performed for the first time in refs [121,122]. The substrate temperature was 700–800 °C, the pressure in the chamber was 50–70 Torr, and the deposition rate was 1–2 μm/h. Work on the deposition of DC on an instrument from WC-Co was also carried out in the USA [35]. In ref [84], the deposit of DC on WC-Co (3 and 6% Co) was also carried out in microwave plasma.

In contrast, the deposition temperature was 900 °C, the proportion of methane in hydrogen was 0.5%, the pressure was 35 Torr, and the deposition time was 15–30 h. In the work cycle [53,56,123], the deposition of DC was carried out at a substrate temperature of 700–750 °C and pressure in the chamber of 80 Torr. The proportion of methane in the gas mixture was 1%, the deposition time was 6 h, and the deposition rate of DC was 1.5 μm/h. In refs [124–126], when nanocrystalline DC was deposited on WC-Co in microwave plasma, nitrogen was added to the gas mixture, and smooth and solid (80 GPa) DC with a thickness of ~10 μm was obtained. These coatings were not inferior to the DC obtained by the hot thread method. However, they did not exceed their operating parameters.

The advantages of the DC deposition method on a WC-Co instrument in microwave plasma were realized only in the last 2–3 years when microwave plasma began being used for the deposition of buffer layers and DC in a single process [127,128]. The variable buffer layers composition WCoB/CoB, W_2CoB_2/WCoB, W_2CoB_2, and others formed on the WC-

Co substrate in microwave plasma when 0.6 vol.% diborane B_2H_6 was added to hydrogen. During DC deposition, the substrate temperature varied from 600 to 1000 °C, the pressure in the chamber was 50 Torr, the proportion of methane and nitrogen in hydrogen was 9 and 1%, respectively, and the deposition time was from 0.5 to 4 h. The hardness and modulus of elasticity of DC were only 15 and 400 GPa, respectively, at low deposition temperatures (600 °C). This hardness is associated with a large proportion of non-diamond carbon. An increase in the substrate temperature during deposition improved the parameters of the DC. The hardness and modulus of the coating's elasticity increased to 32 and 600 GPa at a deposition temperature of 800 °C and at 1000 °C, up to 600 and 700 GPa, respectively. Due to the addition of nitrogen to the gas mixture and a sufficiently high content of methane within it (9%), a smooth nanocrystalline DC with a noticeable proportion of non-diamond carbon was deposited in the microwave plasma. In ref [95], WC-Co substrates with buffer layers of molybdenum dicarbide Mo_2C were used to precipitate DC from microwave plasma, and in ref [96], diffusion layers of HfC/HfC-SiC were used.

In the works [16–20], samples of a new type of superhard tool with significantly higher wear resistance were created based on DC, intended for use in the production of parts and structures made of hard-to-process composite materials. The main problem of DC deposition in microwave plasma was solved due to the creation of inhomogeneities in the working chamber of methane concentration in the plasma volume and temperature along the perimeter of the substrate, the so-called edge effect.

In refs [16,17], an original design of the substrate holder was proposed, which makes it possible to implement a deposition mode of homogeneous DC properties and eliminates direct heating of high-aspect (height-to-diameter ratio) substrates in a microwave plasma reactor. The transition to indirect heating made it possible to reduce the edge effect and provided protection for the edges of the substrates from overheating. The coatings with uniform morphology were obtained both in the center and in the periphery. In refs [17,18], it established that the temperature affecting the diamond growth rate and the grain size of polycrystalline DC significantly depends on the distance between the plasma-forming surface of the substrate holder, the growth surface of the substrate, and the temperature range for the microwave reactor "ARDIS-100" with a power of 2.9 kW (determined with an accuracy of 20 °C). This technology enables the application of DC to cutters made of WC-6% Co, with an aspect ratio of 0.45. As a result, the roughness in the center and periphery of the coating differed by no more than 10% (Figure 7) [17]. A practical method of regulating the temperature of the growth surface and controlling the structure of polycrystalline DC is an operational change in the distance Δh between the substrate and the substrate holder during synthesis in microwave plasma, with the position of the cooled substrate holder unchanged.

In refs [19,20], the processes stimulating the effect of the secondary nucleation and reducing the size of diamond crystallites were used when nitrogen was added to the gas mixture during DC deposition in the plasma chemical reactor ARDIS-100 (2.45 GHz, 5 kW) with a variable supply of the gas mixture CH_4/H_2 and $CH_4/H_2/N_2$ at different stages of growth. In the ARDIS-100, the gas mixture is supplied from above. Typical gas flow values are 0.2–1 L/min. The DC deposition process was carried out at a total gas consumption of 500 std cm^3/min and a chamber pressure of 65 Torr at a microwave power of 3.0 kW. The substrates in the synthesis process were at a temperature of 750 °C, and the deposition rate was 2–3 μm/h. The methane concentration at the time of plasma ignition was 0% and gradually increased to 4% in ~2 min, with a simultaneous increase in the microwave radiation power and substrate temperature (Figure 8).

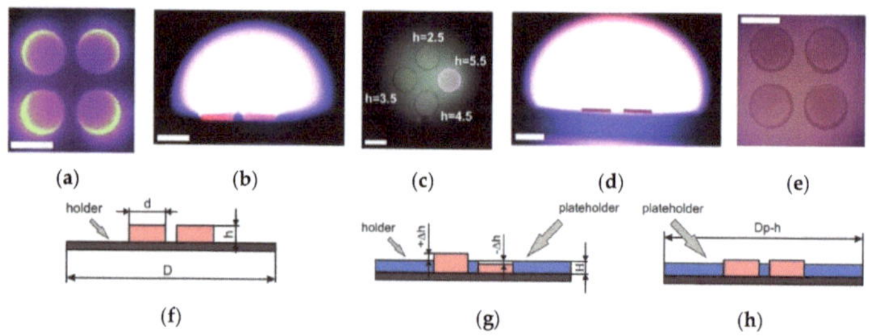

Figure 7. Top view (**a**) and side view (**b**) of the samples in an open substrate holder during the first series of growth of CVD diamond; top view (**c**) of the substrates with different heights in the holder holes during the second series of growth of CVD diamond; and side view (**d**) and top view (**e**) of substrates with equal heights in the holder during the third series of growth of CVD diamond. Schematic views of the arrangement of the substrates in the open holder (h/d = 0.45) during the first series of growth of CVD diamond (**f**); the same for the close-type holder with substrates of different heights during the second series of growth of CVD diamond (**g**), and the same for the open holder (h/d = 0.45) during the third series of growth of CVD diamond (**h**). The plate holder diameter is Dp–h = 57 mm, and the height is H = 3.5 mm. White scale bars size in the lower left corner of sub-figures (**a**–**e**) is 10 mm [17].

Figure 8. SEM images of diamond films deposited at different temperatures and distances between the substrate and plate holder surfaces: (**a**) Δh = −1 mm, T = 700 °C, half of the radius; (**b**) Δh = 0 mm, T = 740 °C, center; (**c**) Δh = 0 mm, T = 740 °C, half of the radius; (**d**) Δh = 0 mm, T = 740 °C, periphery; (**e**) Δh = 1 mm, T = 760 °C, half of the radius; and (**f**) Δh = 2 mm, T = 870 °C, half of the radius. Micrographs (**a**,**e**,**f**) relate to the second CVD series, and micrographs (**b**–**d**) relate to the third CVD series [17].

The temperature range ensuring the uniform deposition of DC on WC-6% Co substrates in the ARDIS-100 reactor is 740–760 °C at a power of 2.9 kW and a distance of $0 \leq \Delta h < 1$ mm. The change in the DC R_a roughness from the center to the periphery does not exceed 15%.

This solution allowed it to preserve the nuclei of diamond particles from burning out and significantly increased their density to ~10^9 cm^{-2} or more. The multi-layer coatings were successive layers of micro- and nanocrystalline diamonds. The synthesis of microcrystalline layers was carried out without the addition of nitrogen. The concentration of N_2 in the total gas flow was maintained at the level of 4% during the deposition of nanocrystalline layers. According to RAMAN spectroscopy data [20], an intermediate layer with a high concentration of the sp_2 phase does not form at the boundary of the micro- and nanocrystalline diamond layers as often as it happens with the growth of gradient layers by the hot thread method, which positively affects the performance characteristics of the tool.

In ref [129], DC deposition in microwave plasma was carried out on buffer layers of a diamond/β-SiC composite grown in the same plant, and the proportion of β-SiC varied from 12 to 68%. The DC layers deposited on the diamond/β-SiC composite demonstrated good adhesion and low-friction coefficients.

11. The Diamond-Coated Carbide Tools for Processing Polymer Composite Materials

In the drilling processes of polymer materials reinforced with carbon fibers (CFRP), conventional cutting tools wear out quickly due to the abrasive effect of carbon fibers on the cutting surfaces of the instrument. Figure 9 shows the wear pattern of the cutting edge, demonstrating the formation of a spout with a larger width than the original tool [84]. The load on the cutting edge is very high in the absence of coolant; therefore, the requirements for the DC are very stringent. The primary defects of the holes processed with a diamond-coated tool are the ejection and destruction of the fiber, cracking of the matrix, and stratification of layers and burrs at the outlet [130–135].

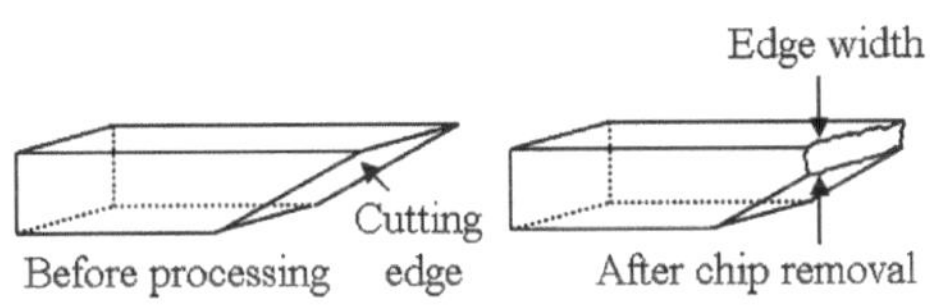

Figure 9. Diagram of the wear of the cutting edge showing the formation of a nose with a wider width than the original tool [84].

When cutting CFRP, a positive effect of DC was noted due to their high hardness and low coefficient of friction [136]. In refs [137,138], the results of tests for cutting the edges of carbon fiber plastics are presented, from which it follows that a diamond-coated tool has better wear resistance than a traditional cutter. However, CFRP delamination occurs when the feed rate of the diamond-coated instrument increases. In ref [4], tool wear was studied when drilling CFRP with various drills, including a DC drill, and it was found that a diamond-coated drill cutting length is 80% longer than a conventional tool.

Boron or silicon-doped DC deposited by the hot thread method on chemically treated WC-Co drills are optimally suited for processing CFRP products [139]. Rockwell hardness measurements have shown that doping with boron and silicon on the surface of a hard alloy can increase the adhesion between DC and WC-Co. The Si-doped drills have better adhesive strength and higher performance and wear resistance than drills without DC when processing CFRP materials with a coated tool. The positive effect of diamond doping with silicon is the reduction of nanoscale diamond crystallites of the coating and the formation of a smoother surface with a low coefficient of friction during cutting [139].

In ref [140], the feeding force and delamination processes were investigated during the operation of drilling carbon plastics with a special geometry tool. It was found that it is necessary to reduce the feed rate to reduce delamination. In ref [141], a mechanical model proposed that it is possible to predict the cutting force when drilling CFRP composite laminates. In ref [142], a physical model was proposed that predicts the critical feed force at the beginning of stratification when drilling CFRP and allows for analyzing the stratification process of drills with different geometries.

In ref [131], the dependence of the stratification processes of the processed material on the cutting speed and feed rate was studied. It shows that an increase in the cutting speed and feed increases the stratification. A comparative analysis of the cutting characteristics for step drills when drilling small diameter holes in CFRP showed that the main factors affecting the service life of a diamond tool are its geometry and feed rate [143]. In ref [144], it was found that when drilling holes in CFRP, the main wear factors are grinding, abrasion, and adhesion of the DC.

Diamond-coated tools with high service properties are the primary tools for processing CFRP [145]. They provide a longer service life and better surface quality [146]. Tools with CVD DC combine record-breaking DC properties with high geometric flexibility. High hardness and wear resistance, low coefficient of friction, and good thermal conductivity reduce cutting forces and heating, which increases the tool's life [147]. However, insufficient adhesion of the DC can reduce its service life. In ref [148], optimal cutting characteristics of polyether ether ketone are composites with a 30% volume fraction of carbon fibers with a CVD.

Figure 10 shows data on the continuity of the tool with and without DC when turning a metal matrix composite A-390 [17]. As can be seen, the DC noticeably increases the resource of the instrument. Figure 11 shows the appearance of the 10th exit hole, treated with uncoated and CVD DC with a grain size of ~3 μm, obtained by the hot thread method [149]. Due to the excellent wear resistance, the tool's durability with DC is much higher.

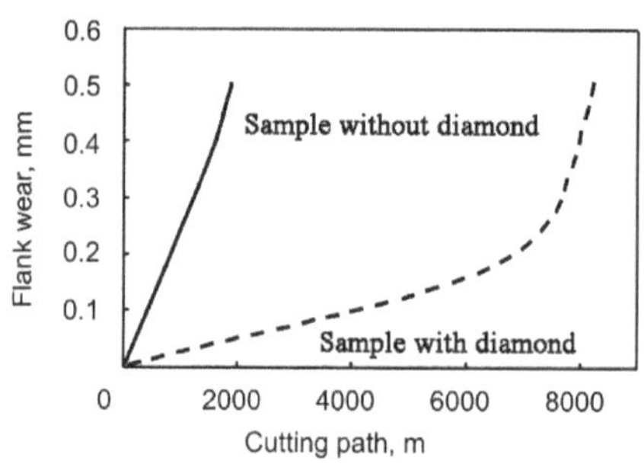

Figure 10. Period of service of the tool with and without diamond coating under the turning of the A-390 type metal-matrix composite with 18% SiC [17].

According to the form of wear of the drill with DC when drilling CFRP, it was established that the observed wear was mainly caused by the abrasive behavior of the reinforcing fiber. The pressure force on the WC-Co drill with DC during the drilling process is less than on the uncoated instrument due to the low coefficient of friction. It shows that the best quality of the holes, especially the inner surface of the hole, is obtained using drills with DC, and it concluded that diamond CVD coatings could solve the problems that arise when processing CFRP [149].

Comparative studies [150] of orthogonal cutting processes with a WC-6% Co tool with and without a CVD nanocrystalline DC with a thickness of 5–6 μm and an average grain size of 50–100 nm showed that at low processing speeds, the cutting force for a tool with and without DC differed slightly, which is due to a large radius of the cutting edge for a tool with a DC, weakening the advantages of a smooth DC. The cutting force decreased compared to a conventional tool with increased cutting speed due to low friction and good wear resistance.

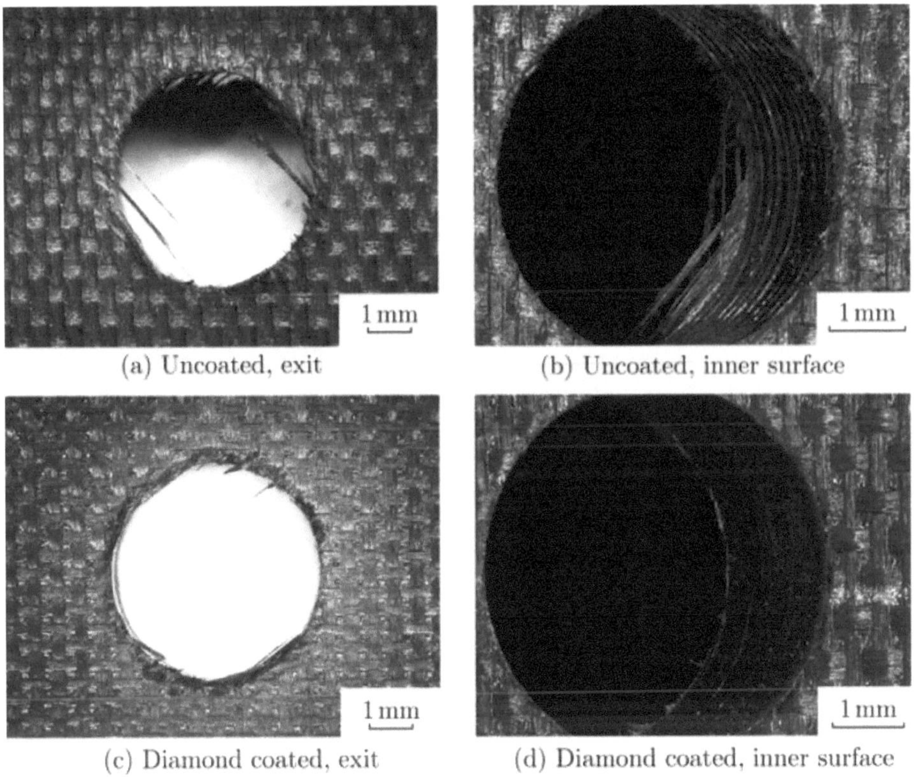

Figure 11. Images for the tenth hole machined by uncoated and diamond-coated drills: (**a**) The outcome made by an uncoated drill, (**b**) the inner surface of the hole made with a drill without coating, (**c**) the outcome made by a diamond-coated drill, and (**d**) the inner surface of the hole made with a drill with diamond coating [149].

12. Conclusions

Despite many experiments on research and development of technology for creating a durable tool with diamond coating, the data obtained should be compared with a great deal of caution since different authors used equipment of varying quality and had other technological qualifications. Therefore, in several cases, the same technological solution is evaluated differently in other works. The literature data on the deposition of diamond coatings on a cutting tool shows that the direct deposition of high-quality DC on WC-Co is a highly complex process. Co catalyzes the formation of a brittle graphite sublayer between diamond and substrate and the mismatch of the CTE of diamond and WC-Co.

Currently, various methods of pretreatment of the substrate surface are used, including chemical etching, mechanical hardening, plasma treatment, laser ablation, application of buffer layers between the substrate and the coating, and various ways to increase the nucleation density of diamond particles.

Tribological characteristics of the tool with DC depend on the microstructure and roughness of the coating. The adhesion, hardness, and wear resistance of the DC increase after several stages of WC-Co pretreatment before deposition of the DC, leading to the creation of a multi-layer, three-dimensional interface.

Murakami reagent and Caro acid are widely used for the chemical etching of WC-Co. The choice of the optimal chemical etching mode for each type of substrate with different concentrations of cobalt, the buffer layer, and DC deposition technology, as well as the operating conditions of the diamond tool, is individually selected. The same applies to the choice of chemical composition, method, mode of deposition, and geometric parameters of buffer layers. The adhesion of the DC is determined by the structure and adhesive characteristics of the buffer layers. So the choice of optimal materials and technology for buffer layers obtaining depends on the type and mode of operation.

The widespread technology of DC deposition by the hot thread method, which provides a low cost of manufacturing a diamond tool, is inferior to the process of plasma deposition using a microwave discharge. The friction coefficient of a diamond-coated tool can be reduced to 0.07 compared with 0.6 for an uncoated tool by reducing the consumption of plasma-forming gas, increasing the growth rate of the coating, as well as significantly expanding the possibilities of nitrogen doping from the gas phase during deposition and formation of two- and multi-layer gradient coatings. The microwave plasma makes it abandon consumable electrodes, increases the purity in the discharge chamber, and contributes to obtaining a perfect structure of diamond coatings. This is especially relevant when creating tools for high-precision mechanical processing of polymer composite materials reinforced with fibers with high elastic properties, which imposes strict requirements on the strength properties of the diamond coating.

Author Contributions: Conceptualization, E.A.; methodology, A.K.; investigation, V.R.; funding acquisition, A.B. visualization, D.S.; conceptualization, V.K.; project administration, S.G.; writing—review and editing S.F. All authors have read and agreed to the published version of the manuscript.

Funding: This research was carried out with the financial support of the Russian Science Foundation within the framework of scientific project No. 22-19-00694.

Data Availability Statement: The data presented in this study are available on request from the corresponding author.

Conflicts of Interest: The authors declare no conflict of interest.

References

1. Faraz, A.; Biermann, D.; Weinet, K. Cutting edge rounding: An innovative tool wear criterion in drilling CFRP composite laminates. *Int. J. Mach. Tools Manuf.* **2009**, *49*, 1185–1196. [CrossRef]
2. Tsao, C.C.; Kuo, K.L.; Hsu, I.C. Evaluation of a novel approach to a delamination factor after drilling composite laminates using a core-saw drill. *Int. J. Adv. Manuf. Technol.* **2012**, *59*, 617–622. [CrossRef]
3. Iliescu, D.; Gethin, D.; Gutierrez, M.E.; Girot, F. Modeling and tool wear in drilling of CFRP. *Int. J. Mach. Tools Manuf.* **2010**, *50*, 204–213. [CrossRef]
4. Tsao, C.C. Experimental study of drilling composite materials with step-core drill. *Mater. Des.* **2008**, *29*, 1740–1744. [CrossRef]
5. Grigoriev, S.N.; Vereschaka, A.A.; Fyodorov, S.V.; Sitnikov, N.N.; Batako, A.D. Comparative analysis of cutting properties and nature of wear of carbide cutting tools with multi-layered nano-structured and gradient coatings produced by using of various deposition methods. *Int. J. Adv. Manuf. Technol.* **2017**, *90*, 3421–3435. [CrossRef]
6. Singla, A.; Singh, N.K.; Singh, Y.; Jangir, D.K. Micro and Nano-Crystalline Diamond Coatings of Co-cemented Tungsten Carbide Tools with Their Characterization. *J. Bio Tribo Corros.* **2021**, *7*, 35. [CrossRef]
7. Tillmann, W. Trends and market perspectives for diamond tools in the construction industry. *Int. J. Refract. Met. Hard Mater.* **2000**, *18*, 301–306. [CrossRef]
8. Kolber, T.; Koepf, A.; Haubner, R.; Hutter, H. WC-Co substrate surface pretreatments with aluminum compounds prior to polycrystalline CVD diamond deposition. *Int. J. Refract. Met. Hard Mater.* **1999**, *17*, 445–452. [CrossRef]
9. Linnik, S.A.; Gaydaychuk, A.V.; Okhotnikov, V.V. Improvem ent to the adhesion of polycrystalline diamond films on WC-Co cemented carbides through ion etching of loosely bound growth centers. *Surf. Coat. Technol.* **2018**, *334*, 227–232. [CrossRef]
10. Dumpala, R.; Chandran, M.; Rao, M.S.R. Engineered CVD diamond coatings for machining and tribological applications. *J. Miner.* **2015**, *67*, 1565–1577. [CrossRef]
11. Park, B.S.; Baik, Y.J.; Lee, K.R.; Eun, K.Y.; Kim, D.H. Behaviour of Co binder phase during diamond deposition on WC-Co substrate. *Diam. Relat. Mater.* **1993**, *2*, 910–917. [CrossRef]
12. Deuerler, F.; Gruner, H.; Pohl, M.; Tikana, L. Wear mechanisms of diamond-coated tools. *Surf. Coat. Technol.* **2001**, *142*, 674–680. [CrossRef]
13. Kubelka, S.; Haubner, R.; Lux, B.; Steiner, R.; Stingeder, G.; Grasserbauer, M. Influences of WC-Co hard metal substrate pre-treatments with boron and silicon on low pressure diamond deposition. *Diam. Relat. Mater.* **1994**, *3*, 1360–1369. [CrossRef]
14. Li, Y.S.; Tang, Y.; Yang, Q.; Xiao, C.; Hirose, A. Diamond deposition on steel substrates with an Al interlayer. *Int. J. Refract. Met. Hard Mater.* **2009**, *27*, 417–420. [CrossRef]

15. Haubner, R.; Kalss, W. Diamond deposition on hardmetal substrates: Comparison of substrate pre-treatments and industrial applications. *Int. J. Refract. Met. Hard Mater.* **2010**, *28*, 75–483. [CrossRef]
16. Ashkinazi, E.E.; Sedov, V.S.; Petrzhik, M.I.; Sovyk, D.N.; Khomich, A.A.; Ralchenko, V.G.; Vinogradov, D.V.; Tsygankov, P.A.; Ushakova, I.N.; Khomich, A.V. Effect of crystal structure on the tribological properties of diamond coatings on hard-alloy cutting tools. *J. Frict. Wear* **2017**, *38*, 252–258. [CrossRef]
17. Ashkihazi, E.E.; Sedov, V.S.; Sovyk, D.N.; Khomich, A.A.; Bolshakov, A.P.; Ryzhkov, S.G.; Khomich, A.V.; Vinogradov, D.V.; Ralchenko, V.G.; Konov, V.I. Plateholder design for deposition of uniform diamond coatings on WC-Co substrates by microwave plasma CVD for efficient turning application. *Diamond Relat. Mater.* **2017**, *75*, 169–175. [CrossRef]
18. Ashkinazi, E.E.; Ralchenko, V.G.; Konov, V.I.; Vinogradov, D.V.; Tsygankov, P.A.; Dryzhak, E.A.; Khomich, A.V. Frictional coefficients between aluminum-silicon alloy and cutting inserts with MPCVD diamond coatings. *Russian Eng. Res.* **2018**, *38*, 457–461. [CrossRef]
19. Khomich, A.V.; Ashkinazi, E.E.; Sedov, V.S.; Sovyk, D.N.; Kozlova, M.V.; Vinogradov, D.V.; Tsygankov, P.A. Microwave CVD deposition and properties of nano/microcrystalline diamond multilayer coatings on tungsten carbide cutting tools. In Proceedings of the International Conference on Mechanical, System and Control Engineering, St. Petersburg, Russia, 19–21 May 2017; pp. 11–15.
20. Khomich, A.A.; Ashkinazi, E.E.; Ralchenko, V.G.; Sedov, V.S.; Khmelnitski, R.A.; Poklonskaya, O.N.; Kozlova, M.V.; Khomich, A.V. Application of Raman spectroscopy for analyzing diamond coatings on a hard alloy. *J. Appl. Spectrosc.* **2017**, *84*, 312–318. [CrossRef]
21. Manaud, J.P.; Poulon, A.; Gomez, S.; Le Petitcorps, Y.A. Comparative study of CrN, ZrN, NbN and TaN layers as cobalt diffusion barriers for CVD diamond deposition on WC-Co tools. *Surf. Coat. Technol.* **2007**, *202*, 222–231. [CrossRef]
22. Kazahaya, K.; Kawai, S.; Matsumoto, Y.; Ishibashi, K.; Imai, T. Diamond Film Coated Tool and Process for Producing the Same. Patent JP, WO2005011902A1, 31 July 2003.
23. Mehlmann, A.K.; Fayer, A.; Dirnfeld, S.F.; Avigal, Y.; Porath, R.; Kochman, A. Nucleation and growth of diamond on cemented carbides by hot-filament chemical vapor deposition. *Diamond Relat. Mater.* **1993**, *2*, 317–322. [CrossRef]
24. Sun, F.H.; Zhang, Z.M.; Chen, M.; Shen, H.S. Improvement of adhesive strength and surface roughness of diamond films on Co-cemented tungsten carbide tools. *Diam. Relat. Mater.* **2003**, *12*, 711–718. [CrossRef]
25. Barletta, M.; Rubino, G.; Gisario, A. Adhesion and wear resistance of CVD diamond coatings on laser treated WC-Co substrates. *Wear* **2011**, *271*, 2016–2024. [CrossRef]
26. Sahoo, B.; Chattopadhyay, A.K. On effectiveness of various surface treatments on adhesion of HF-CVD diamond coating to tungsten carbide inserts. *Diamond Relat. Mater.* **2002**, *11*, 1660–1669. [CrossRef]
27. Gomez, H.; Durham, D.; Xiao, X.; Lukitsch, M.; Luc, P.; Chou, K.; Sachdev, A.; Kumar, A. Adhesion analysis and dry machining performance of CVD diamond coatings deposited on surface modified WC-Co turning inserts. *J. Mater. Process. Technol.* **2012**, *212*, 523–533. [CrossRef]
28. Xu, F.; Xu, J.H.; Yuen, M.F.; Zheng, L.; Lu, W.Z.; Zuo, D.W. Adhesion improvement of diamond coatings on cemented carbide with high cobalt content using PVD interlayer. *Diamond Relat. Mater.* **2013**, *34*, 70–75. [CrossRef]
29. Lu, F.X.; Tang, W.Z.; Tong, Y.M.; Miao, J.Q.; He, L.F.; Li, C.M.; Chen, G.C. Novel pretreatment of hard metal substrate for better performance of diamond coated cutting tools. *Diam. Relat. Mater.* **2006**, *15*, 2039–2045. [CrossRef]
30. Polini, R. Chemically vapor deposited diamond coatings on cemented tungsten carbides: Substrate pretreatments, adhesion and cutting performance. *Thin Solid Film.* **2006**, *515*, 4–13. [CrossRef]
31. Bouzakis, K.D.; Michailidis, N.; Skordaris, G.; Bouzakis, E.; Biermann, D.; M'Saoubi, R. Cutting with coated tools: Coating technologies, characterization methods and performance optimization. *CIRP Annals Manuf. Technol.* **2012**, *6*, 703–723. [CrossRef]
32. Geng, C.L.; Tang, W.Z.; Hei, L.F.; Liu, S.T.; Lu, F.X. Fracture strength of two-step pretreated and diamond coated cemented carbide microdrills. *Int. J. Refract. Met. Hard Mater.* **2007**, *25*, 159–165. [CrossRef]
33. Lei, X.; Shen, B.; Cheng, L.; Sun, F.; Ming, C. Influence of pretreatment and deposition parameters on the properties and cutting performance of NCD coated PCB micro drills. *Int. J. Refract. Met. Hard Mater.* **2014**, *43*, 30–41. [CrossRef]
34. Settineri, L.; Bucciotti, F.; Cecano, F.; Faga, M.G. Surface properties of diamond coatings for cutting tools. *CIRP Annals Manuf. Technol.* **2007**, *56*, 573–576. [CrossRef]
35. Mallika, K.; Komanduri, R. Diamond coatings on cemented tungsten carbide tools by low-pressure microwave CVD. *Wear* **1999**, *224*, 245–266. [CrossRef]
36. Inspektor, A.; Oles, E.J.; Bauer, C.E. Theory and practice in diamond coated metal-cutting tools. *Int. J. Refract. Met. Hard Mater.* **1997**, *15*, 49–56. [CrossRef]
37. Haubner, R.; Schubert, W.D.; Lux, B. Interactions of hard metal substrates during diamond deposition. *Int. J. Refract. Met. Hard Mater.* **1998**, *16*, 177–185. [CrossRef]
38. Haubner, R.; Lux, B. On the formation of diamond coatings on WC/Co hard metal tools. *Int. J. Refract. Met. Hard Mater.* **1996**, *14*, 111–118. [CrossRef]
39. Polini, R.; Bravi, F.; Casadei, F.; D'Antonio, P.; Traversa, E. Effect of substrate grain size and surface treatments on the cutting properties of diamond coated Co-cemented tungsten carbide tools. *Diamond Relat. Mater.* **2002**, *11*, 726–730. [CrossRef]
40. Yang, T.; Wei, Q.; Qi, Y.; Wang, Y.; Xie, Y.; Luo, J.; Yu, Z. Microstructure evolution of thermal spray WC-Co interlayer during hot. *J. Alloy. Compd.* **2015**, *639*, 659–668. [CrossRef]
41. Kim, S.; Seo, B.; Son, S.-H. Dissolution behavior of cobalt from WC-Co hard metal scraps by oxidation and wet milling process. *Hydrometallurgy* **2014**, *143*, 28–33. [CrossRef]

42. Norafifah, H.; Noordina, M.Y.; Izman, S.; Kurniawan, D. Acid pretreatment of WC-Co for surface roughening and cobalt removal prior to CVD diamond coating. *Appl. Mech. Mater.* **2013**, *315*, 592–596. [CrossRef]
43. Kim, J.S.; Park, Y.M.; Bae, M.K.; Kim, C.W.; Kim, D.W.; Shin, D.C.; Kim, T.G. Cutting performance of tungsten carbide tools coated with diamond thin films after etching for various times. *Mod. Phys. Lett. B* **2018**, *32*, 1850236. [CrossRef]
44. Lin, C.-R.; Bendao, M.-K.; Liou, C.-W. Effects of surface modification on improvement of diamond coating on tungsten carbide cutting tool. *Int. J. Mech. Eng. Robot. Res.* **2016**, *5*, 168–171. [CrossRef]
45. Rugoczky, P.; Lassy, G.; Lakatos, J. Characterization of the structure of WC-Co composite support layer of diamond coatings. *Int. J. Adv. Sci. Eng. Technol.* **2018**, *6*, 60–65.
46. Zhou, H.; Li, M.; Yuan, B. Mechanical property and characterization on diamond films deposited on WC-Co substrates. *Mater. Sci. Forum* **2016**, *874*, 333–338. [CrossRef]
47. Vieira da Silva Neto, J.; Rodríguez, L.A.A.; Fraga Mariana, A.C.; André, C.; Raonei, A.C.; Evaldo, J.; Airoldi, V.J.T. WC-Co substrate preparation and deposition conditions for high adhesion of CVD diamond coating. *Rev. Bras. Apl. Vácuo* **2016**, *35*, 53–57. [CrossRef]
48. Wang, X.; Wang, C.; He, W.; Sun, F. Co evolutions for WC-Co with different Co contents during pretreatment and HFCVD diamond film growth processes. *Trans. Nonfer. Met. Soc. China* **2018**, *28*, 469–486. [CrossRef]
49. Tönshoff, H.K.; Mohlfeld, A.; Gey, C.; Winkler, J. Mechanical pretreatment for improved adhesion of diamond coatings. *Surf. Coat. Technol.* **1999**, *116–119*, 440–446. [CrossRef]
50. Neves, D.; Diniz, A.E.; Lima, M.S.F. Microstructural analyses and wear behavior of the cemented carbide tools after laser surface treatment and PVD coating. *Appl. Surf. Sci.* **2013**, *282*, 680–688. [CrossRef]
51. Deuerler, F.; Woehrl, N.; Buck, V. Characterization of nanostructured diamond coatings on various hard metal surfaces. *Int. J. Refract. Met. Hard Mater.* **2006**, *24*, 392–398. [CrossRef]
52. Barshilia, H.C.; Ananth, A.; Khan, J.; Srinivas, G. Ar+H_2 plasma etching for improved adhesion of PVD coatings on steel substrates. *Vacuum* **2012**, *86*, 1165–1173. [CrossRef]
53. Xu, Z.; Lev, L.; Lukitsch, M.; Kumar, A. Effects of surface pretreatments on the deposition of adherent diamond coatings on cemented tungsten carbide substrates. *Diam. Relat. Mater.* **2007**, *16*, 461–466. [CrossRef]
54. Arroyo, J.M.; Diniz, A.E.; Lima, M.S.F. Cemented carbide surface modifications using laser treatment and its effects on hard coating adhesion. *Surf. Coat. Technol.* **2010**, *204*, 2410–2416. [CrossRef]
55. Neves, D.; Diniz, A.E.; Lima, M.S.F. Efficiency of the laser texturing on the adhesion of the coated twist drills. *J. Mater. Process. Technol.* **2006**, *179*, 139–145. [CrossRef]
56. Xu, Z.; Kumar, A.; Lev, L.; Lukitsch, M.; Sikder, A. Adhesion improvement of CVD diamond coatings on WC-Co substrates for machining applications. *Mater. Res. Soc. Symp.* **2005**, *843*, T3.33. [CrossRef]
57. Veillere, A.; Guillemet, T.; Xie, Z.Q.; Zuhlke, C.A.; Alexander, D.R.; Silvain, J.F.; Heintz, J.-M.; Chandra, N.; Lu, Y.F. Influence of WC-Co substrate pretreatment on diamond film deposition by laser-assisted combustion synthesis. *ACS Appl. Mater. Interfaces* **2011**, *3*, 1134–1139. [CrossRef] [PubMed]
58. Creasey, S.; Lewis, D.B.; Smith, I.J.; Munz, W.D. SEM image analysis of droplet formation during metal ion etching by a steered arc discharge. *Surf. Coat. Technol.* **1997**, *97*, 163–175. [CrossRef]
59. Wei, R.; Vajo, J.J.; Matossian, J.N.; Gardos, M.N. Aspects of plasma-enhanced magnetron-sputtered deposition of hard coatings on cutting tools. *Surf. Coat. Technol.* **2002**, *158–159*, 465–472. [CrossRef]
60. Alves, S.M.; Albano, W.; de Oliveira, A.J. Improvement of coating adhesion on cemented carbide tools by plasma etching. *J. Braz. Soc. Mech. Sci. Eng.* **2016**, *39*, 845–856. [CrossRef]
61. Shen, X.; Wang, X.; Sun, F.; Ding, C. Sandblasting pretreatment for deposition of diamond films on WCCo hard metal substrates. *Diam. Relat. Mater.* **2017**, *73*, 7–14. [CrossRef]
62. Ahmed, W.; Sein, H.; Ali, N.; Gracio, J.; Woodwards, R. Diamond films grown on cemented WC-Co dental burs using an improved CVD method. *Diamond Relat. Mater.* **2003**, *12*, 1300–1306. [CrossRef]
63. Deuerler, F.; Vandenberg, H.; Tabersky, R.; Freundlieb, A.; Pies, M.; Buck, V. Pretreatment of substrate surface for improved adhesion of diamond films on hard metal cutting tools. *Diamond Relat. Mater.* **1996**, *5*, 1478–1489. [CrossRef]
64. Tonshoff, H.K.; Mohlfeld, A.; Gey, C.; Winkler, J. Surface modification of cemented carbide cutting tools for improved adhesion of diamond coatings. *Surf. Coat. Technol.* **1998**, *108*, 543–550. [CrossRef]
65. Buck, V.; Deuerler, F. Enhanced nucleation of diamond films on pretreated substrates. *Diamond Relat. Mater.* **1998**, *7*, 1544–1552. [CrossRef]
66. Sarangi, S.K.; Chattopadhyay, A.; Chattopadhyay, A.K. Effect of pretreatment methods and chamber pressure on morphology, quality and adhesion of HFCVD diamond coating on cemented carbide inserts. *Appl. Surf. Sci.* **2008**, *254*, 3721–3733. [CrossRef]
67. Tang, Y.; Li, Y.S.; Yang, Q.; Hirose, A. Deposition and characterization of diamond coatings on WC-Co cutting tools with W/Al interlayer. *Diamond Relat. Mater.* **2010**, *19*, 496–499. [CrossRef]
68. Tang, Y.; Li, Y.S.; Yang, Q.; Hirose, A. Study of carbideforming element interlayers for diamond nucleation and growth on silicon and WC-Co substrates. *Thin Solid Film.* **2010**, *519*, 1606–1610. [CrossRef]
69. Polini, R.; Barletta, M.; Cristofanilli, G. Wear resistance of nano- and micro-crystalline diamond coatings onto WC-Co with Cr/CrN interlayers. *Thin Solid Film.* **2010**, *519*, 1629–1635. [CrossRef]
70. Hei, H.; Shen, Y.; Ma, J.; Li, X.; Yu, S.; Tang, B.; Tang, W. Effect of substrate temperature on SiC interlayers for diamond coatings deposition on WC-Co substrates. *Vacuum* **2014**, *109*, 15–20. [CrossRef]

71. Endler, I.; Leonhardt, A.; Scheibe, H.J.; Born, R. Interlayers for diamond deposition on tool materials. *Diamond Relat. Mater.* **1996**, *5*, 299–303. [CrossRef]
72. Baldus, H.; Jansen, M. Novel high-performance ceramicsamorphous inorganic networks from molecular precursors. *Angew. Chem. Int. Edit.* **1997**, *36*, 328–343. [CrossRef]
73. Claudia, F. The toxicological mode of action and the safety of synthetic amorphous silica-a nanostructured material. *Toxicology* **2012**, *294*, 61–79.
74. Cui, Y.X.; Shen, B.; Sun, F.H. Diamond deposition on WC-Co substrate with amorphous SiC interlayer. *Surf. Eng.* **2014**, *30*, 237–243. [CrossRef]
75. Cui, Y.X.; Shen, B.; Sun, F.H. Influence of amorphous ceramic interlayers on tribological properties of CVD diamond films. *Appl. Surf. Sci.* **2014**, *313*, 918–925. [CrossRef]
76. Fan, F.; Tang, W.; Liu, S.; Hei, L.; Li, C.; Chen, G.; Lu, F. An effort to enhance adhesion of diamond coatings to cemented carbide substrates by introducing Si onto the interface. *Surf. Coat. Technol.* **2006**, *200*, 6727–6732. [CrossRef]
77. Fischer, M.; Chandran, M.; Akhvlediani, R.; Hoffman, A. Interplay between adhesion and interfacial properties of diamond films deposited on WC-10%Co substrates using a CrN interlayer. *Diamond Relat. Mater.* **2016**, *70*, 167–172. [CrossRef]
78. Wei, Q.; Yu, Z.M.; Ashfold, M.N.R.; Ye, J.; Ma, L. Synthesis of micro- or nano-crystalline diamond films on WC-Co substrates with various pretreatments by hot filament chemical vapor deposition. *Appl. Surf. Sci.* **2010**, *256*, 4357–4364. [CrossRef]
79. Sun, X.; Li, Y.; Wan, B.; Yang, L.; Yang, Q. Deposition of diamond coatings on Fe-based substrates with Al and Al/AlN interlayers. *Surf. Coat. Technol.* **2015**, *284*, 139–144. [CrossRef]
80. Xiao, X.; Sheldon, B.W.; Konca, E.; Lev, L.C.; Lukitsch, M.J. The failure mechanism of chromium as the interlayer to enhance the adhesion of nanocrystalline diamond coatings on cemented carbide. *Diamond Relat. Mater.* **2009**, *18*, 1114–1117. [CrossRef]
81. Petrikowski, K.; Fenker, M.; Gaebler, J.; Hagemann, A.; Pleger, S.; Schaefer, L. Study of CrNx and NbC interlayers for HFCVD diamond deposition onto WC-Co substrates. *Diam. Relat. Mater.* **2013**, *33*, 38–44. [CrossRef]
82. Xu, F.; Yuen, M.F.; He, B.; Wang, C.D.; Zhao, X.R.; Tang, X.L.; Zuo, D.W.; Zhang, W.J. Microstructure and tribological properties of cubic boron nitride films on Si_3N_4 inserts via boron-doped diamond buffer layers. *Diam. Relat. Mater.* **2014**, *49*, 9–13. [CrossRef]
83. Gowri, M.; Li, H.; Kacsich, T.; Schermer, J.J.; van Enckevort, W.J.P.; ter Meulen, J.J. Critical parameters in hot filament chemical vapor deposition of diamond films on tool steel substrates with CrN interlayers. *Surf. Coat. Technol.* **2007**, *201*, 4601–4608. [CrossRef]
84. Raghuveer, M.S.; Yoganand, S.N.; Jagannadham, K.; Lemaster, R.L.; Bailey, J. Improved CVD diamond coatings on WC-Co tool substrates. *Wear* **2001**, *253*, 1194–1206. [CrossRef]
85. Hojman, E.; Akhvlediani, R.; Layyous, A.; Hoffman, A. Diamond CVD film formation onto WC-Co substrates using a thermally nitrided Cr diffusion-barrier. *Diam. Relat. Mater.* **2013**, *39*, 65–72. [CrossRef]
86. Hojman, E.; Akhvlediani, R.; Alagem, E.; Hoffman, A. Cobalt out-diffusion and carbon phase composition at the WC-10% Co/diamond film interface investigated by XPS, SEM, Raman and SIMS. *Phys. Stat. Sol. A* **2012**, *209*, 1726–1731. [CrossRef]
87. Chandran, M.; Hoffman, A. Diamond film deposition on WC-Co and steel substrates with a CrN interlayer for tribological applications. *J. Phys. D Appl. Phys.* **2016**, *49*, 13002. [CrossRef]
88. Fischer, M.; Chandran, M.; Akhvlediani, R.; Hoffman, A. The influence of deposition temperature on the adhesion of diamond films deposited on WC-Co substrates using a Cr-N interlayer. *Phys. Stat. Sol. A* **2015**, *212*, 2628–2635. [CrossRef]
89. Linnik, S.A.; Gaydaychuk, A.V.; Barishnikov, E.Y. The feasibility of usage TiN and CrN barrier sublayers for improving the adhesion of polycrystalline diamond films on WC-Co hard alloys. *Key Eng. Mater.* **2016**, *685*, 583–586. [CrossRef]
90. Liu, M.N.; Bian, Y.B.; Zheng, S.J.; Zhu, T.; Chen, Y.G.; Chen, Y.L.; Wang, J.S. Growth and mechanical properties of diamond films on cemented carbide with buffer layers. *Thin Solid Film.* **2015**, *584*, 165–169. [CrossRef]
91. Ye, F.; Mohammadtaheri, M.; Li, Y.; Shiri, S.; Yang, Q.; Chen, N. Diamond nucleation and growth on WC-Co inserts with Cr2O3-Cr interlayer. *Surf. Coat. Technol.* **2018**, *340*, 190–198. [CrossRef]
92. Liu, W.; Man, W.; Cao, Y.; Lyu, J.; Sun, J.; Yu, J.; Dai, D. The influence on the performance of CVD diamond coatings by the treatment of hydrogen peroxide. *Vacuum* **2018**, *152*, 91–96. [CrossRef]
93. Spear, K.E. Diamond-ceramic coating of the future. *J. Amer. Ceram. Soc.* **1989**, *72*, 171–191. [CrossRef]
94. Friedrich, C.; Berg, G.; Broszeit, E.; Berger, C. Datensammlung zu Hartstoffeigenschaften. *Mater. Und Werkst.* **1997**, *28*, 59–76. [CrossRef]
95. Gao, J.; Hei, H.; Zheng, K.; Gao, X.; Liu, X.; Tang, B.; He, Z.; Yu, S. On the use of Mo/Mo_2C gradient interlayers in diamond deposition onto cemented carbide substrates. *Surf. Rev. Lett.* **2016**, *23*, 1550109. [CrossRef]
96. Gao, J.; Hei, H.; Zheng, K.; Wang, R.; Shen, Y.; Liu, X.; Tang, B.; He, Z.; Yu, S. Design and synthesis of diffusionmodified HfC/HfC-SiC bilayer system onto WC-Co substrate for adherent diamond deposition. *J. Alloys Comp.* **2016**, *705*, 376–383. [CrossRef]
97. Yu, X.; Liu, Y.; Ma, L.; Yang, G.; Wang, C. Investigation of thick CVD diamond film with SiC interlayer on tungsten carbide for possible usage in geologic explorations. *Vacuum* **2013**, *94*, 53–56. [CrossRef]
98. Hei, H.; Ma, J.; Li, X.; Yu, S.; Tang, B.; Shen, Y.; Tang, W. Preparation and performance of chemical vapor deposition diamond coatings synthesized onto the cemented carbide micro-end mills with a SiC interlayer. *Surf. Coat. Technol.* **2015**, *261*, 272–277. [CrossRef]
99. Yu, S.W.; Fan, P.W.; Tang, W.Z.; Li, X.J.; Hu, H.L.; Hei, H.J.; Zhang, S.K.; Lu, F.X. Pressure dependence of morphology and phase composition of SiC films deposited by microwave plasma chemical vapor deposition on cemented carbide substrates. *Thin Solid Film.* **2011**, *520*, 828–832. [CrossRef]

100. Cabral, G.; Gäbler, J.; Lindner, J.; Grácio, J.; Polini, R. A study of diamond film deposition on WC-Co inserts for graphite machining: Effectiveness of SiC interlayers prepared by HFCVD. *Diam. Relat. Mater.* **2008**, *17*, 1008–1014. [CrossRef]
101. Srikanth, V.V.S.S.; Staedler, T.; Jiang, X. Structural and compositional analyses of nanocrystallinediamond/β-SiC composite films. *Appl. Phys. A Mater. Sci. Process.* **2008**, *91*, 149–155. [CrossRef]
102. Tian, Q.; Huang, N.; Yang, B.; Zhuang, H.; Wang, C.; Zhai, Z.; Li, J.; Jia, X.; Liu, L.; Jiang, X. Diamond/β-SiC film as adhesion-enhanced interlayer for top diamond coatings on cemented tungsten carbide substrate. *J. Mater. Sci. Technol.* **2017**, *33*, 1097–1106. [CrossRef]
103. Ren, Y.; Yu, X.; Liu, F.; Wang, Y.-H. Low-temperature β-SiC interlayer for diamond film on cemented carbide. *Surf. Eng.* **2017**, *33/1*, 20–26. [CrossRef]
104. Cui, Y.; Wang, W.; Shen, B.; Sun, F. A study of CVD diamond deposition on cemented carbide ball-end milling tools with high cobalt content using amorphous ceramic interlayers. *Diam. Relat. Mater.* **2015**, *59*, 21–29. [CrossRef]
105. Poklonskaya, O.N.; Khomich, A.A. Raman scattering in a diamond crystal implanted by high-energy nickel ions. *J. Appl. Spectrosc.* **2013**, *80/5*, 715–720. [CrossRef]
106. Haubner, R.; Lindlbauer, A.; Lux, B. Diamond nucleation and growth on refractory metals using microwave plasma deposition. *Int. J. Refract. Met. Hard Mater.* **1996**, *14*, 119–125. [CrossRef]
107. Faure, C.; Teulé-Gay, L.; Manaud, J.; Poulon-Quintin, A. Mechanisms of time-modulated polarized nano-crystalline diamond growth. *Surf. Coat. Technol.* **2013**, *222*, 97–103. [CrossRef]
108. Rong, W.; Hei, H.; Zhong, Q.; Shen, Y.; Liu, X.; Wang, X.; Zhou, B.; He, Z.; Yu, S. Nanostructured TaxC interlayer synthesized via double glow plasma surface alloying process for diamond deposition on cemented carbide. *Appl. Surf. Sci.* **2015**, *359*, 41–47. [CrossRef]
109. Yu, S.; Gao, J.; Li, X.; Ma, D.; Hei, H.; Shen, Y.; Rong, W.; Liu, X.; Ne, Z.; Tang, B. Effect of substrate temperature on tantalum carbides interlayers synthesized onto WC-Co substrates for adherent diamond deposition. *Surf. Rev. Lett.* **2019**, *26/2*, 1850138. [CrossRef]
110. Poulon-Quintin, A.; Faure, C.; Teulé-Gay, L.; Manaud, J.P. A multilayer innovative solution to improve the adhesion of nanocrystalline diamond coatings. *Appl. Surf. Sci.* **2015**, *331*, 27–34. [CrossRef]
111. Wei, Q.; Yu, Z.; Ashfold, M.; Chen, Z. Effects of thickness and cycle parameters on fretting wear behavior of CVD diamond coatings on steel substrates. *Surf. Coat. Technol.* **2010**, *205*, 158–167. [CrossRef]
112. Dumpala, R.; Kumar, N.; Kumaran, C.R.; Dash, S.; Ramamoorthy, B.; Rao, M.S.R. Adhesion characteristics of nano- and micro-crystalline diamond coatings: Raman stress mapping of the scratch tracks. *Diam. Relat. Mater.* **2014**, *44*, 71–77. [CrossRef]
113. Dumpala, R.; Ramamoorthy, B.; Rao, M.S.R. Graded composite diamond coatings with top-layer nanocrystallinityand interfacial integrity: Crosssectional Raman mapping. *Appl. Surf. Sci.* **2014**, *289*, 545–550. [CrossRef]
114. Fraga, M.A.; Contin, A.; Rodríguez, L.A.A.; Vieira, J.; Campos, R.A.; Corat, E.J.; Airoldi, V.J.T. Nano- and microcrystalline diamond deposition on pretreated WC-Co substrates: Structural properties and adhesion. *Mater. Res. Express* **2016**, *3*, 025601. [CrossRef]
115. Liu, C.; Wang, J.; Liu, S.; Xiong, L.; Weng, J.; Cui, X. Effects of surface pretreatment on nucleation and growth of ultra-nanocrystalline diamond films. *Plasma Sci. Technol.* **2015**, *17*, 496–501. [CrossRef]
116. Liu, Y.; Du, M.; Zhang, M.; Wei, Q.; Yu, Z.; Li, X. Growth of diamond coatings on functionally graded cemented carbides. *Int. J. Refract. Met. Hard Mater.* **2015**, *49*, 307–313. [CrossRef]
117. Liu, Z.; Xu, F.; Xu, J.; Tang, X.; Liu, Y.; Zuo, D. Preparation and cutting performance of boron-doped diamond coating on cemented carbide cutting tools with high cobalt content. *Int. J. Chem. Molecul. Nucl. Mater. Metallurg. Eng.* **2015**, *9*, 577–580.
118. Srinivasan, B.; Rao, M.S.R.; Rao, B.C. On the development of a dual-layered diamond-coated tool for the effective machining of titanium Ti-6Al-4V alloy. *J. Phys. D Appl. Phys.* **2017**, *50*, 015302. [CrossRef]
119. Xu, Y.; Chen, K.; Wang, S.; Chen, S.; Chen, X. Influence of pretreatment on diamond-coated tool nucleation and mashining performance. *Surf. Rev. Lett.* **2018**, *25*, 1850020.
120. Zhou, H.; Yuan, B.; Lyu, J.; Jiang, N. A novel approach of deposition for uniform diamond films on circular saw blades. *Plasma Sci. Technol.* **2017**, *19*, 115502. [CrossRef]
121. Amirhaghi, S.; Reehal, H.S.; Plappert, E.; Bajic, Z.; Wood, R.J.K.; Wheeler, D.W. Growth and erosive wear performance of diamond coatings on WC substrates. *Diam. Relat. Mater.* **1999**, *8*, 845–849. [CrossRef]
122. Amirhaghi, S.; Reehal, H.S.; Wood, R.J.K.; Wheeler, D.W. Diamond coatings on tungsten carbide and their erosive wear properties. *Surf. Coat. Technol.* **2001**, *135*, 126–138. [CrossRef]
123. Xu, Z.; Lev, L.; Lukitsch, M.; Kumar, A. Deposition of adherent diamond coating on WC-Co substrate. *Mater. Res. Soc. Symp. Proc.* **2006**, *890*, 6. [CrossRef]
124. Hu, J.; Chou, Y.K.; Thompson, R.G.; Burgess, J.; Street, S. Characterizations of nano-crystalline diamond coating cutting tools. *Surf. Coat. Technol.* **2007**, *202*, 1113–1117. [CrossRef]
125. Hu, J.; Chou, Y.K.; Thompson, R.G. Nanocrystalline diamond coating tools for machining high-strength Al alloys. *Int. J. Refract. Met. Hard Mater.* **2008**, *26*, 135–144. [CrossRef]
126. Chou, Y.K.; Thompson, R.G.; Kumar, A. CVD-diamond technologies for dry drilling applications. *Thin Solid Film.* **2010**, *518*, 7487–7491. [CrossRef]
127. Johnston, J.M.; Baker, P.; Catledge, S.A. Improved nanostructured diamond adhesion on cemented tungsten carbide with boride interlayers. *Diam. Relat. Mater.* **2016**, *69*, 114–120. [CrossRef]
128. Johnston, J.M.; Catledge, S.A. Metal-boride phase formation on tungsten carbide (WC-Co) during microwave plasma chemical vapor deposition. *Appl. Surf. Sci.* **2016**, *364*, 315–321. [CrossRef]

129. Ziyao, Y.; Guo, Y.; Giang, X.; Li, C.; Liu, L.; Yang, B.; Song, H.; Zhai, Z.; Lu, Z.; Li, H.; et al. New multilayered diamond/β-SiC composite architectures for high-performance hard coating. *Mater. Design.* **2020**, *186*, 108207.
130. Tsao, C.C.; Hocheng, H. Taguchi analysis of delamination associated with various drill bits in drilling of composite material. *Int. J. Mach. Tools Manuf.* **2004**, *44*, 1085–1090. [CrossRef]
131. Davim, J.P.; Reis, P. Study of delamination in drilling carbon fiber reinforced plastics (CFRP) using design experiments. *Compos. Struct.* **2003**, *59*, 481–487. [CrossRef]
132. Shyha, I.; Soo, S.L.; Aspinwall, D.; Bradley, S. Effect of laminate configuration and feed rate on cutting performance when drilling holes in carbon fibre reinforced plastic composites. *J. Mater. Process. Technol.* **2010**, *210*, 1023–1034. [CrossRef]
133. Karpat, Y.; Değer, B.; Bahtiyar, O. Drilling thick fabric woven CFRP laminates with double point angle drills. *J. Mater. Process. Technol.* **2012**, *212*, 2117–2127. [CrossRef]
134. Abrão, A.M.; Faria, P.E.; Rubio, J.C.C.; Reis, P.; Davim, J.P. Drilling of fiber reinforced plastics. A Review. *J. Mater. Process. Technol.* **2007**, *186*, 1–7. [CrossRef]
135. Tsao, C.C.; Hocheng, H. Effect of eccentricity of twist drill and candle stick drill on delamination in drilling composite materials. *Int. J. Mach. Tools Manuf.* **2005**, *45*, 125–130. [CrossRef]
136. Köpf, A.; Feistritzer, S.; Udier, K. Diamond coated cutting tools for machining of non-ferrous metals and fibre reinforced polymers. *Int. J. Refract. Met. Hard Mater.* **2006**, *24*, 354–359. [CrossRef]
137. Ahmad, J.S.; Sridhar, G. Edge trimming of CFRP composites with diamond coated tools: Edge wear and surface characteristics. In Proceedings of the General Aviation Technology Conference and Exhibit, Wichita, KS, USA, 16–28 April 2002. SAE Technical Papers. [CrossRef]
138. Duboust, N.; Melis, D.; Pinna, C.; Ghadbeig, H.; Collis, A.; Ayvar-Soberanis, S.; Kerrigan, K. Machining of carbon fiber: Optical surface damage characterization and tool wear study. *Procedia CIRP* **2016**, *45*, 71–74. [CrossRef]
139. Guo, Z.J.; Chang, W.X.; Bin, S.; Hong, S.F. Effect of boron and silicon doping on improving the cutting performance of CVD diamond coated cutting tools in machining CFRP. *Int. J. Refract. Met. Hard Mater.* **2013**, *41*, 285–292.
140. Tsao, C.C. Thrust force and delamination of core-saw drill during drilling of carbon fiber reinforced plastics (CFRP). *Int. J. Adv. Manufactur. Technol.* **2008**, *37*, 23–28. [CrossRef]
141. Zhang, L.B.; Wang, L.J.; Liu, X.Y. A mechanical model for predicting critical thrust forces in drilling composite laminates. *J. Eng. Manuf.* **2001**, *215*, 135–146. [CrossRef]
142. Hocheng, H.; Tsao, C.C. Comprehensive analysis of delamination in drilling of composite materials with various drill bits. *J. Mater. Process. Technol.* **2003**, *140*, 335–339. [CrossRef]
143. Shyha, I.S.; Aspinwall, D.K.; Soo, S.L.; Bradley, S. Drill geometry and operating effects when cutting small diameter holes in CFRP. *Int. J. Mach. Tools Manuf.* **2009**, *49*, 1008–1014. [CrossRef]
144. Rawat, S.; Attia, H. Wear mechanisms and tool life management of WC-Co drills during dry high speed drilling of woven carbon fibre composites. *Wear* **2009**, *267*, 1022–1030. [CrossRef]
145. Mu, J.; Xu, J.H.; Chen, Y.; Fu, Y.C. CFRP drilling with brazed diamond core drill. *Sol. Stat. Phenom.* **2011**, *175*, 27–32. [CrossRef]
146. Wang, Y.G.; Yan, X.P.; Chen, X.G.; Sun, C.Y.; Liu, G. Cutting performance of carbon fiber reinforced plastics using PCD tool. *Adv. Mater. Res.* **2011**, *215*, 14–18. [CrossRef]
147. Heaney, P.J.; Sumant, A.V.; Torres, C.D.; Carpick, R.W.; Pfefferkorn, F.E. Diamond coatings for micro end mills: Enabling the dry machining of aluminum at the micro-scale. *Diam. Relat. Mater.* **2008**, *17*, 223–233. [CrossRef]
148. Davim, J.P.; Mata, F. Chemical vapour deposition (CVD) diamond coated tools performance in machining of PEEK composites. *Mater. Des.* **2008**, *29*, 1568–1574. [CrossRef]
149. Zhang, J.; Shen, B.; Sun, F.-H. Fabrication and drilling tests of chemical vapor deposition diamond coated drills in machining carbon fiber reinforced plastics. *J. Shanghai Jiaotong Univ. Sci.* **2013**, *18*, 394–400. [CrossRef]
150. An, Q.; Ming, W.; Cai, X.; Chen, M. Effects of tool parameters on cutting force in orthogonal machining of T700/LT03A unidirectional carbon fiber reinforced polymer laminates. *J. Reinforc. Plast. Compos.* **2015**, *34*, 591–602. [CrossRef]

C Journal of Carbon Research

Article

Detonation Synthesis Nanodiamond Soot as a Promising Filler for Polymer Composites

Oleg V. Lebedev [1,2,*], Tikhon S. Kurkin [1], Evgeny K. Golubev [1], Alexander L. Vasiliev [2,3], Andrey K. Gatin [4], Galina P. Goncharuk [1] and Alexander N. Ozerin [1]

1 Enikolopov Institute of Synthetic Polymer Materials Russian Academy of Sciences (ISPM RAS), Profsoyuznaya Street 70, 117393 Moscow, Russia
2 Moscow Institute of Physics and Technology, Institutsky lane 9, Dolgoprudny, 141700 Moscow, Russia
3 Federal Scientific Research Center "Crystallography and Photonics" Russian Academy of Sciences, Leninsky Ave 59, 119333 Moscow, Russia
4 N. N. Semenov Federal Research Center for Chemical Physics Russian Academy of Sciences (FRCCP RAS), Kosygina Street 4, 119991 Moscow, Russia
* Correspondence: oleg.lebedev@phystech.edu

Abstract: In this work, the results of a complex investigation of structure and properties of nanodiamond soot (NDS) of detonation synthesis are presented. Size distribution of NDS particles, dispersed in different liquid media, was investigated using dynamic light scattering and laser diffraction analysis methods. The results of the investigation, as well as the results of zeta-potential measurements, allowed us to characterize the agglomeration process of the NDS particles as independent of the medium, making NDS a good model filler for research of composite-modified nanosized particles. Additional data obtained using scanning electron microscopy, scanning tunneling microscopy, atomic force microscopy, X-ray diffraction, and Raman spectroscopy, demonstrated that in NDS the spherical nanodiamond (ND) particles with diameter ~5 nm are densely packed into strong-coupled aggregates with diameter ~300 nm, surrounded by graphite nanoribbons. X-ray diffraction analysis estimated the volume fraction of NDs in NDS as ~45 vol.%, simultaneously showing that the graphite is not defective, which was confirmed with the electron diffraction method. It was demonstrated that this structure of NDS allows to efficiently use NDS as a filler for polymer composites to increase polymer characteristics such as electrical conductivity or tribological characteristics, similarly to conventionally applied fillers such as carbon black.

Keywords: nanodiamond soot; polymer nanocomposites; carbon black; conductivity; tribology

Citation: Lebedev, O.V.; Kurkin, T.S.; Golubev, E.K.; Vasiliev, A.L.; Gatin, A.K.; Goncharuk, G.P.; Ozerin, A.N. Detonation Synthesis Nanodiamond Soot as a Promising Filler for Polymer Composites. *C* **2022**, *8*, 69. https://doi.org/10.3390/c8040069

Academic Editor: Gil Gonçalves

Received: 30 September 2022
Accepted: 21 November 2022
Published: 27 November 2022

1. Introduction

Nanodiamonds (NDs) have been considered as one of the most promising fillers for nanocomposites in the last few decades [1–4]. NDs can combine properties of a diamond such as extremely high hardness, chemical and bio-stability, high thermal conductivity, etc., with an ability to use them as an effective filler for improvement of polymer characteristics [5–7].

ND particles can be obtained by the detonation of strong explosives and ablation of carbon target by laser pulses [2,8]. A significant advantage of the detonation method is the formation of diamond nanoparticles with a typical size of 4–5 nm. The structures of ND particles are of a special interest for investigation [4]. In the work of Ginés et al. [9], the zeta potential of NDs was investigated. Positive zeta potentials in nano-structured carbons were explained by the presence of graphitic planes at the surface of ND particles, which leave oxygen-free Lewis sites and promote the suppression of acidic functional groups. Electron microscopy and Raman spectroscopy have been used to show that positive zeta potential of nanodiamond is only exhibited in the presence of sp^2 carbon at the surface. In the review by Shrestha et al. [10] it was concluded that exploiting the surface characteristics of the nanoparticles, including surface charge and functionality, is an effective method of size control of aggregate/agglomerate nanoparticles.

Along with detonation synthesis NDs, polymers modified with NDs are also an attractive subject for research [11]. In the review by Karami et al. [12], it was noted that exceptional properties of detonation NDs and its inherent rich surface chemistry make them suitable for the production of multifunctional materials. In the work of Kidalov et al. [13], it was shown that the thermal conductivity of nanocomposites based on nanodiamonds can reach up to 10 W/(m·K). In the work of Houshyar et al. [14], using conventional methods of processing such as melt mixing, the authors obtained filaments based on polypropylene (PP) filled with NDs that were characterized by exceptional long-term stability performance, mechanical and biological properties.

A wide range of methods for the controlled modification of diamond powder opens up the possibilities of increasing such characteristics of polymer composites as adhesive strength [3,6]. In the work of Kurkin et al. [6], it was shown that the adhesive strength of oriented poly(vinyl alcohol) (PVA) fibers in an epoxy binder can be significantly improved by simultaneously modifying the polymer fiber and the polymer matrix with fine particles of detonation-synthesized ND powder with oppositely charged surfaces.

Unfortunately, using NDs for the modification of polymers can be very expensive, especially if high filler content is required. This is due to the complex purification process required to remove residue metals and non-diamond forms of carbon [3]. Additionally, while purified ND powder can be characterized by a relatively high level of thermal conductivity, it cannot be used efficiently to increase electrical conductivity of polymers [5].

Nanodiamond soot (NDS) is a more affordable type of filler with characteristics close to those of NDs [11,15]. NDS is obtained at the post-detonation stage of an ND synthesis process, preceding the expensive final stage of NDs purification from non-diamond carbon [8]. This reduces the cost of NDS fillers, making them a less expensive alternative to the pure NDs. Additionally, since NDS can contain a high-volume fraction of non-diamond carbon, such as graphite, it is potentially possible to obtain electrically conductive polymer composites modified with NDS.

In the works of Satonkina et al. [4] and Chen et al. [15], structure, composition, and surface properties of detonation synthesis NDS were characterized using elemental analysis, high resolution transmission electron microscopy, X-ray diffraction (XRD), Raman spectroscopy, Fourier-IR, X-ray photoelectron spectroscopy, and small angle X-ray scattering. It was shown that the condensed carbon in the NDS of choice had a complex composition, comprising clusters of diamond, graphite and amorphous carbon, all of which were in the nanometer scale. Two types of graphite phases were present in the ND. The first phase was tangled and highly curved graphite nanoribbons; the other was carbon spheres and carbon mantles graphitized from the NDs. Such nanoribbons were effectively used previously in the work of Alaferdov et al. [16] to obtain polymer composites with excellent conductivity (up to 40 S/cm) and very low percolation threshold (0.42 vol.%).

In the work of Kurkin et al. [17], the structure and properties of the oriented PVA fibers modified with NDs and NDS of detonation synthesis were investigated using XRD, electron microscopy, and mechanical testing. It was shown that NDs and NDS are effective modifiers of the physical–mechanical properties of the oriented PVA fibers. NDS particles in the PVA matrix retained a high level of dispersion without aggregation up to a filling degree of 3 vol.%. Maximum values of the longitudinal elastic modulus and the energy stored in the modified PVA fiber upon stretching up to the moment of the fracture was achieved at a small (1 vol.%) NDS filling value. The adhesion strength of the oriented modified by NDS PVA fibers in the epoxy matrix substantially exceeded the adhesion strength of the nonmodified fibers.

In the works of Neitzel et al. [18,19] it was demonstrated that an addition of ND particles in a polymer matrix (epoxy) is an efficient way of decreasing the friction coefficient of the material. In the work of Lebedev et al. [20], tribological and percolation properties of composites based on PP modified with non-conductive NDS particles were investigated.

By measuring tribological characteristics such as wear rate, it was shown that there is a possibility of formation of a surface layer enriched with NDS. This effect was additionally studied in [21] for electrically conductive such fillers as carbon black (CB).

It should be mentioned that most of the currently published works are either focused on the research of NDs or polymer nanocomposites filled with NDs. There are very few works that investigate the properties of specific NDS particles, and no works that connect the structure of such particles with the properties of polymer composites based on the same filler. At the same time, researching possible areas of application of such a unique type of filler can be very useful for the polymer composite industry and are of a great scientific and practical interest.

The performance of various types of nanosized and submicron polydisperse fillers (including inorganic and carbon particles, as well as more complex organic/inorganic hybrid particles) in polymer composites is generally associated with the structural properties of these fillers [10,22]. Among the most common methods of obtaining information about the phase composition, structural hierarchy and colloidal behavior of polydisperse fillers are XRD, dynamic light scattering (DLS), laser diffraction (LD) and static light scattering, along with various types of microscopy techniques. While each of these techniques has its intrinsic limitations, their combination often provides complementary information [23,24], which is necessary to determine the most optimal procedure of composite material design and processing [25].

The novelty of our work is in a comprehensive study of the properties of NDS of detonation synthesis, characterized by high-volume fractions of graphite in the form of graphite nanoribbons. The study is based on the thorough analysis of the structure of such NDS powder and dispersions under various conditions. The structural properties of NDS were characterized by a wide variety of techniques, including XRD, DLS, LD, Raman spectroscopy, atomic force microscopy (AFM), scanning and transmission electron microscopy (SEM and TEM, respectively), scanning tunnelling microscopy (STM), and electron diffraction. By using the NDS as a filler for a polymer, investigated earlier in our previous works, and comparing the effect of the filler structure on the migration of nanoparticles to the boundaries of polymer composite melts, it makes it possible to draw conclusions on the overall possibility of NDS acting as a more versatile substitute of such conventionally used fillers, as CB. The obtained results can provide a solid foundation for the future design of new carbon-based polymer composite materials.

2. Materials and Methods

As the main object of the investigation, NDS of detonation synthesis was used. NDS was obtained from 2,4,6-trinitrotoluene, providing the high graphite/nanodiamond ratio required to ensure high levels of electrical conductivity. NDS was provided by the Special Design and Technological Bureau "Technolog" (St. Petersburg, Russia). The details on the synthesis process can be found in the works of Dolmatov et al. [3,11].

Commercially available amorphous–crystalline PP grade H030 GP/3 (SIBUR International GmbH, Vienna, Austria) was used as the base polymer for the composite materials studied in this work [26]. As the reference electrically conductive filler, commercially available electrically conductive CB of brand P267-E (equivalent of N 472, Institute of Hydro-carbons Processing of the Siberian Branch of the Russian Academy of Sciences IHCP SB RAS, Saint Petersburg, Russia) with a specific surface area of 144 m^2/g [27] was chosen. Previously, this filler was used in the investigation of nanoparticles migration in PP-based composite melts [21].

Samples of composites were obtained via melt mixing of the polymer granules and nanoparticle powders in a micro-compounder (DACA Instruments, Goleta, CA, USA) designed to prepare compositions in laboratory quantities (up to 4 cm^3). The quantity of the material in the micro-compounder chamber was calculated so the chamber was filled fully. The mixing procedure was carried out at 200 °C and 500 rpm for 10 min after the addition of the components into the compounder.

DC electrical conductivity of the filler powders was measured using a 34401A multimeter (Agilent, Santa Clara, CA, USA) and the four-probe method. A special appliance was used to change volume fraction of the powder between electrodes.

To carry out measurements of electrical resistance of the composite melts, granules of the prepared composites of various compositions were placed in a heated mold between two flat electrodes inside a circular hole in a spacer made of polytetrafluoroethylene [21]. The composite granules were quickly heated to 205 °C, after which a pressure of 5 MPa was gradually applied to the composite melt. The temperature and pressure values were maintained for the duration of the experiments. Constant electrical voltage was applied between the electrodes during the entire experiment time. The voltage was provided by a highly stable AKIP 1147/1 power supply (Prist, Moscow, Russia). The standard preset value of the electric voltage was 2 V. A 6485/E (Keithley, Cleveland, OH, USA) picoammeter was used for the current measurements. During the experiments, the values of the electric current between the electrodes were continuously recorded using a multichannel analog-to-digital converter E14-440 (L-Card, Moscow, Russia) at frequency of 100 Hz.

Hydrodynamic diameters of NDS in various liquid media (acetone, DMSO, ethanol, isopropanol) were determined via DLS at 25 °C using a Microtrac Nanotrac model Zetatrac instrument (Microtrac Inc., Montgomeryville, PA, USA) equipped with a 3 mW solid-state laser operating at 780 nm. Scattered light was detected at 180° in a controlled reference self-beating mode, and the particle size was calculated from the power spectrum analysis of Doppler-shifted signal over several runs. For every tested sample, 5 runs each of 60 s duration were performed. The sphere-equivalent hydrodynamic diameter of the NDS structures was calculated from the particle diffusion coefficient via the Stokes–Einstein equation, using solution viscosities for liquid media at 25 °C. Electrophoretic mobility of NDS structures was measured in liquid media using a Microtrac Nanotrac model Zetatrac instrument by applying AC/DC voltage to built-in sample cell electrodes to determine the magnitude and sign of particle zeta-potential. Zeta-potential was calculated from electrophoretic mobility via the Smoluchowski equation. All measurements and data analysis were performed using NanoFLEX Software. NDS concentrations of 0.5% (*w/v*) were utilized in all DLS and zeta-potential measurements. NDS dispersions were preliminarily treated using a CUD-500 ultrasonic disperser from Criamid, Moscow, Russia (2 min, 250 W, 30 kHz).

Particle size analysis by LD was performed using Wintrac 3000 instrument equipped with a He-Ne laser (λ = 632.8 nm) and auxiliary solid-state laser operating at λ = 405 nm, allowing for particle characterization in 0.1–2000 μm size range. Prior to the measurements, NDS powder was dispersed in a liquid medium using the built-in circulation system equipped ultrasound probe (60 W, 40 kHz). Scattering pattern measurements were performed with sample circulating through a flow-cell at pumping speed of 2000 rotations per minute. For every sample, 60 scans of 2 s duration each were performed. To obtain particle size distributions, each scattering pattern was processed according to Fraunhofer diffraction and Mie scattering models using WINTRAC Software. The reported results are the product of averaging of the 60 scans for every measured sample.

XRD patterns (2Θ = 10–105°) were obtained using a Bruker D8 Advance powder diffractometer ($\lambda(CuK_{\alpha})$ = 1.5418 Å) equipped with focusing germanium crystal monochromator, Ni-filter and a LYNXEYE 1D scintillating detector. XRD patterns were recorded in the transmission mode, with NDS powder placed between two layers of amorphous poly(ethylene terephthalate) films. Processing of diffraction patterns (peak separation, corrections for non-coherent background scattering and scattering from poly(ethylene terephalate)) was performed using Origin Pro software.

AFM measurements were performed using a Solver HV atomic-force microscope (NT-MDT, Moscow, Russia) with standard HA-NC cantilevers (chip size 3.6 × 1.6 × 0.45 mm, cantilever size 87 × 32 × 1.75 μm, curvature radius of the tip 10 nm, force constant 5.8 N/m). The measurements were carried out in a tapping mode with the acquisition of surface topography and phase contrast.

STM measurements were performed using a scanning tunneling microscope (STM, Omicron NanoTechnology, Taunusstein, Germany) at 300 K and a residual gas pressure of 2×10^{-10} Torr. According to the standard procedure, tips made of tungsten wires were used as STM probes. STM topography measurements were carried out in the constant current mode (U_{bias} = 2.0 V, I_{tunnel} = 3.1 nA).

Prior to the AFM and STM measurements, NDS dispersions in 2-propanol (concentration of solid phase 0.5 wt.%) were deposited on a highly oriented pyrolytic graphite substrate and dried at the ambient conditions (air, T = 22 °C).

Raman spectra were obtained using a RamanStation-400 (Perkin-Elmer, Waltham, MA, USA) dispersive echelle Raman spectrometer equipped with a single-mode 785 nm diode laser excitation (1–100 mW), thermoelectrically cooled (−52 °C) CCD matrix detector (Andor Inc., Belfast, North Ireland), XYZ-motorized sample stage and an ×30 microscope objective used for delivering laser excitation and collecting Raman photons. Raman spectra were acquired in 180° backscattering geometry. The solid samples were placed onto the standard Au-coated slide on the XYZ-table and focused under the instrument objective. The Raman spectra were obtained within 3400–100 cm^{-1} Raman shifts range at 1,2,4 or 8 cm^{-1} optical resolution. In different experiments, depending on the luminescence activity of the samples, the laser excitation power of 2–25 mW and signal accumulation times of 20–180 s were used for the spectra recording. A dedicated Bruker OPUS software package was used for routine processing (baseline correction, smoothing, etc.) of the Raman spectra.

Tribological tests were conducted using a pin-on-disk-type testing machine (cylindrical stainless-steel pin) in the dry friction mode according to ASTM G99-05 procedure. Test parameters values can be found in [20]. For each test piece, 6 identical test runs were performed, measuring friction coefficient dependence on time. The obtained results were averaged.

SEM studies of the NDS powder were performed using a JSM-6000 (Neoscope II) scanning microscope (Jeol, Japan). TEM studies were performed using a fully digital 200 kV S/TEM FEI Tecnai Osiris system (FEI Company, Hillsboro, OR, USA). Samples were prepared by deposition of the NDS solutions on holey C film on Au grid.

3. Results

Firstly, the NDS powder was investigated "as it is" using SEM. Selected results of the microscopy study are presented in Figure 1.

(a) (b)

Figure 1. (**a**,**b**) Scanning electron microscopy images of NDS powder.

It can be seen that the powder contains soft-looking (the packing of the carbon phase looks not very dense) agglomerates of diameter >~3 μm (Figure 1a). In turn, these relatively large agglomerates are seemingly formed from aggregates of diameter >100 nm (Figure 1b) [22].

The results of SEM data analysis for the powder of NDS were in some agreement with the measurements of NDS structures (primary particles, aggregates, agglomerates [22]) size distributions in acetone using DLS and LD analysis (LDA), as can be seen from Figure 2.

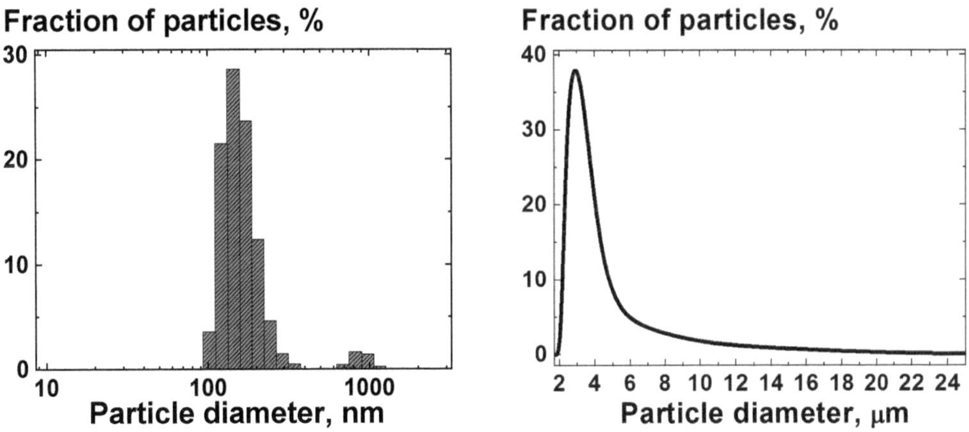

(a) (b)

Figure 2. Distribution of diameter of the NDS particles in a suspension of NDS powder in acetone, obtained with (**a**) DLS and (**b**) LD methods.

The size distribution for NDS aggregates/agglomerates consisted of peaks at ~250 nm and ~950 nm, obtained using DLS (Figure 2a), and at ~3.2 μm, obtained using LDA (Figure 2b). The difference in the size distributions determined by the DLS and LDA can be attributed to the additional pretreatment using ultrasound with higher power when the size distributions were measured by the DLS method, compared to the treatment of the NDS in the circulation system when the size distributions was measured by LD.

NDS size distributions were also obtained for additional liquid media, such as DMSO, ethanol, and isopropanol, as well as for different time periods after ultrasonication: 0 h, 24 h, 120 h, 1 week, 3 weeks, and 4 weeks. The results were very close to the ones obtained for NDS dispersion in acetone soon after ultrasonication (in the range of an error). The results of the DLS measurements for different liquid media and time periods after ultrasonication are presented in Supplementary Materials.

For all the investigated dispersions in different media (acetone, DMSO, ethanol, isopropanol) zeta-potentials were measured after 0 h, 24 h, 120 h, 1 week, 3 weeks, and 4 weeks after the initial ultrasonication. The value of zeta-potential for NDS dispersions were in the range of uncertainty for all the investigated media even after 4 weeks: 20 ± 4 mV for dispersion in acetone soon after the ultrasonication. As was mentioned in the introduction section, according to Ginés et al. [9], the positive zeta potential is only exhibited on nanodiamond with non-zero concentrations of sp^2 bonding, specifically attributed to graphene-like shells. Although it is well known in colloid physics that zeta-potential above +30 mV is expected for stable dispersions, particle dispersions with zeta-potential in a range of +20–30 mV can also maintain the size distribution over a long period of time, depending on the particle size [28].

The next step was the investigation of the NDS powder using TEM. Selected results of the microscopy study are presented in Figure 3.

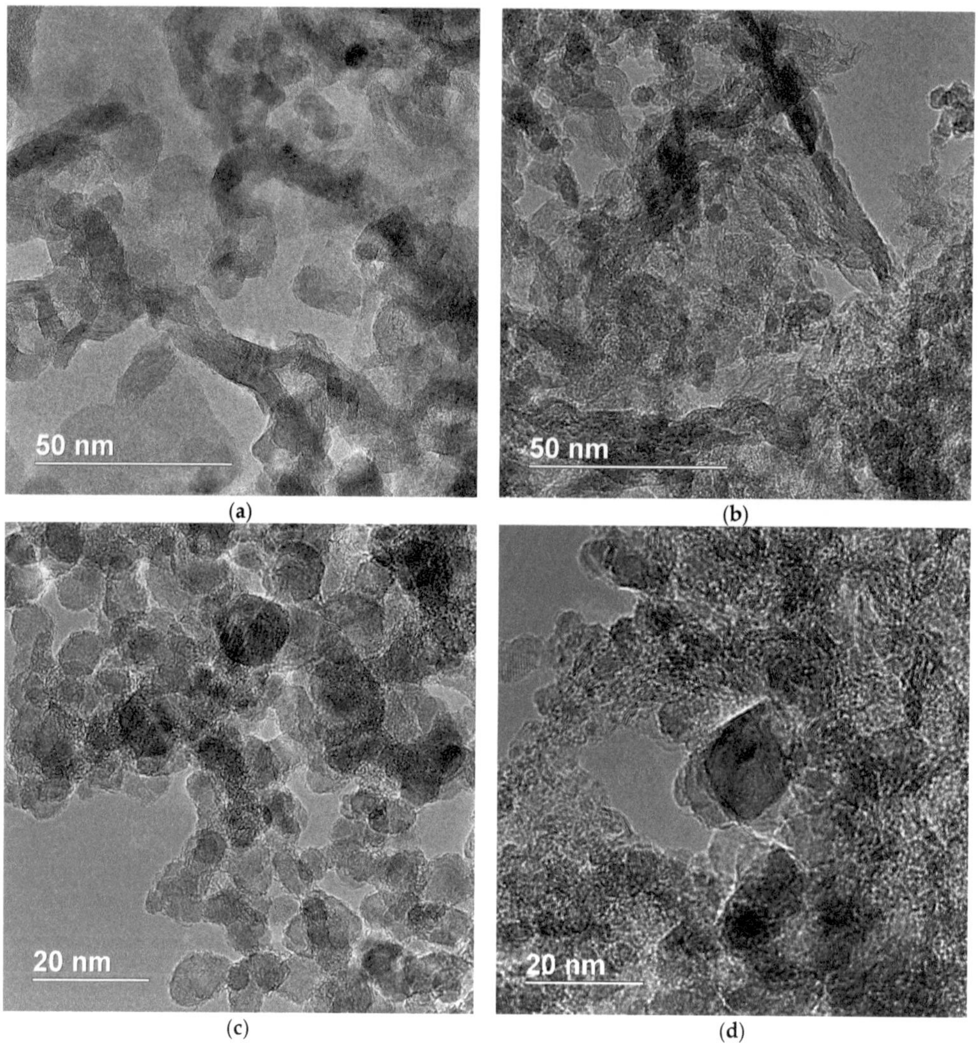

Figure 3. (**a**–**d**) Transmission electron microscopy images of NDS powder.

As it can be seen from the TEM results, especially in Figure 3a,b, the graphite phase is partially presented in the form of graphite nanoribbons/nanobelts. The primary NDS particles themselves seem to have an average diameter of ~5 nm. The primary NDS particles are considered to be ND crystallites [12]. Additionally, relatively large flakes of graphite (~10–15 nm) can be observed (Figure 3c,d).

To be able to investigate the composition of NDS particles as well as their structure, XRD and Raman spectroscopy methods were used. The results of the study are presented in Figure 4.

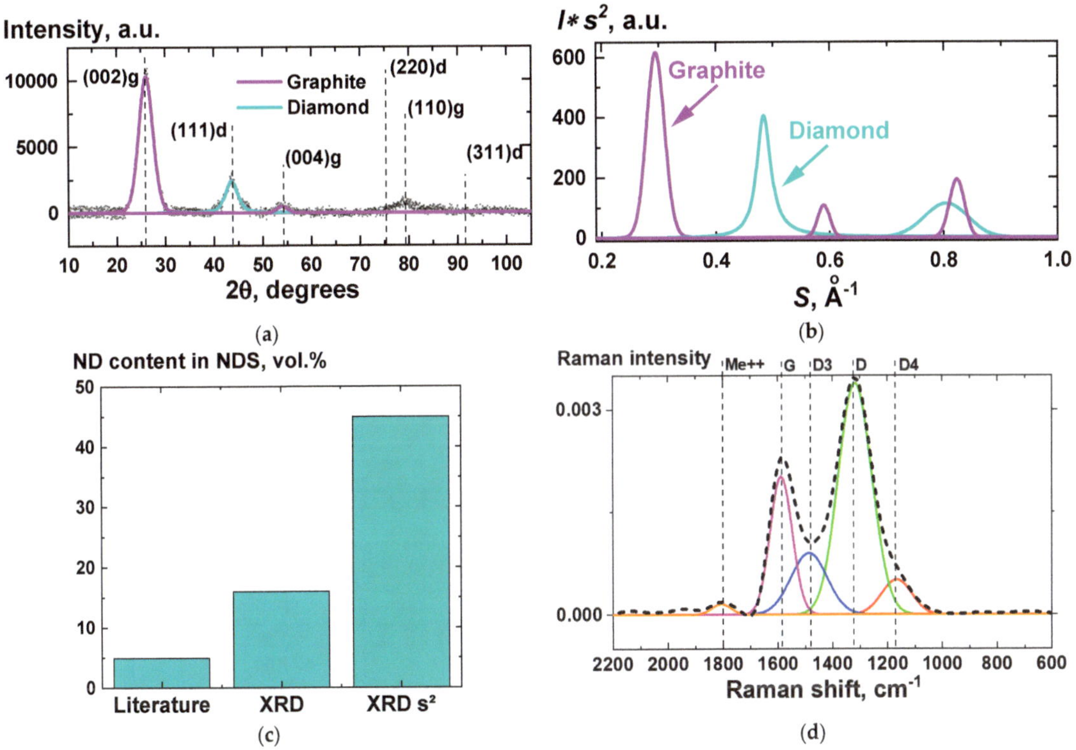

Figure 4. (**a**,**b**)—XRD patterns without (**a**) and with (**b**) implementation of Ruland's method [29]; (**c**)—results of estimation of ND in NDS content value from XRD patterns without ("XRD") and with implementation of Ruland's method ("XRD s2"), and also the value provided by the manufacturer [11] ("Literature"); (**d**)—Raman spectroscopy results for NDS powder. Dotted line (**a**,**d**) represents the obtained experimental data, solid lines (**a**,**b**,**d**)—peaks from decomposition of the experimental data. Dashed vertical lines (**a**,**d**) indicate the centers of the peaks.

From the XRD data, it is possible to estimate the interlayer distance between 002 planes of atoms in the graphite phase. According to the literature data [30], the distance is ~0.35 nm. The experimental value, obtained from the XRD data (Figure 4a,b) turned out to be very close (0.348 ± 0.05) nm.

From XRD data it is also possible to estimate volume fraction of graphite in the NDS powder. The results of the comparison of the obtained value with the literature data [11] are given in Figure 4c. According to the literature, the volume fraction of the NDs in NDS is 4.9 vol.%. Taking into account that this value was obtained using the weighting procedure after purification of NDs, which can lead to a significant decrease of the initial NDs weight value, the real ND content in NDS can be much higher. According to the XRD analysis, the volume fraction of ND in NDS can be estimated as either 16 vol.%, or 45 vol.%, depending if the Ruland's method [29] is implemented (Figure 4b) or not (Figure 4a). No non-carbon impurities were observed according to the XRD data.

As is known from the publications on highly deformed carbon materials [31,32], the Raman scattering bands of the NDS cannot be described by only two bands, G and D. There are always two to three additional bands, which have been labeled according to the literature [31,32] in Figure 4d. The physical meaning of these additional bands is discussed in the relevant publications [31–33].

As the additional methods of NDS agglomerate size distribution and NDS particle structure analysis, AFM, along with STM and the electron diffraction method, were used. Selected results of these methods' implementation are presented in Figure 5.

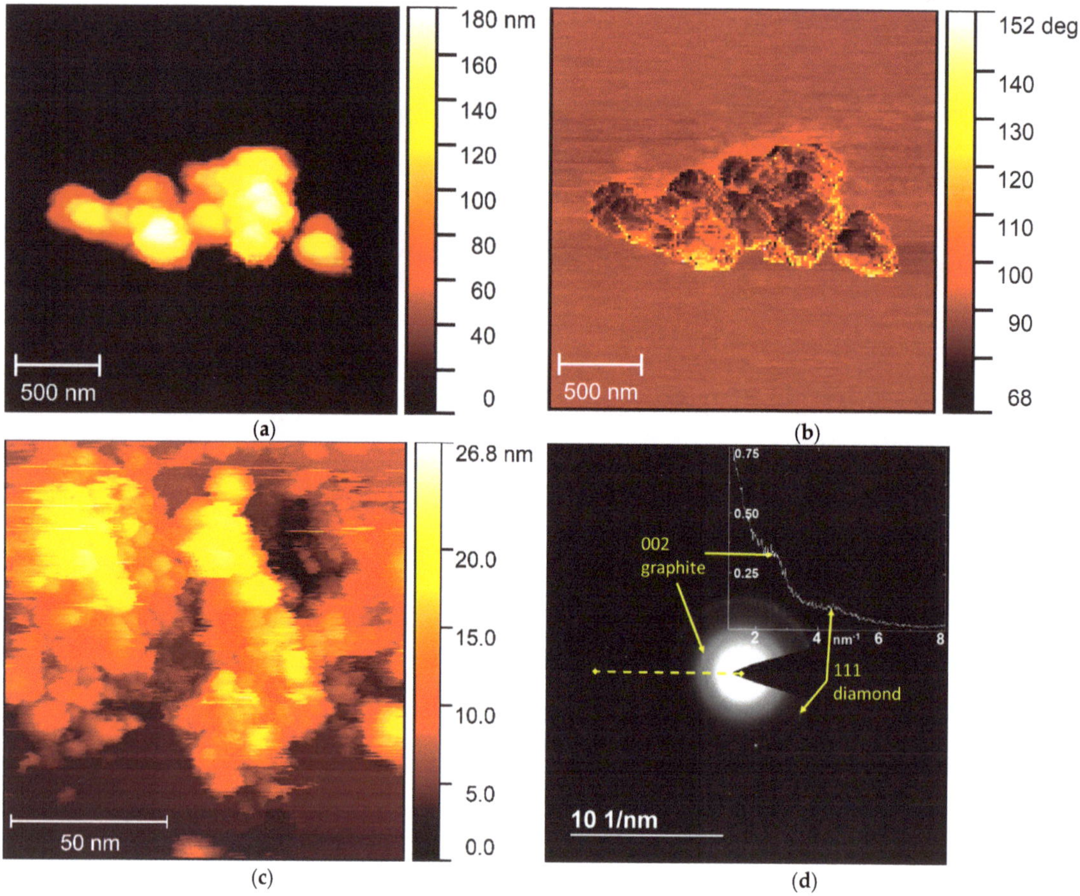

Figure 5. (**a**) Height distribution; and (**b**) phase contrast mode images obtained for NDS powder using AFM in the semi-contact mode; (**c**)—STM results for the NDS powder; (**d**)—electron diffraction results for the NDS powder. The inset (**d**) shows distribution of normalized intensity along the dashed yellow line. The yellow arrows point at the scattering peaks, corresponding to the diamond 111 and graphite 002 planes of carbon atoms.

It can be seen in Figure 5a,b that a large NDS agglomerate with diameter ~1.5 μm comprises small strong-coupled [22] aggregates with diameter of ~250 nm. Thus, it can be said that the AFM results correlate in some way with the SEM (Figure 1), DLS (Figure 2a), and LDA (Figure 2b) data. At the same time, in the STM images (Figure 5c) isolated ND particles can be observed with the diameter ~4–5 nm, which correlates well with the TEM data (Figure 3).

In the electron diffraction pattern (Figure 5d), two characteristic reflexes can be observed, corresponded to the graphite and diamond phases. From the analysis of the graphite reflex, the distance between 002 planes of carbon atoms in graphite can also be evaluated. The obtained value (3.45 $\pm$ 0.1) nm was in the range of error from the value taken from the literature [30] and obtained experimentally from the XRD data (Figure 4a,b).

To evaluate the potential of the NDS powder as a filler for electroconductive polymer composites, the NDS powder resistivity was measured depending on the volume fraction of the nanoparticles. The results of the comparison of resistivity of NDS and CB powders for various volume fraction values are presented in Figure 6a. Additionally, composites, based on PP and containing 10 wt.% NDS and 10 wt.% CB, were subjected to annealing at 205 °C, while the resistivity was constantly measured between the two electrodes in contact with the composite melts. The resulting dependencies of resistivity of the composite melts versus time are presented in Figure 6b.

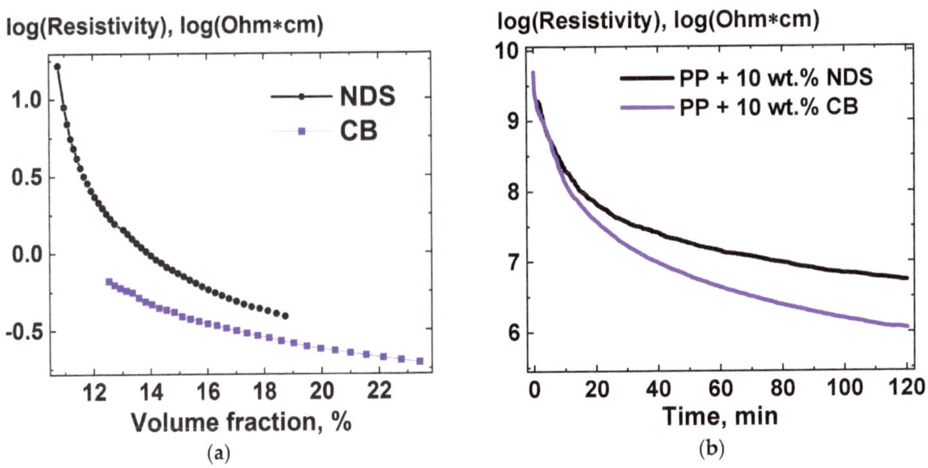

Figure 6. (**a**)—Resistivity of NDS and CB powders versus volume fraction of the powder particles in a cylindric tube, measured between two steel pistons acting as electrodes; (**b**)—resistivity of melts of the composites PP + 10 wt.% NDS and PP + 10 wt.% CB [21], measured between two aluminum electrodes located at the bottom and top of a closed cylindric mold (radius—1.5 cm, height—0.2 cm) at temperature of 205 °C, versus time.

As can be seen in Figure 6, both results for resistivity of the NDS powder and the NDS-containing composite demonstrate similar behavior with volume fraction change or with time if to compare with the corresponding dependencies for CB.

For PP-based composites containing 2.5 and 10 wt.% NDS, dependencies of friction coefficient on time were obtained before and after annealing the composites at 205 °C for 2 h. The averaged friction coefficient values versus time curves for six independent runs are presented in Figure 7a, while the results of averaging of the friction coefficient values versus time of annealing at 205 °C are presented in Figure 7b. The averaging was performed for the values of friction coefficient in the (2–8) h range of tribological test run time.

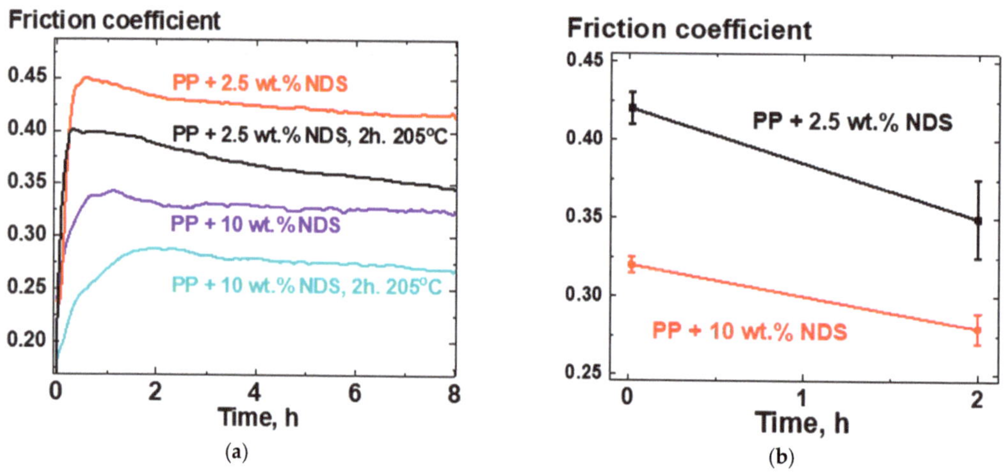

Figure 7. (**a**)—Friction coefficient versus time of a tribological test (steel pin on a rotating composite disk) curves, averaged over six independent test results for each curve, for composites PP + 2.5 wt.% NDS and PP + 10 wt.% NDS before and after keeping the composites for 2 h in a melted state at 205 °C; (**b**)—friction coefficient versus time of heat treatment at 205 °C for composites PP + 2.5 wt.% NDS and PP + 10 wt.% NDS.

4. Discussion

It is known that bonding strength between subunits (primary particles) in an aggregate/agglomerate strongly affects the size of the aggregate/agglomerate determined by different methods, as well the conditions required for the deagglomeration [10,22]. Based on the bonding strength between the subunits, agglomerated particles are known to be roughly categorized into "soft" and "hard" agglomerates. It is not expected that nanoparticles agglomerate size distribution, obtained via DLS method, completely corresponds to the agglomerate size distribution in the polymer matrix in a polymer composite. Moreover, size distribution of nanoparticle agglomerates can be affected by the method of processing used to obtain the polymer composite. However, it is believed that certain structural levels of the NDS, e.g., their hard aggregates comprising of strongly bonded individual nanoparticles, are structurally determined rather by fundamental characteristics and intrinsic properties of the primary particles (material, surface charge and functionality) [10,22] than by an external medium. In this case, specifics of the size distributions of different structural elements for different states of the nanoparticles (powder, dispersion in a liquid medium, filler in a polymer matrix) can be correlated with each other. That is why we believe that the methods used are sufficiently informative for characterization of the NDS aggregates/agglomerates.

From the data obtained using SEM (Figure 1), DLS (Figure 2a), LDA (Figure 2b), and AFM (Figure 5a,b), it was concluded that NDS primary particles tend to aggregate/agglomerate into aggregates/agglomerates of a certain size relatively independently of the medium they are dispersed in, or for how long they are kept in that medium. This feature makes NDS a good model object for the research of properties of composite materials to compare the effect of different polymer matrices on the overall properties of the composites.

The results of TEM (Figure 3), XRD (Figure 4a,b), Raman spectroscopy (Figure 4d), STM (Figure 5c), and electron diffraction (Figure 5d) show that, along with the NDs, a high amount of graphite (Figure 4c) is present in NDS. While ND crystallites of ~4–5 nm gather up in relatively small strong-coupled agglomerates, they are surrounded by graphite in a nanoribbon form. Although from the microscopy data it is possible to roughly estimate the transverse dimensions of the nanoribbons (5–10 nm), estimation of longitudinal dimensions can be quite difficult. At the same time, the microscopy data are sufficient to draw conclusions about the high aspect ratio of such graphite nanoribbons. Due to the high aspect ratio of such a form of graphite, it is possible to achieve very low percolation thresholds for electrical conductivity, as was demonstrated in [16]. At the same time, as the results of XRD and electron diffraction indicate, the graphite phase in NDS is not defective, meaning it can provide decent levels of electrical conductivity even in NDS contain a high fraction of non-conductive NDs. This data agrees with the results of comparison of resistivity for NDS and CB powders for various volume fraction values (Figure 6a), as well as with the results of resistivity measurements with time for composite melts containing NDS and CB. The increase in electrical conductivity of the composite material based on PP is caused by an increase in the concentration of NDS in the near-surface layer when the composite is subjected to thermal treatment at melting temperature [21], as is shown in Figure 6b. High electric conductivity of NDS is a consequence of its structural specifics, such as the presence of graphite phase in the form of nanoribbons.

Results of the tribological tests (Figure 7) allow us to say that using the effect of nanoparticles migration to the boundaries of nanocomposite melt, as was investigated in [21], can be an effective way to improve tribological properties of composite materials filled with NDS particles, specifically due to the presence in NDS of the extremely hard ND cores [12,18,19]. Particularly, it can be seen in Figure 7 that annealing time effectively acted as an increase of the filler content in the composite, which agrees with the conclusions made in [21] on the process of enrichment of a composite melt surface layer with time.

Altogether the results presented in this work demonstrate the versatility and utility of such easily obtainable and cheap filler as NDS. While containing simultaneously hard and chemically resistant ND particles and flexible electrically conductive graphite nanoribbons with a high aspect ratio, NDS has a potential to help meet multiple requirements for a polymer composite by improving different characteristics of the composite simultaneously. This, along with affordability of the filler, can be a very appealing factor for many industries working with polymer composites and looking to find a way to improve the product by substituting such conventional fillers such as CB.

As for future research, implementation of the NDS investigated in this work as a filler used to improve characteristics of ultra-high-molecular-weight polyethylene-based laminates is planned. Additionally, we plan to investigate NDS of different synthesis and study how it affects polymer composites filled with NDS of different types.

Supplementary Materials: The following supporting information can be downloaded at: https://www.mdpi.com/article/10.3390/c8040069/s1.

Author Contributions: Conceptualization, O.V.L. and A.N.O.; methodology, O.V.L., T.S.K., E.K.G., A.L.V., A.K.G., G.P.G. and A.N.O.; validation, O.V.L., T.S.K., E.K.G., A.L.V., A.K.G., G.P.G. and A.N.O.; formal analysis, O.V.L. and A.N.O.; investigation, O.V.L., G.P.G., T.S.K., E.K.G., A.L.V., A.K.G.; resources, A.N.O.; data curation, O.V.L.; writing—original draft preparation, O.V.L.; writing—review and editing, O.V.L. and A.N.O.; visualization, O.V.L.; supervision, A.N.O.; project administration, O.V.L. and A.N.O.; funding acquisition, A.N.O. All authors have read and agreed to the published version of the manuscript.

Funding: The reported study was funded by RSF according to the research project No. 22-13-00359.

Institutional Review Board Statement: Not applicable.

Informed Consent Statement: Not applicable.

Acknowledgments: The authors express their deep gratitude to the team of the Laboratory of the Structure of Polymeric Materials of the ISPM RAS for their attention, cooperation and support at all stages of the work. The authors are also grateful to Dolmatov Valeriy Yurievich for providing them with the NDS powder for the research.

Conflicts of Interest: The authors declare no conflict of interest.

References

1. Dolmatov, V.Y.; Rudenko, D.V.; Dorokhov, A.O.; Malygin, A.A.; Kozlov, A.S.; Marchukov, V.A. Obtaining of Detonation Diamonds from Individual Explosives. *Combust. Explos. Shock Waves* **2021**, *57*, 232–237. [CrossRef]
2. Shvidchenko, A.V.; Eidelman, E.D.; Vul', A.Y.; Kuznetsov, N.M.; Stolyarova, D.Y.; Belousov, S.I.; Chvalun, S.N. Colloids of Detonation Nanodiamond Particles for Advanced Applications. *Adv. Colloid Interface Sci.* **2019**, *268*, 64–81. [CrossRef] [PubMed]
3. Dolmatov, V.Y. Detonation-Synthesis Nanodiamonds: Synthesis, Structure, Properties and Applications. *Russ. Chem. Rev.* **2007**, *76*, 339–360. [CrossRef]
4. Satonkina, N.P.; Ershov, A.P.; Kashkarov, A.O.; Rubtsov, I.A. Elongated Conductive Structures in Detonation Soot of High Explosives. *RSC Adv.* **2020**, *10*, 17620–17626. [CrossRef] [PubMed]
5. Kidalov, S.V.; Shakhov, F.M.; Vul', A.Y.; Ozerin, A.N. Grain-Boundary Heat Conductance in Nanodiamond Composites. *Diam. Relat. Mater.* **2010**, *19*, 976–980. [CrossRef]
6. Kurkin, T.S.; Tikunova, E.P.; Yablokova, M.Y.; Kechek'yan, A.S.; Beshenko, M.A.; Dolmatov, V.Y.; Ozerin, A.N. Effect of a Detonation-Synthesized Nanodiamond Powder on the Adhesive Strength of a Polymer Fiber to an Epoxy Binder. *Dokl. Chem.* **2014**, *457*, 118–121. [CrossRef]
7. Zhang, Y.; Rhee, K.Y.; Hui, D.; Park, S.J. A Critical Review of Nanodiamond Based Nanocomposites: Synthesis, Properties and Applications. *Compos. Part B Eng.* **2018**, *143*, 19–27. [CrossRef]
8. Dolmatov, V.Y. Assessment of Applicability of Explosive Charges for Synthesis of Detonation Nanodiamonds. *J. Superhard Mater.* **2016**, *38*, 373–376. [CrossRef]
9. Ginés, L.; Mandal, S.; Ashek-I-Ahmed; Cheng, C.-L.; Sow, M.; Williams, O.A. Positive Zeta Potential of Nanodiamonds. *Nanoscale* **2017**, *9*, 12549–12555. [CrossRef]
10. Shrestha, S.; Wang, B.; Dutta, P. Nanoparticle Processing: Understanding and Controlling Aggregation. *Adv. Colloid Interface Sci.* **2020**, *279*, 102162. [CrossRef]
11. Dolmatov, V.Y.; Ozerin, A.N.; Kulakova, I.I.; Bochechka, O.O.; Lapchuk, N.M.; Myllymäki, V.; Vehanen, A. Detonation Nanodiamonds: New Aspects in the Theory and Practice of Synthesis, Properties and Applications. *Russ. Chem. Rev.* **2020**, *89*, 1428–1462. [CrossRef]
12. Karami, P.; Salkhi Khasraghi, S.; Hashemi, M.; Rabiei, S.; Shojaei, A. Polymer/Nanodiamond Composites—A Comprehensive Review from Synthesis and Fabrication to Properties and Applications. *Adv. Colloid Interface Sci.* **2019**, *269*, 122–151. [CrossRef] [PubMed]
13. Kidalov, S.V.; Shakhov, F.M.; Vul', A.Y. Thermal Conductivity of Nanocomposites Based on Diamonds and Nanodiamonds. *Diam. Relat. Mater.* **2007**, *16*, 2063–2066. [CrossRef]
14. Houshyar, S.; Sarker, A.; Jadhav, A.; Kumar, G.S.; Bhattacharyya, A.; Nayak, R.; Shanks, R.A.; Saha, T.; Rifai, A.; Padhye, R.; et al. Polypropylene-Nanodiamond Composite for Hernia Mesh. *Mater. Sci. Eng. C* **2020**, *111*, 110780. [CrossRef] [PubMed]
15. Chen, P.; Huang, F.; Yun, S. Characterization of the Condensed Carbon in Detonation Soot. *Carbon N. Y.* **2003**, *41*, 2093–2099. [CrossRef]
16. Alaferdov, A.V.; Lebedev, O.V.; Roggero, U.F.S.; Hernandez-Figueroa, H.E.; Nista, S.V.G.; Trindade, G.M.; Danilov, Y.A.; Ozerin, A.N.; Moshkalev, S.A. Highly Conductive Nanographite/Ultra-High-Molecular-Weight Polyethylene Composite. *Results Mater.* **2022**, *15*, 100298. [CrossRef]
17. Kurkin, T.S.; Ozerin, A.N.; Kechek'yan, A.S.; Gritsenko, O.T.; Ozerina, L.A.; Alkhanishvili, G.G.; Sushchev, V.G.; Dolmatov, V.Y. The Structure and Properties of Polymer Composite Fibers Based on Poly(Vinyl Alcohol) and Nanodiamond of Detonation Synthesis. *Nanotechnologies Russ.* **2010**, *5*, 340–351. [CrossRef]
18. Neitzel, I.; Mochalin, V.; Bares, J.A.; Carpick, R.W.; Erdemir, A.; Gogotsi, Y. Tribological Properties of Nanodiamond-Epoxy Composites. *Tribol. Lett.* **2012**, *47*, 195–202. [CrossRef]
19. Neitzel, I.; Mochalin, V.; Knoke, I.; Palmese, G.R.; Gogotsi, Y. Mechanical Properties of Epoxy Composites with High Contents of Nanodiamond. *Compos. Sci. Technol.* **2011**, *71*, 710–716. [CrossRef]
20. Lebedev, O.V.; Bogdanova, O.I.; Goncharuk, G.P.; Ozerin, A.N. Tribological and Percolation Properties of Polypropylene/Nanodiamond Soot Composites. *Polym. Polym. Compos.* **2020**, *28*, 369–377. [CrossRef]
21. Lebedev, O.V.; Goncharuk, G.P.; Ozerin, A.N. Changes in Electrical Conductance of Polymer Composites Melts Due to Carbon Nanofiller Particles Migration. *Polymers* **2021**, *13*, 1030. [CrossRef] [PubMed]
22. Schaefer, D.W.; Justice, R.S. How Nano Are Nanocomposites? *Macromolecules* **2007**, *40*, 8501–8517. [CrossRef]
23. Pabisch, S.; Feichtenschlager, B.; Kickelbick, G.; Peterlik, H. Effect of Interparticle Interactions on Size Determination of Zirconia and Silica Based Systems—A Comparison of SAXS, DLS, BET, XRD and TEM. *Chem. Phys. Lett.* **2012**, *521*, 91–97. [CrossRef]
24. Ali, H.E.; Radwan, R.R. Synthesis, Characterization and Evaluation of Resveratrol-Loaded Functionalized Carbon Nanotubes as a Novel Delivery System in Radiation Enteropathy. *Eur. J. Pharm. Sci.* **2021**, *167*, 106002. [CrossRef]
25. Sanusi, O.M.; Benelfellah, A.; Papadopoulos, L.; Terzopoulou, Z.; Malletzidou, L.; Vasileiadis, I.G.; Chrissafis, K.; Bikiaris, D.N.; Aït Hocine, N. Influence of Montmorillonite/Carbon Nanotube Hybrid Nanofillers on the Properties of Poly(Lactic Acid). *Appl. Clay Sci.* **2021**, *201*, 105925. [CrossRef]
26. PP H030 GP/3. Available online: http://www.sibur-int.com/upload/documents/TDSH030GP3v14.pdf (accessed on 29 December 2020).

27. P267-E. Available online: https://studylib.ru/doc/2305578/e-lektroprovodnyj-tehnicheskij-uglerod-p-267-e---p-268 (accessed on 29 December 2020).
28. Napper, D.H. *Polymeric Stabilization of Colloidal Dispersions*; Colloid Science: A Series of Monographs. 2; Academic Press: Cambridge, MA, USA, 1983; ISBN 9780125139809.
29. Mogilevskaya, E.L.; Akopova, T.A.; Zelenetskii, A.N.; Ozerin, A.N. The Crystal Structure of Chitin and Chitosan. *Polym. Sci. Ser. A* **2006**, *48*, 116–123. [CrossRef]
30. Calderon, H.A.; Okonkwo, A.; Estrada-Guel, I.; Hadjiev, V.G.; Alvarez-Ramírez, F.; Robles Hernández, F.C. HRTEM Low Dose: The Unfold of the Morphed Graphene, from Amorphous Carbon to Morphed Graphenes. *Adv. Struct. Chem. Imaging* **2016**, *2*, 10. [CrossRef] [PubMed]
31. Sadezky, A.; Muckenhuber, H.; Grothe, H.; Niessner, R.; Pöschl, U. Raman Microspectroscopy of Soot and Related Carbonaceous Materials: Spectral Analysis and Structural Information. *Carbon N. Y.* **2005**, *43*, 1731–1742. [CrossRef]
32. Mermoux, M.; Chang, S.; Girard, H.A.; Arnault, J.-C. Raman Spectroscopy Study of Detonation Nanodiamond. *Diam. Relat. Mater.* **2018**, *87*, 248–260. [CrossRef]
33. Zou, Q.; Li, Y.G.; Zou, L.H.; Wang, M.Z. Characterization of Structures and Surface States of the Nanodiamond Synthesized by Detonation. *Mater. Charact.* **2009**, *60*, 1257–1262. [CrossRef]

C Journal of Carbon Research

MDPI

Article

Laser-Assisted Growth of Carbon-Based Materials by Chemical Vapor Deposition

Abiodun Odusanya [1], Imteaz Rahaman [2], Pallab Kumar Sarkar [2], Abdelrahman Zkria [3], Kartik Ghosh [1] and Ariful Haque [2,4,*]

1 Department of Physics, Astronomy, and Materials Science, Missouri State University, Springfield, MO 65897, USA; abiodun234@live.missouristate.edu (A.O.); kartikghosh@missouristate.edu (K.G.)
2 Electrical Engineering, Ingram School of Engineering, Texas State University, San Marcos, TX 78666, USA; imteaz.rahaman@txstate.edu (I.R.); vlw57@txstate.edu (P.K.S.)
3 Department of Applied Science for Electronics and Materials, Kyushu University, Fukuoka 816-8580, Japan; abdelrahman_zkria@kyudai.jp
4 Materials Science, Engineering, and Commercialization Program, Texas State University, San Marcos, TX 78666, USA
* Correspondence: ahaque@txstate.edu

Citation: Odusanya, A.; Rahaman, I.; Sarkar, P.K.; Zkria, A.; Ghosh, K.; Haque, A. Laser-Assisted Growth of Carbon-Based Materials by Chemical Vapor Deposition. C 2022, 8, 24. https://doi.org/10.3390/c8020024

Academic Editor: Gil Goncalves

Received: 20 February 2022
Accepted: 7 April 2022
Published: 26 April 2022

Abstract: Carbon-based materials (CBMs) such as graphene, carbon nanotubes (CNT), highly ordered pyrolytic graphite (HOPG), and pyrolytic carbon (PyC) have received a great deal of attention in recent years due to their unique electronic, optical, thermal, and mechanical properties. CBMs have been grown using a variety of processes, including mechanical exfoliation, pulsed laser deposition (PLD), and chemical vapor deposition (CVD). Mechanical exfoliation creates materials that are irregularly formed and tiny in size. On the other hand, the practicality of the PLD approach for large-area high-quality CMB deposition is quite difficult. Thus, CVD is considered as the most effective method for growing CBMs. In this paper, a novel pulsed laser-assisted chemical vapor deposition (LCVD) technique was explored to determine ways to reduce the energy requirements to produce high quality CBMs. Different growth parameters, such as gas flow rate, temperature, laser energy, and deposition time were considered and studied thoroughly to analyze the growth pattern. CBMs are grown on Si and Cu substrates, where we find better quality CBM films on Cu as it aids the surface solubility of carbon. Raman spectroscopy confirms the presence of high-quality PyC which is grown at a temperature of 750 °C, CH_4 gas flow rate of 20 sccm, a laser frequency of 10 Hz, and an energy density of 0.116 J/cm^2 per pulse. It is found that the local pulsed-laser bombardment helps in breaking the carbon-hydrogen bonds of CH_4 at a much lower substrate temperature than its thermal decomposition temperature. There is no significant change in the 2D peak intensity in the Raman spectrum with the further increase in temperature which is the indicator of the number of the graphene layer. The intertwined graphene flakes of the PyC are observed due to the surface roughness, which is responsible for the quenching in the Raman 2D signal. These results will provide the platform to fabricate a large area single layer of graphene, including the other 2D materials, on different substrates using the LCVD technique.

Keywords: CBMs; laser; graphene; 2D materials; LCVD

1. Introduction

Over the last few decades, CBMs have been studied thoroughly in order to gain a better understanding of their adsorption, processes, and isotherms [1]. They are the most widely used electrodes due to their low cost, variety of forms (powders, fibers, aerogels, composites, sheets, monoliths, tubes, etc.), ease of processing, relatively inert electrochemistry, and controllable porosity [2,3]. Activated carbon (AC), graphene, carbon nanotubes (CNTs), carbon nanofibers (CNFs), biochar (BC), and carbon aerogels (CAs) are all examples of carbon-based materials. Graphene nanomaterials (GFNs) are less toxic than other carbon-based nanomaterials. Their highly flexible physicochemical properties allow the molecule's various functional groups to be altered and perform specific functions [4]. Its sp^2 hybridized atoms are held together in a hexagonal honeycomb lattice, and each carbon atom is tightly bound to its three neighboring atoms which can be wrapped into a CNT. The authors present a summary of the latest studies on laser-assisted synthesis of graphene-based materials, along with their development and use as electrodes in supercapacitor and battery applications [5].

Graphene can be synthesized by mechanical exfoliation, liquid-phase exfoliation, reduction of graphene oxide, epitaxial growth and chemical vapor deposition techniques [6–10]. Sample preparation, substrate, reactor dimension, reactor temperature, reactor pressure, gas flow rates, sample position, annealing condition, growing condition, and cooling condition are the critical factors for the growth of graphene by CVD technique [11,12]. Conventionally the hydrothermal and heat treatment methods are suitable for nanoparticle synthesis. The hydrothermal method also takes a lot of time, and the process usually consists of multistage processing steps which is not suitable for mass-scale production [13,14]. The LCVD is a unique CVD technique where the global heat source for the furnace is replaced with a localized laser-heated spot. There are mainly two types of LCVD: pyrolytic and photolytic [15]. The pyrolytic approach is very popular nowadays for producing carbon nanomaterials (CNMs). The fabrication of advanced application materials such as CNTs is the most common use of the pyrolysis process. Usually, it involves a two-stage process: firstly, the precursors for nanomaterials are generated by pyrolysis, while in the second stage, these precursors are usually deposited on nickel-iron and cobalt [16]. The LCVD differs from conventional CVD where the growth area can be limited to the area through which the laser beam passes. Low temperature, shorter reaction times, environmental friendliness, energy savings, catalyst-free growth on insulating substrates, high productivity, improved reproducibility, scalability, and excellent control over experimental parameters are the advantages of laser-based techniques compared to the other conventional methods [17]. The CVD technique is commonly used to make pyrolytic carbons. The PyC structure is a graphite-like, poorly crystalline pure elemental carbon structure. Over the last few decades, pyrolytic carbon has opened a wide scope of new applications including electrochemical measurements, conductive vias through dielectric, carbon-silicon Schottky-boundary diodes and safeguarding of electromagnetic radiation in a wide spectral range.

PyCs are made using the CVD process by heating a hydrocarbon at a high temperature range between 1200 and 1400 °C in the absence of oxygen [18]. The fluidized bed chemical vapor deposition (FBCVD) technique is used to produce PyC even at a higher temperature range between 1250 and 1450 °C. As the deposition temperature increases, the density of PyC drops. In these studies, the researchers were unable to reduce the high temperature requirement in the production of PyCs, and they also failed to account for the effect of deposition time [19]. PyC thin films were produced on SiO_2 substrates by CVD at 950 °C under 20 Torr pressure for 30 min. Temperature, pressure, and deposition time were tuned to adjust the thickness and roughness, but this research was conducted only on one type of substrate. Moreover, the deposition time is long which does not help with the reduction in thermal budget. The effect of different substrates was not analyzed [20]. In another study, Hu et al. made PyC at a very high temperature of around 1100 °C. However, the impact of varying deposition time and CH_4 flow rate is not addressed during the formation of CBMs [21]. To investigate the effect of temperature, CBMs were grown at three different system temperatures, i.e., 927, 1527, and 2127 °C. High temperatures obviously hasten the formation of carbon structures. However, the temperature of deposited material is exceptionally high [22]. Barberio et al. presented a laser-plasma-driven approach for producing carbon-based nanomaterials, specifically bi- and few-layer graphene. After 10 s of laser irradiation, the number of layers gradually increases from one to two, and after 40 s, the material transforms into graphite. However, the effect of deposition temperature is not considered here [23]. The pyrolysis of hydrocarbons is a complex process with a large number of different reaction pathways. Ali et al. produced PyC on Si by CVD from gaseous hydrocarbon precursors at about 1000 °C which consists of small, lamellar graphitic ribbons with the size of several nanometers and relatively small amounts of amorphous carbon [24]. Another study shows that Ni–Cu alloy compared to Si is more suitable at low temperatures due to its catalytic nature, low carbon solubility, and more uniform grain size [25]. The film obtained by the proposed method exhibited excellent uniformity and a high monolayer ratio [26]. Only a few studies have been undertaken to reduce the deposition temperature of the CVD process for the growth of carbon-based materials such as graphene. Using thermal CVD, graphene is typically grown between 1000 and 1050 °C [27]. Laser CVD (LCVD) has a lot of promise for local graphene manufacturing using carbon-based materials [28]. The use of a pulsed laser as a secondary source of energy could be a viable option for lowering the deposition temperature and thereby cost to grow CBMs using the CVD technique [29].

To address the aforementioned problems of high energy requirements for CBM production, we studied the growth of CBMs at a low thermal budget by constructing the LCVD system that includes a laser source, mass flow controllers (MFCs), a gas source, a resistant heater, and a vacuum system. Si and Cu substrates are considered in this research to determine the effect of the substrate on PyC growth and to obtain better quality CBM films. In a word, the effects of the deposition temperature, CH_4 flow rate, and laser frequency on the growth of CBMs were studied in this research paper to determine the optimum growth conditions.

2. Experimental Setup

2.1. *Nd:YAG Laser*

A Nd:YAG laser was used for this research work. The Nd:YAG laser system has four energy levels and can be built in both pulsed and continuous modes. A Xenon or Krypton flash tube is employed as a pumping source [30]. The pulse duration of this laser is 5–7 ns. The principal components of this system are: a traditional pulse laser deposition (PLD) chamber, a solid-state Nd:YAG laser of wavelength 266 nm, mass flow controllers (MFC), a roughing and turbomolecular pump, a baratron gauge, a bourdon gauge, a heater, a temperature controller, a UV mirror, a network of gas supply lines, and the precursor gases (mixture of H_2 and Ar gas). For this experiment, two systems were designed as the primary sources of heating: the first one is the use of resistive heating in the chamber and the second one is inductive heating using a quartz tube in an induction furnace. To initiate a deposition, the substrates are placed in the deposition chambers, and it is pumped down from atmospheric pressure to about 5×10^{-5} Torr. Pulsed laser beam incidents on the copper and silicon substrates occurred with the aid of a 248 nm UV mirror for different deposition times of 1–5 min. Figure 1 depicts the deposition process in the chamber.

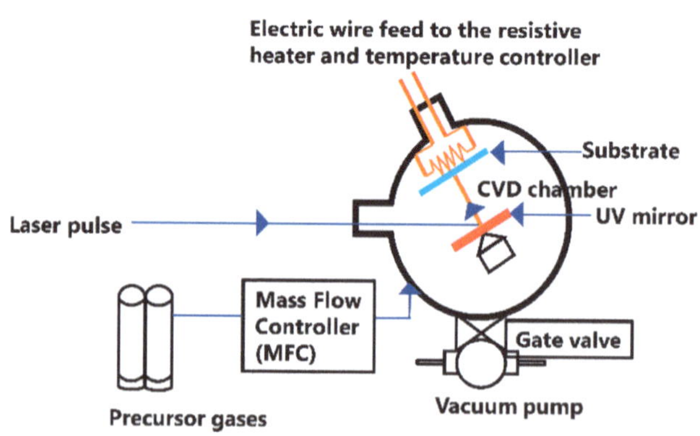

Figure 1. Schematic diagram of the deposition process in the chamber.

2.2. Characterization: Raman Spectroscopy

In this research, the Raman spectrum was produced using a Horiba Labram Raman-PL system at room temperature. As an excitation source, a green laser with a wavelength of 532 nm was used on the sample.

3. Result Analysis

In this study, silicon dioxide wafer ($Si_{x,y}$) and copper foil ($Cu_{x,y}$) were used as substrates where x represents the sample number. Here, y is '0' when there is no laser irradiation and '1' when there is laser irradiation. The deposition pressure was maintained at 550 Torr, Ar/H_2 flow rate was set to 10 sccm, and the energy density was calculated to be 0.116 J/cm^2 per pulse. In this research, the CH_4 flow rate, temperature, substrate, and time were mainly varied to observe the overall effects of the process parameter.

3.1. Variation in CH_4 Flow Rate and Temperature

The deposition is undertaken with two different flow rates of CH_4: 10 and 20 sccm. Here, the deposition temperature is varied between 650 and 850 °C and the effect of temperature change is observed. Before we proceed, the Raman spectra for commercial grade graphene obtained from the Airforce Research Laboratory are shown in Figure 2.

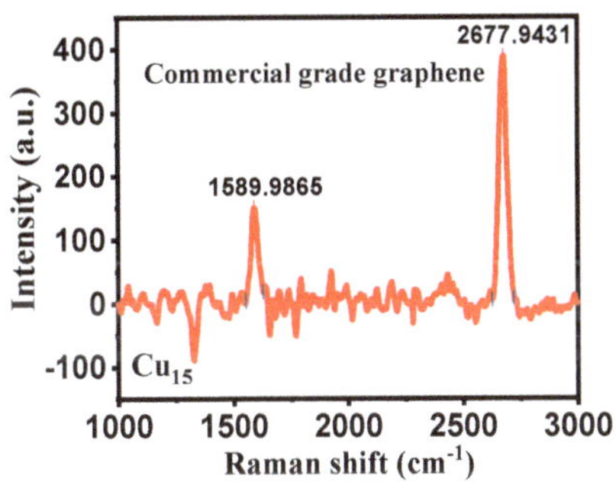

Figure 2. Raman spectra of pristine high−quality graphene.

3.1.1. Constant CH_4 Flow Rate at 10 sccm and Varied Temperature

At first, we kept a constant CH_4 flow rate of 10 sccm and varied the temperature between 650 and 850 °C using copper substrates. The laser frequency was likewise changed between 5 and 10 Hz, with a 5 min deposition duration. Figure 3 shows the Raman spectra considering all these parameters.

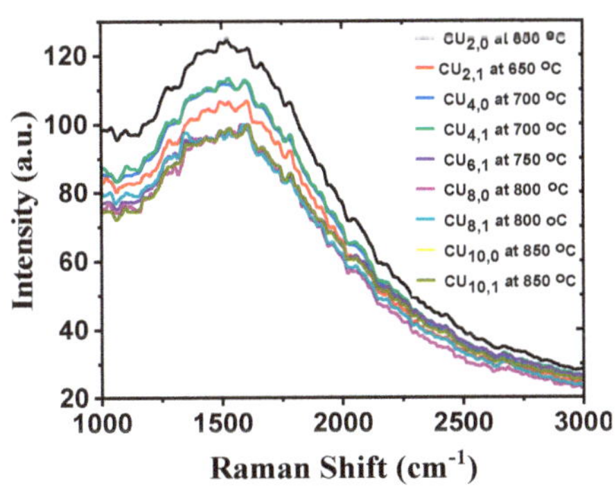

Figure 3. Raman spectra at a constant CH_4 flow rate at 10 sccm and different deposition temperatures.

It is observed that there are no distinguishable peaks to indicate crystalline carbon growth. According to Grove's model, the mass transfer of reactant gaseous species is solely dependent on mass diffusion. As a result, there is a concentration gradient of gaseous species, and the flux of mass transport from the gas phase is inadequate to activate the necessary reaction at the substrate surface. This indicates that the given flow rate of 10 sccm is insufficient for deposition to occur. Furthermore, the requisite temperature to thermally break the carbon–hydrogen bonds (413 kJ/mol) at the required dehydrogenation energy is not met.

3.1.2. Constant CH_4 Flow Rate at 20 sccm and at Different Temperatures

Figure 4 shows the Raman spectra graphs for copper and silicon substrates, respectively. At the same time, we have adjusted the temperature while maintaining a constant CH_4 flow rate of 20 sccm and a laser frequency of 10 Hz.

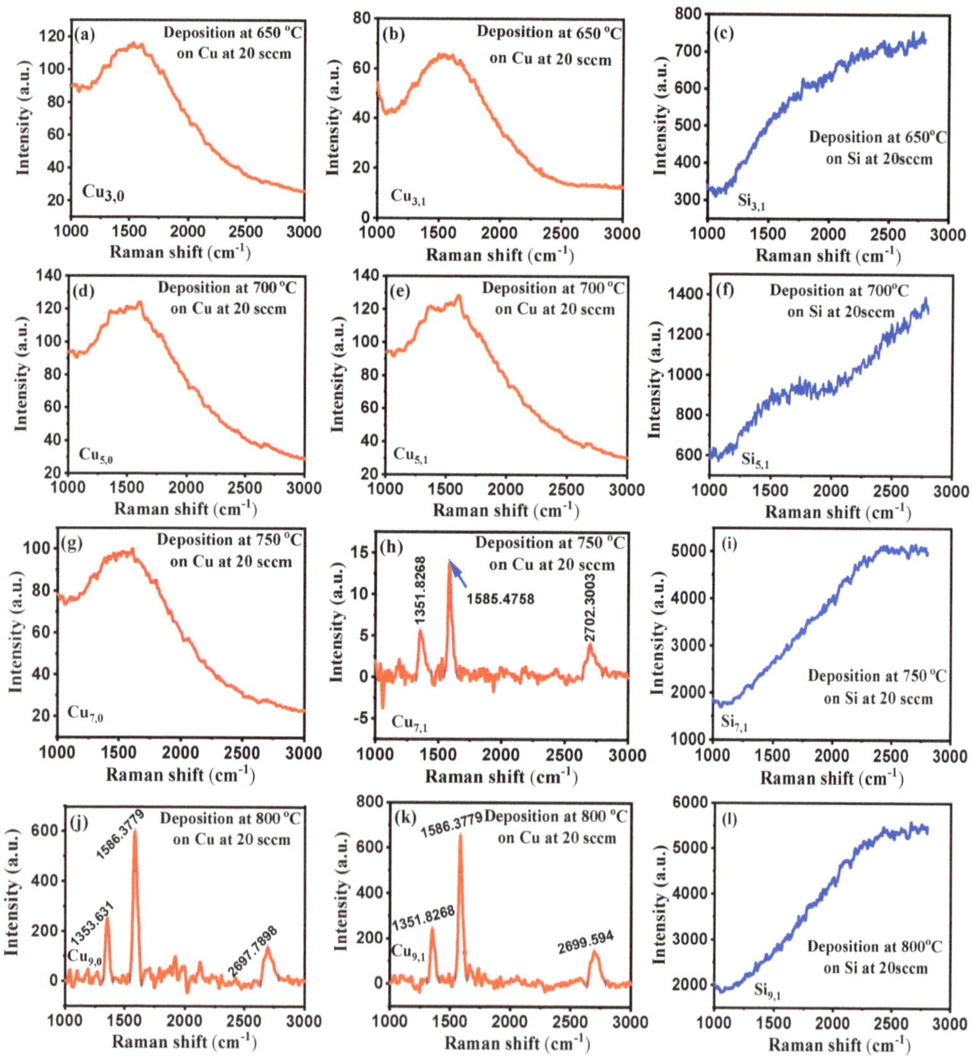

Figure 4. *Cont.*

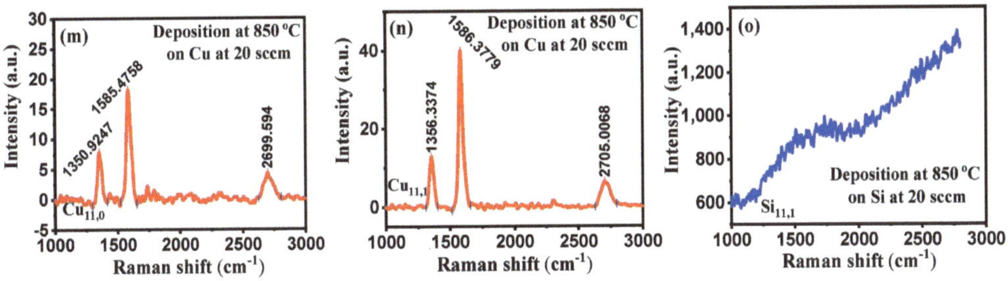

Figure 4. (**a–o**) Raman spectra at constant CH_4 flow rate at 20 sccm with different deposition temperatures on Cu and Si substrates.

As the CH_4 flow rate of 10 sccm is too low for the deposition, we increased the CH_4 flow rate to 20 sccm and repeated the experiment. From Figure 4, it is observed that SiO_2 substrate has no distinguishable peak to ensure the deposition of any type of CBM in crystalline form. Similarly, there are no identifiable peaks for the copper substrate at 650 °C, but a very slight one is observed on the film grown at 700 °C. However, we found three peaks at 1351.8, 1585.5, and 2702.3 cm^{-1} on the laser incident part of our sample at 750 °C, which are D, G, and 2D peaks, respectively. These are the characteristic peaks for carbon-based materials such as graphene. We further raised the temperature of the resistive heater to 850 °C at an increment rate of 50 °C and found CBMs in both laser incident and non-incident areas. The Raman shifts for the D, G, and 2D peaks were all in the same ballpark. This means that at 20 sccm, we were able to commence the reaction by generating enough mass diffusion and flux from the gas phase and the substrate surface. Furthermore, the pulsed laser photons effectively raised the local temperature of the bombarded region. Under these deposition conditions, our CBMs were highly defective due to the predominance of the D peak intensity compared to pristine graphene.

3.2. Variation in Deposition Time

Here, we varied the deposition time and kept other parameters constant as shown in Table 1 and Figure 5a–f.

Table 1. Experimental parameters of different deposition times with constant CH_4 flow rate at 20 sccm and 850 °C deposition temperature.

Sample ID	Process Parameters			
	Temp (°C)	Laser Pulse Frequency (Hz)	CH_4 Flow Rate (sccm)	Deposition Time (min)
$Cu_{11,0}$	850	10	20	5
$Cu_{11,1}$	850	10	20	5
$Cu_{12,0}$	850	10	20	3
$Cu_{12,1}$	850	10	20	3
$Cu_{13,0}$	850	10	20	1
$Cu_{13,1}$	850	10	20	1

Our main goal was to shorten the CBM deposition time and sharpen the 2D peak by reducing the number of layers. The deposition time was reduced from 5 to 3 and then to 1 min, but there was no significant improvement in the 2D peak. From this, it is concluded that it might not be graphene but rather a graphene-like material. It was observed that the deposited substrate was PyC. It is produced by heating a hydrocarbon (CH_4) nearly to its thermal decomposition temperature of around 1000–1200 °C. This thermal decomposition temperature breaks its bonds, releasing carbon-free radicals and allowing graphite to crystalize in the absence of oxygen (pyrolysis) [31]. The surface roughness of PyC films does not vary significantly with film thickness [32], and it is also responsible for quenching the 2D signal in samples $Cu_{11,1}$, $Cu_{12,1}$, and $Cu_{13,1}$, regardless of the deposition time.

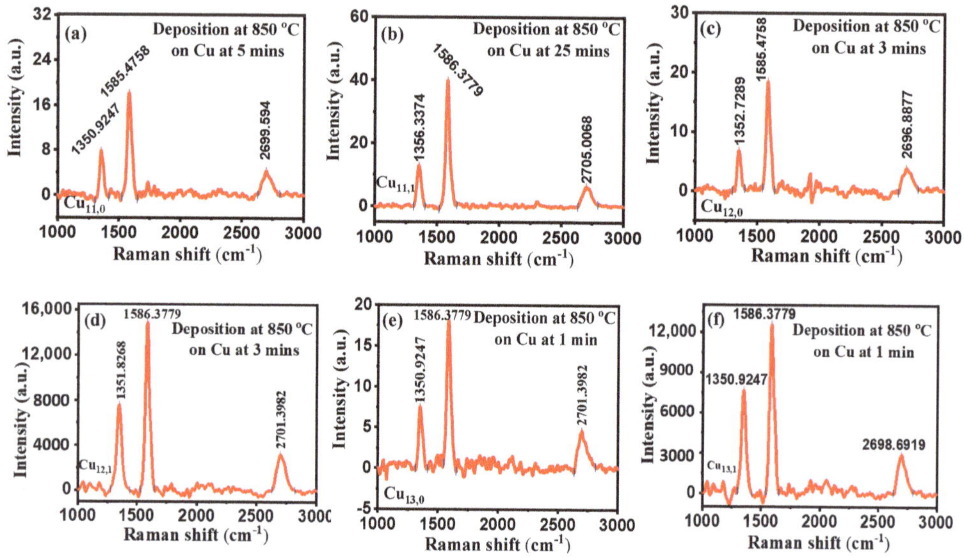

Figure 5. (**a**–**f**) Raman spectra at different deposition times with a constant CH_4 flow rate of 20 sccm and 850 °C deposition temperature.

It was observed that our CBMs were highly defective because of the prominence of the intensity of the D peak when compared with high quality graphene in Figure 2. Sample Cu_{15} is the commercial-grade graphene that we purchased and used as the gold standard for our deposition. Here, we can see that it is a single layer graphene with zero defects. It can be seen from the D peak that sample $Cu_{13,1}$ (Figure 5f) has the highest defect concentration, the defect data looks rather inconclusive, and more research needs to be undertaken on this; the same goes for the carbon concentration shown from the G peak.

3.3. Effects of Laser Energy on PyC Deposition

Here, we compared two depositions, one grown by purely thermal CVD and the other by LCVD. The process parameters are shown in Table 2, and the Raman data plots are shown in Figure 6a–c below.

Table 2. Effect of laser energy on PyC deposition.

Sample ID		Process Parameters			
		Temp (°C)	Laser Pulse Frequency (Hz)	CH_4 Flow Rate (sccm)	Deposition Time (min)
$Si_{11,1}$	$Cu_{11,0}$	850	10	20	5
	$Cu_{11,1}$	850	10	20	5
	$Cu_{14,0}$	850	0	20	5

Using samples Cu_{11} and Cu_{14}, we evaluated the effect of a pulsed laser on the formation of PyC. In both the laser and non-laser bombarded regions of sample Cu_{11}, we found the D, G, and 2D peaks. This implies that the laser shots generate a local laser bombardment region and act as a complementary energy source, which raises the local temperature of the laser spot. This raise in temperature provides the additional energy required to initiate the reaction and increases the deposition area. On the other hand, the thermal energy supplied by the resistive heater was insufficient to initiate the PyC deposition on Cu_{14} sample.

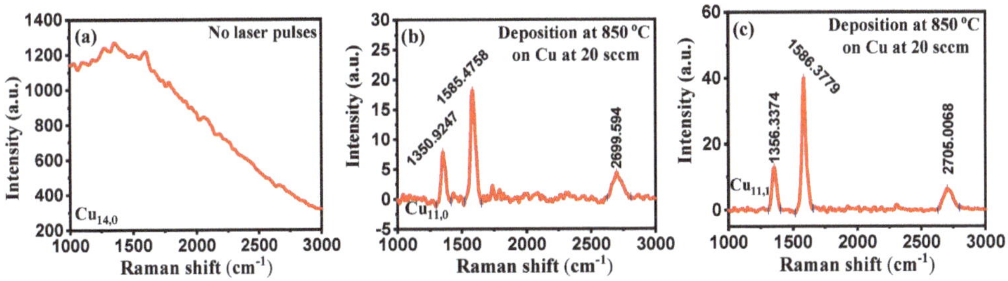

Figure 6. (**a–c**) Raman data plots showing the effect of: (**a**) no laser pulses; (**b**) laser incident spot; (**c**) laser non-incident spot on PyC deposition.

3.4. Effects of the Different Substrates on PyC Deposition

The performance of silicon and copper substrates in PyC deposition was compared at 850 °C with process parameters set to 10 Hz, CH_4 flow rate of 20 sccm, and deposition time of 5 min. The Raman plots are shown, considering all these conditions, in Figure 4 (n, o). Although there is no discernible peak in the Raman spectra for the silicon substrate, which indicates the presence of PyC, the bump-like shape of the graph indicates the presence of amorphous carbon. On the other hand, copper substrate exhibits crystalline sp^2 hybridization in the form of PyC which is ensured by the D, G, and 2D peaks in the Raman spectrum. As a result, the copper substrate acts as a catalyst for PyC deposition by aiding the surface solubility of carbon on copper.

3.5. Reaction Mechanism for PyC Deposition

Taking a cue from Gajewski, Grzegorz, and Pao, Chun-Wei [33], we investigated the chemical pathways for CH_4 decomposition and dehydrogenation on the copper substrate surface. Multiple ground and transition states were used in this study since the dehydrogenation process is a multi-step process with a series of intermediates. This reaction had six ground states (GS) and four transitional stages (TS), as indicated below:

GS1–CH_4
GS2–CH_3 + H
GS3–CH_2 + 2H
GS4–CH + 3H
GS5–C + 4H
GS6–1/2C_2 + 4H
(TS1): $CH_4 \rightarrow CH_3 + H$ (ΔH = +0.75 eV; E_{act} = +1.57 eV)
(TS2): $CH_3 \rightarrow CH_2 + H$ (ΔH = +0.83 eV; E_{act} = +1.36 eV)
(TS3): $CH_2 \rightarrow CH + H$ (ΔH = +0.41 eV; E_{act} = +0.94 eV)
(TS4): $CH \rightarrow C + H$ (ΔH = +1.22 eV; E_{act} = +1.84 eV)

In the gas phase, the activation barriers for CH_4 decomposition are substantially larger than on the surface of the copper substrate. All the reactions are endothermic except for the formation of carbon. The entire reaction is described as follows:

$CH_4(g) \rightarrow C(s) + 4H$ (ΔH = +3.20 eV)

3.6. Quantitative Analysis of Results

The number of layers can be estimated by dividing the peak intensities of D and G or 2D and G [34]. The number of layers can also be calculated using the ratio of the areas of the individual peaks. The estimated ratio of intensities by the number of graphene layers is shown in Table 3.

Table 3. The ratio of peak intensities and number of layers.

Number of Graphene Layers	Approximate Peak Intensities	
Single layer	$\frac{I_D}{I_G} = 0$	$\frac{I_{2D}}{I_G} = 2$
Double layer	$\frac{I_D}{I_G} = 0.05$	$\frac{I_{2D}}{I_G} = 1$
Few layers	$\frac{I_D}{I_G} = 0.1$	$\frac{I_{2D}}{I_G} = 0.8$
Multi-layer	$\frac{I_D}{I_G} = 0.18$	$\frac{I_{2D}}{I_G} = 0.5$

Table 4 below shows the peak positions, intensity, and the number of layers obtained from each deposition from Raman analyses.

Table 4. Peak positions, intensities, and the number of layers from the Raman signal.

ID	Peak Position			Peak Intensity			$\frac{I_{2D}}{I_G}$	Peak Area			$\frac{A_{2D}}{A_G}$
	D	G	2D	D	G	2D		D	G	2D	
$Cu_{7,1}$	1351	1585	2702	5.5	14.3	4.2	0.3	267.1	548.4	377.4	0.7
$Cu_{9,0}$	1352	1586	2697	237.3	612.9	142.3	0.2	8828.0	27,855.0	11,766.2	0.4
$Cu_{9,1}$	1351	1586	2699	234.2	621.5	155.2	0.3	9723.4	24,434.8	13,616.0	0.6
$Cu_{11,0}$	1350	1585	2699	7.8	18.3	4.2	0.2	294.9	822.5	330.7	0.4
$Cu_{11,1}$	1354	1586	2704	9.1	26.0	5.1	0.2	451.4	1157.6	401.9	0.4
$Cu_{12,0}$	1351	1585	2697	6.9	18.3	3.8	0.2	263.6	752.8	350.1	0.5
$Cu_{12,1}$	1352	1586	2701	6858	14,769	3103	0.2	320,096.0	644,120.2	249,714.2	0.4
$Cu_{13,0}$	1350	1586	2701	8.6	18.1	4.5	0.3	427.3	792.3	365.5	0.5
$Cu_{13,1}$	1350	1586	2698	7607	12,899	3008	0.2	366,476	633,137	250,835	0.4
Cu_{15}	0	1589	2677	0	140.6	375.4	2.7	0	5129.9	15,718.5	3.1

We can conclude from Table 4 that all grown samples had many graphene flakes based on the intensity ratios. Instead of looking at individual peaks, the same information can be obtained by looking at the ratio of the areas. The Cu_{15} is a commercial-grade graphene sample that we bought and used as a gold standard for deposition. It is identified that graphene is a single layer with no imperfections. The D peak shows that sample $Cu_{13,1}$ has the highest defect concentration; however, the defect data appears to be equivocal, and more investigation is needed. Additionally, the carbon concentration represented by the G peak is also the same.

4. Conclusions

In this paper, the impacts of laser irradiation on the deposition of pyrolytic carbon using optimal deposition parameters on the copper and silicon dioxide substrates were systematically investigated. Copper foil was found to be the more suitable substrate to grow PyC than silicon dioxide wafer for our experimental setup because copper aids the surface solubility of carbon. From Raman data, we concluded that the PyC shares similar characteristics with graphene because of the coinciding Raman shifts of the D, G, and 2D peaks at 1350, 1600, and 2700 cm^{-1}, respectively. Furthermore, the PyC is randomly oriented and intertwined with graphene flakes produced by heating a hydrocarbon nearly to its thermal decomposition temperature. Pyrolytic carbon is relatively more defect-ridden than high-quality graphene because of the intertwined nature of its constituent graphene flakes. Despite having graphene as its fundamental building block, the intertwining of the graphene flakes causes some surface roughness that does not vary significantly with film thickness, and it is also responsible for quenching the 2D peak on the Raman spectra regardless of deposition time, as seen in samples $Cu_{11,1}$, $Cu_{12,1}$, and $Cu_{13,1}$.

Author Contributions: A.O., I.R. and P.K.S. contributed equally to this work. Data curation A.O., I.R. and P.K.S. Conceptualization, A.O. and I.R. Formal analysis I.R. and P.K.S. Methodology A.Z., Project Administration A.H. and K.G. Resources and supervision A.H. Validation A.Z. Writing—original draft A.O., I.R. and P.K.S. All authors have read and agreed to the published version of the manuscript.

Funding: This work was supported by the NSF Grant (DMR-08211593).

Institutional Review Board Statement: Not applicable.

Informed Consent Statement: Not applicable.

Data Availability Statement: Not applicable.

Acknowledgments: We would like to acknowledge Rishi Patel, senior research scientist, Jordan Valley Innovation Center (JVIC), Springfield, MO 65806, USA, for his feedback and the useful discussions.

Conflicts of Interest: On behalf of all authors, the corresponding author states that there is no conflict of interest.

References

1. Sabzehmeidani, M.M.; Mahnaee, S.; Ghaedi, M.; Heidari, H.; Roy, V.A.L. Carbon based materials: A review of adsorbents for inorganic and organic compounds. *Mater. Adv.* **2021**, *2*, 598–627. [CrossRef]
2. Frackowiak, E. Carbon materials for supercapacitor application. *Phys. Chem. Chem. Phys.* **2007**, *9*, 1774–1785. [CrossRef]
3. Pandolfo, A.; Hollenkamp, A.F. Carbon properties and their role in supercapacitors. *J. Power Sources* **2006**, *157*, 11–27. [CrossRef]
4. McCallion, C.; Burthem, J.; Rees-Unwin, K.; Golovanov, A.; Pluen, A. Graphene in therapeutics delivery: Problems, solutions and future opportunities. *Eur. J. Pharm. Biopharm.* **2016**, *104*, 235–250. [CrossRef]
5. Kumar, R.; del Pino, A.P.; Sahoo, S.; Singh, R.K.; Tan, W.K.; Kar, K.K.; Matsuda, A.; Joanni, E. Laser processing of graphene and related materials for energy storage: State of the art and future prospects. *Prog. Energy Combust. Sci.* **2022**, 100981. [CrossRef]
6. Yazdi, G.R.; Iakimov, T.; Yakimova, R. Epitaxial Graphene on SiC: A Review of Growth and Characterization. *Crystals* **2016**, *6*, 53. [CrossRef]
7. Bourlinos, A.B.; Georgakilas, V.; Zboril, R.; Steriotis, T.A.; Stubos, A.K. Liquid-Phase Exfoliation of Graphite Towards Solubilized Graphenes. *Small* **2009**, *5*, 1841–1845. [CrossRef]
8. Xu, Y.; Cao, H.; Xue, Y.; Li, B.; Cai, W. Liquid-Phase Exfoliation of Graphene: An Overview on Exfoliation Media, Techniques, and Challenges. *Nanomaterials* **2018**, *8*, 942. [CrossRef]
9. Zhang, J.; Yang, H.; Shen, G.; Cheng, P.; Zhang, J.; Guo, S. Reduction of graphene oxide vial-ascorbic acid. *Chem. Commun.* **2010**, *46*, 1112–1114. [CrossRef]
10. Bhuyan, A.; Uddin, N.; Islam, M.; Bipasha, A.; Hossain, S. Synthesis of graphene. *Int. Nano Lett.* **2016**, *6*, 65–83. [CrossRef]
11. Fidanova, T.; Petrov, S.; Napoleonov, B.; Marinova, V.; Petrova, D.; Rafailov, P.; Lin, S.H.; Dimitrov, D. Single and multilayer graphene grown by CVD technique: Characterization for electro-optical applications. In Proceedings of the AIP Conference Proceedings, Sofia, Bulgaria, 26 February 2019; p. 020017. [CrossRef]
12. Wang, J.; Ren, Z.; Hou, Y.; Yan, X.-L.; Liu, P.-Z.; Zhang, H.; Zhang, H.; Guo, J. A review of graphene synthesis at low temperatures by CVD methods. *New Carbon Mater.* **2020**, *35*, 193–208. [CrossRef]
13. Ren, J.; Zhao, D.; Qi, F.; Liu, W.; Chen, Y. Heat and hydrothermal treatments on the microstructure evolution and mechanical properties of plasma sprayed hydroxyapatite coatings reinforced with graphene nanoplatelets. *J. Mech. Behav. Biomed. Mater.* **2020**, *101*, 103418. [CrossRef]
14. Li, C.; Yang, Z.; Tang, Z.; Guo, B.; Tian, M.; Zhang, L. A scalable strategy for constructing three-dimensional segregated graphene network in polymer via hydrothermal self-assembly. *Chem. Eng. J.* **2019**, *363*, 300–308. [CrossRef]
15. Duty, C.; Jean, D.; Lackey, W. Laser chemical vapour deposition: Materials, modelling, and process control. *Int. Mater. Rev.* **2001**, *46*, 271–287. [CrossRef]
16. Zahid, M.U.; Pervaiz, E.; Hussain, A.; Shahzad, M.I.; Niazi, M.B.K. Synthesis of carbon nanomaterials from different pyrolysis techniques: A review. *Mater. Res. Express* **2018**, *5*, 052002. [CrossRef]
17. Collins, P.; Avouris, P. Nanotubes for electronics. *Sci. Am.* **2000**, *283*, 62–69. [CrossRef]
18. Ross, M.; James, C.; Couzens, G.; Klawitter, J. Pyrocarbon small joint arthroplasty of the extremities. In *Joint Replacement Technology*; Elsevier: Amsterdam, The Netherlands, 2014; pp. 628–673.
19. López-Honorato, E.; Meadows, P.J.; Xiao, P. Fluidized bed chemical vapor deposition of pyrolytic carbon–I. Effect of deposition conditions on microstructure. *Carbon* **2009**, *47*, 396–410. [CrossRef]
20. McEvoy, N.; Peltekis, N.; Kumar, S.; Rezvani, E.; Nolan, H.; Keeley, G.P.; Blau, W.; Duesberg, G.S. Synthesis and analysis of thin conducting pyrolytic carbon films. *Carbon* **2012**, *50*, 1216–1226. [CrossRef]
21. Hu, Z.J.; Zhang, W.G.; Hüttinger, K.J.; Reznik, B.; Gerthsen, D. Influence of pressure, temperature and surface area/volume ratio on the texture of pyrolytic carbon deposited from methane. *Carbon* **2003**, *41*, 749–758. [CrossRef]

22. Chen, M.; Zhu, Y.; Xia, J.; Wu, H. Molecular insights into the initial formation of pyrolytic carbon upon carbon fiber sur-face. *Carbon* **2019**, *148*, 307–316. [CrossRef]
23. Saeed, M.; Alshammari, Y.; Majeed, S.A.; Al-Nasrallah, E. Chemical vapour deposition of graphene—Synthesis, charac-terisation, and applications: A review. *Molecules* **2020**, *25*, 3856. [CrossRef]
24. Shah, A.; Stenberg, P.; Karvonen, L.; Ali, R.; Honkanen, S.; Lipsanen, H.; Peyghambarian, N.; Kuittinen, M.; Svirko, Y.; Kaplas, T. Pyrolytic carbon coated black silicon. *Sci. Rep.* **2016**, *6*, 25922. [CrossRef]
25. Lobo, L.S.; Carabineiro, S.A.C. Carbon Formation at High Temperatures (550–1400 °C): Kinetics, Alternative Mechanisms and Growth Modes. *Catalysts* **2020**, *10*, 465. [CrossRef]
26. Wu, Y.; Wang, S.; Komvopoulos, K. A review of graphene synthesis by indirect and direct deposition methods. *J. Mater. Res.* **2020**, *35*, 76–89. [CrossRef]
27. Kumar, R.; Singh, R.K.; Singh, D.; Joanni, E.; Yadav, R.M.; Moshkalev, S.A. Laser-assisted synthesis, reduction and micro-patterning of graphene: Recent progress and applications. *Coord. Chem. Rev.* **2017**, *342*, 34–79. [CrossRef]
28. Barberio, M.; Antici, P. Laser-Plasma Driven Synthesis of Carbon-Based Nanomaterials. *Sci. Rep.* **2017**, *7*, 12009. [CrossRef]
29. Kim, B.J.; Nasir, T.; Choi, J.-Y. Direct Growth of Graphene at Low Temperature for Future Device Applications. *J. Korean Ceram. Soc.* **2018**, *55*, 203–223. [CrossRef]
30. Diode Laser Pumped Nd: YAG Laser. Available online: https://www.tau.ac.il/~{}lab3/72_laser_res/yag_manual.pdf (accessed on 20 March 2021).
31. Hu, S.; Lee, C.-Y.; Chiu, H.-T. Chemical Vapor Deposition of Carbon Nanocoils Three-Dimensionally in Carbon Fiber Cloth for All-Carbon Supercapacitors. *ACS Omega* **2019**, *4*, 195–202. [CrossRef]
32. Dovbeshko, G.I.; Romanyuk, V.R.; Pidgirnyi, D.V.; Cherepanov, V.V.; Andreev, E.O.; Levin, V.M.; Kuzhir, P.P.; Kaplas, T.; Svirko, Y.P. Optical Properties of Pyrolytic Carbon Films Versus Graphite and Graphene. *Nanoscale Res. Lett.* **2016**, *10*, 234. [CrossRef]
33. Gajewski, G.; Pao, C.-W. Ab initio calculations of the reaction pathways for methane decomposition over the Cu (111) surface. *J. Chem. Phys.* **2011**, *135*, 064707. [CrossRef]
34. Faggio, G.; Grillo, R.; Foti, A.; Agnello, S.; Messina, F.; Messina, G. Micro-photoluminescence of Carbon Dots Deposited on Twisted Double-Layer Graphene Grown by Chemical Vapor Deposition. *ACS Appl. Mater. Interfaces* **2021**, *13*, 7324–7333. [CrossRef]

C Journal of Carbon Research

MDPI

Article

Chemical Reduction of GO: Comparing Hydroiodic Acid and Sodium Borohydride Chemical Approaches by X-ray Photoelectron Spectroscopy

Wei Liu [1] and Giorgio Speranza [2,3,4,*]

1 MERLin, School of Chemistry, Edgeworth David Building, Level 2, The University of Sydney, Sydney, NSW 2006, Australia; w.liu@unsw.edu.au
2 Fondazione Bruno Kessler, Sommarive Str. 18, 38123 Trento, Italy
3 Department of Industrial Engineering, University of Trento, Sommarive Str. 9, 38123 Trento, Italy
4 IFN—CNR, CSMFO Lab., Alla Cascata Str. 56/C Povo, 38123 Trento, Italy
* Correspondence: speranza@fbk.eu

Abstract: The efficiency of two wet chemical processes based on hydroiodic acid (HI) and sodium borohydride ($NaBH_4$) used to reduce graphene oxide (GO) have been studied. At this aim, the oxygen abundance of reduced graphene oxide (rGO) was studied as a function of the reductant concentration. A number of rGO samples were produced and their chemical compositions were studied using X-ray photoelectron spectroscopy. The analyses show that the reduction of the oxygen concentration proceeds non-linearly. At the beginning, when pristine GO is utilized a higher extent of reduction is obtained. The oxygen concentration decreases from ~32% to 10.5% by increasing the HI concentration to 0.24 M. A steeper reduction was observed for $NaBH_4$, where the oxygen concentration lowers to ~13.6% using just 50 mg of $NaBH_4$. Next, reduction reactions performed with increasing amounts of reductants in aqueous suspensions show a progressive saturation effect, indicating a limit in the final oxygen concentration. We obtained a residual oxygen concentration of 5.3% using 7.58 M of HI and 8.6% with 1200 mg of $NaBH_4$. The chemical analysis highlights that the reduction of the oxygen concentration in rGO samples is mainly derived from the cleavage of C-OH bonds and the next reconstruction of C-C bonds.

Keywords: reduced graphene oxide; hydroiodic acid; sodium borohydride; XPS; chemical analysis

Citation: Liu, W.; Speranza, G. Chemical Reduction of GO: Comparing Hydroiodic Acid and Sodium Borohydride Chemical Approaches by X-ray Photoelectron Spectroscopy. C 2022, 8, 20. https://doi.org/10.3390/c8020020

Academic Editors: Carlo Mariani and Gil Goncalves

Received: 10 February 2022
Accepted: 18 March 2022
Published: 22 March 2022

1. Introduction

Graphene is a material that has raised the attention of the scientific community for its exceptional properties such as its outstanding electron conductivity, excellent mechanical strength, superior thermal conductivity, and high surface area [1,2]. More than two decades of research on graphene led to a number of patents regarding both its application and production methods. However, the use of graphene on a large scale is strictly dependent on the possibility of producing sufficient amounts of material to satisfy the industrial needs.

At present, the production of graphene flakes has reached thousands of tones/year and the fabrication of continuous graphene sheets with dimension of meters are available [3], allowing for its exploitation in various industrial applications. The rich list of chemical reactions utilized to functionalize the GO sheets enables a large variety of uses in different areas [4,5]. The covalent functionalization of GO sheets includes the transformation of hydroxyl-based functional groups to increase the density of more reactive carboxyl groups, or to produce esters, involving reactions with amide or isocyanate. This permits the further functional modification of the GO surface chemistry using organic molecules [6]. The non-covalent functionalization of GO offers a rich selection of functional groups that can interact via hydrogen, electrostatic, and π-π interactions with the GO basal plane. This opens up the possibility to attach complex molecules such as proteins, DNA, drugs, and amine-terminated polymers. GO may also be considered as a precursor for the fabrication of complex systems by adding nanoparticles, organic compounds, or biomolecules. Metallic nanoparticles, such as Au, Pt, Ru, TiO_2, and ZnO, etc., are added to improve the catalytic and electrochemical properties of GO [4]. Organic molecules, such as porphyrins, aromatic dyes, alkylamines, ionic liquids, pyrene, perylenediimide, and cyclodextrins, etc., are added to improve the solubility of inorganic solvents, processability in composites, and photoactive or electroactive properties [7]. Biomolecules are added to improve biocompatibility, to fabricate bioelectrodes and biosensors, and for general bioapplications [4,8,9]. Another important application of GO is the fabrication of composites to enhance the mechanical, electrical, and electrochemical properties of materials [10,11]. GO is also utilized in protective coatings against high energy radiation and corrosion [12,13]. Environment and energy are hot research areas due to the present emergency caused by climate change. GO is utilized for environmental remediation by absorbing toxic substances [14–16]. As for energy applications, GO and graphene are emerging as important components [17–19]. In this respect, the doping of graphene, for example, with N atoms [20] is used to modify its electric properties, thus allowing a higher efficiency in energy storage devices such as supercapacitors, batteries, and fuel cells [21].

A significant quantity of graphene particulates is produced using chemical exfoliation processes. However, the exfoliation routes have, as a side effect, the production of highly oxidized, micro/nano sized sheets. The high degree of surface oxidation ensures an easy dispersivity in water, but the high degree of oxidation is also linked to a loss of the original graphene properties. GO sheets are transparent, mirroring the presence of a bandgap whose amplitude varies from 0.02 to 2 eV, depending on the oxygen concentration [22,23] and on the kind of carbon–oxygen bonds [24]. Crystalline graphene, is formed by the dense packing of carbon atoms into a honeycomb lattice. In this lattice structure, the carbon configuration is described by the sp^2 hybridization. The reaction with oxygen atoms leads to a change of the hybridization from sp^2 to sp^3. Electrons involved in the C–O bonds lose their mobility and the C–O bond becomes a scattering site. This is mirrored by a reduction of the carrier mobility from 10^5 cm^2/Vs of a single, pure graphene sheet [25] down to 0.05–200 cm^2/Vs in graphene oxides [26]. In addition, the mechanical properties are affected by the presence of surface oxidation. Pristine graphene sheets possess a Young modulus as high as 1 TPa [27], which decreases to 380–470 GPa in the case of GO sheets [28].

The promise of recovering the original graphene properties by reducing graphene oxide led to intense research. The reduction of GO results in two main effects that are strictly related each other: the loss of material transparency and the recovery of a certain degree of conductivity. The increasing extinction coefficient with the removal of oxygen atoms reflects the recovery of the graphene structure. The peculiar electronic configuration described by the Dirac cones at the K and K′ points of the hexagonal Brillouin zone results in a high light absorption. Despite graphene is just one atom thick, it absorbs a significant fraction of incident white light ($\pi\alpha = 2.3\%$) [29]. In rGO, the density of the functional groups is found to deeply affect the absorption coefficient [30]. The decrease of the bandgap with the extent of reduction may change from 1.69 eV to a fraction of eV [31,32]. As a consequence, if functional groups are retained on the surface, rGO becomes a semiconductor thanks to the regeneration of the planar structure [33]. As seen above, the carrier mobility in the presence of oxygen functional groups may be of the order of 100–200 cm^2/Vs due to electron delocalization [34,35]. This electron mobility is two orders of magnitude bigger than that of organic semiconductors, which typically is less than 1 cm^2/Vs, preventing their utilization despite the low cost and flexibility. With the recovery of the carrier mobility in rGO, a correspondent increase in the material conductivity is obtained [36], thus making rGO interesting for applications in electronics and conductive composites [37].

Therefore, an efficient elimination of oxygen species from GO is important for any application involving the electronic properties of rGO [11,37,38]. Aiming at obtaining an efficient reduction, a variety of reducing agents have been explored [39]. rGO was synthesized using hydrazine [40], ascorbic acid [41], H_2S [42], $NaBH_4$ [43], hydrohalic acid [44], metal iodide [45], and amino acids [46], etc. The final rGO products showed significantly different properties depending on their reducing potential and reaction mechanisms. As a result, the C/O ratio ranges from 30 to 2.2, and the conductivities changed from 3.2×10^{-5} to 55,088 S m^{-1} [41,45–47].

The need to produce rGO with programmed amounts of oxygen calls for the definition of efficient reduction protocols. Here, we compare two chemical processes, based on hydroiodic acid and sodium borohydride, and the rGO samples produced by varying the reductant concentration. The surface chemistry was analyzed using X-ray photoelectron spectroscopy and correlated to the reductant's concentration. The comparison between the two reduction chemical routes will enable the definition of optimal operation conditions and the limits of the processes.

2. Experimental

2.1. Chemicals

A water suspension of GO with concentration of 4 mg/mL was purchased from Graphenea (San Sebastian, Spain) with a monolayer content higher than 95% (measured in 0.05 wt% suspensions) and was used for the reduction process. HI (57 wt%, 99.999%), $NaBH_4$ (98%), NaOH (97%), and ethanol (99.5%) were purchased from Sigma-Aldrich-Merck (Darmstadt, Germany) and used as received. Milli-Q water was used in the whole reduction process.

2.2. Reduction of Graphene Oxide with HI

rGO samples obtained using HI were prepared following the procedure in our previous report [48]. Eight samples were obtained by using 0.06 M, 0.12 M, 0.24 M, 0.47 M, 0.95 M, 1.89 M, 3.79 M, and 7.58 M HI concentrations. In a typical synthesis, 50 mL of 0.8 mg/mL GO solution was first prepared by diluting the Graphenea GO suspension and ultrasonicating for 5 min. Next, 50 mL of HI solution was added into above solution and ultrasonicated for another 5 min. The flask with above mixture was transferred into an oil bath at 90 °C and the reduction reaction was kept at this temperature for 4h. When the reaction finished, the flask was cooled down with an ice bath. The product in the flask was collected by using a cellulose filtration paper. The rGO product was then repeatedly washed with water and ethanol and vacuum dried overnight.

2.3. Reduction of Graphene Oxide with $NaBH_4$

The $NaBH_4$-reduced rGO samples were synthesized with a previously reported procedure, although with some modifications [49]. First, 12.5 mL of GO suspension was diluted with Milli-Q water to 1.0 mg/mL under ultrasonication. Next, a $NaBH_4$ solution was prepared by dissolving varied amounts (50 mg, 300 mg, 600 mg, 900 mg, and 1200 mg) of $NaBH_4$ in 50 mL of NaOH solution (pH = 9). The $NaBH_4$—NaOH solution was slowly added into the GO solution. The mixture was then heated to 80 °C in an oil bath and the reaction proceeded at this temperature under continuous stirring for 1h. The rGO product was collected by filtration and washed repeatedly with water and ethanol before vacuum drying overnight.

2.4. Characterization

X-ray photoelectron spectroscopy was utilized to study the surface chemical composition of the samples. A Kratos Axis DLD Ultra (Manchester, UK) instrument was used to carry out the spectral investigation. Thin rGO films, obtained by the two chemical routes described above, were dried under vacuum overnight for degassing and drying. For all the HI and $NaBH_4$ concentrations, the rGO samples' conductivity was high enough to ensure the acquisition of high-quality spectra without charge compensation. For each sample, a wide spectrum was acquired at a pass energy of 150 eV to control the presence of residuals from the reduction processes. The high-resolution C 1s and O 1s core-lines were acquired using a 20 eV pass energy and an energy step of 0.05 eV. These acquisition conditions were sufficient to study the chemical bonds and the associated concentrations. The obtained spectra were analyzed using home-made RxpsG software based on the R platform [50].

3. Results and Discussion

Information about the reductant concentration for each rGO sample is summarized in Table 1. rGO samples from the HI reduction and NaBH4 reduction processes are denoted as S_Hx and S_Nx, respectively.

Table 1. Summary of the rGO produced using HI and $NaBH_4$. The correspondent reductant concentrations are indicated.

Sample Name	HI Concentration	Sample Name	$NaBH_4$ Amount
GO	//		
S_H1	0.06 M	S_N1	50 mg
S_H2	0.12 M	S_N2	300 mg
S_H3	0.24 M	S_N3	600 mg
S_H4	0.47 M	S_N4	900 mg
S_H5	0.95 M	S_N5	1200 mg
S_H6	1.89 M		
S_H7	3.79 M		
S_H8	7.58 M		

The extent of reduction was controlled by XPS. Figure 1A shows the wide spectra obtained from the rGO samples obtained using HI as a reductant.

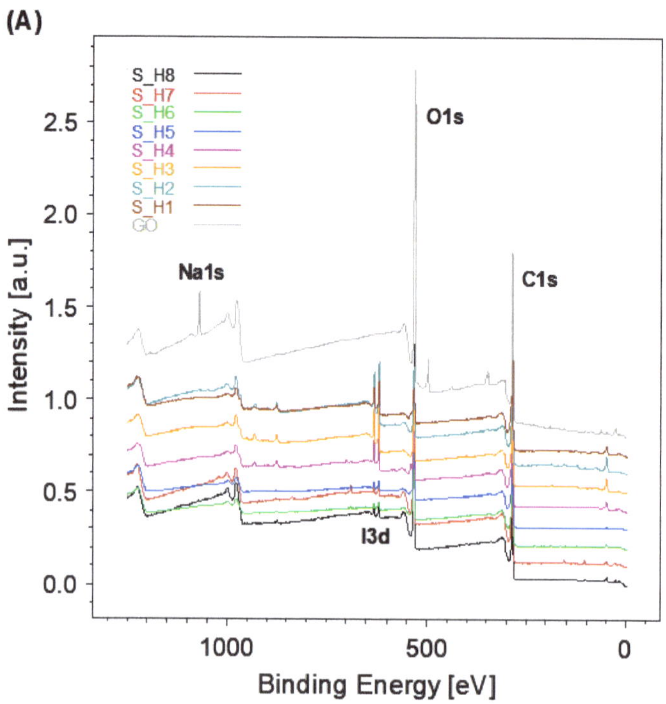

Figure 1. *Cont.*

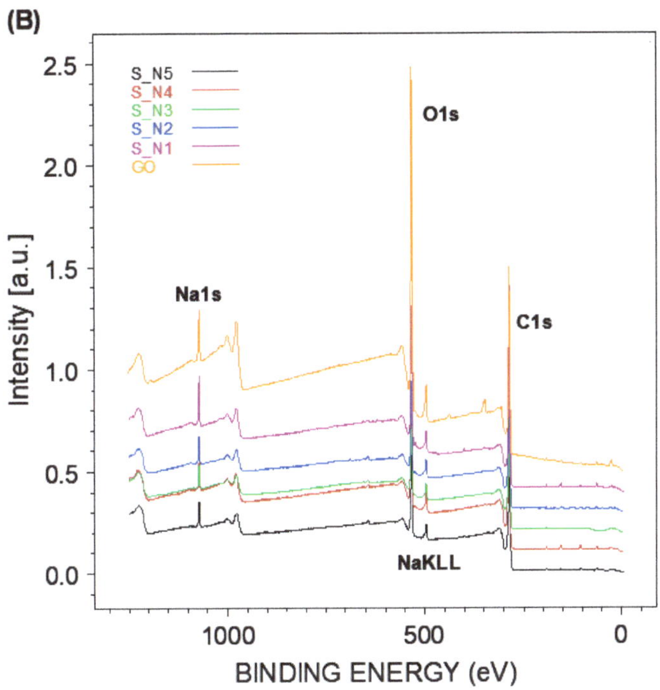

Figure 1. (**A**) Wide spectra of rGO samples synthesized using HI at increasing concentrations. The wide spectrum of GO is plotted as a reference. (**B**) Wide spectra of rGO samples synthesized using $NaBH_4$ at increasing concentrations. All the spectra are normalized to the C 1s intensities to highlight O 1s variations and shifted along the *Y* axis to better show the spectral features. The main spectral components are indicated in both figures.

The wide spectra show that the rGO samples are composed by only carbon and oxygen, with residuals of iodine derived from the HI. In GO, iodine is absent, while the wide spectrum reveals a small Na impurity derived from the synthesis process. To compare the spectra, the intensities were normalized with respect to the C 1s. As can be seen from the spectra, by increasing the concentration of the HI in the solution, the GO undergoes a progressive reduction. This is clearly visible following the intensity of the O 1s, which is expectedly higher in the GO sample and gradually decreases with increasing HI concentration (see also Table 2). A similar behavior is found when analyzing the GO reduced using $NaBH_4$, as shown in Figure 1B. In addition, in this case, all the spectra are normalized with respect to the intensity of the C 1s core-line. This highlights the rapid decrease of the oxygen intensity corresponding to the initial values of the reductant concentrations. Next, there is a much slower decrease of the oxygen concentration with an increasing concentration of $NaBH_4$ (see O concentration trend in Table 2).

In addition to carbon and oxygen, the wide spectra also display the presence of other elements derived from the chemical processing. In particular, at a high BE, a sharp peak due to Na is visible. This also introduces the Auger KLL component near the oxygen core-line. Boron 1s is hardly visible in some of the wide spectra at ~190 eV due to the very low cross section of this element. Regarding the core-line spectra, the high resolution C 1s and O 1s spectra were then fitted to follow the changes of the surface chemistry mirrored by variations of the bonding. Examples of high resolution C1s and O1s core-lines, together with the correspondent peak fitting, are illustrated in Figure 2A,B. Figure 2A shows a main peak assigned to the sp^2 graphitic phase of the S_H2 sample. The fitting component is characterized by a certain asymmetry, mirroring the semimetallic property associated with this phase. The degree of asymmetry may be associated with the reconstruction of the C–C aromatic bonds typical of the graphene structure. This causes the occurrence of energy loss processes described by the pronounced intensity of the C 1s tail at high binding energy. The semimetallic character of the graphitic C 1s component is the reason for selecting a Shirley base-line for the background subtraction. A Gaussian lineshape was used for the other fitting components needed to describe the other chemical bonds. The first component at ~285 eV is derived from hydrocarbon contaminations. Component 3 at 286.2 eV is assigned to the hydroxyl and epoxy-like bonds, while the component at ~287.5 eV is assigned to the carbonyl bond and carbon atoms bonded to two oxygen atoms in a –O–C–O– bond [51,52].

Table 2. Oxygen and carbon–oxygen concentrations.

Sample	O conc. (%)	CO_x conc. (%)	Sample	O conc. (%)	CO_x conc. (%)
GO	32.34	30.02			
S_H1	21.86	22.24	S_N1	30.02	32.34
S_H2	16.94	16.96	S_N2	13.59	13.07
S_H3	10.50	10.98	S_N3	9.78	9.97
S_H4	10.29	10.21	S_N4	8.91	8.91
S_H5	9.09	9.01	S_N5	8.62	8.63
S_H6	8.06	8.52			
S_H7	7.72	7.76			
S_H8	5.32	5.77			

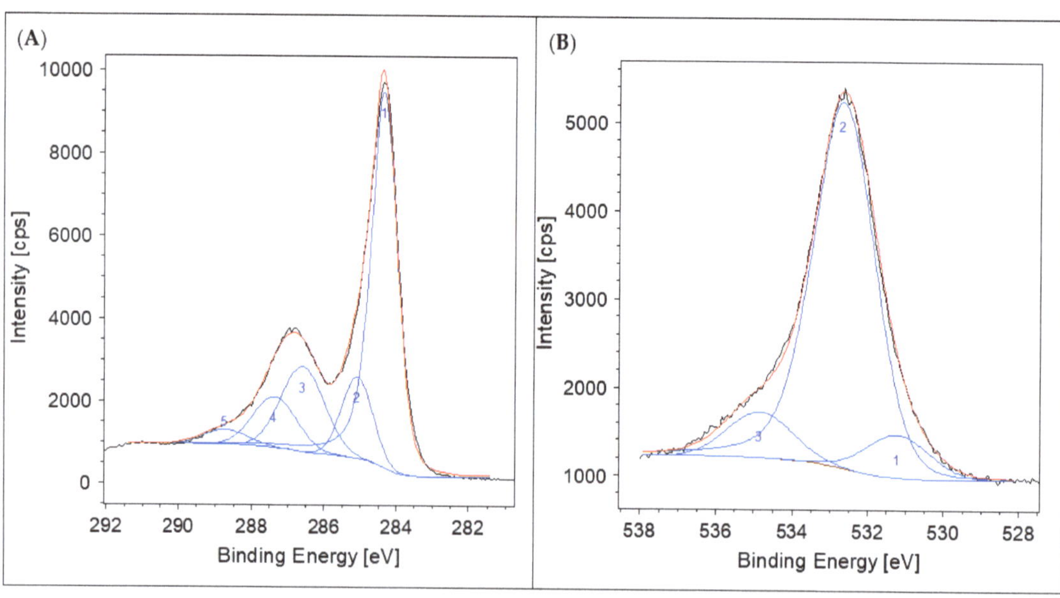

Figure 2. (**A**) C 1s and (**B**) O 1s core-lines of the S_H2 sample reported together with their peak fitting.

Finally, the carboxyl bond is found at a higher BE. Similar peak fitting was performed on the C 1s from the other HI-reduced rGO samples. In the case of oxygen, the core-lines were fitted using three Gaussian components, as reported in Figure 2B. The first two components are assigned to carbon–oxygen bonds and, specifically, to C=O bonds at lower BE and C–O single bonds at the higher BE. The remaining third fitting component at ~534 eV (or higher) is derived from oxygen in H_2O molecules adsorbed on the surfaces of rGO-HI samples [51,52]. A similar interpretation also holds for the C 1s and O 1s core-lines from the rGO-$NaHB_4$ samples. Figure 3A,B reports an example of the peak fitting of the two core-lines. As seen in Figure 1B, residuals of Na are present in all the S_N samples. One final note regarding the complete overlap of the Auger KL_1L_{23} component falling at 532 eV: in all the peak fittings of the oxygen core-lines, the contribution of this Auger component is negligible and, for this reason, it was ignored. As in the previous case, a Shirley baseline was utilized for the background subtraction of the high-resolution C 1s spectrum. In addition, the main peak corresponding to the sp^2 phase is described by an asymmetric lineshape to account for the energy losses according to the notes above. The remaining components are Gaussian and are assigned to the chemical bonds previously described. Component 2 at ~285.0 eV is assigned to hydrocarbon contamination. At increasing BE, hydroxyl and epoxy, carbonyl, O–C–O bonds, and carboxyl groups can be found, respectively. There are differences in the peak fitting of the oxygen core-line. The components at lower BE are assigned to contributions derived from NaOH and $NaHBO_3$. The two components at ~532 and ~532.7 eV are assigned to carbon–oxygen bonds, while the component at the higher BE accounts for the oxygen derived from adsorbed H_2O molecules.

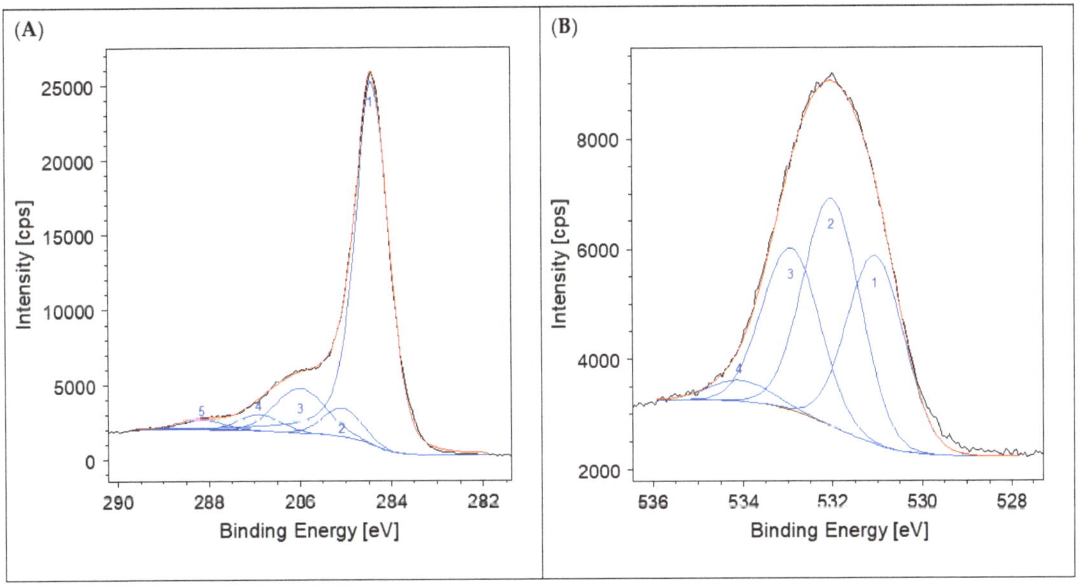

Figure 3. (**A**) C 1s and (**B**) O 1s core-lines of the S_N3 sample reported together with their peak fitting.

Figure 4A,B shows the trend of oxygen concentrations in S_H and S_N samples.

Increasing the amounts of HI and $NaBH_4$ reductants in the GO water suspension results in a reduction of the concentration of oxygen in both cases. Interestingly, an excellent agreement between the oxygen concentration and the abundance of the oxidized carbon components is obtained. The latter was estimated considering the stoichiometry of the carbon–oxygen bond described by the various fit components.

As illustrated in Figure 4, for both HI and $NaBH_4$, a higher effect is obtained at the beginning, where a lower reductant concentration corresponds to a higher decrease of oxygen. Consequently, the abundances of carbon atoms involved in bonds with oxygen show the same behavior, having a tendency to display a flattening of the trend. Table 2 summarizes the values of the oxygen and C–O_x concentrations for the various reductant concentrations.

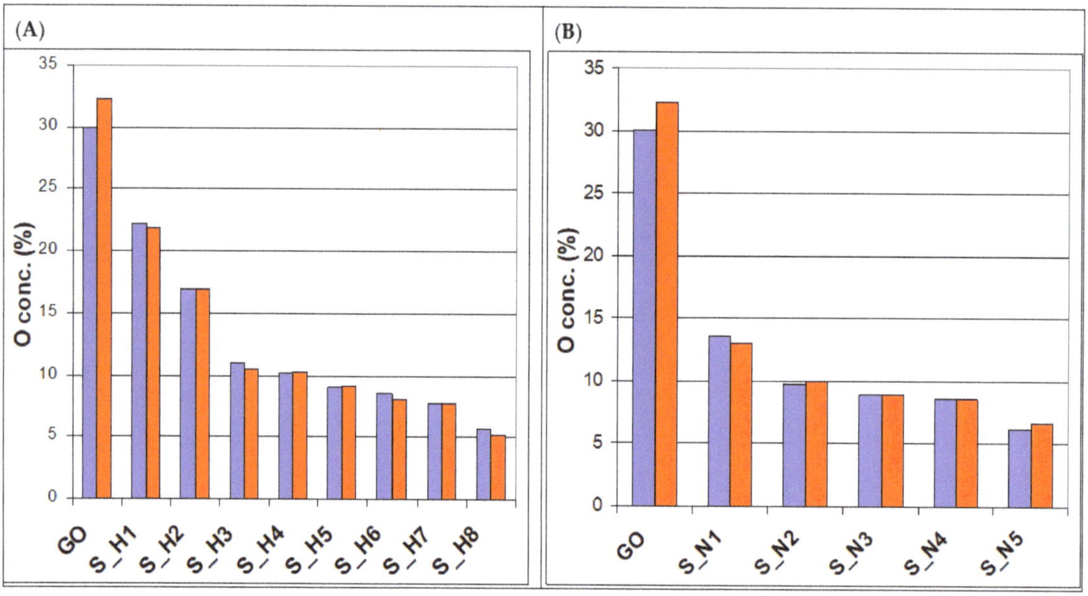

Figure 4. Trend of the total oxygen concentrations (blue) and of the concentration of oxygen derived from the C1s fit and carbon oxidized fit components of carbon (red) in rGO reduced using HI (**A**) or $NaBH_4$ (**B**).

HI, as well as other hydrohalic acids such as HBr and HCl, is used in the electrophilic addition and nucleophilic substitution chemical reactions. Considering the nucleophilic nature of halides, I^- possesses a higher nucleophilicity with respect to Cl^- and Br^-. In addition, HI is very efficient in attacking epoxide C–O bonds [47]. This renders HI the reductant of choice to remove oxygen from GO hydroxyl and epoxy bonds, which constitute the main portion of the carbon–oxygen bonds. In Figure 5A, a representative scheme of the possible chemical reactions that can occur between HI and GO is presented. In agreement with these reaction models, the cleavage of ether and iodination, iodination of alcohols, reduction of aromatic iodides, and partial reduction of the carbonyl groups may occur [44]. This could also explain why the reaction of GO with an increasing concentration of HI leads to a progressive decrease of the reduction efficiency. When hydroxyl and epoxy groups are largely removed, the other kind of bonds, such as carboxyl, lactone, –O–C–O–, and carbonyl groups, are likely still present on the rGO surface, as can be seen in the high-resolution C 1s spectra. As a matter of fact, in our GO sample, the hydroxyl and epoxy groups amount to ~24% of the total oxygen concentration, which is reported as 30% in Table 2. The remaining part of the carbon–oxygen bonds, which is around 6%, corresponds to the percentage of the oxygen left on the sample surface after the reduction with higher HI concentrations (see the CO_x concentration in Table 2), which is in good agreement with the residual oxygen abundance found in the S_H8 sample.

(A)

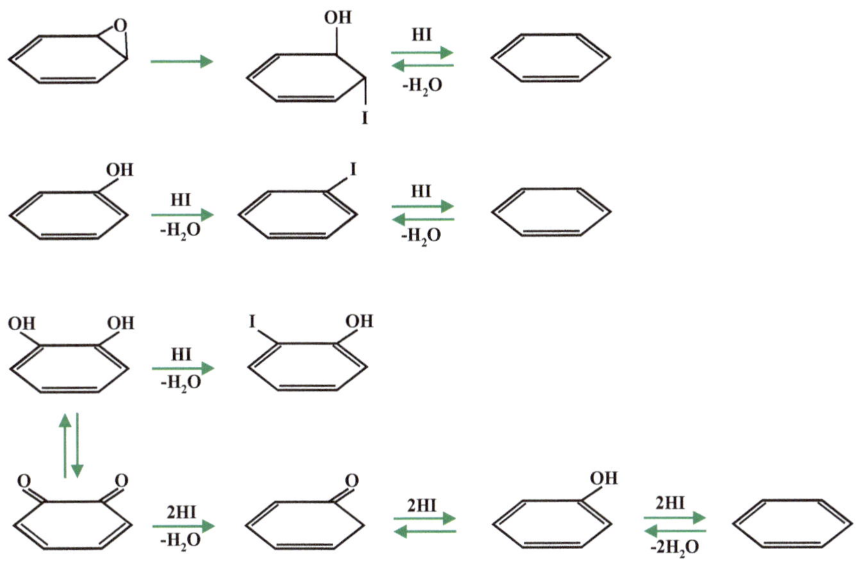

(B)

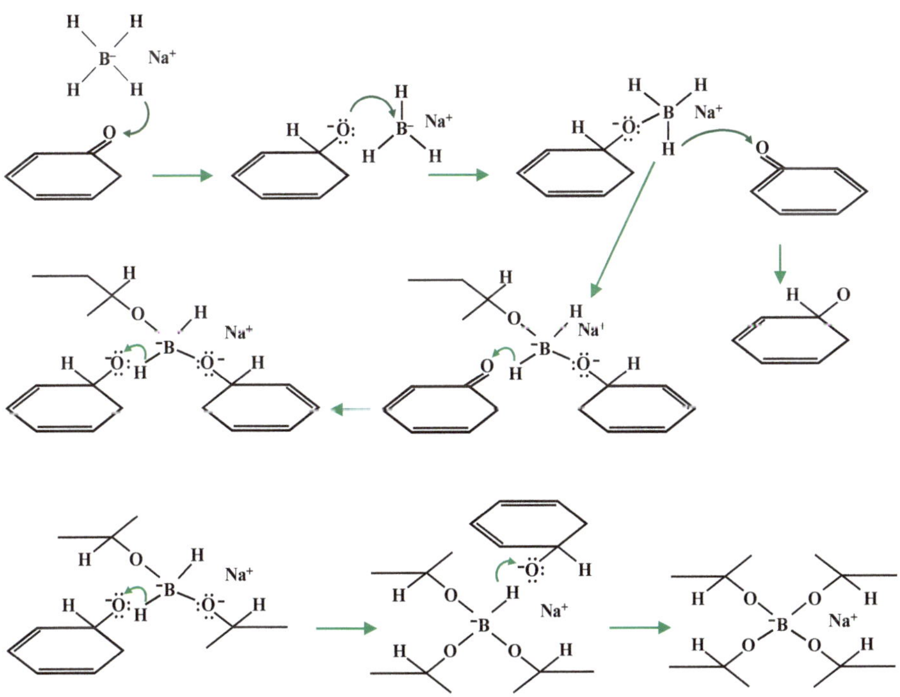

Figure 5. (**A**) Model of the chemical reaction occurring between oxygen-containing aromatic groups and HI; (**B**) progression of carbonyl group reduction until the complete depletion of B–H bonds and boron oxidation.

Regarding the use of $NaBH_4$, this is a compound containing a BH_4^- anion with a tetrahedral structure. In aqueous solution, this anion reacts easily with electrophile species, such as carbonyl groups, as illustrated in Figure 5B. The borohydride anion induces a hydride transfer reaction, resulting in an oxyanion and an electron-deficient BH_3 molecule The oxyanion can then react with BH_3, restoring the borohydride. This reaction proceeds until B–H bonds have been depleted. In reality, this ideal reaction scheme does not occur in common reduction processes, where the reduction efficiency is related to the kind of original compound. $NaBH_4$ is able to react with aldehyde and ketone carbonyl groups, while carbonyl compounds, such as esters or amides, are not reducible. However, a simulation report showed that, in a NaOH solution, the OH^- anion can interact with a hydroxyl group in GO to generate negatively charged GO. Due to this electron transfer, it is easy for the epoxy ring to be opened with a small energy barrier. These oxygen atoms on negatively charged GO can be removed with the assistance of Na^+ and a water molecule. In this process, NaOH can be considered as a catalyst [53]. The reduction reactions are listed below:

$$\text{GO-2OH} + 2\text{NaOH} = \text{GO-2O}^- + 2\text{Na}^+ + 2\text{H}_2\text{O} \quad (1)$$

$$\text{GO-2O}^- + 2\text{Na}^+ + \text{H}_2\text{O} = \text{rGO} + 2\text{NaOH} \quad (2)$$

As apparent from Figure 4B, the trend of the oxygen concentration is similar to that obtained using HI. Additionally, the lower value of oxygen corresponding to the higher value of $NaBH_4$ concentration is comparable to that from the HI reduction process. We can then hypothesize that, in this case, $NaBH_4$ primarily removes hydroxyl and epoxy groups. This hypothesis is confirmed by the results reported previously [54], where the authors studied the trend of carbon–oxygen single and double bonds as a function of the reduction reaction time. The authors found that the concentration of carbonyl bonds remains more or less constant with time, while C–O single bonds undergo a decrease when increasing the reaction time. These findings are also in agreement with the work of other authors, where $NaBH_4$ preserved carboxyl groups useful for the further functionalization of the rGO sheets [55]. These results are partially in agreement with other results showing that $NaBH_4$ transforms the carbonyl in hydroxyl groups, which is subsequently removed along with epoxy bonds [43]. The decreasing efficiency of the reductant already found for HI at increasing concentrations is also found in the case of $NaBH_4$. In addition, in this case, the reduction of the oxygen abundance when increasing the concentration of $NaBH_4$ deviates from a linear trend and shows saturation. This is in agreement with other studies reporting a similar trend of the residual oxygen in rGO vs $NaBH_4$ concentration [43] or vs increasing reduction time processing [54]. Interestingly, the residual oxygen concentration left on the S_N5 sample is much lower than that obtained in [43,54] after a reaction time of 24 h.

4. Conclusions

The increasing demand of producing high-quality reduced graphene oxide possessing high conductivity and low defect density has motivated this study, which was aimed at comparing the efficiency of hydroiodic acid and sodium borohydride. These chemical compounds are known to be active towards oxidized species, but their reactivity varies upon the nature of the bond formed by oxygen. XPS was then used to shed light on the reaction mechanisms at the basis of the GO reduction. Characterization analyses were performed at increasing reductant concentration. Results show that, for both HI and $NaBH_4$, the initial reduction process proceeds efficiently. However, when the residual oxygen concentration lowers, the reaction with oxygen atoms, and their successive removal, becomes more complex. As such, results show that, at the higher HI and $NaBH_4$ concentration, the residual oxygen concentration is minimal and lower with respect to that found by other studies where these reductant agents were utilized. As a matter of fact, results show that the final oxygen concentrations are 5.3% and 8.6%, making HI more efficient than $NaBH_4$–NaOH. The improved sheet degree of GO reduction may open up more practical uses of this material in all of the applicative sectors where electrical properties of the material play an important role.

Author Contributions: Conceptualization, G.S.; Data curation, G.S. and W.L.; Formal analysis, G.S. and W.L.; Writing—original draft, G.S.; Writing—review & editing, W.L. All authors have read and agreed to the published version of the manuscript.

Funding: This research received no external funding.

Data Availability Statement: The data presented in this study are available from the corresponding author upon request.

Conflicts of Interest: The authors declare no conflict of interest.

References

1. Georgakilas, V.; Perman, J.A.; Tucek, J.; Zboril, R. Broad Family of Carbon Nanoallotropes: Classification, Chemistry, and Applications of Fullerenes, Carbon Dots, Nanotubes, Graphene, Nanodiamonds, and Combined Superstructures. *Chem. Rev.* **2015**, *115*, 4744–4822. [CrossRef] [PubMed]
2. Allen, M.J.; Tung, V.C.; Kaner, R.B. Honeycomb carbon: A review of graphene. *Chem. Comm.* **2010**, *110*, 132–145. [CrossRef] [PubMed]
3. Kong, W.; Kum, H.; Bae, S.-H.; Shim, J.; Kim, H.; Kong, L.; Meng, K.; Wang, K.; Kim, C.; Kim, J. Path towards graphene commercialization from lab to market. *Nat. Nanotechnol.* **2019**, *14*, 927–938. [CrossRef] [PubMed]
4. Chen, D.; Feng, H.; Li, J. Graphene Oxide: Preparation, Functionalization, and Electrochemical Applications. *Chem. Rev.* **2012**, *112*, 6027–6053. [CrossRef]
5. Wang, Y.; Li, S.; Yang, H.; Luo, J. Progress in the functional modification of graphene/graphene oxide: A review. *RSC Adv.* **2020**, *10*, 15328.
6. Xiong, Y.; Xie, Y.; Zhang, F.; Ou, E.; Jiang, Z.; Ke, L.; Hu, D.; Xu, W. Reduced graphene oxide/hydroxylated styrene-butadiene-styrene tri-block copolymer electroconductive nanocomposites: Preparation and properties. *Mater. Sci. Eng. B* **2012**, *177*, 1157–1163. [CrossRef]
7. Joshi, D.J.; Koduru, J.R.; Malek, N.I.; Hussain, C.M.; Kailasa, S.K. Surface modifications and analytical applications of graphene oxide: A review. *Trends Anal. Chem.* **2021**, *144*, 116448. [CrossRef]
8. Shareena, T.P.D.; McShan, D.; Dasmahapatra, A.K.; Tchounwou, P.B. A Review on Graphene-Based Nanomaterials in Biomedical Applications and Risks in Environment and Health. *Nano-Micro Lett.* **2018**, *10*, 53. [CrossRef]
9. Liu, Y.; Dong, X.; Chen, P. Biological and chemical sensors based on graphene materials. *Chem. Soc. Rev.* **2012**, *41*, 2283–2307. [CrossRef]
10. Ahmad, H.; Fan, M.; Hui, D. Graphene oxide incorporated functional materials: A review. *Compos. B* **2018**, *145*, 270–280. [CrossRef]
11. Wang, Z.; Nelson, J.K.; Hillborg, H.; Zhao, S.; Schadler, L.S. Graphene Oxide Filled Nanocomposite with Novel Electrical and Dielectric Properties. *Adv. Mater.* **2012**, *24*, 3134–3137. [CrossRef] [PubMed]
12. Türkaslan, S.S.; Ugur, S.S.; Türkaslan, B.E.; Fantuzzi, N. Evaluating the X-ray-Shielding Performance of Graphene-Oxide-Coated Nanocomposite Fabric. *Materials* **2022**, *15*, 1441. [CrossRef] [PubMed]
13. Chauhan, D.S.; Quraishi, M.A.; Ansari, K.R.; Saleh, T.A. Graphene and graphene oxide as new class of materials for corrosion control and protection: Present status and future scenario. *Progr. Org. Coat.* **2020**, *147*, 105741. [CrossRef]
14. Obayomi, K.S.; Lau, S.Y.; Danquah, M.; Chiong, T.; Takeo, M. Advances in graphene oxide based nanobiocatalytic technology for wastewater treatment. *Environ. Nanotechnol. Monit. Manag.* **2022**, *17*, 100647. [CrossRef]
15. Siwinska-Stefanska, K.; Fluder, M.; Tylus, W.; Jesionowski, T. Investigation of amino-grafted TiO_2/reduced graphene oxide hybrids as a novel photocatalyst used for decomposition of selected organic dyes. *J. Environ. Manag.* **2018**, *212*, 395–404. [CrossRef]
16. Perreault, F.; de Faria, A.F.; Elimelech, M. Environmental applications of graphene-based nanomaterials. *Chem. Soc. Rev.* **2015**, *44*, 5861–5896. [CrossRef] [PubMed]
17. Tian, Y.; Yu, Z.; Cao, L.; Zhang, L.X.; Sun, C.; Wang, D.W. Graphene oxide: An emerging electromaterial for energy storage and conversion. *J. Energy Chem.* **2021**, *55*, 323–344. [CrossRef]
18. Chen, K.; Song, S.; Liu, F.; Xue, D. Structural design of graphene for use in electrochemical energy storage devices. *Chem. Soc. Rev.* **2015**, *44*, 6230–6257. [CrossRef]
19. Guo, S.; Dong, S. Graphene nanosheet: Synthesis, molecular engineering, thin film, hybrids, and energy and analytical applications. *Chem. Soc. Rev.* **2011**, *40*, 2644–2672. [CrossRef]
20. Kamedulski, P.; Truszkowski, S.; Lukaszewicz, J.P. Highly Effective Methods of Obtaining N-Doped Graphene by Gamma Irradiation. *Materials* **2020**, *13*, 4975. [CrossRef]
21. Mankge, N.S.; Madito, M.J.; Hlongwa, N.W.; Kuvarega, A.T. Review of electrochemical production of doped graphene for energy storage applications. *J. Energy Storage* **2022**, *46*, 103527. [CrossRef]

22. Velasco-Soto, M.A.; Perez-Garcia, S.A.; Alvarez-Quintana, J.; Cao, Y.; Nyborg, L.; Licea-jimenez, L. Selective band gap manipulation of graphene oxide by its reduction with mild reagents. *Carbon* **2015**, *93*, 967–973. [CrossRef]
23. Shen, Y.; Yang, S.; Zhou, P.; Sun, Q.; Wang, P.; Wan, L.; Li, J.; Chen, L.; Wang, X.; Ding, S.; et al. Evolution of the band-gap and optical properties of graphene oxide with controllable reduction level. *Carbon* **2013**, *62*, 157–164. [CrossRef]
24. Hasan, M.T.; Senger, B.J.; Ryan, C.; Culp, M.; Gonzalez-Rodriguez, R.; Coffer, J.L.; Naumov, A.V. Optical Band Gap Alteration of Graphene Oxide via Ozone Treatment. *Sci. Rep.* **2017**, *7*, 6411. [CrossRef]
25. Weiss, N.O.; Zhou, H.; Liao, L.; Liu, Y.; Jiang, S.; Huang, Y.; Duan, X. Graphene: An emerging electronic material. *Adv. Mater.* **2012**, *24*, 5782–5825. [CrossRef]
26. Gupta, S.; Joshi, P.; Narayan, J. Electron mobility modulation in graphene oxide by controlling carbon melt lifetime. *Carbon* **2020**, *170*, 327–3337. [CrossRef]
27. Lee, C.; Wei, X.; Kysar, J.W.; Hone, J. Measurement of the elastic properties and intrinsic strength of monolayer graphene. *Science* **2008**, *321*, 385–388. [CrossRef]
28. Liu, L.; Zhang, J.; Zhao, J.; Liu, F. Mechanical properties of graphene oxides. *Nanoscale* **2012**, *4*, 5910–5916. [CrossRef] [PubMed]
29. Nair, R.R.; Blake, P.; Grigorenko, A.N.; Novosdelov, K.S.; Booth, T.J.; Stauber, T.; Peres, N.M.R.; Geim, A.K. Fine Structure Constant Defines Visual Transparency of Graphene. *Science* **2008**, *320*, 1308. [CrossRef]
30. Su, R.; Lin, S.F.; Chen, D.Q.; Chen, G.H. Study on the Absorption Coefficient of Reduced Graphene Oxide Dispersion. *J. Phys. Chem. C* **2014**, *118*, 12520–12525. [CrossRef]
31. Jin, Y.; Zheng, Y.; Podkolzin, S.G.; Lee, W. Band gap of reduced graphene oxide tuned by controlling functional groups. *J. Mater. Chem. C* **2020**, *8*, 4885–4894. [CrossRef]
32. Sehrawat, A.P.; Islam, S.S.; Mishra, P.; Ahmad, S. Reduced graphene oxide (rGO) based wideband optical sensor and the role of Temperature, Defect States and Quantum Efficiency. *Sci. Rep.* **2018**, *8*, 3537.
33. Wang, Y.; Wang, L.; Wang, H.Y.; Chen, Q.D.; Sun, H.B. Ultrafast Spectroscopic Study of Insulator-Semiconductor-Semimetal Transitions in Graphene Oxide and Its Reduced Derivatives. *J. Phys. Chem. C* **2019**, *123*, 22550–22555. [CrossRef]
34. Bhaumik, A.; Haque, A.; Taufique, M.; Karnati, P.; Patel, R.; Nath, M.; Ghosh, K. Reduced Graphene Oxide Thin Films with Very Large Charge Carrier Mobility Using Pulsed Laser Deposition. *J. Mater. Sci. Eng.* **2017**, *6*, 4. [CrossRef]
35. Wang, Y.; Chen, Y.; Lacey, S.D.; Xu, L.; Xie, H.; Li, T.; Danner, V.A.; Hu, L. Reduced graphene oxide film with record-high conductivity and mobility. *Mater. Today* **2018**, *21*, 186–192. [CrossRef]
36. Mohan, V.B.; Brown, R.; Jayaraman, K.; Bhattacharyya, D. Characterisation of reduced graphene oxide: Effects of reduction variables on electrical conductivity. *Mater. Sci. Eng. B* **2015**, *193*, 49–60. [CrossRef]
37. Tarcan, R.; Todor-Boer, O.; Petrovai, I.; Leordean, C.; Astilean, S.; Botiz, I. Reduced graphene oxide today. *J. Mater. Chem. C* **2020**, *8*, 1198–1224. [CrossRef]
38. Smith, A.T.; LaChance, A.M.; Zeng, S.; Liu, B.; Sun, L. Synthesis, properties, and applications of graphene oxide/reduced graphene oxide and their nanocomposites. *Nano Mater. Sci.* **2019**, *1*, 31–47. [CrossRef]
39. Agarwal, V.; Zetterlund, P.B. Strategies for reduction of graphene oxide—A comprehensive review. *Chem. Eng. J.* **2021**, *405*, 127018. [CrossRef]
40. Park, S.; An, J.; Potts, J.R.; Velamakanni, A.; Murali, S.; Ruoff, R.S. Hydrazine-reduction of graphite- and graphene oxide. *Carbon* **2011**, *49*, 3019–3023. [CrossRef]
41. Zhang, J.; Yang, H.; Shen, G.; Cheng, P.; Zhang, J.; Guo, S. Reduction of graphene oxide via ascorbic acid. *Chem. Commun.* **2010**, *46*, 1112–1114. [CrossRef] [PubMed]
42. Zhang, C.; Lv, W.; Zhang, W.; Zheng, X.; Wu, M.-B.; Wei, W.; Tao, Y.; Li, Z.; Yang, Q. Reduction of Graphene Oxide by Hydrogen Sulfide: A Promising Strategy for Pollutant Control and as an Electrode for Li-S Batteries. *Adv. Energy Mater.* **2014**, *4*, 1301565. [CrossRef]
43. Shin, H.-J.; Kim, K.K.; Benayad, A.; Yoon, S.-M.; Park, H.K.; Jung, I.-S.; Jin, M.H.; Jeong, H.-K.; Kim, J.M.; Choi, J.-Y.; et al. Efficient Reduction of Graphite Oxide by Sodium Borohydride and Its Effect on Electrical Conductance. *Adv. Funct. Mater.* **2009**, *19*, 1987–1992. [CrossRef]
44. Moon, I.K.; Lee, J.; Ruoff, R.S.; Lee, H. Reduced graphene oxide by chemical graphitization. *Nat. Comm.* **2010**, *1*, 73. [CrossRef] [PubMed]
45. Liu, C.; Hao, F.; Zhao, Q.; Luo, S.; Lin, H. Low temperature reduction of free-standing graphene oxide papers with metal iodides for ultrahigh bulk conductivity. *Sci. Rep.* **2014**, *4*, 3965. [CrossRef] [PubMed]
46. Chen, D.; Li, L.; Guo, L. An environment-friendly preparation of reduced graphene oxide nanosheets via amino acid. *Nanotechnology* **2011**, *22*, 325601. [CrossRef]
47. Chua, C.K.; Pumera, M. Chemical reduction of graphene oxide: A synthetic chemistry viewpoint. *Chem. Soc. Rev.* **2014**, *43*, 291–312. [CrossRef]
48. Liu, W.; Speranza, G. Tuning the Oxygen Content of Reduced Graphene Oxide and Effects on Its Properties. *ACS Omega* **2021**, *6*, 6195–6205. [CrossRef]
49. Tan, S.M.; Ambrosi, A.; Chua, C.K.; Pumera, M. Electron transfer properties of chemically reduced graphene materials with different oxygen contents. *J. Mater. Chem. A* **2014**, *2*, 10668–10675. [CrossRef]
50. Speranza, G.; Canteri, R. RxpsG a new open project for Photoelectron and Electron Spectroscopy data processing. *SoftwareX* **2019**, *10*, 100282. [CrossRef]

51. Moulder, J.F.; Stickle, W.F.; Sobol, P.E.; Bomben, K.D. *Handbook of X-ray Photoelectron Spectroscopy: A Reference Book of Standard Spectra for Identification and Interpretation of Xps Data*; EdenPraire, Physical Electronics: Minnesota, MN, USA, 1995.
52. Beamson, G.; Briggs, D. *High Resolution XPS of Organic Polymers: The Scienta ESCA300 Database*; Surface Spectra: Chichester, UK, 1992.
53. Chen, C.; Kong, W.; Duana, H.-M.; Zhang, J. Theoretical simulation of reduction mechanism of graphene oxide in sodium hydroxide solution. *Phys. Chem. Chem. Phys.* **2014**, *16*, 12858–12864. [CrossRef] [PubMed]
54. Guex, L.G.; Sacchi, B.; Peuvot, K.F.; Andersson, R.L.; Pourrahimi, A.M.; Strom, V.; Farris, S.; Olsson, R.T. Experimental review: Chemical reduction of graphene oxide (GO) to reduced graphene oxide (rGO) by aqueous chemistry. *Nanoscale* **2017**, *9*, 9562. [CrossRef] [PubMed]
55. Tien, H.W.; Huang, Y.L.; Yang, S.Y.; Hsiao, S.T.; Liao, W.H.; Li, H.M.; Wang, Y.S.; Wang, J.Y.; Ma, C.C.M. Preparation of transparent, conductive films by graphene nanosheet deposition on hydrophilic or hydrophobic surfaces through control of the pH value. *J. Mater. Chem.* **2012**, *22*, 2545–2552. [CrossRef]

C Journal of Carbon Research

MDPI

Article

Carbon Composites—Graphene-Oxide-Catalyzed Sugar Graphitization

Madhu Singh and Randy L. Vander Wal *

John and Willie Leone Family, Department of Energy and Mineral Engineering and The EMS Energy Institute, Penn State University, University Park, PA 16802, USA; madhusingh@alumni.psu.edu
* Correspondence: ruv12@psu.edu

Abstract: Utilization of biopolymers to form graphitic carbons is challenged by their high oxygen content and resulting curved and defective carbon lamellae upon high-temperature heat-treatment. Two composites, one with graphene-oxide (GO) and the other with reduced graphene-oxide (rGO) as fillers, respectively, in a matrix of sugar, each for the same added 2.5 wt.%, exhibited different degrees of graphitization compared to pure sugar on its own. Reactive oxygen groups on GO contribute to reactive templating and crystallite formation. Under high-temperature heat-treatment, sugar, a well-known non-graphitizing precursor, is converted to graphitic carbon in the presence of GO. Possessing fewer oxygen groups, rGO forms two phases in the sugar matrix—a non-graphitic phase and a graphitic phase. The latter is attributed to the remaining oxygen on the rGO.

Keywords: biopolymer; graphitization; sugar; templating; graphene oxide

Citation: Singh, M.; Vander Wal, R.L. Carbon Composites—Graphene-Oxide-Catalyzed Sugar Graphitization. *C* **2022**, *8*, 15. https://doi.org/10.3390/c8010015

Academic Editor: Gil Goncalves

Received: 15 January 2022
Accepted: 9 February 2022
Published: 14 February 2022

1. Introduction

Graphitic or "soft" carbons are ideally suited as electrode materials for energy storage systems due to low cost, high electrical conductivity, stable physicochemical property, and long cycle life in Li batteries [1–3]. Synergistically, the extended and contiguous lamellae provide high electrical conductivity (Battery University https://batteryuniversity.com/learn/article/bu_309_graphite (accessed on 4 December 2021); Properties and Characteristics of Graphite—Poco Graphite http://poco.com/Portals/0/Literature/Semiconductor/IND-109441-0115.pdf (accessed on 4 December 2021)).

Presently, petroleum-derived cokes (Asbury Carbons, PetCoke—https://asbury.com/materials/coke/ (accessed on 3 December 2021)), coal tar pitches and mined (i.e., "natural") graphite are the primary precursors for graphitic carbons [4,5], while oil-based synthetic resins and polymers are preferred precursors for manufactured carbon materials for their graphitic quality and uniformity [6–8]. Unfortunately, petroleum-derived cokes and coal tar pitches are nonrenewable resources that require intensive energy for their production (American Energy Reliance (AER)—Introduction to Petroleum Coke https://www.americanenergyalliance.org/2014/02/introduction-to-petroleum-coke/ (accessed on 3 December 2021); Repsol, Petcoke production https://www.repsol.com/en/products-and-services/green-fuel-grade petcoke/production/index.cshtml (accessed on 4 December 2021)). Moreover, they are accompanied by high organic volatile emissions. Meanwhile, natural graphite reserves are limited and the mining process is environmentally poor (Whoriskey, P., In your phone, in their air. A trace of graphite is in consumer tech. In these Chinese villages, it's everywhere. 2 October 2016. https://www.washingtonpost.com/graphics/business/batteries/graphite-mining-pollution-in-china/ (accessed on 3 December 2021); The unseen side of graphite mining. 6 December 2011. http://rukshanmaliq.blogspot.com/2011/12/unseen-side-of-graphite-mining.html (accessed on 3 December 2021)).

Bio-based, renewable precursors such as cellulose would displace the oil and mined graphite as presently used for the manufacture of carbons for energy storage or composites [9–11]. Water usage, as used in oil production and processing, and CO_2 emissions would be reduced. Yet, when lignocellulosic biomass and other carbohydrate-based or oxygen-rich precursors are carbonized, the formation of non-hexagonal carbon ring structures (pentagons, heptagons, and octagons) within graphene-like sheets prevents graphitization [12,13]. The resulting disordered carbon lacks the galleries formed between parallel and stacked lamellae. Moreover, the electrical conductivity of these carbons is also poor when compared to graphitic carbons [14–16]. Templating offers a path to convert a non-graphitizing bio-based precursor, such as sugar, into a graphitizing carbon.

Relevant to interaction with the matrix, distinctive chemical features of graphene-oxide (GO) and reduced graphene-oxide (rGO) are the sp^2 framework and oxygen content. Physical templating is defined as the interaction between the emerging aromatic domains within the matrix and the sp^2 framework of the GO leading to alignment and orientation. Reactive templating is defined as chemical bond formation between the matrix and GO. Bonding would be favored by either or both components being in radical form, namely, matrix decomposition products or radical sites on GO (or rGO) generated by loss of CO or CO_2. The different oxygen contents for GO versus rGO affect both its physical and chemical templating ability. High oxygen content will lead to extensive radical sites at elevated temperatures. Oppositely, high oxygen content disrupts the sp^2 framework with the conversion of many sp^2 basal sites into sp^3 carbons supporting phenolic and epoxide groups [17,18] and the size of contiguous sp^2 domains is reduced.

These definitions are supported by reactive force field (ReaxFF)-based molecular dynamics (MD) simulations. These simulations reveal that the addition of graphene to polyacrylonitrile (PAN) precursor enhances the alignments of 6-member carbon rings in carbonization [19,20], with π-π interactions presumably being the origin of the alignment force. Related simulations for double-walled nanotubes and graphene within graphitizing PAN showed that the alignment between the template surface and newly formed all-carbon rings improves continuously during the carbonization process by providing a stable sp^2 basal surface [21]. MD simulation trajectories have also revealed that graphene surface carbon atoms are relatively inactive and do not form bonds with the polymer matrix, even at high temperatures, due to their conjugated sp^2 electronic configuration [20,22]. Instead, C-H bonds at the graphene edges break and the undercoordinated carbon atoms at the graphene edges then form bonds with PAN chains. Here, we test oxygen groups as the labile entity to create reactive edge sites.

These concepts are supported by experimental studies. For example, addition of a small amount of GO into PAN increased graphitic order and preferred orientation in the vicinity of visible GO [23]. A separate study found that, during the thermal stabilization of PAN/rGO, rGO confined the cyclization and dehydrogenation reaction of PAN to facilitate the formation of large, conjugated structures [24]. In a related study, graphene was shown to induce carbonization of polyimide [25]. Films doped with 2 wt.% GO were found to have higher crystallization with lamellae orientation and interlayer distance closer to that of ideal graphene compared to that doped with 2 wt.% rGO. This was attributed to bonding between the GO hydroxyl and carboxylic groups and PAN. Elsewhere, PAN/GO composite films spin-coated on substrates were stabilized in air and then carbonized under inert atmosphere at 1000 °C to form carbon films [26]. Cyclization of PAN and partial reduction of GO occurred simultaneously during stabilization. High electrical conductivity appeared due to the π-π interaction between the graphene planes and the cyclized PAN structures [27]. Using SU-8, a photoresist polymer, as host matrix, at the early stages of carbonization, graphene induced the carbon atoms in its near vicinity to form oriented layers parallel to the graphene layer [28]. Upon graphitization, graphene narrowed the distribution of graphite grain orientations.

Here, in contrast to these graphitizable matrix precursors, sugar was chosen as the precursor for its well-known non-graphitizable char as an exacting test of templating to form graphitic carbon—our first objective. As a uniform mixture, the potential for crystallite formation and growth could be probed beyond the additive-matrix interface—the second objective. Last, as a simple disaccharide, results with sugar serve as a reference for more complex, but chemically equivalent, biopolymers, such as starch and cellulose—the third objective of this study.

2. Materials and Methods

2.1. Materials

GO and rGO were used as received from Cheap Tubes Inc. (Grafton, VT, US). GO has 45~55 atomic % oxygen while rGO exhibits ~5 atomic % oxygen, as specified by the manufacturer from XPS measurements. X-Y for both materials is in the 300–800 nm range, each with 2–4 layers stacked for a thickness of ~1.2 nm. rGO is produced from the same graphene oxide, hence, the nominal sheet sizes and stacking are equivalent. XPS analysis by the vendor did not report metal contamination, and our own results did not reveal any Mn or Fe contamination to the 0.1 at.% level. TGA-MS studies have shown similar profiles, i.e., CO and CO_2 gasification products form at similar temperature ranges and in amounts traceable to functional group identity [29,30]. From ambient to 550 °C, there is a progressive loss of oxygen in the form of CO and CO_2. Based on these studies ~50–70% of the oxygen has been transformed into CO and CO_2 for both graphene oxide forms. Each form (rGO and GO) will lose roughly the same fraction of oxygen, albeit different absolute amounts. Food grade sugar (chemically, sucrose) was purchased from Sigma-Aldrich (St. Louis, MO, US).

2.2. Formulation

To synthesize the composites, 2.5 wt.% of a graphene form is thoroughly mixed with the matrix material, both in powdered form. These are mixed to a visible uniform dispersion of the filler in the matrix. The mixture is transferred to an aluminum mold, which is placed inside a closed reactor under autogenic pressure conditions of ¾″ internal diameter and 25 mL internal volume. The mold fits into the inner dimensions of the tubing reactor. The reactor is then sealed and purged of any oxygen with an ultra-high purity nitrogen manifold. Pressurized nitrogen helps check for air leaks from the reactor tube and helps purge it of any air/oxygen. The reactor is purged 4–5 times; a slight nitrogen overpressure of ~5 psi is maintained in the reactor to ensure an inert atmosphere during the heating process.

Carbonization: Once sealed and purged, the tubing reactors were plunged into a pre-heated and aerated sand bath. The temperature of the sand bath was set at 550 °C and was maintained isothermally for a duration of ~ 5 h. This temperature (550 °C) may vary, based on the rates of carbonization of the matrix material used. Most matrices used in this work carbonize in the range of 550–650 °C for 5–6 h. The reactors oscillated to provide vertical agitation over a 2.5 cm amplitude and at a frequency of 100 oscillations per minute for the entire duration of the heat treatment process. Carbonization occurs under autogenous pressure; pressures of up to 1500–2000 psi were reached at the end of the process. At the end of carbonization, the reactors were quenched in cold water for ~5 s. Once cool, the intermediate carbonized form of the composite was retrieved after venting out released gases in a fume hood. The material was washed with acetone and left to dry overnight in a fume hood before graphitization.

Graphitization: This is the next step in the formulation process. Here, the carbonized product was held in graphite crucibles and heated to ~2700 °C in Argon (under a 2 psi overpressure) in a Centorr Vacuum Industries (Nashua, NH, US) series 45 graphitization furnace. The furnace was purged of any oxygen by flushing the furnace chamber with Argon down to 100 millitorrs at least three times before heating. Temperatures up to 1400 °C were measured with a K-type thermocouple, and a Mirage Ircon pyrometer was used to measure higher temperatures up to 2700 °C using a viewport. Samples were held at 2700 °C for ~1 h, after which the furnace and samples were cooled overnight before retrieving them for further characterization.

2.3. Material Characterization

Transmission electron microscopy: Transmission electron microscopy (TEM) and its high-resolution version (HRTEM) was performed using a FEI (Hillsboro, OR, US) Talos F200X equipped with a 200 keV FEG source with a resolution of 0.12 Å. C-C composite samples were ground and dispersed in methanol. The sample was sonicated using a sonication horn for about ~5–10 min before being dropped onto 300 mesh C/Cu lacey TEM grids. Samples were also characterized by selected area electron diffraction (SAED) patterns performed with the same electron microscope.

X-ray Diffraction: X-ray diffraction (XRD) measurements of the samples were carried out in a PANalytical (Malvern, UK) Empyrean X-ray diffractometer using a Cu source (λ = 1.54 Å), para-focusing optics, and a PIXcel 3D detector. Spectrum was collected from 2θ = 5–95°. Measurements were performed after correcting for instrument broadening using an internal silicon standard. Background subtraction and peak deconvolution of the XRD pattern was performed using JADE® (Christchurch, New Zealand) software, which also calculated the d_{002} layered plane spacing as per the Bragg equation. For the sugar (matrix only) and rGO + sugar composite, the two d_{002} values correspond to peaks obtained by deconvolution. The carbon layer plane lateral extent (La) and stack height (Lc) were calculated by applying the Scherrer equation [31].

Several dozen TEM images were acquired across the samples. Multiple samples were prepared and 2–3 widely separated areas on each grid were examined. Given that the material was formed and processed as a powder, there was no control of particle (and crystallite) orientation. Images were selected as representative but not intended to be definitive for graphitic structure formation. XRD was performed on a ZBI silicon plate holding 50 mg of powder. With powder at ~+50 mesh, the material was deemed sufficiently "fine" to uniformly present crystallites in all orientations for unbiased spectra by our prior experience with such powder forms of graphitized carbons. The XRD instrument, an Empyrean III, did not allow for sample rotation

TEM and XRD are complementary techniques, where TEM/HRTEM is useful for illustrating the development of extended, stacked lamellae, while XRD offers a "bulk" measurement over a larger sample. One must exercise caution while discerning lattice spacing from bright field TEM images. This approach is risky given focusing adjustments at any given magnification. SAED patterns are preferred for lattice measurements, as they are not subject to focusing. As shown further, XRD and HRTEM measurements and characterization were consistent and complimentary.

3. Results

Figure 1 shows an HRTEM image of the matrix, sugar, along with its corresponding strong ring-like diffraction pattern, when it was subjected to high-temperature heat treatment by itself, without any added fillers. Sugar is a non-graphitizing (Here, in the context of the TEM images, "non-graphitizing" refers to ribbons, curved lamellae and voids observed, along with absence of ordered stacking.) carbon and its nanostructure exhibits this with multiple intertwined few layer ribbons visible corresponding to the strong ring-like SAED pattern. It indicates the presence of many randomly oriented crystals in the material, with no long-range order, characteristic of a non-graphitizing material.

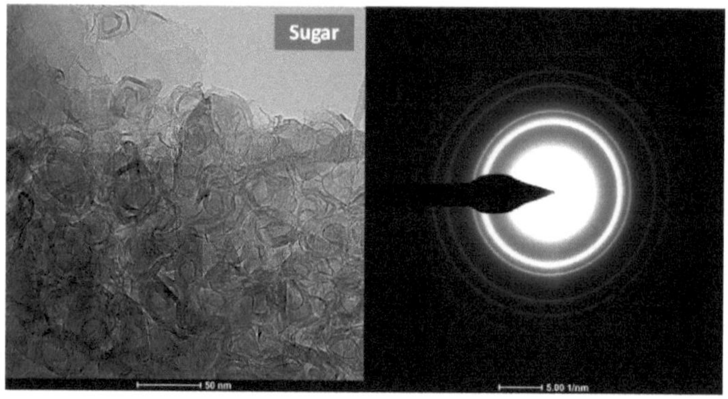

Figure 1. HRTEM image and SAED pattern of the matrix, sugar, after high-temperature heat treatment.

The following panel in Figure 2 shows GO and rGO when heat treated by themselves, as well as when their respective composite materials were formed with GO and rGO as fillers, respectively, in sugar as the surrounding matrix. GO and rGO possess similar starting sheet-like nanostructure, with rGO derived from GO, its oxidized counterpart. On being added as fillers within the sugar matrix and subjected to high-temperature heat treatment, both materials showed very different nanostructures. GO shows a sheet-like nanostructure, while rGO contained a mix of inter-locked ribbons and graphitic sheets, as discussed and shown further.

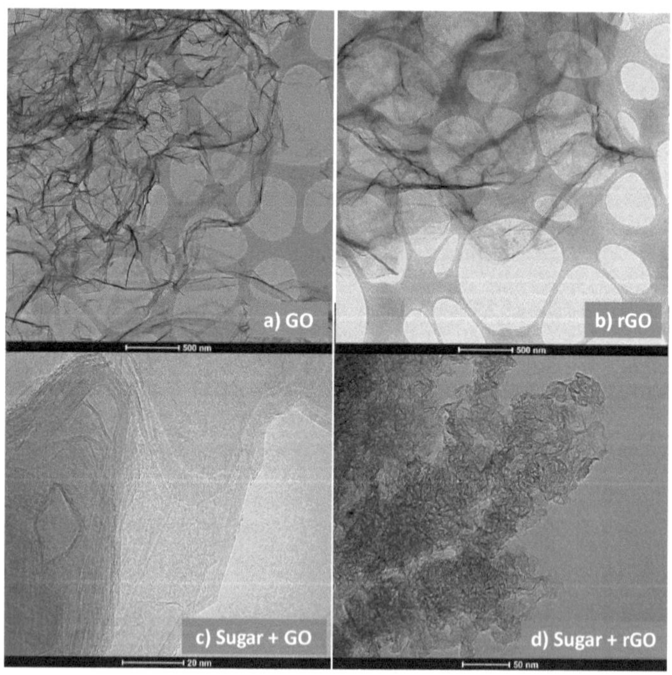

Figure 2. (**a**) GO and (**b**) rGO heat treated by themselves; compared to their respective composites nanostructures with sugar—(**c**) sugar + GO and (**d**) sugar + rGO.

Figure 3 shows comparative XRD spectra for sugar, a well-known non-graphitizing carbon, and two composites, one with rGO + sugar, the other with GO + sugar, each for the same added 2.5 wt.%. The key to this comparison is that the rGO was produced from the same graphene oxide; hence, the nominal sheet sizes and stacking are equivalent. The relatively smaller diffraction angle of the pure sugar is characteristic for the curved lamellae, possessing larger separation, as illustrated by the HRTEM image below. The XRD breadth reflects the low stacking order in the material.

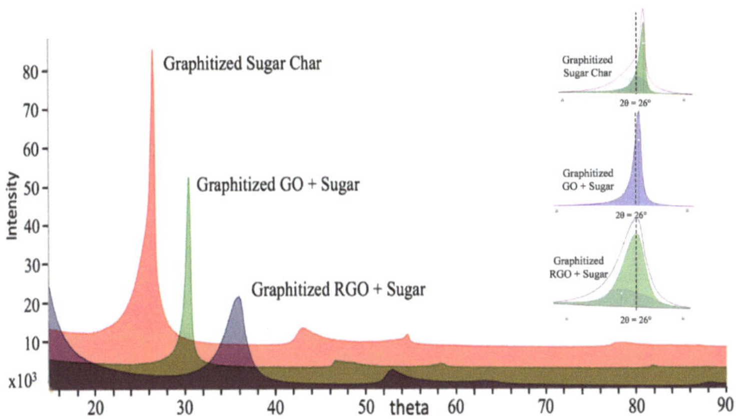

Figure 3. X-ray diffraction spectra for three composites with sugar as matrix precursor, pure and with GO or rGO as additives, illustrating different graphitizing behavior. Note that the X-axis has been shifted to show the three peaks as distinct. However, the peaks occur at 2θ = 26°. The inset shows XRD-peak deconvolution for the three materials.

Lattice and crystallite parameters for the sugar and GO composites are summarized in Table 1. Overall results are quantified by the graphitization index, a metric based on the lattice d-spacing, as extracted from XRD spectra. The limits are 1 for a graphitic material and 0 for a non-graphitic material. (There can be a slight negative value for certain materials. This is due to the limits of the definition). The result is telling—the lower amount of graphitic additive (GO) produces a more graphitic composite.

Table 1. Summary of the crystallite parameters of the listed composites after high-temperature heat treatment.

Filler/ Lattice Parameter	No filler. Only Graphitized Sugar	Graphene Oxide	Reduced GO
d_{002} [Å]	3.48; 3.344	3.39	3.6; 3.43
L_c (using 002) [nm]; 2θ = 26°	2; 16	13	2; 4
L_c (using 004) [nm]; 2θ = 44°	10	7	4
L_a (using 110) [nm]; 2θ = 54°	16	25	10
g (Degree of graphitization)	~1	0.6	0.14

Redefined as: g = (3.440-d_002)/(3.440–3.345) (Originally defined by Franklin, 1951, [32]).

As shown in the inset of Figures 3 and 4, deconvolution of the main d_{002} peak for the rGO + sugar composite reveals two d_{002} peaks, one broader, appearing at a lower (mean) angle of 24° and a much sharper peak, appearing at 26°, representing a smaller d_{002} spacing and corresponding to higher stacking order than the former by Scherrer's analysis. The broader peak is consistent with a non-graphitized, sugar-derived carbon, while the latter is indicative of a very graphitic phase in the composite.

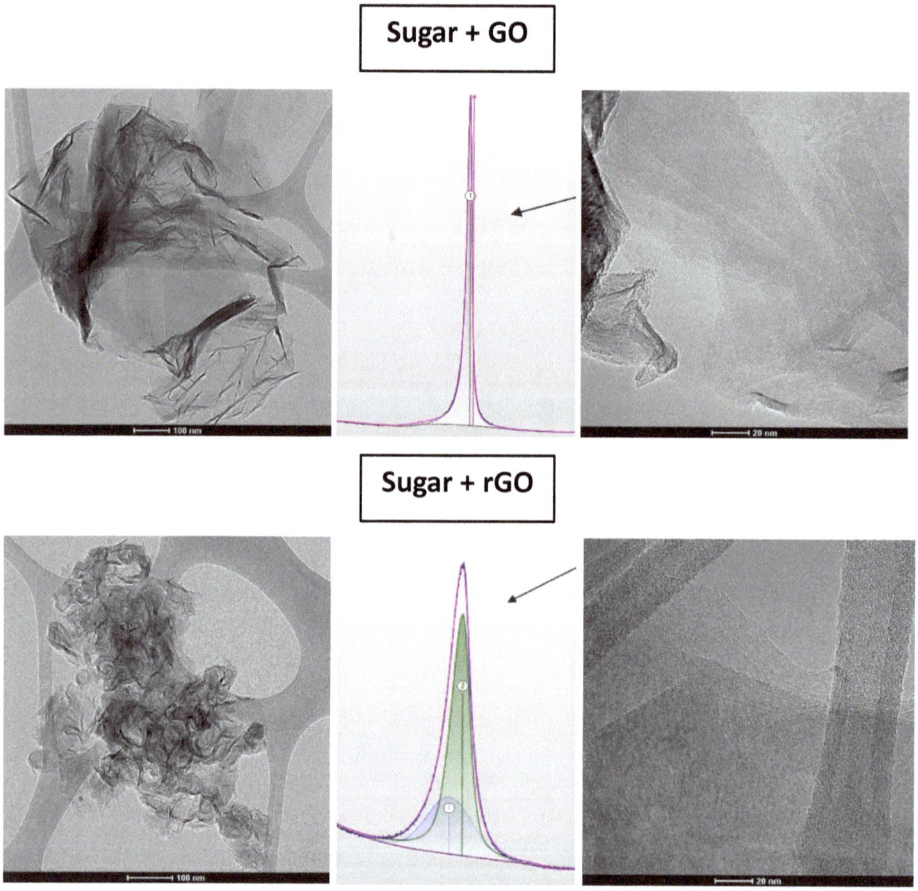

Figure 4. HRTEM images for C-C composites fabricated from the indicated precursors. Center panels show the XRD d_{002} peaks for each, illustrating the single or two components for the two composites.

The existence of the two phases is also supported by HRTEM images and their corresponding SAED patterns shown in Figure 5. A clear spot pattern is indicative of long-range order and crystallinity in one direction, while an intermediate spot-ring pattern indicates many randomly oriented crystal structures, resulting in their respective spot patterns merging into a ring-like pattern. This is consistent with reactive templating, as discussed further. GO + sugar, in contrast, showed only one uniform phase throughout, as shown in Figure 6, as well as in prior HRTEM images of the composite in this manuscript.

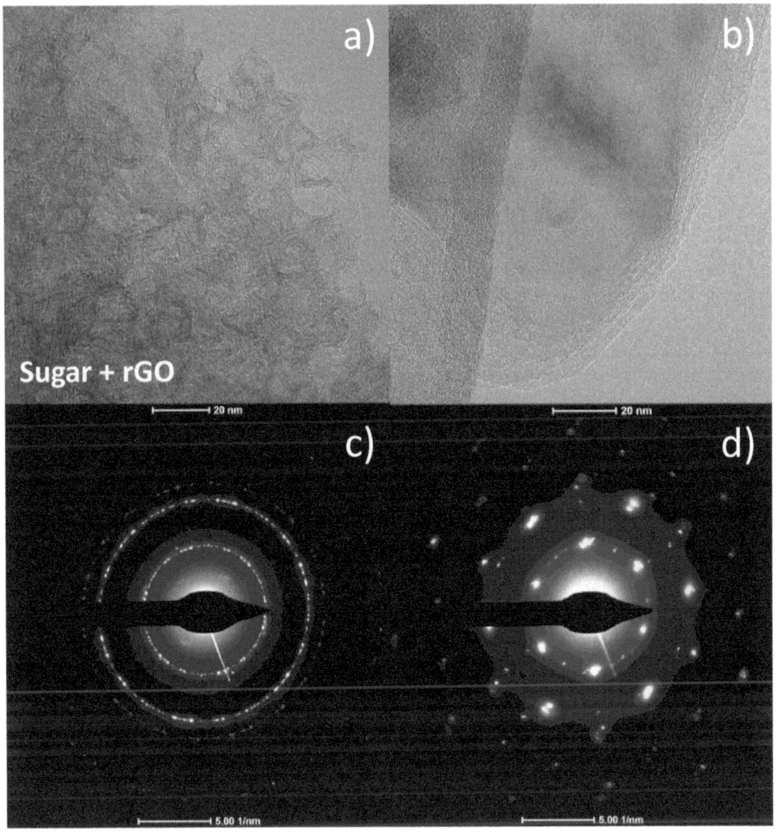

Figure 5. (**a**,**b**) HRTEM images and (**c**,**d**) SAED patterns for the phases observed in the sugar + rGO C-C composite. (**a**,**c**) correspond to the phase with randomly oriented crystals, while (**b**,**d**) correspond to the phase with sheet-like graphitic unidirectional crystallinity.

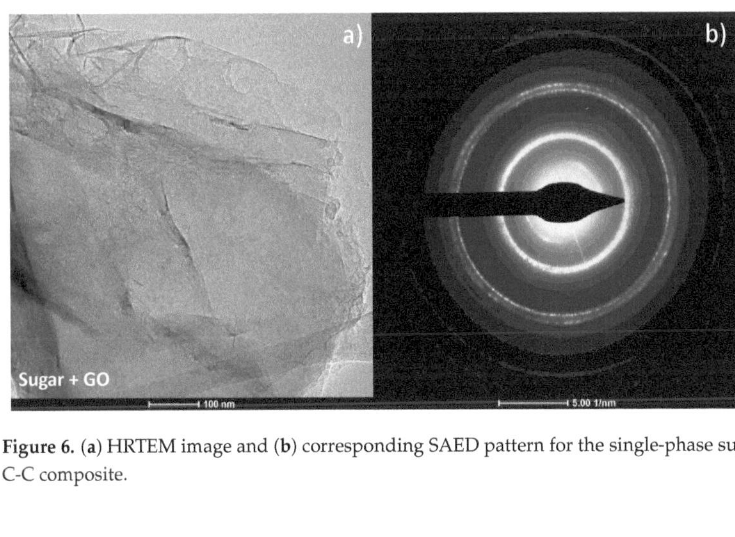

Figure 6. (**a**) HRTEM image and (**b**) corresponding SAED pattern for the single-phase sugar + GO C-C composite.

4. Discussion

The graphitic phase component for the rGO + sugar is some fraction of the composite that is highly graphitic whereas the rest is not, being normal, sugar-derived char as is historically known. In contrast, the sugar char with GO exhibited a single (002) diffraction peak, signaling a single graphitic phase. Comparing rGO and its precursor, GO, in a non-graphitizing matrix, demonstrates reactive templating. The reduced form with largely restored basal plane carbons contained few remaining oxygen groups to form free radical sites during carbonization. As a result, rGO is largely unable to reactively bond with the matrix and exert reactive templating. Given the weak interaction between an extended aromatic π network and a non-aromatic (sp^3) precursor (sugar), the possibility for physical templating appears to be minimal.

In contrast, GO forms many radical sites that will seek termination by reactive bonding with radical sites within the decomposing matrix. These comparative results demonstrate reactive bonding between "like" additive and matrix with similar chemistries and oppositely, little templating when allotrope additive and matrix are dissimilar i.e., mismatched in their chemistry. By this comparison, the limiting "factor" for rGO as an additive to induce graphitization (through reactive templating) is its residual oxygen content after reduction from the GO form.

Therein, the graphitic component for the rGO composite and the GO composite both arise through, and are illustrations of, reactive templating, wherein radical sites (mainly on the sheet edges) formed by oxygen group loss during carbonization bond with the numerous free radicals formed from the sugar matrix during thermal decomposition. Figures 5 and 6 show HRTEM images for the rGO and GO composites, in which the graphitic and uniform phase for the GO composite is apparent, whereas rGO, by comparison, reveals both graphitized and non-graphitized carbon regions. Notably, the two phases visible for the rGO composite via HRTEM are qualitatively consistent with the 2-peak deconvolution of its corresponding XRD pattern, as shown in the middle panel. Still, a contribution of the sp^2 network to aromatic ring alignment cannot be ruled out.

5. Summary

In summary, comparing GO and its oxygen-depleted reduced form, rGO, in a non-graphitizing matrix demonstrates reactive templating. For rGO, given the incompatibility of an extended aromatic π network to physically template a non-aromatic precursor, the possibility for physical templating is negligible; we see a composite that behaves similarly to its dominant matrix phase, sugar. With GO as a filler, on the other hand, there is a creation of multiple reactive sites on carbonization, and the general absence of any significant π network further removes any potential contribution to such bonding with the matrix. This shows reactive bonding and, therefore, templating to form a graphitizing carbon when the filler and matrix are alike (as in GO + sugar), rather than when they are dissimilar (as in rGO + sugar), in which case the allotrope possesses stable sp^2-bound carbons.

Author Contributions: Conceptualization, R.L.V.W.; Data curation, M.S.; Formal analysis, M.S.; Funding acquisition, R.L.V.W.; Investigation, M.S.; Methodology, M.S.; Supervision, R.L.V.W.; Visualization, R.L.V.W.; Writing—original draft, M.S.; Writing—review and editing, M.S. and R.L.V.W. All authors have read and agreed to the published version of the manuscript.

Funding: This work was supported under Grant W911NF-17-1-0513, under the ARO STIR program (2017–2018).

Data Availability Statement: Data is contained within the article.

Acknowledgments: Characterizations (TEM, XRD, EELS) were performed using the facilities of the Materials Research Institute at The Pennsylvania State University. HRTEM guidance from Jennifer Gray and Ke Wang is gratefully acknowledged.

Conflicts of Interest: The authors declare no conflict of interest.

References

1. Yoshino, A. Development of the Lithium-Ion Battery and Recent Technological Trends. In *Lithium-Ion Batteries. Advances and Applications*; Pistoia, G., Ed.; Elsevier: Amsterdam, The Netherlands, 2014; pp. 1–20. [CrossRef]
2. Matthé, R.; Eberle, U. The Voltec System—Energy Storage and Electric Propulsion. In *Lithium-Ion Batteries. Advances and Applications*; Pistoia, G., Ed.; Elsevier: Amsterdam, The Netherlands, 2014; pp. 151–176. [CrossRef]
3. To Engage with Chemical engineers and Researchers Working in the Field of Battery Technologies. In Proceedings of the Battery and Energy Storage Workshop Proceedings, New York, NY, USA, 21–22 October 2019.
4. Krauss, U.; Schmidt, H.; Taylor, H.; Sutphin, D. International Strategic Minerals Inventory Summary Report; Natural Graphite. In *Circular*; U.S. Department of the Interior, Geological Survey: Washington, DC, USA, 1989.
5. Lee, S.M.; Kang, C.S.; Roh, J.S. Bulk graphite: Materials and manufacturing process. *Carbon Lett.* **2015**, *16*, 135–146. [CrossRef]
6. Fitzer, E. The future of carbon-carbon composites. *Carbon* **1987**, *25*, 163–190. [CrossRef]
7. Savage, E. *Carbon-Carbon Composites*; Springer: Berlin/Heidelberg, Germany, 2012. [CrossRef]
8. Newcomb, B.A. Processing, structure, and properties of carbon fibers. *Compos. Part A Appl. Sci. Manuf.* **2016**, *91*, 262–282. [CrossRef]
9. Gao, Z.; Zhang, Y.; Song, N.; Li, X. Biomass-derived renewable carbon materials for electrochemical energy storage. *Mater. Res. Lett.* **2017**, *5*, 69–88. [CrossRef]
10. Correa, C.R.; Kruse, A. Biobased Functional Carbon Materials: Production, Characterization, and Applications—A Review. *Materials* **2018**, *11*, 1568. [CrossRef]
11. Xie, X.; Goodell, B. Thermal Degradation and Conversion of Plant Biomass into High Value Carbon Products. In *Deterioration and Protection of Sustainable Biomaterials*; American Chemical Society: Washington, DC, USA, 2014; Volume 1158, pp. 147–158.
12. Jenkins, G.M.; Kawamura, K. *Polymeric Carbons—Carbon Fibre, Glass and Char*; Cambridge University Press: Cambridge, UK, 1976.
13. McDonald-Wharry, J.; Manley-Harris, M.; Pickering, K. Reviewing, Combining, and Updating the Models for the Nanostructure of Non-Graphitizing Carbons Produced from Oxygen-Containing Precursors. *Energy Fuels* **2016**, *30*, 7811–7826. [CrossRef]
14. Han, J.; Kwon, J.H.; Lee, J.-W.; Lee, J.H.; Roh, K.C. An effective approach to preparing partially graphitic activated carbon derived from structurally separated pitch pine biomass. *Carbon* **2017**, *118*, 431–437. [CrossRef]
15. Stephan, A.M.; Kumar, P.; Ramesh, R.; Thomas, S.; Jeong, S.K.; Nahm, K.S. Pyrolitic carbon from biomass precursors as anode materials for lithium batteries. *Mater. Sci. Eng. A* **2006**, *430*, 132–137. [CrossRef]
16. Miao, M.; Zuo, S.; Zhao, Y.; Wang, Y.; Xia, H.; Tan, C.; Gao, H. Selective oxidation rapidly decomposes biomass-based activated carbons into graphite-like crystallites. *Carbon* **2018**, *140*, 504–507. [CrossRef]
17. Song, J.; Wang, X.; Chang, C.-T. Preparation and Characterization of Graphene Oxide. *J. Nanomater.* **2014**, *2014*, 276143. [CrossRef]
18. Zhang, Z.; Schniepp, H.C.; Adamson, D.H. Characterization of graphene oxide: Variations in reported approaches. *Carbon* **2019**, *154*, 510–521. [CrossRef]
19. Rajabpour, S.; Mao, Q.; Gao, Z.; Talkhoncheh, M.K.; Zhu, J.; Schwab, Y.; Kowalik, M.; Li, X.; Van Duin, A.C. Polyacrylonitrile/Graphene Nanocomposite: Towards the Next Generation of Carbon Fibers. *arXiv* **2020**, arXiv:2006.11985.
20. Saha, B.; Furmanchuk, A.; Dzenis, Y.; Schatz, G.C. Multi-step mechanism of carbonization in templated polyacrylonitrile derived fibers: ReaxFF model uncovers origins of graphite alignment. *Carbon* **2015**, *94*, 694–704. [CrossRef]
21. Papkov, D.; Beese, A.M.; Goponenko, A.; Zou, Y.; Naraghi, M.; Espinosa, H.D.; Saha, B.; Schatz, G.C.; Moravsky, A.; Loutfy, R.; et al. Extraordinary Improvement of the Graphitic Structure of Continuous Carbon Nanofibers Templated with Double Wall Carbon Nanotubes. *ACS Nano* **2013**, *7*, 126–142. [CrossRef]
22. Rajabpour, S.; Mao, Q.; Gao, Z.; Talkhoncheh, M.K.; Zhu, J.; Schwab, Y.; Kowalik, M.; Li, X.; Van Duin, A.C. Low-temperature carbonization of polyacrylonitrile/graphene carbon fibers: A combined ReaxFF molecular dynamics and experimental study. *Carbon* **2021**, *174*, 345–356.
23. Papkov, D.; Goponenko, A.; Compton, O.C.; An, Z.; Moravsky, A.; Li, X.-Z.; Nguyen, S.T.; Dzenis, Y.A. Improved Graphitic Structure of Continuous Carbon Nanofibers via Graphene Oxide Templating. *Adv. Funct. Mater.* **2013**, *23*, 5763–5770. [CrossRef]
24. Lei, S.; Zhong, S.; Wang, Y.; Tong, Y.; Xu, L. Preparation of monodisperse reduced graphene oxide/polyacrylonitrile composite and its thermal-induced structural transformation. *Mater. Lett.* **2015**, *161*, 108–111. [CrossRef]
25. Ma, L.; Wang, Y.; Wang, Y.; Wang, C.; Gao, X. Graphene induced carbonization of polyimide films to prepared flexible carbon films with improving-thermal conductivity. *Ceram. Int.* **2020**, *46*, 3332–3338. [CrossRef]
26. Baek, G.Y.; Lee, H.S.; Jung, J.-M.; Hwang, I.-T.; Shin, J.; Jung, C.-H.; Choi, J.-H. Preparation of conductive carbon films from polyacrylonitrile/graphene oxide composite films by thermal treatment. *J. Ind. Eng. Chem.* **2018**, *58*, 87–91. [CrossRef]
27. Zhu, D.; Xu, C.; Nakura, N.; Matsuo, M. Study of carbon films from PAN/VGCF composites by gelation/crystallization from solution. *Carbon* **2002**, *40*, 363–373. [CrossRef]
28. Cunning, B.V.; Wang, B.; Shin, T.J.; Ruoff, R.S. Structure-directing effect of single crystal graphene film on polymer carbonization and graphitization. *Mater. Horiz.* **2019**, *6*, 796–801. [CrossRef]
29. Figueiredo, J.L.; Pereira, M.F.R.; Freitas, M.M.A.; Órfão, J.J.M. Characterization of Active Sites on Carbon Catalysts. *Ind. Eng. Chem. Res.* **2007**, *46*, 4110–4115. [CrossRef]

30. Kundu, S.; Wang, Y.; Xia, W.; Muhler, M. Thermal Stability and Reducibility of Oxygen-Containing Functional Groups on Multiwalled Carbon Nanotube Surfaces: A Quantitative High-Resolution XPS and TPD/TPR Study. *J. Phys. Chem. C* **2008**, *112*, 16869–16878. [CrossRef]
31. Patterson, A.L. The Scherrer Formula for X-Ray Particle Size Determination. *Phys. Rev.* **1939**, *56*, 978–982. [CrossRef]
32. Franklin, R.E. The structure of graphitic carbons. *Acta Crystallogr.* **1951**, *4*, 253–261. [CrossRef]

C Journal of Carbon Research

MDPI

Article

Structural and Electrochemical Characteristics of Platinum Nanoparticles Supported on Various Carbon Carriers

Margarita Kozlova [1,2], Sergey Butrim [1,2], Maksim Solovyev [1,2], Artem Pushkarev [1,2], Irina Pushkareva [1,2], Valery Kalinichenko [2,3,4], Svetlana Akelkina [2] and Sergey Grigoriev [1,2,5,6,*]

1 National Research University "Moscow Power Engineering Institute", 14, Krasnokazarmennaya St., 111250 Moscow, Russia; KozlovaMarV@mpei.ru (M.K.); ButrimSI@mpei.ru (S.B.); SolovyevMaxA@mpei.ru (M.S.); PushkarevAS@mpei.ru (A.P.); PushkarevaIV@mpei.ru (I.P.)
2 National Research Centre, "Kurchatov Institute", 1, Akademika Kurchatova Sq., Moscow 123182, Russia; kalinval47@mail.ru (V.K.); akelkina_sv@nrcki.ru (S.A.)
3 N.M. Emanuel Institute of Biochemical Physics, Russian Academy of Sciences, 4, Kosygina St., 119991 Moscow, Russia
4 N.N. Semenov Federal Research Center for Chemical Physics, Russian Academy of Sciences, 4, Kosygina St., 119991 Moscow, Russia
5 HySA Infrastructure Center of Competence, Faculty of Engineering, North-West University, Potchefstroom 2531, South Africa
6 A.N. Nesmeyanov Institute of Organoelement Compounds, Russian Academy of Sciences, 28, Vavilova St., 119991 Moscow, Russia
* Correspondence: GrigoryevSA@mpei.ru; Tel.: +7-495-362-7206

Abstract: Graphene-like materials have attracted significant attention as alternative catalyst carriers due to the broad possibilities of changing their shape, composition, and properties. In this study we investigated the structural and electrochemical characteristics of platinum electrocatalysts supported on reduced graphene oxide (rGO), including those modified with amine functionalities, nitrogen heteroatoms (rGO-Am), and oxygen enriched (rGO-O). Synthesis of Pt nanoparticles (20 wt.%) on the graphene-like nanomaterials surface was carried out using a modified polyol procedure. The Pt^{20}/rGO-Am showed a lower Pt nanoparticles size together with high Pt utilization and EASA values compared to rGO-supported catalysts and the Pt/C reference sample due to the uniform distribution of nucleation centers on the surface of graphene nanoparticles, and the greater ability of these centers to electrically bond with platinum.

Keywords: reduced graphene oxide; aminated graphene; ozonized graphene; functionalized catalyst carrier; carbon-supported platinum nanoparticle; platinum utilization; functional group; polyol reduction; polymer electrolyte membrane fuel cell; PEMFC

Citation: Kozlova, M.; Butrim, S.; Solovyev, M.; Pushkarev, A.; Pushkareva, I.; Kalinichenko, V.; Akelkina, S.; Grigoriev, S. Structural and Electrochemical Characteristics of Platinum Nanoparticles Supported on Various Carbon Carriers. C **2022**, *8*, 14. https://doi.org/10.3390/c8010014

Academic Editor: Gil Goncalves

Received: 24 December 2021
Accepted: 4 February 2022
Published: 14 February 2022

1. Introduction

Polymer electrolyte membrane fuel cells (PEMFC) are being developed as power sources for transport, stationary and portable systems as an alternative to conventional internal combustion engines, secondary batteries, and other conventional power sources [1]. The catalyst for the oxygen reduction reaction (ORR) (usually carbon-supported Pt nanoparticles [2]) is the main component of the membrane-electrode assembly (MEA) of PEMFC, which determines its performance, efficiency and durability [3]. The high specific surface area of catalysts based on Pt nanoparticles is achieved by using carbon carries [4]. Carbon carries must have such properties as: a high specific surface area, high electrical conductivity, developed pore structure and corrosion resistance [5]. In particular, in order to ensure the transport of reagents/reaction products to/from the active centers, the carbon supports must have a large number of mesopores and macropores, and an appropriate morphology [6]. The most commonly used carbon supports are carbon blacks (for example, Vulcan XC-72), which do not completely fulfil catalyst support requirements, especially on the corrosion resistance [7,8]. The disadvantage of carbon black consists of its microporous structure. The metal particles may sink inside micropores, decreasing the electrochemically active surface area (EASA). This leads to the development of mesoporous carbons with high mesopores (2 < pore sizes < 50 nm) degree. These materials have attracted a lot of attention due to their regular structure, high surface area, large pore volume and a narrow pore size distribution [9].

The challenging task is the development of alternative supports for Pt nanoparticles that increase the activity and stability of the catalyst, reduce the noble metal loading, and provide optimal properties of the active layer (hydrophobicity/hydrophilicity, porous structure, etc.) that allow the mass transfer facilitation [10,11]. Therefore, graphene-like materials and their derivatives are considered as promising catalyst supports.

Graphene first obtained by mechanical exfoliation [12] is significantly exploited for energy storage systems because of its excellent physical-chemical properties: high specific surface area (up to 2600 $m^2 g^{-1}$), high thermal and electrical conductivity, corrosion resistance and mechanical strength (up to 130 GPa) [13]. In graphene, each carbon atom is covalently bonded with neighboring carbon atoms by three in-plane σ-bonds (from sp^2 orbitals) and one π-bond (from the remaining 2p orbital) perpendicular to the basal plane comprising the electron delocalization network [14,15]. For many electrochemical systems, including PEMFC, dispersed derivatives of graphene, such as three-dimensional reduced graphene oxide (rGO) are of practical interest [16,17]. Due to the sp2-hybridized carbon atoms the chemical and electrochemical stability of graphene-like materials are quite high [18,19], and these materials are considered as promising supports of Pt nanoparticles. However, the graphene basal plane is inert and does not have enough defects that could act as the active centers of electrochemical reactions or nucleation of nanoparticles [20]. Carbon materials modified with heteroatoms (B, N, S or P) allow for breaking the surface electroneutrality and redistributing the electronic density, providing the appearance of active centers for the deposition of Pt nanoparticles, as well as controlling their properties [21]. This allows for achieving higher activity and stability of electrocatalysts in ORR and other reactions compared with counterparts based on carbon black or graphene-like materials without doping [22]. In particular, aminated graphene exhibits complicated morphology with a tendency to form a wrinkled and corrugated structure, which facilitates the formation of porous films and aerogels, making it promising for catalyst and sensing applications [23].

The proposed study is focused on the structural, morphological and electrochemical properties of electrocatalysts supported on rGO modified with amine functionalities and nitrogen heteroatoms (rGO-Am), and an oxygen enriched one (rGO-O) obtained by ozone treatment. The synthesized catalysts (with 20 wt.% of Pt) were characterized using different techniques, including X-ray photoemission spectroscopy (XPS), X-ray diffraction (XRD), transmission electron microscopy (TEM), and cyclic voltammetry.

2. Materials and Methods

Graphene oxide (GO) was obtained by modified Hummers' method, as described in [24,25].

The rGO-O was obtained by ozone–oxygen mixture treatment which was performed at 20 °C by barrier discharge in the ozone generator "MEDOZONE" UOTA-60-01. The ozone concentration was 30–50 mg L^{-1}, and the processing was carried out for 40–50 min until the ozone absorption stopped [26].

The rGO-Am was synthesized via the following three-stage approach. At the first stage, GO was treated with hydrobromic acid. At the second stage the brominated graphene oxide was subjected to ammonolysis with alcoholic ammonia solution. At the third stage the aminated graphene oxide was reduced with hydrazine.

Synthesis of rGO-supported Pt nanoparticles (20 wt.%) was carried out by a modified polyol reduction approach [27]. Briefly, 11.2 mL $H_2PtCl_6 \times 2H_2O$ was used as the platinum precursor, and 180 mL of ethylene glycol was used as the reducing agent. The synthesis was carried out in a Drexel flask with argon flowing in the flask, which was kept at 75 °C for 2 h, and then heated to a reduction temperature of 180 °C, and kept for 1 h. Washing was performed by decantation six times. Pt supported on Vulcan XC-72 (Pt/C) was used as the reference catalyst (Prometheus R&D Ltd., Rostov-on-Don, Russia).

XRD (SmartLab SE X-ray Diffractometer, Rigaku SmartLab, Tokyo, Japan) was carried out using CuK_α radiation. The diffraction data for Pt^{20}/C, Pt^{20}/rGO, Pt^{20}/rGO-O and Pt^{20}/rGO-Am catalyst were recorded for 2θ angles between 5° and 120°. The crystallite size of Pt nanoparticles (d) was determined by Debye–Scherrer's equation:

$$d = \frac{K\lambda}{\beta cos\theta} \quad (1)$$

where K is the Scherrer constant (for spherical particles with cubic symmetry is equal to 0.94), λ is the wavelength of the X-ray used (0.15406 nm), β is the full width of the diffraction peak at half maximum in radians (FWHM), and θ is the Bragg angle.

The morphological features of the catalysts were studied by transmission electron microscopy (TEM) using a Titan^{TM} 80–300 S/TEM ("FEI") (80–300 kV, space resolution by points is equal to 0.07–0.08 nm).

The XPS studies were performed using a PHOIBOS 150 (SPECS Surface Nano Analysis GmbH, Berlin, Germany) hemispherical analyzer and monochromatic Al Kα radiation (Kurchatov Complex of Synchrotron and Neutron Investigation, NRC "Kurchatov Institute", Moscow, Russia). Obtained spectrums (C 1*s*, O 1*s*, Pt 4*f*) were measured and processed according to the approach described in [28,29].

Electrochemical studies were performed using a potentiostat Solartron 1285 (AMETEK Inc., Berwyn, PA, USA) in a three-electrode cell using Ag/AgCl/sat. KCl solution reference electrode (Ag/AgCl), polished glassy carbon working electrode, and Pt wire counter electrode [27,28]. Cyclic voltammograms were measured using Ar-purged 0.5 M H_2SO_4 solution in the potential range of –0.15 to 1.2 V at the sweep rate of 20 mV s^{-1}.

Electrocatalytic ink consisting of 16 mg of catalyst and 8 mL of water was prepared by ultrasonic treatment using a probe ultrasonic homogenizer (Cole Parmer, Vernon Hills, IL, USA) for 20 s with the following treatment in an ultrasonic bath for 40 min. Then, a 10 µL inks aliquot was dropped onto the working electrode surface. The pretreatment of the working electrode was performed via cyclic voltammetry at 100 mV s^{-1} until the stable curve was obtained.

Electrochemical surface area was calculated from cyclic voltammetry analysis by using the following equation:

$$EASA = \left(\frac{Q}{Q_H * L_{pt}}\right) * 10^{-1} \quad (2)$$

where Q is the charge associated with the adsorption/desorption of hydrogen monolayers (average value between anodic and cathodic sweep was taken) (µC cm^{-2}), L_{Pt} is the Pt loading on the working electrode (mg cm^{-2}), and Q_H is the charge required to oxidize the monolayer of hydrogen on a Pt surface (µC cm^{-2}) [30].

The total Pt surface area was calculated using the equation below, considering particles have spherical shape [28]:

$$Spt = \frac{6000}{p * d} \quad (3)$$

where *Spt*—the total surface area of Pt particles (m^2 g^{-1}), *d*—the Pt particle diameter (nm) taken as the mean value between d_{XRD} and d_{TEM}, a *p*—the density of Pt (21.4 g m^{-3}).

The accessibility of the Pt nanoparticles and the catalytic layer connectivity could be analyzed using the Pt utilization. Pt utilization of 100% means that all Pt nanoparticles are integrated into the entire electron conductivity network and are fully accessible by the electrolyte. The Pt utilization was calculated using the equation below:

$$Upt = \left(\frac{EASA}{S_{pt}}\right) * 100 \quad (4)$$

where U_{pt} is the Pt utilization of the catalyst (%) [28,31].

3. Results and Discussion

TEM-micrographs of Pt^{20}/rGO, Pt^{20}/rGO-O and Pt^{20}/rGO-Am catalysts are shown on Figure 1. All catalysts exhibit a layered structure which is characterized by some extent of Pt nanoparticles agglomeration. Using well-known ImageJ software [32], an analysis of the size distribution of Pt nanoparticles were performed (no less than 100 particles for each catalyst have been taken into account). Pt nanoparticles are well-dispersed with a narrow size distribution in the range of 2–3 nm with an average size of 2.8, 2.5 and 2.4 nm for Pt^{20}/rGO, Pt^{20}/rGO-O and Pt^{20}/rGO-Am, respectively. However, rGO-supported Pt nanoparticles are characterized by agglomeration, which leads to a decrease in catalysts EASA [28,33]. Pt^{20}/rGO-Am has the smallest average nanoparticle size because the nitrogen doping of rGO allows for the formation of additional defects on the rGO basal surface, which provides additional active centers for stabilizing Pt nanoparticles during their deposition. This contributes to a more uniform distribution of smaller Pt nanoparticles [34,35]. It should be noted that the average sizes of Pt nanoparticles in all obtained rGO-based catalysts were lower compared to the commercial catalyst Pt^{20}/C (average size of 3.1 nm) which is taken as a well-known reference catalyst [36].

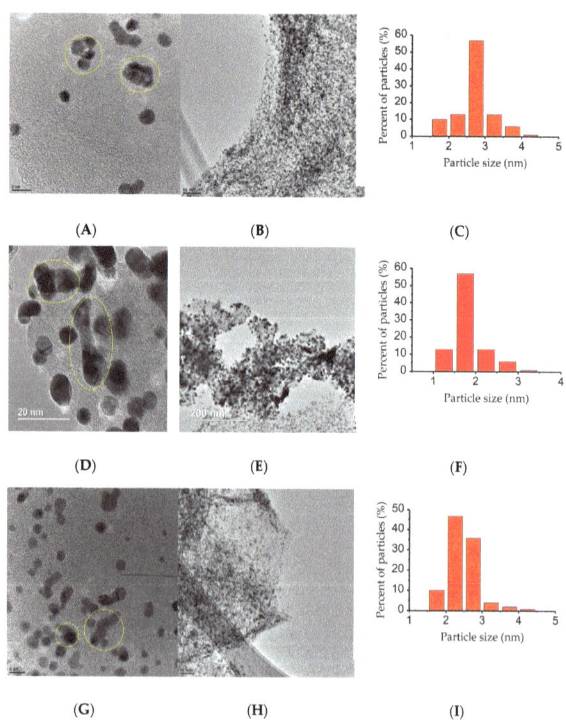

Figure 1. TEM images of Pt^{20}/rGO (**A**,**B**), Pt^{20}/rGO-O (**D**,**E**) and Pt^{20}/rGO-Am (**G**,**H**), the corresponding Pt nanoparticles size distribution (**C**,**F**,**I**).

In addition, the XRD investigations were performed to confirm the crystalline structures of the Pt on different supports (Figure 2). The characteristic peaks at $2\theta \approx 23.3°$ with respect to the carbon (002) plane are detected for all studied catalysts. The diffraction peaks at 2θ of 39°, 46°, 67°, 81° and 85° correspond to the (111), (200), (220), (311) and (222) planes of the face-centered cubic structure of Pt crystallites, respectively [37]. The average crystallite size was calculated for the XRD peak assigned to Pt(220), as this is far from the background signal of carbon support [38]. The Pt crystallite size of Pt^{20}/C, Pt^{20}/rGO, Pt^{20}/rGO-O and Pt^{20}/rGO-Am was found to be ca. 2.83, 2.70, 2.85 and 2.69 nm, respectively. The obtained average crystallite sizes are given in Table 1 together with the particle sizes obtained by TEM. The mean Pt(220) crystallite size values are in a narrow range of 2.69–2.85 nm, which generally correlates well with the Pt nanoparticles mean size values (2.4–3.1 nm).

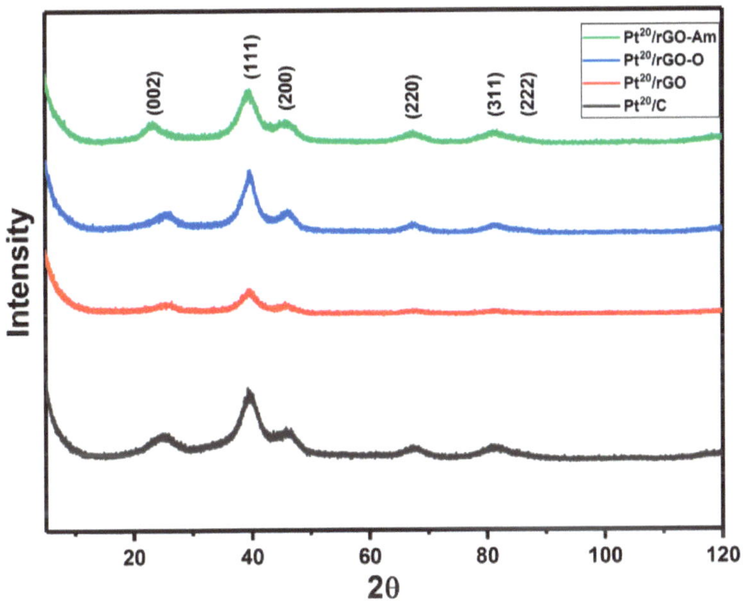

Figure 2. X-ray diffraction patterns for Pt^{20}/C, Pt^{20}/rGO, Pt^{20}/rGO-O and Pt^{20}/rGO-Am.

Table 1. Comparative parameters of the obtained catalysts.

Sample	d_{TEM}, nm	d_{XRD}, nm	EASA, $m^2\ g^{-1}$	S_{Pt}, $m^2\ g^{-1}$	U_{Pt}, %
Pt^{20}/C	3.1	2.83	70.5	94.7	77.4
Pt^{20}/rGO	2.8	2.70	67.1	101.9	65.8
Pt^{20}/rGO-O	2.5	2.85	40.6	105.0	25.8
Pt^{20}/rGO-Am	2.4	2.69	87.5	110.3	79.3

The elemental composition of Pt^{20}/rGO, Pt^{20}/rGO-O and Pt^{20}/rGO-Am catalysts was studied using X-ray photoelectron spectroscopy. According to the C 1*s* spectra (Figure 3A,C,E), rGO-supported catalysts primarily consist of sp^2–carbon (~284.4 eV). The sp^2/sp^3 ratios of Pt^{20}/rGO, Pt^{20}/rGO-O and Pt^{20}/rGO-Am are 38.69, 23.4 and 1.23.

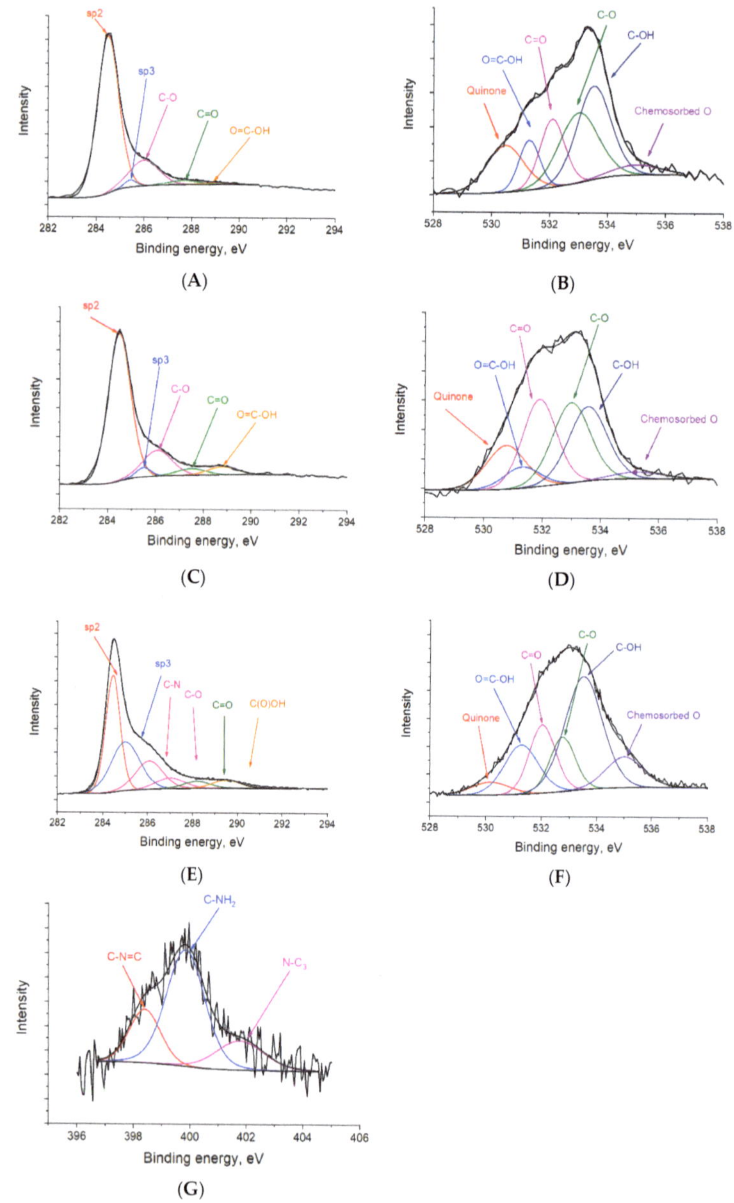

Figure 3. C 1s (**A**,**C**,**E**), O 1s (**B**,**D**,**F**) and N 1s (**G**) spectrums of catalysts: Pt^{20}/rGO (**A**,**B**), Pt^{20}/rGO-O (**C**,**D**) and Pt^{20}/rGO-Am (**E**,**F**,**G**).

According to the C 1s spectra, Pt^{20}/rGO and Pt^{20}/rGO-O are rather similar to each other and differ mainly in the relative intensity ratios of O=C-OH, C=O, C–O, C-OH and quinone groups [29]. The existence of C–N bonds (286.07 eV) [39], as well as relatively high sp^3 carbon atoms ratios, should be noticed in Pt^{20}/rGO-Am.

The O 1s spectra of catalysts are characterized by several peaks and allow for a somewhat more detailed assessment of the oxygen-containing groups. In particular, quinone (~530.3–530.5 eV), carboxyl (~531.1 eV), carbonyl (~532.1 eV), C–O (~532.7–533.0 eV), and hydroxyl (~533.7 eV) functional groups [29,40] are detected. Lower binding energy peaks in the O 1s spectra could be ascribed to the quinone [41] or Me-O. The existence of oxygen functional groups in rGO is due to the incomplete thermal reduction of graphene oxide at temperatures below 2000 °C [42]. The C/O value of each catalyst shows oxygen content and equals to 15.57, 9.35 and 8.87 at.% for Pt^{20}/rGO, Pt^{20}/rGO-O and Pt^{20}/rGO-Am, respectively. As such, the ozone treatment allows for significantly increasing the concentration of oxygen functional groups in comparison with the bare rGO, which is in good agreement with the literature. Ozone treatment of carbon black at 140 °C was responsible for a C/O decrease from 22.4 to 18.4 and increase in the oxygen functionalities content, especially in the carboxyl groups [43].

It is worth noting that the N 1s peak in Pt^{20}/rGO-Am confirms the presence of N in the rGO structure, according to XPS spectrum fitting the nitrogen content in rGO is ca. 1.41 at.% (Table 2). Thus, the peaks observed in Pt^{20}/rGO-Am spectra (Figure 3G) show the presence of amino groups (399.83 eV) [23], and demonstrate that the significant part of the N atoms being embedded in the rGO structure in the form of quaternary N (401.78 eV) and pyridine N (398.4 eV) [29,44]. It is important to note that the amine functionality with its peak area percentage of ca. 60% appears to be a dominant type of nitrogen species. The dominancy of amine functionality was reported for bare aminated rGO as well [23]. It is well known that amino, carbonyl and carboxyl groups can replace chlorine in the coordination sphere of hexachloroplatinates [45–47]. The introduction of functional groups into the rGO allows for the immobilization of platinum precursor molecules (H_2PtCl_6) on the support surface. The features of these groups determine the Pt nanoparticles morphology. Moreover, the nitrogen heteroatoms presumably contribute to better Pt nanoparticles distribution, as described recently [48–50].

Table 2. Some results of XPS spectrums fitting of Pt^{20}/rGO, Pt^{20}/rGO-O and Pt^{20}/rGO-Am.

Sample	C/O	C sp2, %	N, at.%	NH_2–, %	N_{pyrr}, %	N_{quat}, %
Pt^{20}/rGO	15.57	76.2	–	–	–	–
Pt^{20}/rGO-O	9.35	71.1	–	–	–	–
Pt^{20}/rGO-Am	8.87	37.0	1.41	59.57	21.98	18.43

Metallic Pt (0) duplet peaks were detected at 71.20/74.55 eV, 71.22/74.57 eV and 71.39/74.72 eV for Pt^{20}/rGO, Pt^{20}/rGO-O and Pt^{20}/rGO-Am (Figure 4). The Pt (0) content of Pt^{20}/rGO, Pt^{20}/rGO-O and Pt^{20}/rGO-Am is ca. 46.18, 52.77 and 55.4%, respectively. The observed binding energy shift (ca. 0.2 eV) of the Pt/rGO-Am catalyst Pt(0) peak toward higher values is associated with a change in the electronic structure of the support (downshifting of the *d*-band center position), which may decrease the specific interaction between the Pt surface and strongly adsorbed intermediates (e.g., OH_{ads}) [51]. Moreover, the decrease in these adsorbed OH_{ads}–Pt interactions could enhance the catalytic activity of Pt towards ORR [52,53]. This is consistent with the apparent absence of Pt (IV) in the Pt 4f spectrum (Figure 4C) [54].

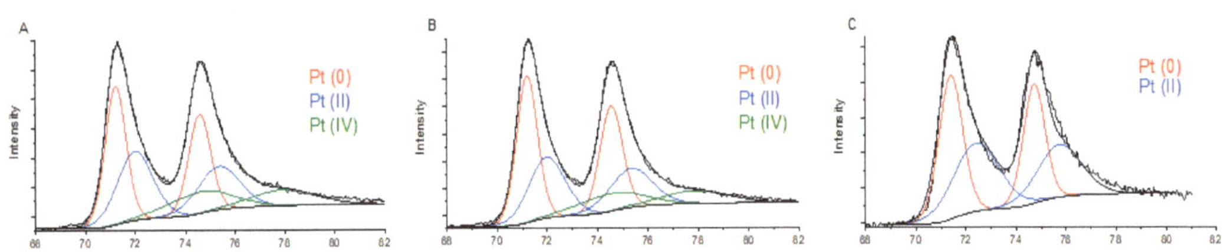

Figure 4. Spectra Pt 4f of Pt^{20}/rGO (**A**), Pt^{20}/rGO-O (**B**) and Pt^{20}/rGO-Am (**C**) catalysts obtained by the XPS method.

The cyclic voltammograms of catalysts of Pt^{20}/C, Pt^{20}/rGO, Pt^{20}/rGO-O, Pt^{20}/rGO-Am are shown in Figure 5. Well-defined peaks, observed in the potential range from −0.15 to 0.2 V vs. Ag/AgCl reference electrode correspond to the desorption/adsorption of the hydrogen monolayer on the surface of Pt nanoparticles [28]. The EASA value for Pt^{20}/C, Pt^{20}/rGO, Pt^{20}/rGO-O and Pt^{20}/rGO-Am is ca. 70.5, 67.1, 40.6 and 87.5 $m^2\ g^{-1}$, respectively. The EASA value for Pt^{20}/rGO is in good agreement with the data reported in the literature [55] for the rGO-supported catalyst (52.1 $m^2\ g^{-1}$) with a similar platinum content and rather close mean Pt nanoparticle size. Pt^{20}/rGO-O has low EASA, presumably due to a high degree of particle agglomeration during synthesis. These areas are unevenly distributed over the rGO surface, and tend to appear near the graphene sheet edges. Therefore, Pt nanoparticles in Pt^{20}/rGO-O tend to be closely formed on the same areas, allowing their aggregation [26]. This could be responsible for lower catalyst EASA and Pt utilization, as discussed below.

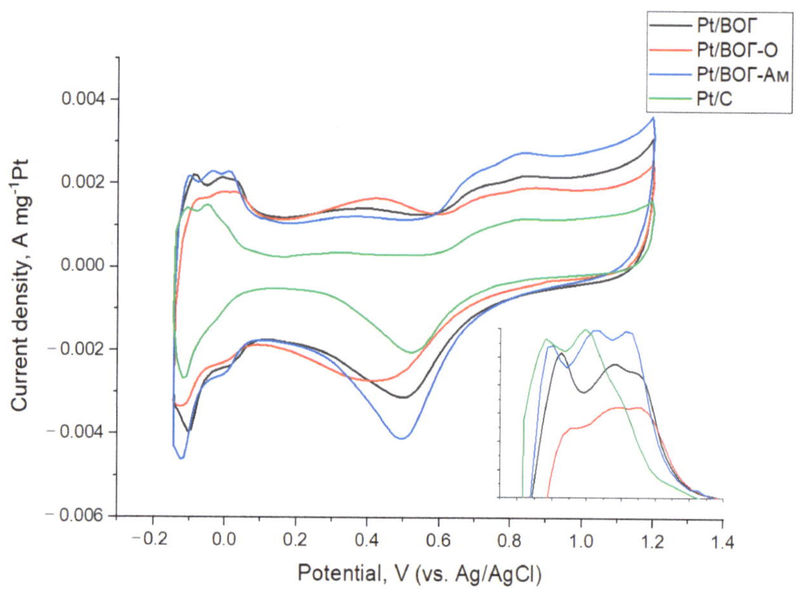

Figure 5. Cyclic voltammograms of Pt^{20}/C, Pt^{20}/rGO, Pt^{20}/rGO-O and Pt^{20}/rGO-Am measured in an Ar-purged 0.5 M H_2SO_4 solution at 20 mV s^{-1}.

The purpose of studying the Pt utilization in the electrocatalyst is to evaluate the electronic bonding of the carbon support with Pt nanoparticles [56]. According to our results, the obtained EASA values are lower than the values of the total Pt surface area calculated from Equation (3), even using a low-loaded thin catalytic film on the working electrode surface. Pt utilization of the various rGO-supported catalyst could vary from 28% to 93% [57,58] depending on synthesis approach and conditions. The obtained results (Table 1) show that the nitrogen doping of rGO allows for increasing the catalyst EASA, reducing the Pt nanoparticles agglomeration degree, and increasing the Pt utilization. This observation is in good agreement with the literature; the interaction between Pt nanoparticles and the N-doped support enhance adhesion of the Pt nanoparticles to the support [35] which could alleviate the aggregation of the nanoparticles. Moreover, first-principles study revealed that Pt nucleation is favorable on pyridinic and pyrrolic N, and the center of the *d*-band can be shifted downward by the nitrogen species [52]. The EASA of Pt^{20}/rGO and Pt^{20}/rGO-O catalysts calculated by CVs exhibited low values compared to Pt^{20}/C, which is probably due to the blocking on the surface of the Pt particles and to the inaccessibility of the catalyst deep inside the micropores of the carbon support [59].

4. Conclusions

In this study, variously modified rGO-based catalytic materials were synthesized and characterized, and their performances were compared with those for a reference Pt/C catalyst. The aminated rGO synthesized in this study could be suggested as a promising PEMFC catalyst support due to the uniform distribution of Pt nanoparticles provided by amino functionalities and embedded nitrogen (pyridine and quaternary) heteroatoms. Amination provides additional sites for Pt nanoparticles nucleation and prevents their further growth during synthesis. The low nanoparticles agglomeration degree ensures a high Pt utilization value, which is higher than the one for the bare rGO-supported catalyst and comparable with those for Pt/C. Pt^{20}/rGO-Am catalyst showed the highest % ECSA among the other catalysts. Ozone treatment of rGO is still a questionable approach, as the Pt nanoparticles morphology is not appropriate for further use as PEMFC catalyst, because of the high degree of catalyst particles agglomeration, which deteriorates the catalyst layer microstructure, thus reducing the MEA performance.

Author Contributions: Conceptualization, M.K., A.P. and I.P.; methodology, I.P., S.B. and M.S.; validation, M.K., A.P. and S.A.; investigation, M.K. and V.K.; resources, S.G.; writing—original draft preparation, M.K. and A.P.; writing—review and editing, I.P. and S.G.; visualization, M.K. and I.P.; supervision, S.G.; project administration, S.G. and I.P.; funding acquisition, S.G. and A.P. All authors have read and agreed to the published version of the manuscript.

Funding: This work has been performed within the framework of the project "Modified carbon nanomaterials for electrodes of fuel cells with solid polymer electrolyte", with the support of a grant from the National Research University "Moscow Power Engineering Institute" for the implementation of scientific research programs "Energy", "Electronics, Radio Engineering and IT", and "Industry 4.0 Technologies for Industry and Robotics in 2020–2022".

Institutional Review Board Statement: Not applicable.

Informed Consent Statement: Not applicable.

Acknowledgments: Thanks to E.V. Kukueva for capturing the TEM images, to R.G. Chumakov for capturing and processing of the XPS spectrums, and to V.V. Podmasterijev for RGO ozone processing.

Conflicts of Interest: The authors declare no conflict of interest. The funders had no role in the design of the study; in the collection, analyses, or interpretation of data; in the writing of the manuscript, or in the decision to publish the results.

References

1. Abdelkareem, M.A.; Elsaid, K.; Wilberforce, T.; Kamil, M.; Sayed, E.T.; Olabi, A. Environmental aspects of fuel cells: A review. *Sci. Total Environ.* **2021**, *752*, 141803. [CrossRef] [PubMed]
2. Wang, Y.; Ruiz Diaz, D.F.; Chen, K.S.; Wang, Z.; Adroher, X.C. Materials, technological status, and fundamentals of PEM fuel cells—A review. *Mater. Today* **2020**, *32*, 178–203. [CrossRef]
3. Hou, J.; Yang, M.; Ke, C.; Wei, G.; Priest, C.; Qiao, Z.; Wu, G.; Zhang, J. Platinum-group-metal catalysts for proton exchange membrane fuel cells: From catalyst design to electrode structure optimization. *EnergyChem* **2020**, *2*, 100023. [CrossRef]
4. Du, L.; Shao, Y.; Sun, J.; Yin, G.; Liu, J.; Wang, Y. Advanced catalyst supports for PEM fuel cell cathodes. *Nano Energy* **2016**, *29*, 314–322. [CrossRef]
5. Lee, J.-H.; Park, S.-J. Recent advances in preparations and applications of carbon aerogels: A review. *Carbon* **2020**, *163*, 1–18. [CrossRef]
6. Grigoriev, S.A.; Millet, P.; Fateev, V.N. Evaluation of carbon-supported Pt and Pd nanoparticles for the hydrogen evolution reaction in PEM water electrolysers. *J. Power Source* **2008**, *177*, 281–285. [CrossRef]
7. Zhang, S.; Yuan, X.-Z.; Hin, J.N.C.; Wang, H.; Friedrich, K.A.; Schulze, M. A review of platinum-based catalyst layer degradation in proton exchange membrane fuel cells. *J. Power Source* **2009**, *194*, 588–600. [CrossRef]
8. Bharti, A.; Cheruvally, G. Influence of various carbon nano-forms as supports for Pt catalyst on proton exchange membrane fuel cell performance. *J. Power Source* **2017**, *360*, 196–205. [CrossRef]
9. Messias, S.; Nunes da Ponte, M.S.; Reis-Machado, A. Carbon Materials as Cathode Constituents for Electrochemical CO_2 Reduction—A Review. *C—J. Carbon Res.* **2019**, *5*, 83. [CrossRef]
10. Pak Hoe, L.; Boaventura, M.; Lagarteira, T.; Kee Shyuan, L.; Mendes, A. Polyol synthesis of reduced graphene oxide supported platinum electrocatalysts for fuel cells: Effect of Pt precursor, support oxidation level and pH. *Int. J. Hydrogen Energy* **2018**, *43*, 16998–17011. [CrossRef]

11. Baranov, I.E.; Grigoriev, S.A.; Ylitalo, D.; Fateev, V.N.; Nikolaev, I.I. Transfer processes in PEM fuel cell: Influence of electrode structure. *Int. J. Hydrogen Energy* **2006**, *31*, 203–210. [CrossRef]
12. Geim, A.K.; Novoselov, K.S. The rise of graphene. *Nanosci. Technol.* **2009**, *6*, 11–19.
13. Jamil, M.F.; Biçer, E.; Yarar Kaplan, B.; Alkan Gürsel, S. One-step fabrication of new generation graphene-based electrodes for polymer electrolyte membrane fuel cells by a novel electrophoretic deposition. *Int. J. Hydrogen Energy* **2021**, *46*, 5653–5663. [CrossRef]
14. Bera, M.; Maji, P.K. Effect of structural disparity of graphene-based materials on thermo-mechanical and surface properties of thermoplastic polyurethane nanocomposites. *Polymer* **2017**, *119*, 118–133. [CrossRef]
15. Armano, A.; Agnello, S. Two-Dimensional Carbon: A Review of Synthesis Methods, and Electronic, Optical, and Vibrational Properties of Single-Layer Graphene. *C—J. Carbon Res.* **2019**, *5*, 67. [CrossRef]
16. Alekseeva, O.K.; Pushkareva, I.V.; Pushkarev, A.S.; Fateev, V.N. Graphene and Graphene-Like Materials for Hydrogen Energy. *Nanotechnol. Russ.* **2020**, *15*, 273–300. [CrossRef]
17. Tkachev, S.V.; Buslaeva, E.Y.; Naumkin, A.V.; Kotova, S.L.; Laure, I.V.; Gubin, S.P. Reduced graphene oxide. *Inorg. Mater.* **2012**, *48*, 796–802. [CrossRef]
18. Erickson, K.; Erni, R.; Lee, Z.; Alem, N.; Gannett, W.; Zettl, A. Determination of the Local Chemical Structure of Graphene Oxide and Reduced Graphene Oxide. *Adv. Mater.* **2010**, *22*, 4467–4472. [CrossRef]
19. Mamat, M.; Grigoriev, S.; Dzhus, K.; Grant, D.; Walker, G. The performance and degradation of Pt electrocatalysts on novel carbon carriers for PEMFC applications. *Int. J. Hydrogen Energy* **2010**, *35*, 7580–7587. [CrossRef]
20. Su, H.; Hu, Y.H. Recent advances in graphene-based materials for fuel cell applications. *Energy Sci. Eng.* **2021**, *9*, 958–983. [CrossRef]
21. Feng, X.; Bai, Y.; Liu, M.; Li, Y.; Yang, H.; Wang, X.; Wu, C. Untangling the respective effects of heteroatom-doped carbon materials in batteries, supercapacitors and the ORR to design high performance materials. *Energy Environ. Sci.* **2021**, *14*, 2036–2089. [CrossRef]
22. Shahgaldi, S.; Hamelin, J. Improved carbon nanostructures as a novel catalyst support in the cathode side of PEMFC: A critical review. *Carbon* **2015**, *94*, 705–728. [CrossRef]
23. Rabchinskii, M.K.; Ryzhkov, S.A.; Kirilenko, D.A.; Ulin, N.V.; Baidakova, M.V.; Shnitov, V.V.; Pavlov, S.I.; Chumakov, R.G.; Stolyarova, D.Y.; Besedina, N.A.; et al. From graphene oxide towards aminated graphene: Facile synthesis, its structure and electronic properties. *Sci. Rep.* **2020**, *10*, 6902. [CrossRef] [PubMed]
24. Pushkarev, A.S.; Pushkareva, I.V.; Grigoriev, S.A.; Kalinichenko, V.N.; Presniakov, M.Y.; Fateev, V.N. Electrocatalytic layers modified by reduced graphene oxide for PEM fuel cells. *Int. J. Hydrogen Energy* **2015**, *40*, 14492–14497. [CrossRef]
25. Grigor'ev, S.A.; Pushkarev, A.S.; Kalinichenko, V.N.; Pushkareva, I.V.; Presnyakov, M.Y.; Fateev, V.N. Electrocatalytic layers based on reduced graphene oxide for fabrication of low-temperature fuel cells. *Kinet. Catal.* **2015**, *56*, 689–693. [CrossRef]
26. Grigoriev, S.A.; Fateev, V.N.; Pushkarev, A.S.; Pushkareva, I.V.; Ivanova, N.A.; Kalinichenko, V.N.; Presnyakov, M.Y.; Wei, X. Reduced Graphene Oxide and Its Modifications as Catalyst Supports and Catalyst Layer Modifiers for PEMFC. *Materials* **2018**, *11*, 1405. [CrossRef]
27. Pushkarev, A.S.; Pushkareva, I.V.; Ivanova, N.A.; du Preez, S.; Bessarabov, D.G.; Chumakov, R.G.; Stankevich, V.G.; Fateev, V.N.; Evdokimov, A.A.; Grigoriev, S.A. Pt/C and Pt/SnOx/C Catalysts for Ethanol Electrooxidation: Rotating Disk Electrode Study. *Catalysts* **2019**, *9*, 271. [CrossRef]
28. Pushkareva, I.V.; Pushkarev, A.S.; Kalinichenko, V.N.; Chumakov, R.G.; Soloviev, M.A.; Liang, Y.; Millet, P.; Grigoriev, S.A. Reduced Graphene Oxide-Supported Pt-Based Catalysts for PEM Fuel Cells with Enhanced Activity and Stability. *Catalysts* **2021**, *11*, 256. [CrossRef]
29. Pushkarev, A.S.; Alekseeva, O.K.; Pushkareva, I.V.; Shapir, B.L.; Chumakov, R.G.; Tishkin, V.V.; Kozlova, M.V.; Kalinichenko, V.N.; Fateev, V.N. Plasma doping of nanostructed reduced graphene oxide. *Nanotechnol. Russ.* **2020**, *15*, 735–740. [CrossRef]
30. Pozio, A.; Francesco, M.; De Cemmi, A.; Cardellini, F.; Giorgi, L.; De Francesco, M.; Cemmi, A.; Cardellini, F.; Giorgi, L. Comparison of high surface Pt/C catalysts by cyclic voltammetry. *J. Power Source* **2002**, *105*, 13–19. [CrossRef]
31. Avcioglu, G.S.; Ficicilar, B.; Bayrakceken, A.; Eroglu, I. High performance PEM fuel cell catalyst layers with hydrophobic channels. *Int. J. Hydrogen Energy* **2015**, *40*, 7720–7731. [CrossRef]
32. Schneider, C.A.; Rasband, W.S.; Eliceiri, K.W. NIH Image to ImageJ: 25 years of image analysis. *Nat. Methods* **2012**, *9*, 671–675. [CrossRef] [PubMed]
33. Spasov, D.D.; Ivanova, N.A.; Pushkarev, A.S.; Pushkareva, I.V.; Presnyakova, N.N.; Chumakov, R.G.; Presnyakov, M.Y.; Grigoriev, S.A.; Fateev, V.N. On the Influence of Composition and Structure of Carbon-Supported Pt-SnO_2 Hetero-Clusters onto Their Electrocatalytic Activity and Durability in PEMFC. *Catalysts* **2019**, *9*, 803. [CrossRef]
34. Puthusseri, D.; Ramaprabhu, S. Oxygen reduction reaction activity of platinum nanoparticles decorated nitrogen doped carbon in proton exchange membrane fuel cell under real operating conditions. *Int. J. Hydrogen Energy* **2016**, *41*, 13163–13170. [CrossRef]
35. Melke, J.; Peter, B.; Habereder, A.; Ziegler, J.; Fasel, C.; Nefedov, A.; Sezen, H.; Wöll, C.; Ehrenberg, H.; Roth, C. Metal–Support Interactions of Platinum Nanoparticles Decorated N-Doped Carbon Nanofibers for the Oxygen Reduction Reaction. *ACS Appl. Mater. Interfaces* **2016**, *8*, 82–90. [CrossRef]
36. Nefedkin, S.I.; Guterman, V.E.; Alekseenko, A.A.; Belenov, S.V.; Ivanenko, A.V.; Klimova, M.A.; Pavlov, V.I.; Panov, S.V.; Paperzh, K.O.; Shubenkov, S.V. Russian Technologies and Nanostructural Materials in High Specific Power Systems Based on Hydrogen–Air Fuel Cells with an Open Cathode. *Nanotechnol. Russ.* **2020**, *15*, 370–378. [CrossRef]

37. Carrera-Cerritos, R.; Baglio, V.; Aricò, A.S.; Ledesma-García, J.; Sgroi, M.F.; Pullini, D.; Pruna, A.J.; Mataix, D.B.; Fuentes-Ramírez, R.; Arriaga, L.G. Improved Pd electro-catalysis for oxygen reduction reaction in direct methanol fuel cell by reduced graphene oxide. *Appl. Catal. B Environ.* **2014**, *144*, 554–560. [CrossRef]
38. Hyun, K.; Lee, J.H.; Yoon, C.W.; Kwon, Y. The Effect of Platinum Based Bimetallic Electrocatalysts on Oxygen Reduction Reaction of Proton Exchange Membrane Fuel Cells. *Int. J. Electrochem. Sci.* **2013**, *8*, 11752–11767.
39. Ma, J.; Habrioux, A.; Luo, Y.; Ramos-Sanchez, G.; Calvillo, L.; Granozzi, G.; Balbuena, P.B.; Alonso-Vante, N. Electronic interaction between platinum nanoparticles and nitrogen-doped reduced graphene oxide: Effect on the oxygen reduction reaction. *J. Mater. Chem. A* **2015**, *3*, 11891–11904. [CrossRef]
40. Stobinski, L.; Lesiak, B.; Malolepszy, A.; Mazurkiewicz, M.; Mierzwa, B.; Zemek, J.; Jiricek, P.; Bieloshapka, I. Graphene oxide and reduced graphene oxide studied by the XRD, TEM and electron spectroscopy methods. *J. Electron Spectros. Relat. Phenom.* **2014**, *195*, 145–154. [CrossRef]
41. Oh, Y.J.; Yoo, J.; Kim, Y.I.; Yoon, J.K.; Na Yoon, H.; Kim, J.-H.; Bin Park, S. Oxygen functional groups and electrochemical capacitive behavior of incompletely reduced graphene oxides as a thin-film electrode of supercapacitor. *Electrochim. Acta* **2014**, *116*, 118–128. [CrossRef]
42. Botas, C.; Álvarez, P.; Blanco, C.; Santamaría, R.; Granda, M.; Gutiérrez, M.D.; Rodríguez-Reinoso, F.; Menéndez, R. Critical temperatures in the synthesis of graphene-like materials by thermal exfoliation–reduction of graphite oxide. *Carbon* **2013**, *52*, 476–485. [CrossRef]
43. Chang, X.; Dong, F.; Yang, S.; Tang, Z.; Zha, F. Well dispersed Pt nanoparticles on commercial carbon black oxidized by ozone possess significantly high electro-catalytic activity for methanol oxidation. *Int. J. Hydrogen Energy* **2019**, *44*, 21559–21568. [CrossRef]
44. Deng, Y.; Xie, Y.; Zou, K.; Ji, X. Review on recent advances in nitrogen-doped carbons: Preparations and applications in supercapacitors. *J. Mater. Chem. A* **2016**, *4*, 1144–1173. [CrossRef]
45. Zou, M.; Du, M.; Zhang, M.; Yang, T.; Zhu, H.; Wang, P.; Bao, S. Synthesis and deposition of ultrafine noble metallic nanoparticles on amino-functionalized halloysite nanotubes and their catalytic application. *Mater. Res. Bull.* **2015**, *61*, 375–382. [CrossRef]
46. Golovizin, V.S.; Levchenko, L.M. Recovery of Platinum Complexes from Chloride Solutions Using Oxidized Carbon Sorbents. *Chem. Sustain. Dev.* **2011**, *19*, 537–540.
47. Kim, K.T.; Kim, Y.G.; Chung, J.S. Adsorption of cationic platinum complex onto carbon support. *Carbon* **1993**, *31*, 1289–1296.
48. Li, Z.; Gao, Q.; Zhang, H.; Tian, W.; Tan, Y.; Qian, W.; Liu, Z. Low content Pt nanoparticles anchored on *N*-doped reduced graphene oxide with high and stable electrocatalytic activity for oxygen reduction reaction. *Sci. Rep.* **2017**, *7*, 43352. [CrossRef]
49. Mao, S.; Wang, C.; Wang, Y. The chemical nature of *N* doping on *N* doped carbon supported noble metal catalysts. *J. Catal.* **2019**, *375*, 456–465. [CrossRef]
50. Zhao, L.; Sui, X.-L.; Li, J.-L.; Zhang, J.-J.; Zhang, L.-M.; Wang, Z.-B. Ultra-fine Pt nanoparticles supported on 3D porous N-doped graphene aerogel as a promising electro-catalyst for methanol electrooxidation. *Catal. Commun.* **2016**, *86*, 46–50.
51. Zhu, J.; Xiao, M.; Zhao, X.; Liu, C.; Ge, J.; Xing, W. Strongly coupled Pt nanotubes/*N*-doped graphene as highly active and durable electrocatalysts for oxygen reduction reaction. *Nano Energy* **2015**, *13*, 318–326. [CrossRef]
52. Tachibana, N.; Yukawa, Y.; Morikawa, K.; Kawaguchi, M.; Shimanoe, K. Pt nanoparticles supported on nitrogen-doped porous carbon as efficient oxygen reduction catalysts synthesized via a simple alcohol reduction method. *SN Appl. Sci.* **2021**, *3*, 338. [CrossRef]
53. Nie, Y.; Wei, Z. Surface-confined Pt-based catalysts for strengthening oxygen reduction performance. *Prog. Nat. Sci. Mater. Int.* **2020**, *30*, 796–806. [CrossRef]
54. Yarar Kaplan, B.; Haghmoradi, N.; Biçer, E.; Merino, C.; Alkan Gürsel, S. High performance electrocatalysts supported on graphene based hybrids for polymer electrolyte membrane fuel cells. *Int. J. Hydrogen Energy* **2018**, *43*, 23221–23230. [CrossRef]
55. Antony, R.P.; Preethi, L.K.; Gupta, B.; Mathews, T.; Dash, S.; Tyagi, A.K. Efficient electrocatalytic performance of thermally exfoliated reduced graphene oxide-Pt hybrid. *Mater. Res. Bull.* **2015**, *70*, 60–67. [CrossRef]
56. Avcioglu, G.S.; Ficicilar, B.; Eroglu, I. Improved PEM fuel cell performance with hydrophobic catalyst layers. *Int. J. Hydrogen Energy* **2018**, *43*, 18632–18641. [CrossRef]
57. Eris, S.; Daşdelen, Z.; Yıldız, Y.; Sen, F. Nanostructured Polyaniline-rGO decorated platinum catalyst with enhanced activity and durability for Methanol oxidation. *Int. J. Hydrogen Energy* **2018**, *43*, 1337–1343. [CrossRef]
58. Şanlı, L.I.; Bayram, V.; Yarar, B.; Ghobadi, S.; Gürsel, S.A. Development of graphene supported platinum nanoparticles for polymer electrolyte membrane fuel cells: Effect of support type and impregnation–reduction methods. *Int. J. Hydrogen Energy* **2016**, *41*, 3414–3427. [CrossRef]
59. Joo, S.H.; Pak, C.; You, D.J.; Lee, S.-A.; Lee, H.I.; Kim, J.M.; Chang, H.; Seung, D. Ordered mesoporous carbons (OMC) as supports of electrocatalysts for direct methanol fuel cells (DMFC): Effect of carbon precursors of OMC on DMFC performances. *Electrochim. Acta* **2006**, *52*, 1618–1626. [CrossRef]

C Journal of Carbon Research

Article

Reduced Graphene Oxide—Polycarbonate Electrodes on Different Supports for Symmetric Supercapacitors

Olena Okhay [1,*], Alexandre Cunha Bastos [2], Kateryna Andreeva [3], Sampo Tuukkanen [4] and Alexander Tkach [2]

1 TEMA—Centre for Mechanical Technology and Automation, Department of Mechanical Engineering, University of Aveiro, 3810-193 Aveiro, Portugal
2 CICECO—Aveiro Institute of Materials, Department of Materials and Ceramic Engineering, University of Aveiro, 3810-193 Aveiro, Portugal; acbastos@ua.pt (A.C.B.); atkach@ua.pt (A.T.)
3 V. Lashkaryov Institute of Semiconductor Physics at NAS of Ukraine, 03028 Kyiv, Ukraine; ev_andreeva322@ukr.net
4 Faculty of Medicine and Health Technology, Tampere University, FI-33101 Tampere, Finland; sampo.tuukkanen@gmail.com
* Correspondence: olena@ua.pt

Abstract: Electrode materials for electrochemical capacitors or supercapacitors (SCs) are widely studied, as they are needed for the development of energy storage devices in electrical vehicles and flexible electronics. In the current work, a self-supported paper of reduced graphene oxide (rGO) with polycarbonate (PC) (as rGO-PC composite) was prepared by simple vacuum filtration and low-temperature annealing. rGO-PC as a freestanding single electrode was studied in a three-electrode system and presented a capacitive energy storage mechanism. To fabricate SCs based on rGO-PC, flexible polyethylene terephthalate (PET) with layers of both Cu tape (Cu tape) and carbon tape (C tape) (PET/Cu/C), as well as PET covered by graphene ink (PET/GrI), were used as supports. Fabricated flexible symmetric SCs have shown similar behavior with a higher areal capacitance value than that on PET/Cu/C substrate.

Keywords: reduced graphene oxide; polymer; vacuum filtration; freestanding electrode; flexible supercapacitors

Citation: Okhay, O.; Bastos, A.C.; Andreeva, K.; Tuukkanen, S.; Tkach, A. Reduced Graphene Oxide—Polycarbonate Electrodes on Different Supports for Symmetric Supercapacitors. *C* **2022**, *8*, 12. https://doi.org/10.3390/c8010012

Academic Editors: Camélia Matei Ghimbeu and Oluwafunmilola Ola

Received: 13 December 2021
Accepted: 31 January 2022
Published: 2 February 2022

1. Introduction

Graphitic carbon monolayers were observed and studied more than several decades ago. At the beginning, in 1859, Sir Benjamin Collins Brodie described the highly lamellar structure of thermally reduced graphite oxide [1]. In 1918, the powder diffraction technique was used to study the structure of graphite, and graphite oxide "paper" was described by Kohlschütter and Haenni [2]. Later, the theory of graphene was prepared and published by Wallace in 1947 [3]. In the following year, Ruess and Vogt first showed just arrived electron microscopy images of few-layer graphite, accompanied by further observation of single graphene layers [4]. In 1986, Boehm et al. referred to a single layer of graphite or a thin layer of sp^2 hybridized carbon atoms arranged in a honeycomb lattice as graphene [5]. However, it was only in 2004 that isolated graphene was reported to be obtained using the simple scotch tape method [6]. The subsequent discoveries of its unusual properties have led to an extraordinary interest amongst researchers across almost all scientific disciplines [7,8]. For instance, graphene has a high mechanical strength and a large specific surface area (2630 m^2/g [9]), which makes it an ideal electrode material for energy storage devices.

Graphene oxide (GO) is a graphene-derived material that includes oxygen functional groups and, thus, has properties rather different from those of graphene. By reduction, these functional groups are removed resulting in reduced graphene oxide (rGO), which is a graphene-related material [10] that can be obtained in a much larger amount by easier ways than a monolayer of graphene. GO can be reduced by various techniques, including chemical and thermal treatments, although the former method typically requires the use of hazardous chemicals. Thus, the thermal reduction technique is the most preferred, being often used to significantly increase the electrical conductivity of rGO material [11]. Moreover, the specific surface area also increases with the reduction temperature but only up to 350 °C [12]. At the same time, the removal of functional groups at a high temperature creates defects in the carbon lattice plane, and the final properties of rGO can significantly depend on the annealing temperature [13]. The lowering of the defect concentration, as well as a significant (~ two times) decrease in the lattice d-spacing, was reported to start at about 200 °C for freestanding GO papers prepared by vacuum filtration [14]. Moreover, freestanding rGO paper heat treated at 200 °C can also show a higher gravimetric or areal capacitance than rGO obtained at much higher temperatures [15].

Although rGO cannot show the same ideal structure as graphene, it can present a low enough sheet resistance from ~1000 Ω/□ to <200 Ω/□ [16] and chemical and thermal stability while not being as expensive as graphene. Thus, rGO is a highly acceptable material to be used in a large amount of sensors, environmental or catalytic applications as well as biological and optoelectronic devices [17]. rGO is especially intensively studied for application in energy storage devices, particularly in supercapacitors (SCs) due to its high enough specific capacitance of ~250 F/g [18] and high specific surface area (SSA) of ~1520 m^2/g [19–22], although a high SSA does not guarantee the rising specific capacitance [23–26]. At the same time, the applications of rGO on flexible substrates are particularly promising [27].

Different methods have been employed for the preparation of rGO electrodes, including suspension-based vacuum filtration; electrophoretic deposition as well as that at microwaves; fiber injection; transferring; and powder-/aerogel-/hydrogel-based deposition by screen printing, gluing, dropping, covering, etc. [28,29]. For example, layer-by-layer (LBL) deposition and solution casting methods present advantages when preparing ultrathin hybrid films with a controllable thickness. However, the LBL technique lacks practical significance for the preparation of thick papers because of the complicated, time-consuming assembly process. Solution casting is another simple and easy method to assemble thin films because it only needs the evaporation of moisture from the cast dispersion on a substrate. The vacuum filtration method involves simple filtration of graphene oxide or hybrid dispersions through a microporous filter membrane with the aid of vacuum pumping, allowing the deposition of highly aligned graphene sheets [30]. Thus, vacuum filtration is also one of the easiest ways to prepare electrodes [31], although the following electrochemical analysis of these materials in supercapacitors can be difficult without additional support.

Polycarbonate (PC) is a thermoplastic polymer with enhanced thermal stability [32,33] and mechanical stiffness [34], and it is used in a wide range of applications [35], such as photonics [36], automotive and aeronautic industries [33,37]. At the same time, the composites of PC with graphene have also been widely studied as promising materials for 3D printing [38], photoactuators [39] and energy storage applications [40]. In addition, based on the studied electrochemical properties of PC [41], it can be used as a supercapacitor membrane [42,43] or as a polymer electrolyte [44,45]. It was also reported to be used after chemical recycling as activated carbon in supercapacitor electrodes [46].

However, to date, no direct use of PC alone or in a composite with rGO as an SC electrode material has been reported to the best of our knowledge. Therefore, freestanding rGO-polycarbonate paper discs were prepared by vacuum filtration and investigated here as freestanding electrodes. In addition, they were assembled onto two different substrates and encapsulated in supercapacitors for electrochemical test analysis and comparison. For that, two substrates were fast and simply prepared in the laboratory from such available and inexpensive materials as flexible polyethylene terephthalate (PET), Cu tape (Cu), carbon tape (C) and graphene ink (GrI).

2. Materials and Methods

2.1. Preparation of Graphene Oxide Solution

A light brown graphene oxide (GO) solution (2 g per 1 L) was prepared based on the Hummer method. Commercially available natural flake graphite, sulfuric acid (99.99%), potassium permanganate (99%), hydrogen peroxide (30 wt% in H_2O) and hydrochloric acid (37%) were used as received from Sigma-Aldrich Chemie GmbH, Taufkirchen, Germany. All aqueous solutions were prepared with deionized distilled water. GO was synthesized using the following process: 50 mL of 98% sulfuric acid was placed in a 250 mL chemical glass flask provided with a magnetic stirrer and fixed in a water bath with a small amount of added ice. Then, 2 g of graphite powder was added to the acid. After stirring for a few minutes, 6 g of dry potassium permanganate was added to the reactant mixture. Later, the mixture was heated to 35 °C and diluted with 100 mL of water for 20 min. The diluted mixture was held for 30 min at a temperature of 70 °C, followed by addition of 100 mL of water and 10 mL of 3% hydrogen peroxide. Immediately thereafter, the mixture was passed through a paper filter under evacuation. The filtered-out material was mixed with 500 mL of distilled water using a magnetic stirrer for 10 min. The GO fraction was separated from the reaction products using a centrifuge. The final product was a light brown GO solution, which did not reveal any signs of settling under further centrifuging for a prolonged time.

2.2. Preparation of rGO-PC Freestanding Electrode

The freestanding rGO-PC paper was obtained by a process similar to that described in our previous work [47] with addition of polycarbonate (PC, $C_{15}H_{16}O_2$) (Sigma-Aldrich Chemie GmbH, Taufkirchen, Germany). Briefly, 0.5 g of PC was dissolved with 30 mL of dichloromethane (CH_2Cl_2, Sigma-Aldrich Chemie GmbH, Taufkirchen, Germany) and mixed with GO solution at a weight ratio of 1:2 (PC:GO). Homogeneous solution was poured into a glass funnel (see Figure 1a) and filtered by laboratory vacuum pump with Millipore membrane filters (pore size = 0.45 μm, Merck Milipore, Darmstadt, Germany). Obtained wet samples (see Figure 1b) were dried at room temperature for 1 day, removed from the membrane as freestanding papers and thermally treated at 180 °C in vacuum oven (Vacutherm, Thermo Fischer Scientific GmbH, Dreieich, Germany) for 1 h, resulting in samples with areal loading of about 0.7 mg/cm^2, referred to here as rGO-PC. Based on the geometrical and theoretical density values, and due to the fact that papers are freestanding, their porosity was assessed as 91%. We assessed the porosity value based on the assumption of the presence of the nominal polymer concentration in the analyzed electrode materials.

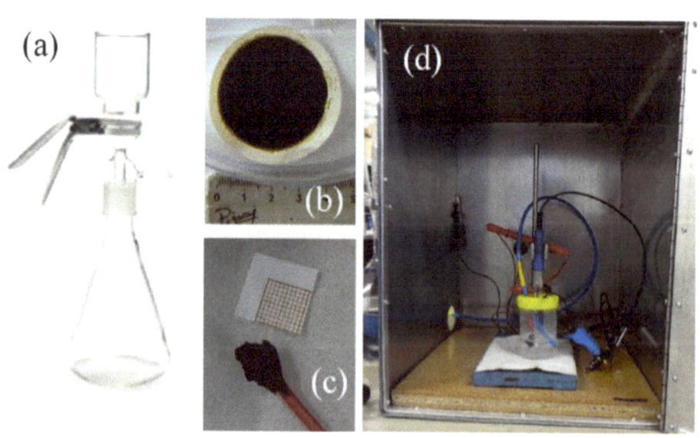

Figure 1. Digital photographs of glass funnel for vacuum filtration (**a**), vet sample after vacuum filtration of solution through PVDF membrane (**b**), rGO-PC paper connected to electrode (**c**) and home-made electrochemical cell used for the measurements of the single electrode in electrolyte (**d**).

2.3. Preparation of rGO-PC-Based Supercapacitors

Two supercapacitors were prepared by laminating two electrodes together with an electrolyte and separator in between according to the scheme shown in Figure 2a. For one of them, PET (Melinex ST506 from Dupont Teijin Films, Dumfries, UK) only covered by adhesive Cu tape (Cu) (Mouser Electronics, Barcelona, Spain) and conductive carbon double-sided tape (C) (Nissin EM Co Ltd., Tokyo, Japan) using a layer-by-layer simple method resulted in PET/Cu/C substrate. rGO-PC freestanding disks were attached to the top C tape (see Figure 2a, right). Two equal PET/Cu/C/rGO-PC electrodes were used to prepare symmetric SC with a TF4050 paper separator (Nippon Kodoshi Corporation, Kochi, Japan) and 1M NaCl aqueous electrolyte. The high electrolyte concentration was selected to ensure an excess of ions such that the risk of electrolyte starvation and additional internal resistance was removed [48]. The electrodes and separator were sandwiched together with the electrolyte and sealed with an adhesive film (UPM Raflatac, Tampere, Finland). The assembly was carried out with the electrodes at 90° angle with respect to each other, forming a square-shaped supercapacitor at the electrode overlap area as shown in Figure 2b. For the electrodes of the second SC, the conductive graphite ink (GrI) (Electrodag PC407C, Acheson Industries Ltd., Cattedown, UK) was blade coated on top of the PET substrate using the mask for patterning and was subsequently cured at 120 °C for 5 min, which resulted in PET/GrI substrate. rGO-PC disk paper was placed onto Gr-ink. Two PET/GrI/rGO-PC electrodes were assembled in full device (see Figure 2c) by the method described above and checked for flexibility (see Figure 2d).

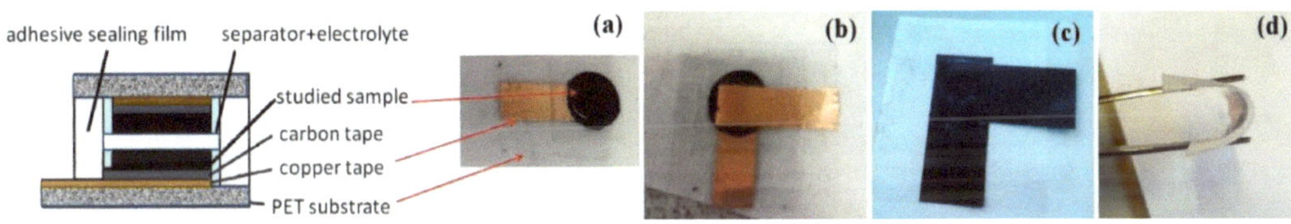

Figure 2. The structure of the assembled supercapacitor (**a**) and digital photographs of SCs on PET with Cu and C tapes (**b**) and PET with graphite ink (**c**). Image of the fabricated SC presenting its flexibility (**d**).

2.4. Characterization

X-ray diffraction (XRD) profiles of the studied samples were collected at room temperature in a continuous scanning mode (step 0.02° and time 10 s) using a diffractometer (SmartLab, Rigaku, Tokyo, Japan) in the 2θ range from 5° to 50°. Morphology of the studied samples was analyzed by scanning electron microscope (SEM, SU70, Hitachi, Tokyo, Japan) under 15 kV acceleration voltage. Current–voltage (I–V) characteristics were measured by the four-point method in a van der Pauw configuration using a home made cell. The single-electrode tests were performed with an AUTOLAB PGSTAT 302N potentiostat (Metrohm AG, Herisau, Switzerland) in a home-made electrochemical cell with the materials to be tested as a working electrode, a platinum wire as a counter electrode and a silver | silver sulphate (K_2SO_4) electrode as a reference ($E_{Ag | Ag2SO4 | K2SO4 sat}$ = 0.444 V vs. saturated calomel electrode (SCE) and 0.70 V vs. normal hydrogen electrode (NHE) at 22 °C). For the electrochemical characterization of the single electrode, a piece of the freestanding rGO-PC paper was directly connected to electrical cable as shown in Figure 1c and immersed in electrolyte in the home-made three-electrode cell presented in Figure 1d.

The supercapacitor properties were measured using a Zennium Electrochemical Workstation (Zahner Elektrik GmbH, Kronach, Germany) in a two-electrode configuration. Cyclic voltammetry (CV) from 0 to 0.9 V was used to obtain a qualitative measure of the supercapacitors. The voltage sweep rate was 100 mV/s. For the testing of supercapacitors, 1 M NaCl electrolyte was used. Galvanostatic charging/discharging (GCD) measurements were performed according to international standard IEC 62391-1:2006, Class 3 [49]. The supercapacitors were charged to 0.9 V in 1 min and held at this potential for 5 min.

3. Results and Discussion

3.1. Structural Characterization of Freestanding rGO-PC Paper

The XRD profile of the vacuum-filtrated rGO-PC freestanding paper is presented in Figure 3, together with the data for the GO, bare rGO paper prepared by the same process and bare polycarbonate film obtained from the corresponding solution on glass substrate for comparison. The strong peak at ~10.4°, which is characteristic for GO, is shown in Figure 3a,b to disappear after vacuum annealing at 180 °C, while a broad rGO peak at ~24° appeared. The final composite XRD pattern includes both weak rGO response of rGO shown in Figure 3b and dominant response of PC shown in Figure 3c without any extra reflections as seen in Figure 3d. The morphology of the rGO-PC film in the plane-view SEM image is presented in Figure 3e, revealing that the components are mixed. According to the cross-section SEM micrograph shown in Figure 3f, the thickness of the composite paper is 7 ± 1 μm.

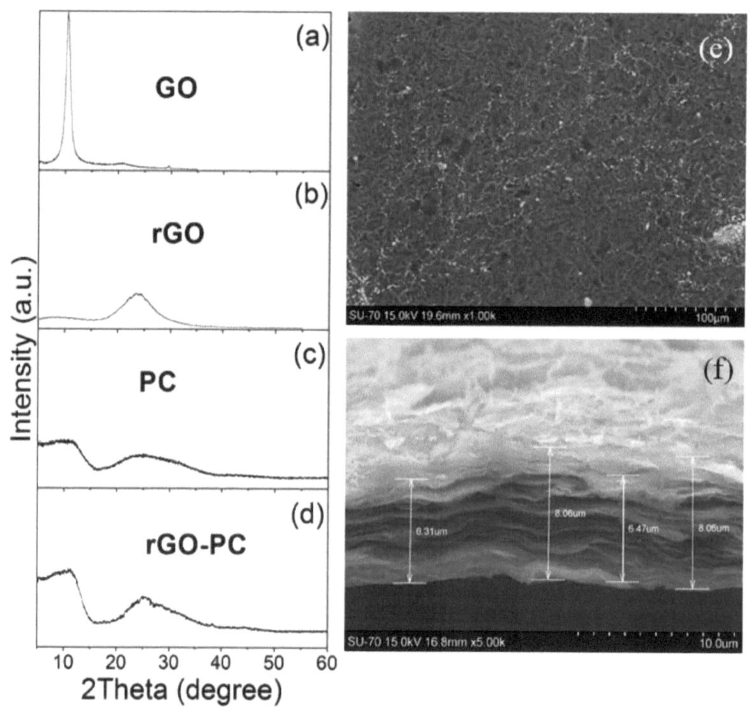

Figure 3. XRD profiles of GO (**a**), rGO after heat treatment at 180 °C (**b**), PC film on glass substrate (**c**) and rGO-PC freestanding paper prepared by vacuum filtration (**d**). SEM images of plan view (**e**) and cross-section (**f**) of rGO-PC paper.

3.2. Electrochemical Characterization of Single Electrode

Before the electrochemical characterization, the I–V test was carried out on the paper as shown in Figure 4. Semimetal electrical behavior can be seen there. A difference of several orders of magnitude (from μA to A) can be found between the current values for GO (see inset of Figure 4) and rGO-PC (see Figure 4). In addition, 33 ± 3 Ω resistance was obtained with a digital multimeter when two probes were spaced at a 1 cm distance, which is low enough for electrical contact during the proposed measurements without additional support.

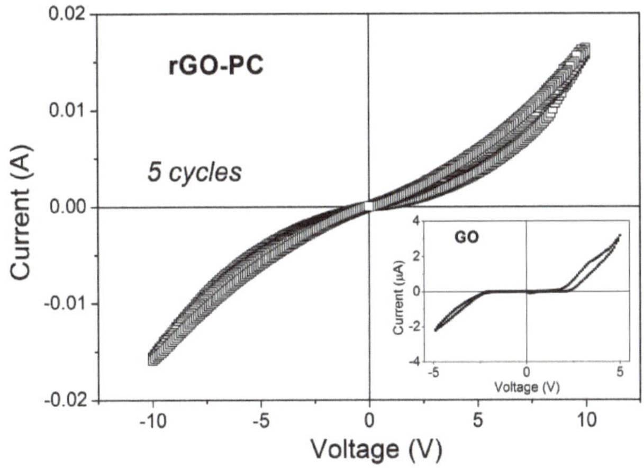

Figure 4. I–V plot of rGO-PC freestanding paper at 5 cycling measurements. Inset shows I–V of GO paper.

For the electrochemical characterization of the single electrode, rGO-PC freestanding paper was prepared as described in the Materials and Methods Section and shown in Figure 1. The self-supported rGO-PC electrode was electrochemically studied at different scan rates from 5 to 100 mV/s. Figure 5a shows the CV curves of the not ideal but quasi-rectangular shape without distinct peaks even at the highest scan rate of 100 mV/s, suggesting the absence of obvious pseudo-capacitive contribution due to oxidation and reduction but a strong capacitive energy storage mechanism typical for an electric double-layer *capacitor* (EDLC). The current change with the scan rate in the CV curve follows the power law shown in Equation (1) [50]:

$$I = a \times v^{b} \tag{1}$$

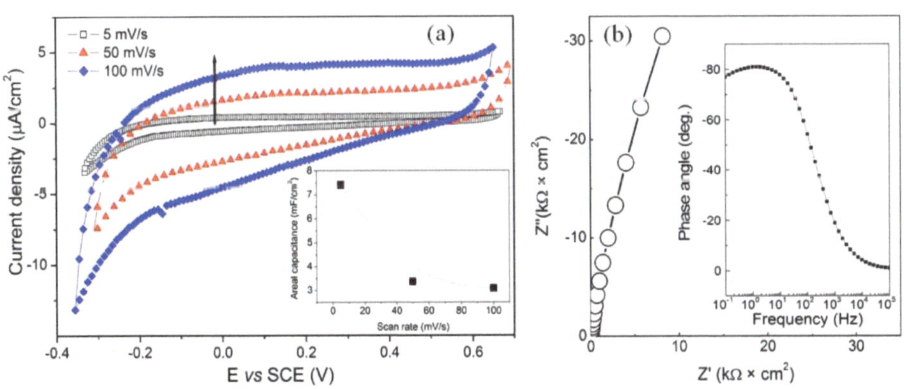

Figure 5. CV of single rGO-PC electrode measured at different scan rates (**a**) and calculated areal capacitances (inset). Nyquist plot of single rGO-PC electrode (**b**) and corresponding Bode plot (inset).

Here, *i* is the response current at 0.1 V, *v* is the scan rate, and *a* and *b* are the variable parameters. The capacitive origin of the energy storage is attributed to the parameter *b* value deduced from the *log(i)* vs. *log(v)* plot close to 1, while battery-type behavior is associated with its value close to 0.5. In this way, parameter *b* was calculated from CV in

Figure 5a (as *log(i)* vs. *log(v)* plot) and found to be higher than 0.8, which supported an EDLC type of storage in the freestanding rGO-PC electrode [51,52].

The areal capacitance of the single rGO-PC electrode C_a can be calculated from CV by Equation (2):

$$C_a = \frac{1}{A \times v \times (V_2 - V_1)} \int_{V_1}^{V_2} I(V)dV, \quad (2)$$

where A is the area of the electrode material; v is the potential scan rate; $V_2 - V_1$ is the potential window; and *I(V)* is the response current at voltage V under CV operation. As can be seen from Figure 1c, the studied material is slightly brittle, and some difficulties appeared during the connection of the material as the working electrode. The area of the electrode was also not easy to determine; however, it was calculated according to the graph paper (also shown in photo of Figure 1c). According to Equation (2), the areal capacitance from CV for the rGO-PC electrode was calculated as ~7.5, 3.4 and 3.1 mF/cm^2 at 5, 50 and 100 mV/s, respectively (see inset of Figure 5a).

In addition, the freestanding rGO-PC was studied by electrochemical impedance spectroscopy (EIS), and the obtained results are presented in Figure 5b as Nyquist plots. The Nyquist plot basically contains two major impedance components: Z' exhibits the resistive and Z" shows the capacitive and/or inductive behavior of the electrode. The slope of the obtained results is almost but not perfectly vertical, which also supports the capacitive energy storage in the studied material [50]. The absence of a perfect vertical impedance line for the electrode could be because of some technical reason, such as penetration of the electrical signal and diffusion of ions in the deep inner parts of the rGO-PC electrodes. A slight and slow response to the electrical signal is also due to the oxygen functionalities, which, on the one hand, increase the resistivity of the electrode but, on the other hand, promote a faster diffusion of ions. The phase angle of the rGO-PC electrode is seen from the Bode plot, shown in the inset of Figure 5b to be close to $-80°$, which can indicate non-ideal capacitor behavior [53,54]. The characteristic frequency f_0 for a phase angle of $-45°$ is 193 Hz. This frequency marks the point at which the resistive and capacitive impedances are equal. The corresponding time constant $t_0 = 1/f_0$ equals 0.005 s, meaning that the studied material has a low frequency response (slow ion transport rate and high enough electrical resistance).

3.3. Electrochemical Characterization of Symmetric SC on Substrates

Furthermore, the prepared rGO-PC papers were assembled onto two PET-based supports to fabricate flexible SCs, and full-cell devices were electrochemically characterized. Two rGO-PC-based supercapacitors on the PET/Cu/C and PET/GrI substrates were prepared by laminating two electrodes together with an electrolyte and separator in between according to the scheme shown in the Materials and Methods Section. The supercapacitors were characterized by CV and GCD in an operating voltage window from 0 to +0.9 V in a two-electrode cell configuration to demonstrate the applicability of the facile preparation method for composite devices on flexible supports. Although the freestanding rGO-PC electrode was characterized with CV in a three-electrode cell to obtain a qualitative view of the device properties, the use of a two-electrode cell configuration should lead to more realistic results, predicting the performance of the materials in an actual application.

The CV curves for rGO-PC on PET/Cu/C and on PET/GrI are shown in Figure 6a. The CV curves for both SCs are not ideally rectangular, indicating also that the material charges and discharges quickly and efficiently. The CV of the SC made on PET/GrI is more rectangular, but its area, as well as corresponding capacitance, is smaller than that of the SC using PET/Cu/C as a support for rGO-PC. The specific capacitance of the full SC was calculated from CV using Equation (3):

$$C_{a\ total} = \frac{1}{2 \times A \times v \times (V_2 - V_1)} \int_{V_1}^{V_2} I(V)dV, \quad (3)$$

where A is the work area of the electrodes (1 cm^2), v is the scan rate, $\int I(V)dt$ is the integral area of the positive and negative sweeps of the CV curve, and V_1 and V_2 represent the voltage window of the charge (from 0 to +0.9 V).

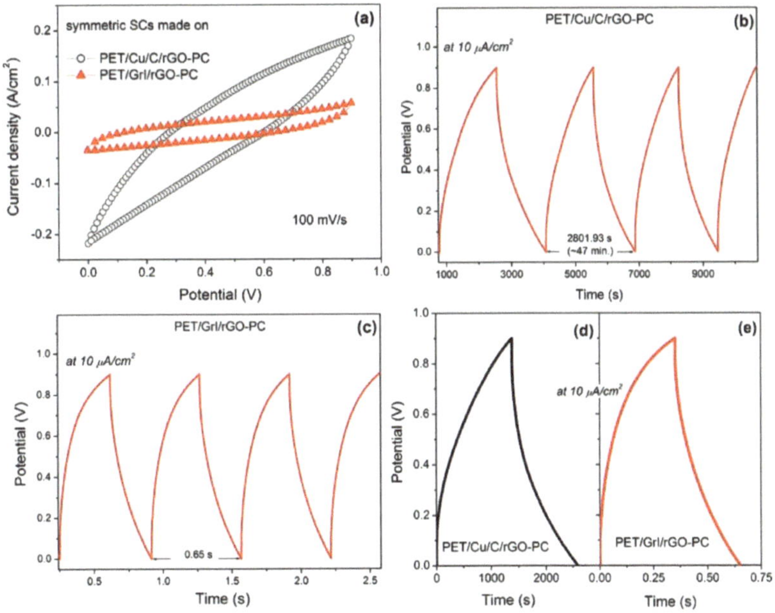

Figure 6. CV measured at 100 mV/s on symmetric supercapacitors made of rGO-PC active material assembled on PET with Cu and C tapes (open symbols) and PET with graphite ink (solid symbols) (**a**). GCD measured at 10 μA/cm^2 on SC made of rGO-PC active material assembled on PET with Cu and C tapes (**b**) and PET with graphite ink (**c**). Comparative figure of single GCD curves at 10 μA/cm^2 for both SCs (**d**,**e**).

Moreover, the more precise areal capacitance of the full SC was calculated from the GCD curves (Figure 6b–e) using Equation (4):

$$C_{u\ total} = \frac{2 \cdot I}{2 \times A \cdot (\Delta V)^2} \int_{t\ (V_{max})}^{t\ (V_{min})} V(t)dt, \quad (4)$$

where $\int V(t)dt$ is the integral area; $t(V_{max})$ is the initial time, at which the voltage has its maximal value and the discharge starts; and $t(V_{min})$ is the final time, at which the discharge finishes (i.e., $V = 0$). Furthermore, for the symmetric SC, the capacitance of a single electrode according to the literature [55] is

$$C_{electrode} = 4 \times C_{\ total} \quad (5)$$

Based on the equations above, the areal capacitances deduced from CV for the single PET/Cu/C/rGO-PC and PET/GrI/ rGO-PC electrodes were found to be ~0.58 and 0.19 mF/cm^2 at 100 mV/s, respectively. The corresponding gravimetric capacitance is 0.83 F/g and 0.27 F/g, respectively. At the same time, the areal capacitances deduced from GCD are 0.93 and 0.25 mF/cm^2 at 10 μA/cm^2, respectively.

Table 1 summarizes the results of this work in comparison with the literature reports on similar rGO-based systems [56–60]. Romero et al. reported that the areal capacitance of 0.23 mF/cm^2 at 10 mV/s for infrared CO_2 laser-induced graphene materials from Kapton® HN polyimide films was higher than that of 0.16 mF/cm^2 for rGO obtained on PET by UV laser [56]. The rGO layers obtained through thermal treatment at 850 °C demonstrated an areal capacitance of ~2.2 mF/cm^2 at 25 mV/s as can be seen in the work by Ghasemi et al. [57]. The areal capacitance of 3.51 mF/cm^2 was reported by Chen et al. for an electrode fabricated by immersion of the carbon cloth into GO solution with further heat treatment [58]. Beidaghi et al. calculated ~3 mF/cm^2 capacitance for rGO interdigital electrodes prepared by electrostatic spray deposition and photolithography lift-off methods [59]. Vacuum-filtrated rGO freestanding films that were chemically reduced by hydrazine hydrate were prepared by Xu et al., and an areal capacitance of ~100 mF/cm^2 was reported at 0.2 mA/cm^2, although it rapidly decreased with the current density increasing to ~10 mF/cm^2 at 10 mA/cm^2 [60]. Moreover, such a decrease, as well as the poor electrical conductivity of the pure rGO film electrode, was ascribed to its compact lamellar structure in the process of vacuum extraction, which could prevent the transmission of electrolyte ions to some extent [60]. As can be seen from Table 1, the obtained values of the areal capacitance for both SCs in the current work are slightly lower than the reported ones for graphene and rGO without PC, thus indicating that the addition of PC does not enhance electrode performance, although the rGO-PC electrodes are applicable to a flexible SC. However, the areal capacitance of rGO papers prepared by vacuum filtration was reported by Zhu et al. to strongly increase with the thickness [61] that can be used for the further study of the current materials.

Table 1. The literature data for areal capacitance of rGO-based SC electrodes for comparison with the results of this work.

Electrode	Electrolyte	Area Capacitance	Reference
rGO-PC, vacuum filtration, placed on PET/Cu/C	NaCl	0.93 mF/cm^2 at 10 μA/cm^2	This work
rGO-PC, vacuum filtration, placed on PET/GrI	NaCl	0.25 mF/cm^2 at 10 μA/cm^2	This work
Graphene, laser synthetized on Kapton	H_3PO_4-PVA	0.23 mF/cm^2 at 10 mV/s	[56]
rGO, placed on Si substrate	KCl	~2.2 mF/cm^2 at 25 mV/s	[57]
rGO, covered on carbon cloth	H_2SO_4	3.5 mF/cm^2 at 0.5 mA/cm^2	[58]
rGO, interdigital microelectrode arrays	KCl	3 mF/cm^2	[59]
rGO, vacuum filtration	H_2SO_4-PVA	~10 mF/cm^2 at 10 mA/cm^2	[60]

4. Conclusions

In the current work, self-supported rGO paper with polycarbonate was prepared by simple vacuum filtration and low-temperature annealing. rGO-PC as a freestanding single electrode was studied in a three-electrode system and represented a capacitive energy storage mechanism. To fabricate flexible supercapacitors, inexpensive PET/Cu/C and PET/GrI were prepared by the facile method and used as flexible supports for the rGO-PC electrode active material. Fabricated flexible symmetric supercapacitors based on rGO-PC have shown similar behavior with a higher areal capacitance value than that on PET/Cu/C substrate. Improved electrochemical results are expected in further work by modifying the thickness and surface of the active material and by using an electrolyte with a wider potential window.

Author Contributions: Conceptualization, data curation and writing—original draft preparation, O.O.; methodology, O.O., A.C.B. and S.T.; validation, O.O., A.C.B. and S.T.; formal analysis, A.T., A.C.B., S.T. and K.A.; resources, O.O., A.C.B. and S.T.; writing—review and editing, A.T. All authors have read and agreed to the published version of the manuscript.

Funding: This article is funded by national funds, through FCT—Fundacao para a Ciencia e a Tecnologia, in the scope of the framework contract foreseen in numbers 4, 5 and 6 of article 23 of the Decree Law 57/2016, of 29 August and UIDB/00481/2020 and UIDP/00481/2020, and CENTRO-01-0145-FEDER-022083—Centro Portugal Regional Operational Programme (Centro2020), under the PORTUGAL 2020 Partnership Agreement, through the European Regional Development Fund. Olena Okhay acknowledges the European Consortium of Innovative Universities (ECIU) for its support with a research mobility grant to the University of Tampere. Alexander Tkach acknowledges the project CICECO Aveiro Institute of Materials (UIDB/50011/2020 and UIDP/50011/2020), financed by national funds through the FCT/MCTES and, when appropriate, co-financed by FEDER under the PT2020 Partnership Agreement.

Institutional Review Board Statement: Not applicable.

Informed Consent Statement: Not applicable.

Data Availability Statement: The data presented in this study are available on request from the corresponding author.

Acknowledgments: The authors acknowledge Jari Keskinen from Tampere University for his valuable contribution to the electrochemical measurements.

Conflicts of Interest: The authors declare no conflict of interest.

References

1. Brodie, B.C. On the Atomic Weight of Graphite. *Philos. Trans. R. Soc. Lond.* **1859**, *149*, 249–259. [CrossRef]
2. Kohlschütter, V.; Haenni, P. Zur Kenntnis des Graphitischen Kohlenstoffs und der Graphitsäure. *Z. Anorg. Allg. Chem.* **1918**, *105*, 121–144. [CrossRef]
3. Wallace, P.R. The band theory of graphite. *Phys. Rev.* **1947**, *71*, 622–634. [CrossRef]
4. Ruess, G.; Vogt, F. Höchstlamellarer Kohlenstoff aus Graphitoxyhydroxyd. *Monatsh. Chem.* **1948**, *78*, 222–242. [CrossRef]
5. Boehm, H.P.; Setton, R.; Stumpp, E. Nomenclature and terminology of graphite intercalation compounds. *Carbon* **1986**, *24*, 241. [CrossRef]
6. Novoselov, K.S.; Geim, A.K.; Morozov, S.V.; Jiang, D.; Zhang, Y.; Dubonos, S.V.; Grigorieva, I.V.; Firsov, A.A. Electric field effect in atomically thin carbon films. *Science* **2004**, *306*, 666. [CrossRef]
7. Novoselov, K.S.; Geim, A.K.; Morozov, S.V.; Jiang, D.; Katsnelson, M.I.; Grigorieva, I.V.; Dubonos, S.V.; Firsov, A.A. Two-dimensional gas of massless Dirac fermions in graphene. *Nature* **2005**, *438*, 197–200. [CrossRef] [PubMed]
8. Zhang, Y.B.; Tan, Y.W.; Stormer, H.L.; Kim, P. Experimental observation of the quantum Hall effect and Berry's phase in graphene. *Nature* **2005**, *438*, 201–204. [CrossRef] [PubMed]
9. Züttel, A.; Sudan, P.; Mauron, P.; Wenger, P. Model for the hydrogen adsorption on carbon nanostructures. *Appl. Phys. A* **2004**, *78*, 941–946. [CrossRef]
10. Sundaram, R.S. Chemically derived graphene. In *Graphene: Properties, Preparation, Characterisation and Devices*; Skakalova, V., Kaiser, A.B., Eds.; Woodhead Publishing: Sawston, UK, 2014; pp. 50–80. [CrossRef]
11. Okhay, O.; Tkach, A.; Hortiguela Gallo, M.J.; Otero-Irurueta, G.; Mikhalev, S.; Staitic, P.; Lufrano, F. Energy storage of supercapacitor electrodes on carbon cloth enhanced by graphene oxide aerogel reducing conditions. *J. Energy Storage* **2020**, *32*, 101839. [CrossRef]
12. Sengupta, I.; Chakraborty, S.; Talukdar, M.; Pal, S.K.; Chakraborty, S. Thermal reduction of graphene oxide: How temperature influences purity. *J. Mater. Res.* **2018**, *33*, 4113. [CrossRef]
13. Ganguly, A.; Sharma, S.; Papakonstantinou, P.; Hamilton, J. Probing the Thermal Deoxygenation of Graphene Oxide Using High-Resolution In situ X-ray-Based Spectroscopies. *J. Phys. Chem. C* **2011**, *115*, 17009–17019. [CrossRef]
14. Ramamoorthy, H.; Buapan, K.; Chiawchan, T.; Thamkrongart, K.; Somphonsane, R. Exploration of the temperature-dependent correlations present in the structural, morphological and electrical properties of thermally reduced free-standing graphene oxide papers. *J. Mater. Sci.* **2021**, *56*, 15134–15150. [CrossRef]
15. Tamboli, S.H.; Seok Kim, B.; Choi, G.; Lee, H.; Lee, D.; Patil, U.M.; Lim, J.; Kulkarni, S.B.; Chan Jun, S.; Hee Cho, H. Post-heating effects on the physical and electrochemical capacitive properties of reduced graphene oxide paper. *Mater. Chem. A* **2014**, *2*, 5077. [CrossRef]
16. Karthick, R.; Brindha, M.; Selvaraj, M.; Ramu, S. Stable colloidal dispersion of functionalized reduced graphene oxide in aqueous medium for transparent conductive film. *J. Colloid Interface Sci.* **2013**, *406*, 69–74. [CrossRef] [PubMed]

17. Tarcan, R.; Todor-Boer, O.; Petrovai, I.; Leordean, C.; Astilean, S.; Botiz, I. Reduced graphene oxide today. *J. Mater. Chem. C* **2020**, *8*, 1198–1224. [CrossRef]
18. Liu, C.; Yu, Z.; Neff, D.; Zhamu, A.; Jang, B.Z. Graphene-based supercapacitor with an ultrahigh energy density. *Nano Lett.* **2010**, *10*, 4863–4868. [CrossRef]
19. El-Kady, M.F.; Strong, V.; Dubin, S.; Kaner, R.B. Laser scribing of high-performance and flexible graphene-based electrochemical capacitors. *Science* **2012**, *335*, 1326–1330. [CrossRef]
20. El-Kady, M.; Shao, Y.; Kaner, R. Graphene for batteries, supercapacitors and beyond. *Nat. Rev. Mater.* **2016**, *1*, 16033. [CrossRef]
21. Ke, Q.; Wang, J. Graphene-based materials for supercapacitor electrodes—A review. *J. Mater.* **2016**, *2*, 37–54. [CrossRef]
22. Lemine, A.S.; Zagho, M.M.; Altahtamouni, T.M.; Bensalah, N. Graphene a promising electrode material for supercapacitors—A review. *Int. J. Energy Res.* **2018**, *42*, 4284–4300. [CrossRef]
23. Zuliani, J.E.; Tong, S.; Jia, C.Q.; Kirk, D.W. Contribution of surface oxygen groups to the measured capacitance of porous carbon supercapacitors. *J. Power Sources* **2018**, *395*, 271–279. [CrossRef]
24. Li, J.; Tang, J.; Yuan, J.; Zhang, K.; Yu, X.; Sun, Y.; Zhang, H.; Qin, L.-C. Porous carbon nanotube/graphene composites for high-performance supercapacitors. *Chem. Phys. Lett.* **2018**, *693*, 60–65. [CrossRef]
25. Huang, Z.-D.; Zhang, B.; Liang, R.; Zheng, Q.-B.; Oh, S.W.; Lin, X.Y.; Yousefi, N.; Kim, J.K. Effects of reduction process and carbon nanotube content on the supercapacitive performance of flexible graphene oxide papers. *Carbon* **2012**, *50*, 4239–4251. [CrossRef]
26. Alazmi, A.; El Tall, O.; Rasul, S.; Hedhili, M.N.; Patole, S.P.; Costa, P.M.F.J. A process to enhance the specific surface area and capacitance of hydrothermally reduced graphene oxide. *Nanoscale* **2016**, *8*, 17782–17787. [CrossRef]
27. Dubal, D.P.; Kim, J.G.; Kim, Y.; Holze, R.; Lokhande, C.D.; Kim, W.B. Supercapacitors based on flexible substrates: An overview. *Energy Technol.* **2014**, *2*, 325–341. [CrossRef]
28. Faruque, M.A.A.; Syduzzaman, M.; Sarkar, J.; Bilisik, K.; Naebe, M. A review on the production methods and applications of graphene materials. *Nanomaterials* **2021**, *11*, 2414. [CrossRef]
29. Li, L.; Zhang, D.; Deng, J.; Fang, J.; Gou, Y. Review—Preparation and application of graphene-based hybrid materials through electrochemical exfoliation. *J. Electrochem. Soc.* **2020**, *167*, 086511. [CrossRef]
30. Liu, S.; Hu, K.; Cerruti, M.; Barthelat, F. Ultra-stiff graphene oxide paper prepared by directed-flow vacuum filtration. *Carbon* **2020**, *158*, 426–434. [CrossRef]
31. Zhao, S.; Li, M.; Wu, X.; Yu, S.H.; Zhang, W.; Luo, J.; Wang, J.; Geng, Y.; Gou, Q.; Sun, K. Graphene-based free-standing bendable films: Designs, fabrications, and applications. *Mater. Today Adv.* **2020**, *6*, 100060. [CrossRef]
32. Li, Z.; Chen, X.; Zhang, C.; Baer, E.; Langhe, D.; Ponting, M.; Zhu, L. High dielectric constant polycarbonate/nylon multilayer films capacitors with self-healing capability. *ACS Appl. Polym. Mater.* **2019**, *1*, 867–875. [CrossRef]
33. Legrand, G.D.; Bendler, G.T. *Handbook of Polycarbonate Science and Technology*; CRC Press: Boca Raton, FL, USA, 1999.
34. Singh, J.; Bansal, S.A.; Singh, A.P. Mechanical Characterization of Polycarbonate-Graphene Oxide (PCG) Nanocomposite. In *Advances in Production and Industrial Engineering*; Pandey, M., Ed.; Springer Nature Singapore Pte Ltd.: Singapore, 2021; pp. 103–111. [CrossRef]
35. Yoon, S.-H.; Jung, H.-T. Grafting polycarbonate onto graphene nanosheets: Synthesis and characterization of high performance polycarbonate–graphene nanocomposites for ESD/EMI applications. *RSC Adv.* **2017**, *7*, 45902–45910. [CrossRef]
36. Li, D.; Del Rio Castillo, A.E.; Jussila, H.; Ye, G.; Ren, Z.; Bai, J.; Chen, X.; Lipsanen, H.; Sun, Z.; Bonaccorso, F. Black phosphorus polycarbonate polymer composite for pulsed fibre lasers. *Appl. Mater. Today* **2016**, *4*, 17–23. [CrossRef]
37. Sasaki, H.; Hamanaka, I.; Takahashi, Y.; Kawaguchi, T. Effect of reinforcement on the flexural properties of injection-molded thermoplastic denture base resins. *J. Prosthodont.* **2015**, *26*, 302–308. [CrossRef]
38. Lago, E.; Toth, P.S.; Pugliese, G.; Pellegrini, V.; Bonaccorso, F. Solution blending preparation of polycarbonate/graphene composite: Boosting the mechanical and electrical properties. *RSC Adv.* **2016**, *6*, 97931–97940. [CrossRef]
39. Leeladhar, P.; Raturi, J.P. Singh. Sunlight-driven eco-friendly smart curtain based on infrared responsive graphene oxide-polymer photoactuators. *Sci. Rep.* **2018**, *8*, 3687. [CrossRef]
40. Yadav, A.; Kumar Yadav, D.; Kumar Mishra, J.; Sahu, R.; Jain, S.K.; Dixit, S.; Agarwal, G.; Jakhar, N.; Tripath, B. Effect of $BaTiO_3$ Nanofillers on the Energy Storage Performance of Polymer Nanocomposites. *Macromol. Symp.* **2021**, *399*, 2100024. [CrossRef]
41. Rogal'skii, S.P.; Pud, A.A.; Shapoval, G.S.; Mel'nik, A.F.; Bryk, M.T. Nature of Initiators for Indirect Electrochemical Reductive Degradation of Polycarbonates in Dimethylformamide. *Theor. Exp. Chem.* **2004**, *40*, 33–38. [CrossRef]
42. Zhu, D.; Hou, J.; Zhang, L.; Gao, Y.; Dai, B.; Lian, Y.; Yan, H.; Zhang, H. Microbial Porous Carbon by Low-Alkali Activation for Flexible Supercapacitors. *J. Electron. Mater.* **2021**, *50*, 6733–6740. [CrossRef]
43. Wu, Y.; Yan, M.; Sun, L.; Shi, W. Flexible yolk-shelled $NiCo_2S_4$ hollow spheres/RGO film electrodes for efficient supercapacitive energy storage. *New J. Chem.* **2018**, *42*, 16174–16182. [CrossRef]
44. Saito, K.; Jehanno, C.; Meabe, L.; Olmedo-Martínez, J.L.; Mecerreyes, D.; Fukushima, K.; Sardon, H. From plastic waste to polymer electrolytes for batteries through chemical upcycling of polycarbonate. *J. Mater. Chem. A* **2020**, *8*, 13921–13926. [CrossRef]
45. Matsumoto, M.; Uno, T.; Kubo, M.; Itoh, T. Polymer electrolytes based on polycarbonates and their electrochemical and thermal properties. *Ionics* **2013**, *19*, 615–622. [CrossRef]
46. Farzana, R.; Rajarao, R.; Bhat, B.R.; Sahajwalla, V. Performance of an activated carbon supercapacitor electrode synthesised from waste Compact Discs (CDs). *J. Ind. Eng. Chem.* **2018**, *65*, 387–396. [CrossRef]

47. Okhay, O.; Gonçalves, G.; Dias, C.; Ventura, J.; Vieira, E.M.F.; Gonçalves, L.M.V.; Tkach, A. Tuning electrical and thermoelectric properties of freestanding graphene oxide papers by carbon nanotubes and heat treatment. *J. Alloys Compd.* **2019**, *781*, 196–200. [CrossRef]
48. Pell, W.; Conway, B.; Marincic, N. Analysis of non-uniform charge/discharge and rate effects in porous carbon capacitors containing sub-optimal electrolyte concentrations. *J. Electroanal. Chem.* **2000**, *491*, 9–21. [CrossRef]
49. IEC-62391-2. Fixed Electric Double Layer Capacitors for Use in Electronic Equipment—Part 2: Sectional Specification—Electric Double Layer Capacitors Fpr Power Application, 1st ed. International Standard, International Electrochemical Commission, ICS Codes 31.060.10. Available online: https://webstore.iec.ch/preview/info_iec62391-1%7Bed1.0%7Den.pdf (accessed on 10 April 2006).
50. Mathis, T.S.; Kurra, N.; Wang, X.; Pinto, D.; Simon, P.; Gogotsi, Y. Energy storage data reporting in perspective-guidelines for interpreting the performance of electrochemical energy storage systems. *Adv. Energy Mater.* **2019**, *9*, 1902007. [CrossRef]
51. Gogotsi, Y.; Penner, R.M. Energy storage in nanomaterials-capacitive, pseudocapacitive, or battery-like? *ACS Nano* **2018**, *12*, 2081–2083. [CrossRef] [PubMed]
52. Okhay, O.; Tkach, A. Graphene/reduced graphene oxide-carbon nanotubes composite electrodes: From capacitive to battery-type behaviour. *Nanomaterials* **2021**, *11*, 1240. [CrossRef] [PubMed]
53. Paleo, A.J.; Staiti, P.; Rocha, A.M.; Squadrito, G.; Lufrano, F. Lifetime assessment of solid-state hybrid supercapacitors based on cotton fabric electrodes. *J. Power Sources* **2019**, *434*, 226735. [CrossRef]
54. Park, H.; Ambade, R.B.; Noh, S.H.; Eom, W.; Koh, K.H.; Ambade, S.B.; Lee, W.J.; Kim, S.H.; Han, T.H. Porous graphene-carbon nanotube scaffolds for fiber supercapacitors. *ACS Appl. Mater. Interfaces* **2019**, *11*, 9011–9022. [CrossRef]
55. Tuukkanen, S.; Krebs, M. Printable power storage: Batteries and supercapacitors. In *Organic and Printed Electronics*, 1st ed.; Nisato, G., Lupo, D., Ganz, S., Eds.; Jenny Stanford Publishing: New York, NY, USA, 2016; pp. 265–291.
56. Romero, F.J.; Gerardo, D.; Romero, R.; Ortiz-Gomez, I.; Salinas-Castillo, A.; Moraila-Martinez, C.L.; Rodriguez, N.; Morales, D.P. Comparison of laser-synthetized nanographene-based electrodes for flexible supercapacitors. *Micromachines* **2020**, *11*, 555. [CrossRef] [PubMed]
57. Ghasemi, F.; Jalali, M.; Abdollahi, A.; Mohammadi, S.; Sanaee, Z.; Mohajerzadeh, S. A high performance supercapacitor based on decoration of MoS_2/reduced graphene oxide with NiO nanoparticles. *RSC Adv.* **2017**, *7*, 52772. [CrossRef]
58. Chen, Y.; Bai, J.; Yang, D.; Sun, P.; Li, X. Excellent performance of flexible supercapacitor based on the ternary composites of reduced graphene oxide/molybdenum disulfide/poly (3,4-ethylenedioxythiophene). *Electrochim. Acta* **2020**, *330*, 135205. [CrossRef]
59. Beidaghi, M.; Wang, C. Micro-supercapacitors based on interdigital electrodes of reduced graphene oxide and carbon nanotube composites with ultrahigh power handling performance. *Adv. Funct. Mater.* **2012**, *22*, 4501–4510. [CrossRef]
60. Xu, L.; Li, Y.; Jia, M.; Zhao, Q.; Jin, X.; Yao, C. Synthesis and characterization of free-standing activated carbon/reduced graphene oxide film electrodes for flexible supercapacitors. *RSC Adv.* **2017**, *7*, 45066–45074. [CrossRef]
61. Zhu, J.; Zu, J.; Liu, J.; Wang, Y.; Pei, M.; Xu, Y. Self-assembled reduced graphene oxide films with different thicknesses as high performance supercapacitor electrodes. *J. Energy Storage* **2020**, *32*, 101795. [CrossRef]

C Journal of Carbon Research

MDPI

Article

The X-ray, Raman and TEM Signatures of Cellulose-Derived Carbons Explained

Petros Kasaira Mubari [1,2,*], Théotime Beguerie [2], Marc Monthioux [1], Elsa Weiss-Hortala [2], Ange Nzihou [2] and Pascal Puech [1,*]

1 Centre d'Elaboration des Matériaux et d'Etudes Structurales (CEMES), CNRS, Université de Toulouse, 29 Rue Jeanne Marvig, CEDEX 04, 31055 Toulouse, France; marc.monthioux@cemes.fr

2 IMT Mines Albi, UMR CNRS 5302, Centre RAPSODEE, Campus Jarlard, Université de Toulouse, CEDEX 09, 81013 Albi, France; theotime.beguerie@mines-albi.fr (T.B.); elsa.weiss@mines-albi.fr (E.W.-H.); ange.nzihou@mines-albi.fr (A.N.)

* Correspondence: petros.mubari@mines-albi.fr (P.K.M.); pascal.puech@cemes.fr (P.P.); Tel.: +33-(0)5-67-52-43-57 (P.P.)

Abstract: Structural properties of carbonized cellulose were explored to conjugate the outcomes from various characterization techniques, namely X-ray diffraction (XRD), Raman spectroscopy, and high-resolution transmission electron microscopy. All these techniques have evidenced the formation of graphene stacks with a size distribution. Cellulose carbonized at 1000 and 1800 °C at a heating rate of 2 °C/min showed meaningful differences in Raman spectroscopy, whereas in XRD, the differences were not well pronounced, which implies that the crystallite sizes calculated by each technique have different significations. In the XRD patterns, the origin of a specific feature at a low scattering angle commonly reported in the literature but poorly explained so far, was identified. The different approaches used in this study were congruous in explaining the observations that were made on the cellulose-derived carbon samples. The remnants of the basic structural unit (BSU) are developed during primary carbonization. Small graphene-based crystallites inherited from the BSUs, which formerly developed during primary carbonization, were found to coexist with larger ones. Even if the three techniques give information on the average size of graphenic domains, they do not see the same characteristics of the domains; hence, they are not identical, nor contradictory but complementary. The arguments developed in the work to explain which characteristics are deduced from the signal obtained by each of the three characterization techniques relate to physics phenomena; hence, they are quite general and, therefore, are valid for all kind of graphenic materials.

Keywords: cellulose; carbonization; disorder; Raman; XRD; crystallite size; HRTEM

Citation: Mubari, P.K.; Beguerie, T.; Monthioux, M.; Weiss-Hortala, E.; Nzihou, A.; Puech, P. The X-ray, Raman and TEM Signatures of Cellulose-Derived Carbons Explained. C **2022**, *8*, 4. https://doi.org/10.3390/c8010004

Academic Editor: Gil Goncalves

Received: 17 December 2021
Accepted: 30 December 2021
Published: 3 January 2022

1. Introduction

Carbonaceous materials derived from biomass have a wide range of applications, ranging from aqueous and gaseous pollutant adsorption [1,2] and catalysis [3–5], to energy storage [6,7]. The carbonization of a non-graphitizable material, such as cellulose, results in the formation of stacked graphenes lacking the long-range near-parallel extension (otherwise designated as the local molecular orientation (LMO)) [8], which is necessary for the solid to crystallize into genuine graphite. In this regard, there is some misinterpretation or confusion in the literature due to the fact that using a graphitization process (i.e., a heat-treatment above 2200 °C up to 3000 °C) does not compulsorily result in obtaining carbon materials with the structure of genuine graphite [9]. In the case of fully non-graphitizable carbons, only randomly oriented turbostratic carbons are obtained.

X-ray diffraction (XRD) patterns are routinely used for carbon characterization and provide parameters, such as L_c (average stack thickness perpendicular to the basal plane in the perfectly coherent domains, i.e., the crystallites) and L_a (average crystallite diameter in the basal plane) which are usually both calculated by the Scherrer equation through the use of relevant peak widths. XRD also provides the lattice constants, a and c, which are usually indirectly calculated from the Bragg equation, which is made possible by calculating the period of a family of atomic planes and correlating it to the geometry of the crystal system. Additional correcting operations are required for small L_a and L_c, although it is nearly never mentioned in the literature [10]. An alternative is to use ad hoc functions obtained from the bottom-up approach [11]. For the turbostratic crystallites, increasing heat-treatment temperatures subsequently improves the stacking which is also accompanied by an increase in L_a, but this does not mean that there is a conversion into graphite. Characteristic features of non-graphitizable carbons include the prevalence of graphene curvatures, the presence of fullerene-like defects, short-distance misorientation between coherent domains, and, in the absence of catalysts, the impossibility to obtain graphite crystallites [12,13]. The aim of graphitization is to obtain graphite, meaning that the turbostratic stacks with inter-graphene spacing equal to or larger than 0.344 nm within a long-distance extended LMO must somehow be brought closer, down to 0.335 nm, as a consequence of the appropriate positioning of graphenes relative to each other within each crystallite. One important traditional way of measuring graphitization is by measuring the value of the spacing between graphene layers using the 2θ position of the $00l$ diffraction peaks in the XRD patterns [14].

With disordered graphenic materials, apparent sizes of the average crystallite determined by XRD, L_a and L_c, are usually used to denote the degree of order [14]. It was demonstrated by Oberlin [8], based on the previous assumption by Franklin [12], that the presence of cross-linking chemical functions in the organic precursor of non-graphitizable carbons hinders the alignment of polyaromatic molecules during primary carbonization (i.e., the early temperature-induced chemical transformation processes which drive the size of the local molecular orientation occurring below 600 °C). As a result, graphene stacks remain with small L_a and misoriented with respect to each other, which leads to graphitization resistance and subsequently determines the pore size.

In some poorly organised carbons, a peak may appear below the inter-graphene diffraction peak at 2θ~20–26°, typically at 2θ~10–13° (values for a Cu X-ray source) [15]. XRD has widely been used in the literature to account for stacking disorder, with stacking faults being defined as the absence of ABAB (Bernal) stacking. The crystallite dimensions in the z direction are reduced accordingly, and in turn, the resulting XRD peaks will broaden.

Raman spectroscopy is another well-established, non-destructive technique which is used to acquire information pertaining to the bulk properties of polycrystalline materials [16–19]. The crystallinity of carbon materials has been increasingly determined by Raman spectroscopy as an alternative to XRD [20]. Graphenic materials exhibit a Raman peak related to the sp^2 C bonding energy, known as the G band, which appears at 1580 cm^{-1}, corresponding to the position for the regular graphite structure. The D band observed around 1350 cm^{-1} in less ordered carbons is mainly associated with point defects and domain edges. While efforts have been made to reconcile XRD and Raman spectroscopy, results have not always been conclusive. During the pyrolysis of wood, Zickler et al. [21] concluded that the estimate of the graphenic crystallite size L_a derived from its correlation with the I_D/I_G ratio, as proposed by Tuinstra and Koenig [19], must be taken with caution. Indeed, not only was it not applicable for $L_a < 2$ nm, as previously stated by Ferrari and Robertson [22], but significant mismatches can also be seen for large L_a due to the ways through which the Raman spectra are exploited [21]. Furthermore, laser power effects [23], which can be material-dependent [24], as well as order/disorder differences between different carbon materials, are expected to also have an impact on Raman spectroscopy results in this first-order region. Besides intensity, both the position and width variations of the Raman D and G bands are also used to determine the degree of structural order [25]. For this purpose, it is a common practice to adopt a five-band decomposition procedure to fit the Raman spectra [20,26–29] and extract tiny shape variations of the Raman spectrum (decomposition is a term definitively better than deconvolution as the number of peaks and the way of proceeding are empirical). The literature dealing with Raman spectroscopy provides other parameters which aim to describe carbon materials. However, this is dependent on the sample type. Therefore, establishing a universal characterisation method is impossible as the texture, nanotexture, and structure (see our definitions of these terms in the "Methods" section) may vary significantly from one carbon material to another.

As a usual complement to Raman spectroscopy and XRD, high-resolution transmission electron microscopy (HRTEM) has been a useful technique for investigating carbon materials at the nanoscale, and tentatively obtaining quantified parameters describing the degree of order, mostly based on lattice fringe imaging [8,30,31]. HRTEM is used to obtain information on the texture, nanotexture, and structure of carbonaceous materials, but is prone to artefacts and intense labour, and is too local for some of the issues to be addressed. Furthermore, when dealing with bulk materials, such as chars, only the graphenes oriented edge-on can be revealed with sufficient contrast, and the fact that HRTEM images are 2D projections of 3D samples is likely to induce misinterpretations of the images.

In this paper, XRD, Raman spectroscopy, and HRTEM are used to study the texture, nanotexture, and crystalline structure of cellulose-derived chars carbonized in the 1000–1800 °C range. It is not of common knowledge that, despite the parameter name being the same, these techniques do not measure the same L_a. It is, therefore, important to compare and discuss what the differences are. First, we outline the techniques used to analyze the samples and the parameters of interest which were used to account for the carbon material organization. Then, we describe the theoretical aspects used to model and calculate the diffraction patterns. Finally, the nanotexture, texture, and structure of the cellulose-derived carbons are described, including the average crystallite sizes, in full agreement with the simulated XRD profiles, the experimental XRD patterns, Raman data, and HRTEM results.

2. Experimental

2.1. Materials, Treatments, and Terminology

Microcrystalline cellulose [CAS: 9004-34-6] was obtained from Sigma Aldrich Chimie S.a.r.l (Saint Quentin Fallavier, France). The carbonized samples were produced following a two-step process. First, 10 g of raw cellulose were pyrolyzed under N_2 with a flow rate of 1 L/min in a Carbolite tubular furnace at a temperature of 800 °C using a heating rate of 2 °C/min. A one-hour isotherm was applied. The resultant chars were collected after cooling and were subjected to further pyrolysis at 1000 °C and 1800 °C in a Nabertherm tubular furnace under a 5 L/min N_2 stream using a heating rate of 2 °C/min. The initial pyrolysis step serves the purpose of removing tars to prevent the clogging of the high-temperature furnace.

We chose 2 samples and annealed them at very different temperatures. 1000 and 1800 °C heat-treatment temperatures were chosen because they are both milestone temperatures, as far as carbonization processes are concerned. The former is in the temperature range where most of the remaining hydrogen atoms (as aromatic -CH) are released, and corresponds to the "coke" stage in industrial processes. The latter is in the range of the end of the secondary carbonization, beyond which graphitization processes may start (regardless whether they will succeed). In our previous studies on graphitizable carbons [11], the differences in terms of crystallite size between these 2 temperatures were really important. In particular, at an annealing temperature of 1800 °C, AB pairs began to form.

To describe the inner organization of graphenic materials, we used a three-term description where all terms were exclusive: the texture describes how graphene stacks are arranged with respect to each other at long distance; the imperfections (curvatures, defects) of the graphene layers determine the nanotexture; and the structure encompasses the diverse ways in which the graphenes pile up within a crystallite, among which the most common configurations are turbostratic, Bernal graphite, and rhombohedral graphite. More details are provided in Monthioux et al. [32].

2.2. X-ray Diffraction

XRD diffraction patterns were recorded from the carbonized cellulose samples on a Bruker RX-D8 Advance X-ray diffractometer with a non-monochromated Cu K_α radiation source (average λ = 1.5406 Å). The domains which scatter the X photons in a coherent manner represent the crystallites. Raw diffractograms exhibit an intense background, presumably due to a variety of X-photon interactions with the matter (Compton effect, absorption, polarization, etc.). Hence, they were treated for removing the background using continuously decreasing functions, which is a routine procedure for X-ray pattern analysis. The resulting diffraction profiles obtained from experiments were fitted using a pseudo-Voigt function for the peaks appearing at $2\theta \approx 13°$ and $2\theta \approx 24°$, and specific parameterized functions were used for the peaks at $2\theta \approx 43°$ and $2\theta \approx 80°$ in MATLAB® [11]. This was completed in accordance with the Rietveld method, and the theoretical line profiles were refined by non-linear least-squares until a measured diffraction line profile was matched [33]. The interplanar spacing was calculated using the $2\theta \approx 24°$ peak position labelled *001*. We decided to use this notation *001* because the material remained turbostratic, whereas both *002* and *003* were only valid for the Bernal and rhombohedral graphite structures, respectively. However, it must be kept in mind that these 3 notations all correspond to the intergraphene spacing. We also proceeded to calculate the periodicity associated with the peak at $2\theta \approx 13°$, which will be named the Non-Identified (NI) peak, as a start. One can note that the NI peak is still present with Co sources (1.789 Å), allowing us to eliminate the artefact hypothesis. The graphenic crystallite sizes L_c were directly obtained from the Scherrer equation applied to the *001*. L_a (pertaining to the *10* asymmetric band for turbostratic carbons) and were obtained as a parameter of the parametrised function. Examples of the functions used are included in the supplementary information of ref. [11]. While a simple Gaussian function accounts well for the *001* peak, more specific functions are necessary for fitting the *10* and *11* bands. In these functions, L_a is the only adjustable parameter (if we do not consider intensity). Thus, using the classical approach, which is based on the Scherrer equation, a complex user-dependent fit, and the shape factor *K*, is no longer necessary for fitting these two peaks with our approach.

2.3. Modelling the X-ray Diffraction Peak at $2\theta \approx 13°$

The various defects (only one will be reported here) and stacking configurations were modelled by first constructing crystallite models with MATLAB®, and then by calculating the related X-ray diffractograms using the highly parallelized XaNSoNS freeware [34] with the ultimate goal to find a diffractogram that matches the experimental profiles.

2.4. Raman Spectroscopy

Raman spectroscopic analysis of carbonized cellulose was performed with an XploRA Plus spectrometer using a low laser power (1.4 mW), an objective of magnification ×50, and laser wavelengths/energies of 638 nm/1.94 eV (red) and 532 nm/2.33 eV (green) considering an acquisition time of 100 s. The laser spot size was larger than 1 µm and no heating was observed. We used two approaches to interpret the Raman spectra. In the first approach, we fitted the experimental spectra with a D band which was decomposed in two contributions and a G band for obtaining accurate values for the respective intensities as well as the linewidths. In the second one, we fitted the spectra with five bands corresponding to a spectral decomposition often used in carbon material characterization.

2.4.1. Fitting Procedure and Parameters When Considering the D and G Bands Only

The usually symmetric D band was fitted with two Voigt functions, a sharp one D_s, and a broad one D_b, located at the same spectral position, and a combination of Gaussian and Fano functions was used for fitting the G band [35]. The parameters of interest are the ratio of the intensities of the D and G bands (I_D/I_G) and the full width at half maximum of the D band ($FWHM_D$), which will be used to estimate L_a. Table 1 outlines the parameters that were of interest after fitting the experimental spectra with two bands. The crystallite size L_a

was calculated in three ways by exploiting Tuinstra–Koenig's correlation for $L_a > 2$ nm [19], the Ferrari's and Robertson's law for $L_a < 2$ nm [22], and the confinement model also for L_a larger than ~2 nm [36].

Table 1. Extracted parameters for the D and G band-fitted Raman spectra.

Extracted Parameters	Significance
(I_D/I_G)	Calculation of L_a [19,22]
D band—$FWHM_D$	Calculation of L_a [36]
D band	Degree of graphenic-sp^2 disorder

Laser energy dependencies were accounted for by taking into consideration the following relation [18]:

$$\frac{I_D}{I_G} = \frac{f(La)}{E_L^{\alpha}} \tag{1}$$

It is expressed as a function of the crystallite size $f(L_a)$, where (I_D/I_G) represents the intensity ratio and E_L is the laser energy whilst α is the exponent to be calculated. $\alpha = 4$ for graphite [37] but for poorly organized carbon, this value decreases [35], although it always stays positive.

After considering the laser energy dependencies, L_a from the Tuinstra–Koenig correlation is given by ($L_a > 2$ nm):

$$L_a = \frac{4.4}{\left(\frac{I_D}{I_G}\right)} \times \left(\frac{2.41}{EL}\right)^{\alpha} \tag{2}$$

L_a from the Ferrari's and Robertson's law is given by ($L_a < 2$ nm):

$$L_a = \sqrt{\left(\frac{\left(\frac{I_D}{I_G}\right)}{0.55 \times \left(\frac{2.41}{EL}\right)^{\alpha}}\right)} \tag{3}$$

while I_D/I_G is expected to be maximum around 2 nm.

L_a from the confinement model is given by ($L_a \geq 2$ nm):

$$L_a = \frac{580}{FWHM_D} \tag{4}$$

with a continuous variation without notable change at $L_a = 2$ nm (for small L_a, the 580 coefficient decreases, allowing to find a higher limit value).

2.4.2. Fitting Procedure and Parameters When Considering Five Bands

The idea of having several bands for the fitting means having correlations which aim to determine the degree of disorder and the amorphous content. Usually, the functions (Gaussian, Lorentzian, Voigt) and the fixed parameters (usually wavenumbers) are selected to reveal the consistent shape variation of the Raman spectrum within a family of samples, while the background is often removed first. The experimental Raman spectra were decomposed into five bands in a manner summarized in Table 2.

Table 2. Consensual attribution from the literature and corresponding functions that were used for the fitting of Raman spectra with five bands [20,28,38,39].

Bands and Usual Position in Graphenic Materials (cm^{-1})	Fitting Function	Representation
$G \approx 1580$	Voigt	sp^2
$D_1 \approx 1350$	Voigt	defects in sp^2 (point defects, edges, curvature)
$D_2 \approx 1620$	Gaussian	defects in sp^2, needed due to fixed G band wavenumber and thus associated to disorder
$D_3 \approx 1500$	Gaussian	amorphous carbon

To quantify both the disorder degree and amorphous content, we used the intensity ratio of the D_3 band to the total intensity of all the bands $\frac{ID_3}{I_{total}}$. As the shapes of our experimental spectra varied with the treatment temperature as well as in function of the excitation energy used, for consistency, we opted for a Voigt shape (a combination of Lorentzian and Gaussian shapes) in some regions and a Gaussian shape in other regions.

2.5. HRTEM Micrographs, Image Filtering Analysis, Fringe Analysis

HRTEM images of the samples were collected using a SACTEM FEI TECNAI F20 microscope. This microscope, equipped with a field emission gun (Schottky type), was Cs-corrected (objective lens) and operated at 100 kV.

The image filtering using Fourier transform was used to extract information relative to the different domains from HRTEM images for some particular spatial frequencies, i.e., corresponding to the intergraphene distance of 3.4 ± 0.2 Å.

No value regarding the average fringe length, either straight or continuous but curved, was extracted from the HRTEM study because values would be biased by the projection effect already mentioned in the introduction section (unless domains are large, which is not the case with our samples). Indeed, projecting lattice fringes from a 3D volume onto a 2D imaging device is likely to generate fringe lengths shorter than they actually are, depending on the actual fringe lengths and the degree of textural and nanotextural anisotropies. One way to tentatively limit this could be to prepare sample TEM specimens as thin as possible, but then the fringe length could be physically altered by the preparation method.

3. Results and Discussion

3.1. X-ray Diffraction

Figure 1a,b present the raw XRD patterns (including the background, abbreviated with bg) for the cellulose carbonized at temperatures 1000 and 1800 °C. An intense background is seen at low scattering angles; hence, it is a routine in XRD analysis to merely subtract it, as mentioned is Section 2.2, in order to enhance the scattered peaks from the material. The bottom profile in Figure 1a,b show the result from the subtraction. Four broad peaks appear at around 13°, 24°, 43°, and 80°, corresponding to the NI peak, the intergraphene distance, and the *10* and *11* Laue indices of the in-plane atomic periodicities in isolated or turbostratically stacked graphenes, respectively.

At first glance, the two XRD profiles of the carbonised cellulose samples in Figure 1 differ in the background intensity, and differ slightly in the intensity, width, and positions of the peaks upon the temperature increase. Employing the Bragg's law ($2.d.\sin(\theta) = n.\lambda$) for the d_{001} peak at ~24°, an intergraphene distance after correction [10] of 3.68 Å was found for the cellulose char at 1000 °C and of 3.55 Å for the cellulose char at 1800 °C. The decrease in the intergraphene distance signifies an improvement in the material nanotexture (Table 3). It is worth noting that the decrease in the I_{bg}/I_{001} ratio from 1.65 to 0.87 is consistent with the assignment of at least part of the background to disorder, as claimed by [40], as it is expected that disorder decreases as the carbonization temperature increases.

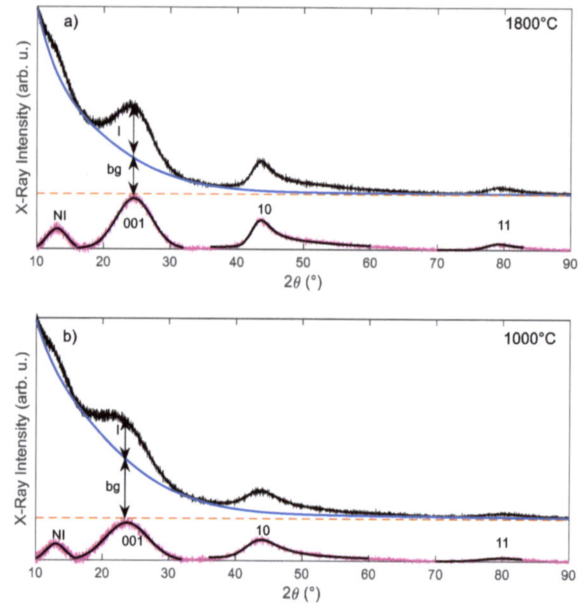

Figure 1. Raw and fitted X-ray diffraction profiles after background removal for cellulose-derived carbons obtained at (**a**) 1800 °C and (**b**) 1000 °C at a heating rate of 2°/min. Black curves for the diffractograms without background are the resulting fitting lines, obtained by the method described in [11].

Table 3. XRD parameters for cellulose-derived carbons at different carbonization temperatures. * From the Bragg's law without correction, corresponding to the apparent intergraphene spacing, ** with correction corresponding to the true intergraphene spacing (see [10]).

Extracted Parameters	Treatment Conditions	
	1000 °C, 2 °C/min	1800 °C, 2 °C/min
L_c(nm) from *001*	1.04	1.26
L_a (nm) from *10*	1.85	3.22
L_a (nm) from *11*	1.76	2.85
d_{001} (Å)	3.75 * (3.68 **)	3.62 * (3.55 **)
$a_{C=C}$ (Å)	1.42	1.42
$FWHM_{NI}$ (°)	3.46	3.46
$\frac{FWHM_{NI}}{FWHM_{001}}$	0.44	0.54
$\frac{A_{NI}}{A_{001}}$	0.20	0.21
$\frac{I_{NI}}{I_{001}}$	0.44	0.39
$\frac{I_{bg}}{I_{001}}$	1.65	0.87

These values are consistent with reports in the literature for biomass-issued carbons [41–43]. While the peak intensities do not increase significantly, the crystallite sizes L_c and L_a increase with increasing the carbonization temperature from 1000 °C to 1800 °C, although moderately for L_c and more extensively for L_a. L_c corresponds mainly in both cases to stacking of three carbon layers (we consider the van der Waals limit). This indicates the expected gradual increase in the dimensions of the turbostratic carbon crystallites with temperature, which mostly occur by gradually merging the crystallites into larger ones thanks to the progressive healing of the peripheral defects and the resulting slight misorientations which affect the crystallites [8].

On the contrary, the width of the peak at 13° remains constant ($FWHM_{NI}$ in Table 3). Despite being intense and single, this peak does not correspond to any periodicity in the graphite structure or its degenerative, turbostratic structure. However, it is striking that, with perfect graphite, the length of the unit cell along c is 6.7 Å, i.e., it could give a position at $2\theta = 13°$. However, of course, it is known that, in the graphite unit cell, the first graphene plane is in phase opposition to the second, leading to destructive interference and no intensity; hence the first scattered intensity appears for twice the 2θ value (corresponding to the *002* peak). Said otherwise, in graphite, the diffraction peak accounting for the (*001*) plane family is not a *001* peak at ~13° but a *002* peak at ~25°, because the unit cell is multiple and involves two equally spaced graphene layers instead of one. This suggests several possible origins for the NI peak: (i) a structural defect for which the middle graphene plane is missing in the cell, or a specific graphene configuration; (ii) the persistence of a structural, periodic feature specific to cellulose; and (iii) a specific feature of the XRD signal, when generated by poorly organized carbon, mostly made of tiny graphene stacks. Modelling was then performed to check two of these hypotheses whereas the others will be just discussed, as follows.

A plausible defect is that of inter-locked graphene sheets exhibiting curvatures as shown in Figure 2a, which is a quite common dislocation figure in graphene-based materials of any kind, including graphite. A peak below the d_{001} peak is present and is called in the following the defect peak (DP) (see Figure 2a). However, should this type of dislocation be the only responsible for the NI peak, the decrease in the intergraphene spacing (upshift of the *001* 2θ position in Figure 1) when carbonizing from 1000 to 1800 °C should also induce an upshift of the NI peak 2θ position, which is not observed. Moreover, the FWHM of the calculated DP peak obtained is not consistent with the experimental diffractograms.

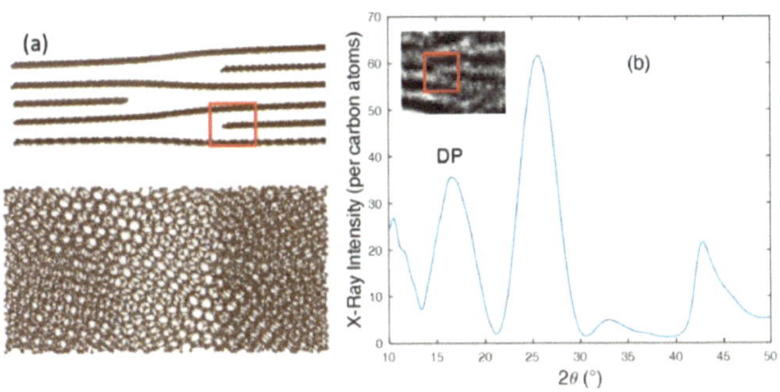

Figure 2. (**a**) Modelled dislocation showing an example of such a defect. (**b**) Simulated diffraction pattern profile obtained from the framed area in the atomistic model in (**a**), with an inset of a corresponding lattice fringe TEM image observed in a defective carbon.

Graphene edges connected or folded while forming loops are able to provide such a constant periodicity whatever the intergraphene distance, as ~6–7 Å corresponds to the diameter of the resulting channel. However, such a graphene termination is not observed in graphenic carbons obtained from the regular carbonization of organic precursors, and corresponds to specific process conditions instead [44].

Another hypothesis to consider could be that the 13° peak accounts for a remaining periodic feature of the cellulose skeleton. Indeed, XRD patterns of pristine (uncarbonized) cellulose do exhibit such a periodicity [45]. Unfortunately, modelling the full carbonization process of cellulose is impossible as it would require unrealistic computer time; hence, only the first steps are reported in the literature [46]. Nevertheless, the removal of atoms through the release of CO, CO_2, and H_2O, during primary carbonisation (i.e., <600 °C) and subsequent polyaromatization during secondary carbonization, should destroy this periodicity. Accordingly, it has been observed that the features corresponding to cellulose disappear [47]. Generally speaking, no such a persistence of an atomic-scale feature from the original organic skeleton upon carbonization was ever observed in the literature, whatever the precursor.

Finally, a peak around 2θ = 13° was also evidenced for poorly organized carbon materials originating from precursors other than cellulose [48,49]. This suggests that the 13° peak is somehow associated with diffraction patterns from poorly organized carbons rather than specific to a cellulose origin. Consequently, we simulated the diffraction patterns for small graphene stacks (turbostratic) with L_a = 0.7, 1, 2, and 3 nm, with the number of layers ranging from 1 to 5 (see Figure 3). From the various modelled spectra reported in Figure 3, we can make several statements:

- Single graphenes generate a large intensity at a low 2θ angle whatever L_a, rapidly decreasing as the scattering angle increases. This is consistent with a recent statement from the literature, which showed that poorly organized carbon is able to contribute significantly to the background, specifically at low angles [40]. Therefore, our modelling supports this statement, and identifies that such a contributing intensity profile at low angles reveals the occurrence of individual graphenes.
- For an odd number (here 3 and 5) of turbostratically stacked layers with L_a in the range 1–2 nm, a peak shows up in the 13° range, resulting from interference effects. The position of this peak is dependent on L_a. Stacks involving an even number of graphenes (here 2 and 4) do not exhibit this peak, meaning that, in this case, interferences are destructive. This destructive effect rapidly vanishes as the number of graphenes in the stacks increases beyond 5.
- The actual *001* peak resulting from the stacking periodicity shift towards lower angles (from ~25.5 to ~22°) while the number of graphenes in the stack decreases (here from 5 to 2), as shown by Fujimoto [10], despite the fact that the intergraphene distance remains the same in our simulation.
- With ultra-small L_a (<1 nm), intensity is generated at a low 2θ angle which can significantly contribute to the background.

From all the possibilities explored in Figure 3 and using the corrected d_{001} values for 1000 and 1800 °C from Table 3, in Figure 4, we reported the calculated diffraction patterns with L_a = 1 nm and three layers. Indeed, the matching between the simulated and experimental 13° peaks is perfect. The $FWHM_{NI}$ of the simulation equals 3.79°, which is very close to the experimental value of 3.46° (see Table 3). This is definitely the only simulation giving such a small value. The ratio of the simulated peak areas A_{NI}/A_{001} equals 0.38. On the other hand, the matching is not good in terms of intensities for the *001* and *10* peaks, which can be easily understood as revealing the presence of a range of crystallites other than the selected one (i.e., three stacked layers with L_a = 1 nm), therefore providing a distribution of L_a size and of layer number per stack.

To summarize, both the precursor types are highly cross-linked as any non-graphitizable material, and the limited carbonization temperatures suggested that the materials developed graphene crystallites with less than five layers (as confirmed by the HRTEM images to be shown below) and L_a < 3 nm. They might contain isolated graphenes, which would contribute to the background mostly at low 2θ angle as well as domains with L_a < 1 nm, whatever the number of layers is. This cannot be ascertained since the related intensity cannot be easily discriminated from the "regular" background (see Section 2.2). On the other hand, nanosized graphene crystallites developed, with 2 to 4 stacked layers. From them, only the crystallites with three layers are specifically seen as generating the 13° peak because, for the crystallites with an even number of layers (two and four), interference effects are destructive. This is why the 13° peak is specific to poorly organised graphenic carbons. We propose to designate this peak as a small graphene crystallite (SGC) peak. Should the carbonization temperature be increased beyond 1800 °C, crystallites would increase in L_a and layer number, resulting in making the 13° peak vanish as the material nanotexture improves.

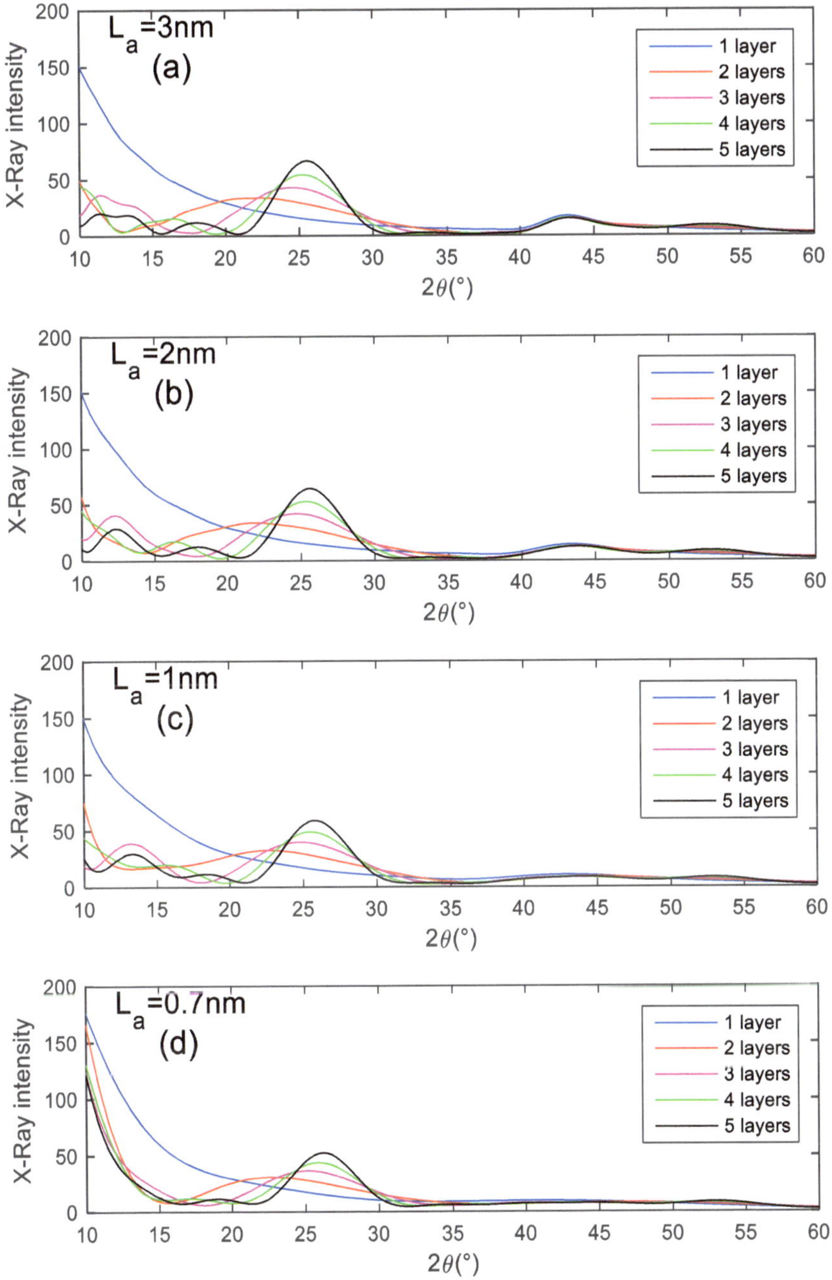

Figure 3. Simulated X-ray diffraction patterns for turbostratic carbon crystallites with the same L_a and a number of graphenes varying from 1 to 5. (**a**–**d**) L_a equals 3, 2, 1, and 0.7 nm, respectively. Each curve corresponds to an average of 10 different configurations (random sequences of plane rotations and positions). The intensity is per carbon atom.

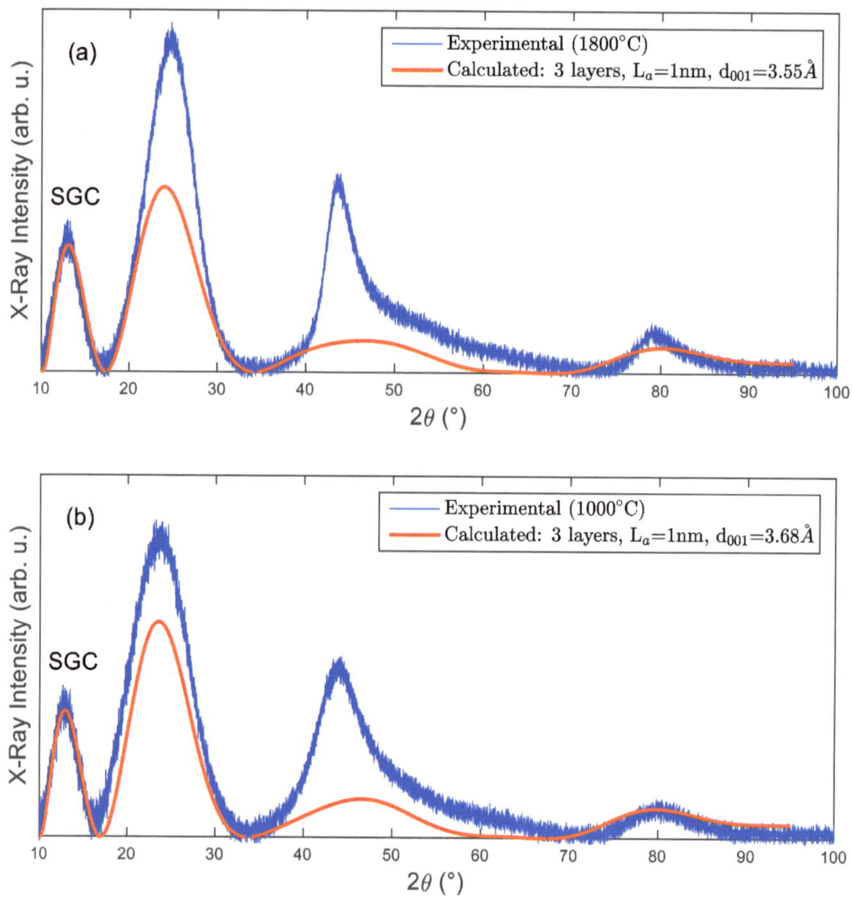

Figure 4. Calculated X-ray diffraction patterns after background removal (red lines) and experimental patterns after background removal normalised to the small graphene crystallite (SGC) peak (blue, noisy lines). The position of the SGC peak is dependent on L_a. (**a**,**b**) correspond to the samples annealed at 1800 and 1000 °C, respectively.

3.2. *Raman Spectroscopy*

Using two wavelengths, we experimentally determined the α value in Equations (2) and (3). We found 0.731 and 1.131 for samples annealed at 1000 and 1800 °C, respectively.

3.2.1. Raman Spectra Fitted with Two Bands (D and G Only)

The first-order (800–1736 cm^{-1} range) Raman spectra of both cellulose chars are shown in Figure 5, exhibiting the regular D and G bands at about 1340 cm^{-1} and 1594 cm^{-1}, respectively, fitted with two bands. Both bands are broad, which is characteristic of disordered graphenic carbons, but sharpen when the carbonisation temperature is increased (from 1000 to 1800 °C). Another notable feature is a corresponding double intensification of the D band whilst the G band increases by ~50% only. The intensity increases due to the increase in graphenic domain size, leading to a more resonant behavior.

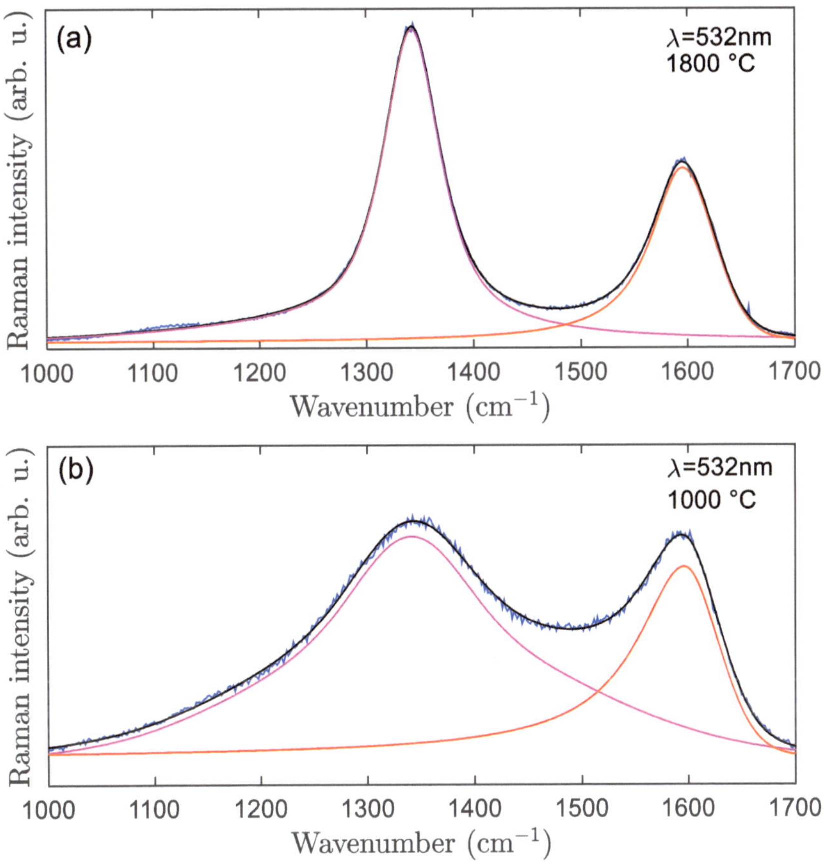

Figure 5. Raman spectra of cellulose carbonized at (**a**) 1800 and (**b**) 1000 °C obtained with an excitation wavelength of 532 nm and fitted with two bands. The magenta lines in both spectra are the result from the fitting of the D band with two components D_s and D_b, as described in Section 2.4.1.

Fitting with two bands, as shown in Figure 5, helps to obtain the intensity of the G and D bands, as well as the full-width at half maximum of the D band ($FWHM_D$). Table 4 lists the average L_a of the crystallites in both cellulose chars calculated from the D and G bands, following the various ways described in Section 2.4. The values obtained from the intensity ratio I_D/I_G are in the order of 1.42 and 2.45 nm at 1000 and 1800 °C, respectively, whereas the L_a values derived from the confinement model using $FWHM_D$ are much higher with a factor of 2 and 3.5, respectively.

Table 4. L_a obtained using the 532 nm excitation wavelength.

Crystallite Size (nm)	1000 °C, 2 °C/min	1800 °C, 2 °C/min
L_a [19]	-	2.45
L_a [22]	1.42	-
L_a [36]	<2.84	8.69

The monotonic increase in L_a for disordered carbons with heat-treatment temperature is obvious and has been reported recurrently in the literature [19,41,50,51]

The D and G bands are significantly broader at low carbonisation temperatures, as a result of the confinement effect at the nanoscale [51]. Because the resonant behavior varies in function of the crystallite size [35], a signal originating from the large crystallites obtained at higher temperatures in a material can dominate the D-band region. Moreover, XRD and Raman do not measure the same characteristics. Indeed, L_a sizes measured by XRD involve distortion-free graphenes (because the XRD peaks result from the X-photons scattered strictly in the same direction of space by the atoms), even though the graphene lattice may be somehow defective (i.e., the graphenes may contain in-plane defects, such as 5–7 ring pairs (so-called Dienes defect [52] or vacancies—which do not alter their planarity)). On the other hand, L_a sizes measured by Raman are not affected by corrugations (long-range angular distortions), providing that the graphene lattice is free of any defect, because each in-plane defect (not only the graphene edges) acts as a scattering site for the electron–hole pairs involved in the double resonance Raman process, travelling across the graphene layers. Thus, complementary information is available with these two techniques.

3.2.2. Raman Spectra Fitted with 5 Bands

Figure 6 shows further analysis of the Raman spectra through fitting them with five bands:

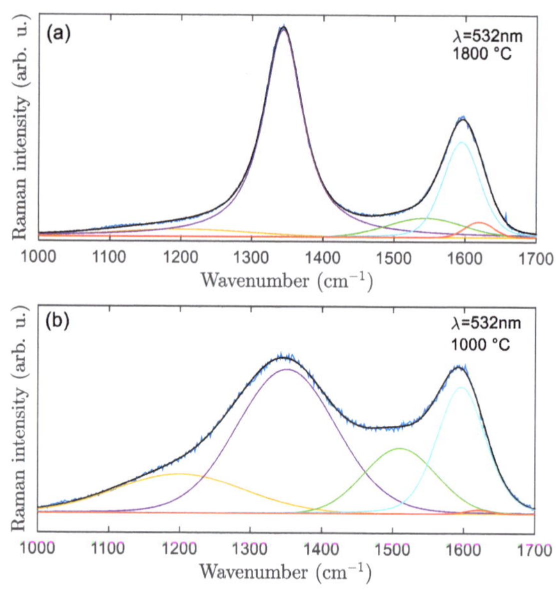

Figure 6. Raman spectra of cellulose carbonized at (**a**) 1800 and (**b**) 1000 °C obtained with an excitation wavelength of 532 nm, and fitted with 5 bands (see Table 2): D_1 (purple) and G (blue), added with D_2 (red), D_3 (green), and D_4 (yellow).

Three additional bands were added to the D_1 and G bands, namely the D_2, D_3, and D_4 bands. Though it is presumably possible to fix the wavenumbers with a set of samples from the same organic origin, it is impossible to do so here as it leads to a less accurate decomposition. The D_3 band at 1510 and 1543 cm^{-1} (for 1000 and 1800 °C carbonisation temperatures, respectively) relates to unstructured (amorphous) carbon. We chose to only discuss the intensity ratio $\frac{ID_3}{I_{total}}$ for quantifying disorders based on our previous studies on biochar materials [29]. From the $\frac{ID_3}{I_{total}}$ ratio in Table 5, we can note that it decreases by a factor of three, accompanied by a proportional decrease in the height of the interband valley, as seen in Figure 6, when the annealing temperature increases from 1000 to 1800 °C. Likewise for the integrated intensity ratio, which decreases by a factor of about 2. The G-band position (1595 cm^{-1}) remains almost unchanged at both temperatures, whilst the D-band is redshifted from 1350 cm^{-1} to 1343 cm^{-1} when the carbonisation temperature increases. The formation of short-ordered turbostratic graphenic stacks was consistent with literature findings [29,53]. In summary, the band position should be adjusted at each temperature and the valley between the D and G band can be accounted for through the D_3 band. By leaving the wavenumber position of the D_3 band free to adjust, the D_3 band may be considered as a good indicator for disorders.

Table 5. Intensity ratios obtained using the 532-nm excitation wavelength.

Excitation Wavelength 532 nm	1000 °C, 2 °C/min	1800 °C, 2 °C/min
Intensities: $\frac{ID_3}{I_{total}}$	0.173	0.055
Integrated intensities: $\frac{AD_3}{A_{total}}$	0.153	0.08

3.3. *HRTEM Image Filtering Analysis, Fringe Analysis*

Figure 7 provides an example of HRTEM images of the carbonized cellulose for both carbonization temperatures and the associated filtered images for a distance of 3.4 ± 0.4 Å.

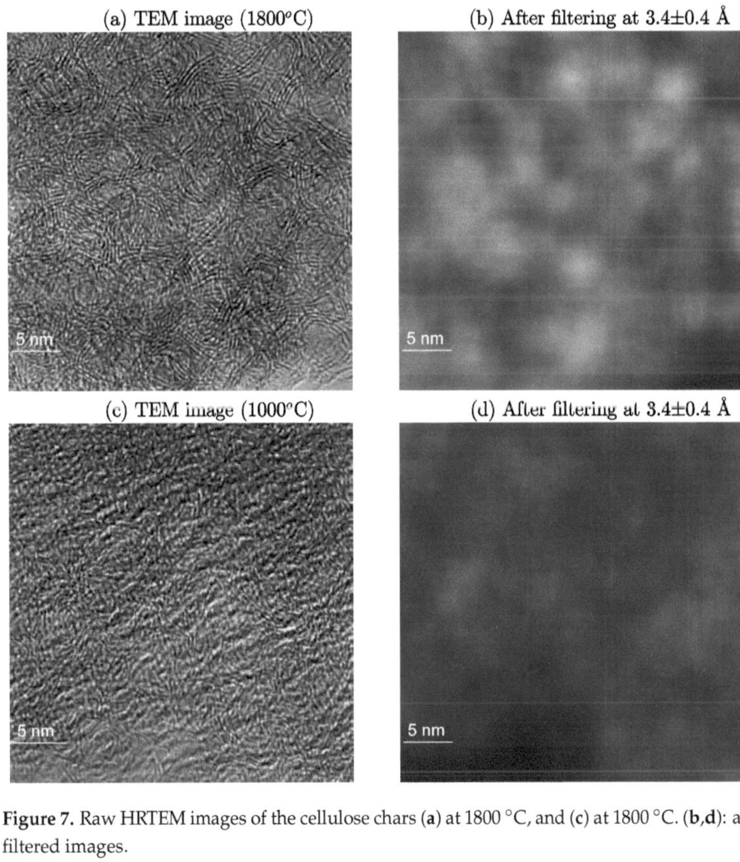

Figure 7. Raw HRTEM images of the cellulose chars (**a**) at 1800 °C, and (**c**) at 1800 °C. (**b**,**d**): associated filtered images.

Comparing Figure 7a,c, it can be seen that, for the cellulose carbonized at 1000 °C, there are fewer graphene sheets. This confirms the transition from a poorly organised material with a short-range ordering of the graphene sheets to a material with improved ordering. When the temperature is increased up to 1800 °C, the graphenes become more apparent because they tend to be larger and their planarity is better; hence, the scattered intensity increases as the number of atoms involved in a single domain increase. The material evolves with a very much larger proportion of curved domains. This is consistent with the merging process of basic structural units [8], which involve straight graphenic planes, but whose merging does not necessarily result in straight domains as it may come with the creation of in-plane defects and local constraints which both affect the planarity. The process is enhanced by the overall effect of increasing the carbonisation temperature which promotes the mobility of atoms in the solid. Although it was chosen not to extract fringe length values from the HRTEM images because the TEM projection of superimposed small domains shorten the associated fringes, which are thus unrepresentative of the sample (see Section 2.5), it is obvious that the lattice fringes increase in length with the increasing temperature, as seen with the TEM images of Figure 7a,c. The average pixel values of the filtered images (Figure 7b,d) are 78 and 114 for the samples annealed at 1000 and 1800 °C, respectively. The increase in the occurrence of 3.4 Å spacing proves that the crystallinity increases, which is fully consistent with the background reduction in XRD.

4. Merging All the Data

Considering all the techniques used in this study, the crystallite size, L_a, as summarized in Table 6, distributes over a range.

Table 6. L_a values obtained directly from XRD (top line), and then from the I_D/I_G obtained by Raman spectroscopy using various correlations available in the literature.

Crystallite Size (nm)	1000 °C, 2 °C/min	1800 °C, 2 °C/min
L_a (XRD)	~1.8	2.8–3.2
L_a [19]	-	~2.5
L_a [22]	1.4	-
L_a [36]	<2.8	8.7

Although L_a is an important indicator of the development of the graphenic domains, none of the L_a obtained should be taken as a universal value because the different techniques do not measure the same feature. L_a obtained from XRD is strictly dependent on the planarity of the graphene. Moreover, we showed that several sizes are visible in the X-ray diffraction patterns. Hence, it is a direct measurement of the average in-plane dimension of the parallel coherent graphene stacks (the genuine crystallites). Should a defect be present in the in-plane graphene lattice while not deforming significantly the orientation of the graphene involved (as for defects such as 5–7 pairs or vacancies for instance), the L_a, as seen by XRD, will not be affected. For the Raman, the "L_a" designation, which is initially a crystallographic term, is not appropriate. This is because the intensity of the D band owes to the amount of both the defects peripheral to the graphenes and the local in-plane defects in the graphene lattice, whereas the G band can manifest even in the absence of graphene because sp^2 carbon chains can also give rise to it. This implies that L_a from Raman not only depends on the in-plane graphene dimension (which is inversely proportional to the number of peripheral defects), but also on the occurrence of local defects dispersed throughout the graphene lattice. The more defective the cellulose-derived carbons are, the more broadened the D band becomes. However, this is not always true in terms of intensity due to the resonance effect and change in behavior around L_a = 2 nm. When the domains are large, the linewidth of the D band is a partial indication of the size of the graphenes involved in the domains, but it can be used more precisely within a sample set to monitor the removal of both the in-plane and peripheral defects with the increasing carbonization temperature. The narrower the D band, the larger the in-plane graphene dimension even with distortion. The higher L_a values obtained from Raman (through the confinement model) with respect to XRD are a result of the overall presence of distortions in the graphenes at 1000 °C, followed by the structural improvement in the graphene sheets due to the healing of local (in-lattice) defects, to which the Raman signal is sensitive whilst the XRD signal is not (or less). Table 6 shows that both the models of Tuinstra and Koenig [19] and Ferrari and Robertson [22] give values closer to that from XRD than the confinement model. This can be understood as follows: their correlations between L_a and I_D/I_G are based on several kinds of samples from various origins and treatments, with L_a determined by X-ray diffraction, leading to an average behavior, eventually not suitable for a specific series of materials. On the contrary, the confinement model [36] does not use XRD data, and was based on studying a homogenous sample series. However, even if other models [25] now also propose to correlate I_D/I_G, and the *FWHMs* of D and G as the confinement model did, no model can account for all the Raman spectra available in the literature so far.

5. Conclusions

Based on HRTEM, cellulose-derived carbons were found to be mainly composed of graphenic crystallites which distribute over a size range, according to XRD and Raman spectroscopy, possibly including ultra-small domains (L_a < 1 nm, contributing to the background of the XRD patterns), but with a prevalence of small domains (revealed by the *10* and 13° peaks of X-ray diffractograms), and some larger distorted domains for the sample annealed at 1800 °C (revealed by Raman spectroscopy and HRTEM). After an annealing at 1000 °C, the crystallites are distributed according to an isotropic texture, even at a short distance, which means that the areas of similar orientation of the graphenes inherited from the local molecular orientation event which took place during the primary carbonization step are very limited in size. Upon further carbonization, the textural isotropy remained because initial misorientations were too large to be healable. At the employed carbonization conditions (a temperature increase from 1000 to 1800 °C), the resulting materials comprise a minor proportion of carbon atoms not involved in the graphenic crystallites, and hence, are probably involved at the grain boundaries between them and assimilated to an amorphous carbon phase, the amount of which decreases with temperature (based on the evolution of the D_3 band from Raman spectroscopy). Meanwhile, the in-plane extension of graphene lattices, yet curved and distorted, increased from ~2.8 to ~8.7 nm with the increasing temperature (based on the pseudo-L_a calculated from the Raman based on the confinement model), mostly because of the healing of some of the defects, both local (in-lattice) and peripheral. On the nanotextural level, the pronounced (short-radius) curvatures impeded the development of large graphene crystallites. Correspondingly, the genuine L_a (according to its crystallographic definition, i.e., the dimension of the average crystallite across the basal plane) only increased from ~2 to 3 nm with the increasing temperature (based on XRD), mostly because of the healing of small-angle distortions. The stacking configuration within the graphene crystallites is random even at 1800 °C, meaning that the crystallographic structure is turbostratic, and from the textural characteristics (unhealable overall isotropy), it can be predicted that it will remain so for any further temperature increase, up to 3000 °C [8,54].

An important conclusion of the work is that the XRD peak corresponding to a scattering angle of ~13° (Cu X-ray source), the origin of which was not understood in the literature so far, is now fully explained. It is specifically due to crystallites involving an odd but limited number (typically three) of carbon layers with L_a in the range 1–2 nm. It does not mean that this presence is exclusive, and crystallites with an even number of layers (two and four) may be present as well, although destructive interference effects do not allow them contributing to the 13° peak. Therefore, the 13° peak, designated in this study as a small graphene–crystallite (SGC) peak, is not a mark for amorphous carbon, but a mark of poorly organized graphenic carbon (i.e., with low nanotexture).

Overall, the study illustrates that technique, such as XRD, Raman, and HRTEM do not quantify the same material characteristics, although, because of the poor suitability of the vocabulary commonly used in the literature, it is often believed so. This work has addressed the reasons for such discrepancies, and figuring them out has allowed seeing these techniques as being highly complementary.

Author Contributions: Writing: P.K.M., P.P. and M.M.; Experiments: P.K.M.; Data treatment: P.P. and P.K.M.; Samples preparation: T.B.; Project management: E.W.-H. and A.N. All authors have read and agreed to the published version of the manuscript.

Funding: P.K.M. is supported by an Université Fédérale de Midi-Pyrénées scholarship. The French "Programme Investissements d'Avenir" (ANR-10-LABEX-22-01 SOLSTICE grant) is acknowledged for its support in the work on cellulose carbonization.

Institutional Review Board Statement: Not applicable.

Data Availability Statement: The data presented in this study are available on request from the corresponding authors.

Acknowledgments: We thank N. Ratel-Ramond for his help in X-ray diffraction experiments.

Conflicts of Interest: The authors declare no conflict of interest.

References

1. Chowdhury, Z.; Abd Hamid, S.B.; Das, R.; Hasan, M.R.; Zain, S.; Khalisanni, K.; Uddin, M. Preparation of Carbonaceous Adsorbents from Lignocellulosic Biomass and Their Use in Removal of Contaminants from Aqueous Solution. *BioResources* **2013**, *8*, 6523–6555. [CrossRef]
2. Mohamad Nor, N.; Lau, L.C.; Lee, K.T.; Mohamed, A.R. Synthesis of activated carbon from lignocellulosic biomass and its applications in air pollution control—A review. *J. Environ. Chem. Eng.* **2013**, *1*, 658–666. [CrossRef]
3. Gan, L.; Zhu, J.; Lv, L. Cellulose hydrolysis catalyzed by highly acidic lignin-derived carbonaceous catalyst synthesized via hydrothermal carbonization. *Cellulose* **2017**, *24*, 5327–5339. [CrossRef]
4. Zhang, P.; Gong, Y.; Li, H.; Chen, Z.; Wang, Y. Solvent-free aerobic oxidation of hydrocarbons and alcohols with Pd@N-doped carbon from glucose. *Nat. Commun.* **2013**, *4*, 1593. [CrossRef]
5. Zhang, Z.; Yang, S.; Li, H.; Zan, Y.; Li, X.; Zhu, Y.; Dou, M.; Wang, F. Sustainable Carbonaceous Materials Derived from Biomass as Metal-Free Electrocatalysts. *Adv. Mater.* **2019**, *31*, e1805718. [CrossRef]
6. Herou, S.; Schlee, P.; Jorge, A.B.; Titirici, M. Biomass-derived electrodes for flexible supercapacitors. *Curr. Opin. Green Sustain. Chem.* **2018**, *9*, 18–24. [CrossRef]
7. Li, R.; Zhou, Y.; Li, W.; Zhu, J.; Huang, W. Structure Engineering in Biomass-Derived Carbon Materials for Electrochemical Energy Storage. *Research* **2020**, *2020*, 8685436. [CrossRef]
8. Oberlin, A. Carbonization and graphitization. *Carbon* **1984**, *22*, 521–541. [CrossRef]
9. Monthioux, M. Comments on: "Structure evolution mechanism of highly ordered graphite during carbonization of cellulose nanocrystals" by Eom et al. [Carbon 150 (2019) 142–152]. *Carbon* **2020**, *160*, 405–406. [CrossRef]
10. Fujimoto, H. Theoretical X-ray scattering intensity of carbons with turbostratic stacking and AB stacking structures. *Carbon* **2003**, *41*, 1585–1592. [CrossRef]
11. Puech, P.; Dabrowska, A.; Ratel-Ramond, N.; Vignoles, G.L.; Monthioux, M. New insight on carbonisation and graphitisation mechanisms as obtained from a bottom-up analytical approach of X-ray diffraction patterns. *Carbon* **2019**, *147*, 602–611. [CrossRef]
12. Franklin, R.E.; Randall, J.T. Crystallite growth in graphitizing and non-graphitizing carbons. *Proc. R. Soc. Lond. Ser. A Math. Phys. Sci.* **1951**, *209*, 196–218.
13. Gutierrez-Pardo, A.; Ramirez-Rico, J.; Cabezas-Rodríguez, R.; Fernández, J. Effect of catalytic graphitization on the electrochemical behavior of wood derived carbons for use in supercapacitors. *J. Power Sources* **2015**, *278*, 18–26. [CrossRef]
14. Maire, J.; Mering, J. Graphitization of soft carbons. *Chem. Phys. Carbon* **1970**, *6*, 125–190.
15. Sembiring, S.; Situmeang, R.; Sembiring, Z. Synthesis and characterization of asphalt composite precursors using amorphous rice husk silica. *Cerâmica* **2019**, *65*, 194–199. [CrossRef]
16. Cançado, L.G.; Takai, K.; Enoki, T.; Endo, M.; Kim, Y.A.; Mizusaki, H.; Jorio, A.; Coelho, L.N.; Magalhães-Paniago, R.; Pimenta, M.A. General equation for the determination of the crystallite size La of nanographite by Raman spectroscopy. *Appl. Phys. Lett.* **2006**, *88*, 163106. [CrossRef]
17. Pimenta, M.A.; Dresselhaus, G.; Dresselhaus, M.S.; Cançado, L.G.; Jorio, A.; Saito, R. Studying disorder in graphite-based systems by Raman spectroscopy. *Phys. Chem. Chem. Phys.* **2007**, *9*, 1276–1290. [CrossRef]
18. Puech, P.; Kandara, M.; Paredes, G.; Moulin, L.; Weiss-Hortala, E.; Kundu, A.; Ratel-Ramond, N.; Plewa, J.-M.; Pellenq, R.; Monthioux, M. Analyzing the Raman Spectra of Graphenic Carbon Materials from Kerogens to Nanotubes: What Type of Information Can Be Extracted from Defect Bands? *C* **2019**, *5*, 69. [CrossRef]

19. Tuinstra, F.; Koenig, J. Raman Spectrum of Graphite. *J. Chem. Phys.* **1970**, *53*, 1126–1130. [CrossRef]
20. Sadezky, A.; Muckenhuber, H.; Grothe, H.; Niessner, R.; Pöschl, U. Raman microspectroscopy of soot and related carbonaceous materials: Spectral analysis and structural information. *Carbon* **2005**, *8*, 1731–1742. [CrossRef]
21. Zickler, G.; Smarsly, B.; Gierlinger, N.; Peterlik, H.; Paris, O. A Reconsideration of the Relationship Between the Crystallite Size La of Carbons Determined by X-ray Diffraction and Raman Spectroscopy. *Carbon* **2006**, *44*, 3239–3246. [CrossRef]
22. Ferrari, A.C.; Robertson, J. Interpretation of Raman spectra of disordered and amorphous carbon. *Phys. Rev. B* **2000**, *61*, 14095–14107. [CrossRef]
23. Ager, J.W.; Veirs, D.K.; Shamir, J.; Rosenblatt, G.M. Laser heating effects in the characterization of carbon fibers by Raman spectroscopy. *J. Appl. Phys.* **1990**, *68*, 3598–3608. [CrossRef]
24. Matthews, M.J.; Pimenta, M.A.; Dresselhaus, G.; Dresselhaus, M.S.; Endo, M. Origin of dispersive effects of the Raman D band in carbon materials. *Phys. Rev. B* **1999**, *59*, R6585–R6588. [CrossRef]
25. Cançado, L.G.; Da Silva, M.G.; Ferreira, E.H.M.; Hof, F.; Kampioti, K.; Huang, K.; Jorio, A. Disentangling contributions of point and line defects in the Raman spectra of graphene-related materials. *2D Mater.* **2017**, *4*, 025039. [CrossRef]
26. Cuesta, A.; Dhamelincourt, P.; Laureyns, J.; Martínez-Alonso, A.; Tascón, J.M.D. Raman microprobe studies on carbon materials. *Carbon* **1994**, *32*, 1523–1532. [CrossRef]
27. Beyssac, O.; Goffé, B.; Petitet, J.-P.; Froigneux, E.; Moreau, M.; Rouzaud, J.-N. On the characterization of disordered and heterogeneous carbonaceous materials by Raman spectroscopy. Georaman 2002, Fifth International Conference on Raman Spectroscopy. *Appl. Earth Sci.* **2003**, *59*, 2267–2276.
28. Couzi, M.; Bruneel, J.-L.; Talaga, D.; Bokobza, L. A multi wavelength Raman scattering study of defective graphitic carbon materials: The first order Raman spectra revisited. *Carbon* **2016**, *107*, 388–394. [CrossRef]
29. Berhanu, S.; Hervy, M.; Weiss-Hortala, E.; Proudhon, H.; Berger, M.-H.; Chesnaud, A.; Faessel, M.; King, A.; Pham Minh, D.; Villot, A.; et al. Advanced characterization unravels the structure and reactivity of wood-based chars. *J. Anal. Appl. Pyrolysis* **2018**, *130*, 79–89. [CrossRef]
30. Harris, P.J.F.; Tsang, S.C. High-resolution electron microscopy studies of non-graphitizing carbons. *Philos. Mag. A* **1997**, *76*, 667–677. [CrossRef]
31. Singh, M.; Vander Wal, R.L. Nanostructure Quantification of Carbon Blacks. *C* **2019**, *5*, 2. [CrossRef]
32. Monthioux, M.; Noé, L.; Kobylko, M.; Wang, Y.; Cazares-Huerta, T.C.; Pénicaud, A. Determining the structure of graphene-based flakes from their morphotype. *Carbon* **2017**, *115*, 128–133. [CrossRef]
33. Lutterotti, L.; Scardi, P. Simultaneous structure and size–strain refinement by the Rietveld method. *J. Appl. Crystallogr.* **1990**, *23*, 246–252. [CrossRef]
34. Neverov, V.S. XaNSoNS: GPU-accelerated simulator of diffraction patterns of nanoparticles. *SoftwareX* **2017**, *6*, 63–68. [CrossRef]
35. Mallet-Ladeira, P.; Puech, P.; Weisbecker, P.; Vignoles, G.L.; Monthioux, M. Behavior of Raman D band for pyrocarbons with crystallite size in the 2–5 nm range. *Appl. Phys. A* **2014**, *114*, 759–763. [CrossRef]
36. Puech, P.; Plewa, J.-M.; Mallet-Ladeira, P.; Monthioux, M. Spatial confinement model applied to phonons in disordered graphene-based carbons. *Carbon* **2016**, *105*, 275–281. [CrossRef]
37. Cançado, L.G.; Jorio, A.; Pimenta, M.A. Measuring the absolute Raman cross section of nanographites as a function of laser energy and crystallite size. *Phys. Rev. B* **2007**, *76*, 064304. [CrossRef]
38. Asadullah, M.; Zhang, S.; Min, Z.; Yimsiri, P.; Li, C.-Z. Effects of biomass char structure on its gasification reactivity. *Bioresour. Technol.* **2010**, *101*, 7935–7943. [CrossRef] [PubMed]
39. Li, X.; Hayashi, J.I.; Li, C.Z. FT-Raman spectroscopic study of the evolution of char structure during the pyrolysis of a Victorian brown coal. *Fuel* **2006**, *85*, 1700–1707. [CrossRef]
40. Kang, D.S.; Lee, S.M.; Lee, S.H.; Roh, J.S. X-ray diffraction analysis of the crystallinity of phenolic resin-derived carbon as a function of the heating rate during the carbonization process. *Carbon Lett.* **2018**, *27*, 108–111.
41. Kercher, A.; Nagle, D. Microstructural Evolution during Charcoal Carbonization by X-ray Diffraction Analysis. *Carbon* **2003**, *41*, 15–27. [CrossRef]
42. Keiluweit, M.; Nico, P.S.; Johnson, M.G.; Kleber, M. Dynamic Molecular Structure of Plant Biomass-Derived Black Carbon (Biochar). *Environ. Sci. Technol.* **2010**, *44*, 1247–1253. [CrossRef] [PubMed]
43. Dahbi, M.; Kiso, M.; Kubota, K.; Horiba, T.; Chafik, T.; Hida, K.; Matsuyama, T.; Komaba, S. Synthesis of hard carbon from argan shells for Na-ion batteries. *J. Mater. Chem. A* **2017**, *5*, 9917–9928. [CrossRef]
44. Campos-Delgado, J.; Kim, Y.A.; Hayashi, T.; Morelos-Gómez, A.; Hofmann, M.; Muramatsu, H.; Endo, M.; Terrones, H.; Shull, R.D.; Dresselhaus, M.S.; et al. Thermal stability studies of CVD-grown graphene nanoribbons: Defect annealing and loop formation. *Chem. Phys. Lett.* **2009**, *469*, 177–182. [CrossRef]
45. Garvey, C.J.; Parker, I.H.; Simon, G.P. On the Interpretation of X-ray Diffraction Powder Patterns in Terms of the Nanostructure of Cellulose I Fibres. *Macromol. Chem. Phys.* **2005**, *206*, 1568–1575. [CrossRef]
46. Qiao, Q.; Li, X.; Huang, L. Crystalline Cellulose under Pyrolysis Conditions: The Structure–Property Evolution via Reactive Molecular Dynamics Simulations. *J. Chem. Eng. Data* **2020**, *65*, 360–372. [CrossRef]
47. Kwon, S.-M.; Kim, N.; Cha, D.-S. An investigation on the transition characteristics of the wood cell walls during carbonization. *Wood Sci. Technol.* **2009**, *43*, 487–498. [CrossRef]

48. Meng, Y.; Contescu, C.; Liu, P.; Wang, S.; Lee, S.H.; Guo, J.; Young, T. Understanding the local structure of disordered carbons from cellulose and lignin. *Wood Sci. Technol.* **2021**, *55*, 587–606. [CrossRef]
49. Noda, T.; Inagaki, M. The Structure of Glassy Carbon. *Bull. Chem. Soc. Jpn.* **1964**, *37*, 1534–1538. [CrossRef]
50. Paris, O.; Zollfrank, C.; Zickler, G.A. Decomposition and carbonisation of wood biopolymers—A microstructural study of softwood pyrolysis. *Carbon* **2005**, *43*, 53–66. [CrossRef]
51. Wu, H.; Gakhar, R.; Chen, A.; Lam, S.; Marshall, C.P.; Scarlat, R. Comparative analysis of microstructure and reactive sites for nuclear graphite IG-110 and graphite matrix A3. *J. Nucl. Mater.* **2019**, *528*, 151802. [CrossRef]
52. Monthioux, M.; Charlier, J.-C. Giving credit where credit is due: The Stone–(Thrower)–Wales designation revisited. *Carbon* **2014**, *75*, 1–4. [CrossRef]
53. McDonald-Wharry, J.; Manley-Harris, M.; Pickering, K.L. Carbonisation of biomass-derived chars and the thermal reduction of a graphene oxide sample studied using Raman spectroscopy. *Carbon* **2013**, *59*, 383–405. [CrossRef]
54. Ouzilleau, P.; Gheribi, A.E.; Chartrand, P.; Soucy, G.; Monthioux, M. Why some carbons may or may not graphitize? The point of view of thermodynamics. *Carbon* **2019**, *149*, 419–435. [CrossRef]

C Journal of Carbon Research

MDPI

Article

Increased Electrical Conductivity of Carbon Nanotube Fibers by Thermal and Voltage Annealing

Varun Shenoy Gangoli [1,2,*], Chris J. Barnett [3,*], James D. McGettrick [4], Alvin Orbaek White [1] and Andrew R. Barron [1,2,5,6,7,*]

1 Energy Safety Research Institute, Bay Campus, Swansea University, Swansea SA1 8EN, UK; alvin.orbaekwhite@swansea.ac.uk
2 Department of Chemistry, Rice University, Houston, TX 77005, USA
3 Department of Physics, Singleton Campus, Swansea University, Swansea SA2 8PP, UK
4 SPECIFIC, Bay Campus, Swansea University, Swansea SA1 8EN, UK; j.d.mcgettrick@swansea.ac.uk
5 Arizona Institutes for Resilience (AIR), University of Arizona, Tucson, AZ 85721, USA
6 Department of Materials Science and Nanoengineering, Rice University, Houston, TX 77005, USA
7 Faculty of Engineering, Universiti Teknologi Brunei, Jalan Tungku Link, Gadong BE1410, Brunei
* Correspondence: v.s.gangoli@swansea.ac.uk (V.S.G.); c.j.barnett@swansea.ac.uk (C.J.B.); a.r.barron@swansea.ac.uk (A.R.B.); Tel.: +44-01792-606930 (A.R.B.)

Abstract: We report the effect of annealing, both electrical and by applied voltage, on the electrical conductivity of fibers spun from carbon nanotubes (CNTs). Commercial CNT fibers were used as part of a larger goal to better understand the factors that go into making a better electrical conductor from CNT fibers. A study of thermal annealing in a vacuum up to 800 °C was performed on smaller fiber sections along with a separate analysis of voltage annealing up to 7 VDC; both exhibited a sweet spot in the process as determined by a combination of a two-point probe measurement with a nanoprobe, resonant Raman spectroscopy, and X-ray photoelectron spectroscopy (XPS). Scaled-up tests were then performed in order to translate these results into bulk samples inside a tube furnace, with similar results that indicate the potential for an optimized method of achieving a better conductor sample made from CNT fibers. The results also help to determine the surface effects that need to be overcome in order to achieve this.

Keywords: carbon; carbon nanotube; electrical conductor; annealing; energy

Citation: Gangoli, V.S.; Barnett, C.J.; McGettrick, J.D.; Orbaek White, A.; Barron, A.R. Increased Electrical Conductivity of Carbon Nanotube Fibers by Thermal and Voltage Annealing. C **2022**, *8*, 1. https://doi.org/10.3390/c8010001

Academic Editor: Gil Goncalves

Received: 6 November 2021
Accepted: 20 December 2021
Published: 23 December 2021

Publisher's Note: MDPI stays neutral with regard to jurisdictional claims in published maps and institutional affiliations.

1. Introduction

Carbon nanotubes (CNTs) have been the subject of many research groups worldwide for nearly three decades now, and their untapped potential is still the subject of newly published articles daily. This is not to say that CNTs have never made it out of the lab, and in fact there are several different applications that are in use already in industries ranging from energy, manufacturing, construction, sports, transportation, and even as art [1–4]. One challenge left unconquered is the replacement of the use of metal, especially aluminum and copper, for electricity transmission. The electric grid varies in material composition from region to region; however, in order to achieve a compromise of weight and electrical conductivity, it tends to consist of an alloy of the two metals more often than not [5]. A compromise it remains, too, given that the weight of copper offsets the practical benefits of its use in long transmission lines, and thermal resistance losses add up in underground cable installations.

CNT-based cables could be a viable alternative, since their electrical conductivity exceeds that of copper while weighing very little, owing to the hollow nature of individual nanotubes [5]. There have been many attempts at this, including spinning out long individual CNT fibers, drawing and extruding acid-doped CNT fibers [6], and even composite fibers of CNTs and the aforementioned copper [7–10]. The latter approach, when not completely made in-house, involves the use of commercial CNT fibers that are often inconsistent when it comes to the uniformity of their physical and chemical properties, which in turn makes them harder to use as a control starting material that is then further optimized for increased electrical conductivity.

One of the approaches used to circumvent this issue with commercially available CNT fibers, and indeed most CNT sources as a whole, is to purify them and remove the impurities, including any residual catalyst and amorphous carbon [11]. Depending on the manufacturing process, this can be a bulk phenomenon or one with a substantial surface-only effect. CNT fibers, when used for electrical transmission, can have different conduction pathways depending on the paths of least available resistance, and the point of contact at the surface of the fibers is a critical step to be accounted for [12–14].

There is a strong need to better understand the contact resistances, and to take steps to address this by purification at the local surface sites. The work described herein was aimed at addressing both points and describes two routes of achieving the same end goal of increased electrical conductivity as a result of lowered contact resistance between the electric probes and the CNT fibers. This study made use of the of thermal and voltage annealing processes with the two-point probe measurements that were also supported by the characterization of spot points, in order to reveal the change in the chemical composition following the annealing. The small-scale experiments were then scaled up on a macro-scale to demonstrate a proof-of-concept for quickly achieving a uniform, more optimized CNT fiber for electrical conductivity applications. Unlike the purification processes that are well known [11], our approach relies on internal electrical heating and not a mixture of chemical treatment and heat. We believe this makes the results described herein a unique methodology.

2. Experimental Section

2.1. Materials

A 100 m yarn of YE-A10 carbon nanotube (CNT) fiber was purchased from Nanocomp Technologies, Inc. (New Hampshire, NH, USA) and handled as previously described [15]. This was a single-ply fiber of single- and few-walled CNTs with a rated average diameter of 130 μm. Several 1 cm-long and 30 cm-long pieces were cut for the experiments as described below.

2.2. Nanoprobe Measurements

A 1 cm-long piece of the CNT fiber was loaded into an Omicron LT nanoprobe (now part of Scienta Omicron, Uppsala, Sweden) equipped with an SEM column (base pressure 1×10^{-10} mbar). Tungsten tips were etched in a 2 M KOH solution and, to ensure that the native probe oxide did not affect the measurements, the tungsten probes were direct-current annealed in the nanoprobe using the method described in [16,17]. The probes were then lowered onto the fiber using the method described in [12,13,18], to ensure the probe pressure did not induce strain in the sample or affect the measured I-V, as well as to ensure consistent measurements, and positioned 20 μm apart using the SEM images, an example of which is shown in Figure 1. I-V measurements between −1 V and +1 V were taken at eight locations on the fiber, and five repeat measurements were taken at each location for statistical accuracy. During the voltage sweeps, the SEM beam was switched off, as it is known that the beam can affect I-V measurements. The I-V measurements here were taken at 25 °C, and this served as a reference sample to use in subsequent experiments.

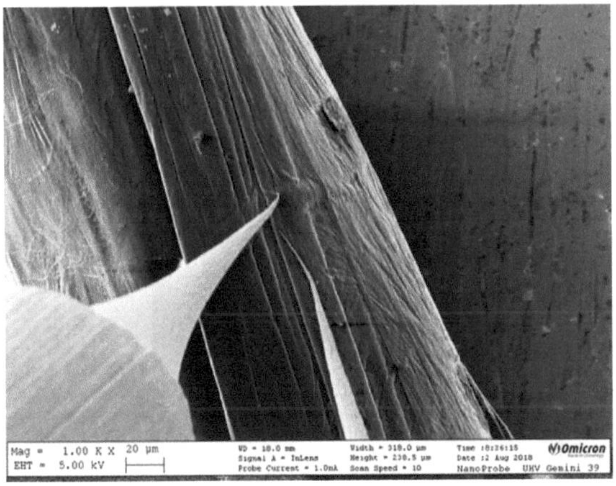

Figure 1. An example SEM image of two tungsten probes 20 μm apart and in contact with the CNT-fiber surface. The probes were used for a two-point-probe I-V measurement in an Omicron nanoprobe between the chosen voltage range as specified in the experimental conditions.

2.3. *Thermal Annealing in Nanoprobe*

A 1 cm-long piece of the Nanocomp YE-A10 CNT fiber was loaded into the nanoprobe's annealing stage and thermally annealed for an hour at 100 °C. The temperature was calculated using the PN heater calibration curve provided by Omicron [12,13]. The fiber was then allowed to cool to 25 °C and then I-V measurements were taken using the same procedure as described above. This piece was then removed from the nanoprobe and marked for subsequent characterization. A separate 1 cm piece of the fiber was treated in the same way, except it was heated to 200 °C. Six more such pieces were used to complete a thermal-analysis study from 100–800 °C in increments of 100 °C, in addition to the reference sample from earlier that was probed at 25 °C without any annealing.

2.4. *Voltage Annealing in Nanoprobe*

A 1 cm-long piece of the CNT fiber was loaded into the nanoprobe's stage and I-V measurements between −2 V and +2 V were taken at eight locations on the fiber, and then five repeat measurements were taken at each location for statistical accuracy. The sample was then removed from the nanoprobe and marked for subsequent characterization. Five more such samples were used, so there was a total of seven samples that were probed from ±1 V to ±7 V, including the reference sample from earlier. The tungsten tips were found to be welded onto the fiber surface at ±8 V, indicating the point of no return under these particular testing conditions.

2.5. *Thermal Annealing in Tube Furnace*

A 30 cm-long piece of the CNT fiber was placed in a quartz tube with a 1″ outer diameter and enclosed inside a vacuum-furnace tube system that uses a Thermo Scientific 54500 (Massachusetts, MA, USA) high-temperature tube furnace. The quartz tube was then pumped down to a vacuum of 9×10^{-3} mbar, followed by thermal annealing at temperatures ranging from 100 °C to 800 °C in 100 °C increments for 1 h. After annealing, the fibers were allowed to cool to room temperature in the vacuum before being removed, and then I-V measurements were taken for a 1 cm section of each fiber between −1 V and +1 V to allow for a direct comparison with the set of samples that were thermally annealed in the nanoprobe itself.

2.6. Characterization

In addition to the I-V measurements in the nanoprobe that helped to generate an average electrical resistance for each sample, resonant Raman spectra were also collected for each sample. These were taken using a Renishaw inVia™ Raman microscope (Wotton-under-Edge, UK) equipped with a 785 nm laser and a Leica PL Fluortar L50x/0.55 long-working-distance objective lens. The Raman laser-excitation powers were determined by increasing the laser irradiance until a spectral transition from baseline noise to a weak signal-to-noise ratio of ~3 at a Raman shift of 1600 cm^{-1} was observed. The actual irradiances are unknown, and not relevant to this discussion. This resulted in a relatively flat background for the 785 nm spectra, with 785 nm chosen to resonate with the majority of the fiber samples as opposed to the other available lasers (633 and 514 nm). The collection times were chosen in order to acquire >25,000 cts at 1600 cm^{-1}, and a minimum of five separate Raman spectrum maps were collected per sample for statistical accuracy. A thermogravimetric analysis was performed on the sample that was thermally annealed in the tube furnace using a TA Instrument SDT Q600 (Delaware, DE, USA), and the samples were run in an open alumina crucible under a continuous air flow of 100 mL/min with a heating ramp rate of 10 °C/min. The samples that were thermally annealed in the tube furnace also had enough material remaining for characterization in a Kratos Axis Supra XPS (Manchester, UK) at a base pressure of ~1×10^{-9} bar. The fibers were mounted free-standing over the relatively large (cm scale) holes on the sample bar to remove any risk of substrate influence on the final data, with each point at least 2 mm apart. The conductive fibers were mounted in an electrical connection with the ground. As such, the charge neutralizer was not used, and no charge correction was applied to the data. A minimum of three X-ray photoelectron spectroscopy (XPS) spectra were collected per fiber sample and analyzed for statistical accuracy using CasaXPS (Version 2.3.23rev1.1K).

3. Results and Discussion

3.1. Thermal Annealing in Nanoprobe

Figure 1 illustrates the reference experiment of an I-V measurement of Nanocomp's YE-A10 CNT fiber (as-is) in the nanoprobe. This involved the use of two tungsten probes that were separated by 20 μm and in contact with the fiber surface, with a voltage sweep from −1 V to +1 V. The current, in amps, was measured at each collected datapoint and plotted as a function of the voltage. One such measurement curve is seen in Figure 2, and each set had 100 datapoints. The electrical resistance was calculated at each point and the same process was repeated for all other such acquisitions at different locations along the CNT fiber surface. This resulted in a lot of data that were used to calculate the reported average electrical resistance and the associated standard deviation for statistical accuracy. These are plotted in Figure 3 for the reference sample measured at 25 °C, along with the subsequent samples generated after thermal annealing under vacuum in the nanoprobe, with annealing temperatures ranging from 100 °C to 800 °C in steps of 100 °C.

Figure 3 is a semi-log plot owing to the multiple orders of magnitude of the change in electrical resistance, with a starting value of 1.28×10^6 Ω, which then remains nearly the same until an annealing temperature of 300 °C, following which it continually decreases with annealing temperature down to a minimum average of 2.98×10^3 Ω at 500 °C, resulting in a three-orders-of-magnitude reduction in electrical resistance under these testing conditions. The electrical resistance then increases in a manner that reverses the trend, going higher than the starting value and reaching 3.17×10^6 Ω at 800 °C. The trend here was strong enough not to merit further study of the annealing conditions past 800 °C. The data, while confirming a sweet spot for minimum electrical resistance for these CNT fibers at an annealing temperature of 500 °C, also show an irreversible performance drop in electrical conductivity with increased annealing temperature. Both of these observations necessitated further investigations.

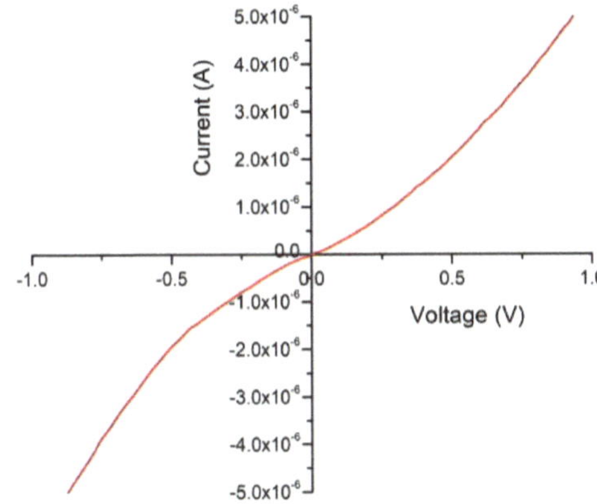

Figure 2. Example of an I-V measurement curve for Nanocomp YE-A10 CNT fiber as measured with two tungsten probes 20 µm apart in an Omicron nanoprobe using a voltage sweep from −1 to +1 V.

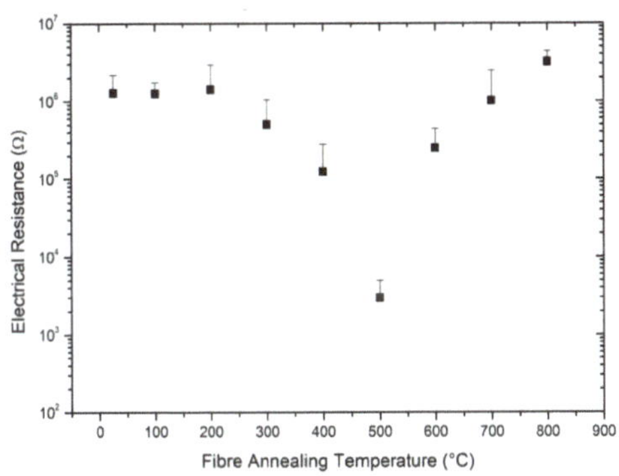

Figure 3. A semi-log plot of CNT fiber average electrical resistance as a function of annealing temperature in the nanoprobe. Electrical resistance drops to a minimum point followed by an irreversible increase.

Given the limited size of the samples (1 cm in length), there was not sufficient material to perform macroscopic analysis, which prompted the scale-up study discussed below; however, resonant Raman microscopy can help to determine any relative change in the quality of the carbon [19–21] in the CNT fibers as a function of the thermal-annealing temperatures in the nanoprobe. Figure 4a shows the average ratio of the intensity of the G- and D-peaks in the spectra collected for each sample, with I_G helping to quantify the amount of graphitic carbon in the sample, which in this case is sp^2-hybridized carbon in the form of graphene and the CNTs themselves. I_D, in contrast, is predominantly a measure of sp^3-hybridized carbon; it is attributed to the content of amorphous carbon that is present in the CNT fibers and has a higher electrical resistance relative to CNTs [22]. Indeed, TEM imaging previously performed on analogous Nanocomp CNT fibers revealed interspersed amorphous carbon [23], although the conclusions drawn from this work are hard to justify given the physical changes to the individual CNTs as part of the TEM sample preparation.

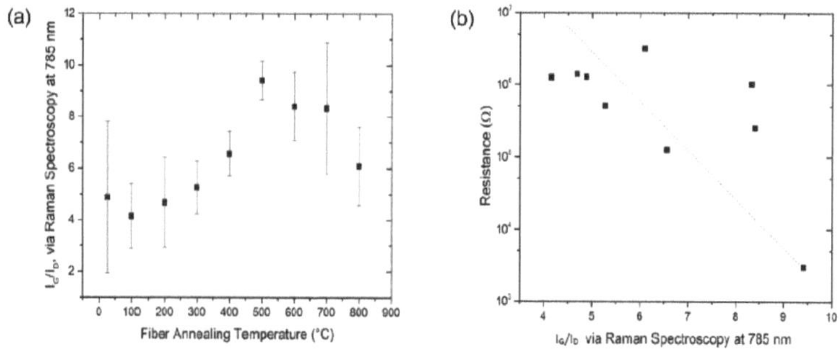

Figure 4. (**a**) Relative intensity of the G- and D-peaks in the resonant Raman spectra for CNT fibers at 785 nm as a function of annealing temperature in the nanoprobe and (**b**) electrical resistance of the CNT fibers as a function of the relative intensity of the G- and D-peaks. There is a general inverse relation of electrical resistance with I_G/I_D.

The relative content of CNTs to amorphous carbon impurities remained essentially constant with annealing temperature up to 300 °C, followed by a definite increase up to a maximum point with the sample annealed at 500 °C, and then it decreased again, although not down to the starting value. The plot in Figure 4a essentially mirrors that in Figure 3, suggesting that the electrical resistance is related to the average I_G/I_D ratio. As seen in Figure 4b, there is generally a downward trend as expected from the two separate trends with annealing temperature. To explain the observed changes with annealing temperature, we hypothesize that with increasing annealing temperature there is removal of amorphous carbon, along with the potential self-repair of graphene and CNTs. Above 600 °C there is both the possibility of the removal of CNTs and of physical damage and disruptions to the CNT conduction pathways themselves [24]. The increased standard deviation at higher temperatures, combined with a decrease in the Raman spectra signal-to-noise ratio, add credence to the loss of materials remaining to resonate at the energy supplied by the 785 nm excitation laser.

3.2. Voltage Annealing in Nanoprobe

Previously reported results have shown an analogous effect of applied voltage on CNTs and CNT fibers compared to thermal annealing, so much so that the former can itself be seen as voltage annealing [12]. This was the drive behind these experiments, wherein 1 cm-long CNT-fiber samples were probed in the nanoprobe as before, except without any thermal annealing and instead using voltage sweeps from ±1 V to ±8 V. This meant that the sample swept from a specific voltage, e.g., ±7 V, only underwent I-V measurements with applied voltage ranging from −7 V to +7 V, rather than seven separate scans from ±1 V, ±2 V, all the way to ±7 V. This decoupled voltage annealing helps to remove any layer-by-layer treatment effects that would otherwise have been possible with the stepwise approach but is more applicable to the investigation of the effect of annealing voltage over electrical resistance, especially when it comes to replicating these results with different samples and sample sizes alike.

Figure 5 shows the reference sample measured at 25 °C and with the voltage sweep from −1 V to +1 V. This remains within the error margins for the sample swept between −2 V and +2 V, but then there is a significant drop at ±3 V with an average electrical resistance of 2.63×10^3 Ω, which is another three-orders-of-magnitude reduction as was observed with thermal annealing. The decrease continues through ±5 V, albeit with a slower rate of decrease, and reaches a minimum value of 1.24×10^3 Ω. There is then a relatively massive jump in resistance immediately afterwards, resulting in an irreversible degradation of those specific samples in terms of their electrical conduction pathways. This consistently occurred during these voltage sweeps, and thus the findings are statistically accurate as well.

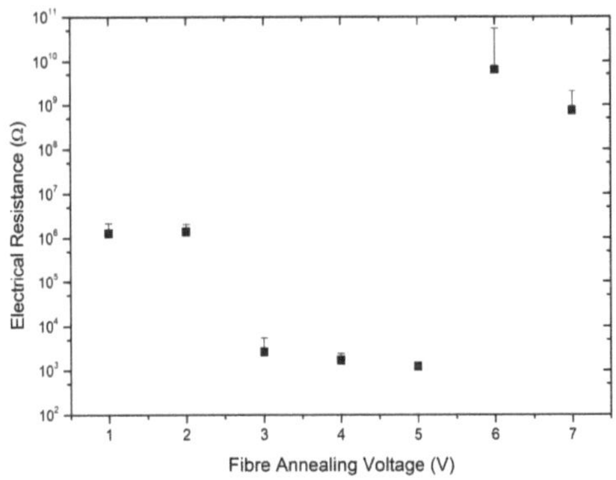

Figure 5. A semi-log plot of CNT fiber average electrical resistance as a function of annealing voltage in the nanoprobe. Electrical resistance drops to a minimum point followed by an irreversible increase.

A similar hypothesis is postulated here, wherein a voltage threshold is required to lower the contact resistance between the probes and the CNT fiber, with the impurities being removed first followed by damage to the conduction pathways as voltage increases.

3.3. Thermal Annealing in Tube Furnace

The experiments until this point involved short CNT-fiber sections, and further characterization would have benefited from scaling up the annealing experiments. Voltage annealing is less trivial to perform with larger sections of fibers, especially given the easier access to tube furnaces of various sizes. Multiple 30 cm-long sections were thus annealed in a tube furnace as described in the methodology section, with annealing temperatures set to match those of the inside of the nanoprobe itself. The operating pressure in the tube furnace was read as 9×10^{-3} mbar, although here the limitation was more the pressure gauge itself, and the actual pressure was likely lower. Since this was not as low a pressure as in the nanoprobe, it was accordingly expected that the annealing effect may be different at different temperatures.

Figure 6 shows the average electrical resistance of the annealed fibers, as measured by the nanoprobe, as a function of the thermal-annealing temperature in the tube furnace. As with the thermal annealing of the microscopic samples (Figure 3), the data for the macroscopic samples (Figure 6) show an initial decrease in resistance (from the reference sample with 1.28×10^6 Ω at 25 °C) with the increased annealing temperature down to 1.63×10^3 Ω at 400 °C. Above this temperature there is a slight increase to 3.63×10^3 Ω at 500 °C before the electrical resistance increases more dramatically at higher temperatures. It could be that the actual sweet spot in both thermal annealing experiments is somewhere between 400–500 °C, but the practical benefits of finer tuning are outweighed by the net improvements already achieved. It should be noted that the two heating systems (controllers and associated temperature measurement) may also cause a difference in the actual temperature versus the indicated temperatures that may be responsible for this offset.

As with the thermal-annealing experiments in the nanoprobe itself, the resonant Raman spectroscopy of the longer CNT-fiber samples that were annealed in a tube furnace shows in Figure 7a the relative content of CNTs to amorphous carbon impurities increasing with annealing temperature, reaching a maximum when the samples were annealed at 400 and 500 °C, and then decreasing again. Once again, this can be hypothesized as a process that begins with the removal of amorphous carbon and is then followed by the removal of CNTs as the annealing temperature continues to increase past 600 °C. The higher standard deviations here can be attributed to the tube furnace not having as uniform a heating field throughout the length of the fiber inside, and this could have also contributed to the slight discrepancy in the comparison of the annealing sweet spot in the nanoprobe.

Figure 7b again relates the change in electrical resistance of the fiber sections to the corresponding average I_G/I_D values, and there is once more a downward trend as expected from the two separate trends of annealing temperature.

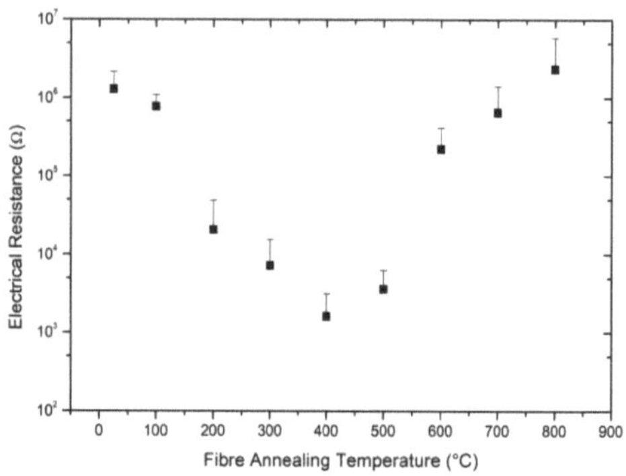

Figure 6. A semi-log plot of CNT fiber average electrical resistance as a function of annealing temperature in a tube furnace. Electrical resistance drops to a minimum point followed by an irreversible increase.

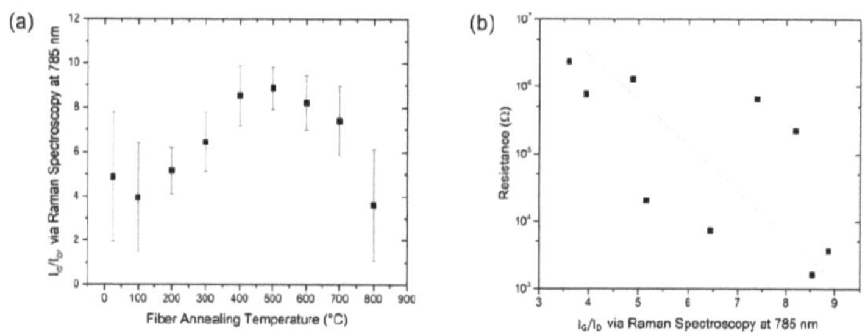

Figure 7. (**a**) Relative intensity of the G- and D-peaks in the resonant Raman spectra for CNT fibers at 785 nm as a function of annealing temperature in the nanoprobe and (**b**) electrical resistance of the CNT fibers as a function of the relative intensity of the G- and D-peaks. There is a general inverse relation of electrical resistance with I_G/I_D.

To better understand any chemical composition changes occurring here, representative sections of the longer tube-furnace-annealed fibers were characterized by XPS, with scans recorded for each annealing temperature. Three areas were scanned on each piece of fiber, with a general survey scan followed by detailed scans of the C 1s, O 1s, Fe 2p, Si 2p, and S 2p representations. The survey scans showed no observable deviation, with all of the chemical elements remaining throughout. Si 2p and S 2p were scanned as these chemicals are used in Nanocomp's manufacturing process of CNT fibers and can potentially leave a residue. Table 1 shows the average percentage composition of the fibers after the respective annealing stage. All the way up to 700 °C, the relative carbon content does not significantly alter from ~95%, and the peaks did not show any change in shape either, with a representative C 1s peak shown in Figure 8; however, annealing at 500 °C reduced the carbon content slightly to 93.8%. It is also noticeable that annealing at 800 °C reduced the carbon content to 85%, although once again the shape of the C 1s peak did not change.

Table 1. Average elemental composition of CNT fibers annealed in tube furnace as measured by XPS.

Annealing Temperature. (°C)	Carbon (%)	Iron (%)	Oxygen (%)	Sulphur (%)	Silicon (%)
100	95.98 ± 0.64	0.24 ± 0.01	3.38 ± 0.58	0.14 ± 0.01	0.22 ± 0.05
200	95.29 ± 0.45	0.26 ± 0.01	3.90 ± 0.37	0.19 ± 0.02	0.32 ± 0.05
300	96.60 ± 0.40	0.24 ± 0.01	2.73 ± 0.39	0.13 ± 0.06	0.26 ± 0.05
400	96.56 ± 0.29	0.33 ± 0.02	2.83 ± 0.28	0.04 ± 0.01	0.22 ± 0.01
500	93.78 ± 0.34	0.68 ± 0.07	4.89 ± 0.38	0.13 ± 0.01	0.50 ± 0.03
600	94.52 ± 0.42	0.24 ± 0.01	4.36 ± 0.32	0.15 ± 0.01	0.70 ± 0.11
700	95.25 ± 0.50	0.24 ± 0.02	3.71 ± 0.37	0.03 ± 0.01	0.74 ± 0.17
800	85.18 ± 2.56	0.42 ± 0.05	10.66 ± 1.47	0.00 ± 0.00	3.73 ± 1.05

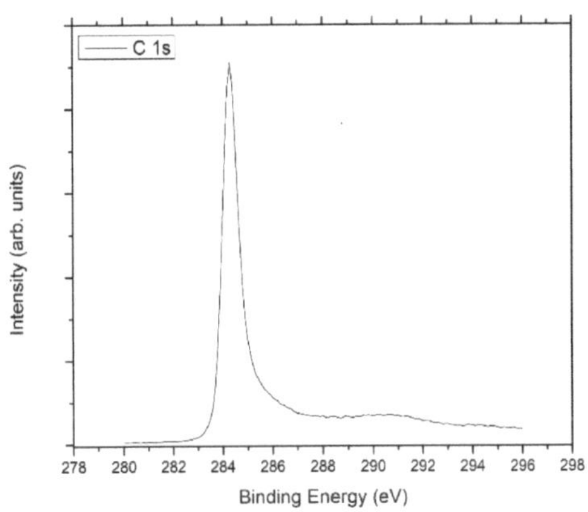

Figure 8. A representative C 1s XPS peak measured for fibers annealed in the tube furnace. The shape of the C 1s peak did not change with annealing temperature all the way to 800 °C.

The O 1s peak was fitted with three component peaks, with the two positioned at 532.1 eV and 533.5 eV associated with oxygen–carbon bonds and the third being a metal-oxide peak positioned at a lower binding energy of 530.3 eV as shown in Figure 9a. Annealing up to 400 °C caused a gradual increase in the metal-oxide component from 8% to 19% of the total oxygen content; however, the overall quantity of oxygen relative to carbon remained constant. Annealing at 500 °C caused a significant increase in the metal-oxide peak to 36% of the total oxygen as seen in Figure 9b, which was then reduced to ~15% after annealing at 600 °C, 700 °C, and 800°C. After annealing at 800 °C, the percentage of oxygen content also increased to 10%.

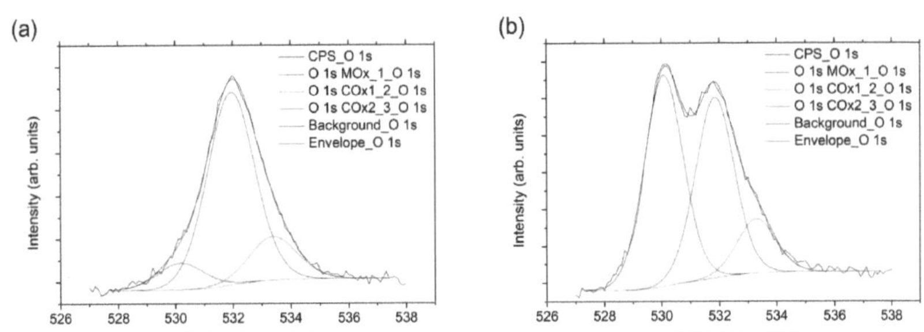

Figure 9. O 1s peak fits for the CNT fibers annealed at (**a**) 100 °C and (**b**) 500 °C showing the relative composition of oxygen content in the fibers varies, indicative of more metal oxide at the higher annealing temperature.

The Fe 2p spectra showed that the iron consisted of Fe (0) and Fe (II/III) in all of the samples, and the quantity measured was low throughout. The percentage Fe was ~0.24% for fibers annealed at up to 300 °C, and this value increased to 0.34% and 0.68% for the fibers annealed at 400 °C and 500 °C, respectively. These increases were seen to be a result of a relative increase in the Fe (II/III) components which, combined with the oxygen peaks from before, heavily suggest the formation of iron oxides near the CNT-fiber surface from the oxidation of residual catalyst. Measurements of the fibers annealed at 600 °C and 700 °C showed that the Fe percentage had dropped to 0.24% again, and this was caused by a reduction in the Fe (II/III) components. The sample annealed at 800 °C had an increased iron content of 0.42% which, combined with the increase in the content of other elements and the decrease of the carbon content here, points towards the loss of CNTs at higher temperatures.

It is likely that the changes are more gradual than these discrete measurements suggest, and a thermogravimetric analysis was performed in order to understand this further. Figure 10 shows an example TGA plot for a piece of Nanocomp's YE-A10 CNT fiber as is, with both weight loss and heat flow recorded as functions of temperature. Four exothermic peaks in heat flow (red) are marked, which are postulated to represent, in order, oxidation of amorphous carbon (1), oxidation of Fe to FeO in addition to the oxidation of any present single-walled CNTs (2), oxidation of multi-walled CNTs (3), and oxidation of FeO to Fe_2O_3/Fe_3O_4 (4) [21,25].

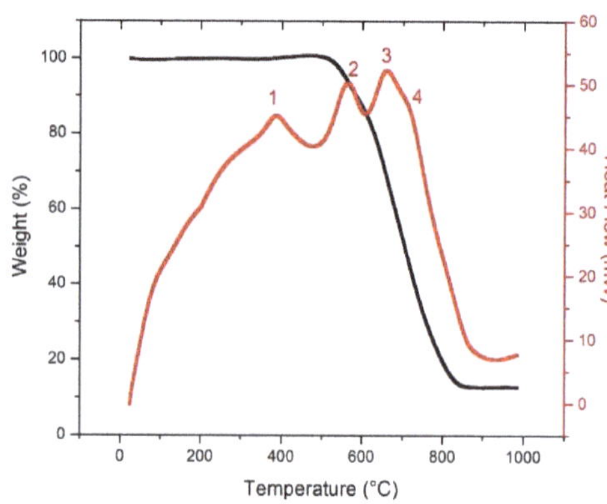

Figure 10. Thermogravimetry of Nanocomp YE-A10 CNT fiber with weight loss (%) and heat flow (mW) reported. There are four exotherms of interest here, which cater to compositional changes in the fiber and can explain the effect of fiber annealing better.

With this basis set, TGA was performed on the fibers annealed in the tube furnace, given that there was enough material remaining to do so. Figure 11a shows the heat-flow exotherm for the fiber annealed at 300 °C, with Figure 11b and 11c doing the same for the fibers annealed at 500 °C and 700 °C, respectively. Given the inherent inhomogeneity of the Nanocomp single-ply YE-A10 fiber, the relative intensities of the exotherms are the takeaway points here, and they show that the fibers annealed to increasing temperatures have already gone through the four stages in the series, including the removal of amorphous carbon, which then yields a fiber with a higher I_G/I_D and a lower electrical resistance, as well as the eventual oxidation of the conductive CNTs, thus leaving behind iron oxide, as seen by XPS, and resulting in a higher electrical resistance due to disruption of the conduction pathways. The actual temperatures of the exotherms need not match the annealing temperatures owing to the different annealing environment, including operating pressure, and it is the relative trends that are relevant here.

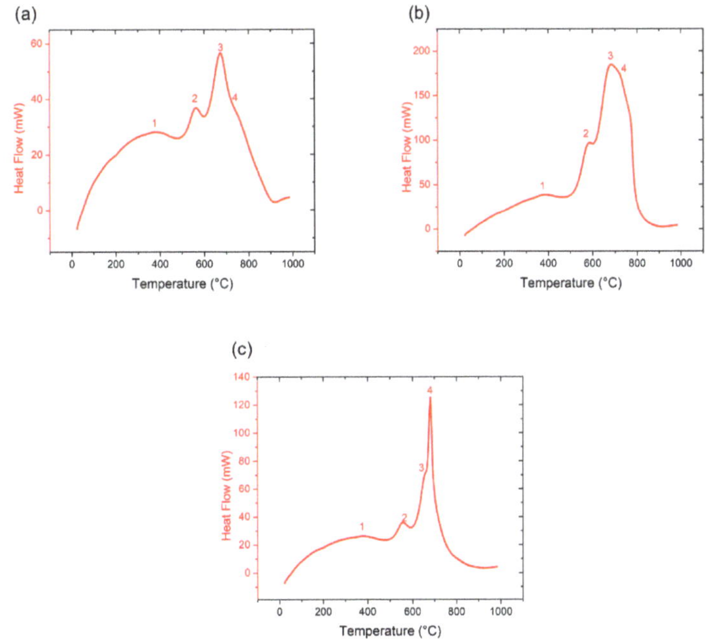

Figure 11. Heat-flow exotherms of the CNT fibers that were previously annealed at (**a**) 300 °C, (**b**) 500 °C, and (**c**) 700 °C, showing the four exotherms of interest here.

4. Conclusions

The scope of this work was initially limited to understanding the effect of thermal annealing on the localized electrical resistance of CNT fibers. A retail spool of Nanocomp's single-ply YE-A10 was purchased accordingly, with the use of an Omicron nanoprobe to both thermally anneal short sections and measure the electrical resistance afterward. There was a clear observation of the Goldilocks principle, i.e., "not too hot, not too cold, just right" [26], with the fibers annealed at 500 °C having a three-orders-of-magnitude reduction in electrical resistance relative to the fiber as is. This led to the performance of a more thorough study, with voltage annealing shown to have a similar effect. Scaled-up tests were performed in a tube furnace which showed similar results, while also providing enough material for further characterization using XPS and TGA in order to indicate the selective removal of non/less-conductive amorphous carbon and CNT self-repair as the primary reasons for the increased conductance. There are also fewer conduction pathways now, with the electrons able to take the path of least resistance with fewer hopping steps in between. Annealing at higher temperatures/voltage results in irreversible changes that cause a significant increase in resistance, often exceeding the starting value. Our results indicate that this is more to do with the loss of conductive CNT materials, with less-conductive catalysts either remaining as is or as oxides. The conduction pathways are also different now, and all of this contributes to fibers that are annealed at 700–800 °C not being viable as starting materials for further processing and electrical conduction applications. This is effectively a purification protocol, although one that has currently only been studied on the localized surface of the CNT fiber and not the bulk fibers. Further research is ongoing to translate these findings to the bulk scale, as well as to better optimize the purification of CNT fibers.

Author Contributions: V.S.G., C.J.B. and J.D.M. conceptualized and performed the experiments, analyzed the results, and contributed to writing the manuscript. A.O.W. contributed towards discussions about the experiments and results. A.R.B. helped design the experiments, provided guidance to data analysis, and helped write the manuscript. All authors have read and agreed to the published version of the manuscript.

Funding: This work was supported by the Office of Naval Research (N00014-15-2717). V.S.G. funded thanks to Salts Healthcare Ltd. A.O.W. is funded through Sêr Cymru II Fellowship by the Welsh Government and the European Regional Development Fund (ERDF). A.O.W. acknowledges funding from Welsh Government Circular Economy Capital Fund FY 2020-21. The authors acknowledge access to the SEM and XPS provided by the Swansea University AIM Facility, funded in part by the EPSRC (EP/M028267/1) and (EP/N020863/1), the European Regional Development Fund through the Welsh Government (80708) and the Welsh Government's Sêr Cymru program.

Data Availability Statement: The data presented in this study are available on request from the corresponding author.

Acknowledgments: We thank Bruce E. Brinson and W. Wade Adams for useful discussions. We would like to acknowledge the assistance provided by Swansea University AIM Facility.

Conflicts of Interest: The authors declare no conflict of interest. The funders had no role in the design of the study; in the collection, analyses, or interpretation of data; in the writing of the manuscript, or in the decision to publish the results.

References

1. Peng, J.; He, Y.; Zhou, C.; Su, S.; Lai, B. The carbon nanotubes-based materials and their applications for organic pollutant removal: A critical review. *Chin. Chem. Lett.* **2021**, *32*, 1626–1636. [CrossRef]
2. Prajapati, S.K.; Malaiya, A.; Kesharwani, P.; Soni, D.; Jain, A. Biomedical applications and toxicities of carbon nanotubes. *Drug Chem. Toxicol.* **2020**, 1–16. [CrossRef] [PubMed]
3. Anzar, N.; Hasan, R.; Tyagi, M.; Yadav, N.; Narang, J. Carbon nanotube—A review on Synthesis, Properties and plethora of applications in the field of biomedical science. *Sens. Int.* **2020**, *1*, 100003. [CrossRef]
4. Jain, N.; Gupta, E.; Kanu, N.J. Plethora of Carbon Nanotubes Applications in Various Fields—A State-of-the-Art-Review. *Smart Sci.* **2021**, 1–24. [CrossRef]
5. Jarosz, P.; Schauerman, C.; Alvarenga, J.; Moses, B.; Mastrangelo, T.; Raffaelle, R.; Ridgley, R.; Landi, B. Carbon nanotube wires and cables: Near-term applications and future perspectives. *Nanoscale* **2011**, *3*, 4542–4553. [CrossRef] [PubMed]
6. Bulmer, J.S.; Kaniyoor, A.; Elliott, J.A. A meta-analysis of conductive and strong carbon nanotube materials. *Adv. Mater.* **2021**, *33*, 2008432. [CrossRef]
7. Hjortstam, O.; Isberg, P.; Soderholm, S.; Dai, H. Can we achieve ultra-low resistivity in carbon nanotube-based metal compo-sites? *Appl. Phys. A Mater. Sci. Process* **2004**, *78*, 1175–1179. [CrossRef]
8. Janas, D.; Liszka, B. Copper matrix nanocomposites based on carbon nanotubes or graphene. *Mater. Chem. Front.* **2018**, *2*, 22–35. [CrossRef]
9. Jordan, M.B.; Feng, Y.; Burkett, S.L. Development of seed layer for electrodeposition of copper on carbon nanotube bundles. *J. Vac. Sci. Technol. B* **2015**, *33*, 021202. [CrossRef]
10. Subramaniam, C.; Sekiguchi, A.; Yamada, T.; Futaba, D.N.; Hata, K. Nano-scale, planar and multi-tiered current pathways from a carbon nanotube–copper composite with high conductivity, ampacity and stability. *Nanoscale* **2016**, *8*, 3888–3894. [CrossRef] [PubMed]
11. Kang, C.S.; Lee, I.J.; Seo, M.S.; Kim, S.H.; Baik, D.H. Effect of purification method on the electrical properties of the carbon nanotube fibers. *Fibers Polym.* **2017**, *18*, 1580–1585. [CrossRef]
12. Barnett, C.J.; Gowenlock, C.E.; Welsby, K.; Orbaek-White, A.; Barron, A.R. Spatial and Contamination-Dependent Electrical Properties of Carbon Nanotubes. *Nano Lett.* **2018**, *18*, 695–700. [CrossRef]
13. Barnett, C.J.; Gowenlock, C.E.; White, A.O.; Barron, A.R. Pressure dependent conduction of individual multi-walled carbon nanotubes: The effect of mechanical distortions. *Nanoscale Adv.* **2021**, *3*, 643–646. [CrossRef]
14. Barnett, C.J.; McGettrick, J.D.; Gangoli, V.S.; Kazimierska, E.; Orbaek White, A.; Barron, A.R. Effect of applied pressure on the electrical resistance of carbon nanotube fibres. *Materials* **2021**, *14*, 2106. [CrossRef] [PubMed]
15. Gangoli, V.S.; Raja, P.M.V.; Esquenazi, G.L.; Barron, A.R. The safe handling of bulk low-density nanomaterials. *SN Appl. Sci.* **2019**, *1*, 644. [CrossRef]
16. Cobley, R.; Brown, R.A.; Barnett, C.J.; Maffeis, T.G.G.; Penny, M.W. Quantitative analysis of annealed scanning probe tips using energy dispersive x-ray spectroscopy. *Appl. Phys. Lett.* **2013**, *102*, 23111. [CrossRef]
17. Barnett, C.J.; Kryvchenkova, O.; Wilson, L.S.J.; Maffeis, T.G.G.; Kalna, K.; Cobley, R. The role of probe oxide in local surface conductivity measurements. *J. Appl. Phys.* **2015**, *117*, 174306. [CrossRef]
18. Smith, N.A.; Lord, A.M.; Evans, J.E.; Barnett, C.J.; Cobley, R.J.; Wilks, S.P. Forming reproducible non-lithographic nano-contacts to assess the effect of contact compressive strain in nanomaterials. *Semicond. Sci. Technol.* **2015**, *30*, 065011. [CrossRef]
19. Dresselhaus, M.; Dresselhaus, G.; Saito, R.; Jorio, A. Raman spectroscopy of carbon nanotubes. *Phys. Rep.* **2005**, *409*, 47–99. [CrossRef]

20. Gangoli, V.S.; Godwin, M.A.; Joshi, S.S.; Allanavar, A.B.; Reddy, G.; Bradley, R.K.; Barron, A.R. The state of HiPco sin-gle-walled carbon nanotubes in 2019. *C* **2019**, *5*, 65.
21. Zhang, K.S.; Pham, D.; Lawal, O.; Ghosh, S.; Gangoli, V.S.; Smalley, P.; Kennedy, K.; Brinson, B.; Billups, W.E.; Hauge, R.; et al. Overcoming catalyst residue inhibition of the functionalization of single-walled carbon nano-tubes via the Billups-Birch reduction. *ACS Appl. Mater. Interfaces* **2017**, *9*, 37972–37980. [CrossRef] [PubMed]
22. Suherman, H.; Sulong, A.B.; Sahari, J. Effect of filler loading concentration, curing temperature and molding pressure on the electrical conductivity of CNTs/graphite/epoxy nanocomposites at high loading of conductive fillers. *Int. J. Mech.* **2010**, *1*, 74–79.
23. Khanbolouki, P.; Tehrani, M. Purification, structural evolutions, and electrical properties of carbon nanotube yarns processed via incandescent annealing. *Carbon* **2020**, *168*, 710–718. [CrossRef]
24. Orbaek, A.W.; Aggarwal, N.; Barron, A.R. The development of a 'process map' for the growth of carbon nanomaterials from ferrocene by injection CVD. *J. Mater. Chem. A* **2013**, *1*, 14122–14132. [CrossRef]
25. Wright, K.D.; Barron, A.R. Catalyst Residue and Oxygen Species Inhibition of the Formation of Hexahapto-Metal Complexes of Group 6 Metals on Single-Walled Carbon Nanotubes. *C* **2017**, *3*, 17. [CrossRef]
26. Obrey, S.J.; Bott, S.G.; Barron, A.R. A Lewis Base Promoted Alkyl/Alkoxide Ligand Redistribution: Reaction of [Me2Al(μ-OCPh3)]$_2$ with THF. *Organometallics* **2001**, *20*, 5119–5124. [CrossRef]

C Journal of Carbon Research

Article

Carbon Nanotubes Use for the Semiconductors ZnSe and ZnS Material Surface Modification via the Laser-Oriented Deposition Technique

Natalia Kamanina [1,2,3,*], Andrey Toikka [1,2,3], Bulat Valeev [4] and Dmitry Kvashnin [4,5,6]

1 Lab for Photophysics of Media with Nanoobjects at Vavilov State Optical Institute, Kadetskaya Liniya V.O., Dom 5, Korp.2/Babushkina Str., Dom 36, Korp.1, 199053 St. Petersburg, Russia; atoikka@obraz.pro
2 Photonica Department, St. Petersburg Electrotechnical University ("LETI"), Ul.Prof.Popova, Dom 5, 197376 St. Petersburg, Russia
3 Petersburg Nuclear Physics Institute, National Research Center "Kurchatov Institute", 1 md. Orlova Roshcha, 188300 Gatchina, Russia
4 Moscow Institute of Physics and Technology (National Research University), 9 Institutskiy per., Dolgoprudny, 141701 Moscow, Russia; bulat_valeev@phystech.edu (B.V.); dgkvashnin@phystech.edu (D.K.)
5 Emanuel Institute of Biochemical Physics RAS, 4 Kosigina St., 119334 Moscow, Russia
6 Physics and Mathematics Department, Pirogov Russian National Research Medical University, Ostrovitianov Str. 1, 117997 Moscow, Russia
* Correspondence: nvkamanina@mail.ru; Tel.: +7-(812)-327-0095

Citation: Kamanina, N.; Toikka, A.; Valeev, B.; Kvashnin, D. Carbon Nanotubes Use for the Semiconductors ZnSe and ZnS Material Surface Modification via the Laser-Oriented Deposition Technique. C **2021**, 7, 84. https://doi.org/10.3390/c7040084

Academic Editor: Gil Goncalves

Received: 16 November 2021
Accepted: 4 December 2021
Published: 7 December 2021

Publisher's Note: MDPI stays neutral with regard to jurisdictional claims in published maps and institutional affiliations.

Abstract: It is known that a material's volume and the surface structuring by the nanoparticles causes a significant change in the material's basic properties. In this aspect, the structuration of the surface of semiconductors is of interest, because their wide potential application in optoelectronics can extend the products' transparency, hardness, wettability, and other important parameters. This paper presents possible methods for the surface modification of zinc selenide and zinc sulfide when carbon nanotubes are deposited on the surface by the application of the laser-oriented technique. It also shows changes of the spectral, mechanical, and wetting characteristics of the considered materials. Using the molecular dynamic simulations, the possible process of the carbon nanotubes penetration into the considered surfaces is presented. The simulation results are partially supported by the obtained experimental data.

Keywords: ZnSe; ZnS; surface; carbon nanotubes; laser-oriented deposition; spectra; microhardness; wetting angle; MD

1. Introduction

Among the different materials involved in general optoelectronics, semiconductor materials occupy a special place that is connected with the unique charge carrier mobility of the semiconductor structures and their operation in the visible and near-IR spectral range. These structures can be involved in the solar energy, gas storage, modulation, optical limiting, etc. schemes as well, and they can dramatically extend the applications of optoelectronics devices [1–4]. ZnSe and ZnS materials can be considered as perspective semiconductors to form the different detectors for the wide spectral range and to develop the mirrors and the optical limiting instruments [5–17]. Researchers have revealed the anti-reflective, luminescent properties of these materials and visualized the process of the nanoparticles' creation based on the ZnSe and ZnS structures and their introduction into the polymer matrices [8–11]. Treatment of these materials by femtosecond lasers has been explained as well [12,16].

Furthermore, it should be remarked that ZnSe and ZnS materials are widely used to make the electrically and optically addressed spatial light modulators (SLM), which operate in the blue and in the visible spectral range. This specific application is connected with the high carrier mobility of these types of the materials. Thus, the response (switch-on and switch-off time) can be revealed with good speed. Indeed, the resolution of the SLM with the photo layers based on the ZnSe and ZnS structures is not so good. However, the compromise between the resolution and the speed can be found via the specific pulsed regime of the laser operations when the delay between the electric pulse and the light one can be varied [17,18].

Different types of classical methods and innovative approaches can be used in order to modify the interface and the surface of the semiconductor materials with the nanoparticles. As promising nano-objects used for this aim, carbon nanotubes (CNTs) [19–24] can be successfully proposed. The basic features of the carbon nanotubes are regarded to be their branched surface, their high conductivity, the strong hardness of their C–C bonds, the little value of the refractive index as well as their complicated and unique mechanisms of the charge carrier moving.

In the current paper, the laser-oriented deposition (LOD) approach, previously shown to treat inorganic crystals and metals, such as KBr [25], BaF_2 [26], and Al [27], is shown as the perspective instrument to modify the properties of the semiconductors based on ZnSe and ZnS. The possibility of the ZnS and ZnSe surfaces' modification via the vertically oriented CNTs deposition is shown. Modification leads to changes in the atomic structure of the surface as well as the mechanical and the optical properties. The obtained modified structures show prospects for an application as a basic element of the optoelectronics schemes, of the solar cells, and of the gas storage systems. Moreover, it can be useful for information technology and displays.

2. Materials and Methods

The matrix ZnSe and ZnS materials have been purchased from the "Alkor Technologies" company in Saint-Petersburg. The diameter of the substrate was 35 mm and the thickness of the ZnSe and ZnS structures was 5 and 3 mm, respectively.

To modify the properties of the semiconductor materials via their surface treatment, the single-wall carbon nanotubes (SWCNTs) type #704121 with the diameter placed in the range of 0.7–1.1 nm purchased from Aldrich Co. have been used. It is important in order to find the coinciding conditions between the dimensions of the CNTs directly with the elementary lattice of the material.

To modify the surface, the IR CO_2 laser with p-polarized irradiation at the wavelength of 10.6 μm and the power of 30 W has been used. The general view of the block scheme has been shown in [25]. It is shown that the laser system was connected with a vacuum hood, which contains the fixing unit samples and the device for depositing substances on the substrate. To explain the LOD technical process more clearly, the scheme is shown in Figure 1.

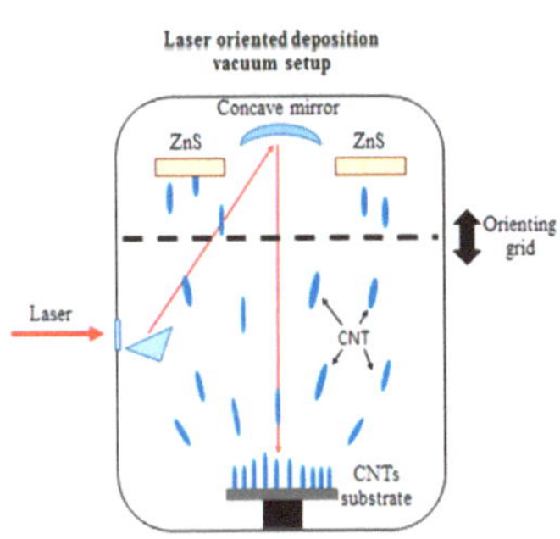

Figure 1. The scheme to show the LOD process used for the ZnS and ZnSe materials more clearly.

CNTs have been placed at the materials interface under the conditions of an additional applied electric field of 100–600 V × cm^{-1} in order to orient the nanotubes in the vertical position during the applied process. It can be possible when, for example, the distance between the grid and the upper substrate can be varied. Some qualitative scheme of this process with the grid incorporated in the vacuum chamber was shown before as well [24]. It permits varying the velocity of the CNTs and forming the covalent bonding between the carbon atoms and the surface matrix materials with good advantage. Using such a scheme, the carbon nanotubes with a given diameter in the interstices of the lattice of the matrix material can be integrated. This procedure has been partially shown in several RU patents [28–30].

The spectra of the CNTs-treated materials have been obtained using the Furrier FSM-1202 instruments as well as using a VIS SF-26 spectrophotometer operated in the range of 250–1200 nm. The Polam-P312 microscope has been applied to make the image of the materials treated. The surface mechanical hardness (abrasive strength) has been revealed using the CM-55 instrument, and the microhardness has been measured via the PMT-3M device using; this device has been produced by "LOMO" (Saint-Petersburg, Russia) with the ability to vary indenter forces as well. The OCA 15EC device has been purchased from LabTech Co. (Saint Petersburg, Russia) and used to control the wetting angle change to predict the Lotus effect.

To describe the implantation of the CNT at an atomic scale, the molecular dynamic (MD) simulations were carried out using the LAMMPS package [31]. Interaction between the ZnS system and CNTs was described via the Tersoff interatomic potential [32], which was parameterized according to [33]. For the description of the C–C interaction, the Tersoff potential was parameterized according to the procedure [34]. To describe the interaction between the CNT and the ZnS substrate, the Lennard–Jones 12-6 potential was used. The interaction parameters were fit as: ε = 0.0113 eV, σ = 1.76 Å. All the simulations were performed at room temperature. The behavior of an atomic under the finite constant temperature was estimated by use of the Nóse–Hoover thermostat [35,36]. The total time of the simulation was 60 ps with the time step of 1 fs. The bottom layer of the ZnS substrate was fixed to avoid the displacement of the whole structure during an implantation process. To avoid the effective energy dissipation during the implantation, the boundary conditions with rescaling of the atomic velocities were applied. The initial velocities of CNTs were set up from 100 to 600 m/s.

3. Results and Discussion

It is well known that the ZnSe and ZnS materials are the most promising key elements in the optical industry. ZnSe is used in X-ray detectors, in the development of the cores for the optical fibers, in the IR-laser techniques for windows of the CO_2 lasers, etc. ZnS is used as a suitable material for the phosphors and for the semiconductor laser technology, etc. The LOD procedure indicates some advantage of the modified properties of such materials. Why? It is connected with the fact that the refractive index of the CNTs is so small and close to 1.1, but the Young modulus is so large and it is placed in the range of some teraPascales. Thus, the Fresnel losses can be decreased, the transparency can be increased, and the mechanical parameters can be improved.

It should be noticed that in order to study the structuration process influence on the physical and the chemical characteristics of the ZnSe and ZnS materials, the substrates with the polished surfaces have been used. The photos of the ZnSe and the ZnS samples studied are presented in Figure 2. It is worth noting that the ring is visible on the ZnS sample in Figure 2b, which was obtained when testing this sample for an abrasion to determine the mechanical strength, which will be discussed later.

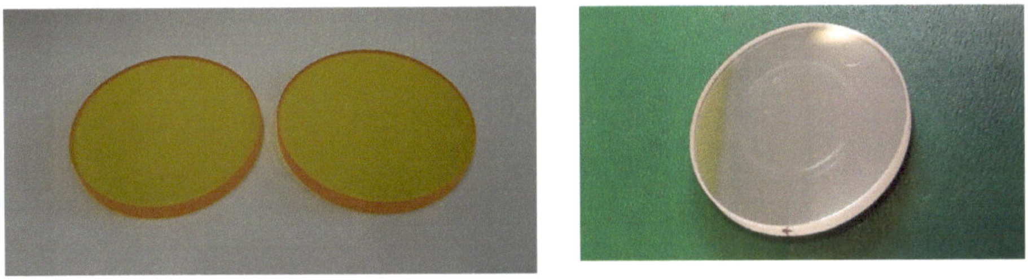

(a) (b)

Figure 2. Photos of the ZnSe (**a**) and ZnS (**b**) samples studied in the current research.

Some change of the spectral characteristics of the ZnSe and the ZnS substrate in the range of 1,4 microns to 2,4 microns is shown in Figure 3. The image shows the influence of the CNTs structuration on the spectral parameters. Figure 3a (curve 1) presents the transmission of the pure ZnSe substrate on the wavelength range indicated above. Figure 3a (curve 2 and curve 3) shows the spectral characteristics after the treatment of the ZnSe surface from one side (Figure 3a; curve 2) and from both sides (Figure 3a; curve 3), respectively. The change of the spectral parameters for the pure ZnS and for the ZnS with one side treated by the CNTs is presented in Figure 3b. The green curve (lower one) shows the transmittance for the pure substrate; the blue curve (upper one) presents the transmittance for the ZnS covered with the CNTs. It is important that the surface treatment of samples with the CNTs can increase the transparency of both materials. It can be coincided with the fact that the CNTs refractive index is close to 1.1 [21,22], which is less than the refractive index of the ZnSe or ZnS materials; thus, the Fresnel losses can be decreased dramatically, which can predict the increase in the transparency.

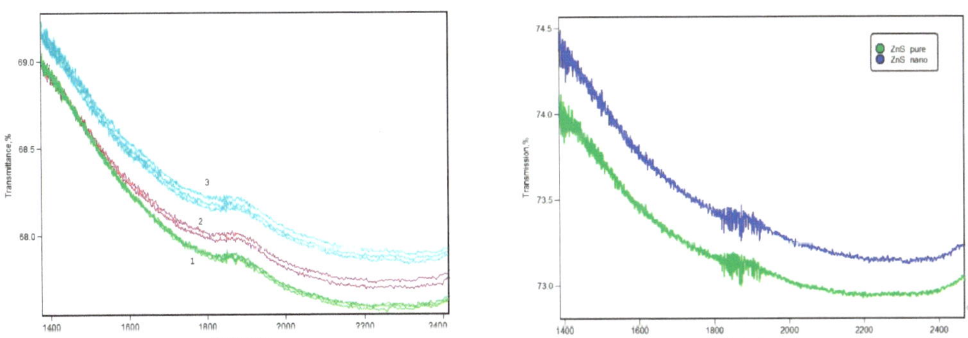

(a) (b)

Figure 3. The ZnSe (**a**) and ZnS (**b**) transmittance in the near-IR-range.

The mechanical properties of the studied materials were investigated by two methods. Experiments were carried out to establish the microhardness of the nanostructured samples as well as to measure the comparative abrasion strength.

The microhardness improvement can be shown in the data presented in Table 1 for the ZnSe and ZnS structures. It was found that an implantation could increase the microhardness of the material up to 5% and 6.7%, respectively. Thus, the same tendency of the microhardness increasing via the CNTs covering by the LOD technique has been obtained for the ZnS and ZnSe substrates.

Table 1. The comparative data of the ZnSe and ZnS microhardness change, when an indentor was 20 g.

Material Studied	Middle Value of the Microhardness, Pa $\times 10^9$
Pure ZnSe	0.1118
ZnSe structured with CNTs	0.1175
Pure ZnS	0.8589
ZnS structured with CNTs	0.91675

Moreover, the surface mechanical hardness (abrasive strength) for the ZnS and ZnSe materials has been established. The data are shown in Table 2. It should be noticed that the abrasion strength was compared with the abrasion of the glass K8 (crown glass), on which, with an indenter of 100 g, 3000 revolutions can be reached until a distinct ring appears. That is, with such a standard, namely crown glass K8, zero strength of the glass is achieved. Analyzing the data from Table 2, one can testify that the tendency to increase not only the microhardness but also the surface strength can be found for the ZnS and ZnSe materials with good advantage.

Table 2. The comparative data of the ZnS and ZnSe surface hardness when an indentor was 100 g.

Material Studied	The Number of Revolutions on the Abrasion Surface before the Appearance of a Distinct Ring
Pure ZnS	100–120
ZnS structured with CNTs	200–220
Pure ZnSe	420
ZneS structured with CNTs	600

It is worth paying attention to the fact that different methods for determining the mechanical strength for the semiconductor materials based on the ZnS and ZnSe systems are not always the same for both. Different methods are more preferable for each of the materials. However, the general tendency of the mechanical strength increasing is maintained always and consistently.

Furthermore, for the ZnS and ZnSe materials, the wetting angle before and after the CNTs deposition has been established. The data of the wetting angle change are shown in Figure 4. When evaluating the averaging by the wetting angle, a sample was taken from the measurements of a batch of samples in the amount of 10 pieces.

One can see the slight change of the wetting angle for the ZnS (Figure 4a,b) and a better increase in the wetting angle for the ZnSe structure after its structuration by the CNTs (Figure 4c,d).

In addition, in order to support the matrix ZnS and ZnSe materials' relief, modification of the AFM images have been made. The data are shown in Figure 5. One can see that the surface relief shown for the pure ZnS and ZnSe materials (Figure 5a,c) is absolutely different from the one that presented for the nanostructured surfaces (Figure 5b,d). Moreover, since the CNTs can be considered as donor–acceptor systems, the formation of the globules at the structured surfaces can be found.

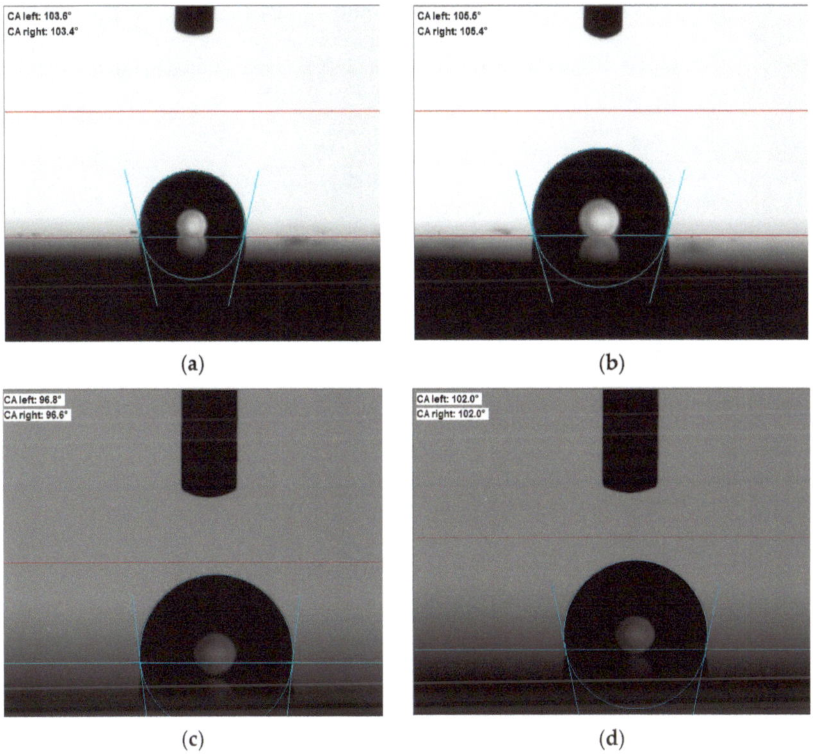

Figure 4. The wetting angle for the ZnS surface, pure (**a**) and with CNTs treatment (**b**). The wetting angle for the ZnSe surface, pure (**c**) and with CNTs treatment (**d**).

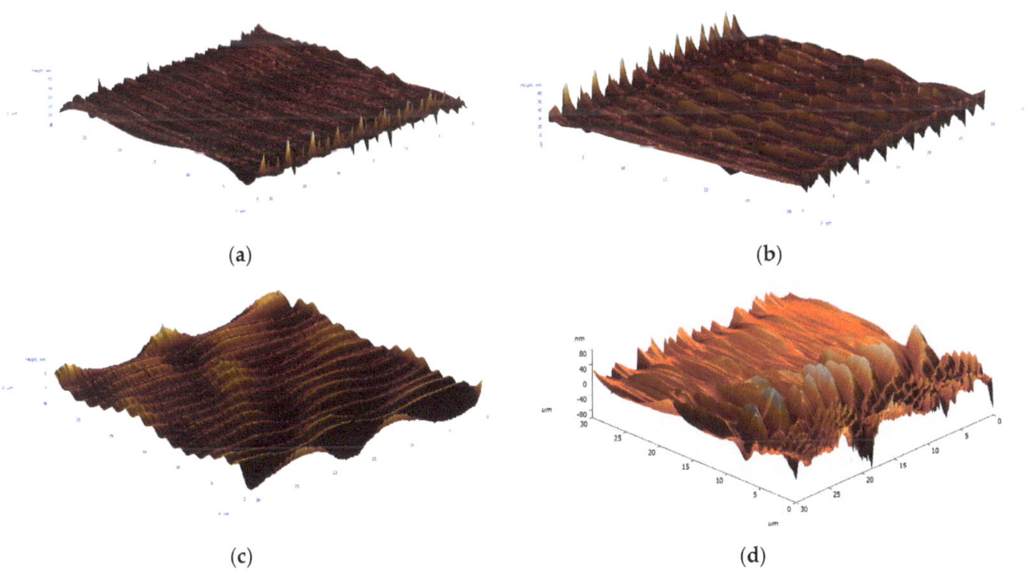

Figure 5. The AFM images of the pure (**a**) and the structured (**b**) ZnS surfaces; the AFM images of the pure (**c**) and structured (**d**) ZnSe surfaces.

Figure 6 shows the results of the MD simulations carried out. We calculated the dependence of the penetration depth of the CNT into the ZnS surface on the CNT diameter and its velocity and an inclination angle. The nanotube diameter was varied from 0.64 to 3.35 nm.

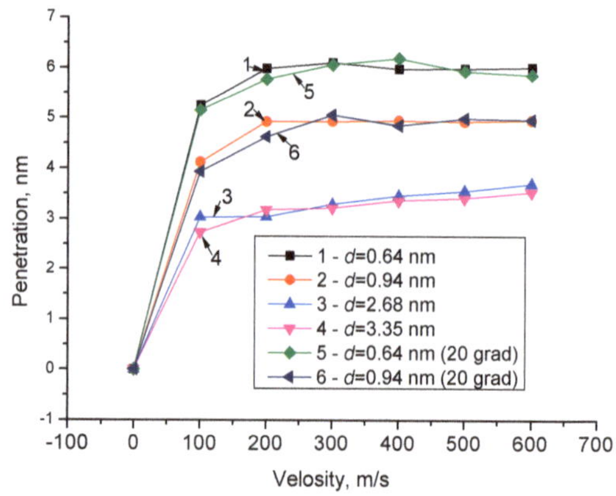

Figure 6. The dependence of the penetration depth of CNTs into ZnS surface on the CNT velocity and its diameter.

It can be seen that the CNT speed and their diameter as well as the angle of inclination on the substrate affects the final values of the penetration depth of the nanoobject into the ZnS surface. The qualitative representation of the process of the penetration of the CNTs with diameters of 0.64 nm (Figure 7a) and 2.68 nm (Figure 7b) into the ZnS surface layers with an initial velocity of 100 m/s is shown. It was found that the CNT with the smaller diameter penetrates more deeply than the CNTs with the wider diameter at the same initial velocity, which reflects its higher stiffness [37]. Figure 6 shows the process at the CNTs speed of 100 m/s.

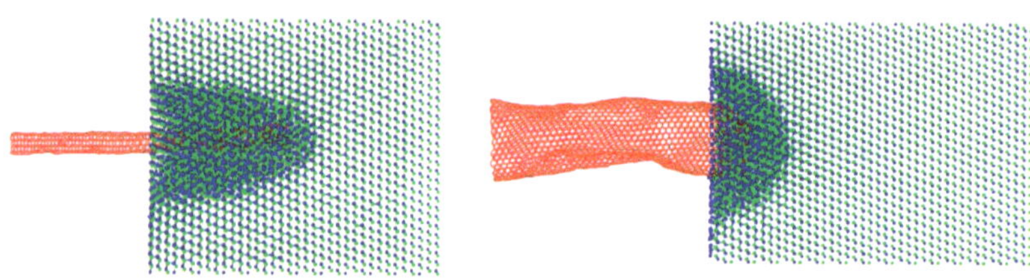

(a) (b)

Figure 7. The qualitative representation of the CNT penetration with the diameter of 0.64 nm (**a**) and 2.68 nm (**b**) into the ZnS surface.

We can plan to make the analogous simulation for the ZnSe materials and will show the results in the future publications.

Thus, the experimental results partially can be added by this simulation; it supports the fact of the effective application of the LOD technique in order to improve the basic properties of such materials with good advantage.

4. Conclusions

Analyzing the obtained and discussed data, the following conclusions should be drawn.

We have compared the properties of pure ZnS and ZnSe materials with those obtained after the modification procedure by the LOD method.

Structuration of the inorganic semiconducting surfaces (based on ZnS and ZnSe) by the CNTs shows the promising way of the surface modification with the aim of their transparency increase. The stronger increasing was observed in the case of treating both sides of the ZnS and ZnSe surfaces. This is due to the low refractive index of the CNTs, which allows reducing the loss of the reflection along the Fresnel law. The structuration of the ZnS and ZnSe surfaces shows the slight increase in the microhardness and the wetting angle. It can be connected with the large value of the Young's module of the CNTs. The formation of the possible link between the carbon atoms and the surface materials permits increasing the hardness of the matrix materials as well. The laser-oriented deposition method can be applied with good advantage in order to improve the properties of the semiconductor materials. The obtained experimental results are partially supported by the molecular dynamic simulations.

It should be noted that all results are repeatedly checked with a multiplicity of the checks once a month. The error is plus or minus 0.5% and less. Moreover, all results of the modified ZnS and ZnSe have testified better data than shown, for example, in the classical books [38,39] for the pure materials.

Established data can show the promising route of the modification of the semiconductors surface with the purpose to form the structured materials with the desired optical properties which make them suitable for the applications as elements of optoelectronics, solar energy, gas storage systems, information technology, displays, and biomedicine. The knowledge accumulated can be used in the education process and for the practices as well.

For the future study, it will be useful to obtain and present the electronic microscopy images; it will be considered for the ZnS and ZnSe structures as well as other semiconducting materials.

Author Contributions: Conceptualization, formal analysis, writing—review and editing N.K.; investigation, A.T.; methodology, software D.K. and B.V. All authors have read and agreed to the published version of the manuscript.

Funding: The calculations were performed using the resources provided by the HybriLIT heterogeneous computing platform of the Joint Institute for Nuclear Research and supercomputer cluster provided by the Materials Modelling and Development Laboratory at NUST "MISIS" (supported via the grant from the Ministry of Education and Science of the Russian Federation No. 14.Y26.31.0005).

Institutional Review Board Statement: This work was carried out as an initiative. Consequently, it did not require any approval and conclusion from the Institute.

Informed Consent Statement: Consent to work; of course, it was obtained from all participants in this study.

Data Availability Statement: Using these primary materials of the article, as well as taking into account earlier developments on other materials, a patent will be created, as well as a message will be published in the media, for example, at LETI University, since the results are useful for teaching students.

Acknowledgments: The authors would like to thank their colleagues from Vavilov State Optical Institute, "LETI" University and Nuclear Physics Institute for the helpful discussions at the scientific seminars. Moreover, we would like to acknowledge our colleagues from Moscow Institutes for the on-line discussions.

Conflicts of Interest: The authors declare no conflict of interest.

References

1. Barbagiovanni, E.G.; Lockwood, D.J.; Simpson, P.J.; Goncharova, L.V. Quantum confinement in Si and Ge nanostructures: Theory and experiment. *Appl. Phys. Rev.* **2014**, *1*, 011302. [CrossRef]
2. Gautier, G.; Leduc, P. Porous silicon for electrical isolation in radio frequency devices: A review. *Appl. Phys. Rev.* **2014**, *1*, 011101. [CrossRef]
3. Yang, D. *Handbook of Photovoltaic Silicon*; Springer: Berlin/Heidelberg, Germany, 2019; p. 967. ISBN 978-3-662-56473-8.
4. Van Stryland, E.W.; Wu, Y.Y.; Hagan, D.J.; Soileau, M.J.; Mansour, K. Optical limiting with semiconductors. *J. Opt. Soc. Am. B* **1988**, *5*, 1980–1988. [CrossRef]

5. Sandeep, A.; Saleem, K.; Parveen, L.; Ishan, G.; Suresh, K. Electrical properties of electrodeposited zinc selenide (ZnSe) nanowires. *J. Mater. Sci. Mater. Electron.* **2014**, *25*, 4150–4155. [CrossRef]
6. Xu, S.; Qi, X.; Xu, X.; Wang, X.; Yang, Z.; Zhang, S.; Lin, T.; He, P. Effects of electroless nickel plating method for low temperature joining ZnS ceramics. *J. Mater. Sci. Mater. Electron.* **2019**, *30*, 15236–15249. [CrossRef]
7. Pushkin, A.V.; Migal, E.A.; Tokita, S.; Korostelin, Y.V.; Potemkin, F.V. Femtosecond graphene mode-locked Fe: ZnSe laser at 4.4 μm. *Opt. Lett.* **2020**, *45*, 738–741. [CrossRef]
8. Chan, L.; DeCuir, E.A.; Fu, R.; Morse, D.E.; Gordon, M.J. Biomimetic nanostructures in ZnS and ZnSe provide broadband anti-reflectivity. *J. Opt.* **2017**, *19*, 114007. [CrossRef]
9. Dmitruk, I.; Berezovska, N.; Degoda, V.; Hrabovskyi, Y.; Kolodka, R.; Podust, G.; Stanovyi, O.; Blonskyi, I. Luminescence of Femtosecond Laser-Processed ZnSe Crystal. *J. Nanomater.* **2021**, *2021*, 6683040. [CrossRef]
10. Cao, J.; Jiang, Z.J. Thickness-dependent Shell Homogeneity of ZnSe/CdSe Core/Shell Nanocrystals and Their Spectroscopic and Electron- and Hole-transfer Dynamics Properties. *J. Phys. Chem. C* **2020**, *124*, 12049–12064. [CrossRef]
11. Kushwaha, K.K.; Kumari, S.; Mahobia, S.K.; Tiwary, S.K.; Sinha, B.K.; Ramrakhiani, M. Synthesis and Characterization of ZnSe/CdSe/ZnS Nano-particles Embedded in Polymer Matrix: A Review. *Int. J. Compos. Const. Mater.* **2018**, *4*, 28–33.
12. Xiao, Y.; Deng, G.; Feng, G.; Ning, S.; Wang, S.; Chen, X.; Yang, H.; Zhou, S. Femtosecond laser induced nano-meter size surface structures on ZnSe film. *AIP Adv.* **2019**, *9*, 015106. [CrossRef]
13. Lin, Z.Q.; Wang, G.G.; Li, L.H.; Wang, H.; Li, L.H.; Wang, H.; Tian, J.L.; Zhang, H.Y.; Han, J.C. Preparation and protection of ZnS surface sub-wavelength structure for infrared window. *Appl. Surf. Sci.* **2019**, *470*, 395–404. [CrossRef]
14. Igweoko, A.E.; Augustine, C.; Idenyi, N.E.; Okorie, B.A.; Anyaegbunam, F.N.C. Influence of processing conditions on the optical properties of chemically deposited zinc sulphide (ZnS) thin film. *Mater. Res. Express* **2018**, *5*, 036413. [CrossRef]
15. Nabesaka, K.; Ishikawa, Y.; Hosokawa, Y.; Uraoka, Y. Photomechanical modification of ZnS microcrystal to enhance electroluminescence by ultrashort-pulse laser processing. *Appl. Phys. Express* **2017**, *10*, 021201. [CrossRef]
16. Li, X.; Li, M.; Liu, H.; Guo, Y. Fabrication of an Anti-Reflective Microstructure on ZnS by Femtosecond Laser Bessel Beams. *Molecules* **2021**, *26*, 4278. [CrossRef]
17. Kamanina, N.V. Using the operator Laplace method to estimate the response time of space-time light modulators. *Tech. Phys.* **1994**, *39*, 83–85.
18. Kamanina, N.V.; Vasilenko, N.A. Influence of operating conditions and of interface properties on dynamic characteristics of liquid-crystal spatial light modulators. *Opt. Quantum Electron.* **1997**, *29*, 1–9. [CrossRef]
19. Zhang, F.; Wang, H.; Wang, C.; Zhao, Y.; Duan, J. Direct femtosecond laser writing of inverted array for broadband antireflection in the far-infrared. *Opt. Lasers Eng.* **2020**, *129*, 106062. [CrossRef]
20. Mensah, N.G.; Nkrumah, G.; Mensah, S.Y.; Allotey, F.K.A. Temperature dependence of the thermal conductivity in chiral carbon nanotubes. *Phys. Lett. A* **2004**, *329*, 369–378. [CrossRef]
21. Robertson, J. Realistic applications of CNTs. *Mater. Today* **2004**, *7*, 46–52. [CrossRef]
22. Namilae, S.; Chandra, N.; Shet, C. Mechanical behavior of functionalized nanotubes. *Chem. Phys. Lett.* **2004**, *387*, 247–252. [CrossRef]
23. Fa, W.; Yang, X.; Chen, J.; Dong, J. Optical properties of the semiconductor carbon nanotube intramolecular junctions. *Phys. Lett.* **2004**, *A323*, 122–131. [CrossRef]
24. Taherpour, A.A.; Aghagolnezhad-Gerdroudbari, A.; Rafiei, S. Theoretical and quantitative structural relationship studies of reorganization energies of [SWCNT(5,5)-Armchair-CnH20] (n=20–310) nanostructures by neural network CFFBP method. *Int. J. Electrochem. Sci.* **2012**, *7*, 2468–2486.
25. Kamanina, N.V.; Likhomanova, S.V.; Kuzhakov, P.V. Advantages of the Surface Structuration of KBr Materials for Spectrometry and Sensors. *Sensors* **2018**, *18*, 3013. [CrossRef] [PubMed]
26. Kamanina, N.; Kuzhakov, P.; Kvashnin, D. Novel Perspective Coatings for the Optoelectronic Elements: Features of the Carbon Nanotubes to Modify the Surface Relief of BaF_2 Materials. *Coatings* **2020**, *10*, 7066. [CrossRef]
27. Kamanina, N.; Borodianskiy, K.; Kvashnin, D. Surface Heterostructure of Aluminum with Carbon Nanotubes Obtained by Laser-Oriented Deposition. *Coatings* **2021**, *11*, 674. [CrossRef]
28. Kamanina, N.V.; Vasilyev, P.Y. Optical Coatings Based on CNTs for the Optical Devises and Nanoelectronics. R.U. Patent 2 355 001 C2, 10 May 2009.
29. Kamanina, N.V.; Vasilyev, P.Y.; Studeonov, V.I. Optical Coating Based on Oriented in the Electric Field CNTs for the Optical Devises, Micro- and Nanoelectronics under the Conditions When the Interface: Solid Substrate-Coating Can Be Eliminated. R.U. Patent 2 405 177 C2, 27 November 2010.
30. Kamanina, N.V.; Kuzhakov, P.V.; Vasilyev, P.Y. Protective Coating for Hygroscopic Optical Materials Based on Laser-Deposited Carbon Nanotubes for the Purpose of Optoelectronics and Medical Equipment. R.U. Patent 2013118962 A, 27 October 2014.
31. Plimpton, S. Fast Parallel Algorithms for Short-Range Molecular Dynamics. *J. Comput. Phys.* **1995**, *117*, 1–19. [CrossRef]
32. Tersoff, J. New Empirical Approach for the Structure and Energy of Covalent Systems. *Phys. Rev. B* **1988**, *37*, 6991–7000. [CrossRef]
33. Benkabou, F.; Aourag, H.; Certier, M. Atomistic Study of Zinc-Blende CdS, CdSe, ZnS, and ZnSe from Molecular Dynamics. *Mater. Chem. Phys.* **2000**, *66*, 10–16. [CrossRef]
34. Tersoff, J. Modeling Solid-State Chemistry: Interatomic Potentials for Multicomponent Systems. *Phys. Rev. B* **1989**, *39*, 5566–5568. [CrossRef]

35. Nosé, S. A Unified Formulation of the Constant Temperature Molecular Dynamics Methods. *J. Chem. Phys.* **1984**, *81*, 511–519. [CrossRef]
36. Hoover, W.G. Canonical Dynamics: Equilibrium Phase-Space Distributions. *Phys. Rev. A* **1985**, *31*, 1695–1697. [CrossRef] [PubMed]
37. Shima, H. Buckling of Carbon Nanotubes: A State of the Art Review. *Materials* **2011**, *5*, 47–84. [CrossRef] [PubMed]
38. Morozova, N.K.; Kuznetsov, V.A.; Ryzhikov, V.D.; Galstyan, V.G. *Zinc Selenide: Preparation and Optical Properties*; Fok, M.V., Ed.; Russian Academy of Sciences, A.V. Shubnikov Institute of Crystallography: St. Petersburg, Russia, 1992; 95p.
39. Bovina, L.A.; Brodin, M.S.; Valakh, M.Y. *Physics of Compounds AIIBVI*; Georgobiani, A.N., Sheinkman, M.K., Eds.; Phys. Mat. Lit.: Nauka, Russia, 1986; 320p.

C Journal of Carbon Research

Article

Premade Nanoparticle Films for the Synthesis of Vertically Aligned Carbon Nanotubes

Abdul Hoque [1], Ahamed Ullah [2], Beth S. Guiton [2] and Noe T. Alvarez [1,*]

[1] Department of Chemistry, University of Cincinnati, Cincinnati, OH 45221, USA; hoqueml@mail.uc.edu
[2] Department of Chemistry, University of Kentucky, Lexington, KY 40506, USA; ahamed.ullah@uky.edu (A.U.); beth.guiton@uky.edu (B.S.G.)
* Correspondence: noe.alvarez@uc.edu

Abstract: Carbon nanotubes (CNTs) offer unique properties that have the potential to address multiple issues in industry and material sciences. Although many synthesis methods have been developed, it remains difficult to control CNT characteristics. Here, with the goal of achieving such control, we report a bottom-up process for CNT synthesis in which monolayers of premade aluminum oxide (Al_2O_3) and iron oxide (Fe_3O_4) nanoparticles were anchored on a flat silicon oxide (SiO_2) substrate. The nanoparticle dispersion and monolayer assembly of the oleic-acid-stabilized Al_2O_3 nanoparticles were achieved using 11-phosphonoundecanoic acid as a bifunctional linker, with the phosphonate group binding to the SiO_2 substrate and the terminal carboxylate group binding to the nanoparticles. Subsequently, an Fe_3O_4 monolayer was formed over the Al_2O_3 layer using the same approach. The assembled Al_2O_3 and Fe_3O_4 nanoparticle monolayers acted as a catalyst support and catalyst, respectively, for the growth of vertically aligned CNTs. The CNTs were successfully synthesized using a conventional atmospheric pressure-chemical vapor deposition method with acetylene as the carbon precursor. Thus, these nanoparticle films provide a facile and inexpensive approach for producing homogenous CNTs.

Keywords: vertically aligned carbon nanotubes; catalyst; catalyst support; nanoparticles; monolayer

Citation: Hoque, A.; Ullah, A.; Guiton, B.S.; Alvarez, N.T. Premade Nanoparticle Films for the Synthesis of Vertically Aligned Carbon Nanotubes. C **2021**, 7, 79. https://doi.org/10.3390/c7040079

Academic Editor: Gil Goncalves

Received: 11 October 2021
Accepted: 16 November 2021
Published: 19 November 2021

Publisher's Note: MDPI stays neutral with regard to jurisdictional claims in published maps and institutional affiliations.

1. Introduction

The synthesis of vertically aligned carbon nanotubes (VA-CNTs) on a planar substrate has been established as a method for growing high-quality CNTs with controlled lengths [1,2]. Conventional CNT synthesis by chemical vapor deposition (CVD) requires a catalyst-support film of aluminum oxide (Al_2O_3) and an active metal catalyst. Furthermore, the interactions between the catalyst and catalyst support are crucial for enhancing the catalyst lifetime and VA-CNT growth [3–5]. Such catalysts are commonly prepared using traditional top-down methods, with films being deposited using physical processes such as e-beam deposition, sputtering, or atomic layer deposition. Unfortunately, these catalysts provide little control over the type of CNTs that are produced, or their diameters, which is largely due to the catalyst and catalyst support preparation method. Typically, a thin film of Al_2O_3 (~10 nm) is deposited as a catalyst support followed by the deposition of a thin catalyst film (0.5–3 nm of Fe) [6,7]. Upon exposure to a high temperature and reduction, the catalyst layer forms densely packed islands, which act as nucleation sites for CNT growth [8]. As a result, it is nearly impossible to control the catalyst nanoparticle size, and consequently, to calculate the CNT diameter ahead of time, as the nanoparticle size generally determines the nanotube diameter [9–11]. In addition, these techniques typically require expensive instrumentation, which may have a significant impact on the large-scale production of CNTs.

The development of less expensive and well-controlled methods for depositing catalysts and catalyst supports has generated considerable research interest. Although controlling the catalyst nanoparticle size is expected to provide control over the CNT diameter, the preparation of monodisperse catalyst nanoparticles and their assembly into thin films is challenging, and few methods have been reported [12]. Premade catalyst nanoparticles with controlled densities and sizes can affect the number of nucleation sites and the CNT diameters, respectively [13]. Efforts in this field have demonstrated the effective growth of VA-CNT arrays [14,15], but a greater control over particle size and the build-up of catalyst monolayer films is required. Investigations of individual catalyst nanoparticles can be used to elucidate their characteristics and their catalytic activity toward CNT growth. The lower temperatures that are required for nanoparticle synthesis should favor diameter control. Moreover, wet chemistry methods for catalyst and catalyst support deposition for CNT synthesis should be less expensive than traditional physical evaporation methods.

The deposition of catalyst nanoparticles on a substrate by drop casting has a limited ability to provide uniform monolayer films [16]. In contrast, the assembly of a uniform nanoparticle monolayer on a chemically functionalized substrate is a suitable technique [14,17,18]. An Al_2O_3 catalyst support layer consisting of premade nanoparticles can prevent catalyst diffusion [19] and increase hydrocarbon decomposition [20]. However, there are many challenges associated with assembling nanoparticles in solution into a continuous thin film, including particle diameter uniformity (monodispersion) and the packing of nanoparticles into high-density two-dimensional films while avoiding the formation of double or multiple layers. The assembly of nanoparticle monolayers on planar substrates has been explored using crosslinking agents as anchoring groups [21–24]. (3-Aminopropyl)triethoxysilane has been applied as an intermediate ligand for the formation of monolayers of nanoparticles with amino ($-NH_2$) groups, which can bind with or replace other organic functional groups [25,26]. Monolayers of iron oxide (Fe_3O_4) nanoparticles on glass plates were successfully assembled using a hydroxypropyl cellulose (HPC) biopolymer thin film, with the HPC intermediate linkers forming hydrogen bonds with both the substrate and nanoparticle surfaces [27]. Therefore, it is understood that nanoparticle assembly as a monolayer depends on the organic or inorganic coating on the nanoparticle surface and the chemical nature of the substrate.

Oleic acid is a carboxylate-containing long-chain organic molecule that is used as a surface stabilizing agent for nanoparticle synthesis [28]. Al_2O_3 or Fe_3O_4 nanoparticles coated with oleic acid can be assembled into a monolayer on a substrate by using organic linker molecules. Bifunctional 11-phosphonoundecanoic acid (PNDA) is a suitable linker that can bind to a metal oxide substrate via its phosphonate group to form a carboxylic acid thin film [21,29]. The carboxylate groups in both oleic acid (on the nanoparticle surfaces) and PNDA (on the metal oxide substrate) can undergo ligand exchange reactions in solution. Chemically bonded carboxylate groups on Fe_3O_4 nanoparticles can facilitate the formation of a uniform monolayer of catalyst particles on a substrate. Monolayer formation on a substrate that is modified with an organic ligand can increase both the uniformity and particle density on the substrate. These features are advantageous, as it has been reported that a high density of catalyst particles is required to grow VA-CNTs [30,31].

Here, we report the bottom-up assembly of premade nanoparticle monolayer films of Al_2O_3 and Fe_3O_4 for the synthesis of VA-CNTs. Using spherical nanoparticles that were synthesized in solution, Al_2O_3 and Fe_3O_4 monolayers were sequentially assembled on an SiO_2 substrate. Nanoparticle separation was performed in order to enhance the homogeneity of the assembled monolayer films. The obtained nanoparticle films were then successfully applied to catalyze the growth of VA-CNTs.

2. Experimental

Materials: Aluminum chloride (Reagents Plus, ≥99%), iron (III) chloride hexahydrate (reagent grade, ≥98%), 1-octadecene (technical grade, ≥90%), oleic acid (technical grade, ≥90%), and PNDA (≥90%) were purchased from Sigma-Aldrich (St. Louis, MO, USA) and used without further purification. Sodium oleate (≥97.0%) was purchased from TCI Chemicals. 1-Hexadecene (technical grade, ≥92%), hexane (ACS grade, ≥99.5%), and ethanol (ACS grade) were purchased from Fischer Scientific. Silicon wafers with

500 nm ± 5% thermal oxide layer on top surface were purchased from University Wafer (South Boston, MA, USA). Ultrathin carbon type-A 400 mesh copper grids with lacey carbon for transmission electron microscopy (TEM) characterization were obtained from Ted Pella (Redding, CA, USA).

Characterization: Fourier transform infrared (FTIR) spectroscopy was used to analyze the aluminum oleate and iron oleate precursors as well as the synthesized nanoparticles. The sizes and shapes of the Al_2O_3 and Fe_3O_4 nanoparticles were characterized by atomic force microscopy (AFM; Bruker Dimension icon ScanAsyst AFM, Billerica, MA, USA), TEM (JEOL JEM-1230, Tokyo, Japan), and scanning electron microscopy (Philips FEI XL30 ESEM, North Billerica, MA, USA). X-ray photoelectron spectroscopy (XPS; K-Alpha, Thermo Scientific, Waltham, MA, USA) using a 400 μm spot area was used to investigate the presence of linker molecules on metal oxide substrates. Raman spectroscopy using an inVia confocal Raman microscope (Renishaw, Gluocestershire, UK) with a 633 nm laser at 10% power was used to evaluate the quality of the grown nanotubes.

Al_2O_3 nanoparticle synthesis: Al_2O_3 nanoparticles were synthesized by the thermal decomposition of an aluminum oleate precursor in 1-octadecene as a high-boiling solvent. The aluminum oleate precursor was prepared by reacting $AlCl_3$ (5 g) and sodium oleate (20 g) in a solvent mixture (70 mL hexane + 40 mL ethanol + 30 mL H_2O) for 4 h under reflux at 70 °C with continuous stirring [32]. Subsequently, 1.5 g (1.7 mmol) of aluminum oleate and 1.5 mL of oleic acid (as a capping agent) were dissolved in 50 mL of 1-octadecene. Using a three-neck boiling flask, this mixture was heated to 320 °C for 2 h with continuous stirring. When the temperature was close to 320 °C, continuous bubbling was observed and the transparent solution became creamy white, indicating the nucleation of Al_2O_3 nanoparticles. After cooling to ambient temperature, the Al_2O_3 nanoparticles were precipitated using a 1:1 mixture of isopropyl alcohol and hexane. Subsequently, Al_2O_3 nanoparticles of different sizes were separated by centrifugation. The large-size oleic-acid-stabilized Al_2O_3 nanoparticles were first separated by centrifugation at 8000 rpm for 60 min, and then smaller Al_2O_3 nanoparticles were separated at 12,000 and 15,000 rpm for 60 min. The Al_2O_3 nanoparticles were then redispersed in hexane or a hexane/isopropyl alcohol mixture (1:1) for further use.

Fe_3O_4 nanoparticle synthesis: Fe_3O_4 nanoparticles with uniform sizes were synthesized from an iron oleate precursor according to the procedure reported by Park et al. with some modifications [33]. Iron oleate (1.5 g, 1.67 mmol) and oleic acid (1.5 mL) were dissolved in 1-hexadecene (50 mL). Using a three-neck boiling flask, the mixture was heated at 284 °C for 60 min with continuous stirring under a nitrogen flow to control oxidation. The black solution was cooled to ambient temperature and 50 mL of ethanol was added to precipitate the Fe_3O_4 nanoparticles. Then, Fe_3O_4 nanoparticles of different sizes were separated by centrifugation and redispersed in hexane for further use [12,34].

Formation of Al_2O_3 nanoparticle monolayer: One side polished SiO_2 layer (500 nm ± 5%) on one side of the silicon wafer was employed as a substrate for the synthesis of VA-CNTs. In order to form a monolayer of Al_2O_3 nanoparticles, the SiO_2 substrate was modified with bifunctional PNDA using the procedure reported by Hanson et al. with some modifications [35]. A 3 mm × 3 mm piece of one-side-polished SiO_2 was cleaned by ultrasonication in acetone for 20 min. The substrate was dried under a nitrogen flow and treated with piranha solution (H_2SO_4/H_2O_2, 3:1) for 30 min to generate hydroxyl groups on the surface [26,35]. After washing with water and drying under a nitrogen flow, the hydroxylated substrate was immersed in a solution of 2 mM PNDA in ethanol/water (1:1, v/v) for 48 h to attach PNDA ligands to the surface. The substrate was washed with ethanol and an ethanol/water mixture (50:50). Subsequently, the substrate was immersed in a 3 mg/mL solution of oleic-acid-coated Al_2O_3 nanoparticles for 3 h to initiate a ligand exchange reaction between the terminal carboxylate group of PNDA and the carboxylate group of the oleic acid on the Al_2O_3 nanoparticle surface, thus forming a nanoparticle monolayer. Any physiosorbed particles were removed from the substrate surface by washing with hexane.

Formation of Fe_3O_4 nanoparticle monolayer: An Fe_3O_4 nanoparticle monolayer was formed on the Al_2O_3 nanoparticle layer using a similar procedure [21]. The substrate with the Al_2O_3 nanoparticle monolayer was annealed at 700 °C for 2 h to remove all of the organic ligands from the surface and to convert any aluminum hydroxide to Al_2O_3. Then, the substrate was treated with piranha solution and immersed in a 2 mM PNDA solution to bind the linker molecules for the Fe_3O_4 monolayer formation. The substrate was immersed in an Fe_3O_4 nanoparticle solution to promote the ligand exchange reaction between the terminal carboxylate group of PNDA and the carboxylate group of oleic acid on the Fe_3O_4 nanoparticle surface. Optimized conditions (3 mg/mL Fe_3O_4 nanoparticle solution for 3 h) gave a compact monolayer of Fe_3O_4 nanoparticles. To remove any physiosorbed particles from the surface, the substrate was washed with hexane. The substrate was then heated at 400 °C for 1 h to remove the oleic acid capping agent and any other organics from the surface.

CNT synthesis: The CNTs were synthesized on the substrate with premade Al_2O_3 and Fe_3O_4 nanoparticle monolayers using CVD. Initially, the substrate was heated at 450 °C for 15 min in the presence of hydrogen in order to thermally reduce the Fe_3O_4 catalyst nanoparticles to metallic iron. Then, VA-CNTs were grown by CVD at 750 °C and atmospheric pressure for 20 min using acetylene as the carbon source with argon as the carrier gas.

3. Results and Discussion

The key aspect of this work was the monolayer assembly of premade Fe_3O_4 catalyst and Al_2O_3 catalyst support nanoparticles for the synthesis of VA-CNTs. It has been reported that the synthesis of CNTs with a specific chirality mostly depends on the interactions between the metal catalyst, catalyst support, and hydrocarbon [36–38]. Figure 1A shows a schematic diagram of the process for assembling the catalyst support and catalyst films for VA-CNT growth. Since Al_2O_3 is a common catalyst support for the large-scale synthesis of VA-CNTs, an Al_2O_3 film was prepared on the SiO_2 surface by using premade Al_2O_3 nanoparticles in solution. The formation of the catalyst support monolayer required the SiO_2 substrate to be activated by introducing hydroxyl groups that could be coupled with linker molecules. Subsequently, the SiO_2 substate was modified with PNDA as an organic linker for binding nanoparticles. Organophosphonic acids have been widely used as ligands for metal oxide (e.g., SiO_2 and Al_2O_3) functionalization in various applications, with the phosphonate group binding to the metal oxide substrate [39–42]. A similar approach using bifunctional PNDA to form a monolayer of iron–platinum bimetallic nanoparticles on an Al_2O_3 substrate indicated that the phosphonate and –COOH groups of PNDA were bound to the Al_2O_3 substrate and the nanoparticle surface, respectively [21]. The sequential formation of Al_2O_3 and Fe_3O_4 nanoparticle monolayers on the functionalized SiO_2 substrate are shown in Figure 1B,C. Oleic acid on the nanoparticle surfaces facilitated the ligand exchange reaction using –COOH as an anchoring group [43].

The oleic-acid-coated Al_2O_3 nanoparticles that were synthesized from the aluminum oleate precursor were spherical, as shown in Figure S1. Larger Al_2O_3 nanoparticles were separated as the decant by centrifugation at 12,000 rpm for 60 min (Figure S1A), whereas comparatively small Al_2O_3 nanoparticles remained in the decanted solution (Figure S1B), which were chosen to form the monolayer on the SiO_2 substrate. FTIR spectroscopy was used to confirm the presence of an oleic acid coating on the surface of the Al_2O_3 nanoparticles (Figure S2). Characteristic bands associated with $–CH_2$, –CH, and C=O groups were observed for oleic acid (Figure S2A), aluminum oleate (Figure S2B), and the oleic-acid-coated Al_2O_3 nanoparticles (Figure S2C). The strong band at 1595 cm^{-1} in the FTIR spectrum of aluminum oleate (Figure S2B) corresponded to a COO-Al species [32], which is typically found at 1600 cm^{-1} [44]. The FTIR spectrum of oleic-acid-capped Al_2O_3 (Figure S2C) showed C–H stretching peaks but no carbonyl peak. The strong peak at ~1465 cm^{-1} and the peak at ~1600 cm^{-1} (Figure S2C), which are characteristic of symmetric and asymmetric carboxylate (COO–) stretching, indicated that oleic acid was bound to the

Al_2O_3 nanoparticle surface [45]. In addition, the appearance of a strong –OH stretching band at 3000–3600 cm^{-1} (Figure S2C) suggested the presence of hydroxylated Al_2O_3 particles. The FTIR spectrum of the Al_2O_3 nanoparticles that were annealed at 700 °C (Figure S2D) indicated that all of the organic functional groups, including oleic acid, were removed from the surface.

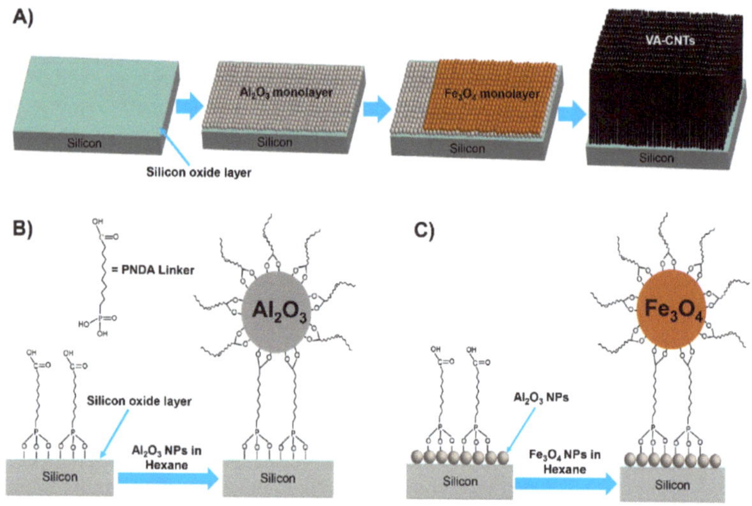

Figure 1. Schematic illustration of catalyst support and catalyst nanoparticle monolayer film assembly for the synthesis of VA-CNTs. (**A**) Step-wise formation of Al_2O_3 and Fe_3O_4 nanoparticle monolayers on a freshly cleaned SiO_2 wafer. Schematic representations of (**B**) Al_2O_3 and (**C**) Fe_3O_4 nanoparticle monolayer assembly.

The sizes and shapes of the Al_2O_3 nanoparticles were characterized using AFM and TEM (Figure 2). Figure 2A shows an AFM image of the Al_2O_3 nanoparticles that were used for monolayer assembly as a catalyst support for VA-CNT growth. As shown by the particle size distribution (Figure 2B), the separated Al_2O_3 nanoparticles had an average size of 6.2 ± 1.4 nm. TEM characterization of the Al_2O_3 nanoparticles was performed before and after annealing (Figure 2C,D). TEM imaging of the oleic-acid-coated Al_2O_3 nanoparticles before annealing was difficult, likely because the organic coating or the hydroxylated nature of the nanoparticles interfered with the electron beam contrast. For the Al_2O_3 nanoparticles before annealing, the TEM sample was prepared by drop casting the nanoparticle solution on the TEM grid and subsequently baking the grid for 2 days at 70 °C. As shown in Figure 2C, the Al_2O_3 nanoparticles before annealing were well distributed. For the Al_2O_3 nanoparticles after annealing at 700 °C for 2 h, the nanoparticles were dispersed in isopropyl alcohol, the large Al_2O_3 nanoparticle aggregates were separated by centrifugation, and a few drops of Al_2O_3 nanoparticle solution were drop cast onto a copper grid for TEM characterization. As shown in Figure 2D, the Al_2O_3 nanoparticles, after annealing, showed some aggregation and distortion. XPS analysis was performed in order to confirm the elemental state of aluminum in the nanoparticles. Figure S3 shows the Al2p and O1s core-level spectra. The Al2p peak at a binding energy of 74.3 eV is consistent with the presence of Al_2O_3 in the nanoparticles [46]. The deconvoluted O1s peaks at 531 and 532.6 eV indicated the presence of two different types of oxygen. The peak at 531 eV corresponded to Al_2O_3, whereas the peak at 532.6 eV could be due to dissolved oxygen (moisture) or hydroxyl groups in the sample [47,48]. Although these results indicate a high content of surface –OH groups on the Al_2O_3 nanoparticles, the high temperature used for CVD will convert them into oxides during the CNT synthesis process.

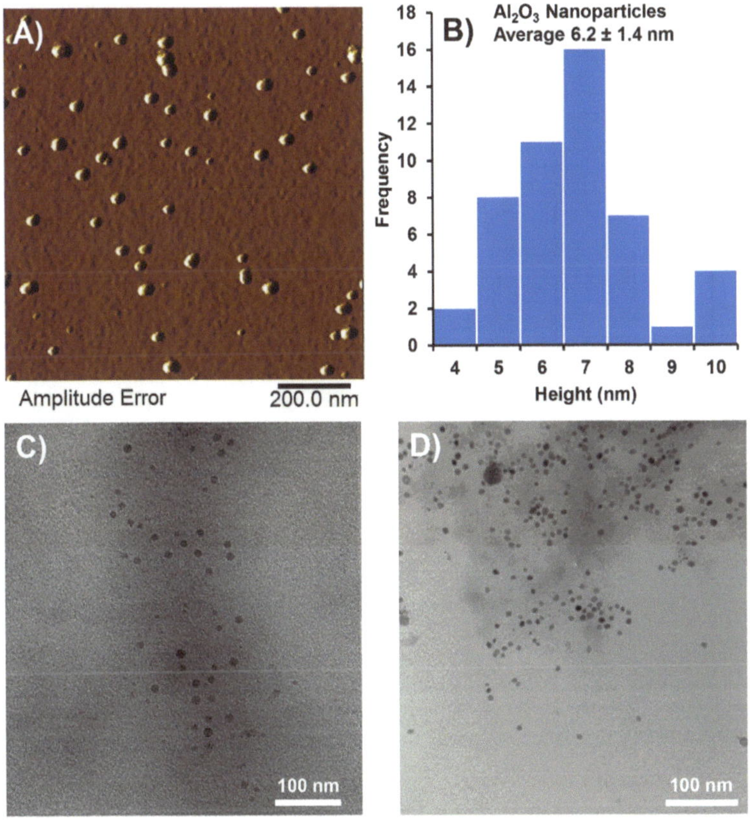

Figure 2. Characterization of synthesized Al_2O_3 nanoparticles. (**A**) AFM image of Al_2O_3 nanoparticles, (**B**) Al_2O_3 nanoparticle size distribution, (**C**) TEM image of Al_2O_3 nanoparticles before annealing, and (**D**) TEM image of Al_2O_3 nanoparticles after annealing at 700 °C.

Fe_3O_4 nanoparticles were synthesized using iron oleate as a precursor and 1-hexadecene as a solvent, similar to the method reported by Park et al. [33]. The obtained oleic-acid-stabilized Fe_3O_4 nanoparticles of different sizes were separated by centrifugation at various speeds. After removing the larger particles, the precipitate that was obtained at 8000 rpm gave a relatively homogenous size distribution, as shown by the AFM image in Figure 3A and the corresponding particle size distribution in Figure 3B. The Fe_3O_4 catalyst nanoparticles with an average size of 9.3 ± 1.2 nm were chosen to form a monolayer on the Al_2O_3 film for VA-CNT growth. The low- and high-magnification TEM images in Figure 3C,D reveal that these Fe_3O_4 nanoparticles had relatively uniform sizes and shapes. FTIR spectral analysis was used to verify the presence of an oleic acid coating on the nanoparticle surface. As shown in Figure S4, the FTIR spectra of oleic acid, iron oleate, and the oleic-acid-coated Fe_3O_4 nanoparticles contained characteristic bands associated with $-CH_2$, –CH, and C=O groups. The characteristics asymmetric and symmetric COO-Al bands at ~1438 and ~1600 cm^{-1} (Figure S4B) confirm the prepared aluminum oleate precursor for the synthesis of Fe_3O_4 catalyst nanoparticles. For the oleic-acid-coated Fe_3O_4 nanoparticles, a strong C–H stretching band appeared at ~2850 cm^{-1}, but the intensity of the carbonyl stretching band at 1710 cm^{-1} was reduced (Figure S4C). The characteristic asymmetric and symmetric stretching bands at ~1438 and ~1600 cm^{-1} (Figure S4C), respectively, confirmed that oleic acid was chemisorbed on the Fe_3O_4 nanoparticles [49]. Although the Fe_3O_4 nanoparticles contained both γ-Fe_2O_3 and Fe_3O_4, the dominant phase was Fe_3O_4 based on the synthetic method that was used [33]. The liquid-phase synthesis of the catalyst nanoparticles provides direct control over particle size because oleic acid acts as a passivation layer to reduce nanoparticle aggregation [50]. In wet chemical processes, an oxygen-free environment and low synthesis temperature reduce the size of the obtained nanoparticles, whereas catalyst formation at high temperatures provides less control over particle size.

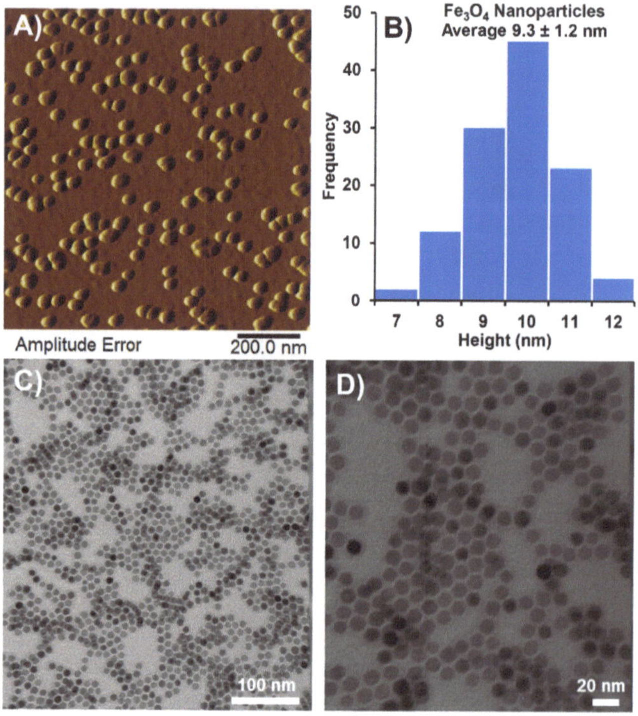

Figure 3. Characterization of synthesized Fe_3O_4 nanoparticles. (**A**) AFM image of Fe_3O_4 nanoparticles, (**B**) Fe_3O_4 nanoparticle size distribution, (**C**) low-magnification TEM image of Fe_3O_4 nanoparticles, and (**D**) high-magnification TEM image of Fe_3O_4 nanoparticles.

The use of catalyst nanoparticle dispersions with uniform sizes for the assembly of monolayer films could provide improved control over CNT characteristics. Importantly, as this approach is less expensive than traditional catalyst preparation processes, scalable VA-CNT growth can be realized. Bifunctional or multifunctional molecules can be employed as linkers on substrates/surfaces depending on the strength of the interaction between the functional group and substrate. Surface hydroxyl (–OH) groups on SiO_2 and Al_2O_3 generated by piranha treatment can enhance the reactivity toward linker molecules [51]. The condensation reaction between the hydroxyl group of a silanol and the phosphonic acid group of PNDA produced a phosphonate group that was bound to the substrate surface and H_2O as a byproduct [35,52]. Phosphonate groups have a higher grafting rate and stronger binding ability with metal oxide substrates than carboxylate groups [53]. The XPS analysis was conducted to confirm the functionalization of silicon oxide substrate with PNDA ligand. Figures S5 and S6 show the XPS survey spectra of the bare SiO_2 substrate and the SiO_2 substrate with bound phosphonate groups (PNDA), which were compared to reference XPS spectra [54]. The core-level Si2p, P2p, O1s, and C1s spectra are shown in Figure S7. The P2p peak at a binding energy of 132.9 eV confirmed the successful binding of phosphonate to the SiO_2 substrate.

An Al_2O_3 nanoparticle monolayer was formed on the PNDA-functionalized SiO_2 substrate via a ligand exchange reaction in an Al_2O_3 nanoparticle solution. The carboxylate (COOH) group of the PNDA linker on the SiO_2 substrate was exchanged with the COOH group of oleic acid on the Al_2O_3 nanoparticle surface [21]. After the successful Al_2O_3 monolayer formation, the substrate was dried at 400 °C for 1 h and characterized by AFM in tapping mode (Figure 4). The surface morphology of the bare SiO_2 substrate (Figure 4A) was very smooth with no particles, whereas the Al_2O_3 nanoparticle assembly (Figure 4B) exhibited well-distributed nanoparticles on the surface. The cross-sectional profile of the bare SiO_2 substrate (Figure 4C) showed that the substrate was almost planar with a height variation of ~0.5 nm. In contrast, the cross-sectional profile of the Al_2O_3 nanoparticle assembly (Figure 4D) showed a nanoparticle layer with a height variation of ~5 nm, which is similar to the average size of the Al_2O_3 nanoparticles that were used for monolayer formation. Thus, the liquid-phase preparation of an Al_2O_3 catalyst support via monolayer assembly is suitable for the development of an easy and inexpensive method for the large-scale synthesis of VA-CNTs. The Al_2O_3 film was annealed at 700 °C for 2 h to remove all of the organic molecules (capping agents and surface ligands) from the surface and convert any aluminum hydroxide to the oxide for application as a catalyst support.

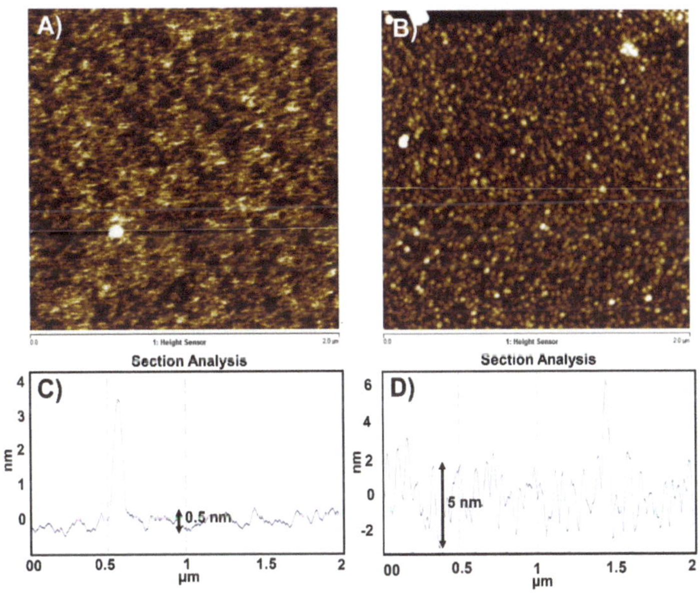

Figure 4. Surface morphology of the SiO_2 substrate before and after assembly of the Al_2O_3 nanoparticle monolayer. AFM images of (**A**) the bare SiO_2 substrate and (**B**) the Al_2O_3 nanoparticle monolayer. Height cross-sectional profiles (surface roughness) of (**C**) the bare SiO_2 substrate and (**D**) the Al_2O_3 nanoparticle monolayer.

Al_2O_3 interacts strongly with iron, which can reduce the aggregation of catalyst particles on the substrate. Thus, an Fe_3O_4 monolayer was formed over the Al_2O_3 layer, which was modified with PNDA using a procedure similar to that for the SiO_2 substrate. The successful binding of PNDA on the Al_2O_3 monolayer was confirmed by XPS. The XPS survey spectrum of the PNDA film on the Al_2O_3 layer (Figure S8) showed all the core-level peaks of Al_2O_3 as well as a phosphorous peak. The Al2p, P2p, O1s, and C1s core-level spectra of PNDA on the Al_2O_3 layer are shown in Figure S9. The three peaks in the C1s spectrum confirmed the presence of carboxylate functionalities on the Al_2O_3 substrate [55]. The C1s peak at 284.5 eV corresponded to regular carbon (CH, CH_2), whereas the peaks at 285.5 and 289 eV corresponded to carbon in the carbonyl (CO) and carboxylate (COOH) groups, respectively. A monolayer of Fe_3O_4 catalyst nanoparticles was formed on the Al_2O_3 layer via the ligand exchange reaction between the COOH groups of the PNDA linker on the functionalized Al_2O_3 surface and the COOH group of oleic acid on the Fe_3O_4 nanoparticles in solution. The low- (15 μm × 15 μm) and high-magnification AFM images (2 μm × 2 μm) of the Fe_3O_4 monolayer (Figure 5A,B) indicated that a compact nanoparticle layer was formed containing nanoparticles of a uniform size. The corresponding cross-sectional profiles of the AFM images (Figure 5C,D) revealed small openings in the monolayer assembly with heights of approximately 10 nm, which is consistent with the average size of the Fe_3O_4 nanoparticles that were used for monolayer formation (9.3 ± 1.2 nm). The surface functionalization of metal oxide substrates with monolayers or multilayers of Fe_3O_4 nanoparticles has been studied for various applications [18]. However, Fe_3O_4 nanoparticle monolayers are expected to be suitable as catalysts for VA-CNT growth because the high density and uniformity of catalyst particles can provide control over the number of nanotubes formed in a specific area, as well as their diameters. The density of the Fe_3O_4 nanoparticles on the Al_2O_3 layer was 1.4×10^{11} nanoparticles/cm^2, which was determined by counting the nanoparticles using a high-magnification (100 nm × 100 nm) AFM image. The number of total Fe_3O_4 nanoparticles were manually counted in a 100 × 100 nm^2 area and their density was extrapolated for a square centimeter area. In addition, the use of a monolayer of uniform catalyst nanoparticles can reduce particle aggregation and catalyst poisoning.

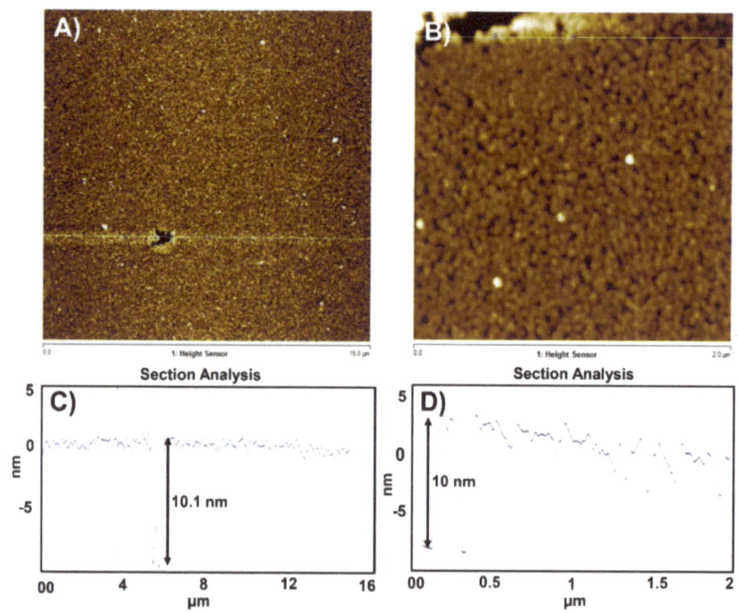

Figure 5. Surface morphology of the Fe_3O_4 nanoparticle monolayer formed over the Al_2O_3 monolayer film. (**A**) Low-magnification (15 μm × 15 μm) and (**B**) high-magnification (2 μm × 2 μm) AFM images of the Fe_3O_4 nanoparticle monolayer. Height cross-sectional profiles corresponding to (**C**) image A and (**D**) image B.

Using the premade monolayer of catalyst nanoparticles, CNTs were synthesized by CVD, which is a suitable technique for large-scale synthesis. First, all of the organics were removed from the catalyst surface by heating at 400 °C for 2 h in the open air. Subsequently, the Fe_3O_4 monolayer was reduced to metallic iron by thermal reduction at 450 °C for 15 min using atomic hydrogen (H) generated from a hydrogen gas (H_2) flow [56]. VA-CNTs were successfully synthesized using acetylene as a carbon precursor at 750 °C for

15 min. A comparative literature study about synthesis methods of vertically aligned CNTs from premade nanoparticles is shown in supplementary information (Table S1). The table compares different methodologies used for assembly of catalyst nanoparticles for CNTs growth. Figure 6A,B show low- and high-magnification SEM images of the grown VA-CNTs where uniform-length of CNTs are observed. Though the density of catalyst nanoparticles in Figure 5B is quite very high, the lower density of CNTs growth can be the result of lower nucleation rates of catalyst particles. Authors think that investigating the suitable synthesis parameters in CVD reaction can increase the nucleation rates which can increase the density of VACNTs. TEM imaging (Figure 6C) indicated that the CNTs that were grown from the catalyst nanoparticle monolayer had relatively similar diameters. The quality of the grown CNTs was assessed using Raman spectral analysis (Figure 6D). The characteristic G peak of CNTs, which corresponds to sp^2 carbon, was more prominent than the D peak, which corresponds to sp^3 carbon, indicating the structural defects of CNTs, which are typical of multilayer CNTs. A small RBM peak was also observed in the Raman spectrum of the VA-CNTs, which indicated the presence of a few single-walled CNTs among the mostly multi-walled CNTs.

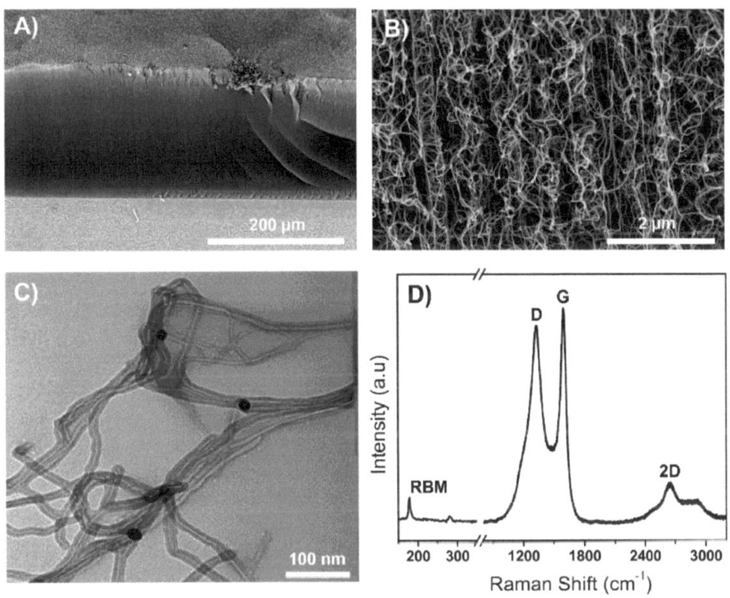

Figure 6. Characterization of VA-CNTs grown from the premade nanoparticle monolayer assembly. (**A**) Low-magnification SEM image, (**B**) high-magnification SEM image, (**C**) high-magnification TEM image, and (**D**) Raman spectrum of VA-CNTs.

4. Conclusions

Spherical Al_2O_3 and Fe_3O_4 nanoparticles were synthesized and separated into different sizes by varying the centrifugation speed, which provided sufficiently homogenous nanoparticles with which to form monolayer films. Using wet chemical processes, continuous monolayer films of Al_2O_3 nanoparticles as a catalyst support and Fe_3O_4 nanoparticles as a catalyst were sequentially assembled on an SiO_2 substrate using PNDA as a linker molecule. This monolayer assembly was successfully applied to catalyze the growth of uniform, high-density VA-CNTs. This approach for preparing a catalyst support monolayer using premade Al_2O_3 nanoparticles provides an inexpensive and easy alternative to the conventional methods for fabricating thin film catalyst supports such as sputtering, e-beam deposition, and other instrumentation that requires ultrahigh vacuum systems.

Furthermore, the use of a catalyst particle monolayer can improve the homogeneity of the obtained CNTs and increase CNT density.

Supplementary Materials: The following are available online at https://www.mdpi.com/article/10.3390/c7040079/s1, Table S1: Current methods of premade catalyst nanoparticles assembly used for CNTs growth; Figure S1: SEM images of spherical shaped aluminum oxide/hydroxide NPs. (A) Large size of aluminum oxide/hydroxide nanoparticles separated by centrifuge technique with 12,000 rpm, (B) Small size of nanoparticles remained in decanted solution. A few drops of nanoparticle solution was drop cast onto a silicon wafer and annealed at 400 °C for an hour before characterization; Figure S2: FTIR spectra of (A) pure oleic acid, (B) aluminum oleate, (C) aluminum oxide nanoparticles without annealing, and (D) aluminum oxide nanoparticles annealed at 700 °C; Figure S3: XPS spectra of aluminum oxide nanoparticles after annealing at 700 °C for 2 h. (A) Survey spectra that represent all the core level peaks, (B) Al 2p, and (C) O 1s; Figure S4: FTIR spectra of (A) pure oleic acid, (B) iron oleate, and (C) oleic acid coated iron oxide nanoparticles; Figure S5: XPS survey spectra of blank silicon oxide substrate that show all the core level peaks; Figure S6: XPS survey spectra of 11-phosphonoundecanoic acid (PNDA) film on silicon substrate showing all core level peaks; Figure S7: XPS spectra of PNDA attached silicon oxide substrate. (A) core level Si 2p peak, (B) core level P 2p peak, (C) core level O 1s peak, and (D) core level C 1s peak; Figure S8: XPS survey spectra of 11-phosphonoundecanoic acid film on alumina monolayer showing all core level peaks; Figure S9: XPS spectra of PNDA attached aluminum oxide substrate. (A) core level Al 2p peak, (B) core level P 2p peak, (C) core level O 1s peak, and (D) core level C 1s peak.

Author Contributions: Conceptualization, methodology, validation, N.T.A. and A.H.; formal analysis, A.H.; investigation, A.H., A.U. and B.S.G.; resources, supervision, project administration, N.T.A.; writing—original draft preparation, A.H.; writing—review and editing, A.H. and N.T.A.; funding acquisition, N.T.A. All authors have read and agreed to the published version of the manuscript.

Funding: This research was funded by NSF, grant number 2016484 under NSF PFI-RP.

Acknowledgments: The authors are thankful for professorship start-up funds from the Department of Chemistry at the University of Cincinnati. Partial salary support for A.U. was provided by an internal seed grant from the University of Kentucky (UK) Energy Research Priority Area program. XPS data were collected at the UK electron microscopy center (EMC), which belongs to the National Science Foundation NNCI Kentucky Multiscale Manufacturing and Nano Integration Node, supported by ECCS-1542174.

Conflicts of Interest: The authors declare no conflict of interest. The funders had no role in the design of the study; in the collection, analyses, or interpretation of data; in the writing of the manuscript, or in the decision to publish the results.

References

1. Alvarez, N.T.; Miller, P.; Haase, M.; Kienzle, N.; Zhang, L.; Schulz, M.J.; Shanov, V. Carbon Nanotube Assembly at Near-Industrial Natural-Fiber Spinning Rates. *Carbon* **2015**, *86*, 350–357. [CrossRef]
2. Alvarez, N.T.; Miller, P.; Haase, M.R.; Lobo, R.; Malik, R.; Shanov, V. Tailoring Physical Properties of Carbon Nanotube Threads during Assembly. *Carbon* **2019**, *144*, 55–62. [CrossRef]
3. Amama, P.B.; Pint, C.L.; Kim, S.M.; Eyink, K.G.; Stach, E.A.; Hauge, R.H.; Maruyama, B. Evolution, Activity, and Lifetime of Alumina-Supported Fe Catalyst during Super Growth of Single-Walled Carbon Nanotube Carpets: Influence of the Type of Alumina. *Mater. Res. Soc. Symp. Proc.* **2010**, *1258*, 217–222. [CrossRef]
4. Modekwe, H.U.; Mamo, M.A.; Moothi, K.; Daramola, M.O. Effect of Different Catalyst Supports on the Quality, Yield and Morphology of Carbon Nanotubes Produced from Waste Polypropylene Plastics. *Catalysts* **2021**, *11*, 692. [CrossRef]
5. Wang, B.; Yang, Y.; Li, L.J.; Chen, Y. Effect of Different Catalyst Supports on the (n,m) Selective Growth of Single-Walled Carbon Nanotube from Co-Mo Catalyst. *J. Mater. Sci.* **2009**, *44*, 3285–3295. [CrossRef]
6. Christen, H.M.; Puretzky, A.A.; Cui, H.; Belay, K.; Fleming, P.H.; Geohegan, D.B.; Lowndes, D.H. Rapid Growth of Long, Vertically Aligned Carbon Nanotubes through Efficient Catalyst Optimization Using Metal Film Gradients. *Nano Lett.* **2004**, *4*, 1939–1942. [CrossRef]
7. Guzmán De Villoria, R.; Figueredo, S.L.; Hart, A.J.; Steiner, S.A.; Slocum, A.H.; Wardle, B.L. High-Yield Growth of Vertically Aligned Carbon Nanotubes on a Continuously Moving Substrate. *Nanotechnology* **2009**, *20*, 1–8. [CrossRef]
8. Pint, C.L.; Kim, S.M.; Stach, E.A.; Hauge, R.H. Rapid and Scalable Reduction of Dense Surface-Supported Metal-Oxide Catalyst with Hydrazine Vapor. *ACS Nano* **2009**, *3*, 1897–1905. [CrossRef]

9. Cheung, C.L.; Kurtz, A.; Park, H.; Lieber, C.M. Diameter-Controlled Synthesis of Carbon Nanotubes. *J. Phys. Chem. B* **2002**, *106*, 2429–2433. [CrossRef]
10. Kukovitsky, E.F.; L'vov, S.G.; Sainov, N.A.; Shustov, V.A.; Chernozatonskii, L.A. Correlation between Metal Catalyst Particle Size and Carbon Nanotube Growth. *Chem. Phys. Lett.* **2002**, *355*, 497–503. [CrossRef]
11. Sinnott, S.B.; Andrews, R.; Qian, D.; Rao, A.M.; Mao, Z.; Dickey, E.C.; Derbyshire, F. Model of Carbon Nanotube Growth through Chemical Vapor Deposition. *Chem. Phys. Lett.* **1999**, *315*, 25–30. [CrossRef]
12. Alvarez, N.T.; Li, F.; Pint, C.L.; Mayo, J.T.; Fisher, E.Z.; Tour, J.M.; Colvin, V.L.; Hauge, R.H. Uniform Large Diameter Carbon Nanotubes in Vertical Arrays from Premade Near-Monodisperse Nanoparticles. *Chem. Mater.* **2011**, *23*, 3466–3475. [CrossRef]
13. Schäffel, F.; Kramberger, C.; Rümmeli, M.H.; Grimm, D.; Mohn, E.; Gemming, T.; Pichler, T.; Rellinghaus, B.; Büchner, B.; Schultz, L. Nanoengineered Catalyst Particles as a Key for Tailor-Made Carbon Nanotubes. *Chem. Mater.* **2007**, *19*, 5006–5009. [CrossRef]
14. Alvarez, N.T.; Orbaek, A.; Barron, A.R.; Tour, J.M.; Hauge, R.H. Dendrimer-Assisted Self-Assembled Monolayer of Iron Nanoparticles for Vertical Array Carbon Nanotube Growth. *ACS Appl. Mater. Interfaces* **2010**, *2*, 15–18. [CrossRef]
15. Alvarez, N.T.; Hamilton, C.E.; Pint, C.L.; Orbaek, A.; Yao, J.; Frosinini, A.L.; Barron, A.R.; Tour, J.M.; Hauge, R.H. Wet Catalyst-Support Films for Production of Vertically Aligned Carbon Nanotubes. *ACS Appl. Mater. Interfaces* **2010**, *2*, 1851–1856. [CrossRef]
16. Yang, Z.; Zhang, S.; Li, L.; Chen, W. Research Progress on Large-Area Perovskite Thin Fi Lms and Solar Modules. *J. Mater.* **2017**, *3*, 231–244. [CrossRef]
17. Chen, J.; Xu, X.; Zhang, L.; Huang, S. Controlling the Diameter of Single-Walled Carbon Nanotubes by Improving the Dispersion of the Uniform Catalyst Nanoparticles on Substrate. *Nano-Micro Lett.* **2015**, *7*, 353–359. [CrossRef]
18. Pauly, M.; Pichon, B.P.; Albouy, P.A.; Fleutot, S.; Leuvrey, C.; Trassin, M.; Gallani, J.L.; Begin-Colin, S. Monolayer and Multilayer Assemblies of Spherically and Cubic-Shaped Iron Oxide Nanoparticles. *J. Mater. Chem.* **2011**, *21*, 16018–16027. [CrossRef]
19. Amama, P.B.; Pint, C.L.; McJilton, L.; Kim, S.M.; Stach, E.A.; Murray, P.T.; Hauge, R.H.; Maruyama, B. Role of Water in Super Growth of Single-Walled Carbon Nanotube Carpets. *Nano Lett.* **2009**, *9*, 44–49. [CrossRef]
20. Hasnan, N.S.N.; Timmiati, S.N.; Lim, K.L.; Yaakob, Z.; Kamaruddin, N.H.N.; Teh, L.P. Recent Developments in Methane Decomposition over Heterogeneous Catalysts: An Overview. *Mater. Renew. Sustain. Energy* **2020**, *9*, 1–18. [CrossRef]
21. Yildirim, O.; Gang, T.; Kinge, S.; Reinhoudt, D.N.; Blank, D.H.A.; van der Wiel, W.G.; Rijnders, G.; Huskens, J. Monolayer-Directed Assembly and Magnetic Properties of FePt Nanoparticles on Patterned Aluminum Oxide. *Int. J. Mol. Sci.* **2010**, *11*, 1162–1179. [CrossRef] [PubMed]
22. Andryszewski, T.; Iwan, M.; Hołdyński, M.; Fiałkowski, M. Synthesis of a Free-Standing Monolayer of Covalently Bonded Gold Nanoparticles. *Chem. Mater.* **2016**, *28*, 5304–5313. [CrossRef]
23. Feichtenschlager, B.; Lomoschitz, C.J.; Kickelbick, G. Tuning the Self-Assembled Monolayer Formation on Nanoparticle Surfaces with Different Curvatures: Investigations on Spherical Silica Particles and Plane-Crystal-Shaped Zirconia Particles. *J. Colloid Interface Sci.* **2011**, *360*, 15–25. [CrossRef]
24. Tizazu, G. A Simple Method for Patterning Nanoparticles on Planar Surfaces. *J. Nanotechnol.* **2019**, *2019*, 1–7. [CrossRef]
25. Karade, V.C.; Sharma, A.; Dhavale, R.P.; Dhavale, R.P.; Shingte, S.R.; Patil, P.S.; Kim, J.H.; Zahn, D.R.T.; Chougale, A.D.; Salvan, G.; et al. APTES Monolayer Coverage on Self-Assembled Magnetic Nanospheres for Controlled Release of Anticancer Drug Nintedanib. *Sci. Rep.* **2021**, *11*, 1–12. [CrossRef]
26. Acres, R.G.; Ellis, A.V.; Alvino, J.; Lenahan, C.E.; Khodakov, D.A.; Metha, G.F.; Andersson, G.G. Molecular Structure of 3-Aminopropyltriethoxysilane Layers Formed on Silanol-Terminated Silicon Surfaces. *J. Phys. Chem. C* **2012**, *116*, 6289–6297. [CrossRef]
27. Zhou, M.; Hedlund, J. Assembly of Oriented Iron Oxide and Zeolite Crystals via Biopolymer Films. *J. Mater. Chem.* **2012**, *22*, 24877–24881. [CrossRef]
28. Wah, F.; Mun, L.; Tai, F.; Bee, S.; Hamid, A. Iron Oxide Nanoparticles Decorated Oleic Acid for High Colloidal Stability. *Adv. Polym. Technol.* **2017**, *37*, 1712–1721. [CrossRef]
29. Tamaki, H.; Abe, S.; Yamagata, S.; Yoshida, Y.; Sato, Y. Self-Assembled Monolayer Formation on a Dental Orthodontic Stainless Steel Wire Surface to Suppress Metal Ion Elution. *Coatings* **2020**, *10*, 367. [CrossRef]
30. Yamada, K.; Kaneko, A.; Kato, H.; Homma, Y. Vertically-Aligned Carbon Nanotube Growth Using Closely Packed Iron Oxide Nanoparticles. *Mater. Express* **2012**, *2*, 257–260. [CrossRef]
31. Signore, M.A.; Rizzo, A.; Rossi, R.; Piscopiello, E.; Di Luccio, T.; Capodieci, L.; Dikonimos, T.; Giorgi, R. Role of Iron Catalyst Particles Density in the Growth of Forest-like Carbon Nanotubes. *Diam. Relat. Mater.* **2008**, *17*, 1936–1942. [CrossRef]
32. Lee, D.H.; Condrate, R.A. FTIR Spectral Characterization of Thin Film Coatings of Oleic Acid on Glasses: I. Coatings on Glasses from Ethyl Alcohol. *J. Mater. Sci.* **1999**, *34*, 139–146. [CrossRef]
33. Park, J.; An, K.; Hwang, Y.; Park, J.E.G.; Noh, H.J.; Kim, J.Y.; Park, J.H.; Hwang, N.M.; Hyeon, T. Ultra-Large-Scale Syntheses of Monodisperse Nanocrystals. *Nat. Mater.* **2004**, *3*, 891–895. [CrossRef] [PubMed]
34. Yu, W.W.; Falkner, J.C.; Yavuz, C.T.; Colvin, V.L. Synthesis of Monodisperse Iron Oxide Nanocrystals by Thermal Decomposition of Iron Carboxylate Salts. *Chem. Commun.* **2004**, 2306–2307. [CrossRef] [PubMed]
35. Hanson, E.L.; Schwartz, J.; Nickel, B.; Koch, N.; Danisman, M.F. Bonding Self-Assembled, Compact Organophosphonate Monolayers to the Native Oxide Surface of Silicon. *J. Am. Chem. Soc.* **2003**, *125*, 16074–16080. [CrossRef] [PubMed]
36. Panic, S.; Bajac, B.; Rakić, S.; Kukovecz; Kónya, Z.; Srdić, V.; Boskovic, G. Molybdenum Anchoring Effect in Fe–Mo/MgO Catalyst for Multiwalled Carbon Nanotube Synthesis. *React. Kinet. Mech. Catal.* **2017**, *122*, 775–791. [CrossRef]

37. He, M.; Zhang, L.; Jiang, H.; Yang, H.; Fossard, F.; Cui, H.; Sun, Z.; Wagner, J.B.; Kauppinen, E.I.; Loiseau, A. FeTiO Based Catalyst for Large-Chiral-Angle Single-Walled Carbon Nanotube Growth. *Carbon* **2016**, *107*, 865–871. [CrossRef]
38. Harutyunyan, A.R.; Chen, G.; Paronyan, T.M.; Pigos, E.M.; Kuznetsov, O.A.; Hewaparakrama, K.; Kim, S.M.; Zakharov, D.; Stach, E.A.; Sumanasekera, G.U. Preferential Growth of Single-Walled Carbon Nanotubes with Metallic Conductivity. *Science* **2009**, *326*, 116–120. [CrossRef]
39. Dubey, M.; Weidner, T.; Gamble, L.J.; Castner, D.G. Structure and Order of Phosphonic Acid-Based Self-Assembled Monolayers on Si(100). *Langmuir* **2010**, *26*, 14747–14754. [CrossRef]
40. Mutin, P.H.; Lafond, V.; Popa, A.F.; Granier, M.; Markey, L.; Dereux, A. Selective Surface Modification of SiO_2-TiO_2 Supports with Phosphonic Acids. *Chem. Mater.* **2004**, *16*, 5670–5675. [CrossRef]
41. Thissen, P.; Valtiner, M.; Grundmeier, G. Stability of Phosphonic Acid Self-Assembled Monolayers on Amorphous and Single-Crystalline Aluminum Oxide Surfaces in Aqueous Solution. *Langmuir* **2010**, *26*, 156–164. [CrossRef]
42. Zhao, R.; Rupper, P.; Gaan, S. Recent Development in Phosphonic Acid-Based Organic Coatings on Aluminum. *Coatings* **2017**, *7*, 133. [CrossRef]
43. Smolensky, E.D.; Park, H.Y.E.; Berquó, T.S.; Pierre, V.C. Surface Functionalization of Magnetic Iron Oxide Nanoparticles for MRI Applications—Effect of Anchoring Group and Ligand Exchange Protocol. *Contrast Media Mol. Imaging* **2011**, *6*, 189–199. [CrossRef] [PubMed]
44. Cross, S.N.W.; Rochester, C.H. Infrared Study of the Adsorption of Ethyl Isocyanate on Silica Immersed in Carbon Tetrachloride. *J. Chem. Soc. Faraday Trans. 1 Phys. Chem. Condens. Phases* **1981**, *77*, 1945–1952. [CrossRef]
45. Lewis, W.K.; Rosenberger, A.T.; Gord, J.R.; Crouse, C.A.; Harruff, B.A.; Fernando, K.A.S.; Smith, M.J.; Phelps, D.K.; Spowart, J.E.; Guliants, E.A.; et al. Multispectroscopic (FTIR, XPS, and TOFMS-TPD) Investigation of the Core-Shell Bonding in Sonochemically Prepared Aluminum Nanoparticles Capped with Oleic Acid. *J. Phys. Chem. C* **2010**, *114*, 6377–6380. [CrossRef]
46. Van den Brand, J.; Snijders, P.C.; Sloof, W.G.; Terryn, H.; De Wit, J.H.W. Acid-Base Characterization of Aluminum Oxide Surfaces with XPS. *J. Phys. Chem. B* **2004**, *108*, 6017–6024. [CrossRef]
47. Alexander, M.R.; Thompson, G.E.; Beamson, G. Characterization of the Oxide/Hydroxide Surface of Aluminum Using X-Ray Photoelectron Spectroscopy: A Procedure for Curve Fitting the O 1s Core Level. *Surf. Interface Anal.* **2000**, *29*, 468–477. [CrossRef]
48. Zähr, J.; Oswald, S.; Türpe, M.; Ullrich, H.J.; Füssel, U. Characterisation of Oxide and Hydroxide Layers on Technical Aluminum Materials Using XPS. *Vacuum* **2012**, *86*, 1216–1219. [CrossRef]
49. Do, B.P.H.; Nguyen, B.D.; Nguyen, H.D.; Nguyen, P.T. Synthesis of Magnetic Composite Nanoparticles Enveloped in Copolymers Specified for Scale Inhibition Application. *Adv. Nat. Sci. Nanosci. Nanotechnol.* **2013**, *4*, 045016. [CrossRef]
50. Gupta, R.; Pancholi, K.; Sa, R.D.E.; Murray, D.; Huo, D.; Droubi, G.; White, M.; Njuguna, J. Effect of Oleic Acid Coating of Iron Oxide Nanoparticles on Properties of Magnetic Polyamide-6 Nanocomposite. *JOM* **2019**, *71*, 3119–3128. [CrossRef]
51. Okhrimenko, D.V.; Nielsen, C.F.; Lakshtanov, L.Z.; Dalby, K.N.; Johansson, D.B.; Solvang, M.; Deubener, J.; Stipp, S.L.S. Surface Reactivity and Dissolution Properties of Alumina-Silica Glasses and Fibers. *ACS Appl. Mater. Interfaces* **2020**, *12*, 36740–36754. [CrossRef] [PubMed]
52. Hoque, E.; DeRose, J.A.; Bhushan, B.; Mathieu, H.J. Self-Assembled Monolayers on Aluminum and Copper Oxide Surfaces: Surface and Interface Characteristics, Nanotribological Properties, and Chemical Stability. *Appl. Scanning Probe Methods IX* **2007**, *111*, 235–281. [CrossRef]
53. Queffélec, C.; Petit, M.; Janvier, P.; Knight, D.A.; Bujoli, B. Surface Modification Using Phosphonic Acids and Esters. *Chem. Rev.* **2012**, *112*, 3777–3807. [CrossRef] [PubMed]
54. Gouzman, I.; Dubey, M.; Carolus, M.D.; Schwartz, J.; Bernasek, S.L. Monolayer vs. Multilayer Self-Assembled Alkylphosphonate Films: X-Ray Photoelectron Spectroscopy Studies. *Surf. Sci.* **2006**, *600*, 773–781. [CrossRef]
55. Hoque, E.; Derose, J.A.; Kulik, G.; Hoffmann, P.; Mathieu, H.J.; Bhushan, B. Alkylphosphonate Modified Aluminum Oxide Surfaces. *J. Phys. Chem. B* **2006**, *110*, 10855–10861. [CrossRef] [PubMed]
56. Bayer, B.C.; Fouquet, M.; Blume, R.; Wirth, C.T.; Weatherup, R.S.; Ogata, K.; Knop-Gericke, A.; Schlögl, R.; Hofmann, S.; Robertson, J. Co-Catalytic Solid-State Reduction Applied to Carbon Nanotube Growth. *J. Phys. Chem. C* **2012**, *116*, 1107–1113. [CrossRef]

C Journal of Carbon Research

MDPI

Article

Influence of Carbon Nanotube Attributes on Carbon Nanotube/Cu Composite Electrical Performances

Rajyashree Sundaram *, Atsuko Sekiguchi *, Guohai Chen, Don Futaba, Takeo Yamada *, Ken Kokubo and Kenji Hata

CNT-Application Research Center, National Institute of Advanced Industrial Science and Technology, Central 5, 1-1-1 Higashi, Tsukuba 3058565, Japan; guohai-chen@aist.go.jp (G.C.); d-futaba@aist.go.jp (D.F.); kokubo.ken@aist.go.jp (K.K.); kenji-hata@aist.go.jp (K.H.)

* Correspondence: Rajyashree.sundaram@cantab.net (R.S.); atsuko-sekiguchi@aist.go.jp (A.S.); takeo-yamada@aist.go.jp (T.Y.)

Abstract: Carbon nanotube (CNT)/copper composites offer promise as lightweight temperature-stable electrical conductors for future electrical and electronic devices substituting copper. However, clarifying how constituent nanotube structures influence CNT/Cu electrical performances has remained a major research challenge. Here, we investigate the correlation between the CNT/Cu electrical performances and nanotube structure by preparing and characterizing composites containing nanotubes of different structural attributes. We prepared three types of composites—single-wall (SW)-CNT/Cu wires, SW-CNT/Cu pillars, and multi-wall (MW)-CNT/Cu wires. The composites were fabricated from the corresponding CNT templates by two-step Cu electrodeposition, which retains template nanotube attributes through the fabrication process. The nanotube characteristics (diameter, G/D, alignment, etc.) in each template as well as the internal structure and electrical performances of the corresponding composites were characterized. SW-CNT/Cu wires and pillars outperformed MW-CNT/Cu wires, showing $\approx 3\times$ higher room-temperature four-probe conductivities (as high as 30–40% Cu-conductivity). SW-CNT/Cu also showed up to $4\times$ lower temperature coefficients of resistances i.e., more temperature-stable conductivities than MW-CNT/Cu. Our results suggest that few-walled small-diameter nanotubes can contribute to superior temperature-stable CNT/Cu conductivities. Better CNT crystallinity (high G/D), fewer nanotube ends/junctions, and nanotube alignment may be additionally beneficial. We believe that these results contribute to strategies for improving CNT/Cu performances to enable the real-world application of these materials as Cu substitutes.

Keywords: carbon nanotube (CNT)/Cu composites; nanotube attributes; electrical conductivity; temperature-coefficient of resistance

Citation: Sundaram, R.; Sekiguchi, A.; Chen, G.; Futaba, D.; Yamada, T.; Kokubo, K.; Hata, K. Influence of Carbon Nanotube Attributes on Carbon Nanotube/Cu Composite Electrical Performances. *C* **2021**, *7*, 78. https://doi.org/10.3390/c7040078

Academic Editor: Gil Goncalves

Received: 2 October 2021
Accepted: 5 November 2021
Published: 15 November 2021

1. Introduction

Carbon nanotube/copper (CNT/Cu) composites are proposed as promising lightweight copper substitutes for next-generation electrical wiring and interconnect applications on account of their rivaling conductivities with superior heat stability, current capacities, and lower densities [1–3]. CNT/Cu up to 2/3rd as light as copper have been fabricated in practically applicable forms, such as planar [4,5] and vertical micro-interconnects [6,7], macroscopic wires [8–15], etc. These composites have been observed to show room-temperature conductivities similar to copper in the range $1.0–4.7 \times 10^5$ S/cm (20–80% of Cu) [4–7,12,13,15–17]. The heat stability of conductivity in CNT/Cu is observed as subdued conductivity reduction with temperature rise vs. copper and is quantified as a reduction in temperature coefficients of resistance (TCR) [5,6,8,11–13,15,16,18]. TCR values as low as 10–50% that of Cu are reported due to nanotube participation (with their inherently low TCR) in electron transport. Such heat-stable conductivities are necessary for electrical wirings/cables/interconnects that reliably operate in high-temperature settings in a variety of applications, including automobiles, aircrafts, and high-power electronics.

In this work, we have attempted to gain an understanding of the effect of constituent nanotube attributes on CNT/Cu electrical performances. The key aspects that are bound to affect CNT/Cu performances are (a) attributes of the constituents—i.e., CNTs and Cu, and (b) the interactions between Cu and nanotubes [2]. Considerable research has been focused on the latter, specifically toward improving Cu-CNT adhesion by including an interfacial additive that can interact with both copper and the CNTs [2]. These are usually oxygen [18] or a metal typically with carbide-forming tendencies (i.e., capability to interact with carbon-based entities), such as Ni, Cr, Al, Ti, etc. [12–14,19]. Studies providing experimental evidence on the influence of CNT/Cu constituent attributes, especially nanotube characteristics (length, diameter, wall number, crystallinity, alignment, etc.) on composite performances are few. Studies in current literature that do study nanotube structure vs. composite properties focus mainly on mechanical performances to which CNTs contribute as reinforcing agents. For instance, small-diameter few-walled CNTs have shown better wear properties, hardness, strength, etc. than large-diameter multi-walled CNTs [20–22]. Very few studies focus on how nanotube attributes affect electrical performances. The experimental studies of Shuai [17] and Zhao et al. [23] indicate that nanotube alignment along the bias direction may be beneficial for electrical conductivities, which is supported by theoretical predictions [24,25]. However, systematic studies on the impact of other parameters besides nanotube orientation, such as crystallinity, diameter, wall number, etc. are yet to be reported. We believe that charting nanotube attributes vs. electrical properties is essential to both tailor CNT/Cu with improved performances and to gain a mechanistic understanding of electron/phonon transport or the stress transfer behaviors in these materials.

A key issue in investigating the effect of nanotube structural parameters on CNT/Cu performances is fabricating composites consisting of nanotubes of pre-determined/regulated structures with uniform CNT–Cu mixing. One major challenge in making such composites lies in preserving the nanotube structure through processing that is involved in fabrication. Dispersion-based CNT/Cu processing strategies, especially metallurgical methods involving ball milling, mixing, sonication, etc. alter nanotube characteristics—mainly, length and crystallinity—by damaging the graphitic lattice and/or slicing CNTs [2,20,21]. These methodologies also preclude CNT orientation control in composites [2,20,21].

Composite fabrication by Cu electrodeposition of CNT templates preserves the nanotube structure during processing. Among electrodeposition methods, the two-step Cu deposition strategy has yielded composites with uniform nanotube–Cu mixing even with high CNT vol % ($\approx$45 vol %) [4–11]. Applying only typical aqueous electrolyte-based Cu electrodeposition on CNT templates leads merely to core–sheath composites consisting of templates coated by Cu with no CNT–Cu mixing in the bulk [8,11]. This is because aqueous electrolytes cannot infiltrate hydrophobic nanotube templates. In contrast, the two-step strategy [4–11] involves deposition from an organic electrolyte capable of wetting and infiltrating CNT templates that first seeds Cu within CNT templates. This is followed by the second seed-growth step (by aqueous electrodeposition) to obtain the final composites exhibiting uniform CNT–Cu mixing throughout.

By employing two-step electrodeposition on CNT templates with well-regulated nanotube attributes, the influence of the nanotube structure on composites with uniform and comparable Cu–CNT mixing can be explored. In this work, we procured three templates with distinct nanotube structural parameters—multi-wall (MW)-CNT wires (CNTWs), single-wall (SW)-CNTWs, and SW-CNT pillars (SW-CNTPs). The nanotube parameters in each template (diameter, G/D, alignment, etc.) were thoroughly characterized. Then, we applied two-step Cu electrodeposition on these templates to fabricate the corresponding composites—i.e., multi-wall MW-CNT/Cu wires, SW-CNT/Cu wires, and SW-CNT/Cu pillars. We characterized the internal structures and electrical performances of these three composite samples and have attempted to correlate CNT/Cu electrical performances with nanotube attributes.

2. Materials and Methods

CNT/Cu samples in this study were fabricated by the two-step Cu electrodeposition of CNT templates with varied nanotube characteristics (as summarized in the schematic; see Figure 1). Details of CNT templates as well as composite fabrication and sample characterization methods are provided below.

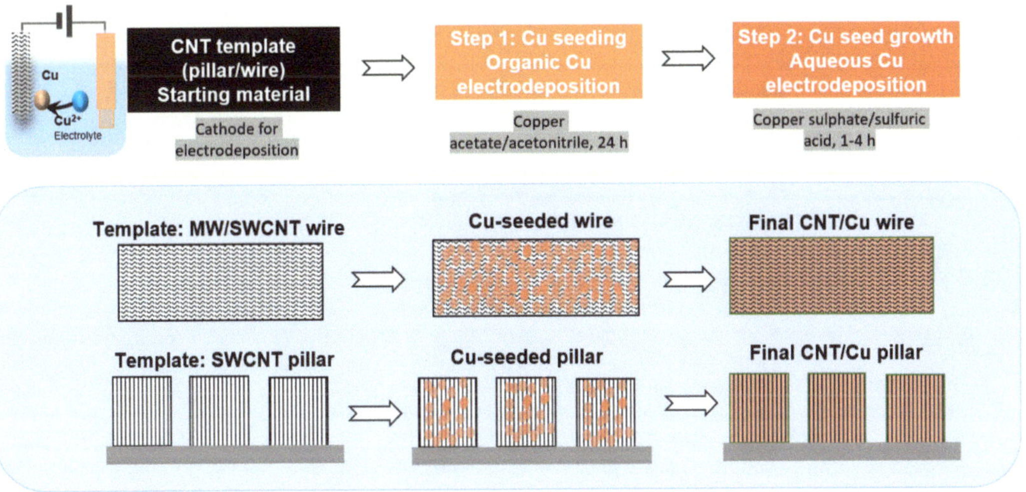

Figure 1. Schematic representing CNT/Cu fabrication by two-step Cu electrodeposition of CNT templates—wires and pillars.

2.1. CNT Templates and Their Characterization

Three CNT templates with distinct nanotube characteristics were used in this study: (i) MW-CNTWs, (ii) SW-CNTWs, and (iii) SW-CNTPs. The MW-CNTWs were industrial wires (Muratec, Murata Machinery Ltd., Kyoto, Japan, 10 m spool) with density ≈ 0.5 g/cm^3 and 40–50 µm in diameter). The wires were continuously twist-spun from vertical MW-CNT arrays (array height and CNT length ≈ 500 µm) grown on substrates by chemical vapor deposition (CVD). The SW-CNTWs (width $\approx$250 µm, density ≈ 0.06 g/cm^3) were tapes directly obtained from a floating catalyst CVD reactor (CNT length 200–500 µm) [26,27]. Typical CNTW linear densities were ≈ 7–9×10^{-4} mg/mm. The SW-CNTPs were synthesized in-house by "super-growth" (water-assisted CVD) on Si substrates, as described previously [7,28,29]. Arrays of pillars (diameter ≈ 100 µm, spaced ≈ 200 µm apart, pillar length and CNT length ≈ 300 µm, 95–98% porosity) were grown upon lithographically patterned CVD catalyst ($\approx$3 nm thick Fe/40 nm thick Al_2O_3 buffer) deposited on patterned W lines (Figure S1, Supplementary Materials). The W lines served as electrical contacts between the pillars and W pads sputtered on Si substrates bearing the pillar array. This enabled contacting of all pillars in the array with the electrochemical workstation to permit Cu electrodeposition. To retain the pillar structure during electrodeposition, before composite fabrication, carbon was deposited from the gas phase amidst nanotubes and nanotube bundles in the SW-CNTPs (as reported in detail previously [7]).

Nanotube bundling and alignment, individual nanotube structure (diameter, wall number, etc.), and CNT graphitic structure in terms of Raman spectroscopy G-band to D-band intensity ratio (G/D) [30,31] were the chief template structural attributes we characterized. CNT bundling and alignment were observed by scanning electron microscopy (SEM, S4800, Hitachi High-Tech, Tokyo, Japan). Each individual nanotube structure was probed by transmission electron microscopy (TEM, Acceleration voltage = 200 kV, EM-002B, Topcon, Tokyo, Japan). TEM samples were prepared by dispersing templates (≈0.1 mg) in absolute ethanol (20 mL) by sonication and placing the dispersions on holey carbon TEM grids, which was followed by allowing the solvent to evaporate overnight. Diameters of 50 nanotubes (for each sample) obtained with ImageJ software were used to calculate average diameters. To obtain G/D ratios, Raman spectra were acquired from the templates (excitation laser wavelength = 532 nm, lateral resolution = 1 μm, Nicolet Almega XR Raman Spectrometer, Thermo-Electron, Tokyo, Japan).

2.2. CNT/Cu Fabrication and Characterization

Composites were prepared from the three different CNT templates using two-step Cu electrodeposition (Figure 1). Routine aqueous electrolytes do not wet and infiltrate hydrophobic CNT templates and typically lead to only surface Cu deposition and core–sheath structured composites [8,11]. To enable internal Cu deposition and fabricate composites with CNTs and Cu uniformly mixed in the bulk [4–11], the two-step strategy was adopted. In the first step, Cu was seeded within templates from an organic electrolyte (anhydrous $Cu(CH_3COO)_2$ (Sigma Aldrich, Tokyo, Japan) in CH_3CN (Wako pure chemicals, Osaka, Japan), 0.25 g/100 mL). To make the final composite samples, in the second step, the Cu seeds were grown by aqueous electrodeposition from commercial acidified aqueous $CuSO_4$ electrolyte (40 g/L $CuSO_4.5H_2O$ and ≈ 10 g/L H_2SO_4, without accelerators/suppressors, ATMI Inc., Danbury, Connecticut, USA). All electrodeposition experiments were carried out on VMP3 electrochemical workstation (Princeton applied research, Oakridge, TN, USA) at constant current mode. The electrodeposition set-up consisted of the CNT template cathode and Cu anode separated by Teflon spacers. In case of the CNTWs, samples (≈2 cm long) were mounted on a stainless-steel (SS) mesh sheet and mechanically clamped using SS clips. This arrangement was connected to the electrochemical station as the working electrode. In case of SW-CNTPs, the Si substrate bearing the pillars was directly used as the cathode (configuration illustrated in schematic shown in Figure S1, Supplementary Materials), which was connected to the electrochemical workstation through the sputtered W pad using crocodile clips.

The seeding condition i.e., current density (and time) for MW-CNTWs was 2.5 mA/cm^2 (24 h), and that for SW-CNTWs and SW-CNTPs was 0.25 mA/cm^2 (24 h). Subsequent aqueous deposition was carried out at 10 mA/cm^2 (1 h) for MW-CNTWs and at 2.5 mA/cm^2 (4 h) for SW-CNTWs and SW-CNTPs. CNT/Cu fabrication parameters were optimized and fixed based on our previous studies [9,10]. Excess electrolyte was removed by rinsing with CH_3CN and deionized water after seeding and aqueous deposition, respectively. After each deposition step, samples were vacuum-dried (100 °C, 1 h) and subjected to reductive annealing (250 °C, 3 h, 150 cm^3/min in H_2 flow) to remove any Cu oxides.

We characterized the composite internal structures to ensure CNT-Cu mixing by cross-sectional SEM (CS-SEM, S4800, Hitachi High-Tech, Tokyo, Japan). In case of wires, cross-sections were obtained by slicing with high-precision scissors. For SW-CNT/Cu pillars, individual pillars were dislodged from the array and sliced using laser cutting (YO_4 laser, Telecentric MD-T1000, Osaka, Japan). CNT/Cu densities were calculated from sample mass (resolution = 0.1 μg, UMX2, Mettler Toledo, Tokyo, Japan), length, and SEM or optical microscopy diameters (average 5 points).

Electrical resistances were acquired using a 4-probe set-up (PS-100 probe station, Lakeshore Cryotronics, Westerville, OH, USA and B1500A analyzer, Keysight, Tokyo, Japan). Contacting was done directly with samples using 25-micron-diameter Cu-Be probes on ≈ 1 cm long wire samples or on individual pillars dislodged from the substrate. Inter-probe distances to calculate conductivities was measured under an optical microscope. Cross-section areas were measured from CS-SEM images using ImageJ software. At least three values were measured for each wire sample, and averages were used for conductivity calculations. High-temperature resistances (up to 373 K) were obtained (at 10^{-3} Pa) with sample temperature regulated by a heating stage (controlled by Lakeshore Model 236 controller, Lakeshore Cryotronics, Westerville, OH, USA) to calculate TCR values. TCR was calculated as the slope of normalized resistance (measured resistance normalized by room temperature resistance, R/R_{298}) vs. temperature difference (difference between measurement temperature and room temperature, $T\text{-}T_{298}$) plot. Average and standard deviation values for electrical performance measurements from (at least) 3 samples of each type are presented. Electrical performances of neat Cu wires (100 µm diameter, 99.9% purity, Nilaco, Tokyo, Japan) and CNT templates were also measured for reference.

3. Results

CNT template SEM, TEM, and Raman spectroscopy characterization reveals key differences in the structure of nanotubes used for CNT/Cu fabrication (Figure 2). Electrical conductivities and their temperature stabilities (TCR) of composites consisting of different nanotube structures show distinct variations even with similar CNT-Cu mixing and composite internal structures (Figures 3 and 4, and Table 1). Differences in the nanotube structural parameters of the three templates as well as the structure and performances of the respective composites are discussed in Sections 3.1 and 3.2.

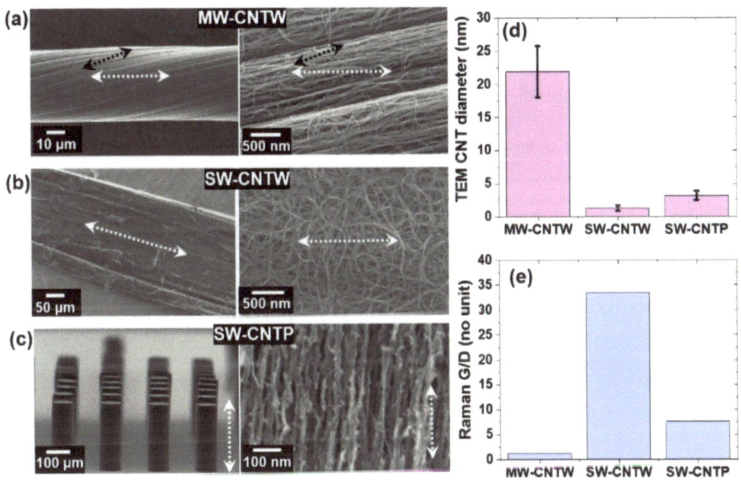

Figure 2. Characterization of CNT templates used for CNT/Cu fabrication. Low- and high-magnification SEM images of (**a**) MW-CNTW, (**b**) SW-CNTW, and (**c**) SW-CNTP samples. White arrows indicate wire/pillar axis. Black arrows in (**a**) indicate twist direction. (**d**) TEM CNT diameter and (**e**) Raman G/D ratios of MW-CNTW, SW-CNTW, and SW-CNTP samples.

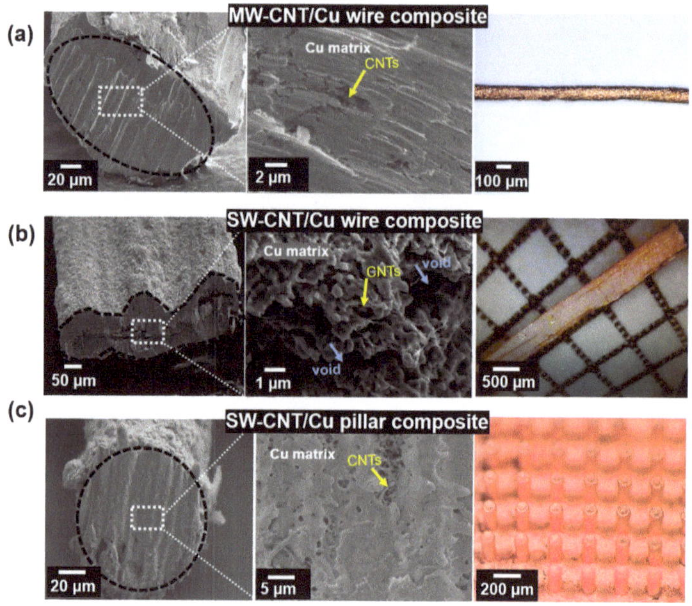

Figure 3. Low- and high-magnification CS-SEM images and optical micrographs of (**a**) MW-CNT/Cu wires, (**b**) SW-CNT/Cu wires, and (**c**) SW-CNT/Cu pillars.

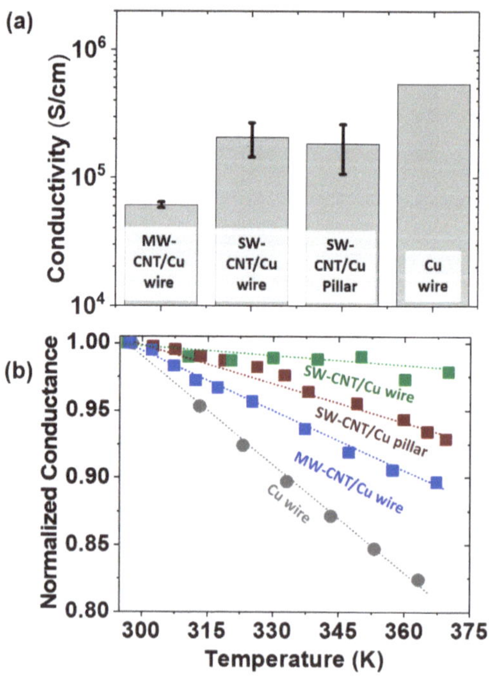

Figure 4. Electrical performances of CNT/Cu composites: (**a**) room-temperature electrical conductivities and (**b**) normalized conductance (to room-temperature conductance) vs. temperature plot.

Table 1. Composite properties (density, room temperature electrical conductivity, and TCR) and parent template nanotube characteristics (CNT diameter and G/D).

Sample	Composite Properties			CNT Attributes		
	Density (g/cm^3)	Conductivity (S/cm)	TCR (/K)	Diameter (nm)	G/D	Remarks
MW-CNT/Cu wire	5.1 ± 0.3	$6.1 \times 10^4 \pm 3.2 \times 10^3$	$1.8 \times 10^{-3} \pm 2.0 \times 10^{-4}$	21.9 ± 3.9	1.3	≈500 μm long MWCNTs with multiple ends and nanotube–nanotube junctions
SW-CNT/Cu wire	2.2 ± 0.2	$2.1 \times 10^5 \pm 6.2 \times 10^4$	$4.4 \times 10^{-4} \pm 1.9 \times 10^{-4}$	1.3 ± 0.4	33.5	≈200–500 μm long SWCNTs with multiple ends and nanotube–nanotube junctions
SW-CNT/Cu pillar	5.0 ± 0.2	$1.8 \times 10^5 \pm 7.6 \times 10^4$	$1.1 \times 10^{-3} \pm 2.1 \times 10^{-4}$	3.2 ± 0.7	7.6	≈300 μm long SWCNTs (aligned) running end-to-end
Cu wire	8.9	$5.4 \times 10^5 \pm 2.8 \times 10^3$	$3.3 \times 10^{-3} \pm 4.3 \times 10^{-4}$	-	-	-

3.1. Nanotube Attributes in the CNT Templates

Nanotubes in MW-CNTWs (Figure 2a) are in general aligned along the twist direction, while in SW-CNTWs, nanotubes are entangled without any alignment (Figure 2b). Both types of wires can be expected to have multiple nanotube–nanotube junctions and ends due to the CNT length obtained at the synthesis stage (<500 μm), which is much shorter than macroscopic wire length scales (i.e., several meters). In SW-CNTPs (Figure 2c, nanotubes are in general aligned along the pillar axis and run end-to-end spanning the entire pillar length. All templates consist of spaces between nanotubes/nanotube bundles into which Cu^{2+} ions infiltrate and deposit copper during composite fabrication. As discussed in a previous work, appropriate template porosity and low densities are vital for optimal internal Cu electrodeposition, especially during the seeding step [10].

TEM analysis (Figure 2d and Figure S2a, Supplementary Materials) confirms the diameter and nanotube type constituting the templates. Nanotube diameters in MW-CNTWs, SW-CNTWs, and SW-CNTPs are 21.9 ± 3.9 nm, 1.3 ± 0.4 nm, and 3.2 ± 0.7 nm, respectively. Extensive SEM and TEM CNT template analysis revealed very little metal CVD catalyst residues. MW-CNTW templates show >96 wt % CNTs and catalyst residues < 1 wt % in air-TGA carried out from room temperature to 1000 °C, as reported in our earlier studies [8,10,11]. As mentioned in the experimental section, SW-CNTPs were obtained by water-assisted CVD, which is known to result in CNT materials of high purity with low catalyst residues [7,28,29]. Gas-phase carbon deposition was carried out on SW-CNTPs after CVD synthesis to include amorphous carbon in between nanotubes and nanotube bundles to retain pillar structural integrity during electrodeposition. Except for post-deposited amorphous carbon in the pillars, the purity of the three templates can be assumed to be comparable.

Raman spectra G peaks are observed at ≈1581/cm for MW-CNTWs and at 1588–1590/cm for SW-CNTWs and SW-CNTPs, which are typical for peak positions of MW- and SW-CNTs, respectively (Figure S2b, Supplementary Materials) [30,31]. In addition, RBM signatures distinctive to single-wall tubes [30,31] are observed from SW-CNTWs and SW-CNTPs, which are absent in the case of MW-CNTWs (Figure S3, Supplementary Materials). The Raman spectroscopy G/D ratios of the templates are also markedly different. The G/D—ratio of the graphitic G-band intensity (1580–1590/cm) and defective D-band (≈1340/cm) intensity—is a figure of merit of graphitic crystallinity of nanocarbon materials [30,31]. Imperfections in the graphitic lattice as well as amorphous impurities lead to lower G/D ratios. MW-CNTWs and SW-CNTPs show lower G/D (<7) than SW-CNTWs (G/D $\approx$ 33.5), indicating the better graphitic lattice quality of nanotubes in SW-CNTWs. To note, the gas-phase deposition of (amorphous) carbon in SW-CNTPs carried out after nanotube pillar CVD synthesis to retain the pillar structure during electrodeposition lowered G/D ratios from $\approx$7.6 to $\approx$1.1 (Figure S4, Supplementary Materials).

To summarize the differences in template nanotube characteristics: the twist-spun MW-CNTWs consist of low-G/D nanotubes with multiple ends and nanotube–nanotube junctions. SW-CNTWs are composed of high-G/D crystalline single-wall nanotubes ($\approx$1.3 nm in diameter), also comprising of multiple ends and junctions. In SW-CNTPs, low-G/D (lower crystallinity) aligned single-wall tubes $\approx$ 3 nm in diameter run end-to-end with additional (amorphous) carbon. These nanotube characteristics can be assumed to be retained in the analogous CNT/Cu samples fabricated by two-step Cu electrodeposition.

3.2. Composite Structure and Performances with Various Nanotube Attributes

CNT/Cu samples prepared from the three templates show nanotubes mixed uniformly in the Cu matrix with a regular Cu surface coating (SEM and optical microscopy images in Figure 3a–c and Figure S5, Supplementary Materials). CS-SEM after organic electrodeposition shows numerous Cu seeds ($\approx$few hundreds of nm in diameter; see Figure S6, Supplementary Materials), which are vital for the subsequent seed growth and fabrication of final composites with uniform CNT–Cu mixing [9,10]. Around 20–40% of Cu mass in the final composites is estimated to be deposited during the Cu-seeding stage (organic deposition), with the rest deposited during aqueous deposition. In terms of composition, all three final composites show comparable Cu wt % of $\approx$98–99 wt % (Table S1) and are lighter than copper, as seen from the CNT/Cu densities (Table 1). In comparison to pure copper (density $\approx$ 8.9 g/cm^3) [32], MW-CNT/Cu wire and SW-CNT/Cu pillar densities are $\approx$ 5.0 g/cm^3, corresponding to $\approx$40–45 vol % nanotubes uniformly mixed in a copper matrix. SW-CNT/Cu wires show even lower densities ($\approx$2 g/cm^3), which can be attributed to voids observed in cross-sections (Figure 3b), in addition to the weight reduction arising from nanotubes uniformly mixed in the Cu matrix. These voids were inherent to these samples, which was possibly due to ultralow densities of the starting CNTW templates (<0.1 g/cm^3). While the presence of voids in SW-CNT/Cu wires makes the nanotube vol % difficult to ascertain, based on sample masses and densities, CNT vol % in the composite wires is estimated to be $\approx$30–40 vol %.

The composite electrical performances in terms of conductivities and their temperature stabilities (TCR) are competitive to Cu, and variations in performances are observed among samples with different CNT attributes (Figure 4 and Table 1). The conductivities we measure for copper are comparable to standard values 5.85 $\times$ 10^5 S/cm [32], confirming the validity of our electrical characterization protocol. The room temperature electrical conductivities (Figure 4a) of MW-CNT/Cu wires are $\approx$10% and an order of magnitude lower than that of Cu. SW-CNT/Cu wire and SW-CNT/Cu pillar conductivities are $\approx$ 3$\times$ higher than that of MW-CNT/Cu wires with similar Cu content (98–99 wt %, Table S1) and of the same order of magnitude as that of pure copper. The values range to $\approx$ 30–40% of Cu conductivity. In terms of specific conductivities (density-normalized values), for SW-CNT/Cu wires, SW-CNT/Cu pillars, and MW-CNT/Cu wires, the values are 1.4$\times$, 0.5$\times$, and 0.2$\times$ that of copper, respectively. The specific conductivity value of SW-CNT/Cu wires exceeding that of copper is in line with observations previously reported for SW-CNT/Cu composites [2,4,5]. Compared to parent CNT templates (Table S2, Supplementary Materials), CNT/Cu samples show two to three orders of magnitude higher conductivities, as may be expected from copper addition to nanotubes. The conductivities increase with Cu inclusion in each step of two-stage electrodeposition, as discussed in our previous studies [8,10]. For example, in MW-CNTWs, the conductivity increases from $\approx$261 S/cm for the neat wires to $\sim$350 S/cm for Cu seeded wires and to $\sim$6.1 $\times$ 10^4 S/cm for the final MW-CNT/Cu wires.

All composites show a decrease in conductance (normalized to room-temperature conductance) with temperature rise i.e., metal-like behavior (Figure 4b). However, composite conductances reduce less rapidly with temperature rise, and lower TCR values are observed vs. pure copper (Table 1, Figure S7). The lower the TCR, higher the temperature stability of the conductivity. The superior temperature stability of the conductivity of CNT/Cu is a widely reported merit of the composites arising from the participation of CNTs in the electron transport [4,6,8,12,13,15,16]. CNTs (and their ensembles) are known to show weak a temperature dependence of conductance (Table S2, Supplementary Materials [33–35]) and low TCR values. Rivalling composite conductivities, especially with high-temperature stability, are crucial for applications requiring conductors in electrical and electronic systems that operate above room temperature—ranging from motor windings, devices in aircrafts/cars, or even as interconnects in high-power electronics. As an aside, the macroscale wire and microscale pillar composites we have fabricated here are in forms that are relevant for real-world application as conducting wiring materials and vertical interconnect structures, respectively.

Comparing TCR values among the composites, SW-CNT/Cu wires show the lowest values (≈13% Cu-TCR), which is followed by the SW-CNT/Cu pillars (≈33% Cu-TCR) and MW-CNT/Cu wires (≈50% Cu-TCR). Compared to MW-CNT/Cu wires, the TCR of the SW-CNT/Cu composites are up to $\approx 4\times$ lower. Considering both electrical conductivities and TCR, the superiority of the electrical performances of the samples in this study can be graded in decreasing order as SW-CNT/Cu wire > SW-CNT/Cu pillar > MW-CNT/Cu wire.

It is surprising that SW-CNT/Cu wires show copper-level electrical conductivities despite voids seen in the cross-section (Figure 3b). It could be construed that the copper outercoating rather than composite bulk containing both CNTs and Cu contributes dominantly to the conductivity. However, if this had been the case, the TCR values should have been closer to that of pure copper, as reported previously [8]. The combination of high conductivity and low TCR observed in SW-CNT/Cu wires despite voids indicate that the single-wall nanotubes in these composites play a significant role in composite electron transport.

4. Discussion

Our work lends experimental evidence to the impact of nanotube attributes on CNT/Cu composite electrical performances. Interpreting the results of our current study, we suggest that few-walled small-diameter nanotubes benefit composite electrical performances. In addition, high crystallinity (high G/D)/fewer nanotube ends and junctions/nanotube alignment may also contribute positively to composite performances.

Among the samples considered here, MW-CNT/Cu wires show lower electrical conductivites and temperature stabilities than SW-CNT/Cu composites (wires and pillars). The poorer performance of MW-CNT/Cu vs. SW-CNT/Cu could boil down to intrinsically lower conductivities of MW-CNTs and their aggregates [36] as well as geometric factors. For comparable nanotube vol %, small-diameter SW-CNTs afford a higher surface to volume ratio. Comparing SW-CNT/Cu pillars and MW-CNT/Cu wires (with ≈45 vol% nanotubes and G/D < 7), it is reasonable to assume the shared CNT-Cu interface area to be higher for SW-CNT/Cu than MW-CNT/Cu. In addition to the dissimilar shared interface areas stemming to large and small nanotube diameters, there are key differences between MW-CNT/Cu wires and SW-CNT/Cu pillars. While MW-CNT/Cu wires have multiple nanotube ends and junctions and CNTs are aligned along the wire twist direction (rather than along applied bias), in SW-CNT/Cu pillars, nanotubes run end-to-end aligned in general along applied bias. These factors may additionally contribute to the better performances of SW-CNT/Cu pillars vs. MW-CNT/Cu wires. Theoretical studies predict that CNTs aligned along the applied bias as well as fewer junctions and ends (that contribute to electron scattering) can lead to higher CNT/Cu electrical conductivities [24,25].

Among the SW-CNT/Cu samples, wires with high-G/D and smaller 1.3 nm diameter tubes show slightly better properties (especially, in terms of TCR) than pillars with low-G/D and 3 nm diameter tubes. This is the case despite SW-CNT/Cu wires containing voids, multiple tube ends and junctions, and a lack of CNT alignment vs. the SW-CNTs in pillars running end-to-end aligned along the applied bias. Based on the CNT vol% of the SW-CNT/Cu samples (pillars $\approx$ 45 vol%, wires estimated as $\approx$30–40 vol%), the higher shared interfacial areas arising from the smaller ($\approx$1.3 nm) diameter tubes in the SW-CNT/Cu wires (vs. $\approx$3-nm tubes in pillars) may be a favorable factor. In addition to diameter, nanotube crystallinity variations seen as different G/D ratios (SW-CNTW G/D $\approx$ 33.5, SW-CNTPs G/D $<$ 7) may also play a role. Higher CNT crystallinity is associated with higher intrinsic nanotube and CNT assembly conductivities [37], which may benefit composite performances. In addition, we hypothesize that better CNT lattice quality can also enhance CNT–Cu interactions and electron transport across the interface. In addition to the innate low G/D of nanotubes in the SW-CNTPs (G/D $\approx$ 7), amorphous carbon deposited after CVD synthesis (to retain pillar structure during electrodeposition; see Figure S4, Supplementary Materials) can perhaps act as additional scattering centers undermining electrical performances. A related effect on CNT/Cu thermal conductivity reduction with amorphous carbon inclusion was observed by Cho et al. [38], which was attributed to carbonaceous impurities at CNT–Cu interfaces acting as thermal barriers.

Our pointers from this study on favorable nanotube attributes i.e., small-diameter, few-wall, high-crystallinity nanotubes with fewer ends and junctions to achieve improved composite electrical properties, are unique in the literature. Studies dedicated to the influence of CNT characteristics on composite performances, especially electrical properties, are few. Most previous studies have attempted to corelate the composite mechanical performances (strength, hardness, etc.) with nanotube wall number, diameter, orientation, etc. [20–23]. Furthermore, a majority of these reports involve fabrication methodologies that damage CNTs/alter nanotube attributes during composite making, involving powder metallurgy techniques such as ball milling or CNT dispersion methods such as sonication [20–23]. In contrast, our composite fabrication methodology, i.e., the two-step electrodeposition of CNT templates, retains template nanotube attributes in the composites. Therefore, the differences in the composite electrical performances we observe in this study are attributable largely to the differences in the nanotube characteristics in the composites (and parent templates).

A drawback of our work is that multiple nanotube attributes (alignment, diameter, G/D, etc.) differ simultaneously in templates and consequently in the composite samples. The influence and impact of an individual CNT structural parameter—e.g., diameter—on composite performances, while all other attributes (G/D, wall number, length, alignment, presence/absence of nanotube junctions, etc.) are held constant, is unclear. In fact, fabricating composites with well-controlled CNT structures as well as synthesizing such regulated nanotubes (and templates) for composite fabrication is a major challenge for the CNT research community. In addition, despite the superior performances observed in SW-CNT/Cu in this work, our samples do not represent the true potential of SW-CNT-based composites because of the presence of amorphous carbon in the pillars and voids in the wires. Nevertheless, we believe that the results presented in this paper offer preliminary cues on the impact constituent nanotubes can exert on composite electrical properties for future research strategies toward performance tailoring.

5. Conclusions

In this work, we have attempted to demonstrate the impact of nanotube structural parameters on CNT/Cu electrical performances. We prepared three types of composites—MW-CNT/Cu wires, SW-CNT/Cu wires, and SW-CNT/Cu pillars—by two-step Cu electrodeposition of MW-CNTW, SW-CNTW, and SW-CNTP templates. The nanotube structural parameters in the respective templates (diameter, G/D, alignment, etc.) were characterized by SEM, TEM, and Raman spectroscopy. Our results indicate that few-walled small-diameter nanotubes are beneficial for composite electrical performances to achieve better temperature-stable conductivities. In addition, our data suggest that high CNT crystallinity (high G/D), fewer nanotube ends and junctions, and nanotube alignment along the bias direction may also positively influence the composite performances. SW-CNT/Cu wires and pillars outperformed MW-CNT/Cu wires with ≈3× higher room-temperature conductivities (values reaching 30–40% of Cu conductivity) even with comparable CNT vol % (≈45 vol %). SW-CNT/Cu also showed better temperature stability of conductivity, with up to 4× lower TCR than MW-CNT/Cu wires. Our results contribute unambiguous experimental evidence for the effect of CNT structural parameters on composite electrical performances enabled by a unique composite-making methodology (two-step Cu electrodeposition of CNT templates) that retains nanotube attributes during the fabrication. We believe that our results can be used as hints for electrical performance tailoring of CNT/Cu to develop these materials as lightweight temperature-stable electrical conductors with performances rivalling copper for real-world applications.

Supplementary Materials: The following are available online at https://www.mdpi.com/article/10.3390/c7040078/s1, Figure S1: (Left) Top-view of SW-CNTP array on W lines that act as electrical contacts for Cu electrodeposition. (Right) Schematic of the SW-CNTP array on Si substrate (top-view) used as cathode for Cu electrodeposition, Figure S2: Example (a) TEM images and (b) Raman spectra of MW-CNTW, SW-CNTW, and SW-CNTP samples used for CNT/Cu fabrication, Figure S3: Radial breathing mode region in the Raman spectra of nanotube templates used for CNT/Cu fabrication, Figure S4: Raman spectra and SEM images of SW-CNTP before and after carbon coating, Figure S5: SEM images of SW-CNT/Cu pillars (array and individual pillar), Figure S6: Cross-section SEM images of the Cu-seeded MW-CNTW, SW-CNTW, and SW-CNTP samples after organic Cu electrodeposition and H_2 reduction, Figure S7: Measured resistance normalized by room-temperature resistance, R/R_{298} vs. difference between measurement temperature and room temperature, $T\text{-}T_{298}$, Table S1: Cu content (wt %) in and cross-section areas of CNT/Cu composites, Table S2: Conductivity and TCR of CNT templates used for CNT/Cu fabrication.

Author Contributions: Conceptualization, methodology, validation and investigation, R.S. and A.S.; Conceptualization, D.F. and K.H.; Investigation, G.C.; Writing—original draft preparation, R.S.; Writing—review and editing, R.S., A.S., G.C., D.F., T.Y., K.K. and K.H.; Project administration, T.Y., D.F., A.S., K.K. and K.H.; Funding acquisition, R.S., A.S. and K.H. All authors have read and agreed to the published version of the manuscript.

Funding: This research was partially funded by the Japan Society for the Promotion of Science (JSPS) by JSPS KAKENHI Grant (Number JP 18F18050) through the JSPS international research fellowship.

Data Availability Statement: Raw data are available from authors on request.

Acknowledgments: We are grateful to M. Nishimura, H. Oosako, R. Shiina, S. Nemoto and M. Irie for their technical support. We thank Muratec, Murata Machinery Ltd. for providing the multi-wall CNT wires used in this work.

Conflicts of Interest: The authors declare no conflict of interest.

References

1. Daneshvar, F.; Chen, H.; Noh, K.; Sue, H.J. Critical challenges and advances in the carbon nanotube–metal interface for next-generation electronics. *Nanoscale Adv.* **2021**, *3*, 942–962. [CrossRef]
2. Sundaram, R.M.; Sekiguchi, A.; Sekiya, M.; Yamada, T.; Hata, K. Copper/carbon nanotube composites: Research trends and outlook. *R. Soc. Open Sci.* **2018**, *5*, 180814. [CrossRef]
3. Tehrani, M. Advanced Electrical Conductors: An Overview and Prospects of Metal Nanocomposite and Nanocarbon Based Conductors. *Phys. Status Solidi (A)* **2021**, *218*, 2000704. [CrossRef]
4. Subramaniam, C.; Yamada, T.; Kobashi, K.; Sekiguchi, A.; Futaba, D.N.; Yumura, M.; Hata, K. One hundred fold increase in current carrying capacity in a carbon nanotube–copper composite. *Nat. Commun.* **2013**, *4*, 2202. [CrossRef] [PubMed]
5. Subramaniam, C.; Sekiguchi, A.; Yamada, T.; Futaba, D.N.; Hata, K. Nano-scale, planar and multi-tiered current pathways from a carbon nanotube–copper composite with high conductivity, ampacity and stability. *Nanoscale* **2016**, *8*, 3888–3894. [CrossRef] [PubMed]
6. Sun, S.; Mu, W.; Edwards, M.; Mencarelli, D.; Pierantoni, L.; Fu, Y.; Jeppson, K.; Liu, J. Vertically aligned CNT-Cu nano-composite material for stacked through-silicon-via interconnects. *Nanotechnology* **2016**, *27*, 335705. [CrossRef]

7. Chen, G.; Sundaram, R.; Sekiguchi, A.; Hata, K.; Futaba, D.N. Through-Silicon-Via Interposers with Cu-Level Electrical Conductivity and Si-Level Thermal Expansion Based on Carbon Nanotube-Cu Composites for Microelectronic Packaging Applications. *ACS Appl. Nano Mater.* **2020**, *4*, 869–876. [CrossRef]
8. Sundaram, R.; Yamada, T.; Hata, K.; Sekiguchi, A. Electrical performance of lightweight CNT-Cu composite wires impacted by surface and internal Cu spatial distribution. *Sci. Rep.* **2017**, *7*, 1–11. [CrossRef]
9. Sundaram, R.; Yamada, T.; Hata, K.; Sekiguchi, A. The influence of Cu electrodeposition parameters on fabricating structurally uniform CNT-Cu composite wires. *Mater. Today Commun.* **2017**, *13*, 119–125. [CrossRef]
10. Sundaram, R.; Yamada, T.; Hata, K.; Sekiguchi, A. The importance of carbon nanotube wire density, structural uniformity, and purity for fabricating homogeneous carbon nanotube–copper wire composites by copper electrodeposition. *Jpn. J. Appl. Phys.* **2018**, *57*, 04FP08. [CrossRef]
11. Sundaram, R.; Yamada, T.; Hata, K.; Sekiguchi, A. Thermal expansion of Cu/carbon nanotube composite wires and the effect of Cu-spatial distribution. *J. Mater. Res. Technol.* **2020**, *9*, 6944–6949. [CrossRef]
12. Leggiero, A.P.; Driess, S.D.; Loughran, E.D.; McIntyre, D.J.; Hailstone, R.K.; Cress, C.D.; Puchades, I.; Landi, B.J. Platinum nanometal interconnection of copper–carbon nanotube hybrid electrical conductors. *Carbon* **2020**, *168*, 290–301. [CrossRef]
13. McIntyre, D.J.; Leggiero, A.P.; Hailstone, R.K.; Puchades, I.; Cress, C.D.; Landi, B.J. Integrated Titanium-Carbon Nanotube Conductors via Joule-Heating Driven Chemical Vapor Deposition. *ECS Trans.* **2020**, *97*, 321–327. [CrossRef]
14. McIntyre, D.J.; Hirschman, R.K.; Puchades, I.; Landi, B.J. Enhanced copper–carbon nanotube hybrid conductors with titanium adhesion layer. *J. Mater. Sci.* **2020**, *55*, 6610–6622. [CrossRef]
15. Leggiero, A.P.; Trettner, K.J.; Ursino, H.L.; McIntyre, D.J.; Schauer, M.; Zeira, E.; Cress, C.D.; Landi, B.J. High conductivity copper–carbon nanotube hybrids via site-specific chemical vapor deposition. *ACS Appl. Nano Mater.* **2018**, *2*, 118–126. [CrossRef]
16. Shuai, J.; Xiong, L.; Zhu, L.; Li, W. Enhanced strength and excellent transport properties of a superaligned carbon nanotubes reinforced copper matrix laminar composite. *Compos. Part. A Appl. Sci. Manuf.* **2016**, *88*, 148–155. [CrossRef]
17. Shuai, J.; Xiong, L.Q.; Lin, Z.H.U.; Li, W.Z. Effects of ply-orientation on microstructure and properties of super-aligned carbon nanotube reinforced copper laminar composites. *Trans. Nonferrous Met. Soc. China* **2017**, *27*, 1747–1758. [CrossRef]
18. Mendoza, M.E.; Solórzano, I.G.; Brocchi, E.A. Mechanical and electrical characterization of Cu-2 wt.% SWCNT nanocomposites synthesized by in situ reduction. *Mater. Sci. Eng. A* **2012**, *544*, 21–26. [CrossRef]
19. Milowska, K.Z.; Ghorbani-Asl, M.; Burda, M.; Wolanicka, L.; Ćatić, N.; Bristowe, P.D.; Koziol, K.K. Breaking the electrical barrier between copper and carbon nanotubes. *Nanoscale* **2017**, *9*, 8458–8469. [CrossRef]
20. Nayan, N.; Shukla, A.K.; Chandran, P.; Bakshi, S.R.; Murty, S.V.S.N.; Pant, B.; Venkitakrishnan, P.V. Processing and characterization of spark plasma sintered copper/carbon nanotube composites. *Mater. Sci. Eng. A* **2017**, *682*, 229–237. [CrossRef]
21. Guiderdoni, C.; Pavlenko, E.; Turq, V.; Weibel, A.; Puech, P.; Estournès, C.; Peigney, A.; Bacsa, W.; Laurent, C. The preparation of carbon nanotube (CNT)/copper composites and the effect of the number of CNT walls on their hardness, friction and wear properties. *Carbon* **2013**, *58*, 185–197. [CrossRef]
22. Sun, Y.; Chen, Q. Diameter dependent strength of carbon nanotube reinforced composite. *Appl. Phys. Lett.* **2009**, *95*, 021901. [CrossRef]
23. Zhao, S.; Zheng, Z.; Huang, Z.; Dong, S.; Luo, P.; Zhang, Z.; Wang, Y. Cu matrix composites reinforced with aligned carbon nanotubes: Mechanical, electrical and thermal properties. *Mater. Sci. Eng. A* **2016**, *675*, 82–91. [CrossRef]
24. Ghorbani-Asl, M.; Bristowe, P.D.; Koziol, K. A computational study of the quantum transport properties of a Cu–CNT composite. *Phys. Chem. Chem. Phys.* **2015**, *17*, 18273–18277. [CrossRef]
25. Hjortstam, O.; Isberg, P.; Söderholm, S.; Dai, H. Can we achieve ultra-low resistivity in carbon nanotube-based metal composites? *Appl. Phys. A* **2004**, *78*, 1175–1179. [CrossRef]
26. Kobayashi, K.; Shukla, B.; Ohmori, S.; Kiyomiya, M.; Hirai, T.; Kuwahara, Y.; Saito, T. Wall-Number Selectivity in Single/Double-Wall Carbon Nanotube Production by Enhanced Direct Injection Pyrolytic Synthesis. *Jpn. J. Appl. Phys.* **2013**, *52(10R)*, 105102. [CrossRef]
27. Saito, T.; Ohshima, S.; Okazaki, T.; Ohmori, S.; Yumura, M.; Iijima, S. Selective diameter control of single-walled carbon nanotubes in the gas-phase synthesis. *J. Nanosci. Nanotechnol.* **2008**, *8*, 6153–6157. [CrossRef] [PubMed]
28. Chen, G.; Dodson, B.; Hedges, D.M.; Steffensen, S.C.; Harb, J.N.; Puleo, C.; Galligan, C.; Ashe, J.; Vanfleet, R.R.; Davis, R.C. Fabrication of high aspect ratio millimeter-tall free-standing carbon nanotube-based microelectrode arrays. *ACS Biomater. Sci. Eng.* **2018**, *4*, 1900–1907. [CrossRef] [PubMed]
29. Hata, K.; Futaba, D.N.; Mizuno, K.; Namai, T.; Yumura, M.; Iijima, S. Water-assisted highly efficient synthesis of impurity-free single-walled carbon nanotubes. *Science* **2004**, *306*, 1362–1364. [CrossRef]
30. Saito, R.; Hofmann, M.; Dresselhaus, G.; Jorio, A.; Dresselhaus, M.S. Raman spectroscopy of graphene and carbon nanotubes. *Adv. Phys.* **2011**, *60*, 413–550. [CrossRef]
31. Reich, S.; Thomsen, C.; Maultzsch, J. *Carbon Nanotubes: Basic Concepts and Physical Properties*, 1st ed.; John Wiley & Sons: Weinheim, Germany, 2004; pp. 142–174.
32. Haynes, W.M. (Ed.) Thermal and physical properties of pure metals. In *CRC Handbook of Chemistry and Physics*, 97th ed.; CRC Press, Taylor and Francis: Boca Raton, FL, USA, 2016; pp. 12–217

33. Behabtu, N.; Young, C.C.; Tsentalovich, D.E.; Kleinerman, O.; Wang, X.; Ma, A.W.; Bengio, E.A.; Waarbeek, R.F.; de Jong, J.J.; Hoogerwerf, R.E.; et al. Strong, light, multifunctional fibers of carbon nanotubes with ultrahigh conductivity. *Science* **2013**, *339*, 182–186. [CrossRef] [PubMed]
34. Zhao, Y.; Wei, J.; Vajtai, R.; Ajayan, P.M.; Barrera, E.V. Iodine doped carbon nanotube cables exceeding specific electrical conductivity of metals. *Sci. Rep.* **2011**, *1*, 1–5. [CrossRef] [PubMed]
35. Niven, J.F.; Johnson, M.B.; Juckes, S.M.; White, M.A.; Alvarez, N.T.; Shanov, V. Influence of annealing on thermal and electrical properties of carbon nanotube yarns. *Carbon* **2016**, *99*, 485–490. [CrossRef]
36. Bulmer, J.S.; Kaniyoor, A.; Elliott, J.A. A Meta-Analysis of Conductive and Strong Carbon Nanotube Materials. *Adv. Mater.* **2021**, *33*, 2008432. [CrossRef] [PubMed]
37. Tsentalovich, D.E.; Headrick, R.J.; Mirri, F.; Hao, J.; Behabtu, N.; Young, C.C.; Pasquali, M. Influence of carbon nanotube characteristics on macroscopic fiber properties. *ACS Appl. Mater. Interfaces* **2017**, *9*, 36189–36198. [CrossRef] [PubMed]
38. Cho, S.; Kikuchi, K.; Kawasaki, A. On the role of amorphous intergranular and interfacial layers in the thermal conductivity of a multi-walled carbon nanotube–copper matrix composite. *Acta Mater.* **2012**, *60*, 726–736. [CrossRef]

C Journal of Carbon Research

MDPI

Review

Graphene Synthesis and Its Recent Advances in Applications—A Review

Anuluxan Santhiran [1], Poobalasuntharam Iyngaran [1], Poobalasingam Abiman [1] and Navaratnarajah Kuganathan [2,3,*]

1 Department of Chemistry, University of Jaffna, Sir. Pon Ramanathan Road, Thirunelvely, Jaffna 40000, Sri Lanka; sanuluxan@gmail.com (A.S.); piyngs7@gmail.com (P.I.); abiman777@gmail.com (P.A.)
2 Department of Materials, Imperial College London, London SW7 2AZ, UK
3 Faculty of Engineering, Environment and Computing, Coventry University, Priory Street, Coventry CV1 5FB, UK
* Correspondence: n.kuganathan@imperial.ac.uk

Abstract: Owing to the remarkable chemical and physical properties, graphene has been widely investigated by researchers over the last 15 years. This review summarizes major synthetic methods such as mechanical exfoliation, liquid phase exfoliation, unzipping of carbon nanotube, oxidation-reduction, arc discharge, chemical vapor deposition, and epitaxial growth of graphene in silicon carbide. Recent advances in the application of graphene in graphene-based lithium ion batteries, supercapacitors, electrochemical sensors, transparent electrodes and environmental based remedies are discussed.

Keywords: graphene; chemical vapor deposition; supercapacitors; batteries; photocatalyst

Citation: Santhiran, A.; Iyngaran, P.; Abiman, P.; Kuganathan, N. Graphene Synthesis and Its Recent Advances in Applications—A Review. C **2021**, 7, 76. https://doi.org/10.3390/c7040076

Academic Editors: Gil Goncalves and Stefano Bellucci

Received: 8 September 2021
Accepted: 8 November 2021
Published: 10 November 2021

1. Introduction

Graphene has been extensively studied by scientific and engineering communities for more than 15 years since its first fabrication reported in 2004 [1]. Graphene is a single layer of two-dimensional carbon atoms in a hexagonal lattice structure and has been widely used in many applications such as electronics [2], energy storing batteries [3], super capacitors [4], fuel cells [5] and solar cells [6,7] owing to its unique high surface area, thermal [8] and electrical conductivity [9] and mechanical strength [10]. Graphene is one of the allotropes of carbon and it is made of hexagons. Other allotropes include fullerenes (0D), nanotubes (1D) and graphite (3D) (see Figure 1). Graphene based nanomaterials include reduced graphene oxide, graphene quantum dots and graphene oxides. Graphene ideally consists of sp^2 carbon atoms but other family members comprise of sp^2 and sp^3 due to the introduction of the functional groups such as hydroxyl, carboxyl, carbonyl and epoxy groups [11].

Graphene consists of two-dimensional layer of carbon atoms with sp^2 hybridization arising from mixing of s, p_x and p_y orbitals. The remaining p_z orbital of each carbon atom forms π bonds with three neighboring carbon atoms, known as valence band and a band of empty π^* orbitals known as conduction band [12]. Carbon has four valence electrons, and three of them form sigma bonds that are back bone of hexagonal structure. The remaining electron forms one third of π bond with the nearest neighbor carbon atom. These out of plane interactions are extremely weak leading to out of plane electrical and thermal conductivities which are ~103 times lower than that of in plane analogues [13]. A single layer of graphene consists of hexagonal layer of carbon atoms. Bi-layer has 2 and few layer graphene has 3 to 10 layers of two-dimensional sheets. Graphene with more than 10 layers is known as thick graphene and has less scientific interest. In Bi- layer and few layer graphene, carbon atoms are stacked in different possible ways, hexagonal or AA stacking, Bernal or AB stacking and rhombohedral way or ABD stacking [14]. In a twisted bilayer graphene, layers are twisted at a small angle [15].

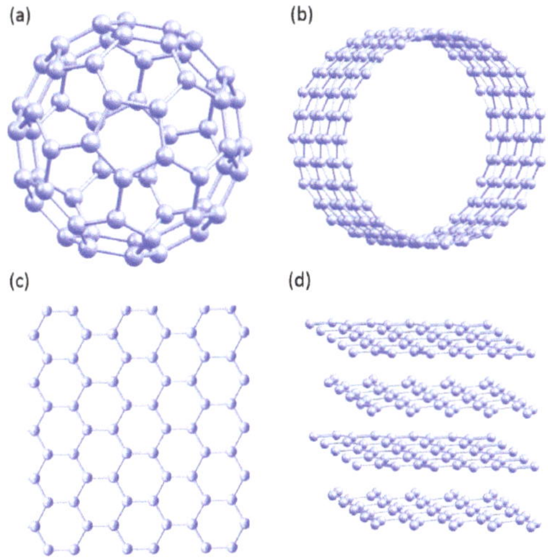

Figure 1. (**a**) Zero-dimensional Fullerene, (**b**) one-dimensional carbon nanotube (**c**) two-dimensional graphene and (**d**) three-dimensional graphite.

Graphene attracted great attention due to its excellent electronic properties. Graphene is a semiconductor with a small band gap [16]. The electron mobility of graphene at room temperature was found to be as high as 15,000 $cm^2V^{-1}s^{-1}$ with small temperature dependency and zero effective mass for the charge carriers [17]. In most of the materials, the electron movement is hindered by phonon scattering. However, in the case of graphene, the movement of electrons is hindered only by defect scattering. As a result, the theoretical limit of resistivity of graphene is estimated to be 10^{-6} Ω cm which is the lowest resistivity measured at room temperature [18]. Absence of interatomic plane coupling provides high thermal conductivity. Graphene with defects exhibits lower thermal conductivity. Average thermal conductivity of high-quality exfoliated graphene is 3000–4000 W/m K and the thermal conductivity of graphene prepared via CVD method is 2500 W/m K [19]. Due to its excellent thermal property, graphene is being used as a temperature sensor, a thermoelectric sensor and a thermal biosensor in energy management systems [20,21]. The graphene has been synthesized using chemical vapor deposition of hydrocarbon on transition metal surfaces [22–25], thermal decomposition of silicon carbide wafer under ultrahigh vacuum [26,27], and chemical and thermal reduction of graphene oxide [28,29]. Among the methods listed above, reduction of graphene oxide is the most economical method [30]. Thus, for electrochemical analysis, graphene has been prepared using this method. In chemical vapor deposition method, there may be impurities of transition metals (Ni or Cu) but the reduction method provides graphene with high purity. Furthermore, this method provides an effective way of studying electrocatalytic effects.

Graphene oxide is a single layer of graphite oxide and is generally produced by the chemical treatment of graphite through oxidation [31]. Graphene oxide comprises various functional groups containing oxygen (see Figure 2). These functional groups are mostly hydroxyl and epoxide groups in the basal planes and consist of trace amount of carbonyl, carboxyl, phenol, lactone and quinone groups at the edges of the sheet [32]. The wide range of these functional groups at the edges and the basal planes of graphene oxide make it functionalized and exfoliated to yield well dispersed solutions of separate graphene oxide sheets in polar and non-polar solutions and therefore it has a wide range of applications in nanocomposites [33], photocatalysis [34], battery [35], capacitors [36] and sensors [37].

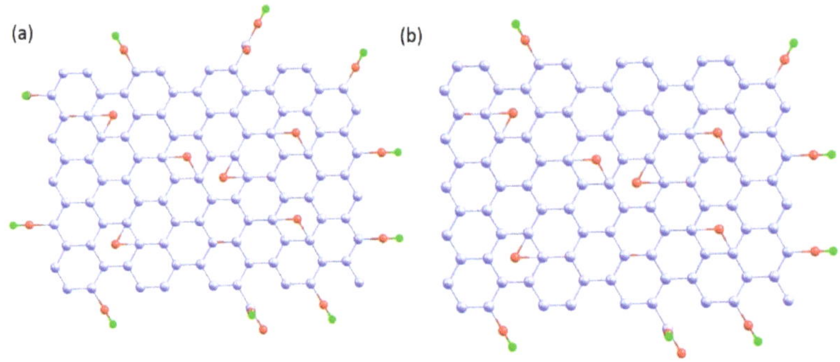

Figure 2. Structures of (**a**) graphene oxide and (**b**) reduced graphene oxide [light purple-carbon, red-oxygen and green-hydrogen].

2. Synthesis

Geim and Novoselov carried out groundbreaking experiments on the two-dimensional graphene using scotch tape method in 2004 [38]. There are bottom-up and top- down approaches available for the synthesis of graphene (see Figure 3). For example, mechanical cleavage is the process where graphite is broken down into graphene (top-down) and on the other hand, in chemical vapor deposition method, graphene is developed in silicon carbide (bottom-up) [39].

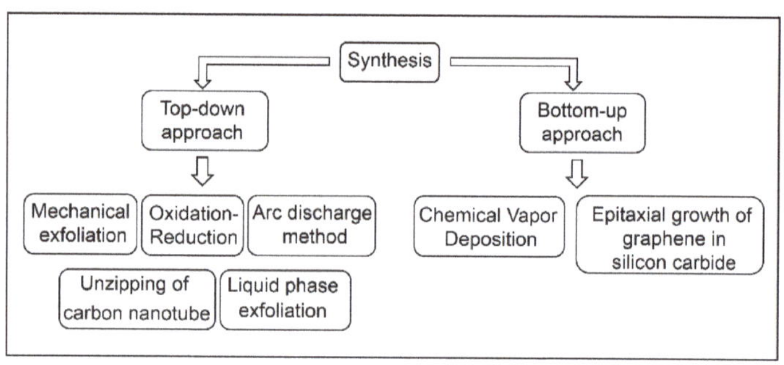

Figure 3. Schematic diagram showing synthesis methods [40].

2.1. Mechanical Exfoilation

As previously mentioned, the graphene sheets are held together by weak van der Waals forces (see Figure 4). If these forces are broken, high purity graphene can be obtained. In mechanical exfoliation method, mechanical energy is used to destroy these weak bonds and separate the individual sheets. Exfoliation is generally a peeling process repeatedly carried out in graphite to obtain layers of graphene. This method was first developed by Geim and Novoselov using highly oriented pyrolytic graphite (HOPG) as a precursor [17]. In this process HOPG sheet of thickness 1 mm is used to dry etching by oxygen plasma to create many 5 μm deep mesas (an isolated flat-topped surface). Then these were used to photoresist and baked to stick on the mesas. Thereafter, scotch tape was used to peel off the graphene layers from graphite. These thin flakes were then released to the acetone and transferred on to a Si substrate. Thus, pure graphene flakes are produced on a Si substrate. The disadvantage of this method is that large industrial production cannot be scaled.

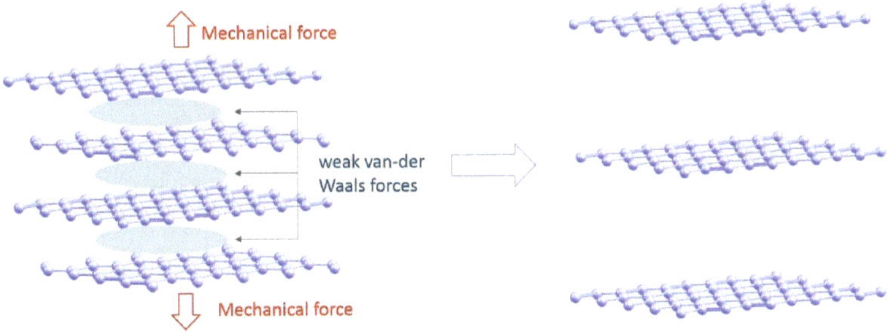

Figure 4. Mechanical exfoliation process in graphite.

2.2. Liquid Phase Exfoliation

Liquid phase exfoliation (LPE) is one of the most commonly used method for the synthesis of graphene and was first reported in 2008 [41]. In this top-down method, a stable dispersion of single layer or few-layer of graphene can be obtained through exfoliation of graphite. LPE includes main steps such as dispersion of graphite in a suitable solvent, exfoliation and the purification of the final products [42]. This method of exfoliation involves separation of graphene layers by overcoming the van der Waals forces. Therefore, the selection of solvent plays a major role. The properties such as surface energy, Hildebrand solubility, Hansen solubility parameters and surface tension have been considered for the selection of the solvent [43,44]. Researchers found that solutions with surface tension (for liquid) within 40–50 mJ/m^2 or the surface energies (for solid) within 70–80 mJ/m^2 are found to be good for the graphene production [44]. Aside for the solvent, the sonification is also an important factor to be considered during this synthesis method. Even though LPE method is an effective method for the graphene production several studies point out that sonification could affect the edges and basal planes [45]. This can be rectified by optimizing the sonification time, temperature and the sonication intensity [44]. One of the major disadvantages of this method is the yield which is not sufficient for industrial applications at macroscopic scale. Other disadvantages include solvents being expensive, toxic and the reduction of the size of the nanosheets. Future studies should be carried to improve the efficiency and the economic feasibility of this method.

2.3. Unzipping of Carbon Nanotube

This process involves cutting of the cylindrical carbon nanotubes (CNT) in axial or longitudinal direction into a flat graphene sheet with single, bi or few layers graphene. Single walled carbon nanotubes (SWCNT) or multiwalled carbon nanotubes (MWCNT) can be used as a starting material. There are several methods which have been widely investigated to unzip CNT such as plasma etching [46], chemical unzipping [47], intercalation and exfoliation [48] and metal catalyzed cutting [49]. Longitudinal unzipping of CNT leads to the formation of nanoribbons.

In chemical unzipping, CNTs were cut longitudinally by the treatment of H_2SO_4 followed by oxidation using $KMnO_4$. Then oxidized graphene will be reduced chemically using NH_4OH and hydrazine monohydrate ($N_2H_4{\cdot}H_2O$) solution [47]. This method is considered as a low significant method because during oxidation the precursor is damaged and graphene becomes electronically inferior due to the presence oxygen defect sites [47,50]. Yield of graphene was improved when intercalation of MWCNT was carried out in oxalic acid (chemical medium) prior to the chemical unzipping. The improvement in the yield was due to the appropriate size of oxalic acid (0.34 nm) which intercalates comfortably between the interlayers of the MWCNT [51]. In plasma etching, polymer film such as poly-methyl methacrylate is being used. In this method, CNT is embedded into the film and then this mixture of film and CNT are separated out in KOH solution. Thereafter CNT is exposed in argon-based plasma. Here the longitudinal C-C bond of CNT is broken down to produce graphene with smooth edges [46]. Intercalation and exfoliation is another method of unzipping CNT. Here lithium and ammonia are allowed to react with CNT. In this treatment, a strong stress is exerted between the layers of CNT resulting in the separation of the graphene layers [48]. Researchers considered other metal nanoparticles such as nickel, cobalt and copper and found that C-C and H-H bonds in MWCNT were dissociated by these nanoparticles [52]. The above-mentioned methods use chemicals which are hazardous and expensive. Unzipping of MWCNT can be achieved by using electric field from tungsten electrode [53]. This method provides defect free and high pure graphene nanoribbons. Therefore, electric field method is more preferable due to feasibility, environmentally friendliness and accuracy.

2.4. Arc Discharge Method

In Arc discharge method, anode and cathode are submerged in a gas or liquid medium of a reaction chamber. During the passage of electricity, the medium is dissociated to produce a high temperature plasma (3727–5725 °C) and this will be sufficient to sublime the precursor [54]. This method is considered to be expensive due to the usage of vacuum. Therefore, the use of air instead of H_2/He medium is preferred to reduce the cost of production of graphene [55]. Li et al. demonstrated a method in which petroleum asphalt is used as a precursor in a water arc discharge system [55]. The use of asphalt which is a carbon rich source and water medium makes this method a cost effective. These improvements in arc discharge method could produce graphene at an affordable price.

2.5. Chemical Vapor Deposition (CVD)

CVD is one of the most promising method for the synthesis of graphene cost effectively and can produce large area of graphene. In this CVD approach, hydrocarbon gaseous species are injected into the reactor and travel through a hot zone where these molecules degrade to form carbon radicals and deposit on the metal surface as single / few layer graphene (see Figure 5). During this process metal surface acts as a catalyst and also influences the deposition mechanism of graphene which plays a key role in the preparation of pure graphene [56]. Researchers have used metals such as Ru, Ir, Pt, Co, Pd and Re. The nickel (Ni) and copper (Cu) are low cost, have better control of layers of graphene and are easier to transfer graphene. Thus, they are widely used as substrates in CVD [56]. Using cold-wall and hot-wall reaction chambers, the CVD growth of graphene has been carried out [57]. In this technique, the growth of graphene is fast, high quality and takes low power consumption. Furthermore, there is an enhancement in charge carrier mobility.

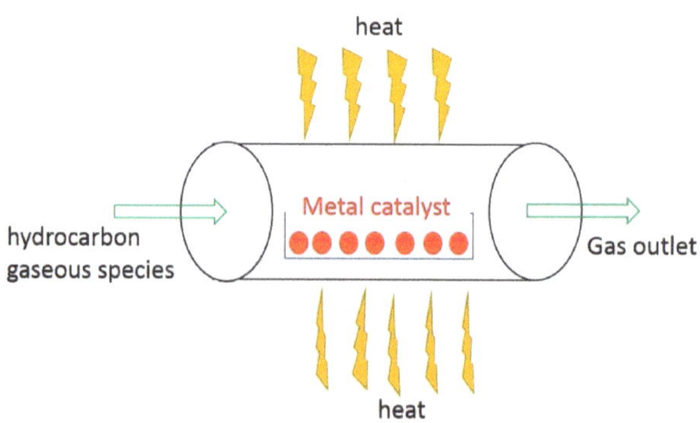

Figure 5. Schematic diagram showing CVD approach.

2.5.1. Chemical Vapor Deposition of Graphene on Nickel

In this synthetic method, Ni films are first annealed at 900 to 1000 °C under Ar/H_2 environment for 20 min to increase the grain size of Ni. Thereafter film is exposed to CH_4/H_2 mixture. Here the methane is degraded to form carbon atoms and dissolves into the film resulting a solid solution. As the last step, these samples are cooled in argon gas.

Ni is highly soluble in carbon at elevated temperature and it reduces with the decreasing temperature [56]. The formation of thickness was dependent on the rate of cooling [58].

2.5.2. Chemical Vapor Synthesis of Graphene on Copper

In this method, Cu foil of thickness 25 μm was heated at 1000 °C in hydrogen atmosphere followed by the introduction of H_2/CH_4 mixture to initiate the formation of graphene. Then the temperature was reduced to room temperature [56]. One of the demerits of this method is that Cu has the very low solubility of carbon even at high concentration of hydrocarbon or longer growth time [56].

2.6. *Epitaxial Growth of Graphene in Silicon Carbide (SiC)*

Epitaxial growth can be made under vacuum or at atmospheric conditions. Different types of SiC such as single crystal SiC wafers, polycrystalline SiC and SiC thin films are used in this method. SiC is heated to 1200–1600 °C in ultra-high vacuum which sublimates silicon (melting point of Si = 1100 °C) leaving carbon atoms in the reaction vessel. Later these carbon atoms aggregate to form graphene [59]. This method produced 1 to 3 layers of graphene and the number of layers dependent on the decomposition temperature [14]. This synthetic technique is capable of producing wafer scale layers of graphene and therefore its potential interest is high in the semiconductor industry [14]. As a development of this method, graphene was synthesized from Ni coated SiC substrate at a lower temperature of 750 °C [60]. This method consists of technical problems such as high cost of SiC substrates (single-crystal SiC) and high temperature. These issues should be properly addressed to extend the practical and economic feasibility.

2.7. *Oxidation-Reduction*

This is one of the widely used methods for the synthesis of graphene from graphite. There are four major routes available for the conversion of graphite to form graphene oxide such as Brodie [61], Staudenmaier [62], Hofmann [63] and Hummers [31] method. All these methods are almost similar to each other except the use of different chemicals and improved steps over the years. All these methods require temperature below 100 °C to maintain low production cost. One of the demerits of these methods is the release of toxic gases such as nitrogen dioxide (NO_2) and dinitrogen tetroxide (N_2O_4) [64,65]. Nowadays the Hummers method is being widely used as this method is fast and safe (see Figure 6). Recent modified version of hummers' method has eliminated the use of sodium nitrate ($NaNO_3$) which reduces the production cost [66]. The oxidation of graphene increases the inter layer spacing of graphite layers from 0.335 to 0.625 nm. This is because of the incorporation of intercalation compounds [67]. This confirms the reduction of van der Waals forces between the inter layers. Therefore, exfoliation by sonication using a suitable solvent can produce single or bilayer or few layer graphene oxide (GO) [67]. Graphene does not contain any functional groups and it is therefore insoluble in water and organic solvents. On the other hand, GO is hydrophilic due to the presence of functional groups containing oxygen that can be dispersed in different solutions such as water, tetrahydrofuran (THF) and ethylene glycol [68]. Reduction methods are used for the restoration of honeycomb lattice from disrupted sp^2 bonding of GO. Mainly chemical, thermal and electrochemical processes are used for the reduction of GO eliminating the oxygen functional groups such as carbonyl, carboxylic and hydroxyl.

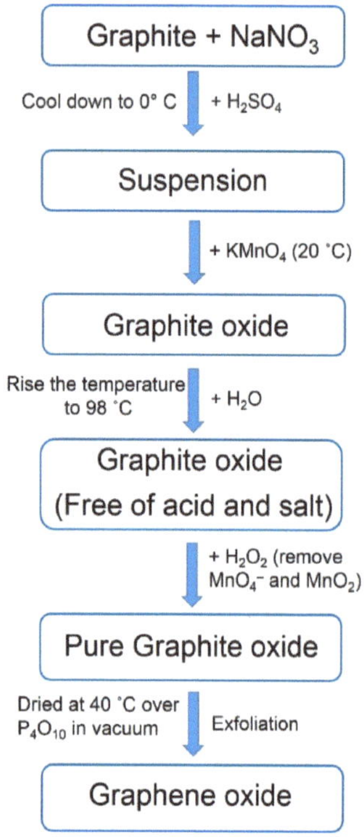

Figure 6. Schematic diagram showing Hummers method [31].

In a chemical reduction of GO, hydrazine (N_2H_4) is commonly used. However, it is a toxic chemical and relatively expensive for the production. Therefore, several reducing agents such as borohydride, aluminum hydride, plant extract, amino acids and microorganism have been widely used [69]. Due to the safety, environmental concerns and synthesis time, thermal reduction [70] and hydrothermal reduction [71] have been investigated by the scientists. Thermal reduction method has demerits such as high temperature, high cost and release of greenhouse gas CO_2 meanwhile hydrothermal reduction requires low temperature there by lower energy consumption [72]. Electrochemical reduction has gained considerable interest owing to its rapid reduction, environmentally benign (fewer toxic reductants) and cost effectiveness [40].

3. Applications

3.1. Lithium Ion Batteries

Rapid decrease of fossil fuels and the environmental pollution necessitated the invention of new energy sources. Lithium ion battery was first introduced in 1999. In 2019, John B. Goodenough, M. Stanley Whittingham and Akira Yoshino were awarded noble prize in chemistry for the invention of Lithium ion batteries (LIBs). Owing to the long cycle and high energy density, LIBs are considered as important energy storage devices. Graphene has been widely investigated for the anode material for LIBs because of its high flexibility, stability and high surface to volume ratio. The speed of Li ion movement determines the power capability of battery [73]. Application of the graphene with transition metal oxides improves the discharge rate and electrochemical stability.

Graphitic carbon has been used as an anode material for LIBs. It possess high crystallinity. Layer arrangement leads to the formation of LiC_6 which is the highest intercalation state of the battery [74]. Zhao et al. reported anodic material prepared by electrochemical exfoliation and it exhibited a capacity of 356 mA h g^{-1} at 1 A g^{-1} with 100% capacity retention after more than 6500 cycles [75]. Ji et al. [76] reported a novel 3D graphene powder synthesized via thermal CVD. Incorporation of this powder in graphite electrode consisted a capacity of 542.8 mA h after 400 cycles and 427.2 mA h after 1000 cycles [76]. Density functional theory calculations reported by Oladipo et al. revealed that the graphene loaded with polypyrroles exhibited a small band gap of 0.39 eV and strong absorbing power [77].

Nanoparticles were allowed to intercalate with graphene to improve the performance of the batteries. Graphene intercalated with Co_3O_4 nanoparticles was analyzed as an anode material for high performance LIBs. Here the nanoparticles were intercalated between sheets to maintain the separation of sheets [73]. This intercalation improved cyclic performance and resulted in high reversible capacity. This combination was found to have specific capacity of 1100 mA h g^{-1} with current density of 74 mA h g^{-1}. After 130 cycles the discharge capacity was observed at 1000 mA h g^{-1} [78]. In another study, SnO_2-graphene intercalation was investigated and charge capacity of 860 mA h g^{-1} with current density of 50 mA g^{-1} was observed. This attainment was due to the nanocrystalline property of SnO_2-graphene. Reversible capacity was found at 810 mA h g^{-1} and 70 % of the reversible capacity was remained after 30 cycles at 570 mA h g^{-1} [79]. Rutile TiO_2-graphene composite have the specific capacity of 87 mA h g^{-1} and anatase TiO_2-graphene have the high current density of 96 mA h g^{-1} [80]. Silicon-boron nitride nanocomposite displayed a reversible capacitance of 785 mA g^{-1} at 450 mA g^{-1} at the end of 800 cycles. The incorporation provides a large Li-ion storage and shorter Li-Li separation [81].

Addition of heteroatoms such as B and N into the graphene frameworks improved the surface wettability due to the presence of more hydrophilic sites and improved reactions between electrolyte and electrode [82]. N-doped graphene was studied for the reversible discharge capacity and it was found that nitrogen atoms improved the Li ion intercalation [83]. In a recent study, N-doped graphene nanosheets were synthesized in ammonia (NH_3) environment by heat treatment of graphene oxide. Reversible capacity of around 250 mA h g^{-1} at 2.1 mA g^{-1} current density and capacity of 900 mA h g^{-1} at 4.2 mA g^{-1} were obtained [84]. Yang et al. reported the fabrication of N-doped porous graphene hybrid nanosheets with induced growth of zeolitic imidazolate framework on graphene [85]. This framework on graphene increases the specific area and the electron transfer within graphene network [85]. Hetero atomic defect increased the distance between sheets and electrolyte wettability enhanced the thermal stability and the electrical conductivity of doped graphene [86]. Graphene has been studied as mostly as anode material instead of cathode material because of its low electrical conductivity, slower electron and Li ion transport and low specific capacity and the agglomeration of the particles while applied as the cathode material [87]. Among the studies reported above, graphene intercalated with Co_3O_4 nanoparticles has high specific capacity and current density. Here the nanoparticles intercalate between graphene sheets and flexibility of graphene increased it performance. There is a strong interaction between graphene and Co_3O_4 preventing volume expansion/contraction and aggregation of nanoparticles during the battery performance. Yusuf et al. [73] reported that this composite effectively shows high surface area, good conductivity, and mechanical flexibility. Thus, this Co_3O_4—graphene composite exhibits a better performance.

3.2. Supercapacitors

Supercapacitors are energy storage devices with good cyclability, high power and fast charge-discharge process [72]. Surface area of the materials used for the fabrication of supercapacitors determines the performance. Several materials have been widely investigated to improve the performance. Among those, graphene and graphene-based composites have received attention owing to their high surface area, distinctive mechanical and electrical properties. Agglomeration of graphene has been mentioned as a major drawback in the performance of supercapacitors. This can be overcome by the incorporation of metal oxide nanoparticles with graphene [72].

Metal oxides such as RuO_2, NiO_2, MnO_2, ZnO and SnO_2 have been investigated to enhance the performance of supercapacitors. Deposition of RuO_2 on the surface of graphene prevents agglomeration and improves electron transfer. High conductivity of RuO_2 nanoparticles and separated graphene sheets provide combined advantage and improved performance with the energy density of 20.1 $Whkg^{-1}$ and an excellent electrochemical stability of 97.9% after 1000 cycles [88]. $Ni(OH)_2$-graphene hydrogel was found to have specific capacitance of ~1212 F g^{-1} at the discharge rate of 2 A g^{-1} [89]. Supercapacitor fabricated with graphene-carbon nanotube (CNT) hybrid attained the highest energy density among the all carbon-based supercapacitors and was found to work at 4V and possess energy density of 460 $Whkg^{-1}$ [90]. This is due to the large surface area and the covalent interaction between graphene and carbon nanotubes. Porous Nickel-graphene-CNT exhibits high diffusion conductivity which is due to the aligned interfacial contact and volume of the 3D nanomaterial [91]. Three-dimensional flower shaped $Li_4Ti_5O_{12}$-graphene nanostructure exhibits a high specific capacitance of 706.52 F g^{-1} with discharge rate of 1 A g^{-1} and pine needles-based carbon nanostructure possess specific capacitance of 314.50 F g^{-1} at the discharge rate of 1 A g^{-1}. Supercapacitor with $Li_4Ti_5O_{12}$-graphene nanostructure as a positive electrode and pine needles-based carbon nanostructure as a negative electrode provides an energy density of 35.06 $Whkg^{-1}$ with power density of 800.08 W kg^{-1} and exhibit cyclic stability of 90.18% after 2000 cycles, respectively [92]. Graphene-MoS_2 composite prepared by Wang et al. [93] exhibits higher specific capacitance when compared with that of graphene. This combination gives the interconnection and interpenetration between nanomaterials which reduced restacking and aggregation. This composite exhibits a capacitance of 58.5 mF/cm^{-2} and showed quick current response to voltage reversal [93].

Chen et al. [94] have reported novel network structure of nitrogen doped porous graphene and this network possessed good conductivity and quick electron transfer. This network has high gravimetric capacitance and volumetric capacitance compared to that of graphene. This supercapacitor exhibits energy density up to 27.6 Wh Kg^{-1} at 600 W kg^{-1} [94]. Graphene based aerogels have been an excellent candidate as electrode materials because of their ultralow density, mechanical stability and flexibility [95,96]. Doping the aerogels with N and S increased the performance of the supercapacitor. Zhang et al. [97] reported the synthesis of N, S co-doped graphene aerogels using 2,5-Dimercapto-1,3,4-thiazole as a precursor. This exhibited a capacitance of 321 F g^{-1} at a current of 1 A g^{-1} which is approximately twice the value of pure graphene aerogel. Here the S and N atoms improved the performance of supercapacitor through synergistic effect [97]. It is generally expected that high specific surface area provides high specific capacitance of carbon-based electrodes. Graphene nanoflakes reported by Arkhipova et al. [98] had specific surface area range of 1720 to 660 m^2 g^{-1} depending on the synthesis time. The highest capacitance of 112 Fg^{-1} was observed for the graphene nanoflakes synthesized in the shortest synthesis time [98]. In supercapacitors, the application of nanoparticles with graphene improves the performance by preventing the agglomeration. Nitrogen doped graphene possess high performance than the graphene made of nanoparticles. This is due to the improvement in the electron transfer via nitrogen doping.

3.3. Electrochemical Sensors

Electrochemical sensing is one of the most important application of graphene. The adsorption of molecules on the surface of graphene influences its electrical conductivity [10]. Carrier concentration of the graphene is dependent on the type of dopant (donor or acceptor) adsorbed on the surface. There are many properties of graphene which increase its efficiency of detecting molecules. Since graphene is a 2D material, whole surface can be exposed to the analyte [99]. Graphene has high conductivity and low noise of distortion. Thus a small change in the concentration can alter electrical conductivity [98]. It consists of very low crystal defects [17,100] which ensure the low noise of distortion due to the thermal switching [101]. Schedin et al. [99] first reported the sensing of graphene in 2007. In this report sensing of gas molecules such as NO_2, NH_3, H_2O and CO is discussed. Re-usage of graphene sensors after the vacuum annealing at temperature of 150 °C or under UV radiation is possible for a short time [99].

Fowler et al. [102] reported the sensing of NO_2, NH_3 and dinitro toluene (DNT, used in explosives) by graphene [102]. Sensing of NO_2 was occurred by the withdrawal of electron from the graphene (hole induced conduction) and in NH_3, electron was donated to the graphene from the gas molecules (electron induced conduction). This explains that the sensing occurs due to the transfer of charge (donation or acceptance) at the surface of graphene [102]. Sundaram et al. [103] have reported the sensing of H_2 on the modified graphene by the deposition of Pd nanoparticles (Nps) and Pd showed high affinity towards H_2. Here the resistance of the bare and Pd-modified graphene sheets has almost the same resistance maximum in the gate dependence curve. During the adsorption of H_2, resistance maximum is strongly shifted to the negative gate voltages. This shift is due to the dissolution and partial dissociation of hydrogen at Pd nanoparticles. The particle's work function decreases as the electrons are transferred from Pd to the graphene sheet. [82]. Gupta et al. [104] reported the ascorbic acid sensor by graphene nanosheets functionalized with polyaniline nanostructures and metal nanoparticles (Au Nps, Ag Nps). Here the nanoparticles are sandwiched between layers of graphene to ensure the good electron transfer. This can be further improved by changing the morphology, size, and the density of particles. Graphene sensor was found to have ultrasensitive up to lower concentration of 1×10^{-12} M and sensor with Au nanoparticles had a low signal-to-noise ratio. [104].

Shan et al. [105] have reported the sensing of glucose using graphene functionalized with glucose enzyme oxidase (GOD). Biosensor was comprised of graphene-polyethyleneimine-functionalized nanocomposite for the detection of glucose. Flavin adenine dinucleotide (FAD) is deeply embedded in a protective protein shell, which makes the direct electron communication difficult with electrodes. As the graphene has the high specific area and extraordinary electron transport, direct electron transfer between GOD and electrode substrate is promoted. Sensor was based on the electron transfer from GOD demonstrating the potential application of graphene in glucose detection. A linear response was recorded between 2 mM to 14 mM in their study [105]. Kang et al. [106] have constructed a glucose sensor employing chitosan. They have sonicated suspension containing graphene and chitosan for 1 h and a stable dark suspension was formed as a result. This results in the formation of C-OH and -COOH functional groups which let the graphene more hydrophilic. Chitosan disperses the graphene and immobilizes the enzyme molecules and this sensor showed a sensitivity of 37.93 μA mM^{-1} cm^{-2} for the concentration ranging between 0.08 mM and 12 mM possessing a long-term stability [106]. As the diabetic glucose concentration is above 7.0 mM and this sensor can be used for the determination of human blood sugar level. The linear range of this sensor is wider than that reported by Liu et al. [107], 0 to 7.8 mM (GOD on MWCNTs-chitosan matrix), Liu and Ju, et al. [108] 0.08 to 0.28 mM (GOD on colloidal gold modified carbon paste electrode) and Huang et al. [109] 0.5 to 11.1 mM (COD at a CdS nanoparticles modified electrode). Wu et al. [110] reported sensor GOD-graphene-PtNps-chitosan which has a detection of 0.6 μM glucose. Performance of this sensor was increased through intercalation of nanoparticles with graphene in order to increase the larger surface area and electrical conductivity. Here the interfering signals from the ascorbic acid and uric acid are negligible compared with the response to glucose [110,111].

Graphene based DNA electrochemical sensors provide a high sensitivity, high selectivity and low cost for the detection of DNA sequences and genes which are mutated and provide a simple and promising diagnosis of patients [112,113]. Sensor is based on the change in conductivity as the DNA molecules are physiosorbed on the graphene surface resulting in the change in the conductivity [114,115]. Zhou et al. [116] have reported DNA electrochemical sensor based on graphene obtained from chemical reduction of graphene oxide. The sensor detects four different current signals of nucleotide bases in DNA (guanine, adenine, cytosine and thymine) [11]. These devices can be separated into two namely, back-gated and liquid-gated [117]. Compared with that of the back-gated sensors, liquid-gated sensors are used widely for the detection of the DNA due to the sensitivity of 0.01 nM and ability to detect single mismatches [118]. This sensor has two electrode sources and a drain which connected by graphene sheet. A drop of buffer with DNA sample is loaded on the graphene surface then the gate voltage is measured so that DNA templates can be identified [119].

Electrochemical sensors based on graphene have been developed for the detection of the heavy metal ions. Li et al. [119,120] have identified a sensor based on Nafion-graphene composite film for the detection of Pb^{2+} and Cd^{2+}. The detection limits of 0.02 μL^{-1} were observed for both heavy metal ions. The linear range is 0.5 μL^{-1} to 50 μL^{-1} and 1.5 μL^{-1} to 30 μL^{-1} for Pb^{2+} and Cd^{2+}, respectively. The sensitivity was found to be higher than Nafion -bismuth electrode [121] and Nafion-CNT coated bismuth film electrode [122]. The enhanced performance is due to the unique properties of graphene such as capacity to adsorb metal ions [119,120].

3.4. Transparent Electrodes

Indium titanium oxide (ITO) and fluorine doped tin oxide (FTO) are commonly used as transparent coating for liquid crystal displays, solar cells and touch panels [123]. Due to the brittle nature and expensiveness of indium, graphene is being studied for the application in the transparent electrodes. The unique chemical, physical and mechanical stability makes graphene perfect material as a transparent electrode for solar cell and displays. Large surface area, inertness towards water and oxygen, high hole transparent mobility make graphene as an ideal material for photovoltaics [14].

Kim et al. [124] reported the growth of graphene films on a 300 nm thick nickel layer and the transmittance was reported to be 80%. Graphene layer is grown on the nickel surface by CVD and then the Nickel can be etched away using $FeCl_3$ leaving the graphene film alone. The reduction in the thickness of the nickel layer and growth time increases the transmittance to 93%. It was mentioned that further reduction of graphene layer thickness by ultraviolet etching increased the sheet resistance. In order to reduce the contact resistance, indium electrodes were deposited on the corner of the square. The minimum sheet resistance is around 280 Ω per square, which is around 30 times smaller than the lowest sheet resistance measured on assembled films [124]. Wang et al. [125] prepared a transparent graphene electrode for dye sensitized solar cells (DSSC) from exfoliated graphene oxide followed by thermal reduction. It was found that it has 70% of transparency over 1000 to 3000 nm and a good conductivity of 550 S/cm. It possesses short circuit photocurrent density of 1.01 mA cm^{-2}, open circuit of 0.7 V and overall power conversion efficiency of 0.26%. Low conversion efficiency was due to the low quality of graphene film [125]. Hong et al. [126] mentioned that composites of graphene with poly (3.4-ethylenedioxythiophene) had the transmittance of more than 80% and the energy conversion efficiency of 4.5%. The thickness of the synthesized film was around 60 nm. When the graphene weight percentage in the composite increased from 0 to 1 %, it was observed that the energy conversion was increased from 2.3 % to 4.5 %. This weight percentage was optimum and further increase of the weight percentage of graphene resulted in the reduction of the energy conversion efficiency [126]. Cu was chosen as a transparent electrode due to its low cost and high electrical conductivity. Thickness of the metal plays a crucial role in the performance because higher thickness gives improved resistance but reduces the transmittance. Kang et al. [127] reported the performance of Cu with different thickness (1, 3, 5, and 7 nm) incorporated with graphene. When the thickness of Cu is higher than 5 nm, the intensity of light entering the electrode is reduced. Cu with thickness of 3 nm exhibits a maximum current density of 12.64 mA/cm^2 [127]. Huang et al [128] reported a sandwich structure made of graphene and silver nanowire where silver nanowire is inserted between two layers of graphene. The stretchable properties of single layer graphene (SLG), double layer graphene (DLG) and sandwiched type structure was analyzed and it was observed that the resistance of the SLG was unmeasurable after 10 cycles, DLG have similar results after 40 cycles but the sand witched type gradually increased to 4 after 20 cycles. Thereafter conductivity structure stabilized up to 100 cycles. This sandwiched type structure exhibits an excellent stretching ability. Under tensile strain cracks and holes were formed in SLG (15.14 %) and DLG (10.28 %) but sandwiched type electrode only experienced fewer cracks and holes (6.4%). This result indicates that the Ag wire suppress the crack formation. This structure effectively reduces the oxidation of the silver wire and withhold the initial resistance of the electrode. Furthermore, it possessed low sheet resistance, high light transmittance (greater than 90 %) and long-term stability when compared to that of SLG and DLG [128].

3.5. Environmental Applications

With the start of industrial revolution in the mid-18th century, many advanced industries were developed with the use of steam, coal and fossil fuels. During this period, environmental pollution started to increase rapidly. Many mining industries released metal ions to the environment. Advancement of motor vehicles and the use of fossil fuels increased the greenhouse gases and other hazardous gases to the environment. Graphene and its derivatives play a crucial role in removing these pollutants from the environment.

3.5.1. Adsorption of Metal Ions

Metal ion pollutants enter into water bodies due to mining and industrial wastes. Generally, activated carbon has been used for the adsorption of metal ion pollutants, however, its production cost and regeneration restrict its applications. Owing to the high surface area, tunable surface chemistry, scalable production and non-corrosive property provide a path for the graphene-based materials for the application in metal ion adsorption [129,130]. Nanoparticles such as iron or iron oxide have been incorporated with the graphene (magnetic graphene nanocomposite) for the adsorption. Here the iron oxide is used to remove the chromium (VI). The graphene and magnetic graphene nanocomposite were compared for the Cr (VI) removal ability. The pure graphene exhibited lower removal efficiency of only 44.6 % (even high concentration of 3 g L^{-1}) while the magnetic graphene nanocomposite showed a higher removal of efficiency of 52.6 % (at lower concentration 0.25 g L^{-1}). The removal efficiency is further improved to 100 % when the 10% nanoparticle loaded with magnetic graphene nanocomposite. This composite completely removes Cr (VI) within 5 min (using 3 g L^{-1}). pH plays an important role for the adsorption of Cr (VI) and adsorption efficiency was higher at lower pH range (pH 1–3). [131,132]. Here the graphene is used as a physical support for iron oxide nanoparticle which improves the surface area and increases the number of binding sites for the metal ions [132]. Magnetically active iron oxide can be efficiently removed from solution using its magnetic properties [133].

3.5.2. Adsorption of Gases

Gosh et al. [134] have studied the adsorption of CO_2 and H_2 on the graphene surface and found that maximum adsorption of 37.93 % takes place by a single layer of graphene. Highest adsorption of H_2 was adsorbed around 3.1 weight % and the CO_2 adsorption was around 21–34 weight % [134]. Density functional theory study on the CO_2 adsorption on defective graphene sheets indicates that CO_2 is adsorbed exothermally on the vacancy defects in the graphene sheets. The energy of physisorption of CO_2 on the mono-vacancy site is around 210 m eV and the chemisorption energy is around 1.72 eV. Further studies revealed that the defective graphene showed four times higher adsorption than defect free graphene and CO_2 molecules could form C-O bond with the carbon atoms of graphene [135]. Liu et al. [136] have studied the influence of the oxygen functional group on the adsorption using CO_2-CH_4, CO_2-N_2, CO_2-H_2O mixtures. The results indicated that the CO_2 which have the oxygen functional group is preferentially adsorbed over CH_4 and N_2. CO_2 is preferentially adsorbed on the functionalized graphene surface with an induced polarity due to the strong quadrupolar moment of CO_2 compared to the species of N_2 and CH_4 where the quadrupole is week. In the CO_2-H_2O mixture, H_2O is adsorbed relatively higher. This study revealed that the surface of the graphene can be tuned for the selective adsorption of gaseous molecules [136].

3.5.3. Graphene Based Photocatalyst

Owing to the low cost and oxidizing ability, TiO_2 is generally used as a semiconductor forming graphene based photocatalyst for the degradation of the organic and biological contaminants. Reduced graphene oxides with 10% of Titanium nanotubes (TNT) showed the photocatalytic degradation of three times higher than the free TNT. TNTs were preferred due to their larger surface area when compared with that of the spherical nanoparticles [137]. As the organic dyes are aromatic, their adsorption on the graphene surface is enhanced by the π-π stacking interaction between sp^2 orbitals of the both systems [138]. Due to the combination of the graphene and photocatalyst, the band gap of the photocatalyst is reduced thus increasing the efficiency of the degradation [138,139]. Graphene improves the electron-hole recombination through the sp^2 hybridized network. Here the graphene acts as an electron acceptor and gives a conductive platform to transport electrons which involves in the oxidation and reduction reaction during degradation [138].

4. Conclusions

Graphene is a two-dimensional carbon network with a considerable research interest. Owing to its unique chemical and physical properties graphene has been investigated and used in many applications such as electronics, energy storing batteries, supercapacitors, solar cells and photocatalysts. Several synthetic methods have been applied to produce graphene such as mechanical exfoliation, liquid phase exfoliation, unzipping of carbon nanotubes, chemical vapor deposition and oxidation and reduction. Among these methods, chemical vapor deposition is the most promising method as this method is cost effective and can produce a large amount of graphene. N-doped graphene nanosheets and nanoparticles incorporated graphene have been investigated to improve the performance of the lithium ion batteries. The incorporation of nitrogen with graphene increases specific area and electron transfer within the network of graphene. Metal oxide nanoparticles such as RuO_2, NiO_2, MnO_2, Co_3O_4, ZnO and SnO_2 and CNT-graphene have been incorporated with graphene to boost the performance of the super capacitors. Among these Co_3O_4 nanoparticles exhibit the best performance. Both graphene and Co_3O_4 have strong interaction and this helps to prevents volume expansion/contraction and the aggregation of nanoparticles. Several electrochemical sensors such as gas sensors, biomolecular sensors and heavy metal ion sensors have been under research and in application as well. Sensors have been developed by the incorporation of metal nanoparticle and used to track NO_2, NH_3, DNT, CO, glucose level, DNA sequences, and metal ions (Pb^{2+}, Cd^{2+}). Biosensors have been developed with chitosan protein. Chitosan protein provides functional groups to make the sensors hydrophilic. Nanoparticles play an important role in these sensors by increasing the performance. Theses biosensors have been utilized for the detection of blood glucose level. Graphene has been studied for the transparent electrode in DSSC and several studies have been undergoing to improve the light conversion efficiencies. Graphene has been developed on Ni (1.01 mA cm^{-2}) and Cu to use as transparent electrodes. Performance of the transparent electrode was depended on the thickness of the metal and the weight percentage of the graphene. As Cu is a low cost and high electrical conductivity material, has an excellent performance of 12.64 mA/cm^2, when compared with the Ni. Unique chemical property of the graphene paved the way for the environmental applications such as adsorption of heavy metals, organic contaminants, gases and applications, where graphene functions as photocatalyst for the degradation of the organic pollutants. Compared to the adsorption of the gases CO_2, H_2O, CH_4, and N_2, CO_2 was preferentially adsorbed by functionalized graphene. As levels of CO_2 is drastically increasing this might be a solution to reduce the levels of CO_2 and to tackle the climate change. Graphene remains as a unique material with exceptional properties that could lead to promising applications.

Author Contributions: Conceptualization, A.S., P.I. and P.A.; data curation, A.S.; writing—original draft preparation, A.S.; writing—review and editing, N.K.; All authors have read and agreed to the published version of the manuscript.

Funding: This research received no funding.

Acknowledgments: We acknowledge the Department of Chemistry, University of Jaffna for providing technical support during the preparation of the manuscript.

Conflicts of Interest: The authors declare no conflict of interest.

References

1. Fang, B.; Chang, D.; Xu, Z.; Gao, C. A Review on Graphene Fibers: Expectations, Advances, and Prospects. *Adv. Mater.* **2020**, *32*, 1902664. [CrossRef]
2. Hass, J.; De Heer, W.A.A.; Conrad, E.H. The growth and morphology of epitaxial multilayer graphene. *J. Phys. Condens. Matter* **2008**, *20*, 323202. [CrossRef]
3. Yoo, E.; Kim, J.; Hosono, E.; Zhou, H.-S.; Kudo, T.; Honma, I. Large Reversible Li Storage of Graphene Nanosheet Families for Use in Rechargeable Lithium Ion Batteries. *Nano Lett.* **2008**, *8*, 2277–2282. [CrossRef]
4. Stoller, M.D.; Park, S.; Zhu, Y.; An, J.; Ruoff, R.S. Graphene-Based Ultracapacitors. *Nano Lett.* **2008**, *8*, 3498–3502. [CrossRef] [PubMed]
5. Kou, R.; Shao, Y.; Wang, D.; Engelhard, M.H.; Kwak, J.H.; Wang, J.; Viswanathan, V.V.; Wang, C.; Lin, Y.; Wang, Y.; et al. Enhanced activity and stability of Pt catalysts on functionalized graphene sheets for electrocatalytic oxygen reduction. *Electrochem. Commun.* **2009**, *11*, 954–957. [CrossRef]
6. Wu, J.; Becerril, H.A.; Bao, Z.; Liu, Z.; Chen, Y.; Peumans, P. Organic solar cells with solution-processed graphene transparent electrodes. *Appl. Phys. Lett.* **2008**, *92*, 263302. [CrossRef]
7. Sun, S.; Gao, L.; Liu, Y. Enhanced dye-sensitized solar cell using graphene-TiO_2 photoanode prepared by heterogeneous coagulation. *Appl. Phys. Lett.* **2010**, *96*, 083113. [CrossRef]
8. Balandin, A.A.; Ghosh, S.; Bao, W.; Calizo, I.; Teweldebrhan, D.; Miao, F.; Lau, C.N. Superior Thermal Conductivity of Single-Layer Graphene. *Nano Lett.* **2008**, *8*, 902–907. [CrossRef]
9. Service, R.F. Materials science. Carbon sheets an atom thick give rise to graphene dreams. *Science* **2009**, *324*, 875–877. [PubMed]
10. Lee, C.; Wei, X.; Kysar, J.W.; Hone, J. Measurement of the Elastic Properties and Intrinsic Strength of Monolayer Graphene. *Science* **2008**, *321*, 385–388. [CrossRef]
11. Zheng, P.; Wu, N. Fluorescence and Sensing Applications of Graphene Oxide and Graphene Quantum Dots: A Review. *Chem.—Asian J.* **2017**, *12*, 2343–2353. [CrossRef] [PubMed]
12. Allen, M.J.; Tung, V.C.; Kaner, R.B. Honeycomb Carbon: A Review of Graphene. *Chem. Rev.* **2009**, *110*, 132–145. [CrossRef]
13. Enoki, T.; Endo, M.; Suzuki, M. *Graphite Intercalation Compounds and Applications*; Oxford University Press: New York, NY, USA, 2003; p. 452.
14. Choi, W.; Lahiri, I.; Seelaboyina, R.; Kang, Y.S. Synthesis of Graphene and Its Applications: A Review. *Crit. Rev. Solid State Mater. Sci.* **2010**, *35*, 52–71. [CrossRef]
15. Nimbalkar, A.; Kim, H. Opportunities and Challenges in Twisted Bilayer Graphene: A Review. *Nano-Micro Lett.* **2020**, *12*, 1–20. [CrossRef]
16. Bacon, M.; Bradley, S.J.; Nann, T. Graphene Quantum Dots. *Part. Part. Syst. Charact.* **2014**, *31*, 415–428. [CrossRef]
17. Geim, A.K.; Novoselov, K.S. The rise of graphene. *Nat. Mater.* **2007**, *6*, 183–191. [CrossRef] [PubMed]
18. Chen, J.-H.; Jang, C.; Xiao, S.; Ishigami, M.; Fuhrer, M.S. Intrinsic and extrinsic performance limits of graphene devices on SiO2. *Nat. Nanotechnol.* **2008**, *3*, 206–209. [CrossRef]
19. Balandin, A.A. Phononics of Graphene and Related Materials. *ACS Nano* **2020**, *14*, 5170–5178. [CrossRef]
20. Rhee, K.Y. Electronic and Thermal Properties of Graphene. *Nanomaterials* **2020**, *10*, 926. [CrossRef]
21. Guo, X.; Cheng, S.; Cai, W.; Zhang, Y.; Zhang, X.A. A review of carbon-based thermal interface materials: Mechanism, thermal measurements and thermal properties. *Mater. Des.* **2021**, *209*, 109936. [CrossRef]
22. Reina, A.; Jia, X.; Ho, J.; Nezich, D.; Son, H.; Bulovic, V.; Dresselhaus, M.S.; Kong, J. Large Area, Few-Layer Graphene Films on Arbitrary Substrates by Chemical Vapor Deposition. *Nano Lett.* **2009**, *9*, 30–35. [CrossRef]
23. Li, X.; Cai, W.; An, J.; Kim, S.; Nah, J.; Yang, D.; Piner, R.; Velamakanni, A.; Jung, I.; Tutuc, E.; et al. Large-Area Synthesis of High-Quality and Uniform Graphene Films on Copper Foils. *Science* **2009**, *324*, 1312–1314. [CrossRef] [PubMed]
24. Shams, S.S.; Zhang, R.; Zhu, J. Graphene synthesis: A Review. *Mater. Sci.-Pol.* **2016**, *33*, 566–578. [CrossRef]
25. Muñoz, R.; Gómez-Aleixandre, C. Review of CVD Synthesis of Graphene. *Chem. Vap. Depos.* **2013**, *19*, 297–322. [CrossRef]
26. Emtsev, K.V.; Bostwick, A.; Horn, K.; Jobst, J.; Kellogg, G.L.; Ley, L.; McChesney, J.; Ohta, T.; Reshanov, S.A.; Röhrl, J.; et al. Towards wafer-size graphene layers by atmospheric pressure graphitization of silicon carbide. *Nat. Mater.* **2009**, *8*, 203–207. [CrossRef] [PubMed]
27. Zhang, R.; Dong, Y.; Kong, W.; Han, W.; Tan, P.-H.; Liao, Z.-M.; Wu, X.; Yu, D. Growth of large domain epitaxial graphene on the C-face of SiC. *J. Appl. Phys.* **2012**, *112*, 104307. [CrossRef]

28. Stankovich, S.; Dikin, D.A.; Piner, R.D.; Kohlhaas, K.A.; Kleinhammes, A.; Jia, Y.; Wu, Y.; Nguyen, S.; Ruoff, R.S. Synthesis of graphene-based nanosheets via chemical reduction of exfoliated graphite oxide. *Carbon* **2007**, *45*, 1558–1565. [CrossRef]
29. Schniepp, H.C.; Li, J.-L.; McAllister, M.J.; Sai, H.; Herrera-Alonso, M.; Adamson, D.H.; Prud'Homme, R.K.; Car, R.; Saville, D.A.; Aksay, I.A. Functionalized Single Graphene Sheets Derived from Splitting Graphite Oxide. *J. Phys. Chem. B* **2006**, *110*, 8535–8539. [CrossRef]
30. Park, S.; Ruoff, R.S. Chemical methods for the production of graphenes. *Nat. Nanotechnol.* **2009**, *4*, 217–224. [CrossRef]
31. Hummers, W.S.; Offeman, R.E. Preparation of Graphitic Oxide. *J. Am. Chem. Soc.* **1958**, *80*, 1339. [CrossRef]
32. Lawal, A.T. Graphene-based nano composites and their applications. A review. *Biosens. Bioelectron.* **2019**, *141*, 111384. [CrossRef] [PubMed]
33. Zhao, W.; Liu, H.; Meng, N.; Jian, M.; Wang, H.; Zhang, X. Graphene oxide incorporated thin film nanocomposite membrane at low concentration monomers. *J. Membr. Sci.* **2018**, *565*, 380–389. [CrossRef]
34. Akyüz, D.; Koca, A. Photocatalytic hydrogen production with reduced graphene oxide (RGO)-CdZnS nano-composites synthesized by solvothermal decomposition of dimethyl sulfoxide as the sulfur source. *J. Photochem. Photobiol. A Chem.* **2018**, *364*, 625–634. [CrossRef]
35. Kumar, N.; Rodriguez, J.R.; Pol, V.G.; Sen, A. Facile synthesis of 2D graphene oxide sheet enveloping ultrafine 1D LiMn2O4 as interconnected framework to enhance cathodic property for Li-ion battery. *Appl. Surf. Sci.* **2019**, *463*, 132–140. [CrossRef]
36. Aghazadeh, M. One-step Electrophoretic/electrochemical Synthesis of Reduced Graphene Oxide/Manganese Oxide (rGO-Mn3O4) Nanocomposite and Study of its Capacitive Performance. *Anal. Bioanal. Electrochem.* **2018**, *10*, 961–973.
37. Jasmi, F.; Azeman, N.H.; Bakar, A.A.A.; Zan, M.S.D.; Badri, K.H.; Su'ait, M.S. Ionic Conductive Polyurethane-Graphene Nanocomposite for Performance Enhancement of Optical Fiber Bragg Grating Temperature Sensor. *IEEE Access* **2018**, *6*, 47355–47363. [CrossRef]
38. Novoselov, K.S.; Geim, A.K.; Morozov, S.V.; Jiang, D.; Zhang, Y.; Dubonos, S.V.; Grigorieva, I.V.; Firsov, A.A. Electric Field Effect in Atomically Thin Carbon Films. *Science* **2004**, *306*, 666–669. [CrossRef]
39. Young, R.; Kinloch, I.A.; Gong, L.; Novoselov, K. The mechanics of graphene nanocomposites: A review. *Compos. Sci. Technol.* **2012**, *72*, 1459–1476. [CrossRef]
40. Lee, X.J.; Hiew, B.Y.Z.; Lai, K.C.; Lee, L.Y.; Gan, S.; Thangalazhy-Gopakumar, S.; Rigby, S. Review on graphene and its derivatives: Synthesis methods and potential industrial impleentation. *J. Taiwan Inst. Chem. Eng.* **2019**, *98*, 163–180. [CrossRef]
41. Hernandez, Y.; Nicolosi, V.; Lotya, M.; Blighe, F.M.; Sun, Z.; De, S.; McGovern, I.T.; Holland, B.; Byrne, M.; Gun'Ko, Y.; et al. High-yield production of graphene by liquid-phase exfoliation of graphite. *Nat. Nanotechnol.* **2008**, *3*, 563–568. [CrossRef]
42. Ciesielski, A.; Samorì, P. Graphene via sonication assisted liquid-phase exfoliation. *Chem. Soc. Rev.* **2014**, *43*, 381–398. [CrossRef]
43. Edwards, R.S.; Coleman, K.S. Graphene synthesis: Relationship to applications. *Nanoscale* **2013**, *5*, 38–51. [CrossRef] [PubMed]
44. Yi, M.; Shen, Z. A review on mechanical exfoliation for the scalable production of graphene. *J. Mater. Chem. A* **2015**, *3*, 11700–11715. [CrossRef]
45. Baig, Z.; Mamat, O.; Mustapha, M.; Mumtaz, A.; Munir, K.; Sarfraz, M. Investigation of tip sonication effects on structural quality of graphene nanoplatelets (GNPs) for superior solvent dispersion. *Ultrason. Sonochem.* **2018**, *45*, 133–149. [CrossRef] [PubMed]
46. Jiao, L.; Zhang, L.; Wang, X.; Diankov, G.; Dai, H. Narrow graphene nanoribbons from carbon nanotubes. *Nature* **2009**, *458*, 877–880. [CrossRef] [PubMed]
47. Kosynkin, D.V.; Higginbotham, A.L.; Sinitskii, A.; Lomeda, J.R.; Dimiev, A.; Price, B.K.; Tour, J. Longitudinal unzipping of carbon nanotubes to form graphene nanoribbons. *Nature* **2009**, *458*, 872–876. [CrossRef]
48. Cano-Márquez, A.G.; Rodriguez-Macias, F.; Campos-Delgado, J.; Espinosa-González, C.G.; Tristán-López, F.; Ramírez-González, D.; Cullen, D.; Smith, D.; Terrones, M.; Vega-Cantu, Y.I. Ex-MWNTs: Graphene Sheets and Ribbons Produced by Lithium Intercalation and Exfoliation of Carbon Nanotubes. *Nano Lett.* **2009**, *9*, 1527–1533. [CrossRef]
49. Thess, A.; Lee, R.; Nikolaev, P.; Dai, H.; Petit, P.; Robert, J.; Xu, C.; Lee, Y.H.; Kim, S.G.; Rinzler, A.G.; et al. Crystalline Ropes of Metallic Carbon Nanotubes. *Science* **1996**, *273*, 483–487. [CrossRef]
50. Liu, J.; Kim, G.-H.; Xue, Y.; Kim, J.Y.; Baek, J.-B.; Durstock, M.; Dai, L. Graphene Oxide Nanoribbon as Hole Extraction Layer to Enhance Efficiency and Stability of Polymer Solar Cells. *Adv. Mater.* **2013**, *26*, 786–790. [CrossRef]
51. Mahmoud, W.E.; Al-Hazmi, F.S.; Al-Harbi, G.H. Wall by wall controllable unzipping of MWCNTs via intercalation with oxalic acid to produce multilayers graphene oxide ribbon. *Chem. Eng. J.* **2015**, *281*, 192–198. [CrossRef]
52. Zhang, G.; Manjooran, N. Nanofabrication and its application in renewable energy. *Johns. Matthey Technol. Rev.* **2014**, *58*, 221–223.
53. Bhuyan, M.S.A.; Uddin, M.N.; Islam, M.M.; Bipasha, F.A.; Hossain, S.S. Synthesis of graphene. *Int. Nano Lett.* **2016**, *6*, 65–83. [CrossRef]
54. Arora, N.; Sharma, N.N. Arc discharge synthesis of carbon nanotubes: Comprehensive review. *Diam. Relat. Mater.* **2014**, *50*, 135–150. [CrossRef]
55. Li, Y.; Chen, Q.; Xu, K.; Kaneko, T.; Hatakeyama, R. Synthesis of graphene nanosheets from petroleum asphalt by pulsed arc discharge in water. *Chem. Eng. J.* **2013**, *215–216*, 45–49. [CrossRef]
56. Zhang, Y.; Zhang, L.; Zhou, C. Review of Chemical Vapor Deposition of Graphene and Related Applications. *Acc. Chem. Res.* **2013**, *46*, 2329–2339. [CrossRef] [PubMed]
57. Arjmandi-Tash, H.; Lebedev, N.; van Deursen, P.M.; Aarts, J.; Schneider, G.F. Hybrid cold and hot-wall reaction chamber for the rapid synthesis of uniform graphene. *Carbon* **2017**, *118*, 438–442. [CrossRef]

58. Cooper, D.R.; D'Anjou, B.; Ghattamaneni, N.; Harack, B.; Hilke, M.; Horth, A.; Majlis, N.; Massicotte, M.; Vandsburger, L.; Whiteway, E.; et al. Experimental Review of Graphene. *ISRN Condens. Matter Phys.* **2012**, *2012*, 501686. [CrossRef]
59. Mishra, N.; Boeckl, J.; Motta, N.; Iacopi, F. Graphene growth on silicon carbide: A review. *Phys. Status Solidi A* **2016**, *213*, 2277–2289. [CrossRef]
60. Cambaz, Z.G.; Yushin, G.; Osswald, S.; Mochalin, V.; Gogotsi, Y. Noncatalytic synthesis of carbon nanotubes, graphene and graphite on SiC. *Carbon* **2008**, *46*, 841–849. [CrossRef]
61. Brodie, B.C. Sur le poids atomique du graphite. *Ann. Chim. Phys.* **1860**, *59*, e472.
62. Staudenmaier, L. Verfahren zur Darstellung der Graphitsäure. *Eur. J. Inorg. Chem.* **1898**, *31*, 1481–1487. [CrossRef]
63. Hofmann, U.; König, E. Untersuchungen über graphitoxyd. *Chemie* **1937**, *234*, 311–336. [CrossRef]
64. Somanathan, T.; Prasad, K.; Ostrikov, K.; Saravanan, A.; Krishna, V.M. Graphene Oxide Synthesis from Agro Waste. *Nanomaterials* **2015**, *5*, 826–834. [CrossRef]
65. Yu, H.; Zhang, B.; Bulin, C.; Li, R.; Xing, R. High-efficient Synthesis of Graphene Oxide Based on Improved Hummers Method. *Sci. Rep.* **2016**, *6*, 36143. [CrossRef]
66. Chen, J.; Yao, B.; Li, C.; Shi, G. An improved Hummers method for eco-friendly synthesis of graphene oxide. *Carbon* **2013**, *64*, 225–229. [CrossRef]
67. Yoon, G.; Seo, D.-H.; Ku, K.; Kim, J.; Jeon, S.; Kang, K. Factors Affecting the Exfoliation of Graphite Intercalation Compounds for Graphene Synthesis. *Chem. Mater.* **2015**, *27*, 2067–2073. [CrossRef]
68. Konios, D.; Stylianakis, M.M.; Stratakis, E.; Kymakis, E. Dispersion behaviour of graphene oxide and reduced graphene oxide. *J. Colloid Interface Sci.* **2014**, *430*, 108–112. [CrossRef] [PubMed]
69. Chua, C.K.; Pumera, M. Chemical reduction of graphene oxide: A synthetic chemistry viewpoint. *Chem. Soc. Rev.* **2013**, *43*, 291–312. [CrossRef] [PubMed]
70. Kumar, P.V.; Bardhan, N.M.; Chen, G.-Y.; Li, Z.; Belcher, A.M.; Grossman, J.C. New insights into the thermal reduction of graphene oxide: Impact of oxygen clustering. *Carbon* **2016**, *100*, 90–98. [CrossRef]
71. Zheng, X.; Peng, Y.; Yang, Y.; Chen, J.; Tian, H.; Cui, X.; Zheng, W. Hydrothermal reduction of graphene oxide; effect on surface-enhanced Raman scattering. *J. Raman Spectrosc.* **2016**, *48*, 97–103. [CrossRef]
72. Chen, W.; Yan, L. Preparation of graphene by a low-temperature thermal reduction at atmosphere pressure. *Nanoscale* **2010**, *2*, 559–563. [CrossRef] [PubMed]
73. Yusuf, M.; Kumar, M.; Khan, M.A.; Sillanpää, M.; Arafat, H. A review on exfoliation, characterization, environmental and energy applications of graphene and graphene-based composites. *Adv. Colloid Interface Sci.* **2019**, *273*, 102036. [CrossRef] [PubMed]
74. Olabi, A.; Abdelkareem, M.A.; Wilberforce, T.; Sayed, E.T. Application of graphene in energy storage device–A review. *Renew. Sustain. Energy Rev.* **2020**, *135*, 110026. [CrossRef]
75. Zhao, X.; Li, H.; Han, F.; Dai, M.; Sun, Y.; Song, Z.; Han, D.; Niu, L. Electrochemical exfoliation of graphene as an anode material for ultra-long cycle lithium ion batteries. *J. Phys. Chem. Solids* **2019**, *139*, 109301. [CrossRef]
76. Ji, X.; Mu, Y.; Liang, J.; Jiang, T.; Zeng, J.; Lin, Z.; Lin, Y.; Yu, J. High yield production of 3D graphene powders by thermal chemical vapor deposition and application as highly efficient conductive additive of lithium ion battery electrodes. *Carbon* **2021**, *176*, 21–30. [CrossRef]
77. Folorunso, O.; Hamam, Y.; Sadiku, R.; Ray, S.S.; Adekoya, G.J. Investigation of graphene loaded polypyrrole for lithium-ion battery. *Mater. Today Proc.* **2020**, *38*, 635–638. [CrossRef]
78. Wu, Z.-S.; Ren, W.; Wen, L.; Gao, L.; Zhao, J.; Chen, Z.; Zhou, G.; Li, F.; Cheng, H.-M. Graphene Anchored with Co3O4 Nanoparticles as Anode of Lithium Ion Batteries with Enhanced Reversible Capacity and Cyclic Performance. *ACS Nano* **2010**, *4*, 3187–3194. [CrossRef]
79. Tian, R.; Zhang, Y.; Chen, Z.; Duan, H.; Xu, B.; Guo, Y.; Kang, H.; Li, H.; Liu, H. The effect of annealing on a 3D SnO2/graphene foam as an advanced lithium-ion battery anode. *Sci. Rep.* **2016**, *6*, 19195. [CrossRef]
80. Wang, D.; Choi, D.; Li, J.; Yang, Z.; Nie, Z.; Kou, R.; Hu, D.; Wang, C.; Saraf, L.V.; Zhang, J.; et al. Self-Assembled TiO_2–Graphene Hybrid Nanostructures for Enhanced Li-Ion Insertion. *ACS Nano* **2009**, *3*, 907–914. [CrossRef]
81. Idrees, M.; Batool, S.; Kong, J.; Zhuang, Q.; Liu, H.; Shao, Q.; Lu, N.; Feng, Y.; Wujcik, E.K.; Gao, Q.; et al. Polyborosilazane derived ceramics-nitrogen sulfur dual doped graphene nanocomposite anode for enhanced lithium ion batteries. *Electrochim. Acta* **2019**, *296*, 925–937. [CrossRef]
82. Lee, S.J.; Theerthagiri, J.; Nithyadharseni, P.; Arunachalam, P.; Balaji, D.; Kumar, A.M.; Madhavan, J.; Mittal, V.; Choi, M.Y. Heteroatom-doped graphene-based materials for sustainable energy applications: A review. *Renew. Sustain. Energy Rev.* **2021**, *143*, 110849. [CrossRef]
83. Reddy, A.L.M.; Srivastava, A.; Gowda, S.R.; Gullapalli, H.; Dubey, M.; Ajayan, P.M. Synthesis of Nitrogen-Doped Graphene Films For Lithium Battery Application. *ACS Nano* **2010**, *4*, 6337–6342. [CrossRef] [PubMed]
84. Wang, H.; Zhang, C.; Liu, Z.; Wang, L.; Han, P.; Xu, H.; Zhang, K.; Dong, S.; Yao, J.; Cui, G. Nitrogen-doped graphene nanosheets with excellent lithium storage properties. *J. Mater. Chem.* **2011**, *21*, 5430–5434. [CrossRef]
85. Yang, J.; Jia, K.; Wang, M.; Liu, S.; Hu, C.; Zhang, K.; Zhang, Y.; Qiu, J. Fabrication of nitrogen-doped porous graphene hybrid nanosheets from metal–organic frameworks for lithium-ion batteries. *Nanotechnology* **2019**, *31*, 145402. [CrossRef]
86. Wu, Z.-S.; Ren, W.; Xu, L.; Li, F.; Cheng, H.-M. Doped Graphene Sheets as Anode Materials with Superhigh Rate and Large Capacity for Lithium Ion Batteries. *ACS Nano* **2011**, *5*, 5463–5471. [CrossRef]

87. Paek, S.-M.; Yoo, E.; Honma, I. Enhanced cyclic performance and lithium storage capacity of SnO_2/graphene nanoporous electrodes with three-dimensionally delaminated flexible structure. *Nano Lett.* **2009**, *9*, 72–75. [CrossRef] [PubMed]
88. Wu, Z.-S.; Wang, D.-W.; Ren, W.; Zhao, J.; Zhou, G.; Li, F.; Cheng, H.-M. Anchoring Hydrous RuO_2 on Graphene Sheets for High-Performance Electrochemical Capacitors. *Adv. Funct. Mater.* **2010**, *20*, 3595–3602. [CrossRef]
89. Zheng, F.; Yang, Y.; Chen, Q. High lithium anodic performance of highly nitrogen-doped porous carbon prepared from a metal-organic framework. *Nat. Commun.* **2014**, *5*, 5261. [CrossRef]
90. Zhu, Y.; Li, L.; Zhang, C.; Casillas, G.; Sun, Z.; Yan, Z.; Ruan, G.; Peng, Z.; Raji, A.-R.; Kittrell, C.; et al. A seamless three-dimensional carbon nanotube graphene hybrid material. *Nat. Commun.* **2012**, *3*, 1225. [CrossRef]
91. Yan, Z.; Ma, L.; Zhu, Y.; Lahiri, I.; Hahm, M.G.; Liu, Z.; Yang, S.; Xiang, C.; Lu, W.; Peng, Z.J.A. Three-dimensional metal–graphene–nanotube multifunctional hybrid materials. *ACS Nano* **2013**, *7*, 58–64. [CrossRef]
92. Xing, L.L.; Wu, X.; Huang, K.J. High-performance supercapacitor based on three-dimensional flower-shaped $Li_4Ti_5O_{12}$-graphene hybrid and pine needles derived honeycomb carbon. *J. Colloid Interface Sci.* **2018**, *529*, 171–179. [CrossRef] [PubMed]
93. Wang, H.; Tran, D.; Moussa, M.; Stanley, N.; Tung, T.T.; Yu, L.; Yap, P.L.; Ding, F.; Qian, J.; Losic, D. Improved preparation of MoS_2/graphene composites and their inks for supercapacitors applications. *Mater. Sci. Eng. B* **2020**, *262*, 114700. [CrossRef]
94. Chen, Z.; Zhao, S.; Zhao, H.; Zou, Y.; Yu, C.; Zhong, W. Nitrogen-doped interpenetrating porous carbon/graphene networks for supercapacitor applications. *Chem. Eng. J.* **2020**, *409*, 127891. [CrossRef]
95. Gopakumar, D.A.; Arumughan, V.; Pottathara, Y.B.; Sisant, S.K.; Pasquini, D.; Bračič, M.; Seantier, B.; Nzihou, A.; Thomas, S.; Rizal, S.; et al. Robust Superhydrophobic Cellulose Nanofiber Aerogel for Multifunctional Environmental Applications. *Polymers* **2019**, *11*, 495. [CrossRef]
96. Pottathara, Y.B.; Tiyyagura, H.R.; Ahmad, Z.; Sadasivuni, K.K. Graphene based aerogels: Fundamentals and applications as supercapacitors. *J. Energy Storage* **2020**, *30*, 101549. [CrossRef]
97. Zhang, L.; Chen, H.; Lu, X.; Wang, Y.; Tan, L.; Sui, D.; Qi, W. Fabrication of N, S co-doped graphene aerogel for high-performance supercapacitors: π-conjugated planar molecules as efficient dopants and pillared agents. *Appl. Surf. Sci.* **2020**, *529*, 147022. [CrossRef]
98. Arkhipova, E.A.; Ivanov, A.S.; Maslakov, K.I.; Egorov, A.V.; Savilov, S.V.; Lunin, V.V. Mesoporous graphene nanoflakes for high performance supercapacitors with ionic liquid electrolyte. *Microporous Mesoporous Mater.* **2019**, *294*, 109851. [CrossRef]
99. Schedin, F.; Geim, A.K.; Morozov, S.V.; Hill, E.W.; Blake, P.; Katsnelson, M.I.; Novoselov, K.S. Detection of individual gas molecules adsorbed on graphene. *Nat. Mater.* **2007**, *6*, 652–655. [CrossRef]
100. Novoselov, K.; Geim, A.K.; Morozov, S.; Jiang, D.; Katsnelson, M.I.; Grigorieva, I.V.; Dubonos, S.V.; Firsov, A.A. Two-dimensional gas of massless Dirac fermions in graphene. *Nature* **2005**, *438*, 197–200. [CrossRef]
101. Novoselov, K.S.; Jiang, Z.; Zhang, Y.; Morozov, S.V.; Stormer, H.L.; Zeitler, U.; Maan, J.C.; Boebinger, G.S.; Kim, P.; Geim, A.K. Room-Temperature Quantum Hall Effect in Graphene. *Science* **2007**, *315*, 1379. [CrossRef]
102. Fowler, J.D.; Allen, M.J.; Tung, V.; Yang, Y.; Kaner, R.B.; Weiller, B.H. Practical Chemical Sensors from Chemically Derived Graphene. *ACS Nano* **2009**, *3*, 301–306. [CrossRef] [PubMed]
103. Sundaram, R.S.; Gómez-Navarro, C.; Balasubramanian, K.; Burghard, M.; Kern, K.J.A.M. Electrochemical modification of graphene. *Adv. Mater.* **2008**, *20*, 3050–3053. [CrossRef]
104. Gupta, S.; Meek, R. Metal nanoparticles-grafted functionalized graphene coated with nanostructured polyaniline 'hybrid' nanocomposites as high-performance biosensors. *Sens. Actuators B Chem.* **2018**, *274*, 85–101. [CrossRef]
105. Shan, C.; Yang, H.; Song, J.; Han, D.; Ivaska, A.; Niu, L. Direct electrochemistry of glucose oxidase and biosensing for glucose based on graphene. *Anal. Chem.* **2009**, *81*, 2378–2382. [CrossRef]
106. Kang, X.; Wang, J.; Wu, H.; Aksay, I.A.; Liu, J.; Lin, Y. Glucose oxidase–graphene–chitosan modified electrode for direct electrochemistry and glucose sensing. *Biosens. Bioelectron.* **2009**, *25*, 901–905. [CrossRef] [PubMed]
107. Liu, Y.; Wang, M.; Zhao, F.; Xu, Z.; Dong, S. The direct electron transfer of glucose oxidase and glucose biosensor based on carbon nanotubes/chitosan matrix. *Biosens. Bioelectron.* **2005**, *21*, 984–988. [CrossRef]
108. Liu, S.; Ju, H. Reagentless glucose biosensor based on direct electron transfer of glucose oxidase immobilized on colloidal gold modified carbon paste electrode. *Biosens. Bioelectron.* **2003**, *19*, 177–183. [CrossRef]
109. Huang, Y.; Zhang, W.; Xiao, H.; Li, G. An electrochemical investigation of glucose oxidase at a CdS nanoparticles modified electrode. *Biosens. Bioelectron.* **2005**, *21*, 817–821. [CrossRef]
110. Wu, H.; Wang, J.; Kang, X.; Wang, C.; Wang, D.; Liu, J.; Aksay, I.A.; Lin, Y.J.T. Glucose biosensor based on immobilization of glucose oxidase in platinum nanoparticles/graphene/chitosan nanocomposite film. *Talanta* **2009**, *80*, 403–406. [CrossRef]
111. Liu, H.; Gao, J.; Xue, M.; Zhu, N.; Zhang, M.; Cao, T.J.L. Processing of graphene for electrochemical application: Noncovalently functionalize graphene sheets with water-soluble electroactive methylene green. *Langmuir* **2009**, *25*, 12006–12010. [CrossRef]
112. Sassolas, A.; Leca-Bouvier, B.; Blum, L.J. DNA Biosensors and Microarrays. *Chem. Rev.* **2007**, *108*, 109–139. [CrossRef]
113. Drummond, T.G.; Hill, M.G.; Barton, J.K. Electrochemical DNA sensors. *Nat. Biotechnol.* **2003**, *21*, 1192–1199. [CrossRef] [PubMed]
114. Ohno, Y.; Maehashi, K.; Matsumoto, K. Label-Free Biosensors Based on Aptamer-Modified Graphene Field-Effect Transistors. *J. Am. Chem. Soc.* **2010**, *132*, 18012–18013. [CrossRef]
115. Kybert, N.J.; Han, G.H.; Lerner, M.B.; Dattoli, E.N.; Esfandiar, A.; Johnson, A.T.C. Scalable arrays of chemical vapor sensors based on DNA-decorated graphene. *Nano Res.* **2013**, *7*, 95–103. [CrossRef]

116. Zhou, M.; Zhai, Y.; Dong, S. Electrochemical Sensing and Biosensing Platform Based on Chemically Reduced Graphene Oxide. *Anal. Chem.* **2009**, *81*, 5603–5613. [CrossRef] [PubMed]
117. Green, N.S.; Norton, M.L. Interactions of DNA with graphene and sensing applications of graphene field-effect transistor devices: A review. *Anal. Chim. Acta* **2015**, *853*, 127–142. [CrossRef] [PubMed]
118. Dong, X.; Shi, Y.; Huang, W.; Chen, P.; Li, L.-J. Electrical Detection of DNA Hybridization with Single-Base Specificity Using Transistors Based on CVD-Grown Graphene Sheets. *Adv. Mater.* **2010**, *22*, 1649–1653. [CrossRef]
119. Li, J.; Guo, S.; Zhai, Y.; Wang, E. High-sensitivity determination of lead and cadmium based on the Nafion-graphene composite film. *Anal. Chim. Acta* **2009**, *649*, 196–201. [CrossRef]
120. Li, J.; Guo, S.; Zhai, Y.; Wang, E.J.E.C. Nafion–graphene nanocomposite film as enhanced sensing platform for ultrasensitive determination of cadmium. *Electrochem. Commun.* **2009**, *11*, 1085–1088. [CrossRef]
121. Kefala, G.; Economou, A.; Voulgaropoulos, A. A study of Nafion-coated bismuth-film electrodes for the determination of trace metals by anodic stripping voltammetry. *Analyst* **2004**, *129*, 1082–1090. [CrossRef]
122. Liu, B.; Lu, L.; Wang, M.; Zi, Y. A Study of Nafion-Coated Bismuth-Film Electrode for the Determination of Zinc, Lead, and Cadmium in Blood Samples. *Electroanalysis* **2008**, *20*, 2363–2369. [CrossRef]
123. Choi, W.-J.; Kwak, D.-J.; Park, C.-S.; Sung, Y.-M. Characterization of Transparent Conductive ITO, ITiO, and FTO Films for Application in Photoelectrochemical Cells. *J. Nanosci. Nanotechnol.* **2012**, *12*, 3394–3397. [CrossRef] [PubMed]
124. Kim, K.S.; Zhao, Y.; Jang, H.; Lee, S.Y.; Kim, J.M.; Kim, K.S.; Ahn, J.-H.; Kim, P.; Choi, J.-Y.; Hong, B.H. Large-scale pattern growth of graphene films for stretchable transparent electrodes. *Nature* **2009**, *457*, 706–710. [CrossRef] [PubMed]
125. Wang, X.; Zhi, L.; Müllen, K. Transparent, Conductive Graphene Electrodes for Dye-Sensitized Solar Cells. *Nano Lett.* **2007**, *8*, 323–327. [CrossRef] [PubMed]
126. Hong, W.; Xu, Y.; Lu, G.; Li, C.; Shi, G. Transparent graphene/PEDOT–PSS composite films as counter electrodes of dye-sensitized solar cells. *Electrochem. Commun.* **2008**, *10*, 1555–1558. [CrossRef]
127. Kang, J.H.; Choi, S.; Park, Y.J.; Park, J.S.; Cho, N.S.; Cho, S.; Walker, B.; Choi, D.S.; Shin, J.-W.; Seo, J.H. Cu/graphene hybrid transparent conducting electrodes for organic photovoltaic devices. *Carbon* **2020**, *171*, 341–349. [CrossRef]
128. Huang, C.-H.; Wu, H.-C.; Chen, B.-F.; Li, Y.-C. Graphene/Silver Nanowires/Graphene Sandwich Composite for Stretchable Transparent Electrodes and Its Fracture Mechanism. *Micromachines* **2021**, *12*, 512. [CrossRef]
129. Ren, X.; Chen, C.; Nagatsu, M.; Wang, X. Carbon nanotubes as adsorbents in environmental pollution management: A review. *Chem. Eng. J.* **2011**, *170*, 395–410. [CrossRef]
130. Sitko, R.; Zawisza, B.; Malicka, E. Graphene as a new sorbent in analytical chemistry. *TrAC Trends Anal. Chem.* **2013**, *51*, 33–43. [CrossRef]
131. Yao, Y.; Yang, Z.; Zhang, D.; Peng, W.; Sun, H.; Wang, S. Magnetic $CoFe_2O_4$–graphene hybrids: Facile synthesis, characterization, and catalytic properties. *Ind. Eng. Chem. Res.* **2012**, *51*, 6044–6051. [CrossRef]
132. Zhu, J.; Wei, S.; Gu, H.; Rapole, S.B.; Wang, Q.; Luo, Z.; Haldolaarachchige, N.; Young, D.P.; Guo, Z. One-Pot Synthesis of Magnetic Graphene Nanocomposites Decorated with Core@Double-shell Nanoparticles for Fast Chromium Removal. *Environ. Sci. Technol.* **2011**, *46*, 977–985. [CrossRef]
133. Liu, M.; Chen, C.; Hu, J.; Wu, X.; Wang, X. Synthesis of magnetite/graphene oxide composite and application for cobalt (II) removal. *J. Phys. Chem. C* **2011**, *115*, 25234–25240. [CrossRef]
134. Ghosh, A.; Subrahmanyam, K.S.; Krishna, K.S.; Datta, S.; Govindaraj, A.; Pati, S.K.; Rao, C.N.R. Uptake of H_2 and CO_2 by Graphene. *J. Phys. Chem. C* **2008**, *112*, 15704–15707. [CrossRef]
135. Liu, Y.; Wilcox, J. CO_2 Adsorption on Carbon Models of Organic Constituents of Gas Shale and Coal. *Environ. Sci. Technol.* **2010**, *45*, 809–814. [CrossRef] [PubMed]
136. Liu, Y.; Wilcox, J. Molecular Simulation Studies of CO_2 Adsorption by Carbon Model Compounds for Carbon Capture and Sequestration Applications. *Environ. Sci. Technol.* **2012**, *47*, 95–101. [CrossRef] [PubMed]
137. Perera, S.; Mariano, R.; Vu, K.; Nour, N.; Seitz, O.; Chabal, Y.; Balkus, K. Hydrothermal synthesis of graphene-TiO_2 nanotube composites with enhanced photocatalytic activity. *ACS Catal.* **2012**, *2*, 949–956. [CrossRef]
138. Zhang, H.; Lv, X.; Li, Y.; Wang, Y.; Li, J. P25-Graphene Composite as a High Performance Photocatalyst. *ACS Nano* **2010**, *4*, 380–386. [CrossRef]
139. Liu, J.; Bai, H.; Wang, Y.; Liu, Z.; Zhang, X.; Sun, D.D. Self-Assembling TiO_2 Nanorods on Large Graphene Oxide Sheets at a Two-Phase Interface and Their Anti-Recombination in Photocatalytic Applications. *Adv. Funct. Mater.* **2010**, *20*, 4175–4181. [CrossRef]

C Journal of Carbon Research

MDPI

Article

Crystallization Kinetics of Poly(lactic acid)–Graphene Nanoscroll Nanocomposites: Role of Tubular, Planar, and Scrolled Carbon Nanoparticles

Oluwakemi Ajala [1], Caroline Werther [2], Rauf Mahmudzade [2], Peyman Nikaeen [3] and Dilip Depan [2,*]

1 Intel Corporation, Hillsboro, OR 97124, USA; kemi.oajala@gmail.com
2 Institute for Materials Research and Innovation, Chemical Engineering Department, Lafayette, LA 70503, USA; caroline.werther@gmail.com (C.W.); rauf.mahmudzade1@louisiana.edu (R.M.)
3 Mechanical Engineering Department, University of Louisiana at Lafayette, P.O. Box 44130, Lafayette, LA 70504, USA; peyman.nikaeen1@louisiana.edu
* Correspondence: ddepan@louisiana.edu

Abstract: Graphene nanoscrolls (GNS) are 1D carbon-based nanoparticles. In this study, they were investigated as a heterogeneous nucleating agent in the poly(lactic acid) (PLA) matrix. The isothermal and non-isothermal melting behavior and crystallization kinetics of PLA-GNS nanocomposites were investigated using a differential scanning calorimeter (DSC). Low GNS content not only accelerated the crystallization rate, but also the degree of crystallinity of PLA. The Avrami model was used to fit raw experimental data, and to evaluate the crystallization kinetics for both isothermal and non-isothermal runs through the nucleation and growth rate. Additionally, the effect of the dimensionality and structure of the nanoparticle on the crystallization behavior and kinetics of PLA is discussed. GNS, having a similar fundamental unit as CNT and GNP, were observed to possess superior mechanical properties when analyzed by the nanoindentation technique. The scrolled architecture of GNS facilitated a better interface and increased energy absorption with PLA compared to CNTs and GNPs, resulting in superior mechanical properties.

Keywords: poly(lactic acid); graphene nanoscrolls; crystallization; kinetics; nanoindentation

Citation: Ajala, O.; Werther, C.; Mahmudzade, R.; Nikaeen, P.; Depan, D. Crystallization Kinetics of Poly(lactic acid)–Graphene Nanoscroll Nanocomposites: Role of Tubular, Planar, and Scrolled Carbon Nanoparticles. C **2021**, 7, 75. https://doi.org/10.3390/c7040075

Academic Editors: Jorge Bedia and Gil Goncalves

Received: 31 August 2021
Accepted: 26 October 2021
Published: 31 October 2021

1. Introduction

PLA, being a bio-based and biodegradable polymer, has attracted significant interest due to its comparably excellent material properties such as high stiffness, high optical transparency, and easy processability [1–4]. PLA is a linear aliphatic thermoplastic polyester prepared by the ring-opening polymerization of lactide [5,6]. PLA typically has a slow crystallization rate, which causes low heat distortion temperature, high gas diffusion, and unstable mechanical properties such as brittleness due to its low glass transition temperature [2]. Therefore, crystallization is enhanced by the addition of nucleating agents that can lower the surface free energy barrier and thus allow crystallization to occur at higher temperatures [2,7]. PLA mostly crystallizes into α and α' crystalline forms, with the α' form being less organized and having a lower Young's modulus than the α form.

Improvements in the mechanical properties and crystallization kinetics were reported due to the inclusion of nanoparticles such as nanoclays, talc, and carbon-based nanoparticles [1,8,9]. Crystallization can also be improved by subjecting PLA to a wide range of thermal treatments. However, these methods yield moderate results and are time consuming, as the crystallization of PLA is slow. The most effective method to achieve superior mechanical properties is the filling of a PLA matrix with inorganic nanoparticles such as clay, carbon nanotubes, graphene, and spherical nanoparticles. Upon good dispersion in the polymer matrix, such nanoparticles have been found to increase both the rate and degree of crystallization, as they act as extra crystallization nuclei. The improvements in crystallization and physical properties of polymers are not only related to the characteristics of the fillers (geometry, aspect ratio, specific surface area), but also to the formed polymer/filler interfacial strength.

Graphene nanoscrolls (GNS) are an increasingly important one-dimensional open ended tubular structure [10–13]. GNS consist of tubular layers of graphene sheets, formed by rolling in a continuous manner, that are distinct from the 1D multiwalled carbon nanotube (MWCNT) [13,14]. The open ends and expansions in the interlayer surface area allows for solvent and molecule accessibility just by facile infiltration [10,11,15]. Non-covalent interactions between overlapping areas provide GNS with structural stability and prevents them from reverting back to initial planar morphology. Li et al. reports that GNS could potentially possess the most superior adhesion property in the carbon family, with values of 0.45, 0.31, and 0.33 J m^{-2} for graphene, graphite, and multiwall carbon nanotubes, respectively [16]. Moreover, in thicker GNS, adhesion energy was observed to also be higher, at 1.72 ± 0.12 J m^{-2}, compared to 0.62 ± 0.02 J m^{-2} in smaller diameter GNS (Figure S2). The synthesis of GNSs can be conducted through a variety of processes, namely, rolling of graphene on a substrate surface [11], well-controlled lyophilization [13], the Langmuir–Blodgett process [15], exfoliation or interaction of graphitic interlayers, and chemical vapor deposition [17]. GNS' favorable properties render it an attractive carbon-based nucleating agent in the crystallization performance of PLA.

To the best of author's knowledge, no study has reported the investigation of the crystallization kinetics of PLA induced by GNS. An in-depth study on the GNS-induced crystallization kinetics of PLA is required to fully realize the potential of PLA. The objective of this study was to investigate the melting behavior and crystallization kinetics of the PLA nanocomposite under isothermal and non-isothermal conditions. It is important to understand the fundamentals of the crystallization kinetics of a polymer affected by crystal structure and morphology induced by a nucleating agent [18]. The crystallization kinetics of PLA-GNS was modeled using the Avrami model, and dimensionality and growth rate were obtained from isothermal and non-isothermal experiments. Another aim was to conduct preliminary studies on the comparative analysis of PLA incorporated with GNS, carbon nanotubes (CNTs), and graphene nanoplatelet (GNPs), which are nanoparticles possessing similar fundamental units.

2. Materials and Methods

2.1. Materials

Poly(lactic acid) (PLA) was procured from Nature-Works (4032D grade). According to the technical data sheet provided by the supplier, its melting point is ~180 °C, glass transition temperature is 55–60 °C, melt flow index range is 190 °C, and density is 1.25 g cm^{-3}. Graphene nanoplatelets (GNPs) having an average of 5–7 atomic layers of 2D graphite were acquired from Sigma-Aldrich, with a lateral size of 0.5–5 µm and BET surface area >500 $m^2 \cdot g^{-1}$. Toluene (≥99.5%) and tetrahydrofuran (THF) were obtained from Sigma-Aldrich. Carbon nanotubes (CNTs) with diameter 1.5 nm and length 1–5 µm were obtained from Sigma Aldrich (St. Louis, MO, USA).

2.2. Preparation of PLA-GNS Polymer Nanocomposite

Graphene nanoplatelets (GNPs) were converted into graphene nanoscrolls (GNS), using our previously reported procedure [19]. Thereafter, 0.5 g of PLA was dissolved in 15 mL THF using a magnetic stirrer for 60 min at 60 °C in an oil bath. Simultaneously, the resultant GNS powder was dispersed in 10 mL THF and sonicated in an SRA TruPower UC-20D (SRA Shops, Walpole, MA, USA) Digital Ultrasonic Cleaner at room temperature. After 60 min of sonication, the GNS-THF dispersion was added to the PLA-THF solution and heated at 60 °C for 120 min using a magnetic stirrer. PLA-GNS samples were prepared with different weight ratios: PLA-GNS-0.1, PLA-GNS-0.5, PLA-GNS-1, and PLA-GNS-2. The solution was subsequently poured into glass petri dishes to allow for solvent evaporation at room temperature under atmospheric pressure overnight. The residual solvent was then removed by placing the nanohybrid membranes in a vacuum oven at 40 °C for 60 min. Similarly, pure PLA was processed under the same conditions by dissolving 0.5 g PLA in 25 mL THF at 60 °C for 3 h. For comparative analysis, PLA-CNT and PLA-GNP samples were prepared following the same solution-mixing method for 2 weight percent of nanoparticle in PLA matrix.

2.3. *Differential Scanning Calorimeter (DSC)*

Thermal analyses of PLA-GNS (~5 mg) were carried out using a Perkin-Elmer 4000 (Perkin Elmer, Houston, TX, USA) Differential Scanning Calorimeter (DSC). The calorimeter was calibrated with indium and experiments were carried out under a constant inert nitrogen gas flow of 19.8 mL min^{-1}. Isothermal crystallization studies were carried out by heating pure PLA and PLA-GNS nanocomposite samples to 180 °C at a heating rate of 10 °C min^{-1} and kept at this temperature for 5 min to completely erase the thermal history. Thereafter, the samples were quenched to the desired crystallization temperature, T_c of 90, 100, and 110 °C, at a cooling rate of 40 °C min^{-1} and kept under isothermal conditions for 60 min. The samples were then cooled to room temperature and heated once again to 180 °C at 10 °C min^{-1}. The degree of crystallinity, X_c, of samples was determined by Equation (1):

$$X_c(\%) = \frac{\Delta H_f - \Delta H_c}{\Delta H_f^0} \times 100 \quad (1)$$

where ΔH_f is the melting enthalpy, ΔH_f^0 is the melting enthalpy of pure crystalline PLA (100% crystalline polylactide = 93 J/g), and ΔH_c is the crystallization enthalpy.

The non-isothermal crystallization measurements were carried out in a Perkin-Elmer 4000 (Perkin Elmer, Houston, TX, USA) calorimeter. The calorimeter was calibrated using an indium standard to ensure reliability of data, and experiments were carried out under a constant flow of nitrogen gas at the rate of 19.8 mL min^{-1}. The sample weight was kept at ~5 mg, sealed hermetically in aluminum pans. For crystallization studies, samples were first heated from 30 to 180 °C at 10 °C min^{-1} and held at 180 °C for 5 min to completely erase the thermal history. Subsequent cooling to 30 °C was undertaken at cooling rates of 2, 5, 7.5, 10, 15, 20, 25, 30, 35, 40, 45, and 50 °C min^{-1}. In order to study the melting behavior after crystallizing, samples underwent a second heating to 180 °C at 10 °C min^{-1} and held for a minute.

2.4. *Nanoindentation*

A Nano Indenter G200 instrument (Agilent, Santa Clara, CA, USA) provided by Agilent-MTS, compliant with ISO 14577, was utilized to measure the elastic modulus and hardness of pure PLA and PLA-GNP, PLA-CNT, and PLA-GNS nanocomposites. Its specifications were as follows: load range of 0.4–500 mN, displacement resolution of less than 0.01 nm, and load resolution of 50 nN. The instrument was equipped with a three-sided pyramidal-shaped Berkovich indenter with a tip radius of about 100 nm. Thin film samples were attached to an aluminum-testing stub using a very thin layer of crystal bond glue. Small indentation loads of 0.5 mN were applied to films with thickness of 50 to 100 μm, to minimize the effect of substrate on the properties of the thin films. The resulting indentation depths were in the range of 200 nm to 1 μm for all the samples. Fifty indentations were conducted at random locations, and average and standard deviation are reported. Poisson's ratio of PLA was 0.33.

3. Results and Discussion

3.1. *Isothermal Crystallization Studies*

Table 1 presents the Avrami parameters, *n* and *k*, for the Avrami exponent and rate constant, respectively. The Avrami equation describes the phase changes during isothermal crystallization [20]. The analysis was performed using the un-subtracted heat flow curve obtained from DSC during isothermal studies and the relative crystallinity, X_t, as a function of time was obtained using the following relation [21,22]:

$$X(t) = \frac{Q_t}{Q_\infty} = \frac{\int_0^t \frac{dH}{dt}\, dt}{\int_0^\infty \frac{dH}{dt}\, dt} \tag{2}$$

where Q_t and Q_∞ are the amounts of enthalpy generated at time t and infinite time, t_∞, respectively, with dH/dt as the rate of heat or enthalpy evolution. The Avrami equation, which is primarily used to model this isothermal crystallization process, is given by [22–24]:

$$1 - X(t) = \exp(-kt^n) \tag{3}$$

where $X(t)$ is the relative crystallinity as a function of time, k is the Avrami rate constant, and n is the Avrami exponent. Relative crystallinity, $X(t)$, is the fractional amount of relative volume fraction of polymer matrix crystallized. The Avrami relation takes into consideration a constant nucleation rate and constant linear growth [23]. The linearized form of the Avrami equation can be expressed as:

$$Log[-\ln\{1 - X(t)\}] = nlog\, t + \log k \tag{4}$$

Table 1. Collated values of heat of fusion (ΔH_f), degree of crystallinity (X_c), induction time (t_0), and half time of crystallization ($t_{1/2}$) and Avrami parameters for pure PLA and PLA-GNS nanocomposites at different isothermal crystallization temperatures, T_c.

Samples	T_c (°C)	ΔH_f (J/g)	X_c (%)	t_0 (min)	$t_{1/2}$ (obs) (min)	$t_{1/2}$ (calc) (min)	n	k (min^{-n})
Pure PLA	90	38.0	40.8	4.06	17.58	17.65	2.37	1.44×10^{-3}
	100	33.8	41.7	1.83	6.46	6.59	2.22	2.16×10^{-2}
	110	33.3	36.0	1.70	6.34	6.49	2.18	2.29×10^{-2}
PLA-GNS-0.1	90	47.2	50.7	2.21	7.06	7.10	2.35	1.66×10^{-2}
	100	39.6	42.5	1.35	3.52	3.44	2.96	7.84×10^{-2}
	110	34.4	36.9	1.70	3.00	2.95	2.65	1.63×10^{-1}
PLA-GNS-0.5	90	56.6	60.7	1.49	6.78	6.64	2.51	1.14×10^{-2}
	100	34.3	36.8	1.03	2.81	2.85	2.13	1.94×10^{-1}
	110	38.5	41.4	1.03	2.60	2.52	2.76	2.32×10^{-1}
PLA-GNS-1	90	51.1	54.9	1.68	6.80	6.51	3.64	2.25×10^{-3}
	100	42.1	45.2	1.35	3.10	3.03	2.79	1.62×10^{-1}
	110	34.3	36.8	1.17	2.43	2.39	2.73	3.99×10^{-1}
PLA-GNS-2	90	50.1	53.8	2.09	7.47	7.38	2.84	6.12×10^{-3}
	100	41.4	44.5	1.37	3.56	3.42	2.59	1.07×10^{-1}
	110	33.6	36.0	1.24	2.24	2.28	2.24	6.37×10^{-1}

The rate constant, k and the exponent, n are determined from the slope and intercept, respectively. The reaction half time of crystallization is denoted by:

$$t_{1/2} = \left(\frac{In\, 2}{k}\right)^{1/n} \tag{5}$$

The Avrami fits were performed for a relative crystallinity range of 0.03–0.2 for an accurate evaluation of primary crystallization in PLA and PLA-GNS nanocomposites. The Avrami equation cannot account for a period where no crystallization occurs; therefore, induction time, t_0, denotes the initial time at which crystallization begins. An induction period in the early stage of crystallization is required for creation of nuclei and is estimated by calculating the first derivative dX/dt, because it should be zero and denote the start of nucleation; therefore, $t = t_0$ (Figure 1a) [25]. From Table 1, the induction time (t_0) decreases for the PLA-GNS nanocomposites; therefore, we can infer that the scrolled conformation of GNS aided the induction period for PLA chains to adsorb and orient on the GNS curved surface. Relative crystallinity plots for pure PLA and PLA-GNS nanocomposites depict the different phases of crystallization occurring as a sigmoidal evolution (Figure 1b). The values of relative crystallinity are in the range of 0–100%, with 0% being the beginning, and 100% the impingement of crystals and crowding that occurs at the end stages of crystallization. The half time of crystallinity is the time at which $X(t)$ is 0.5 or 50%, and in all cases the time it takes for the evolution to plateau is shortest in PLA-GNS when compared with pure PLA. GNS plays an active role in the crystal growth and nucleation by reducing the activation energy and acting as the surface on which PLA can orient its chains and assist in helical chain packing. The log-log plots in Figure 1c depict the linear portions of the relative crystallinity plots, and the Avrami exponent and rates are obtained from the slope and intercept, respectively.

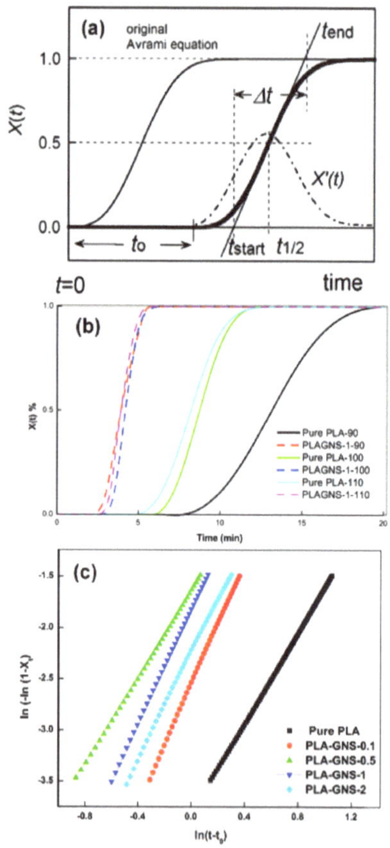

Figure 1. (**a**) Schematic illustration of extraction of the Avrami exponent, (**b**) relative crystallinity vs. time as obtained from the Avrami equation, and (**c**) log plots of PLA and its nanohybrids.

The n values obtained for pure PLA and PLA-GNS nanocomposites are in the ranges of 2.1–2.3 and 2.1–3.6, respectively (Table 1). This indicates that GNS has an evident nucleating effect on PLA, because the n depicts the dimensionality of inner crystal growth and structure. From the n values obtained, PLA-GNS nanocomposites followed three-dimensional heterogeneous nucleation and sporadic growth [26].

3.2. Non-Isothermal Crystallization Studies

Nonisothermal crystallization of pure PLA and PLA-GNS nanocomposites was analyzed with scan rates ranging from 2 to 50 °C min^{-1}. It is essential to understand the kinetics of nonisothermal crystallization because most polymer processing applications are carried out under nonisothermal conditions. PLA is known to crystallize very slowly; therefore, we quantified the role of GNS in the polymer matrix during nonisothermal conditions.

Figure 2 presents second melting curves for pure PLA and PLA-GNS-0.5, and thermal parameters are summarized in Table 2. Firstly, pure PLA melting endotherms possessed single peaks, T_{m1}, and cold crystallization (T_{cc}), which is marked by an increased intensity and a broad area with increasing cooling rates (Figure 2a).

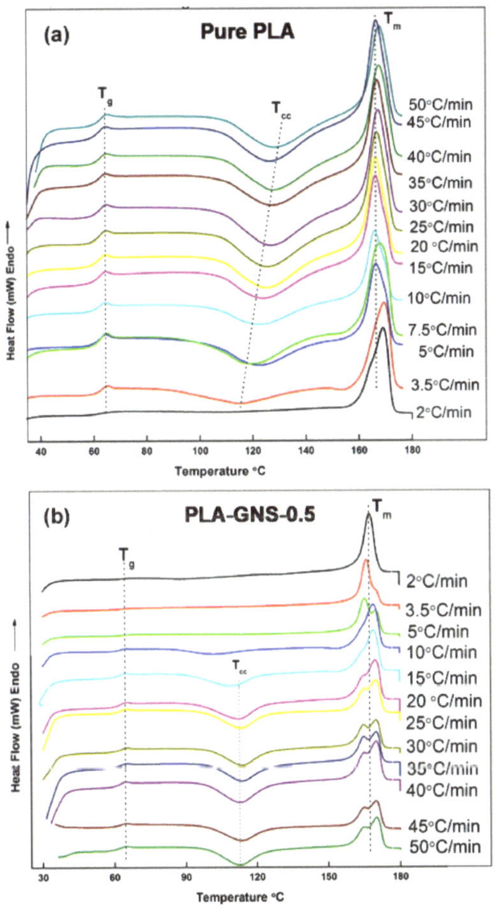

Figure 2. (**a**) DSC thermograms of pure PLA crystallized at various cooling rates, while (**b**) represents DSC thermograms of PLA-GNS nanohybrid at various cooling rates.

By comparison, the presence of double melting peaks (at higher cooling rates) in PLA-GNS-0.5, as seen in Figure 2b, is due to the presence of disordered α′-crystals. The formation of these crystals is kinetically favored compared to the ordered and stable α-crystals. Lower temperatures favor the disordered α′-crystals, whereas higher temperatures favor the ordered α-crystal [27]. Moreover, for PLA-GNS-0.5 there is an absence of cold crystallization peaks, T_{cc}, until the cooling rate increases to 10 °C min^{-1}, which is due to the phase transition occurring from α′ to α-crystals during the melting process. This suggests that GNS enhanced nucleation and crystal growth in PLA as cooling rates increased. Maximum crystallinity was observed at about 49% for PLA-GNS-0.1 at a cooling rate of 2 °C min^{-1} compared to pure PLA at 33%. The glass transition temperatures, T_g, listed in Table 2 for pure PLA and PLA-GNS nanocomposites are 62.7 ± 1.7 °C. Results indicate that the presence of GNS in the polymer matrix has little to no effect on the glass transition temperature. Previous research reports similar observations [28].

Table 2. Thermal parameters of pure PLA and PLA-GNS nanocomposites at different cooling rates.

Sample	ϕ (°C min^{-1})	T_g (°C)	T_{m1} (°C)	T_{m2} (°C)	T_{cc} (°C)	ΔH_f (J g^{-1})	ΔH_c (J g^{-1})	X_c (%)
Pure PLA	2	62.7	168.9	-		30.7		33.02
	3.5	62.7	169.3	-	115.2	35.7	−16.5	20.58
	5	62.3	167.9	-	119.3	40.2	−31.7	9.15
	7.5	62.2	166.5	-	123.6	30.2	−29.3	0.98
	10	61.8	166.1	-	123.0	24.0	−21.2	2.99
	15	61.9	166.4	-	124.3	29.6	−29.6	−0.09
	20	61.9	166.5	-	125.5	31.4	−32.1	−0.72
	25	62.4	166.9	-	125.9	33.2	−30.8	2.58
	30	61.5	166.7	-	126.6	33.1	−32.5	0.63
	35	62.2	167.4	-	127.1	35.4	−33.7	1.81
	40	62.0	167.0	-	127.4	32.5	−31.0	1.61
	45	62.6	167.6	-	128.0	30.3	−32.8	−2.78
	50	62.6	168.0	-	128.4	29.7	−24.3	5.77
PLA-GNS-0.1	2	62.9	167.6	-	-	45.3	-	48.61
	3.5	63.7	166.2	-	-	33.2	-	35.62
	5	62.6	165.1	169.9	-	29.3	-	31.52
	7.5	62.4	168.9	-	105.4	40.8	−11.8	31.17
	10	61.9	168.3	-	101.0	37.1	−9.9	29.23
	15	62.2	165.3	169.1	115.4	32.5	−30.2	2.52
	20	62.1	165.0	169.5	117.0	34.5	−33.0	1.63
	25	62.3	165.1	169.6	116.4	33.4	−35.9	−2.75
	30	61.8	164.6	169.8	114.1	29.5	−30.7	−1.29
	35	62.0	164.3	169.9	113.9	31.3	−32.7	−1.49
	40	63.0	164.8	170.3	114.4	30.6	−30.9	−0.27
	45	62.9	165.8	170.3	117.3	36.6	−41.2	−4.98
	50	61.9	165.1	170.2	114.8	34.5	−37.4	−3.14
PLA-GNS-0.5	2	63.6	167.3	-	-	39.9	-	42.87
	3.5	62.7	166.0	-	-	32.2	-	34.58
	5	63.5	165.4	170.0	-	32.6	-	35.00
	10	62.7	168.9	-	101.9	39.5	−10.3	31.38
	15	62.4	168.9	-	108.9	35.7	−24.1	12.52
	20	62.4	164.5	169.6	112.2	38.4	−32.1	6.74
	25	62.3	164.9	169.8	113.5	32.0	−30.7	1.41
	30	62.3	164.7	169.8	113.7	28.8	−29.9	−1.20
	35	62.5	164.7	170.0	113.7	30.1	−29.1	1.14
	40	62.5	164.8	169.8	113.0	37.8	−36.7	1.17
	45	62.8	164.8	169.8	113.5	27.0	−26.0	1.05
	50	62.6	164.6	170.1	112.8	29.6	−26.0	3.89
PLA-GNS-1	2	62.4	166.1	-	-	33.5	-	36.02
	3.5	62.2	165.5	-	-	36.0	-	38.71
	5	63.5	164.8	169.8	-	23.5	-	25.25
	7.5	61.0	163.6	169.0	-	31.2	−3.0	30.29
	10	62.4	164.3	169.2	111.3	35.0	−23.5	12.37
	15	61.4	163.4	168.5	111.1	33.5	−13.8	21.20
	20	62.7	164.2	170.0	111.6	31.6	−25.9	6.06
	25	62.3	164.2	169.6	113.4	29.2	−23.2	6.46
	30	62.6	165.0	170.3	111.9	48.0	−47.0	1.07
	35	62.4	164.6	170.3	113.8	35.1	−37.3	−2.33
	40	62.7	165.0	170.4	114.7	29.9	−29.5	0.45
	45	62.8	165.0	170.3	115.2	36.1	−37.1	−1.11
	50	62.5	163.9	170.0	112.6	36.9	−38.2	−1.39

Table 2. *Cont.*

Sample	ϕ (°C min^{-1})	T_g (°C)	T_{m1} (°C)	T_{m2} (°C)	T_{cc} (°C)	ΔH_f (J g^{-1})	ΔH_c (J g^{-1})	X_c (%)
PLA-GNS-2	2	64.4	166.9	-	-	33.2	-	35.70
	3.5	63.0	165.3	169.8	-	28.2	-	30.31
	5	62.9	165.0	170.1	-	31.4	-	33.77
	7.5	62.0	164.6	169.4	-	24.5	−2.1	24.07
	10	62.3	-	168.6	105.8	35.5	−15.4	21.55
	15	62.3	-	169.2	109.5	37.2	−22.6	15.66
	20	61.9	-	169.7	110.4	36.6	−30.6	6.42
	25	62.1	163.8	169.8	112.0	32.0	−27.8	4.49
	30	62.1	163.8	169.8	110.8	25.9	−23.9	2.14
	35	62.3	164.7	170.3	113.9	45.1	−41.4	3.89
	40	61.7	162.8	169.3	109.0	22.7	−19.8	3.18
	45	62.6	163.3	169.8	110.5	27.8	−27.0	0.86
	50	62.6	163.5	170.4	109.7	38.7	−35.7	3.16

The Avrami model describes the free growth of crystals prior to impingement, which were obtained for relative crystallinity, $X(t)$, at 10–50%. The rate constant, k, and exponent, n, determined from the slope and intercept, respectively, are listed in Table 3. A relationship between the crystallization temperatures T extracted from DSC and time t is given by Equation (5) during non-isothermal crystallization. The relationship between crystallization time, t, and corresponding temperature during the nonisothermal crystallization is represented as:

$$t = \frac{T_0 - T}{\phi} \tag{6}$$

where T_0 is the onset of crystallization, T is the temperature at crystallization time t, and ϕ is the cooling rate [28].

Table 3. Avrami parameters, n and k, of pure PLA and PLA-GNS nanocomposites at selected cooling rates.

ϕ (°C min^{-1})	Sample	n	k	R^2	$t_{1/2}$
2	Pure PLA	1.81	1.54×10^{-3}	0.9999	30.25
	PLA-GNS-0.1	3.76	5.19×10^{-3}	1	28.82
	PLA-GNS-0.5	4.32	2.56×10^{-3}	0.9998	29.2
	PLA-GNS-1	3.46	4.00×10^{-3}	1	30.32
	PLA-GNS-2	2.72	8.30×10^{-3}	0.9935	30.6
3.5	Pure PLA	2.79	5.37×10^{-2}	0.999	23.05
	PLA-GNS-0.1	4.04	7.23×10^{-3}	1	17.77
	PLA-GNS-0.5	3.56	1.02×10^{-1}	1	17.97
	PLA-GNS-1	3.45	8.32×10^{-2}	0.9999	18.42
	PLA-GNS-2	3.47	6.99×10^{-2}	1	18.42
5	Pure PLA	1.24	7.07×10^{-2}	0.9985	15.73
	PLA-GNS-0.1	3.1	8.70×10^{-1}	0.9999	13.33
	PLA-GNS-0.5	2.94	2.40×10^{-1}	0.9931	14.32
	PLA-GNS-1	4.4	9.30×10^{-2}	0.998	13.78
	PLA-GNS-2	4.11	3.55×10^{-1}	0.9996	13.32
7.5	Pure PLA	1.55	1.85×10^{-1}	0.9975	12.73
	PLA-GNS-0.1	2.63	6.43×10^{-2}	0.9999	10.62
	PLA-GNS-0.5	2.14	3.87×10^{-1}	1	10.07
	PLA-GNS-1	2.89	7.29×10^{-2}	0.9981	9.72
	PLA-GNS-2	2.97	8.52×10^{-2}	0.9994	9.6
10	Pure PLA	2.34	3.08×10^{0}	0.9983	11.5
	PLA-GNS-0.1	2.74	8.97×10^{0}	1	8.33
	PLA-GNS-0.5	2.79	1.12×10^{1}	0.9997	8.25
	PLA-GNS-1	2.52	6.81×10^{0}	0.9977	8.57
	PLA-GNS-2	2.57	1.09×10^{1}	0.9991	8.55

For pure PLA, the dimensionality of growth for the nonisothermal process is seen in n values in the range of 1.24–2.79, as seen in Table 3, and PLA-GNS nanocomposites, with n values in the range of 1.96–4.4. These values also decreased at faster cooling rates due to higher undercooling. PLA has a very slow crystallization rate, however, at a cooling rate of 2 °C min^{-1}, and the half time of crystallization, $t_{1/2}$, was 30.25 min for pure PLA and 28.2 min for PLA-GNS-0.1. The enhanced crystallization rate was even more evident at higher scan rates, such as at 3.5 °C min^{-1}, where the $t_{1/2}$ of pure PLA was 23.05 min and of PLA-GNS-0.1 was 17.77 min. GNS plays a heterogeneous nucleating role by reducing the activation energy required for surface nucleation to occur, and aiding polymer diffusion and packing on the nanoparticle.

3.3. *Comparative Analysis of PLA-GNS, PLA-CNT, PLA-GNP*

Structurally, GNPs, CNTs, and GNS are made up of graphite layers, and therefore have the same fundamental structural unit [29]. GNPs have a two-dimensional surface made up of several atom layers of graphene sheets, whereas CNTs consist of curled graphite layers with two half fullerenes capping either end of the tube [30]. A GNS, which is very similar to a CNT, is a one-dimensional nanoparticle made up of rolled-or-scrolled GNPs with open ended tubular topology. In order to investigate the influence of the varying dimensional features of these nanoparticles on the crystallization behavior of PLA, isothermal crystallization was quantified to compare the crystallization kinetics of PLA induced by these three nanoparticles having different dimensions.

In order to compare the crystallization kinetics, the primary crystallization of PLA incorporated with GNPs, CNTs, and GNS was modelled using the Avrami equation. Figure 3a,b depicts the relative crystallinity curves and log-log plots of pure PLA, PLA-GNP, PLA-CNT, and PLA-GNS. As indicated in Table 4 and Figure 3a,b, crystallization occurs much faster in PLA-CNT with a half time of crystallization, $t_{1/2}$, = 5 min. This may be due to the large surface curvature of CNT having a diameter of ~2 nm, which allows polymer chains to orient along the tube axis. This leads to a phenomenon called soft epitaxy, in which strict lattice matching is not required [31].

The planar geometry and scrolled architecture of GNPs and GNS are shown in the Supplementary Figure S1a,b, respectively. Similarly, the morphology of pure PLA and PLA-GNS is shown in Figure S1c,d. Geometric confinement plays a major role because it dictates the polymer chain orientation on the nanoparticle with high surface curvature. Compared to the GNS and CNT, the GNP is a 2D flat surface with no surface curvature; hence, polymer chains would require strict lattice matching and epitaxy as the main growth mechanism. The epitaxial growth of polymer crystals leads to different orientations of the polymer chains, thereby increasing the rate of crystallization [31]. The Avrami exponent, n, depicts the multidimensionality of crystal growth with the addition of the heterogeneous nucleating agents. The requirement for strict lattice matching may have resulted in the decline in crystal nucleation and growth in the GNP due to low surface curvature. This is because of the confined nucleation and growth of PLA lamella between the GNP layers [32]. PLA chains absorbed on the surface of GNP needed more time to adjust their conformation, making the nucleation process more complex and decreasing the rate of crystallization; thus, $t_{1/2}$ was larger in PLA-GNP at ~13 min. Further, due to the variation in the diameters of the GNS, geometric confinement would have played a role in areas of smaller diameter, d_2, and strict lattice matching and size-dependent epitaxy in areas of larger diameter, d_1. It has previously been found that the overall kinetics of crystallization was accelerated by both graphene and CNTs; however, the ability to accelerate was much stronger for CNTs than that for graphene [29,30]. Our results indicated that the crystallinity is affected by the better interfacial adhesion between the PLA and GNS, whereas the crystallization half time ($t_{1/2}$) is controlled by epitaxy; the scrolled GNS surface helps in the crystallization, whereas the time to crystallize was intermediate in the CNTs and GNPs.

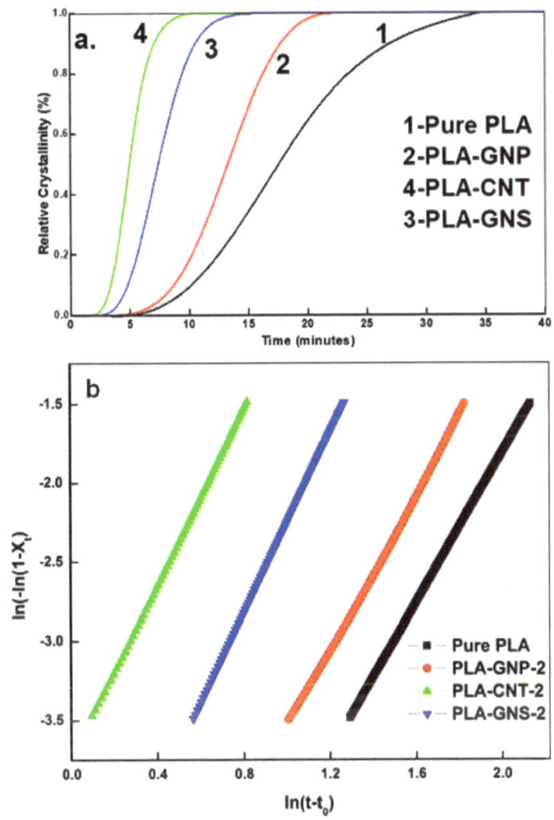

Figure 3. (**a**) Relative crystallinity vs. time plots of pure PLA and its hybrids with various nanoparticles; (**b**) log plots of PLA and its nanohybrids.

Table 4. Thermal parameters for isothermal crystallization at T_c = 90 of pure PLA, PLA-GNS, PLA-GNP, and PLA-CNT.

Scheme	Parameters	PLA-GNS	PLA-CNT	PLA-GNP
	n	2.84	2.76	2.46
	k (min^{-n})	6.12×10^{-3}	2.34×10^{-2}	2.43×10^{-3}
	R^2	1	0.9998	0.9989
	t_0 (minutes)	5.4	3.4	9.2
	$t_{1/2}$ (minutes)	7.5	5	13.3
	ΔH_f (J g^{-1})	51.3433	50.9566	48.3337
	X_c (%)	55.149	54.733	51.916
	T_m (°C)	168.35	168.35	167.97
	T_g (°C)	66.13	64.4	65.5

A mechanical comparison using the load control nanoindentation technique was undertaken to test the elastic modulus and hardness of pure PLA and PLA-GNS-0.5 using the Oliver and Phar method (Figure 4). Incorporating GNS in PLA increases the elastic modulus by 81% to 4.7 GPa and hardness by 90% in the polymer nanocomposite. CNT improved elastic modulus by ~42% and hardness by 70%, whereas GNP increased the elastic modulus and hardness by ~69% and 95%, respectively.

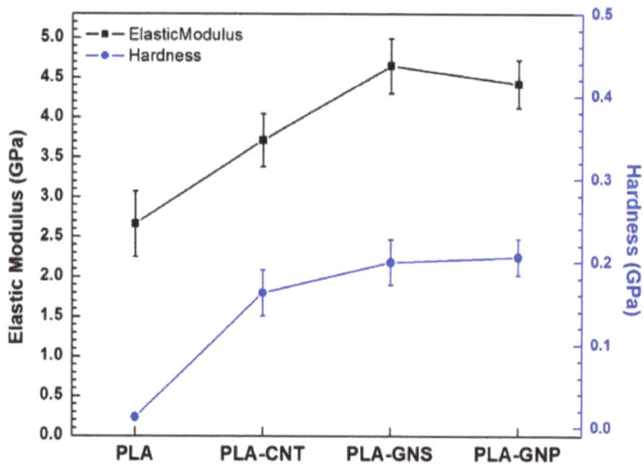

Figure 4. Elastic modulus and hardness curves of pure PLA and its hybrids with various nanoparticles via the nanoindentation test.

GNS possesses higher structural stability, which provides it with the capability to overcome non-uniform dispersion and poor adhesion of their planar and tubular counterparts. Unlike the bamboo-like compartments of CNTs, the continuity of a rolled-up graphene sheet provides a uniform response with superior mechanical, thermal, and electrical properties. PLA is known to fail by crazing and has low resistance to crack initiation and propagation. Moreover, in PLA nanohybrids, crystallites can be stress concentration initiators and decrease impact strength. Cracks propagate through each uniformly distributed nanofiller, resulting in dissipation of energy throughout the polymer matrix, because the cracks must pass through the nanofillers while adhering well to the PLA matrix. Because strong interface favors the increase in energy absorption by nanofillers, the greatest improvement was observed in PLA-GNS because GNS presented a better interface with PLA than CNTs and GNPs. Similar to crazes, defects at interfaces dissipate energy. We previously reported that GNS provides a scrolled surface with higher strained surface, which not only provides a better interface, but also facilitates trans-crystallization, resulting in a trans-crystalline layer of PLA on GNS [19]. This is evident by the planar and wrapped surface morphology of crystallized PLA on GNS (Figure S2d). Trans-crystallization is known to avoid crazing propagation [33]. Thus, the presence of GNS diminishes the apparition of crazes, defects, and voids, which normally dissipate crack energy and increase impact strength.

He et al. suggests that GNS possesses superior adhesion due to the strain generated during scrolling within graphene monolayers [34]. Its well-defined radius of curvature builds up strain in the curved walls and, compared to planar pristine graphene, the adhesion force is 2.5-fold stronger and allows enhanced bonding interfaces between PLA and graphene. These obtained results strongly suggest GNS not only enhanced the elastic modulus and hardness of PLA, but its overall mechanical properties surpassed that of its PLA-CNT and PLA-GNP counterparts.

4. Conclusions

The incorporation of GNS as a nucleating agent accelerated the crystallization rate of the PLA under isothermal and non-isothermal conditions. The increase in the crystallization rates was reflected in the decreased induction time and half times of crystallization. The melting behavior of PLA-GNS nanocomposites suggested that GNS did not alter the polymorphic nature of PLA. During non-isothermal conditions, however, the inclusion of GNS resulted in the formation of imperfect α'-crystals. Overall crystallinity was enhanced in the polymer nanocomposite, suggesting that GNS assisted in the ordering of PLA chains. GNS, CNT, and GNP served as heterogeneous nucleating agents by shortening the induction time and accelerating the crystallization rate in the PLA matrix.

Supplementary Materials: The following are available online at https://www.mdpi.com/article/10.3390/c7040075/s1. Figure S1: Schematic illustration of the structure of graphene nanoscrolls indicating the smaller diameter (d_1) and larger diameter (d_2), Figure S2: Transmission electron micrographs of (a) graphene nanoplatelets (GNP), (b) graphene nanoscrolls (GNS) and scanning electron micrographs of (c) pure PLA and (d) PLA crystallized on GNS, Figure S3: FTIR spectra of pure PLA and its various nanohybrids with carbon nanoparticles, Figure S4: X-ray diffraction (XRD) patterns of pure PLA and its various nanohybrids with carbon nanoparticles.

Author Contributions: Conceptualization, O.A. and D.D.; methodology, O.A.; software, O.A., C.W., R.M., P.N.; validation, O.A., D.D., P.N. and R.M.; formal analysis, O.A.; investigation, O.A.; resources, D.D.; data curation, O.A.; writing—original draft preparation, O.A.; writing—review and editing, D.D.; visualization, O.A. and C.W.; supervision, D.D.; project administration, D.D.; funding acquisition, D.D. All authors have read and agreed to the published version of the manuscript.

Funding: This research was funded by Louisiana Board of Regents, grant number LEQSF (2020-21)-ENH-DE-20.

Conflicts of Interest: The authors declare no conflict of interest. The funders had no role in the design of the study; in the collection, analyses, or interpretation of data; in the writing of the manuscript, or in the decision to publish the results.

References

1. Murariu, M.; Dubois, P. PLA composites: From production to properties. *Adv. Drug Deliv. Rev.* **2016**, *107*, 17–46. [CrossRef]
2. Feng, Y.; Ma, P.; Xu, P.; Wang, R.; Dong, W.; Chen, M.; Joziasse, C. The crystallization behavior of poly(lactic acid) with different types of nucleating agents. *Int. J. Biol. Macromol.* **2018**, *106*, 955–962. [CrossRef]
3. Farah, S.; Anderson, D.G.; Langer, R. Physical and Mechanical Properties of PLA, and Their Functions in Widespread Applications—A Comprehensive Review. *Adv. Drug. Deliv. Rev.* **2016**, *107*, 367–392. [CrossRef] [PubMed]
4. Hamad, K.; Kaseem, M.; Yang, H.W.; Deri, F.; Ko, Y.G. Properties and medical applications of poly(lactic acid): A review. *Express Polym. Lett.* **2015**, *9*, 435–455. [CrossRef]
5. Wachirahuttapong, S.; Thongpin, C.; Sombatsompop, N. Effect of PCL and compatibility contents on the morphology, crystallization and mechanical properties of PLA/PCL blends. *Energy Procedia* **2016**, *89*, 198–206. [CrossRef]
6. Saeidlou, S.; Huneault, M.A.; Li, H.; Park, C.B. Poly(lactic acid) crystallization. *Prog. Polym. Sci.* **2012**, *37*, 1657–1677. [CrossRef]
7. Li, H.; Huneault, M.A. Effect of nucleation and plasticization on the crystallization of poly(lactic acid). *Polymer* **2007**, *48*, 6855–6866. [CrossRef]
8. Nampoothiri, K.M.; Nair, N.R.; John, R.P. An overview of the recent developments in polylactide (PLA) research. *Bioresour. Technol.* **2010**, *101*, 8493–8501. [CrossRef]
9. Castro-Aguirre, E.; Iniguez-Franco, F.; Samsudin, H.; Fang, X.; Auras, R. Poly(lactic acid)—Mass production, processing, industrial applications, and end of life. *Adv. Drug Deliv. Rev.* **2016**, *107*, 333–366. [CrossRef] [PubMed]
10. Zheng, B.; Xu, Z.; Gao, C. Mass production of graphene nanoscrolls and their application in high rate performance supercapacitors. *Nanoscale* **2016**, *8*, 1413–1420. [CrossRef]
11. Dhar, P.; Gaur, S.S.; Kumar, A.; Katiyar, V. Cellulose nanocrystal templated graphene nanoscrolls for high performance supercapacitors and hydrogen storage: An experimental and molecular simulation study. *Sci. Rep.* **2018**, *8*, 1–15. [CrossRef]
12. Zheng, B.N.; Gao, C. Preparation of graphene nanoscrolls/poly(aniline) composites and their use in high performance super capacitors. *New Carbon Mater.* **2016**, *31*, 315–320. [CrossRef]
13. Xu, Z.; Zheng, B.; Chen, J.; Gao, C. Highly efficient synthesis of neat graphene nanoscrolls from graphene oxide by well-controlled lyophilization. *Chem. Mater.* **2014**, *26*, 6811–6818. [CrossRef]
14. Zhao, J.; Yang, B.; Yang, Z.; Zhang, P.; Zheng, Z.; Ren, W.; Yan, X. Facile preparation of large-scale graphene nanoscrolls from graphene oxide sheets by cold quenching in liquid nitrogen. *Carbon* **2014**, *79*, 470–477. [CrossRef]
15. Gao, Y.; Chen, X.; Xu, H.; Zou, Y.; Gu, R.; Xu, M.; Jen, A.K.Y.; Chen, H. Highly efficient fabrication of nanoscrolls from functionalized graphene oxide by Langmuir Blodgett method. *Carbon* **2010**, *48*, 4475–4482. [CrossRef]
16. Li, H.; Papadakis, R.; Jafri, S.H.M.; Thersleff, T.; Michler, J.; Ottosson, H.; Leifer, K. Superior adhesion of graphene nanoscrolls. *Commun. Phys.* **2018**, *1*, 44. [CrossRef]
17. Chen, X.; Li, L.; Sun, X.; Kia, H.G.; Peng, H. A novel synthesis of graphene nanoscrolls with tunable dimension at a large scale. *Nanotechnology* **2012**, *23*, 055603. [CrossRef]
18. Liu, M.; Zhao, Q.; Wang, Y.; Zhang, C.; Mo, Z.; Cao, S. Melting behaviors, isothermal and non isothermal crystallization kinetics of nylon 1212. *Polymer* **2003**, *44*, 2537–2545. [CrossRef]

19. Ajala, O.; Werther, C.; Nikaeen, P.; Singh, R.P.; Depan, D. Influence of graphene nanoscrolls on the crystallization behavior and nano-mechanical properties of poly(lactic acid). *Polym. Adv. Technol.* **2019**, *30*, 1825–1835. [CrossRef]
20. Bianchi, O.; Oliveira, R.V.B.; Fiorio, R.; Martins, J.D.N.; Zattera, A.J.; Canto, L.B. Assessment of Avrami, Ozawa and Avrami–Ozawa equations for determination of EVA crosslinking kinetics from DSC measurements. *Polym. Test.* **2008**, *27*, 722–729. [CrossRef]
21. Jiang, X.L.; Luo, S.J.; Sun, K.; Chen, X.D. Effect of nucleating agents on crystallization kinetics of PET. *Express Polym. Lett.* **2007**, *1*, 245–251. [CrossRef]
22. Wu, D.; Cheng, Y.; Feng, S.; Yao, Z.; Zhang, M. Crystallization behavior of polylactide/graphene composites. *Ind. Eng. Chem. Res.* **2013**, *52*, 6731–6739. [CrossRef]
23. Lorenzo, A.T.; Arnal, M.L.; Albuerne, J.; Müller, A.J. DSC isothermal polymer crystallization kinetics measurements and the use of the Avrami equation to fit the data: Guidelines to avoid common problems. *Polym. Test.* **2007**, *26*, 222–231. [CrossRef]
24. Barrau, S.; Vanmansart, C.; Moreau, M.; Addad, A.; Stoclet, G.; Lefebvre, J.M.; Seguela, R. Crystallization behavior of carbon nanotube-polylactide nanocomposites. *Macromolecules* **2011**, *44*, 6496–6502. [CrossRef]
25. Kummara, S.; Tashiro, K.; Monma, T.; Horita, K. Isotope Effect on the Melt–Isothermal Crystallization of Polyoxymethylene D/H Random Copolymers and D/H Blend Samples. *Macromolecules* **2015**, *48*, 8070–8081. [CrossRef]
26. Sperling, L.H. *Introduction to Physical Polymer Science*, 4th ed.; John Wiley & Sons, Inc.: Hoboken, NJ, USA, 2005.
27. Kawai, T.; Rahman, N.; Matsuba, G.; Nishida, K.; Kanaya, T.; Nakano, M.; Okamoto, H.; Kawada, J.; Usuki, A.; Honma, N.; et al. Crystallization and melting behavior of poly(L-lactic acid). *Macromolecules* **2007**, *40*, 9463–9469. [CrossRef]
28. Wang, H.; Qiu, Z. Crystallization behaviors of biodegradable poly(L-lactic acid)/graphene oxide nanocomposites from the amorphous state. *Thermochim. Acta* **2011**, *526*, 229–236. [CrossRef]
29. Xu, Z.; Niu, Y.; Yang, L.; Xie, W.; Li, H.; Gan, Z.; Wang, Z. Morphology, rheology and crystallization behavior of polylactide composites prepared through addition of five-armed star polylactide grafted multiwalled carbon nanotubes. *Polymer* **2010**, *51*, 730–737. [CrossRef]
30. Xu, J.Z.; Chen, T.; Yang, C.L.; Li, Z.M.; Mao, Y.M.; Zeng, B.Q.; Hsiao, B.S. Isothermal crystallization of poly(L-lactide) induced by graphene nanosheets and carbon nanotubes: A comparative study. *Macromolecules* **2010**, *43*, 5000–5008. [CrossRef]
31. Li, L.; Li, C.Y.; Ni, C. Polymer crystallization-driven, periodic patterning on carbon nanotubes. *J. Am. Chem. Soc.* **2006**, *128*, 1692–1699. [CrossRef]
32. Xin, S.; Li, Y.; Zhao, H.; Bian, Y.; Li, W.; Han, C.; Dong, Q.; Ning, Z.; Dong, L. Confinement crystallization of poly(lactide) induced by mutiwalled carbon nanotubes and graphene nanosheets. *J. Therm. Anal. Calor.* **2015**, *122*, 379–391. [CrossRef]
33. Zafeiropoulos, N.E.; Baillie, C.A.; Matthews, F.L. A study of transcrystallinity and its effect on the interface in flax fiber reinforced composite materials. *Composite A Appl. Sci. Manufact.* **2015**, *32*, 525–543. [CrossRef]
34. He, Y.; Chen, W.F.; Yu, W.B.; Ouyang, G.; Yang, G.W. Anomalous interface adhesion of graphene membranes. *Sci. Rep.* **2013**, *3*, 2660. [CrossRef] [PubMed]

C Journal of Carbon Research

MDPI

Article

Applicability of Atmospheric Pressure Plasma Jet (APPJ) Discharge for the Reduction in Graphene Oxide Films and Synthesis of Carbon Nanomaterials

Sri Hari Bharath Vinoth Kumar [1,2], Josefa Ibaceta-Jaña [1], Natalia Maticuic [2], Krystian Kowiorski [3], Matthias Zelt [2], Ulrich Gernert [4], Ludwika Lipińska [3], Bernd Szyszka [1], Rutger Schlatmann [2], Uwe Hartmann [5] and Ruslan Muydinov [1,2,*]

1 Institute for High-Frequency and Semiconductor-System Technologies, Technische Universität Berlin, Einsteinufer 25, 10587 Berlin, Germany; s.vinothkumar@tu-berlin.de (S.H.B.V.K.); josefa.ibaceta@gmail.com (J.I.-J.); bernd.szyszka@tu-berlin.de (B.S.)
2 PvcomB, Helmholtz-Zentrum Berlin für Materialien und Energie GmbH, Schwarzschildstraße 3, 12489 Berlin, Germany; natalia.maticiuc@helmholtz-berlin.de (N.M.); matthias.zelt@helmholtz-berlin.de (M.Z.); rutger.schlatmann@helmholtz-berlin.de (R.S.)
3 Graphene and Composites Research Group, Łukasiewicz Research Network—Institute of Microelectronics and Photonics, Aleja Lotników 32/46, 02-668 Warsaw, Poland; krystian.kowiorski@itme.edu.pl (K.K.); ludwika.lipinska@itme.edu.pl (L.L.)
4 Zentraleinrichtung Elektronenmikroskopie (ZELMI), Straße des 17. Juni 135, 10623 Berlin, Germany; ulrich.gernert@tu-berlin.de
5 Impellis GmbH & Co., KG, Am Försterberg 12, 32805 Horn-Bad Meinberg, Germany; u.hartmann@impellis.de
* Correspondence: ruslan.muydinov@helmholtz-berlin.de

Citation: Vinoth Kumar, S.H.B.; Ibaceta-Jaña, J.; Maticuic, N.; Kowiorski, K.; Zelt, M.; Gernert, U.; Lipińska, L.; Szyszka, B.; Schlatmann, R.; Hartmann, U.; et al. Applicability of Atmospheric Pressure Plasma Jet (APPJ) Discharge for the Reduction in Graphene Oxide Films and Synthesis of Carbon Nanomaterials. C **2021**, 7, 71. https://doi.org/10.3390/c7040071

Academic Editor: Gil Goncalves

Received: 16 September 2021
Accepted: 11 October 2021
Published: 14 October 2021

Abstract: Atmospheric pressure plasma jets (APPJ) are widely used in industry for surface cleaning and chemical modification. In the recent past, they have gained more scientific attention especially in the processing of carbon nanomaterials. In this work, a novel power generation technique was applied to realize the stable discharge in N_2 (10 vol.% H_2) forming gas in ambient conditions. This APPJ was used to reduce solution-processed graphene oxide (GO) thin films and the result was compared with an established and optimized reduction process in a low–pressure capacitively coupled (CCP) radiofrequency (RF) hydrogen (H_2) plasma. The reduced GO (rGO) films were investigated by Raman spectroscopy and X-ray photoelectron spectroscopy (XPS). Effective deoxygenation of GO was observed after a quick 2 s treatment by AAPJ. Further deoxygenation at longer exposure times was found to proceed with the expense of GO–structure integrity. By adding acetylene gas into the same APPJ, carbon nanomaterials on various substrates were synthesized. The carbon materials were characterized by Raman spectroscopy, scanning electron microscopy (SEM), and energy-dispersive X-ray (EDX) analyses. Fullerene-like particles and graphitic carbon with short carbon nanotubes were detected on Si and Ag surfaces, respectively. We demonstrate that the APPJ tool has obvious potential for the versatile processing of carbon nanomaterials.

Keywords: atmospheric plasma; carbon nanomaterials; graphene oxide; plasma treatment

1. Introduction

Plasma techniques have been part of the material science and technology domain for the past several decades [1]. They are well integrated with today's research and industries in diverse applications due to high versatility, flexibility, and scalability. Recently, there has been a growing interest in the use of atmospheric plasma systems, and the APPJ systems, in particular [2–5]. The fundamentals of this technology have been reviewed in the literature [6–14]. APPJ can generate a high flux of various metastable but chemically very active species. Its use in ambient conditions allows setting up smaller, modular, and low-cost plasma sources for surface treatment. Since APPJs work with a gas flow, the ignition of plasma is spatially separated from the treatment surface [15]. This attribute has the following advantages: (i) the surface to be treated is safe from damage caused by arcs or discharge currents; and (ii) surfaces of objects with various shapes, geometries, and sizes can be treated. A limitation of the APPJs is the small area of a treatment restricted to a few millimeters. This limitation can be successfully overcome by scanning [16], using robotic arms [17], or an array of APPJs [18,19].

The applications of APPJs are numerous and can be broadly summarized into (i) surface modification; (ii) deposition of coatings, and (iii) biomedical applications [2,3,5]. This article relates to the former two themes. The crossover of the APPJ applications with the synthesis and processing of carbon nanomaterials is particularly interesting [20–23].

In the last four decades, the field of carbon materials has been extended with fullerenes, carbon dots, nanodiamonds, nanofibers, nanotubes, and graphene [24]. From these carbon-allotropic forms, the growth and prospects of graphene have been compelling [25]. In our previous work, we highlighted the importance and scope of atmospheric plasma such as APPJ in the reduction in GO, which is one of the strategies to produce graphene films [26].

There is also an emerging interest in the production of free-standing graphene (gas-phase synthesized graphene, GSG) at atmospheric pressure. The GSG is formed in the afterglow region of plasma discharge where nucleation and growth take place [27,28]. Microwave (MW)-induced plasma is predominantly utilized for such synthesis [29–31]. Jašek et al. [31] deposited GSG layers from ethanol vapors in argon in this way. Such plasma heated the Si/SiO_2 substrate to ~650–1150 °C, depending on plasma conditions. The results are commendable, but the approach needs further development to operate at lower temperatures suited for sensitive (e.g., flexible) substrates. Researchers have also synthesized GSG using high-power APPJ (30–40 kW power, 300–400 A) fixed and enclosed in a reactor with graphitic walls [24,32–34]. The reactor was further enclosed in a vacuum chamber with helium or argon feed-through. As a graphene forming admixture, propane/butane or methane, or acetylene were used. A drastic cooling of the discharge flow resulted in the formation of a solid form of carbon. Crumbled graphene sheets of various sizes (up to a few hundred nanometers) with the number of layers in the range of 1–10 have been obtained. Although this technique is attractive, some of the downsides are [32–34]: (i) a very high power consumption; (ii) the necessity of special auxiliary systems including a vacuum chamber with cooling; and (iii) the crumbled nature of graphene, and the need for a post-synthesis solution process to realize particular applications such as transparent conductive layers, sensors, and other functional coatings.

Motivated by the need for an atmospheric plasma system to be able to deposit graphene films in ambient conditions, we developed a low-power APPJ system. Although the glowing gas jet is named "Plasma" in the literature, it is not strictly correct in our case since most gas species are not ionized but excited. Our free jet system provides a lower temperature of neutrals as compared to the known APPJs. Due to the discontinuity of power coupling, not all the flowing gas is electrically affected. The potential applications of this system are demonstrated for: (i) reduction in GO films and (ii) deposition of carbon nanomaterials on various substrates (described in Sections 3.1 and 3.2, respectively).

2. Materials and Methods

As substrates, an Eagle XG® (Corning Inc., Corning, NY, USA) glass of 1.1 mm thickness and p-type (100)–oriented Si (Microchemicals GmbH, Ulm, Germany) were used. Before deposition, the substrates were sonicated at 50 °C in organic solvents (acetone, isopropanol) and washed in a glass-washer using a universal detergent (Hellmanex®) and deionized water.

All GO samples were prepared only on Si substrates. The aqueous dispersion of GO was prepared with flakes obtained from graphite by a modified Hummers method. Further experimental details are described in the literature [35–37]. The 2 mg/mL GO dispersion in deionized water (DI) (flake size 1–100 μm) was spin-coated at 2000 rpm for 30 s on 1.5 × 1.5 cm^2 substrates. The films were dried at room temperature (RT) overnight in a N_2–filled glovebox. As found by SEM and confirmed by ellipsometry these films had a thickness of ~20 nm and a multilayered porous microstructure.

The commercial PlasmaJet® (Raantec GmbH & Co., KG, Borgholzhausen, Germany) system was specially redesigned for our purposes. Since molecular H_2 inhibits plasma generation, the electronic components of the power generator have been modified. Power generation was also optimized for the gas mixtures used and thus operated at an MF-induced (31.8 kHz) direct current (DC) power (480 W). The free jet nozzle was constructed to provide a vortex-like flow of gas. The nozzle head design details are available in the patent DE 195 32 412 C 2. The internal glowing chamber was equipped with a ring-electrode, which also worked as an outcoupling window. Another electrode was placed in the middle of the chamber on the back side. Fine-tuning of the power generation provided the following gas excitation conditions:

- high enough excitation degree;
- gas leaving the nozzle electrically unchanged;
- gas species remaining excited outside the nozzle;
- the relaxation of excited species producing energy in the infrared (IR) range.

So far, no detailed optical emission spectroscopy (OES) analysis of our free jets has been performed, however, it is planned for future work. The GO–reduction treatment was performed in a flow (30 L/h) of forming gas (10 vol.% H_2 in N_2). The substrate to nozzle distance was 8 mm, and the plasma irradiation time was varied at 2 s, 5 s, and 10 s. To avoid rGO reoxidation in air, the freshly treated surfaces were exposed to the same flow for 20 s after the plasma power supply was turned off. For all treatments/deposition with the APPJ, no external sample heating was used. The GO reduced with APPJ in ambient conditions was compared to rGO reduced in a H_2-plasma environment. For this, an RF–plasma treatment was carried out in the AKT 1600 system (Applied Materials, Santa Clara, CA, USA) described elsewhere [38]. After comprehensive optimization where RF-power, temperature, and treatment time were varied, the conditions chosen were: 100 W RF power, 30 s treatment, and 200 °C substrate temperature. The heating was performed in hydrogen for 10 min and the cooling in the load lock chamber for 5 min under vacuum. A pure H_2 (99.9999%) environment at a process pressure of 133.3 Pa was used during the treatment.

The in-house prepared RF-sputtered Corning®-glass/~20 nm *i*-ZnO/~20 nm Ag film stack (referred to as ZnO/Ag from here onwards) and Si were used as substrates for carbon APPJ deposition. A Corning®-glass of size 2.5 × 2.5 cm^2 was used for the sputtering of ZnO/Ag. The sputtering process for i-ZnO was described in previous work [39]. Silver films were deposited in the DC magnetron sputtering system A600V7 (Leybold Optics Dresden GmbH, Dresden, Germany) at a 75-mm target–to–substrate distance. The base pressure was 5.37×10^{-1} Pa, the process was performed in Ar (99.999%) at a pressure of 5.16×10^{-1} Pa. The power density was 1.33 W/cm^2, no intentional heating of a substrate was used. The substrate holder oscillated with a 4.8 m/min speed. No special pretreatment such as etching of SiO_x from silicon wafer was applied before carbon depositions. The metallic silver film grew (111)-textured on (0002)-ZnO seed layer (Figure S1a), which was in good agreement with the literature [40]. Herewith, the Ag-film was not textured in-plane (Figure S1). For carbon synthesis, the APPJ discharge contained a mixture of C_2H_2 (feedstock gas, 0.5 L/h flow) and forming gas (10 vol.% H_2 in N_2, 30 L/h flow). The substrate to nozzle distance varied between 3 and 6 mm. The 1 × 1 cm^2 cut ZnO/Ag films, optimized based on resistance measurements in Van der Pauw geometry (Rsh ≈ 10 Ω), were chosen for the carbon depositions. The photographs of the APPJ working with two considered gas mixtures are shown in Figure S2. All freshly treated or deposited samples were immediately packed into vacuum-sealed plastic bags for transport or storage until the next processing step or analysis.

Scanning electron microscopy (SEM) and energy dispersive X-ray analyses (EDX) were performed on a Hitachi SU8030 SEM (Tokyo, Japan) with a CFEG (cold field emitter gun) and EDX analyzer EDAX TEAM system with a 30 mm^2 Octane SDD (silicon drift detector). XRD measurements were completed on the Bruker Discovery D8 (Karlsruhe, Germany) system using Cu K_{α} radiation and the high-resolution mode of the 1D-LYNXEYE XE-T detector. For the acquisition of the pole figure, Polycap optics were installed.

XPS measurements were carried out using nonmonochromatic Mg K_{α} radiation (1254 eV) [41]. The raw XPS spectra of the various GO were fit using the XPSPEAK 4.1 software by non-linear least-squares fittings with a Gaussian/Lorentzian ratio of 80:20 [42]. Micro-Raman spectroscopy was performed in a high-resolution Horiba Jobin Yvon LabRAM HR800 (Bensheim, Germany) spectrometer. Samples deposited with the C_2H_2 plasma on Si and ZnO/Ag substrates were measured with an excitation source of 633 nm and a power of 2.49 mW. The GO and rGO (reduced with APPJ) on Si were measured with an excitation wavelength of 532 nm with a power of 2.60 mW, while the set of GO and rGO prepared with the RF H_2–plasma was measured with 633 nm excitation source and power of 5 mW. Different Raman excitation wavelengths used during the investigation were solely due to lab availability. The acquisition parameters were grating of 600 lines per millimeter, an aperture of 400 μm, and an objective of 100×. The Raman spectra were fit using Fityk software (version 1.3.1).

3. Results and Discussion

3.1. Reduction in GO

The fitted Raman spectra for the GO and various rGO are presented in Figure 1 where a combination of three pseudo-Voigt and two Gaussian functions were used [43]. The fit data such as peak centers, intensity, and full width at half maximum (FWHM) have been summarized in the supplementary information, Table S1. In pure graphite, the peak at ~1580 cm^{-1} corresponded to the G-band arising from the E_{2g}-symmetry of the carbon lattice [44]. In the case of GO/rGO, this peak either blueshifted or redshifted based on the excitation laser frequency. The spectra also exhibited peaks at ~1350 cm^{-1} (D–band) and ~1620 cm^{-1} (D′–band) corresponding to the disorder-induced phonon mode due to the finite size of crystals and defects [45]. The D′–band has also been attributed to vacancy defects in pentagonal and octagonal rings in the literature [46]. The shoulder at ~1200 cm^{-1} was the D*–band also arising from disordered graphitic lattice provided by the sp^2–sp^3 bonds at the edges of the network [44]. The attribution of the D″–band at ~1480 cm^{-1} was ambiguous [43] as it has been assigned to (i) phonon density of states in finite-sized crystals [47], (ii) amorphous lattices, since the intensity of the D″ band was found to decrease with an increase in crystallinity [48], (iii) amorphous sp^2–bonded forms, including functionalized small molecules [44,49], and (iv) a double resonance process with the iTO phonon branch [50].

In the literature, the peak at ~1447 cm^{-1} has been attributed to either the scissoring mode of –CH_2 [51–53] or the C=C ring stretching in thiophene (containing nitrogen) [54,55]. Moreover, a peak ~1450 cm^{-1} in nanocrystalline diamond is said to arise from the sum modes of sp^2 C=C and C–H vibrations of trans-polyacetylene type segments at grain boundaries [42,56]. In the case of the rGO samples here, this peak is likely to be from the mode of –CH_2 as it was confirmed with XPS analysis that foreign atoms were not incorporated into the rGO. Therefore, the emergence of these peaks indicated that the sample structure deviated from the typical GO structure [57]. With a 10 s APPJ-reduction, the structure of rGO seemed to further deviate from the 5 s APPJ-reduced sample (see supplementary information, Figure S3). The overtone bands (2D or G′, D + G, and 2G [58,59]) were found in the region 2500–3300 cm^{-1} for the GO and they decreased in intensity with increasing free-jet exposure time (supplementary information, Figure S3). This can be attributed to the formation of defects and diminishing sp^2–C domains [59–61].

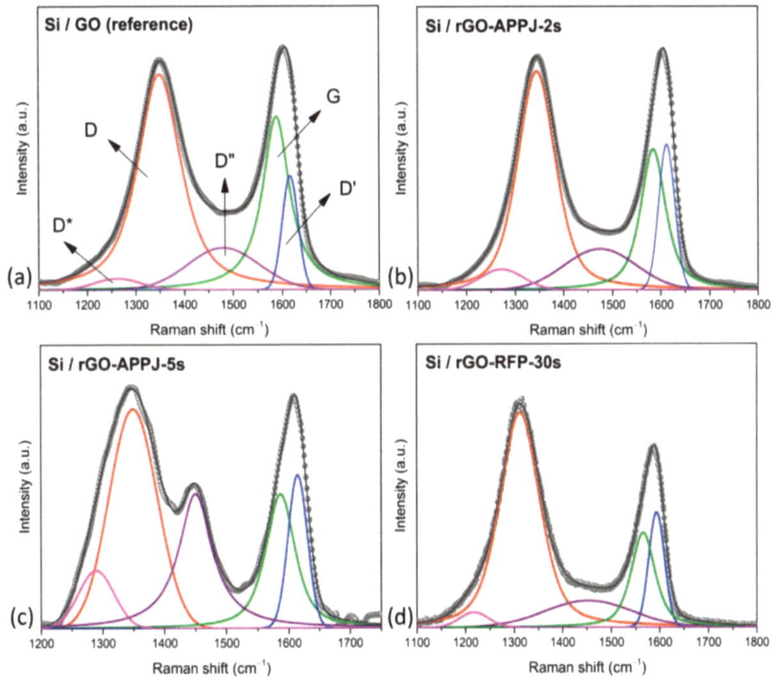

Figure 1. Fitted Raman spectra of GO (**a**), 2 s APPJ reduced GO (**b**), 5 s APPJ reduced GO (**c**), and 30 s RF plasma reduced GO (**d**) on Si substrates.

The XPS survey of the various GO is shown in the supplementary information, Figure S4. As expected, peaks arising from carbon (~284 eV) and oxygen (~531 eV) were observed along with silicon (99–169 eV) from the substrate. Interestingly, nitrogen signals were not found in the APPJ–processed GO samples, which were exposed to plasma containing nitrogen species. This suggested that nitrogen doping did not occur for the rGO. For samples treated with APPJ for 5 and 10 s, traces of iron (Fe) were found originating from the sample holder (supplementary information, Figure S4). Figure 2 presents the XPS core levels of the GO and various rGO. The chemical components identified from the spectra were sp^2–hybridized carbon (sp^2–C), sp^3–hybridized carbon (sp^3–C, C–C/C–H), hydroxyl-(C–OH), epoxide-(C–O–C), carbonyl-(C=O), and carboxyl-(CO(OH)) groups, respectively [60,62–67]. For the sample rGO–APPJ–5s, a feature at binding energy >290 eV was distinctively observed. In the literature, this has been assigned to energy losses due to π–π * excitations, which is typical for carbonaceous systems with delocalized π–electrons [68], as well as to carbon atoms bonded to metals [69].

The increasing trend of the $I_D/(I_G + I_{D'})$ ratio from Raman spectroscopy was an indicator of deoxygenation in GO (Table 1). The increasing ratio arose from the increasing quantity of disorder, which in turn was associated with the appearance of new sp^2–C domains with the loss of carbon–oxygen bonds [70]. The loss of oxygen bonds was also verified from the XPS quantification of the components' fit for the C1s spectra. The crystallite size was calculated from the areas of the D band and G band in Raman spectra using the Tuinstra and Koenig relation [71–73]. The calculated average crystallite size (La) for GO was 10.8 nm when the Raman signals were measured with a 532-nm laser. It reduced to 9.6 nm and 8.4 nm when the GO was reduced with APPJ plasma for 2 and 5 s, respectively. In the case of RF plasma treatment, the crystallite was reduced to 9.2 nm from 10.2 nm (Raman signals recorded with 633 nm laser). The average distance between the defects reduced to 7.9 nm and 7.4 nm from 8.4 nm for an APPJ treatment time of 2 and 5 s, respectively. With the 30 s RF plasma treatment, the distance between defects reduced to 7.9 nm from 8.2 nm. This suggests that with plasma irradiation defects were created in the rGO which is already well established in the literature. The 30 s RF plasma irradiation produced similar defects as the 2 s APPJ exposure, indicating that the APPJ had higher energetic plasma species than the RF-plasma used.

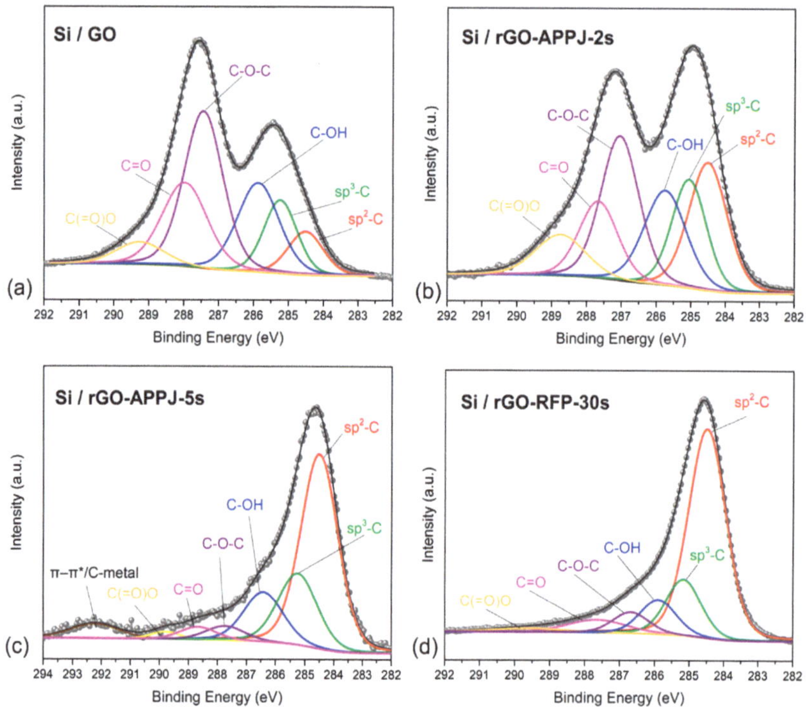

Figure 2. High-resolution XPS spectra in the C1s region for GO (**a**), 2 s APPJ reduced GO (**b**), 5 s APPJ reduced GO (**c**), and 30 s RF plasma reduced GO (**d**) on Si substrates.

Table 1. The calculated values of $I_D/(I_G + I_{D'})$, the average crystallite size (L_a), and the average distance between defects (L_D) for GO and various rGO from Raman spectroscopy (Figure 1 and Table S1). rGO-RF-30s and its reference GO sample (indicated by †) were measured with 633 nm laser excitation, while the remaining samples were measured with 532 nm laser excitation. Quantification (area distribution in %) of the components fitted in the C1s spectra (Figure 2). For the rGO-APPJ-5s sample, a contribution of 5.2% from the feature at binding energy >290 eV (discussed earlier) was not tabulated.

Sample	Raman Analysis			XPS Quantification Analysis					
	$I_D/(I_G + I_{D'})$	L_a (nm)	L_D (nm)	sp^2–C	sp^3–C	C–OH	C–O–C	C=O	C(O)OH
GO	0.75	10.8	8.4	8.2	12.9	20	32.8	20.3	5.8
rGO-APPJ-2s	0.76	9.6	7.9	21.3	16.1	17.1	23.8	12.8	8.9
rGO-APPJ-5s	0.79	8.4	7.4	52.0	20.1	14.0	3.9	3.3	1.5
GO †	1.06	10.2	8.2	41.7	23.1	9.1	7.7	5.6	9.2
rGO-RF-30s †	1.08	9.7	7.9	61.3	15.4	9.5	5.8	5.8	2.1

The hydrogen RF plasma typically constitutes free electrons, neutrals (molecular H_2, atomic H), and charged ions (H^-, H^+, H_2^+, and H_3^+) [74]. The plasma composition is relatively stable and homogeneous stipulating nearly equilibrium conditions for chemical interactions. At the same time, the electrodeless RF-coupling powers the high temperature of neutrals. Hydrogen-containing APPJ reveals mostly nitrogen second positive system bands and NH [75]; typical species from N_2 discharge are N, NO•, OH•, N_2, N_2^+, and CN [76]. As pointed above, the gas in our free jet was expected to be negligibly ionized, especially at the distance of the surface reaction event. In GO, the carbon bonded to oxygen accounted for nearly 79% of the total carbon bonds. In the case of carbon-to-carbon bonds (Table 1), the fraction of sp^3-hybridized content (12.9%) was found to be higher than sp^2-hybridized bonds (8.2%). Even with a short 2 s APPJ-treatment, the oxygen-containing group loss was observed, and the amount of sp^2-carbon content (~21%) increased relative to sp^3-carbon (~16%). With a 5 s APPJ-treatment, the net oxygen-containing group fraction reduced to ~23%, and the restoration of sp^2-C increased to 52%. To compare, the optimized hydrogen RF-plasma reduced net oxygen-content in the GO film down to ~24%, while the sp^2-C content increased to ~61%. Compared to the 5 s APPJ-treatment, the RF plasma treatment yielded a lower content of C-OH and O=C-OH groups, whereas the APPJ-treatment yielded a lower content of C-O-C and C=O groups.

3.2. *Synthesis of Carbon Nanomaterials*

3.2.1. Deposition on ZnO/Ag Substrates

It is known, that the (111)-facet of fcc-metals such as Cu and Ni serve as a template for the honeycomb carbon network [77,78]. The lattice mismatch with graphene in these cases exceeds −10%. According to our estimation, the 111-Ag lattice would have less than +2% mismatch at normal conditions. That is why we have chosen it as a substrate, which is a standard for low-emissivity coating technology. Figure 3 presents the Raman spectra of carbon deposited on ZnO/Ag substrates, and they were fitted with a combination of Lorentzian and Gaussian (only for D3 band) functions [44]. The parameters extracted from the fitted peak are summarized in the supplementary information, Table S2 and the broad Raman spectrum is presented in the supplementary information, Figure S5. In the 250–750 cm^{-1} region (Figure S5), several peaks and bands were strongly observed for samples deposited at a distance of 3 mm for 1 s (and weakly when deposited at a distance of 6 mm from the substrate). These features are attributed to the underlying glass substrate [79], ZnO [80], and Ag–O [81]. Except for the sample deposited at 3 mm and 1 s, the samples portrayed well-separated D and G bands suggesting relatively low amorphous content or high crystalline (graphitic) content [82].

The G band split into G^- band and G^+ band (Figure 3), which is usually observed in carbon nanotubes (CNT) as a result of curvature [83]. This splitting has also been observed when graphene sheets are strained [84], wrinkled [85], or folded and scrolled [86].

A weak 2D band (~2700 cm^{-1}) was detected in all the samples (supplementary information, Figure S5). The 2D band is a well-known characteristic feature of all the sp^2–C based materials [87–89]. The films deposited for 2 s revealed relatively prominent 2D peaks, as well as the G + D band (~2900 cm^{-1}) (supplementary information, Figure S5). The existence of these features is assessed as a lower disorder in the films [60]. The D3 band (~1500 cm^{-1}) is assigned to the presence of amorphous carbon [44,90]. The D4 band (~1200 cm^{-1}) is assigned to the aliphatic compounds grafted on the disordered carbon network [90,91] or defective graphite [90,92]. In some samples (Figure 3b,c), a weaker D″ band (1076–1095 cm^{-1}) is also additionally observed originating from defects [90,92,93].

The estimated crystallite size (L_a) from Raman spectroscopy were 8.7 nm, 10.2 nm, 10 nm, and 11.6 nm for the samples Ag–C–D3–1S, Ag–C–D3–2S, Ag–C–D6–1S, and Ag–C–D6–2S, respectively [71–73]. Similarly, these samples had an estimated defect between the point defects (L_D) as 7.5 nm, 8.2 nm, 8 nm, and 8.7 nm, in the same order. The SEM images (supplementary information, Figures S6 and S7) evidenced several microstructural changes as compared to the initial silver surface (Figure S7a,b). Silver recrystallized and probably shrank, which led to a formation of pores (dark regions ranged in size from a few 10s of nm to ~100 nm). Point EDX analysis of the largest dark areas revealed greater Zn and lesser Ag-content as compared to the flat particle-free areas. These relatively homogeneous areas were notably smoother in treated samples than the surface of the initial silver film. One can also distinguish in a few places an increase in the size of Ag grains, but the surface appears to be coated with carbon. Nano needle-like structures (or short nanotubes) were also observed, whose diameter was ~50 nm, and length was in the range of 100s of nm. Similar to the pores, the number of short nanotubes also decreased at a higher nozzle distance and longer deposition time. Bright dots observed in the SEM images with higher magnification represent silver particles according to EDX.

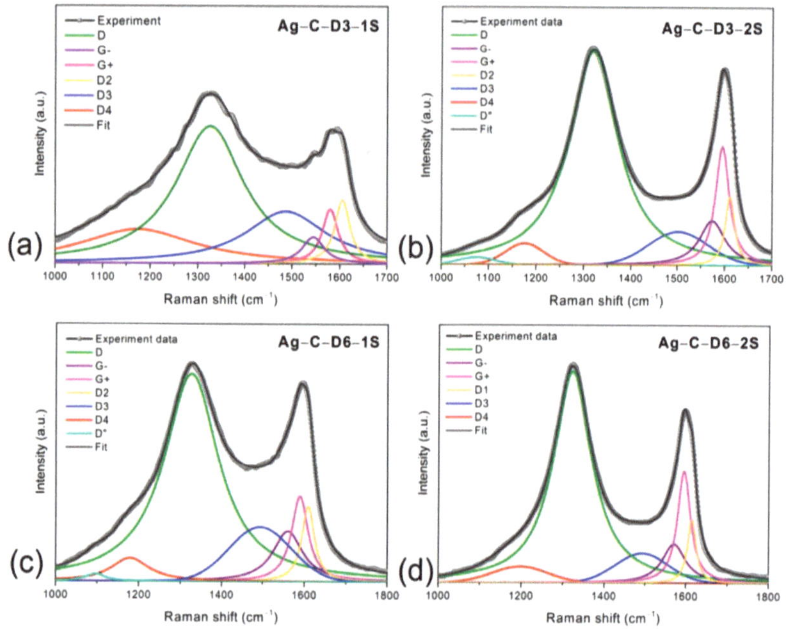

Figure 3. Fitted Raman spectra for carbon deposited on ZnO/Ag substrates at various nozzle-to-substrate distances and durations: (**a**) at 3 mm for 1 s, (**b**) at 3 mm for 2 s, (**c**) at 6 mm for 1 s, and (**d**) at 6 mm for 2 s.

3.2.2. Deposition on Si Substrates

It is known that native Si-wafers are oxidized from the surface. This oxide is very smooth and dense but not crystalline. We took this substrate for the next experiments to discard the template impact of a substrate. The dynamic linearly oscillating deposition was undertaken at the nozzle-to-substrate distance of 6 mm to avoid overheating the substrate. The Raman spectra of the carbon deposited on Si substrates are shown in Figure 4 and the peak assignments are summarized in Table 2. The carbon deposited for 2–20 s displayed a similar spectral feature, and their interpretation was not straightforward. Some prominently observed features in the spectra were the bands in the range 1050–1100 cm^{-1} coinciding with C–C stretching observed in hydrocarbon chains [94,95] and several peaks up to 1450 cm^{-1} which are observed in C60/C70-fullerene materials (see Table 2) [96,97].

It should be also noted that the region 1100–1150 cm^{-1} and 1430–1470 cm^{-1} were related to the trans-polyacetylene segments at the grain boundaries [98]. The peaks ~1060 cm^{-1} and ~1180 cm^{-1} also have been reported for the nanocrystalline phase of carbon [99]. The sp^2–C phase or the G band (1560 cm^{-1}) from the in-plane stretching mode of carbon is observed only when the deposition time is high at 30 s [100,101]. The carbon with the broad G and D bands indicates the existence of an amorphous structure [82]. However, features of the fullerene-like structure were also superimposed into the dominant broad G and D bands. This suggests a possible composite structure of fullerene-like matter and amorphous films. The SEM images of the carbon samples are presented in the supplementary information, Figure S8. For up to a deposition time of 20 s, the surface coverage was nonuniform and low with clusters of spherical particles. In the sample deposited for 30 s, spherical particles of various sizes (several 10 s of nm) were observed with some in aggregated form. These aggregates (supplementary information, Figure S5d) were comparable to onion-like fullerene/nanocarbon (fullerenoid) structures reported in the literature [20,102–105].

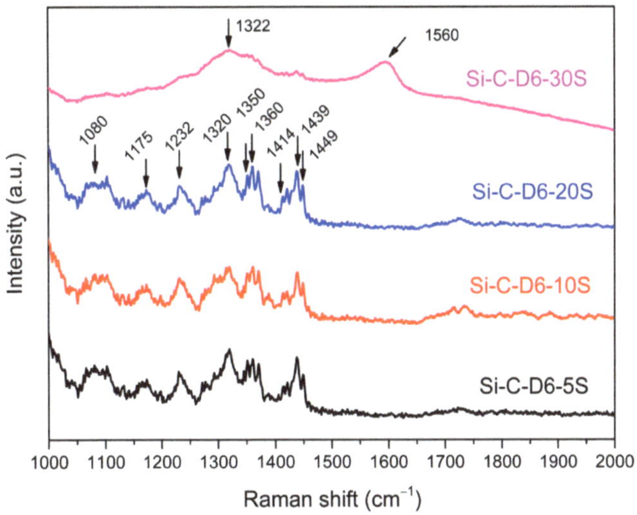

Figure 4. Raman spectra of the carbon deposited on Si substrates for various time and nozzle distances.

The idea of using acetylene to synthesize or to deposit the sp^2-carbon network was based on the corresponding catalytic synthesis of benzene. This can proceed for instance via adsorption on Pd-surface [106], or be promoted by transition metals, which are known to catalyze the so-called [2 + 2 + 2]-polymerization [107]. Silicon atoms may also form intermediates driving the polymerization process [108]. In combination with admixtures increasing the Lewis acid character of the promoter (e.g., Si^{+4}), the reaction in question can proceed at much lower temperatures than the established 500–600 °C [109]. The entire mechanism of polymerization in plasma from a single carbon atom to fullerenes was proposed and described in detail elsewhere [110]. In the case of a plasma torch process at atmospheric pressure, the formation of a similar carbon-onion-like structure passes through dissociation, nucleation, and agglomeration events in the carbon feedstock gas [102]. The spherical agglomerates form owing to defects in fullerene molecules, which fulfill the role of reactive centers [111]. Further investigation of our free jet must provide the information on existing reactive species and give an idea of polymerization mechanism.

Table 2. Features observed in the Raman spectra of Carbon deposited on Si substrates.

Peak	Vibrational Assignment	Reference
1080 cm^{-1}	C–H bend + C–C stretch	[112,113]
1175 cm^{-1}	C–H bend or E_1'' (C70) or $H_g(5)$ (C60)	[98,113,114]
1232 cm^{-1}	A_1' (C70)	[96]
1320 cm^{-1}	$G_u(5)$ (C60) or E_1' (C70) or the D-band	[82,97]
1350 cm^{-1}	$T_{2g}(4)$ (C60) or E_2' (C70)	[97]

Table 2. *Cont.*

Peak	Vibrational Assignment	Reference
1360 cm^{-1}	D band related sp^2-C	[114]
1370 cm^{-1}	$H_g(7)$ (C60) or E_2' (C70)	[97]
1414 cm^{-1}	$H_u(6)$ (C60) or E_1' (C70)	[97]
1439 cm^{-1}	$T_{1u}(4)$ (C60) or E_1' (C70)	[96,97]
1449 cm^{-1}	$H_g(7)$ (C60) or E_1'' (C70)	[97]
1560 cm^{-1}	$H_u(7)$ (C60) or the G band	[115,116]

4. Conclusions

In this work, a novel free jet system was evaluated in ambient conditions. The optimized MF-power generation could produce excited species in hydrogen-containing gas mixtures making them chemically active. Using 10% vol. H_2/N_2 forming gas, ~20 nm GO-films were successfully reduced within a processing time of 2 s according to XPS and Raman analyses. The rGO films thus obtained were comparable in quality with the rGO films reduced in RF hydrogen plasma under optimized and established conditions. A longer free jet treatment resulted in further deoxygenation with the appearance of a strong Raman peak at 1447 cm^{-1}, which was attributed to trans-polyacetylene (–CH_2) chains. The appearance of such chains reflects the breaking of the honeycomb carbon network in GO. A closer analytical look into the free-jet constituents and their energies is planned to control and improve the process, e.g., to not damage the GO surface at longer treatment.

By introducing the acetylene admixture into the forming gas, we realized the deposition of carbon films by our free jet system. Multiple activations of the depositing surface via dynamic oscillating treatment resulted according to Raman spectroscopy analysis in a formation of the fullerene C70-like matter. SEM revealed the presence of aggregated spherical nanoparticles from a few-10s to over a few-100s of nm. These particles resembled the onion-like nanostructure, which contains fullerene-like particles, as reported in the literature. The carbon films deposited on (111)-textured Ag surface had nano needle-like structures (a few 10s nm in diameter and a few 100s nm long). Raman spectroscopy also revealed the presence of graphitic carbon (estimated crystallite size, $L_a \approx 9$–12 nm). The number of nanocrystalline carbon needles reduced with increasing deposition time and nozzle-to-substrate distance.

The preliminary results based on our free jet system demonstrated unique simplicity in the synthesis of various carbon nanostructures at atmospheric pressure. The presented method opens an inexpensive, scalable, and entirely vacuum-free way to produce rGO coatings. The low gas temperature of the jet enables using any sensitive substrates including flexible ones. Furthermore, we believe that a much faster (seconds vs. tens of minutes) and vacuum-free reduction process has more potential to be realized on an industrial scale, e.g., in reel-to-reel production concepts. A deep scientific understanding of the excitation processes occurring in gases coupled electrically in this way is of imperative importance.

Supplementary Materials: The following are available online at https://www.mdpi.com/article/10.3390/c7040071/s1: Figure S1: XRD pattern and pole-figure of the ZnO/Ag substrate, Figure S2: Photographs of the APPJ used in this work, Figure S3: Raman spectra of GO and rGO prepared with APPJ, Figure S4: XPS survey spectra for Si, GO, and various rGO, Figure S5: Wide Raman spectra of carbon deposited on ZnO/Ag substrates. Figures S6 and S7: SEM images of carbon deposited on ZnO/Ag substrates, Figure S8: SEM images of carbon deposited on Si substrates, Table S1: Parameters from Raman spectroscopy of GO and various rGO, Table S2: Parameters from Raman spectroscopy of various carbon deposited on ZnO/Ag substrates.

Author Contributions: The conceptualization, organization, supervision, XRD measurements, review, and editing of the draft, R.M.; development of the special APPJ system, U.H.; experimental realization, analyses of data, and writing of the original draft, S.H.B.V.K.; Raman measurements and support with data analyses, J.I.-J.; XPS measurements and support with data analyses, N.M.; synthesis of GO and rGO powders and preparation of their stable suspensions, K.K., RF-plasma treatments in hydrogen, M.Z.; SEM and EDX analyses, U.G.; scientific and administrative support on GO and rGO precursors, L.L.; project administration and funding acquisition, B.S. and R.S. All authors have read and agreed to the published version of the manuscript.

Funding: The authors acknowledge support from the German Federal Ministry for Economic Affairs and Energy (BMWi) in the framework of the project "speedCIGS" (grant no. 0324095H) and the German Federal Ministry of Education and Research (BMBF) in the framework of the project "Catlab" (grant no. 03EW0015A/B). The authors also acknowledge the support from the German Research Foundation and the Open Access Publication Fund of TU Berlin. Josefa Ibaceta-Jaña acknowledges the financial support provided by the Scholarship Becas Chile-DAAD 2017/91645541.

Data Availability Statement: The data presented in this study are available on request from the corresponding author.

Acknowledgments: We sincerely thank Ansgar Jahn for the first basic technical support provided by RAANTEC GmbH & Co. KG (Borgholzhausen, Germany). We gratefully acknowledge our colleagues Alexander Steigert from the Institute for Nanospectroscopy, HZB (Berlin, Germany) who provided the *i*-ZnO films on the glass and Katja Mayer-Stillrich and Manuel Hartig from PVcomB, HZB (Berlin, Germany) for the deposition of Ag-films. We thank Markus R. Wagner for providing access and support for the Raman spectroscopy measurements. The experimenters thank Iver Lauermann from PVcomB, HZB (Berlin, Germany) for providing chemical infrastructure. In a few cases, Christoph Fahrenson from ZELMi, TU Berlin (Germany) made SEM images.

Conflicts of Interest: The authors declare no conflict of interest. The funders had no role in the design of the study; in the collection, analyses, or interpretation of data; in the writing of the manuscript, or in the decision to publish the results.

References

1. Adamovich, I.; Baalrud, S.D.; Bogaerts, A.; Bruggeman, P.J.; Cappelli, M.; Colombo, V.; Czarnetzki, U.; Ebert, U.; Eden, J.G.; Favia, P.; et al. The 2017 Plasma Roadmap: Low temperature plasma science and technology. *J. Phys. D Appl. Phys.* **2017**, *50*, 323001. [CrossRef]
2. Penkov, O.V.; Khadem, M.; Lim, W.-S.; Kim, D.-E. A review of recent applications of atmospheric pressure plasma jets for materials processing. *J. Coat. Technol. Res.* **2015**, *12*, 225–235. [CrossRef]
3. Belmonte, T.; Henrion, G.; Gries, T. Nonequilibrium Atmospheric Plasma Deposition. *J. Therm. Spray Technol.* **2011**, *20*, 744–759. [CrossRef]
4. Di, L.; Zhang, J.; Zhang, X. A review on the recent progress, challenges, and perspectives of atmospheric-pressure cold plasma for preparation of supported metal catalysts. *Plasma Process. Polym.* **2018**, *15*, 1700234. [CrossRef]
5. Braný, D.; Dvorská, D.; Halašová, E.; Škovierová, H. Cold atmospheric plasma: A powerful tool for modern medicine. *Int. J. Mol. Sci.* **2020**, *21*, 2932. [CrossRef] [PubMed]
6. Eliasson, B.; Kogelschatz, U. Nonequilibrium Volume Plasma Chemical Processing. *IEEE Trans. Plasma Sci.* **1991**, *19*, 1063–1077. [CrossRef]
7. Schütze, A.; Jeong, J.Y.; Babayan, S.E.; Park, J.; Selwyn, G.S.; Hicks, R.F. The atmospheric-pressure plasma jet: A review and comparison to other plasma sources. *IEEE Trans. Plasma Sci.* **1998**, *26*, 1685–1694. [CrossRef]
8. Conrads, H.; Schmidt, M. Plasma generation and plasma sources. *Plasma Sources Sci. Technol.* **2000**, *9*, 441–454. [CrossRef]
9. Merche, D.; Vandencasteele, N.; Reniers, F. Atmospheric plasmas for thin film deposition: A critical review. *Thin Solid Film.* **2012**, *520*, 4219–4236. [CrossRef]
10. Bogaerts, A.; Neyts, E.; Gijbels, R.; Van der Mullen, J. Gas discharge plasmas and their applications. *Spectrochim. Acta Part B At. Spectrosc.* **2002**, *57*, 609–658. [CrossRef]
11. Tanaka, H.; Ishikawa, K.; Mizuno, M.; Toyokuni, S.; Kajiyama, H.; Kikkawa, F.; Metelmann, H.-R.; Hori, M. State of the art in medical applications using non-thermal atmospheric pressure plasma. *Rev. Mod. Plasma Phys.* **2017**, *1*, 3. [CrossRef]
12. Lu, X.; Laroussi, M.; Puech, V. On atmospheric-pressure non-equilibrium plasma jets and plasma bullets. *Plasma Sources Sci. Technol.* **2012**, *21*, 034005. [CrossRef]
13. Winter, J.; Brandenburg, R.; Weltmann, K.-D. Atmospheric pressure plasma jets: An overview of devices and new directions. *Plasma Sources Sci. Technol.* **2015**, *24*, 064001. [CrossRef]
14. Lu, X.; Naidis, G.V.; Laroussi, M.; Ostrikov, K. Guided ionization waves: Theory and experiments. *Phys. Rep.* **2014**, *540*, 123–166. [CrossRef]
15. De Geyter, N.; Morent, R. Cold plasma surface modification of biodegradable polymer biomaterials. In *Biomaterials for Bone Regeneration. Novel Techniques and Applications*; Woodhead Publishing: Sawston, UK, 2014; pp. 202–224.
16. Alotaibi, F.; Tung, T.T.; Nine, M.J.; Kabiri, S.; Moussa, M.; Tran, D.N.H.; Losic, D. Scanning atmospheric plasma for ultrafast reduction of graphene oxide and fabrication of highly conductive graphene films and patterns. *Carbon* **2018**, *127*, 113–121. [CrossRef]

17. Chemin, J.-B.; Bulou, S.; Baba, K.; Fontaine, C.; Sindzingre, T.; Boscher, N.D.; Choquet, P. Transparent anti-fogging and self-cleaning TiO_2/SiO_2 thin films on polymer substrates using atmospheric plasma. *Sci. Rep.* **2018**, *8*, 9603. [CrossRef]
18. Nie, Q.; Cao, Z.; Ren, C.S.; Wang, D.Z.; Kong, M.G. A two-dimensional cold atmospheric plasma jet array for uniform treatment of large-area surfaces for plasma medicine. *New J. Phys.* **2009**, *11*, 115015. [CrossRef]
19. Qian, C.; Fang, Z.; Yang, J.; Kang, M. Investigation on Atmospheric Pressure Plasma Jet Array in Ar. *IEEE Trans. Plasma Sci.* **2014**, *42*, 2438–2439. [CrossRef]
20. Mostofizadeh, A.; Li, Y.; Song, B.; Huang, Y. Synthesis, Properties, and Applications of Low-Dimensional Carbon-Related Nanomaterials. *J. Nanomater.* **2011**, *2011*, 685081. [CrossRef]
21. Gonzalez-Aguilar, J.; Moreno, M.; Fulcheri, L. Carbon nanostructures production by gas-phase plasma processes at atmospheric pressure. *J. Phys. D Appl. Phys.* **2007**, *40*, 2361–2374. [CrossRef]
22. Vollath, D. Plasma synthesis of nanopowders. *J. Nanopart. Res.* **2008**, *10*, 39–57. [CrossRef]
23. Sun, X.; Bao, J.; Li, K.; Argyle, M.D.; Tan, G.; Adidharma, H.; Zhang, K.; Fan, M.; Ning, P. Advance in Using Plasma Technology for Modification or Fabrication of Carbon-Based Materials and Their Applications in Environmental, Material, and Energy Fields. *Adv. Funct. Mater.* **2020**, *31*, 2006287. [CrossRef]
24. Antonova, I.V.; Shavelkina, M.B.; Ivanov, A.I.; Soots, R.A.; Ivanov, P.P.; Bocharov, A.N. Graphene Flakes for Electronic Applications: DC Plasma Jet-Assisted Synthesis. *Nanomaterials* **2020**, *10*, 2050. [CrossRef] [PubMed]
25. Kong, W.; Kum, H.; Bae, S.H.; Shim, J.; Kim, H.; Kong, L.; Meng, Y.; Wang, K.; Kim, C.; Kim, J. Path towards graphene commercialization from lab to market. *Nat. Nanotechnol.* **2019**, *14*, 927–938. [CrossRef]
26. Vinoth Kumar, S.H.B.; Muydinov, R.; Szyszka, B. Plasma assisted reduction of graphene oxide films. *Nanomaterials* **2021**, *11*, 382. [CrossRef]
27. Dato, A.; Radmilovic, V.; Lee, Z.; Phillips, J.; Frenklach, M. Substrate-free gas-phase synthesis of graphene sheets. *Nano Lett.* **2008**, *8*, 2012–2016. [CrossRef]
28. Tatarova, E.; Henriques, J.; Luhrs, C.C.; Dias, A.; Phillips, J.; Abrashev, M.V.; Ferreira, C.M. Microwave plasma based single step method for free standing graphene synthesis at atmospheric conditions. *Appl. Phys. Lett.* **2013**, *103*, 134101. [CrossRef]
29. Dato, A. Graphene synthesized in atmospheric plasmas—A review. *J. Mater. Res.* **2019**, *34*, 214. [CrossRef]
30. Toman, J.; Jasek, O.; Snirer, M.; Kudrle, V.; Jurmanova, J. On the interplay between plasma discharge instability and formation of free-standing graphene nanosheets in a dual-channel microwave plasma torch at atmospheric pressure. *J. Phys. D Appl. Phys.* **2019**, *52*, 265205. [CrossRef]
31. Jašek, O.; Toman, J.; Jurmanová, J.; Šnírer, M.; Kudrle, V.; Buršíková, V. Study of graphene layer growth on dielectric substrate in microwave plasma torch at atmospheric pressure. *Diam. Relat. Mater.* **2020**, *105*, 107798. [CrossRef]
32. Shavelkina, M.B.; Ivanov, P.P.; Bocharov, A.N.; Amirov, R.K. 1D modeling of the equilibrium plasma flow in the scope of direct current plasma torch assisted graphene synthesis. *J. Phys. D Appl. Phys.* **2019**, *52*, 495202. [CrossRef]
33. Shavelkina, M.B.; Amirov, R.K.; Kavyrshin, D.I.; Yusupov, D.I. Plasma Jet-Assisted Synthesis of Graphene Using a DC Plasma Torch. *Plasma Phys. Rep.* **2019**, *45*, 1080–1086. [CrossRef]
34. Shavelkina, M.; Ivanov, P.; Bocharov, A.; Amirov, R. Distinctive Features of Graphene Synthesized in a Plasma Jet Created by a DC Plasma Torch. *Materials* **2020**, *13*, 1728. [CrossRef] [PubMed]
35. Hummers, W.S.; Offeman, R.E. Preparation of Graphitic Oxide. *J. Am. Chem. Soc.* **1958**, *80*, 1339. [CrossRef]
36. Nasiłowska, B.; Bogdanowicz, Z.; Hińcza, K.; Mierczyk, Z.; Góźdź, S.; Djas, M.; Kowiorski, K.; Bombalska, A.; Kowalik, A. Graphene oxide aerosol deposition and its influence on cancer cells. Preliminary results. *Materials* **2020**, *13*, 4464. [CrossRef]
37. Chlanda, A.; Kowiorski, K.; Małek, M.; Kijeńska-Gawrońska, E.; Bil, M.; Djas, M.; Strachowski, T.; Swieszkowski, W.; Lipińska, L. Morphology and Chemical Purity of Water Suspension of Graphene Oxide FLAKES Aged for 14 Months in Ambient Conditions. A Preliminary Study. *Materials* **2021**, *14*, 4108. [CrossRef]
38. Rohde, M.; Zelt, M.; Gabriel, O.; Neubert, S.; Kirner, S.; Severin, D.; Stolley, T.; Rau, B.; Stannowski, B.; Schlatmann, R. Plasma enhanced chemical vapor deposition process optimization for thin film silicon tandem junction solar cells. *Thin Solid Film.* **2014**, *558*, 337–343. [CrossRef]
39. Muydinov, R.; Ruske, F.; Neubert, S.; Steigert, A.; Klaus, M.; Selve, S.; Köppel, G.; Szyszka, B. Combination of nitrogen mediated crystallisation with post-deposition annealing—Towards ultra-thin ZnO:Al contacts. *Thin Solid Film.* **2015**, *589*, 750–754. [CrossRef]
40. Arbab, M. The base layer effect on the d.c. conductivity and structure of direct current magnetron sputtered thin films of silver. *Thin Solid Film.* **2001**, *381*, 15–21. [CrossRef]
41. Lauermann, I.; Steigert, A. CISSY: A station for preparation and surface/interface analysis of thin film materials and devices. *J. Large-Scale Res. Facil. JLSRF* **2016**, *2*, A67. [CrossRef]
42. Díez-Betriu, X.; Álvarez-García, S.; Botas, C.; Álvarez, P.; Sánchez-Marcos, J.; Prieto, C.; Menéndez, R.; de Andrés, A. Raman spectroscopy for the study of reduction mechanisms and optimization of conductivity in graphene oxide thin films. *J. Mater. Chem. C* **2013**, *1*, 6905. [CrossRef]
43. Claramunt, S.; Varea, A.; López-Díaz, D.; Velázquez, M.M.; Cornet, A.; Cirera, A. The importance of interbands on the interpretation of the raman spectrum of graphene oxide. *J. Phys. Chem. C* **2015**, *119*, 10123–10129. [CrossRef]
44. Sadezky, A.; Muckenhuber, H.; Grothe, H.; Niessner, R.; Pöschl, U. Raman microspectroscopy of soot and related carbonaceous materials: Spectral analysis and structural information. *Carbon* **2005**, *43*, 1731–1742. [CrossRef]

45. Zhao, X.; Ando, Y. Raman Spectra and X-Ray Diffraction Patterns of Carbon Nanotubes Prepared by Hydrogen Arc Discharge. *Jpn. J. Appl. Phys.* **1998**, *37*, 4846–4849. [CrossRef]
46. Kudin, K.N.; Ozbas, B.; Schniepp, H.C.; Prud'Homme, R.K.; Aksay, I.A.; Car, R. Raman Spectra of Graphite Oxide and Functionalized Graphene Sheets. *Nano Lett.* **2008**, *8*, 36–41. [CrossRef]
47. Nemanich, R.J.; Solin, S.A. First- and second-order Raman scattering from finite-size crystals of graphite. *Phys. Rev. B* **1979**, *20*, 392–401. [CrossRef]
48. Vollebregt, S.; Ishihara, R.; Tichelaar, F.D.; Hou, Y.; Beenakker, C.I.M. Influence of the growth temperature on the first and second-order Raman band ratios and widths of carbon nanotubes and fibers. *Carbon* **2012**, *50*, 3542–3554. [CrossRef]
49. Goodman, P.A.; Li, H.; Gao, Y.; Lu, Y.F.; Stenger-Smith, J.D.; Redepenning, J. Preparation and characterization of high surface area, high porosity carbon monoliths from pyrolyzed bovine bone and their performance as supercapacitor electrodes. *Carbon* **2013**, *55*, 291–298. [CrossRef]
50. Saito, R.; Hofmann, M.; Dresselhaus, G.; Jorio, A.; Dresselhaus, M.S. Raman spectroscopy of graphene and carbon nanotubes. *Adv. Phys.* **2011**, *60*, 413–550. [CrossRef]
51. Liu, Y.; Zhou, J.; Zhang, X.; Liu, Z.; Wan, X.; Tian, J.; Wang, T.; Chen, Y. Synthesis, characterization and optical limiting property of covalently oligothiophene-functionalized graphene material. *Carbon* **2009**, *47*, 3113–3121. [CrossRef]
52. Jenrette, E.; Pradhan, S.K.; Rutherford, G.; Flowers, J.; Ha, D.; Pradhan, A.K. Quantum-dot-conjugated graphene oxide as an optical tool for biosensor. *Opt. Express* **2015**, *23*, 25017. [CrossRef] [PubMed]
53. Chaturvedi, D.; Balaji, S.; Bn, V.; Ariese, F.; Umapathy, S.; Rangarajan, A. Different Phases of Breast Cancer Cells: Raman Study of Immortalized, Transformed, and Invasive Cells. *Biosensors* **2016**, *6*, 57. [CrossRef]
54. Baibarac, M.; Lapkowski, M.; Pron, A.; Lefrant, S.; Baltog, I. SERS spectra of poly(3-hexylthiophene) in oxidized and unoxidized states. *J. Raman Spectrosc.* **1998**, *29*, 825–832. [CrossRef]
55. Nyamekye, C.K.A.; Bobbitt, J.M.; Zhu, Q.; Smith, E.A. The evolution of total internal reflection Raman spectroscopy for the chemical characterization of thin films and interfaces. *Anal. Bioanal. Chem.* **2020**, *412*, 6009–6022. [CrossRef]
56. Ferrari, A.C.; Robertson, J. Origin of the 1150-cm^{-1} Raman mode in nanocrystalline diamond. *Phys. Rev. B* **2001**, *63*, 121405. [CrossRef]
57. Wu, J.B.; Lin, M.L.; Cong, X.; Liu, H.N.; Tan, P.H. Raman spectroscopy of graphene-based materials and its applications in related devices. *Chem. Soc. Rev.* **2018**, *47*, 1822–1873. [CrossRef] [PubMed]
58. Krishnamoorthy, K.; Veerapandian, M.; Yun, K.; Kim, S.-J. The chemical and structural analysis of graphene oxide with different degrees of oxidation. *Carbon* **2013**, *53*, 38–49. [CrossRef]
59. Ma, B.; Rodriguez, R.D.; Ruban, A.; Pavlov, S.; Sheremet, E. The correlation between electrical conductivity and second-order Raman modes of laser-reduced graphene oxide. *Phys. Chem. Chem. Phys.* **2019**, *21*, 10125–10134. [CrossRef]
60. Ganguly, A.; Sharma, S.; Papakonstantinou, P.; Hamilton, J. Probing the Thermal Deoxygenation of Graphene Oxide Using High-Resolution In Situ X-ray-Based Spectroscopies. *J. Phys. Chem. C* **2011**, *115*, 17009–17019. [CrossRef]
61. Martins Ferreira, E.H.; Moutinho, M.V.O.; Stavale, F.; Lucchese, M.M.; Capaz, R.B.; Achete, C.A.; Jorio, A. Evolution of the Raman spectra from single-, few-, and many-layer graphene with increasing disorder. *Phys. Rev. B* **2010**, *82*, 125429. [CrossRef]
62. Wei, G.; Yu, J.; Gu, M.; Tang, T.B. Dielectric relaxation and hopping conduction in reduced graphite oxide. *J. Appl. Phys.* **2016**, *119*, 224102. [CrossRef]
63. Kamedulski, P.; Truszkowski, S.; Lukaszewicz, J.P. Highly effective methods of obtaining n-doped graphene by gamma irradiation. *Materials* **2020**, *13*, 4975. [CrossRef] [PubMed]
64. Chen, X.; Wang, X.; Fang, D. A review on C1s XPS-spectra for some kinds of carbon materials. *Fuller. Nanotub. Carbon Nanostruct.* **2020**, *28*, 1048–1058. [CrossRef]
65. Ahn, M.; Liu, R.; Lee, C.; Lee, W. Designing Carbon/Oxygen Ratios of Graphene Oxide Membranes for Proton Exchange Membrane Fuel Cells. *J. Nanomater.* **2019**, *2019*, 6464713. [CrossRef]
66. Teng, C.C.; Ma, C.C.M.; Lu, C.H.; Yang, S.Y.; Lee, S.H.; Hsiao, M.C.; Yen, M.Y.; Chiou, K.C.; Lee, T.M. Thermal conductivity and structure of non-covalent functionalized graphene/epoxy composites. *Carbon* **2011**, *49*, 5107–5116. [CrossRef]
67. Sygellou, L.; Paterakis, G.; Galiotis, C.; Tasis, D. Work Function Tuning of Reduced Graphene Oxide Thin Films. *J. Phys. Chem. C* **2016**, *120*, 281–290. [CrossRef]
68. Tatarova, E.; Dias, A.; Henriques, J.; Abrashev, M.; Bundaleska, N.; Kovacevic, E.; Bundaleski, N.; Cvelbar, U.; Valcheva, E.; Arnaudov, B.; et al. Towards large-scale in free-standing graphene and N-graphene sheets. *Sci. Rep.* **2017**, *7*, 10175. [CrossRef]
69. Chowdhury, F.A.; Hossain, M.A.; Uchida, K.; Tamura, T.; Sugawa, K.; Mochida, T.; Otsuki, J.; Mohiuddin, T.; Boby, M.A.; Alam, M.S. Graphene oxide/carbon nanoparticle thin film based IR detector: Surface properties and device characterization. *AIP Adv.* **2015**, *5*, 107228. [CrossRef]
70. Marrani, A.G.; Motta, A.; Palmieri, V.; Perini, G.; Papi, M.; Dalchiele, E.A.; Schrebler, R.; Zanoni, R. A comparative experimental and theoretical study of the mechanism of graphene oxide mild reduction by ascorbic acid and N -acetyl cysteine for biomedical applications. *Mater. Adv.* **2020**, *1*, 2745–2754. [CrossRef]
71. Tuinstra, F.; Koenig, J.L. Raman Spectrum of Graphite. *J. Chem. Phys.* **1970**, *53*, 1126–1130. [CrossRef]
72. Cançado, L.G.; Takai, K.; Enoki, T.; Endo, M.; Kim, Y.A.; Mizusaki, H.; Jorio, A.; Coelho, L.N.; Magalhães-Paniago, R.; Pimenta, M.A. General equation for the determination of the crystallite size La of nanographite by Raman spectroscopy. *Appl. Phys. Lett.* **2006**, *88*, 163106. [CrossRef]

73. Puech, P.; Kandara, M.; Paredes, G.; Moulin, L.; Weiss-Hortala, E.; Kundu, A.; Ratel-Ramond, N.; Plewa, J.-M.; Pellenq, R.; Monthioux, M. Analyzing the Raman Spectra of Graphenic Carbon Materials from Kerogens to Nanotubes: What Type of Information Can Be Extracted from Defect Bands? *C J. Carbon Res.* **2019**, *5*, 69. [CrossRef]
74. Kalache, B.; Novikova, T.; Fontcuberta i Morral, A.; Roca i Cabarrocas, P.; Morscheidt, W.; Hassouni, K. Investigation of coupling between chemistry and discharge dynamics in radio frequency hydrogen plasmas in the Torr regime. *J. Phys. D Appl. Phys.* **2004**, *37*, 1765–1773. [CrossRef]
75. Debrabandere, D.; Eynde, X.V.; Reniers, F. Effects of hydrogen addition in nitrogen atmospheric pressure plasma on its optical and electrical properties and silicon-based deposits compositions. *J. Phys. Conf. Ser.* **2011**, *275*, 012011. [CrossRef]
76. Ilik, E.; Durmuş, Ç.; Akan, T. Optical Investigations of N2 Atmospheric Pressure Plasma Jets. *Adıyaman Univ. J. Sci.* **2020**, *10*, 326–338. [CrossRef]
77. Parashar, U.K.; Bhandari, S.; Srivastava, R.K.; Jariwala, D.; Srivastava, A. Single step synthesis of graphene nanoribbons by catalyst particle size dependent cutting of multiwalled carbon nanotubes. *Nanoscale* **2011**, *3*, 3876. [CrossRef] [PubMed]
78. Niu, T.; Zhou, M.; Zhang, J.; Feng, Y.; Chen, W. Growth Intermediates for CVD Graphene on Cu(111): Carbon Clusters and Defective Graphene. *J. Am. Chem. Soc.* **2013**, *135*, 8409–8414. [CrossRef] [PubMed]
79. Fernandez, T.T.; Gross, S.; Arriola, A.; Privat, K.; Withford, M.J. Revisiting ultrafast laser inscribed waveguide formation in commercial alkali-free borosilicate glasses. *Opt. Express* **2020**, *28*, 10153. [CrossRef] [PubMed]
80. Montenegro, D.N.; Hortelano, V.; Martínez, O.; Martínez-Tomas, M.C.; Sallet, V.; Muñoz-Sanjosé, V.; Jiménez, J. Non-radiative recombination centres in catalyst-free ZnO nanorods grown by atmospheric-metal organic chemical vapour deposition. *J. Phys. D Appl. Phys.* **2013**, *46*, 235302. [CrossRef]
81. Martina, I.; Wiesinger, R.; Schreiner, M.; Martína, I.; Wiesinger, R.; Jembrih-Simbürger, D.; Schreiner, M.; Martina, I.; Wiesinger, R.; Schreiner, M. Micro-Raman Characterisation of Silver Corrosion Products: Instrumental Set Up and Reference Database. *e-Preserv. Sci.* **2012**, *9*, 1–8.
82. Scheibe, H.-J.; Drescher, D.; Alers, P. Raman characterization of amorphous carbon films. *Fresenius J. Anal. Chem.* **1995**, *353*, 695–697. [CrossRef]
83. Jorio, A.; Saito, R. Raman spectroscopy for carbon nanotube applications. *J. Appl. Phys.* **2021**, *129*, 021102. [CrossRef]
84. Mohiuddin, T.M.G.; Lombardo, A.; Nair, R.R.; Bonetti, A.; Savini, G.; Jalil, R.; Bonini, N.; Basko, D.M.; Galiotis, C.; Marzari, N.; et al. Uniaxial strain in graphene by Raman spectroscopy: Gpeak splitting, Grüneisen parameters, and sample orientation. *Phys. Rev. B* **2009**, *79*, 205433. [CrossRef]
85. Mikael, S.; Seo, J.-H.; Javadi, A.; Gong, S.; Ma, Z. Wrinkled bilayer graphene with wafer scale mechanical strain. *Appl. Phys. Lett.* **2016**, *108*, 183101. [CrossRef]
86. Podila, R.; Rao, R.; Tsuchikawa, R.; Ishigami, M.; Rao, A.M. Raman Spectroscopy of Folded and Scrolled Graphene. *ACS Nano* **2012**, *6*, 5784–5790. [CrossRef] [PubMed]
87. Vidano, R.P.; Fischbach, D.B.; Willis, L.J.; Loehr, T.M. Observation of Raman band shifting with excitation wavelength for carbons and graphites. *Solid State Commun.* **1981**, *39*, 341–344. [CrossRef]
88. Ferrari, A.C.; Meyer, J.C.; Scardaci, V.; Casiraghi, C.; Lazzeri, M.; Mauri, F.; Piscanec, S.; Jiang, D.; Novoselov, K.S.; Roth, S.; et al. Raman Spectrum of Graphene and Graphene Layers. *Phys. Rev. Lett.* **2006**, *97*, 187401. [CrossRef]
89. Ferrari, A.C. Raman spectroscopy of graphene and graphite: Disorder, electron-phonon coupling, doping and nonadiabatic effects. *Solid State Commun.* **2007**, *143*, 47–57. [CrossRef]
90. Couzi, M.; Bruneel, J.-L.; Talaga, D.; Bokobza, L. A multi wavelength Raman scattering study of defective graphitic carbon materials: The first order Raman spectra revisited. *Carbon* **2016**, *107*, 388–394. [CrossRef]
91. Pawlyta, M.; Rouzaud, J.N.; Duber, S. Raman microspectroscopy characterization of carbon blacks: Spectral analysis and structural information. *Carbon* **2015**, *84*, 479–490. [CrossRef]
92. Niwase, K.; Tanaka, T.; Kakimoto, Y.; Ishihara, K.N.; Shingu, P.H. Raman Spectra of Graphite and Diamond Mechanically Milled with Agate or Stainless Steel Ball-Mill. *Mater. Trans. JIM* **1995**, *36*, 282–288. [CrossRef]
93. Herziger, F.; Tyborski, C.; Ochedowski, O.; Schleberger, M.; Maultzsch, J. Double-resonant la phonon scattering in defective graphene and carbon nanotubes. *Phys. Rev. B* **2014**, *90*, 245431. [CrossRef]
94. Czamara, K.; Majzner, K.; Pacia, M.Z.; Kochan, K.; Kaczor, A.; Baranska, M. Raman spectroscopy of lipids: A review. *J. Raman Spectrosc.* **2015**, *46*, 4–20. [CrossRef]
95. Kochan, K.; Peng, H.; Gwee, E.S.H.; Izgorodina, E.; Haritos, V.; Wood, B.R. Raman spectroscopy as a tool for tracking cyclopropane fatty acids in genetically engineered *Saccharomyces cerevisiae*. *Analyst* **2019**, *144*, 901–912. [CrossRef] [PubMed]
96. Meilunas, R.; Chang, R.P.H.; Liu, S.; Jensen, M.; Kappes, M.M. Infrared and Raman spectra of C 60 and C 70 solid films at room temperature. *J. Appl. Phys.* **1991**, *70*, 5128–5130. [CrossRef]
97. Schettino, V.; Pagliai, M.; Cardini, G. The Infrared and Raman Spectra of Fullerene C 70. DFT Calculations and Correlation with C 60. *J. Phys. Chem. A* **2002**, *106*, 1815–1823. [CrossRef]
98. Dychalska, A.; Popielarski, P.; Franków, W.; Fabisiak, K.; Paprocki, K.; Szybowicz, M. Study of CVD diamond layers with amorphous carbon admixture by Raman scattering spectroscopy. *Mater. Sci.* **2015**, *33*, 799–805. [CrossRef]
99. Marcinauskas, L.; Černauskas, M.; Milieška, M.; Reso, D. The effect of hydrogen addition in argon-acetylene plasma on the structure of amorphous carbon films. *J. Non-Cryst. Solids* **2017**, *466–467*, 1–7. [CrossRef]

100. Birrell, J.; Gerbi, J.E.; Auciello, O.; Gibson, J.M.; Johnson, J.; Carlisle, J.A. Interpretation of the Raman spectra of ultrananocrystalline diamond. *Diam. Relat. Mater.* **2005**, *14*, 86–92. [CrossRef]
101. Hasdeo, E.H.; Nugraha, A.R.T.; Dresselhaus, M.S.; Saito, R. Breit-Wigner-Fano line shapes in Raman spectra of graphene. *Phys. Rev. B* **2014**, *90*, 245140. [CrossRef]
102. Zhao, S.; Hong, R.; Luo, Z.; Lu, H.; Yan, B. Carbon Nanostructures Production by AC Arc Discharge Plasma Process at Atmospheric Pressure. *J. Nanomater.* **2011**, *2011*, 1–6. [CrossRef]
103. Chen, L.S.; Wang, C.J. Low Temperature Large Scale CVD Synthesis of Nano Onion-Like Fullerenes. *Adv. Mater. Res.* **2012**, *490–495*, 3211–3214. [CrossRef]
104. Yang, Y.; Liu, X.; Guo, X.; Wen, H.; Xu, B. Synthesis of nano onion-like fullerenes by chemical vapor deposition using an iron catalyst supported on sodium chloride. *J. Nanopart. Res.* **2011**, *13*, 1979–1986. [CrossRef]
105. Du, A.B.; Liu, X.G.; Fu, D.J.; Han, P.D.; Xu, B.S. Onion-like fullerenes synthesis from coal. *Fuel* **2007**, *86*, 294–298. [CrossRef]
106. Kaltchev, M.; Stacchiola, D.; Molero, H.; Wu, G.; Blumenfeld, A.; Tysoe, W.T. On the reaction pathway for the formation of benzene from acetylene catalyzed by palladium. *Catal. Lett.* **1999**, *60*, 11–14. [CrossRef]
107. Broere, D.L.J.; Ruijter, E. Recent advances in transition-metal-catalyzed [2+2+2]-cyclo(co) trimerization reactions. *Synthesis* **2012**, *44*, 2639–2672. [CrossRef]
108. Sakurai, H.; Nakadaira, Y.; Hosomi, A.; Eriyama, Y.; Hirama, K.; Kabuto, C. Intramolecular cyclotrimerization of macrocylic and acyclic triynes with Group 6 metal carbonyls. The formation of fulvene and benzene. *J. Am. Chem. Soc.* **1984**, *106*, 8315–8316. [CrossRef]
109. Shapiro, I.; Weiss, H.G. Cyclization of acetylene-d2 to benzene-d6. *J. Am. Chem. Soc.* **1957**, *79*, 3294. [CrossRef]
110. Cota-Sanchez, G.; Soucy, G.; Huczko, A.; Beauvais, J.; Drouin, D. Effect of Iron Catalyst on the Synthesis of Fullerenes and Carbon Nanotubes in Induction Plasma. *J. Phys. Chem. B* **2004**, *108*, 19210–19217. [CrossRef]
111. Berezkin, V.I. Fullerenes as nuclei of carbon black particles. *Phys. Solid State* **2000**, *42*, 580–585. [CrossRef]
112. Boyer, M.-I.; Quillard, S.; Rebourt, E.; Louarn, G.; Buisson, J.P.; Monkman, A.; Lefrant, S. Vibrational Analysis of Polyaniline: A Model Compound Approach. *J. Phys. Chem. B* **1998**, *102*, 7382–7392. [CrossRef]
113. Morávková, Z.; Šeděnková, I.; Bober, P. The First Stages of Chemical and Electrochemical Aniline Oxidation—Spectroscopic Comparative Study. *Appl. Sci.* **2020**, *10*, 2091. [CrossRef]
114. Capote, G.; Mastrapa, G.C.; Trava-Airoldi, V.J. Influence of acetylene precursor diluted with argon on the microstructure and the mechanical and tribological properties of a-C:H films deposited via the modified pulsed-DC PECVD method. *Surf. Coat. Technol.* **2015**, *284*, 145–152. [CrossRef]
115. Kuzmany, H.; Pfeiffer, R.; Hulman, M.; Kramberger, C. Raman spectroscopy of fullerenes and fullerene–nanotube composites. *Philos. Trans. R. Soc. Lond. Ser. A Math. Phys. Eng. Sci.* **2004**, *362*, 2375–2406. [CrossRef]
116. Frezzotti, M.L. Diamond growth from organic compounds in hydrous fluids deep within the Earth. *Nat. Commun.* **2019**, *10*, 4952. [CrossRef] [PubMed]

C Journal of Carbon Research

MDPI

Review

Fabrication and Supercapacitor Applications of Multiwall Carbon Nanotube Thin Films

Kyle Jiang and Rosario A. Gerhardt *

School of Materials Science and Engineering, Georgia Institute of Technology, Atlanta, GA 30332, USA; ksjiang@mit.edu
* Correspondence: rosario.gerhardt@mse.gatech.edu; Tel.: +1-404-894-6886

Abstract: Multiwalled carbon nanotubes (MWCNTs) are a one-dimensional nanomaterial with several desirable material properties, including high mechanical tensile modulus and strength, high electrical conductivity, and good thermal conductivity. A wide variety of techniques have been optimized to synthesize MWCNTs and to fabricate thin films of MWCNTs. These synthesis and fabrication methods vary based on precursor materials, process parameters, and physical and chemical principals, and have a strong influence on the properties of the nanotubes and films. Thus, the fabrication methods determine the performance of devices that can exploit the advantageous material properties of MWCNTs. Techniques for the fabrication of carbon nanotubes and carbon nanotube thin films are reviewed, followed by a discussion of the use of MWCNTs as an electrode material for electrochemical double-layer supercapacitors (EDLCs). EDLCs feature high power density, excellent reversibility and lifetime, and improved energy density over electrolytic capacitors. Beyond surveying fabrication techniques previously explored for MWCNT electrodes, an alternative approach based on inkjet printing capable of depositing a small amount of active material is discussed. Such an approach allows for a high degree of control over electrode properties and can potentially reduce cost and active material waste, which are essential components to the gradual conversion to green energy.

Keywords: MWCNT; mechanical properties; electrical properties; synthesis; fabrication

Citation: Jiang, K.; Gerhardt, R.A. Fabrication and Supercapacitor Applications of Multiwall Carbon Nanotube Thin Films. *C* **2021**, *7*, 70. https://doi.org/10.3390/c7040070

Academic Editor: Gil Goncalves

Received: 5 September 2021
Accepted: 24 September 2021
Published: 30 September 2021

1. Introduction

Carbon nanotubes are a one-dimensional nanomaterial and have been extensively studied since their initial discovery in the soot of evaporated graphite by Iijima [1] in the early 1990s. There are two primary variants of carbon nanotubes: single-walled carbon nanotubes (SWCNTs) and multi-walled carbon nanotubes (MWCNTs). SWCNTs consist of a single cylinder of graphitic carbon, while MWCNTs are composed of several concentric layers of graphene. While SWCNTs have outer diameters in the range of around 1–2 nm [2], MWCNTs have larger outer diameters, often on the order of 10–100 nm [3]. Lengths of nanotubes can reach the order of micro- or millimeters. Because nanotubes are synthesized from readily available carbon precursor materials, their cost can be low compared to other nanomaterials. In general, MWCNTs can be synthesized via a wide variety of physical and chemical processes with less stringent requirements on materials and synthesis parameters compared to SWCNTs. For example, with deposition-based synthesis techniques such as CVD, catalyst particle sizes can be larger for MWCNTs than for SWCNTs. Catalysts are not required at all to synthesize MWCNTs from methods such as arc discharge and laser ablation [4].

The thin film device topology, which is defined as a layer of material whose thickness is on the order of nanometers to microns [5], allows for a high surface area-to-volume ratio, low material usage, and features enhanced material properties in the normal direction. These are advantageous for several applications of carbon nanotubes, including sensors, gas separators, and electrically conductive devices. Several methods have been used to fabricate thin films of multiwall nanotubes including direct synthesis on the application substrate, transfer from an initial substrate to the application substrate, and a variety of wet deposition techniques that deposit previously synthesized nanotubes onto the application substrate. The adhesion between nanotube materials and substrate, properties of the thin film, and performance in various applications are determined by the film fabrication method and parameters.

A crucial application for carbon nanotube thin films is in energy storage materials. In 2003, energy derived from fossil fuels, which are nonrenewable and known to increase atmospheric concentration of CO_2 and other "greenhouse gases" [6], made up 95% of the global energy consumption [7]. Renewable forms of energy, such as hydroelectric, wind, and solar, typically generate power intermittently and therefore require energy storage technologies such as fuel cells, batteries, and supercapacitors to supply the electrical grid during downtimes [8]. Compared to fuel cells and batteries, supercapacitors store less energy per active material mass because they operate based on the principle of electrostatic energy storage as opposed to chemical storage [9]; however, supercapacitors can deliver more power per active material mass because of the more facile and reversible charge transfer through ionic transport and electronic conduction at the electrodes [10]. Due to its high electrical conductivity and compatibility with electrolytes, carbon is a desirable electrode material. In lithium batteries, graphite is a common commercial anode material. In double layer supercapacitors, electrodes are typically composed of high-specific surface area carbons [11]. Multiwalled carbon nanotube thin films therefore provide enhanced energy storage due to higher available surface area onto which ions can adsorb.

In this article, previous investigations into the synthesis and supercapacitor applications of MWCNTs and their thin films are discussed, with an emphasis on the use of nanotube materials in supercapacitor electrodes for electrical energy storage. An experimental section describes the fabrication, simulation, electrical, and electrochemical characterization of MWCNT electrodes spin-coated and inkjet-printed onto photopaper. Following is an outline of the sections of the paper.

Section 2 overviews MWCNTs as a nanomaterial. The most prominent techniques for their synthesis are highlighted. MWCNTs can be synthesized via physical routes, i.e., the evaporation of graphite or other carbon precursors by electric current or laser sources, or via chemical routes, i.e., the growth of nanotubes on metallic catalyst particles from gas precursors. This is followed by a brief overview of their mechanical and electrical properties.

Section 3 focuses on MWCNT thin films. The broad classes of MWCNT thin film synthesis techniques—direct growth, substrate transfer, and wet deposition—are discussed, with emphasis on wet deposition. Wet deposition techniques are of interest in thin film fabrication because they can be performed with commercial nanotubes, which lowers cost, shortens fabrication time, and offers flexibility in substrate selection. Following fabrication techniques, the mechanical, electrical, and other properties of MWCNT thin films are presented. MWCNT thin films possess enhanced properties because of the intrinsic material properties of MWCNTs.

Section 4 discusses the application of MWCNT thin films in supercapacitor electrodes. The double-layer capacitor device is introduced. Some methods to improve the electrochemical performance of electrodes, including acid functionalization, compositing with pseudocapacitive components, is overviewed. An experimental section follows that details the fabrication of a double-layer supercapacitor using spin-coating and inkjet-printing methods, which have not been extensively studied in the literature.

2. Multiwalled Carbon Nanotubes

2.1. Synthesis

Several techniques have been used to synthesize MWCNTs. These techniques have been previously reviewed [12–14]. Among these methods, arc discharge, laser ablation, and chemical vapor deposition are representative of the synthesis of MWCNTs in the laboratory.

Other methods include flame synthesis, electrolysis, and high-pressure carbon monoxide (HiPCO) synthesis.

2.1.1. Arc Discharge

MWCNTs were first discovered after synthesis by arc discharge [1]. In this technique, a graphite anode is sublimated by an applied dc current in a low-pressure, high-temperature atmosphere, and nanotubes are extracted from the negative electrode (cathode). A schematic of a typical arc discharge synthesis is shown in Figure 1. Several parameters affect the quality and morphology of nanotubes, including the atmospheric gas, pressure, gas flow rate, input current, and electrode configuration. The effects of various parameters have previously been reviewed [15].

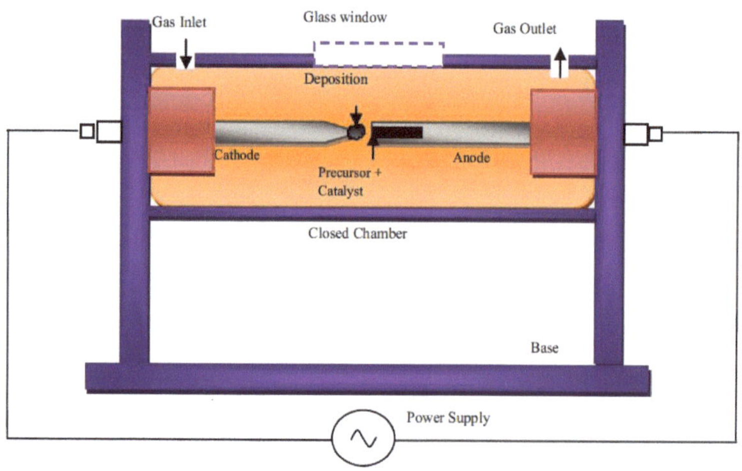

Figure 1. Schematic of arc discharge synthesis of MWCNTs [15]. Reproduced with permission from Elsevier.

An important parameter in nanotube growth with arc discharge is the choice of atmosphere. In previous works, nanotubes have been synthesized in helium, argon, nitrogen, hydrogen, and mixtures between these and other gases. At high temperatures, gas excited by the applied electric current ionizes, allowing for vaporized carbon to mobilize from the anode toward the cathode. Gases with higher ionization energies, such as helium, require a larger electrical potential between the electrodes to ionize. The potential, in turn, depends on the applied current and electrical conductivity of the induced plasma.

Besides the choice of atmosphere, gas pressure affects the morphology of nanotubes, as it determines the concentration of ionized gas in the discharge chamber. Cadek et al. [16] have determined that in helium atmosphere, both yield (characterized by electron paramagnetic resonance and thermogravimetric analysis of nanotubes purified by a PmPV/toluene suspension) and purity of nanotubes increased increasing pressure up to 500 Torr. Zhao [17] determined that 500 Torr was also an optimal pressure for yield in arc- discharge nanotube synthesis in hydrogen atmospheres. On the other hand, Kim and Kim [18] have found that in air atmosphere, decreasing the air pressure from ambient to 300 Torr increased the MWCNT yield.

Another important parameter in arc discharge is the applied dc current. Electrical current causes the vaporization of the anode; the current density determines the concentration of carbon precursor that is vaporized from the graphitic anode. Furthermore, electrical current induces excitation of the atmosphere to cause ionization of gas into plasma, which carries the carbon vapor to the cathode, where it is deposited. As assessed by their I_d/I_g defect intensity ratio, nanotubes synthesized via arc discharge with higher currents up to 70 A were found to have a reduced isotropic effect, resulting in more ordered nanotubes [19].

2.1.2. Laser Ablation

Another technique for MWCNT synthesis is laser ablation, which is similar to arc discharge in that both methods involve a graphitic precursor. However, instead of electrical current, a laser (either continuous or pulsed) vaporizes the graphite target in an inert atmosphere. As the vapor condenses on the walls of the reactor, nanotubes are formed. Laser ablation is a high-yield process, and nanotubes produced by laser ablation are generally of high quality [14]. Interestingly, however, while single-walled nanotubes have been frequently synthesized by this method, to the best of the authors' knowledge, no previous works have synthesized MWCNTs with laser ablation. Possible reasons for this include the large amount of energy required by laser methods, low yield compared to chemical methods, and the similarity of nanotube properties to those synthesized by arc discharge, which requires less specialized equipment and may thus be lower cost.

2.1.3. Chemical Vapor Deposition

Unlike arc discharge and laser ablation, chemical vapor deposition (CVD) promotes growth of nanotubes on metallic catalysts rather than graphite. In CVD, an inert carrier gas brings hydrocarbon gases such as benzene and acetylene into the reaction chamber, which is heated at elevated temperatures to decompose the reactants, forming nanotubes on the catalysts. The morphology and diameters of the nanotubes depend heavily on the catalysts used. Due to the lower temperature and power requirements of CVD, it is a scalable synthesis technique for MWCNTs. Synthesizing nanotubes via thermal CVD in oxidizing conditions, such as in air, was found to yield nanotubes free of amorphous carbon [20]. Other process parameters, such as the catalytic growth rate and secondary pyrolysis rates, also govern the structure and quality of nanotubes.

2.2. *Properties of Individual Nanotubes*

The geometry of SWCNTs can be described by the chiral vector, which points in the direction that a graphene sheet (hexagonal tessellation of sp^2-hybridized carbon) would have to be rolled to form the same helical structure as the nanotube [21]. Since the graphene sheet is two-dimensional, the chiral vector has two components and is expressed with respect to a basis consisting of two vectors pointing at 30 and 90 degrees relative to the edge of one hexagon in the graphene sheet. The three main classes of SWCNT geometries are armchair (with coordinates (n, n)), zigzag (with coordinates $(n, 0)$), and chiral (all coordinates that are not reduceable to armchair or zigzag). These geometries and the conventional basis elements (a_1, a_2) used to describe chiral vectors are illustrated in Figure 2. The angle between the locus of zigzag vectors and that of armchair vectors is 30 degrees. Vectors that form angles outside the range of [0, 30] degrees with respect to the locus of zigzag vectors are always congruent to one in this range with an appropriate change of basis. For example, a chiral vector $(0, n)$ is congruent to $(n, 0)$—a zigzag nanotube—when the basis elements are rotated clockwise by 60 degrees.

2.2.1. Mechanical Properties

The mechanical properties of MWCNTs and their composites have been reviewed [22]. Due to the strength of other carbon compounds, carbon nanotubes were proposed to have a high tensile modulus and mechanical strength. Indeed, Treacy et al. [23] obtained a Young's modulus of 1.8 TPa (standard deviation of 1.4 TPa) for a series of 11 individual MWCNTs, which were found to be much higher than for other carbon fibers. Yu et al. [24] have determined moduli for 19 individual MWCNTs in the range of 270 GPa–950 GPa, and strengths in the range of 11 GPa–63 GPa. Furthermore, strains at failure were as high as 12%. Nanotubes were loaded in a scanning electron microscope (SEM) and attached to the tips of two AFM cantilevers by the deposition of a carbonaceous material [24]. Stress–strain curves for the individual nanotubes are shown in Figure 3. The discrepancy in the tensile strengths obtained in the two experiments [23,24] may be explained primarily by structural defects on the nanotubes, loading mode, and strain rate, and less importantly by experimental uncertainties in the estimation of nanotube lengths. It has been shown theoretically that structural defects are known to reduce the tensile modulus [25,26]. Yu, et al. [24] used a tensile-testing approach to find the elastic modulus, which was found to apply the most stress on the outermost layer of the MWCNTs as evidenced by the "sword-in-sheath" failure where the inner layers remained mostly intact after the nanotube broke. In contrast, Treacy et al. [23] deduced the modulus by observing the variation in cantilever vibration amplitude with temperature with the nanotube fixed at the base. Differences in the testing mode results in loading of different regions of the nanotube, which is radially anisotropic. This can result in different measured values of the modulus.

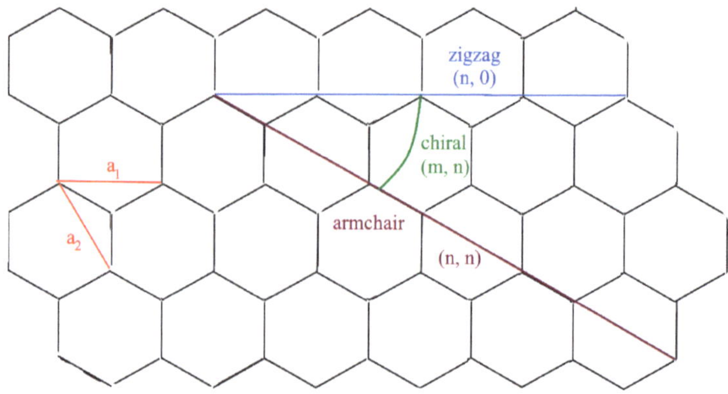

Figure 2. Illustration of the typical basis (vectors a_1 and a_2) and the chiral vectors of zigzag, armchair, and general chiral single-walled nanotubes on a sheet of graphene.

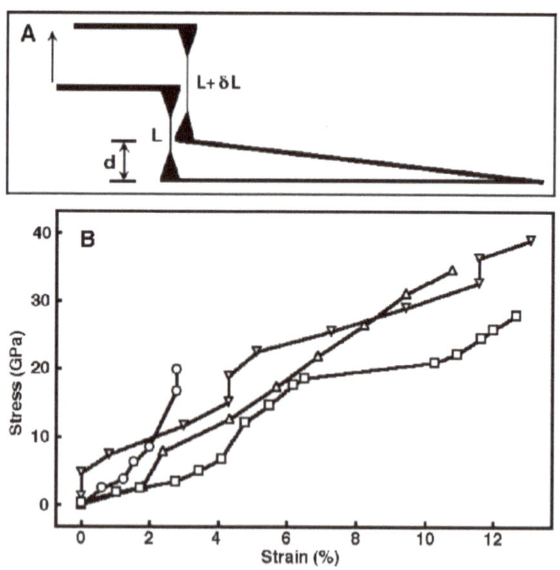

Figure 3. (**A**) Loading configuration and (**B**) stress–strain curves for four individual MWCNTs showing high tensile moduli and strengths [24]. Used with permission from The American Association for the Advancement of Science.

Salvetat et al. [27] found a relationship between tensile modulus and disorder in MWCNTs. The disorder in MWCNTs is controlled by synthesis process, with arc-discharge producing the most uniform nanotubes with the smallest number of defects, and catalytic CVDs producing nanotubes with a high concentration of defects and impurities. With a higher degree of disorder, the modulus decreases by several orders of magnitude, and the variation in modulus increases. In addition to structural defects, chirality and nanotube size have also been shown to affect the tensile strengths of nanotubes [28]. A tensile test of several nanotubes with different chiralities and diameters revealed that armchair nanotubes and nanotubes with smaller diameter had higher strengths [28]. As shown in Figure 4, the tensile strength was found to depend strongly on the chiral angle, which was attributed to the variation in the stress experienced by carbon–carbon bonds due to the orientation of those bonds relative to the axis [28]. Nanotubes with bonds oriented at a greater angle relative to the tensile axis experience a smaller effective stress; this angle is maximized at 30 degrees for armchair nanotubes. The dependence of strength on diameter followed an inverse-square root relationship [28], suggesting that nanotubes with smaller diameters have higher strengths.

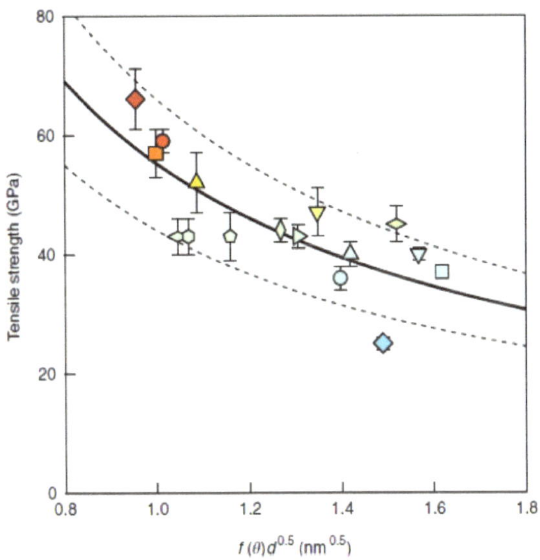

Figure 4. Relationship between tensile strength and nanotube geometry, showing the dependence of strength on the chiral angle and diameter [28]. Used with permission from Springer Nature.

2.2.2. Electrical Properties

The electrical properties of MWCNTs are unique because of their one-dimensional carbon structure. Single SWCNTs are either semiconducting or metallic, and the behavior solely depends on their geometry as determined by the chiral vector. The definition of the chiral vector was shown in Figure 2. All armchair nanotubes are conducting, as well as zigzag nanotubes with n evenly divisible by 3, and generally, chiral nanotubes with a difference between the coordinates that is divisible by 3. This is due to the periodicity of carbon atoms about the circumferential direction of the nanotube, resulting in only a few allowable wavevectors in that dimension, which contrasts with graphene, an infinite 2-dimensional plane. The periodicity determines the conduction and valence bands of the nanotube and the resulting electronic density of states [29]. Therefore, assuming a random distribution of chiral vectors, one out of three nanotubes will be metallic, while two-thirds will exhibit semiconducting behavior. While the preferential synthesis of metallic or semiconducting nanotubes is difficult, it is possible by tuning the shape and size of catalyst particles [30], substrate [31], and other growth conditions.

MWCNTs are composed of several concentric layers of SWCNTs and therefore allow for electronic conduction in the radial direction as well as along the nanotube axis. As such, at sufficiently high temperatures (above ~10 K), the electrical behavior of MWCNTs is very similar to that of bulk graphite, a mix between metallic and semiconducting [32].

3. MWCNT Thin Films

3.1. Fabrication

Several thin film fabrication methods exist for MWCNTs. Fabrication techniques for carbon nanotube thin films have previously been reviewed [13]. There are three general categories for MWCNT thin film synthesis methods. In the first, nanotubes are directly grown (via methods discussed in Section 2) on the final substrate. In another method, known as substrate transfer, nanotubes are grown on one substrate, which is subsequently removed and replaced with the final substrate. Finally, films can be created by depositing a slurry or a suspension that is composed of nanotubes onto the final substrate. Table 1 summarizes different techniques for MWCNT film synthesis that have been previously investigated.

3.1.1. Direct Growth

MWCNTs can be directly grown on the substrate used for the final application. Methods for nanotube growth were outlined in Section 2. The advantages of directly growing nanotubes on the final substrate include stronger adhesion, fewer processing steps leading to faster overall production from nanotube precursor materials to the final device, and more uniform films. Furthermore, the geometry of nanotubes themselves can be more precisely controlled by the synthesis technique, whereas other techniques typically cannot control individual nanotube geometries. This is important for applications that require specific material properties at the nanoscale. However, direct growth methods can be expensive because they require specialized equipment used in nanotube synthesis, such as CVD furnaces or arc discharge apparatuses. Furthermore, nanotube synthesis can be time-consuming and is generally less scalable for high-throughput production compared to other techniques.

3.1.2. Substrate Transfer

Films grown on one substrate can be subsequently transferred to another substrate. The transfer occurs by exploiting the surface energy differences between nanotubes and substrate. In several cases, carbon nanotubes were synthesized by CVD on a suitable catalyst substrate, then transferred to another substrate more suitable for the final application. Cole et al. [33] synthesized vertically-aligned MWCNT by thermal CVD on Si/SiO_2 substrate sputter-coated with Al_2O_3/Fe catalyst. A schematic of the process is shown in Figure 5. To transfer the MWCNT to the final PET substrate, a quartz roller compressed aligned MWCNTs against the PET substrate forming a dense aligned film of MWCNTs [33]. Honda et al. [34] prepared a thin film of MWCNTs by transferring from an Si substrate initially coated with Fe CVD catalyst to an aluminum substrate required for the final electrode application. The alignment of the MWCNTs on the initial substrate was maintained following the transfer by applying a homogeneous coating of phenolic glue containing graphitic carbon [34]. SWCNTs have been similarly transferred from an aluminum oxide filter to PET using a wet-transfer process that utilizes the difference in wettability between the substrates [35]. It was also observed that nanotubes could also be transferred to other substrates including glass, PMMA, and PVDF, which are uncommon substrates for direct growth by first reducing their surface energies through techniques such as octadecyl trichlorosilane treatment [35].

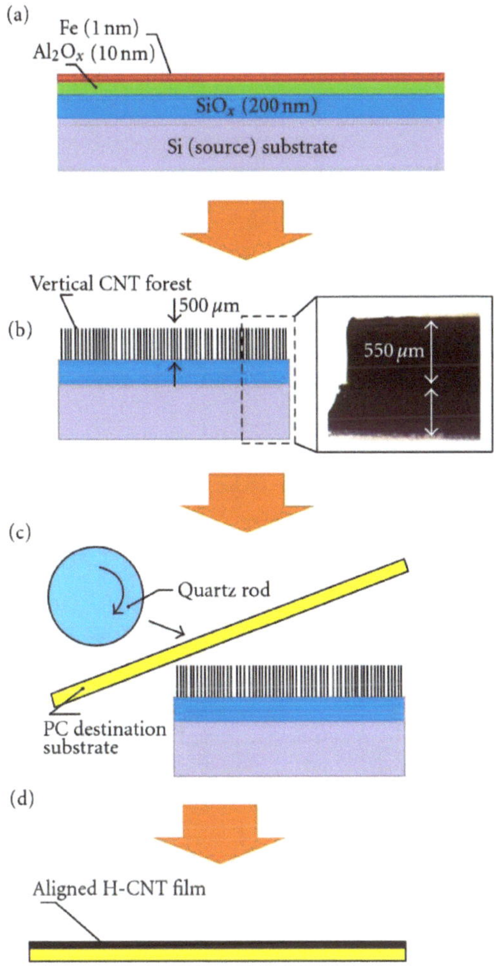

Figure 5. Schematic of the substrate transfer technique of MWCNT film synthesis [33]. Starting substrate sputtered with Fe and Al_2O_X (**a**). MWCNTs synthesized on the starting substrate with CVD (**b**). Final substrate aligned and pressed against grown MWCNTs (**c**). Final CNT film on target substrate (**d**). Used with permission from Hindawi.

3.1.3. Wet Deposition

Wet deposition methods, illustrated in Table 1, involve first the synthesis of MWCNTs followed by the dispersion in a suspension or the mixture into a slurry. Suspensions have lower concentrations of MWCNTs and thus tend to have lower viscosities and densities than slurries. As such, deposition methods are usually only applicable to either suspensions or slurries. Wet deposition techniques are advantageous because they can be used to form films from commercial, pre-synthesized nanotubes; no specialized equipment to directly synthesize nanotubes from precursor materials is required. Furthermore, wet deposition methods are generally faster to perform than direct growth or substrate transfer. Therefore, these methods are typically more scalable than other thin film fabrication methods.

However, challenges still exist for wet deposition. For suspension-based depositions, nanotubes should be uniformly dispersed in the solvent to form uniform thin films. Uniform dispersion of nanotubes is difficult due to their one-dimensional geometry and high aspect ratio, which results in significant van der Waals attractive interactions. To address this issue, nanotubes can be charged via functionalization, resulting in electrostatic repulsion that counters the van der Waals attraction, or dispersed in a surfactant, which breaks apart adjacent nanotubes and lowers the surface tension between the nanotubes and solvent. Slurry-based methods are also faced with the challenge of removing agglomeration of solid particles. Slurry-based methods are also sensitive to the specific fabrication steps and result in films with greater variation in local properties, which may not be desirable for certain applications. Finally, because the nanotube wet mixture is deposited onto the substrate, interactions between the dispersion and the substrate are critical. Deposition parameters must be tuned to achieve good uniformity and utilization of active material.

Dip coating is a popular technique thin film synthesis from sol–gels [36], but it has also been applied to thin-film fabrication from MWCNT dispersions. In this technique, the substrate is initially submerged in the dispersion then withdrawn at a specified velocity. As the substrate leaves the dispersion, the solvent is removed from the substrate via two physical processes: evaporation due to contact with the air, and vertical draining due to gravity.

Like dip coating, drop casting deposits material through the evaporation of solvent from dispersion on a substrate. In this technique, a small quantity of solution is first dropped on the substrate. Optionally, the droplet can be flowed over the substrate by orienting the substrate at an angle with respect to the vertical or flowing gas across the substrate to propel the droplet. By patterning the substrate, a preferred nanotube orientation can also be achieved.

Spin coating is another dispersion deposition technique. However, unlike in dip coating and drop casting, the relative motion between the dispersion and the substrate is not linear. Rather, a small volume of liquid mixture is dropped at the center of a substrate that is rotating at a specified speed. The high shear rate of the substrate simultaneously spreads a volume of liquid in the lateral direction and reduces its thickness. This results in a highly uniform thin film whose thickness can be controlled by the rotational speed and the spinning time [37,38].

In inkjet printing, a small volume of dispersion is ejected by a piezoelectric excitation along the nozzle [39]. The droplet size and velocity are controlled by the properties of the ink and the nozzle voltage profile. Inkjet printing has the potential to be very cost effective since commercial inkjet printers can simply be used after modifying the cartridge to be compatible with MWCNT dispersions. Although similar in principle, inkjet printing is distinct from drop casting because of the very small volume of liquid in each droplet. Furthermore, unlike other thin film fabrication techniques, inkjet printing also allows for high precision in the droplet location. Thin films with highly specific geometries can be printed, which is ideal for applications such as thin-film transistors and patterned sensors and electrodes. Nonetheless, several challenges remain for the effective use of inkjet printing in nanotube deposition. Because of the very small nozzle orifice diameter, the ink cannot contain agglomerates that are larger than about 10% of the orifice [39]. As-made dispersions must be filtered before printing, and depending on the dispersion technique, the filtering process can remove a significant number of nanotubes, resulting in low material utilization.

Screen printing is a method similar to inkjet printing in the sense that it is capable of controlling the location of the deposited inks. Instead of depositing the ink drop-by-drop, the ink is pressed using a squeegee against a screen containing an open mesh with the desired pattern. The substrate below the screen receives the ink in the open areas and after evaporation, the film is formed.

Meyer rod coating, also known as metering rod coating, creates a thin film by spreading a slurry over a surface [40]. A suspension or dispersion cannot be used because the viscosity is too low. The rod is threaded with a uniform diameter, leaving a small space between adjacent threads. As the rod is pulled over the substrate, a film with uniform thickness on the order of the thread diameter is formed behind the rod.

A similar technique to Meyer rod coating for slurry-based thin film fabrication is doctor blading, otherwise known as tape casting [41]. Instead of a threaded rod, however, a doctor blade is used that consists of a plane set at a uniform distance from the substrate. Optionally, a reservoir fixed to the back of the blade supplies the dispersion. The blade mechanism can be pulled over the substrate, or the substrate can be moved relative to the blade. The gap distance between the blade and substrate is the primary factor in controlling the thickness of the film.

Table 1 summarizes MWCNT thin-film wet deposition fabrication methods. The operating principles are described, and schematics display the general configuration of the substrate and dispersion. Common applications of films produced via the methods are provided in the final column [42–71]. While the primary process parameters controlling film properties are listed, the mechanisms by which the thin film is formed in each case are often dependent on dispersion properties as well. For instance, after the dispersion droplet has been ejected from the nozzle of an inkjet printer, the dynamics of the droplet as it impacts the substrate and the solvent begins to evaporate. These quantities are characterized by the Reynolds, Weber, and Ohnesorge numbers [42], where the Reynolds number represents the dispersion viscosity, the Weber number represents the surface tension of the droplet's contact with the substrate and the air, and the Ohnesorge number is a function of the Reynolds and Weber numbers. Likewise, the qualities of doctor bladed films are dependent on several factors, including the blading mechanism and the dispersion's fluid properties. While the thickness of the film is primarily controlled by the gap between the substrate and blade, the wet thickness of the cast is different from the film thickness because of solvent evaporation and dry shrinkage, which are affected by casting speed, pressure, and several other parameters [41].

Table 1. Schematics and applications of thin-film fabrication techniques for MWCNT dispersions and slurries.

Fabrication Technique	Description	Schematic	Applications
Dip coating	Substrate extracted from dispersion Thin film formed by evaporation and gravitational draining	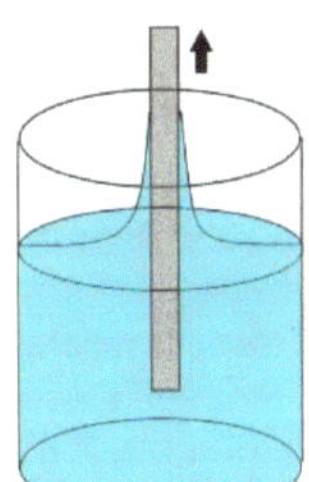	Chemical sensing [43,44] Corrosion resistance [45] Biomedical [46] Supercapacitors [47,48]
Drop casting/Drying	Dispersion droplet placed on substrate Droplet optionally moved across substrate Evaporation removes solvent	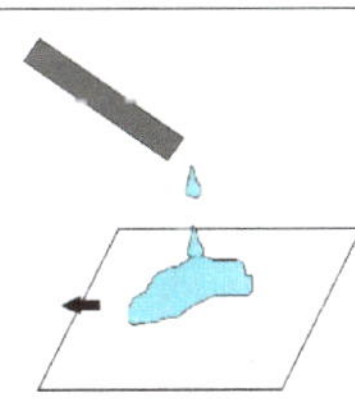	Chemical sensing [49] Supercapacitors [50,51] Electrically conductive films [52] Gas separation [53] Photovoltaics [54]
Spin coating	Dispersion dropped onto rotating substrate Shearing force from rotation spreads dispersion over surface Evaporation removes solvent		Transparent films Photovoltaics [55,56] Electromagnetic shielding [57]

Table 1. *Cont.*

Fabrication Technique	Description	Schematic	Applications
Inkjet printing	Electrical potential applied across piezoelectric nozzle, ejecting droplets on substrate Evaporation removes solvent	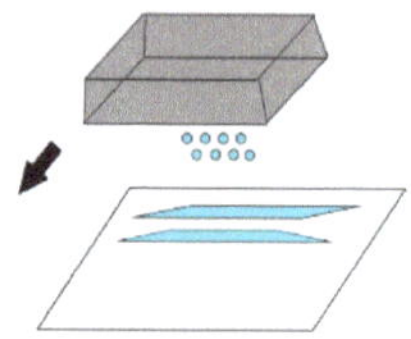	Electrically conductive films [58–61] Transistor devices [62–65]
Screen printing	Ink squeezed through open areas of a screen Ink is deposited on substrate Evaporation removes solvent	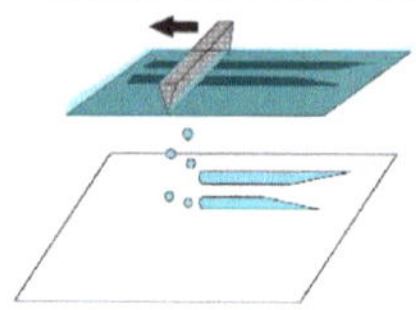	Piezoresistive pressure sensors [66–68] Bio sensing [69,70] Chemical sensing [43] Photovoltaics [55]
Meyer rod coating	Slurry is cast behind threaded rod Rod is translated without rotation over substrate	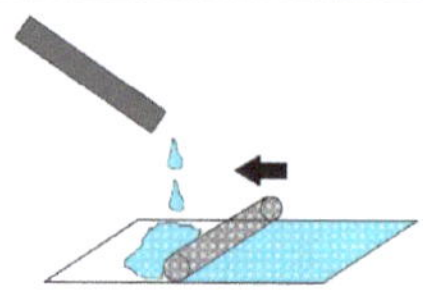	Supercapacitors [71,72]
Doctor blading	Slurry is cast in front of blade or in a reservoir Doctor blade is drawn over substrate Thickness of film is related to distance between blade and carrier	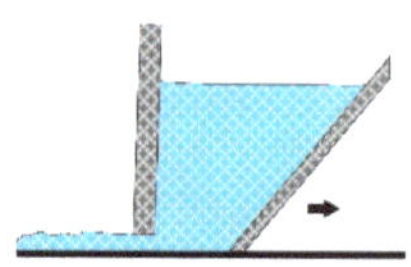	Supercapacitors

3.2. *Properties of Multiwalled Nanotube Films*

The properties of MWCNT films fabricated from different techniques vary widely, and therefore have a significant bearing on the specific applications for the film. For instance, inkjet printing and screen printing allow for patterned deposits of nanotubes, which is essential for sensors and other devices requiring specific placement of nanotubes. For other applications, such as electrochemical supercapacitors, much larger quantities of active material are needed than can be efficiently deposited by printing techniques. As such, slurry-based methods such as lamination, Meyer rod coating, and doctor blading have been preferred [51,71]. In this section, different properties of MWCNT thin films synthesized through various techniques are reviewed and connected to potential applications.

3.2.1. Mechanical Properties

A study by Ma et al. [73] determined the mechanical properties of pure MWCNT films using the digital speckle correlation method (DSCM). In this approach, paint with known mechanical properties was sprayed on the surface of the film. Tensile tests were performed using the paint speckles as indicators of the displacement field and correcting for the effect of paint on the mechanical properties. MWCNT thin film was fabricated by filtration of a MWCNT:PVA dispersion and subsequent removal of the microporous substrate. The stress–strain curve is shown in Figure 6. It was determined that the corrected elastic modulus of the film was ~1.04 GPa and the tensile strength was ~26.6 MPa, showing good agreement with theoretical analyses [73].

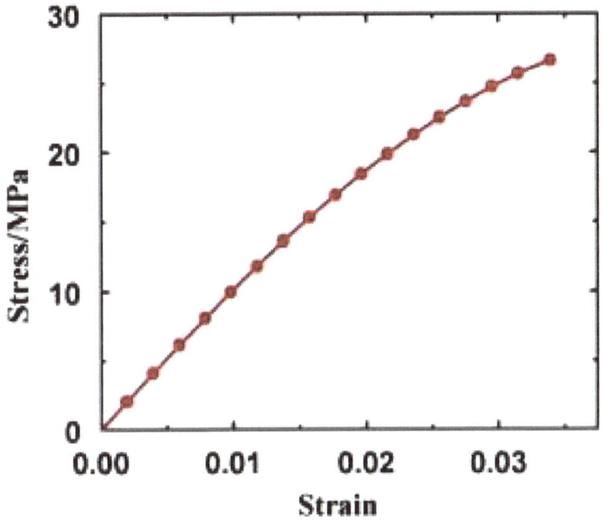

Figure 6. Stress–strain curve of an MWCNT film obtained by the digital speckle correlation method [73]. Used with permission from Elsevier.

MWCNTs have been used to increase the strength of polymer thin films. In a study by Xu et al. [74], spin coating a dispersion of 0.1 wt% MWCNT in chloroform with EPON SU-8 photoresist epoxy increased the elastic modulus of the thin film over the pure epoxy by 20%, from 4.2 GPa to 5 GPa. The authors attribute the strong interfacial adhesion between the MWCNTs and epoxy to (1) nanomechanical locking between twisted MWCNTs and epoxy, partially caused by a mismatch in the thermal expansion coefficient and (2) level bond and nonbond interactions [74]. The spin coating process realigned the MWCNTs to be parallel to the thin film plane and perpendicular to the crack direction through the centrifugal force, but was limited by the concentration of solution that could be deposited. Koval'chuk et al. [75] fabricated nanocomposite thin films by hot pressing powders of either unidecyl-functionalized nanotubes (C_{11}–MWCNT) or unfunctionalized nanotubes (MWCNT). They used eiher isotactic polypropylene (iPP) or syndiotactic polypropylene (sPP) and found an increase in the Young's modulus by 37%, 66%, 30%, and 84% for iPP/MWCNT, sPP/MWCNT, iPP/C_{11}–MWCNT, and sPP/C_{11}–MWCNT composites, respectively, compared to uncomposited iPP and sPP [75], shown in Figure 7. It was observed that the Young's modulus increases more significantly at low MWCNT loading (<~1 wt%) but plateaus upon higher loading amounts. Tensile strength was also affected by MWCNT loading, although less significantly. At high loading, strength was adversely affected by MWCNTs, which was attributed to nanotube agglomeration [75]. Sun et al. [76] investigated the mechanical properties of multilayered chitosan/MWCNT composite films prepared using the sublimation and compression method (SAC) and casting-evaporation (CE) method. The precursor dispersion was a mixture of acid treated MWCNT dispersion and a 1.0 wt% aqueous chitosan solution. Films produced by both SAC and CE showed increased modulus and ultimate tensile strength after compositing with MWCNTs, from 484 to 502 MPa and 1288 to 1613 MPa, respectively. Tensile strength was similarly improved from 4.43 to 9.82 MPa for films produced by SAC and from 8.86 to 12.36 MPa for films produced by CE. The improved mechanical properties are attributed to the suppression of cracks caused by crazing by the added MWCNTs [76].

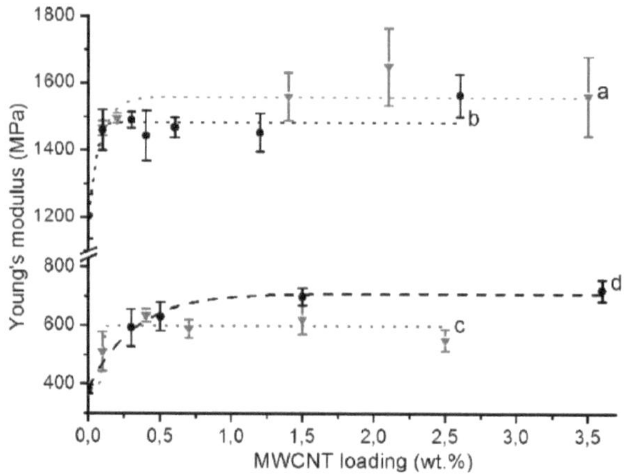

Figure 7. Elastic moduli as a function of unfunctionalized and functionalized MWCNT loading in polypropylene composites: (**a**) iPP/MWCNT, (**b**) iPP/C_{11}–MWCNT, (**c**) sPP/MWCNT, and (**d**) sPP/C_{11}–MWCNT [75]. Reprinted with permission from A.A. Koval'chuk, V.G. Shevchenko, A.N. Shchegolkhin, P.M. Nedorezova, A.N. Klyamkina, A.M. Aladyshev, Effect of Carbon Nanotube Functionalization on the Structural and Mechanical Properties of Polypropylene/MWCNT Composites, *Macromolecules* **2008**, *41*, 7536–7542. Copyright 2008 American Chemical Society.

3.2.2. Electrical Properties

The electrical properties of MWCNT thin films are important for applications in transistors and electrodes. Lau et al. [77] investigated the effect of nanotube functionalization by thermal treatment, acid treatment, and UV–ozone treatment on electrical conductivity. The MWCNT mat was fabricated on glass filter paper by filtration and drying. The electrical conductivity of the unmodified nanotube mat was ~49.5 S/cm [77]. In the case of thermal treatment, increasing the treatment time with temperature held fixed at 500 °C increased the thermal conductivity to ~4000 S/cm at 6 h [77]. Similarly, increasing the temperature of the thermal treatment with time held fixed at 1 h increased the electrical conductivity to ~2800 S/cm at just above 500 °C [77]. Similar trends in the conductivity were observed for increasing treatment time and temperature for acid treatment as well as increasing treatment time and ozone flow rates for UV–ozonolysis. The maximum conductivity observed was over 16,000 S/cm for ozone treatment at 300 °C for 30 min and an ozone flow rate of 5 L/min [77].

Because of its high electrical conductivity, MWCNTs have been used to improve the conductivity of polymer films. Chen et al. [78] dispersed acid-treated MWCNTs in DMAc and reacted with sorbitol and PEDOT:PSS. The mixture was spin coated onto silicon wafers. With increasing reaction time from 0 to 6 h, the resistivity of the film decreased from 8.3×10^5 Ω·cm to 260 Ω·cm due to increased homogeneity of the distribution of MWCNTs in the PEDOT:PSS matrix [78]. Ram et al. [79] composited MWCNTs with PVDF in DMF and solution-casted the mixture. The DC conductivity of the film increased to ~10^{-3} S/cm (resistivity ~1000 Ω·cm) with 5 wt% MWCNT, with a percolation threshold at ~1 wt% MWCNT. Furthermore, the AC conductivity for composites less than 1 wt% MWCNTs were found to increase monotonically with frequency, which was attributed to insulating behavior and the contribution of "electron hopping" to conductivity [79]. Chen and Tao et al. [80] determined the conductivity of PDMS-backfilled MWCNT-graphene aerogels fabricated by vacuum filtration to be ~6.3 S/cm (resistivity of 0.16 Ω·cm), and this was irrespective of the number of tensile or bending cycles the films were subjected to. The film also showed more stable conductivity with applied strain compared to graphene-PDMS or MWCNT-PDMS composites, with a ~30% decrease in conductivity at 30% tensile strain [80].

Kim et al. [81] prepared MWCNT-PMMA films using a solution casting technique from hydrochloric acid-treated MWCNTs sonicated with PMMA and PVP in DMF, with a PET substrate. The sheet resistance of the sheets, as measured by 2-probe multimeter, was observed to decrease from 4×10^4 kΩ/□ for 1.0 wt% MWCNTs to ~250 kΩ/□ for 4.0 wt% MWCNTs [81]. Vacuum filtration of polyurethane–MWCNT composites has been explored by Fernandez-d'Arlas et al. [82], and the DC conductivity was found to follow a percolation pattern with a percolation threshold of 2.5% volume ratio MWCNTs and a maximum observed conductivity on the order of 10^{-2} S/cm (resistivity ~100 Ω·cm) at 10.5% volume ratio MWCNTs. Spray-coated, and melt-mixed thin films of MWCNT composites with various polymers were investigated by Staudinger et al. [83]. The electrical resistivities of melt-mixed PC–MWCNT, PMMA–MWCNT, and PVDF–MWCNT all drastically decreased with MWCNT concentration below 1 wt% and more gradually at higher concentrations, and beyond 2 wt% MWCNT loading, the resistivity for each composite was on the order of 10 Ω·cm [83].

Okimoto et al. [62] prepared SWCNTs by dispersion in DMF. When deposited at varying thicknesses via inkjet printing, the nanotubes changed properties, from a semiconducting material at low thicknesses to metallic at high thicknesses, an effect attributed to the emergent interactions between nanotubes when the coverage is uniform and complete. The IV response is shown in Figure 8. The electrical characteristics were found to depend strongly on areal coverage, which is explained by the variation in carrier mobility due to coverage [62]. Mustonen et al. [63] also investigated inkjet-printed carbon nanotube TFTs. The charge mobility in the TFTs was between 1.6 and 4.2 cm^2/V-s. The on–off current ratio for the TFTs was on the order of 105. Similar results were obtained by another group investigating flexible TFTs using a top-gate geometry and silver electrodes. Kim et al. [64] developed an inkjet-printed hybrid TFT using zinc oxide as the n-channel material and SWCNT as the p-channel material on both glass (SiO_2) and plastic (polyimide) substrate. The performance of the TFT was rigorously tested using a 5-stage complementary-ring oscillator, revealing a propagation delay of only 140 ns, as shown in Figure 9. The voltage behavior for the CNTs printed on SiO_2 is evidently more stable than for those on the polyimide substrate, but on both substrates, the static power consumption of the TFTs is exceptional (<0.25 W at 5 V) [64].

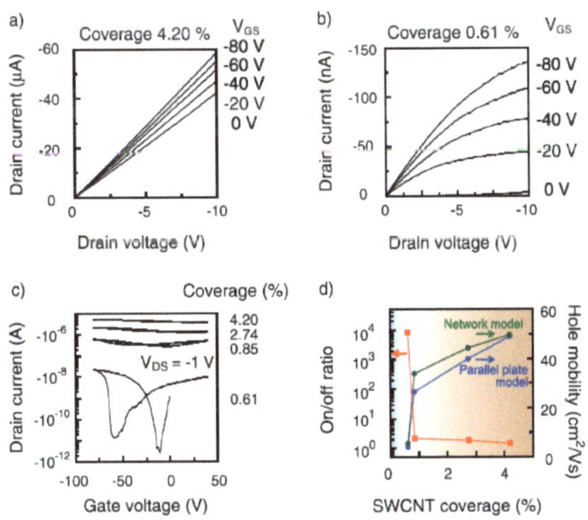

Figure 8. Inkjet-printed carbon nanotube thin-film transistor characteristics [62]. Figure (**a**) shows the output characteristics of the print with higher areal coverage, while figure (**b**) shows the output characteristics of the print with lower coverage. Figures (**c**,**d**) show the transfer characteristics of the prints and the on-off ratio dependence on coverage, respectively. Used with permission from John Wiley and Sons.

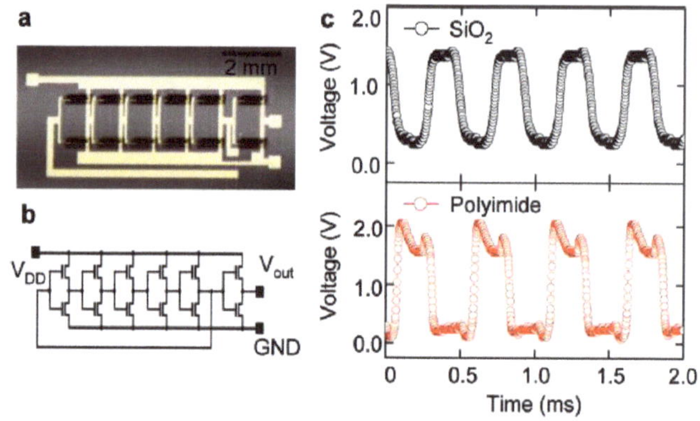

Figure 9. Five-stage ring oscillator network built with inkjet-printed CNT thin-film transistors with a drain voltage of 2.5 V [64]. (**a**) image of device (**b**) equivalent circuit (**c**) Voltage profile for devices built on SiO_2 and polyimide. Reprinted with permission from B. Kim, S. Jang, M.L. Geier, P.L. Prabhumirashi, M.C. Hersam, A. Dodabalapur, High-Speed, Inkjet-Printed Carbon Nanotube/Zinc Tin Oxide Hybrid Complementary Ring Oscillators, Nano Letters 14 (6), **2014**, 3683–3687. Copyright 2014 American Chemical Society.

Multi-walled carbon-nanotubes printed on paper with different pore sizes have been demonstrated by Muhlbauer et al. [61] to have a wide range of electrical properties, especially in response to ac inputs. The effect of different number of deposition layers was also explored, and it was concluded that four layers provided a uniform, consistent, and interconnected network of nanotubes. Depositions on papers with smaller pores were found to have a lower impedance, attributed to higher material retention during filtration. This is reflected in Figure 10, which shows a much lower impedance magnitude for MWCNTs deposited on 410 and 413 filter papers than on 415 and 417 papers. The study demonstrated that samples produced by drying carbon nanotubes all displayed an RL equivalent network in series with another resistor as part of an RC segment. This interesting behavior of high-frequency inductive behavior was attributed to the wrapping of the nanotubes around pores in the paper substrate [61]. For vacuum-filtered carbon nanotubes deposited on paper, a constant-phase element parallel to resistor branch was observed to explain the region of the Nyquist plot that contains a negative imaginary part of the impedance. Vacuum-filtered depositions on paper were characterized to be more insulating in nature (i.e., resistive more than reactive) due to the large pore size and more inhomogeneous surface [61].

3.2.3. Other Properties

While mechanical and electrical properties of MWCNT thin films are of primary interest for general film applications, other material properties are also influenced by the fabrication technique. Optical transmittance is important for optoelectronic films and transparent films. Many investigations into the electrical properties of MWCNT thin films have also measured the optical transmittance, since a common application for thin films is as transparent conductors. It has been found that increasing the MWCNT concentration tends to decrease the transmittance because of the high absorbance and scattering of light by carbon nanotubes. Chen et al. found a decrease in the optical transmittance by more than 10% across the visible spectrum after compositing with nanotubes, but the transmittance could be improved ~5% by increasing the MWCNT/sorbitol/PEDOT:PSS reaction time from 2 h to 4 h, and by ~10% by increasing the reaction time to 6 h [78]. Ram et al. [79] discovered a more drastic decrease in transmittance with nanotube concentration in PVDF composites, down to 0% transmittance across the visible spectrum for 5.0 wt% MWCNT.

For PET and PMMA composites, Kim et al. [81] found a gradually decreasing transmittance with increased nanotube concentration, in the range of 64% to 89% at 550 nm when increasing the MWCNT weight ratio from 1.0 wt% to 4.0 wt%. Staudinger et al. [83] observed lower transmittances ranging from 37% to 50% for 8 layers of 2 wt% MWCNTs spray coated on PVDF, 22% to 30% for 8 layers of 4 wt% MWCNTs on PC, and 6% to 10% for 8 layers of 3 wt% MWCNTs on PMMA. The differences were attributed to varying qualities of the MWCNT dispersions, since nanotube agglomerates in PMMA with sizes on the order of the wavelength of visible light can adversely affect the transmittance, in addition to differences in the surface roughness of the composites [83]. The optical transmittance of MWCNT thin films generally decreases as the electrical conductivity increases, which is reasonable since carbon nanotubes tend to absorb visible light but have higher electrical conductivity than the substrate [84–86]. Treatments of the device with alcohol spray [84] could simultaneously increase both the transmittance and electrical conductivity.

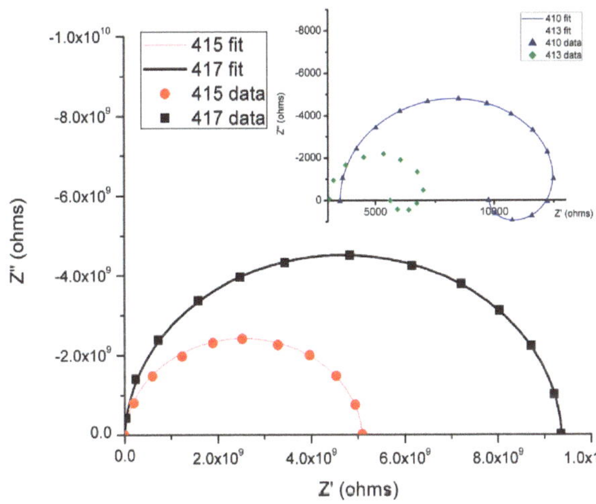

Figure 10. Nyquist plots for frequency response of multi-walled carbon nanotubes deposited on 410 and 413 (inset), and 415 and 417 filter papers [61]. Used with permission from Springer Nature.

The thermal properties of carbon nanotube films have also been previously investigated. Hone et al. [87] measured the variation of thermal conductivity with temperature of a magnetically-aligned SWCNT film. At 300 K, the thermal conductivity was almost 220 W/m-K, which was much higher than a film with unaligned SWCNTs, with thermal conductivity of less than 50 W/m-K [87]. Yang et al. [88] investigated mechanical and thermal properties of MWCNT/epoxy composites and found a nearly linear variation of the thermal conductivity with nanotube volume fraction, up to 0.9 W/m-K for commercial MWCNTs and 1.3 W/m-K for TETA-treated MWCNTs. Pradhan et al. [89] synthesized MWCNTs with CVD on anodized aluminum oxide and removed the template with NaOH. The effective thermal conductivity of aligned MWCNTs was nearly an order of magnitude higher than randomly oriented MWCNTs and had a less significant temperature dependence.

4. Multiwall Carbon Nanotube Thin Films for Supercapacitor Applications

Carbon nanotubes are ideal for use in energy storage devices due to excellent conductivity that improves the power density; their one-dimensional geometry; mechanical strength that allows them to support the electrode and withstand stresses during electrochemical cycling; high specific surface area that increases the contact between electrode and electrolyte and promotes charge transfer; and low mass density [90]. Several reviews have discussed the use of carbon materials in supercapacitors and micro-supercapacitors [91].

Table 2 summarizes some of the previous publications that have investigated carbonaceous materials in supercapacitor electrodes, as well as properties obtained in our group with MWCNTs. Previous research efforts have used a wide variety of electrode fabrication techniques (e.g., pressing, lamination, and direct synthesis on the current collector), electrolyte systems (e.g., aqueous, organic), and substrates (both conducting and insulating), so direct comparison of the active material performance is difficult. Furthermore, few investigations have explored spin coating and inkjet printing techniques for fabricating electrodes because overall capacitance generally improves with increased loading of the active material and a limited interest in extremely lightweight energy storage. However, as shown in Figure 11, the specific capacitance is overall not heavily influenced by the electrode mass. Figure 11 also reveals that among different carbon materials, single-walled carbon nanotubes (SWCNTs) generally feature the highest specific capacitances, while MWCNTs have similar or slightly better performances compared to other carbon materials. However, the merit of using MWCNTs as supercapacitor electrode materials arises from their lower cost relative to single-walled nanotubes [92].

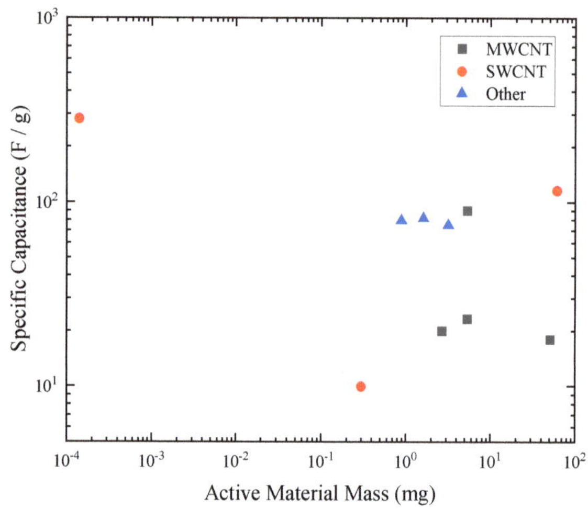

Figure 11. Relationship between specific capacitance and the mass of active material per electrode (either directly measured or estimated based on other measurements) from previous investigations of carbon-based supercapacitor electrodes. Data taken from references: [50,71,93–100].

4.1. Electrochemical Double-Layer Capacitors

Electrochemical double-layer supercapacitors (ECDL) are energy storage devices that store electrical energy in the double-layer interface that forms between the solid electrode and electrolyte. As depicted in Figure 12, the Helmholtz layer is composed of the outer surface of the charged electrode and a layer of oppositely charged ions in the electrolyte solution electrostatically attracted to the electrode [101,102]. Depending on the concentration and size of the ions in solution and the surface area of the Helmholtz layer, the capacitance of the Helmholtz layers on both electrodes in the capacitor dominates the ECDL capacitance. The diffuse layer forms outside of the Helmholtz layer and consists of a concentration gradient of solvated electrolyte ions [103]. The concentration gradient of charge carriers results in a gradual potential gradient directed towards the electrode, which also contributes to the capacitance, although generally not as significantly as the Helmholtz capacitance.

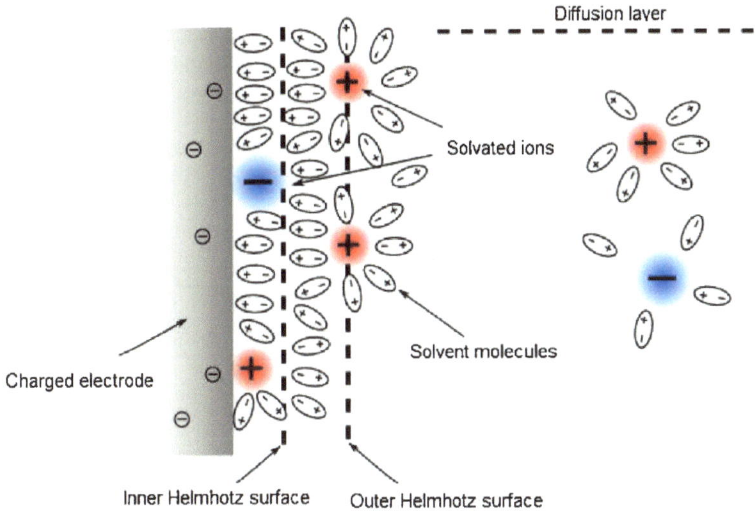

Figure 12. Helmholtz layer comprising the inner and outer Helmholtz surfacers and diffuse layer developed on the interface between the electrolyte and a charged electrode [102]. Used with permission from Elsevier.

An et al. [104] fabricated carbon nanotube thin film pellet capacitor electrodes by pressing CNTs synthesized by arc discharge with PVC binder in THF and subsequent high-temperature heat treatment. They achieved a high specific capacitance of 180 F/g, high energy density of 6.5 Wh/kg, and low equivalent series resistance (ESR) of 105 mΩ using a polished Ni foam current collector. Li et al. [105] also treated carbon nanotubes with sulfuric and nitric acid, resulting in high specific surface area of 279 m^2/g and specific capacitance of 56.5 F/g. Chen et al. [106] grew carbon nanotubes directly on graphite foil substrate, followed by acid treatment. The nanotubes were found to be densely packed on the surface of the graphite. Highly rectangular CV curves are obtained, and a specific capacitance of 147 F/g was achieved at a scan rate of 25 mV/s, affirming the excellent capacitive properties of carbon nanotubes. Adding even a very small weight percentage (0.15 wt%) of MWCNTs to activated carbon (AC) electrode resulted in a significant decrease in supercapacitor impedance and increased the specific capacitance [107]. The improved electrochemical performance was explained by the ability of a small amount of MWCNTs to reduce the size of macropores and prevent the blocking of micropores in AC after the binder is added in the slurry-based electrode fabrication process. The increased micropore volume facilitates charge transfer.

4.2. Compositing Carbon Nanotubes with Conducting Polymer and Metal Oxides

Although carbon nanotubes alone have shown exceptional performance as electrical double-layer capacitor electrodes, capacitances and specific energies can be further increased by introducing conducting polymers or metal oxides on the nanotubes. This resulted in the so-called pseudocapacitance effect, wherein ions in the electrolyte desolve and adsorb onto the electrode via Faradaic charge transfer [108]. For example, Frackowiak et al. [101] deposited polypyrrole (PPy), a conducting polymer, onto nanotubes to increase the specific capacitance from 80 F/g to 170 F/g, with a clear two-step charging mechanism. The potentiostat scans of the supercapacitors are shown in Figure 13, which shows much higher currents for the composite electrode. In addition to increasing the capacitance, the compositing of polypyrrole increased the operable voltage range for the charge and discharge, which is important for practical applications of the supercapacitor [109]. Kong et al. [110] synthesized MWNT/PANI composite via an in-situ chemical oxidation polymerization technique. The composite electrodes had a specific capacitance of 224 F/g. Other groups [111–113] have investigated the performance of carbon nanotube-metal oxide composites and have found that the addition of metal oxides to carbon nanotubes improves the specific capacitance, specific energy, and stability of the electrochemical capacitor. Teresawa, et al. [114] found that compositing MWCNT capacitor electrodes with RuO_2 using a solution evaporation fabrication process increased the capacitance beyond that of SWCNT electrodes due to the pseudocapacitive redox reactions between the electrolyte ions and the mesoporous texture of the MWCNTs. The enhanced capacitance allowed the electrode to support higher mechanical strains and stresses [115].

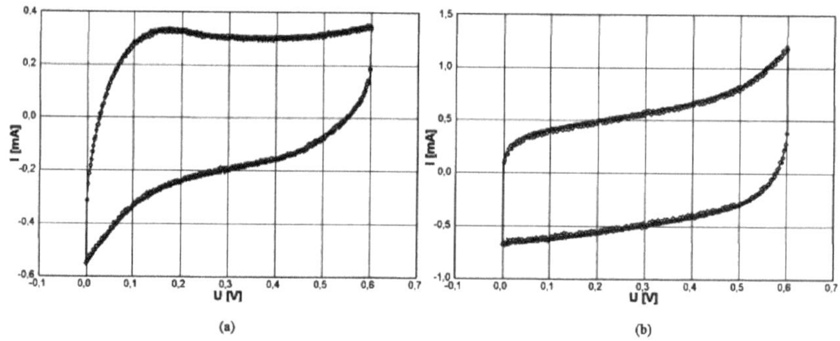

Figure 13. Potentiostats of MWCNT electrodes without (**a**) and with (**b**) polypyrrole (Ppy) in 1 M H_2SO_4 electrolyte [109]. Reprinted with permission from Elsevier.

4.3. Other Improvements to the Carbon Capacitor Electrode

While compositing nanotubes with conducting polymer increases the capacitance through the pseudocapacitive effect, it has the side effect of increasing the resistance of the electrode and thus decreasing the power density of the capacitor [115]. To avoid this, one method previously explored in the literature is the purposeful functionalization of surface groups on nanotubes, such as carboxyl (–COOH), hydroxyl (–OH), and carbonyl (=C=O:) groups [116]. Functionalization is conducted through treatment with a mixture of sulfuric acid and nitric acid [117] and is typically performed at elevated temperatures. For instance, Du, Yeh, and Pan [116] produced MWCNT electrodes by electrophoretic deposition (EPD), and the functionalized groups not only improved the capacitance but also increased the uniformity of the EPD suspension by increasing the mutual repulsion of nanotubes in dispersion. Ma et al. [118] found that functionalizing the electrode with sulfuric and nitric acid removed impurities, and more importantly increased the pore size, allowing better penetration of the electrolyte. Yun et al. [118] fabricated monolayer graphene/MWCNT electrodes via spray coating and functionalization with significantly increased surface area (~160 times) and lowered internal resistance compared to unmodified monolayer graphene electrodes. As shown in Figure 14, monolayer graphene/functionalized MWCNT electrodes exhibited much higher specific capacitance and energy and power densities than monolayer graphene electrodes.

A distinct process from functionalization that can be used to introduce pseudocapacitive redox reactions on MWCNT electrodes is ammoxidation in ammonia and air [92]. While functionalization alters the physical structure of nanotubes by breaking the walls and introducing mesopores to improve the specific surface, ammoxidation was observed to preserve the structure of nanotubes but modify the chemical composition of nanotubes [92], which allows for pseudo-faradaic charge transfer like the type that occurs in conducting polymer and metal oxide compounds. In another investigation, Geng et al. [119] observed that ammoxidation decreases the surface area and may not be as effective as functionalization and pretreatment by ball milling for the improvement of capacitance.

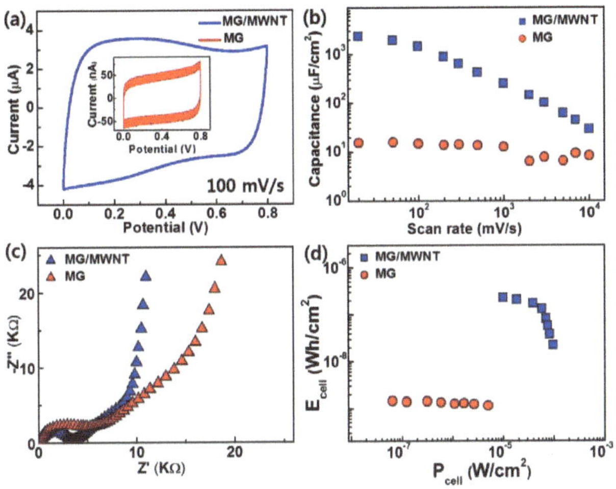

Figure 14. Electrochemical performance of supercapacitors with MWCNT and monolayer graphene composite electrodes in all-solid-state supercapacitors (PVA-H_3PO_4 gel electrolyte) [118]. Cyclic voltammetry is shown in (**a**). The variation of area-specific capacitance with scan rate is shown in (**b**). Impedance spectra are shown in (**c**). Specific energy and specific power are plotted in (**d**). Used with permission from Elsevier.

Table 2 below reviews carbon-based double-layer supercapacitors from the literature. The most common deposition processes for electrodes are pressing and lamination, but less conventional methods such as dip coating, direct growth on the substrate, and evaporation have also been explored. Depending on the deposition method, various amounts of active material are incorporated into the electrode. As discussed previously and shown in Figure 11, though the specific capacitance varies by several orders of magnitude across studies from ~1 F/g to ~200 F/g, it is not significantly affected by the mass of the electrode but does vary significantly with the material composition and electrolyte. Although the thickness of the electrode is not commonly reported by the literature, this suggests that the specific capacitance does not depend heavily on electrode thickness. Other design considerations, such as the choice of the substrate, additives, and electrolyte, in addition to post-processing treatments, should have a greater impact on device performance. Additives such as conducting polymers allow for charge storage through adsorbed ions and can significantly increase the specific capacitance. The substrate choice is also important, as it affects the impedance of the overall electrode. The electrolyte determines the ionic conductivity and stability of the supercapacitor.

4.4. Spin-Coated and Inkjet-Printed MWCNT Capacitor Electrodes

To further investigate their conductivity and performance in electrochemical devices, multiwall carbon nanotubes were dispersed in aqueous solvent with a surfactant. The dispersion was spin-coated and inkjet-printed onto commercial photo paper substrate. While these techniques have not been extensively used for supercapacitor electrode fabrication in previous investigations, they offer alternative methods for low-cost fabrication of miniature charge storage devices.

4.4.1. Experimental

Materials

Multiwall carbon nanotubes (Cheaptubes, purity >95%) synthesized by combustion CVD and purified using acid treatments were used without further modification. Deionized (DI) water (Thermo Scientific purifier) was used as solvent. Sodium dodecylbenzene sulfonate (SDBS) and sodium chloride (NaCl) were used as surfactant and electrolyte salt, respectively. The substrate is inkjet-compatible Epson glossy photo paper with a thickness of 230 μm (9 mil) and sheet density of 183 g/m^2. Photo paper was cut into desired sizes, treated with UV–ozone before deposition to eliminate contaminants, activate polymers on the surface [120], and improve the hydrophilicity of the paper, which is essential for depositing a water-based ink [121]. Filter paper (410, VWR) with particle retention of 1 μm was used as the separator material for capacitive devices.

Multiwall Nanotube Ink Dispersion Preparation

Nanotube ink was prepared using a formulation described previously [122]. A mixture of 1 mg/mL MWCNTs added to DI water is ultrasonicated for 20 min. Sodium dodecyl benzenesulfonate (SDBS) is added to the suspension at a concentration of 5 mg/mL to disperse the nanotubes by reducing aggregation caused by the van der Waals interactions between adjacent nanotubes [123]. The resulting dispersion is magnetically stirred for 12 h, after which the inks appeared visually uniform.

Spin Coating

MWCNT dispersion was spin coated (CEE/Brewer Science 100) onto photo paper substrate (1.5 cm × 4 cm) at a rotational speed of 6000 RPM. 25 μL of the MWCNT dispersion was pipetted onto the substrate after accelerating to the steady-state speed. After deposition, the substrate was spun for 30 s to promote evaporation of the solvent. Additional layers were casted immediately on top of previous layers. The final depositions were dried in atmospheric conditions for 24 h before electrical measurements and further processing.

Inkjet Printing

MWCNT dispersion was inkjet-printed (Microfab Jetlab II, nozzle orifice diameter 50 μm) onto photo paper substrate. A 5 cm (2 in.) square of photo paper was placed flat on an unheated platen. To prevent flocculation of the nozzle, the ink was filtered using a 5 μm nylon Luer-lock syringe filter, which removed all visible nanotube agglomerates from the ink. The supernatant ink was subsequently loaded into an inkjet vial. Steady drop formation and ejection velocity of ~2.5 m/s was achieved using a dwell voltage of 50 V, dwell time of 30 μs, and no echo pulse. Uniform rectangular grid patterns with an inter-droplet spacing of 50 μm were printed using on-the-fly deposition technique, which consists of simultaneously ejecting droplets and continuously moving the platen. Additional layers are printed immediately on top of previous layers. The final depositions are let to rest in atmospheric conditions for 24 h before electrical measurements and further processing.

Device Fabrication

Capacitive devices were constructed from as-deposited MWCNT inks on photo paper substrate. Because photo paper is non-conducting—the AC impedance at 0.1 Hz was measured to be ~1011 Ω—the active nanotube material was used as the current collector. Two identical rectangular electrodes (1.5 cm × 3 cm) were either spin-coated or inkjet-printed. The electrodes were overlapped with an area of 1.5 cm × 2 cm, leaving 1.5 cm × 1 cm of active material exposed at either end for interfacing with electrochemical characterization devices. A small sheet of 410 filter paper (separator) was spin coated on both sides with a total of 60 μL of NaCl: DI water electrolyte solution. The wetted filter paper was inserted between the electrodes, and the paper stack was clamped using alligator contact clips to avoid evaporation of electrolyte.

Material and Sample Characterization

The diameter distribution of the multiwall nanotube starting material was estimated using transmission electron microscopy (TEM) and the ImageJ image processing software. Due to the limited number of discernable nanotubes from TEM, which results from the clustering and transparency of the nanotubes, statistical bootstrapping was used to approximate the diameter distribution of the nanotube population used in the study.

Deposited MWCNT films were observed using optical microscopy (Pentaview). Electrical properties of the films were measured using four-point probe (Keithley 6221 DC current source with Keithley 2182-A nanovoltmeter) operated in the delta mode and impedance spectroscopy (Solartron 1260 frequency response analyzer with Solartron 1296 dielectric interface) with AC impedance magnitude of 100 mV and frequency range from MHz to 10 mHz. Electrochemical characterization of the capacitor device was performed with a Gamry IFC5000 potentiostat operated at a scan rate of 100 mV/s and a potential window of −0.5 V to 1.0 V.

Computational Simulation of Thin Films

The electric currents (ec) physics interface of COMSOL Multiphysics 5.6 was used to simulate the electrical resistance of MWCNT thin films with different layer counts assuming isotropic metallic conduction. The thin film was simulated as a thin cylindrical disk with diameter of $d_f = 0.692$ cm, which roughly corresponds to the area of spin-coated and inkjet-printed films in the experimental study. The thickness was approximated to be proportional to the number of layers at a per-layer thickness of $t_l = 2.5 \times 10^{-8}$ m, based on a previous report of an inkjet-printed MWCNT film [124]. The governing system of equations is shown in Equations (1)–(3).

$$\nabla \cdot \vec{J} = 0 \text{ Current Conservation} \tag{1}$$

$$\vec{J} = \sigma \vec{E} + \vec{J_e} \text{ Ohm's Law} \tag{2}$$

$$\vec{E} = -\nabla V \text{ Definition of Electric Potential} \tag{3}$$

where J is the current density, J_e is the external current density, E is the electric field, and V is the electric potential. The conductivity of the film material is specified as $\sigma = 143$ S/m, reported previously as the bulk conductivity of MWCNTs synthesized via CVD for 12 h [125].

Four point-terminals are defined along the x-axis diameter on the top surface of the model with spacing of $s = 0.155$ cm, the inter probe spacing on the 4 point probe used in experimental measurements. Viewed from the top, the left outer terminal (Terminal 1) is a current source I_0 (set to 100 μA and 1 μA), and the right outer terminal (Terminal 4) is a grounded boundary ($V_0 = 0$ V). The remaining terminals (Terminals 2 and 3) are for sense measurements. All other boundaries are insulating.

4.4.2. Results and Discussion

Material Structure and Deposition Morphology

The TEM images of the MWCNT material (Figure 15) show randomly oriented nanotubes with variation in both inner and outer tube diameters. To determine the mean outer diameters of the MWCNTs, several measurements were made along nanotubes in the micrograph using the ImageJ image processing software. The mean diameter of the nanotubes is 15.5 nm, with a standard deviation of ~4 nm. This is comparable to the MWCNT dimensions reported in previous investigations of electrical properties [126,127]. The random orientation and curling of the nanotubes are desirable for creating dispersions; however, agglomeration of the nanotubes, as seen in the leftmost region of Figure 15, are undesirable for inkjet printing due to the possibility of clogging the nozzle [128]. This justifies use of surfactants to disassemble agglomerates and uniformly disperse the active material [123,129,130].

It is important to note that due to the low viscosity of the MWCNT ink, some of the deposited ink was centrifuged away from the substrate during the deposition of each layer by spin-coating. As a result, the active material loading deposited on the substrate is less than the nominal amount contained in the dispersion. Nonetheless, the depositions appear uniform across the surface of the substrate. Individual film layers dry quickly, allowing for layers to be deposited continuously.

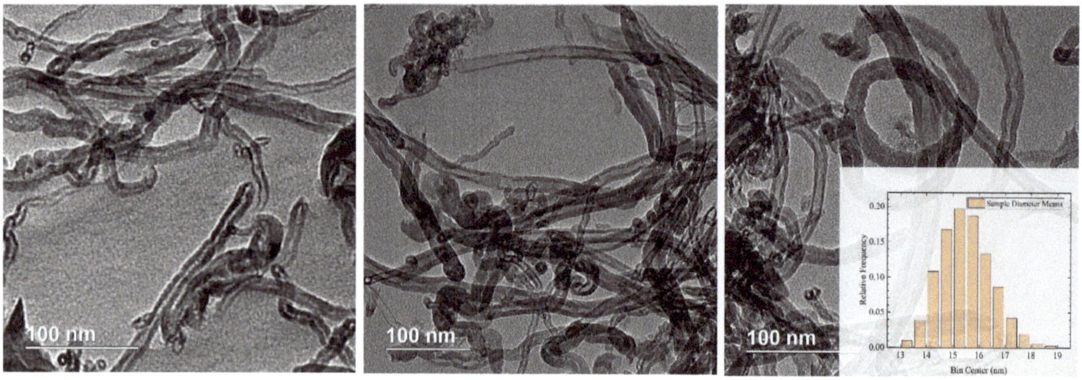

Figure 15. Transmission electron microscope images of multiwall nanotubes used to make electrodes. Inset shows the CNT diameter distribution as determined by statistical bootstrapping from the nanotubes in the images.

Figure 16 shows optical microscope images of MWCNT thin films with various numbers of layers produced by spin-coating. Even though nanotubes can be observed to cover the substrate area at each layer count, there are several particulates on each sample that suggest agglomeration still exists to some degree in the ink dispersion, despite the use of surfactant. These regions become more prevalent as the number of layers is increased, which supports the assumption that layers are deposited independently of one another on the existing surface. Agglomerates are proposed to contribute to electrical conductivity of the spin-coated samples, since connected regions can form conductive paths. In addition to the agglomerates, the dispersion can also be observed to wet the paper, as indicated by the grey regions behind the agglomerates. The degree of wetting increases with higher layer counts. However, the presence of non-wetted, lighter regions on the substrate indicate that the spin-coated films are mostly non-homogeneous.

Due to the nature of inkjet printing, which deposits a small volume of ink at each location, no active material was lost during the deposition process. As in spin-coating, the short drying time of the dispersion compared to the layer deposition time allowed layers of MWCNT to be printed continuously. Figure 17 shows optical images of the inkjet-printed films with various number of layers. Unlike the spin-coated films, the inkjet-printed films show no visible agglomerates, even at high layer counts. This is because inks for inkjet-printing were filtered before printing to prevent flocculation and subsequent clogging of the nozzle. Despite the lack of agglomerates, the optical images of the highly uniform films produced by inkjet printing confirm that the ink has absorbed into the substrate, which contributes to the electrical conductivity of the films.

The inkjet printing technique used in this study, known as on-the-fly, continuously deposits ink while moving the platen. Another technique, known as drop-on-demand, halts the platen at every print location and ejects a single droplet of ink before moving to the next position. Drop-on-demand is not investigated in the present study because of the associated long printing times, which may not be scalable for practical applications.

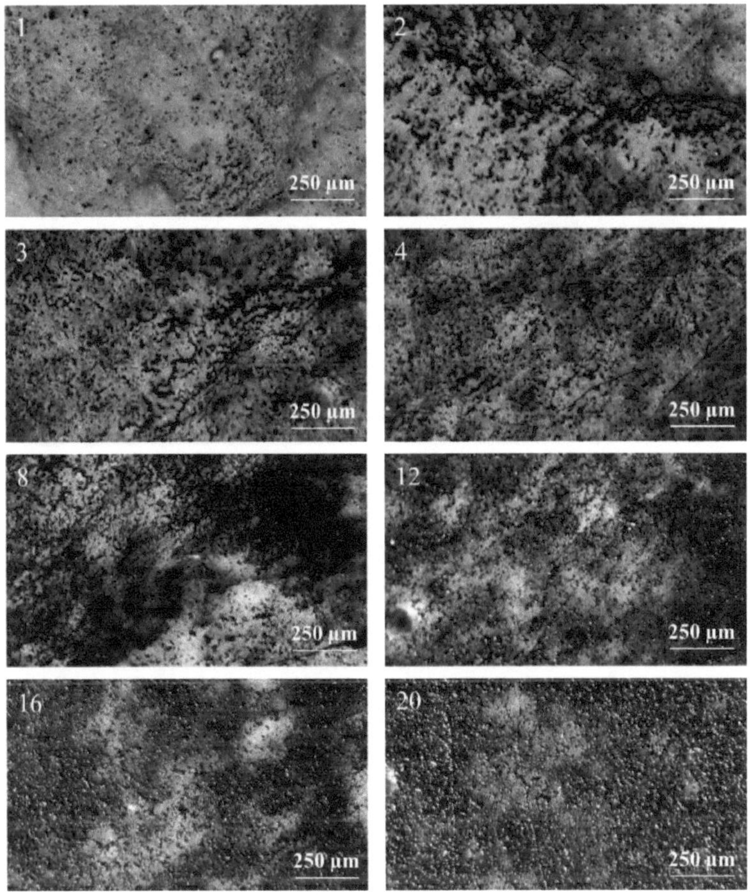

Figure 16. Optical microscope images of spin-coated MWCNT films on photo paper with varying layer counts (1–20 as labeled in each panel), showing visible agglomerates.

DC Electrical Characterization and Simulation of MWCNT Films

Figure 18 shows the sheet resistances of spin-coated and inkjet-printed films as measured by dc 4-point probe. Film resistance is strongly dependent on the number of layers, which correlates with the total amount of conductive material deposited onto the photo paper. As the number of layers increases, the sheet resistance falls rapidly, indicating the formation of conductive pathways. With 20 layers of spin coating, a sheet resistance of 1.64 kΩ/□ is achieved. With 20 overwritings of inkjet printing, the sheet resistance of 2.69 kΩ/□ is slightly more than one order of magnitude larger than that of the analogous spin-coated sample. The higher sheet resistance is attributed to the increased uniformity and lack of agglomerates, which are proposed to increase the conductivity in spin coated films. Increasing the number of overwritings of inkjet printing, a minimum sheet resistance of 166 Ω/□ is achieved. These resistances are comparable to sheet resistances obtained through spin coating and vacuum filtration on several filter papers (~1 kΩ for 20 layers) [131].

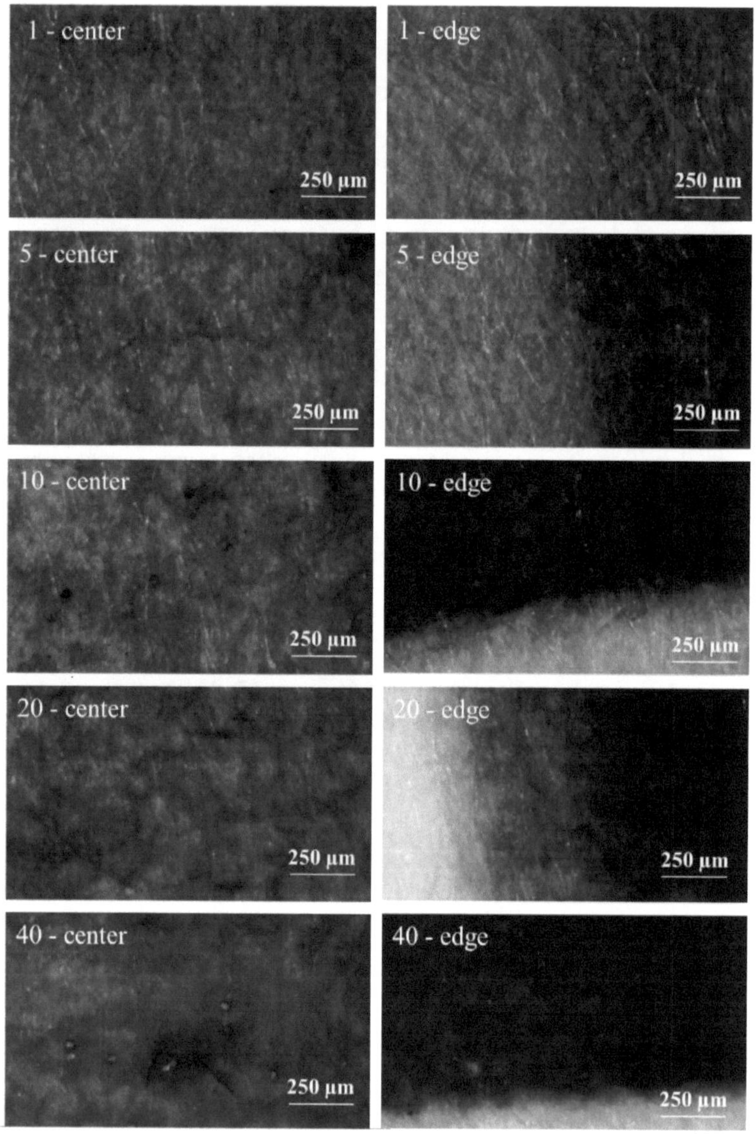

Figure 17. Optical microscope images of inkjet-printed MWCNT films on photo paper at 50 μm drop spacing and varying layer counts (1–40 as labeled in each panel), showing both center and edge regions of films.

Potential distributions and normalized electric field lines within the simulated MWCNT films with 10, 40, and 100 layers are shown in Figure 19. As expected, the profiles are symmetric about the x-axis diameter, and the potential along the y-axis diameter is halfway between the potentials at the working terminals. The effect of the finite sample size is clearly indicated by the electric field lines, which are nearly tangent to the circular sample boundary at radial positions near the edge of the sample. It is important to note that the similarity between the color distributions in the three cases suggests a linear current-voltage response at the terminals, which is necessary for the characterization of sheet resistance with 4-point probe. This agrees with the linearity of the governing equations and boundary conditions. Even though the potential can be quite high (on the order of ~5 V) in some areas of the low-layer count models when excited with 100 μA, which are not encountered in experimental 4-point probe measurements, the calculated film resistances are independent of the input current. Simulating the system with a reduced current of 1 μA results in reduced but nonetheless dimensionally similar potential distributions, as discussed further below.

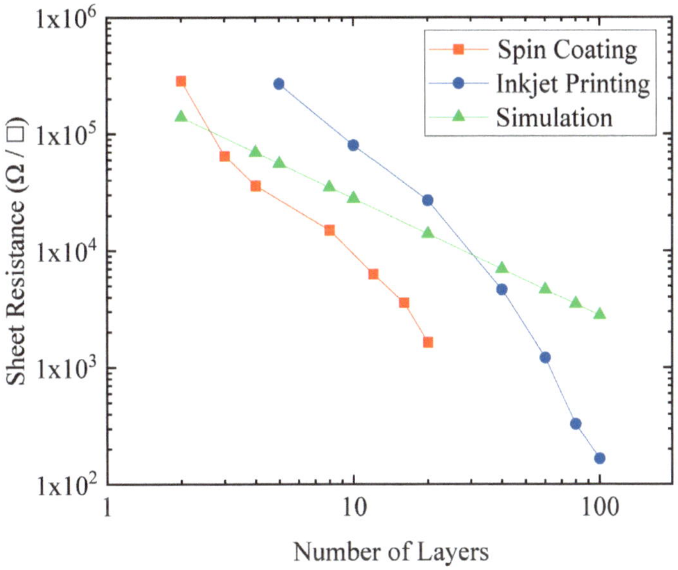

Figure 18. DC four-point probe measurements of sheet resistance for spin-coated and inkjet-printed MWCNT thin films, and the simulated sheet resistance.

Figure 20 shows the potentials along the diameter of the simulated film at a depth of half of the film thickness for films with 20, 40, 60, 80, and 100 layers when the input current was set to 1 μA. The boundaries of the plot represent the edges of the film. The working (1, 4) and sense (2, 3) terminals are labeled. The potential difference between the sense terminals is used to compute the sheet resistance of the simulated film, which were plotted alongside the experimental measurements in Figure 18. Because the resistivity of the film material is held constant, the simulated sheet resistance is inversely proportional to the thickness. Furthermore, because the thickness in the simulation is proportional to the number of layers, the sheet resistance is also inversely proportional to the layer count. This contrasts with the inverse-square relationship between sheet resistance and layer count for spin-coated and in inkjet-printed films. If the thicknesses of MWCNT films scale linearly with layer counts, which is applicable if layers adsorb independently and identically to the existing surface, then the resistivity of the material varies inversely with the layer count. Such a behavior is comparable to the variation of resistivity with thickness reported for thin films of metallic conductors [132,133]. However, the thicknesses of deposited films are not expected to scale exactly linearly with the number of layers because of the relatively low active material loading in the ink and a correspondingly low surface layer density of nanotubes. The reduced sheet resistance likely arises from an increased number of interlayer conductive pathways formed on the substrate. Film thicknesses will be measured in future work and will help elucidate the resistive and nature of charge-transfer in MWCNT thin films.t

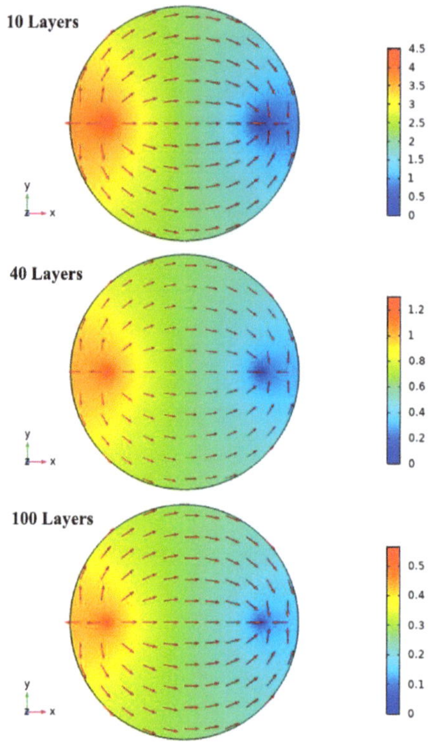

Figure 19. Potential and electric field color plots of simulated thin films.

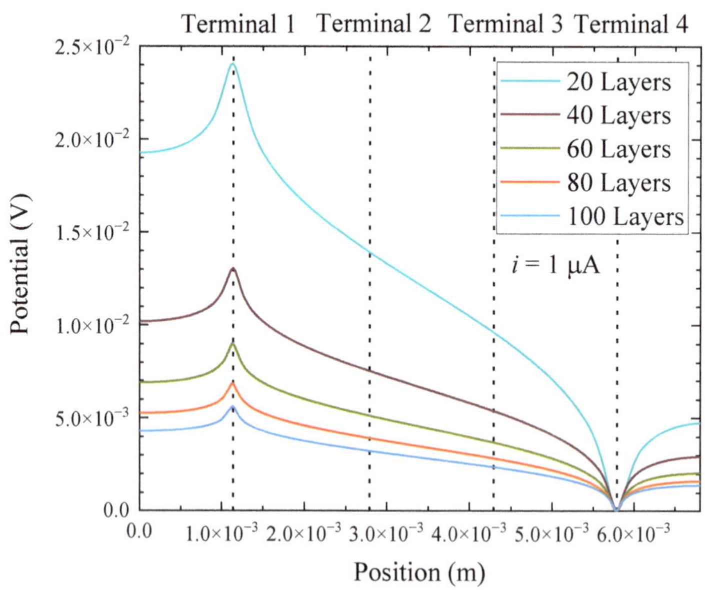

Figure 20. Electric potentials along the half-thickness of the simulated film. Terminal 1 is the current input. Terminal 4 is grounded. Terminals 2 and 3 are the sense terminals.

AC Characterization of MWCNT Films

Impedance spectroscopy data for the spin-coated and inkjet-printed MWCNT films (Figure 21) show that all films exhibit purely resistive behavior at low frequencies, which is indicated by a constant impedance magnitude and a phase angle of nearly 0, and a transition to a capacitive reactance beyond the corner frequency. The capacitive reactance is indicated by (1) the magnitude decreasing by one order of magnitude for each decade in frequency and (2) the phase angle approaching −90 degrees. Near the corner frequency, the impedance angles for all films become positive. This indicates that some inductive charge transfer effects are occurring within the film, which can likely be attributed to the tubular nanostructure of MWCNTs, resulting in coiled conductive pathways [59].

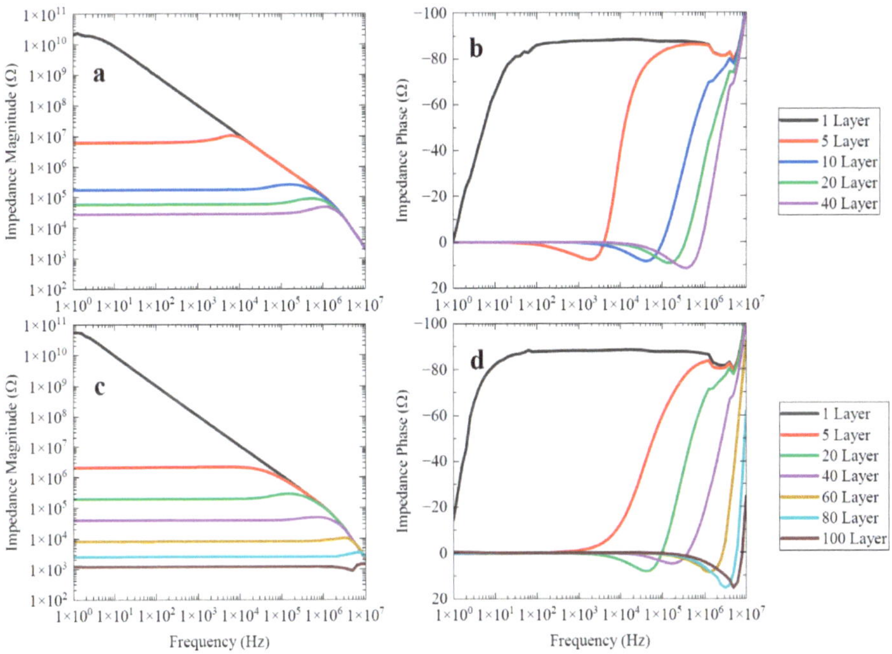

Figure 21. Electrical impedance spectroscopy for spin-coated samples (magnitude in (**a**) and phase in (**b**)) and inkjet-printed samples (magnitude in (**c**) and phase in (**d**)).

For both spin coating and inkjet printing, films with only one layer of deposition appeared to have insulating behavior. This is evidenced by the high impedance magnitude at 0.1 Hz (which approximates DC conditions) low corner frequency, and predominantly capacitive behavior across all measured frequencies. Films with only one layer do not contain enough conducting pathways to provide acceptable conductivity. Increasing the number of layers—depositions in the case of spin coating or number of overwritings in the case of inkjet printing—introduces additional conducting pathways via the random positioning and alignment of nanotubes across different layers throughout the film. When the number of layers becomes large, critical conductive networks have already been formed, but the film impedance continues to decrease because of the formation of additional conductive paths. The effect diminishes with increasing layers, which agrees well with the general theory of percolation. For a spin-coated film with 20 layers, the impedance at DC approaches 1.2×10^4 Ω. For an inkjet-printed film with 40 layers, the impedance at DC approaches 3.5×10^4 Ω. The spin-coated films exhibit lower impedances for the same number of layers compared to the inkjet-printed films, which agrees with the trends observed for the DC sheet resistances. This is because, the amount of material deposited for spin coating is significantly larger compared to inkjet printing. Furthermore, as discussed previously, agglomerates in spin-coated films are absent in inkjet-printed films, and agglomerates may contribute to enhanced in-plane conductivity.

Electrochemical Capacitor

To demonstrate one application of inkjet-printed MWCNT thin films, a simple capacitive device was fabricated using only two inkjet-printed electrodes and an electrolyte-soaked separator. A schematic and picture of the device are shown in Figure 22. Unlike typical capacitor electrodes with rigid metallic current collectors, the active material in the present work is deposited on photopaper, allowing for flexible operation. The active area with liquid electrolyte is clamped but not thoroughly sealed, as shown in Figure 22, and measurements are performed immediately after the device is constructed.

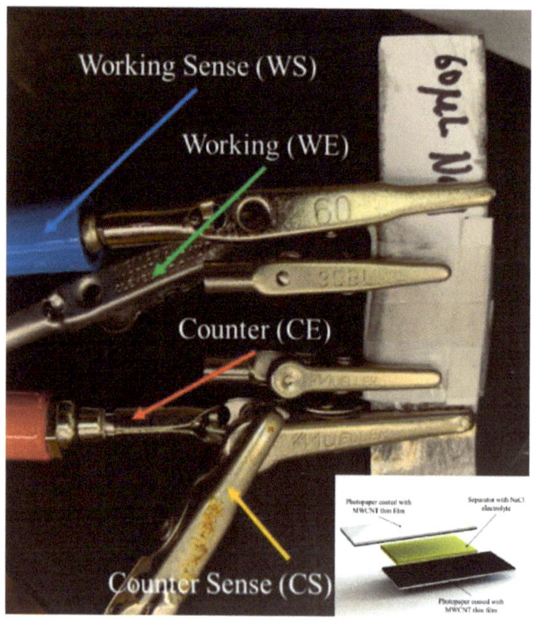

Figure 22. Photo and schematic of the capacitive device connected to the electrochemical measurement workstation. Thicknesses of layers in the schematic inset are exaggerated for visibility.

The cyclic voltammogram for the device obtained with only 10 inkjet-printed layers is shown in Figure 23. Compared to previous works utilizing carbon nanotubes as active material (Table 2), the devices fabricated in the present work utilize far less weight of active material due to the lower active material concentration in dispersion, the use of thin film deposition techniques, and low layer counts. For the 10-layer electrode used here, the active mass of each electrode is 1.6×10^{-5} g, compared to typical masses on the order of several milligrams for previous works shown in Table 2. Additionally, no additives are used in the synthesis of the electrode film, other than the SDBS surfactant during ink preparation. Due to low material utilization and high resistivity of electrical contacts, the specific capacitances obtained for the device (0.22 F/g) are significantly lower than electrodes fabricated using conventional techniques, such as pressing and lamination. The high resistance of the device is indicated by the sloped profile of the CV curve, which contrasts with the box-shaped profiles typical of electric double-layer supercapacitors [134] Nonetheless, the device demonstrates the application of conducting films in flexible charge storage. To enhance the electrochemical performance of inkjet-printed thin film devices, the contact between the active material and electrolyte can be improved, and substrates with improved electrical conductivity and external electrode contacts will be investigated in the future.

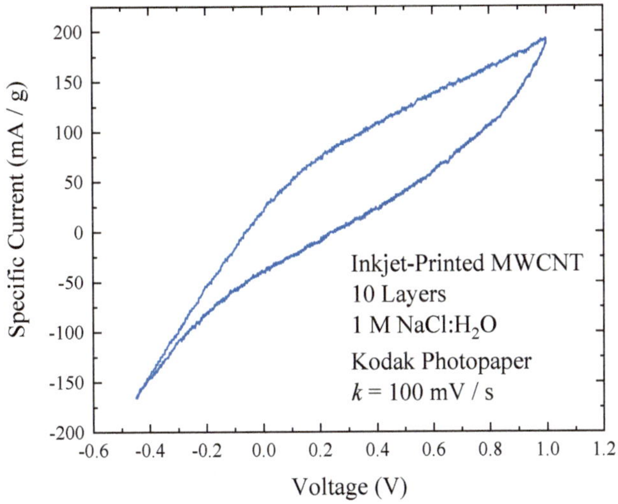

Figure 23. Cyclic voltammetry of the capacitive device built with inkjet-printed MWCNTs.

Table 2. Summary of previous investigations of carbon material-based electric double-layer supercapacitors.

Active Material	Additives	Deposition Process	Active Material Details	Substrate Material/Surface Treatments	Post-Proessing	Electrolyte	Specific Capacitance [F/g]	Ref.
MWCNT	SDBS (surf.)	Spin coating/Inkjet printing	Spin coating: 400 μg (25 μg/layer, 8 layers) Inkjet printing: 60 μg (6.0 μg/layer, 10 layers)	Photo paper/UV ozone	N/A	1 M NaCl in H_2O	0.02	Current Work
MWCNT (acet. decomp.)	PVDF (bind.); acetylene black (cond.)	Pressing	3.4–8.5 mg (85 wt% MWCNT, 4–10 mg total electrode mass)	Stainless steel	N/A	6 M KOH	4–135	[93]
MWCNT (CCVD)	PTFE (bind.); carbon black (cond.)	Lamination	51 mg (85 wt% active material, 15 mg/cm^2, 4 cm^2)	Aluminum/polyurethane paint (cond.)	N/A	1.5 M NEt_4BF_4 in acetonitrile	18	[71]
MWCNT	PVDF (bind.); acetylene black (cond.) Activation with KOH	Pressing	1.7–17 mg (85 wt% active material, 2–20 mg total)	Gold	N/A	1 M H_2SO_4	90	[94]
MWCNT	$Mg(NO_3)_2$ (EPD electrolyte)	Electrophoretic deposition	0.8 cm × 0.8 cm (electrode area)	Nickel foil	Ar/H_2 gas treatment	6 N KOH	21	[135]
MWCNT	N/A	Colloid deposition and evaporation	2.67 mg (26.7 mg/mL, 0.1 mL)	Nickel foil	Drying at room temperature	6 M KOH	20	[50]
MWCNT	Nitric acid treatment	Drying	25 μm (thickness), 800 mg/mL (MWCNT density)	N/A	Drying; thermally cross-linking	38 wt% H_2SO_4	102	[51]
MWCNT	Phenolic resin (PF) (bind.), nitric acid treatment	Molding	1 g/cm^3 (bulk density)	Graphite	Carbonization; Immersion in sulfuric/nitric acid	38 wt% H_2SO_4	15–25	[109]
MWCNT (aligned)	PPy (polymerization on MWCNTs)	Direct growth via pyrolysis of toluene and ferrocene	100 μm (aligned MWCNT length); 0.16 cm^2 (electrode area)	Gold-coated Mylar	N/A	0.5 M KCl	2.55 [F/cm^2]	[136]
MWCNT/PPy	PPy (polymerization on MWCNTs)	Direct growth with Ni catalyst	8–10 μm (aligned MWCNT length)	Titanium	Soaking in distilled water; bubbling in nitrogen (eliminate oxygen)	0.1 M $LiClO_4$	N/A	[137]
MWCNT (in-plane aligned)/PPy	PPy (polymerization on MWCNTs)	MWCNT solution as polymerization	0.2–2.5 μm (MWCNT length); 60 wt% MWCNTs	Graphite	N/A	0.5 M KCl	192	[138]
SWCNT	Sulfuric acid/nitric acid treatment	Dip coating	61.2 mg (by weighing)	Non-woven cotton paper	Drying at room temperature	PVA/phosphoric acid (solid)	116	[95]

Table 2. *Cont.*

Active Material	Additives	Deposition Process	Active Material Details	Substrate Material/Surface Treatments	Post-Proessing	Electrolyte	Specific Capacitance [F/g]	Ref.
SWCNT	SDBS (surf.)	Inkjet printing	300 µg (0.3 mg/cm^2, 1 cm^2)	Newspaper; Kodak photo paper/PVDF coating	N/A	1 M $LiPF_6$ in EC: DEC	3–33	[96]
SWCNT	PVDC (bind.); THF (solv.)	Pressing	150 µm (thickness of pellet), 750 mg/mL (SWCNT density)	Nickel foil/foam; Polishing	N/A	38 wt% H_2SO_4	180	[96]
SWCNT	SDS (surf.)	Suspension deposition and evaporation	141 ng (by QCM)	Platinum	Heating in oven	0.1 M $TBAPF_6$	283	[97]
SWCNT	N/A	Dip coating	0.85 mg/cm^2	NTP	N/A	3 M NaCl	37.3	[139]
SWCNT/PPy	PPy(polymerization on SWCNTs), PTFE (bind.), super P (cond.)	Pressing	250 µm (electrode thickness)	Nickel foam	Vacuum drying	7.5 M KOH	265	[140]
Activated Carbon (AC)/MWNT	PVDA or PVA (bind.)	Film application	120 µm (thickness of AC/MWNT)	Aluminum foil	N/A	1 M $TEABF_4$	120	[72]
Nanostructured Carbon	N/A	Cluster beam deposition	3.17 mg (1 g/cm^3 density, 6 cm × 6 cm × 0.8 µm)	Aluminum	N/A	Quartenary ammonium salt in PC	75	[98]
Graphene	N/A	Inkjet printing	890 µg (8.9 µg/layer, 100 layers)	Titanium foil	Reduction of GO in flowing N_2	1 M H_2SO_4	48–132	[99]
Graphene/MWNT	Functionalized with nitric/sulfuric acid	Spray coating of MWCNTs onto CVD-grown graphene	500 nm (thickness of MWNT film)	Polyethylene terephthalate (PET)	N/A	PVA-H_3PO_4 (gel)	2.54 [mF/cm^2]	[110]
Graphene/MWNT foam	Activation with KOH	Foam grown via one-step ambient pressure CVD		Nickel foam	Cooling to room temperature	6 M KOH	286	[141]
Graphene/PANI	SDBS (surf.)	Inkjet printing	1.62 mg (800 × 800 array, 13 µm separation, 1.5 mg/cm^2) 5 layers	Quartz; carbon fabric	N/A	1 M H_2SO_4	82	[100]

5. Summary and Outlook

In this review, the major synthesis techniques for generating MWCNTs and depositing MWCNT thin films were discussed. Due to their unique geometry and mechanical and electrical properties, multiwall carbon nanotubes exhibit excellent performance when used as active components in electrochemical energy storage. Spin coating and inkjet printing were investigated as techniques for MWCNT supercapacitor electrode fabrication. Electrical measurements revealed that nanotube films were conductive, and that the film conductivity increases with increasing material loading. Multiwall nanotube thin films were used as electrodes in symmetric double-layer flexible capacitors. The thin film electrodes exhibited acceptable charge storage, achieving similar performance as electrodes formed from other fabrication techniques, such as pressing and casting, which utilize a much higher amount of active material as well as additives. This suggests that spin coating and inkjet printing are viable techniques for producing high-performance flexible capacitors for small-scale energy storage. To further improve the performance of carbon nanotubes in electrochemical capacitors, different electrolyte systems with higher working potential ranges, as well as different substrate materials and device configurations, will be further investigated in the future. The quality of the ink dispersion, which greatly influences the quality of deposited films, can be improved by using different surfactants and mixing techniques. Fine control over thin film properties enabled by low-volume deposition techniques will become increasingly important as devices are diversified and miniaturized.

Funding: This research received no external funding.

Data Availability Statement: Data available by sending an email to Gerhardt at rosario.gerhardt @mse.gatech.edu.

Acknowledgments: This work was conducted by K.J. as an undergraduate research project at the Georgia Institute of Technology. R.A.G. acknowledges use of the Goizueta Foundation Faculty Chair to cover the costs of accessing Institute for Electronics and Nanotechnology (IEN) facilities to conduct the inkjet printing experiments as well as the able assistance from David Gottfried. The authors would like to thank Siva Sethuraman for preparation and microscope images of the high-layer count inkjet-printed films. Morgan Watt is acknowledged for acquiring the TEM images of the multiwall carbon nanotube precursor material.

Conflicts of Interest: The authors declare no conflict of interest.

References

1. Iijima, S. Helical microtubules of graphitic carbon. *Nature* **1991**, *354*, 56–58. [CrossRef]
2. Moisala, A.; Nasibulin, A.G.; Kauppinen, E.I. The role of metal nanoparticles in the catalytic production of single-walled carbon nanotubes—A review. *J. Phys. Condens. Matter* **2003**, *15*, S3011–S3035. [CrossRef]
3. Siegal, M.P.; Overmyer, D.L.; Provencio, P.P. Precise control of multiwall carbon nanotube diameters using thermal chemical vapor deposition. *Appl. Phys. Lett.* **2002**, *80*, 2171–2173. [CrossRef]
4. Terrones, M. Carbon nanotubes: Synthesis and properties, electronic devices and other emerging applications. *Int. Mater. Rev.* **2004**, *49*, 325–377. [CrossRef]
5. Jilani, A.; Abdel-Wahab, M.S.; Hammad, A.H. Advance Deposition Techniques for Thin Film and Coating. In *Modern Technologies for Creating the Thin-film Systems and Coatings*; IntechOpen: London, UK, 2017. [CrossRef]
6. Dincer, I. Renewable energy and sustainable development: A crucial review. *Renew. Sustain. Energy Rev.* **2000**, *4*, 157–175. [CrossRef]
7. Chow, J.; Kopp, R.J.; Portney, P.R. Energy Resources and Global Development. *Science* **2003**, *302*, 1528–1531. [CrossRef]
8. Lehtola, T.; Zahedi, A. Solar energy and wind power supply supported by storage technology: A review. *Sustain. Energy Technol. Assess.* **2019**, *35*, 25–31. [CrossRef]
9. Hall, P.J.; Bain, E. Energy-storage technologies and electricity generation. *Energy Policy* **2008**, *36*, 4352–4355. [CrossRef]
10. Miller, J.R.; Simon, P. MATERIALS SCIENCE: Electrochemical Capacitors for Energy Management. *Science* **2008**, *321*, 651–652. [CrossRef]
11. Qu, D. Studies of the activated carbons used in double-layer supercapacitors. *J. Power Sources* **2002**, *109*, 403–411. [CrossRef]
12. Baddour, C.E.; Briens, C. Carbon Nanotube Synthesis: A Review. *Int. J. Chem. React. Eng.* **2005**, *3*. [CrossRef]
13. Muhlbauer, R.L.; Gerhardt, R.A. A review on the synthesis of carbon nanotube thin films. In *Carbon Nanotubes: Synthesis and Properties*; Nova Science Publishers: New York, NY, USA, 2012; pp. 107–156.

14. Rafique, M.M.A.; Iqbal, J. Production of Carbon Nanotubes by Different Routes-A Review. *J. Encapsulation Adsorpt. Sci.* **2011**, *1*, 29–34. [CrossRef]
15. Arora, N.; Sharma, N. Arc discharge synthesis of carbon nanotubes: Comprehensive review. *Diam. Relat. Mater.* **2014**, *50*, 135–150. [CrossRef]
16. Cadek, M.; Murphy, R.; McCarthy, B.; Drury, A.; Lahr, B.; Barklie, R.; Panhuis, M.I.H.; Coleman, J.; Blau, W. Optimisation of the arc-discharge production of multi-walled carbon nanotubes. *Carbon* **2002**, *40*, 923–928. [CrossRef]
17. Zhao, T. Gas and pressure effects on the synthesis of amorphous carbon nanotubes. *Chin. Sci. Bull.* **2004**, *49*, 2569–2571. [CrossRef]
18. Kim, H.H.; Kim, H.J. Preparation of carbon nanotubes by DC arc discharge process under reduced pressure in an air atmosphere. *Mater. Sci. Eng. B* **2006**, *133*, 241–244. [CrossRef]
19. Bagiante, S.; Scalese, S.; Scuderi, V.; D'Urso, L.; Messina, E.; Compagnini, G.; Privitera, V. Role of the growth parameters on the structural order of MWCNTs produced by arc discharge in liquid nitrogen. *Phys. Status Solidi (B)* **2010**, *247*, 884–887. [CrossRef]
20. Wienecke, M.; Bunescu, M.-C.; Deistung, K.; Fedtke, P.; Borchartd, E. MWCNT coatings obtained by thermal CVD using ethanol decomposition. *Carbon* **2006**, *44*, 718–723. [CrossRef]
21. Rahmandoust, M.; Öchsner, A. Buckling Behaviour and Natural Frequency of Zigzag and Armchair Single-Walled Carbon Nanotubes. *J. Nano Res.* **2012**, *16*, 153–160. [CrossRef]
22. Coleman, J.N.; Khan, U.; Blau, W.J.; Gun'Ko, Y. Small but strong: A review of the mechanical properties of carbon nanotube–polymer composites. *Carbon* **2006**, *44*, 1624–1652. [CrossRef]
23. Treacy, M.M.J.; Ebbesen, T.W.; Gibson, J.M. Exceptionally high Young's modulus observed for individual carbon nanotubes. *Nature* **1996**, *381*, 678–680. [CrossRef]
24. Yu, M.-F.; Lourie, O.; Dyer, M.J.; Moloni, K.; Kelly, T.F.; Ruoff, R.S. Strength and Breaking Mechanism of Multiwalled Carbon Nanotubes Under Tensile Load. *Science* **2000**, *287*, 637–640. [CrossRef] [PubMed]
25. Sammalkorpi, M.; Krasheninnikov, A.; Kuronen, A.; Nordlund, K.; Kaski, K. Mechanical properties of carbon nanotubes with vacancies and related defects. *Phys. Rev. B* **2004**, *70*, 245416. [CrossRef]
26. Mielke, S.L.; Troya, D.; Zhang, S.; Li, J.-L.; Xiao, S.; Car, R.; Ruoff, R.S.; Schatz, G.C.; Belytschko, T. The role of vacancy defects and holes in the fracture of carbon nanotubes. *Chem. Phys. Lett.* **2004**, *390*, 413–420. [CrossRef]
27. Salvetat, J.-P.; Bonard, J.-M.; Thomson, N.; Kulik, A.; Forró, L.; Benoit, W.; Zuppiroli, L. Mechanical properties of carbon nanotubes. *Appl. Phys. A* **1999**, *69*, 255–260. [CrossRef]
28. Takakura, A.; Beppu, K.; Nishihara, T.; Fukui, A.; Kozeki, T.; Namazu, T.; Miyauchi, Y.; Itami, K. Strength of carbon nanotubes depends on their chemical structures. *Nat. Commun.* **2019**, *10*, 1–7. [CrossRef] [PubMed]
29. Charlier, J.-C.; Issi, J.-P. Electronic structure and quantum transport in carbon nanotubes. *Appl. Phys. A* **1998**, *67*, 79–87. [CrossRef]
30. Harutyunyan, A.R.; Chen, G.; Paroyan, T.M.; Pigos, E.M.; Kuznetsov, O.A.; Hewparakrama, K.; Kim, S.M.; Zakharov, D.; Stach, E.A.; Sumanasekera, G.U. Preferential Growth of Single-Walled Carbon Nanotubes with Metallic Conductivity. *Science* **2009**, *36*, 116–120. [CrossRef]
31. Ding, L.; Tselev, A.; Wang, J.; Yuan, D.; Chu, H.; McNicholas, T.P.; Li, Y.; Liu, J. Selective Growth of Well-Aligned Semiconducting Single-Walled Carbon Nanotubes. *Nano Lett.* **2009**, *9*, 800–805. [CrossRef]
32. Langer, L.; Stockman, L.; Heremans, J.P.; Bayot, V.; Olk, C.H.; Van Haesendonck, C.; Bruynseraede, Y.; Issi, J.-P. Electrical resistance of a carbon nanotube bundle. *J. Mater. Res.* **1994**, *9*, 927–932. [CrossRef]
33. Cole, M.; Hiralal, P.; Ying, K.; Li, C.; Zhang, Y.; Teo, K.; Ferrari, A.; Milne, W. Dry-Transfer of Aligned Multiwalled Carbon Nanotubes for Flexible Transparent Thin Films. *J. Nanomater.* **2012**, *2012*, 1–8. [CrossRef]
34. Honda, Y.; Haramoto, T.; Takeshige, M.; Shiozaki, H.; Kitamura, T.; Ishikawa, M. Aligned MWCNT Sheet Electrodes Prepared by Transfer Methodology Providing High-Power Capacitor Performance. *Electrochem. Solid-State Lett.* **2007**, *10*, A106–A110. [CrossRef]
35. Shin, J.-H.; Shin, D.-W.; Patole, S.P.; Lee, J.H.; Park, S.M.; Yoo, J.B. Smooth, transparent, conducting and flexible SWCNT films by filtration–wet transfer processes. *J. Phys. D Appl. Phys.* **2009**, *42*, 045305. [CrossRef]
36. Brinker, C.; Frye, G.; Hurd, A.; Ashley, C. Fundamentals of sol-gel dip coating. *Thin Solid Films* **1991**, *201*, 97–108. [CrossRef]
37. Scriven, L.E. Physics and Applications of DIP Coating and Spin Coating. *MRS Proc.* **1988**, *121*, 717. [CrossRef]
38. Sahu, N.; Parija, B.; Panigrahi, S. Fundamental understanding and modeling of spin coating process: A review. *Indian J. Phys.* **2009**, *83*, 493–502. [CrossRef]
39. Derby, B. Inkjet Printing of Functional and Structural Materials: Fluid Property Requirements, Feature Stability, and Resolution. *Annu. Rev. Mater. Res.* **2010**, *40*, 395–414. [CrossRef]
40. MacLeod, D. Wire-Wound Rod Coating. In *Coatings Technology Handbook*, 3rd ed.; Taylor and Francis: Boca Raton, FL, USA, 2006.
41. Mistler, R.; Twiname, E.R. The Tape Casting Process. In *Tape Casting: Theory and Practice*; The American Ceramic Society: Westerville, OH, USA, 2000; pp. 83–186.
42. Shin, P.; Sung, J.; Lee, M.H. Control of droplet formation for low viscosity fluid by double waveforms applied to a piezoelectric inkjet nozzle. *Microelectron. Reliab.* **2011**, *51*, 797–804. [CrossRef]
43. Sánchez, M.; Rincón, M. Sensor response of sol–gel multiwalled carbon nanotubes-TiO2 composites deposited by screen-printing and dip-coating techniques. *Sens. Actuators B Chem.* **2009**, *140*, 17–23. [CrossRef]
44. Choi, J.; Park, E.J.; Park, D.W.; Shim, S.E. MWCNT–OH adsorbed electrospun nylon 6,6 nanofibers chemiresistor and their application in low molecular weight alcohol vapours sensing. *Synth. Met.* **2010**, *160*, 2664–2669. [CrossRef]

45. Bakhsheshi-Rad, H.; Hamzah, E.; Dias, G.J.; Saud, S.; Yaghoubidoust, F.; Hadisi, Z. Fabrication and characterisation of novel ZnO/MWCNT duplex coating deposited on Mg alloy by PVD coupled with dip-coating techniques. *J. Alloys Compd.* **2017**, *728*, 159–168. [CrossRef]
46. Abrishamchian, A.; Hooshmand, T.; Mohammadi, M.; Najafi, F. Preparation and characterization of multi-walled carbon nanotube/hydroxyapatite nanocomposite film dip coated on Ti–6Al–4V by sol–gel method for biomedical applications: An in vitro study. *Mater. Sci. Eng. C* **2013**, *33*, 2002–2010. [CrossRef] [PubMed]
47. Raut, S.S.; Sankapal, B.R. Comparative studies on MWCNTs, Fe2O3 and Fe2O3/MWCNTs thin films towards supercapacitor application. *New J. Chem.* **2016**, *40*, 2619–2627. [CrossRef]
48. Karade, S.S.; Sankapal, B.R. Room temperature PEDOT: PSS encapsulated MWCNTs thin film for electrochemical supercapacitor. *J. Electroanal. Chem.* **2016**, *771*, 80–86. [CrossRef]
49. Kalaivani, A.; Narayanan, S.S. Cadmium Selenide Quantum Dots—MWCNTs Nanocomposite Modified Electrode for the Determination of Epinephrine. *Adv. Mater. Res.* **2014**, *938*, 176–181. [CrossRef]
50. Du, C.; Yeh, J.; Pan, N. High power density supercapacitors using locally aligned carbon nanotube electrodes. *Nanotechnology* **2005**, *16*, 350–353. [CrossRef]
51. Niu, C.; Sichel, E.K.; Hoch, R.; Moy, D.; Tennent, H. Supercapacitors using carbon nanotubes films by electrophoretic deposition. *Appl. Phys. Lett.* **2006**, *160*, 1480–1482. [CrossRef]
52. Sanli, A.; Benchirouf, A.; Ramalingame, R.; Palaniyappan, S.; Sharma, R.; Muller, C.; Kanoun, O. Characterization of the dielectric properties of multiwalled carbon nanotubes (MWCNTs)/PEDOT:PSS nanocomposites. In Proceedings of the 2014 IEEE 11th International Multi-Conference on Systems, Signals Devices (SSD14), Barcelona, Spain, 11–14 February 2014; pp. 1–5. [CrossRef]
53. Wu, B.; Li, X.; An, D.; Zhao, S.; Wang, Y. Electro-casting aligned MWCNTs/polystyrene composite membranes for enhanced gas separation performance. *J. Membr. Sci.* **2014**, *462*, 62–68. [CrossRef]
54. Peng, S.; Wu, Y.; Zhu, P.; Thavasi, V.; Mhaisalkar, S.G.; Ramakrishna, S. Facile fabrication of polypyrrole/functionalized multiwalled carbon nanotubes composite as counter electrodes in low-cost dye-sensitized solar cells. *J. Photochem. Photobiol. A Chem.* **2011**, *223*, 97–102. [CrossRef]
55. Munkhbayar, B.; Hwang, S.; Kim, J.; Bae, K.; Ji, M.; Chung, H.; Jeong, H. Photovoltaic performance of dye-sensitized solar cells with various MWCNT counter electrode structures produced by different coating methods. *Electrochim. Acta* **2012**, *80*, 100–107. [CrossRef]
56. Chen, H.; Wang, J.; Jia, C.; Mou, J.; Zhu, L. Highly efficient dye-sensitized solar cell with a novel nanohybrid film of Cu2ZnSnS4-MWCNTs as counter electrode. *Appl. Surf. Sci.* **2017**, *422*, 591–596. [CrossRef]
57. Zhou, J.; Lubineau, G. Improving Electrical Conductivity in Polycarbonate Nanocomposites Using Highly Conductive PEDOT/PSS Coated MWCNTs. *ACS Appl. Mater. Interfaces* **2013**, *5*, 6189–6200. [CrossRef] [PubMed]
58. Kordas, K.; Mustonen, T.; Toth, G.; Jantunen, H.; Lajunen, M.; Soldano, C.; Talapatra, S.; Kar, S.; Vajtai, R.; Ajayan, P.M. Inkjet Printing of Electrically Conductive Patterns of Carbon Nanotubes. *Small* **2006**, *2*, 1021–1025. [CrossRef] [PubMed]
59. Song, J.-W.; Kim, J.; Yoon, Y.-H.; Choi, B.-S.; Kim, J.-H.; Han, C.-S. Inkjet printing of single-walled carbon nanotubes and electrical characterization of the line pattern. *Nanotechnology* **2008**, *19*, 095702. [CrossRef] [PubMed]
60. Small, W.R.; Panhuis, M.I.H. Inkjet Printing of Transparent, Electrically Conducting Single-Walled Carbon-Nanotube Composites. *Small* **2007**, *3*, 1500–1503. [CrossRef]
61. Muhlbauer, R.L.; Joshi, S.; Gerhardt, R. The effect of substrate pore size on the network interconnectivity and electrical properties of dropcasted multiwalled carbon nanotube thin films. *J. Mater. Res.* **2013**, *28*, 1617–1624. [CrossRef]
62. Okimoto, H.; Takenobu, T.; Yanagi, K.; Miyata, Y.; Shimotani, H.; Kataura, H.; Iwasa, Y. Tunable Carbon Nanotube Thin-Film Transistors Produced Exclusively via Inkjet Printing. *Adv. Mater.* **2010**, *22*, 3981–3986. [CrossRef]
63. Mustonen, T.; Mäklin, J.; Kordas, K.; Halonen, N.; Toth, G.; Saukko, S.; Vähäkangas, J.; Jantunen, H.; Kar, S.; Ajayan, P.M.; et al. Controlled Ohmic and nonlinear electrical transport in inkjet-printed single-wall carbon nanotube films. *Phys. Rev. B* **2008**, *77*. [CrossRef]
64. Kim, B.; Jang, S.; Geier, M.L.; Prabhumirashi, P.L.; Hersam, M.C.; Dodabalapur, A. High-Speed, Inkjet-Printed Carbon Nanotube/Zinc Tin Oxide Hybrid Complementary Ring Oscillators. *Nano Lett.* **2014**, *14*, 3683–3687. [CrossRef]
65. Ha, M.; Xia, Y.; Green, A.; Zhang, W.; Renn, M.J.; Kim, C.H.; Hersam, M.C.; Frisbie, C.D. Printed, Sub-3V Digital Circuits on Plastic from Aqueous Carbon Nanotube Inks. *ACS Nano* **2010**, *4*, 4388–4395. [CrossRef]
66. Khan, S.; Dahiya, R.; Tinku, S.; Lorenzelli, L. Conformable tactile sensing using screen printed P(VDF-TrFE) and MWCNT-PDMS composites. *Sensors* **2014**, 862–865. [CrossRef]
67. Khan, S.; Lorenzelli, L.; Dahiya, R.S. Bendable piezoresistive sensors by screen printing MWCNT/PDMS composites on flexible substrates. In Proceedings of the 2014 10th Conference on Ph.D. Research in Microelectronics and Electronics (PRIME), Grenoble, France, 30 June–3 July 2014; pp. 1–4. [CrossRef]
68. Khan, S.; Dang, W.; Lorenzelli, L.; Dahiya, R. Flexible Pressure Sensors Based on Screen-Printed P(VDF-TrFE) and P(VDF-TrFE)/MWCNTs. *IEEE Trans. Semicond. Manuf.* **2015**, *28*, 486–493. [CrossRef]
69. Viswanathan, S.; Rani, C.; Anand, A.V.; Ho, J.-A.A. Disposable electrochemical immunosensor for carcinoembryonic antigen using ferrocene liposomes and MWCNT screen-printed electrode. *Biosens. Bioelectron.* **2009**, *24*, 1984–1989. [CrossRef]
70. Viswanathan, S.; Rani, C.; Ho, A. Electrochemical immunosensor for multiplexed detection of food-borne pathogens using nanocrystal bioconjugates and MWCNT screen-printed electrode. *Talanta* **2012**, *94*, 315–319. [CrossRef] [PubMed]

71. Portet, C.; Yushin, G.; Gogotsi, Y. Electrochemical performance of carbon onions, nanodiamonds, carbon black and multiwalled nanotubes in electrical double layer capacitors. *Carbon* **2007**, *45*, 2511–2518. [CrossRef]
72. Markoulidis, F.; Lei, C.; Lekakou, C.; Figgemeier, E.; Duff, D.; Khalil, S.; Martorana, B.; Cannavaro, I. High-performance Supercapacitor cells with Activated Carbon/MWNT nanocomposite electrodes. *IOP Conf. Ser. Mater. Sci. Eng.* **2012**, *40*, 012021. [CrossRef]
73. Ma, Y.J.; Yao, X.F.; Wang, D. Experimental investigation on mechanical properties of CNT film using digital speckle correlation method. *Opt. Lasers Eng.* **2012**, *50*, 1575–1581. [CrossRef]
74. Xu, X.; Thwe, M.M.; Shearwood, C.; Liao, K. Mechanical properties and interfacial characteristics of carbon-nanotube-reinforced epoxy thin films. *Appl. Phys. Lett.* **2002**, *81*, 2833–2835. [CrossRef]
75. Koval'Chuk, A.A.; Shevchenko, V.G.; Shchegolikhin, A.N.; Nedorezova, P.M.; Klyamkina, A.N.; Aladyshev, A.M. Effect of Carbon Nanotube Functionalization on the Structural and Mechanical Properties of Polypropylene/MWCNT Composites. *Macromolecules* **2008**, *41*, 7536–7542. [CrossRef]
76. Sun, F.; Cha, H.-R.; Bae, K.; Hong, S.; Kim, J.-M.; Kim, S.H.; Lee, J.; Lee, D. Mechanical properties of multilayered chitosan/CNT nanocomposite films. *Mater. Sci. Eng. A* **2011**, *528*, 6636–6641. [CrossRef]
77. Lau, C.H.; Cervini, R.; Clarke, S.R.; Markovic, M.G.; Matisons, J.G.; Hawkins, S.C.; Huynh, C.P.; Simon, G.P. The effect of functionalization on structure and electrical conductivity of multi-walled carbon nanotubes. *J. Nanoparticle Res.* **2008**, *10*, 77–88. [CrossRef]
78. Chen, Y.; Kang, K.; Han, K.; Yoo, K.; Kim, J. Enhanced optical and electrical properties of PEDOT: PSS films by the addition of MWCNT-sorbitol. *Synth. Met.* **2009**, *159*, 1701–1704. [CrossRef]
79. Ram, R.; Rahaman, M.; Khastgir, D. Electrical properties of polyvinylidene fluoride (PVDF)/multi-walled carbon nanotube (MWCNT) semi-transparent composites: Modelling of DC conductivity. *Compos. Part A Appl. Sci. Manuf.* **2015**, *69*, 30–39. [CrossRef]
80. Chen, M.; Tao, T.; Zhang, L.; Gao, W.; Li, C. Highly conductive and stretchable polymer composites based on graphene/MWCNT network. *Chem. Commun.* **2013**, *49*, 1612–1614. [CrossRef]
81. Kim, K.-S.; Park, S.-J. Effect of silver doped MWCNTs on the electrical properties of conductive MWCNTs/PMMA thin films. *Synth. Met.* **2010**, *160*, 123–126. [CrossRef]
82. Fernandez-d'Arlas, B.; Khan, U.; Rueda, L.; Martin, L.; Ramos, J.A.; Coleman, J.N.; Gozalez, M.L.; Valea, A.; Mondragon, I.; Corcuera, M.A.; et al. Study of the mechanical, electrical and morphological properties of PU/MWCNT composites obtained by two different processing routes. *Compos. Sci. Technol.* **2012**, *72*, 235–242. [CrossRef]
83. Staudinger, U.; Thoma, P.; Lüttich, F.; Janke, A.; Kobsch, O.; Gordan, O.; Pötschke, P.; Voit, B.; Zahn, D.R. Properties of thin layers of electrically conductive polymer/MWCNT composites prepared by spray coating. *Compos. Sci. Technol.* **2017**, *138*, 134–143. [CrossRef]
84. Jung, D.; Kim, D.; Lee, K.H.; Overzet, L.J.; Lee, G.S. Transparent film heaters using multi-walled carbon nanotube sheets. *Sens. Actuators A Phys.* **2013**, *199*, 176–180. [CrossRef]
85. Saw, L.; Mariatti, M.; Azura, A.; Azizan, A.; Kim, J. Transparent, electrically conductive, and flexible films made from multiwalled carbon nanotube/epoxy composites. *Compos. Part B Eng.* **2012**, *43*, 2973–2979. [CrossRef]
86. Li, J.; Liu, J.; Gao, C.; Zhang, J.; Sun, H. Influence of MWCNTs Doping on the Structure and Properties of PEDOT:PSS Films. *Int. J. Photoenergy* **2009**, *2009*, 1–5. [CrossRef]
87. Hone, J.; Llaguno, M.; Nemes, N.M.; Johnson, A.T.; Fischer, J.E.; Walters, D.A.; Casavant, M.J.; Schmidt, J.; Smalley, R.E. Electrical and thermal transport properties of magnetically aligned single wall carbon nanotube films. *Appl. Phys. Lett.* **2000**, *77*, 666–668. [CrossRef]
88. Yang, K.; Gu, M.; Guo, Y.; Pan, X.; Mu, G. Effects of carbon nanotube functionalization on the mechanical and thermal properties of epoxy composites. *Carbon* **2009**, *47*, 1723–1737. [CrossRef]
89. Pradhan, N.R.; Duan, H.; Liang, J.; Iannacchione, G. The specific heat and effective thermal conductivity of composites containing single-wall and multi-wall carbon nanotubes. *Nanotechnology* **2009**, *20*, 245705. [CrossRef] [PubMed]
90. Antiohos, D.; Romano, M.; Chen, J.; Razal, J.M. Carbon nanotubes for energy applications. In *Syntheses and Applications of Carbon Nanotubes and Their Composites*; IntechOpen: London, UK, 2013. [CrossRef]
91. Bu, F.; Zhou, W.; Xu, Y.; Du, Y.; Guan, C.; Huang, W. Recent developments of advanced micro-supercapacitors: Design, fabrication and applications. *NPJ Flex. Electron.* **2020**, *4*, 1–16. [CrossRef]
92. Jurewicz, K.; Babeł, K.; Pietrzak, R.; Delpeux, S.; Wachowska, H. Capacitance properties of multi-walled carbon nanotubes modified by activation and ammoxidation. *Carbon* **2006**, *44*, 2368–2375. [CrossRef]
93. Frackowiak, E.; Metenier, K.; Bertagna, V.; Beguin, F. Supercapacitor electrodes from multiwalled carbon nanotubes. *Appl. Phys. Lett.* **2000**, *77*, 2421–2423. [CrossRef]
94. Frackowiak, E.; Jurewicz, K.; Szostak, K.; Delpeux, S.; Béguin, F. Nanotubular materials as electrodes for supercapacitors. *Fuel Process. Technol.* **2002**, *77–78*, 213–219. [CrossRef]
95. Hu, S.; Rajamani, R.; Yu, X. Flexible solid-state paper based carbon nanotube supercapacitor. *Appl. Phys. Lett.* **2012**, *100*, 104103. [CrossRef]
96. Hu, L.; Wu, H.; Cui, Y. Printed energy storage devices by integration of electrodes and separators into single sheets of paper. *Appl. Phys. Lett.* **2010**, *96*, 183502. [CrossRef]

97. Liu, C.-Y.; Bard, A.J.; Wudl, F.; Weitz, I.; Heath, J.R. Electrochemical Characterization of Films of Single-Walled Carbon Nanotubes and Their Possible Application in Supercapacitors. *Electrochem. Solid-State Lett.* **1999**, *2*, 577–578. [CrossRef]
98. Diederich, L.; Barborini, E.; Piseri, P.; Podestà, A.; Milani, P.; Schneuwly, A.; Gallay, R. Supercapacitors based on nanostructured carbon electrodes grown by cluster-beam deposition. *Appl. Phys. Lett.* **1999**, *75*, 2662–2664. [CrossRef]
99. Le, L.T.; Ervin, M.H.; Qiu, H.; Fuchs, B.E.; Lee, W.Y. Graphene supercapacitor electrodes fabricated by inkjet printing and thermal reduction of graphene oxide. *Electrochem. Commun.* **2011**, *13*, 355–358. [CrossRef]
100. Xu, Y.; Hennig, I.; Freyberg, D.; Strudwick, A.J.; Schwab, M.G.; Weitz, T.; Cha, K.C.-P. Inkjet-printed energy storage device using graphene/polyaniline inks. *J. Power Sources* **2014**, *248*, 483–488. [CrossRef]
101. Sharma, P.; Bhatti, T. A review on electrochemical double-layer capacitors. *Energy Convers. Manag.* **2010**, *51*, 2901–2912. [CrossRef]
102. Libich, J.; Máca, J.; Vondrák, J.; Čech, O.; Sedlaříková, M. Supercapacitors: Properties and applications. *J. Energy Storage* **2018**, *17*, 224–227. [CrossRef]
103. Kisza, A. The capacitance of the diffuse layer of electric double layer of electrodes in molten salts. *Electrochim. Acta* **2005**, *51*, 2315–2321. [CrossRef]
104. An, K.H.; Kim, W.S.; Park, Y.S.; Choi, Y.C.; Lee, S.M.; Chung, D.C.; Bae, D.J.; Lim, S.C.; Lee, Y.H. Supercapacitors Using Single-Walled Carbon Nanotube Electrodes. *Adv. Mater.* **2001**, *13*, 497–500. [CrossRef]
105. Li, C.; Wang, D.; Liang, T.; Wang, X.; Ji, L. A study of activated carbon nanotubes as double-layer capacitors electrode materials. *Mater. Lett.* **2004**, *58*, 3774–3777. [CrossRef]
106. Chen, J.; Li, W.; Wang, D.; Yang, S.; Wen, J.; Ren, Z. Electrochemical characterization of carbon nanotubes as electrode in electrochemical double-layer capacitors. *Carbon* **2002**, *40*, 1193–1197. [CrossRef]
107. Markoulidis, F.; Lei, C.; Lekakou, C.; Duff, D.; Khalil, S.; Martorana, B.; Cannavaro, I. A method to increase the energy density of supercapacitor cells by the addition of multiwall carbon nanotubes into activated carbon electrodes. *Carbon* **2014**, *68*, 58–66. [CrossRef]
108. Conway, B.; Birss, V.; Wojtowicz, J. The role and utilization of pseudocapacitance for energy storage by supercapacitors. *J. Power Sources* **1997**, *66*, 1–14. [CrossRef]
109. Frackowiak, E.; Jurewicz, K.; Delpeux, S.; Beguin, F. Nanotubular materials for supercapacitors. *J. Power Sources* **2001**, *97–98*, 822–825. [CrossRef]
110. Kong, L.-B.; Zhang, J.; An, J.-J.; Luo, Y.-C.; Kang, L. MWNTs/PANI composite materials prepared by in-situ chemical oxidative polymerization for supercapacitor electrode. *J. Mater. Sci.* **2008**, *43*, 3664–3669. [CrossRef]
111. Gao, B.; Hao, L.; Fu, Q.; Su, L.; Yuan, C.; Zhang, X. Hydrothermal synthesis and electrochemical capacitance of $RuO_2 \cdot xH_2O$ loaded on benzenesulfonic functionalized MWCNTs. *Electrochim. Acta* **2010**, *55*, 3681–3686. [CrossRef]
112. Rakhi, R.; Alshareef, H. Enhancement of the energy storage properties of supercapacitors using graphene nanosheets dispersed with metal oxide-loaded carbon nanotubes. *J. Power Sources* **2011**, *196*, 8858–8865. [CrossRef]
113. Zhi, M.; Xiang, C.; Li, J.; Li, M.; Wu, N. Nanostructured carbon–metal oxide composite electrodes for supercapacitors: A review. *Nanoscale* **2013**, *5*, 72–88. [CrossRef]
114. Terasawa, N.; Mukai, K.; Yamato, K.; Asaka, K. Superior performance of non-activated multi-walled carbon nanotube polymer actuator containing ruthenium oxide over a single-walled carbon nanotube. *Carbon* **2012**, *50*, 1888–1896. [CrossRef]
115. Lewandowski, A.; Galiński, M. Practical and theoretical limits for electrochemical double-layer capacitors. *J. Power Sources* **2007**, *173*, 822–828. [CrossRef]
116. Du, C.; Yeh, J.; Pan, N. High power density supercapacitors electrodes of carbon nanotube films by electrophoretic deposition. *Nanotechnology* **2006**, *17*, 5314–5318. [CrossRef]
117. Ma, R.; Liang, J.; Wei, B.; Zhang, B.; Xu, C.; Wu, D. Study of electrochemical capacitors utilizing carbon nanotube electrodes. *J. Power Sources* **1999**, *84*, 126–129. [CrossRef]
118. Yun, J.; Kim, D.; Lee, G.; Ha, J.S. All-solid-state flexible micro-supercapacitor arrays with patterned graphene/MWNT electrodes. *Carbon* **2014**, *79*, 156–164. [CrossRef]
119. Geng, X.; Li, F.; Wang, D.-W.; Cheng, H.-M. The influence of pretreatment on multi-wall carbon nanotubes for electrochemical capacitors. *Carbon* **2013**, *60*, 564. [CrossRef]
120. Sham, M.L.; Li, J.; Ma, P.C.; Kim, J.-K. Cleaning and Functionalization of Polymer Surfaces and Nanoscale Carbon Fillers by UV/Ozone Treatment: A Review. *J. Compos. Mater.* **2009**, *43*, 1537–1564. [CrossRef]
121. Tsao, C.W.; Hromada, L.; Liu, J.; Kumar, P.; DeVoe, D. Low temperature bonding of PMMA and COC microfluidic substrates using UV/ozone surface treatment. *Lab Chip* **2007**, *7*, 499–505. [CrossRef] [PubMed]
122. Huang, L.; Chen, K.; Peng, C.; Gerhardt, R.A. Highly conductive paper fabricated with multiwalled carbon nanotubes and poly(3,4-ethylenedioxythiophene)-poly(styrenesulfonate) by unidirectional drying. *J. Mater. Sci.* **2011**, *46*, 6648–6655. [CrossRef]
123. Uddin, N.M.; Capaldi, F.M.; Farouk, B. Molecular dynamics simulations of carbon nanotube dispersions in water: Effects of nanotube length, diameter, chirality and surfactant structures. *Comput. Mater. Sci.* **2012**, *53*, 133–144. [CrossRef]
124. Chen, P.; Chen, H.; Qiu, J.; Zhou, C. Inkjet printing of single-walled carbon nanotube/RuO2 nanowire supercapacitors on cloth fabrics and flexible substrates. *Nano Res.* **2010**, *3*, 594–603. [CrossRef]
125. Singjai, P.; Changsarn, S.; Thongtem, S. Electrical resistivity of bulk multi-walled carbon nanotubes synthesized by an infusion chemical vapor deposition method. *Mater. Sci. Eng. A* **2007**, *443*, 42–46. [CrossRef]

126. Safadi, B.; Andrews, R.; Grulke, E.A. Multiwalled carbon nanotube polymer composites: Synthesis and characterization of thin films. *J. Appl. Polym. Sci.* **2002**, *84*, 2660–2669. [CrossRef]
127. Castillo, F.Y.; Socher, R.; Krause, B.; Headrick, R.; Grady, B.P.; Prada-Silvy, R.; Pötschke, P. Electrical, mechanical, and glass transition behavior of polycarbonate-based nanocomposites with different multi-walled carbon nanotubes. *Polymer* **2011**, *52*, 3835–3845. [CrossRef]
128. Lee, A.; Sudau, K.; Ahn, K.H.; Lee, S.J.; Willenbacher, N. Optimization of Experimental Parameters to Suppress Nozzle Clogging in Inkjet Printing. *Ind. Eng. Chem. Res.* **2012**, *51*, 13195–13204. [CrossRef]
129. Ju, L.; Zhang, W.; Wang, X.; Hu, J.; Zhang, Y. Aggregation kinetics of SDBS-dispersed carbon nanotubes in different aqueous suspensions. *Colloids Surf. A Physicochem. Eng. Asp.* **2012**, *409*, 159–166. [CrossRef]
130. Lin, L.; Peng, H.; Ding, G. Dispersion stability of multi-walled carbon nanotubes in refrigerant with addition of surfactant. *Appl. Therm. Eng.* **2015**, *91*, 163–171. [CrossRef]
131. Zhang, H.; Gerhardt, R.A. AAMC Conference Presentation Given by Prof. Gerhardt in Miami, December 2016 Which Includes Data from Mulhbauer. *Ph.D. Thesis*. 2014. Available online: smartech-gatech.edu (accessed on 5 September 2021).
132. Lacy, F. Developing a theoretical relationship between electrical resistivity, temperature, and film thickness for conductors. *Nanoscale Res. Lett.* **2011**, *6*, 636. [CrossRef]
133. Karoui, A. Aluminum Ultra Thin Film Grown by Physical Vapor Deposition for Solar Cell Electric Nanocontacts. *ECS Trans.* **2011**, *41*, 21–28. [CrossRef]
134. Patil, S.S.; Bhat, T.S.; Teli, A.M.; Beknalkar, S.A.; Dhavale, S.B.; Faras, M.M.; Karanjkar, M.M.; Patil, P.S. Hybrid solid state supercapacitors (HSSC's) for high energy & power density: An overview. *Eng. Sci.* **2020**. [CrossRef]
135. Du, C.; Pan, N. Supercapacitors using carbon nanotubes films by electrophoretic deposition. *J. Power Sources* **2006**, *160*, 1487–1494. [CrossRef]
136. Hughes, M.; Shaffer, M.S.P.; Renouf, A.C.; Singh, C.; Chen, G.Z.; Fray, D.J.; Windle, A.H. Electrochemical Capacitance of Nanocomposite Films Formed by Coating Aligned Arrays of Carbon Nanotubes with Polypyrrole. *Adv. Mater.* **2002**, *14*, 382–385. [CrossRef]
137. Chen, J.; Huang, Z.; Wang, D.; Yang, S.; Li, W.; Wen, J.; Ren, Z. Electrochemical synthesis of polypyrrole films over each of well-aligned carbon nanotubes. *Synth. Met.* **2001**, *125*, 289–294. [CrossRef]
138. Hughes, M.; Chen, G.; Shaffer, M.S.P.; Fray, D.J.; Windle, A.H. Electrochemical Capacitance of a Nanoporous Composite of Carbon Nanotubes and Polypyrrole. *Chem. Mater.* **2002**, *14*, 1610–1613. [CrossRef]
139. Barisci, J.N.; Wallace, G.; Baughman, R.H. Electrochemical studies of single-wall carbon nanotubes in aqueous solutions. *J. Electroanal. Chem.* **2000**, *488*, 92–98. [CrossRef]
140. An, K.H.; Jeon, K.K.; Heo, J.K.; Lim, S.C.; Bae, D.J.; Lee, Y.H. High-Capacitance Supercapacitor Using a Nanocomposite Electrode of Single-Walled Carbon Nanotube and Polypyrrole. *J. Electrochem. Soc.* **2002**, *149*, A1058–A1062. [CrossRef]
141. Wang, W.; Guo, S.; Penchev, M.; Ruiz, I.; Bozhilov, K.N.; Yan, D.; Ozkan, M.; Ozkan, C.S. Three dimensional few layer graphene and carbon nanotube foam architectures for high fidelity supercapacitors. *Nano Energy* **2012**, *2*, 294–303. [CrossRef]

C Journal of Carbon Research

MDPI

Article

Detonation Synthesis Nanodiamond Soot as a Promising Filler for Polymer Composites

Oleg V. Lebedev [1,2,*], Tikhon S. Kurkin [1], Evgeny K. Golubev [1], Alexander L. Vasiliev [2,3], Andrey K. Gatin [4], Galina P. Goncharuk [1] and Alexander N. Ozerin [1]

1 Enikolopov Institute of Synthetic Polymer Materials Russian Academy of Sciences (ISPM RAS), Profsoyuznaya Street 70, 117393 Moscow, Russia
2 Moscow Institute of Physics and Technology, Institutsky lane 9, Dolgoprudny, 141700 Moscow, Russia
3 Federal Scientific Research Center "Crystallography and Photonics" Russian Academy of Sciences, Leninsky Ave 59, 119333 Moscow, Russia
4 N. N. Semenov Federal Research Center for Chemical Physics Russian Academy of Sciences (FRCCP RAS), Kosygina Street 4, 119991 Moscow, Russia
* Correspondence: oleg.lebedev@phystech.edu

Abstract: In this work, the results of a complex investigation of structure and properties of nanodiamond soot (NDS) of detonation synthesis are presented. Size distribution of NDS particles, dispersed in different liquid media, was investigated using dynamic light scattering and laser diffraction analysis methods. The results of the investigation, as well as the results of zeta-potential measurements, allowed us to characterize the agglomeration process of the NDS particles as independent of the medium, making NDS a good model filler for research of composite-modified nanosized particles. Additional data obtained using scanning electron microscopy, scanning tunneling microscopy, atomic force microscopy, X-ray diffraction, and Raman spectroscopy, demonstrated that in NDS the spherical nanodiamond (ND) particles with diameter ~5 nm are densely packed into strong-coupled aggregates with diameter ~300 nm, surrounded by graphite nanoribbons. X-ray diffraction analysis estimated the volume fraction of NDs in NDS as ~45 vol.%, simultaneously showing that the graphite is not defective, which was confirmed with the electron diffraction method. It was demonstrated that this structure of NDS allows to efficiently use NDS as a filler for polymer composites to increase polymer characteristics such as electrical conductivity or tribological characteristics, similarly to conventionally applied fillers such as carbon black.

Keywords: nanodiamond soot; polymer nanocomposites; carbon black; conductivity; tribology

Citation: Lebedev, O.V.; Kurkin, T.S.; Golubev, E.K.; Vasiliev, A.L.; Gatin, A.K.; Goncharuk, G.P.; Ozerin, A.N. Detonation Synthesis Nanodiamond Soot as a Promising Filler for Polymer Composites. *C* **2022**, *8*, 69. https://doi.org/10.3390/c8040069

Academic Editor: Gil Gonçalves

Received: 30 September 2022
Accepted: 21 November 2022
Published: 27 November 2022

1. Introduction

Nanodiamonds (NDs) have been considered as one of the most promising fillers for nanocomposites in the last few decades [1–4]. NDs can combine properties of a diamond such as extremely high hardness, chemical and bio-stability, high thermal conductivity, etc., with an ability to use them as an effective filler for improvement of polymer characteristics [5–7].

ND particles can be obtained by the detonation of strong explosives and ablation of carbon target by laser pulses [2,8]. A significant advantage of the detonation method is the formation of diamond nanoparticles with a typical size of 4–5 nm. The structures of ND particles are of a special interest for investigation [4]. In the work of Ginés et al. [9], the zeta potential of NDs was investigated. Positive zeta potentials in nano-structured carbons were explained by the presence of graphitic planes at the surface of ND particles, which leave oxygen-free Lewis sites and promote the suppression of acidic functional groups. Electron microscopy and Raman spectroscopy have been used to show that positive zeta potential of nanodiamond is only exhibited in the presence of sp^2 carbon at the surface. In the review by Shrestha et al. [10] it was concluded that exploiting the surface characteristics of the nanoparticles, including surface charge and functionality, is an effective method of size control of aggregate/agglomerate nanoparticles.

Along with detonation synthesis NDs, polymers modified with NDs are also an attractive subject for research [11]. In the review by Karami et al. [12], it was noted that exceptional properties of detonation NDs and its inherent rich surface chemistry make them suitable for the production of multifunctional materials. In the work of Kidalov et al. [13], it was shown that the thermal conductivity of nanocomposites based on nanodiamonds can reach up to 10 W/(m·K). In the work of Houshyar et al. [14], using conventional methods of processing such as melt mixing, the authors obtained filaments based on polypropylene (PP) filled with NDs that were characterized by exceptional long-term stability performance, mechanical and biological properties.

A wide range of methods for the controlled modification of diamond powder opens up the possibilities of increasing such characteristics of polymer composites as adhesive strength [3,6]. In the work of Kurkin et al. [6], it was shown that the adhesive strength of oriented poly(vinyl alcohol) (PVA) fibers in an epoxy binder can be significantly improved by simultaneously modifying the polymer fiber and the polymer matrix with fine particles of detonation-synthesized ND powder with oppositely charged surfaces.

Unfortunately, using NDs for the modification of polymers can be very expensive, especially if high filler content is required. This is due to the complex purification process required to remove residue metals and non-diamond forms of carbon [3]. Additionally, while purified ND powder can be characterized by a relatively high level of thermal conductivity, it cannot be used efficiently to increase electrical conductivity of polymers [5].

Nanodiamond soot (NDS) is a more affordable type of filler with characteristics close to those of NDs [11,15]. NDS is obtained at the post-detonation stage of an ND synthesis process, preceding the expensive final stage of NDs purification from non-diamond carbon [8]. This reduces the cost of NDS fillers, making them a less expensive alternative to the pure NDs. Additionally, since NDS can contain a high-volume fraction of non-diamond carbon, such as graphite, it is potentially possible to obtain electrically conductive polymer composites modified with NDS.

In the works of Satonkina et al. [4] and Chen et al. [15], structure, composition, and surface properties of detonation synthesis NDS were characterized using elemental analysis, high resolution transmission electron microscopy, X-ray diffraction (XRD), Raman spectroscopy, Fourier-IR, X-ray photoelectron spectroscopy, and small angle X-ray scattering. It was shown that the condensed carbon in the NDS of choice had a complex composition, comprising clusters of diamond, graphite and amorphous carbon, all of which were in the nanometer scale. Two types of graphite phases were present in the ND. The first phase was tangled and highly curved graphite nanoribbons; the other was carbon spheres and carbon mantles graphitized from the NDs. Such nanoribbons were effectively used previously in the work of Alaferdov et al. [16] to obtain polymer composites with excellent conductivity (up to 40 S/cm) and very low percolation threshold (0.42 vol.%).

In the work of Kurkin et al. [17], the structure and properties of the oriented PVA fibers modified with NDs and NDS of detonation synthesis were investigated using XRD, electron microscopy, and mechanical testing. It was shown that NDs and NDS are effective modifiers of the physical–mechanical properties of the oriented PVA fibers. NDS particles in the PVA matrix retained a high level of dispersion without aggregation up to a filling degree of 3 vol.%. Maximum values of the longitudinal elastic modulus and the energy stored in the modified PVA fiber upon stretching up to the moment of the fracture was achieved at a small (1 vol.%) NDS filling value. The adhesion strength of the oriented modified by NDS PVA fibers in the epoxy matrix substantially exceeded the adhesion strength of the nonmodified fibers.

In the works of Neitzel et al. [18,19] it was demonstrated that an addition of ND particles in a polymer matrix (epoxy) is an efficient way of decreasing the friction coefficient of the material. In the work of Lebedev et al. [20], tribological and percolation properties of composites based on PP modified with non-conductive NDS particles were investigated.

By measuring tribological characteristics such as wear rate, it was shown that there is a possibility of formation of a surface layer enriched with NDS. This effect was additionally studied in [21] for electrically conductive such fillers as carbon black (CB).

It should be mentioned that most of the currently published works are either focused on the research of NDs or polymer nanocomposites filled with NDs. There are very few works that investigate the properties of specific NDS particles, and no works that connect the structure of such particles with the properties of polymer composites based on the same filler. At the same time, researching possible areas of application of such a unique type of filler can be very useful for the polymer composite industry and are of a great scientific and practical interest.

The performance of various types of nanosized and submicron polydisperse fillers (including inorganic and carbon particles, as well as more complex organic/inorganic hybrid particles) in polymer composites is generally associated with the structural properties of these fillers [10,22]. Among the most common methods of obtaining information about the phase composition, structural hierarchy and colloidal behavior of polydisperse fillers are XRD, dynamic light scattering (DLS), laser diffraction (LD) and static light scattering, along with various types of microscopy techniques. While each of these techniques has its intrinsic limitations, their combination often provides complementary information [23,24], which is necessary to determine the most optimal procedure of composite material design and processing [25].

The novelty of our work is in a comprehensive study of the properties of NDS of detonation synthesis, characterized by high-volume fractions of graphite in the form of graphite nanoribbons. The study is based on the thorough analysis of the structure of such NDS powder and dispersions under various conditions. The structural properties of NDS were characterized by a wide variety of techniques, including XRD, DLS, LD, Raman spectroscopy, atomic force microscopy (AFM), scanning and transmission electron microscopy (SEM and TEM, respectively), scanning tunnelling microscopy (STM), and electron diffraction. By using the NDS as a filler for a polymer, investigated earlier in our previous works, and comparing the effect of the filler structure on the migration of nanoparticles to the boundaries of polymer composite melts, it makes it possible to draw conclusions on the overall possibility of NDS acting as a more versatile substitute of such conventionally used fillers, as CB. The obtained results can provide a solid foundation for the future design of new carbon-based polymer composite materials.

2. Materials and Methods

As the main object of the investigation, NDS of detonation synthesis was used. NDS was obtained from 2,4,6-trinitrotoluene, providing the high graphite/nanodiamond ratio required to ensure high levels of electrical conductivity. NDS was provided by the Special Design and Technological Bureau "Technolog" (St. Petersburg, Russia). The details on the synthesis process can be found in the works of Dolmatov et al. [3,11].

Commercially available amorphous–crystalline PP grade H030 GP/3 (SIBUR International GmbH, Vienna, Austria) was used as the base polymer for the composite materials studied in this work [26]. As the reference electrically conductive filler, commercially available electrically conductive CB of brand P267-E (equivalent of N 472, Institute of Hydro-carbons Processing of the Siberian Branch of the Russian Academy of Sciences IHCP SB RAS, Saint Petersburg, Russia) with a specific surface area of 144 m^2/g [27] was chosen. Previously, this filler was used in the investigation of nanoparticles migration in PP-based composite melts [21].

Samples of composites were obtained via melt mixing of the polymer granules and nanoparticle powders in a micro-compounder (DACA Instruments, Goleta, CA, USA) designed to prepare compositions in laboratory quantities (up to 4 cm^3). The quantity of the material in the micro-compounder chamber was calculated so the chamber was filled fully. The mixing procedure was carried out at 200 °C and 500 rpm for 10 min after the addition of the components into the compounder.

DC electrical conductivity of the filler powders was measured using a 34401A multimeter (Agilent, Santa Clara, CA, USA) and the four-probe method. A special appliance was used to change volume fraction of the powder between electrodes.

To carry out measurements of electrical resistance of the composite melts, granules of the prepared composites of various compositions were placed in a heated mold between two flat electrodes inside a circular hole in a spacer made of polytetrafluoroethylene [21]. The composite granules were quickly heated to 205 °C, after which a pressure of 5 MPa was gradually applied to the composite melt. The temperature and pressure values were maintained for the duration of the experiments. Constant electrical voltage was applied between the electrodes during the entire experiment time. The voltage was provided by a highly stable AKIP 1147/1 power supply (Prist, Moscow, Russia). The standard preset value of the electric voltage was 2 V. A 6485/E (Keithley, Cleveland, OH, USA) picoammeter was used for the current measurements. During the experiments, the values of the electric current between the electrodes were continuously recorded using a multichannel analog-to-digital converter E14-440 (L-Card, Moscow, Russia) at frequency of 100 Hz.

Hydrodynamic diameters of NDS in various liquid media (acetone, DMSO, ethanol, isopropanol) were determined via DLS at 25 °C using a Microtrac Nanotrac model Zetatrac instrument (Microtrac Inc., Montgomeryville, PA, USA) equipped with a 3 mW solid-state laser operating at 780 nm. Scattered light was detected at 180° in a controlled reference self-beating mode, and the particle size was calculated from the power spectrum analysis of Doppler-shifted signal over several runs. For every tested sample, 5 runs each of 60 s duration were performed. The sphere-equivalent hydrodynamic diameter of the NDS structures was calculated from the particle diffusion coefficient via the Stokes−Einstein equation, using solution viscosities for liquid media at 25 °C. Electrophoretic mobility of NDS structures was measured in liquid media using a Microtrac Nanotrac model Zetatrac instrument by applying AC/DC voltage to built-in sample cell electrodes to determine the magnitude and sign of particle zeta-potential. Zeta-potential was calculated from electrophoretic mobility via the Smoluchowski equation. All measurements and data analysis were performed using NanoFLEX Software. NDS concentrations of 0.5% (*w/v*) were utilized in all DLS and zeta-potential measurements. NDS dispersions were preliminarily treated using a CUD-500 ultrasonic disperser from Criamid, Moscow, Russia (2 min, 250 W, 30 kHz).

Particle size analysis by LD was performed using Wintrac 3000 instrument equipped with a He-Ne laser (λ = 632.8 nm) and auxiliary solid-state laser operating at λ = 405 nm, allowing for particle characterization in 0.1–2000 μm size range. Prior to the measurements, NDS powder was dispersed in a liquid medium using the built-in circulation system equipped ultrasound probe (60 W, 40 kHz). Scattering pattern measurements were performed with sample circulating through a flow-cell at pumping speed of 2000 rotations per minute. For every sample, 60 scans of 2 s duration each were performed. To obtain particle size distributions, each scattering pattern was processed according to Fraunhofer diffraction and Mie scattering models using WINTRAC Software. The reported results are the product of averaging of the 60 scans for every measured sample.

XRD patterns (2Θ = 10–105°) were obtained using a Bruker D8 Advance powder diffractometer ($\lambda(CuK_{\alpha})$ = 1.5418 Å) equipped with focusing germanium crystal monochromator, Ni-filter and a LYNXEYE 1D scintillating detector. XRD patterns were recorded in the transmission mode, with NDS powder placed between two layers of amorphous poly(ethylene terephthalate) films. Processing of diffraction patterns (peak separation, corrections for non-coherent background scattering and scattering from poly(ethylene terephalate)) was performed using Origin Pro software.

AFM measurements were performed using a Solver HV atomic-force microscope (NT-MDT, Moscow, Russia) with standard HA-NC cantilevers (chip size 3.6 × 1.6 × 0.45 mm, cantilever size 87 × 32 × 1.75 μm, curvature radius of the tip 10 nm, force constant 5.8 N/m). The measurements were carried out in a tapping mode with the acquisition of surface topography and phase contrast.

STM measurements were performed using a scanning tunneling microscope (STM, Omicron NanoTechnology, Taunusstein, Germany) at 300 K and a residual gas pressure of 2×10^{-10} Torr. According to the standard procedure, tips made of tungsten wires were used as STM probes. STM topography measurements were carried out in the constant current mode (U_{bias} = 2.0 V, I_{tunnel} = 3.1 nA).

Prior to the AFM and STM measurements, NDS dispersions in 2-propanol (concentration of solid phase 0.5 wt.%) were deposited on a highly oriented pyrolytic graphite substrate and dried at the ambient conditions (air, T = 22 °C).

Raman spectra were obtained using a RamanStation-400 (Perkin-Elmer, Waltham, MA, USA) dispersive echelle Raman spectrometer equipped with a single-mode 785 nm diode laser excitation (1–100 mW), thermoelectrically cooled (−52 °C) CCD matrix detector (Andor Inc., Belfast, North Ireland), XYZ-motorized sample stage and an ×30 microscope objective used for delivering laser excitation and collecting Raman photons. Raman spectra were acquired in 180° backscattering geometry. The solid samples were placed onto the standard Au-coated slide on the XYZ-table and focused under the instrument objective. The Raman spectra were obtained within 3400–100 cm^{-1} Raman shifts range at 1,2,4 or 8 cm^{-1} optical resolution. In different experiments, depending on the luminescence activity of the samples, the laser excitation power of 2–25 mW and signal accumulation times of 20–180 s were used for the spectra recording. A dedicated Bruker OPUS software package was used for routine processing (baseline correction, smoothing, etc.) of the Raman spectra.

Tribological tests were conducted using a pin-on-disk-type testing machine (cylindrical stainless-steel pin) in the dry friction mode according to ASTM G99-05 procedure. Test parameters values can be found in [20]. For each test piece, 6 identical test runs were performed, measuring friction coefficient dependence on time. The obtained results were averaged.

SEM studies of the NDS powder were performed using a JSM-6000 (Neoscope II) scanning microscope (Jeol, Japan). TEM studies were performed using a fully digital 200 kV S/TEM FEI Tecnai Osiris system (FEI Company, Hillsboro, OR, USA). Samples were prepared by deposition of the NDS solutions on holey C film on Au grid.

3. Results

Firstly, the NDS powder was investigated "as it is" using SEM. Selected results of the microscopy study are presented in Figure 1.

(a) (b)

Figure 1. (**a**,**b**) Scanning electron microscopy images of NDS powder.

It can be seen that the powder contains soft-looking (the packing of the carbon phase looks not very dense) agglomerates of diameter >~3 μm (Figure 1a). In turn, these relatively large agglomerates are seemingly formed from aggregates of diameter >100 nm (Figure 1b) [22].

The results of SEM data analysis for the powder of NDS were in some agreement with the measurements of NDS structures (primary particles, aggregates, agglomerates [22]) size distributions in acetone using DLS and LD analysis (LDA), as can be seen from Figure 2.

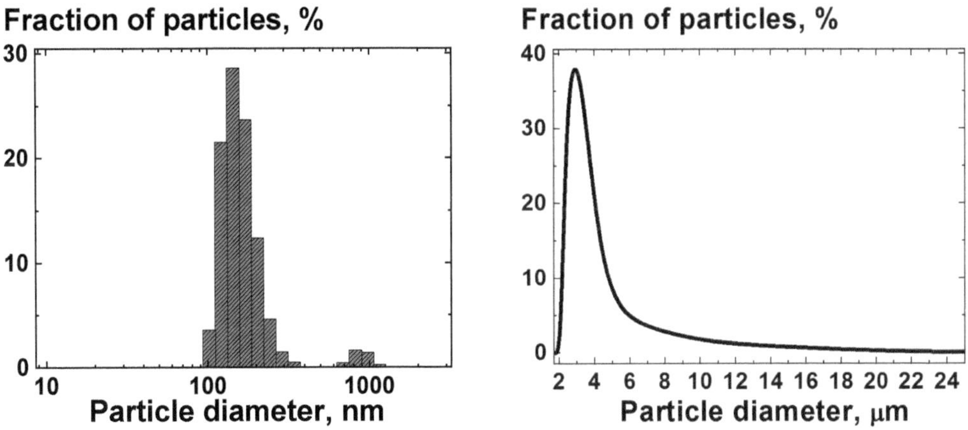

(a) (b)

Figure 2. Distribution of diameter of the NDS particles in a suspension of NDS powder in acetone, obtained with (**a**) DLS and (**b**) LD methods.

The size distribution for NDS aggregates/agglomerates consisted of peaks at ~250 nm and ~950 nm, obtained using DLS (Figure 2a), and at ~3.2 μm, obtained using LDA (Figure 2b). The difference in the size distributions determined by the DLS and LDA can be attributed to the additional pretreatment using ultrasound with higher power when the size distributions were measured by the DLS method, compared to the treatment of the NDS in the circulation system when the size distributions was measured by LD.

NDS size distributions were also obtained for additional liquid media, such as DMSO, ethanol, and isopropanol, as well as for different time periods after ultrasonication: 0 h, 24 h, 120 h, 1 week, 3 weeks, and 4 weeks. The results were very close to the ones obtained for NDS dispersion in acetone soon after ultrasonication (in the range of an error). The results of the DLS measurements for different liquid media and time periods after ultrasonication are presented in Supplementary Materials.

For all the investigated dispersions in different media (acetone, DMSO, ethanol, isopropanol) zeta-potentials were measured after 0 h, 24 h, 120 h, 1 week, 3 weeks, and 4 weeks after the initial ultrasonication. The value of zeta-potential for NDS dispersions were in the range of uncertainty for all the investigated media even after 4 weeks: 20 ± 4 mV for dispersion in acetone soon after the ultrasonication. As was mentioned in the introduction section, according to Ginés et al. [9], the positive zeta potential is only exhibited on nanodiamond with non-zero concentrations of sp^2 bonding, specifically attributed to graphene-like shells. Although it is well known in colloid physics that zeta-potential above +30 mV is expected for stable dispersions, particle dispersions with zeta-potential in a range of +20–30 mV can also maintain the size distribution over a long period of time, depending on the particle size [28].

The next step was the investigation of the NDS powder using TEM. Selected results of the microscopy study are presented in Figure 3.

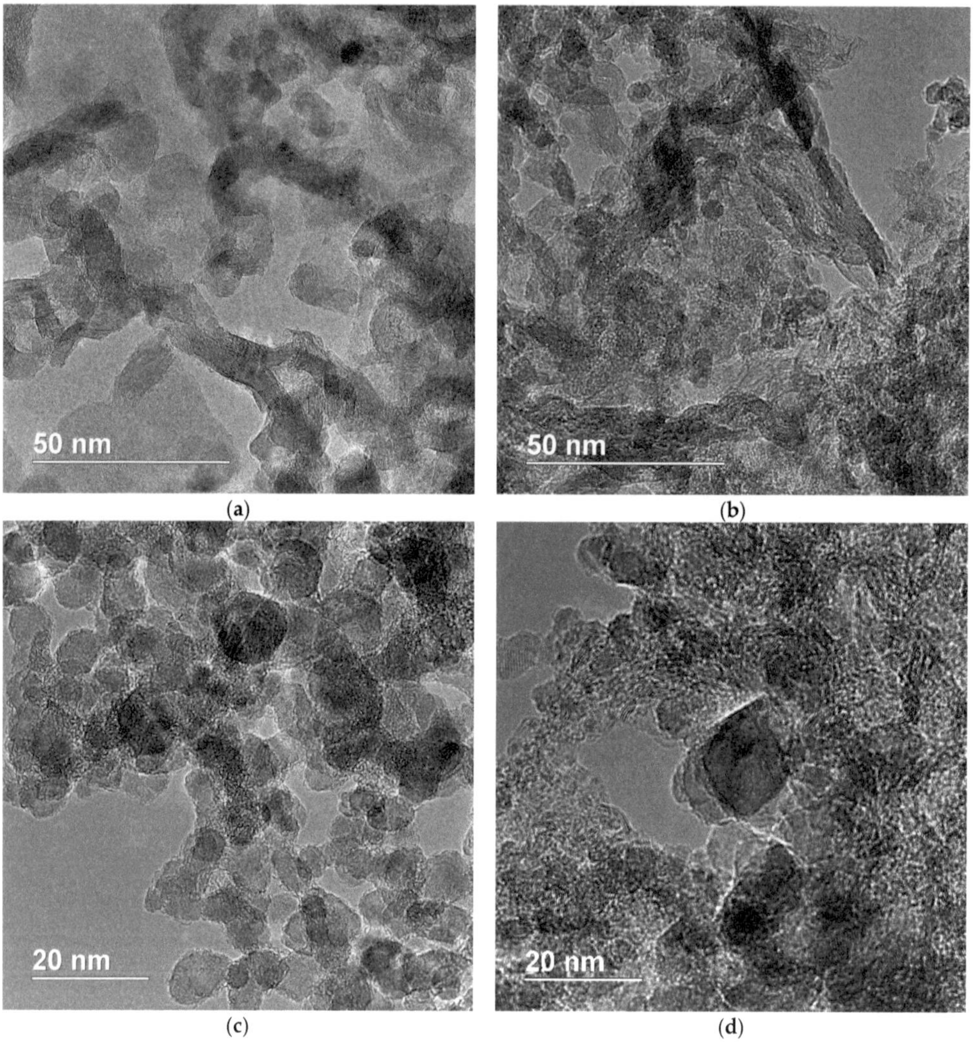

Figure 3. (**a**–**d**) Transmission electron microscopy images of NDS powder.

As it can be seen from the TEM results, especially in Figure 3a,b, the graphite phase is partially presented in the form of graphite nanoribbons/nanobelts. The primary NDS particles themselves seem to have an average diameter of ~5 nm. The primary NDS particles are considered to be ND crystallites [12]. Additionally, relatively large flakes of graphite (~10–15 nm) can be observed (Figure 3c,d).

To be able to investigate the composition of NDS particles as well as their structure, XRD and Raman spectroscopy methods were used. The results of the study are presented in Figure 4.

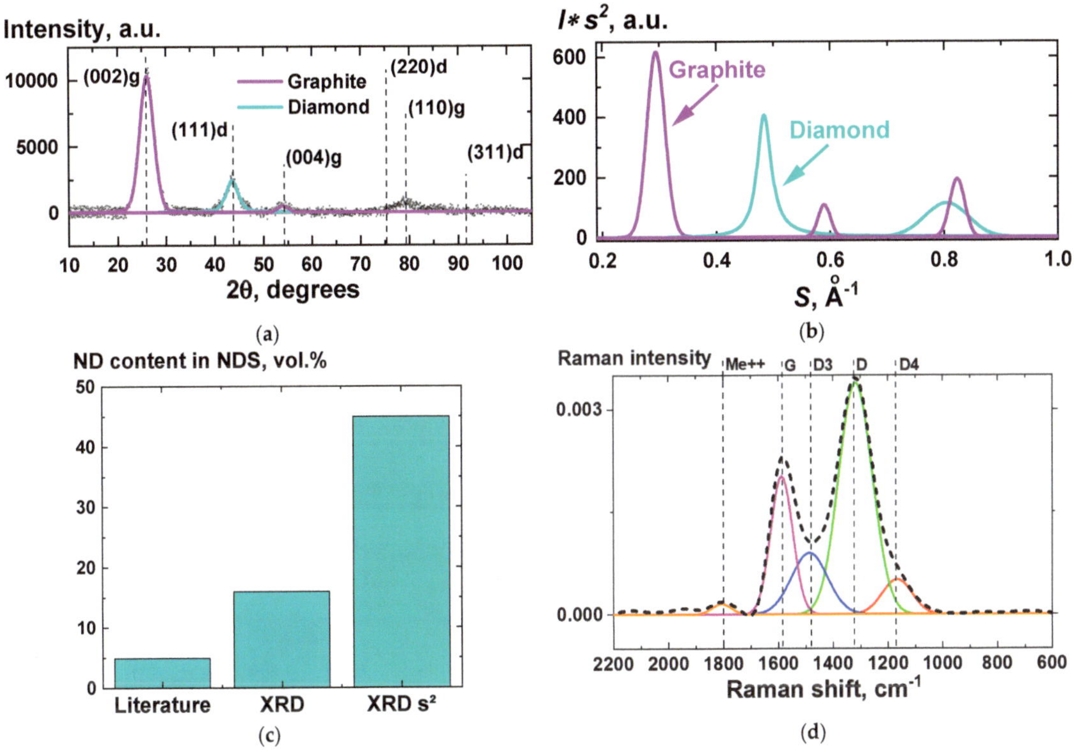

Figure 4. (**a**,**b**)—XRD patterns without (**a**) and with (**b**) implementation of Ruland's method [29]; (**c**)—results of estimation of ND in NDS content value from XRD patterns without ("XRD") and with implementation of Ruland's method ("XRD s2"), and also the value provided by the manufacturer [11] ("Literature"); (**d**)—Raman spectroscopy results for NDS powder. Dotted line (**a**,**d**) represents the obtained experimental data, solid lines (**a**,**b**,**d**)—peaks from decomposition of the experimental data. Dashed vertical lines (**a**,**d**) indicate the centers of the peaks.

From the XRD data, it is possible to estimate the interlayer distance between 002 planes of atoms in the graphite phase. According to the literature data [30], the distance is ~0.35 nm. The experimental value, obtained from the XRD data (Figure 4a,b) turned out to be very close (0.348 ± 0.05) nm.

From XRD data it is also possible to estimate volume fraction of graphite in the NDS powder. The results of the comparison of the obtained value with the literature data [11] are given in Figure 4c. According to the literature, the volume fraction of the NDs in NDS is 4.9 vol.%. Taking into account that this value was obtained using the weighting procedure after purification of NDs, which can lead to a significant decrease of the initial NDs weight value, the real ND content in NDS can be much higher. According to the XRD analysis, the volume fraction of ND in NDS can be estimated as either 16 vol.%, or 45 vol.%, depending if the Ruland's method [29] is implemented (Figure 4b) or not (Figure 4a). No non-carbon impurities were observed according to the XRD data.

As is known from the publications on highly deformed carbon materials [31,32], the Raman scattering bands of the NDS cannot be described by only two bands, G and D. There are always two to three additional bands, which have been labeled according to the literature [31,32] in Figure 4d. The physical meaning of these additional bands is discussed in the relevant publications [31–33].

As the additional methods of NDS agglomerate size distribution and NDS particle structure analysis, AFM, along with STM and the electron diffraction method, were used. Selected results of these methods' implementation are presented in Figure 5.

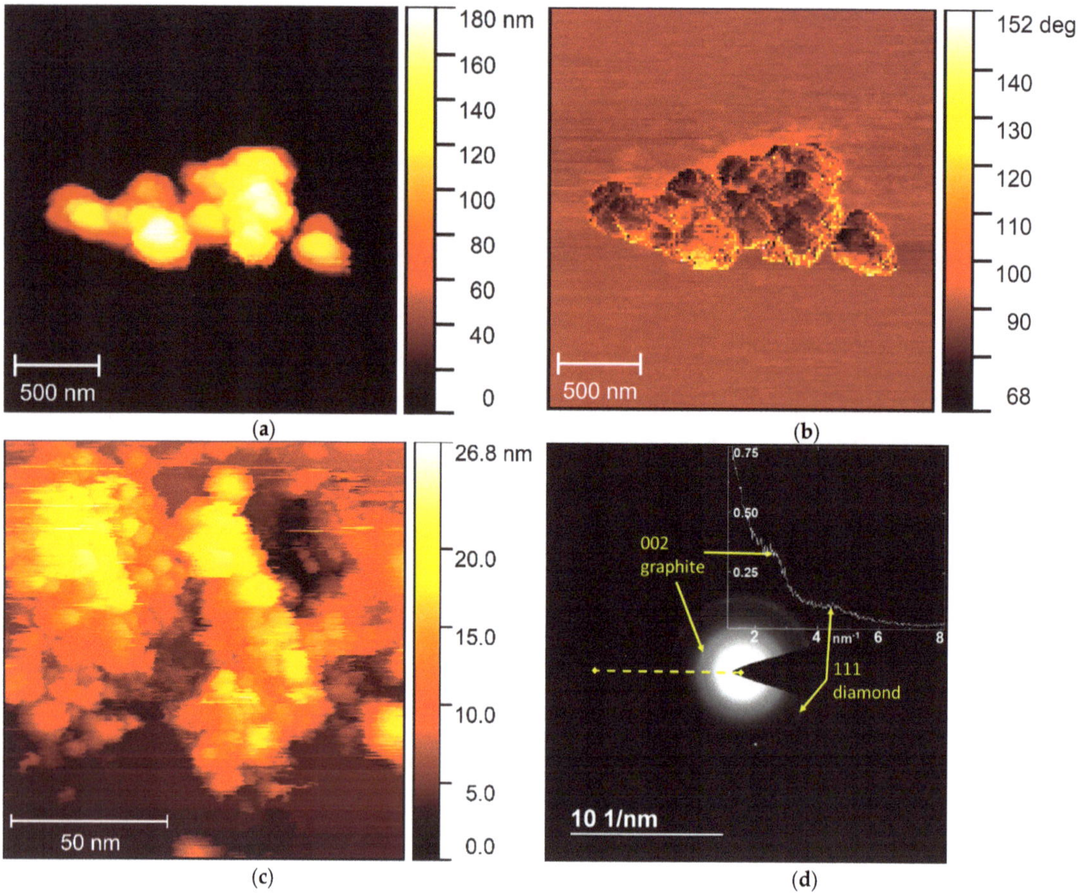

Figure 5. (**a**) Height distribution; and (**b**) phase contrast mode images obtained for NDS powder using AFM in the semi-contact mode; (**c**)—STM results for the NDS powder; (**d**)—electron diffraction results for the NDS powder. The inset (**d**) shows distribution of normalized intensity along the dashed yellow line. The yellow arrows point at the scattering peaks, corresponding to the diamond 111 and graphite 002 planes of carbon atoms.

It can be seen in Figure 5a,b that a large NDS agglomerate with diameter ~1.5 μm comprises small strong-coupled [22] aggregates with diameter of ~250 nm. Thus, it can be said that the AFM results correlate in some way with the SEM (Figure 1), DLS (Figure 2a), and LDA (Figure 2b) data. At the same time, in the STM images (Figure 5c) isolated ND particles can be observed with the diameter ~4–5 nm, which correlates well with the TEM data (Figure 3).

In the electron diffraction pattern (Figure 5d), two characteristic reflexes can be observed, corresponded to the graphite and diamond phases. From the analysis of the graphite reflex, the distance between 002 planes of carbon atoms in graphite can also be evaluated. The obtained value (3.45 ± 0.1) nm was in the range of error from the value taken from the literature [30] and obtained experimentally from the XRD data (Figure 4a,b).

To evaluate the potential of the NDS powder as a filler for electroconductive polymer composites, the NDS powder resistivity was measured depending on the volume fraction of the nanoparticles. The results of the comparison of resistivity of NDS and CB powders for various volume fraction values are presented in Figure 6a. Additionally, composites, based on PP and containing 10 wt.% NDS and 10 wt.% CB, were subjected to annealing at 205 °C, while the resistivity was constantly measured between the two electrodes in contact with the composite melts. The resulting dependencies of resistivity of the composite melts versus time are presented in Figure 6b.

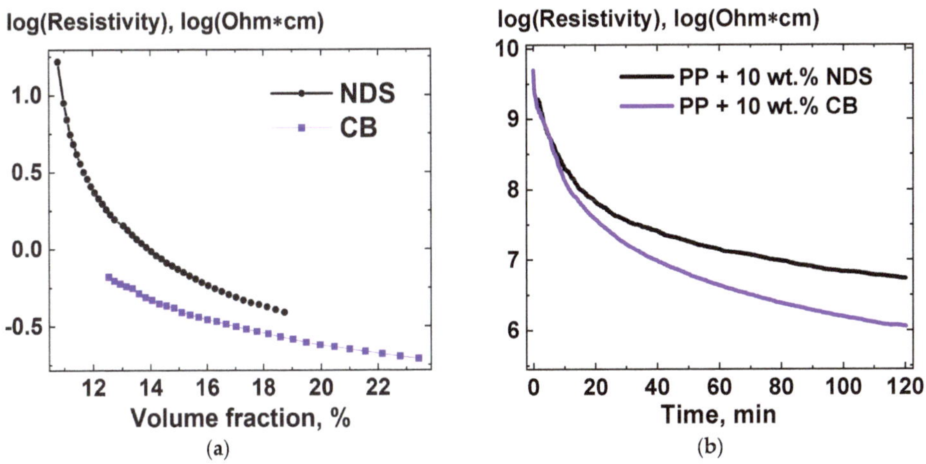

Figure 6. (**a**)—Resistivity of NDS and CB powders versus volume fraction of the powder particles in a cylindric tube, measured between two steel pistons acting as electrodes; (**b**)—resistivity of melts of the composites PP + 10 wt.% NDS and PP + 10 wt.% CB [21], measured between two aluminum electrodes located at the bottom and top of a closed cylindric mold (radius—1.5 cm, height—0.2 cm) at temperature of 205 °C, versus time.

As can be seen in Figure 6, both results for resistivity of the NDS powder and the NDS-containing composite demonstrate similar behavior with volume fraction change or with time if to compare with the corresponding dependencies for CB.

For PP-based composites containing 2.5 and 10 wt.% NDS, dependencies of friction coefficient on time were obtained before and after annealing the composites at 205 °C for 2 h. The averaged friction coefficient values versus time curves for six independent runs are presented in Figure 7a, while the results of averaging of the friction coefficient values versus time of annealing at 205 °C are presented in Figure 7b. The averaging was performed for the values of friction coefficient in the (2–8) h range of tribological test run time.

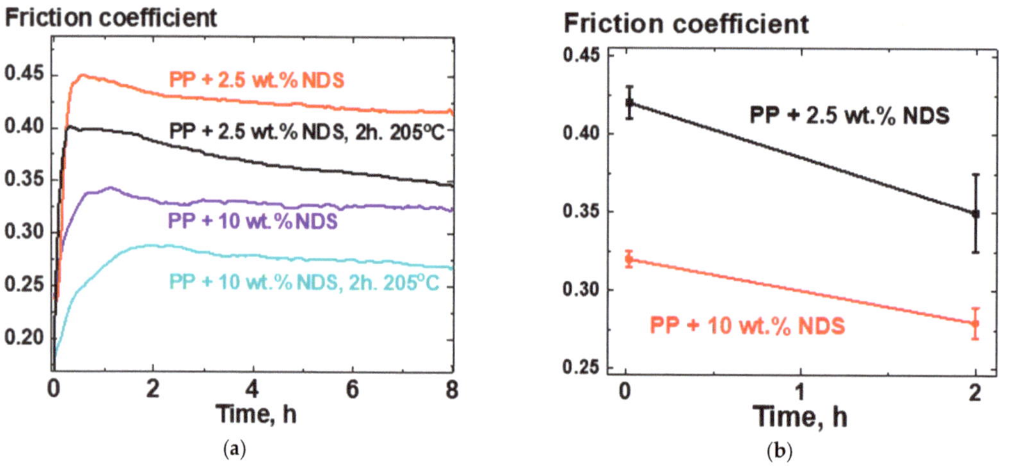

Figure 7. (**a**)—Friction coefficient versus time of a tribological test (steel pin on a rotating composite disk) curves, averaged over six independent test results for each curve, for composites PP + 2.5 wt.% NDS and PP + 10 wt.% NDS before and after keeping the composites for 2 h in a melted state at 205 °C; (**b**)—friction coefficient versus time of heat treatment at 205 °C for composites PP + 2.5 wt.% NDS and PP + 10 wt.% NDS.

4. Discussion

It is known that bonding strength between subunits (primary particles) in an aggregate/agglomerate strongly affects the size of the aggregate/agglomerate determined by different methods, as well the conditions required for the deagglomeration [10,22]. Based on the bonding strength between the subunits, agglomerated particles are known to be roughly categorized into "soft" and "hard" agglomerates. It is not expected that nanoparticles agglomerate size distribution, obtained via DLS method, completely corresponds to the agglomerate size distribution in the polymer matrix in a polymer composite. Moreover, size distribution of nanoparticle agglomerates can be affected by the method of processing used to obtain the polymer composite. However, it is believed that certain structural levels of the NDS, e.g., their hard aggregates comprising of strongly bonded individual nanoparticles, are structurally determined rather by fundamental characteristics and intrinsic properties of the primary particles (material, surface charge and functionality) [10,22] than by an external medium. In this case, specifics of the size distributions of different structural elements for different states of the nanoparticles (powder, dispersion in a liquid medium, filler in a polymer matrix) can be correlated with each other. That is why we believe that the methods used are sufficiently informative for characterization of the NDS aggregates/agglomerates.

From the data obtained using SEM (Figure 1), DLS (Figure 2a), LDA (Figure 2b), and AFM (Figure 5a,b), it was concluded that NDS primary particles tend to aggregate/agglomerate into aggregates/agglomerates of a certain size relatively independently of the medium they are dispersed in, or for how long they are kept in that medium. This feature makes NDS a good model object for the research of properties of composite materials to compare the effect of different polymer matrices on the overall properties of the composites.

The results of TEM (Figure 3), XRD (Figure 4a,b), Raman spectroscopy (Figure 4d), STM (Figure 5c), and electron diffraction (Figure 5d) show that, along with the NDs, a high amount of graphite (Figure 4c) is present in NDS. While ND crystallites of ~4–5 nm gather up in relatively small strong-coupled agglomerates, they are surrounded by graphite in a nanoribbon form. Although from the microscopy data it is possible to roughly estimate the transverse dimensions of the nanoribbons (5–10 nm), estimation of longitudinal dimensions can be quite difficult. At the same time, the microscopy data are sufficient to draw conclusions about the high aspect ratio of such graphite nanoribbons. Due to the high aspect ratio of such a form of graphite, it is possible to achieve very low percolation thresholds for electrical conductivity, as was demonstrated in [16]. At the same time, as the results of XRD and electron diffraction indicate, the graphite phase in NDS is not defective, meaning it can provide decent levels of electrical conductivity even in NDS contain a high fraction of non-conductive NDs. This data agrees with the results of comparison of resistivity for NDS and CB powders for various volume fraction values (Figure 6a), as well as with the results of resistivity measurements with time for composite melts containing NDS and CB. The increase in electrical conductivity of the composite material based on PP is caused by an increase in the concentration of NDS in the near-surface layer when the composite is subjected to thermal treatment at melting temperature [21], as is shown in Figure 6b. High electric conductivity of NDS is a consequence of its structural specifics, such as the presence of graphite phase in the form of nanoribbons.

Results of the tribological tests (Figure 7) allow us to say that using the effect of nanoparticles migration to the boundaries of nanocomposite melt, as was investigated in [21], can be an effective way to improve tribological properties of composite materials filled with NDS particles, specifically due to the presence in NDS of the extremely hard ND cores [12,18,19]. Particularly, it can be seen in Figure 7 that annealing time effectively acted as an increase of the filler content in the composite, which agrees with the conclusions made in [21] on the process of enrichment of a composite melt surface layer with time.

Altogether the results presented in this work demonstrate the versatility and utility of such easily obtainable and cheap filler as NDS. While containing simultaneously hard and chemically resistant ND particles and flexible electrically conductive graphite nanoribbons with a high aspect ratio, NDS has a potential to help meet multiple requirements for a polymer composite by improving different characteristics of the composite simultaneously. This, along with affordability of the filler, can be a very appealing factor for many industries working with polymer composites and looking to find a way to improve the product by substituting such conventional fillers such as CB.

As for future research, implementation of the NDS investigated in this work as a filler used to improve characteristics of ultra-high-molecular-weight polyethylene-based laminates is planned. Additionally, we plan to investigate NDS of different synthesis and study how it affects polymer composites filled with NDS of different types.

Supplementary Materials: The following supporting information can be downloaded at: https://www.mdpi.com/article/10.3390/c8040069/s1.

Author Contributions: Conceptualization, O.V.L. and A.N.O.; methodology, O.V.L., T.S.K., E.K.G., A.L.V., A.K.G., G.P.G. and A.N.O.; validation, O.V.L., T.S.K., E.K.G., A.L.V., A.K.G., G.P.G. and A.N.O.; formal analysis, O.V.L. and A.N.O.; investigation, O.V.L., G.P.G., T.S.K., E.K.G., A.L.V., A.K.G.; resources, A.N.O.; data curation, O.V.L.; writing—original draft preparation, O.V.L.; writing—review and editing, O.V.L. and A.N.O., visualization, O.V.L.; supervision, A.N.O.; project administration, O.V.L. and A.N.O.; funding acquisition, A.N.O. All authors have read and agreed to the published version of the manuscript.

Funding: The reported study was funded by RSF according to the research project No. 22-13-00359.

Institutional Review Board Statement: Not applicable.

Informed Consent Statement: Not applicable.

Acknowledgments: The authors express their deep gratitude to the team of the Laboratory of the Structure of Polymeric Materials of the ISPM RAS for their attention, cooperation and support at all stages of the work. The authors are also grateful to Dolmatov Valeriy Yurievich for providing them with the NDS powder for the research.

Conflicts of Interest: The authors declare no conflict of interest.

References

1. Dolmatov, V.Y.; Rudenko, D.V.; Dorokhov, A.O.; Malygin, A.A.; Kozlov, A.S.; Marchukov, V.A. Obtaining of Detonation Diamonds from Individual Explosives. *Combust. Explos. Shock Waves* **2021**, *57*, 232–237. [CrossRef]
2. Shvidchenko, A.V.; Eidelman, E.D.; Vul', A.Y.; Kuznetsov, N.M.; Stolyarova, D.Y.; Belousov, S.I.; Chvalun, S.N. Colloids of Detonation Nanodiamond Particles for Advanced Applications. *Adv. Colloid Interface Sci.* **2019**, *268*, 64–81. [CrossRef] [PubMed]
3. Dolmatov, V.Y. Detonation-Synthesis Nanodiamonds: Synthesis, Structure, Properties and Applications. *Russ. Chem. Rev.* **2007**, *76*, 339–360. [CrossRef]
4. Satonkina, N.P.; Ershov, A.P.; Kashkarov, A.O.; Rubtsov, I.A. Elongated Conductive Structures in Detonation Soot of High Explosives. *RSC Adv.* **2020**, *10*, 17620–17626. [CrossRef] [PubMed]
5. Kidalov, S.V.; Shakhov, F.M.; Vul', A.Y.; Ozerin, A.N. Grain-Boundary Heat Conductance in Nanodiamond Composites. *Diam. Relat. Mater.* **2010**, *19*, 976–980. [CrossRef]
6. Kurkin, T.S.; Tikunova, E.P.; Yablokova, M.Y.; Kechek'yan, A.S.; Beshenko, M.A.; Dolmatov, V.Y.; Ozerin, A.N. Effect of a Detonation-Synthesized Nanodiamond Powder on the Adhesive Strength of a Polymer Fiber to an Epoxy Binder. *Dokl. Chem.* **2014**, *457*, 118–121. [CrossRef]
7. Zhang, Y.; Rhee, K.Y.; Hui, D.; Park, S.J. A Critical Review of Nanodiamond Based Nanocomposites: Synthesis, Properties and Applications. *Compos. Part B Eng.* **2018**, *143*, 19–27. [CrossRef]
8. Dolmatov, V.Y. Assessment of Applicability of Explosive Charges for Synthesis of Detonation Nanodiamonds. *J. Superhard Mater.* **2016**, *38*, 373–376. [CrossRef]
9. Ginés, L.; Mandal, S.; Ashek-I-Ahmed; Cheng, C.-L.; Sow, M.; Williams, O.A. Positive Zeta Potential of Nanodiamonds. *Nanoscale* **2017**, *9*, 12549–12555. [CrossRef]
10. Shrestha, S.; Wang, B.; Dutta, P. Nanoparticle Processing: Understanding and Controlling Aggregation. *Adv. Colloid Interface Sci.* **2020**, *279*, 102162. [CrossRef]
11. Dolmatov, V.Y.; Ozerin, A.N.; Kulakova, I.I.; Bochechka, O.O.; Lapchuk, N.M.; Myllymäki, V.; Vehanen, A. Detonation Nanodiamonds: New Aspects in the Theory and Practice of Synthesis, Properties and Applications. *Russ. Chem. Rev.* **2020**, *89*, 1428–1462. [CrossRef]
12. Karami, P.; Salkhi Khasraghi, S.; Hashemi, M.; Rabiei, S.; Shojaei, A. Polymer/Nanodiamond Composites—A Comprehensive Review from Synthesis and Fabrication to Properties and Applications. *Adv. Colloid Interface Sci.* **2019**, *269*, 122–151. [CrossRef] [PubMed]
13. Kidalov, S.V.; Shakhov, F.M.; Vul', A.Y. Thermal Conductivity of Nanocomposites Based on Diamonds and Nanodiamonds. *Diam. Relat. Mater.* **2007**, *16*, 2063–2066. [CrossRef]
14. Houshyar, S.; Sarker, A.; Jadhav, A.; Kumar, G.S.; Bhattacharyya, A.; Nayak, R.; Shanks, R.A.; Saha, T.; Rifai, A.; Padhye, R.; et al. Polypropylene-Nanodiamond Composite for Hernia Mesh. *Mater. Sci. Eng. C* **2020**, *111*, 110780. [CrossRef] [PubMed]
15. Chen, P.; Huang, F.; Yun, S. Characterization of the Condensed Carbon in Detonation Soot. *Carbon N. Y.* **2003**, *41*, 2093–2099. [CrossRef]
16. Alaferdov, A.V.; Lebedev, O.V.; Roggero, U.F.S.; Hernandez-Figueroa, H.E.; Nista, S.V.G.; Trindade, G.M.; Danilov, Y.A.; Ozerin, A.N.; Moshkalev, S.A. Highly Conductive Nanographite/Ultra-High-Molecular-Weight Polyethylene Composite. *Results Mater.* **2022**, *15*, 100298. [CrossRef]
17. Kurkin, T.S.; Ozerin, A.N.; Kechek'yan, A.S.; Gritsenko, O.T.; Ozerina, L.A.; Alkhanishvili, G.G.; Sushchev, V.G.; Dolmatov, V.Y. The Structure and Properties of Polymer Composite Fibers Based on Poly(Vinyl Alcohol) and Nanodiamond of Detonation Synthesis. *Nanotechnologies Russ.* **2010**, *5*, 340–351. [CrossRef]
18. Neitzel, I.; Mochalin, V.; Bares, J.A.; Carpick, R.W.; Erdemir, A.; Gogotsi, Y. Tribological Properties of Nanodiamond-Epoxy Composites. *Tribol. Lett.* **2012**, *47*, 195–202. [CrossRef]
19. Neitzel, I.; Mochalin, V.; Knoke, I.; Palmese, G.R.; Gogotsi, Y. Mechanical Properties of Epoxy Composites with High Contents of Nanodiamond. *Compos. Sci. Technol.* **2011**, *71*, 710–716. [CrossRef]
20. Lebedev, O.V.; Bogdanova, O.I.; Goncharuk, G.P.; Ozerin, A.N. Tribological and Percolation Properties of Polypropylene/Nanodiamond Soot Composites. *Polym. Polym. Compos.* **2020**, *28*, 369–377. [CrossRef]
21. Lebedev, O.V.; Goncharuk, G.P.; Ozerin, A.N. Changes in Electrical Conductance of Polymer Composites Melts Due to Carbon Nanofiller Particles Migration. *Polymers* **2021**, *13*, 1030. [CrossRef] [PubMed]
22. Schaefer, D.W.; Justice, R.S. How Nano Are Nanocomposites? *Macromolecules* **2007**, *40*, 8501–8517. [CrossRef]
23. Pabisch, S.; Feichtenschlager, B.; Kickelbick, G.; Peterlik, H. Effect of Interparticle Interactions on Size Determination of Zirconia and Silica Based Systems—A Comparison of SAXS, DLS, BET, XRD and TEM. *Chem. Phys. Lett.* **2012**, *521*, 91–97. [CrossRef]
24. Ali, H.E.; Radwan, R.R. Synthesis, Characterization and Evaluation of Resveratrol-Loaded Functionalized Carbon Nanotubes as a Novel Delivery System in Radiation Enteropathy. *Eur. J. Pharm. Sci.* **2021**, *167*, 106002. [CrossRef]
25. Sanusi, O.M.; Benelfellah, A.; Papadopoulos, L.; Terzopoulou, Z.; Malletzidou, L.; Vasileiadis, I.G.; Chrissafis, K.; Bikiaris, D.N.; Aït Hocine, N. Influence of Montmorillonite/Carbon Nanotube Hybrid Nanofillers on the Properties of Poly(Lactic Acid). *Appl. Clay Sci.* **2021**, *201*, 105925. [CrossRef]
26. PP H030 GP/3. Available online: http://www.sibur-int.com/upload/documents/TDSH030GP3v14.pdf (accessed on 29 December 2020).

27. P267-E. Available online: https://studylib.ru/doc/2305578/e-lektroprovodnyj-tehnicheskij-uglerod-p-267-e---p-268 (accessed on 29 December 2020).
28. Napper, D.H. *Polymeric Stabilization of Colloidal Dispersions*; Colloid Science: A Series of Monographs. 2; Academic Press: Cambridge, MA, USA, 1983; ISBN 9780125139809.
29. Mogilevskaya, E.L.; Akopova, T.A.; Zelenetskii, A.N.; Ozerin, A.N. The Crystal Structure of Chitin and Chitosan. *Polym. Sci. Ser. A* **2006**, *48*, 116–123. [CrossRef]
30. Calderon, H.A.; Okonkwo, A.; Estrada-Guel, I.; Hadjiev, V.G.; Alvarez-Ramírez, F.; Robles Hernández, F.C. HRTEM Low Dose: The Unfold of the Morphed Graphene, from Amorphous Carbon to Morphed Graphenes. *Adv. Struct. Chem. Imaging* **2016**, *2*, 10. [CrossRef] [PubMed]
31. Sadezky, A.; Muckenhuber, H.; Grothe, H.; Niessner, R.; Pöschl, U. Raman Microspectroscopy of Soot and Related Carbonaceous Materials: Spectral Analysis and Structural Information. *Carbon N. Y.* **2005**, *43*, 1731–1742. [CrossRef]
32. Mermoux, M.; Chang, S.; Girard, H.A.; Arnault, J.-C. Raman Spectroscopy Study of Detonation Nanodiamond. *Diam. Relat. Mater.* **2018**, *87*, 248–260. [CrossRef]
33. Zou, Q.; Li, Y.G.; Zou, L.H.; Wang, M.Z. Characterization of Structures and Surface States of the Nanodiamond Synthesized by Detonation. *Mater. Charact.* **2009**, *60*, 1257–1262. [CrossRef]

C Journal of Carbon Research

MDPI

Article

Atomic Layer Deposition of Nanolayered Carbon Films

Zhigang Xiao [1,*], Kim Kisslinger [2] and Rebhadevi Monikandan [3]

[1] Department of Electrical Engineering and Computer Science, Alabama A&M University, Normal, AL 35762, USA
[2] Center for Functional Nanomaterials, Brookhaven National Laboratory, Upton, NY 11973, USA; kisslinger@bnl.gov
[3] Materials Characterization Facility, Georgia Institute of Technology, Atlanta, GA 30332, USA; rjeevagan3@mail.gatech.edu
* Correspondence: zhigang.xiao@aamu.edu; Tel.: +1-256-372-5679; Fax: +1-256-372-5855

Abstract: In this paper, carbon thin films were grown using the plasma-enhanced atomic layer deposition (PE-ALD). Methane (CH_4) was used as the carbon precursor to grow the carbon thin film. The grown film was analyzed by the high-resolution transmission electron micrograph (TEM), X-ray photoelectron spectroscopy (XPS) analysis, and Raman spectrum analysis. The analyses show that the PE-ALD-grown carbon film has an amorphous structure. It was found that the existence of defective sites (nanoscale holes or cracks) on the substrate of copper foil could facilitate the formation of nanolayered carbon films. The mechanism for the formation of nanolayered carbon film in the nanoscale holes was discussed. This finding could be used for the controlled growth of nanolayered carbon films or other two-dimensional nanomaterials while combining with modern nanopatterning techniques.

Keywords: nanolayered graphene thin film; two-dimensional nanomaterial; localized growth; plasma-enhanced atomic layer deposition

Citation: Xiao, Z.; Kisslinger, K.; Monikandan, R. Atomic Layer Deposition of Nanolayered Carbon Films. C **2021**, *7*, 67. https://doi.org/10.3390/c7040067

Academic Editor: Gil Goncalves

Received: 26 August 2021
Accepted: 24 September 2021
Published: 27 September 2021

1. Introduction

Since the discovery of mechanically exfoliated graphene in 2004 [1], research on two-dimensional (2D) nanomaterials has grown exponentially due to their compelling physical, chemical, and electrical properties such as high current-carrying capacity, high mobility, and ballistic quantum transport [2–15]. Graphene, a single layer of carbon atoms bonded together in a hexagonal lattice, has a perfect 2D structure, while other 2D nanomaterials such as 2D boron nitride and transition metal dichalcogenide nanomaterials have 2D structures similar to graphene. The unique 2D structure with a lateral size of about 100 nm to a few micrometers and a thickness of only a single atom or few atoms confines electrons and holes to move in a 2D plane, resulting in a strong quantum effect, which produce novel physical, chemical, and electrical properties in the 2D nanomaterials and devices [16–19]. Their strong in-plane covalent bond and atomic thickness also make the 2D nanomaterials have excellent mechanical strength and flexibility [20,21]. Electrons and holes undergo ballistic transport on the sub-micrometer scale without collision in 2D nanomaterials and do not suffer from the scale limitations of current metal-oxide-semiconductor field-effect transistors (MOSFET); therefore, nanoelectronic devices built on 2D materials offer many benefits for further miniaturization in the device fabrication and the possibility to revolutionize the silicon-based semiconductor technology in the future [22–28], allowing for the Moore's Law to continue on its path [29].

Current methods for the preparation of 2D nanomaterials are divided into the top-down method and the bottom-up method [17]. The top-down method relies on the exfoliation of thin layer 2D crystals from their parent layered bulk crystals using processes such as mechanical cleavage and mechanical force-assisted liquid exfoliation, while the bottom-up method is based on the chemical reactions such as chemical vapor deposition (CVD) growth and wet-chemical syntheses. The top-down method has numerous limitations for the application of device fabrication, while the bottom-up method is more promising and could grow 2D nanomaterials with desired compositions and crystal phases and enhanced physical, chemical, and electrical properties. The CVD method is the most popular and effective method for the large-scale synthesis of 2D nanomaterials [30–35].

Atomic layer deposition (ALD) is a special modification of conventional chemical vapor deposition (CVD) for growing thin films [36]. In conventional chemical vapor deposition (CVD), precursors are simultaneously supplied to a substrate, but in atomic layer deposition (ALD), the precursors are alternately exposed onto the substrate and subsequently purged for each precursor. One ALD cycle generally consists of two half-reactions, which are achieved by repeating four steps (the first precursor exposure and purging steps for the first half-reaction, and the second precursor exposure and purging steps for the second half-reaction) [37–40]. Plasma-enhanced atomic layer deposition (PE-ALD) is an extension of thermal ALD in which gas-phase plasma electron impact is substituted for thermal agitation, thereby achieving much lower substrate temperatures [41]. Sequential self-limiting surface reactions allow for precise thickness control, high conformality, and scalability to large surface areas, making ALD an ideal candidate for growing 2D nanomaterials.

In this research, we report the growth of carbon thin films using the plasma-enhanced atomic layer deposition (PE-ALD) and the finding that the existence of defective sites (nanoscale holes or cracks) on the substrate of copper foil could facilitate the formation of nanolayered carbon films. This finding could be used for the controlled growth of nanolayered carbon films or other two-dimensional nanomaterials while combining with modern nanopatterning techniques.

2. Experimental Details (Methods)

A plasma-enhanced atomic layer deposition system (ALD-150LX, Kurt J. Lesker Company, Frankfurt am Main, Germany) was used to grow the carbon film in this research [42]. Methane (CH_4) was used as the precursor for the carbon source, and Ar/H_2 were used for both plasma and purge gases during the deposition. The process chamber had a base pressure of 5×10^{-7} Torr. Carrier gas mass flow controllers (MFC), high-speed ALD valves, and shut-off valves were used to control the flow of carrier gas, precursor, and reactant gas into the process chamber. The inductively coupled remote plasma source was set at 1000 W to generate plasma and the substrate holder, which is electrically grounded was heated at 500 °C during the growth of carbon films. Two types of substrates were used for the growth of carbon films. One substrate was a 25 mm thick Cu foil (Alfa Aesar, Havrier, MA, USA, 99.8%), the other one was a 3-inch diameter <100> silicon wafer (n-type, with ~3 Ω cm resistivity from Virginia Semi-conductor, Inc., Fredericksburg, VA, USA) covered by a 100 nm thick copper (Cu) thin film. The Cu thin film was grown on the silicon substrate with e beam evaporation using an e beam/thermal evaporation system (PVD 75 from Kurt J. Lesker), and copper (Cu) slug (3.175 mm diameter × 3.175 mm length; 99.99% pure from Alfa Aesar) was used as the Cu source material. The Cu foil was dipped in buffered hydrofluoric acid (BHF) (Transene Company, Danvers, MA, USA) for 30 s to etch away any possible oxides such as copper oxide (CuO) on the surface before being transformed to the deposition chamber. The Cu-thin-film-covered silicon substrate was transformed to the deposition chamber right after the deposition of the Cu thin film on the silicon substrate. After being transferred onto the substrate holder, the Cu foil substrate was cleaned by H_2 (65 sccm)/Ar (65 sccm) plasma gas at 1 kW for 10 min, immediately followed by the growth of the carbon film, while the Cu-thin-film-covered silicon substrate was cleaned by H_2 (65 sccm)/Ar (65 sccm) plasma gas at 1 kW for 1 min. The total chamber pressure was about 1.2 Torr. The time for each pulse cycle during the growth of the carbon film was set as follows: 100 ms for CH4, 1000 ms for Ar purge, 6000 ms for H_2 (65 sccm)/Ar (65 sccm) plasma, and 1000 ms for Ar purge. The carbon films were grown for 200 cycles.

2.1. X-ray Photoelectron Spectroscopy (XPS) for Analysis of the Composition and Atomic Concentration of the Carbon Film

X-ray photoelectron spectroscopy (XPS) was performed using the Thermo K-alpha XPS system. The samples were cleaned with IPA and nitrogen gun before running on the XPS. Surface analysis by XPS was achieved by irradiating the sample with a monochromatic source of Al K Alpha X Rays (1.486 KeV). The tool was calibrated with Cu2p, Ag3d, and Au4f internal standards. Low-energy ionized Argon gas was used as a flood gun source for charge neutralization.

2.2. Scanning Electron Micrograph (SEM) for the Imaging and Analysis of the Surface Morphology of Carbon Films

A scanning electron microscope/focused ion beam (SEM/FIB) dual-beam system (Helios 600 from Thermo Fisher Scientific, Waltham, MA, USA) was used to image and analyze the surface of the grown carbon films.

2.3. High-Resolution Tunneling Electron Micrograph (HRTEM) Imaging and Energy Dispersive Spectroscopy (EDS) Mapping of Carbon Films

To characterize the carbon film structure and composition in a transmission electron microscope (TEM), electron transparent cross-sectional samples (lamella) of the grown carbon films were extracted using the standard in situ lift-out method [43] in a scanning electron microscope/focused ion beam (SEM/FIB) dual-beam system (Helios 600 from Thermo Fisher Scientific). High-resolution images were acquired with 200 keV e-beam energy in a high-resolution analytical transmission electron microscope (JEOL JEM-2100F, Showa, Tokyo, Japan). High-resolution images and energy dispersive spectroscopy (EDS) maps were acquired with 200 keV e-beam energy in a scanning TEM (STEM)/TEM (Talos F200X from Thermo Fisher Scientific) equipped with a four-quadrant 0.9 sr X-ray detector.

2.4. Raman Spectroscopy Analysis of Carbon Films

Raman spectroscopy was obtained using a Renishaw Invia Qontor Confocal Raman Microscope. A 488 nm optically pumped semiconductor laser (OPSL, Lagos, Nigeria) was used. An automated 2D Raman mapping was performed to scan the sample surface for the analysis of Raman shift. A continuous overlay of Raman over optical image (white light) map is achieved using a small step size of 1 micron and a 2-micron laser spot size.

3. Results and Discussion

Table A1 summarizes the compositions and atomic concentrations, which were obtained by the XPS analysis for the carbon film grown by the remote plasma-enhanced atomic layer deposition. The carbon film has atomic concentrations of 72% C, 20% O, 3% Cu, and 5% others (F, Ca, and Cl). Figure A1 shows the high-resolution XPS scan of the C sp^2 in the carbon thin film, indicating that the PE-ALD-grown carbon film is an amorphous film. The C 1s core-level XPS peaks of highly ordered graphene layers exhibit asymmetric shapes with higher intensity at higher binding energy from the center position of the peak due to the high electron densities at Fermi levels [44–48]. Figure A2a is an SEM image of the surface of the carbon film grown on the copper (Cu) foil substrate, and Figure A2b is the SEM image of an enlarged view of the carbon film, while Figure A2c is an SEM image of the surface of the carbon film grown on the Cu-thin-film covered silicon substrate, and Figure A2d is the SEM image of an enlarged view of the carbon film. Figure A3 is the cross-sectional TEM image of the copper (Cu) foil substrate with a ~50 nm thick carbon thin film grown on it, showing a rough substrate surface with some nanoscale holes or nanoscale cracks on the surface. After the further cross-sectional analysis of the grown carbon thin-film sample by high-resolution transmission electron micrograph (TEM), it was found that the film grown in the nanoscale holes had the structure of nanolayered graphene, while the film grown on the hole-free flat smooth surface was an amorphous carbon film without any nanolayered structure. Figure A4a is the high-resolution cross-sectional TEM image of the carbon thin film grown in a nanoscale hole, clearly showing some nanolayered structures; Figure A4b is an enlarged view of the film inside the hole, showing nanolayered structures, while Figure A4c is an enlarged view of the film on the bottom surface of the Cu nanohole, showing nanolayered carbon films, which have a difference of about 0.22 nm between layers [49]. Figure A5a–c shows the high-resolution cross-sectional TEM images of the carbon thin film grown in another Cu nanoscale hole, presenting similar nanolayered structures in Figure A4a–c. Figure A6a is the high-resolution cross-sectional TEM image of the carbon film grown on a hole-free flat smooth surface of the same Cu foil substrate, showing a homogenous amorphous carbon film without any nanolayered structures on the interface or anywhere else; Figure A6b is an enlarged view of the film. Figure A7a is the high-resolution cross-sectional TEM image of the carbon film grown on the Cu-thin-film-covered silicon substrate, showing a similar amorphous structure as Figure A6a, without any nanolayered structures; Figure A7b,c shows the enlarged views of the film. The TEM image of the carbon film grown on the Cu thin film further confirms that the carbon film grown on a hole-free surface of copper is an amorphous film in the experimental conditions.

The carbon film grown on the chromium (Cr)/nickel (Ni) film was also investigated. A Ni film was grown on the silicon wafer using the e-beam evaporation, followed by a thin Cr layer. The carbon film was then grown using the PE-ALD with the same experimental condition for the growth of the carbon films on copper foil and a copper-covered silicon substrate. Figure A8a shows the high-resolution cross-sectional TEM image of the carbon film on the Cr/Ni film. The film is an amorphous carbon film and is similar to those in Figures A6 and A7, without any nanolayered structures on the interface or anywhere else. Figure A8b–f summarizes the energy dispersive spectroscopy (EDS) elemental map analysis in the interface region of the sample: Figure A8b is the color-coded EDS elemental map of carbon (C), where the green color represents the C K-alpha X-ray signal; Figure A8c is the EDS elemental map of Cr (blue); Figure A8d is the EDS elemental map of O (red), which overlaps with the Cr in Figure A8c, indicating that the Cr layer was oxidized; Figure A8e is the EDS elemental map of Ni (yellow); Figure A8f is the color-coded EDS elemental map of C, Cr, O, Ni combined together. Figure A8g shows the high-angle annular dark-field (HAADF) STEM micrograph of the same region mapped in Figure A8f, clearly showing the high atomic number (Z) Cr and Ni layers.

Figure A9 shows the two typical Raman spectra measured from the carbon film grown on the Cu foil substrate, where the red line has the characteristic D and G peaks, while the black line does not have the D and G peaks. The Raman analysis was performed with the 2D mapping measurement on the surface of carbon film, where the areas for the measurement were selected randomly. The analysis from 100 measurement scans shows that about 60% of the measurements have a Raman spectrum similar to the red line, while about 40% have a Raman spectrum similar to the black line. The 2D mapping measurement did not find the 2D peaks, indicating that the PE-ALD-grown carbon film is an amorphous film [50–52].

Mechanism of chemical vapor deposition (CVD) or atomic layer deposition (ALD) includes diffusion of gas toward the substrate surface, absorption to the surface, reaction, desorption, and diffusion of by-product. The possible reason why the nanolayered carbon film was formed in the nanoscale Cu hole is that the hole selected the neutral species (methane molecules (CH_4)) or ionic species (CH_3) with a certain moving direction and only allowed the species (CH_4 or CH_3) that had a parallel or close parallel moving direction to the wall of nanohole to enter the hole, which facilitated the nanolayered structure. Figure A10a,b shows the possible mechanism for the growth of nanolayered carbon film in the nanoscale Cu hole and the growth of homogeneous amorphous carbon film on a flat smooth Cu surface, respectively. The nanoscale hole in Figure A10a functions similar to a filter that blocks most methane molecules (CH_4) or other ionized groups such as methyl group (CH_3) with random moving directions, and only those with a moving direction parallel or close parallel to the wall of nanohole can enter the hole. The species with the same moving direction arrive on the Cu interface in order, facilitating the formation of the nanolayers while the hole-free flat Cu surface in Figure A10b is subjected to methane molecules (CH_4) or ionized groups from every direction simultaneously, resulting in a uniform amorphous carbon film without forming any nanolayered structure. Researchers found that electric fields could be manipulated to grow nanostructured carbon films [53,54] or graphene [55]. The electric field made the ionic species flow in a certain direction, and the ionic species moving in order facilitated the growth of the nanostructured carbon films or graphene. Nanoscale holes could be precisely patterned and fabricated with the hole diameter, down to a few nanometers, using the nanofabrication patterning techniques [56]; further research on the growth of carbon films on a substrate with uniform-patterned nanoscale holes may confirm the mechanism or suggest a different explanation about the phenomenon that the formation of the amorphous carbon layer could be varied by the existence of defective sites (nanoscale holes or cracks) on the substrate.

4. Summary

In summary, carbon thin films were grown using the plasma-enhanced atomic layer deposition (PE-ALD). The carbon film was analyzed using the high-resolution transmission electron micrograph (TEM), X-ray photoelectron spectroscopy (XPS) analysis, and Raman spectrum analysis, indicating that the PE-ALD-grown carbon film has an amorphous structure. The existence of defective sites (nanoscale holes or cracks) on the substrate was found to facilitate the formation of nanolayered carbon films in the plasma-enhanced atomic layer deposition of carbon films. This finding could be used for the controlled growth of nanolayered carbon films or other two-dimensional nanomaterials while combining with modern nanopatterning techniques.

Author Contributions: Z.X. designed the experiments, fabricated the materials, analyzed the data, and wrote the manuscript. K.K. performed the HRTEM experiment and contributed to analyzing the data and writing the manuscript. R.M. performed the Raman and XPS experiments and contributed to analyzing the data and writing the manuscript. All authors have read and agreed to the published version of the manuscript.

Funding: U.S. Department of Energy, Office of Basic Energy Sciences, under Contract No. DE-SC00112704; National Science Foundation under Grant No. CBET-1740687 and Grant No. ECCS-1542174; Department of Defense under Grant No. W911NF-16-1-0554.

Acknowledgments: Research carried out in part at the Center for Functional Nanomaterials, Brookhaven National Laboratory, which is supported by the U.S. Department of Energy, Office of Basic Energy Sciences, under Contract No. DE-SC00112704; XPS and Raman measurements were carried out at the IEN/IMAT Material Characterization Facility, in part at the Georgia Tech Institute for Electronics and Nanotechnology, a member of the National Nanotechnology Coordinated Infrastructure, which is supported by the National Science Foundation (Grant ECCS-1542174); the research was supported by National Science Foundation under Grant No. CBET-1740687 and Department of Defense under Grant No. W911NF-16-1-0554. The authors gratefully thank Fernando Camino at Brookhaven National Laboratory for help in obtaining the SEM pictures.

Conflicts of Interest: The authors declare no conflict of interest.

Appendix A

Table A1. Compositions of the carbon thin film measured by XPS.

Atomic Concentration (%) of Carbon Film			
C	**O**	**Cu**	**Others (F, Ca, Cl)**
72	20	3	5

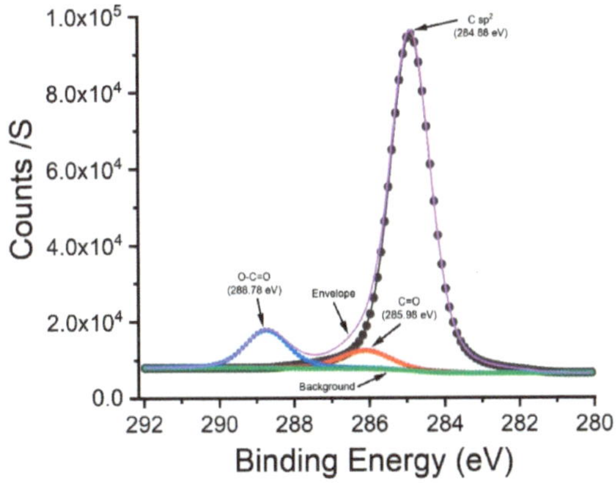

Figure A1. High-resolution XPS scan of C sp^2 in the carbon film.

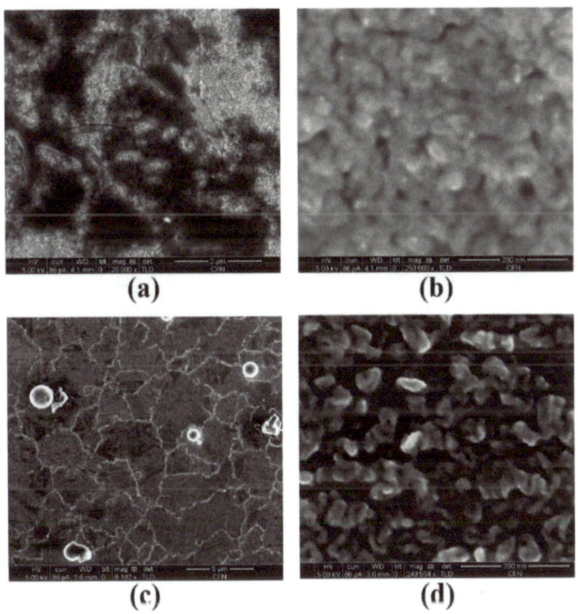

Figure A2. (**a**) SEM image of a carbon film grown on the copper (Cu) foil substrate; (**b**) SEM image of an enlarged view of the carbon film grown on the copper (Cu) foil substrate; (**c**) SEM image of a carbon film grown on the Cu-thin-film-covered silicon substrate; (**d**) SEM image of an enlarged view of the carbon film grown on the Cu-thin-film-covered silicon substrate.

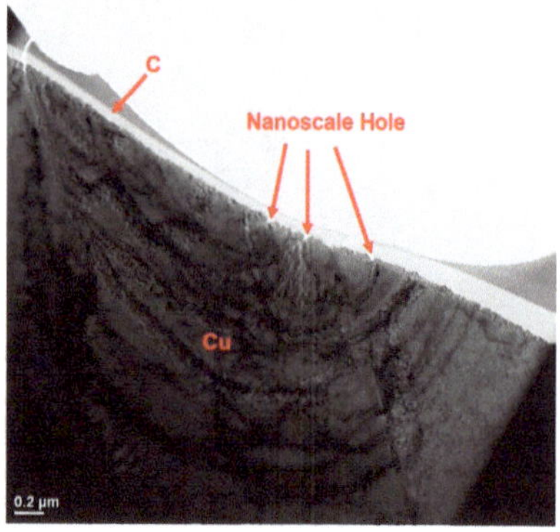

Figure A3. TEM micrograph of the TEM sample. The carbon film was grown on the copper foil substrate, which has some nanoscale holes on the surface. Scale bar in the figure is 0.2 µm.

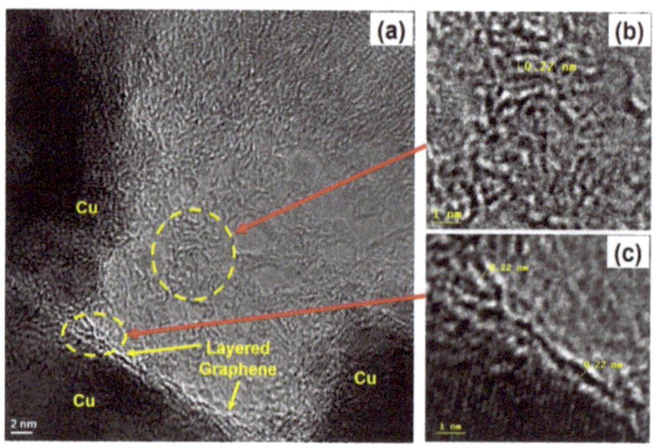

Figure A4. (**a**) High-resolution cross-sectional TEM image of the nanolayered carbon film in a nanoscale hole; (**b**) an enlarged view of the film inside the hole, showing nanolayered structures; (**c**) an enlarged view of the film on the interface of Cu substrate, showing nanolayered structures. Scale bar in (**a**): 2 nm; scale bar in (**b**,**c**): 1 nm.

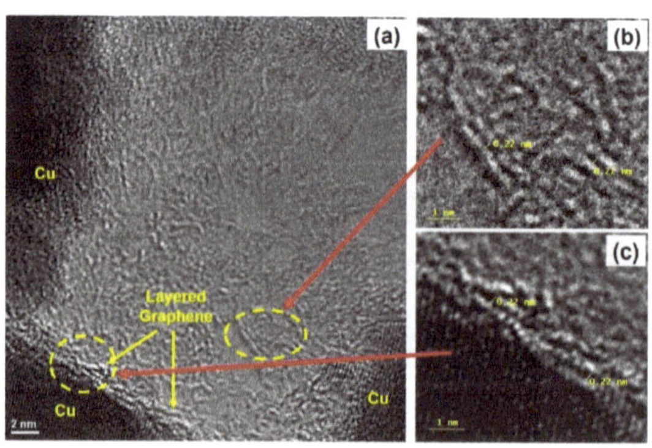

Figure A5. (**a**) High-resolution cross-sectional TEM image of the nanolayered carbon film in another nanoscale hole; (**b**) an enlarged view of the film inside the hole, showing nanolayered structures; (**c**) an enlarged view of the film on the surface of Cu substrate, showing nanolayered structures. Scale bar in (**a**): 2 nm; scale bar in (**b**,**c**): 1 nm.

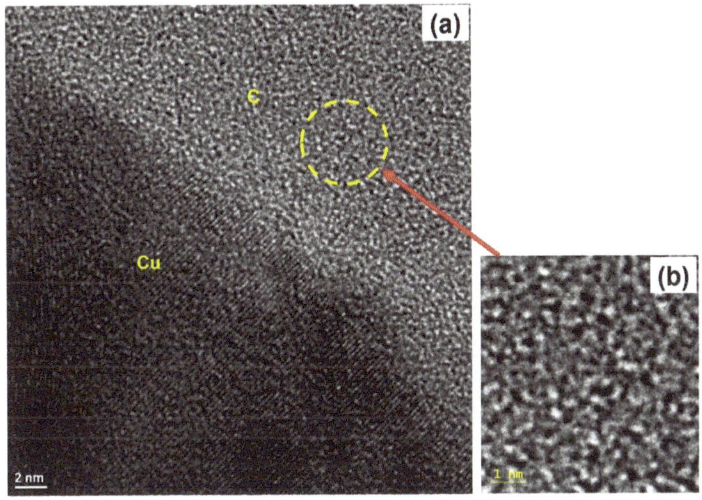

Figure A6. (**a**) High-resolution cross-sectional TEM image of the carbon film on a flat smooth surface of the same copper foil substrate, showing a homogenous amorphous carbon film without any nanolayered structures on the interface or anywhere else; (**b**) an enlarged view of the film. Scale bar in (**a**): 2 nm; scale bar in (**b**): 1 nm.

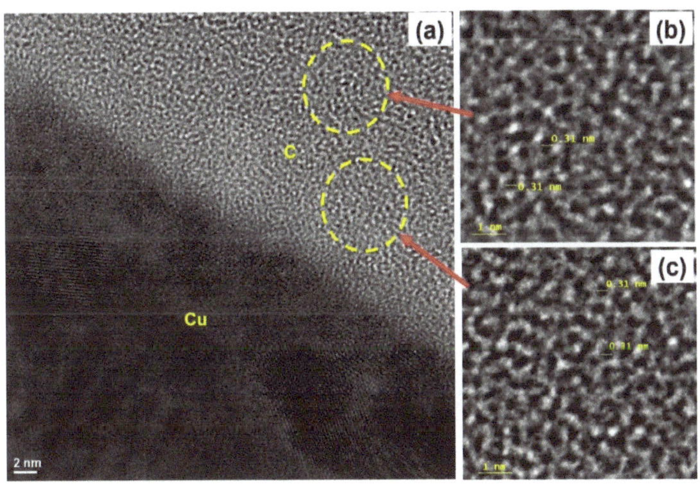

Figure A7. (**a**) High-resolution cross-sectional TEM image of the carbon film on a copper film which was grown on a silicon wafer, showing a homogenous amorphous carbon film without any nanolayered structures on the interface or anywhere else; (**b**,**c**) enlarged views of the film. Scale bar in (**a**): 2 nm; Scale bar in (**b**,**c**): 1 nm.

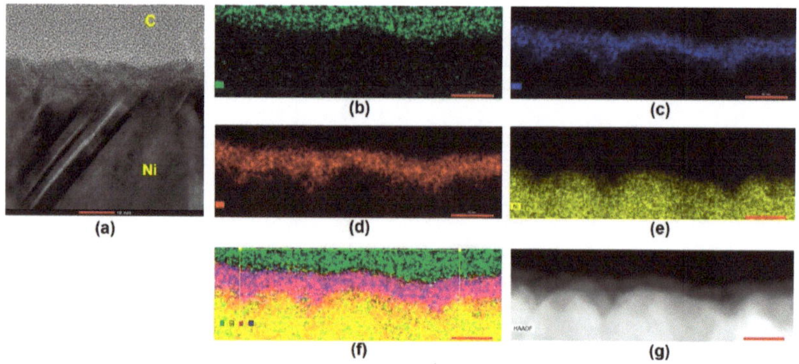

Figure A8. (**a**) High-resolution cross-sectional TEM image of the carbon film on a Ni film that was covered by a thin Cr film and grown on a silicon wafer, showing a homogenous amorphous carbon film without any nanolayered structures on the interface or anywhere else; (**b**) color-coded EDS elemental map of carbon (**c**) in the interface region, where the green color represents the C K-alpha x-ray signal); (**c**) color-coded EDS elemental map of Cr (blue); (**d**) color-coded EDS elemental map of O (red), which overlaps with the Cr in (**c**), indicating that the Cr layer was oxidized; (**e**) color-coded EDS elemental map of Ni (yellow); (**f**) color-coded EDS elemental map of C, Cr, O, Ni combined together; (**g**) HAADF STEM micrograph of the same region mapped in (**f**). Scale bar in (**a–g**): 10 nm.

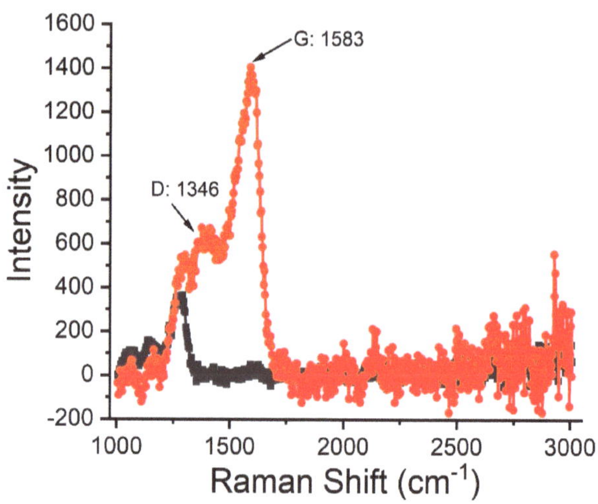

Figure A9. Raman spectrum of the carbon film grown on the Cu foil substrate, where the red line shows that the measurement has the D and G peaks, while the black line shows that the measurement does not have the D and G peaks.

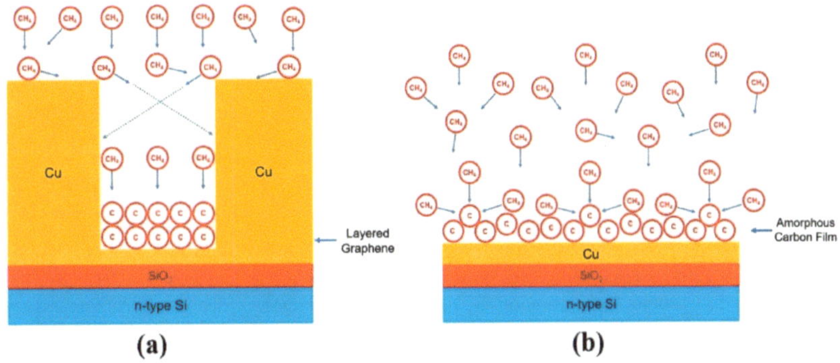

Figure A10. (**a**) Diagram for describing the mechanism about the formation of nanolayered carbon film in a nanoscale Cu hole, where the hole functions as a filter and select only the species moving in the direction parallel to the hole wall to enter the hole and reach the bottom; (**b**) diagram for describing the mechanism about the formation of the amorphous carbon film on a flat smooth Cu surface due to the random moving direction of species.

References

1. Novoselov, K.S.; Geim, A.K.; Morozov, S.V.; Jiang, D.; Zhang, Y.; Dubonos, S.V.; Grigorieva, I.V.; Firsov, A.A. Electric field effect in atomically thin carbon films. *Science* **2004**, *306*, 666–669. [CrossRef]
2. Stoller, M.D.; Park, S.; Zhu, Y.; An, J.; Ruoff, R.S. Graphene-based ultracapacitors. *Nano Lett.* **2008**, *8*, 3498–3502. [CrossRef]
3. Lee, C.; Wei, X.D.; Kysar, J.W.; Hone, J. Measurement of the elastic properties and intrinsic strength of monolayer graphene. *Science* **2008**, *321*, 385–388. [CrossRef] [PubMed]
4. Nair, R.R.; Blake, P.; Grigorenko, A.N.; Novoselov, K.S.; Booth, T.J.; Stauber, T.; Peres, N.M.R.; Geim, A.K. Fine structure constant defines visual transparency of graphene. *Science* **2008**, *320*, 1308. [CrossRef]
5. Xu, M.S.; Liang, T.; Shi, M.M.; Chen, H.Z. Graphene-like two-dimensional materials. *Chem. Rev.* **2013**, *113*, 3766–3798. [CrossRef] [PubMed]
6. Gupta, A.; Sakthivel, T.; Seal, S. Recent development in 2D materials beyond graphene. *Prog. Mater. Sci.* **2015**, *73*, 44–126. [CrossRef]
7. Ferrari, A.C.; Bonaccorso, F.; Fal'Ko, V.; Novoselov, K.S.; Roche, S.; Bøggild, P.; Borini, S.; Koppens, F.H.L.; Palermo, V.; Pugno, N.; et al. Science and technology roadmap for graphene, related two-dimensional crystals, and hybrid systems. *Nanoscale* **2015**, *7*, 4598–4810. [CrossRef] [PubMed]
8. Novoselov, K.S.; Geim, A.K.; Morozov, S.V.; Jiang, D.; Katsnelson, M.I.; Grigorieva, I.V.; Dubonos, S.V.; Firsov, A.A. Two-dimensional gas of massless Dirac fermions in graphene. *Nature* **2005**, *438*, 197–200. [CrossRef]
9. Schedin, F.; Geim, A.K.; Morozov, S.V.; Hill, E.W.; Blake, P.; Katsnelson, M.I.; Novoselov, K.S. Detection of individual gas molecules adsorbed on graphene. *Nat. Mater.* **2007**, *6*, 652–655. [CrossRef] [PubMed]
10. Weng, Q.; Wang, X.; Bando, Y.; Golberg, D. Functionalized hexagonal boron nitride nanomaterials: Emerging properties and applications. *Chem. Soc. Rev.* **2016**, *45*, 3989–4012. [CrossRef]
11. Li, L.; Chen, Y. Atomically thin boron nitride: Unique properties and applications. *Adv. Funct. Mater.* **2016**, *26*, 2594–2608. [CrossRef]
12. Tan, C.; Zhang, H. Two-dimensional transition metal dichalcogenide nanosheet-based composites. *Chem. Soc. Rev.* **2015**, *44*, 2713–2731. [CrossRef]
13. Chhowalla, M.; Shin, H.; Eda, G.; Li, L.; Loh, K.; Zhang, H. The chemistry of two-dimensional transition metal dichalcogenide nanosheets. *Nat. Chem.* **2013**, *5*, 263–275. [CrossRef]
14. Huang, X.; Zeng, Z.; Zhang, H. Metal dichalcogenide nanosheets: Preparation, properties and applications. *Chem. Soc. Rev.* **2013**, *42*, 1934–1946. [CrossRef]
15. Lv, R.; Robinson, J.A.; Schaak, R.E.; Sun, D.; Mallouk, T.T.; Terrones, M. Transition metal dichalcogenides and beyond: Synthesis, properties, and applications of single- and few-layer nanosheets. *Acc. Chem. Res.* **2015**, *48*, 56–64. [CrossRef]
16. Zhang, H. Ultrathin two-dimensional nanomaterials. *ACS Nano* **2015**, *9*, 9451–9469. [CrossRef]
17. Tan, C.; Cao, X.; Wu, X.; He, Q.; Yang, J.; Zhang, X.; Chen, J.; Zhao, W.; Han, S.; Nam, G.; et al. Recent advances in ultrathin two-dimensional nanomaterials. *Chem. Rev.* **2017**, *117*, 6225–6331. [CrossRef] [PubMed]
18. Geim, A.K.; Novoselov, K.S. The rise of graphene. *Nat. Mater.* **2007**, *6*, 183–191. [CrossRef]
19. Zhang, Y.; Tan, Y.; Stormer, H.L.; Kim, P. Experimental observation of the quantum hall effect and berry's phase in graphene. *Nature* **2005**, *438*, 201–204. [CrossRef] [PubMed]
20. Bertolazzi, S.; Brivio, J.; Kis, A. Stretching and breaking of ultrathin MoS_2. *ACS Nano* **2011**, *5*, 9703–9709. [CrossRef] [PubMed]
21. Akinwande, D.; Petrone, N.; Hone, J. Two-dimensional flexible nanoelectronics. *Nat. Commun.* **2014**, *5*, 5678. [CrossRef] [PubMed]

22. Conti, S.; Pimpolari, L.; Calabrese, G.; Worsley, R.; Majee, S.; Polyushkin, D.; Paur, M.; Pace, S.; Keum, D.; Fabbri, F.; et al. Low-voltage 2D materials-based printed field-effect transistors for integrated digital and analog electronics on paper. *Nat. Commun.* **2020**, *11*, 3566. [CrossRef] [PubMed]
23. Liu, Y.; Halim, U.; Ding, M.; Liu, Y.; Wang, Y.; Jia, C.; Chen, P.; Duan, X.; Wang, C.; Song, F.; et al. Solution-processable 2D semiconductors for high performance large-area electronics. *Nature* **2018**, *562*, 254–258.
24. Liu, C.; Chen, H.; Wang, S.; Liu, Q.; Jiang, Y.; Zhang, D.; Liu, M.; Zhou, P. Two-dimensional materials for next-generation computing technologies. *Nat. Nanotechnol.* **2020**, *15*, 545–557. [CrossRef]
25. Li, L.; Shao, L.; Liu, X.; Gao, A.; Wang, H.; Zheng, B.; Hou, G.; Shehzad, K.; Yu, L.; Miao, F. Room-temperature valleytronic transistor. *Nat. Nanotechnol.* **2020**, *15*, 743–749. [CrossRef]
26. Qian, Q.; Lei, J.; Wei, J.; Zhang, Z.; Tang, G.; Zhong, K.; Zheng, Z.; Chen, K. 2D materials as semiconducting gate for field-effect transistors with inherent over-voltage protection and boosted ON-current. *NPJ 2D Mater. Appl.* **2019**, *3*, 24. [CrossRef]
27. Fiori, G.; Bonaccorso, F.; Iannaccone, G.; Palacios, T.; Neumaier, D.; Seabaugh, A.; Banerjee, S.K.; Colombo, L. Electronics based on two-dimensional materials. *Nat. Nanotechnol.* **2014**, *7*, 768–779. [CrossRef]
28. Chhowalla, M.; Jena, D.; Zhang, H. 2D semiconductors for transistors. *Nat. Rev. Mater.* **2016**, *1*, 16052. [CrossRef]
29. Moore, G.E. Cramming more components onto integrated circuits. *Electronics* **1965**, *38*, 114–117. [CrossRef]
30. Yu, J.; Li, J.; Zhang, W.; Chang, H. Synthesis of high quality two-dimensional materials via chemical vapor deposition. *Chem. Sci.* **2015**, *6*, 6705–6716. [CrossRef] [PubMed]
31. Najmaei, S.; Liu, Z.; Zhou, W.; Zou, X.; Shi, G.; Lei, S.; Yakobson, B.I.; Idrobo, J.-C.; Ajayan, P.M.; Lou, J. Vapour phase growth and grain boundary structure of molybdenum disulphide atomic layers. *Nat. Mater.* **2013**, *12*, 754–759. [CrossRef]
32. Van der Zande, A.M.; Huang, P.Y.; Chenet, D.A.; Berkelbach, T.C.; You, Y.; Lee, G.; Heinz, T.F.; Reichman, D.R.; Muller, D.; Hone, J.C. Grains and grain boundaries in highly crystalline monolayer molybdenum disulphide. *Nat. Mater.* **2013**, *12*, 554–561. [CrossRef] [PubMed]
33. Schmidt, H.; Wang, S.; Chu, L.; Toh, M.; Kumar, R.; Zhao, W.; Neto, A.H.C.; Martin, J.; Adam, S.; Özyilmaz, B.; et al. Transport properties of monolayer MoS_2 grown by chemical vapor deposition. *Nano Lett.* **2014**, *14*, 1909–1913. [CrossRef] [PubMed]
34. Chang, Y.; Zhang, W.; Zhu, Y.; Han, Y.; Pu, J.; Chang, J.; Hsu, W.; Huang, J.; Hsu, C.; Chiu, M.; et al. Monolayer $MoSe_2$ grown by chemical vapor deposition for fast photodetection. *ACS Nano* **2014**, *8*, 8582–8590. [CrossRef] [PubMed]
35. Xia, J.; Huang, X.; Liu, L.; Wang, M.; Wang, L.; Huang, B.; Zhu, D.; Li, J.; Gu, C.; Meng, X. CVD synthesis of large-area, highly crystalline $MoSe_2$ atomic layers on diverse substrates and application to photodetectors. *Nanoscale* **2014**, *6*, 8949–8955. [CrossRef] [PubMed]
36. George, S.M. Atomic layer deposition: An overview. *Chem. Rev.* **2010**, *110*, 111–131. [CrossRef] [PubMed]
37. Valdivia, A.; Tweet, D.J.; Conley, J.F., Jr. Atomic layer deposition of two dimensional MoS_2 on 150 mm substrates. *J. Vac. Sci. Technol.* **2016**, *34*, 021515. [CrossRef]
38. Jin, Z.; Shin, S.; Kwon, D.; Hana, S.; Min, Y. Novel chemical route for atomic layer deposition of MoS_2 thin film on SiO_2/Si substrate. *Nanoscale* **2014**, *6*, 14453–14458. [CrossRef]
39. Dai, T.; Liu, Y.; Fan, X.; Liu, X.; Xie, D.; Li, Y. Synthesis of few-layer 2H-$MoSe_2$ thin films with wafer-level homogeneity for high-performance photodetector. *Nanophotonics* **2018**, *7*, 1959–1969. [CrossRef]
40. Dai, T.; Liu, Y.; Liu, X.; Xie, D.; Li, Y. High performance photodetectors constructed on atomically thin few-layer MoSe2 synthesized using atomic layer deposition and a chemical vapor deposition chamber. *J. Alloys Compd.* **2019**, *785*, 951–957. [CrossRef]
41. Zhang, Y.; Ren, W.; Jiang, Z.; Yang, S.; Jing, W.; Shi, P.; Wu, X.; Ye, Z. Low-temperature remote plasma-enhanced atomic layer deposition of graphene and characterization of its atomic-level structure. *J. Mater. Chem. C* **2014**, *2*, 7570–7574. [CrossRef]
42. Xiao, Z.; Kisslinger, K.; Chance, S.; Banks, S. Comparison of hafnium dioxide and zirconium dioxide grown by plasma-enhanced atomic layer deposition for the application of electronic materials. *Crystals* **2020**, *10*, 136. [CrossRef]
43. Stevie, F.A.; Gianuzzi, L.A.; Prenitzer, B.I. *Introduction to Focused Ion Beams: Instrumentation, Theory, Techniques and Practice*; Springer International Publishing: New York, NY, USA, 2005.
44. Bulusheva, L.G.; Kanygin, M.A.; Arkhipov, V.E.; Popov, K.M.; Fedoseeva, Y.V.; Smirnov, D.A.; Okotrub, A.V. In situ X-ray photoelectron spectroscopy study of lithium interaction with graphene and nitrogen-doped graphene films produced by chemical vapor deposition. *J. Phys. Chem. C* **2017**, *121*, 5108. [CrossRef]
45. Patil, U.V.; Pawbake, A.S.; Machuno, L.G.B.; Gelamo, R.V.; Jadkark, S.R.; Rout, C.S.; Late, D.J. Effect of plasma treatment on multilayer graphene: X-ray photoelectron spectroscopy, surface morphology investigations and work function measurements. *RSC Adv.* **2016**, *6*, 48843. [CrossRef]
46. Matei, D.G.; Weber, N.; Kurasch, S.; Wundrack, S.; Woszczyna, M.; Grothe, M.; Weimann, T.; Ahlers, F.; Stosch, R.; Kaiser, U.; et al. Functional single-layer graphene sheets from aromatic monolayers. *Adv. Mater.* **2013**, *25*, 4146. [CrossRef] [PubMed]
47. Yamada, Y.; Kim, J.; Matsuo, S.; Sato, S. Nitrogen-containing graphene analyzed by X-ray photoelectron spectroscopy. *Carbon* **2014**, *70*, 59–74. [CrossRef]
48. Kovtun, A.; Jones, D.; Dell'Elce, S.; Treossi, E.; Liscio, A.; Palermo, V. Accurate chemical analysis of oxygenated graphene-based materials using X-ray photoelectron spectroscopy. *Carbon* **2019**, *143*, 268–275. [CrossRef]
49. Mao, S.; Yu, K.; Chang, J.; Steeber, D.A.; Ocola, L.E.; Chen, J. Direct growth of vertically-oriented graphene for field-effect transistor biosensor. *Sci. Rep.* **2013**, *3*, 1696. [CrossRef]

50. Yang, Z.; Bhowmick, S.; Sen, F.G.; Banerji, A.; Alpas, A.T. Roles of sliding-induced defects and dissociated water molecules on low friction of graphene. *Sci. Rep.* **2018**, *8*, 121. [CrossRef] [PubMed]
51. Tian, H.; Yang, Y.; Xie, D.; Cui, Y.; Mi, W.; Zhang, Y.; Ren, T. Wafer-scale integration of graphene-based electronic, optoelectronic and electroacoustic devices. *Sci. Rep.* **2014**, *4*, 3598. [CrossRef]
52. Li, M.; Liu, D.; Wei, D.; Song, X.; We, D.; Wee, A. Controllable synthesis of graphene by plasma-enhanced chemical vapor deposition and its related applications. *Adv. Sci.* **2016**, *3*, 1600003. [CrossRef]
53. Shakerzadeh, M.; Teo, E.H.T.; Sorkin, A.; Bosman, M.; Tay, B.K.; Su, H. Plasma density induced formation of nanocrystals in physical vapor deposited carbon films. *Carbon* **2011**, *49*, 1733. [CrossRef]
54. Shakerzadeh, M.; Loh, G.C.; Xu, N.; Chow, W.L.; Tan, C.W.; Lu, C.; Yap, R.C.C.; Tan, D.; Tsang, S.H.; Teo, E.H.T.; et al. Re-ordering chaotic carbon: Origins and application of textured carbon. *Adv. Mater.* **2012**, *24*, 4112–4123. [CrossRef] [PubMed]
55. Zhang, Y.; Tang, S.; Deng, D.; Deng, S.; Chen, J.; Xu, N. Growth direction manipulation of few-layer graphene in the vertical plane with parallel arrangement. *Carbon* **2013**, *56*, 103–108. [CrossRef]
56. Manfrinato, V.R.; Camino, F.E.; Stein, A.; Zhang, L.; Lu, M.; Stach, E.A.; Black, C.T. Patterning Si at the 1 nm length scale with aberration corrected electron-beam lithography: Tuning of plasmonic properties by design. *Adv. Funct. Mater.* **2019**, *29*, 1903429. [CrossRef]

C Journal of Carbon Research

Article

Conformational Analysis of [60]PCBM from DFT Simulations of Electronic Energies, Bond Strain and the ^{13}C NMR Spectrum: Input Geometry Determination and Ester Bond Rotation Dynamics

Tong Liu [1] and T. John S. Dennis [2,3,*]

1 School of Physics and Astronomy, Queen Mary University of London, London E1 4NS, UK; t.liu@qmul.ac.uk
2 MOE Key Laboratory of Macromolecular Synthesis and Functionalization, Department of Polymer Science and Engineering, Zhejiang University, Hangzhou 310027, China
3 Haina-Carbon Nanostructure Research Center, Yangtze Delta Region Institute of Tsinghua University, Jiaxing 314006, China
* Correspondence: j.dennis@zju.edu.cn

Abstract: With the aim of determining the best input geometry for DFT calculations of [60]PCBM, the geometry of 24 chemically possible [60]PCBM conformers were optimised and their electronic energies and average bond strains were determined. A DFT analysis of the relevant dihedral angles provided insights into the dynamical behaviour of the ester group through sterically restricted bond rotations. In addition, the ^{13}C NMR spectra of the six better performing conformers were simulated and compared with an experiment. There is a close correlation between average bond strain, total electronic energy and mean absolute error of the simulated ^{13}C NMR spectra of the ester carbons. The best overall candidate conformer for the input geometry had the C61-C4, C4-C3 and C3-C2 single bonds of the alkyl chain in *syn*, *anti* and *anti* arrangements, respectively, and had the C2-C1 and C1-O single bonds of the ester in *syn* and *anti* arrangements, respectively. This contrasts strikingly with most representations of PCBM in the literature, which depict all relevant bonds in *anti* arrangements.

Keywords: PCBM; conformer; structure

Citation: Liu, T.; Dennis, T.J.S. Conformational Analysis of [60]PCBM from DFT Simulations of Electronic Energies, Bond Strain and the ^{13}C NMR Spectrum: Input Geometry Determination and Ester Bond Rotation Dynamics. *C* **2021**, *7*, 66. https://doi.org/10.3390/c7030066

Academic Editor: Gil Goncalves

Received: 16 August 2021
Accepted: 16 September 2021
Published: 21 September 2021

1. Introduction

[60]PCBM, or, correctly, methyl 4-[3′*H*-phenyl cyclopropa(C_{60}-I_h) [5,6] fulleren-1,9-yl]butanoate, was first synthesised in 1995 as one of six novel homo- and methano-bridged cyclo-additions to C_{60} [1]. Since then, [60]PCBM has become one of the most important n-type molecules in organic electronics. The initial application was as the archetypal electron acceptor in bulk heterojunction photovoltaics [2]. However, it has gone on to be used in many other nano-electronic applications, including perovskite photovoltaics [3], field effect transistors [4–6], light emitting diodes [7,8] and photodetectors [9–11].

To aid in the understanding of these applications experimental research has been supported by hundreds of quantum chemical ab initio calculations and simulations. Indeed, a Google search carried out on the day of submission for "PCBM" AND "DFT" (where AND is logical) returned over 10^5 internet matches, and a Web of Science search indicated that about 10% of primary sources on [60]PCBM involve quantum chemical calculations. However, to perform accurate calculations/simulations, an accurate input geometry needs to be used. Solution ^{13}C NMR spectroscopy indicates that [60]PCBM has C_S molecular point group symmetry in which the phenyl group bisects the mirror plane [1]. In addition, ^{1}H NMR spectroscopy shows well-defined second-order spin–spin coupling effects indicative of substantially hindered rotation about the single bonds between heavy atoms of the ester group [12]. With this, [60]PCBM has potential to exist as 32 conformers that are consistent with [60]PCBM's C_S symmetry, of which 24 are chemically possible.

We trialled all 24 conformers, aiming to determining the best input geometry for simulations of non-solid [60]PCBM. Although the results were obtained in relation to simulating the ^{13}C NMR spectrum, they are generally applicable. In addition, although they were obtained in relation to a liquid phase experiment, they may provide insights into the mechanism of packing of [60]PCBM in the solid phase. This is because solid samples used in solar cells and other organic electronics applications are almost invariably obtained from solution (e.g., via spin coating), where the molecule would initially have the conformational structure, and need to conform during condensation to the restraints on the dynamical behaviour of the ester group, obtained here.

2. Materials and Methods

All ab initio calculations were performed using the Gaussian 16 package [13]. The gauge independent atomic orbitals (GIAO) DFT simulations of the ^{13}C NMR spectrum employed the ωB97XD method coupled with the cc-pVTZ basis set. The range separation parameter for electron exchange, ω, was set to zero bohr^{-1}, while the other parameters were kept at their default values, and the SCRF solvation model was used. This is a method we previously found [14] to substantially outperform the very commonly used B3LYP method when specifically applied to PCBM. All other calculations were carried out at a B3LYP/cc-pVTZ level together with Grimme-D3 empirical dispersion with Becke–Johnson damping.

An initial input geometry for PCBM generated using Gaussview software [15] and subsequently freely optimised to a minimum with tight cut-offs. A frequency calculation test of this optimised geometry revealed no vibrational frequencies, indicating that the optimisation converged to a minimum. The optimised geometry did not, however, conform to any of the 24 C_S-symmetry conformers. Instead, it had C_1 symmetry with the ester lying slightly off the mirror plane. To ensure than the conformers had the experimental C_S symmetry, starting with the optimised geometry, the five relevant dihedral angles of each conformer were set and constrained to either 0° or 180°. Each of the conformers was then re-optimised subject to the 5 constrained dihedral angles (all other bond dihedral angles, bond angles and bond lengths were freely varied).

The [60]PCBM used in this study was synthesised following an adaption of the original method of Hummelen et al., [1] purified by HPLC on a silica semi-preparative column and characterised at room temperature by ^{13}C NMR spectroscopy at 600 MHz (125 MHz for C) in carbon disulphide solution. A two-page set-by-step description of our synthesis method and an HPLC chromatogram of our purification method was recently published elsewhere [12].

3. Results and Discussion

[60]PCBM, as its trivial name suggests, may be considered as consisting of three parts: a phenyl group, a C_{61} cyclopropafullerenyl group and a butyric acid methyl ester group. The conformational structure of the conjoined phenyl cyclopropafullerenyl substituent to the ester is trivial. This is because there is only one conformation of that grouping that is consistent with the molecule's C_s symmetry. This conformer has phenyl carbons C1 and C4 and cyclopropafullerene carbons C20, C30, C31, C40 and C61 on the mirror plane while all other atoms exist as mirrored pairs bisecting the mirror plane.

For the ester group to retain PCBM's C_S symmetry in solution, all heavy elements of the ester group (C4, C3, C2, C1, both oxygens and the methyl carbon) must lie on the mirror plane. In addition, one of the methyl hydrogens must also be on the mirror plane and the other eight hydrogens of the ester group must exit in mirrored pairs bisecting the mirror plane. To lie on the mirror plane, the dihedral angles of the heavy elements of the ester must be either 0° or 180°. There are no other possible dihedral angles that are consistent with C_S symmetry. These dihedral angles are those of the C61-C4, C4-C3, C3-C2, C2-C1 and C1-O single bonds. With each of the five dihedral angles having one of two values (0° or 180°), there are $2^5 = 32$ C_S-consistent conformers. However, of these 32 possibilities, eight are chemically absurd. This is because for four of them, the ester group would penetrate the fullerene cage, and for the other four conformers, the ester group would penetrate the phenyl ring. This leaves 24 chemically possible conformers.

3.1. Proposed Conformer Nomenclature

To differentiate the 24 remaining conformers while keeping a reasonably short trivial name, we propose the following nomenclature. In parentheses, prefix [60]PCBM with the five relative orientations about the bonds in the ester group, which are designated "s" for *syn* and "a" for *anti* with respect to the relevant single bonded heavy elements of the addend that lie on the mirror plane. We present them in the following order: C61-C4, C4-C3, C3–C2, C2–C1 and C1–O, and place a hyphen after the third bond to differentiate the conformational behaviour of the alkyl chain from that of the ester end.

An example is (*sas*-sa)[60]PCBM, which has C61-C4 *syn*, C4-C3 *anti* and C3-C2 *syn* for the alkyl chain part, and C2-C1 *syn* and C1-O *anti* for the COOMe part. A diagram of (*sas-sa*)[60]PCBM is given in Figure 1, which was randomly chosen to illustrate this naming system. For example, the C61-C4 bond has the attached carbons C3 of the ester and C1 of the phenyl ring in a *syn* conformation on the mirror plane; hence, the conformation about this bond is denoted as *s*. On the other hand, the arrangement of the attached carbons to the C3-C4 bond (C2 and C61, respectively) are in an *anti a* conformation. It must be kept in mind that the *s* or *a* designation refers to the heavy elements (C and O).

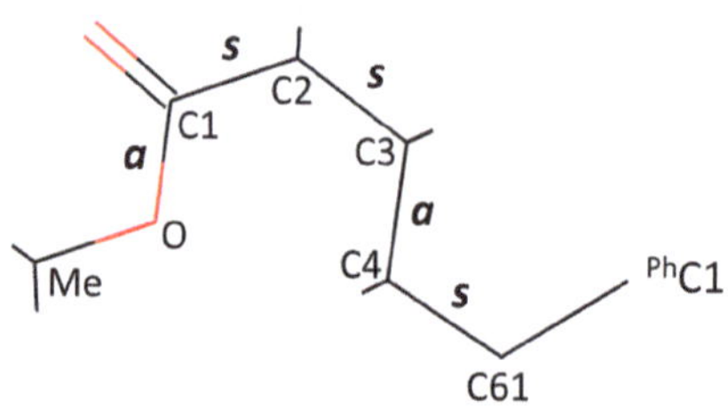

sas-sa

Figure 1. A diagram of the ester part for (*sas-sa*)[60]PCBM illustrating the nomenclature for the conformers used in this communication. It should be understood that C61 also attaches to carbons C1 and C9 of the fullerene (which is omitted for clarity).

All 24 possible ester conformations are shown in Figure 2 at their B3LYP/cc-pVTZ optimised geometries. For clarity, as with Figure 1, the conformationally invariant phenyl and cyclopropafullerenyl groups are omitted and only the relevant atoms are shown.

3.2. Bond Strain and Electronic Energy Analysis

In this part of the work, the initial freely optimised geometry (as detailed in Section 2) had its five relevant dihedral set to angles to either 0° (*syn*) or 180° (*anti*) to generate the 24 C_S symmetric conformers. The geometries of the 24 conformers were then re-optimised subject to the five dihedral angles, each being constrained as appropriate, while all bond lengths, bond angles and other dihedral angles could freely vary. The initial geometry from which the 24 conformers were generated was at a minimum on the B3LYP/cc-pVTZ potential energy surface. As such, with all 24 conformers being generated from this one optimised structure by varying it along only one coordinate (one dihedral angle), the 24 conformers should be near the minima.

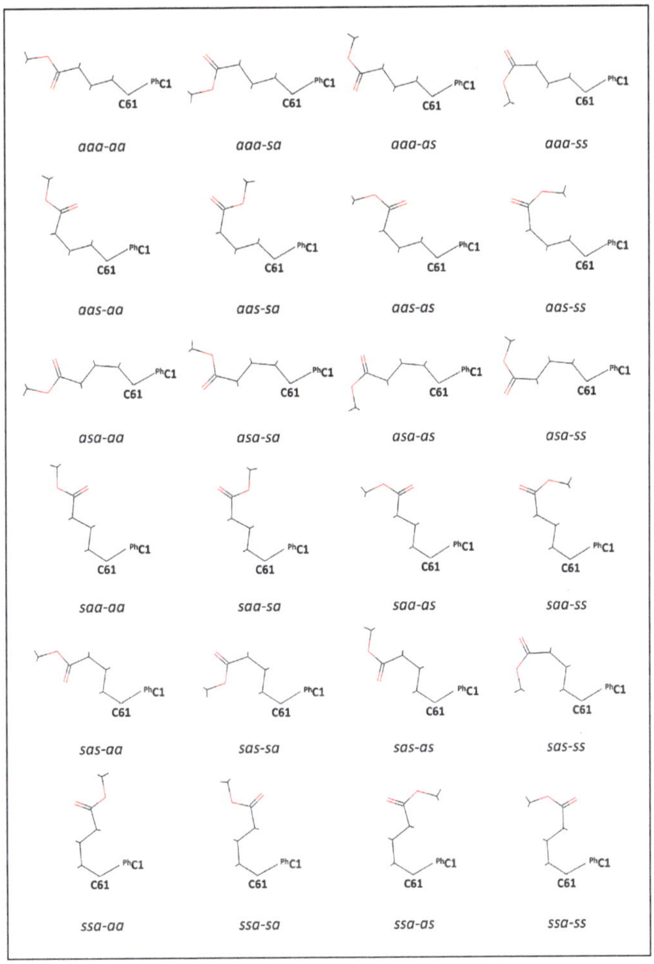

Figure 2. The 24 possible conformers of [60]PCBM. For clarity, the structures of the phenyl and cyclopropafullerenyl groups are not shown as they are invariant for all conformers. The ester group is shown at its C_S-constrained B3LYP/cc-pVTZ optimised geometry for each conformer.

It is apparent from inspecting Figure 2 that many of the bond angles are unrealistic. For example, the C1-C2-C3 bond angle in (*aas-ss*)[60]PCBM at the end of the second row is exceptionally wide at 138° where about 112° is expected. As further examples of the unrealistic bond angles, the six *xxx-ss* conformers have their C1-O-Me bond angles ranging from 126.6° to 129.7°. These are all very wide compared to experimentally known values from powder x-ray diffraction studies, −116.0° [16]. Hence, even without simulating the ^{13}C NMR spectra of the six *xxx-ss* conformers, these may be discounted as candidates for the proper input geometry. Similarly, all 12 *axx-xx* and four *ssa-xx* conformers have extremely large C61-C4-C3 bond angles, ranging from 120.7°–135.9° (*cf.* experiment [16] at 114.4°); conversely, the remaining eight *sxx-xx* conformers (*saa-xx* and *sas-xx*) have more realistic angles (ranging from 113.9°–115.4°). This suggests that the 12 *axx-xx* and four *ssa-xx* conformers may be also discounted. A full table of every relevant bond angle is given as supporting information (Table S1).

To give a quantitative measure of the bond strain for each of the 24 conformers, all the relevant bond angles for each conformer were measured using Gaussview software [17].

Following this, the absolute differences from the experimental solid state [16] bond angle were obtained, and finally, from these, the mean absolute deviation (MAD) was obtained for each conformer. These 24 MADs are plotted against their relevant conformer in Figure 3. The MADs range from 0.6 to 12.3 degrees per bond angle. Within each of the six *xxx*-combinations, it is the *xxx-ss* that have the highest MADs (which were qualitatively rejected earlier in the discussion), validating this method.

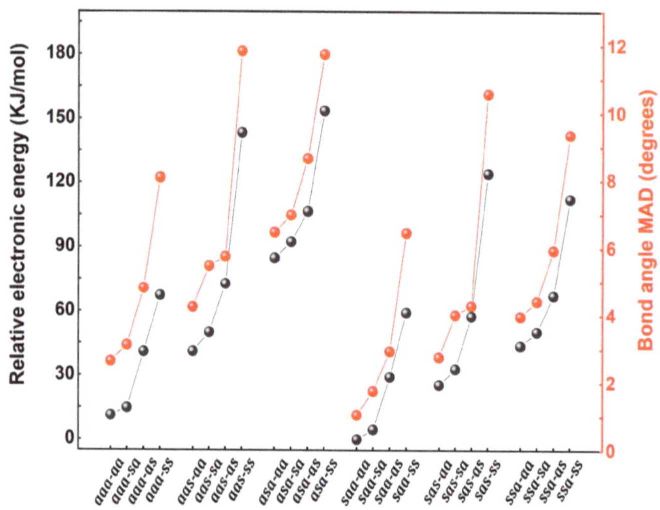

Figure 3. The C–C–C bond angle MADs and relative total electronic energies of the 24 conformers. The lowest 6 conformers were shortlisted for simulation of their ^{13}C NMR spectra.

The degree of bond strain in the conformers is also likely to be reflected in their total electronic energies. As such, the total electronic energy of each of the 24 conformers is also given in Figure 3. These are presented as relative to the lowest energy conformer, which is set to zero. These energies range up to 153 kJ/mol (1.59 eV) above that of the most stable conformer.

Two initial deductions can be drawn from Figure 3. The first is that there is a very close correlation between MAD of the bond angles and the relative total electronic energy, with remarkably similar trends in both. The second is that entire pattern breaks into two separate series of three separate four-membered groups. One series involves the 12 *axx-xx* conformers and another involves the 12 *sxx-xx* conformers, within each series, there are three groups of four conformers (with each group consisting of an *xxx-aa*, *-sa*, *-as*, *-ss* set).

The members of each of the six groups all follow the same trend. That is, the *xxx-aa* member is the most stable (having the lowest total electronic energies and bond strain MADs), closely followed by the *-sa* member. There is then a much larger step to the *-as* member followed by an even larger step to the *-ss* member. This indicates that the -COOMe part is quite unstable having the C1-O bond *syn* and is even more unstable if the C2-C1 bond is also *syn*. It also indicated that provided the C1-O bond is *anti*, there is little difference in terms of energy or bond strain on whether the C2-C1 bond is *anti* or *syn*.

Just as the members of the groups all follow the same trend, the groups themselves within their respective series also follow a matching trend in that there is an increase in instability and/or bond strain in the order *xaa-xx* < *xas-xx* < *xsa-xx*.

The final trend compares the two series and indicates that members of the *sxx-xx* series are shifted downwards by about 20 kJ/mol (200 meV) in electronic energy and about 2° in bond angle MAD relative to their respective analogs in the *axx-xx* series. If the C61-C4 bond is *anti*, then C3C4 bond is oriented slightly towards the fullerene. Therefore, to alleviate steric issues with the ester group becoming too close to the fullerene, the ^{Ph}C1-C61-C4 bond angle becomes strained in that it decreases from the known value of 115.5° [16] (^{Ph}C1 is used to avoid ambiguity as C1 of the phenyl and C1 of the C_{60} are both bonded to C61). It decreases to 107.6° for all *aaa-xx* conformers, 106.9–106.4° for the *aas-xx* conformers and 103.5–102.7° for the *asa-xx* conformers. Conversely, if the C61-C4 bond is *syn*, then the C3-C4 bond is oriented almost vertically away from the fullerene. However, this then brings it closer to the phenyl group. To alleviate similar steric issues with the ester group becoming too close to the phenyl, the ^{Ph}C1-C61-C4 bond angle opens relatively slightly to 115.9–116.3° for the *saa-xx* conformers, 117.1–117.3° for the *aas-xx* conformers and 120.0–120.4° for the *asa-xx* conformers. The average deviation for each group is slightly larger for the *axx* series than it is for the *sxx* series, which may explain why the members of the former series have slightly decreasing electronic energies and bond strain MADs. It is also noticed that both these trends in increasing bond strain follow the same *xaa-xx* < *xas-xx* < *xsa-xx* trend for series members, as identified above. Hence, it seems likely that the C61-C4 bond plays a significant role in the conformational behaviour of [60]PCBM, with a strong preference for it being carbon-*syn*.

3.3. Simulation of the ^{13}C NMR Spectra (sp^3 Hybridised Carbons)

Based on these analyses, the conformers with the six lowest bond angle strains, as measured by their mean absolute deviations of the relevant bond angles from experiment, were shortlisted to have their ^{13}C NMR spectra simulated. These six conformers also had the six lowest electronic energies, and are in order of increasing bond angle MAD, *saa-aa*, *saa-sa*, *aaa-aa*, *sas-aa*, *aaa-as* and *aaa-sa*.

^{13}C NMR spectra of the six candidate conformers were simulated. These are presented for the sp^3 region, together with the experimental spectrum, in Figure 4. Inspection of these spectra gives a clear indication that only the (b) *saa-aa* and (c) *saa-sa* conformations have simulated spectra that can be compared favourably to that of the experimental spectrum. Both follow the experimental line pattern very closely. It is also indicated that *saa-sa* is the better of the two on groups, having the spacings closer to the experiment. Conversely, both (d) *saa-as* and (g) *sas-aa* have the two lines near 50 ppm, much wider spacing than the experiment, and the other three lines are quite evenly spaced rather than having the experimental groupings. Finally, (e) *aaa-aa* and (f) *aaa-sa* are the worst of the six shortlisted conformers, with each having the first five lines with reasonably equal spacings.

The observation from the six simulated sp^3 NMR spectra that *saa-aa* and *saa-sa* are very similar to each other and are the two closest to experiment agrees with the conclusion made earlier from the bond strain and electronic energy analysis. This is because these two conformers have the lowest two bond strains and electronic energies of all the conformers, but their values are very similar. Conversely, of the six simulated spectra, *aaa-aa* and *aaa-sa* also appear very similar but give the worst agreement with the experiment. This also accords with the previous analysis, as these have the highest two bond strains and electronic energies of the six shortlisted conformers but are very close to each other in value. That the bond angle MADs (average bond strains), the total electronic energies and the quality of the simulated ^{13}C NMR spectra all follow the same trends provides compelling evidence that *saa-aa* and *saa-sa* are the two most stable structures for the conformational arrangements of the ester group and are thus the two most occupied conformers.

The discussion above is limited to the sp^3 region for two reasons: (1) it is a simple area of the spectrum where comparisons between the different conformers are easy to make, and (2) it involves the very carbons that are relevant to the conformations of the ester (nb., the sp^2 fullerene and phenyl carbons do not conformationally vary in any of the 24 conformation). Analysis of the far more complicated sp^2-hybridised carbons also indicates that *saa-sa* performed slightly better than *saa-aa*; for this reason, the *saa-as* conformer was used in our recent publication on the full spectrum [14].

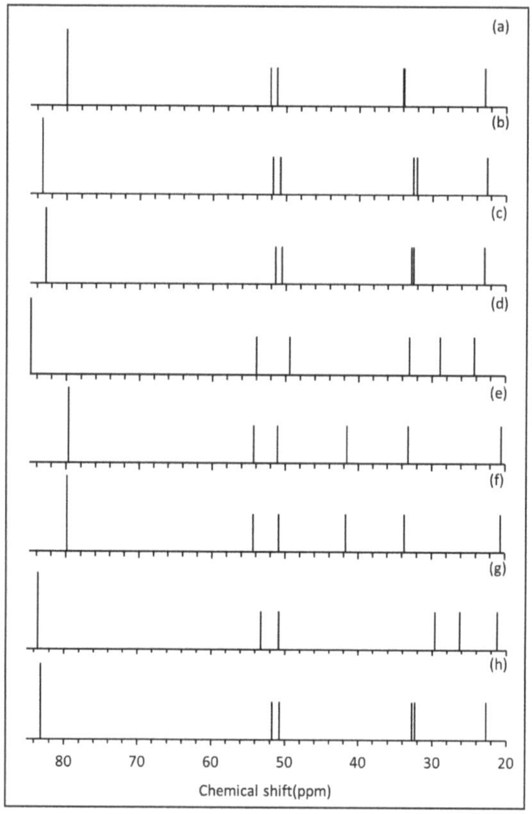

Figure 4. The experimental and simulated spectra of the 6 potential input geometries. (**a**) Experiment; (**b**) *saa-aa*; (**c**) *saa-sa*; (**d**) *saa-as*; (**e**) *aaa-aa*; (**f**) *aaa-sa*; (**g**) *sas-aa*. Only the *saa-aa* and *eaa-sa* conformers have simulated spectra that resemble the experimental spectrum, with *saa-sa* being slightly the closer. (**h**) A 60:40 weighted average of (**b**,**c**), as discussed in Section 3.4.

3.4. *Conformational Analysis of the Ester Group*

Insights into the dynamical behaviour of the ester group may be obtained with reference to our recent experimental study of second-order spin–spin couplings in the ^{1}H NMR spectrum. The observed very strong second-order spin–spin couplings at the C4 carbon resulted from restricted rotation about the ester C61-C4 and C4-C3 bonds [12]. For these two bonds, there is *gauche-syn/anti-gauche′* and back again restricted rotation. That is, for these two bonds, there is no continued rotation from *gauche* to *gauche′* for steric reasons [12]. Hence, it seems that *axx-xx* conformers cannot be obtained owing to this restricted rotation. For the same reasons, the *xsx-xx* conformers also cannot be obtained. This restricted rotation explanation is consistent with the NMR simulations in Figure 4, as these give relatively poor results when the ester is symmetry-constrained into these unfavourable conformations. The poor agreement of the *sas-aa* conformer's NMR spectrum with the experiment may also be explained in terms of dynamical behaviour. Although these conformers may be obtained as there is some degree of full rotation about the C2–C3 bond [12], occupation of the *syn* conformation in this bond should be relatively short lived.

To further investigate these conclusions and extend them to all heavy atom bonds of the ester groups, we conducted a DFT investigation. To this end, we considered the optimised conformer with the lowest total electronic energy and bond strain, (*saa*-*a*a)[60]PCBM, to be a "standard" to compare other conformers. This involved re-optimising the geometry of the standard (*saa*-aa)[60]PCBM so that all bond lengths, bond angles and dihedral angles were fixed, but then the following dihedral angles were each in turn varied from 0° to 180° in 10° steps: (a) ^{Ph}C1-C61-C4-C3, (b) C61-C4-C3-C2, (c) C4-C3-C2-C1, (d) C3-C2-C1-O and (e) C2-C1-O-Me. In this way, plots of relative energy against dihedral angle may be obtained for each of the five heavy-atom dihedral angles of the ester group (Figure 5). The fixing of the bond lengths and angles may be justified under the Born–Oppenheimer approximation. This is because bond length/angle variations are vibrational motion, and these are extremely slow relative to the rotational motion of dihedral angle variation. Hence, bond lengths and bond angles may, to a good approximation, be considered fixed on the time scale of the rotations. These five plots of relative energy against dihedral angle are considered, in turn, below in the abovementioned order.

C61-C4 Bond

A plot of the ^{Ph}C1-C61-C4-C3 dihedral angle (which represents rotation about the C61-C4 bond) against relative energy is presented in Figure 5a. The blue graph is shown at full scale, whereas for clarity, the orange graph is an expansion in energy. This figure reveals two minima at near-gauche and near-gauche' dihedral angles of +68° and 292° (−68°). The slight variation away from the ideal *gauche* dihedral angle of 60° is likely to be due to the large asymmetry in the barriers on either side of these minima. The barrier to rotation from *gauche* to *gauche' via* the *syn* conformer (0°) is 15.3 kJ/mol (176.7 meV), which corresponds to 6.18 × kT at room temperature (RT). A Boltzmann population analysis indicates that about 1 in 500 molecules (0.2%) have enough thermal energy to overcome this barrier. As the barrier is the same height in both directions, the *gauche* and *gauche'* conformers are equally populated.

On the other hand, rotation from *gauche* to *gauche' via* the *anti* (180°) conformer is impossible, as the barrier is 177 kJ/mol (1.84 eV) or 71.5 × kT RT. A Boltsmann analysis suggests that only 10^{-29}% of molecules could overcome this barrier. However, the energy difference between (*aaa-aa*)[60]PCBM and (*saa-aa*)[60]PCBM from Figure 3, at 11.6 kJ/mol (120 meV), is much less that that from Figure 5a. A Boltzmann/Arrhenius analysis based on this energy difference suggests that population of the *aaa-aa* conformer would be slightly less than 1% of that of the *saa-aa* conformer, and as will be shown for the remaining dihrdral angles, the populations would be far below 1% for the other 11 *axx-xx* conformers. However, even a 1% population is unlikely to be achieved. This is because the 11.6 kJ/mol energy difference is that of a vibrationally relaxed conformer. As mentioned earlier, with vibrational motions of the bonds (length and angle changes) being about 2 orders of magnitude slower than rotations about the bonds, it can be concluded that the relative population of the *aaa-aa* conformer is effectively zero. This is because with rotational rapid ascent up the barrier towards *aaa-aa* along the coordinate of Figure 5a, the configuration would be rapidly repelled from further assent long before it had any opportunity to even start vibrationally relaxing.

Within the *gauche* and *gauche'* conformers about the C61-C4 bond, the ester group undergoes librational behaviour whereby the ^{Ph}C1-C61-C4-C3 dihedral varies between −17° and +14° (*gauche*) and −14° and +17° (*gauche'*) either side of the minima, the extremes of which correspond to kT at room temperature (2.48 kJ/mol or 25.7 meV) above the minima in energy.

There are two shoulders on the central peak near 130 and 230 degrees. These are at elevated instability as they occur at dihedral angles about C61-C4 where one or the other of the two hydrogens on C4 is eclipsed with the hydrogen-less ^{Ph}C1 carbon.

It should be noted that the *aaa-aa* conformer and the *saa-aa* conformer are related to each other by a rotation of 180° about the C61-C4 bond. As such, they lie on a line through the potential energy surface (PES). Considering the PES along this line while retaining C_S symmetry should show that the *aaa-aa* conformer is a maximum and *saa-sa* is a minimum. The *saa-aa* conformer does not, however, represent the global minimum on the B3LYP/cc-pVTZ PES of (*xaa*-aa)[60]PCBM. Inspection of Figure 5a shows that the global minimum regarding the relative orientation about the C61-C4 bond has a dihedral angle of about 70°. Confirmation of this comes from an OPT-FREQ calculation that reveals this dihedral angle to be 68.14° with all other relevant dihedrals angles being within a degree of 180°. This structure is lower in energy than the similarly optimised (*saa-aa*)[60]PCBM structure. In addition, it has no imaginary vibrational frequencies (indicative of a minimum on the PES). However, this structure has C_1 point group symmetry rather than the experimentally observed C_S symmetry. As such, simulations based on this structure in solution would not, in general, resemble the experiment. For example, simulated ^{13}C NMR spectra of PCBM would split into 72 resonances instead of the experimentally seen 42 resonances; degenerate energy levels would also be split. As such, it seems likely that the barrier for *gauche*(′) to *syn* may be lower than that calculated, and that in reality, the librational motion goes back and forth between 90° and −90° (270°), with the *syn* conformation (0°) being the centre of the wide motion.

From the analysis of the orientational behaviour about the C61-C4 bond, it can be concluded that none of the 12 *axx-xx* conformers of [60]PCBM has any significant population. As such, (*aaa-aa*)[60]PCBM, the most commonly used representation of [60]PCBM in the literature, is essentially energetically forbidden, which suggests that at least some of the many ab initio calculations based on it may have unreliable conclusions.

C4-C3 Bond

The plot of dihedral angle rotation about the C4-C3 bond against relative energy, shown in Figure 5b, indicates that there are three relatively stable conformers, *gauche* (92°), *anti* (180°) and *gauche*′ (272°), and that the *syn* conformer (0°) at about 75 kJ/mol (~800 meV) represents an unsurmountable barrier (over 30 × kT at RT) to rotation from *gauche*′ to *syn* to *gauche*. This barrier would be much higher (at about 4.2 eV) if the neighbouring C3–C2 dihedral angle was not able to change to partially alleviate the bond strain.

The greater variation away from the ideal *gauche* angle than that seen for the C61–C4 bond above is due to the even greater asymmetry in the barriers. On this occasion, the barrier for rotation from *gauche*(′) to *anti* is 12.18 kJ/mol (126.3 meV, 4.91 × kT at RT), and that to rotate from *anti* to *gauche*(′) is 16.32 kJ/mol (169.2 meV, 6.58 × kT at RT).

A Boltzmann analysis based on these barriers indicates that 0.734% of molecules have sufficient thermal energy to overcome the barrier and rotate from the *gauche*(′) conformers to the *anti* conformer, and that 0.138% of molecules may similarly rotate from *anti* to *gauche*(′). An Arrhenius analysis indicates that there is a 16:84 population ratio of the two conformers, with the *anti* (180°) conformer being the more favoured. As the barrier to rotation from *anti* to *gauche* and that *anti* to *gauche*′ are identical, the *gauche* and *gauche*′ conformers would be equally populated (at *ca.* 8% each). These findings are consistent with the conclusion we recently drew on the populations of these three conformers based on well-resolved second-order spin–spin coupling splitting of the hydrogens attached to Carbon C4 by those attached to C3. Based on a CJG conformer analysis, a 12½:75:12½ population ratio for the *gauche*, *anti* and *gauche*′ conformers was estimated [12]. The slight discrepancy might come from imperfect assumptions made in the CJG analysis, in this work, or both. Nevertheless, both this work and our recent CJG estimate from second-order ^{1}H NMR spin–spin couplings have the general conclusion that the *syn* conformer is extremely unstable, the *anti* conformer is the most stable with *gauche*(′) islands of stability either side of the *anti* conformer, and that the population ratio of the *gauche*, *anti* and *gauche*′ conformers is about 1:8:1. With the *syn* conformation being unstable, all *xsx-xx* conformers have essentially zero population.

Like the ^{Ph}C1-C61-C4-C3 dihedral angle discussed above, there is librational motion about the minima of the *gauche*, *anti* and *gauche*′ conformers. In this case, the librations occur over a 30° range, ±15° either side of the *anti* conformer (i.e., between C61-C4-C3-C2 dihedral angles of 165°–195°), and over an 18° range, ±9° either side of the *gauche*(′) conformers (i.e., between dihedral angles of 79°–97° and 263°–181°, respectively).

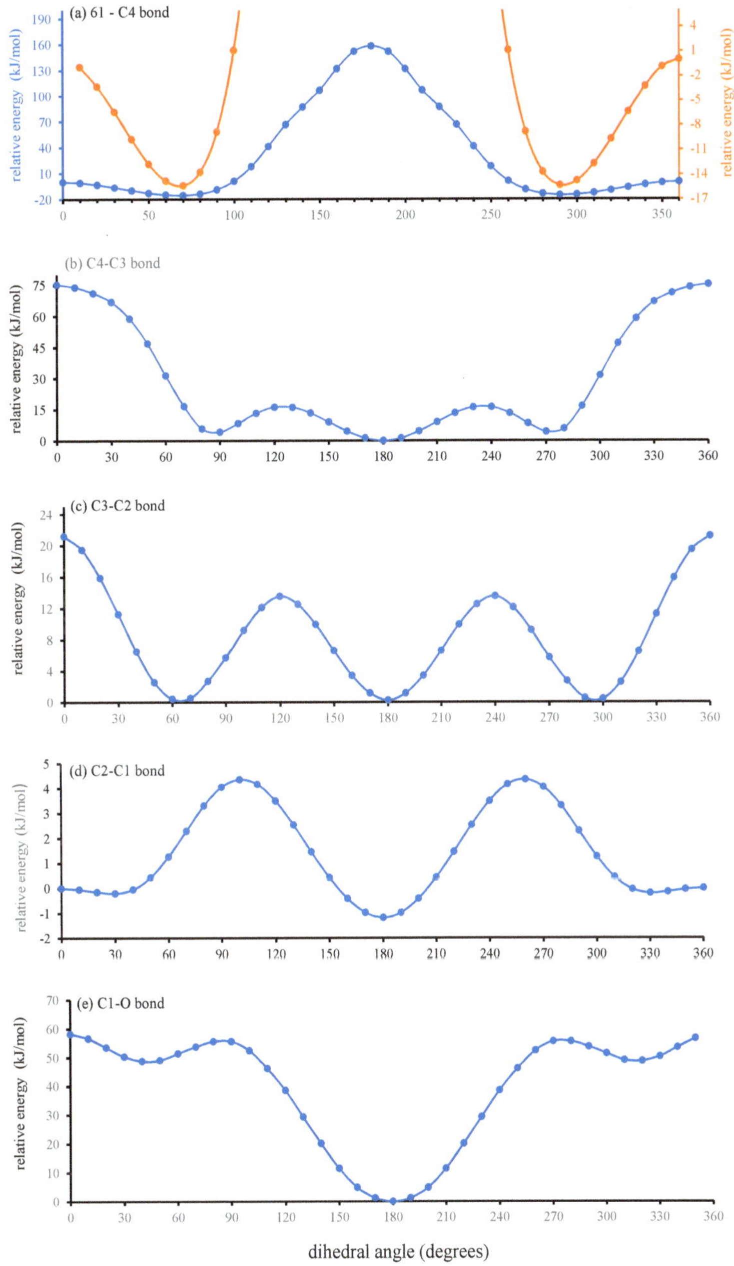

Figure 5. Plots of relative electronic energy as a function of the (**a**) ^{Ph}C1-C61-C4-C3 (**b**) C61-C4-C3-C2, (**c**) C4-C3-C2-C1, (**d**) C3-C2-C1-O and (**e**) C2-C1-O-Me dihedral angles (i.e., for rotation about the C61C4, C4-C3, C3-C2, C2-C1 and C1-O bonds, respectively).

C3-C2 Bond

A plot of relative electronic energy vs the C4-C3-C2-C1 dihedral angle is given in Figure 5c. Three equal minima are seen at the *gauche* dihedral angles of −68°, −68° (292°) and at the *anti* dihedral angle of 180°, with equal *gauche-anti* and *gauche'-anti* barriers of 13.4 kJ/mol (140 meV, 5.47 × *kT* at RT). The more stable 68°, 180° and 292° dihedrals have the hydrogens of C3 and C4 staggered, whereas the unstable 0°, 120° and 240° have those hydrogens eclipsed.

The barrier to rotation through the *syn* conformer is higher at 21 kJ/mol (220 meV). Like the previously discussed bond, this barrier would be much higher, 155 kJ/mol (1.6 eV) this time, if it wasn't for the ability of the neighbouring C2–C1 dihedral angle to change and thereby partially alleviate the bond strain. Again, like the previous bond, the *syn* conformation is unstable compared to the *anti* conformation, meaning that all *xxs-xx* conformers also have essentially zero population. An Arrhenius analysis of these barriers suggests that the *gauche-anti-gauche'* transition occurs 20 times more often than that of *gauche-syn-gauche'*. This situation is like that for the C4-C3 bond, where the rest of the ester group "wags" back and forth from *gauche* to *gauche'* via the stable middle conformer (*anti* in this case) and very rarely by the unstable conformer. However, unlike the ^{1}H NMR resonances from the hydrogen on carbon C4 being split by those of C3, the NMR spectrum does not indicate significant second-order spin–spin couplings for the hydrogens of carbon C2 being split by those of C3, which we recently interpreted as resulting from near free rotation about the C3-C2 bond [12]. However, this interpretation is not correct. This is because the *gauche-anti-gauche'* transition occurring 20 times more often than the *gauche-syn-gauche'* transition is inconsistent with free rotation. Therefore, a new explanation is needed of the experimental ^{1}H NMR observation of a first-order 1:2:1 triplet for the two symmetrically equivalent hydrogens of carbon C2 by those of carbon C3. Instead of "free rotation", the lack of second-order effects is better explained by the near equal energy barriers for the *gauche(')* to *anti* and the *anti* to *gauche(')* transitions, which would result in near equal populations of the three conformers. This is because at second order, the usual first-order 1:2:1 triplet occurs when the populations of the three staggered conformers about this bond are equal. Hence, is it the effectively equal *gauche(')-anti* and *anti-gauche(')* barriers, rather than free rotation about the C3-C2 bond, that gives the near equal populations that account for the lack of second-order NMR effects ordinarily expected from non-free rotation.

C2-C1 Bond

From Figure 5d, it is apparent that there are two stable conformers about this bond, *syn* at 0° and *anti* at 180°, with the *anti* conformer being slightly more stable by 0.974 kJ/mol (10.1 meV). The two wells near +30° and −30° (330°) are so shallow, at only 0.19 kJ/mol (2 meV, 0.078 × *kT* at RT), that they are insignificant in relation to ester dynamics at room temperature.

About the *syn* conformer, there is librational motion over a very wide range of about 150°. That is, between +75° and −75° (285°). Meanwhile, about the more stable *anti* conformer, the librational range is only 70°, at ±35° (i.e., between C3-C2-C1-O dihedral angles of 155° and 215°). However, the energy barrier for rotation from *anti* to *syn* is far lower than those discussed so far at only 2.2 *kT* at RT (5.53 kJ/mol, 57.3 meV), and that for *syn* to *anti* is even lower at 1.8 *kT* at RT (4.56 kJ/mol, 47.3 meV). As such, these barriers suggest that there is almost free rotation about this bond. Indeed, a Boltzmann indicates that about 16% of molecules have sufficient thermal energy to rotate from *syn* to *anti* at room temperature, and that about 11% of molecules have sufficient energy to rotate the other way. An Arrhenius analysis based on these barriers indicates that there is an approximate 60:40 population ratio of the two conformers with the *anti* conformer being the more favoured. The much low barriers to full rotation about this bond account for the observation made from Figure 3 that the *xxx-aa* and *xxx-sa* conformers, which involve this bond, have the lowest bond strain differences. With this, in simulating molecular spectra of [60]PCBM, it may be advantageous to use a 60:40 weighted average of relevant spectra of (saa-aa)[60]PCBM and (saa-sa)[60]PCBM. For example, Figure 4 h shows a weighted ^{13}C NMR spectrum of the sp^3-hybridised carbons of [60]PCBM, which is remarkably close to the experimental spectrum (Figure 4a).

C1-O Bond

The final ester bond to have a dihedral angle for heavy elements is the C1-O single bond. The plot of relative electronic energy as a function of the C2-C1-O–Me dihedral angle (Figure 5e) shows three minima near *gauche* (45°), *anti* (180°) and *gauche'* (−45° or 315°) dihedral angles. However, only the *anti* conformer is stable, and thereby has any significant population. This is because there is a relatively small barrier of 2.8 kT at RT (6.9 kJ/mol, 72 meV) to *gauche(')* to anti rotation, but an essentially unsurmountable barrier of 22.5 kT at RT (53.7 kJ/mol, 577 meV) for the reverse rotation, giving an approximate 1:10^8:1 population ratio of the *gauche, anti* and *gauche'* conformers. Within the stable *anti* conformer, there is librational motion about the C1-O bond between dihedral angles of 166° and 194°; i.e., ±14° either side of *anti.*

4. Conclusions

In summary, the conformational behaviour of the organic electronics n-type material, [60]PCBM, was examined via DFT techniques. This was informed by total electronic energy and average bond strain calculations. With the PhC61 group being conformationally invariant, the relative orientations about the ester bonds C61-C4, C4-C3, C3-C2, C2-C1 and C1-O (which could be either *syn* or *anti* to maintain C_S symmetry) gave rise to conformers. There are 24 chemically possible conformers. However, the conformational analysis suggests that only two of them may exit. These had (1) the C61-C4 bond in a *syn* configuration with all other relevant bonds *anti*, and (2) both the C61-C4 and C2-C1 bonds in *syn* configurations with all other relevant bonds being *anti* (with a respective 60:40 population ratio). With this, there is only reasonably free rotation about the C2-C1 bond of the ester group, with all other bonds suffering from strongly restricted bond rotations owing to steric hindrance from the nearby and bulky fullerene and phenyl groups.

These findings were supported by computational analyses that gave these two conformers, designated (*saa-aa*)[60]PCBM and (saa-sa)[60]PCBM, as those with the two lowest total electronic energies and the two lowest bond strains. In addition, DFT simulations of the ^{13}C spectrum of the ester sp^3-hybridised carbons based on these two conformations show a striking resemblance to the experiment, whereas those of other relatively low-energy conformers are barely recognisable in comparison to the experiment. A 60:40 weighted average of ^{13}C NMR spectra of these two conformers (reflecting the populations of the two conformers) also gave a spectrum remarkably close to that of the experiment.

A possibly important implication of these findings is they suggest that most representations of [60]PCBM in the literature are unrealistic. This is because they depict the relative orientations about the C61-C4 bond in the particularly unstable *anti* (180°) conformation. This very commonly depicted *anti* conformation has the ester group oriented roughly parallel and close to the surface of the fullerene. The much better representation found here is the opposite *syn* conformation about this bond that has the ester group oriented about perpendicular to the surface. This *syn* conformation is still consistent with C_S symmetry while being substantially more stable. As such, many published ab initio calculations on this highly important molecule to the field of organic electronics may have employed an improper input geometry with varying degrees of potential repercussions on the validity of the resulting conclusions.

Although [60]PCBM would not maintain C_S symmetry when packed into a solid, the solution-state structure is, nevertheless, highly important to the solid-state structure, as solid [60]PCBM is almost invariably formed from solutions. As such, the conformational conclusions presented here may also provide insights to aid dynamical modelling of the mechanisms of the formation of solution-processed [60]PCBM films.

Supplementary Materials: The following are available online at https://www.mdpi.com/article/10.3390/c7030066/s1, Table S1: A full table of every relevant bond angle for each of the 24 [60]PCBM conformers together with their absolute and relative electronic energies.

Author Contributions: Conceptualisation, T.L. and T.J.S.D.; methodology, T.L. and T.J.S.D.; validation, T.L. and T.J.S.D.; formal analysis, T.L. and T.J.S.D.; investigation, T.L. and T.J.S.D.; resources, T.J.S.D.; data curation, T.L. and T.J.S.D.; writing—original draft preparation, T.L.; writing—review and editing, T.J.S.D.; visualisation, T.L.; supervision, T.J.S.D.; project administration, T.J.S.D.; funding acquisition, T.J.S.D. All authors have read and agreed to the published version of the manuscript.

Funding: This research received no external funding.

Institutional Review Board Statement: Not applicable.

Data Availability Statement: The raw data (including all Gaussian 16 output files) and analysis spreadsheets are available from the authors on request.

Acknowledgments: We thank the China Scholarship Council for a Ph.D. studentship to T.L. Sections 3.2 and 3.3 of this research utilised Queen Mary's Apocrita HPC facility, supported by QMUL Research-IT.

Conflicts of Interest: The authors declare no conflict of interest.

References

1. Hummelen, J.C.; Knight, B.W.; LePeq, F.; Wudl, F.; Yao, J.; Wilkins, C.L. Preparation and Characterization of Fulleroid and Methanofullerene Derivatives. *J. Org. Chem.* **1995**, *60*, 532–538. [CrossRef]
2. Yu, Y.; Jin, B.; Peng, R.F.; Fan, L.S.; Cai, L.H.; Fan, B.; Chu, S.J. Photovoltaic performance of PCBM analogs with different ester groups as acceptor in the polymer solar cells. *Synth. Met.* **2016**, *212*, 44–50. [CrossRef]
3. Abrusci, A.; Stranks, S.D.; Docampo, P.; Yip, H.L.; Jen, A.K.Y.; Snaith, H.J. High-Performance Perovskite-Polymer Hybrid Solar Cells via Electronic Coupling with Fullerene Monolayers. *Nano Lett.* **2013**, *13*, 3124–3128. [CrossRef] [PubMed]
4. Waldauf, C.; Schilinsky, P.; Perisutti, M.; Hauch, J.; Brabec, C.J. Solution-processed organic n-type thin-film transistors. *Adv. Mater.* **2003**, *15*, 2084–2088. [CrossRef]
5. Seo, J.H.; Gutacker, A.; Walker, B.; Cho, S.N.; Garcia, A.; Yang, R.Q.; Nguyen, T.Q.; Heeger, A.J.; Bazan, G.C. Improved Injection in n-Type Organic Transistors with Conjugated Polyelectrolytes. *J. Am. Chem. Soc.* **2009**, *131*, 18220–18221. [CrossRef] [PubMed]
6. Zhao, X.; Liu, T.; Shi, W.; Hou, X.; Liu, Z.; Dennis, T.J.S. Understanding Charge Transport in Endohedral Fullerene Single-Crystal Field-Effect Transistors. *J. Phys. Chem. C* **2018**, *122*, 8822–8828. [CrossRef]
7. Winder, C.; Muhlbacher, D.; Neugebauer, H.; Sariciftci, N.S.; Brabec, C.; Janssen, R.A.J.; Hummelen, J.K. Polymer solar cells and infrared light emitting diodes: Dual function low bandgap polymer. *Mol. Cryst. Liq. Cryst.* **2002**, *385*, 213–220. [CrossRef]
8. Zhang, Q.M.; Chen, L.J.; Jia, W.Y.; Lei, Y.L.; Xiong, Z.H. Traps as interaction sites for hyperfine mixing: The origin of magnetoconductance in organic light-emitting diodes. *Org. Electron.* **2016**, *39*, 318–322. [CrossRef]
9. Wang, X.H.; Hofmann, O.; Das, R.; Barrett, E.M.; Demello, A.J.; Demello, J.C.; Bradley, D.D.C. Integrated thin-film polymer/fullerene photodetectors for on-chip microfluidic chemiluminescence detection. *Lab Chip* **2007**, *7*, 58–63. [CrossRef] [PubMed]
10. Tang, Z.; Ma, Z.F.; Sanchez-Diaz, A.; Ullbrich, S.; Liu, Y.; Siegmund, B.; Mischok, A.; Leo, K.; Campoy-Quiles, M.; Li, W.; et al. Polymer: Fullerene Bimolecular Crystals for Near-Infrared Spectroscopic Photodetectors. *Adv. Mater.* **2017**, *29*, 1702184. [CrossRef] [PubMed]
11. Zhao, X.; Liu, T.; Liu, H.; Wang, S.; Li, X.; Zhang, Y.; Hou, X.; Liu, Z.; Shi, W.; Dennis, T.J.S. Organic Single-Crystalline p-n Heterojunctions for High-Performance Ambipolar Field-Effect Transistors and Broadband Photodetectors. *ACS Appl. Mater. Interfaces* **2018**, *10*, 42715–42722. [CrossRef] [PubMed]
12. Liu, T.; Abrahams, I.; Dennis, T.J.S. Conformational Analysis of [60]PCBM via Second-Order Proton NMR Spin–Spin Coupling Effects. *J. Phys. Chem. Lett.* **2020**, *11*, 5397–5401. [CrossRef] [PubMed]
13. Frisch, M.J.; Trucks, G.W.; Schlegel, H.B.; Scuseria, G.E.; Robb, M.A.; Cheeseman, J.R.; Scalmani, G.; Barone, V.; Petersson, G.A.; Nakatsuji, H.; et al. *Gaussian 16, Revision, C.01*; Gaussian Inc.: Wallingford, CT, USA, 2016.
14. Liu, T.; Misquitta, A.J.; Abrahams, I.; Dennis, T.J.S. Characterization of the Fullerene Derivative [60]PCBM, by High-Field Carbon, and Two-Dimensional NMR Spectroscopy, Coupled with DFT Simulations. *Carbon* **2021**, *173*, 891–900. [CrossRef]
15. Hamprecht, F.A.; Cohen, A.J.; Tozer, D.J.; Handy, N.C. Development and assessment of new exchange-correlation functionals. *J. Chem. Phys.* **1998**, *109*, 6264–6271. [CrossRef]
16. Casalegno, M.; Zanardi, S.; Frigerio, F.; Po, R.; Carbonera, C.; Marra, G.; Nicolini, T.; Raos, G.; Meille, S.V. Solvent-free phenyl-C61-butyric acid methyl ester (PCBM) from clathrates: Insights for organic photovoltaics from crystal structures and molecular dynamics. *Chem. Commun.* **2013**, *49*, 4525–4527. [CrossRef] [PubMed]
17. Dennington, R.; Keith, T.A.; Millam, J.M. *GaussView, Version 6*; Semichem Inc.: Shawnee Mission, KS, USA, 2019.

Journal of Carbon Research

MDPI

Article

Applicability and Limitations of Simplified Elastic Shell Theories for Vibration Modelling of Double-Walled Carbon Nanotubes

Matteo Strozzi [1,2,*], Oleg V. Gendelman [2], Isaac E. Elishakoff [3] and Francesco Pellicano [4]

1 Department of Sciences and Methods for Engineering, University of Modena and Reggio Emilia, 42122 Reggio Emilia, Italy
2 Faculty of Mechanical Engineering, Technion, Israel Institute of Technology, Haifa 32000, Israel; ovgend@me.technion.ac.il
3 Department of Ocean and Mechanical Engineering, Florida Atlantic University, Boca Raton, FL 33431, USA; elishako@fau.edu
4 Department of Engineering "Enzo Ferrari", Centre InterMech Mo.Re., University of Modena and Reggio Emilia, 41125 Modena, Italy; francesco.pellicano@unimore.it
* Correspondence: matteo.strozzi@unimore.it

Abstract: The applicability and limitations of simplified models of thin elastic circular cylindrical shells for linear vibrations of double-walled carbon nanotubes (DWCNTs) are considered. The simplified models, which are based on the assumptions of membrane and moment approximate thin-shell theories, are compared with the extended Sanders–Koiter shell theory. Actual discrete DWCNTs are modelled by means of couples of concentric equivalent continuous thin, circular cylindrical shells. Van der Waals interaction forces between the layers are taken into account by adopting He's model. Simply supported and free–free boundary conditions are applied. The Rayleigh–Ritz method is considered to obtain approximate natural frequencies and mode shapes. Different aspect and thickness ratios, and numbers of waves along longitudinal and circumferential directions, are analysed. In the cases of axisymmetric and beam-like modes, it is proven that membrane shell theory, differently from moment shell theory, provides results with excellent agreement with the extended Sanders–Koiter shell theory. On the other hand, in the case of shell-like modes, it is found that both membrane and moment shell theories provide results reporting acceptable agreement with the extended Sanders–Koiter shell theory only for very limited ranges of geometries and wavenumbers. Conversely, for shell-like modes it is found that a newly developed, simplified shell model, based on the combination of membrane and semi-moment theories, provides results in satisfactory agreement with the extended Sanders–Koiter shell theory in all ranges.

Keywords: carbon nanotubes; double-walled; linear vibrations; Sanders–Koiter theory; simplified shell models

Citation: Strozzi, M.; Gendelman, O.V.; Elishakoff, I.E.; Pellicano, F. Applicability and Limitations of Simplified Elastic Shell Theories for Vibration Modelling of Double-Walled Carbon Nanotubes. C **2021**, 7, 61. https://doi.org/10.3390/c7030061

Academic Editor: Gil Goncalves

Received: 2 July 2021
Accepted: 4 August 2021
Published: 9 August 2021

1. Introduction

Thin-walled mechanical components, such as beams, plates and shells, are adopted as structural elements in several engineering fields, in particular mechanical, aeronautical and aerospace, due to their specific mechanical properties, such as high strength and stiffness, low weight and light inertia.

As an example, monolithic rectangular tanks are used as engineering structures in sewage and water treatment plants, and recreational facilities. In most cases they are made with constant thickness walls, but adoption of variable thickness (e.g., trapezoidal cross-section) walls is justified in structures where hydrostatic pressure acts on the walls [1,2].

On the other hand, examples of thin shell structures adopted in mechanical systems are piping components, turbine disks, pressure vessels and liquid-retaining tanks, while aircrafts, missiles and rockets represent some examples of thin shell structure applications in aeronautical and aerospace field [3].

The wide application of thin shell structures in engineering field is due to their very high load-carrying capability, strength-to-weight ratio and structural stiffness. In particular, thin shell structures are able to efficiently support applied external forces by virtue of their characteristic spatial curvature, and, as a result, thin shells are much stronger and stiffer than all other structural forms [4].

The simplest, and therefore most adopted theories, to model shell behaviour are the elastic shell theories. These theories predict, adequately, both stresses and deformations for shells exhibiting elastic deformations, i.e., deformations for which Hooke's law holds into the equilibrium equations [5]. On the other hand, these theories are limited to elastic, i.e., reversible, deformations of shells and therefore they cannot be adopted in the case of structural fracture or failure [6].

An actual shell is a three-dimensional body; by using Kirchhoff–Love's assumptions [7], the three-dimensional problem of shell equilibrium and straining reduces to a two-dimensional problem at the shell middle surface. Therefore, in thin elastic shell theories, the shell problem is reduced to the study of the middle surface deformations, i.e., strains and changes in curvature and torsion [8].

Several thin elastic shell theories with varying degrees of accuracy were derived in literature, depending on the degree of simplification of the elasticity equilibrium equations. The approximations necessary for the development of a simplified but adequate thin elastic shell theory have been subject of relevant studies, performed in the past, by eminent elastic continuum mechanics scientists.

Love was the first to develop an approximate but accurate thin shell theory based on classical linear elasticity [9]. In order to simplify the strain–displacement relationships, and therefore the expression of the elastic strain energy, Love applied to the shell theory the same hypotheses that were originally developed by Kirchhoff for the plate bending theory, together with the small deflection and thin shell hypotheses. These assumptions are commonly called "Kirchhoff–Love's assumptions", and Love's theory of thin elastic shells is also referred to as "first-order approximation shell theory" [10].

After Love, many researchers tried to eliminate some inadequacies of his approximate theory, with particular regard to its inconsistent treatment of small terms into the equilibrium equations. The need for a mathematically rigorous two-dimensional set of shell equations adopting Kirchhoff–Love's assumptions led to different versions of the first-order approximation shell theory.

Reissner [11] developed a first-order approximation theory by deriving equilibrium, strain–displacement and stress resultant equations for thin shells directly from the three-dimensional elasticity theory. Reissner's linear theory of thin elastic shells considers Kirchhoff–Love's assumptions and neglects the distance of the arbitrary point of the shell from the middle surface with respect to the radii of curvature of the shell middle surface.

Sanders [12] also developed a first-order approximation theory starting from the principle of virtual work and applying Kirchhoff–Love's assumptions. Sanders linear theory of thin elastic shells, which also removes inconsistencies of Love theory in small terms, was later modified and refined by Koiter, giving rise to Sanders–Koiter shell theory [13].

Donnell [14], Vlasov [15] and Mushtari [16], independently, developed a simplified thin shell theory assuming, besides Kirchhoff–Love's hypotheses, infinitesimal tangential displacements along longitudinal and circumferential directions into the strain–displacement relationships and equilibrium equations. Due to their simplicity, the governing equations of Donnell–Vlasov–Mushtari shell theory were found to be extremely convenient for solving many engineering problems related to shells.

Other first-order approximation shell theories, close to Love theory and considering Kirchhoff–Love's assumptions, were developed by Timoshenko [17] and Naghdi [18].

Besides first-order approximation shell theories, a second class of thin elastic shell theories, which are commonly referred to as higher-order approximation shell theories, was developed. To this second class they are assigned all thin elastic shell theories in which at least one of Kirchhoff–Love's assumptions is not respected.

Lur'ye [19], Flugge [20] and Byrne [21], independently of each other, developed a second-order approximation shell theory deriving equations of equilibrium, strain–displacement relationships and stress resultants from the application of Kirchhoff's hypotheses and small deflection assumption to the equations of the three-dimensional theory of elasticity, while removing the thin shell assumption.

Novozhilov [22] developed another version of the second-order approximation shell theory by deriving the strain–displacement relationships from the three-dimensional theory of elasticity by applying Kirchhoff's assumptions and the stress resultant equations from the strain energy expression, again removing, similarly to Lur'ye–Flugge–Byrne theory, thin shell assumption.

Other second-order approximation theories, obtained applying Kirchhoff–Love's assumptions but removing the thinness one, were developed also by Gol'denveizer [23] and Reissner [24].

Since the governing equations of the general thin elastic shell theories are eight-order partial differential equations, then the computation of the shell stress and strain components is very difficult. As a consequence, researchers involved in thin shell stress-strain analysis were focused on the simplification of the governing partial differential equations of shells.

The aforementioned thin elastic shell theories assumed that the bending (flexural) stresses were of the same order of the stretching (extensional) stresses. If one of the above-mentioned types of stresses is negligible, in comparison with the other one, then it is possible to obtain a considerable simplification in the characteristic equations and relationships of the shell theory, in the form of a simplified (reduced) elastic shell model.

The first possible type of stress state for thin elastic shells is the membrane (momentless) state, which is present if the bending stresses are negligible compared with the stretching ones [25,26]. The governing equations of the membrane shell theory can be derived directly from the equations of the general shell theory by neglecting bending and torsional moments on the stress state of the thin shell. The membrane shell theory is able to describe the state of stress with reasonable accuracy in the case of relatively short and very thin shells, when bending and torsional deformations are very small, while stretching deformations are preeminent, i.e., pure membrane state of stress.

The second possible type of stress state for thin elastic shells is the moment (forceless) state, which is present if the stretching stresses are negligible compared with the bending ones [27,28]. The governing equations of the moment shell theory can be obtained directly from the equations of the general shell theory by neglecting stretching (extensional) forces on the stress state of the thin shell. Contrary to membrane shell theory, moment shell theory is able to describe the state of stress with reasonable accuracy in the case of very long and relatively thick shells, when stretching deformations are very small, while bending and torsional deformations are preeminent, i.e., pure moment state of stress.

Research in nanotechnology has attracted numerous investigators in recent years. Recent status is reported in the monographs written by Elishakoff et al. [29] and Liew et al. [30], as well as in the volumes edited by Tserpes and Silvestre [31] and Harik [32].

Carbon nanotubes (CNTs) are used as ultrahigh frequency resonators in a large number of nano-electro-mechanical devices such as sensors, oscillators, charge detectors and field emission devices [33,34]. The reduction of the size and the increment of the stiffness of a nano-mechanical resonator magnify its resonant frequencies and decrease its energy consumption, therefore improving its sensitivity.

A relevant issue of CNTs is represented by their geometry, in particular their cross-section, which is important for the obtained results. Actually, for the practical CNT based composites, most of CNTs are compressed or collapsed to achieve a very dense packing. This specific CNT configuration, which presents a non-circular cross section, is beneficial for improving performance: in fact, collapsed carbon nanotubes represent ultralight structural materials with very high specific strength and stiffness comparable to current carbon fibre reinforced polymer composites, and with ultrahigh thermal conductivity [35–37].

A large number of experiments were conducted on single-walled carbon nanotubes (SWCNTs) [38–40]. In these works, the linear vibrations of free–free SWCNTs are analysed by adopting resonant Raman spectroscopy with laser excitation wavelengths in the range of nanometres. Several Raman peaks are observed, which correspond to the vibrational modes of the CNTs, and the CNT chirality is univocally assigned by measuring one radial breathing mode (RBM) frequency and applying the theory of the resonant transitions.

Because of their nanoscale size, it is very difficult to investigate the mechanical properties of the nanotubes adopting experimental techniques, which require the use of high resolution transmission electron microscopes and do not allow the natural frequencies of the different vibration modes to be easily separated within the frequency spectrum.

On the other hand, it was found that molecular dynamics (MD) simulations provide good predictions of the mechanical behaviour of SWCNTs and multi-walled carbon nanotubes (MWCNTs) under external forces, with results close to the experiments [41–43]. In these works, natural frequencies of RBMs and beam-like modes (BLMs) considering CNTs with different geometries, boundary conditions and chirality are computed adopting the molecular mechanics (MM) potential.

However, it has been demonstrated that MD simulations are computationally inefficient, especially when dealing with MWCNTs incorporating a large number of atoms. Therefore, more efficient analytical and numerical methods have been recently developed in order to analyse classes of CNTs in a more general and efficient way; such methods are generally based on continuous models for the nanotube and allow a strong reduction of the number of degrees of freedom. In particular, it was found that continuous thin elastic shell models can correctly predict static buckling and free vibration behaviour of CNTs obtained from experiments and MD simulations [44–48].

The analogies between thin elastic shell model and CNT structure led to an extensive adoption of thin elastic shell theories for CNT mechanical deformation analysis: applicability and limitations of thin elastic shell theories for vibration modelling of CNTs were studied by considering Love [49], Donnell–Mushtari [50–52], Sanders–Koiter [53,54] and Flügge–Byrne–Lur'ye [55,56] theories.

Refs. [57–61] demonstrated the excellent agreement of the natural frequencies of SWCNTs and MWCNTs obtained by considering Sanders–Koiter shell theory with the results of experimental studies and MD simulations, for a large range of geometries and wavenumbers. Therefore, in the present work, the Sanders–Koiter shell theory is assumed as the reference theory for the investigation of applicability and limitations of the simplified thin elastic shell models.

The starting point of the present paper is represented by the simplified thin elastic shell theory developed by Manevitch for vibration modelling of SWCNTs and reported in Ref. [62]. Specifically, Manevitch proposed a reduced version of Sanders–Koiter shell theory based on the hypotheses of the membrane shell theory (negligible middle surface changes in curvature and torsion) for axisymmetric modes and a combination of the hypotheses of semi-membrane and moment shell theories (negligible middle surface circumferential and shear strains) for beam-like and shell-like modes. Comparisons with the complete (extended) Sanders–Koiter shell theory proved the applicability of reduced (simplified) Sanders–Koiter shell theory for relatively long and thick SWCNTs. Later, Manevitch and Kaplunov [63,64] refined the previously adopted assumptions on the vanishing of circumferential and shear middle surface deformations by means of asymptotic analysis in the vicinity of the lowest cut-off frequency.

In the present paper, the investigation on the applicability and limitations of simplified models of thin elastic shells for SWCNT linear vibrations of Ref. [62] is extended to DWCNTs, where the effect of the van der Waals interaction forces between the adjacent layers completely changes the scenario. The Sanders–Koiter shell theory, retaining three middle surface strains and three middle surface changes in curvature and torsion, is applied as reference model. Three different simplified models are compared with Sanders–Koiter shell theory: (i) a shell model retaining only the three middle surface strains, based on the assumptions of membrane shell theory, (ii) a shell model retaining only the three middle surface changes in curvature and torsion, based on the assumptions of moment shell theory, (iii) a shell model retaining the three middle surface strains and the middle surface change in curvature along the circumferential direction, obtained combining membrane and semi-moment shell theories.

The aim of the present study is to propose a new simplified elastic shell model applicable with satisfactory accuracy for linear vibrations of DWCNTs in a large range of geometries and wavenumbers, including cases in which membrane and moment shell models result in large relative errors. This new simplified shell model, as a reduced version of Sanders–Koiter shell theory, by presenting a lower number of middle surface components within the expression of the elastic strain energy, gives rise to a lower computational effort into the numerical simulations with respect to extended Sanders–Koiter shell theory.

2. Extended Sanders–Koiter Shell Theory

In the present paper, the actual discrete DWCNT of Figure 1a is modelled by means of a couple of concentric equivalent continuous thin elastic cylindrical shells with van der Waals interaction forces. In Figure 1b,c, a continuous thin elastic cylindrical shell with radius R, length L and thickness h is shown; a cylindrical coordinate system $(O; x, \theta, z)$ is considered, where the origin O of the reference system is located at the centre of one end of the cylindrical shell. Three different displacements are present: longitudinal $u\ (x, \theta, t)$, circumferential $v\ (x, \theta, t)$ and radial $w\ (x, \theta, t)$; the radial displacement w is assumed positive outward; (x, θ) are the longitudinal and angular coordinates of an arbitrary point on the middle surface of the shell; z is the radial coordinate along the thickness h; t is the time.

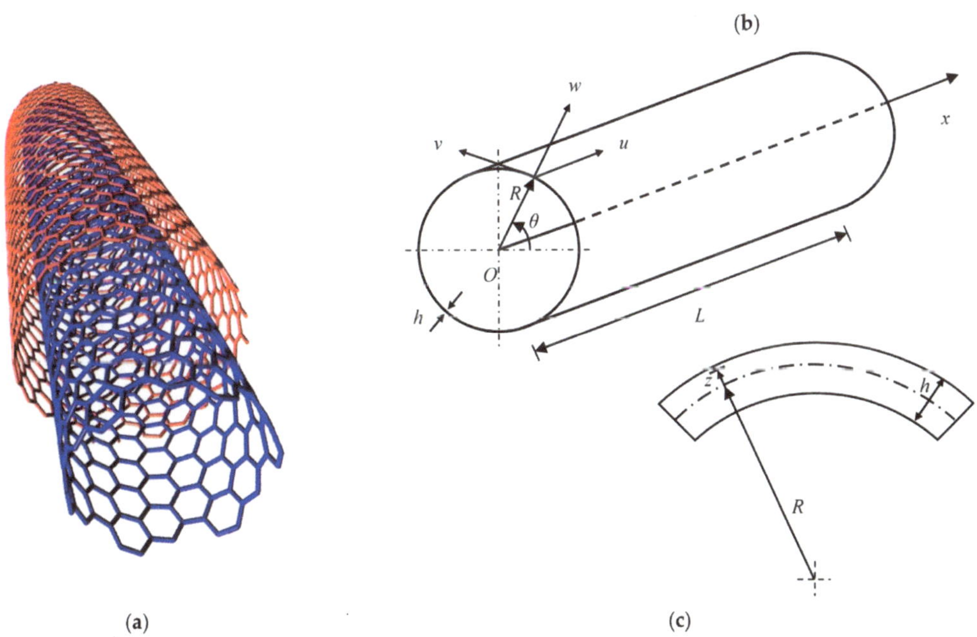

Figure 1. Continuum modelling of a DWCNT. (**a**) Actual discrete DWCNT. (**b**) Geometry of the equivalent continuous single cylindrical shell. (**c**) Cross-section of the surface of the equivalent continuous single cylindrical shell.

2.1. Strain–Displacement Relationships

The dimensionless displacement field of the i-th cylindrical shell is written in the form [61]:

$$\tilde{u}_i = \frac{u_i}{R_i} \quad \tilde{v}_i = \frac{v_i}{R_i} \quad \tilde{w}_i = \frac{w_i}{R_i} \quad i = 1,2 \tag{1}$$

where (u_i, v_i, w_i) is the dimensional displacement field and R_i is the radius of the i-th shell.

In this paper, the Sanders–Koiter shell theory is used to model the DWCNT dynamics. The linear relationships between strains and displacements in Sanders–Koiter shell theory are based on "Kirchhoff–Love's assumptions" [7]: (i) the thickness of the shell h is small if compared with the other dimensions of the shell, for example the radius of curvature of the middle surface R and the length L, (ii) strains and displacements are sufficiently small so that the quantities of second and higher-order magnitude into the strain–displacement relationships may be neglected in comparison with the first-order terms, (iii) the transverse normal stress is small with respect to the other normal stress components and may be neglected (i.e., $\sigma_z = 0$), (iv) normals to the undeformed middle surface remain straight and normal to the deformed middle surface, suffering no extension ("Kirchhoff's kinematic hypothesis").

As for the first geometric assumption, with regard to the thickness ratio R/h, it was demonstrated that the range of validity of thin shell theories, in the present work Sanders–Koiter thin shell theory, is given by $20 \leq R/h \leq 500$, see Ref. [16] for more details. Moreover, the consequence of the last geometric assumption is that the transverse shear strains of the thin circular cylindrical shell may be neglected ($\gamma_{xz} = \gamma_{\theta z} = 0$); in addition, also the rotary inertia of the shell is neglected.

According to the Sanders–Koiter shell theory, the dimensionless strain components at an arbitrary point of the i-th shell are related to the dimensionless middle surface strains and changes in curvature and torsion of the i-th shell by the following relationships [61]:

$$\tilde{\varepsilon}_{x,i} = \tilde{\varepsilon}_{x,0,i} + \zeta_i \tilde{k}_{x,i} \quad \tilde{\varepsilon}_{\theta,i} = \tilde{\varepsilon}_{\theta,0,i} + \zeta_i \tilde{k}_{\theta,i} \quad \tilde{\gamma}_{x\theta,i} = \tilde{\gamma}_{x\theta,0,i} + \zeta_i \tilde{k}_{x\theta,i} \quad i = 1,2 \tag{2}$$

where $\zeta_i = z_i/R_i$ is the dimensionless radial coordinate of the i-th cylindrical shell.

The dimensionless middle surface linear strains of the i-th cylindrical shell are [61]:

$$\tilde{\varepsilon}_{x,0,i} = \alpha_i \frac{\partial \tilde{u}_i}{\partial \eta} \quad \tilde{\varepsilon}_{\theta,0,i} = \frac{\partial \tilde{v}_i}{\partial \theta} + \tilde{w}_i \quad \tilde{\gamma}_{x\theta,0,i} = \frac{\partial \tilde{u}_i}{\partial \theta} + \alpha_i \frac{\partial \tilde{v}_i}{\partial \eta} \quad i = 1,2 \tag{3}$$

where $\eta = x/L$ is the dimensionless longitudinal coordinate of the shell and $\alpha_i = R_i/L$.

The dimensionless middle surface changes in curvature and torsion of the i-th shell are [61]:

$$\tilde{k}_{x,i} = -\alpha_i^2 \frac{\partial^2 \tilde{w}_i}{\partial \eta^2} \quad \tilde{k}_{\theta,i} = \frac{\partial \tilde{v}_i}{\partial \theta} - \frac{\partial^2 \tilde{w}_i}{\partial \theta^2} \quad \tilde{k}_{x\theta,i} = -2\alpha_i \frac{\partial^2 \tilde{w}_i}{\partial \eta \partial \theta} + \frac{1}{2}\left(3\alpha_i \frac{\partial \tilde{v}_i}{\partial \eta} - \frac{\partial \tilde{u}_i}{\partial \theta}\right) \quad i = 1,2 \tag{4}$$

2.2. Force and Moment Resultants

The dimensionless force and moment resultants can be written as functions of the dimensionless middle surface strains and changes in curvature and torsion of the i-th cylindrical shell [61]:

$$\begin{aligned}
\tilde{N}_{x,i} &= \frac{1}{(1-\nu^2)}(\tilde{\varepsilon}_{x,0,i} + \nu \tilde{\varepsilon}_{\theta,0,i}) & \tilde{N}_{\theta,i} &= \frac{1}{(1-\nu^2)}(\tilde{\varepsilon}_{\theta,0,i} + \nu \tilde{\varepsilon}_{x,0,i}) & \tilde{N}_{x\theta,i} &= \frac{1}{2(1+\nu)}\tilde{\gamma}_{x\theta,0,i} \\
\tilde{M}_{x,i} &= \frac{\beta_i}{12(1-\nu^2)}(\tilde{k}_{x,i} + \nu \tilde{k}_{\theta,i}) & \tilde{M}_{\theta,i} &= \frac{\beta_i}{12(1-\nu^2)}(\tilde{k}_{\theta,i} + \nu \tilde{k}_{x,i}) & \tilde{M}_{x\theta,i} &= \frac{\beta_i}{24(1+\nu)}\tilde{k}_{x\theta,i} \quad i = 1,2 \\
\tilde{Q}_{x,i} &= \beta_i \left(\alpha_i \frac{\partial \tilde{M}_{x,i}}{\partial \eta} + \frac{\partial \tilde{M}_{x\theta,i}}{\partial \theta}\right) & \tilde{Q}_{\theta,i} &= \beta_i \left(\alpha_i \frac{\partial \tilde{M}_{x\theta,i}}{\partial \eta} + \frac{\partial \tilde{M}_{\theta,i}}{\partial \theta}\right)
\end{aligned} \tag{5}$$

where ν is the Poisson's ratio of the cylindrical shell and $\beta_i = h/R_i$.

2.3. Membrane Approximate Shell Theory

A thin elastic shell supports an arbitrary external loading by means of stress resultants (internal forces) and stress couples (bending and twisting moments), see dimensionless force and moment resultants (5).

However, under appropriate loading and boundary conditions, the resulting bending and twisting moments are either zero, or so small that they may be neglected [3].

Such a state of stress is referred to as *membrane state of stress* because of the analogy to membranes that cannot support bending and twisting moments. The corresponding theory of thin shells that deals with this state of stress is called *membrane theory of thin shells* [3].

It follows from the above that the membrane theory neglects all moment resultants, i.e.,

$$\widetilde{M}_{x,i} = \widetilde{M}_{\theta,i} = \widetilde{M}_{x\theta,i} = 0 \quad i = 1,2 \tag{6}$$

From the corresponding moment relations (5), it follows that neglecting the moments can be justified by assuming negligible changes in curvature and torsion of the i-th shell middle surface:

$$\widetilde{k}_{x,i} = \widetilde{k}_{\theta,i} = \widetilde{k}_{x\theta,i} = 0 \quad i = 1,2 \tag{7}$$

and therefore we deal with the membrane or *momentless state of stress* of the shell [4].

2.4. Moment Approximate Shell Theory

Similarly, under appropriate loading and boundary conditions, the resulting internal forces are either zero, or so small that they may be neglected [3].

Such a state of stress is referred to as *moment state of stress*, where the corresponding theory of thin shells is called *moment theory of thin shells* [3].

It follows from the above that the moment theory neglects all force resultants, i.e.,

$$\widetilde{N}_{x,i} = \widetilde{N}_{\theta,i} = \widetilde{N}_{x\theta,i} = 0 \quad i = 1,2 \tag{8}$$

From the corresponding force relations (5), it follows that neglecting the forces can be justified by assuming very small, at least negligible strains of the i-th shell middle surface:

$$\widetilde{\varepsilon}_{x,0,i} = \widetilde{\varepsilon}_{\theta,0,i} = \widetilde{\gamma}_{x\theta,0,i} = 0 \quad i = 1,2 \tag{9}$$

and therefore we deal with the moment or *forceless state of stress* of the shell [4].

2.5. Stress-Strain Relationships

Considering plane stress hypothesis ($\sigma_z = 0$, Kirchhoff–Love's assumptions), the dimensionless stresses of the i-th thin circular cylindrical shell are related to the corresponding dimensionless strains, in case of homogeneous isotropic elastic material, by the following local constitutive equations [61]:

$$\widetilde{\sigma}_{x,i} = \frac{1}{(1-\nu^2)}(\widetilde{\varepsilon}_{x,i} + \nu\widetilde{\varepsilon}_{\theta,i}) \quad \widetilde{\sigma}_{\theta,i} = \frac{1}{(1-\nu^2)}(\widetilde{\varepsilon}_{\theta,i} + \nu\widetilde{\varepsilon}_{x,i}) \quad \widetilde{\tau}_{x\theta,i} = \frac{1}{2(1+\nu)}\widetilde{\gamma}_{x\theta,i} \quad i = 1,2 \tag{10}$$

2.6. Elastic Strain Energy

The dimensionless elastic strain energy of the i-th thin elastic circular cylindrical shell, which models a SWCNT, under plane stress hypothesis ($\sigma_z = 0$), is expressed as [61]:

$$\widetilde{U}_i = \frac{1}{2\beta_i}\int_0^1\int_0^{2\pi}\int_{-\beta_i/2}^{\beta_i/2} (\widetilde{\sigma}_{x,i}\widetilde{\varepsilon}_{x,i} + \widetilde{\sigma}_{\theta,i}\widetilde{\varepsilon}_{\theta,i} + \widetilde{\tau}_{x\theta,i}\widetilde{\gamma}_{x\theta,i})d\eta d\theta d\zeta \quad i = 1,2 \tag{11}$$

By inserting Equations (2) and (10) into Equation (11), the following expression for the elastic strain energy of the i-th circular cylindrical shell is obtained [61]:

$$\begin{aligned}\widetilde{U}_i =& \frac{1}{2}\frac{1}{(1-\nu^2)}\int_0^1\int_0^{2\pi}\left(\widetilde{\varepsilon}^2_{x,0,i}+\widetilde{\varepsilon}^2_{\theta,0,i}+2\nu\widetilde{\varepsilon}_{x,0,i}\widetilde{\varepsilon}_{\theta,0,i}+\frac{(1-\nu)}{2}\widetilde{\gamma}^2_{x\theta,0,i}\right)d\eta d\theta+ \\ &+\frac{1}{2}\frac{\beta_i^2}{12(1-\nu^2)}\int_0^1\int_0^{2\pi}\left(\widetilde{k}^2_{x,i}+\widetilde{k}^2_{\theta,i}+2\nu\widetilde{k}_{x,i}\widetilde{k}_{\theta,i}+\frac{(1-\nu)}{2}\widetilde{k}^2_{x\theta,i}\right)d\eta d\theta\end{aligned} \quad i = 1,2 \quad (12)$$

In Equation (12), the first term on the right-hand side, which is related to shell middle surface strains, is referred to as membrane (i.e., stretching) energy, while the second one, which is related to shell middle surface changes in curvature and torsion, is referred to as bending (i.e., flexural) energy.

The dimensionless elastic strain energy of a DWCNT, given by two concentric SWCNTs, is:

$$\widetilde{U} = \sum_{i=1}^{2}\delta_i\widetilde{U}_i \quad (13)$$

where $\delta_i = R_i/R_1$ and R_1 is the radius of the inner shell.

2.7. Kinetic Energy

The dimensional time variable t is made dimensionless by adopting a reference frequency ω_0, which is the lowest extensional circular frequency of a ring under plane strain hypothesis, in the form [61]:

$$\omega_0 = \sqrt{\frac{E}{\rho(1-\nu^2)R_1^2}} \quad \tau = \omega_0 t \quad (14)$$

where E is the Young's modulus, ν is the Poisson's ratio and ρ is the mass density of the two layers of the DWCNT, and τ is the dimensionless time variable.

The dimensionless velocity field of the i-th cylindrical shell is written in the form [61]:

$$\widetilde{u}'_i = \frac{d\widetilde{u}_i}{d\tau} = \frac{\dot{u}_i}{R_i\omega_0} \quad \widetilde{v}'_i = \frac{d\widetilde{v}_i}{d\tau} = \frac{\dot{v}_i}{R_i\omega_0} \quad \widetilde{w}'_i = \frac{d\widetilde{w}_i}{d\tau} = \frac{\dot{w}_i}{R_i\omega_0} \quad i = 1,2 \quad (15)$$

where $(\dot{u}_i, \dot{v}_i, \dot{w}_i)$ is the corresponding dimensional velocity field.

The dimensionless kinetic energy of the i-th thin cylindrical shell, which models a SWCNT, by neglecting the rotary inertia effect, is given by [61]:

$$\widetilde{T}_i = \frac{1}{2}\frac{1}{(1-\nu^2)}\delta_i^2\int_0^1\int_0^{2\pi}\left(\widetilde{u}'^2_i+\widetilde{v}'^2_i+\widetilde{w}'^2_i\right)d\eta d\theta \quad i = 1,2 \quad (16)$$

The dimensionless kinetic energy of a DWCNT, given by two concentric SWCNTs, is:

$$\widetilde{T} = \sum_{i=1}^{2}\delta_i\widetilde{T}_i \quad (17)$$

2.8. Van der Waals Interaction Energy

The dimensionless pressure exerted on the i-th circular cylindrical shell due to van der Waals interactions between the two layers (i, j) of a DWCNT is written as a function of the dimensionless radial displacements $(\widetilde{w}_i, \widetilde{w}_j)$ of the two layers in the form [61]:

$$\widetilde{p}_i(\eta,\theta) = \widetilde{c}_{ij}(\delta_i\widetilde{w}_i - \delta_j\widetilde{w}_j) \quad i,j = 1,2 \quad i \neq j \quad (18)$$

where $\widetilde{c}_{ij}$ is the dimensionless van der Waals interaction coefficient between layers (*i*, *j*), which can be expressed, by adopting the formulation proposed by He et al. [51], as:

$$\widetilde{c}_{ij} = -\left(\frac{1001\pi\widetilde{\varepsilon}\widetilde{\sigma}^{12}}{3\widetilde{a}^4}\widetilde{E}_{ij}^{13} - \frac{1120\pi\widetilde{\varepsilon}\widetilde{\sigma}^{6}}{9\widetilde{a}^4}\widetilde{E}_{ij}^{7}\right)\delta_j \quad i,j = 1,2 \quad i \neq j \tag{19}$$

where $(\widetilde{\sigma}, \widetilde{\varepsilon}, \widetilde{a})$ are dimensionless van der Waals interaction parameters, see [61] for more details.

The dimensionless elliptical integrals of Equation (19) are given by [51]:

$$\widetilde{E}_{ij}^{m} = (\delta_j + \delta_i)^{-m} \int_0^{\pi/2} \frac{d\theta}{\left(1-\widetilde{k}_{ij}\cos^2\theta\right)^{m/2}} \quad i,j = 1,2 \quad i \neq j \quad m = 7,13 \tag{20}$$

and the dimensionless geometric coefficient of Equation (20) is expressed as [51]:

$$\widetilde{k}_{ij} = \frac{4\delta_j\delta_i}{(\delta_j+\delta_i)^2} \quad i,j = 1,2 \quad i \neq j \tag{21}$$

The dimensionless van der Waals interaction energy of the *i*-th thin cylindrical shell, which models a SWCNT, is expressed as follows [61]:

$$\widetilde{V}_i = -\frac{1}{2}\frac{1}{\beta_i}\int_0^1\int_0^{2\pi} \widetilde{p}_i(\eta,\theta)\,\widetilde{w}_i\,d\eta d\theta \quad i = 1,2 \tag{22}$$

and therefore the dimensionless van der Waals interaction energy of a DWCNT is:

$$\widetilde{V} = \sum_{i=1}^{2} \delta_i \widetilde{V}_i \tag{23}$$

3. Linear Vibration Modelling of DWCNTs

A modal vibration, i.e., a synchronous motion, of a DWCNT can be formally written as [61]:

$$\begin{aligned} \widetilde{u}_i(\eta,\theta,\tau) &= \widetilde{U}_i(\eta,\theta)\widetilde{f}_i(\tau) \qquad \widetilde{v}_i(\eta,0,\tau) = \widetilde{V}_i(\eta,\theta)\widetilde{f}_i(\tau) \\ \widetilde{w}_i(\eta,\theta,\tau) &= \widetilde{W}_i(\eta,\theta)\widetilde{f}_i(\tau) \end{aligned} \quad i = 1,2 \tag{24}$$

where $(\widetilde{U}_i, \widetilde{V}_i, \widetilde{W}_i)$ are the three components of the modal shape of the *i*-th circular cylindrical shell and $\widetilde{f}_i$ is the corresponding dimensionless time law, which is supposed to be the same for the three dimensionless displacements $(\widetilde{u}_i, \widetilde{v}_i, \widetilde{w}_i)$ (modal vibration hypothesis).

The modal shape components $(\widetilde{U}_i, \widetilde{V}_i, \widetilde{W}_i)$ are expanded by means of a double mixed series in terms of *m*-th degree Chebyshev orthogonal polynomials $T_m^*(\eta)$ along the longitudinal direction η and harmonic functions ($\cos n\theta$, $\sin n\theta$) along the circumferential direction θ, in the form [61]:

$$\begin{aligned} \widetilde{U}_i(\eta,\theta) &= \sum_{m=0}^{M_u}\sum_{n=0}^{N} \widetilde{U}_{i,m,n}\, T_m^*(\eta)\cos n\theta \qquad \widetilde{V}_i(\eta,\theta) = \sum_{m=0}^{M_v}\sum_{n=0}^{N} \widetilde{V}_{i,m,n}\, T_m^*(\eta)\sin n\theta \\ \widetilde{W}_i(\eta,\theta) &= \sum_{m=0}^{M_w}\sum_{n=0}^{N} \widetilde{W}_{i,m,n}\, T_m^*(\eta)\cos n\theta \end{aligned} \quad i = 1,2 \tag{25}$$

where $T_m^* = T_m(2\eta - 1)$, *m* denotes the number of longitudinal half-waves, *n* represents the number of circumferential waves and $(\widetilde{U}_{i,m,n}, \widetilde{V}_{i,m,n}, \widetilde{W}_{i,m,n})$ are unknown coefficients, which can be calculated by imposing boundary conditions.

3.1. Boundary Conditions

In this paper, the linear vibrations of simply supported and free–free DWCNTs are investigated.

3.1.1. Simply Supported DWCNTs

In case of simply supported DWCNTs, the boundary conditions are written as [61]:

$$\tilde{v}_i = 0 \quad \tilde{w}_i = 0 \quad \tilde{N}_{x,i} = 0 \quad \tilde{M}_{x,i} = 0 \quad \eta = 0,1 \quad i = 1,2 \tag{26}$$

Starting from conditions (26), taking into consideration Equation (5) for dimensionless force and moment resultants and Equation (24) for dimensionless displacements, together with expansions (25) for the corresponding modal shape components, the following relationships are derived:

$$\tilde{V}_i(\eta,\theta) = \sum_{m=0}^{M_v}\sum_{n=0}^{N} \tilde{V}_{i,m,n}\, T^*_m(\eta)\sin n\theta = 0 \qquad \eta = 0,1 \quad i = 1,2 \tag{27}$$

$$\tilde{W}_i(\eta,\theta) = \sum_{m=0}^{M_w}\sum_{n=0}^{N} \tilde{W}_{i,m,n}\, T^*_m(\eta)\cos n\theta = 0 \qquad \eta = 0,1 \quad i = 1,2 \tag{28}$$

$$\tilde{U}_{i,\eta}(\eta,\theta) = \sum_{m=0}^{M_u}\sum_{n=0}^{N} \tilde{U}_{i,m,n}\, T^*_{m,\eta}(\eta)\cos n\theta = 0 \qquad \eta = 0,1 \quad i = 1,2 \tag{29}$$

$$\tilde{W}_{i,\eta\eta}(\eta,\theta) = \sum_{m=0}^{M_w}\sum_{n=0}^{N} \tilde{W}_{i,m,n}\, T^*_{m,\eta\eta}(\eta)\cos n\theta = 0 \qquad \eta = 0,1 \quad i = 1,2 \tag{30}$$

where $(\cdot)_{,\eta} = \partial(\cdot)/\partial\eta$ and $(\cdot)_{,\eta\eta} = \partial^2(\cdot)/\partial\eta^2$.

The linear algebraic system given by Equations (27)–(30) can be solved analytically in terms of coefficients $(\tilde{U}_{i,1,n}, \tilde{U}_{i,2,n}, \tilde{V}_{i,0,n}, \tilde{V}_{i,1,n}, \tilde{W}_{i,0,n}, \tilde{W}_{i,1,n}, \tilde{W}_{i,2,n}, \tilde{W}_{i,3,n})$, for $n \in [0, N]$. Therefore, in the case of simply supported DWCNTs, eight different coefficients for each cylindrical shell are obtained.

3.1.2. Free–Free DWCNTs

In case of free–free DWCNTs, the boundary conditions are written as [61]:

$$\tilde{N}_{x,i} = 0 \quad \tilde{N}_{x\theta,i} + \tilde{M}_{x\theta,i} = 0 \quad \tilde{Q}_{x,i} + \frac{\partial \tilde{M}_{x\theta,i}}{\partial\theta} = 0 \quad \tilde{M}_{x,i} = 0 \quad \eta = 0,1 \quad i = 1,2 \tag{31}$$

It can be observed that the boundary conditions (31), applied at the free edges of the DWCNTs, are of "natural type" (forces and moments) and not of "geometric type" (displacements and velocities); since the Rayleigh–Ritz method is used to solve the equations, then only the "geometric type" boundary conditions (see simply supported ones) have to be exactly satisfied, while the "natural type" boundary conditions will be satisfied by the minimization of the total energy of the system.

3.2. Rayleigh–Ritz Method

In the case of MWCNTs, the maximum number of variables needed to describe a general vibration mode with n circumferential waves is given by $N_p = N_s \times (M_u + M_v + M_w + 3 - p)$, where N_s is the number of concentric SWCNTs, $M_u = M_v = M_w$ is the maximum number of longitudinal half-waves considered and p is the number of equations needed to satisfy the boundary conditions.

By means of a specific convergence analysis it was obtained that $M_u = M_v = M_w = 11$ provides accurate results with relatively reduced computational effort. Therefore, for the DWCNTs ($N_s = 2$), in the case of simply supported edges, it is imposed $p = 8$ and it is found $N_p = 56$, while in the case of free–free edges, it is imposed $p = 0$ and it is found $N_p = 72$.

For a multi-mode analysis including different values of circumferential waves n, the number of degrees of freedom of the system is computed by the relation $N_{max} = N_p \times (N + 1)$, where N is the maximum number of circumferential waves considered.

Equations (24) are inserted in the expressions of elastic strain energy (12), kinetic energy (16) and van der Waals interaction energy (22) in order to compute the value of Rayleigh quotient $R(\widetilde{q})$, where $\widetilde{q}$ is a vector containing all the unknown coefficients of expansions (25), which is defined in the form [61]:

$$\widetilde{q} = \begin{bmatrix} \vdots \\ \widetilde{U}_{i,m,n} \\ \widetilde{V}_{i,m,n} \\ \widetilde{W}_{i,m,n} \\ \vdots \end{bmatrix} \quad i = 1,2 \tag{32}$$

After imposing stationarity to Rayleigh quotient, the following classic eigenvalue problem in dimensionless form is obtained [61]:

$$(-\widetilde{\omega}^2\widetilde{\boldsymbol{M}} + \widetilde{\boldsymbol{K}})\widetilde{q} = 0 \tag{33}$$

which provides approximate dimensionless circular frequencies (eigenvalues $\widetilde{\omega}_j$) and modal shapes (eigenvectors $\widetilde{q}_j$), with $j = (1,2 \ldots ,N_{\max})$, where $\widetilde{\boldsymbol{M}}$ and $\widetilde{\boldsymbol{K}}$ denote the dimensionless mass and stiffness matrices, respectively.

The approximate modal shape of the j-th mode of the i-th thin circular cylindrical shell is given by Equation (25), where the coefficients $(\widetilde{U}_{i,m,n}, \widetilde{V}_{i,m,n}, \widetilde{W}_{i,m,n})$ are replaced with $(\widetilde{U}^{(j)}_{i,m,n}, \widetilde{V}^{(j)}_{i,m,n}, \widetilde{W}^{(j)}_{i,m,n})$, which are the components of the j-th eigenvector $\widetilde{q}$ of Equation (33).

The vector function [61]:

$$\widetilde{\boldsymbol{W}}^{(j)}(\eta,\theta) = \begin{bmatrix} \widetilde{U}_i^{(j)}(\eta,\theta) \\ \widetilde{V}_i^{(j)}(\eta,\theta) \\ \widetilde{W}_i^{(j)}(\eta,\theta) \end{bmatrix} \quad i = 1,2 \tag{34}$$

is the approximation of the j-th eigenfunction vector of the original problem.

The components of the j-th eigenfunction Vector (34) are normalised imposing [61]:

$$\max[\max[\widetilde{U}_i^{(j)}(\eta,\theta), \widetilde{V}_i^{(j)}(\eta,\theta), \widetilde{W}_i^{(j)}(\eta,\theta)]] = 1 \quad i = 1,2 \tag{35}$$

4. Numerical Results

In the present work, natural frequencies of DWCNTs obtained by using simplified (reduced) thin shell models are compared with the results of the extended (complete) Sanders–Koiter shell theory. The mechanical parameters reported in Table 1 are adopted. DWCNTs with different values of aspect ratio L/R_2 and thickness ratio R_2/h are considered. Vibration modes with different numbers of longitudinal half-waves m and circumferential waves n are investigated.

Table 1. Mechanical parameters of the considered DWCNTs [46].

Young's Modulus E [TPa]	5.5
Poisson's ratio ν	0.19
Mass density ρ [kg/m^3]	11,700
Layer thickness h [nm]	0.066
Interlayer distance d [nm]	0.34

4.1. Experimental Validation of the Extended Sanders–Koiter Shell Theory for SWCNTs

In this section, the model based on the extended Sanders–Koiter shell theory is validated in the linear field by means of comparisons with experimental data available in the literature in terms of natural frequencies of the radial breathing modes (RBMs).

The RBM is a specific vibrational mode that is often used to identify experimentally the CNTs by resonant Raman spectroscopy (RRS). This mode corresponds to the "vibrational" numbers ($m = 0$, $n = 0$) and it appears only in case of free–free boundary conditions.

The existing experimental data for the CNT vibrations are mainly focused on the RBMs of SWNTs because the spectrum of the nanotubes is quite complex. Moreover, the natural frequency associated with the RBM of a SWNT is inversely proportional to the diameter of, and is independent from, the length of the carbon nanotube [38].

Resonant Raman spectroscopy (RRS) represents a powerful technique to study the quantum properties of electrons and phonons in carbon nanotubes and to determine the atomic structure of an isolated SWNT. The RBM exhibits strong resonant characteristics in the Raman spectra, because it corresponds to the symmetric in-phase motion of all the carbon atoms in the radial direction of the tube. In this configuration all the carbon atoms undergo the same radial displacement [39].

For the RBMs, the radial displacement $\widetilde{w}(\eta, \theta, \tau)$ is spatially uniform, i.e., it is independent of η and $\theta(\widetilde{w} = \widetilde{w}(\tau))$. In this special case, the bending stiffness of the SWNTs does not appear, because the radial breathing vibration does not involve the bending deformation and it corresponds to an uniaxial stress state of the graphene sheet [40].

The comparisons reported in Table 2 show that the extended Sanders–Koiter shell theory gives natural frequencies with a good agreement with the experiments in terms of radial breathing modes (RBMs) of the free–free SWCNT under different diameters, where the relative errors are always less than 5%, and it appears a satisfactory accuracy.

Table 2. Natural frequencies of the radial breathing mode ($m = 0$, $n = 0$). Comparisons between resonant Raman spectroscopy and extended Sanders–Koiter shell theory. Free–free SWCNT with mechanical parameters of Table 1.

	Natural Frequency [10^{12} Hz]		
SWCNT Diameter D [nm]	**Resonant Raman Spectroscopy [40]**	**Extended Sanders–Koiter Shell Theory**	**Difference %**
1.018	7.165	6.905	3.63
1.036	7.105	6.785	4.50
1.053	6.865	6.669	2.85
1.137	6.295	6.177	1.87
1.409	5.276	5.025	4.76
1.416	5.216	4.964	4.83
1.435	5.066	4.895	3.37
1.467	4.947	4.788	3.21
1.492	4.917	4.711	4.19
1.528	4.797	4.594	4.23
1.540	4.737	4.559	3.76
1.564	4.677	4.494	3.91
1.599	4.617	4.393	4.85
1.627	4.527	4.318	4.62
1.644	4.437	4.271	3.74

4.2. Comparisons between Extended Sanders–Koiter Shell Theory and Simplified Thin Shell Models

In the numerical simulations performed to compare simplified and extended shell models, the aspect ratio L/R_2 assumes values from 10 (relatively short DWCNT) to 50 (long DWCNT) and 100 (very long DWCNT), while the thickness ratio R_2/h assumes values from 25 (relatively thick DWCNT) to 50 (thin DWCNT) and 250 (very thin DWCNT).

It must be observed that, since the thickness ratio R_2/h considered in the numerical simulations is within the range of validity $20 \leq R/h \leq 500$ of Kirchhoff–Love's assumptions, then the DWCNT can be modelled by means of two concentric thin elastic circular cylindrical shells, and the Sanders–Koiter thin shell theory can be correctly applied, see ref. [7] for more details.

As previously stated, in the present paper, the actual discrete DWCNTs are modelled by means of couples of concentric equivalent continuous thin cylindrical shells with van der Waals interactions.

In the case of cylindrical shells, the modal shapes associated with each natural frequency can be classified as primarily radial (flexural), longitudinal (axial) or circumferential (torsional). It was demonstrated that the lowest natural frequency is usually associated with the modal shape where the transverse (i.e., radial) displacement w is prevalent, while the other two natural frequencies are usually higher by one order of magnitude and they are associated with modal shapes where the displacements in the tangent plane (i.e., longitudinal u and circumferential v) dominate, see Refs. [4,7,8] for more details.

Moreover, since a DWCNT is constituted by two concentric cylindrical shells (i.e., SWCNTs), then each modal shape of the DWCNT is present twice, once referred to the inner cylindrical shell (with prevalent radius R_1) and once referred to the outer cylindrical shell (with prevalent radius R_2). As reported in literature [61], between the two corresponding natural frequencies, the one associated with the modal shape referred to the inner cylindrical shell (inner radius R_1) is usually lower.

Therefore, in the following numerical simulations, only the natural frequencies corresponding to modal shapes with prevalent radial displacement w and prevalent inner radius R_1 will be analysed, which, according to the previous observations, represent the fundamental (i.e., lowest) frequencies of DWCNTs.

In this paper, the extended Sanders–Koiter shell theory, which retains three middle surface strains $(\tilde{\varepsilon}_{x,0,i}, \tilde{\varepsilon}_{\theta,0,i}, \tilde{\gamma}_{x\theta,0,i})$ and three middle surface changes in curvature and torsion $\left(\tilde{k}_{x,i}, \tilde{k}_{\theta,i}, \tilde{k}_{x\theta,i}\right)$ in the expression of the elastic strain energy (12), is taken as reference model. Three different simplified elastic shell models are compared with the extended Sanders–Koiter shell theory to investigate their applicability and limitations for vibration modelling of DWCNTs in case of axisymmetric ($n = 0$), beam-like ($n = 1$) and shell-like ($n \geq 2$) modes.

The first simplified elastic shell model studied is a reduced version of the Sanders–Koiter shell theory developed on the basis of the assumptions of the membrane shell theory (6) and (7), retaining only the three middle surface strains $(\tilde{\varepsilon}_{x,0,i}, \tilde{\varepsilon}_{\theta,0,i}, \tilde{\gamma}_{x\theta,0,i})$ and neglecting the three middle surface changes in curvature and torsion $\left(\tilde{k}_{x,i}, \tilde{k}_{\theta,i}, \tilde{k}_{x\theta,i}\right)$.

The second simplified elastic shell model analysed is a reduced version of the Sanders–Koiter shell theory developed on the basis of the assumptions of the moment shell theory (8) and (9), retaining only the three middle surface changes in curvature and torsion $\left(\tilde{k}_{x,i}, \tilde{k}_{\theta,i}, \tilde{k}_{x\theta,i}\right)$ and neglecting the three middle surface strains $(\tilde{\varepsilon}_{x,0,i}, \tilde{\varepsilon}_{\theta,0,i}, \tilde{\gamma}_{x\theta,0,i})$.

It must be stressed that, for shell-like modes, as reported in literature [3], the membrane shell theory is accurate for relatively short and thin shells, when the effect of the middle surface changes in curvature and torsion $\left(\tilde{k}_{x,i}, \tilde{k}_{\theta,i}, \tilde{k}_{x\theta,i}\right)$ is negligible (very small bending/torsional deformations); conversely, the moment shell theory is accurate for relatively long and thick shells, when the effect of the middle surface strains $(\tilde{\varepsilon}_{x,0,i}, \tilde{\varepsilon}_{\theta,0,i}, \tilde{\gamma}_{x\theta,0,i})$ is negligible (very small stretching deformations).

Therefore, for shell-like modes, a third simplified elastic shell model is introduced, which is a new approximate model, developed on the basis of a combination of the hypotheses of the membrane and semi-moment theories of thin shells, retaining the middle surface terms $\left(\tilde{\varepsilon}_{x,0,i}, \tilde{\varepsilon}_{\theta,0,i}, \tilde{\gamma}_{x\theta,0,i}, \tilde{k}_{\theta,i}\right)$ and neglecting the middle surface changes in curvature and torsion $\left(\tilde{k}_{x,i}, \tilde{k}_{x\theta,i}\right)$.

4.2.1. Axisymmetric Vibration Modes

In this section, natural frequencies of simply supported DWCNTs, in case of axisymmetric vibration modes ($n = 0$), are obtained. Different numbers of longitudinal half-waves m are considered. Aspect ratios $L/R_2 = (10,50,100)$ and thickness ratios $R_2/h = (25,50,250)$ are investigated. Different elastic shell theories are compared.

In Tables 3–5, comparisons between extended Sanders–Koiter and simplified membrane shell theories are reported. A negligible percentage difference (≈0%) is obtained for every number of longitudinal half-waves m and every value of aspect L/R_2 and thickness R_2/h ratios.

In Tables 6–8, comparisons between extended Sanders–Koiter and simplified moment shell theories are reported. A very large percentage difference (≈100%) is obtained for every number of longitudinal half-waves m and every value of aspect L/R_2 and thickness R_2/h ratios.

Therefore, in case of axisymmetric vibration modes, membrane shell theory, which neglects the middle surface changes in curvature and torsion $\left(\widetilde{k}_{x,i}, \widetilde{k}_{\theta,i}, \widetilde{k}_{x\theta,i}\right)$ and retains the middle surface strains $(\widetilde{\varepsilon}_{x,0,i}, \widetilde{\varepsilon}_{\theta,0,i}, \widetilde{\gamma}_{x\theta,0,i})$, gives an excellent agreement, with respect to the extended Sanders–Koiter shell theory, and can be adopted to properly compute the natural frequencies of DWCNTs.

These results, obtained for DWCNTs, confirm the applicability of membrane and limitations of moment thin shell theories for the axisymmetric vibration modelling of simply supported thin circular cylindrical shells [4]. In fact, in this case, membrane (stretching) energy is prevalent, while bending (flexural) energy is negligible, in the expression of the elastic strain energy of the cylindrical shell.

Table 3. Simply supported DWCNT with aspect ratio $L/R_2 = 10$ and thickness ratios $R_2/h = (25,50,250)$. Axisymmetric modes ($n = 0$). Comparisons between Sanders–Koiter and membrane shell theories.

	Natural Frequency [10^{12} Hz]		
Mode (*m*,*n*)	**Sanders–Koiter Shell Theory**	**Membrane Shell Theory**	**Difference %**
Aspect ratio $L/R_2 = 10$	Thickness ratio $R_2/h = 25$		
(1,0)	2.34634	2.34634	0.00
(2,0)	2.36193	2.36191	0.00
(3,0)	2.42883	2.42875	0.00
(4,0)	2.21080	2.21053	0.01
(5,0)	2.26693	2.26615	0.03
Aspect ratio $L/R_2 = 10$	Thickness ratio $R_2/h = 50$		
(1,0)	1.12536	1.12536	0.00
(2,0)	1.13416	1.13416	0.00
(3,0)	1.17636	1.17635	0.00
(4,0)	1.06656	1.06652	0.00
(5,0)	1.08789	1.08779	0.01
Aspect ratio $L/R_2 = 10$	Thickness ratio $R_2/h = 250$		
(1,0)	0.21566	0.21566	0.00
(2,0)	0.21759	0.21759	0.00
(3,0)	0.22817	0.22817	0.00
(4,0)	0.20556	0.20556	0.00
(5,0)	0.20879	0.20879	0.00

Table 4. Simply supported DWCNT with aspect ratio $L/R_2 = 50$ and thickness ratios $R_2/h = (25,50,250)$. Axisymmetric modes ($n = 0$). Comparisons between Sanders–Koiter and membrane shell theories.

	Natural Frequency [10^{12} Hz]		**Difference %**
Mode (m,n)	**Sanders–Koiter Shell Theory**	**Membrane Shell Theory**	
Aspect ratio $L/R_2 = 50$	Thickness ratio $R_2/h = 25$		
(1,0)	2.34288	2.34288	0.00
(2,0)	2.34328	2.34328	0.00
(3,0)	2.34397	2.34397	0.00
(4,0)	2.34498	2.34498	0.00
(5,0)	2.34636	2.34636	0.00
Aspect ratio $L/R_2 = 50$	Thickness ratio $R_2/h = 50$		
(1,0)	1.12347	1.12347	0.00
(2,0)	1.12369	1.12369	0.00
(3,0)	1.12407	1.12407	0.00
(4,0)	1.12462	1.12462	0.00
(5,0)	1.12537	1.12537	0.00
Aspect ratio $L/R_2 = 50$	Thickness ratio $R_2/h = 250$		
(1,0)	0.21526	0.21526	0.00
(2,0)	0.21530	0.21530	0.00
(3,0)	0.21538	0.21538	0.00
(4,0)	0.21550	0.21550	0.00
(5,0)	0.21566	0.21566	0.00

Table 5. Simply supported DWCNT with aspect ratio $L/R_2 = 100$ and thickness ratios $R_2/h = (25,50,250)$. Axisymmetric modes ($n = 0$). Comparisons between Sanders–Koiter and membrane shell theories.

	Natural Frequency [10^{12} Hz]		**Difference %**
Mode (m,n)	**Sanders–Koiter Shell Theory**	**Membrane Shell Theory**	
Aspect ratio $L/R_2 = 100$	Thickness ratio $R_2/h = 25$		
(1,0)	2.34278	2.34278	0.00
(2,0)	2.34288	2.34288	0.00
(3,0)	2.34304	2.34304	0.00
(4,0)	2.34328	2.34328	0.00
(5,0)	2.34359	2.34359	0.00
Aspect ratio $L/R_2 = 100$	Thickness ratio $R_2/h = 50$		
(1,0)	1.12342	1.12342	0.00
(2,0)	1.12347	1.12347	0.00
(3,0)	1.12356	1.12356	0.00
(4,0)	1.12369	2.12369	0.00
(5,0)	1.12386	1.12386	0.00
Aspect ratio $L/R_2 = 100$	Thickness ratio $R_2/h = 250$		
(1,0)	0.21525	0.21525	0.00
(2,0)	0.21526	0.21526	0.00
(3,0)	0.21528	0.21528	0.00
(4,0)	0.21530	0.21530	0.00
(5,0)	0.21534	0.21534	0.00

Table 6. Simply supported DWCNT with aspect ratio $L/R_2 = 10$ and thickness ratios R_2/h = (25,50,250). Axisymmetric modes ($n = 0$). Comparisons between Sanders–Koiter and moment shell theories.

Mode (m,n)	Natural Frequency [10^{12} Hz] Sanders–Koiter Shell Theory	Moment Shell Theory	Difference %
Aspect ratio $L/R_2 = 10$	Thickness ratio $R_2/h = 25$		
(1,0)	2.34634	0.00243	99.9
(2,0)	2.36193	0.00971	99.6
(3,0)	2.42883	0.02185	99.1
(4,0)	2.21080	0.03884	98.2
(5,0)	2.26693	0.06105	97.3
Aspect ratio $L/R_2 = 10$	Thickness ratio $R_2/h = 50$		
(1,0)	1.12536	0.00061	99.9
(2,0)	1.13416	0.00243	99.8
(3,0)	1.17636	0.00546	99.5
(4,0)	1.06656	0.00971	99.1
(5,0)	1.08789	0.01526	98.6
Aspect ratio $L/R_2 = 10$	Thickness ratio $R_2/h = 250$		
(1,0)	0.21566	0.00002	100
(2,0)	0.21759	0.00010	99.9
(3,0)	0.22817	0.00022	99.9
(4,0)	0.20556	0.00039	99.8
(5,0)	0.20879	0.00061	99.7

Table 7. Simply supported DWCNT with aspect ratio $L/R_2 = 50$ and thickness ratios R_2/h = (25,50,250). Axisymmetric modes ($n = 0$). Comparisons between Sanders–Koiter and moment shell theories.

Mode (m,n)	Natural Frequency [10^{12} Hz] Sanders–Koiter Shell Theory	Moment Shell Theory	Difference %
Aspect ratio $L/R_2 = 50$	Thickness ratio $R_2/h = 25$		
(1,0)	2.34288	0.00010	100
(2,0)	2.34328	0.00039	100
(3,0)	2.34397	0.00087	100
(4,0)	2.34498	0.00155	99.9
(5,0)	2.34636	0.00244	99.9
Aspect ratio $L/R_2 = 50$	Thickness ratio $R_2/h = 50$		
(1,0)	1.12347	0.00002	100
(2,0)	1.12369	0.00010	100
(3,0)	1.12407	0.00022	100
(4,0)	1.12462	0.00039	100
(5,0)	1.12537	0.00061	99.9
Aspect ratio $L/R_2 = 50$	Thickness ratio $R_2/h = 250$		
(1,0)	0.21526	0.00000	100
(2,0)	0.21530	0.00000	100
(3,0)	0.21538	0.00001	100
(4,0)	0.21550	0.00001	100
(5,0)	0.21566	0.00002	100

Table 8. Simply supported DWCNT with aspect ratio L/R_2 = 100 and thickness ratios R_2/h = (25,50,250). Axisymmetric modes (n = 0). Comparisons between Sanders–Koiter and moment shell theories.

Mode (m,n)	Natural Frequency [10^{12} Hz] Sanders–Koiter Shell Theory	Natural Frequency [10^{12} Hz] Moment Shell Theory	Difference %
Aspect ratio L/R_2 = 100	Thickness ratio R_2/h = 25		
(1,0)	2.34278	0.00002	100
(2,0)	2.34288	0.00010	100
(3,0)	2.34304	0.00022	100
(4,0)	2.34328	0.00039	100
(5,0)	2.34359	0.00061	100
Aspect ratio L/R_2 = 100	Thickness ratio R_2/h = 50		
(1,0)	1.12342	0.00001	100
(2,0)	1.12347	0.00002	100
(3,0)	1.12356	0.00005	100
(4,0)	1.12369	0.00010	100
(5,0)	1.12386	0.00015	100
Aspect ratio L/R_2 = 100	Thickness ratio R_2/h = 250		
(1,0)	0.21525	0.00000	100
(2,0)	0.21526	0.00000	100
(3,0)	0.21528	0.00000	100
(4,0)	0.21530	0.00000	100
(5,0)	0.21534	0.00001	100

Moreover, since, from the computations, it was found that, for axisymmetric vibration modes, the middle surface tangential shear strain is equal to zero $(\widetilde{\gamma}_{x\theta,0,i} = 0)$, then this last component can be neglected. Therefore, a further simplified model, based on the hypotheses of the semi-membrane shell theory $\left(\widetilde{\gamma}_{x\theta,0,i} = \widetilde{k}_{x,i} = \widetilde{k}_{\theta,i} = \widetilde{k}_{x\theta,i} = 0\right)$ and retaining the middle surface strains $(\widetilde{\varepsilon}_{x,0,i}, \widetilde{\varepsilon}_{\theta,0,i})$ (two terms instead of six) in the expression of the elastic strain energy (12), can be used. This model again provides an excellent agreement, but with a much lower computational effort, than the extended Sanders–Koiter shell theory. These last comparisons are omitted for the sake of brevity.

In particular, from the performed simulations it was found that, for sufficiently long and thin DWCNTs ($L/R_2 \geq 50$, $R_2/h \geq 50$), the contribution of the circumferential normal strain $\widetilde{\varepsilon}_{\theta,0,i}$ is strongly prevalent with respect to the one of the longitudinal normal strain $\widetilde{\varepsilon}_{x,0,i}$ $(\widetilde{\varepsilon}_{\theta,0,i} \gg \widetilde{\varepsilon}_{x,0,i})$.

4.2.2. Beam-Like Vibration Modes

In this section, natural frequencies of simply supported DWCNTs in the case of beam-like modes (n = 1) are obtained. Different numbers of longitudinal half-waves m are considered. Aspect ratios L/R_2 = (10,50,100) and thickness ratios R_2/h = (25,50,250) are investigated. Different elastic shell theories are compared.

In Tables 9–11, comparisons between extended Sanders–Koiter and simplified membrane shell theories are reported. Again, a negligible percentage difference (≈0%) is obtained for every number of longitudinal half-waves m and every value of aspect L/R_2 and thickness R_2/h ratios.

In Tables 12–14, comparisons between extended Sanders–Koiter and simplified moment shell theories are reported. Again, a very large percentage difference (≈100%) is found for every number of longitudinal half-waves m and every value of aspect L/R_2 and thickness R_2/h ratios.

Therefore, also in case of beam-like vibration modes, membrane shell theory, which neglects the middle surface changes in curvature and torsion $\left(\widetilde{k}_{x,i}, \widetilde{k}_{\theta,i}, \widetilde{k}_{x\theta,i}\right)$ and retains the middle surface strains $(\widetilde{\varepsilon}_{x,0,i}, \widetilde{\varepsilon}_{\theta,0,i}, \widetilde{\gamma}_{x\theta,0,i})$, gives an excellent agreement with respect to the extended Sanders–Koiter shell theory and can be adopted to properly compute the natural frequencies of DWCNTs.

These results, obtained for DWCNTs, confirm applicability of membrane and limitations of moment thin shell theories also for beam-like vibration modelling of simply supported thin circular cylindrical shells [4]. In fact, also in this case, membrane energy is prevalent, while bending energy is negligible, in the expression of the elastic strain energy of the cylindrical shell.

From the computations it was obtained that, for beam-like vibration modes, all middle surface strains are not negligible. Therefore, a simplified elastic shell model, based on the hypotheses of the membrane shell theory $\left(\widetilde{k}_{x,i} = \widetilde{k}_{\theta,i} = \widetilde{k}_{x\theta,i} = 0\right)$ and taking into account all the middle surface strains $(\widetilde{\varepsilon}_{x,0,i}, \widetilde{\varepsilon}_{\theta,0,i}, \widetilde{\gamma}_{x\theta,0,i})$ (three terms instead of six) in the expression of the elastic strain energy (12), can be adopted. This model provides an excellent agreement but with a lower computational effort than the extended Sanders–Koiter shell theory.

In particular, from the performed simulations, similarly to the axisymmetric modes, it was found that, for sufficiently long and thin DWCNTs ($L/R_2 \geq 50$, $R_2/h \geq 50$), the contribution of the circumferential normal strain $\widetilde{\varepsilon}_{\theta,0,i}$ is strongly prevalent with respect to the longitudinal normal $\widetilde{\varepsilon}_{x,0,i}$ and tangential shear $\widetilde{\gamma}_{x\theta,0,i}$ strains $(\widetilde{\varepsilon}_{\theta,0,i} \gg (\widetilde{\varepsilon}_{x,0,i}, \widetilde{\gamma}_{x\theta,0,i}))$.

Therefore, from the results reported in the present and in the previous section, it can be derived that membrane shell theory, as a simplified elastic shell model, provides natural frequencies with excellent agreement with respect to the extended Sanders–Koiter shell theory with regard to axisymmetric and beam-like modes. The goal of the next section is to verify if the membrane shell theory can be rightly applied also to shell-like modes or of it is necessary to integrate/complete it with moment shell theory in order to obtain an acceptable agreement with the results of the extended Sanders–Koiter shell theory.

Table 9. Simply supported DWCNT with aspect ratio $L/R_2 = 10$ and thickness ratios $R_2/h = (25,50,250)$. Beam-like vibration modes ($n = 1$). Comparisons between Sanders–Koiter and membrane shell theories.

Mode (m,n)	Natural Frequency [10^{12} Hz]		Difference %
	Sanders–Koiter Shell Theory	Membrane Shell Theory	
Aspect ratio $L/R_2 = 10$	Thickness ratio $R_2/h = 25$		
(1,1)	0.12064	0.12063	0.01
(2,1)	0.39066	0.39060	0.01
(3,1)	0.70059	0.70040	0.03
(4,1)	1.00182	1.00134	0.05
(5,1)	1.27305	1.27195	0.09
Aspect ratio $L/R_2 = 10$	Thickness ratio $R_2/h = 50$		
(1,1)	0.06251	0.06251	0.00
(2,1)	0.20074	0.20073	0.00
(3,1)	0.35710	0.35707	0.01
(4,1)	0.50633	0.50626	0.01
(5,1)	0.63743	0.63729	0.02
Aspect ratio $L/R_2 = 10$	Thickness ratio $R_2/h = 250$		
(1,1)	0.01290	0.01290	0.00
(2,1)	0.04102	0.04102	0.00
(3,1)	0.07235	0.07235	0.00
(4,1)	0.10179	0.10179	0.00
(5,1)	0.12716	0.12716	0.00

Table 10. Simply supported DWCNT with aspect ratio $L/R_2 = 50$ and thickness ratios $R_2/h = (25,50,250)$. Beam-like modes ($n = 1$). Comparisons between Sanders–Koiter and membrane shell theories.

	Natural frequency [10^{12} Hz]		**Difference %**
Mode (*m*,*n*)	**Sanders–Koiter Shell Theory**	**Membrane Shell Theory**	
Aspect ratio $L/R_2 = 50$	Thickness ratio $R_2/h = 25$		
(1,1)	0.00531	0.00531	0.00
(2,1)	0.02098	0.02098	0.00
(3,1)	0.04620	0.04620	0.00
(4,1)	0.07985	0.07984	0.01
(5,1)	0.12111	0.12109	0.02
Aspect ratio $L/R_2 = 50$	Thickness ratio $R_2/h = 50$		
(1,1)	0.00277	0.00277	0.00
(2,1)	0.01091	0.01091	0.00
(3,1)	0.02401	0.02401	0.00
(4,1)	0.04144	0.04144	0.00
(5,1)	0.06275	0.06274	0.02
Aspect ratio $L/R_2 = 50$	Thickness ratio $R_2/h = 250$		
(1,1)	0.00057	0.00057	0.00
(2,1)	0.00226	0.00226	0.00
(3,1)	0.00498	0.00498	0.00
(4,1)	0.00857	0.00857	0.00
(5,1)	0.01295	0.01295	0.00

Table 11. Simply supported DWCNT with aspect ratio $L/R_2 = 100$ and thickness ratios $R_2/h = (25,50,250)$. Beam-like modes ($n = 1$). Comparisons between Sanders–Koiter and membrane shell theories.

	Natural Frequency [10^{12} Hz]		**Difference %**
Mode (*m*,*n*)	**Sanders–Koiter Shell Theory**	**Membrane Shell Theory**	
Aspect ratio $L/R_2 = 100$	Thickness ratio $R_2/h = 25$		
(1,1)	0.00133	0.00133	0.00
(2,1)	0.00531	0.00531	0.00
(3,1)	0.01189	0.01189	0.00
(4,1)	0.02098	0.02098	0.00
(5,1)	0.03263	0.03263	0.00
Aspect ratio $L/R_2 = 100$	Thickness ratio $R_2/h = 50$		
(1,1)	0.00069	0.00069	0.00
(2,1)	0.00277	0.00277	0.00
(3,1)	0.00619	0.00619	0.00
(4,1)	0.01091	0.01091	0.00
(5,1)	0.01697	0.01697	0.00
Aspect ratio $L/R_2 = 100$	Thickness ratio $R_2/h = 250$		
(1,1)	0.00014	0.00014	0.00
(2,1)	0.00057	0.00057	0.00
(3,1)	0.00128	0.00128	0.00
(4,1)	0.00226	0.00226	0.00
(5,1)	0.00352	0.00352	0.00

Table 12. Simply supported DWCNT with aspect ratio $L/R_2 = 10$ and thickness ratios $R_2/h = (25,50,250)$. Beam-like modes ($n = 1$). Comparisons between Sanders–Koiter and moment shell theories.

Mode (m,n)	Natural Frequency [10^{12} Hz] Sanders–Koiter Shell Theory	Moment Shell Theory	Difference %
Aspect ratio $L/R_2 = 10$	Thickness ratio $R_2/h = 25$		
(1,1)	0.12064	0.05079	57.9
(2,1)	0.39066	0.05734	85.3
(3,1)	0.70059	0.06802	90.3
(4,1)	1.00182	0.08297	91.7
(5,1)	1.27305	0.10238	92.0
Aspect ratio $L/R_2 = 10$	Thickness ratio $R_2/h = 50$		
(1,1)	0.06251	0.01057	83.1
(2,1)	0.20074	0.01232	93.9
(3,1)	0.35710	0.01512	95.8
(4,1)	0.50633	0.01899	96.2
(5,1)	0.63743	0.02407	96.2
Aspect ratio $L/R_2 = 10$	Thickness ratio $R_2/h = 250$		
(1,1)	0.01290	0.00038	97.0
(2,1)	0.04102	0.00045	98.9
(3,1)	0.07235	0.00057	99.2
(4,1)	0.10179	0.00072	99.3
(5,1)	0.12716	0.00093	99.3

Table 13. Simply supported DWCNT with aspect ratio $L/R_2 = 50$ and thickness ratios $R_2/h = (25,50,250)$. Beam-like modes ($n = 1$). Comparisons between Sanders–Koiter and moment shell theories.

Mode (m,n)	Natural Frequency [10^{12} Hz] Sanders–Koiter Shell Theory	Moment Shell Theory	Difference %
Aspect ratio $L/R_2 = 50$	Thickness ratio $R_2/h = 25$		
(1,1)	0.00531	0.00007	98.7
(2,1)	0.02098	0.00027	98.7
(3,1)	0.04620	0.00060	98.7
(4,1)	0.07985	0.00106	98.7
(5,1)	0.12111	0.00165	98.6
Aspect ratio $L/R_2 = 50$	Thickness ratio $R_2/h = 50$		
(1,1)	0.00277	0.00002	99.4
(2,1)	0.01092	0.00007	99.4
(3,1)	0.02401	0.00015	99.4
(4,1)	0.04144	0.00026	99.4
(5,1)	0.06275	0.00041	99.3
Aspect ratio $L/R_2 = 50$	Thickness ratio $R_2/h = 250$		
(1,1)	0.00057	0.00000	100
(2,1)	0.00226	0.00000	100
(3,1)	0.00498	0.00001	99.9
(4,1)	0.00857	0.00001	99.9
(5,1)	0.01295	0.00002	99.9

Table 14. Simply supported DWCNT with aspect ratio $L/R_2 = 100$ and thickness ratios R_2/h = (25,50,250). Beam-like modes ($n = 1$). Comparisons between Sanders–Koiter and moment shell theories.

	Natural Frequency [10^{12} Hz]		Difference %
Mode (m,n)	Sanders–Koiter Shell Theory	Moment Shell Theory	
Aspect ratio $L/R_2 = 100$	Thickness ratio $R_2/h = 25$		
(1,1)	0.00133	0.00002	98.7
(2,1)	0.00531	0.00007	98.7
(3,1)	0.01189	0.00015	98.7
(4,1)	0.02098	0.00027	98.7
(5,1)	0.03263	0.00042	98.7
Aspect ratio $L/R_2 = 100$	Thickness ratio $R_2/h = 50$		
(1,1)	0.00069	0.00000	100
(2,1)	0.00277	0.00002	99.4
(3,1)	0.00619	0.00004	99.4
(4,1)	0.01091	0.00007	99.4
(5,1)	0.01697	0.00010	99.4
Aspect ratio $L/R_2 = 100$	Thickness ratio $R_2/h = 250$		
(1,1)	0.00014	0.00000	100
(2,1)	0.00057	0.00000	100
(3,1)	0.00128	0.00000	100
(4,1)	0.00226	0.00000	100
(5,1)	0.00352	0.00000	100

4.2.3. Shell-Like Vibration Modes

In this section, natural frequencies of simply supported DWCNTs in the case of shell-like vibration modes ($n \geq 2$) are obtained. Different numbers of longitudinal half-waves m are considered. Aspect ratios L/R_2 = (10,50,100) and thickness ratios R_2/h = (25,50,250) are investigated. Different elastic shell theories are compared.

In Figures 2–6, comparisons between extended Sanders–Koiter and simplified membrane shell theories are reported. From these comparisons, the following observations can be made:

- by increasing the number of longitudinal half-waves m the percentage difference decreases (the contribution of the middle surface changes $\left(\tilde{k}_{x,i}, \tilde{k}_{\theta,i}, \tilde{k}_{x\theta,i}\right)$ decreases);
- by increasing the number of circumferential waves n the percentage difference increases (the contribution of the middle surface changes $\left(\tilde{k}_{x,i}, \tilde{k}_{\theta,i}, \tilde{k}_{x\theta,i}\right)$ increases);
- by increasing the aspect ratio L/R_2 the percentage difference increases (the contribution of the middle surface changes $\left(\tilde{k}_{x,i}, \tilde{k}_{\theta,i}, \tilde{k}_{x\theta,i}\right)$ increases);
- by increasing the thickness ratio R_2/h the percentage difference decreases (the contribution of the middle surface changes $\left(\tilde{k}_{x,i}, \tilde{k}_{\theta,i}, \tilde{k}_{x\theta,i}\right)$ decreases).

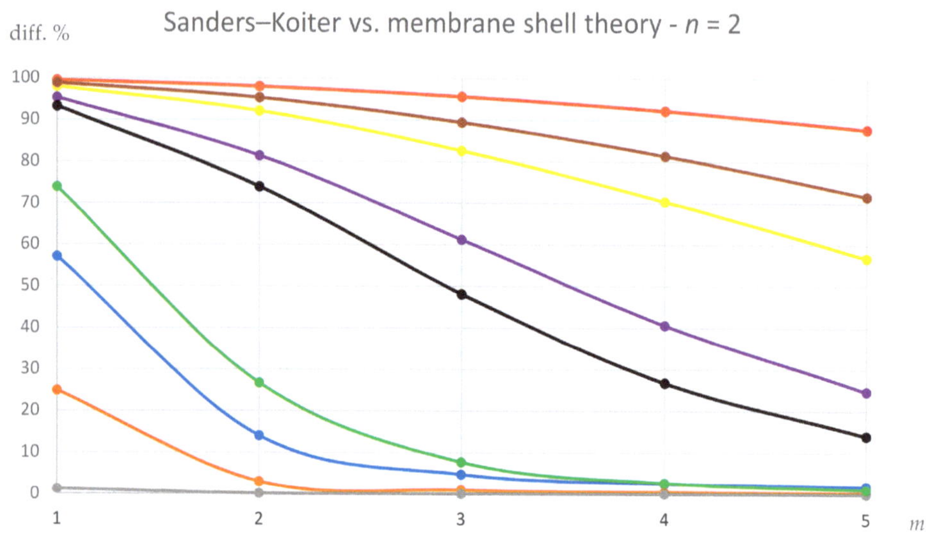

Figure 2. Simply supported DWCNT of Table 1 with different values of aspect ratio L/R_2 and thickness ratio R_2/h. Shell-like modes (n = 2). Comparisons between extended Sanders–Koiter and simplified membrane shell theories, where ▬ L/R_2 = 10, R_2/h = 25; ▬ L/R_2 = 10, R_2/h = 50; ▬ L/R_2 = 10, R_2/h = 250; ▬ L/R_2 = 50, R_2/h = 25; ▬ L/R_2 = 50, R_2/h = 50; ▬ L/R_2 = 50, R_2/h = 250; ▬ L/R_2 = 100, R_2/h = 25; ▬ L/R_2 = 100, R_2/h = 50; ▬ L/R_2 = 100, R_2/h = 250.

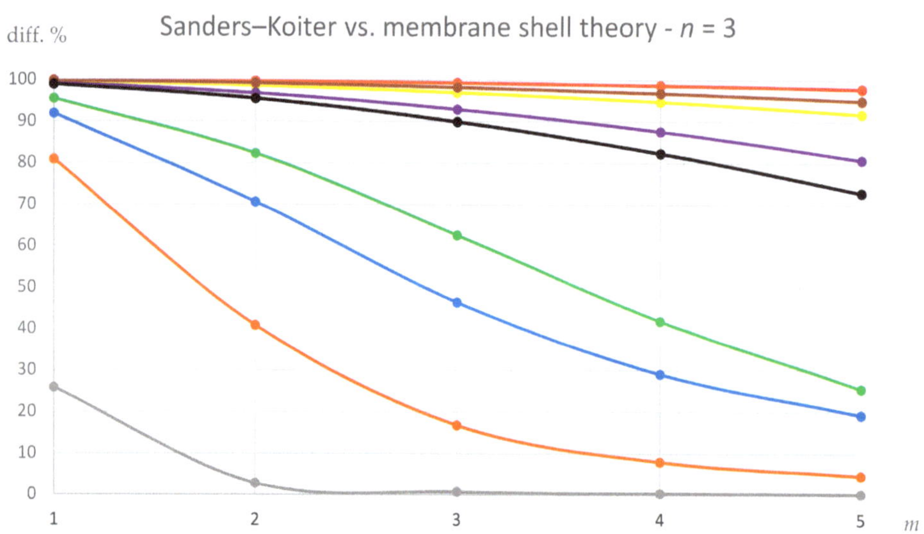

Figure 3. Simply supported DWCNT of Table 1 with different values of aspect ratio L/R_2 and thickness ratio R_2/h. Shell-like modes (n = 3). Comparisons between extended Sanders–Koiter and simplified membrane shell theories, where ▬ L/R_2 = 10, R_2/h = 25; ▬ L/R_2 = 10, R_2/h = 50; ▬ L/R_2 = 10, R_2/h = 250; ▬ L/R_2 = 50, R_2/h = 25; ▬ L/R_2 = 50, R_2/h = 50; ▬ L/R_2 = 50, R_2/h = 250; ▬ L/R_2 = 100, R_2/h = 25; ▬ L/R_2 = 100, R_2/h = 50; ▬ L/R_2 = 100, R_2/h = 250.

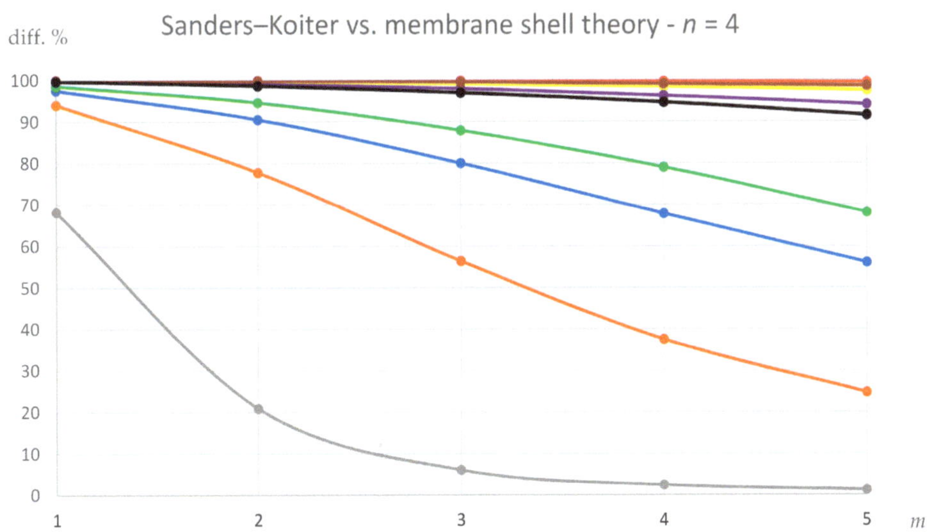

Figure 4. Simply supported DWCNT of Table 1 with different values of aspect ratio L/R_2 and thickness ratio R_2/h. Shell-like modes (n = 4). Comparisons between extended Sanders–Koiter and simplified membrane shell theories, where ▬ L/R_2 = 10, R_2/h = 25; ▬ L/R_2 = 10, R_2/h = 50; ▬ L/R_2 = 10, R_2/h = 250; ▬ L/R_2 = 50, R_2/h = 25; ▬ L/R_2 = 50, R_2/h = 50; ▬ L/R_2 = 50, R_2/h = 250; ▬ L/R_2 = 100, R_2/h = 25; ▬ L/R_2 = 100, R_2/h = 50; ▬ L/R_2 = 100, R_2/h = 250.

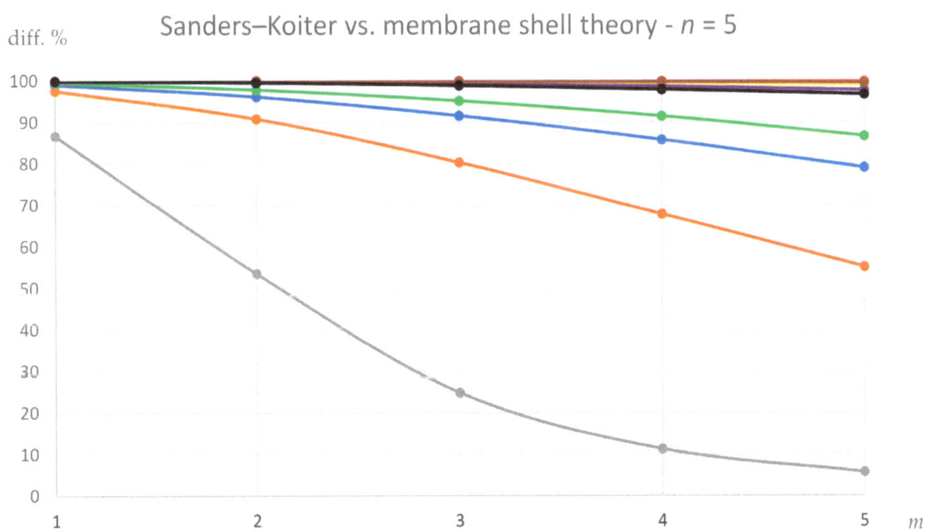

Figure 5. Simply supported DWCNT of Table 1 with different values of aspect ratio L/R_2 and thickness ratio R_2/h. Shell-like modes (n = 5). Comparisons between extended Sanders–Koiter and simplified membrane shell theories, where ▬ L/R_2 = 10, R_2/h = 25; ▬ L/R_2 = 10, R_2/h = 50; ▬ L/R_2 = 10, R_2/h = 250; ▬ L/R_2 = 50, R_2/h = 25; ▬ L/R_2 = 50, R_2/h = 50; ▬ L/R_2 = 50, R_2/h = 250; ▬ L/R_2 = 100, R_2/h = 25; ▬ L/R_2 = 100, R_2/h = 50; ▬ L/R_2 = 100, R_2/h = 250.

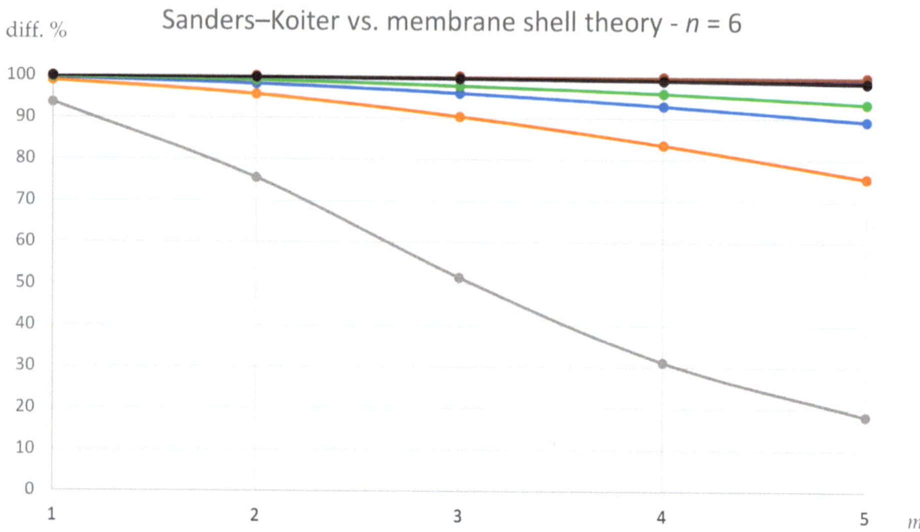

Figure 6. Simply supported DWCNT of Table 1 with different values of aspect ratio L/R_2 and thickness ratio R_2/h. Shell-like modes ($n = 6$). Comparisons between extended Sanders–Koiter and simplified membrane shell theories, where ▬ $L/R_2 = 10$, $R_2/h = 25$; ▬ $L/R_2 = 10$, $R_2/h = 50$; ▬ $L/R_2 = 10$, $R_2/h = 250$; ▬ $L/R_2 = 50$, $R_2/h = 25$; ▬ $L/R_2 = 50$, $R_2/h = 50$; ▬ $L/R_2 = 50$, $R_2/h = 250$; ▬ $L/R_2 = 100$, $R_2/h = 25$; ▬ $L/R_2 = 100$, $R_2/h = 50$; ▬ $L/R_2 = 100$, $R_2/h = 250$.

Therefore, in case of shell-like modes, membrane shell theory yields satisfactory results only for relatively short and very thin DWCNTs, when the effect of middle surface changes in curvature and torsion $\left(\widetilde{k}_{x,i}, \widetilde{k}_{\theta,i}, \widetilde{k}_{x\theta,i}\right)$ is negligible (very small bending and torsional deformations) and the effect of middle surface strains $(\widetilde{\varepsilon}_{x,0,i}, \widetilde{\varepsilon}_{\theta,0,i}, \widetilde{\gamma}_{x\theta,0,i})$ results prevalent (very high stretching deformations).

Starting from the previous observations, the following results can be obtained:

- the percentage difference decreases as the number of longitudinal half-waves m increases, the number of circumferential waves n decreases, the aspect ratio L/R_2 decreases and the thickness ratio R_2/h increases;
- the minimum percentage difference is obtained at ($m = 5$, $n = 2$, $L/R_2 = 10$, $R_2/h = 250$) and is equal to 0.01;
- the percentage difference increases as the number of longitudinal half-waves m decreases, the number of circumferential waves n increases, the aspect ratio L/R_2 increases and the thickness ratio R_2/h decreases;
- the maximum percentage difference is obtained at ($m = 1$, $n = 6$, $L/R_2 = 100$, $R_2/h = 25$) and is equal to 100.

In Figures 7–11, comparisons between extended Sanders–Koiter and simplified moment shell theories are reported. From these comparisons, the following observations can be made:

- by increasing the aspect ratio L/R_2 the percentage difference decreases (the contribution of the middle surface strains $(\widetilde{\varepsilon}_{x,0,i}, \widetilde{\varepsilon}_{\theta,0,i}, \widetilde{\gamma}_{x\theta,0,i})$ decreases);
- by increasing the thickness ratio R_2/h the percentage difference increases (the contribution of middle surface strains $(\widetilde{\varepsilon}_{x,0,i}, \widetilde{\varepsilon}_{\theta,0,i}, \widetilde{\gamma}_{x\theta,0,i})$ increases);
- the curve showing maximum percentage difference corresponds to ($L/R_2 = 10$, $R_2/h = 250$);
- the curve showing minimum percentage difference corresponds to ($L/R_2 = 100$, $R_2/h = 25$).

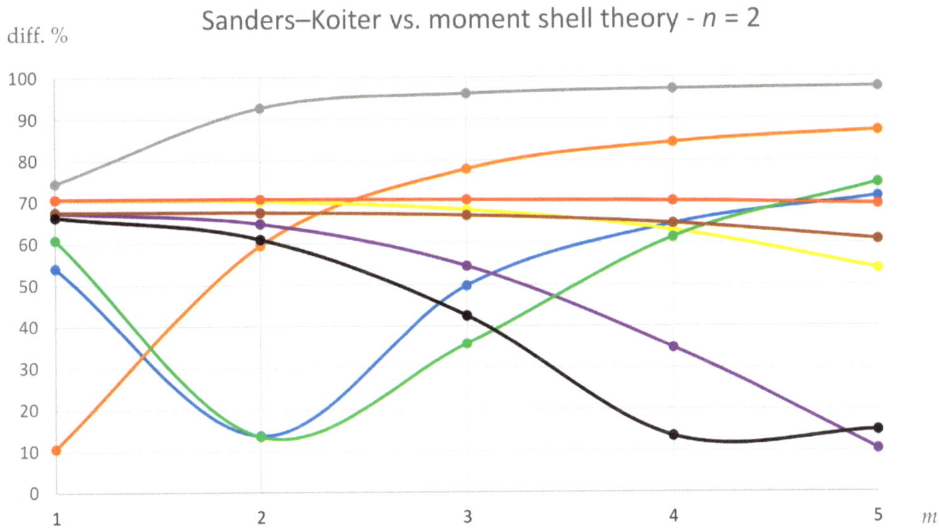

Figure 7. Simply supported DWCNT of Table 1 with different values of aspect ratio L/R_2 and thickness ratio R_2/h. Shell-like modes ($n = 2$). Comparisons between extended Sanders–Koiter and simplified moment shell theories, where ▬ $L/R_2 = 10$, $R_2/h = 25$; ▬ $L/R_2 = 10$, $R_2/h = 50$; ▬ $L/R_2 = 10$, $R_2/h = 250$; ▬ $L/R_2 = 50$, $R_2/h = 25$; ▬ $L/R_2 = 50$, $R_2/h = 50$; ▬ $L/R_2 = 50$, $R_2/h = 250$; ▬ $L/R_2 = 100$, $R_2/h = 25$; ▬ $L/R_2 = 100$, $R_2/h = 50$; ▬ $L/R_2 = 100$, $R_2/h = 250$.

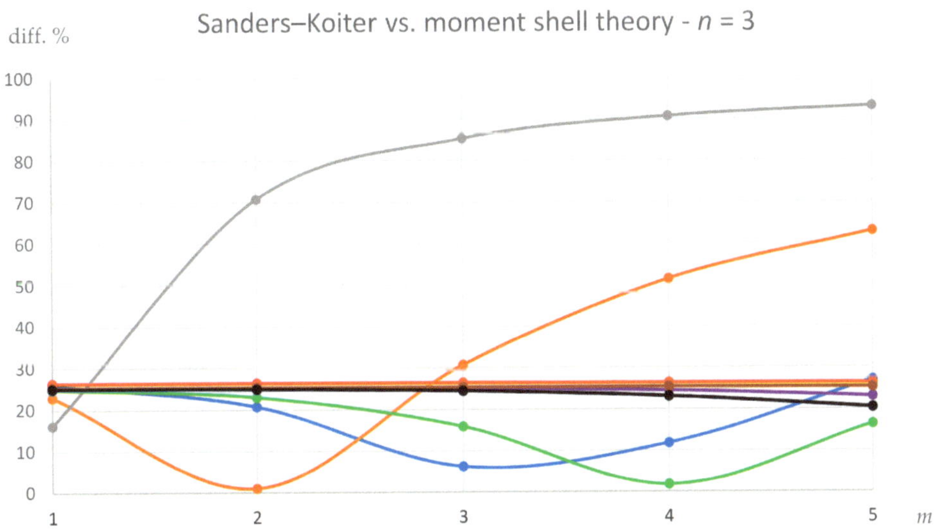

Figure 8. Simply supported DWCNT of Table 1 with different values of aspect ratio L/R_2 and thickness ratio R_2/h. Shell-like modes ($n = 3$). Comparisons between extended Sanders–Koiter and simplified moment shell theories, where ▬ $L/R_2 = 10$, $R_2/h = 25$; ▬ $L/R_2 = 10$, $R_2/h = 50$; ▬ $L/R_2 = 10$, $R_2/h = 250$; ▬ $L/R_2 = 50$, $R_2/h = 25$; ▬ $L/R_2 = 50$, $R_2/h = 50$; ▬ $L/R_2 = 50$, $R_2/h = 250$; ▬ $L/R_2 = 100$, $R_2/h = 25$; ▬ $L/R_2 = 100$, $R_2/h = 50$; ▬ $L/R_2 = 100$, $R_2/h = 250$.

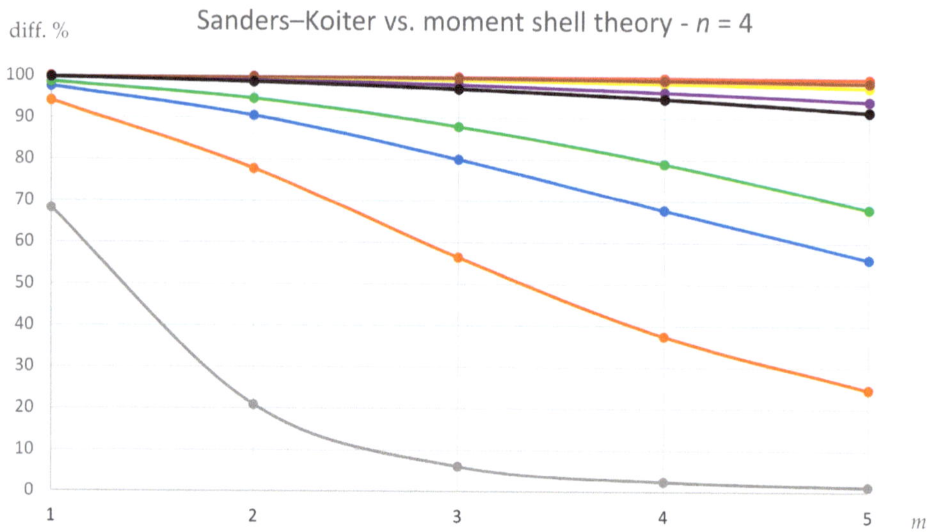

Figure 9. Simply supported DWCNT of Table 1 with different values of aspect ratio L/R_2 and thickness ratio R_2/h. Shell-like modes (n = 4). Comparisons between extended Sanders–Koiter and simplified moment shell theories, where ▬ $L/R_2 = 10$, $R_2/h = 25$; ▬ $L/R_2 = 10$, $R_2/h = 50$; ▬ $L/R_2 = 10$, $R_2/h = 250$; ▬ $L/R_2 = 50$, $R_2/h = 25$; ▬ $L/R_2 = 50$, $R_2/h = 50$; ▬ $L/R_2 = 50$, $R_2/h = 250$; ▬ $L/R_2 = 100$, $R_2/h = 25$; ▬ $L/R_2 = 100$, $R_2/h = 50$; ▬ $L/R_2 = 100$, $R_2/h = 250$.

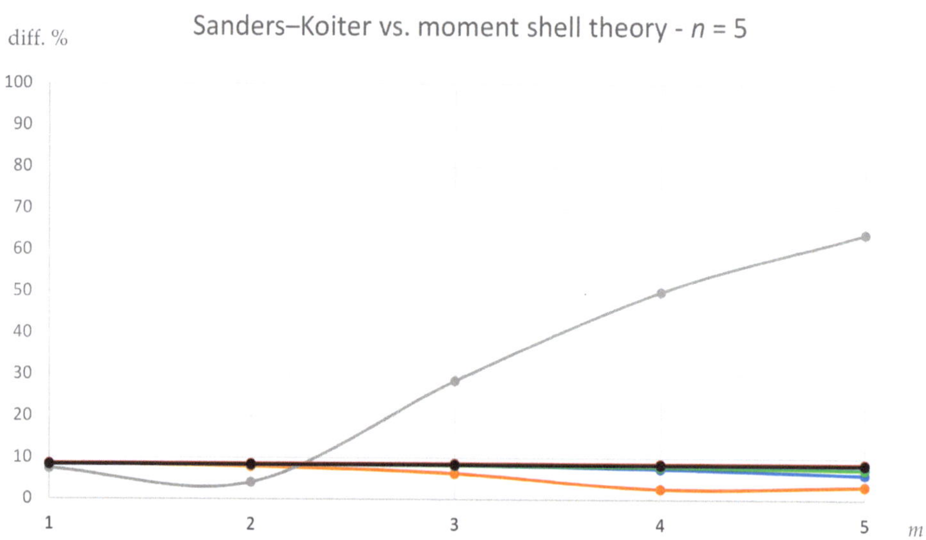

Figure 10. Simply supported DWCNT of Table 1 with different values of aspect ratio L/R_2 and thickness ratio R_2/h. Shell-like modes (n = 5). Comparisons between extended Sanders–Koiter and simplified moment shell theories, where ▬ $L/R_2 = 10$, $R_2/h = 25$; ▬ $L/R_2 = 10$, $R_2/h = 50$; ▬ $L/R_2 = 10$, $R_2/h = 250$; ▬ $L/R_2 = 50$, $R_2/h = 25$; ▬ $L/R_2 = 50$, $R_2/h = 50$; ▬ $L/R_2 = 50$, $R_2/h = 250$; ▬ $L/R_2 = 100$, $R_2/h = 25$; ▬ $L/R_2 = 100$, $R_2/h = 50$; ▬ $L/R_2 = 100$, $R_2/h = 250$.

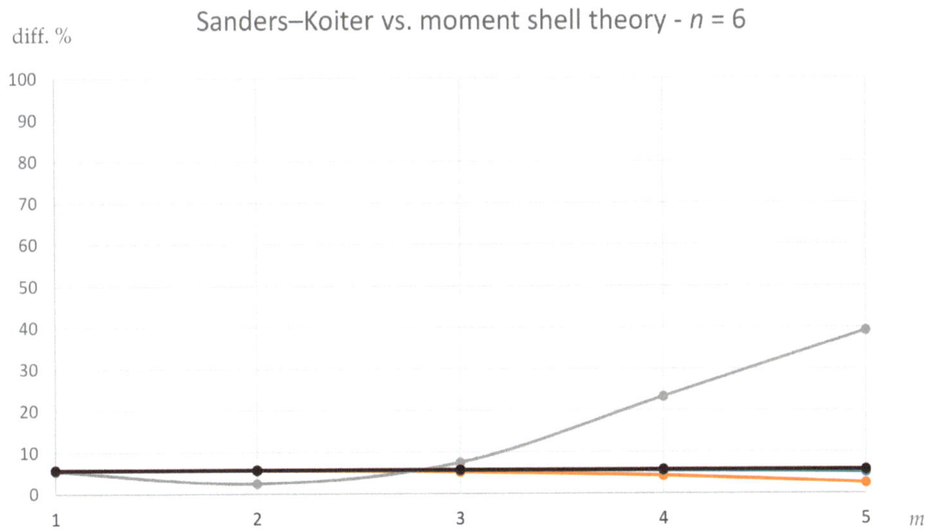

Figure 11. Simply supported DWCNT of Table 1 with different values of aspect ratio L/R_2 and thickness ratio R_2/h. Shell-like modes ($n = 6$). Comparisons between extended Sanders–Koiter and simplified moment shell theories, where $L/R_2 = 10$, $R_2/h = 25$; $L/R_2 = 10$, $R_2/h = 50$; $L/R_2 = 10$, $R_2/h = 250$; $L/R_2 = 50$, $R_2/h = 25$; $L/R_2 = 50$, $R_2/h = 50$; $L/R_2 = 50$, $R_2/h = 250$; $L/R_2 = 100$, $R_2/h = 25$; $L/R_2 = 100$, $R_2/h = 50$; $L/R_2 = 100$, $R_2/h = 250$.

Therefore, in case of shell-like modes, moment shell theory yields satisfactory results only for very long and relatively thick DWCNTs, when the effect of middle surface strains $(\tilde{\varepsilon}_{x,0,i}, \tilde{\varepsilon}_{\theta,0,i}, \tilde{\gamma}_{x\theta,0,i})$ is negligible (very small stretching deformations) and therefore the effect of middle surface changes in curvature and torsion is prevalent $\left(\tilde{k}_{x,i}, \tilde{k}_{\theta,i}, \tilde{k}_{x\theta,i}\right)$ (very high bending/torsional deformations).

Starting from the previous observations, the following results can be obtained:

- by increasing the number of longitudinal half-waves m the percentage difference of the curve ($L/R_2 = 10$, $R_2/h = 250$) showing maximum percentage difference increases (the contribution of the middle surface strains $(\tilde{\varepsilon}_{x,0,i}, \tilde{\varepsilon}_{\theta,0,i}, \tilde{\gamma}_{x\theta,0,i})$ increases);
- by increasing the number of circumferential waves n the percentage difference of the curve ($L/R_2 = 10$, $R_2/h = 250$) showing maximum percentage difference decreases (the contribution of the middle surface strains $(\tilde{\varepsilon}_{x,0,i}, \tilde{\varepsilon}_{\theta,0,i}, \tilde{\gamma}_{x\theta,0,i})$ decreases);
- the maximum percentage difference is obtained at ($m = 5$, $n = 2$, $L/R_2 = 10$, $R_2/h = 250$) and is equal to 97.6;
- by increasing the number of longitudinal half-waves m the percentage difference of the curve ($L/R_2 = 100$, $R_2/h = 25$) showing maximum percentage difference is approximately constant (the contribution of the middle surface strains $(\tilde{\varepsilon}_{x,0,i}, \tilde{\varepsilon}_{\theta,0,i}, \tilde{\gamma}_{x\theta,0,i})$ does not vary);
- by increasing the number of circumferential waves n the percentage difference of the curve ($L/R_2 = 100$, $R_2/h = 25$) showing minimum percentage difference decreases (the contribution of the middle surface strains $(\tilde{\varepsilon}_{x,0,i}, \tilde{\varepsilon}_{\theta,0,i}, \tilde{\gamma}_{x\theta,0,i})$ decreases);
- the minimum percentage difference is obtained at ($m = 1 \div 5$, $n = 6$, $L/R_2 = 100$, $R_2/h = 25$) and is equal to 5.84.

These results clearly confirm the limitations of membrane and moment shell theories in shell-like vibration modelling of simply supported thin circular cylindrical shells reported in the literature. In particular, in Ventsel [3] it is stated that "the area of application of the membrane theory of cylindrical shells is restricted by fairly short shells ($L/R \ll \sqrt{R/h}$)" and "the accuracy of the membrane solution decreases with an increase in the shell length". Moreover, in Soedel [4] it is written that the membrane approximation "has some justification for shells and arches vibrating in shapes where the stretching of the middle surface is a dominating contributor to motion resistance" (extensional approximation) and "for thickness-to-radius ratios up to approximately $h/R = 0.01$ (i.e., for $R/h \geq 100$), the membrane approximation gives very good results. Only when the shell starts to become a "thick" shell we see a pronounced effect of bending on the transverse natural frequencies. This is shown for $h/R = 0.1$ (i.e., for $R/h \leq 10$)" (inextensional approximation).

From the computations it was found that, for shell-like modes, similarly to beam-like modes, all middle surface strains are not negligible. Moreover, it was obtained that, in the case of shell-like modes, the middle surface change in curvature along the circumferential direction $\widetilde{k}_{\theta,i}$ increases with increasing the aspect L/R_2 and thickness R_2/h ratios, and so also this term must be taken into account within the model.

Therefore, starting from these results, a new simplified elastic shell model is proposed, based on a combination of the hypotheses of membrane and semi-moment shell theories, neglecting the middle surface terms $\left(\widetilde{k}_{x,i}, \widetilde{k}_{x\theta,i}\right)$ and retaining the middle surface terms $\left(\widetilde{\varepsilon}_{x,0,i}, \widetilde{\varepsilon}_{\theta,0,i}, \widetilde{\gamma}_{x\theta,0,i}, \widetilde{k}_{\theta,i}\right)$ (four terms instead of six) within the expression of the elastic strain energy (12).

In Figures 12–16, comparisons between the extended Sanders–Koiter shell theory and the proposed simplified elastic shell model are reported. It can be observed that:

- for a number of circumferential waves $n = 2$, the value of percentage difference is lower than 2 for every DWCNT geometry (L/R_2, R_2/h) and number of longitudinal half-waves m;
- for $n > 2$, the percentage difference of relatively short DWCNTs ($L/R_2 = 10$) increases with increasing m and n, while the percentage difference of sufficiently long DWCNTs ($L/R_2 \geq 50$) is quasi-constant with increasing m and decreases with increasing n, becoming negligible;
- the maximum percentage difference of a relatively short ($L/R_2 = 10$) and thick ($R_2/h = 25$) DWCNT is obtained at ($m = 5$, $n = 4$) and is equal to 8.51;
- the maximum percentage difference of a relatively short ($L/R_2 = 10$) and thin ($R_2/h = 50$) DWCNT is obtained at ($m = 5$, $n = 5$) and is equal to 6.47.

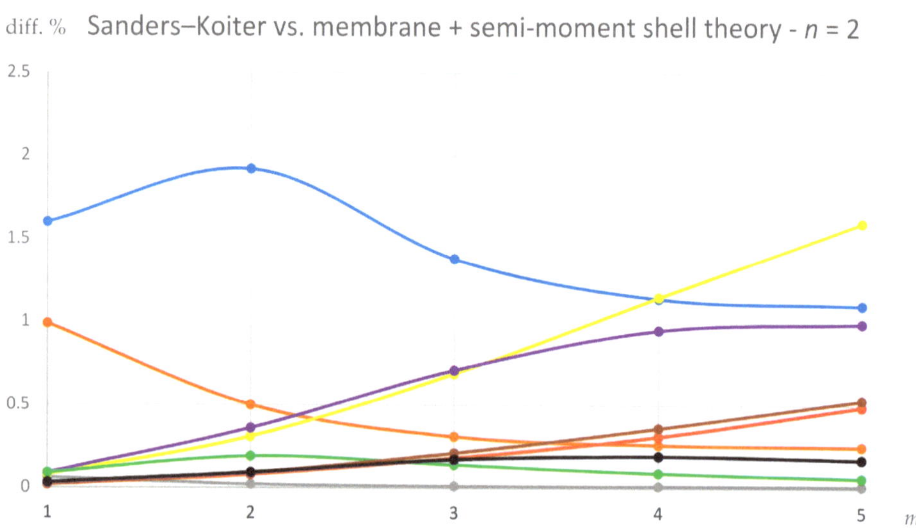

Figure 12. Simply supported DWCNT of Table 1 with different values of aspect ratio L/R_2 and thickness ratio R_2/h. Shell-like modes ($n = 2$). Comparisons between extended Sanders–Koiter and membrane + semi-moment shell theory, where $L/R_2 = 10$, $R_2/h = 25$; $L/R_2 = 10$, $R_2/h = 50$; $L/R_2 = 10$, $R_2/h = 250$; $L/R_2 = 50$, $R_2/h = 25$; $L/R_2 = 50$, $R_2/h = 50$; $L/R_2 = 50$, $R_2/h = 250$; $L/R_2 = 100$, $R_2/h = 25$; $L/R_2 = 100$, $R_2/h = 50$; $L/R_2 = 100$, $R_2/h = 250$.

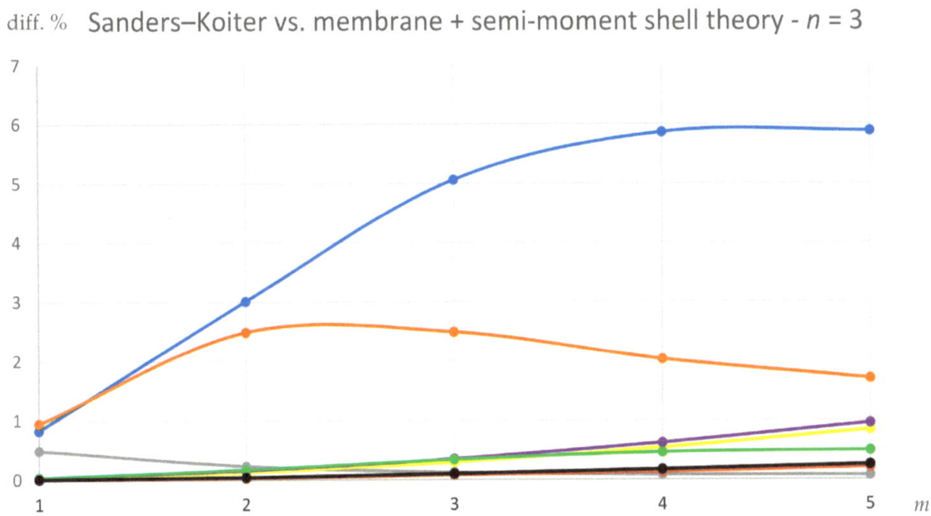

Figure 13. Simply supported DWCNT of Table 1 with different values of aspect ratio L/R_2 and thickness ratio R_2/h. Shell-like modes ($n = 3$). Comparisons between extended Sanders–Koiter and membrane + semi-moment shell theory, where $L/R_2 = 10$, $R_2/h = 25$; $L/R_2 = 10$, $R_2/h = 50$; $L/R_2 = 10$, $R_2/h = 250$; $L/R_2 = 50$, $R_2/h = 25$; $L/R_2 = 50$, $R_2/h = 50$; $L/R_2 = 50$, $R_2/h = 250$; $L/R_2 = 100$, $R_2/h = 25$; $L/R_2 = 100$, $R_2/h = 50$; $L/R_2 = 100$, $R_2/h = 250$.

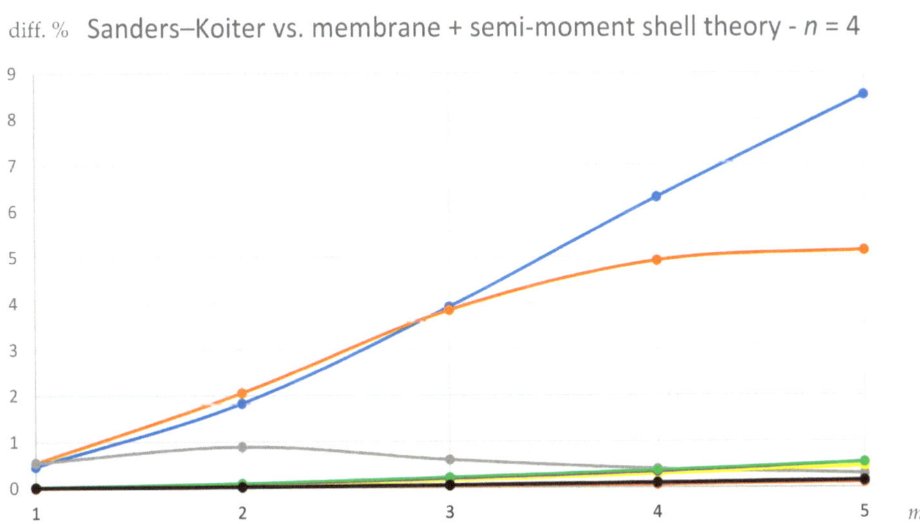

Figure 14. Simply supported DWCNT of Table 1 with different values of aspect ratio L/R_2 and thickness ratio R_2/h. Shell-like modes ($n = 4$). Comparisons between extended Sanders–Koiter and membrane + semi-moment shell theory, where $L/R_2 = 10$, $R_2/h = 25$; $L/R_2 = 10$, $R_2/h = 50$; $L/R_2 = 10$, $R_2/h = 250$; $L/R_2 = 50$, $R_2/h = 25$; $L/R_2 = 50$, $R_2/h = 50$; $L/R_2 = 50$, $R_2/h = 250$; $L/R_2 = 100$, $R_2/h = 25$; $L/R_2 = 100$, $R_2/h = 50$; $L/R_2 = 100$, $R_2/h = 250$.

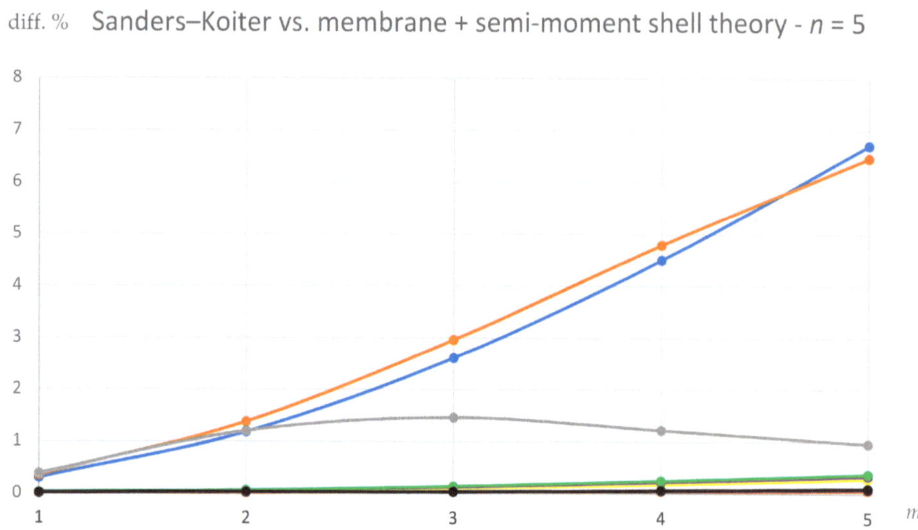

Figure 15. Simply supported DWCNT of Table 1 with different values of aspect ratio L/R_2 and thickness ratio R_2/h. Shell-like modes ($n = 5$). Comparisons between extended Sanders–Koiter and membrane + semi-moment shell theory, where $L/R_2 = 10$, $R_2/h = 25$; $L/R_2 = 10$, $R_2/h = 50$; $L/R_2 = 10$, $R_2/h = 250$; $L/R_2 = 50$, $R_2/h = 25$; $L/R_2 = 50$, $R_2/h = 50$; $L/R_2 = 50$, $R_2/h = 250$; $L/R_2 = 100$, $R_2/h = 25$; $L/R_2 = 100$, $R_2/h = 50$; $L/R_2 = 100$, $R_2/h = 250$.

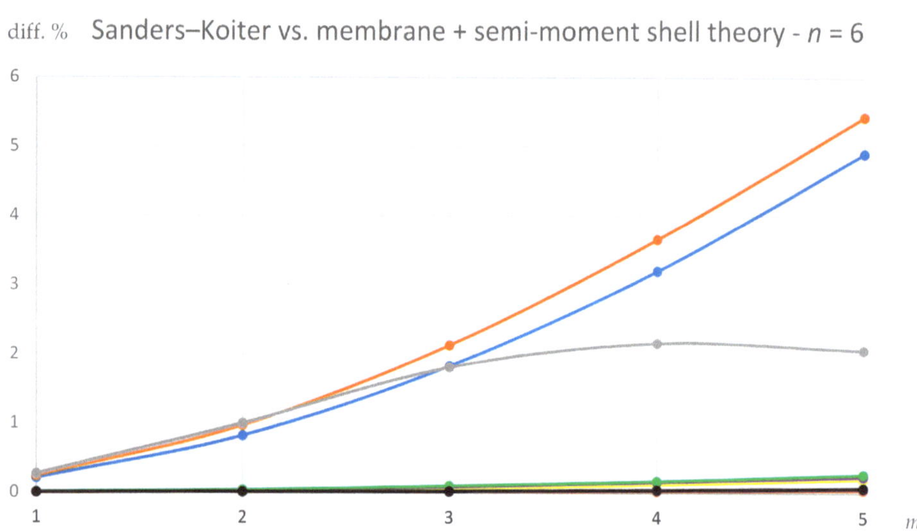

Figure 16. Simply supported DWCNT of Table 1 with different values of aspect ratio L/R_2 and thickness ratio R_2/h. Shell-like modes ($n = 6$). Comparisons between extended Sanders–Koiter and membrane + semi-moment shell theory, where $L/R_2 = 10$, $R_2/h = 25$; $L/R_2 = 10$, $R_2/h = 50$; $L/R_2 = 10$, $R_2/h = 250$; $L/R_2 = 50$, $R_2/h = 25$; $L/R_2 = 50$, $R_2/h = 50$; $L/R_2 = 50$, $R_2/h = 250$; $L/R_2 = 100$, $R_2/h = 25$; $L/R_2 = 100$, $R_2/h = 50$; $L/R_2 = 100$, $R_2/h = 250$.

In all considered ranges, the new developed simplified elastic shell model, based on a combination of membrane and semi-moment shell theories, provides results in satisfactory agreement and lower computational effort than the extended Sanders–Koiter theory.

In particular, from the performed simulations, it was found that, for sufficiently long and thin DWCNTs ($L/R_2 \geq 50$, $R_2/h \geq 50$), the contribution of the middle surface change in curvature along the circumferential direction $\widetilde{k}_{\theta,i}$ is strongly prevalent with respect to the ones of the middle surface strains ($\widetilde{k}_{\theta,i} \gg (\widetilde{\varepsilon}_{x,0,i}, \widetilde{\varepsilon}_{\theta,0,i}, \widetilde{\gamma}_{x\theta,0,i})$).

In addition, from the performed simulations, it was found that, for relatively thick DWCNTs ($R_2/h = 25$), when the effect of the middle surface changes in curvature and torsion is prevalent, 80% of the difference between extended Sanders–Koiter shell theory and new developed simplified elastic shell model is provided by the middle surface change in torsion $\widetilde{k}_{x\theta,i}$ with respect to the change in curvature along the longitudinal direction $\widetilde{k}_{x,i}$.

4.3. Characteristic Assumptions of the New Developed Reduced Sanders–Koiter Shell Theory

Starting from the simplified elastic shell models previously treated, in this section a reduced version of the extended Sanders–Koiter shell theory developed on the basis of the assumptions of the membrane and moment shell theories is reported.

For the axisymmetric modes ($n = 0$), a simplified elastic thin shell model based on the assumptions of the semi-membrane shell theory, which neglects the middle surface terms ($\widetilde{\gamma}_{x\theta,0,i}, \widetilde{k}_{x,i}, \widetilde{k}_{\theta,i}, \widetilde{k}_{x\theta,i}$) and retains the middle surface strains ($\widetilde{\varepsilon}_{x,0,i}, \widetilde{\varepsilon}_{\theta,0,i}$), is proposed, see Section 4.2.1:

$$n = 0 \quad \left(\widetilde{\gamma}_{x\theta,0,i}, \widetilde{k}_{x,i}, \widetilde{k}_{\theta,i}, \widetilde{k}_{x\theta,i}\right) = 0 \quad (\widetilde{\varepsilon}_{x,0,i}, \widetilde{\varepsilon}_{\theta,0,i}) \neq 0 \tag{36}$$

For the beam-like modes ($n = 1$), a simplified elastic thin shell model based on the assumptions of the membrane shell theory, which neglects the middle surface changes in curvature and torsion of the shell $\left(\widetilde{k}_{x,i}, \widetilde{k}_{\theta,i}, \widetilde{k}_{x\theta,i}\right)$ and retains the middle surface strains ($\widetilde{\varepsilon}_{x,0,i}, \widetilde{\varepsilon}_{\theta,0,i}, \widetilde{\gamma}_{x\theta,0,i}$), is proposed, see Section 4.2.2:

$$n = 1 \quad \left(\widetilde{k}_{x,i}, \widetilde{k}_{\theta,i}, \widetilde{k}_{x\theta,i}\right) = 0 \quad (\widetilde{\varepsilon}_{x,0,i}, \widetilde{\varepsilon}_{\theta,0,i}, \widetilde{\gamma}_{x\theta,0,i}) \neq 0 \tag{37}$$

For the shell-like modes ($n \geq 2$), a new developed simplified elastic thin shell model based on a combination of the assumptions of the membrane and semi-moment shell theory, which neglects the middle surface changes in curvature and torsion $\left(\widetilde{k}_{x,i}, \widetilde{k}_{x\theta,i}\right)$ and retains the middle surface terms ($\widetilde{\varepsilon}_{x,0,i}, \widetilde{\varepsilon}_{\theta,0,i}, \widetilde{\gamma}_{x\theta,0,i}, \widetilde{k}_{\theta,i}$), is proposed, see Section 4.2.3:

$$n \geq 2 \quad \left(\widetilde{k}_{x,i}, \widetilde{k}_{x\theta,i}\right) = 0 \quad \left(\widetilde{\varepsilon}_{x,0,i}, \widetilde{\varepsilon}_{\theta,0,i}, \widetilde{\gamma}_{x\theta,0,i}, \widetilde{k}_{\theta,i}\right) \neq 0 \tag{38}$$

The reduced Sanders–Koiter shell theory reported is a simplified elastic shell model that is applicable for linear vibrations of DWCNTs (with maximum relative error around 8%) in a very large range of geometries and numbers of waves, including some cases in which membrane and moment shell models result in much larger relative errors.

These results are very significant for further applications to CNT vibrations because simplified elastic shell models, which are characterized by a lower number of middle surface components in the expression of the elastic strain energy, and therefore a lower computational effort into the numerical simulations, with respect to the extended elastic shell models (e.g., the extended Sanders–Koiter shell theory), result particularly useful for linear (and even more nonlinear) vibrations of MWCNTs with large number of layers.

On the other hand, it must be stressed that these simplified elastic shell models are applicable only for radial deflection of thin cylindrical shells in the presence of simple boundary conditions (e.g., simply supported edges). When longitudinal or circumferential deflections become essential, or more complicated boundary conditions are applied, then more accurate elastic shell models (e.g., extended Sanders–Koiter shell theory) should be adopted, see Refs. [3,4,7] for more details.

5. Conclusions

Applicability and limitations of simplified models of thin elastic circular cylindrical shells for linear vibrations of double-walled carbon nanotubes (DWCNTs) with comparison to the extended Sanders–Koiter shell theory are studied. Different values of aspect and thickness ratios, and different numbers of waves along the longitudinal and circumferential directions, are analysed. In case of axisymmetric modes it is found that a reduced form of the membrane shell theory, also neglecting the tangential shear strain, gives excellent agreement with the extended Sanders–Koiter shell theory. Again, for beam-like modes, it is demonstrated that membrane shell theory gives excellent agreement with the extended Sanders–Koiter shell theory. As for shell-like modes it is found that both membrane and moment shell theories are not in acceptable agreement with the extended Sanders–Koiter shell theory for a large range of DWCNT aspect and thickness ratios. Therefore, in case of shell-like modes, a new simplified elastic shell model, based on the combination of membrane and semi-moment theories and neglecting middle surface changes in curvature along the axial direction and torsion, is developed, providing satisfactory agreement with the extended Sanders–Koiter shell theory. By considering the semi-membrane shell theory for axisymmetric modes, the membrane shell theory for beam-like modes and the combination of membrane and semi-moment theories for shell-like modes, it is obtained a reduced version of the Sanders–Koiter shell theory that presents a significantly enlarged range of applicability compared with both membrane and moment shell theories, giving results with excellent, or at least acceptable, agreement for all geometries and numbers of waves with respect to the extended Sanders–Koiter shell theory.

Author Contributions: Conceptualization, M.S. and I.E.E.; methodology, M.S. and I.E.E.; supervision: O.V.G. and F.P.; writing—original draft preparation, M.S. and I.E.E.; writing—review and editing, O.V.G. and F.P. All authors have read and agreed to the published version of the manuscript.

Funding: This research was funded by Israel Science Foundation, grant number 1696/17 (authors Matteo Strozzi and Oleg V. Gendelman) and Edmund Safra Foundation, grant number 2018 (author Isaac E. Elishakoff).

Institutional Review Board Statement: Not applicable.

Acknowledgments: This work is dedicated to the blessed memory of Leonid I. Manevitch, outstanding educator, respected teacher, admired scholar, who died, after severe illness, on 20 August 2020, at the age of 82. Manevitch followed the research presented in this paper with great attention to the end, dispensing useful suggestions on mathematical modelling and pointing out interesting comparative works in literature. In previously published works Manevitch investigated applicability and limitations of simplified elastic shell theories for vibration modelling of SWCNTS: the Authors of the present work extended these relevant studies to DWCNTs with the perspective of investigate linear and nonlinear vibrations of MWCNTs.

Conflicts of Interest: The authors declare no conflict of interest.

References

1. Buczkowski, W.; Szymczak-Graczyk, A.; Walczak, Z. Experimental validation of numerical static calculations for a monolithic rectangular tank with walls of trapezoidal cross-section. *Bull. Pol. Acad. Sci. Tech. Sci.* **2017**, *65*, 799–804. [CrossRef]
2. Szymczak-Graczyk, A. The Effect of Subgrade Coefficient on Static Work of a Pontoon Made as a Monolithic Closed Tank. *Appl. Sci.* **2021**, *11*, 4259. [CrossRef]
3. Ventsel, E.; Krauthammer, T. Thin Plates and Shells. In *Theory, Analysis, and Applications*; The Pennsylvania State University, Marcel Dekker: New York, NY, USA, 2001.
4. Soedel, W. *Vibrations of Shells and Plates*, 3rd ed.; Marcel Dekker: New York, NY, USA, 2004.
5. Calladine, C. *Theory of Shell Structures*; Cambridge University Press: New York, NY, USA, 1983.
6. Yamaki, N. *Elastic Stability of Circular Cylindrical Shells*; Elsevier: Amsterdam, The Netherlands, 1984.
7. Leissa, A.W. *Vibration of Shells*; Acoustical Society of America: Columbus, OH, USA, 1993.
8. Amabili, M. *Nonlinear Vibrations and Stability of Shells and Plates*; Cambridge University Press: New York, NY, USA, 2008.
9. Love, A.E.H. *A Treatise on the Mathematical Theory of Elasticity*, 4th ed.; Dover: New York, NY, USA, 1944.
10. Green, A.E. On the Linear Theory of Thin Elastic Shells. *Proc. R. Soc. Lond. Ser. A Math. Phys. Sci.* **1962**, *266*, 143–160.
11. Reissner, E. Stress strain relations in the theory of thin elastic shells. *J. Math. Phys.* **1952**, *31*, 109–119. [CrossRef]

12. Sanders, J.L. *An Improved First Approximation Theory for Thin Shells*; NASA TR-R24; U.S. Government Printing Office: Washington, DC, USA, 1959.
13. Koiter, W.T. A consistent First Approximation in the General Theory of Thin Elastic Shells. In Proceedings of the Symposium on Theory of Thin Elastic Shells, Delft, The Netherlands, 24–28 August 1959.
14. Donnell, L.H. *Beams, Plates, and Shells*; McGraw-Hill: New York, NY, USA, 1976.
15. Vlasov, V.Z. *General Theory of Shells and Its Applications in Engineering*; NASA Technical Translation F-99: Washington, DC, USA, 1964.
16. Mushtari, K.M. Certain generalizations of the theory of thin shells. *Izv. Fiz. Mat. Ob-Va. Pri Kaz. Un-Te* **1938**, *11*, 28–56. (In Russian)
17. Timoshenko, S.; Woinowsky-Krieger, S. *Theory of Plates and Shells*; McGraw-Hill: New York, NY, USA, 1959.
18. Naghdi, P.M. A survey on resent progress in the theory of thin elastic shells. *Appl. Mech. Rev.* **1956**, *9*, 365–368.
19. Lur'ye, A.I. General theory of elastic shells. *Prikl. Mat. Mekh.* **1940**, *4*, 7–34. (In Russian)
20. Flugge, W. *Stresses in Shells*, 2nd ed.; Springer: Berlin/Heidelberg, Germany, 1962.
21. Byrne, R. *Theory of Small Deformations of a Thin Elastic Shell*; Seminar Reports in Mathematics; University of California: Berkeley, CA, USA, 1944; Volume 2, pp. 103–152.
22. Novozhilov, V.V. *Thin Shell Theory*; Springer: Groningen, The Netherlands, 1964. (In Russian)
23. Gol'denveizer, A.L. *Theory of Elastic Thin Shells*; Pergamon Press: New York, NY, USA, 1961.
24. Reissner, E. A new derivation of the equations of the deformation of elastic shells. *Am. J. Math.* **1941**, *63*, 177–184. [CrossRef]
25. Sokolovskii, V.V. Momentless shells of revolution. *Prikl. Mat. Mekh.* **1938**, *1*, 81–88. (In Russian)
26. Truesdell, C. The Membrane Theory of Shells of Revolution. *Trans. Am. Math. Soc.* **1945**, *58*, 96–166. [CrossRef]
27. Gibson, J.E. *Thin Shells. Computing and Theory*; Pergamon Press: New York, NY, USA, 1980.
28. Gibson, J.E. *Linear Elastic Theory of Thin Shells*; Pergamon Press: New York, NY, USA, 1965.
29. Elishakoff, I.; Pentaras, D.; Dujat, K.; Versaci, C.; Muscolino, G.; Wang, C.M.; Ghyselinck, G. *Carbon Nanotubes and Nano Sensors: Vibrations, Buckling, and Ballistic Impact*; John Wiley & Sons: London, UK, 2012.
30. Liew, K.M.; Jianwei, Y.; Zhang, L. *Mechanical Behaviors of Carbon Nanotubes: Theoretical and Numerical Approaches*; Elsevier: Amsterdam, The Netherlands, 2017.
31. Tserpes, K.I.; Silvestre, N. *Modeling of Carbon Nanotubes, Graphene and Their Composites*; Springer International Publishing: Cham, Switzerland, 2014.
32. Harik, V. *Trends in Nanoscale Mechanics: Mechanics of Carbon Nanotubes, Graphene, Nanocomposites and Molecular Dynamics*; Springer: Dordrecht, The Netherlands, 2014.
33. Rahman, A.; Ali, I.; Al Zahrani, S.M.; Eleithy, R.H. A review of the applications of nanocarbon polymer composites. *NANO Brief Rep. Rev.* **2011**, *6*, 185–203. [CrossRef]
34. Thostenson, E.T.; Ren, Z.; Chou, T.W. Advances in the science and technology of carbon nanotubes and their composites: A review. *Compos. Sci. Technol.* **2001**, *61*, 1899–1912. [CrossRef]
35. Jensen, B.D.; Kim, J.W.; Sauti, G.; Wise, K.E.; Dong, L.; Wadley, H.N.G.; Park, J.G.; Liang, R.; Siochi, E.J. Toward ultralight high-strength structural materials via collapsed carbon nanotube bonding. *Carbon* **2020**, *156*, 538–548. [CrossRef]
36. Al-Ghalith, J.; Xu, H.; Dumitrica, T. Collapsed carbon nanotubes as building blocks for high-performance thermal materials. *Phys. Rev. Mater.* **2017**, *1*, 056001. [CrossRef]
37. Zhang, S.; Ma, Y.; Suresh, L.; Hao, A.; Bick, M.; Tan, S.C.; Chen, J. Carbon Nanotube Reinforced Strong Carbon Matrix Composites. *Am. Chem. Soc. ACS Nano* **2020**, *14*, 9282–9319. [CrossRef]
38. Rao, A.M.; Richter, E.; Bandow, S.; Chase, B.; Eklund, P.C.; Williams, K.A.; Fang, S.; Subbaswamy, K.R.; Menon, M.; Thess, A.; et al. Diameter-selective Raman scattering from vibrational modes in carbon nanotubes. *Science* **1997**, *275*, 187–191. [CrossRef]
39. Bandow, S.; Asaka, S.; Saito, Y.; Rao, A.M.; Grigorian, L.; Richter, E.; Eklund, P.C. Effect of the growth temperature on the diameter distribution and chirality of single-wall carbon nanotubes. *Phys. Rev. Lett.* **1998**, *80*, 3779–3782. [CrossRef]
40. Jorio, A.; Saito, R.; Hafner, J.H.; Lieber, C.M.; Hunter, M.; McClure, T.; Dresselhaus, G.; Dresselhaus, M.S. Structural (n,m) determination of isolated single-wall carbon nanotubes by resonant Raman scattering. *Phys. Rev. Lett.* **2001**, *86*, 1118–1121. [CrossRef]
41. Gupta, S.S.; Bosco, F.G.; Batra, R.C. Breakdown of structural models for vibrations of single-wall zigzag carbon nanotubes. *J. Appl. Phys.* **2009**, *106*, 063527. [CrossRef]
42. Cheng, H.C.; Liu, Y.L.; Wu, C.H.; Chen, W.H. On radial breathing vibration of carbon nanotubes. *Comput. Methods Appl. Mech. Eng.* **2010**, *199*, 2820–2827. [CrossRef]
43. Gupta, S.S.; Bosco, F.G.; Batra, R.C. Wall thickness and elastic moduli of single-walled carbon nanotubes from frequencies of axial, torsional and inextensional modes of vibration. *Comput. Mater. Sci.* **2010**, *47*, 1049–1059. [CrossRef]
44. Qian, D.; Wagner, G.J.; Liu, W.K.; Yu, M.F.; Ruoff, R.S. Mechanics of Carbon Nanotubes. *Appl. Mech. Rev.* **2002**, *55*, 495–533. [CrossRef]
45. Huang, Y.; Wang, Z. *Mechanics of Nanotubes, Comprehensive Structural Integrity Handbook*; Elsevier: New York, NY, USA, 2003; Volume 8, pp. 551–579.
46. Yakobson, B.I.; Brabec, C.J.; Bernholc, J. Nanomechanics of Carbon Tubes: Instability Beyond Linear Response. *Phys. Rev. Lett.* **1996**, *76*, 2511–2514. [CrossRef]
47. Ru, C.Q. Effective Bending Stiffness of Carbon Nanotubes. *Phys. Rev. B* **2000**, *62*, 9973–9976. [CrossRef]

48. Ru, C.Q. Degraded Axial Buckling Strain of Multiwalled Carbon Nanotubes Due to Interlayer Slips. *J. Appl. Phys.* **2001**, *89*, 3426–3433. [CrossRef]
49. Liew, K.; Wang, Q. Analysis of wave propagation in carbon nanotubes via elastic shell theories. *Int. J. Eng. Sci.* **2007**, *45*, 227–241. [CrossRef]
50. Ru, C.Q. Axially compressed buckling of a double-walled carbon nanotube embedded in an elastic medium. *J. Mech. Phys. Solids* **2001**, *49*, 1265–1279. [CrossRef]
51. He, X.Q.; Eisenberger, M.; Liew, K.M. The effect of Van der Waals interaction modeling on the vibration characteristics of multiwalled carbon nanotubes. *J. Appl. Phys.* **2006**, *100*, 124317. [CrossRef]
52. Elishakoff, I.; Pentaras, D. Fundamental natural frequencies of double-walled carbon nanotubes. *J. Sound Vib.* **2009**, *322*, 652–664. [CrossRef]
53. Silvestre, N.; Wang, C.M.; Zhang, Y.Y.; Xiang, Y. Sanders shell model for buckling of single-walled carbon nanotubes with small aspect ratio. *Compos. Struct.* **2011**, *93*, 1683–1691. [CrossRef]
54. Silvestre, N. On the accuracy of shell models for torsional buckling of carbon nanotubes. *Eur. J. Mech. A/Solids* **2012**, *32*, 103–108. [CrossRef]
55. Wang, C.Y.; Ru, C.Q.; Mioduchowski, A. Applicability and Limitations of Simplified Elastic Shell Equations for Carbon Nanotubes. *J. Appl. Mech.* **2004**, *71*, 622–631. [CrossRef]
56. Ghavanloo, E.; Fazelzadeh, S.A. Vibration characteristics of single-walled carbon nanotubes based on anisotropic elastic shell model including chirality effect. *Appl. Math. Model.* **2012**, *36*, 4988–5000. [CrossRef]
57. Strozzi, M.; Smirnov, V.V.; Manevitch, L.I.; Pellicano, F. Nonlinear vibrations and energy exchange of single-walled carbon nanotubes. Radial breathing modes. *Compos. Struct.* **2018**, *184*, 613–632. [CrossRef]
58. Strozzi, M.; Smirnov, V.V.; Manevitch, L.I.; Milani, M.; Pellicano, F. Nonlinear vibrations and energy exchange of single-walled carbon nanotubes. Circumferential flexural modes. *J. Sound Vib.* **2016**, *381*, 156–178. [CrossRef]
59. Strozzi, M.; Pellicano, F. Nonlinear Resonance Interaction between Conjugate Circumferential Flexural Modes in Single-Walled Carbon Nanotubes. *Shock Vib.* **2019**, *2019*, 3241698. [CrossRef]
60. Strozzi, M.; Smirnov, V.V.; Manevitch, L.I.; Pellicano, F. Nonlinear normal modes, resonances and energy exchange in single-walled carbon nanotubes. *Int. J. Non-Linear Mech.* **2020**, *120*, 103398. [CrossRef]
61. Strozzi, M.; Pellicano, F. Linear vibrations of triple-walled carbon nanotubes. *Math. Mech. Solids* **2018**, *23*, 1456–1481. [CrossRef]
62. Strozzi, M.; Manevitch, L.I.; Pellicano, F.; Smirnov, V.V.; Shepelev, D.S. Low-frequency linear vibrations of single-walled carbon nanotubes: Analytical and numerical models. *J. Sound Vib.* **2014**, *333*, 2936–2957. [CrossRef]
63. Kaplunov, J.; Manevitch, L.I.; Smirnov, V.V. Vibrations of an elastic cylindrical shell near the lowest cut-off frequency. *Proc. R. Soc. A* **2018**, *472*, 20150753. [CrossRef]
64. Andrianov, I.; Kaplunov, J.; Kudaibergenov, A.; Manevitch, L. The effect of a weak nonlinearity on the lowest cut-off frequencies of a cylindrical shell. *Z. Fur Angew. Math. Und Phys.* **2018**, *9*, 1–12. [CrossRef]

C Journal of Carbon Research

MDPI

Article

The Growth Behavior of Amorphous Hydrogenated Carbon a-C:H Layers on Industrial Polycarbonates—A Weak Interlayer and a Distinct Dehydrogenation Zone

Torben Schlebrowski [1,*], Melanie Fritz [1], Lucas Beucher [1], Yongxin Wang [2], Stefan Wehner [1] and Christian B. Fischer [1,3,*]

1 Department of Physics, University Koblenz-Landau, 56070 Koblenz, Germany; melfritz@uni-koblenz.de (M.F.); lucasbeucher@uni-koblenz.de (L.B.); wehner@uni-koblenz.de (S.W.)
2 Key Laboratory of Marine New Materials and Related Technology, Zhejiang Key Laboratory of Marine Materials and Protection Technology, Ningbo Institute of Materials Technology and Engineering, Chinese Academy of Sciences, Ningbo 315201, China; yxwang@nimte.ac.cn
3 Materials Science, Energy and Nano-Engineering Department, Mohammed VI Polytechnic University, Ben Guerir 43150, Morocco
* Correspondence: schlebrowski@uni-koblenz.de (T.S.); chrbfischer@uni-koblenz.de (C.B.F.); Tel.: +49-261-287-2365 (T.S.); +49-261-287-2345 (C.B.F.)

Abstract: Polycarbonate (PC) is a material that is used in many areas: automotive, aerospace engineering and data storage industries. Its hardness is of particular importance, but some applications are affected by its low wettability or scratch susceptibility. This can be changed either by blending with other polymers, or by surface modifications, such as the application of an amorphous hydrogenated carbon layer (a-C:H). In this study, individual a-C:H layers of different thicknesses (10–2000 nm) were deposited on PC by RF PECVD. Both the layer morphology with AFM and SEM and the bonding states of the carbon on the surface with synchrotron-assisted XPS and NEXAFS were studied. The aim was to investigate the coatability of PC and the stability of the a-C:H. Special attention was paid to the interlayer region from 0 to 100 nm, since this is responsible for the layer to base material bonding, and to the zone of dehydrogenation (from about 1000 nm), since this changes the surface composition considerably. For PC, the interlayer was relatively small with a thickness of only 20 nm. Additionally, a correlation was found between the evolving grain structure and the development of the C-H peak according to NEXAFS C K-edge measurements.

Keywords: a-C:H grain development; PECVD plasma coating; gradual film deposition; synchrotron radiation; a-C:H-polymer interlayer; surface morphology

Citation: Schlebrowski, T.; Fritz, M.; Beucher, L.; Wang, Y.; Wehner, S.; Fischer, C.B. The Growth Behavior of Amorphous Hydrogenated Carbon a-C:H Layers on Industrial Polycarbonates—A Weak Interlayer and a Distinct Dehydrogenation Zone. C **2021**, 7, 59. https://doi.org/10.3390/c7030059

Academic Editor: Gil Goncalves

Received: 29 June 2021
Accepted: 26 July 2021
Published: 29 July 2021

1. Introduction

Polycarbonate (PC) is a transparent polymer and is used in many technical applications due to its advantageous mechanical properties and its processability. For example, it is utilized not only in aerospace and automotive engineering, but also in the construction industry and optical lenses [1–3]. However, PC itself has disadvantages, such as high hydrophobicity and notch sensitivity. These properties can be modified—for example, by blending with other thermoplastic polymers [2]. Another way to influence these properties is to deposit a (thin) layer of amorphous carbon (a-C) or amorphous hydrogenated carbon (a-C:H) on its surface [4–8]. These carbon layers consist of a network of π–sp^2, σ–sp^2 and sp^3 hybridized carbon atoms in which the hydrogen is embedded [6,9,10].

The wettability of these layers is influenced by the morphology [11,12] and the chemical bonds on the surface [13,14], as well as by the carbon bonds present on the surface (sp^2 or sp^3 hybridized) [15–18]. Coatings with a-C:H are suitable for altering the wettability of the surface sustainably [5,8,19,20]. The ratio of the carbon hybridization states can be influenced by the plasma parameters during the coating process, the gas used, the amount of hydrogen [6,9,17,21,22] and the coating thickness itself [19,20,23]. This carbon state ratio also affects the scratch resistance of the coating obtained. If the sp^2/sp^3 ratio is very low, a very hard, diamond-like coating is obtained (high hardness, chemical inertness and high electrical resistance) [21]. If the proportion of sp^3 bonds on the surface is reduced in favor of the sp^2 bonds, the layer has graphite-like properties, such as improved electrical conductivity on the surface of the coated material [21].

Various techniques can be used to deposit such a-C:H layers: e.g., ion beam deposition, magnetron sputtering and radio frequency plasma enhanced chemical vapor deposition (RF-PECVD) [6,21,23,24]. The latter was used in this study to deposit coatings on PC, as its low process temperatures make it ideal for coating non-conductive polymers [21,25].

However, before such a layer can be used to specifically influence the properties of such a material, the basic coatability must first be investigated. The important question is whether an interlayer forms and how pronounced it is. The interlayer is a mixed phase of the base material and the a-C:H layer deposited by the plasma. During the plasma process, the species which are generated interact with the substrate via impinging and surface reorganization [26]; additionally, material from the substrate may be eroded and incorporated into the forming layer. These processes lead to better crosslinking of the two systems, and the more pronounced the interlayer is, the better the adhesion of the a-C:H layer [24]. Polyoxymethylene (POM), for example, does not form a measurable interlayer; the layer can simply be detached [27]. High-density polyethylene (HDPE), on the other hand, has a very pronounced interlayer that is 40 nm thick and has very good connectivity to the polymer [8,28]. As soon as the plasma has hardly any influence on the base material, the interlayer formation is completed and regular layer growth begins. Furthermore, it must be determined whether the so-called dehydrogenation of the layers takes place as soon as a certain layer thickness is exceeded [5,8,27–29]. Here, hydrogen is released from the a-C:H layer, whereby the proportion of sp^2-bonded carbon increased at the expense of the sp^3 bonds, which ultimately influenced the layers' properties. The layer stability itself and the detection of non-stable layer thicknesses must also be checked in advance.

In the current study, these investigations were carried out for the transparent polymer PC. By RF-PECVD technique and acetylene plasma, a-C:H layers of various thicknesses (0–2000 nm) were deposited and examined. Special attention was paid to the a-C:H layer from 10 to 50 nm. Fine subdivisions were used here, since the interlayer manifests itself in this range. The layer stability was investigated ex situ by scanning electron microscopy (SEM). A more microscopic view of the surface morphology was possible with atomic force microscopy (AFM). Finally, the carbon bond states at the film's surface were analyzed by X-ray photoelectron spectroscopy (XPS) and near-edge X-ray absorption fine structure (NEXAFS).

2. Materials and Methods

2.1. Sample Preparation

PC material was supplied as 2 mm thick sheets by Goodfellow (Bad Nauheim, Germany). It was of industrial quality. Samples with Ø 10 mm were stamped out and subsequently cleaned with isopropanol in an ultrasonic bath. For the coating, they were mounted on a vacuum-compatible aluminum holder and placed in a vacuum chamber at a distance of 275 mm in front of the plasma source. The high frequency plasma source used was a Copra DN 400 (CCR GmbH, Troisdorf, Germany) operating at 13.65 MHz [24]. The coating of the sample surface was processed in two separate plasma processes: first, the samples were cleaned and activated for 10 min in oxygen plasma. Second, an acetylene plasma was used for the a-C:H coating. The process duration depended on the desired coating thickness [5,8,27]. The processes can be read about in detail elsewhere [23]. During the plasma processes, the sample temperature never exceeded 40 °C [25]. Different layer thicknesses from 0 (O_2-treated PC sample) to 2000 nm were realized this way. The thickness intervals were variable and depended on the thickness range investigated: 0–100 nm in 10 nm steps in the interlayer region, 100–1000 nm in 100 nm steps and 1000–2000 nm in

250 nm steps. The coating rates were, apart from small variations in initial growth, constant over time and were about 10 nm/min [23]. In order to check the coating thicknesses subsequently, silicon wafers partially covered with aluminum foil were coated together with the PC samples. The thickness of each deposited layer was determined using a profilometer (Dektak 3, Veeco Instruments Inc., Plainview, NY, USA) [23].

2.2. Sample Surface Morphology

The analysis of each sample's surface structure was carried out in two steps. A scanning electron microscope (SEM515, Phillips, 7 kV, WD 20 mm, FEI Company, Amsterdam, The Netherlands) was first used to perform more macroscopic examinations. The SEM images provided information on the stability of the film and whether it was intact. However, for this technique, a 7–10 nm thick gold layer must first be applied to the sample's surface to increase the conductivity. This is to avoid charging effects or even beam damage by the electron beam. A more detailed analysis of the surface was obtained with an atomic force microscope (AFM, Asylum Research, Oxford Instruments MFP-3D-BIO). The AFM was operated in contact mode with AC240-TS tips from Oxford Instruments in ambient conditions. For both microscopy techniques the measurements were repeated in at least three locations to ensure homogeneity of the surface. The AFM data obtained were then analyzed using Gwyddion (open source, version 2.58). First, a plane correction was performed. Subsequently, a mean and a median filter were used, followed by a conservative noise reduction. These steps are the same as those which were used by Catena et al. [5,8,27]. Finally, the zero level was shifted to the minimum high value of the sample surface for a unified representation. Roughness, through R_q and R_a, was evaluated, and images were exported accordingly. Due to the small measuring range, however, the layer integrity cannot be determined, which is why the SEM and AFM techniques complement each other.

2.3. Sample Surface Chemical Composition

The chemical composition of the surface and the near-surface layers were studied by performing X-ray photoelectron spectroscopy (XPS) and near-edge X-ray absorption fine structure (NEXAFS) measurements at the synchrotron BESSY II, Helmholtz Zentrum Berlin. The measurements were performed during the low-alpha phase to avoid beam damage and strong charging at the sample surface. The HE-SGM beamline was used, which is described in detail elsewhere [30]. The beamline is equipped with an extra electron gun to compensate for beam-induced charging effects on the sample surfaces, which are still present during the low-alpha phase. Both techniques are used to investigate the carbon bond states present at the surface/near-surface region.

The XPS investigations were divided into two steps. First, a full spectrum (survey) from 0 to 700 eV was measured to study the overall chemical composition of each sample's surface. Subsequently, the C1s peaks were analyzed in detail and evaluated with CasaXPS (software version 2.3.18, Casa Software Ltd., Teignmouth, UK). For the analysis, the C1s peaks were placed at 284.8 eV to compensate for any static charge present [31]. The fractions of sp^2, sp^3, C-O and C=O bonds in the total C1s peak were determined via peak deconvolution. When plotting in the final diagram, the C=O bonds were evaluated as sp^2 bonds and the C-O fractions as sp^3 bonds. To ensure the homogeneity of the sample surfaces and to check correctness, the measurements were repeated in at least two locations on the surface.

During the NEXAFS studies, both the C K-edge and the O K-edge were measured on the sample surfaces. The O K-edge was used to check the samples for the presence or absence of oxygen on the surface. The measurement of the C K-edge was also repeated in at least two locations to ensure correctness. The angle between the sample and the detector was 55° during the measurements. The measurements were realized in partial electron yield (PEY). A counter voltage was applied to the detector and not all electrons emerging from the sample surface during the X-ray irradiation reached the detector. Only electrons that have exited close to the surface have the necessary energy to reach the detector, and the measurement is therefore very surface-sensitive [30]. The C K-edge was evaluated according to Watt et al. [32] using the commercial software Origin 8.1. Further investigations of the spectra were then performed with a self-written peak-evaluation program.

3. Results and Discussion

SEM measurements were performed to verify the condition of the coatings. The results for the range of thin films (up to 100 nm) are displayed in Figure 1. Shown are the oxygen plasma-treated sample (a), the films of 10–50 nm a-C:H thickness in 10 nm steps (b-f) and the 100 nm a-C:H film (g). There were no changes in the state of the deposited layers between 50 and 100 nm.

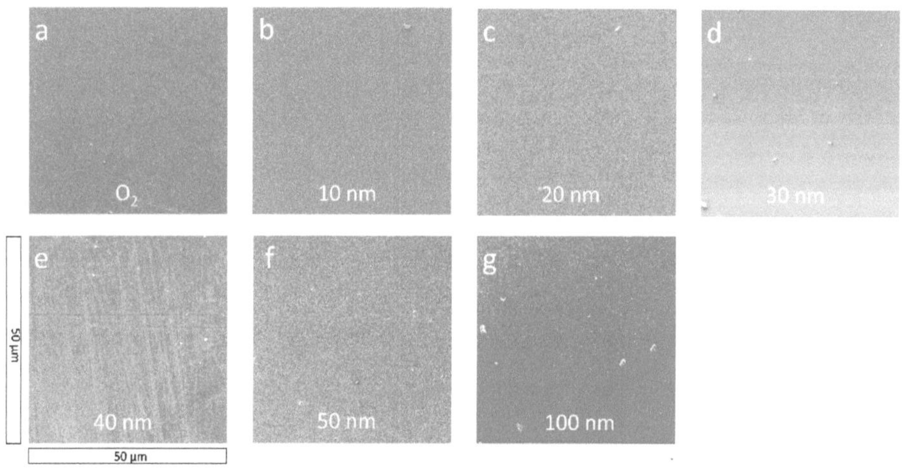

Figure 1. SEM images for the thin a-C:H layers on PC. Shown are the O_2-treated PC sample (**a**); a-C:H deposited in thicknesses from 10 to 50 nm (**b–f**) and 100 nm (**g**).

The O_2-treated sample in Figure 1a shows a smooth, clean surface. Only a few particles are visible on the surface. When a 10 nm a-C:H layer was applied (Figure 1b), a homogeneous, closed layer was formed. With increasing layer thickness, this behavior remained unchanged (Figure 1c,d). When the a-C:H layer reached 40 nm, the sample showed many trace-like structures in the SEM. This was probably due to a scratched surface of the sample during production. A reorganization of the layer would also be conceivable (e.g., stress-induced [33]), but the structures were probably too directional for this. Both the 50 and 100 nm samples (Figure 1f,g) had homogeneous, closed layers.

The SEM measurements of the thicker layers are shown in Figure 2, starting with 200 nm a-C:H layer thickness (Figure 2a). The homogeneous, closed surface already observed for the 100 nm sample continued for the 200 and 300 nm samples (Figure 2b). However, the deposition of a 400 nm thick a-C:H layer led to a layer failure on PC. In Figure 2c, the cracks that run through the film are clearly visible. This was most likely due to excessive stress in the film. These stresses occur when there is a high proportion of sp^3 bonds in the system [34,35]. When they reach a critical value, the system ruptures and the stress is released. For 500 nm a-C:H on PC (Figure 2d), the layer was again homogeneous and closed (=stable). This held up to a film thickness of 800 nm (not shown here). At 900 nm (Figure 2e), however, the system was again unstable. That layer also cracked due to stress. From a film thickness of 1000 nm (Figure 2f), the film remained closed and homogeneous up to 2000 nm (Figure 2j).

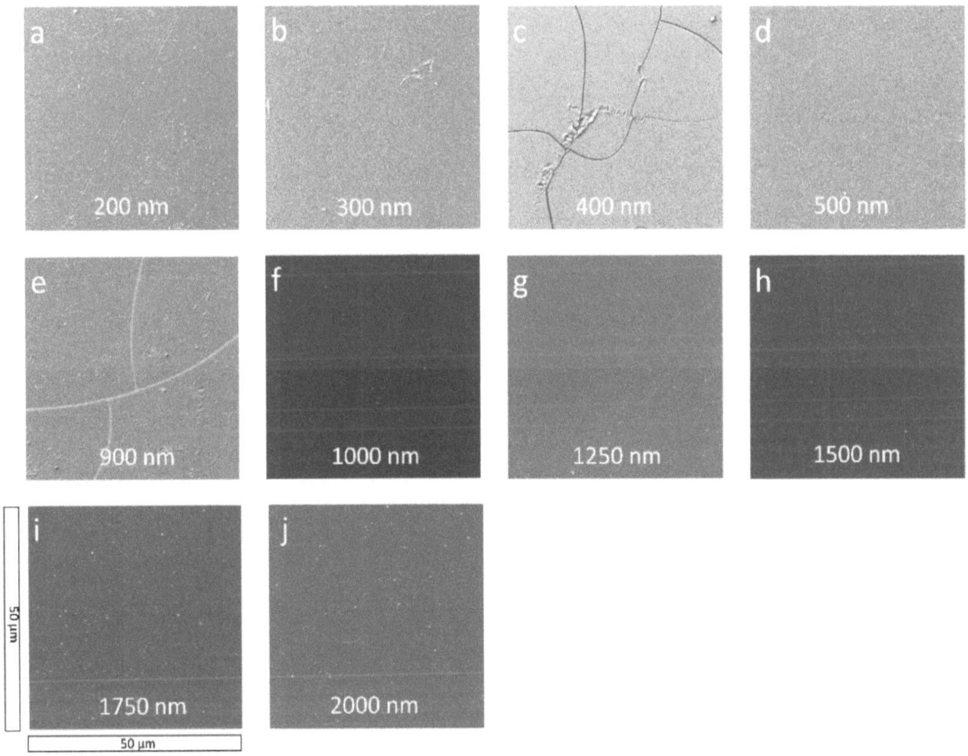

Figure 2. SEM results for thick a-C:H layers deposited on PC. Shown are the film thicknesses of 200 nm (**a**) to 2000 nm (**j**).

Figure 3 shows the three-dimensional images of the AFM measurements. The respective values for the roughness are given in Table 1. The O_2 treated sample (a) was relatively smooth with a roughness R_a of 2.069 nm, but the surface contained particles already visible in the SEM. With the application of a 10 nm a-C:H layer, the surface remained homogeneous. At 20 nm (c) a granular structure was formed. The formation of these grains is a sign that the interlayer phase is completed and normal a-C:H growth begins [5,8]. At 30 nm thickness (d) the most homogeneous surface was present (R_a of 0.789), and only single, small particles were visible. One a-C:H layer was most likely just completed and the next one was beginning to form. The same was applicable to the sample with 50 nm layer thickness (f), but the process was already more advanced.

Figure 4 shows the results of the AFM measurements for layer thicknesses from 200 nm (a) to 2000 nm (i). The corresponding roughness values are listed in Table 1. For 200 nm a-C:H film thickness (Figure 4a), multiple but small grains are visible. At the thickness of 300 nm (b), the number of grains decreased just slightly, but they increased strongly in size. The situation was similar for the 400 nm layer thickness (c). Relatively fewer and smaller grains were on the 500 nm layer (d). The 1000 nm sample (e), like the 2000 nm sample (i), was hardly occupied by grains. The 1250, 1500 and 1750 nm layers showed pronounced grain structures, although of varying intensity (f–h).

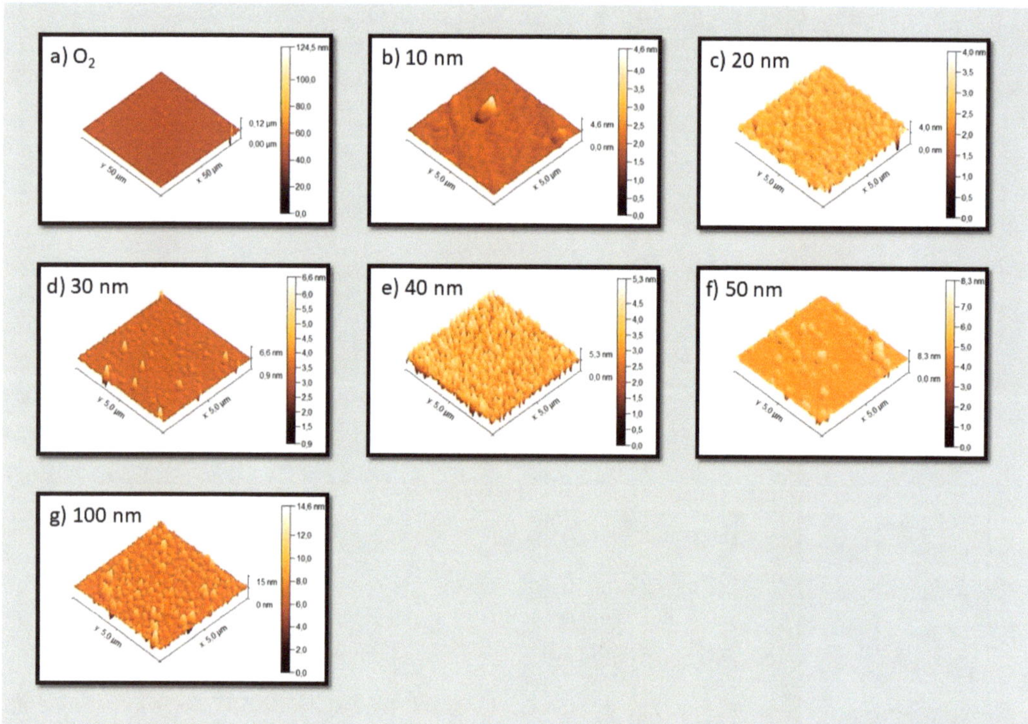

Figure 3. AFM 3D-images for the O_2 plasma-treated PC sample (**a**) and PC coated with thin a-C:H layers from 10 nm (**b**) to 100 nm (**g**).

Table 1. AFM roughness values R_a and R_q: analyzed are the O_2 plasma-treated PC sample and PC samples coated with a-C:H layers of various thickness in ascending order.

Sample	R_a [nm]	R_q [nm]
O_2	2.069 ± 1.578	5.256 ± 15.320
10	2.759 ± 2.095	4.775 ± 5.086
20	2.363 ± 0.401	2.999 ± 0.649
30	0.789 ± 0.240	1.350 ± 0.581
40	3.812 ± 0.719	4.932 ± 1.091
50	2.940 ± 1.744	5.155 ± 2.925
100	5.161 ± 1.383	7.140 ± 2.134
200	4.978 ± 1.381	6.946 ± 2.162
300	5.729 ± 3.585	9.890 ± 6.357
400	3.850 ± 1.968	6.783 ± 3.564
500	2.048 ± 1.395	3.886 ± 3.613
1000	7.443 ± 7.519	15.250 ± 16.670
1250	4.546 ± 1.454	6.432 ± 2.398
1500	3.790 ± 1.080	5.161 ± 2.387
1750	3.460 ± 1.274	4.813 ± 2.868
2000	2.139 ± 1.189	4.331 ± 2.745

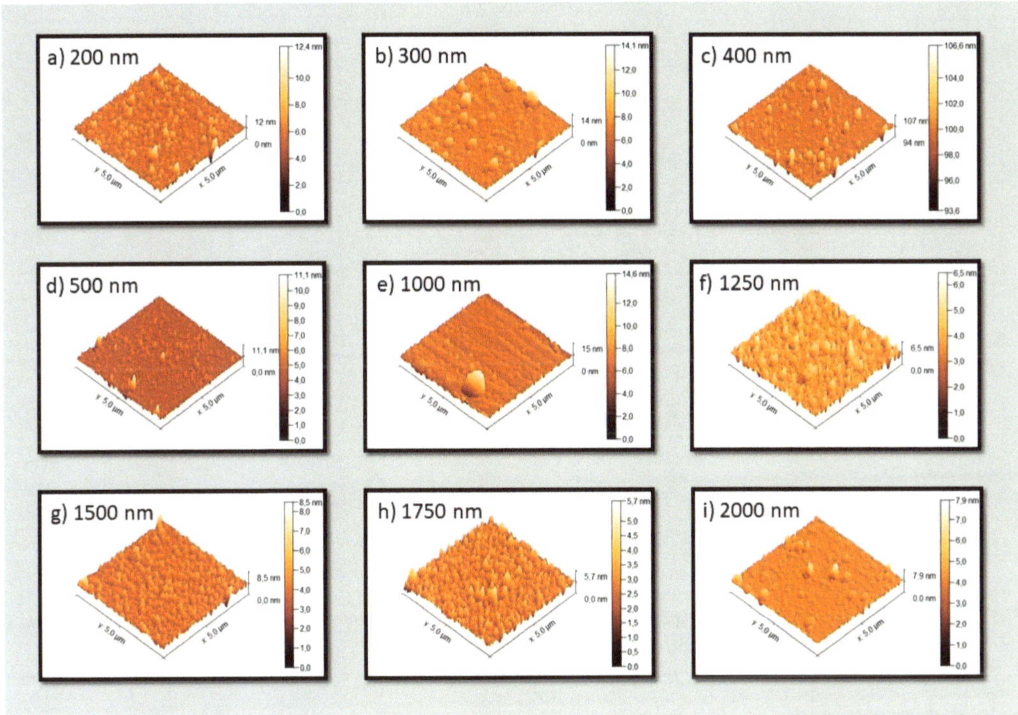

Figure 4. AFM 3D-images for PC samples coated with thick a-C:H layers from 200 nm (**a**) to 2000 nm (**i**).

The XPS results for thin a-C:H layers on PC are shown in Figure 5. The percentages of sp^2 and sp^3 binding after deconvolution of the C1s peak with CasaXPS are plotted against the layer thickness in nanometers. The positions of the peaks were determined with the help of the NIST database and from other sources [19,20,36–38]. For the plot, the C-O bonds were counted as sp^3 bonds and the C=O bonds as sp^2 bonds. The respective values for the whole XPS measurements are shown in Table 2. The relatively high oxygen content of some of these layer surfaces is probably due to saturation of still active bonds on the surface, which occurs after plasma treatment when the vacuum chamber is vented with ambient air. Due to the relatively high surface sensitivity, the saturation of the remaining surface bonds has a great influence on the measurement results obtained. The saturation of broken bonds by internal stress in the layers is also possible. The analysis of the thin layers in 10 nm steps is essential to determining the interlayer thickness. For the O_2 plasma-treated raw polymer, 15% sp^3 bonded and 85% sp^2 bonded carbon were deduced. By depositing an a-C:H layer on PC, the sp^3 fraction increased and dominated the carbon network with a fraction of 55%. When a layer thickness of 20 nm was reached, the sp^2 and sp^3 bonds were balanced. As the a-C:H thickness increased further, the layer behavior changed drastically. The network was strongly sp^2 dominated (95%), and the sp^3 bonding disappeared almost completely. In the AFM images, the pure layer growth on the surface was also visible (Figure 3c,d). The fraction of sp^2 bonds was low with a layer thickness of 40 nm, but remained highly a-C:H dominated (70–80%) up to a layer thickness of 100 nm. Based on the strongly changing layer behavior in the XPS measurements, the interlayer thickness was estimated to be about 20 nm.

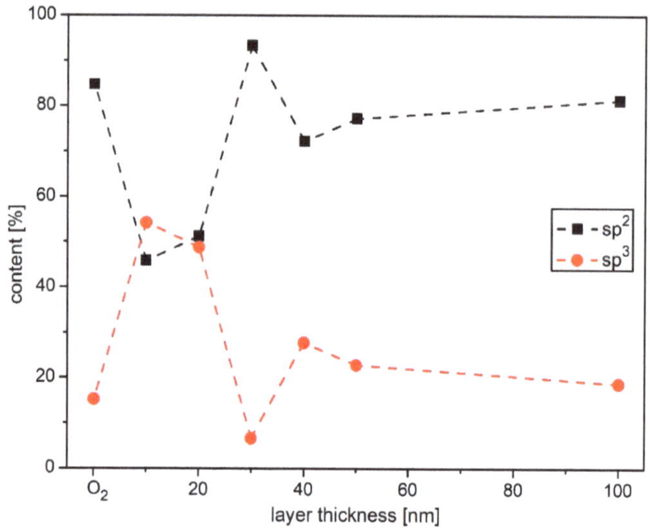

Figure 5. XPS results for thin a-C:H layers on PC. The sp^2 content (black squares) and sp^3 (red circles) bound carbon are plotted against layer thickness (dashed lines show only a trend).

Table 2. Results of XPS measurements separated into the percentages of sp^2, sp^3, C-O and C=O binding to the C1s peak after deconvolution.

Sample	sp^2 [%]	sp^3 [%]	C-O [%]	C=O [%]
O_2	83.49	4.55	10.70	1.26
10	42.55	38.59	15.65	0.37
20	48.58	43.42	5.38	2.63
30	90.14	4.24	2.39	3.24
40	72.19	7.02	20.79	0.00
50	74.58	12.97	9.84	2.62
100	79.39	11.43	7.29	1.90
200	34.02	48.81	14.04	3.12
300	51.77	35.44	12.26	0.53
400	80.51	4.05	13.76	1.69
500	68.10	22.99	0.00	8.91
1000	23.63	50.68	14.96	10.73
1250	84.95	14.40	0.00	0.65
1500	83.12	16.88	0.00	0.00
1750	82.94	10.70	5.20	1.17
2000	82.58	0.31	16.06	1.05

The results of the XPS measurements for the thick a-C:H layers deposited on PC begin at a thickness of 100 nm and are plotted in Figure 6. The dominant bond changed briefly to sp^3 (63%) at a film thickness of 200 nm. As the film thickness increased, the proportion of sp^2 bonds increased again at the expense of the sp^3 bonds, and again reached over 80%. In the course of further coating, the proportion of sp^2 bonds decreased again until at a film thickness of 1000 nm, the sp^3 bond again dominated. The proportions retained were similar to those at 10 and 200 nm. As before, the quantity of sp^3 bonds decreased again with increasing film thickness. After reaching an a-C:H layer of 1250 nm, the two bonds remained almost constant up to a film thickness of 2000 nm. After 1250 nm, the a-C:H growth of the film changed compared to the previous behavior. The strongly reduced proportion of sp^3 bonds was due to dehydration of the layer. In this process, hydrogen leaves the layer and lowers the amount of sp^3-bonded carbons. The free bonds are converted into sp^2 hybridized carbons. Thus, re-hybridization of the carbon atom orbitals occurs. This behavior has already been observed for several polymers coated with a-C:H [8,28,29]. Figure 7 presents exemplary deconvolutions of the 10, 100 and 1000 nm a-C:H layer thicknesses for the C1s peak using CasaXPS.

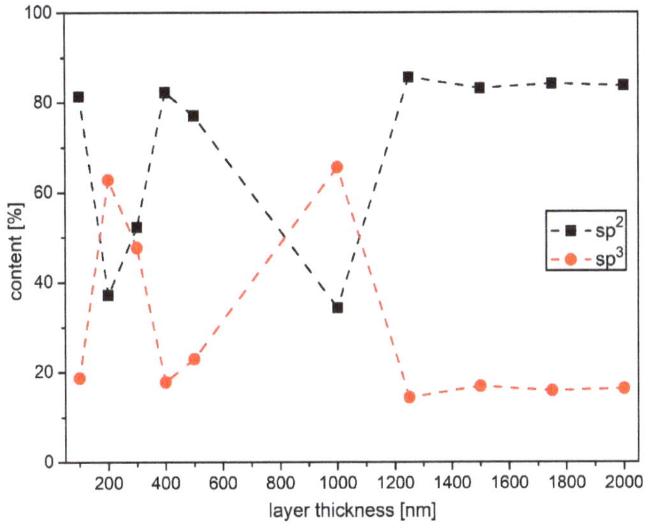

Figure 6. XPS results for thick a-C:H layers deposited on PC, up to 2000 nm. The black squares are the sp^2 content and the red circles are the sp^3 bound carbons plotted against layer thickness (dashed lines only show a trend).

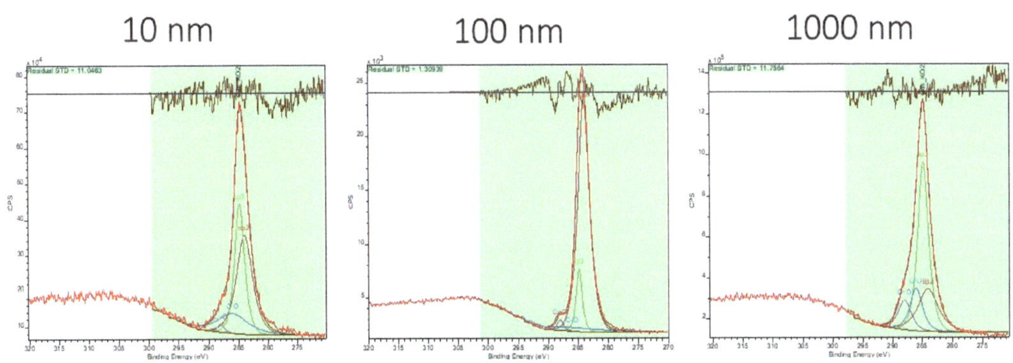

Figure 7. Exemplary XPS peak deconvolution for the a-C:H layer thicknesses of 10, 100 and 1000 nm. The analysis was performed on the sp^2, sp^3, C-O and C=O fractions using CasaXPS software.

A striking feature is the almost periodic behavior of the carbon bonds over the layer thickness. The sp^2 bond dominated, and in a short film thickness interval the sp^3 bond was dominant. With each step, however, the proportion of sp^3 bonds increased slightly (55%, 63% and 65%). After the sp^3 bond was dominant for a short time, the bond ratio suddenly changed to a strongly sp^2 dominated carbon network. A comparison with SEM images shows that this change of the dominant bond was accompanied by layer failure. The first change occurred at 20 nm film thickness (Figure 5). With a short delay, no layer failure occurred at 40 nm, but the sample surface was covered with fine, thread-like structures

(Figure 1e). It is possible that the stress caused by the sp^3 bonds was not yet strong enough to completely crack the layer. However, it is noticeable that there was a strong C-O content of 20.79 % (Table 2). It is possible that individual, broken C-C bonds were saturated with oxygen from the environment. As the layer grew from 300 to 400 nm, the dominant bond changed again from sp^3 to sp^2. The SEM images for the 400 nm sample (Figure 2c) showed a clear cracking of the layer, which took place shortly after the change of the dominant bond. The same applied to the change to the 900 nm sample (Figure 2e). In the XPS the change took place while reaching 1000 nm a-C:H on the sample—thereby only after the layer failure was visible in the SEM. Most likely the fraction of sp^3 bonds had already reached its highest value at 900 nm and at the step to 1000 nm it decreased, which was accompanied with a stress release (=breakup of the layer). Unfortunately, the 900 nm sample was not measurable in the XPS. This change from sp^3 to sp^2 as the dominant bond with a subsequent layer failure has already been observed for other polymers [19,20,23,29]. The more pronounced the interlayer, the greater the delay. In the case of PC, it is very weak with a thickness of 20 nm, and the crack in the layer follows more or less directly after the change in the dominant bond (XPS: change in bond at 20 nm, SEM striking layer behavior at 40 nm). As revealed for polyhydroxy butyrate, the interlayer is of average strength, ending in a delay of 100 nm [20]; for polybutylene adipate terephthalate the interlayer is very pronounced with a delay of 200 nm between the change and the layer failure. Even after failure of the layer, it continues to adhere to the polymer instead of rolling up [23]. Both polyamide 6 (PA6) and HDPE have very pronounced interlayers (40 nm) [28,29]. In both cases, the bonding ratio is very stable, and the layers only break down at very high layer thicknesses. In the case of PA6, no change from sp^3 to sp^2 as the dominant bond could be detected. However, at the step to 1500 nm film thickness, there was a very strong drop in sp^3 bonds (zone of dehydrogenation), followed by film failure at 1750 nm [29]. For HDPE, there was a change from sp^3 to sp^2 at 50 nm, but layer failure was only observable at 1000 nm [8,28]. As in previous studies, the change from sp^2 to sp^3 as the dominant bond had no effect on the integrity of the layer [19,20,23,29].

For further analysis of the chemical composition of the coating surface, NEXAFS investigations were performed. The results of the C K-edge are plotted in Figure 8. The determination of the peak location was based on the work of other groups and specified as follows: C=C π (284.85 eV), C-H (286.15 eV), C-C (288.35 eV) and C=C σ (292.55 eV) [39–43]. The spectra were analyzed as described in Section 2.3. The diagram is divided into three different thickness ranges: A includes the 0–50 nm a-C:H layer thicknesses; B includes the layers from 100 to 500 nm; and finally, C layers from 1000 to 2000 nm. For better clarification, there is a zoom of the region from 284 to 289 eV to the right of each. In Figure 8A, it can be seen that the ratio of the C=C π peak to the C-C peak tended to tilt in favor of the C=C π peak with increasing film thickness. Thus, the sp^2 bonding was more pronounced with increasing film thickness at the expense of the sp^3 bonding. At 50 nm film thickness, the C–C peak was barely visible at 288.15 eV, and the C=C π was relatively emphasized at 284.85 eV. This is also consistent with the results already observed in the XPS (Figure 5). A new feature is that the C K-edge formed a strong C-H slope for the layer thicknesses of 20 and 40 nm. These were the thicknesses with the largest numbers of grains on the surface just before the end of the layer. A correlation is clearly visible. When the layer was complete and grains were barely visible, this peak disappeared (Figure 3b,d,f and Figure 8A). The 100 nm a-C:H layer thickness on PC again showed the peak for the C-H bond (Figure 8B). Again, many grains are visible on the surface in the AFM (Figure 3g). The C=C π peak is strongly emphasized, the C-C peak only weakly developed. This is again in agreement with the XPS measurements. At 200 and 300 nm the C=C π peak decreased, and the C-C peak at 288.15 is more pronounced. Already in the XPS measurements, it could be seen that the fraction of sp^3 was higher (Figure 6). The flanks of the C-H bond were also recognizable. The AFM images (Figure 4a,b) also showed some grains on the surface of the layer. When reaching the 500 nm layer thickness, the C-H peak almost disappeared and was hardly recognizable. The AFM image showed here (Figure 4d) only isolated and mostly very small grains on the surface. The C-K edge of the 1000 nm a-C:H layer on PC is plotted in Figure 8C. Unfortunately, the measurement in NEXAFS proved to be extremely difficult, so the curve was plotted here more for the purpose of completeness. The C-C peak seems to be superimposed or broadened by a possible charge, which would make an evaluation very error-prone, so it will not be discussed further, except for the non-existent C-H peak. As expected from the previous results, the AFM image (Figure 4e) hardly shows any grain structures. For the thicknesses of 1250 and 1750 nm (1500 nm was not measurable), the ratios of the C=C π and the C-C peak were about the same, although both peaks are visible in a defined way. The peak of the C-H bond is also observable, although it is weaker for the larger of the two layer thicknesses (1750 nm). Compared with the AFM images (Figure 4f,h), a decrease in the density of the grains with increasing layer thickness can be seen. Finally, when the film thickness increased to 2000 nm, the ratio again tipped in favor of the C=C π peak at 284.85 eV. In the XPS, on the other hand, the ratio of sp^2 to sp^3 was almost constant from 1000 nm layer thickness, so the C K-edge result for the 2000 nm a-C:H layer differs here from that of the XPS measurements. In the AFM images hardly any grains are visible for this layer thickness (Figure 4i), and also the C-H peak is no longer present in the NEXAFS measurements.

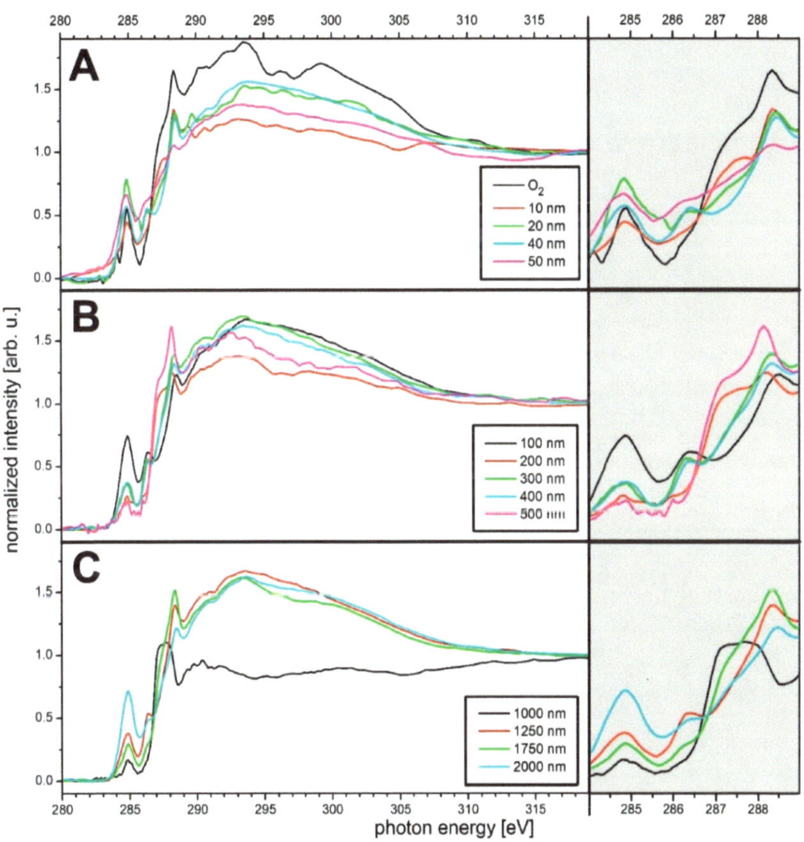

Figure 8. NEXAFS C K-edge spectra for a-C:H layers of various thicknesses deposited on PC. (**A**) The a-C:H thicknesses 0–50 nm, (**B**) 100–500 nm and (**C**) 1000–2000 nm. On the left are the full spectra (280–320 eV); on the right is a magnification of the region from 284 to 289 eV. Here are the peaks located for C=C π (284.85 eV), C-H (286.15 eV) and C-C (288.15 eV).

Overall, however, the two methods correspond to each other and a correlation between grain density on the layer surface and the formation of a C-H peak was found on the basis of the AFM measurements. Apart from minor deviations (for example, a somewhat more pronounced C-H peak would have been expected with a film thickness of 200 nm, Figure 4a), this pattern is confirmed. Thus, with the help of the NEXAFS measurement, conclusions can be drawn about the growth status of the current layer. A possible explanation for the more defined C-H peak with many grains is the size of the surface that is measured. If the layer is smooth, the total surface area is smaller than if the layer is covered with grains. Since the hydrogens represent the ends of the carbon chains, a larger total surface area means that there are also more ends that are detected.

4. Conclusions

Depositions of a-C:H layers were successfully arranged on the polymer PC in various thicknesses (10–2000 nm). The coating was applied by RF-PECVD with acetylene plasma in a vacuum chamber. The obtained layers were subsequently investigated by SEM, AFM, and synchrotron assisted XPS and NEXAFS. Particular attention was paid to the region from 0–100 nm, where the interlayer is formed, which is responsible for the connection of the layer to the base material. It was determined to be 20 nm thick by XPS and AFM measurements. In the AFM, grains are visible from this thickness onwards, which indicate normal a-C:H growth. The XPS shows a completely changing bonding behavior of the carbon atoms on the surface of the layer, which indicates a change in the growth process.

With the SEM measurements, unstable layer thicknesses could be found, where the layer cracks due to internal stresses (400 nm, 900 nm). If the XPS measurements are considered, it is evident that these layer stabilities immediately follow the change from sp^3 to sp^2 as the dominant bond in the carbon network. Rigid sp^3 bonds break due to internal stress and the carbon atoms reorganize to sp^2 bonds. The range of dehydrogenation for the a-C:H layers on PC could be determined by XPS to the thickness range of 1250–2000 nm. Here, hydrogen leaves the layers and formerly sp^3 bonded carbon atoms reorganize into a strong sp^2 dominated network. The NEXAFS investigations confirm the overall results of the XPS. In addition, a correlation between the number and size of grains on the layer surface and the formation of the C-H peak in the C K-edge was found. The more the surface is maximized by the grains, the more accentuated is the C-H peak. Since the hydrogen atoms terminate the carbon chains and towards the outside, it is obvious that the larger the surface area, the more C-H bonds are present.

The coatability of PC with a-C:H layers was examined and areas of unstable layer thickness were found. In addition, effects were detected that influence the surface bonding ratio. With these fundamentals, targeted modification of the surface properties, such as the application of a thin, scratch-resistant a-C:H layer for e.g., optical components, is now possible. However, further studies must be performed for respective applications.

Author Contributions: T.S., Y.W. and C.B.F. conceptualized the study, and developed and organized the processes. T.S. prepared the samples and contributed the AFM measurements. T.S. also performed the data acquisition and evaluation of XPS and NEXAFS, evaluation of SEM and AFM results, overall data curation, visualization and wrote the manuscript draft. M.F. and L.B. contributed to the acquisition of synchrotron measurements, XPS and NEXAFS. T.S. and C.B.F. validated the results, and reviewed and edited the manuscript. Both S.W. and C.B.F. contributed in terms of funding acquisition, project administration, supervision and overall guidance in the study. All authors have read and agreed to the published version of the manuscript.

Funding: T.S., L.B. and C.B.F. gratefully acknowledge the financial support (funding agreements BESSY 2: 62 and 145) provided by the Helmholtz-Zentrum Berlin (HZB), Germany.

Data Availability Statement: The data presented in this study are available on request from the corresponding author.

Acknowledgments: The authors thank Heinz Busch (NTTF Coatings GmbH, Rheinbreitbach, Germany) for the possibility of industrial coating of the samples. Additionally, we thank Barbara Hahn, Youssef Benjillali and Mohammed Bentaouit from the University of Applied Sciences Koblenz (Department of Material Analysis, RheinAhrCampus) for the SEM equipment and measurements. We also thank Alexei Nefedov from KIT (Karlsruhe Institute for Technology, Group: Chemistry of oxidic and organic interfaces; X-ray and Electron Spectroscopy at Interfaces) for beamtime support. We thank HZB for the allocation of synchrotron radiation beamtime. The authors thankfully acknowledge the financial support by the Helmholtz-Zentrum Berlin (HZB), Germany.

Conflicts of Interest: The authors declare no conflict of interest.

References

1. Ibrahim, S.S.; Al Jaafari, A.A.; Ayesh, A.S. Physical characterizations of three phase polycarbonate nanocomposites. *J. Plast. Film Sheet.* **2011**, *27*, 275–291. [CrossRef]
2. Kausar, A. A review of filled and pristine polycarbonate blends and their applications. *J. Plast. Film Sheet.* **2018**, *34*, 60–97. [CrossRef]
3. Krishnan, A.V.; Stathis, P.; Permuth, S.F.; Tokes, L.; Feldman, D. Bisphenol-A: An estrogenic substance is released from polycarbonate flasks during autoclaving. *Endocrinology* **1993**, *132*, 2279–2286. [CrossRef]
4. Wang, Y.; Guan, W.; Fischer, C.B.; Wehner, S.; Dang, R.; Li, J.; Wang, C.; Guo, W. Microstructures, mechanical properties and tribological behaviors of amorphous carbon coatings in-situ grown on polycarbonate surfaces. *Appl. Surf. Sci.* **2021**, *563*, 150309. [CrossRef]
5. Catena, A.; Guo, Q.; Kunze, M.R.; Agnello, S.; Gelardi, F.M.; Wehner, S.; Fischer, C.B. Morphological and Chemical Evolution of Gradually Deposited Diamond-Like Carbon Films on Polyethylene Terephthalate: From Subplantation Processes to Structural Reorganization by Intrinsic Stress Release Phenomena. *ACS Appl. Mater. Interfaces* **2016**, *8*, 10636–10646. [CrossRef] [PubMed]
6. Grill, A. Plasma-deposited diamondlike carbon and related materials. *IBM J. Res. Dev.* **1999**, *43*, 147–162. [CrossRef]
7. Couderc, P.; Catherine, Y. Structure and physical properties of plasma-grown amorphous hydrogenated carbon films. *Thin Solid Film.* **1987**, *146*, 93–107. [CrossRef]
8. Catena, A.; Agnello, S.; Rösken, L.M.; Bergen, H.; Recktenwald, E.; Bernsmann, F.; Busch, H.; Cannas, M.; Gelardi, F.M.; Hahn, B.; et al. Characteristics of industrially manufactured amorphous hydrogenated carbon (a-C:H) depositions on high-density polyethylene. *Carbon* **2016**, *96*, 661–671. [CrossRef]
9. Mohagheghpour, E.; Rajabi, M.; Gholamipour, R.; Larijani, M.M.; Sheibani, S. Ion beam energy dependence of surface and structural properties of amorphous carbon films deposited by IBSD method on Ni–Cu alloy. *J. Mater. Res.* **2017**, *32*, 1258–1266. [CrossRef]
10. Paul, R.; Das, S.N.; Dalui, S.; Gayen, R.N.; Roy, R.K.; Bhar, R.; Pal, A.K. Synthesis of DLC films with different sp2/sp3 ratios and their hydrophobic behaviour. *Phys. Rev. B* **2008**, *41*, 055309. [CrossRef]
11. Feng, L.; Li, S.; Li, Y.; Li, H.; Zhang, L.; Zhai, J.; Song, Y.; Liu, B.; Jiang, L.; Zhu, D. Super-Hydrophobic Surfaces: From Natural to Artificial. *Adv. Mater.* **2002**, *14*, 1857–1860. [CrossRef]
12. Banerjee, D.; Mukherjee, S.; Chattopadhyay, K.K. Controlling the surface topology and hence the hydrophobicity of amorphous carbon thin films. *Carbon* **2010**, *48*, 1025–1031. [CrossRef]
13. Ostrovskaya, L.; Perevertailo, V.; Ralchenko, V.; Dementjev, A.; Loginova, O. Wettability and surface energy of oxidized and hydrogen plasma-treated diamond films. *Diam. Relat. Mater.* **2002**, *11*, 845–850. [CrossRef]
14. Piazza, F.; Morell, G. Wettability of hydrogenated tetrahedral amorphous carbon. *Diam. Relat. Mater.* **2009**, *18*, 43–50. [CrossRef]
15. Bismarck, A.; Brostow, W.; Chiu, R.; Hagg Lobland, H.E.; Ho, K.K.C. Effects of surface plasma treatment on tribology of thermoplastic polymers. *Polym. Eng. Sci.* **2008**, *48*, 1971–1976. [CrossRef]
16. Koidl, P.; Wild, C.; Dischler, B.; Wagner, J.; Ramsteiner, M. Plasma Deposition, Properties and Structure of Amorphous Hydrogenated Carbon Films. *MSF* **1991**, *52–53*, 41–70. [CrossRef]
17. Robertson, J. Diamond-like amorphous carbon. *Mater. Sci. Eng. R Rep.* **2002**, *37*, 129–281. [CrossRef]
18. Zhou, Y.; Wang, B.; Song, X.; Li, E.; Li, G.; Zhao, S.; Yan, H. Control over the wettability of amorphous carbon films in a large range from hydrophilicity to super-hydrophobicity. *Appl. Surf. Sci.* **2006**, *253*, 2690–2694. [CrossRef]
19. Schlebrowski, T.; Beucher, L.; Bazzi, H.; Hahn, B.; Wehner, S.; Fischer, C.B. Prediction of a-C:H layer failure on industrial relevant biopolymer polylactide acide (PLA) foils based on the sp2/sp3 ratio. *Surf. Coat. Technol.* **2019**, *368*, 79–87. [CrossRef]
20. Schlebrowski, T.; Beucher, L.; Bazzi, H.; Hahn, B.; Wehner, S.; Fischer, C.B. Changing Contents of Carbon Hybridizations in Amorphous Hydrogenated Carbon Layers (a-C:H) on Sustainable Polyhydroxybutyrate (PHB) Exhibit a Significant Deterioration in Stability, Depending on Thickness. *C* **2019**, *5*, 52. [CrossRef]
21. Robertson, J. Plasma Deposition of Diamond-Like Carbon. *Jpn. J. Appl. Phys.* **2011**, *50*, 01AF01. [CrossRef]
22. Lee, S.-C.; Tai, F.-C.; Wei, C.-H. Correlation between sp2/sp3 Ratio or Hydrogen Content and Water Contact Angle in Hydrogenated DLC Film. *Mater. Trans.* **2007**, *48*, 2534–2538. [CrossRef]

23. Schlebrowski, T.; Acharchi, H.; Hahn, B.; Wehner, S.; Fischer, C.B. Refinement of Sustainable Polybutylene Adipate Terephthalate (PBAT) with Amorphous Hydrogenated Carbon Films (a-C:H) Revealing Film Instabilities Influenced by a Thickness-Dependent Change of sp2/sp3 Ratio. *Materials* **2020**, *13*, 1077. [CrossRef]
24. Fischer, C.B.; Rohrbeck, M.; Wehner, S.; Richter, M.; Schmeißer, D. Interlayer formation of diamond-like carbon coatings on industrial polyethylene: Thickness dependent surface characterization by SEM, AFM and NEXAFS. *Appl. Surf. Sci.* **2013**, *271*, 381–389. [CrossRef]
25. Rohrbeck, M.; Körsten, S.; Fischer, C.B.; Wehner, S.; Kessler, B. Diamond-like carbon coating of a pure bioplastic foil. *Thin Solid Film.* **2013**, *545*, 558–563. [CrossRef]
26. Neyts, E.; Bogaerts, A.; Van de Sanden, M.C.M. Reaction mechanisms and thin a-C: H film growth from low energy hydrocarbon radicals. *J. Phys. Conf. Ser.* **2007**, *86*, 012020. [CrossRef]
27. Catena, A.; Kunze, M.R.; Agnello, S.; Gelardi, F.M.; Wehner, S.; Fischer, C.B. Amorphous hydrogenated carbon (a-C:H) depositions on polyoxymethylene: Substrate influence on the characteristics of the developing coatings. *Surf. Coat. Technol.* **2016**, *307*, 658–665. [CrossRef]
28. Schlebrowski, T.; Rouabeh, W.; Wehner, S.; Fischer, C.B. Specifying the interlayer turning point and dehydrogenation in a-C:H layers plasma deposited on high-density polyethylene with X-ray synchrotron techniques. *Thin Solid Film.* **2019**, *691*, 137617. [CrossRef]
29. Schlebrowski, T.; Lüber, H.; Beucher, L.; Fritz, M.; Benjillali, Y.; Bentaouit, M.; Hahn, B.; Wehner, S.; Fischer, C.B. Plasma Supported Deposition of Amorphous Hydrogenated Carbon (a-C:H) on Polyamide 6: Determining Interlayer Completion and Dehydrogenation Effects during Layer Growth. *Polymers* **2021**, *13*, 1886. [CrossRef] [PubMed]
30. Nefedov, A.; Wöll, C. Advanced Applications of NEXAFS Spectroscopy for Functionalized Surfaces. In *Surface Science Techniques*; Bracco, G., Holst, B., Eds.; Springer Berlin Heidelberg: Berlin/Heidelberg, Germany, 2013; pp. 277–303. ISBN 978-3-642-34242-4.
31. Fang, D.; He, F.; Xie, J.; Xue, L. Calibration of Binding Energy Positions with C1s for XPS Results. *J. Wuhan Univ. Technol. Mat. Sci. Ed.* **2020**, *35*, 711–718. [CrossRef]
32. Watts, B.; Thomsen, L.; Dastoor, P.C. Methods in carbon K-edge NEXAFS: Experiment and analysis. *J. Electron. Spectrosc. Relat. Phenom.* **2006**, *151*, 105–120. [CrossRef]
33. Schlebrowski, T.; Ouali, R.; Hahn, B.; Wehner, S.; Fischer, C.B. Comparing the Influence of Residual Stress on Composite Materials Made of Polyhydroxybutyrate (PHB) and Amorphous Hydrogenated Carbon (a-C:H) Layers: Differences Caused by Single Side and Full Substrate Film Attachment during Plasma Coating. *Polymers* **2021**, *13*, 184. [CrossRef]
34. Lacerda, R.G.; Marques, F.C. Hard hydrogenated carbon films with low stress. *Appl. Phys. Lett.* **1998**, *73*, 617–619. [CrossRef]
35. Marques, F.C.; Lacerda, R.G.; Odo, G.Y.; Lepienski, C.M. On the hardness of a-C:H films prepared by methane plasma decomposition. *Thin Solid Film.* **1998**, *332*, 113–117. [CrossRef]
36. Tai, F.C.; Lee, S.C.; Wei, C.H.; Tyan, S.L. Correlation between ID/IG Ratio from Visible Raman Spectra and sp2/sp3 Ratio from XPS Spectra of Annealed Hydrogenated DLC Film. *Mater. Trans.* **2006**, *47*, 1847–1852. [CrossRef]
37. Solomon, J.L.; Madix, R.J.; Stöhr, J. Orientation and absolute coverage of benzene, aniline, and phenol on Ag(110) determined by NEXAFS and XPS. *Surf. Sci.* **1991**, *255*, 12–30. [CrossRef]
38. Dietrich, P.M.; Horlacher, T.; Girard-Lauriault, P.-L.; Gross, T.; Lippitz, A.; Min, H.; Wirth, T.; Castelli, R.; Seeberger, P.H.; Unger, W.E.S. Adlayers of dimannoside thiols on gold: Surface chemical analysis. *Langmuir* **2011**, *27*, 4808–4815. [CrossRef]
39. Stöhr, J. *NEXAFS Spectroscopy*; Springer: Berlin/Heidelberg, Germany, 1992; ISBN 978-3-662-02853-7.
40. Brüster, B.; Amozoqueño, C.; Grysan, P.; Peral, I.; Watts, B.; Raquez, J.-M.; Dubois, P.; Addiego, F. Resolving Inclusion Structure and Deformation Mechanisms in Polylactide Plasticized by Reactive Extrusion. *Macromol. Mater. Eng.* **2017**, *302*, 1700326. [CrossRef]
41. Díaz, J.; Anders, S.; Zhou, X.; Moler, E.J.; Kellar, S.A.; Hussain, Z. Analysis of the π^* and σ^* bands of the x-ray absorption spectrum of amorphous carbon. *Phys. Rev. B* **2001**, *64*, 125204. [CrossRef]
42. Dhez, O.; Ade, H.; Urquhart, S.G. Calibrated NEXAFS spectra of some common polymers. *J. Electron Spectrosc. Relat. Phenom.* **2003**, *128*, 85–96. [CrossRef]
43. Diaz, J.; Monteiro, O.R.; Hussain, Z. Structure of amorphous carbon from near-edge and extended x-ray absorption spectroscopy. *Surf. Coat. Technol.* **2007**, *76*, 094201. [CrossRef]

C Journal of Carbon Research

MDPI

Article

Reduction of Device Operating Temperatures with Graphene-Filled Thermal Interface Materials

Jacob S. Lewis †

Phonon Optimized Engineered Materials (POEM) Center, Bourns College of Engineering, University of California-Riverside, Riverside, CA 92521, USA; jlewi014@ucr.edu
† Present address: Intel Corporation, Hillsboro, OR 97124, USA.

Abstract: The majority of research into few layer graphene (FLG) thermal interface materials (TIM) concerns the direct quantification of innate composite properties with much less direct analysis of these materials in realistic applications. In this study, equilibrium temperatures of engineered device substitutes fixed to passive heat sink solutions with varying FLG concentration TIMs are experimentally measured at varying heat dissipation rates. A custom, precisely-controlled heat source's temperature is continually measured to determine equilibrium temperature at a particular heat dissipation. It is found that altering the used FLG TIM concentrations from 0 vol.% to as little as 7.3 vol.% resulted in a decrease of combined TIM and passively-cooled heat sink thermal resistance from 4.23 °C/W to 2.93 °C/W, amounting to a reduction in operating temperature of ≈108 °C down to ≈85 °C at a heat dissipation rate of 20 W. The results confirm FLG TIMs' promising use in the application of device heat dissipation in a novel, controllable experimental technique.

Keywords: graphene; thermal interface materials; temperature; passive cooling; power density; heat sink; VLSI cooling; thermal resistance

Citation: Lewis, J.S. Reduction of Device Operating Temperatures with Graphene-Filled Thermal Interface Materials. C **2021**, 7, 53. https://doi.org/10.3390/c7030053

Academic Editor: Gil Goncalves

Received: 16 June 2021
Accepted: 15 July 2021
Published: 21 July 2021

1. Introduction

The continual progress of semiconductor electronics technology very often comes at the cost of increased power consumption levels to be dissipated as waste heat [1]. The progressive densification of very large-scale integration circuits and advent of "5G" telecommunications technologies have rendered thermal management to be an ever greater device design consideration due to unintended device behavior changes and reduction in reliability at higher temperatures [2–7]. Of increasing importance to society with the adoption of renewable energy sources, photovoltaic cells suffer reduced efficiency at elevated operating temperatures [8,9]. The most common strategy to address device waste heat, due to its relative ease compared to improving device efficiency, is to manage the operating temperature of electronic devices by dissipating generated thermal energy into the environment as easily as possible. The most common method to increase heat dissipation to ambient from small devices is to fix a heat sink into contact in order to increase the exchange rate with the environment. However, between any two solid surfaces in contact, such as a device and heat sink, there is a substantial portion of the total contacting surfaces air-gapped from one another due to unavoidable mating surface imperfections. Industry mitigates the problem of extremely thermally insulating air gaps with the introduction of a thermal interface material (TIM) in the junction to take the space of air [10]. Alternatively, curing polymers are often used to encapsulate heat-producing chips to protect them from environmental contaminants, resulting in an unintentional thermal insulation. Polymers used in these two applications—particularly the former—are often filled with conductive particles such as silver, copper, Al_2O_3, AlN, boron nitride, ZnO, diamond, graphite, carbon nanotubes, and randomly oriented few-layer graphene (FLG) [11–24]. Though these polymer materials remain substantially superior in thermal conduction than the air that they replace, there exists substantial room for improvement in these materials considering they have typically two orders of magnitude lower thermal conductivity than the semiconductors and metals on either side of the thermal junction.

The initial FLG TIM research effort was inspired by the extraordinary thermal conductivity of graphene ranging from 2000 to 5300 W/m°C [25–33]. Early works on FLG TIMs reported impressive thermal conductivities up to 5 W/m°C at room temperature with as low as 10 vol.% of FLG [34,35]. Recent research has produced randomly oriented FLG TIMs with thermal conductivities of ≈12 W/m°C and ≈7 W/m°C for curing and non-curing composites, respectively [36–39]. Such high composite thermal conductivities were achieved with high FLG load levels of over 20 vol.%, resulting in a high composite electrical conductivity as a result of FLG's intrinsic properties. When a composite is loaded with a critical concentration of discrete electrically conductive particles then the composite is said to have reached the electrical percolation threshold and its overall composite electrical conductivity rises precipitously [40–45]. Using multiple filler materials has long been an active area of research due to synergistic enhancements in composite thermal conductivity [13,46–53]. The advent of composite electrical conductivity resultant from electrical percolation is problematic for electronic devices that would employ such materials because it can cause unintentional electrical shorting phenomena. A recent work in hybrid composites showed at least 11 orders of magnitude of composite electrical conductivity by varying the constituent and total load fraction of FLG and hexagonal boron nitride in cured polymer composites [54].

The ultimate goal of every TIM is to achieve as low of a device operating temperature as possible with a given device's power consumption and heat sink's dissipation rate to ambient environment. The flow of heat from electronic device source to ambient environment can be and often is thought of with an analogy to an electronic circuit, where a TIM is a comparatively large resistor in the circuit impeding the flow of heat from high to low potential. Lowering the resistance of this component affords a lower operating temperature for any given device heat dissipation, allowing for the use of increasingly powerful electronics within a similar operating temperature window.

It is very common in TIM research to characterize material properties with techniques such as Laser Flash Analysis, Transient Plane Source, and ASTM D5470 and then determine their performance in a thermal junction from those measured values [10,35,55–68]. These measured values can often end up as parameters in a finite element simulation to predict a device's operating temperature. Studies to directly experimentally determine the operating temperatures of CPU devices with the application of a particular TIM of interest are certainly worthy of consideration but are unfortunately beholden to the dynamic, somewhat uncontrollable power outputs of such complex systems [69–72]. Additionally, it is typical in these studies to use integrated CPU thermistors to measure operating temperatures. This practice is particularly problematic as a result of localized hot spots on the chip leaving unclear which of the numerous local thermistors to consider representative. In this work, a direct comparison of electronic device operating temperatures, by way of a highly controllable experiment quite challenging when using a live CPU, is analyzed with TIMs of different FLG concentrations as a translation between the substantial research in these composites and real world applications.

2. Materials and Methods

2.1. Sample Preparation

In place of a more complicated central processing unit with an at least somewhat variable heat output running Prime95 software, a custom coiled wire heater was used, shown in Figure 1a. This heating element employed a Nickel and Chromium alloy wire often used in electronics heating applications sandwiched between two pieces of Kapton tape. The wire used had a diameter of 72 (±2.5) μm, as measured by micrometer. Electrical current was driven through the coil by a programmable power supply (Chroma 62000P, Foothill Ranch, CA, USA) to produce Joule heating. The wire was hand-coiled by pinning a pre-cut length of wire down with tape onto a hard surface and carefully curling each end around the pinned central point until a roughly uniform coil was achieved. As small of a length of resistor wire extending out of the edge of the coil as possible was sought—typically around 1 cm—to reduce the fraction of heat that is generated outside of the coil. After, the tape with the coil stuck to its bottom side were carefully peeled from the hard surface and joined by another similar piece of tape to produce a completed heating element sandwich.

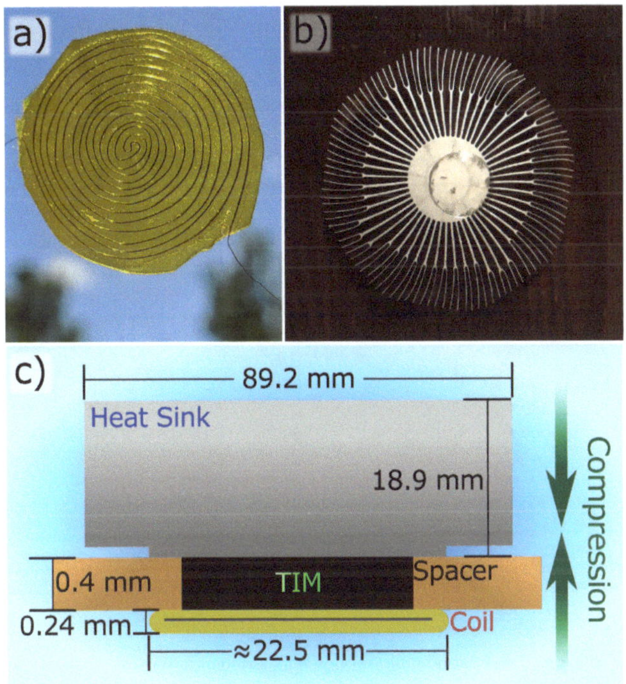

Figure 1. (**a**) Close-up picture of a representative custom heating coil used for experimentation. (**b**) A top-down picture of a CPU heat sink, with thermal interface mating surface face down, with imperfections from previous use. (**c**) A schematic of the sample fabrication process with relevant dimensions marked.

The heat sink was a stock unit sold in tandem with LGA-1155 socket processors, shown in Figure 1b, sourced directly from industry (Intel Corp. E97379-001, Santa Clara, CA, USA). To reduce experimental complexity and improve repeatability, the attached fan was removed. The heat sink was attached to TIM and heating coil in a passive cooling arrangement, resulting in increased thermal resistance, generating higher device temperatures than in an actively cooled setup, as intended by the product designers.

The polymer matrix used in the composite samples is a Diglycidyl ether of Bisphenol A (DGEBA) resin with a Triethylenetetramine hardening agent (Allied High Tech Products Inc., Rancho Dominguez, CA, USA) [73]. A curing polymer was chosen as a matrix for experimental ease and because a singular test would be conducted on each sample, rendering reliability concerns of a cured TIM negligible [74,75]. The FLG used in the composites has a vendor-specified lateral dimension of 25 μm (XG Sciences, Lansing, MI, USA). First, an amount of resin is dispensed into a cup and subjected to a rough vacuum to draw out air bubbles and is then weighed on a scale. This weight is then used to calculate the amount of hardening agent to be used—12% of the weight of resin—and the FLG for a particular targeted filler level. Next, the FLG is added to the resin and the two are mixed in a bladeless, two axis planetary mixer (Flacktek Inc., Landrum, SC, USA) between 1500 and

2000 rpm, depending on difficulty mixing resultant from FLG load level, for 90 s. Then the hardening agent is added and mixed once more with the FLG-polymer mixture under similar conditions. Different TIM samples of DGEBA with FLG filler levels of 0 vol.%, 3.6 vol.%, 7.3 vol.%, 25.5 vol.%, and 43.6 vol.% were prepared. The composites of FLG filler concentrations above 7.3 vol.% were not further evaluated for experimentation, aside from SEM analysis, because of an inability to control their TIM bond line thickness as a result of the apparent increase in viscosity with increasing filler level, as quantitatively reported previously [76]. Work on similar composites has shown electrical conductivity of composites with total FLG-specific loading at around 7.3 vol.% should be on the order of magnitude of 10^{-8} S/cm [54,77].

The mixed but still un-cured polymer composite is then placed on top of a heating coil that is pinned on two sides at the edge by spacers—copper metal sheets of 400 μm thickness—that are themselves pinned by clamps, as shown in Figure 1c. The heat sink is then lowered onto the composite and heating coil and clamped down to compress the TIM and guarantee the eventual TIM thickness is defined by the spacers. It is this step in the process that closely mimics real-world industrial applications of TIMs that rendered the high load level FLG TIMs unusable in the present study. If the TIM apparent viscosity is low enough, excess composite material spills out the remaining two directions that are not occluded by the spacers. Special care was taken to ensure that there would be excess material so complete heater coverage is ensured. The clamped stack is then left to cure over night. After this, the spacers are carefully removed by pulling directly away from the cured TIM. The small resistance wire leads are soldered to a larger, less resistive wire with a length long enough to clear the overhang of the heat sink assembly, and finally that is alligator clipped to test cables connected to the power supply.

2.2. Experimental Procedure

The heating coil, cured TIM, and heat sink assembly test vehicle is placed on top of a thermally insulating teflon plate with the coil side down in ambient atmosphere, shown in Figure 2. Between the teflon plate and the heating coil is a Type-J thermocouple (OMEGA Engineering, Inc., Norwalk, CT, USA) that probes the temperature of the heating coil itself, getting a direct measurement of coil temperature which is analogous to measuring the direct temperature of an electronic device. This entire vertical stack is lightly compressed in a clamp just enough to ensure the light thermocouple remains in place and in good contact. The clamp is imperfectly thermally isolated from the top of the heat sink with a small silicone pad insulation. The thermocouple was read with a voltmeter (Keithley 2182A, Solon, OH, USA). A computer running a python script retrieved temperature measurements from the voltmeter by passing standard commands for programmable instruments (SCPI) to it. In a similar fashion, the computer changed the coil heat dissipation rate via SCPI commands to the programmable power supply.

The experimental goal was to determine the equilibrium temperature of the heating coil at every investigated heat dissipation rate. This was done by applying a current through the heater, continually measuring the thermocouple temperature, and waiting for the time derivative of temperature to approach zero, accounting for measurement noise. Equilibrium was determined by comparing the most recent thermocouple temperature reading ($\phi_{(0)}$), with previous recordings ($\phi_{(x)}$), for likeness; the specific logical conditions are shown in Table 1. Additionally, a check to see if any of the previous four recordings were higher than the most recent—which could not occur without noise—was done to see if signal noise was a considerable factor in the most recent measurements so as to co-opt this unavoidable artifact as an indication of proximity to thermal equilibrium.

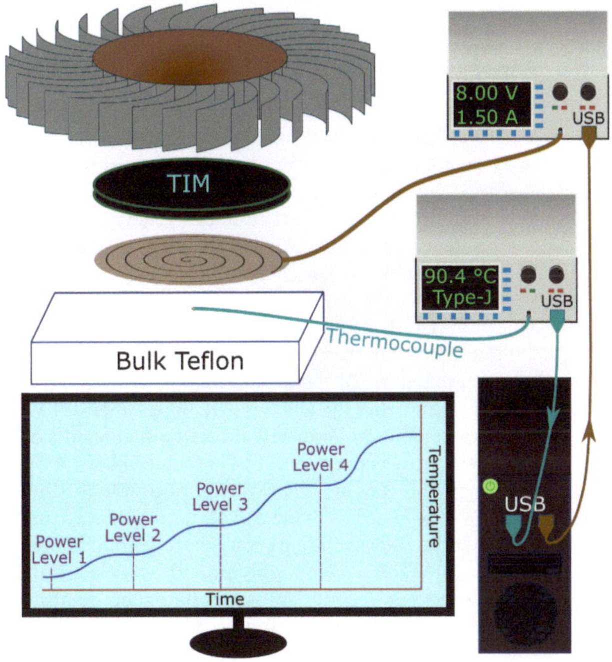

Figure 2. Schematic of the experimental procedure. The heating coil that is bonded to the heat sink by the TIM is placed on top of a thermocouple and teflon insulator. The computer continually reads thermocouple and changes the power supply's current to the heating coil when certain conditions are met.

Table 1. The set of conditions that define equilibrium temperature at a given heat dissipation level, where $\phi_{(x)}$ is the temperature recording x iterations back from the most recent measurement, $\phi_{(0)}$.

Measurement Position	Condition
1–4 previous	$\phi_{(4)} > \phi_{(0)}$, or $\phi_{(3)} > \phi_{(0)}$, or $\phi_{(2)} > \phi_{(0)}$, or $\phi_{(1)} > \phi_{(0)}$
5 previous	$0.996 \times \phi_{(0)} \leq \phi_{(5)} \leq 1.004 \times \phi_{(0)}$
10 previous	$0.99 \times \phi_{(0)} \leq \phi_{(10)} \leq 1.01 \times \phi_{(0)}$
20 previous	$0.99 \times \phi_{(0)} \leq \phi_{(20)} \leq 1.01 \times \phi_{(0)}$
30 previous	$0.98 \times \phi_{(0)} \leq \phi_{(30)} \leq 1.02 \times \phi_{(0)}$

The heat at each step was calculated by the product of the pre-measured heating coil resistance and the square of the applied voltage. In this fashion, the power supply's output current readings aren't depended on for the determination of power output and is valid for invariant electrical loads. The heating coil's resistance change is negligible because of Nickel-Chromium alloy wire's very low resistivity dependence on the considered temperature range [78,79]. The areal heat density equivalent is trivially calculated by the heat power divided by the internal heating coil surface area, approximated as a true circle, disregarding Kapton tape borders that extend beyond the confines of the internal planar coil diameter. All prepared test vehicles' coil circumferences were smaller than that of the heat sink mating surface.

Scanning Electron Microscopy (Carl Zeiss AG, Oberkochen, BW, DE, Gemini SEM) with an accelerating voltage of 2 kV, a working distance of 5 mm, and a secondary electron detector is conducted upon a composite's post-cure, fractured surface. Raman experiments

(Renishaw plc, Wotton-under-Edge, England, UK, InVia Raman) were conducted with an excitation wavelength of 633 in a backscattering configuration [80].

3. Results

TIMs with high and low FLG concentration regimes were prepared. However, the high load level composites—25.5 and 43.6 vol.%—did not have a reliable TIM bond line thickness as a result of difficulty in compressing the composite to the spacer's own thickness due to a vast apparent increase in composite viscosity. To apply enough pressure to clear excess high FLG concentration material inevitably removed the spacers from position. The 43.6 vol.% composite material was removed from the experimental vehicle to act as a representative SEM sample, albeit at a higher load level than the composites analyzed further. Figure 3a shows a SEM of this sample at high magnification in which either individual graphene sheets or stacks of FLG sheets thin enough to not be resolved by SEM are apparent in the middle of the micrograph, pointed to with an arrow. It is suggested from previous research that popular composite mixing processes could serve to further exfoliate FLG materials to smaller thicknesses [81]. Supplementary Figure S1 shows a SEM micrograph and Supplementary Figure S2 shows Raman results of a previous, identically prepared composite with a comparable FLG filling level of 5.4 vol.%.

Figure 3b contains the raw thermocouple readings over time of an experiment on the 7.3 vol.% FLG TIM. One can see in the data a continual, repeating process of rising temperature, then stabilizing temperature, followed by another period of quickly rising temperature. The power supply provides an electrical current and holds at that amount. The thermocouple continually reports its temperature and when the conditions set in Table 1 are met, the power supply increments to a higher current, repeating the process. One can particularly see from the inset plot in Figure 3b that the temperature does not fully reach equilibrium by the time the power level is increased once more. A polynomial fit of the readings prior to the next ramping cycle is extrapolated forward to give a qualitative idea how minor the under-estimation of equilibrium temperature and how much time is saved compared to getting to true equilibrium temperature. However, due to the predominately relative conditions the thermal equilibrium definition algorithm uses, this underestimation does increase at higher power levels. Put another way, $\phi_{(500)} - (0.99 \times \phi_{(500)}) < \phi_{(0)} - (0.99 \times \phi_{(0)})$, meaning that the condition will modestly loosen as the scale of temperature increases, as later recordings will. Supplementary Figure S3 shows differential scanning calorimetry results of pure DGEBA epoxy over typical electronic devices' operating temperatures.

All temperature measurement techniques have a sensor time constant (τ), which is the time that must elapse for the sensor to read 63.2% of a step temperature change [82,83]. Typical sensor time constants for similar fine thermocouples of the vendor is $\approx$5 s in flowing water. Given that the thermocouple is pressed into a highly mechanically compliant, and thus great surface contact, polyimide film, τ in this application should be somewhat greater. The thermocouple reads 99.3% of a step change temperature at 5τ, giving a minimum time to read true temperature of $\approx$25 s. As a result, the raw data shown in Figure 3b is actually the temperature of the heating coil over time convoluted by the response characteristic of the thermocouple. Given the time scales between successive current ramping events ($\approx$8 min average), it is clear that the majority of the observed behavior in Figure 3b is in fact a delayed temperature response of the heating coil itself and not purely the sensor heating up.

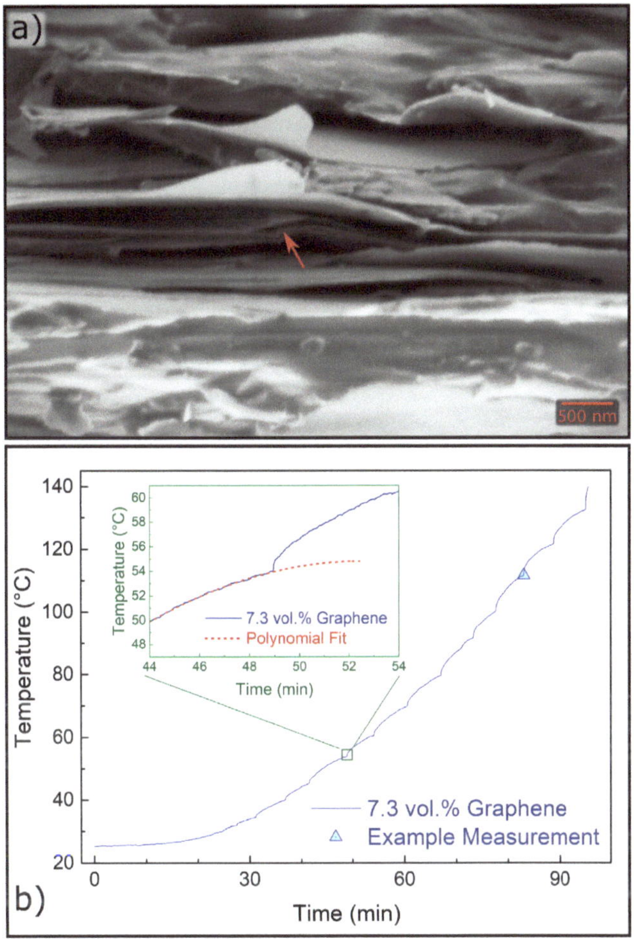

Figure 3. (**a**) Pseudo-colored SEM of a representative FLG composite with an arrow pointing to a definite FLG flake. (**b**) Thermocouple temperature readings over time in the course of an experiment.

Recording the equilibrium temperature for the experimental vehicles with TIMs of DGEBA with FLG filler levels of 0 vol.%, 3.6 vol.%, and 7.3 vol.% showed that for comparable heating coil power levels increasing concentrations of FLG resulted in lower operating temperatures for the heating coil, as seen in Figure 4. At a total heat dissipation rate of 20 W, the operating temperatures were ≈108 °C, ≈91 °C, and ≈85 °C for the 0 vol.%, 3.6 vol.%, and 7.3 vol.% samples, respectively. These operating temperatures correspond to a reduction in the increase of operating temperatures of 20% for the 3.6 vol.% TIM and 29% for the 7.3 vol.% TIM versus pure DGEBA. The modest non-linear behavior at the high power dissipation in the 7.3 vol.% data is attributed to the loosening of the conditions that define thermal equilibrium at higher temperatures, as discussed previously.

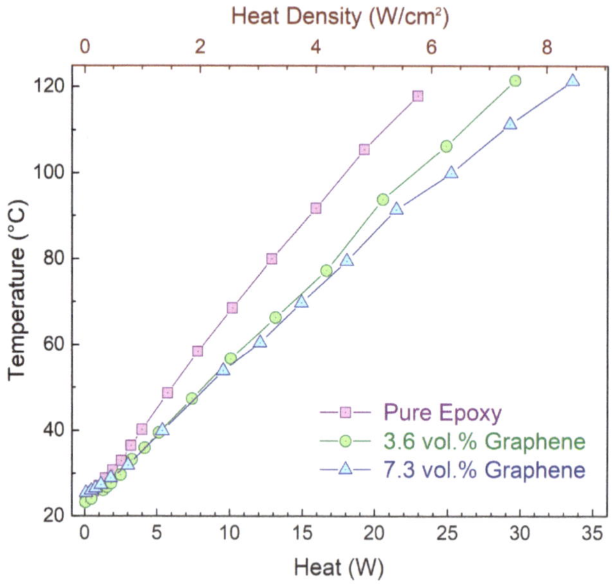

Figure 4. Measured thermal equilibria for each experimental vehicle with different TIM FLG concentrations at different heating coil power levels.

Most large logic chips will throttle down to a lower power state if their operating temperature reaches ≈100 °C to prevent damage and unintended behavior. The passively-cooled test vehicles were able to dissipate ≈4.5 W, ≈5.7 W, and ≈6.4 W for 0 vol.%, 3.6 vol.%, and 7.3 vol.% TIMs, respectively, at typical throttling temperature. Table 2 shows the full least squares fits of equilibrium temperature (Φ) with respect to heat dissipation (q) as well as coefficient of determination for each TIM sample with intercepts fixed at the ambient room temperature at the time of each experiment.

Table 2. Linear dependence of equilibrium temperature (Φ) on heat dissipation (q) with a fixed intercept at the ambient room temperature at the time of experiment for each TIM loading level

TIM Load Level	Least Squares Fit	r^2
0 vol.%	$\Phi(q) = 4.23q + 23.6$	0.9996
3.6 vol.%	$\Phi(q) = 3.37q + 22.4$	0.9998
7.3 vol.%	$\Phi(q) = 2.93q + 25.5$	0.9995

The slopes of the least squares fitted lines corresponds to the combined thermal resistances of the TIM composites and passively cooled heat sinks. Increasing the FLG concentration from 0 vol.% to 7.3 vol.% decreased the thermal resistance from 4.23 °C/W to 2.93 °C/W. Alternatively, switching from 400 µm of pure DGEBA to 7.3 vol.% of FLG in series with the heat sink decreased the thermal resistance by 1.3 °C/W.

4. Discussion

The vast majority of research into TIMs consider the composite material properties which then provide the ability to calculate the junction thermal resistance [10]. In contrast, this work directly experimentally determines device equilibrium temperatures for different TIM materials inside of identical heat sink conditions at different device heat dissipation rates. Previous works have analyzed equilibrium temperatures of CPUs with different applied TIMs at similar computational loads [69–72]. However, this method has a few largely unaddressed limitations. CPU systems are complex, dynamic, and are difficult to simplify without a proprietary knowledge of design. A particular processor's power consumption can vary even under identical computational loads due to gained efficiencies doing identical tasks in succession [84–87]. In addition, different architectures can compute tasks substantially differently leading to diminished repeatability among similar processors of different generations. Although a very minor factor, higher performance random access memory can lead to less processor idle time and more work done over time as a result. Moreover, heat generation takes place in predominately localized places that too also alter substantially over the course of architectural improvements, leaving which integrated CPU thermistor to trust as representative as overall processor temperature ambiguous [88]. This technique does indeed provide rough comparisons between different TIMs to determine relative performance, but no two seemingly identical CPU loads are in fact precisely so.

The results obtained are consistent with previous studies of both TIM material characterizations and comparative TIM CPU operating temperature studies in that increased FLG content, at least below a certain loading level at which point, improves TIM performance [31,34,37,39,54,69,70,72,89,90]. The present increase of TIM performance with additional FLG filler materials is a result of increasing the effective medium thermal conductivity at low filler levels. The FLG acts as discrete pathways for heat to flow through to a greater degree than it would through pure polymer. When the load level increases beyond a particular amount—termed the percolation threshold—the composite thermal conductivity increases precipitously as a result of developed direct pathways through the composite from filler to filler [91–95]. It should be noted that the percolation threshold was found in similar composites at higher load levels–above 20 vol.%–than in composites analyzed in the present study [37]. Recently, research has been conducted into thermal conductivity alterations over power cycles, very similar to the test conducted at present [96]. An increase in composite thermal conductivity—with accompanying reduction in thermal resistance—less than 5% could be expected in this composite as a result of the experiments.

These results provide the operating temperature of each test vehicle at a given, trustworthy heat dissipation level, made possible by the novel experimental technique. By removing the dependence on such a dynamic and complicated heating element as a CPU, greater control of output power is achieved. Real CPUs and similar graphics processing units vary in power output over an order of magnitude. Because of this, the ability to determine what an operating temperature should be anywhere in that heat dissipation range is useful for CPU designers and consumer electronics overall.

This study's experimental design was focused on maximal control of variables. Using a TIM of a bond line thickness of 400 μm is larger than those typically found in industry. This large bondline thickness makes the composite thermal conductivity of ever increasing importance and can minimize the contributions to thermal resistance of TIM-junction surface contact resistance. Recently, the trade-offs of TIM thickness and load level with subsequent composite thermal conductivity alterations was investigated in great detail using very similar composites finding lower thermal junction resistances in higher FLG concentration composites of the same thickness [97]. A thick bond line thickness was chosen in this study to render negligible on a relative basis any thickness measurement error. Additionally, CPU device heat dissipation solutions come with a fan to circulate air through the heat sink fins. To minimize complexity of reverse engineering multiple industrial CPU vendor fan angular velocities and conclude upon a representative fan speed versus operating temperature curve, a passive heat sink cooling set-up was used. Future research along a similar vein should focus on both smaller bond line thicknesses, $<$150 μm, adding an active cooling fan with a realistic speed versus operating temperature behavior, and varying conditions of the mating surfaces.

5. Conclusions

This study provides a more direct, concrete understanding of the benefits that FLG TIMs provide to electronics products by way of a more streamlined and controllable experimental method than those often used in this vein of research. The results obtained were consistent with previous studies using different experimental methods. Increasing the FLG concentration from 0 vol.% to 7.3 vol.% decreased the TIM and passively-cooled heat sink thermal resistance from 4.23 °C/W to 2.93 °C/W. At a heat dissipation rate of 20 W, this reduced thermal resistance results in a reduction of operating temperature from ≈108 °C down to ≈85 °C. These results provide important context between research of FLG composite properties and industrial application.

Supplementary Materials: The following are available at https://www.mdpi.com/article/10.3390/c7030053/s1, Figure S1: 5.4 vol.% SEM micrograph, Figure S2: 5.4 vol.% Raman spectra, Figure S3: Pure DGEBA Differential Scanning Calorimetry results.

Funding: This research was conducted under financial assistance from the Graduate Assistance in Areas of National Need (GAANN) Fellowship.

Institutional Review Board Statement: Not applicable.

Informed Consent Statement: Not applicable.

Data Availability Statement: Not applicable.

Acknowledgments: Facilities and equipment for this study were generously provided by Alexander A. Balandin. Timothy Perrier's assistance in setting up the experimental apparatus was appreciated greatly. The author is thankful towards Zahra Barani for useful discussions.

Conflicts of Interest: The authors declare no conflict of interest.

Abbreviations

The following abbreviations are used in this manuscript:

TIM	Thermal interface material
FLG	few layer graphene
DGEBA	Diglycidyl ether of Bisphenol A
SCPI	standard commands for programmable instruments
τ	Sensor time constant
$\phi_{(x)}$	The experimental temperature recording x iterations previous to the most recent
Φ	Least squares fit of equilibrium temperature as a function of applied heat
q	Applied heat

References

1. Lewis, J.S. Thermal and Electrical Performance Control and Lifespan Progression of Graphene-Based Polymer Composites. Ph.D. Thesis, University of California-Riverside, Riverside, CA, USA, 2020.
2. McCune, E. Fundamentals for Energy-Efficient Massive MIMO. In Proceedings of the 2017 IEEE Wireless Communications and Networking Conference Workshops (WCNCW), San Francisco, CA, USA, 19–22 March 2017; pp. 1–6. [CrossRef]
3. Guo, A.; Del Alamo, J.A. Unified Mechanism for Positive-A nd Negative-Bias Temperature Instability in GaN MOSFETs. *IEEE Trans. Electron Devices* **2017**, *64*, 2142–2147. [CrossRef]
4. Yu, Z.; Wang, R.; Hao, P.; Guo, S.; Ren, P.; Huang, R. Non-Universal Temperature Dependence of Hot Carrier Degradation (HCD) in FinFET: New Observations and Physical Understandings. In Proceedings of the 2018 IEEE Electron Devices Technology and Manufacturing Conference, EDTM 2018-Proceedings, Kobe, Japan, 13–16 March 2018; Institute of Electrical and Electronics Engineers Inc.: Manhattan, NY, USA, 2018; pp. 34–36. [CrossRef]
5. Bury, E.; Chasin, A.; Kaczer, B.; Chuang, K.H.; Franco, J.; Simicic, M.; Weckx, P.; Linten, D. Self-heating-aware CMOS reliability characterization using degradation maps. In Proceedings of the IEEE International Reliability Physics Symposium Proceedings, Burlingame, CA, USA, 1–15 March 2018; Institute of Electrical and Electronics Engineers Inc.: Manhattan, NY, USA, 2018; Volume 2018, pp. 2A.31–2A.36. [CrossRef]
6. Aslan, Y.; Kiper, C.E.; Johannes van den Biggelaar, A.; Johannsen, U.; Yarovoy, A. Passive Cooling of mm-Wave Active Integrated 5G Base Station Antennas Using CPU Heatsinks. In Proceedings of the 2019 16th European Radar Conference (EuRAD), Paris, France, 2–4 October 2019; pp. 121–124.
7. Smoyer, J.L.; Norris, P.M. Brief Historical Perspective in Thermal Management and the Shift Toward Management at the Nanoscale. *Heat Transf. Eng.* **2019**, *40*, 269–282. [CrossRef]
8. Saadah, M.; Hernandez, E.; Balandin, A. Thermal Management of Concentrated Multi-Junction Solar Cells with Graphene-Enhanced Thermal Interface Materials. *Appl. Sci.* **2017**, *7*, 589. [CrossRef]

9. Mahadevan, B.K.; Naghibi, S.; Kargar, F.; Balandin, A.A. Non-Curing Thermal Interface Materials with Graphene Fillers for Thermal Management of Concentrated Photovoltaic Solar Cells. *C J. Carbon Res.* **2019**, *6*, 2. [CrossRef]
10. Lewis, J.S.; Perrier, T.; Barani, Z.; Kargar, F.; Balandin, A.A. Thermal interface materials with graphene fillers: Review of the state of the art and outlook for future applications. *Nanotechnology* **2021**, *32*, 142003. [CrossRef] [PubMed]
11. Wong, C.P.; Bollampally, R.S. Thermal conductivity, elastic modulus, and coefficient of thermal expansion of polymer composites filled with ceramic particles for electronic packaging. *J. Appl. Polym. Sci.* **1999**, *74*, 3396–3403. [CrossRef]
12. Murshed, S.; Leong, K.; Yang, C. Enhanced thermal conductivity of TiO_2—Water based nanofluids. *Int. J. Therm. Sci.* **2005**, *44*, 367–373. [CrossRef]
13. Sim, L.; Ramanan, S.; Ismail, H.; Seetharamu, K.; Goh, T. Thermal characterization of Al_2O_3 and ZnO reinforced silicone rubber as thermal pads for heat dissipation purposes. *Thermochim. Acta* **2005**, *430*, 155–165. [CrossRef]
14. Cola, B.A.; Xu, X.; Fisher, T.S. Increased real contact in thermal interfaces: A carbon nanotube/foil material. *Appl. Phys. Lett.* **2007**, *90*, 093513. [CrossRef]
15. Zeng, J.; Cao, Z.; Yang, D.; Sun, L.; Zhang, L. Thermal conductivity enhancement of Ag nanowires on an organic phase change material. *J. Therm. Anal. Calorim.* **2009**, *101*, 385–389. [CrossRef]
16. Yu, W.; Xie, H.; Li, Y.; Chen, L. Experimental investigation on thermal conductivity and viscosity of aluminum nitride nanofluid. *Particuology* **2011**, *9*, 187–191. [CrossRef]
17. Gao, Z.; Zhao, L. Effect of nano-fillers on the thermal conductivity of epoxy composites with micro-Al_2O_3 particles. *Mater. Des.* **2015**, *66*, 176–182. [CrossRef]
18. Burger, N.; Laachachi, A.; Mortazavi, B.; Ferriol, M.; Lutz, M.; Toniazzo, V.; Ruch, D. Alignments and network of graphite fillers to improve thermal conductivity of epoxy-based composites. *Int. J. Heat Mass Transf.* **2015**, *89*, 505–513. [CrossRef]
19. Yu, W.; Zhao, J.; Wang, M.; Hu, Y.; Chen, L.; Xie, H. Thermal conductivity enhancement in thermal grease containing different CuO structures. *Nanoscale Res. Lett.* **2015**, *10*, 1–8. [CrossRef] [PubMed]
20. Du, H.; Qi, Y.; Yu, W.; Yin, J.; Xie, H. T-shape ZnO whisker: A more effective thermal conductive filler than spherical particles for the thermal grease. *Int. J. Heat Mass Transf.* **2017**, *112*, 1052–1056. [CrossRef]
21. Quinton, B.; Elston, L.; Scofield, J.; Mukhopadhyay, S. Aligned Carbon Nanotube Arrays Bonded to Solid Graphite Substrates: Thermal Analysis for Future Device Cooling Applications. *C J. Carbon Res.* **2018**, *4*, 28. [CrossRef]
22. NarayanPrabhu, K. The effect of load and addition of MWCNTs on silicone based TIMs on thermal contact heat transfer across Cu/Cu interface. *Mater. Res. Express* **2019**, *6*, 1165h9. [CrossRef]
23. Theerthagiri, J.; Salla, S.; Senthil, R.A.; Nithyadharseni, P.; Madankumar, A.; Arunachalam, P.; Maiyalagan, T.; Kim, H.S. A review on ZnO nanostructured materials: Energy, environmental and biological applications. *Nanotechnology* **2019**, *30*, 392001. [CrossRef] [PubMed]
24. Sharma, V.; Kagdada, H.L.; Jha, P.K.; Śpiewak, P.; Kurzydłowski, K.J. Thermal transport properties of boron nitride based materials: A review. *Renew. Sustain. Energy Rev.* **2020**, *120*, 109622. [CrossRef]
25. Balandin, A.A.; Ghosh, S.; Bao, W.; Calizo, I.; Teweldebrhan, D.; Miao, F.; Lau, C.N. Superior Thermal Conductivity of Single-Layer Graphene. *Nano Lett.* **2008**, *8*, 902–907. [CrossRef] [PubMed]
26. Balandin, A.A. Thermal properties of graphene and nanostructured carbon materials. *Nat. Mater.* **2011**, *10*, 569–581. [CrossRef]
27. Ghosh, S.; Calizo, I.; Teweldebrhan, D.; Pokatilov, E.P.; Nika, D.L.; Balandin, A.A.; Bao, W.; Miao, F.; Lau, C.N. Extremely high thermal conductivity of graphene: Prospects for thermal management applications in nanoelectronic circuits. *Appl. Phys. Lett.* **2008**, *92*, 151911. [CrossRef]
28. Seol, J.H.; Jo, I.; Moore, A.L.; Lindsay, L.; Aitken, Z.H.; Pettes, M.T.; Li, X.; Yao, Z.; Huang, R.; Broido, D.; et al. Two-Dimensional Phonon Transport in Supported Graphene. *Science* **2010**, *328*, 213–216. [CrossRef] [PubMed]
29. Cai, W.; Moore, A.L.; Zhu, Y.; Li, X.; Chen, S.; Shi, L.; Ruoff, R.S. Thermal transport in suspended and supported monolayer graphene grown by chemical vapor deposition. *Nano Lett.* **2010**, *10*, 1645–1651. [CrossRef] [PubMed]
30. Wang, H.; Kurata, K.; Fukunaga, T.; Ago, H.; Takamatsu, H.; Zhang, X.; Ikuta, T.; Takahashi, K.; Nishiyama, T.; Takata, Y. Simultaneous measurement of electrical and thermal conductivities of suspended monolayer graphene. *J. Appl. Phys.* **2016**, *119*, 244306. [CrossRef]
31. Zhang, P.; Zeng, J.; Zhai, S.; Xian, Y.; Yang, D.; Li, Q. Thermal Properties of Graphene Filled Polymer Composite Thermal Interface Materials. *Macromol. Mater. Eng.* **2017**, *302*, 1700068. [CrossRef]
32. Li, A.; Zhang, C.; Zhang, Y.F. Thermal Conductivity of Graphene-Polymer Composites: Mechanisms, Properties, and Applications. *Polymers* **2017**, *9*, 437. [CrossRef]
33. Balandin, A.A. Phononics of graphene and related materials. *ACS Nano* **2020**, *14*, 5170–5178. [CrossRef] [PubMed]
34. Shahil, K.M.; Balandin, A.A. Graphene-multilayer graphene nanocomposites as highly efficient thermal interface materials. *Nano Lett.* **2012**, *12*, 861–867. [CrossRef] [PubMed]
35. Fu, Y.X.; He, Z.X.; Mo, D.C.; Lu, S.S. Thermal conductivity enhancement of epoxy adhesive using graphene sheets as additives. *Int. J. Therm. Sci.* **2014**, *86*, 276–283. [CrossRef]
36. Shtein, M.; Nadiv, R.; Buzaglo, M.; Regev, O. Graphene-Based Hybrid Composites for Efficient Thermal Management of Electronic Devices. *Acs Appl. Mater. Interfaces* **2015**, *7*, 23725–23730. [CrossRef]

37. Kargar, F.; Barani, Z.; Salgado, R.; Debnath, B.; Lewis, J.S.; Aytan, E.; Lake, R.K.; Balandin, A.A. Thermal Percolation Threshold and Thermal Properties of Composites with High Loading of Graphene and Boron Nitride Fillers. *ACS Appl. Mater. Interfaces* **2018**, *10*, 37555–37565. [CrossRef] [PubMed]
38. Kargar, F.; Barani, Z.; Balinskiy, M.; Magana, A.S.; Lewis, J.S.; Balandin, A.A. Dual-Functional Graphene Composites for Electromagnetic Shielding and Thermal Management. *Adv. Electron. Mater.* **2019**, *5*, 1800558. [CrossRef]
39. Naghibi, S.; Kargar, F.; Wright, D.; Huang, C.Y.T.; Mohammadzadeh, A.; Barani, Z.; Salgado, R.; Balandin, A.A. Noncuring Graphene Thermal Interface Materials for Advanced Electronics. *Adv. Electron. Mater.* **2020**, 1901303. [CrossRef]
40. Biercuk, M.J.; Llaguno, M.C.; Radosavljevic, M.; Hyun, J.K.; Johnson, A.T.; Fischer, J.E. Carbon nanotube composites for thermal management. *Appl. Phys. Lett.* **2002**, *80*, 2767–2769. [CrossRef]
41. Martin, C.; Sandler, J.; Shaffer, M.; Schwarz, M.K.; Bauhofer, W.; Schulte, K.; Windle, A. Formation of percolating networks in multi-wall carbon-nanotube—Epoxy composites. *Compos. Sci. Technol.* **2004**, *64*, 2309–2316. [CrossRef]
42. Stankovich, S.; Dikin, D.A.; Dommett, G.H.; Kohlhaas, K.M.; Zimney, E.J.; Stach, E.A.; Piner, R.D.; Nguyen, S.B.T.; Ruoff, R.S. Graphene-based composite materials. *Nature* **2006**, *442*, 282–286. [CrossRef] [PubMed]
43. Pang, H.; Chen, T.; Zhang, G.; Zeng, B.; Li, Z.M. An electrically conducting polymer/graphene composite with a very low percolation threshold. *Mater. Lett.* **2010**, *64*, 2226–2229. [CrossRef]
44. Zhang, H.B.; Zheng, W.G.; Yan, Q.; Yang, Y.; Wang, J.W.; Lu, Z.H.; Ji, G.Y.; Yu, Z.Z. Electrically conductive polyethylene terephthalate/graphene nanocomposites prepared by melt compounding. *Polymer* **2010**, *51*, 1191–1196. [CrossRef]
45. Potts, J.R.; Dreyer, D.R.; Bielawski, C.W.; Ruoff, R.S. Graphene-based polymer nanocomposites. *Polymer* **2011**, *52*, 5–25. [CrossRef]
46. Wang, Q.; Gao, W.; Xie, Z. Highly thermally conductive room-temperature-vulcanized silicone rubber and silicone grease. *J. Appl. Polym. Sci.* **2003**, *89*, 2397–2399. [CrossRef]
47. Yu, A.; Ramesh, P.; Sun, X.; Bekyarova, E.; Itkis, M.E.; Haddon, R.C. Enhanced Thermal Conductivity in a Hybrid Graphite Nanoplatelet—Carbon Nanotube Filler for Epoxy Composites. *Adv. Mater.* **2008**, *20*, 4740–4744. [CrossRef]
48. Zhou, T.; Wang, X.; Liu, X.; Xiong, D. Improved thermal conductivity of epoxy composites using a hybrid multi-walled carbon nanotube/micro-SiC filler. *Carbon* **2010**, *48*, 1171–1176. [CrossRef]
49. Li, T.L.; Hsu, S.L.C. Enhanced Thermal Conductivity of Polyimide Films via a Hybrid of Micro- and Nano-Sized Boron Nitride. *J. Phys. Chem. B* **2010**, *114*, 6825–6829. [CrossRef]
50. Kemaloglu, S.; Ozkoc, G.; Aytac, A. Properties of thermally conductive micro and nano size boron nitride reinforced silicon rubber composites. *Thermochim. Acta* **2010**, *499*, 40–47. [CrossRef]
51. Yang, K.; Gu, M. Enhanced thermal conductivity of epoxy nanocomposites filled with hybrid filler system of triethylenetetramine-functionalized multi-walled carbon nanotube/silane-modified nano-sized silicon carbide. *Compos. Part A Appl. Sci. Manuf.* **2010**, *41*, 215–221. [CrossRef]
52. Pak, S.Y.; Kim, H.M.; Kim, S.Y.; Youn, J.R. Synergistic improvement of thermal conductivity of thermoplastic composites with mixed boron nitride and multi-walled carbon nanotube fillers. *Carbon* **2012**, *50*, 4830–4838. [CrossRef]
53. Teng, C.C.; Ma, C.C.M.; Chiou, K.C.; Lee, T.M. Synergetic effect of thermal conductive properties of epoxy composites containing functionalized multi-walled carbon nanotubes and aluminum nitride. *Compos. Part B Eng.* **2012**, *43*, 265–271. [CrossRef]
54. Lewis, J.S.; Barani, Z.; Magana, A.S.; Kargar, F.; Balandin, A.A. Thermal and electrical conductivity control in hybrid composites with graphene and boron nitride fillers. *Mater. Res. Express* **2019**, *6*, 085325. [CrossRef]
55. Yang, S.Y.; Lin, W.N.; Huang, Y.L.; Tien, H.W.; Wang, J.Y.; Ma, C.C.M.; Li, S.M.; Wang, Y.S. Synergetic effects of graphene platelets and carbon nanotubes on the mechanical and thermal properties of epoxy composites. *Carbon* **2011**, *49*, 793–803. [CrossRef]
56. Ma, A.-j.; Chen, W.; Hou, Y. Enhanced Thermal Conductivity of Epoxy Composites with MWCNTs/AlN Hybrid Filler. *Polym. Plast. Technol. Eng.* **2012**, *51*, 1578–1582. [CrossRef]
57. Yu, L.; Park, J.S.; Lim, Y.S.; Lee, C.S.; Shin, K.; Moon, H.J.; Yang, C.M.; Lee, Y.S.; Han, J.H. Carbon hybrid fillers composed of carbon nanotubes directly grown on graphene nanoplatelets for effective thermal conductivity in epoxy composites. *Nanotechnology* **2013**, *24*, 155604. [CrossRef]
58. Fang, X.; Fan, L.W.; Ding, Q.; Wang, X.; Yao, X.L.; Hou, J.F.; Yu, Z.T.; Cheng, G.H.; Hu, Y.C.; Cen, K.F. Increased thermal conductivity of eicosane-based composite phase change materials in the presence of graphene nanoplatelets. *Energy Fuels* **2013**, *27*, 4041–4047. [CrossRef]
59. Goli, P.; Legedza, S.; Dhar, A.; Salgado, R.; Renteria, J.; Balandin, A.A. Graphene-enhanced hybrid phase change materials for thermal management of Li-ion batteries. *J. Power Sources* **2014**, *248*, 37–43. [CrossRef]
60. Prolongo, S.G.; Moriche, R.; Jiménez-Suárez, A.; Sánchez, M.; Ureña, A. Epoxy Adhesives Modified with Graphene for Thermal Interface Materials. *J. Adhes.* **2014**, *90*, 835–847. [CrossRef]
61. Zhang, P.; Li, Q.; Xuan, Y. Thermal contact resistance of epoxy composites incorporated with nano-copper particles and the multi-walled carbon nanotubes. *Compos. Part A Appl. Sci. Manuf.* **2014**, *57*, 1–7. [CrossRef]
62. Olowojoba, G.B.; Eslava, S.; Gutierrez, E.S.; Kinloch, A.J.; Mattevi, C.; Rocha, V.G.; Taylor, A.C. In situ thermally reduced graphene oxide/epoxy composites: Thermal and mechanical properties. *Appl. Nanosci.* **2016**, *6*, 1015–1022. [CrossRef] [PubMed]
63. Lian, G.; Tuan, C.C.; Li, L.; Jiao, S.; Wang, Q.; Moon, K.S.; Cui, D.; Wong, C.P. Vertically Aligned and Interconnected Graphene Networks for High Thermal Conductivity of Epoxy Composites with Ultralow Loading. *Chem. Mater.* **2016**, *28*, 6096–6104. [CrossRef]

64. Moriche, R.; Prolongo, S.; Sánchez, M.; Jiménez-Suárez, A.; Chamizo, F.; Ureña, A. Thermal conductivity and lap shear strength of GNP/epoxy nanocomposites adhesives. *Int. J. Adhes. Adhes.* **2016**, *68*, 407–410. [CrossRef]
65. Huang, T.; Zeng, X.; Yao, Y.; Sun, R.; Meng, F.; Xu, J.; Wong, C. Boron nitride@graphene oxide hybrids for epoxy composites with enhanced thermal conductivity. *RSC Adv.* **2016**, *6*, 35847–35854. [CrossRef]
66. Yuan, C.; Xie, B.; Huang, M.; Wu, R.; Luo, X. Thermal conductivity enhancement of platelets aligned composites with volume fraction from 10% to 20%. *Int. J. Heat Mass Transf.* **2016**, *94*, 20–28. [CrossRef]
67. Zhao, Y.H.; Zhang, Y.F.; Wu, Z.K.; Bai, S.L. Synergic enhancement of thermal properties of polymer composites by graphene foam and carbon black. *Compos. Part B Eng.* **2016**, *84*, 52–58. [CrossRef]
68. Dmitriev, A.A.; Dmitriev, A.S.; Makarov, P.; Mikhailova, I. New nanocomposite surfaces and thermal interface materials based on mesoscopic microspheres, polymers and graphene flakes. *AIP Conf. Proc.* **2018**, *1957*, 020003. [CrossRef]
69. Renteria, J.; Legedza, S.; Salgado, R.; Balandin, M.; Ramirez, S.; Saadah, M.; Kargar, F.; Balandin, A. Magnetically-functionalized self-aligning graphene fillers for high-efficiency thermal management applications. *Mater. Des.* **2015**, *88*, 214–221. [CrossRef]
70. Suh, D.; Moon, C.M.; Kim, D.; Baik, S. Ultrahigh Thermal Conductivity of Interface Materials by Silver-Functionalized Carbon Nanotube Phonon Conduits. *Adv. Mater.* **2016**, *28*, 7220–7227. [CrossRef] [PubMed]
71. Dmitriev, A.S.; Valeev, A.R. Graphene nanocomposites as thermal interface materials for cooling energy devices. *J. Phys. Conf. Ser.* **2017**, *891*, 012359. [CrossRef]
72. Mai, P.T.; Bui, T.A.; Van Tran, H.; Van Pham, T.; Nguyen, D.N.; Phan, M.N.; Bui, T.H. Application of Graphene Silicone Grease in heat dissipation for the Intel Core i5 Processor. *JOIV Int. J. Informatics Vis.* **2019**, *3*, 222–226. [CrossRef]
73. Plazek, D.J.; Choy, I.C. The physical properties of bisphenol-a-based epoxy resins during and after curing. II. Creep behavior above and below the glass transition temperature. *J. Polym. Sci. Part B Polym. Phys.* **1989**, *27*, 307–324. [CrossRef]
74. Prasher, R.; Matayabas, J. Thermal contact resistance of cured gel polymeric thermal interface material. *IEEE Trans. Compon. Packag. Technol.* **2004**, *27*, 702–709. [CrossRef]
75. Due, J.; Robinson, A.J. Reliability of thermal interface materials: A review. *Appl. Therm. Eng.* **2013**, *50*, 455–463. [CrossRef]
76. Naghibi, S. Noncuring Graphene Thermal Interface Materials for Advanced Electronics. Ph.D. Thesis, University of California-Riverside, Riverside, CA, USA, 2020.
77. Nouri-Borujerdi, A.; Kazemi-Ranjbar, S. Thermal and electrical conductivity of a graphene-based hybrid filler epoxy composite. *J. Mater. Sci.* **2021**, *56*, 15151–1516 . [CrossRef]
78. Arnold, A. Nickel-chromium-aluminium-copper resistance wire. *Proc. IEE Part B Radio Electron. Eng.* **1956**, *103*, 439–447. [CrossRef]
79. Yao, Y.; Arajs, S.; Anderson, E. Electrical resistivity of nickel-rich nickel-chromium alloys between 4 and 300 K. *J. Low Temp. Phys.* **1975**, *21*, 369–376. [CrossRef]
80. Shahil, K.M.F.; Hossain, M.Z.; Goyal, V.; Balandin, A.A. Micro-Raman spectroscopy of mechanically exfoliated few-quintuple layers of Bi2Te3, Bi2Se3, and Sb2Te3 materials. *J. Appl. Phys.* **2012**, *111*, 054305. [CrossRef]
81. Paton, K.R.; Varrla, E.; Backes, C.; Smith, R.J.; Khan, U.; O'Neill, A.; Boland, C.; Lotya, M.; Istrate, O.M.; King, P.; et al. Scalable production of large quantities of defect-free few-layer graphene by shear exfoliation in liquids. *Nat. Mater.* **2014**, *13*, 624–630. [CrossRef] [PubMed]
82. Henning, C.; Parker, R. Transient response of an intrinsic thermocouple. *J. Heat Transf.* **1967**, *89*, 146–152. [CrossRef]
83. Farahmand, K.; Kaufman, J.W. Experimental measurement of fine thermocouple response time in air. *Exp. Heat Transf.* **2001**, *14*, 107–118. [CrossRef]
84. Yang, L.; Dick, R.P.; Memik, G.; Dinda, P. HAPPE: Human and Application-Driven Frequency Scaling for Processor Power Efficiency. *IEEE Trans. Mob. Comput.* **2013**, *12*, 1546–1557. [CrossRef]
85. Podzimek, A.; Bulej, L.; Chen, L.Y.; Binder, W.; Tuma, P. Analyzing the Impact of CPU Pinning and Partial CPU Loads on Performance and Energy Efficiency. In Proceedings of the 2015 15th IEEE/ACM International Symposium on Cluster, Cloud and Grid Computing, Shenzhen, China, 4–7 May 2015; pp. 1–10. [CrossRef]
86. Travers, M. *Cpu Power Consumption Experiments and Results Analysis of Intel i7-4820k*; Technical Report; Newcastle University: Newcastle upon Tyne, UK, 2015.
87. Rumi, M.A.; Hasibul Hasan, D.M. CPU power consumption reduction in android smartphone. In Proceedings of the 2015 3rd International Conference on Green Energy and Technology (ICGET), Dhaka, Bangladesh, 11 September 2015; pp. 1–6. [CrossRef]
88. Zajac, P.; Janicki, M.; Szermer, M.; Napieralski, A. Evaluating the impact of scaling on temperature in FinFET-technology multicore processors. *Microelectron. J.* **2014**, *45*, 1806–1813. [CrossRef]
89. Tang, B.; Hu, G.; Gao, H.; Hai, L. Application of graphene as filler to improve thermal transport property of epoxy resin for thermal interface materials. *Int. J. Heat Mass Transf.* **2015**, *85*, 420–429. [CrossRef]
90. Zhang, Y.; Zhang, M. Behavior of a graphene/epoxy composite used as thermal interface material for LED heat dissipation. *Mater. Test.* **2017**, *59*, 1037–1042. [CrossRef]
91. Shenogina, N.; Shenogin, S.; Xue, L.; Keblinski, P. On the lack of thermal percolation in carbon nanotube composites. *Appl. Phys. Lett.* **2005**, *87*, 133106. [CrossRef]
92. Bonnet, P.; Sireude, D.; Garnier, B.; Chauvet, O. Thermal properties and percolation in carbon nanotube-polymer composites. *Appl. Phys. Lett.* **2007**, *91*, 201910. [CrossRef]

93. Zheng, R.; Gao, J.; Wang, J.; Feng, S.P.; Ohtani, H.; Wang, J.; Chen, G. Thermal percolation in stable graphite suspensions. *Nano Lett.* **2012**, *12*, 188–192. [CrossRef] [PubMed]
94. Gu, J.; Xie, C.; Li, H.; Dang, J.; Geng, W.; Zhang, Q. Thermal percolation behavior of graphene nanoplatelets/polyphenylene sulfide thermal conductivity composites. *Polym. Compos.* **2014**, *35*, 1087–1092. [CrossRef]
95. Shtein, M.; Nadiv, R.; Buzaglo, M.; Kahil, K.; Regev, O. Thermally conductive graphene-polymer composites: Size, percolation, and synergy effects. *Chem. Mater.* **2015**, *27*, 2100–2106. [CrossRef]
96. Lewis, J.S.; Perrier, T.; Mohammadzadeh, A.; Kargar, F.; Balandin, A.A. Power Cycling and Reliability Testing of Epoxy-Based Graphene Thermal Interface Materials. *C J. Carbon Res.* **2020**, *6*, 26. [CrossRef]
97. Sudhindra, S.; Kargar, F.; Balandin, A.A. Noncured Graphene Thermal Interface Materials for High-Power Electronics: Minimizing the Thermal Contact Resistance. *Nanomaterials* **2021**, *11*, 1699. [CrossRef] [PubMed]

C Journal of Carbon Research

MDPI

Article

Fused Filament Fabrication Three-Dimensional Printing Multi-Functional of Polylactic Acid/Carbon Black Nanocomposites

Nectarios Vidakis [1], Markos Petousis [1,*], Emmanuel Velidakis [1], Nikolaos Mountakis [1], Peder Erik Fischer-Griffiths [2], Sotirios Grammatikos [2] and Lazaros Tzounis [3]

[1] Mechanical Engineering Department, Hellenic Mediterranean University, Estavromenos, 71004 Heraklion, Greece; vidakis@hmu.gr (N.V.); mvelidakis@hmu.gr (E.V.); mh90@edu.hmu.gr (N.M.)

[2] Group of Sustainable Composites, Department of Manufacturing and Civil Engineering, Norwegian University of Science and Technology, 2815 Gjøvik, Norway; pederef@stud.ntnu.no (P.E.F.-G.); sotirios.grammatikos@ntnu.no (S.G.)

[3] Department of Materials Science & Engineering, University of Ioannina, 45110 Ioannina, Greece; latzounis@uoi.gr

* Correspondence: markospetousis@hmu.gr; Tel.: +30-281-037-9227

Citation: Vidakis, N.; Petousis, M.; Velidakis, E.; Mountakis, N.; Fischer-Griffiths, P.E.; Grammatikos, S.; Tzounis, L. Fused Filament Fabrication Three-Dimensional Printing Multi-Functional of Polylactic Acid/Carbon Black Nanocomposites. C **2021**, 7, 52. https://doi.org/10.3390/c7030052

Academic Editor: Gil Goncalves

Received: 23 June 2021
Accepted: 15 July 2021
Published: 17 July 2021

Abstract: Conductive Polymer Composites (CPCs) have recently gained an extensive scientific interest as feedstock materials in Fused Filament Fabrication (FFF) Three-dimensional (3D) printing. Polylactic Acid (PLA), widely used in FFF 3D printing, as well as its Carbon Black (CB) nanocomposites at different weight percentage (wt.%) filler loadings (0.5, 1.0, 2.5 and 5.0 wt.%), were prepared via a melt mixing filament extrusion process in this study and utilized to manufacture FFF 3D printed specimens. The nanocomposites were examined for their electrical conductivity. The highest loaded 3D printed CPC (5.0 wt.%) was tested as an electrothermal Joule heating device. Static tensile, flexural, Charpy's impact and Vickers microhardness mechanical properties were investigated for the neat and PLA/CB 3D printed nanocomposites. Dynamic Mechanical Analysis (DMA) revealed a stiffening mechanism for the PLA/CB nanocomposites. Scanning Electron Microscopy (SEM) elucidated the samples' internal and external microstructural characteristics. The PLA/CB 5.0 wt.% nanocomposite demonstrated also antibacterial properties, when examined with a screening process, against *Escherichia coli* (*E. coli*) and *Staphylococcus aureus* (*S. aureus*). It can be envisaged that the 3D printed PLA/CB CPCs exhibited a multi-functional performance, and could open new avenues towards low-cost personalized biomedical objects with complex geometry, amongst others, i.e., surgery tools, splints, wearables, etc.

Keywords: fused filament fabrication (FFF); three-dimensional (3D) printing; conductive polymer composites (CPCs); nanocomposites; electrothermal effect; antibacterial properties

1. Introduction

Additive Manufacturing (AM) is in the forefront of research, receiving a continuous industrial interest for potential applications in the fields of engineered polymer composites [1], printed electronics [2], personalized health and biomedical equipment [3], civil engineering infrastructure i.e., buildings, bridges, etc. [4], amongst others. AM enables the manufacturing of components and end-use products with customizable geometries. It is an efficient technology, especially when small batches are required, which are implausible for conventional manufacturing processes that present several limitations and freedom to fabricate parts with complex shapes [5]. Three-dimensional (3D) printing, which is one of the most famous AM technologies, has recently attracted interest on both academia, as well as various industrial sectors for objects' manufacturing and/or prototyping purposes [6]. To that end, it is worth mentioning that 3D printing is a cutting-edge technology allowing the production of 3D objects with more complex geometry compared to conventionally machined parts [5]. Among the increasing number of 3D printing processes, the most well-known ones are: (i) Fused Filament Fabrication (FFF), (ii) Stereo-lithography, (iii) Selective Laser Sintering (SLS), (iv) Binder Jetting, (v) Material Jetting and Drop on Demand (DOD) and (vi) Digital Light processing (DLP) [7].

FFF is the most famous, widely employed and extremely emerging 3D printing technique for applications ranging from scientific development, home 3D printing and Rapid Prototyping, up to manufacturing parts at industrial scale (especially for high performance advanced 3D components). Moreover, FFF has recently penetrated various industrial sectors and became a mature technology, replacing conventional ones, i.e., Plastic Forming (PF), Computer Numerical Control (CNC) machining, (plastic) joining and injection molding [6]. In FFF, a thermoplastic polymeric material is used as the feedstock in filament form. It is being heated above its Melting Point (Tm) and extruded with a movable nozzle in the X–Y direction to create a 3D structure via a layer-by-layer addition manner [8]. FFF could offer thus numerous advantages and a direct solution to: (i) the acute shortage of goods, as for instance in the COVID-19 pandemic situation [9], (ii) freedom of design of 3D objects, (iii) mass customization and (iv) reduced material's consumption leading to waste minimization [10].

Great achievements have been made so far regarding the FFF 3D printing technology, as for instance: (i) the 3D printer's hardware parts i.e., the heated bed, the preheated chamber of the filament, etc., (ii) the FFF 3D printers' automation and accuracy and (iii) the extruder head (metallic materials' properties towards optimum heat flow for homogeneous heat transfer to the polymeric filament and melt properties), allowing high quality AM filamentous 3D printing, that could significantly affect the 3D printed final object "bulk" properties. Nevertheless, FFF is limited by the availability of functional thermoplastic materials, with most of the existing stock filament materials properties being investigated and reported in literature so far [11–13] and a lot of research presented for the development of new polymeric materials for AM in general [14].

Functional thermoplastic filaments can endow additional functionalities to the final 3D printed object, as for instance enhanced physicochemical properties e.g., electrical, thermal, magnetic, mechanical, melt-rheological properties (allowing better processing), etc., giving an added value to the 3D printed final end-product [15]. To that end, is of great interest to investigate the thermoplastic polymer engineering part, by tuning the filament's melt rheological properties via additives, such as viscosity modifiers, polymer blends, nanoparticulate additives, etc., resulting into higher quality 3D printed parts (extensive interlayer fusion), high production rate and production yield of the FFF printing process [16].

To date, there is an increasing demand for FFF 3D printable materials exhibiting multi-functional properties, i.e., materials with enhanced physical and mechanical properties [5]. Namely, functional thermoplastic materials as filaments i.e., electrically conductive and thermoelectric [17], magnetically active [18], capacitive and piezoresistive [19], flexible and stretchable [20], mechanically reinforced [21], antimicrobial [22], etc., have been used for FFF 3D printing, yielding multifunctional 3D objects with tunable properties. Therefore, polymer nanocomposites could offer a unique strategy towards novel filaments and by extension FFF 3D printed derived objects. Specifically, for the polymer nanocomposites, one should consider the nanoparticle filler geometry, density and surface chemistry. Additionally, the nanoscale interactions should be assessed, e.g., the well-known filler-filler interaction and the filler-matrix interactions that have profound implications on the macroscopic behavior of the 3D printed nanocomposite objects as well as to the 3D printing process [23].

Polylactic acid (PLA) is a high performance biocompatible and biodegradable thermoplastic in nature engineering polymeric material, already being widely used and several times reported in FFF 3D printing [24]. In general, polymer/carbon nanoparticulate nanocomposites have attracted a significant attention over the last decade, as lightweight and flexible electrically conductive materials [25,26]. The resulting Conductive Polymer

Composites (CPCs) could find numerous applications, due to the material's conductive behavior, which can be tuned by the nanoparticle filler loading, the particle aspect ratio, surface area and surface chemistry, as well as the processing/mixing parameters. Carbon black (CB) is an allotrope of carbon, while it has been widely utilized as a nanoadditive in various polymer matrices to increase specific properties, such as fracture toughness [27], tensile strength [28], electrical [29] and thermal conductivity [30]. Polymer/CB CPCs have demonstrated sensing functionalities i.e., (i) gas or liquid chemical and/or electrochemical sensors, (ii) strain sensors for wearables or touch/pressure sensors via a piezoresistive or a piezoelectric under-lying mechanism and (iii) temperature sensors [31]. PLA/CB conductive nanocomposite thermoplastic materials could be ideal feedstock materials for AM FFF 3D printing processes, and therefore systematic investigations are required to highlight the fundamental process-structure-property relationship.

CPC multifunctional filaments with electrical properties have been fabricated and utilized for FFF 3D printing towards biomedical sensors [32], piezoresistive strain sensors [33], actuators [34] and resistors for Joule-heating [35]. Specifically, Joule resistive heating or Ohmic heating is the effect when an applied electric current pass through an electrical conductor, resulting in heat generation. In principle, when voltage difference is applied between two points of a conductive network, an electric field is being generated, which accelerates charge carriers in the field direction, resulting in higher kinetic energy. Then, electrons collide within the conductor's ions and the particles are being scattered. Under this atomic scale electronic charge carrier transport mechanism, electrical energy is converted to thermal energy [36]. From Joule's first law, it arises that the consumed power of heating, which is generated by an electrical conductor, is proportional to the electrical resistance and the square of the current: $P = I^2 \cdot R$, where P: Power (W), I: Current (A) and R: Resistance (Ohm). According to the Ohmic conductor model, the transduction of electrical power into Joule-heating is given by the relation $P = V^2/R$, where V (V) denotes the applied input voltage, or otherwise known as voltage bias (Vbias). Nowadays, there are several applications of resistive heating with numerous solutions already existing in the market [37], while 3D printed CPCs could be ideal lightweight candidates for advanced applications.

Aumnate et al. reported on reinforcing Polypropylene (PP) with graphene-polylactic acid microcapsules for FFF 3D printed objects. At low graphene loading (0.75 wt.%), the 3D printed construct showed neither shrinkage nor warping. Graphene nanocomposite 3D printed constructs with 70% infill exhibited higher mechanical performance, when compared to neat PP [38]. In another study related to isotactic polypropylene/graphene 3D printed nanocomposites, the authors reported that the presence of graphene induced shear thinning during extrusion (improved processability aspect), while at 5% and 10% of graphene loading, the storage modulus decreased considerably; namely nearly 50% for the 10% loading [39]. Recently, Lei et al. reported on FFF 3D printed carbon black (CB)/polypropylene composites with excellent microwave absorption performance [40], while Kwok et al. reported on PP/CB electrically conductive nanocomposite filaments for FFF 3D-printed circuits and sensors [41], both focusing only onto the electrical properties of the 3D printed composites. CPC filaments for FFF 3D printing and electrically conductive 3D objects have been reported so far, especially focusing on the electrical and thermoelectric properties of 3D printed nanocomposites, as reported elsewhere [42]. In the 3D printing technology, Polyethylene/graphene nanocomposites have been investigated for heat diffusion applications and advanced electronic devices [43], Low Density Polyethylene (LDPE) composite filled with copper particles and Sn95Ag4Cu1 has been studied for compressive sensing applications [44], while Acrylonitrile Butadiene Styrene/graphite has been examined for electrochemical sensor development [45]. PLA/CB studied in this work has been employed in electrochemical sensors [46], electronic components and circuits [47] and electrodes [48], with none of these studies focusing on the characterization of the material, which was commercial in some cases [47]. Additionally, the material developed did not have multifunctional performance, with research focusing to cope with the requirements of specific applications in 3D printing.

Although there is a lot of interest in literature for the development of conductive materials in AM in general, the effect of specific fillers on properties (mechanical, thermal, electrothermal, antibacterial, etc.) of various matrices is not yet reported. Additionally, the methodology for the preparation of conductive nanocomposites suitable for 3D printing has many technological aspects that need to be investigated, in order to be optimized. Finally, in literature the electrical conductivity is studied in different composites, but the development of nanocomposites suitable for 3D printing with multifunctional performance is still not available in literature. To the best of the author's knowledge, a detailed study of PLA/CB conductive nanocomposites focusing on the electrical percolation threshold as a function of the filler loading, as well as a detailed static mechanical and thermomechanical analyses of the 3D printed samples has not been reported yet.

In this work PLA/CB nanocomposites at 0.5, 1.0, 2.5 and 5.0 wt.% filler loadings were produced by a thermomechanical melt mixing process, to produce 1.75 mm, suitable for 3D printing. Neat PLA and PLA/CB nanocomposites in filament form were then utilized as feedstock material for the specimens FFF 3D printing manufacturing step. FFF process is based on the extrusion principle. All 3D printed specimens were then fully investigated for their static and thermomechanical properties, according to international standards. The highest reinforcement regarding the tensile properties was found for the PLA/CB of 0.5 wt.%, while in the flexural, impact and microhardness tests, the highest values were observed for the PLA/CB 2.5 wt.% nanocomposites.

The PLA/CB nanocomposites were then fully characterized in terms of their electrical conductivity. The PLA/CB 5.0 wt.% also performed as a Joule heater device, as well as a conductive circuitry resistor, evaluating the multi-functional nature of the 3D printed nanocomposites prepared and investigated in this work. The PLA/CB nanocomposites exhibited an electrically conductive behavior from the 2.5 wt.% loading. The electrical conductivity increased with the increase of the filler loading up to the highest loading of 5.0 wt.% prepared in this work. The highest electrical conductivity (σ) was obtained at the "cross-layer" direction for the PLA/CB 5.0 wt.% nanocomposite (~0.9 S/m), which also exhibited an electrothermal "Joule-heating" function upon being exposed to a voltage bias (Vbias) of 15 V. Scanning Electron Microscopy (SEM) fractographic analyses revealed the underlying fracture mechanisms of the tensile tested specimens, while Raman spectroscopy revealed the graphitic nature of CB and the PLA matrix characteristic spectroscopic fingerprints. Thermogravimetric Analysis (TGA) showed the effect of filler loading on the thermal properties of the fabricated nanocomposites. Enhanced thermal stability was observed for all the PLA/CB nanocomposites prepared in this work. Finally, the antimicrobial properties of the 3D printed PLA/CB nanocomposites of this work were examined with a screening process against *E. coli* and *S. aureus* bacteria. A mild antibacterial performance was found only at the nanocomposite with the highest filler loading. Overall, the FFF 3D printed PLA/CB CPCs manufactured herein verified their multifunctional nature and could be employed in various applications as flexible conductors, resistors for circuitry, as well as piezoresistive sensors for strain sensing, etc., while their antibacterial performance is an additional asset in biomedical, hygienic and other health-related applications.

2. Materials and Methods

2.1. Materials

Polylactic Acid (PLA), the matrix material in this study, was of 3052D grade, with molecular weight of 116,000 g/mol. It was procured from Plastika Kritis SA (Heraklion, Greece) in the form of coarse powder. A highly conductive Carbon Black (CB) grade (C-NERGY Super P Conductive Carbon Black) was received from Nanografi Nanotechnology AS (Tallinn, Estonia) as a mechanical reinforcing and electrically conductive agent.

2.2. Fabrication of Filaments and FFF 3D Printing Process of PLA and PLA/CB Nanocomposites

PLA and CB filler mixtures were prepared under five (5) different filler ratios. Mixing percentages were of 0.1%, 0.5%, 1.0%, 2.5% and 5.0% in a weight to weight (*w*/*w*) percentage ratio (wt.%). Pure PLA was also produced as a reference material for all nanocomposite materials. Adequate quantity of each material was weighted and mixed using mechanical mixing equipment. To ensure the lack of humidity in all material mixtures/formulations, drying procedures were implemented throughout the process steps. Namely, PLA coarse powder was dried for 24 h at 50 °C using a laboratory oven. Further drying was applied after the mixing process for all materials, in this case for 5 h at 50 °C.

Extruded filaments with 1.75 mm in diameter suitable for FFF 3D printing were produced, with a melt mixing thermomechanical process, using a 3D Evo (3D Evo B.V., Utrecht, the Netherlands) single screw extruder. Specifically, 3D Evo Composer 450 was employed, which features an optimized and appropriate screw for mixing/compounding purposes during the consecutive extrusion procedure. Its chamber has four (4) heating zones, and the device also features a filament diameter real-time measurement system. In this way, produced filament quality control is implemented and monitored in real-time with the extrusion/production process, determining the filament diameter accuracy and adjusting in real-time the extrusion parameters automatically, to maintain the required accuracy tolerances. The extruded filament diameter was measured to be in all cases 1.68 mm $\pm$ 0.07 mm, which is an adequate accuracy for consistent and high-quality 3D printing. All the produced materials, i.e., pure PLA and the five different PLA/CB nanocomposites were processed under the same extrusion parameters. Temperatures were set to each heating zone; 175 °C at heat zone 4 (closer to hopper), 205 °C at heat zone 3 and 2 (middle stage) and 195 °C at heat zone 1 (closer to extruder's nozzle). Screw rotational speed was set to 7.4 rpm (extruder's range is 2.5–15 rpm) and the built-in winder rotational speeds were automatically adjusted to achieve the requested diameter. Additional quality control tests were also manually conducted with random diameter measurements to the filament diameter, using a high-quality electronic caliper and optical quality control for defects. The produced filaments were then dried at the same conditions, before being used to produce the various 3D printed specimens. The filament diameter was inserted in the 3D printer slicer software, to calculate the necessary feed rate during the entire 3D printing process.

Fused Filament Fabrication (FFF) was selected to manufacture the neat PLA as well as the PLA/CB nanocomposite specimens. An Intamsys (Shanghai, China) 3D Printer was chosen, using an extruder's nozzle of 0.4 mm. Specifically, the Intamsys HT 3D printer was used for specimens' fabrication through AM. Initial trails were performed to obtain the optimum FFF 3D printing parameters (set of parameters) for high quality 3D printing for the pure PLA and the PLA nanocomposites. For the mechanical characterization, the electrical conductivity and the antibacterial test measurements, all 3D geometric models were designed employing the 3D design Autodesk® Fusion 360™ (Autodesk®, Inc., San Rafael, CA, USA) software platform. 3D geometry was finally exported to Standard Tessellation Language (STL) files from the Computer Aided Design (CAD) software platform.

Figure 1 shows the flow chart illustrating the methodology followed in this study to produce and characterize the PLA and the PLA/CB nanocomposites. Figure 2 summarizes the complete list of the 3D printing parameters used to fabricate the specimens in the 3D printer, after an initial screening optimization approach. All other parameters, not mentioned in Figure 2 were set to their default values, in the Intamsuite software tool (after choosing PLA as the reference material in the 3D printer software tool), which was used as the slicer software tool for this study.

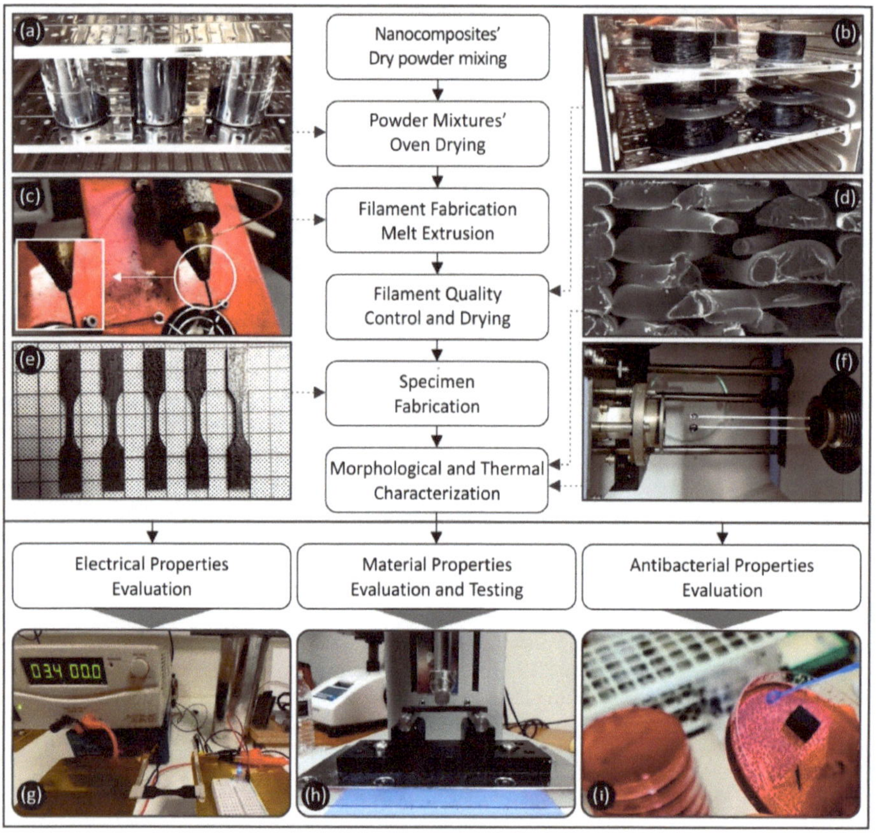

Figure 1. Flow chart illustrating the different steps to manufacture the filaments, as well as the 3D printed specimens and the characterization of the nanocomposites fabricated in this work: (**a**) powders' drying process, (**b**) filaments' drying process, (**c**) filaments' extrusion process, (**d**) tensile test specimens' morphological characterization on SEM, (**e**) tensile test specimens, (**f**) thermogravimetric analysis on the nanocomposites, (**g**) electrical conductivity measurements on the specimens, (**h**) flexural test on specimen, (**i**) antibacterial screening process on the nanocomposites.

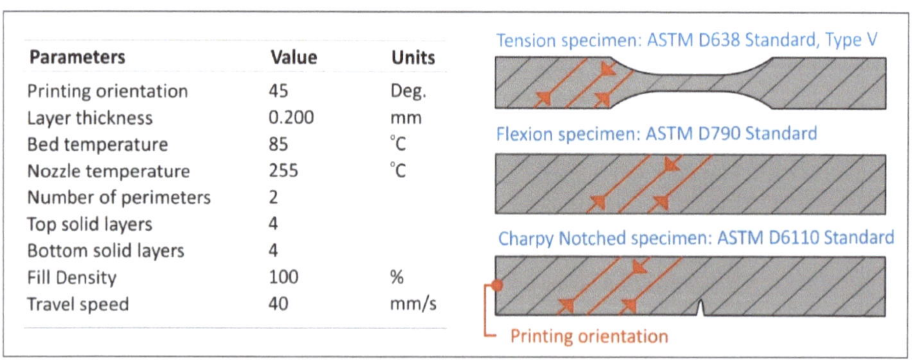

Parameters	Value	Units
Printing orientation	45	Deg.
Layer thickness	0.200	mm
Bed temperature	85	°C
Nozzle temperature	255	°C
Number of perimeters	2	
Top solid layers	4	
Bottom solid layers	4	
Fill Density	100	%
Travel speed	40	mm/s

Figure 2. The optimum FFF 3D printing parameters followed in this study and set up to slicer software to manufacture the neat PLA, as well as the PLA/CB different nanocomposite specimens.

2.3. Characterization Techniques

Raman spectroscopy was performed for the pure PLA and CB, as well as the PLA/CB nanocomposite 3D printed specimens, using a Labram HR-Horiba (Kyoto, Japan) scientific micro-Raman system. An optical microscope equipped with a 50× long working distance objective was utilized for delivering the excitation light and collecting the back-scattering Raman activity. An Ar^+ ion laser line (514.5 nm) at 1.5 mW power at the focal plane was utilized for the Raman excitation. All acquired spectra shown in this work have been treated with a baseline correction through subtraction of a linear or polynomial fit of the baseline from the raw spectra, to remove tilted baseline variation caused by various noises, i.e., fluorescent background, etc.

Thermogravimetric Analysis (TGA) was carried out for the 3D printed samples, namely the neat PLA, as well as the PLA/CB nanocomposites in oxygen atmosphere. The measurements were performed with a Perkin Elmer Diamond TG/TDA (Waltham, MA, USA) with a heating cycle from 30 °C to 550 °C with a heating step of 10 °C/min.

SEM characterization was used to analyze the nano/micro structuring of both the 3D printed specimens' side surface, as well as the fractured surfaces of 3D printed tensile specimens. SEM microstructural analyses were performed using a JEOL JSM 6362LV (Jeol Ltd., Norwood, MA, USA) electron microscope in high-vacuum mode at 20 kV acceleration voltage. Prior to SEM investigations, samples were sputter-coated with gold (Au) to avoid charging effects.

The through-thickness (through-layer) as well as in-plane (cross-layer) electrical conductivity of 3D printed PLA/CB nanocomposite samples was determined using square-shaped samples of 10.0 by 10.0 mm^2 and 3.2 mm thickness. Ag paste (room temperature fast curing curing/drying silver paste received from Agar scientific, Stansted, Essex, United Kingdom) was applied in the appropriate surfaces to contact the samples for the respective through-layer and cross-layer conductivity measurements. Afterwards, 2-point probe electrical resistance measurements were performed with an Agilent Multimeter (Agilent 34401A6$\frac{1}{2}$, Agilent, Santa Clara, CA, USA) to derive the electrical resistivity (ρ) and conductivity (σ), respectively, considering the specific sample dimensions.

Thermal images demonstrating the nanocomposite samples capability to function as electrothermal Joule heating materials were captured with a Seek Thermal Compact PRO Compact XR Imaging Camera IR Infrared Imager (Santa Barbara, CA, USA) with a temperature working window of −40 to 330 °C.

Specimens were tested for the determination of their mechanical properties in room temperature (23 °C). Specifically, quasi-static tensile test experiments were carried out at room temperature (23 °C) following the ASTM D638-02a international standard. According to the standard, a type V specimen of 3.2 mm thickness was chosen, and a total of six (6) specimens were manufactured and tested for each case. An Imada MX2 (Imada Inc., Northbrook, IL, USA) tension/flexure test apparatus in tensile mode using standardized grips was utilized to carry out the tensile test experiments at an elongation rate of 10 mm/min.

Flexural (three-point bending) tests were carried out also at room temperature conditions on 3D printed specimens (64.0 mm length, 12.4 mm width and 3.2 mm thickness), according to the ASTM D790-10 international standard (three-point bending test with 52.0 mm support span). Flexural test specimens were manufactured with the same 3D printing parameters as for the tensile test specimens. An Imada MX2 machine in flexural mode setup was employed for the three-point bending tests. The same speed of 10 mm/min was set for the testing procedure, and a total of six (6) specimens were manufactured and tested for the neat PLA, as well as PLA/CB nanocomposites.

Impact tests were performed according to the ASTM D6110-04 international standard. Specimens were 3D printed with the following dimensions: 80.0 mm (length) by 8.0 mm (width) by 10.0 mm (thickness). Six (6) notched specimens in total were tested using a Terco MT 220 (Terco, Kungens Kurva, Sweden) Charpy's impact apparatus. Release height of the apparatus hammer was the same for all the experiments (367 mm).

Microhardness measurements were conducted according to the ASTM E384-17 international standard. The specimens' surface was fully polished before each set of measurements. An Innova Test 300-Vickers (Innovatest Europe BV, Maastricht, The Netherlands) apparatus was employed, while the applied force was set to 100 gF and duration of 10s was selected for indentation. Imprints were measured under six (6) different specimens for each one of the PLA and PLA/CB nanocomposites.

Dynamic Mechanical Analysis (DMA) was performed using a TA Instruments DMA850 instrument (New Castle, DE, USA). Samples were 3D printed in dimensions of length 58.0–60.0 mm, width 14.0–15.0 mm and thickness 2.7–3.2 mm. Due to the samples having rough side edges from the FFF manufacturing process, all samples were polished in two steps using 240 and 400 grain sandpaper under water flow. Prior to testing, samples were dried at a temperature of 35 °C for a minimum of 48 h. The DMA testing procedure consisted of a temperature ramp from room temperature to 130 °C (and in some cases up to 135 °C), at a rate of 3 °C min^{-1}. Testing was conducted using the three-point bending fixture. Samples were preloaded to 0.1 N. A sinusoidal displacement was applied to the samples with a constant amplitude of 30.0 μm and a frequency of 1.0 Hz throughout the tests. Data was collected by the instrument at a sampling rate of 0.33 Hz. The recorded parameters were the storage modulus, loss modulus, tan δ, temperature, time and oscillation angular frequency.

The antibacterial performance of the nanocomposites of this work, was investigated with the agar well diffusion method [49] implemented in a microbiological lab, for two different bacteria, i.e., *E. coli* and *S. aureus*. Petri dishes (850 mm in diameter) with suitable bacterium growth material (MC.2, C.010066 for the *E. coli* and Chapman, C.010068 for the *S. aureus*), were employed for the tests. Two rectangular specimens (one for each bacterium) of 12.4 mm by 12.4 mm and 3.2 mm height were 3D printed with the same parameters described above, from pure PLA and each one of the different nanocomposite materials prepared in this work. The bacteria were inoculated in the corresponding Petri dishes and the specimens were placed near the center of the Petri dishes, with one specimen placed in each Petri dish. The Petri dishes were placed in an oven at 37 °C for 24 h for the diffusion of the antimicrobial agent into the nanocomposite and inhibit germination and growth of the test microorganism. Then, the width of the inhibition growth zone was measured.

3. Results and Discussion

3.1. Raman Analysis of PLA, CB and PLA/CB Nanocomposites

Figure 3 shows the Raman spectrum of PLA, CB and PLA/CB nanocomposite (at 2.5 wt.% of filler loading) at the spectral region 300–3300 cm^{-1}. It is worth mentioning that the Raman spectra of PLA/CB nanocomposites at higher filler loadings were governed by the carbon nanoadditive "characteristic" spectroscopic responses, which are highly resonant to the 532 nm laser excitation source, without any visible peaks arising due to the PLA polymeric matrix. It can be very clearly observed in the CB and in the PLA/CB spectra the characteristic D-band and G-band peaks of CB graphitic carbon allotrope materials, centered at ca. 1345 cm^{-1} and 1565 cm^{-1}, respectively [50].

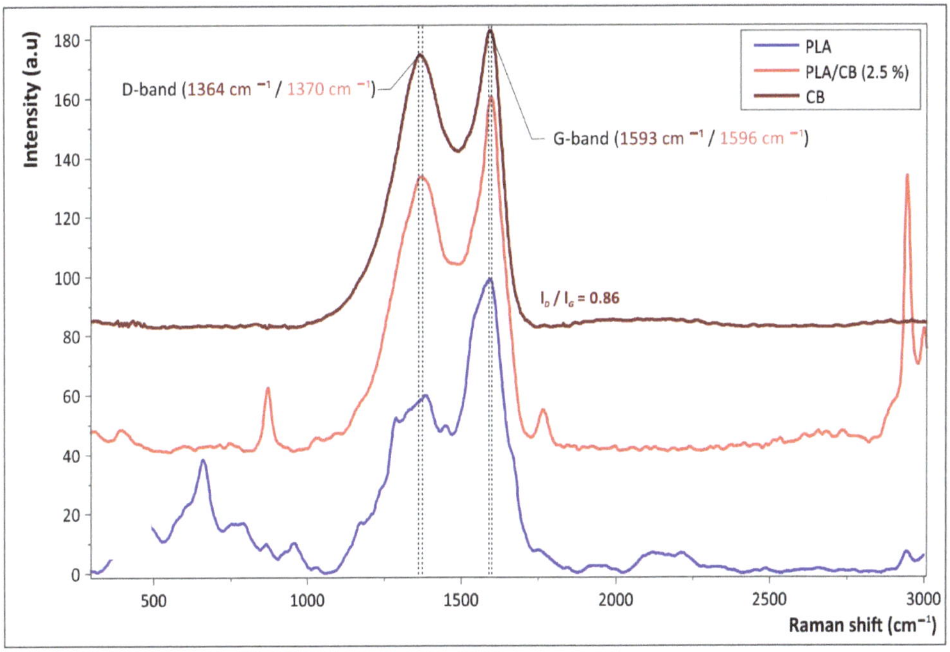

Figure 3. Raman spectra of PLA, CB and PLA/CB nanocomposite (at 2.5 wt.% of filler loading) at the spectral region 300–3300 cm^{-1}.

The graphitization and degree of crystallinity of carbon nanoallotrope materials have been previously correlated to the relative intensity ratio of the corresponding D and G-bands (I_D/I_G) [51]. Specifically, the I_D/I_G intensity band ratio of CB utilized as electrically conductive and mechanical reinforcing filler in this study is 0.86 (indicated also as inset in the corresponding Raman spectrum in Figure 3). For the PLA/CB spectrum, there are characteristic peaks attributed to the PLA matrix, as well as the graphitic fingerprints of CB (indicated with dashed lines). Specifically, the PLA characteristic fingerprints attributed to the PLA macromolecular chains' chemistry (backbone chain, side and end-terminal groups) are the following: 465 and 589 cm^{-1} (C–O–C vibration), 655 cm^{-1} (C=O stretching vibration), 936 cm^{-1} (C–COO vibration), 1249 and 1320 cm^{-1} (CH deformation vibration), 1361–1385 cm^{-1} (CH_3 deformation vibrations of PLA), 1587–1596 cm^{-1} (asymmetric C=O stretching vibrations of carboxylate groups of PLA) and 2947 cm^{-1} (CH_3 symmetric and asymmetric stretching vibration) [21], while only few of them are visible in the PLA/CB nanocomposite, due to the intense peaks of CB.

3.2. Thermogravimetric Analysis (TGA) of 3D Printed Neat PLA and PLA/CB Nanocomposites

TGA investigations revealed the critical degradation temperature of the neat PLA, as well as the PLA/CB nanocomposites under oxygen, simulating thus any plausible degradation may occur during the melt mixing filament extrusion process and the 3D printing consecutive process for the specimen's manufacturing/preparation.

Figure 4 shows the TGA (Figure 4a), and the DTG (Figure 4b) corresponding graphs of the 3D printed neat PLA, as well as the PLA/CB nanocomposites in the temperature window of 50–550 °C. Three distinct thermal windows can be observed. The first one is up to ~270 °C that all materials exhibited thermal stability without any observed weight loss (%). From 270 °C up to 374 °C, there is a second temperature window, corresponding to the materials' thermal degradation and decomposition with the onset temperature of decomposition (T_{on}) at 273 °C. In the third temperature window (>370–380 °C up to 550 °C), all the PLA substance for all samples' formulation has been fully decomposed.

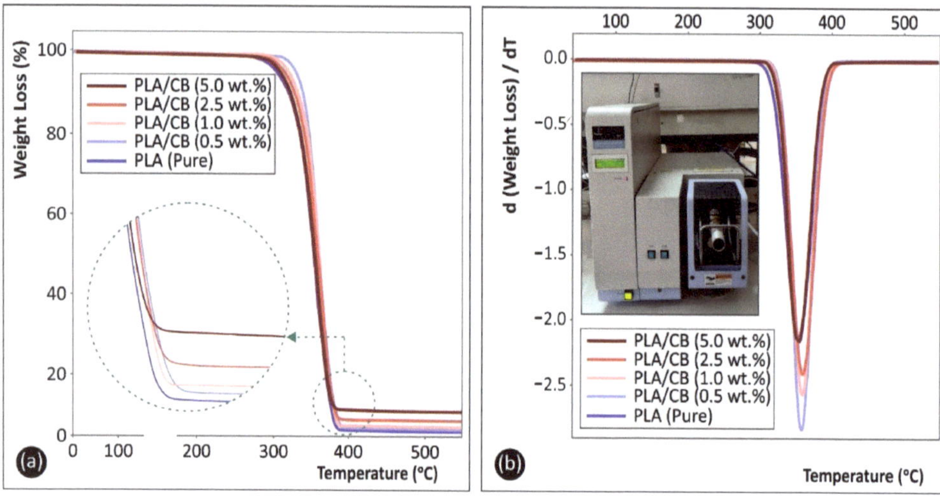

Figure 4. (**a**) Thermogravimetric Analysis (TGA) and (**b**) Differential Thermogravimetry (DTG) plots of 3D printed neat PLA and PLA/CB nanocomposites (inset TGA scan in Figure 4a shows the nanocomposites' remnant due to CB, as well as the increased thermal stability induced by the CB nanofiller at all particle loadings).

The inset TGA scan in Figure 4a shows more precisely (i) the remnant material attributed to the CB nanofiller above 370–380 °C, as well as (ii) the slightly increased materials' (PLA) thermal stability induced by the CB nanofiller at all particle loadings, due to the CB thermal properties. It is worth mentioning that all materials exhibited similar decomposition characteristics, especially for the behavior of the polymer substance used as the matrix, corroborating the extremely high control of the filament extrusion and 3D printing process upon applying the same parameters and resulting, thus, in the same thermal treatment history for all the 3D printed/manufactured samples. The overall thermal stability of the 3D printed specimens does not seem to be impacted; however, it is even enhanced by the addition of the CB, which nanoparticles are well-known for their thermal conductivity; endowing an enhanced thermal degradation performance of the PLA/CB nanocomposites.

3.3. Scanning Electron Microscopy of 3D Printed Sample's Side Surface

SEM microstructural analyses of the 3D printed tensile test samples' side surface morphology (at two different magnifications) were performed to examine the 3D printed specimens' external structure, revealing more precisely the 3D printed samples' interlayer fusion. Figure 5 depicts the side surface morphology of neat PLA and PLA/CB 3D printed nanocomposites at two different magnifications: namely neat PLA (Figure 5a,b), as well as 0.5 wt.% (Figure 5c,d), 2.5 wt.% (Figure 5e,f) and 5.0 wt.% (Figure 5g,h) nanocomposites. For all samples, an excellent interlayer fusion can be observed between the different AM layers, apart from the PLA/CB at 5.0 wt.% (Figure 5g,h) where some defects and discontinuities can be seen. For 3D printed AM specimens, the fusion between the layers affects the stress transfer between the 3D printed filaments' layers, resulting in an optimum interfacial shear strength of the adjacent layers and filaments of the 3D printed parts. This is a prerequisite towards high mechanical performance 3D printed nanocomposite parts, exhibiting an enhancement and reinforcement mechanism due to the presence of the nanofillers.

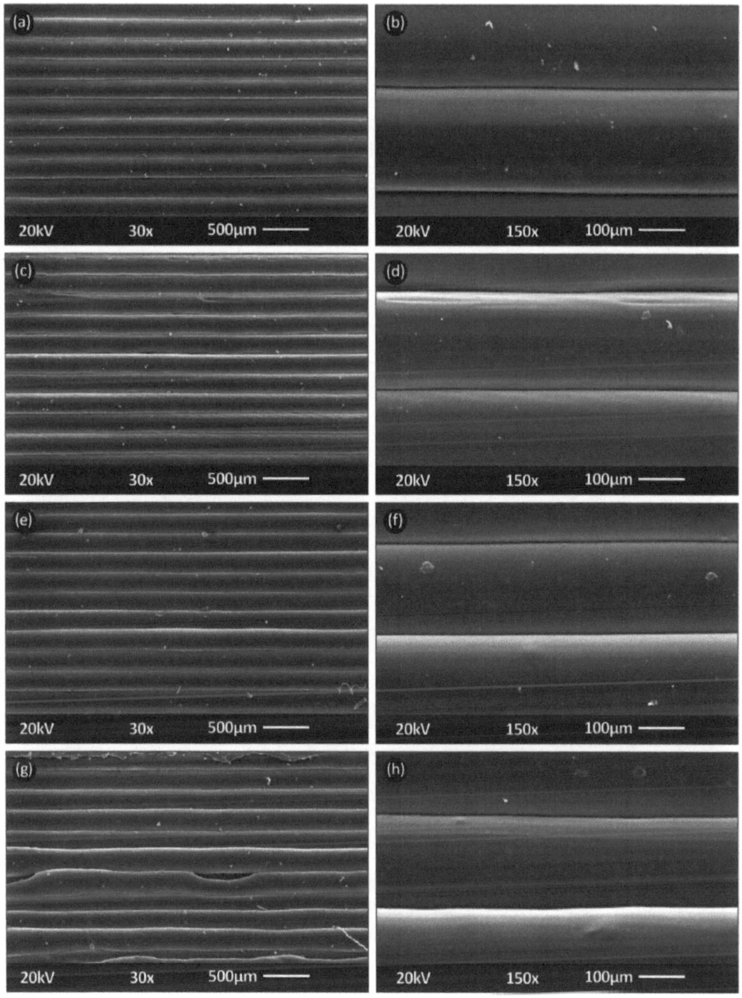

Figure 5. Side surface morphology of 3D printed tensile test specimens of neat PLA (**a**,**b**), as well as PLA/CB with 0.5 wt.% (**c**,**d**), 2.5 wt.% (**e**,**f**) and 5.0 wt.% (**g**,**h**) CB filler loading.

The high quality of the interlayer fusion for the majority of the samples in this study indirectly proves: (i) the optimum 3D printing parameters initially determined and applied in this study, as well as (ii) homogeneously dispersed CB nanofiller in the PLA extruded filaments, since nanoparticle aggregation in the feedstock filament could significantly affect the 3D printing quality introducing inhomogeneities, defects, discontinuities, etc. It is well-known that the existence of micro-aggregates in the filament feedstock could impart structural defects in the final 3D printed samples, since it can affect the polymer melt rheological properties, causing thus nozzle clogging phenomena. In the 3D printed nanocomposites herein, it is worth mentioning that only in the case of PLA/CB (5.0 wt.%) nanocomposite the SEM side surface microstructure reveals some inhomogeneity in the 3D printed layer thickness, as well as some discontinuities and defects/holes between the layers. This is most likely attributed to the increased polymer melt viscosity hampering during the 3D printing filamentous extrusion process.

3.4. Electrical Conductivity of 3D Printed PLA/CB Nanocomposites

Figure 6a illustrates schematically the two-probe electrical resistance measurements performed in this study to determine the electrical conductivity of the 3D printed PLA/CB nanocomposites. The electrical conductivity has been derived in two different directions of the 3D printing process, to examine any plausible anisotropic electrical conductivity behavior, typically known in additive manufacturing produced specimens.

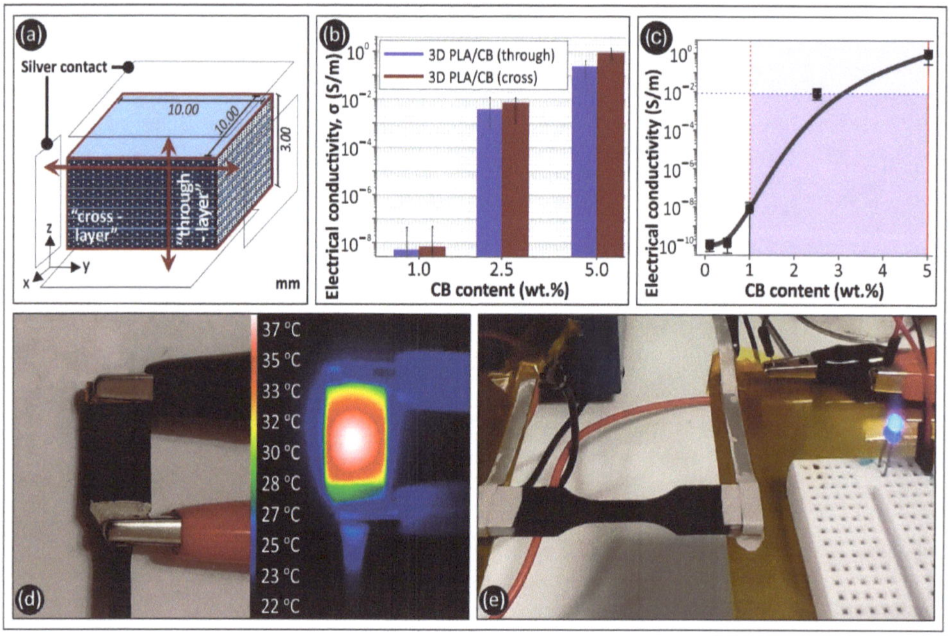

Figure 6. (**a**) schematic illustration of the electrical conductivity measurements for the 3D printed PLA/CB CPC specimens at two different directions; (**b**) electrical conductivity of the 3D printed conductive nanocomposites at 1.0, 2.5 and 5.0 wt.% CB content (at two different measurement directions; namely the "through layer" and the "cross-layer"), as well as (**c**) a detailed plot of the PLA/CB electrical conductivity at various filler loadings (0.1, 0.5, 1.0, 2.5 and 5.0 wt.% CB filler content), demonstrating the PLA/CB CPC percolation curve; (**d**) a real sample digital image (with the two deposited electrodes), as well as an IR-T image showing the electrothermal Joule heating performance of the PLA/CB (5.0 wt.%) nanocomposite at an applied DC Vbias of 15 V; (**e**) LED in operation when current is passing through a PLA/CB (5.0 wt.%) nanocomposite in a closed circuit applying a Vbias of 15 V.

Figure 6b shows the electrical conductivity of the PLA/CB nanocomposites at 1.0, 2.5 and 5.0 wt.% CB content, all of which are above the electrical percolation threshold, exhibiting typical "conductor" characteristics (the 1.0 wt.% is slightly above the percolation threshold; the samples are not showing a fully insulating behavior, which is expected when measuring a polymeric matrix). Conductivity values of the percolated samples are given for the two different measurement directions: namely the "through layer" or through thickness, as well as the "cross-layer" or in-plane direction. It can be easily observed that for all 3D printed PLA/CB CPCs, the cross-layer conductivity values are a bit higher than the through thickness ones. This is a direct proof that the 3D printed nanocomposite samples exhibit "anisotropic" electrical transport properties induced by the FFF process. This finding is in good agreement with other 3D printed CPCs reported elsewhere [32]. Namely, the highest conductivities were determined for the 5.0 wt.% PLA CPC nanocomposites, i.e., 0.3 ± 0.2 S/m and 0.9 ± 0.5 S/m for the "through layer" and "cross-layer" directions, respectively. Nevertheless, the slight difference in the electrical conductivity values at the two directions reveal a high quality and extent of interlayer fusion between the 3D printed consecutive layers: resulting in an almost homogeneous and identical "bulk" electrical conductivity behavior.

Figure 6c represents the electrical conductivity percolation curve for the 3D printed PLA/CB nanocomposites at different filler loadings; namely 0.1, 0.5, 1.0, 2.5 and 5.0 wt.% CB filler content (all values correspond to the "cross-layer" conductivity). As it can be observed, a conductive behavior for the PLA/CB nanocomposites is achieved at CB filler loading of 1.0 wt.%, increasing further with the increase of the filler content. The percolation curve exhibits a typical "S-shaped", otherwise known as sigmoidal curve shape, which is typical for randomly dispersed conductive fillers in an insulating polymer matrix [52].

Figure 6d depicts the Joule heating electrothermal performance of the highest electrically conductive PLA/CB nanocomposite sample (5.0 wt.%), upon applying a DC voltage bias (V_{bias}) of 15 V (using a conventional DC power supply). The infrared thermography (IR-T) image has been captured after a time duration of t = 30 s that the DC voltage has been applied at the two terminals of the specimen. As it is shown in the figure, the sample's surface temperature raises at approximately ~35 °C.

The exploitation of the electrothermal, otherwise known as "Joule-heating" effect, has been widely studied by the research community for different scopes and in various sectors utilizing conductive nanoparticle nanocomposites [53]. Figure 6e represents the operation of an LED device when current is passing through a 5.0 wt.% filled PLA/CB nanocomposite in a closed circuit, upon applying a voltage bias (V_{bias}) of 15 V, demonstrating the potential use of the PLA/CB nanocomposites as lightweight resistors. Figure 6d,e represent two significant demonstration scenarios for various practical applications of the 3D printed PLA/CB nanocomposites, arising from the electrically conductive nature of the 3D printed PLA/CB samples induced by the dispersed CB conductive phase. These two representative experiments reveal the multifunctional character of the 3D printed PLA conductive nanocomposites fabricated in this study, which could enable practical future applications such as flexible circuitry, flexible wires and flexible and wearable sensors, expanding further the well-known and emerging field of 3D printed electronics.

3.5. Mechanical Properties

3.5.1. Tensile Properties of 3D Printed Neat PLA and PLA/CB Nanocomposites

In Figure 7, the tensile properties of 3D printed neat PLA, as well as PLA/CB nanocomposites with 0.1, 0.5, 1.0, 2.5 and 5.0 wt.% CB filler loadings are presented. More specifically, Figure 7a illustrates the comparative and representative stress–strain curves for the different specimens. Figure 7b shows the average tensile strength values, and Figure 7c the tensile modulus of elasticity average values, along with the corresponding standard deviations, all as a function of the filler loading. The CB imparts a negligible knock down effect at the lowest filler loading, while it yields enhanced tensile strength and modulus values for the rest i.e., 0.5, 1.0, 2.5 and 5.0 wt.% filler loadings. Accordingly, the PLA/CB with 0.5 wt.% CB exhibits the best performance and the highest reinforcing mechanism; namely ~17.1% increase for the tensile strength and ~14.1% increase for the tensile modulus, respectively.

By increasing further, the CB content in the PLA matrix, the CB efficiently enhances the tensile properties as compared to the reference neat PLA; however not as much prominently as the 0.5 wt.% loading. This could be explained by decreased filamentous interfacial strength between adjacent filaments in the bulk 3D printed sample and/or limited interdiffusion of the polymeric chains from one filament to another by the presence of CB fillers, preventing an extended polymer chain mobility in the melt state at higher filler loadings. These findings also corroborate some small defects, inhomogeneities and interlayer voids observed by the SEM side surface images at the highest filler loading of 5.0 wt.%. It can be deduced then that 0.5 wt.% of CB content is the mechanical percolation threshold of the 3D printed PLA/CB nanocomposites, endowing the optimum mechanical properties of the 3D printed specimens in quasi-static tension mode.

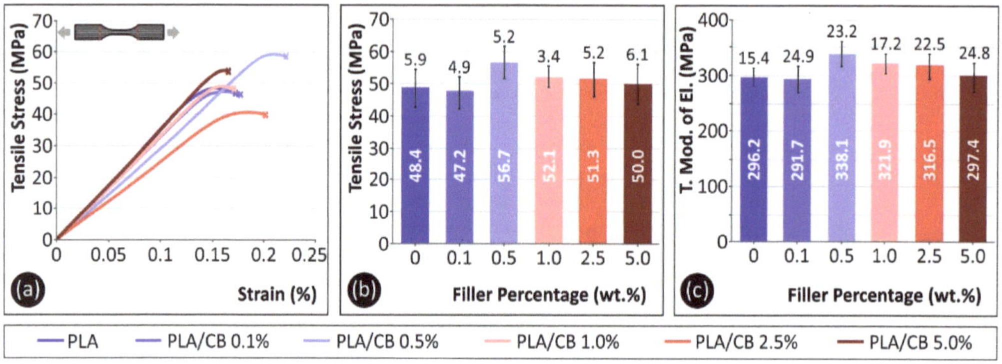

Figure 7. (**a**) Tensile stress (MPa) vs. strain (%) representative curves for neat PLA and PLA/CB nanocomposites. (**b**) average tensile strength, and (**c**) average tensile modulus of elasticity values along with the calculated standard deviation values, both as a function of CB filler loading (wt.%).

3.5.2. SEM Fractography Microstructural Analysis

Figure 8 depicts the fractography microstructural investigations of FFF 3D printed PLA nanocomposites after the tensile test experiments. The SEM images reveal both the specimens' fractured area characteristics and the specific fracture mechanism ("ductile" or "brittle" fracture mechanism), as well as the 3D printed sample's internal microstructure i.e., existence of voids, extent of the adjacent filament fusion, etc. Specifically, neat PLA (Figure 8a,b) compared to 0.5 wt.% (Figure 8c,d), 2.5 wt.% (Figure 8e,f) and 5.0 wt.% (Figure 8g,h) CB filler loading SEM fractographic images are shown at two different magnifications, to elucidate the different underlying fracture mechanisms.

An excellent interdiffusion between the filaments, with no visible voids in the fractured surface can be seen only in the case of neat PLA. On the other hand, for all nanocomposites, some discontinuities and defects in the specimen's internal structure can be observed, with a general trend the internal structure to become worse with the increased filler loading. Furthermore, a relatively "brittle" fracture mechanism occurs for all the samples, seen from the low magnification images, since no filaments have been found to be pulled-out from the specimens' fractured surfaces.

At high magnification images, all samples exhibited almost similar characteristics with the surface microstructure PLA/CB nanocomposites exhibiting a slightly increased micro-roughness compared to the neat PLA. From the SEM fractographic analyses, some important findings and speculations could be deduced i.e., (i) no microaggregates could be observed in the nanocomposites' fractured surfaces even up to 5.0 wt.% filler loading, indirectly proving a sufficient CB nanodispersion (especially seen by the high magnification images), and (ii) a high-quality internal structure can be observed only for neat PLA, whereas PLA/CB nanocomposites exhibit some structural defects e.g., voids, etc., with the worst case at 5.0 wt.% filler loading. The fact that the neat PLA exhibited an excellent internal structure suggest that optimum set of printing parameters have been used in this study.

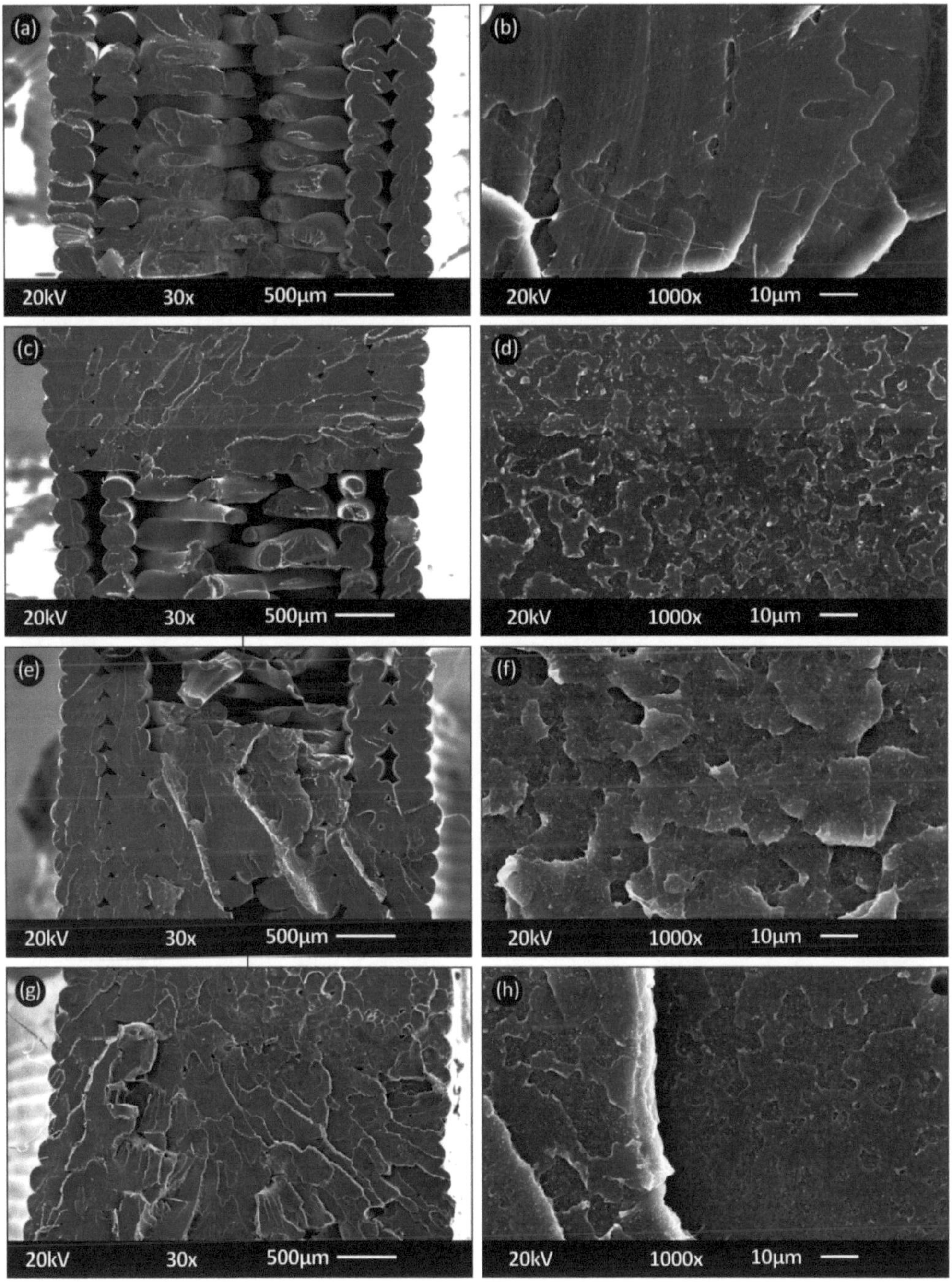

Figure 8. Fractured surface morphology of PLA/CB 3D printed nanocomposites after tensile tests: neat PLA (**a**,**b**) compared to 2.5 wt.% (**c**,**d**), 5.0 wt.% (**e**,**f**) and 10.0 wt.% (**g**,**h**) CB filler loading, respectively.

3.5.3. Flexural Properties of 3D Printed Neat PLA and PLA/CB Nanocomposites

Figure 9 summarizes the flexural properties of 3D printed neat PLA and PLA/CB nanocomposite specimens with 0.1, 0.5, 1.0, 2.5 and 5.0 wt.% CB filler loadings. Figure 9a shows representative flexural stress–strain curves for the different tested samples. Figure 9b illustrates the average flexural strength response, while Figure 9c, shows the average flexural modulus of elasticity of the 3D printed specimens. In all cases, the average values along with the calculated standard deviations are presented. Regarding the CB reinforcement in the flexion mode, interestingly enough, the 0.1 and 0.5 wt.% of CB impart a slight knock down effect compared to the reference neat PLA, both in the flexural strength and modulus. However, at 1.0, 2.5 and 5.0 wt.% the flexural properties are promising and a clear enhancement trend was observed, with the 2.5 wt.% PLA/CB exhibiting the best performance and the highest reinforcing mechanism; namely ~38.3% increase for the flexural strength and ~31.2% increase for the flexural modulus, respectively. By increasing further, the CB content in the PLA matrix and reaching 5.0 wt.%, CB efficiently enhances the flexural properties as compared to the reference neat PLA; however not as much prominently as the 2.5 wt.% loading. Overall, it can be presumed that at 2.5 wt.% of CB content, the mechanical percolation threshold is reached for the applied flexural strain field of the 3D printed PLA/CB nanocomposites.

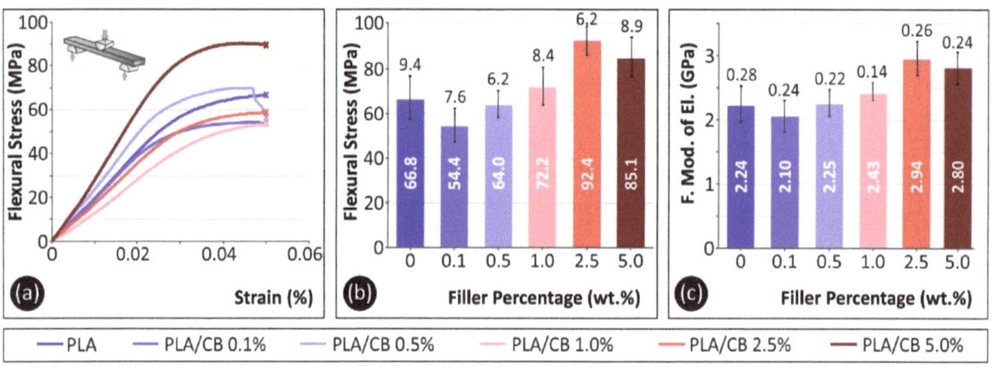

Figure 9. (**a**) flexural stress (MPa) vs. strain (%) representative curves for neat PLA and PLA/CB nanocomposites. (**b**) flexural strength and (**c**) flexural modulus of elasticity average values, along with the calculated standard deviation values, both as a function of the CB different filler loadings (wt.%).

Although the highest calculated flexural strength was not developed on the same filler loading with the tensile tests, flexural tests results follow a similar trend to the tensile tests results. This verifies the reliability of the specimens' preparation process, since no significant differences or large deviations were observed between the tests results.

3.5.4. Impact Properties and Micro-Hardness of 3D Printed Neat PLA and PLA/CB Nanocomposites

Figure 10a shows the average values of the Charpy's notched impact strength (kJ/m^2) results for neat PLA and PLA/CB nanocomposites at different filler loadings. Figure 10b presents the respective micro-hardness (Vickers (HV)) response of the neat PLA and PLA/CB nanocomposites. The impact strength mean values are presented for the different 3D printed specimens in Figure 10a. As it can be observed, the impact strength is increasing by the presence of CB in the polymer matrix for all the filler loadings, while a very slight and negligible decrease could be observed for the 5.0 wt.% nanocomposite. Specifically, the highest impact strength improvement was found for the 1.0 wt.% nanocomposite (~24.6%). The slight decrease of the impact properties at 5.0 wt.% might be attributed to (i) some "critical" number of interlayer voids in the samples' side and internal structure at this CB nanoparticle loading, discussed in more detail in the SEM fractography analyses section, as well as (ii) a possible hampered polymer chain interdiffusion between the different layers and the adjacent filaments during melt deposition, due to the increase of the melt viscosity with the increased particle filler loading, accompanied with hindered polymer chain mobility.

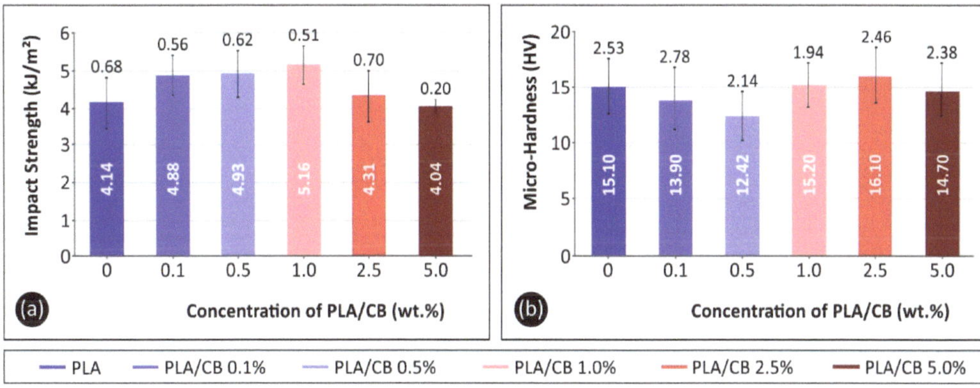

Figure 10. (**a**) Charpy's notched impact strength test results (kJ/m^2), and (**b**) micro-hardness (Vickers (HV)) performance of the neat PLA compared to the PLA/CB nanocomposites (both properties are shown as the calculated average values together with their corresponding standard deviation).

Regarding the micro-hardness (Figure 10b), the PLA/CB nanocomposites follow exactly the trend of the flexural properties with respect to the different filler loadings. Especially, the biggest improvement was observed for the 2.5 wt.% nanocomposite (~6.6%). The enhancement in the microhardness values is directly associated with the stiffening of the material induced by the CB nanoparticles in the PLA polymer matrix. Namely, micro-hardness slightly decreases for the 0.1 and 0.5 wt.% nanocomposites, while it slightly increases for the 1.0 and 2.5 wt.%, with a slight decrease finally at the highest filled PLA at 5.0 wt.%.

3.6. Thermomechanical Analysis of 3D Printed Neat PLA and PLA/CB Nanocomposites

Figure 11 illustrates the 3D printed specimens' thermomechanical response, i.e., the storage modulus (E', MPa), loss modulus (E", MPa) and Tan δ, as a function of temperature, derived from the DMA thermomechanical experiments. Storage moduli values at low temperatures (~30 °C) coincide with the flexural moduli of the tested material, with the highest increase and reinforcing mechanism for the 2.5 wt.% nanocomposite. However, the 5.0 wt.% nanocomposite show also a promising thermomechanical reinforcement mechanism, which is an opposite behavior with regards to the flexural and the tensile properties. This interesting finding is attributed to the nature of the DMA set-up conditions and the samples' significantly lower dimensions, where statistically a plausible much smaller number of defects and discontinuities is present in the tested specimens affecting the bulk mechanical property.

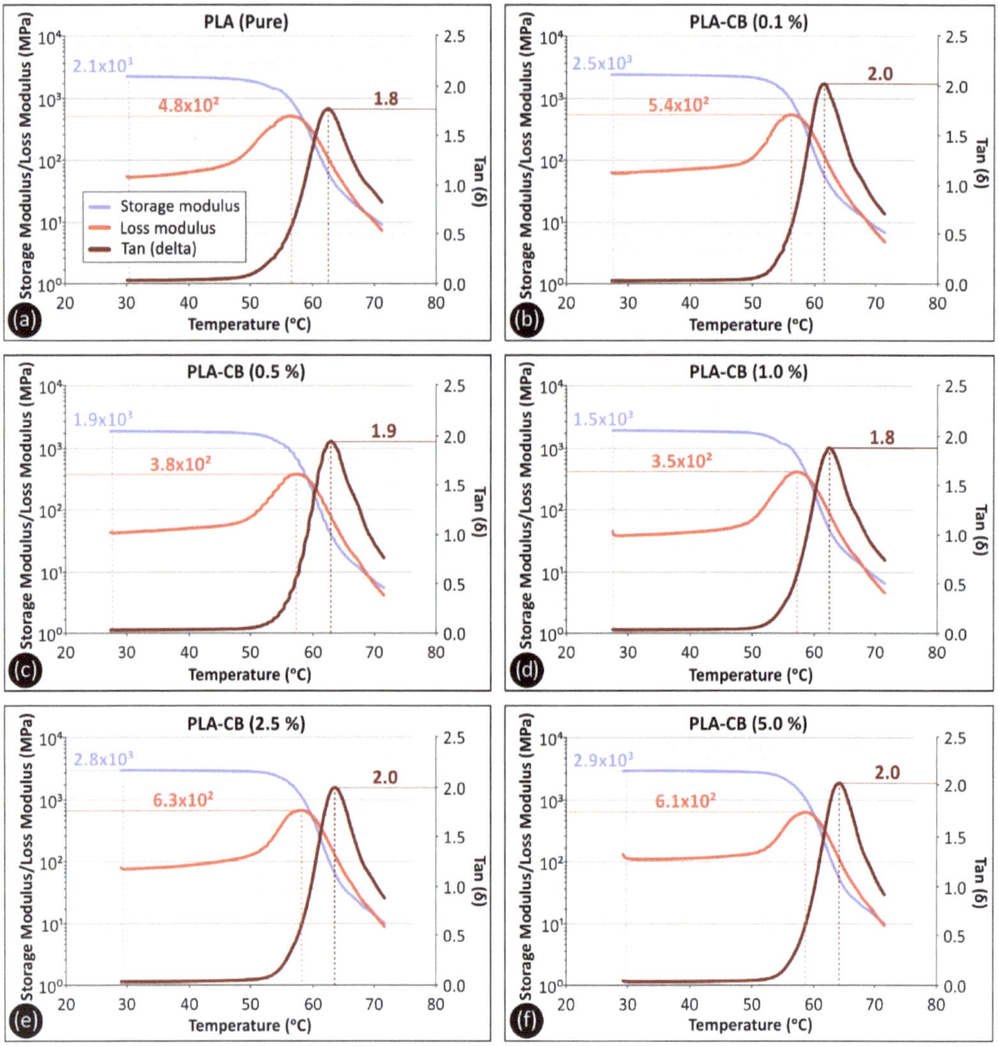

Figure 11. Storage modulus, loss modulus and Tan δ plots of the neat PLA and PLA/CB nanocomposites at different filler loadings: (**a**) pure PLA, (**b**) PLA/CB wt.% 0.1, (**c**) PLA/CB wt.% 0.5, (**d**) PLA/CB wt.% 1.0, (**e**) PLA/CB wt.% 2.5, (**f**) PLA/CB wt.% 5.0.

The storage modulus reveals a decreasing tendency with the increase in temperature. The highest damping is recorded up to 0.1 wt.% doped samples, and marginally decreased for the 0.5, 1.0, 2.5 and 5.0 wt.% nanocomposites. The Tan δ values for all samples exhibit an increasing trend up to the temperature range of 62 °C, indicating that up to this temperature range the mechanical energy is stored elastically in the polymeric macromolecular chains; however, with a decreased ability as the temperature increases, manifested by the decreased storage moduli values. On the other hand, >60–65 °C the Tan δ values show a drop, revealing the softening of the material before fusion, and the transition to the viscoelastic regime. It is worth mentioning that the increase in the storage moduli for all the nanocomposite samples found by the DMA experiments is associated with the stiffening of the polymeric chains in the presence of CB nanofillers. The highest storage modulus value was observed for the 2.5 wt.%, which was also almost identical with the 5.0 wt.% nanocomposite, exhibiting an increase of approximately 40% when compared to the neat PLA.

3.7. Antibacterial Performance of the PLA/CB 3D Printed Nanocomposites

Figure 12 demonstrates the antibacterial results after the 24 h bacteria culture for the highest concentration PLA/CB nanocomposite studied in this research work (5.0 wt.%). A mild clearly visible antibacterial activity was observed, while non-measurable antimicrobial activity has been found for the PLA/CB at lower filler loadings and no antibacterial performance was found for the neat PLA, with the process followed. The developed Inhibition Zone (IZ) was almost the same for both *Staphylococcus aureus* and *Escherichia coli* bacteria. The antibacterial results presented in Figure 12 show that the addition of CB in the PLA matrix can introduce antibacterial properties to the polymer, at least for the two bacteria tested in this study. It should be mentioned that the agar well diffusion method employed in this work exposes the specimens in a severe bacteria environment and it is suitable mainly as a screening process, providing information regarding the potential of the material's antibacterial performance. Still, this is an important finding, considering that the developed nanocomposites were 3D printed, making it possible to build on-demand parts with FFF 3D printing process exhibiting antibacterial properties.

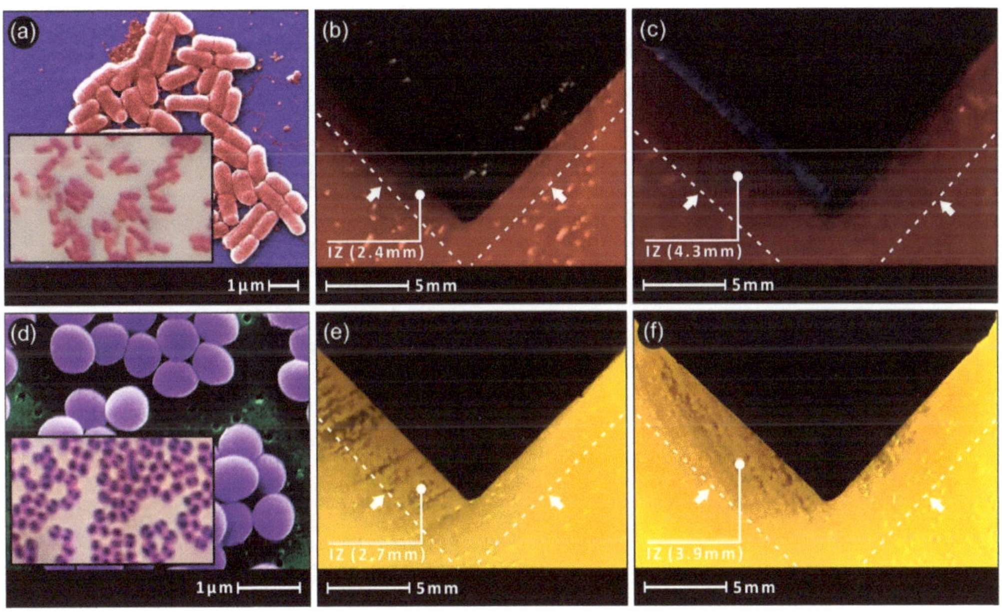

Figure 12. Agar well diffusion method pictures after 24 h culture showing the Inhibition Zone (IZ) of (**a**) *Escherichia coli* bacteria against PLA/CB (5.0 wt.%) (**b–d**) and *Staphylococcus aureus* against PLA/CB (5.0 wt.%) (**e**,**f**).

4. Conclusions

PLA/CB nanocomposites at 0.5, 1.0, 2.5 and 5.0 wt.% filler loadings were produced by melt mixing followed by a filament extrusion process, to produce the feedstock material utilized for the FFF 3D printing consecutive manufacturing of PLA/CB. All filament extrusion and 3D printing parameters have been initially optimized, yielding high quality specimens and an efficient 3D printing process. Raman spectroscopy revealed the graphitic nature of CB, as well as the corresponding PLA matrix characteristic fingerprints, and the PLA/CB nanocomposites (at 2.5 wt.% filler loading) peaks, due to the constituent matrix and dispersed filler phases. TGA analysis proved that the temperatures employed during the extrusion and the 3D printing process are appropriate for the materials prepared and tested in this work and the overall thermal stability of the samples does not seem to be impacted. The PLA/CB 3D printed nanocomposites have been fully investigated for their static and thermomechanical properties; while they have been fully analyzed in terms of their electrically conductivity and demonstrated as (i) Joule heater devices and (ii) conductive circuitry resistor multi-functional nanocomposites.

The electrical percolation threshold was studied, with the nanocomposites exhibiting an electrically conductive behavior from the 2.5 wt.% loading and increasing by increasing the filler loading up to 5.0 wt.%. The nanocomposites showed an anisotropic electrically conductive behavior in the "through" thickness and "cross-layer" 3D printing specimens' measurement directions. The highest electrical conductivity (σ) was obtained at the "cross-layer" direction for the PLA/CB (5.0 wt.%) nanocomposite (~0.9 S/m), while it exhibited an electrothermal "Joule-heating" function with its surface temperature raising at ~35 °C within 30 sec. upon being to a voltage bias (V_{bias}) of 15 V. It is worth to be mentioned that PLA/CB (10.0 wt.%) electrically conductive filaments were also produced; however, they were not processable towards a continuous FFF 3D printing process, hence they are not included this work.

Detailed mechanical investigations i.e., tensile, flexural, impact, microhardness and Dynamic Mechanical Analysis (DMA) tests revealed the structure-property relationship of the fabricated 3D printed nanocomposite materials, as well as elucidating the mechanical reinforcement mechanisms under static and dynamic mechanical loadings for the PLA/CB nanocomposites. Scanning Electron Microscopy (SEM) fractographic analyses revealed the underlying fracture mechanisms of the tensile tested specimens. Additionally, the agar well-known diffusion method revealed a mild antibacterial performance of the nanocomposite at the highest filler loading of 5.0 wt.% prepared in this work. The overall results especially related to the mechanical performance of the 3D printed nanocomposites manufactured and tested in this study are summarized in Figure 13. As it is shown, 0.5 wt.% PLA/CB CPC had the highest tensile strength enhancement, although higher loadings had similar tensile strength results. In the remaining mechanical properties, the 2.5 wt.% PLA/CB CPC had the most enhanced performance, making it overall the best filler loading, regarding the mechanical response of the nanocomposite studied in this work.

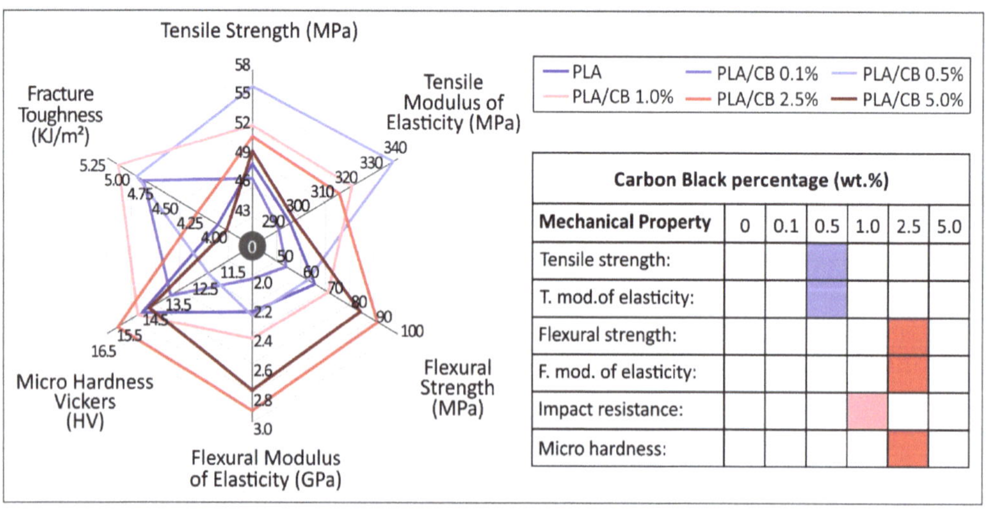

Figure 13. Overall results on the mechanical properties and performance of neat PLA compared to PLA/CB 3D printed nanocomposites.

It could be envisaged that the proposed multi-functional 3D printed PLA/CB CPC manufactured nanocomposites with enhanced mechanical and electrical properties, Joule-heating capability and antibacterial properties could create the seeds for future 3D shaped multifunctional objects where complex 3D architectures and customizability are required; i.e., low-power demanding applications such as wearables, flexible conductors, resistors for circuitry, IoT sensors and personalized biomedical components (wearable strain sensors, antimicrobial splints, etc.).

Author Contributions: Conceptualization, N.V. and L.T.; methodology, M.P.; software, N.M.; validation, L.T., E.V. and S.G.; formal analysis, N.M., P.E.F.-G. and M.P.; investigation, N.M. and E.V.; resources, N.V., data curation, P.E.F.-G. and N.M.; writing—original draft preparation, E.V., L.T. and S.G.; writing—review and editing, E.V., M.P. and S.G.; visualization, P.E.F.-G. and N.V.; supervision, N.V.; project administration, M.P.; funding acquisition, N.V. The manuscript was written through contributions of all authors. All authors have read and agreed to the published version of the manuscript.

Funding: This research received no external funding.

Institutional Review Board Statement: Not applicable.

Informed Consent Statement: Not applicable.

Data Availability Statement: The data presented in this study are available on request from the corresponding author.

Acknowledgments: Authors would like to thank Aleka Manousaki from the University of Crete for taking the SEM images presented in this work.

Conflicts of Interest: The authors declare no conflict of interest.

References

1. Tzounis, L.; Petousis, M.; Grammatikos, S.; Vidakis, N. 3D Printed Thermoelectric Polyurethane/Multiwalled Carbon Nanotube Nanocomposites: A Novel Approach towards the Fabrication of Flexible and Stretchable Organic Thermoelectrics. *Materials* **2020**, *13*, 2879. [CrossRef] [PubMed]
2. Kapnopoulos, C.; Mekeridis, E.D.; Tzounis, L.; Polyzoidis, C.; Zachariadis, A.; Tsimikli, S.; Gravalidis, C.; Laskarakis, A.; Vouroutzis, N.; Logothetidis, S. Fully gravure printed organic photovoltaic modules: A straightforward process with a high potential for large scale production. *Sol. Energy Mater. Sol. Cells* **2016**, *144*, 724–731. [CrossRef]
3. Tzounis, L.; Bangeas, P.I.; Exadaktylos, A.; Petousis, M.; Vidakis, N. Three-Dimensional Printed Polylactic Acid (PLA) Surgical Retractors with Sonochemically Immobilized Silver Nanoparticles: The Next Generation of Low-Cost Antimicrobial Surgery Equipment. *Nanomaterials* **2020**, *10*, 985. [CrossRef]
4. Zhang, J.; Wang, J.; Dong, S.; Yu, X.; Han, B. A review of the current progress and application of 3D printed concrete. *Compos. Part A Appl. Sci. Manuf.* **2019**, *125*, 105533. [CrossRef]
5. Gnanasekaran, K.; Heijmans, T.; van Bennekom, S.; Woldhuis, H.; Wijnia, S.; de With, G.; Friedrich, H. 3D printing of CNT- and graphene-based conductive polymer nanocomposites by fused deposition modeling. *Appl. Mater. Today* **2017**, *9*, 21–28. [CrossRef]
6. Takagishi, K.; Umezu, S. Development of the Improving Process for the 3D Printed Structure. *Sci. Rep.* **2017**, *7*, 39852. [CrossRef]
7. Velu, R.; Raspall, F.; Singamneni, S. Chapter 8—3D printing technologies and composite materials for structural applications. In *Green Composites for Automotive Applications*; Koronis, G., Silva, A., Eds.; Woodhead Publishing: Sawston, UK, 2019; pp. 171–196.
8. Tian, X.; Liu, T.; Yang, C.; Wang, Q.; Li, D. Interface and performance of 3D printed continuous carbon fiber reinforced PLA composites. *Compos. Part A Appl. Sci. Manuf.* **2016**, *88*, 198–205. [CrossRef]
9. Vidakis, N.; Petousis, M.; Tzounis, L.; Maniadi, A.; Velidakis, E.; Mountakis, N.; Kechagias, J.D. Sustainable Additive Manufacturing: Mechanical Response of Polyamide 12 over Multiple Recycling Processes. *Materials* **2021**, *14*, 466. [CrossRef] [PubMed]
10. Ngo, T.D.; Kashani, A.; Imbalzano, G.; Nguyen, K.T.Q.; Hui, D. Additive manufacturing (3D printing): A review of materials, methods, applications and challenges. *Compos. Part B Eng.* **2018**, *143*, 172–196. [CrossRef]
11. Savvakis, K.; Petousis, M.; Vairis, A.; Vidakis, N.; Bikmeyev, A.T. Experimental Determination of the Tensile Strength of Fused Deposition Modeling Parts. In Proceedings of the ASME 2014 International Mechanical Engineering Congress and Exposition, Montreal, QC, Canada, 14–20 November 2014.
12. Vidakis, N.; Petousis, M.; Vairis, A.; Savvakis, K.; Maniadi, A. On the compressive behavior of an FDM Steward Platform part. *J. Comput. Des. Eng.* **2017**, *4*, 339–346. [CrossRef]
13. Vidakis, N.; Petousis, M.; Vairis, A.; Savvakis, K.; Maniadi, A. A parametric determination of bending and Charpy's impact strength of ABS and ABS-plus fused deposition modeling specimens. *Prog. Addit. Manuf.* **2019**, *4*, 323–330. [CrossRef]

14. Tan, L.J.; Zhu, W.; Zhou, K. Recent Progress on Polymer Materials for Additive Manufacturing. *Adv. Funct. Mater.* **2020**, *30*, 2003062. [CrossRef]
15. Vidakis, N.; Petousis, M.; Savvakis, K.; Maniadi, A.; Koudoumas, E. A comprehensive investigation of the mechanical behavior and the dielectrics of pure polylactic acid (PLA) and PLA with graphene (GnP) in fused deposition modeling (FDM). *Int. J. Plast. Technol.* **2019**, *23*, 195–206. [CrossRef]
16. Chacón, J.M.; Caminero, M.A.; García-Plaza, E.; Núñez, P.J. Additive manufacturing of PLA structures using fused deposition modelling: Effect of process parameters on mechanical properties and their optimal selection. *Mater. Des.* **2017**, *124*, 143–157. [CrossRef]
17. Vidakis, N.; Maniadi, A.; Petousis, M.; Vamvakaki, M.; Kenanakis, G.; Koudoumas, E. Mechanical and Electrical Properties Investigation of 3D-Printed Acrylonitrile–Butadiene–Styrene Graphene and Carbon Nanocomposites. *J. Mater. Eng. Perform.* **2020**, *29*, 1909–1918. [CrossRef]
18. Taylor, A.P.; Cuervo, C.V.; Arnold, D.P.; Velásquez-García, L.F. Fully 3D-Printed, Monolithic, Mini Magnetic Actuators for Low-Cost, Compact Systems. *J. Microelectromech. Syst.* **2019**, *28*, 481–493. [CrossRef]
19. Hohimer, C.J.; Petrossian, G.; Ameli, A.; Mo, C.; Pötschke, P. 3D printed conductive thermoplastic polyurethane/carbon nanotube composites for capacitive and piezoresistive sensing in soft pneumatic actuators. *Addit. Manuf.* **2020**, *34*, 101281.
20. Le, T.; Song, B.; Liu, Q.; Bahr, R.A.; Moscato, S.; Wong, C.; Tentzeris, M.M. A novel strain sensor based on 3D printing technology and 3D antenna design. In Proceedings of the 2015 IEEE 65th Electronic Components and Technology Conference (ECTC), San Diego, CA, USA, 26–29 May 2015; pp. 981–986.
21. Vidakis, N.; Petousis, M.; Maniadi, A.; Koudoumas, E.; Liebscher, M.; Tzounis, L. Mechanical Properties of 3D-Printed Acrylonitrile–Butadiene–Styrene TiO_2 and ATO Nanocomposites. *Polymers* **2020**, *12*, 1589. [CrossRef]
22. Vidakis, N.; Petousis, M.; Velidakis, E.; Mountakis, N.; Tzounis, L.; Liebscher, M.; Grammatikos, S.A. Enhanced Mechanical, Thermal and Antimicrobial Properties of Additively Manufactured Polylactic Acid with Optimized Nano Silica Content. *Nanomaterials* **2021**, *11*, 1012. [CrossRef] [PubMed]
23. Thaler, D.; Aliheidari, N.; Ameli, A. Mechanical, electrical, and piezoresistivity behaviors of additively manufactured acrylonitrile butadiene styrene/carbon nanotube nanocomposites. *Smart Mater. Struct.* **2019**, *28*, 084004. [CrossRef]
24. Vidakis, N.; Petousis, M.; Velidakis, E.; Liebscher, M.; Tzounis, L. Three-Dimensional Printed Antimicrobial Objects of Polylactic Acid (PLA)-Silver Nanoparticle Nanocomposite Filaments Produced by an In-Situ Reduction Reactive Melt Mixing Process. *Biomimetics* **2020**, *5*, 42. [CrossRef] [PubMed]
25. Tzounis, L.; Liebscher, M.; Fuge, R.; Leonhardt, A.; Mechtcherine, V. P- and n-type thermoelectric cement composites with CVD grown p- and n-doped carbon nanotubes: Demonstration of a structural thermoelectric generator. *Energy Build.* **2019**, *191*, 151–163. [CrossRef]
26. Tzounis, L. Chapter 9—Synthesis and Processing of Thermoelectric Nanomaterials, Nanocomposites, and Devices. In *Nanomaterials Synthesis*; Pottathara, Y.B., Thomas, S., Kalarikkal, N., Grohens, Y., Kokol, V., Eds.; Elsevier: Amsterdam, The Netherlands, 2019; pp. 295–336.
27. Foteinidis, G.; Tsirka, K.; Tzounis, L.; Baltzis, D.; Paipetis, A.S. The Role of Synergies of MWCNTs and Carbon Black in the Enhancement of the Electrical and Mechanical Response of Modified Epoxy Resins. *Appl. Sci.* **2019**, *9*, 3757. [CrossRef]
28. Papageorgiou, D.G.; Tzounis, L.; Papageorgiou, G.Z.; Bikiaris, D.N.; Chrissafis, K. β-nucleated propylene–ethylene random copolymer filled with multi-walled carbon nanotubes: Mechanical, thermal and rheological properties. *Polymer* **2014**, *55*, 3758–3769. [CrossRef]
29. Tzounis, L.; Pegel, S.; Zafeiropoulos, N.E.; Avgeropoulos, A.; Paipetis, A.S.; Stamm, M. Shear alignment of a poly(styrene-butadiene-styrene) triblock copolymer/MWCNT nanocomposite. *Polymer* **2017**, *131*, 1–9. [CrossRef]
30. Zhou, B.; Luo, W.; Yang, J.; Duan, X.; Wen, Y.; Zhou, H.; Chen, R.; Shan, B. Thermal conductivity of aligned CNT/polymer composites using mesoscopic simulation. *Compos. Part A Appl. Sci. Manuf.* **2016**, *90*, 410–416. [CrossRef]
31. Tzounis, L.; Zappalorto, M.; Panozzo, F.; Tsirka, K.; Maragoni, L.; Paipetis, A.S.; Quaresimin, M. Highly conductive ultra-sensitive SWCNT-coated glass fiber reinforcements for laminate composites structural health monitoring. *Compos. Part B Eng.* **2019**, *169*, 37–44. [CrossRef]
32. Han, T.; Kundu, S.; Nag, A.; Xu, Y. 3D Printed Sensors for Biomedical Applications: A Review. *Sensors* **2019**, *19*, 1706. [CrossRef]
33. Christ, J.F.; Aliheidari, N.; Ameli, A.; Pötschke, P. 3D printed highly elastic strain sensors of multiwalled carbon nanotube/thermoplastic polyurethane nanocomposites. *Mater. Des.* **2017**, *131*, 394–401. [CrossRef]
34. Haghiashtiani, G.; Habtour, E.; Park, S.-H.; Gardea, F.; McAlpine, M.C. 3D printed electrically-driven soft actuators. *Extrem. Mech. Lett.* **2018**, *21*, 1–8. [CrossRef]
35. Jaksic, N.I.; Desai, P.D. Characterization of resistors created by fused filament fabrication using electrically-conductive filament. *Procedia Manuf.* **2018**, *17*, 37–44. [CrossRef]
36. Liu, S.; Li, Y.; Shen, Y.; Lu, Y. Mechanical performance of carbon fiber/epoxy composites cured by self-resistance electric heating method. *Int. J. Adv. Manuf. Technol.* **2019**, *103*, 3479–3493. [CrossRef]
37. Ibrahim, Y.; Kempers, R.; Amirfazli, A. 3D printed electro-thermal anti- or de-icing system for composite panels. *Cold Reg. Sci. Technol.* **2019**, *166*, 102844. [CrossRef]
38. Aumnate, C.; Potiyaraj, P.; Saengow, C.; Giacomin, A.J. Reinforcing polypropylene with graphene-polylactic acid microcapsules for fused-filament fabrication. *Mater. Des.* **2021**, *198*, 109329. [CrossRef]

39. Shmueli, Y.; Lin, Y.-C.; Zuo, X.; Guo, Y.; Lee, S.; Freychet, G.; Zhernenkov, M.; Kim, T.; Tannenbaum, R.; Marom, G.; et al. In-situ X-ray scattering study of isotactic polypropylene/graphene nanocomposites under shear during fused deposition modeling 3D printing. *Compos. Sci. Technol.* **2020**, *196*, 108227. [CrossRef]
40. Lei, L.; Yao, Z.; Zhou, J.; Wei, B.; Fan, H. 3D printing of carbon black/polypropylene composites with excellent microwave absorption performance. *Compos. Sci. Technol.* **2020**, *200*, 108479. [CrossRef]
41. Kwok, S.W.; Goh, K.H.H.; Tan, Z.D.; Tan, S.T.M.; Tjiu, W.W.; Soh, J.Y.; Ng, Z.J.G.; Chan, Y.Z.; Hui, H.K.; Goh, K.E.J. Electrically conductive filament for 3D-printed circuits and sensors. *Appl. Mater. Today* **2017**, *9*, 167–175. [CrossRef]
42. Vidakis, N.; Petousis, M.; Tzounis, L.; Velidakis, E.; Mountakis, N.; Grammatikos, S.A. Polyamide 12/Multiwalled Carbon Nanotube and Carbon Black Nanocomposites Manufactured by 3D Printing Fused Filament Fabrication: A Comparison of the Electrical, Thermoelectric, and Mechanical Properties. *C* **2021**, *7*, 38.
43. Jing, J.; Chen, Y.; Shi, S.; Yang, L.; Lambin, P. Facile and scalable fabrication of highly thermal conductive polyethylene/graphene nanocomposites by combining solid-state shear milling and FDM 3D-printing aligning methods. *Chem. Eng. J.* **2020**, *402*, 126218. [CrossRef]
44. Tan, J.C.; Low, H.Y. Multi-materials fused filament printing with embedded highly conductive suspended structures for compressive sensing. *Addit. Manuf.* **2020**, *36*, 101551. [CrossRef]
45. Petroni, J.M.; Neves, M.M.; de Moraes, N.C.; Bezerra da Silva, R.A.; Ferreira, V.S.; Lucca, B.G. Development of highly sensitive electrochemical sensor using new graphite/acrylonitrile butadiene styrene conductive composite and 3D printing-based alternative fabrication protocol. *Anal. Chim. Acta* **2021**, *1167*, 338566. [CrossRef]
46. João, A.F.; Castro, S.V.F.; Cardoso, R.M.; Gamela, R.R.; Rocha, D.P.; Richter, E.M.; Muñoz, R.A.A. 3D printing pen using conductive filaments to fabricate affordable electrochemical sensors for trace metal monitoring. *J. Electroanal. Chem.* **2020**, *876*, 114701. [CrossRef]
47. Flowers, P.F.; Reyes, C.; Ye, S.; Kim, M.J.; Wiley, B.J. 3D printing electronic components and circuits with conductive thermoplastic filament. *Addit. Manuf.* **2017**, *18*, 156–163. [CrossRef]
48. Abdalla, A.; Hamzah, H.H.; Keattch, O.; Covill, D.; Patel, B.A. Augmentation of conductive pathways in carbon black/PLA 3D-printed electrodes achieved through varying printing parameters. *Electrochim. Acta.* **2020**, *354*, 136618. [CrossRef]
49. Balouiri, M.; Sadiki, M.; Ibnsouda, S.K. Methods for in vitro evaluating antimicrobial activity: A review. *J. Pharm. Anal.* **2016**, *6*, 71–79. [CrossRef] [PubMed]
50. Li, Z.; Zabihi, O.; Wang, J.; Li, Q.; Wang, J.; Lei, W.; Naebe, M. Hydrophilic PAN based carbon nanofibres with improved graphitic structure and enhanced mechanical performance using ethylenediamine functionalized graphene. *RSC Adv.* **2017**, *7*, 2621–2628. [CrossRef]
51. Felisberto, M.; Tzounis, L.; Sacco, L.; Stamm, M.; Candal, R.; Rubiolo, G.H.; Goyanes, S. Carbon nanotubes grown on carbon fiber yarns by a low temperature CVD method: A significant enhancement of the interfacial adhesion between carbon fiber/epoxy matrix hierarchical composites. *Compos. Commun.* **2017**, *3*, 33–37. [CrossRef]
52. Tzounis, L.; Petousis, M.; Liebscher, M.; Grammatikos, S.; Vidakis, N. Three-Dimensional (3D) Conductive Network of CNT-Modified Short Jute Fiber-Reinforced Natural Rubber: Hierarchical CNT-Enabled Thermoelectric and Electrically Conductive Composite Interfaces. *Materials* **2020**, *13*, 2668. [CrossRef] [PubMed]
53. Liebscher, M.; Tzounis, L.; Junger, D.; Dinh, T.T.; Mechtcherine, V. Electrical Joule heating of cementitious nanocomposites filled with multi-walled carbon nanotubes: Role of filler concentration, water content, and cement age. *Smart Mater. Struct.* **2020**, *29*, 125019. [CrossRef]

C Journal of Carbon Research

Article

Comments on the XPS Analysis of Carbon Materials

David J. Morgan [1,2]

[1] Cardiff Catalysis Institute, School of Chemistry, Cardiff University, Park Place, Cardiff CF10 3AT, UK; morganDJ3@cardiff.ac.uk
[2] HarwellXPS: ESPRC National Facility for X-ray Photoelectron Spectroscopy, Research Campus at Harwell (RCaH), Didcot OX11 0FA, UK

Abstract: The surface chemistry of carbon materials is predominantly explored using x-ray photoelectron spectroscopy (XPS). However, many published papers have critical failures in the published analysis, stemming from an ill-informed approach to analyzing the spectroscopic data. Herein, a discussion on lineshapes and changes in the spectral envelope of predominantly graphitic materials are explored, together with the use of the D-parameter, to ascertain graphitic content, using this information to highlight a simple and logical approach to strengthen confidence in the functionalization derived from the carbon core-level spectra.

Keywords: carbon; XPS; oxygen; fitting; quantification

Citation: Morgan, D.J. Comments on the XPS Analysis of Carbon Materials. *C* **2021**, *7*, 51. https://doi.org/10.3390/c7030051

Academic Editor: Gil Goncalves

Received: 21 June 2021
Accepted: 5 July 2021
Published: 6 July 2021

1. Introduction

The chemistry of carbon materials is extremely important in fields such as catalysis, energy storage, composite materials, and sensor technology to name but a few examples [1]. In the analysis of such materials, to obtain an understanding of the surface properties the use of x-ray photoelectron spectroscopy (XPS) has been ubiquitous due to its inherent surface sensitivity. Given the truly heterogenous nature of most carbons, with their bonds to hetero atoms, different chemical species, and varying amounts of sp^2 and sp^3 carbon, the carbon spectral envelope can become quite convoluted.

Many researchers approach the analysis of carbon from a single point of view, specifically analysis of solely the C (1s) region. Complementary analysis of the corresponding heteroatom spectral regions (e.g., O (1s), S (2p), and N (1s)) is required to support the conclusions drawn by the C (1s) analysis [2–5].

Erroneous analysis also stems from failure to apply an asymmetric shape to graphitic carbon, which arises from final-state effects due to the photoelectron ejection [6–9], while a failure to consider the presence of defects or changes in the electronic structure of the carbon can also lead to mistaken assignments of peaks in a fitting model.

This paper reviews and addresses some of the key points that are commonly made in error in the published literature, including photoemission line shapes, the use of the carbon Auger and simple checks a researcher can perform to move towards a better understanding of their carbon material and have confidence in their chemical state assignments from peak fitting.

2. Experimental

Original data presented in this paper were collected on one of two XPS systems. The first is a Thermo Scientific K-Alpha$^+$ photoelectron spectrometer which utilizes microfocused monochromatic Alkα radiation over typically a 400 × 600 μm elliptical analysis area (400-micron spot mode). Data were collected at a pass energy of 40 eV with a step size of 0.1 eV.

The second system is a Kratos Axis Ultra DLD photoelectron spectrometer, again utilizing monochromatic Alkα radiation but operating in its hybrid spectroscopy mode, with the slot aperture giving a rectangular analysis area of 700 × 300 μm. Unless otherwise stated the pass energy was 20 eV for all high-resolution data collected with a step size of 0.1 eV.

Highly-oriented pyrolytic graphite (HOPG) samples were of ZYH quality, measuring 10 × 10 mm and acquired from Scanwel (Gwynedd, UK). All HOPG samples were cleaved using adhesive tape in the entry lock of the spectrometer.

All graphitic samples, including HOPG, are taken to have the C (1s) peak for sp^2 carbon at a binding energy of 284.5 eV, while sp^3 samples were taken to have a C (1s) energy of 285 eV. All data were processed using CasaXPS v2.3.24 [10].

3. Results

3.1. The C (1s) Spectrum of Diamond, HOPG and Graphitic Carbons

This paper will not consider the fitting of adventitious or polymeric carbon species but will focus on predominantly graphitic and diamond-like materials. For discussion on the former, the reader is directed to [11].

Diamond, with its sp^3 hybridization, typically exhibits a Voigt-like peak shape and is taken to have a binding energy of 285 eV [12,13]. The C (1s) core-level spectrum of graphitic carbon, such as that of HOPG, is characterized by an asymmetric peak shape with a binding energy typically reported as 284.5 eV (although values between 284.3 and 284.6 eV have been reported [12,14]) and a characteristic π-π^* shake-up structure with structure centered around 291 and 294.5 eV (Figure 1).

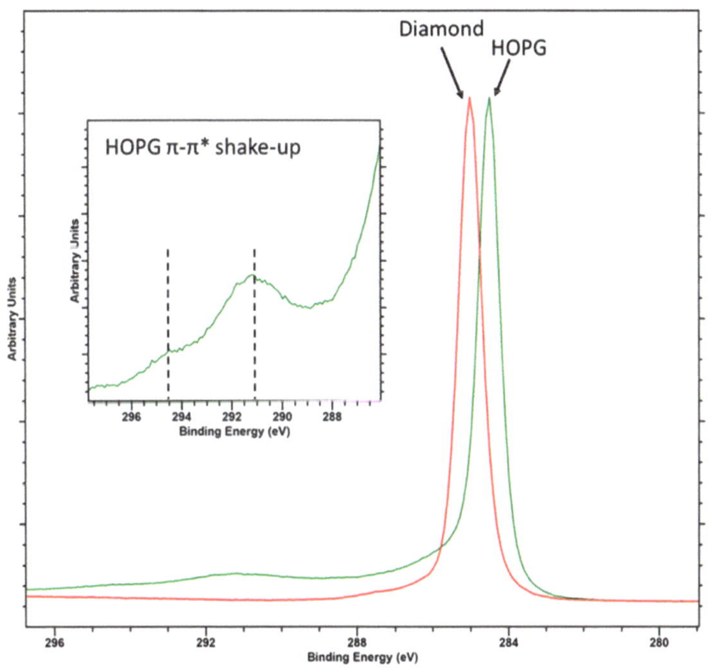

Figure 1. Overlayed C (1s) core level for diamond (red spectra) and HOPG (green spectra) with inset of the graphitic satellite structure.

The fitting of the graphitic C (1s) envelope is one of the most common errors encountered in the literature [15], and the asymmetric shape is something commonly ignored in may published papers. Recently, Linford et al. [11] reported a method for fitting the C (1s) envelope of graphitic carbon materials, using a reference spectrum of clean graphite/HOPG as a model for the graphitic backbone. This method is useful and is indeed a method the current author has used for many years. However, as shown in Figure 2, an apparently pure sp^2 graphitic carbon (here a "nano-onion" [3]) cannot be fitted with the parameters taken from HOPG and likely stems from a combination of hydrogenation and the curvature of the carbon surface [14].

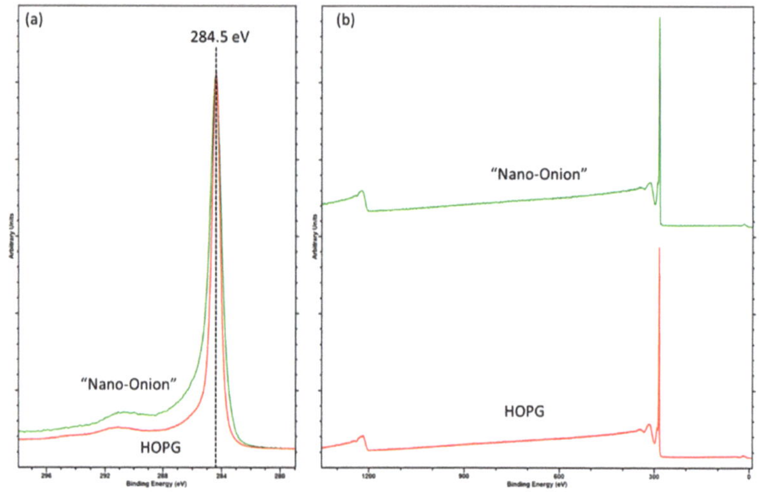

Figure 2. (**a**) C (1s) core-level and (**b**) survey spectra for HOPG, and hydrogenated "nano-onion" carbons illustrating the influence of hydrogenation on the C (1s) spectra.

Hydrogenation undoubtedly introduces some sp^3 character given the broadening around 285 eV and accompanied by a reduction in the D-parameter (discussed in the following section) to 20.5. However, note the broadening with respect to the narrow HOPG spectrum, taking the difference spectra (not shown) shows some perturbation of the π-π* shake-up, which may be expected if there is a lattice disruption, accompanied with some broadening centered at 284.8 and 284 eV and are values consistent with Blume et al. [14] and assigned as disordered (thought to be due to random orientations of dangling bonds with respect to the carbon atoms) and defective carbon, respectively. Such defective induced broadening will cause a change in the main peak asymmetry due to screening of localized areas of charge and the need to adjust any asymmetric fitting parameters for the graphitic peak.

As will be shown in the following sections, analysis, especially fitting, of the C (1s) envelope for graphitic materials is complex and should not be fitted unless a rigorous and self-consistent model is used.

3.2. The C (KVV) Auger Peak and the D-Parameter

Not to be confused with the D-band in Raman spectroscopy of carbon materials, which is a measure of disorder of the graphitic material [16–18], use of the XPS-derived D-parameter to discriminate carbon states was demonstrated by Lascovich and co-workers [19–22], who showed a linear relationship existed between the extremes of pure sp^2 and sp^3 carbons by taking the first differential of the carbon Auger peak, where depending upon the relative concentrations of sp^2 and sp^3 carbon, the separation (D, hence D-parameter) of the differential maxima and minima differ greatly. Figure 3 shows a plot of the values of Lascovich et al. together with the first differential of the Auger peaks for diamond and HOPG. This approach has successfully been used to distinguish sp^2 and sp^3 rich carbon rich domains through XPS mapping experiments [23,24].

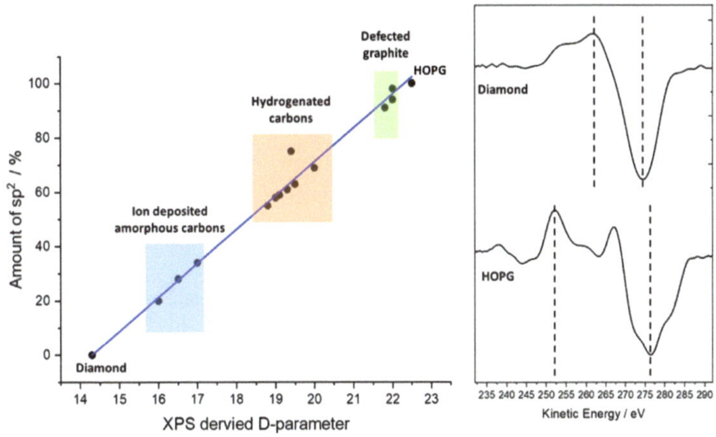

Figure 3. Plot of XPS-derived D-parameter (values taken from [19–22]) and the first differential of the C (KVV) Auger for diamond and clean HOPG indicating the XPS-derived D-parameter.

Consideration should be given in taking the D-parameter as an exact value, however. The electron kinetic energy of the Auger peak is lower than that of the C (1s) photoelectron kinetic energy (ca. 260 and 1200 eV, respectively), yielding an electron inelastic mean free path (IMFP) of 3.3 nm for the core-level photoelectron and 1.1 nm for the Auger electron—a threefold difference in electron depth and therefore a similar magnitude of difference in the XPS information depth [25]. The Auger electrons are more surface sensitive, hence sensitive to contamination. This is demonstrated by the data in Figure 4, where a freshly cleaved HOPG sample has been analyzed and then gently cleaned using an argon cluster source (operating at 4 kV, 2000 atoms); note how the C (1s) data are almost indistinguishable, while the Auger peaks differ significantly. The D-parameters here are 22 eV (cleaved) and 23.5 eV (cluster cleaned [26]), the latter being higher than that previously reported, while the former is in the range typically reported for graphite [12,13,19–21,23,27]; note none of these references mention in situ cleaning.

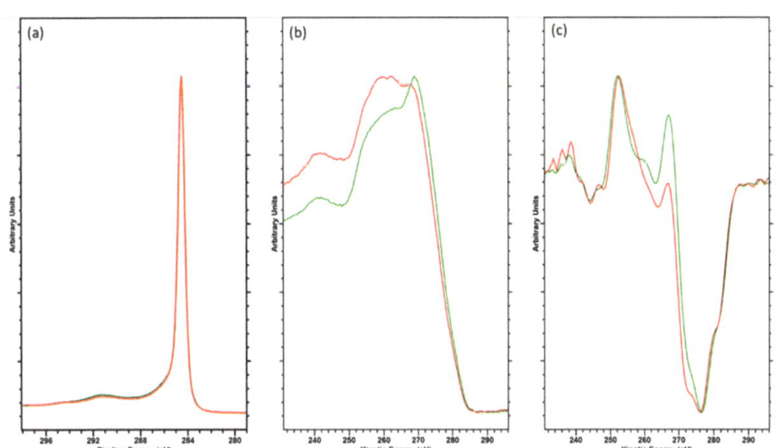

Figure 4. Overlayed data for (**a**) C (1s) core-level, (**b**) C (KVV) Auger and (**c**) first derivative C (KVV) spectra for freshly cleaved (red) and argon cluster cleaned (green) HOPG.

Irrespective of these values, there will be an uncertainty in the determination of the D-parameter depending on the method and parameters used. Table 1 shows calculated values of the D-parameter from two common XPS analysis software packages using the default available options illustrating how, depending on the method used, the D-parameter can vary by as much as 1.5 eV, corresponding to a variance of ca. 20% sp^2 content using the data plotted in Figure 3.

Table 1. D-parameter values for cluster cleaned HOPG determined by different software and methods.

Software	Method	Parameters Used	D-Parameter/eV
Thermo Scientific Avantage v5.9925	D-parameter in-built routine	Gaussian, Smooth width 2, Differentiation width 1	24.0
		Savitsky–Golay, Smooth width 2, Differentiation width 1	24.0
		Moving Average, Smooth width 2, Differentiation width 1	24.0
		Wiener, Smooth width 2, Differentiation width 1	22.5
		Gaussian, Smooth width 2, Differentiation width 1, Quartic Fit	23.5
		Savitsky–Golay, Smooth width 2, Differentiation width 1, Quartic Fit	23.5
		Moving Average, Smooth width 2, Differentiation width 1, Quartic Fit	23.5
		Wiener, Smooth width 2, Differentiation width 1, Quartic Fit	23.0
CasaXPS v2.3.24	Polynomial regression followed by "SP" background type	"PR" background with value of 35, Differentiation of synthetic envelope	24.5
		"PR" background with value of 32, Differentiation of synthetic envelope	24.0
	Differentiation followed by "SP" background type	Savitsky–Golay Quadratic smoothing width 5	24.5
		Savitsky–Golay Quadratic smoothing width 7	24.0
		Savitsky–Golay Quadratic smoothing width 9	23.5
		Savitsky–Golay Quadratic smoothing width 15	23.5

It is worth noting the C (KVV) Auger is typically recorded at a high pass energy, a step size of 0.5 eV and sufficient scans for a good signal-to-noise ratio (typically at least ×50) [28].

3.3. Complementary Use of the C (1s) and O (1s) Spectra

The presence of oxygen functionalities on the carbon surface is expected to have a notable influence on the shape of the C (1s) spectra due to the formation of different carbon–oxygen bonds. However, these changes are not always clear, especially if the moieties are in low concentration or poorly dispersed. The C-O signals themselves are generally found between ca. 285 and 290 eV, i.e., between the C-C peak and the satellite structure for graphitic carbon, and are fitted with the assumption that each C-O species will have equal FWHMs. Table 2 gives some typical ranges for carbon species relative to graphitic carbon; note how some fall in the region of the carbon satellite structure.

A meaningful fit can only be made by appreciation of the asymmetric tail of the sp^2 carbon which forms a pseudo-background on which these signals are superimposed. A simple check of the fit can be made by comparing the corrected integrated photoelectron peak areas of the C-O-containing functions (while ignoring any C-C and π-π^* structure) with that of the total oxygen content. Ideally, these should be equal but given uncertainty in fitting given appropriate line shape selection, etc., then a small mismatch in the range of ca. 1 at % can be considered good. An example of this approach is shown in Figure 5, the data are of a commercially available plasma-oxidized few-layer graphene sample, where the total oxygen content (ca. 17 at%) is approximately equal with all four oxygen-containing functionalities; the difference being <0.3 at%.

Table 2. Typical ranges for C (1s) species relative to graphitic carbon at 284.5 eV [29].

Speciation	Binding Energy Range Relative to 284.5 eV
C-C/C-H	284.8–285.5
C-O-C/C-OH	286.0–287.0
R-C = O/O-C-O	287.5–288.5
O-C = O/CO3	288.5–289.0
R-(CO)-O-(CO)-R	289.0–290.0
R-O-(CO)-O-R	290.0–295.5
CF/Cl	286.5–287.5
CF_2	290.0–291.0
CF_3	291.0–292.0

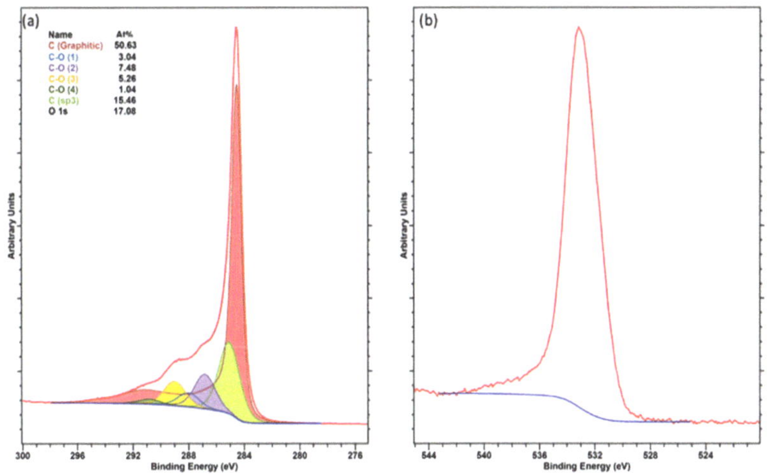

Figure 5. Fitted (**a**) C(1s) and unfitted (**b**) O (1s) spectra for an oxidized graphitic carbon.

Note that in this example, the O (1s) has not been fitted, but by having increased confidence in our C (1s) fit, we could now begin to assign chemistry to the O (1s) region, but this is not always necessary, nor easy, especially as some functionalities can have very similar binding energies [29]. The oxygen region can also be used diagnostically, with the tail to higher binding energy indicating the satellite structure of carbonyl-containing fragments and should be used to inform the fit of the C (1s) envelope.

Using this informed approach, the fit yields a sp^3 carbon content of ca. 15.5%, in excellent agreement with the measured D-parameter of this sample (value 21 ($\pm$0.5 eV), which is ca. 15% sp^3 based on data in Figure 3), whereas significant deviation is found when fitting the C (1s) spectrum alone, leading to some larger disparity between the sp^3 content derived from the D-parameter and peak fit [30].

Of course, the data in Figure 5 are a well-defined case, but serve to illustrate that appreciation of these spectra features can influence our fitting and understanding of the C (1s) spectrum, especially if the C (1s) spectra is not as clearly oxidized as that shown.

Such assignments can be confirmed by chemical derivatization [31,32] using fluorine- or nitrogen-containing reagents with emphasis on changes observed in the O(1s) region. For example, Burgess et al. have shown for acid-treated HOPG, alcohol, ketone and ether functions can be selectively reacted and identified; -COOH functionalities can also be derivatized, but no evidence of these were found in that work [33]. Unpublished work from the same laboratory shows such derivatization also works for carbon powders.

4. Conclusions

A brief overview has been presented on a systematic approach to extracting chemical information for the functionalities on graphitic carbons. Using the logical approach and, where applicable, parameters discussed herein, using the O (1s) spectrum in a complimentary manner, then an algorithmicized approach to fitting, such as that shown in Figure 6, will lead to more reliable and meaningful chemical speciation being extracted from the C (1s) spectrum of the carbon material for functionalized carbon such as graphitic and fullerene materials [34,35].

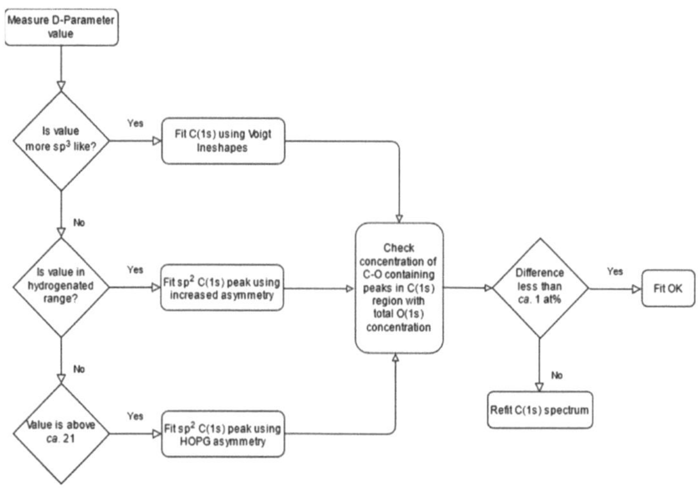

Figure 6. Methodology for the fitting of carbon materials described as a flowchart.

Funding: The EPSRC is acknowledged for funding the National Facility for XPS (HarwellXPS), operated by Cardiff University and UCL, under contract number PR16195.

Institutional Review Board Statement: Not applicable.

Data Availability Statement: Data sharing is not applicable to this article as no new data were created or analyzed in this study.

Conflicts of Interest: The author declares no conflict of interest.

References

1. Cazorla-Amorós, D. Grand challenges in carbon-based materials research. *Front. Mater.* **2014**, *1*. [CrossRef]
2. Blanchard, N.P.; Hatton, R.A.; Silva, S.R.P. Tuning the work function of surface oxidised multi-wall carbon nanotubes via cation exchange. *Chem. Phys. Lett.* **2007**, *434*, 92–95. [CrossRef]
3. Butenko, Y.V.; Krishnamurthy, S.; Chakraborty, A.K.; Kuznetsov, V.L.; Dhanak, V.R.; Hunt, M.R.C.; Šiller, L. Photoemission study of onionlike carbons produced by annealing nanodiamonds. *Phys. Rev. BCondens. Matter Mater. Phys.* **2005**, *71*. [CrossRef]
4. Raffa, V.; Ciofani, G.; Nitodas, S.; Karachalios, T.; D'Alessandro, D.; Masini, M.; Cuschieri, A. Can the properties of carbon nanotubes influence their internalization by living cells? *Carbon* **2008**, *46*, 1600–1610. [CrossRef]
5. Jeong, H.K.; Yun, P.L.; Lahaye, R.J.W.E.; Park, M.H.; Kay, H.A.; Ick, J.K.; Yang, C.W.; Chong, Y.P.; Ruoff, R.S.; Young, H.L. Evidence of graphitic AB stacking order of graphite oxides. *J. Am. Chem. Soc.* **2008**, *130*, 1362–1366. [CrossRef]
6. An, K.H.; Yang, C.M.; Seo, K.; Park, K.A.; Lee, Y.H. A diameter-dependent separation of semiconducting from metallic single-wall carbon nanotubes by using nitronium ions. *Curr. Appl. Phys.* **2006**, *6*. [CrossRef]
7. Cheung, T.T.P. X-ray photoemission of carbon: Lineshape analysis and application to studies of coals. *J. Appl. Phys.* **1982**, *53*, 6857–6862. [CrossRef]
8. Ouyang, Y.; Peng, J.C.; Wang, H.; Peng, Z.H. The rehybridization of electronic orbitals in carbon nanotubes. *Chinese Phys. B* **2008**, *17*, 3123–3129. [CrossRef]
9. Chen, X.; Wang, X.; Fang, D. A review on C1s XPS-spectra for some kinds of carbon materials. *Full- Nanotub. Carbon Nanostructures* **2020**, *28*, 1048–1058. [CrossRef]

10. Fairley, N.; Fernandez, V.; Richard-Plouet, M.; Guillot-Deudon, C.; Walton, J.; Smith, E.; Flahaut, D.; Greiner, M.; Biesinger, M.; Tougaard, S.; et al. Systematic and collaborative approach to problem solving using X-ray photoelectron spectroscopy. *Appl. Surf. Sci. Adv.* **2021**, *5*, 100112. [CrossRef]
11. Gengenbach, T.R.; Major, G.H.; Linford, M.R.; Easton, C.D. Practical guides for x-ray photoelectron spectroscopy (XPS): Interpreting the carbon 1s spectrum. *J. Vac. Sci. Technol. A* **2021**, *39*, 013204. [CrossRef]
12. Kaciulis, S. Spectroscopy of carbon: From diamond to nitride films. *Surf. Interface Anal.* **2012**, *44*, 1155–1161. [CrossRef]
13. Kaciulis, S.; Mezzi, A.; Calvani, P.; Trucchi, D.M. Electron spectroscopy of the main allotropes of carbon. *Surf. Interface Anal.* **2014**, *46*, 966–969. [CrossRef]
14. Blume, R.; Rosenthal, D.; Tessonnier, J.P.; Li, H.; Knop-Gericke, A.; Schlögl, R. Characterizing Graphitic Carbon with X-ray Photoelectron Spectroscopy: A Step-by-Step Approach. *Chem Cat Chem* **2015**, *7*, 2871–2881. [CrossRef]
15. Major, G.H.; Avval, T.G.; Moeini, B.; Pinto, G.; Shah, D.; Jain, V.; Carver, V.; Skinner, W.; Gengenbach, T.R.; Easton, C.D.; et al. Assessment of the frequency and nature of erroneous x-ray photoelectron spectroscopy analyses in the scientific literature. *J. Vac. Sci. Technol. A* **2020**, *38*, 061204. [CrossRef]
16. Merlen, A.; Buijnsters, J.G.; Pardanaud, C. A guide to and review of the use of multiwavelength Raman spectroscopy for characterizing defective aromatic carbon solids: From graphene to amorphous carbons. *Coatings* **2017**, *7*, 153. [CrossRef]
17. Wu, J.B.; Lin, M.L.; Cong, X.; Liu, H.N.; Tan, P.H. Raman spectroscopy of graphene-based materials and its applications in related devices. *Chem. Soc. Rev.* **2018**, *47*, 1822–1873. [CrossRef]
18. Ferrari, A.C.; Robertson, J. Raman spectroscopy of amorphous, nanostructured, diamond-like carbon, and nanodiamond. *Philos. Trans. R. Soc. A Math. Phys. Eng. Sci.* **2004**, *362*, 2477–2512. [CrossRef]
19. Lascovich, J.C.; Scaglione, S. Comparison among XAES, PELS and XPS techniques for evaluation of Sp^2 percentage in a-C:H. *Appl. Surf. Sci.* **1994**, *78*, 17–23. [CrossRef]
20. Lascovich, J.C.; Giorgi, R.; Scaglione, S. Evaluation of the sp^2/sp^3 ratio in amorphous carbon structure by XPS and XAES. *Appl. Surf. Sci.* **1991**, *47*, 17–21. [CrossRef]
21. Lascovich, J.C.; Santoni, A. Study of the occupied electronic density of states of carbon samples by using second derivative carbon KVV Auger spectra. *Appl. Surf. Sci.* **1996**, *103*, 245–253. [CrossRef]
22. Lascovich, J.C.; Rosato, V. Analysis of the electronic structure of hydrogenated amorphous carbon via Auger spectroscopy. *Appl. Surf. Sci.* **1999**, *152*, 10–18. [CrossRef]
23. Barlow, A.J.; Scott, O.; Sano, N.; Cumpson, P.J. Multivariate auger feature imaging (MAFI): A new approach towards chemical state identification of novel carbons in XPS imaging. *Surf. Interface Anal.* **2015**, *47*, 173–175. [CrossRef]
24. Barlow, A.J.; Popescu, S.; Artyushkova, K.; Scott, O.; Sano, N.; Hedley, J.; Cumpson, P.J. Chemically specific identification of carbon in XPS imaging using Multivariate Auger Feature Imaging (MAFI). *Carbon* **2016**, *107*, 190–197. [CrossRef]
25. Tanuma, S.; Powell, C.J.; Penn, D.R. Calculations of electron inelastic mean free paths. V. Data for 14 organic compounds over the 50–2000 eV range. *Surf. Interface Anal.* **1994**, *21*, 165–176. [CrossRef]
26. Morgan, D.J. Cluster cleaned HOPG by XPS. *Surf. Sci. Spectra* **2017**, *24*, 024003. [CrossRef]
27. Lesiak, B.; Kövér, L.; Tóth, J.; Zemek, J.; Jiricek, P.; Kromka, A.; Rangam, N. C sp^2 /sp^3 hybridisations in carbon nanomaterials—XPS and (X)AES study. *Appl. Surf. Sci.* **2018**, *452*, 223–231. [CrossRef]
28. Xpssimplified. Available online: https://xpssimplified.com/elements/carbon.php (accessed on 21 June 2021).
29. Beamson, G.; Briggs, D. (Eds.) *High Resolution XPS of Organic Polymers: The Scienta ESCA300 Database*; Wiley & Sons: Chichester, UK, 1992; p. 267.
30. Theodosiou, A.; Spencer, B.F.; Counsell, J.; Jones, A.N. An XPS/UPS study of the surface/near-surface bonding in nuclear grade graphites: A comparison of monatomic and cluster depth-profiling techniques. *Appl. Surf. Sci.* **2020**, *508*. [CrossRef]
31. Malitesta, C.; Losito, I.; Sabbatini, L.; Zambonin, P.G. Applicability of chemical derivatization—X-ray photoelectron spectroscopy (CD-XPS) to the characterization of complex matrices: Case of electrosynthesized polypyrroles. *J. Electron Spectros. Relat. Phenom.* **1998**, *97*, 199–208. [CrossRef]
32. Pippig, F.; Sarghini, S.; Holländer, A.; Paulussen, S.; Terrvn, H. TFAA chemical derivatization and XPS. Analysis of OH and NH_x polymers. *Surf. Interface Anal.* **2009**, *41*, 421–429. [CrossRef]
33. Burgess, R.; Buono, C.; Davies, P.R.; Davies, R.J.; Legge, T.; Lai, A.; Lewis, R.; Morgan, D.J.; Robinson, N.; Willock, D.J. The functionalisation of graphite surfaces with nitric acid: Identification of functional groups and their effects on gold deposition. *J. Catal.* **2015**, *323*, 10–18. [CrossRef]
34. Fortner, J.D.; Kim, D.I.; Boyd, A.M.; Falkner, J.C.; Moran, S.; Colvin, V.L.; Hughes, J.B.; Kim, J.H. Reaction of water-stable C60 aggregates with ozone. *Environ. Sci. Technol.* **2007**, *41*, 7497–7502. [CrossRef] [PubMed]
35. Weippert, J.; Ulas, S.; Waldt, E.; Amati, M.; Gregoratti, L.; Kiskinova, M.; Böttcher, A. C 68: A non-IPR fullerene capable of binding extraordinary amounts of Cs atoms. *Fuller. Nanotub. Carbon Nanostructures* **2019**, *27*, 206–214. [CrossRef]

C Journal of Carbon Research

MDPI

Review

Carbons Formed in Methane Thermal and Thermocatalytic Decomposition Processes: Properties and Applications

Emmi Välimäki [1], Lasse Yli-Varo [1], Henrik Romar [2] and Ulla Lassi [1,*]

[1] Research Unit of Sustainable Chemistry, University of Oulu, P.O. Box 4300, FI-90014 Oulu, Finland; emmi.valimaki@oulu.fi (E.V.); lasse.yli-varo@oulu.fi (L.Y.-V.)
[2] Hycamite TCD Technologies Ltd., FI-67100 Kokkola, Finland; henrik.romar@hycamite.com
* Correspondence: ulla.lassi@oulu.fi; Tel.: +358-400294090

Abstract: The hydrogen economy will play a key role in future energy systems. Several thermal and catalytic methods for hydrogen production have been presented. In this review, methane thermocatalytic and thermal decomposition into hydrogen gas and solid carbon are considered. These processes, known as the thermal decomposition of methane (TDM) and thermocatalytic decomposition (TCD) of methane, respectively, appear to have the greatest potential for hydrogen production. In particular, the focus is on the different types and properties of carbons formed during the decomposition processes. The applications for carbons are also investigated.

Keywords: methane; hydrogen; carbon; catalytic decomposition; wet decomposition; dry decomposition; hydrogen economy

Citation: Välimäki, E.; Yli-Varo, L.; Romar, H.; Lassi, U. Carbons Formed in Methane Thermal and Thermocatalytic Decomposition Processes: Properties and Applications. C **2021**, 7, 50. https://doi.org/10.3390/c7030050

Academic Editor: Gil Goncalves

Received: 4 June 2021
Accepted: 23 June 2021
Published: 25 June 2021

1. Introduction

The hydrogen economy has attracted a significant amount of interest since the European Union (EU) revealed a new strategy to reach carbon neutrality by 2050. The increased interest in hydrogen is due to its potential as a fuel and as an energy carrier, its capability for energy storage, and its use as a carbon-neutral feedstock without carbon dioxide (CO_2) emissions. At present, hydrogen is mainly produced from fossil fuels, and hydrogen production in the EU releases about 70 to 100 million tons of CO_2 per annum [1]. If a means of production can be developed that does not release CO_2, hydrogen will play a key role in creating a climate-neutral Europe. One major issue to be resolved is how to make the CO_2-free hydrogen production process cost competitive compared with fossil-based hydrogen processes.

Production of hydrogen from methane is considered to be the best option compared to the use of other hydrocarbons. This is due to the fact that methane is abundant, and can be easily transported and stored when needed. Methane also has a high hydrogen to carbon ratio of 4:1 [2]. In addition to methane's significant importance in hydrogen production, methane is also used in power generation and methanol production [3]. Hydrogen can be produced from methane in several ways. In this article, we present the most important production processes and note some advantages and disadvantages for each process. At present, the most commonly used processes for hydrogen production from methane, and especially renewable biomethane, are steam methane reforming (SMR), dry methane reforming (DMR), and partial oxidation (PO) [4]. None of these processes are CO_2-neutral; CO_2 is emitted, or a separate carbon capture process is required after the main process. The addition of the extra step increases the price of the hydrogen produced. In the thermal decomposition of methane (TDM) and thermocatalytic decomposition (TCD) of methane, no oxygen is involved in the reaction, and solid carbon is formed. Compared to other methane conversion methods in which carbon is released to the atmosphere, TDM and TCD processes create carbon capture and therefore have a significant effect on carbon's neutrality and footprint. Most of the recently published papers consider the formation and use of the hydrogen formed in the decomposition of methane, whereas the carbons formed in the process are considered as less important. However, the carbon formed has its own markets and possible applications, such as usage in water purification and energy storage applications. A full use of the carbons formed in the TDM and TCD processes in potential applications would result in increased value, and increase the profitability of these decomposition processes and the overall hydrogen economy. In this paper, we focus on the carbon produced from the TCD process, its properties, and possible applications. We also briefly examine other methane conversion processes to highlight the benefits of TCD.

2. Production Methods for Hydrogen and Carbon from Methane

Various methods exist for producing hydrogen from methane, such as SMR, DRM, PO, TDM, and TCD. The latter two processes are considered more environmentally friendly because they do not create CO_2 or carbon monoxide (CO) emissions [5]. One important difference between these methods is the end use of the carbon. In the first three reactions (SMR, DRM, and PO), carbon is emitted as CO_2 or CO, whereas in TDM and TCD, solid deposits of carbon are formed. Instead of producing CO_2 gas, TDM/TCD processes produce carbon as a potentially valuable solid. The reactions for these processes and corresponding reaction enthalpies are presented in Table 1.

Table 1. Chemical reactions for hydrogen production from methane and corresponding reaction enthalpies. In the first three reactions (SMR, DRM, and PO), carbon is emitted as CO_2 or CO, whereas in TDM and TCD, solid deposits of carbon are formed.

Production Method	Chemical Equation	Reaction Enthalpies	Equation Number
Steam methane reforming (SMR)	$CH_4 + H_2O \rightarrow CO + 3H_2$	$\Delta H_{298K} = 206$ kJ/mol	(1)
Water gas shift reaction (WGS)	$CO + H_2O \rightarrow CO_2 + H_2$	$\Delta H_{298K} = -41$ kJ/mol	(2)
Dry reforming of methane (DRM)	$CH_4 + CO_2 \rightarrow 2CO + 2H_2$	$\Delta H_{298K} = 247$ kJ/mol	(3)
Partial oxidation (PO)	$CH_4 + 0.5O_2 \rightarrow CO + 2H_2$	$\Delta H_{298K} = -23$ kJ/mol	(4)
Thermal decomposition of methane (TDM) Thermocatalytic decomposition of methane (TCD)	$CH_4 \rightarrow C + 2H_2$	$\Delta H_{298K} = 75$ kJ/mol	(5)

SMR (Equation (1)) is one of the most mature processes in hydrogen production, starting from methane. The SMR reaction is endothermic and therefore has a high energy consumption. However, the SMR process has some severe drawbacks and challenges that should be considered. The process is usually carried out at relatively high temperatures, 973–1273 K, and at reaction pressures ranging from 0.3 to 2.5 MPa [6]. Another drawback is that SMR generates large amounts of CO_2 [5]. According to Soltani et al. [7], SMR generates approximately 7 kg of CO_2 to every 1 kg of H_2 produced. Although SMR is one of the least expensive processes in large-scale hydrogen production [6], a harsh reaction environment can create corrosion risks, especially if sulfur is present, a condition that can appear if the gas is not adequately purified [6,8]. In SMR, the catalyst plays a key role in the process. The use of a catalyst increases the yield of hydrogen and lowers the temperature, and no pressure is used. The cost of the catalysts, possible toxicity of the active metals used, and availability and deactivation of the catalyst by the formation of coke, should be considered [4,9]. To gain additional hydrogen from the SMR process, the water gas shift (WGS) reaction (Equation (2)) is usually introduced. In the WGS, CO is transformed into CO_2 and hydrogen, usually after the SMR reaction [9]. WGS is an exothermic reaction that is normally undertaken in two reactors: one with a high temperature in which thermal equilibrium is reached and one with a lower temperature in which the catalyst provides higher conversions; that latter process is much slower. In addition to SMR, WGS is also affected by process parameters, and therefore the variation in performance can vary. Another drawback of WGS is that the activity of the reaction decreases at lower temperatures, which creates additional carbon. This can be an issue, especially in fuel cell applications where lower CO is required.

Another well-known hydrogen production process is dry reforming (Equation (3)). One of the main benefits of DRM is that it consumes methane and CO_2 to create hydrogen, and it is also cheaper when compared with SMR and PO [6]. However, like SMR, it produces CO [9]. The operating temperature ranges from 923 to 1123 K, and the pressure is normally 0.1 MPa [6].

PO is an exothermic reaction (Equation (4)) carried out at a pressure of 10 MPa and a temperature range of 1223 to 1373 K [4,10]. Due to the exothermic reaction, the process is difficult to control because of the induced hotspots on the catalyst [8]. If air is used as an oxygen source, nitrogen separation is required from the gas prior to the reaction, a process that increases operational costs. It is therefore recommended to use pure oxygen and/or separate the oxygen from the air [10]. Oxygen separation from air makes the process costly [4]. The benefits of the process are short residence time, high conversion rates, high selectivity of syngas, good economic conversion, and compactness [8,11–13].

Thermal decomposition of methane has two alternating routes (Equation (5)): catalytic (TCD) and non-catalytic (TDM) [14]. The latter has been used for several decades for the production of carbon black. Methane is one of the most stable hydrocarbons due to its very strong carbon–hydrogen bond and high molecular structure symmetry [15]. In TDM, the reaction temperature is relatively high, and may reach 1473 K for a reasonable hydrogen yield [14,15]. By using a catalyst, this temperature can be significantly lowered, and the catalyst can influence the formed carbon. Both processes are CO_2-free, and therefore they have gained increased interest. In this paper, however, we focus on the TCD catalytic reaction, because the carbon formed can have different morphologies and greater value than with the use of TDM. TCD has remained in the laboratory phase mostly because of the deactivation of the catalyst, the possibility of CO_2 formation during reactivation of the catalyst, and the unreacted methane in hydrogen [15]. Several catalysts have been tested for the reaction, and these can be divided into two separate groups: noble metals and transition metals, such as nickel and iron; and carbon-based materials, such as activated carbon [16]. Metal catalysts can be supported by high surface area materials, such as Al_2O_3 and SiO_2, and carbon can be doped with some metals. Different catalysts have varying effects on the process, and so it is important to select the best catalyst to produce carbon. The effect of catalysts is discussed later in this paper. Economic and environmental issues should also be considered.

In the TCD process, the methane molecules decompose on the surface of the catalytic metals, and in the crevices or cracks of the catalyst. The reaction generates hydrogen molecules and carbon atoms. Carbon diffuses to the metal and accumulates to form carbon deposits [17,18]. In TCD, amorphous, graphitic, and carbon nanotube (CNT) morphologies of carbon can be found, and the process parameters affect the morphology of the carbon. The reaction temperature is above 1473 K, and the carbon formed is mainly amorphous [16]. The CNTs' growth mechanism can be divided into two types, tip growth and base growth, in which one or the other is active, depending on the interaction between catalyst metal and support weakness. Weak interaction leads to tip growth, in which the catalyst particle is lifted by the growing CNT. Strong supports lead to base growth, in which the growth mechanism proceeds through an open tip. Regardless of the manner in which the carbon grows on the catalyst, one of the main problems to be resolved is the separation of the carbon and catalyst used without generating any CO_2 emissions [19]. The size of the catalyst particle can affect the growth of the carbon in TCD process. This is schematically shown in Figure 1, in which the possible effect of large and small nickel catalyst particles on the growth of carbon is shown. Large nickel particles induce more tip growth of carbon, whereas, in the case of small nickel particles, carbon grows on the catalyst particle and encapsulates it.

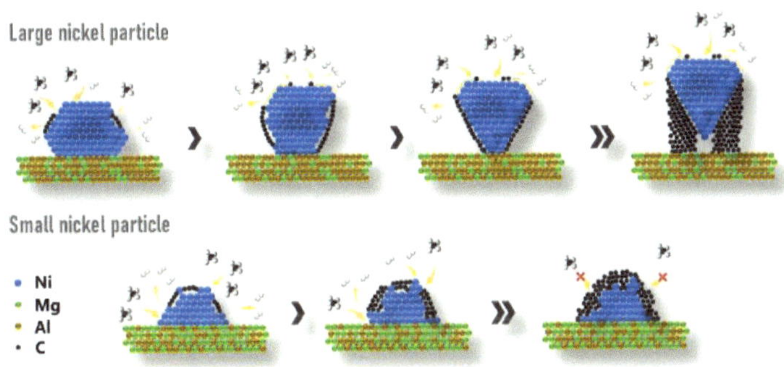

Figure 1. Possible mechanisms of carbon growth in the presence of large and small nickel particles [18]. Large nickel particles induce tip growth of carbon, whereas, in the case of small nickel particles, carbon grows on the catalyst particle and encapsulates it.

3. Carbon Morphologies

Carbon is a polymorphic material; it can exist in more than one form, such as diamond, graphite, and fullerene (also known as carbon nano-onions (CNOs) and CNTs). Its morphologies are shown in Figure 2. In addition, the level of crystallinity varies; carbon can have amorphous, turbostratic, and graphitic crystallinity [20]. Graphite has a crystal structure that is composed of carbon atoms hexagonally arranged in layers. Each carbon atom has a strong covalent bond among three neighboring atoms, and the fourth electron can form a weak bond between layers, which allows these layers to be separated. These increase, for example, the lubricant properties of graphite. In addition, the hexagonal sheets have relatively high electrical conductivity.

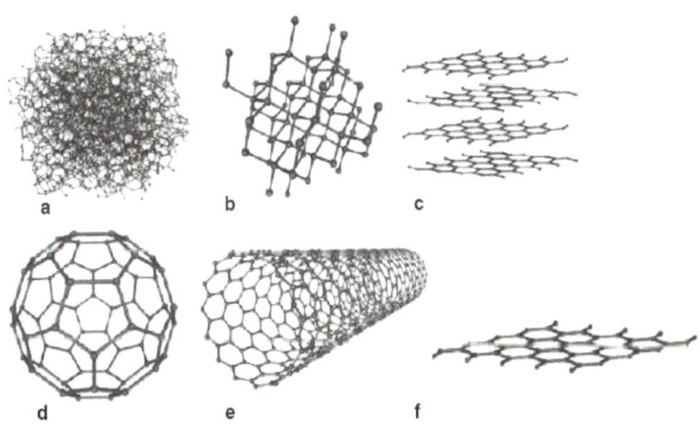

Figure 2. Structures of various carbon morphologies: (**a**) amorphous carbon, (**b**) diamond, (**c**) graphite, (**d**) fullerene, (**e**) carbon nanotube (CNT), and (**f**) graphene [21]. Adapted with permission from [21]. Springer Nature, 2011.

The structure, texture, and morphologies of the carbons formed in the TCD reaction depend on the reaction conditions, such as pressure, temperature, catalysts, and gas phase composition. These properties make them suitable for many applications. The properties that affect applicability, and are thus of common interest, are surface area, pore volume, pore size, porosity, and surface chemistry, in addition to the morphology [22]. Transmission electron microscopy (TEM) and scanning electron microscopy (SEM) are used to study the length, thickness, morphology, impurities, and topography of carbon deposits. Thermogravimetric analysis (TGA) is used to quantify the amount of the deposits and the thermal stability of the specimen. Specific surface area is measured according to the theory of Brunauer–Emmett–Teller, commonly called BET analysis. X-ray diffraction (XRD) is used to create diffractograms of the deposits, and the crystallinity, phase structure, and purity of the specimen can be obtained from XRD. Figure 3 shows XRD patterns for carbons produced by the TCD process using three different nickel catalysts. XRD peaks indicate the crystal structure of materials. In Figure 3, the pink circle indicates carbon morphology, which, in this case, is graphite for all produced carbons. Raman spectra can also be used to determine purity, crystallinity, and degree of structural disorder, for example, the defects and tube alignment of the CNT [23–26].

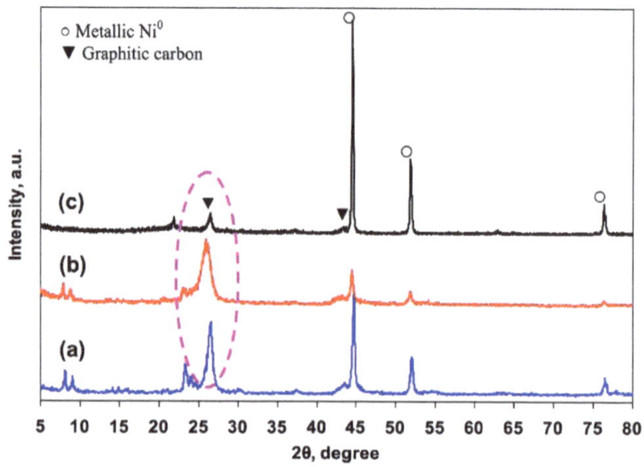

Figure 3. XRD patterns for carbon materials produced by the TCD process over reduced (**a**) Ni/Z-25, (**b**) Ni/Z-400, and (**c**) Ni/AS catalysts [27]. In all cases, a graphite-like carbon crystal structure is seen. Adapted with permission from [27]. Elsevier, 2016.

One of the most studied components in the TCD process to produce carbon are CNTs. These are rolled graphene layers that can form single- or multiwalled tubes, which can differ in thickness, diameter, and layer amount [28]. Typical properties of CNTs are high electrical conductivity and high tensile strength. Physically and chemically, CNTs are stable [29]. They are used in polymers, wastewater/pollution treatment, conductive paints/coatings, biosensors, electronics, energy storage, and environmental applications [29,30]. A major application is to reinforce polymers to improve thermal, electrical, and mechanical properties, and interest in using CNTs in the production of lithium-ion batteries has increased, because they improve the efficiency of the batteries [30].

Carbon fibers (CNFs) are polycrystalline; they have both graphitic and amorphous regions. The amorphous regions lack hexagonal carbon networks that are characteristic of graphite [20]. CNFs have high tensile strength and stiffness, and low density and coefficient of thermal expansion. They are used in the aerospace, vehicle manufacturing, energy, sports equipment, and chemical and textile industries, and as a reinforcement in composite materials [30].

Carbon fullerenes resemble the shape of a sphere [20]. One form of carbon fullerenes is commonly referred to as a CNO, which are multiwalled fullerenes [31]. Carbon fullerenes have not been found in the TCD reaction, but CNOs have. Fullerenes have shown significant promise in several applications, such as solid lubrication, electromagnetic shielding, fuel cells, heterogeneous catalysis, energy storage, electro-optical devices, and supercapacitors [32]. CNOs can have a crucial effect on the TCD process. According to the investigations of Zhou et al. [19], formed CNOs encapsulated the catalyst and, therefore, the catalyst was deactivated.

4. Process Parameters Affecting the Properties of Carbon in TDM and TCD Processes

As mentioned previously, the TCD of methane produces hydrogen and carbon. The yields and carbon morphology are dependent on various process parameters, such as the kind of catalyst used, size of the catalyst's metal particle, promoter material, reaction temperature, pressure, and gas hourly space velocity (GHSV) [33–40]. According to Li et al. [40], the carbon yield is mostly affected by the particle size of the metallic catalyst, and also affects the catalytic life [41]. It has also been mentioned that the carbon morphology affects the efficiency of thermocatalytic decomposition when carbon is used as a catalyst support [35]. The factors affecting catalytic life have been widely studied, and it is valuable to understand how this occurs and the kind of carbon that is formed. In the TDM process, temperatures are higher than those in TCD, and are typically above 1200 °C. The main products in TDM are carbon black graphite-like carbon and coke [42].

4.1. *Effect of Catalyst and Promoters on the Carbon Amount and Quality*

Guil-Lopez et al. [43] compared metal and carbon catalysts in the TCD process, and their differences in terms of hydrogen production and carbon morphology (size and shape). They used Ni and Fe metal catalysts with different supports and six different carbon catalysts, which included carbon black, activated carbon, graphene, and CNTs. Interest in carbon catalysts has risen due to their lower cost, higher temperatures, which affect the methane conversion rate, and the possibility that there is no need to separate catalyst carbon and produced carbon [16] According to the results of a study by Guil-Lopez et al. [43], metal catalysts are more prone to produce CNTs when carbon catalysts produced more graphene-like carbon. Carbon catalyst activity depends on the chemical structure, such as defects, BET surface area, and pore volume. By designing and preparing carbon materials with specific pore systems, catalytic activity and stability can be improved [22]. When comparing amorphous carbon with more ordered carbon, Muradov found that amorphous carbon had more catalytic activity than ordered carbon [44].

Nickel-based catalysts have gained interest due to their capability to produce high-value carbon, such as CNTs and CNFs, and because of their high catalytic activity [45]. The drawbacks of using Ni catalysts are their sensitivity to the operation temperature and quick deactivation [46]. Torres et al. [36] studied the effect of numerous catalysts on TCD and found that the Cu-doped Ni/Al_2O_3 catalyst had a higher yield in carbon formation and a longer activity time at a temperature range of 823–873 K. In addition, the bimetallic Ni-Cu catalyst produced shorter carbon filaments, and the diameter of the filament was more constant compared with the metallic catalyst. Saraswat and Pant [38] studied Cu/Zn metals as promoters with a Ni aluminosilicate catalyst in TCD. Their results indicate that the yield and type of carbon depends on the added amount of Cu and Zn promoters, in addition to the process temperature. Saraswat et al. [47] studied the effect of the process parameters on the methane conversion and carbon amount. First, they studied the effect of temperature on yield. They examined the process at five different temperatures, ranging from 823 to 1023 K in 323 degree steps. The best methane conversion rate was found at 1023 K. In further studies, they used this temperature. The highest carbon yields were at 1023 K and with the catalyst combination of 50% Ni–10% Cu with the support of SiO_2. The carbon yield using this 50% Ni–10% Cu/SiO_2 catalyst was 710%. The second highest carbon yield (610%) was obtained using the 50% Ni–5% Cu catalyst. Table 2 presents the results of Saraswat et al. [46], indicating the catalyst and support used, temperature and GHSV of the TCD process, and the produced carbon amounts and yields. Bai et al. [33] tested active CNFs as catalysts that were loaded with Ni. They stated that the surface structure and textural properties of the CNFs can affect the stability and activity of the catalyst. Their produced carbon was consistent in diameter and relatively long. Nickel particle size has been proven to have a significant effect on carbon formation and catalyst deactivation [18,39]. Smaller metal particle sizes increase the deactivation rate of the catalysts. This can be explained by the fact that CNTs are prone to form on a large particle when a smaller size is encapsulated by carbon [35].

Table 2. Different Ni and Ni/Cu catalysts on SiO_2 support and their effect on the carbon amounts and yields in the TCD reaction (at 1023 K and with GHSV of 1800).

Catalyst	Support	Operating Temperature [K]	GHSV [mL/g_{cat} h]	g_C/g_{CH4} Fed	Carbon Yield %	Source
50 wt% Ni	SiO_2	1023	1800	0.504	600%	[47]
50 wt% Ni/5 wt% Cu	SiO_2	1023	1800	0.531	610%	
50 wt% Ni/10 wt% Cu	SiO_2	1023	1800	0.619	710%	
50 wt% Ni/15 wt% Cu	SiO_2	1023	1800	0.467	550%	
50 wt% Ni/20 wt% Cu	SiO_2	1023	1800	0.374	500%	

Fe catalysts are cheaper and more environmentally friendly compared to Ni, although they require a somewhat higher temperature range [43]. Torres et al. [24] tested Fe catalysts with and without cobalt doping. Results indicated that cobalt-doped catalysts had higher carbon formation and longer activity. Both catalysts had an average diameter of around 12 nm. The difference was that the cobalt-doped catalyst had straighter CNTs than the iron catalyst. In addition, less metal was infused into the carbon with the cobalt-doped catalyst at 18.1 wt%, when the iron catalyst had 32.5 wt%. Carbon morphology can vary with the catalyst, as shown in Figure 4, which illustrates the carbons produced in the TCD process with iron and nickel catalysts. It can be seen that more nanotube-like carbon (a–c) is formed using the nickel catalyst, whereas the use of iron catalyst leads to the formation of more laminal (d–e) carbon.

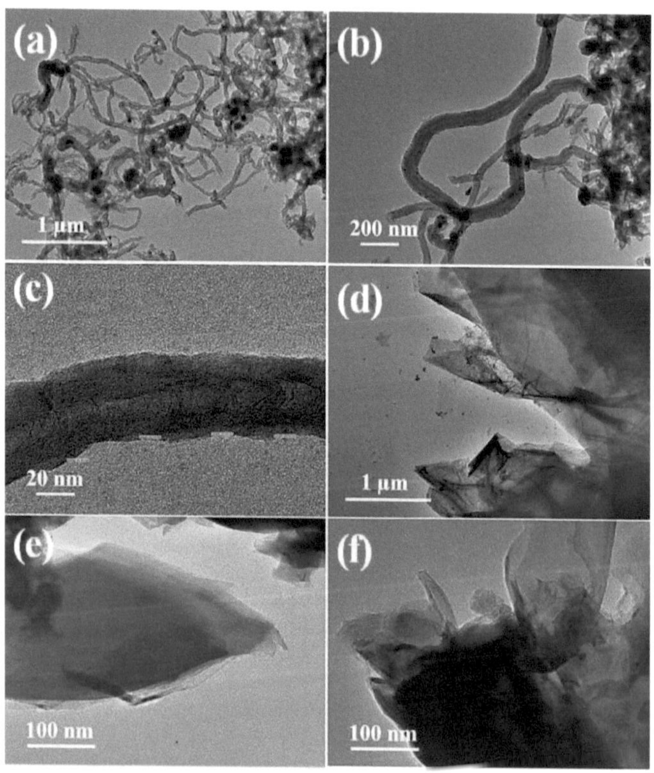

Figure 4. TEM images of the nanocarbon deposits using (**a–c**) Ni/MgO and (**d–f**) Fe/MgO catalysts [48]. Use of a nickel catalyst leads to nanotube-like carbon, whereas the use of an iron catalyst leads to the formation of more laminal carbon. Adapted with permission from [48]. Elsevier, 2017.

Other catalysts that have been investigated for TCD are molten metal catalysts and noble catalysts. The high cost of noble catalysts (Pt, Rh, Pd) prevents their use. Therefore, they are mostly used as a promoter to improve catalyst capability to produce carbon and thermal stability [45]. Takenaka et al. [49] studied the effect of different metals (Cu, Rh, Pd, Ir, and Pt) with nickel catalysts. The results indicated that a nickel catalyst with the addition of Pd significantly improved not only the catalytic life and hydrogen yield but also the carbon yield. Using a molten metal catalyst is one option to confront the issue of how to separate the produced carbon from the catalyst [50]. The idea is that solid carbon floats to the surface where it can be collected [37]. Guo et al. [51] studied the effect of support on the TCD process with mixed metal oxides (MMOs) and Al_2O_3. The results indicated that the MMOs resulted in a better hydrogen yield than Al_2O_3, but a lower carbon yield. The highest carbon yield, of 4.55 g_C/g_{Cat}, was obtained with a Ni3Al catalyst. Ermakova and Ermakov [52] studied different types of support and their effect on the carbon yield. The highest yield of carbon (45 g_C/g_{Cat}) was obtained with SiO_2 support. Without any support, the carbon yield was 16.5 g_C/g_{Cat}, and with common Al_2O_3 support the carbon yield was 14 g_C/g_{Cat}. Awadallah et al. [27] studied the type of carbon that is formed on the catalyst with different supports. They used a Ni catalyst supported by zeolite samples with Si/Al ratios of 25 and 400 and amorphous silica. The Ni catalyst with zeolite 400 support yielded 372% of carbon, whereas with amorphous silica support the carbon yield was 9%. Zeolite samples formatted multiwalled CNTs in which amorphous silica formed graphene layers. They stated that this could be explained by the crystallite size of the Ni species and dispersion. Small crystallite size and high metal dispersion promote the growth of multiwalled CNTs. By comparison, lower dispersion and agglomerated Ni particles promote more horizontal growth to form graphene layers. Details of these studies can be found in Table 3, in which the type of catalyst and support, temperature, GHSV, and carbon yields are given.

Table 3. Different catalysts and supports, GHSV, and their effect on the carbon amounts and yields in the TCD reaction at 973 K.

Catalyst	Support	Operating Temperature [K]	GHSV [mL/g_{cat} h]	g_C/g_{Cat}	Carbon Yield %	Source
Ni0.5Al (Molar ratio)	MMOs	973	60,000 *	0.11		[51]
Ni1Al	MMOs	973	60,000 *	0.49		
Ni2Al	MMOs	973	60,000 *	2.02		
Ni3Al	MMOs	973	60,000 *	4.55		
Ni0.5	Al_2O_3	973	60,000 *	0.67		
Ni1	Al_2O_3	973	60,000 *	1.15		
Ni2	Al_2O_3	973	60,000 *	1.36		
Ni3	Al_2O_3	973	60,000 *	1.29		
85wt% Fe	ZrO_2	973	8000	13.5		[52]
85wt% Fe	Al_2O_3	973	8000	14		
85wt% Fe	TiO_2	973	8000	17.4		
85wt% Fe	SiO_2	973	8000	45		
85wt% Fe	–	973	8000	16.5		
40 wt% Ni	Z-25	973	6000 *		176%	[27]
40 wt% Ni	Z-400	973	6000 *		372%	
40 wt% Ni	Amorphous silica	973	6000 *		9%	

* Calculated from flow rate to GHSV.

4.2. Effect of Process Temperature on Carbon Amount and Quality

Temperature can affect the BET surface and, according to Torres et al. [23], the BET value increases with the decrease in temperature. An increase in temperature can affect the amount of carbon formed [53]. It is also suggested that, to obtain higher crystallinity of carbon, higher reaction temperatures could be beneficial, but only up to a certain point.

Above 1473 K, carbon morphology is more amorphous [16,45]. In Table 4, two different studies are shown in which temperature was the variable in the TCD process. Table also shows the catalyst used and the corresponding process conditions. Carbon yields for each catalyst/support are also presented.

Table 4. The effect of temperature on carbon yield based on two different studies using Ni and Fe catalysts. As can be seen, the carbon yield increased with the increase in temperature.

Catalyst	Support	Operating Temperature [K]	GHSV [mL/g_{cat} h]	Carbon Yield %	Source
Ni	MgO	1173	9000 *	927%	[48]
Ni	MgO	1073	9000 *	863%	
Ni	MgO	973	9000 *	608%	
Fe	MgO	1173	9000 *	1205%	
Fe	MgO	1073	9000 *	1055%	
Fe	MgO	973	9000 *	810%	
NiO		1173	9000 *	815%	[54]
NiO		1073	9000 *	691%	
NiO		973	9000 *	341%	
Fe_2O_3		1173	9000 *	663%	
Fe_2O_3		1073	9000 *	419%	
Fe_2O_3		973	9000 *	196%	

* Calculated from flow rate to GHSV.

In the research of Pudukudy et al. [48], the catalysts Ni and Fe were studied, based on MgO support at different temperatures. The Fe catalyst had a higher carbon yield at all temperatures than the Ni catalyst. With both catalysts, the carbon yield increased with the increase in temperature as can be seen in Table 4. In another study, Pudukudy et al. [54] investigated the effect of temperature on NiO and Fe_2O_3 catalysts, and its effect on the carbon yield. As the temperature increased, the carbon yield increased, and NiO had overall higher yields compared with Fe_2O_3. At 973 K, the carbon yields were 341% and 196% using NiO and Fe_2O_3 catalysts, respectively. At the highest reaction temperature, 1173 K, the carbon yields were 815% and 663% for NiO and Fe_2O_3 catalysts, respectively. In the studies of Takenaka et al. [41], temperature had the opposite effect. Carbon and hydrogen yields both decreased with the increase in the reaction temperature. They also found that at 773 K fish-bone type carbon fibers formed, whereas at 973 K they were mostly multiwalled CNTs (MWCNTs). Saraswat and Pant [47] showed that the carbon yield increased with increasing temperatures from 823 to 1023 K but decreased from 1023 to 1073 K. They used Ni and bimetallic catalysts with the support of zeolite MCM-22.

4.3. *Effect of GHSV on Carbon Amount and Quality*

Increasing the methane flow rate affects the deactivation rate and the carbon. According to Gao et al. [23], the carbon fiber is thicker and the BET decreases when the methane flow rate is increased.

Saraswat et al. [47] studied the effect of GHSV on methane conversion and carbon yield. They previously determined that the optimal catalyst and temperature were 50%Ni-10%Cu/SiO_2 at 1023 K. This combination resulted in the highest conversion rate. They then studied the influence of GHSV with different velocities ranging from 600 to 6000 mL/g_{cat}h. They found that the catalyst had the highest activity at 600 mL/g_{cat}h and, as the GHSV increased, both methane and carbon yields decreased. This suggests that, at a higher flow rate, the catalyst is covered with carbon deposits; therefore, by deactivating the catalyst and using lower flow rates, the decomposition reaction is more effective on the catalyst surface.

4.4. *Summary*

It is important to note that the reaction is affected by many parameters, and the results can vary significantly. It is crucial to identify the parameters that result in good carbon yields without compromising the hydrogen yield. Table 5 summarizes TCD reactions with different process parameters, catalysts, and supports used, and their effect on the carbon amount and morphology. It should be noted that although researchers may report in their experiments that only one type of carbon is formed, all of the other carbon morphologies may also be present in the process; often, the dominant carbon or the form of carbon that is of most interest to the researcher is the only one reported.

Table 5. Other catalysts and supports used in the TCD process, operation conditions, and their effect on the produced carbon yield and morphology.

Catalyst	Support	Operating Temperature [K]	GHSV [$mL/g_{cat}h$]	Carbon Deposits [g_C/g_{cat}]	Carbon Morphology	Source
Fe (porous)		1173	9000	6.62	Layered graphene sheets	[55]
Ni		773	9000	354–398	CNT	[56]
40 wt% Ni	SiO_2	773	90,000 *	491	Fish-bone carbon nanofibers	[41]
40 wt% Ni	SiO_2	973	90,000 *		MWCNT	
75 wt% Ni	SiO_2	823	30,000 *		Longer and thicker than at 923 K	[57]
35 wt% Ni/40 wt% Fe	SiO_2	823	30,000 *		Thinner and shorter than without Fe	
65 wt% NI/10 wt% Fe	SiO_2	823	30,000 *			
75 wt% Ni	SiO_2	923	18,000 *		MWCNT	
35 wt% Ni/40 wt% Fe	SiO_2	923	18,000 *		bamboo shaped	
65 wt% NI/10 wt% Fe	SiO_2	923	18,000 *			
50 wt% Ni/25 wt% Fe	Al_2O_3	973	4200 *	562	CNT	[58]
50 wt% Ni	MCM-22	1023	1800	3.63	CNT	[38]
50 wt% Ni/5 wt% Cu	MCM-22	1023	1800	4.26	CNT	
50 wt% Ni/5 wt% Cu/5 wt% Zn	MCM-22	1023	1800	5.68	CNT	
50 wt% Ni/10 wt% Cu	MCM-22	1023	1800	5.5	CNT	
50 wt% Ni/10 wt% Zn	MCM-22	1023	1800	5.45	CNT	
5 wt% Ni	SiO_2	873	2440	27.16 **		[59]
10 wt% Ni	SiO_2	873	2440	39.73 **		
20 wt% Ni	SiO_2	873	2440	57.94 **		
30 wt% Ni	SiO_2	873	2440	62.05 **	CNT	
40 wt% Ni	SiO_2	873	2440	35.99 **		
50 wt% Ni	SiO_2	873	2440	23.91 **		
60 wt% Ni	SiO_2	873	2440	16.54 **		
70 wt% Ni	SiO_2	873	2440	9.01 **		
90 wt% Ni	SiO_2	873	2440	1.00 **	CNT	
69 wt% Fe	Al_2O_3	973	6000	2.28		[60]
69 wt% Fe	Al_2O_3	1073	6000	4.25		
69 wt% Fe	Al_2O_3	1123	6000	2.3		
69 wt% Fe	Al_2O_3	1173	6000	3.02		
69 wt% Fe	Al_2O_3	1073	3000	3.6		
69 wt% Fe	Al_2O_3	1073	8000	4.8		
12.3Fe/1Mo/6.15Al_2O_3(molar ratio)	Al_2O_3	1023	1500	1.92	bamboo shaped	[28]
50 wt% 12.3Fe/1Mo/6.15MgO(molar ratio)	MgO	1023	1500	8.26	tubular	
8.6 wt% Ni	AC	1123	1620	7.01		[61]
16.6 wt% Ni	AC	1123	1620	9.23		
23.3 wt% Ni	AC	1123	1620	7.92		
30 wt% Ni	AC	1123	1620	6.10		
Graphitized carbon black (Carbopack C)		1123	3800	0.08		[62]
Graphitized carbon black (Carbopack B)		1123	3800	0.12		
Carbon black (Fluke 05120)		1123	3800	0.65		
Carbon black (Fluka 05120)		1123	9500	0.69		
Carbon black (Fluka 03866)		1123	3800	0.212		
Carbon black (Black pearls 2000)		1123	3800	0.22		
Industrial carbon black (HS-50)		1123	3800	0.28		
Commercial activated carbon (CGNorit)		1123	3800	0.45		
Commercial activated carbon (CGNorit)		1123	9500	0.6		
Commercial activated carbon (CGNorit)		1123	19,000	0.66		
26.5 wt% Fe	SiO_2	1073	105,000	7.5	CNT	[63]
26.5 wt% Fe	SiO_2	1073	105,000	22.5	CNT	
90 wt% Fe	Al_2O_3	898	45,000	5.5		[64]
85 wt% Fe/5 wt% Co	Al_2O_3	823	45,000	16		
50 wt% Fe	Al_2O_3	823	45,000	26.5		
50 wt% Fe/6 wt% Co	Al_2O_3	823	45,000	52.4		

* Calculated from flow rate to GHSV. ** Calculated from mol_C/mol_{Ni} to g_C/g_{Ni}.

5. Possible Applications for Produced Solid Carbon

Demand for carbon products is expected to grow in the future [24]. Currently, the most common production methods are plasma arc discharge, pulsed laser gasification, and catalytic chemical vapor deposition. The first of these has a high operation temperature of 4000–6000 K, and the latter has normal pressure and a temperature of 800–1200 K. In TCD, both products' hydrogen and carbon may be valuable in the future; therefore, there is significant interest in the process and applications for produced carbon. These applications vary depending on the type of carbon produced; for example, CNOs have been applied in lubrication oils and supercapacitors. CNTs have been used in water purification treatment and in conductive polymers [25,65].

In this article, we introduce possible applications in which carbon materials are currently being used or could be used. Not all of the studies relate to carbon materials manufactured by TCD. Nonetheless, this process has the potential to become economically feasible because, in many applications, increasing the scale of manufacturing is one of the biggest challenges [66]. It should be noted that, although many of these applications have been tested in the laboratory, they may not be ethically viable or safe for humans in practice; a good example is water purification, because carbon nanotubes are considered toxic to humans [67].

5.1. *Batteries*

Battery manufacturers continuously attempt to improve charging rate, energy density, operating temperature, power density, safety, energy density, and sufficient electrochemical cycling characteristics. Electrode materials play a key role in battery development. CNTs have been studied as an anode material for Li-ion batteries [68]. Traditionally, in Li-ion batteries, graphite has been used as the anode material because of its good electric conductivity. In high power applications, graphene is replaced with materials that have higher capacity, energy, and power density. CNT morphology, high conductivity and tensile strength, and inertness to chemical degradation can improve battery capacity and decrease the risk of pulverization as an anode material, and can therefore be considered as a replacement material for graphene. If CNTs are used in batteries, they should be free from contaminants, such as metal catalysts, amorphous carbon, fullerenes, and nanotubes with divergent morphology. Large-scale production is still under development and the prices remain relatively high.

5.2. *Supercapacitors*

TCD-produced carbon provides a less-expensive, environmentally friendly choice for supercapacitor materials, without compromising the material properties required [68]. Porous carbon materials are attractive materials for the production of electrodes due to their high thermal and chemical stability, good electrical conductivity, and porous structure. Optimal supercapacitor materials should have an accessible and high specific surface area in the microporous region. A certain balance should be obtained between the energy storage-related pores and interconnecting pores to achieve the best properties; therefore, the ability to control the porosity of the carbon by production parameters is crucial [31].

5.3. *Water Treatment Application*

Because CNTs have a relatively large specific surface area and a porous structure, they may be used as an absorbent of organic pollutants from wastewater [69,70]. Although CNTs show promise in this regard, a number critical obstacles must first be overcome. First is the poor separability of the carbon and pollutant. Second is that studies have shown that CNTs' only have the ability to absorb a single organic pollutant, whereas wastewater can have several pollutants, such as metal ions, anions, humic acid, and organic solvents. Third is the economic issue. Thus, in the more practical applications of CNTs, more attention should paid to finding means to make the operation more feasible for water treatment [71].

CNTs can also be toxic to the human body [72,73], and therefore, it should be carefully considered whether they should be applied to wastewater treatment.

5.4. Composite Materials

MWCNTs may be used as a composite material. Combined with polymers, several applications are possible [74]. This composite material may be used, for example, in wearable electronics in textile sectors [75], antennas [76], sensors [77], solar cells [78,79], and EMI shielding [80–83]. The use of these materials can also overcome existing challenges and allow industrialization of new applications, such as healthcare devices, packaging, and shield protection for humans. Previous studies have also noted that this type of composite material may replace steel in various appliances due to its superior electric conductivity.

5.5. Cosmetics and Medical Applications

Carbon-based nanoparticles are being used in cosmetics and medical applications. For example, fullerenes have potential in biomedicine because of their antioxidant activity. They show promise in dermatological and skin care products, such as sunscreen, skin whitening, and antiaging products [83]. Fullerenes can act as a vehicle to improve drug delivery and hair growth. Carbon black can be used as a colorant in cosmetic products, and activated carbon is used in toothpastes [80,81]. Charcoal has been used in medicine dating back to ancient Egyptian civilizations. Currently, charcoal is used for teeth whitening, smoothing the skin, facial cleansing, pore striping, facial masks, and soaps. It has been noted that it can treat acne, bites or cuts, minor infections, itchy scalp, and seborrheic dermatitis. Activated charcoal removes dirt, oil, and impurities, absorbs toxins, has anti-bacterial properties, tightens pores, smooths skin, reduces inflammation, and decongests and brightens the skin [82].

6. Conclusions

CO_2-free hydrogen production is a key area in the effort to achieve the EU's ambitious environmental and carbon neutrality goals. TCD, which is a CO_2-free process, may be used in the future to produce both hydrogen and solid carbon. In this article, we focused on the production of solid carbon, including its properties, the effect of the TCD process on its quality, and its possible applications. Carbon produced using TCD could have a significant effect on the hydrogen economy and increase the value added by the process. Compared to other methane-based hydrogen production processes, this carbon is bound in a solid form and can be used in carbon capture with zero CO_2 emissions.

Carbon can take several forms, such as graphene, diamond, fullerenes, and nanotubes. These different morphologies can have various properties and applications. Therefore, understanding the manufacturing process and parameter effects is crucial to obtain the quality that is desired for the application. In the TCD catalyst, the promoter, temperature, and methane gas flow can influence the carbon type and quality.

Because carbon materials are currently widely used or studied for various applications, carbon manufactured by TCD could provide the impetus for its use in several application areas, such as batteries, supercapacitors, and composite material. This would increase their potential uses, and make them more economically friendly and feasible. However, in some applications, such as water purification, CNT toxicity to humans remains an issue that must be considered.

Author Contributions: Conceptualization, U.L., H.R., E.V. and L.Y.-V.; Writing—original draft preparation, E.V.; Writing—review and editing, U.L, H.R., E.V. and L.Y.-V.; visualization, U.L., H.R.; supervision, U.L.; funding acquisition, E.V., U.L. and H.R. All authors have read and agreed to the published version of the manuscript.

Funding: This research was funded by the Fortum and Neste Foundation, grant number 20200148.

Institutional Review Board Statement: Not applicable.

Informed Consent Statement: Not applicable.

Conflicts of Interest: The authors declare no conflict of interest in regard to this work.

References

1. European Comission. *A Hydrogen Strategy for a Climate-Neutral Europe*; European Commission: Brussels, Belgium, 2020.
2. Yang, L.; Liu, F.; He, J. Natural Sand as a Non-Conventional Catalyst for Hydrogen Production by Methane Thermo-Catalytic Decomposition. *Int. J. Hydrogen Energy* **2019**, *44*, 11625–11633. [CrossRef]
3. Gamal, A.; Eid, K.; El-Naas, M.H.; Kumar, D.; Kumar, A. Catalytic Methane Decomposition to Carbon Nanostructures and Cox-Free Hydrogen: A Mini-Review. *Nanomaterials* **2021**, *11*, 1226. [CrossRef] [PubMed]
4. Chen, L.; Qi, Z.; Zhang, S.; Su, J.; Somorjai, G.A. Catalytic Hydrogen Production from Methane: A Review on Recent Progress and Prospect. *Catalysts* **2020**, *10*, 858. [CrossRef]
5. Abbas, H.F.; Wan Daud, W.M.A. Hydrogen Production by Methane Decomposition: A Review. *Int. J. Hydrogen Energy* **2010**, *35*, 1160–1190. [CrossRef]
6. Abdullah, B.; Abd Ghani, N.A.; Vo, D.V.N. Recent Advances in Dry Reforming of Methane over Ni-Based Catalysts. *J. Clean. Prod.* **2017**, *162*, 170–185. [CrossRef]
7. Soltani, R.; Rosen, M.A.; Dincer, I. Assessment of CO_2 Capture Options from Various Points in Steam Methane Reforming for Hydrogen Production. *Int. J. Hydrogen Energy* **2014**, *39*, 20266–20275. [CrossRef]
8. Usman, M.; Wan Daud, W.M.A.; Abbas, H.F. Dry Reforming of Methane: Influence of Process Parameters—A Review. *Renew. Sustain. Energy Rev.* **2015**, *45*, 710–744. [CrossRef]
9. Levalley, T.L.; Richard, A.R.; Fan, M. The Progress in Water Gas Shift and Steam Reforming Hydrogen Production Technologies—A Review. *Int. J. Hydrogen Energy* **2014**, *39*, 16983–17000. [CrossRef]
10. Kumar, S.; Kumar, S.; Prajapati, J.K. Hydrogen Production by Partial Oxidation of Methane: Modeling and Simulation. *Int. J. Hydrogen Energy* **2009**, *34*, 6655–6668. [CrossRef]
11. De, C.; Roseno, K.T.; Schmal, M.; Brackmann, R.; Alves, R.M.B.; Giudici, R. Partial Oxidation of Methane on Neodymium and Lanthanium Chromate Based Perovskites for Hydrogen Production. *Int. J. Hydrogen Energy* **2019**, *44*, 8166–8177. [CrossRef]
12. Larimi, A.S.; Alavi, S.M. Ceria-Zirconia Supported Ni Catalysts for Partial Oxidation of Methane to Synthesis Gas. *Fuel* **2012**, *102*, 366–371. [CrossRef]
13. Ruckenstein, E.; Hu, Y.H. Methane Partial Oxidation over NiO/MgO Solid Solution Catalysts. *Appl. Catal. A Gen.* **1999**, *183*, 85–92. [CrossRef]
14. Ibrahim, A.A.; Al-Fatesh, A.S.; Khan, W.U.; Soliman, M.A.; Al Otaibi, R.L.; Fakeeha, A.H. Thermo-Catalytic Methane Decomposition: A Review of State of the Art of Catalysts. *J. Chem. Soc. Pakistan* **2015**, *37*, 1269–1297.
15. Ashik, U.P.M.; Wan Daud, W.M.A.; Abbas, H.F. Production of Greenhouse Gas Free Hydrogen by Thermocatalytic Decomposition of Methane—A Review. *Renew. Sustain. Energy Rev.* **2015**, *44*, 221–256. [CrossRef]
16. Keipi, T.; Tolvanen, K.E.S.; Tolvanen, H.; Konttinen, J. Thermo-Catalytic Decomposition of Methane: The Effect of Reaction Parameters on Process Design and the Utilization Possibilities of the Produced Carbon. *Energy Convers. Manag.* **2016**, *126*, 923–934. [CrossRef]
17. Pudukudy, M.; Yaakob, Z.; Akmal, Z.S. Direct Decomposition of Methane over SBA-15 Supported Ni, Co and Fe Based Bimetallic Catalysts. *Appl. Surf. Sci.* **2015**, *330*, 418–430. [CrossRef]
18. Liang, W.; Yan, H.; Chen, C.; Lin, D.; Tan, K.; Feng, X.; Liu, Y.; Chen, X.; Yang, C.; Shan, H. Revealing the Effect of Nickel Particle Size on Carbon Formation Type in the Methane Decomposition Reaction. *Catalysts* **2020**, *10*, 890. [CrossRef]
19. Zhou, L.; Enakonda, L.R.; Harb, M.; Saih, Y.; Aguilar-Tapia, A.; Ould-Chikh, S.; Hazemann, J.-l.; Li, J.; Wei, N.; Gary, D.; et al. Fe Catalysts for Methane Decomposition to Produce Hydrogen and Carbon Nano Materials. *Appl. Catal. B Environ.* **2017**, *208*, 44–59. [CrossRef]
20. Callister, W.D. *Materials Science and Engineering: An Introduction*, 7th ed.; John Wiley & Sons: New York, NY, USA, 2007.
21. Noked, M.; Soffer, A.; Arubach, D. The Electrochemistry of Activated Carbonaceous Materials: Past, Present, and Future. *J. Solid State Electrochem.* **2011**, *15*, 1563–1578. [CrossRef]
22. Qian, J.X.; Chen, T.W.; Enakonda, L.R.; Liu, D.B.; Basset, J.M.; Zhou, L. Methane Decomposition to Pure Hydrogen and Carbon Nano Materials: State-of-the-Art and Future Perspectives. *Int. J. Hydrogen Energy* **2020**, *45*, 15721–15743. [CrossRef]
23. Gao, L.Z.; Kiwi-Minsker, L.; Renken, A. Growth of Carbon Nanotubes and Microfibers over Stainless Steel Mesh by Cracking of Methane. *Surf. Coatings Technol.* **2008**, *202*, 3029–3042. [CrossRef]
24. Torres, D.; Pinilla, J.L.; Suelves, I. Cobalt Doping of α-Fe/Al2O3 Catalysts for the Production of Hydrogen and High-Quality Carbon Nanotubes by Thermal Decomposition of Methane. *Int. J. Hydrogen Energy* **2020**, *45*, 19313–19323. [CrossRef]
25. Liu, Q.; Wu, P.; He, J.; Liu, C.; Jiang, W. Catalytic Decomposition of Methane by Two-Step Cascade Catalytic Process: Simultaneous Production of Hydrogen and Carbon Nanotubes. *Chem. Eng. Res. Des.* **2020**, *163*, 96–106. [CrossRef]
26. Soni, S.K.; Thomas, B.; Kar, V.R. A Comprehensive Review on CNTs and CNT-Reinforced Composites: Syntheses, Characteristics and Applications. *Mater. Today Commun.* **2020**, *25*, 101546. [CrossRef]
27. Awadallah, A.E.; El-Desouki, D.S.; Aboul-Gheit, N.A.K.; Ibrahim, A.H.; Aboul-Gheit, A.K. Effect of Crystalline Structure and Pore Geometry of Silica Based Supported Materials on the Catalytic Behavior of Metallic Nickel Particles during Methane Decomposition to COx-Free Hydrogen and Carbon Nanomaterials. *Int. J. Hydrogen Energy* **2016**, *41*, 16890–16902. [CrossRef]

28. Torres, D.; Pinilla, J.L.; Lázaro, M.J.; Moliner, R.; Suelves, I. Hydrogen and Multiwall Carbon Nanotubes Production by Catalytic Decomposition of Methane: Thermogravimetric Analysis and Scaling-up of Fe-Mo Catalysts. *Int. J. Hydrogen Energy* **2014**, *39*, 3698–3709. [CrossRef]
29. Williams, P.T. Hydrogen and Carbon Nanotubes from Pyrolysis-Catalysis of Waste Plastics: A Review. *Waste Biomass Valorization* **2020**, *12*, 1–28. [CrossRef]
30. Dagle, R.; Dagle, V.; Bearden, M.; Holladay, J.; Krause, T.; Ahmed, S. *R & D Opportunities for Development of Natural Gas Conversion Technologies for Co-Production of Hydrogen and Value-Added Solid Carbon Products*; U.S. Department of Energy Fuel Cell Technologies Office: Chicago, IL, USA, 2017.
31. Gu, W.; Yushin, G. Review of Nanostructured Carbon Materials for Electrochemical Capacitor Applications: Advantages and Limitations of Activated Carbon, Carbide-Derived Carbon, Zeolite-Templated Carbon, Carbon Aerogels, Carbon Nanotubes, Onion-like Carbon, and Graphene. *Wiley Interdiscip. Rev. Energy Environ.* **2014**, *3*, 424–473. [CrossRef]
32. Han, F.D.; Yao, B.; Bai, Y.J. Preparation of Carbon Nano-Onions and Their Application as Anode Materials for Rechargeable Lithium-Ion Batteries. *J. Phys. Chem. C* **2011**, *115*, 8923–8927. [CrossRef]
33. Bai, Z.; Chen, H.; Li, B.; Li, W. Methane Decomposition over Ni Loaded Activated Carbon for Hydrogen Production and the Formation of Filamentous Carbon. *Int. J. Hydrogen Energy* **2007**, *32*, 32–37. [CrossRef]
34. Lobo, L.S.; Carabineiro, S.A.C. Carbon Formation at High Temperatures. *Catalysts* **2020**, *10*, 465. [CrossRef]
35. Guizani, C.; Escudero Sanz, F.J.; Salvador, S. The Nature of the Deposited Carbon at Methane Cracking over a Nickel Loaded Wood-Char. *Comptes Rendus Chim.* **2016**, *19*, 423–432. [CrossRef]
36. Torres, D.; Pinilla, J.L.; Suelves, I. Co-, Cu-and Fe-Doped Ni/Al2o3 Catalysts for the Catalytic Decomposition of Methane into Hydrogen and Carbon Nanofibers. *Catalysts* **2018**, *8*, 300. [CrossRef]
37. Upham, D.C.; Agarwal, V.; Khechfe, A.; Snodgrass, Z.R.; Gordon, M.J.; Metiu, H.; McFarland, E.W. Catalytic Molten Metals for the Direct Conversion of Methane to Hydrogen and Separable Carbon. *Science* **2017**, *358*, 917–921. [CrossRef]
38. Saraswat, S.K.; Pant, K.K. Ni-Cu-Zn/MCM-22 Catalysts for Simultaneous Production of Hydrogen and Multiwall Carbon Nanotubes via Thermo-Catalytic Decomposition of Methane. *Int. J. Hydrogen Energy* **2011**, *36*, 13352–13360. [CrossRef]
39. Christensen, K.O.; Chen, D.; Lødeng, R.; Holmen, A. Effect of Supports and Ni Crystal Size on Carbon Formation and Sintering during Steam Methane Reforming. *Appl. Catal. A Gen.* **2006**, *314*, 9–22. [CrossRef]
40. Hasnan, N.S.N.; Timmiati, S.N.; Lim, K.L.; Yaakob, Z.; Kamaruddin, N.H.N.; Teh, L.P. Recent Developments in Methane Decomposition over Heterogeneous Catalysts: An Overview. *Mater. Renew. Sustain. Energy* **2020**, *9*, 3. [CrossRef]
41. Takenaka, S.; Kobayashi, S.; Ogihara, H.; Otsuka, K. Ni/SiO_2 Catalyst Effective for Methane Decomposition into Hydrogen and Carbon Nanofiber. *J. Catal.* **2003**, *217*, 79–87. [CrossRef]
42. Marquardt, T.; Bode, A.; Kabelac, S. Hydrogen Production by Methane Decomposition: Analysis of Thermodynamic Carbon Properties and Process Evaluation. *Energy Convers. Manag.* **2020**, *221*. [CrossRef]
43. Guil-Lopez, R.; Botas, J.A.; Fierro, J.L.G.; Serrano, D.P. Comparison of Metal and Carbon Catalysts for Hydrogen Production by Methane Decomposition. *Appl. Catal. A Gen.* **2011**, *396*, 40–51. [CrossRef]
44. Muradov, N.; Smith, F.; T-Raissi, A. Catalytic Activity of Carbons for Methane Decomposition Reaction. *Catal. Today* **2005**, *102–103*, 225–233. [CrossRef]
45. Ahmed, S.; Aitani, A.; Rahman, F.; Al-Dawood, A.; Al-Muhaish, F. Decomposition of Hydrocarbons to Hydrogen and Carbon. *Appl. Catal. A Gen.* **2009**, *359*, 1–24. [CrossRef]
46. Qian, J.X.; Chen, T.W.; Enakonda, L.R.; Liu, D.B.; Mignani, G.; Basset, J.M.; Zhou, L. Methane Decomposition to Produce COx-Free Hydrogen and Nano-Carbon over Metal Catalysts: A Review. *Int. J. Hydrogen Energy* **2020**, *45*, 7981–8001. [CrossRef]
47. Saraswat, S.K.; Pant, K.K. Synthesis of Hydrogen and Carbon Nanotubes over Copper Promoted Ni/SiO2 Catalyst by Thermocatalytic Decomposition of Methane. *J. Nat. Gas Sci. Eng.* **2013**, *13*, 52–59. [CrossRef]
48. Pudukudy, M.; Yaakob, Z.; Mazuki, M.Z.; Takriff, M.S.; Jahaya, S.S. One-Pot Sol-Gel Synthesis of MgO Nanoparticles Supported Nickel and Iron Catalysts for Undiluted Methane Decomposition into COx Free Hydrogen and Nanocarbon. *Appl. Catal. B Environ.* **2017**, *218*, 298–316. [CrossRef]
49. Takenaka, S.; Shigeta, Y.; Tanabe, E.; Otsuka, K. Methane Decomposition into Hydrogen and Carbon Nanofibers over Supported Pd-Ni Catalysts. *J. Catal.* **2003**, *220*, 468–477. [CrossRef]
50. Munera Parra, A.A.; Agar, D.W. Molten Metal Capillary Reactor for the High-Temperature Pyrolysis of Methane. *Int. J. Hydrogen Energy* **2017**, *42*, 13641–13648. [CrossRef]
51. Guo, Z.; Zheng, J.E.; Liu, Y.; Chu, W. Insight into the Role of Metal/Oxide Interaction and Ni Availabilities on NiAl Mixed Metal Oxide Catalysts for Methane Decomposition. *Appl. Catal. A Gen.* **2018**, *555*, 1–11. [CrossRef]
52. Ermakova, M.A.; Ermakov, D.Y. Ni/SiO_2 and Fe/SiO_2 Catalysts for Production of Hydrogen and Filamentous Carbon via Methane Decomposition. *Catal. Today* **2002**, *77*, 225–235. [CrossRef]
53. Inaba, M.; Zhang, Z.; Matsuoka, K.; Soneda, Y. Optimization of the Reaction Conditions for Fe-Catalyzed Decomposition of Methane and Characterization of the Produced Nanocarbon Fibers. *Catal. Today* **2019**, *332*, 11–19. [CrossRef]
54. Pudukudy, M.; Yaakob, Z.; Dahani, N.; Takriff, M.S.; Hassan, N.S.M. Production of COx Free Hydrogen and Nanocarbon via Methane Decomposition Over Unsupported Porous Nickel and Iron Catalysts. *J. Clust. Sci.* **2017**, *28*, 1579–1594. [CrossRef]

55. Pudukudy, M.; Yaakob, Z.; Kadier, A.; Takriff, M.S.; Hassan, N.S.M. One-Pot Sol–Gel Synthesis of Ni/TiO2 Catalysts for Methane Decomposition into COx Free Hydrogen and Multiwalled Carbon Nanotubes. *Int. J. Hydrogen Energy* **2017**, *42*, 16495–16513. [CrossRef]
56. Li, Y.; Zhang, B.; Xie, X.; Liu, J.; Xu, Y.; Shen, W. Novel Ni Catalysts for Methane Decomposition to Hydrogen and Carbon Nanofibers. *J. Catal.* **2006**, *238*, 412–424. [CrossRef]
57. Wang, W.; Wang, H.; Yang, Y.; Jiang, S. Ni-SiO2 and Ni-Fe-SiO2 Catalysts for Methane Decomposition to Prepare Hydrogen and Carbon Filaments. *Int. J. Hydrogen Energy* **2012**, *37*, 9058–9066. [CrossRef]
58. Wang, G.; Jin, Y.; Liu, G.; Li, Y. Production of Hydrogen and Nanocarbon from Catalytic Decomposition of Methane over a Ni-Fe/Al_2O_3 Catalyst. *Energy Fuels* **2013**, *27*, 4448–4456. [CrossRef]
59. Venugopal, A.; Naveen Kumar, S.; Ashok, J.; Hari Prasad, D.; Durga Kumari, V.; Prasad, K.B.S.; Subrahmanyam, M. Hydrogen Production by Catalytic Decomposition of Methane over Ni/SiO_2. *Int. J. Hydrogen Energy* **2007**, *32*, 1782–1788. [CrossRef]
60. Torres, D.; De Llobet, S.; Pinilla, J.L.; Lázaro, M.J.; Suelves, I.; Moliner, R. Hydrogen Production by Catalytic Decomposition of Methane Using a Fe-Based Catalyst in a Fluidized Bed Reactor. *J. Nat. Gas Chem.* **2012**, *21*, 367–373. [CrossRef]
61. Sarada Prasad, J.; Dhand, V.; Himabindu, V.; Anjaneyulu, Y. Production of Hydrogen and Carbon Nanofibers through the Decomposition of Methane over Activated Carbon Supported Ni Catalysts. *Int. J. Hydrogen Energy* **2011**, *36*, 11702–11711. [CrossRef]
62. Suelves, I.; Lázaro, M.J.; Moliner, R.; Pinilla, J.L.; Cubero, H. Hydrogen Production by Methane Decarbonization: Carbonaceous Catalysts. *Int. J. Hydrogen Energy* **2007**, *32*, 3320–3326. [CrossRef]
63. Takenaka, S.; Serizawa, M.; Otsuka, K. Formation of Filamentous Carbons over Supported Fe Catalysts through Methane Decomposition. *J. Catal.* **2004**, *222*, 520–531. [CrossRef]
64. Avdeeva, L.B.; Reshetenko, T.V.; Ismagilov, Z.R.; Likholobov, V.A. Iron-Containing Catalysts of Methane Decomposition: Accumulation of Filamentous Carbon. *Appl. Catal. A Gen.* **2002**, *228*, 53–63. [CrossRef]
65. Fau, G.; Gascoin, N.; Gillard, P.; Steelant, J. Methane Pyrolysis: Literature Survey and Comparisons of Available Data for Use in Numerical Simulations. *J. Anal. Appl. Pyrolysis* **2013**, *104*, 1–9. [CrossRef]
66. Rashed, A.O.; Merenda, A.; Kondo, T.; Lima, M.; Razal, J.; Kong, L.; Huynh, C.; Dumée, L.F. Carbon Nanotube Membranes—Strategies and Challenges towards Scalable Manufacturing and Practical Separation Applications. *Sep. Purif. Technol.* **2021**, *257*, 117929. [CrossRef]
67. Ahmad, J.; Naeem, S.; Ahmad, M.; Usman, A.R.A.; Al-Wabel, M.I. A Critical Review on Organic Micropollutants Contamination in Wastewater and Removal through Carbon Nanotubes. *J. Environ. Manag.* **2019**, *246*, 214–228. [CrossRef] [PubMed]
68. Maziarka, P.; Sommersacher, P.; Wang, X.; Kienzl, N.; Retschitzegger, S.; Prins, W.; Hedin, N.; Ronsse, F. Tailoring of the Pore Structures of Wood Pyrolysis Chars for Potential Use in Energy Storage Applications. *Appl. Energy* **2021**, *286*, 116431. [CrossRef]
69. Peng, X.; Li, Y.; Luan, Z.; Di, Z.; Wang, H.; Tian, B.; Jia, Z. Adsorption of 1,2-Dichlorobenzene from Water to Carbon Nanotubes. *Chem. Phys. Lett.* **2003**, *376*, 154–158. [CrossRef]
70. Avcı, A.; İnci, İ.; Baylan, N. Adsorption of Ciprofloxacin Hydrochloride on Multiwall Carbon Nanotube. *J. Mol. Struct.* **2020**, *1206*, 127711. [CrossRef]
71. Peng, J.; He, Y.; Zhou, C.; Su, S.; Lai, B. The Carbon Nanotubes-Based Materials and Their Applications for Organic Pollutant Removal: A Critical Review. *Chin. Chem. Lett.* **2021**. [CrossRef]
72. Mohanta, D.; Patnaik, S.; Sood, S.; Das, N. Carbon Nanotubes: Evaluation of Toxicity at Biointerfaces. *J. Pharm. Anal.* **2019**, *9*, 293–300. [CrossRef]
73. Keipi, T.; Hankalin, V.; Nummelin, J.; Raiko, R. Techno-Economic Analysis of Four Concepts for Thermal Decomposition of Methane: Reduction of CO_2 Emissions in Natural Gas Combustion. *Energy Convers. Manag.* **2016**, *110*, 1–12. [CrossRef]
74. Dhineshbabu, N.R.; Mahadevi, N.; Assein, D. Electronic Applications of Multi-Walled Carbon Nanotubes in Polymers: A Short Review. *Mater. Today Proc.* **2020**. [CrossRef]
75. Wang, X.; Yang, B.; Liu, J.; Zhu, Y.; Yang, C.; He, Q. A Flexible Triboelectric-Piezoelectric Hybrid Nanogenerator Based on P(VDF-TrFE) Nanofibers and PDMS/MWCNT for Wearable Devices. *Sci. Rep.* **2016**, *6*. [CrossRef]
76. Zahir, H.; Wojkiewicz, J.; Alexander, P.; Kone, L.; Belkacem, B.; Bergheul, S.; Lasri, T. Design Fabrication and Characterisation of Polyaniline and Multiwall Carbon Nanotubes Composites-based Patch Antenna. *IET Microw. Antennas Propag.* **2016**, *10*, 88–93. [CrossRef]
77. Turkani, V.S.; Maddipatla, D.; Narakathu, B.B.; Saeed, T.S.; Obare, S.O.; Bazuin, B.J.; Atashbar, M.Z. A Highly Sensitive Printed Humidity Sensor Based on a Functionalized MWCNT/HEC Composite for Flexible Electronics Application. *Nanoscale Adv.* **2019**, *1*, 2311–2322. [CrossRef]
78. Kim, D.H.; Dudem, B.; Yu, J.S. High-Performance Flexible Piezoelectric-Assisted Triboelectric Hybrid Nanogenerator via Polydimethylsiloxane-Encapsulated Nanoflower-like ZnO Composite Films for Scavenging Energy from Daily Human Activities. *ACS Sustain. Chem. Eng.* **2018**, *6*, 8525–8535. [CrossRef]
79. Kar, E.; Bose, N.; Dutta, B.; Mukherjee, N.; Mukherjee, S. MWCNT@SiO2 Heterogeneous Nanofiller-Based Polymer Composites: A Single Key to the High-Performance Piezoelectric Nanogenerator and X-Band Microwave Shield. *ACS Appl. Nano Mater.* **2018**, *1*, 4005–4018. [CrossRef]
80. Rincon, A.M. Chapter 6—Presence of nanomaterials on consumer products: Food, cosmetics, and drugs. In *Exposure to Engineered Nanomaterials in the Environment*; Elsevier: Amsterdam, The Netherlands, 2019; pp. 165–181, ISBN 9780128148365.

81. Viana, Í.E.L.; Weiss, G.S.; Sakae, L.O.; Niemeyer, S.H.; Borges, A.B.; Scaramucci, T. Activated Charcoal Toothpastes Do Not Increase Erosive Tooth Wear. *J. Dent.* **2021**, *109*, 103677. [CrossRef]
82. Sanchez, N.; Fayne, R.; Burroway, B. Charcoal: An Ancient Material with a New Face. *Clin. Dermatol.* **2020**, *38*, 262–264. [CrossRef] [PubMed]
83. Nafisi, S.; Maibach, H.I. Nanotechnology in cosmetics. In *Cosmetic Science and Technology: Theoretical Principles and Applications*; Elsevier: Amsterdam, The Netherlands, 2017; pp. 337–361, ISBN 9780128020548.

C Journal of Carbon Research

MDPI

Article

Fluorescent Nanodiamonds Synthesized in One-Step by Pulsed Laser Ablation of Graphite in Liquid-Nitrogen

Massimo Cazzanelli [1,*], Luca Basso [2], Claudio Cestari [1], Nicola Bazzanella [1], Enrico Moser [1], Michele Orlandi [1], Alessandro Piccoli [1] and Antonio Miotello [1]

[1] Department of Physics, University of Trento, Via Sommarive 14, 38123 Povo, Italy; claudio.cestari@unitn.it (C.C.); nicola.bazzanella@unitn.it (N.B.); enrico.moser@unitn.it (E.M.); michele.orlandi@unitn.it (M.O.); alessandro.piccoli@unitn.it (A.P.); antonio.miotello@unitn.it (A.M.)

[2] Center for Integrated Nanotechnologies, Sandia National Laboratories, Albuquerque, NM 87123, USA; lbasso@sandia.gov

* Correspondence: massimo.cazzanelli@unitn.it

Abstract: In this work, we present a relevant upgrade to the technique of pulsed laser ablation of fluorescent nanodiamonds (NDs), relying on an automatized graphite-target movement maintaining a constant level of liquid nitrogen over its surface during hours of deposition. Around 60 mg of NDs nanopowder was synthesized and optomagnetically characterized to assess its optical quality. Chemical purification of the ablated nanopowders, which removes the graphitic byproducts, permits to obtain pure fluorescent NDs with an efficiency of 7 ± 1% with respect to the total nanopowder mass. This value compares positively with the efficiency of other commercial NDs synthesis techniques such as detonation, cavitation, and high pressure–high temperature.

Keywords: pulsed laser ablation; nanodiamonds; NV-centers

Citation: Cazzanelli, M.; Basso, L.; Cestari, C.; Bazzanella, N.; Moser, E.; Orlandi, M.; Piccoli, A.; Miotello, A. Fluorescent Nanodiamonds Synthesized in One-Step by Pulsed Laser Ablation of Graphite in Liquid-Nitrogen. C **2021**, 7, 49. https://doi.org/10.3390/c7020049

Academic Editor: Gil Goncalves

Received: 31 May 2021
Accepted: 19 June 2021
Published: 21 June 2021

1. Introduction

The unique characteristics of nanodiamonds (NDs), for example, their excellent optical and mechanical properties [1] and the fluorescence properties of their nitrogen-vacancy (NV) centers, dependent on external physical fields [2], opened the possibility of producing devices with nanoscale quantum sensing capabilities. NV-fluorescent NDs in fact are almost unique nanoprobes, allowing for the detection of changes in environmental parameters such as temperature [3,4] and magnetic [5] and electric fields [6]. In addition, the biocompatibility of NDs allows *in-vivo* imaging [7].

Recently, we demonstrated the synthesis of NV fluorescent NDs through pulsed laser ablation (PLA) of graphite in a nitrogen-containing environment. We carried out the synthesis in both a controlled nitrogen [8] atmosphere and in liquid nitrogen [9]. We also demonstrated the possibility to produce NDs within different confining media and starting targets [10].

The most effective condition, in terms of number of fluorescent NDs and fluorescence intensity, is PLA in liquid nitrogen (LN_2), as reported in [9]. The experimental drawback of this new technique is related to the yield in terms of material produced during the synthesis process. In fact, for biological application such as imaging or sensing, a relevant (~mg mass, not obtained in our previous work) amount of highly fluorescent pure ND material is required in order to obtain a good contrast with low laser pumping powers. In addition, a significant amount of NDs permits the employment of strong, industry-compatible, chemical cleaning techniques that allow for the removal of almost all the optically parasitic graphitic byproducts. The aim of this paper is to present our upgraded automatized experimental setup where laser ablation of graphitic targets in LN_2 produces significant quantities of nanopowders (pure NDs with NV centers + graphite) allowing for cleaning out the embedded NV-containing NDs, and to precisely estimate the liquid-nitrogen laser-ablated fluorescent nanodiamond production yield.

2. Materials and Methods

The experimental apparatus (general view, details, and schematics) is shown in Figure 1. Pulsed laser irradiation is performed by using a KrF excimer laser (Lambda-Physik Coherent LPX220i) having wavelength of 248 nm and pulse duration of 20 ns. Laser energy is set to 690 mJ per pulse, and the laser is operated at a pulse rate of 10 Hz. The laser beam is deflected by a motor-driven oscillating mirror and consequently focused by a 400 mm-focal-length lens on the target surface. The pyrolytic graphite target is placed in a glass vial ("internal LN_2 vial" in Figure 1a, where the top surface of the target is covered by a ~5 mm-thick liquid nitrogen layer. To limit the unavoidable liquid nitrogen evaporation that occurs during the high-power excimer laser irradiation, the glass box was placed in a polystyrene box ("external LN_2 dewar" in Figure 1a) and filled with liquid nitrogen. The ablation process was automatized as much as possible with the aim to perform ~hours of deposition. The system was then equipped with three additional step-motor driven movements (Figure 1b): the Z-movement, the target rotation around its central cylindrical axis, and the laser beam rastering along the graphite cylinder height. These movements are controlled with Arduino© and more thoroughly described in Figure 1. The movement along the Z-axis allows movement of the target up and down along the vertical direction during the deposition process. It is also employed before the ablation procedure starts, to put the graphite target inside the internal LN_2 vial, in a position in which the top surface is covered by a ~5 mm thick LN_2 layer. We have found empirically that this LN_2 layer thickness is a good compromise between the confining requisites needed for NDs to nucleate from the graphite and the minimization of UV laser fluence losses due to liquid absorption/scattering. During the ablation process and the consequent LN_2 evaporation, this motor-driven, Arduino-controlled movement allows the target to downshift, with a specifically calibrated constant speed, in order to maintain the same LN_2 layer thickness above the target surface.

The other two movements allow the homogeneous ablation of the whole graphite cylindrical surface and also help to avoid the formation of detrimental macroscopic craters in the target. The first movement is the target rotation; the cylindrical graphite target rotates with a specific angular velocity. The second movement is the laser beam rastering along the longitudinal cylinder height. This movement is obtained by putting the mirror in a holder that can oscillate. The maximum oscillation angle of the mirror is set by observing where the laser beam is impinging on the target surface, and avoiding that the laser beam goes beyond the target length. These three movements are a crucial upgrade in the experimental apparatus, increasing the laser-deposition from ~µg to ~mg of material. In fact, the movements also minimize the manual operations required during the ablation, thus allowing a more efficient use of the deposition time, and avoiding continuous manual refills of the LN_2 level above the graphite target. This upgrade allowed us to perform more than 6 h of deposition (but longer depositions can be done and repetition rate of the laser can be safely increased up to 100 Hz), with partial liquid nitrogen refills at 45 minute interdelays, thus reaching ~57 mg of synthesized material (pure NDs with NV centers + graphite) as opposed to previous fractions of µg. In fact now, with all the procedures adopted and the consequent technical development of the apparatus, we can make around 200.000 laser shots per deposition session, as opposed to the 20.000 laser shots previously.

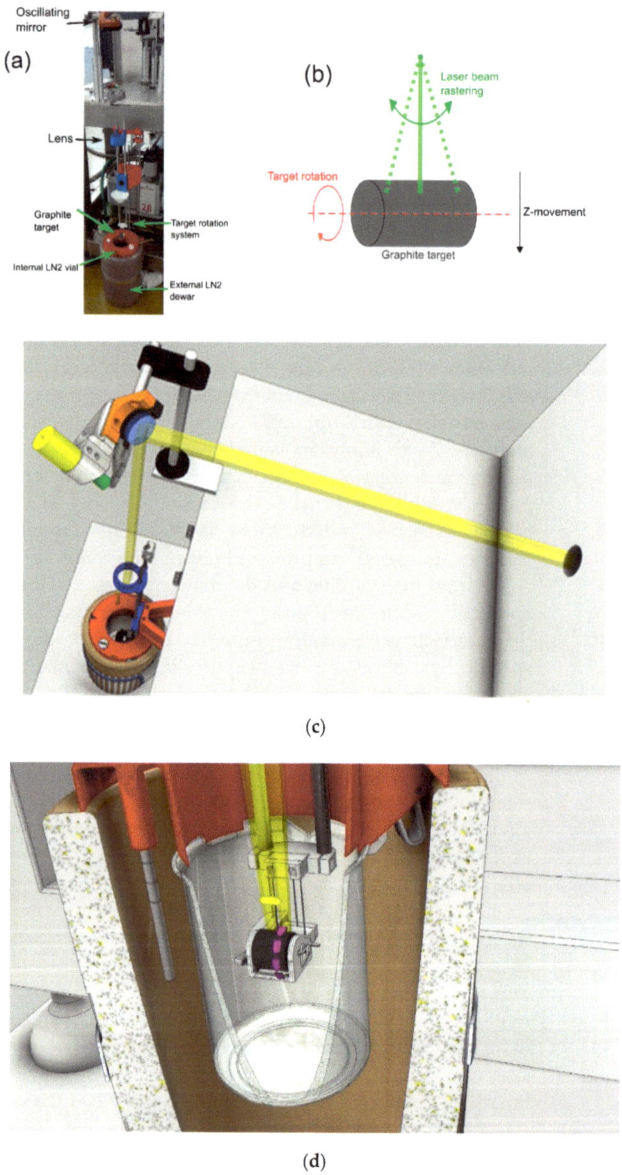

Figure 1. (**a**) Experimental apparatus. (**b**) Schematics of the oscillating mirror automatized system. The laser beam is made to impinge the cylinder surface at 45°. (**c**) Details of the experimental apparatus. The yellow transparent line is the 248 nm high fluence laser beam. The steering mirror reflects it through a 400 mm focal lens (fixed inside the blue circular holder). The graphitic target is fully immersed inside a LN_2 bath. (**d**) Graphite cylindrical target with a view of the roto-step-translational system, which lets the target rotate when the laser impinges on the sample.

The photoluminescence spectra and the optically detected magnetic resonance spectra (ODMR) of our samples were measured employing a confocal microRaman Jobin–Yvon LabRam apparatus, where the confocal microscope section is represented schematically in Figure 2. The excitation and the collection of the light from the micrometrically sized region of our samples were accomplished with a 100× objective. The sample was fixed on an epoxy-glass sample holder equipped with a Hertzian RF antenna (when measuring ODMR), which provided the RF pumping of the NV ground state energy levels, |g>, to excited state, |e> (see the energetic levels scheme in Figure 2). Under tunable radiofrequency irradiation, we probed the microphotoluminescence spectrum coming from a ~2 μm × 2 μm × 5 μm volume containing an ensemble of NDs and performed an optical imaging of the emitting NDs. In particular, we equipped such a micro-Raman/microphotoluminescence apparatus with a tunable radiofrequency (RF) source, capable of furnishing up to 1 W of RF power in the 2.6–3.2 GHz range. The RF irradiation served as the ground |g> level pump to suitably populate the $m_S = \pm 1$ sublevels, which are essential to activate the RF-dependent fluorescence mechanism related to the NV^- defect center in (nano)diamonds [2]. We then recorded the microphotoluminescence spectra on a ~2 μm × 2 μm × 5 μm region (using an Olympus 50×, NA = 0.5, and working distance = 10.7 mm objective) where a cluster assembly of NDs was present at the optical imaging, while varying the RF frequency in the 2.6–3.2 GHz spectral region at constant 0.5 W power.

The bare photoluminescence spectra were recorded when no RF irradiation was superimposed on the NDs samples.

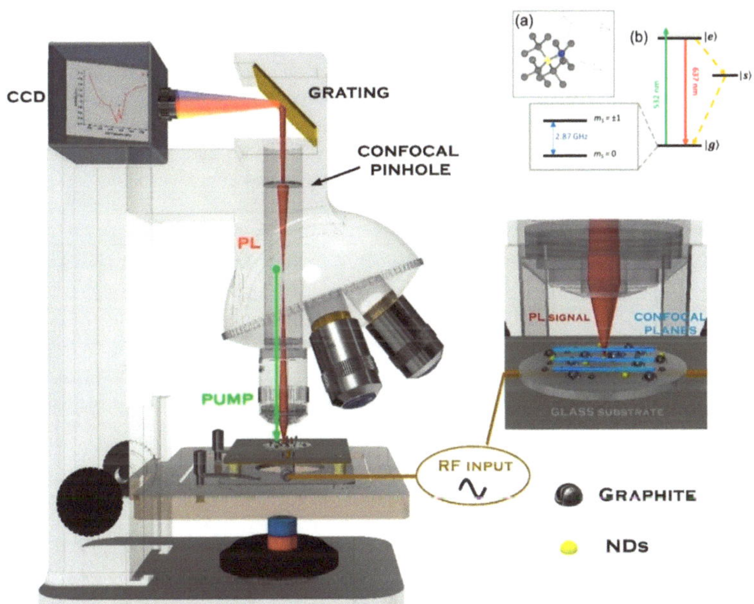

Figure 2. The confocal microscope allowed for the selection of the focal plane of the NDs clusters. The light was spectrally dispersed and finally post-processed in order to get an optically detected magnetic resonance spectrum combined with an optical image of the very region of the sample under investigation. The figure reports the general case when residual graphite is still residually present. Energy levels scheme of NDs are also reported in the top right inset. Section (**a**) shows the NV center crystal structure, and (**b**) the energy levels of the NV color center.

3. Results and Discussion

After the deposition, we followed two graphite cleaning procedures: the first relying on long-lasting thermal annealing of the nanopowders in order to remove as much graphite as possible, and the second employing a sulfonitric solution with a selective removal of graphite that achieves the preservation of the embedded NDs.

The thermal annealing at 425 °C and atmospheric condition was performed [11] to remove selectively the sp^2 carbon atoms (the sp^3 carbon atoms require a temperature higher than 500 °C to be oxidized). Assuming an efficiency of the graphite–diamond transition of 1% (but Yang et al. [12] reported efficiency for ablation of graphite in water up to 5%), the amount of pure NDs was estimated to be on the order of ~mg. The morphological analysis of the sample was done via scanning electron microscopy (SEM). Fluorescence and optically detected magnetic resonance of the samples were also analyzed. Typical SEM images for the as-deposited (as dep) sample and for those that were annealed (5 h ann and 77 h ann) are reported in Figure 3. As it can be seen, nanostructures having a size in the order of hundreds nm were formed. From comparison of the SEM images ("as dep" vs. "ann" samples), one can easily see that the NDs are completely embedded within the parasitic graphite matrix. When the sp^2 carbon atoms are etched under thermal oxidation (425 °C), the morphology of the sample significantly changes. Figure 3 (5 h ann) shows how the nanoparticles become more detached from the surrounding graphitic environment. The visible surface of the NDs is partially etched and it is not as smooth as in the "as dep" sample, indicating that the oxygen atoms act as efficient catalyzers for the removal of the NDs graphitic surface layers. Worth noting is the presence of submicrometric "holes" in the graphitic matrix (Figure 3, 5 h ann)—those are likely due to a complete removal of the sp^2 carbon atoms "gluing" the NDs clusters to the surrounding graphite. After 77 h of additional annealing (Figure 3, 77 h ann), this very situation is even more clear. The separation between the NDs borders and the graphite matrix is clearly observed—the amorphous carbon atoms and the NDs no longer form a continuous film as in the "as dep" sample. The etching process is digging (i.e., removing atoms) around and below the nanoparticles. This proves further that the nanoparticle core is not made by sp^2 atoms, but by sp^3 hybridized carbon atoms.

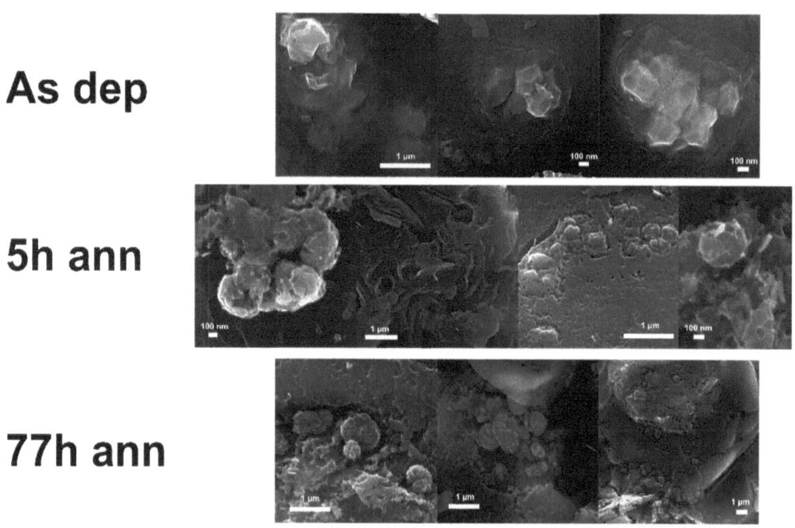

Figure 3. Typical SEM images for the as-deposited sample and after annealing at 425 °C for 5 h and 77 h.

As seen from SEM, this method is efficient only in partially removing the NDs-surrounding graphite. For this reason, we also followed the alternative cleaning protocol that consists of treating ablation soot in acid. The methodology, adapted from a detonation NDs industrial treatment [13], starts with dispersion of the soot in 30 mL of sulfonitric solution (three parts sulfuric acid at 98% concentration, one part nitric acid at 65% concentration). The mixture is then heated to 190 °C and kept under reflux for 3 h using an oil bath and a recondensation column. The resulting solution is then cooled and neutralized with NaOH. In order to deposit the NDs, the neutralized solution is then centrifuged for 30 min at 3000 rpm. Subsequently, the supernatant liquid is decanted and the sediment washed with distilled water in order to remove the salts formed during the neutralization, and centrifuged again. After drying, the samples were characterized via SEM imaging as reported in Figure 4.

Figure 4. Typical SEM images for the chemically cleaned sample. The NDs are clearly visible in agglomerated form and the graphite is virtually absent.

The yield of the chemical purification process, quantified as mass of the final product over the mass of the total ablation soot, is 7 ± 1%. This result is in line with other research based on laser ablation techniques that report yields between 5 and 10% [13,14]. HPHT shows yields around 15% [15] and cavitation-assisted techniques around 10% [16]. Detonation falls in the 4–10% range of the mass of the initial explosive charge mass [16,17].

NV-center optical properties were characterized by studying the NDs fluorescence spectra of our nanopowders under continuous 532 nm laser irradiation (Figure 5) using the confocal apparatus described in the previous section. A broad red emission was detected, consistent with the well-established [18] NV^- radiation pattern. An additional photoluminescence tail, probably due to graphitic surface layers on the NDs, was observed in the red region of the spectrum (>700 nm). The most important message is that the photoluminescence after the thermal and chemical treatments, performed as described, was still compatible with that observed in our previous work [8,9]. The sharp peaks at <600 nm are the D and G Raman peaks related to residual NDs-surrounding graphite shells. ODMR measurements are sensitive only to NV^- center fluorescence [2]. The observable effect is that the NV^- fluorescence undergoes a significant intensity change when radiofrequency coirradiation is superimposed to the samples and tuned to around 2870 MHz (transition of NV ground state energy level ǀg> to excited state ǀe>, as illustrated in Figure 2)

In fact, the ODMR curve (Figure 6) confirms the presence of NV^- centers as testified by the ~2870 MHz fluorescence dip. In particular, its spectral broadness is compatible with previous observation [8] and it is attributed to local strain fields acting on the observed micrometric-sized NDs clusters.

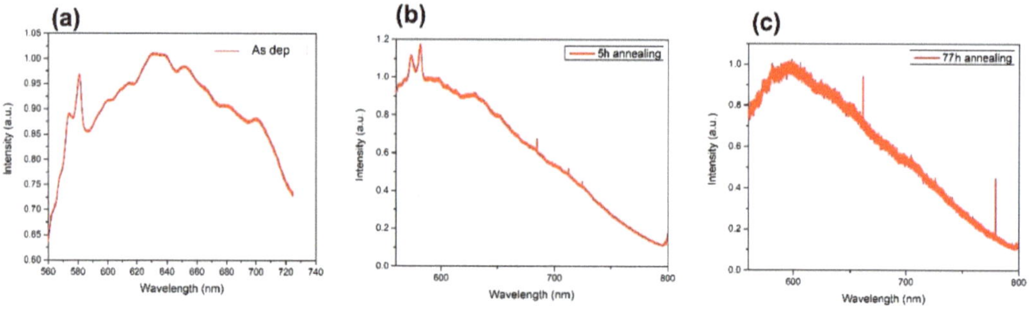

Figure 5. PL spectra for the as-deposited (**a**), the 425 °C 5 h annealed (**b**), and (**c**) the 425 °C 77 h annealed sample.

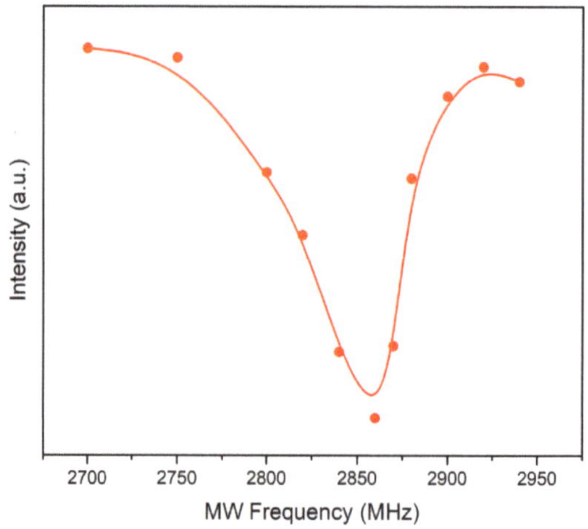

Figure 6. Representative ODMR spectrum of the PLA NDs after the cleaning procedure. The line is a guide for the eye.

4. Conclusions

In conclusion, we designed and built an efficient automatized liquid-nitrogen pulsed laser ablation apparatus capable of synthesizing milligrams of fluorescent NDs. In fact, within a single 6 h experiment more than 50 mg ablation powder (pure NDs with NV centers + graphite) were obtained, one order of magnitude larger than reported in [8,9]. This represents an important step forward with respect to previous existing literature [8,9], allowing to chemically clean out almost all the graphitic byproducts, preserving their optical and magneto-optical features, and letting us estimate the PLA-yield around values competitive with other production techniques. PLA fluorescent NDs are therefore closer to sensing applications [19].

Author Contributions: Conceptualization, M.C., L.B., M.O., and A.M.; methodology, L.B., M.C., C.C., N.B., M.O., A.P., and E.M.; writing, all authors; SEM imaging, N.B.; design of the experimental apparatus, C.C., N.B., and M.C.; optical characterization, E.M. and M.C.; supervision, M.C. and A.M. All authors have read and agreed to the published version of the manuscript.

Funding: The research activity is partially supported by the NaDia project within the research framework of Q@TN. This work was performed, in part, at the Center for Integrated Nanotechnologies, an Office of Science User Facility operated for the U.S. Department of Energy (DOE) Office of Science. Sandia National Laboratories is a multimission laboratory managed and operated by National Technology and Engineering Solutions of Sandia, LLC, a wholly owned subsidiary of Honeywell

International, Inc., for the U.S. DOE's National Nuclear Security Administration, under contract DE-NA-0003525. The views expressed in the article do not necessarily represent the views of the U.S. DOE or the United States Government.

Institutional Review Board Statement: Not applicable.

Informed Consent Statement: Not applicable.

Data Availability Statement: The data are available upon request to massimo.cazzanelli@unitn.it.

Acknowledgments: In this section, you can acknowledge any support given which is not covered by the author contribution or funding sections. This may include administrative and technical support, or donations in kind (e.g., materials used for experiments).

Conflicts of Interest: The authors declare no conflict of interest.

References

1. Mochalin, V.N.; Shenderova, O.; Ho, D.; Gogotsi, Y. The properties and applications of nanodiamonds. *Nat. Nanotechnol.* **2012**, *7*, 11–23. [CrossRef] [PubMed]
2. Doherty, M.W.; Manson, N.B.; Delaney, P.; Jelezko, F.; Wrachtrup, J.; Hollenberg, L.C. The nitrogen-vacancy colour centre in diamond. *Phys. Rep.* **2013**, *528*, 1–45. [CrossRef]
3. Sotoma, S.; Terada, D.; Segawa, T.F.; Igarashi, R.; Harada, Y.; Shirakawa, M. Enrichment of ODMR-active nitrogen-vacancy centres in five-nanometre-sized detonation-synthesized nanodiamonds: Nanoprobes for temperature, angle and position. *Sci. Rep.* **2018**, *8*, 1–8. [CrossRef] [PubMed]
4. Simpson, D.A.; Morrisroe, E.; McCoey, J.M.; Lombard, A.H.; Mendis, D.C.; Treussart, F.; Hollenberg, L.C. Non-neurotoxic nanodiamond probes for intraneuronal temperature mapping. *ACS Nano* **2017**, *11*, 12077–12086. [CrossRef] [PubMed]
5. Maze, J.R.; Stanwix, P.L.; Hodges, J.S.; Hong, S.; Taylor, J.M.; Cappellaro, P.; Lukin, M.D. Nanoscale magnetic sensing with an individual electronic spin in diamond. *Nature* **2008**, *455*, 644–647. [CrossRef] [PubMed]
6. Chen, M.; Pierstorff, E.D.; Lam, R.; Li, S.Y.; Huang, H.; Osawa, E.; Ho, D. Nanodiamond-mediated delivery of water-insoluble therapeutics. *ACS Nano* **2009**, *3*, 2016–2022. [CrossRef] [PubMed]
7. Kucsko, G.; Maurer, P.C.; Yao, N.Y.; Kubo, M.; Noh, H.J.; Lo, P.K.; Lukin, M.D. Nanometre-scale thermometry in a living cell. *Nature* **2013**, *500*, 54–58. [CrossRef] [PubMed]
8. Basso, L.; Gorrini, F.; Cazzanelli, M.; Bazzanella, N.; Bifone, A.; Miotello, A. An all-optical single-step process for production of nanometric-sized fluorescent diamonds. *Nanoscale* **2018**, *10*, 5738–5744. [CrossRef] [PubMed]
9. Basso, L.; Bazzanella, N.; Cazzanelli, M.; Miotello, A. On the route towards a facile fluorescent nanodiamonds laser-synthesis. *Carbon* **2019**, *153*, 148–155. [CrossRef]
10. Basso, L.; Gorrini, F.; Bazzanella, N.; Cazzanelli, M.; Dorigoni, C.; Bifone, A.; Miotello, A. The modeling and synthesis of nanodiamonds by laser ablation of graphite and diamond-like carbon in liquid-confined ambient. *Appl. Phys. A* **2018**, *124*, 72. [CrossRef]
11. Osswald, S.; Yushin, G.; Mochalin, V.; Kucheyev, S.O.; Gogotsi, Y. Control of sp2/sp3 carbon ratio and surface chemistry of nanodiamond powders by selective oxidation in air. *J. Am. Chem. Soc.* **2006**, *128*, 11635–11642. [CrossRef] [PubMed]
12. Aleksenskii, A.E. Technology of preparation of detonation Nanodiamonds. In *Detonation Nanodiamonds: Science and Applications*; Vul', Y.A., Shenderova, O.A., Eds.; Pan Stanford Publishing Pte. Ltd.: Boca Raton, FL, USA, 2014; pp. 51–52.
13. Yang, L.; May, P.W.; Yin, L.; Smith, J.A.; Rosser, K.N. Growth of diamond nanocrystals by pulsed laser ablation of graphite in liquid. *Diam. Relat. Mater.* **2007**, *16*, 725–729. [CrossRef]
14. Xiao, J.; Liu, P.; Yang, G.W. Nanodiamonds from coal under ambient conditions. *Nanoscale* **2015**, *7*, 6114–6125. [CrossRef] [PubMed]
15. Boudou, J.P.; Curmi, P.A.; Jelezko, F.; Wrachtrup, J.; Aubert, P.; Sennour, M.; Gaffet, E. High yield fabrication of fluorescent nanodiamonds. *Nanotechnology* **2009**, *20*, 235602. [CrossRef] [PubMed]
16. Arnault, J.C. (Ed.) *Nanodiamonds: Advanced Material Analysis, Properties and Applications*, 1st ed.; Elsevier: Amsterdam, The Netherlands, 2017; pp. 29–40.
17. Schrand, A.M.; Hens, S.A.C.; Shenderova, O.A. Nanodiamond particles: Properties and perspectives for bioapplications. *Crit. Rev. Solid State* **2009**, *34*, 18–74. [CrossRef]
18. Chung, P.H.; Perevedentseva, E.; Cheng, C.L. The particle size-dependent photoluminescence of nanodiamonds. *Surf. Sci.* **2007**, *601*, 3866–3870. [CrossRef]
19. Basso, L.; Cazzanelli, M.; Orlandi, M.; Miotello, A. Nanodiamonds: Synthesis and application in sensing, catalysis, and the possible connection with some processes occurring in space. *Appl. Sci.* **2020**, *10*, 4094. [CrossRef]

C Journal of Carbon Research

MDPI

Article

Energy Absorption in Carbon Fiber Composites with Holes under Quasi-Static Loading

Omar Alhyari and Golam Newaz *

Mechanical Engineering Department, Wayne State University, Detroit, MI 48202, USA; omar.alhyari@wayne.edu
* Correspondence: gnewaz@eng.wayne.edu

Abstract: Composite tubular structures have shown promise as energy absorbers in the automobile industry. This paper investigates the energy absorption characteristics of carbon fiber reinforced plastic (CFRP) tubes with pre-existing holes. Holes may represent an extreme case of impact damage that perforates the tube, e.g., stones from road surface impacting the tubes. Tubes with holes represent more conservative performance characteristics, since impact damage of the same size will have residual material, which may carry some load. Tubes with holes can provide the lower limit of CFRP tube performance under axial crushing relative to impact damaged tubes with perforation diameter close to the hole diameter. In this study, tubes with lay-up of [05/902/04] with one and two holes in defined locations and different diameters are experimentally studied under quasi-static loading. It was found that specific energy absorption (SEA) reduces by 50% with one or two holes of 15 mm size, 100 mm from top of the tube. The SEA reduction is about 60% lower than the regular tube when the diameter of the hole is 20 mm located at 100 mm from top. The most severe reduction occurs if the location of single or double holes are 75 mm from the top. In this case, a SEA reduction of 75% can be expected. Results indicate that holes can significantly alter the energy absorption capability of the tubes. It is also clear that in axial crushing of composite tubes, the location of the hole (100 to 75 mm) appears to create more pronounced effect than the size of the hole itself (15 vs. 20 mm) for the cases investigated. The failure modes for tubes with holes seem to preserve similar damage modes with delamination, frond creation, and brittle fracture, which is typically observed in regular composite tubes under axial crushing load. This is due to primarily front end crushing, which dominates the failure modes, while hole induced damage occurs later.

Keywords: carbon fiber reinforced plastics (CFRP); energy absorption; composite materials; thin walled tubes; damage; quasi-static loading

Citation: Alhyari, O.; Newaz, G. Energy Absorption in Carbon Fiber Composites with Holes under Quasi-Static Loading. *C* **2021**, *7*, 16. https://doi.org/10.3390/c7010016

Received: 5 January 2021
Accepted: 22 January 2021
Published: 1 February 2021

1. Introduction

The Corporate Average Fuel Economy (CAFE) standards and the strong safety standards have forced auto manufacturers to search for materials that can achieve the high strength to weight ratio without manufacturing complexity or increasing the cost. It was well known that traditional materials like steel will not meet the auto industry's requirements, which has led manufacturers to investigate the possibility of using lighter weight composite materials. Recent developments in all aluminum vehicles such as Ford 150 pick-up trucks show a trend to use lightweight materials.

Automakers see an increase in fuel efficiency each time they can reduce the vehicle mass. Auto manufacturers continued looking for ways to achieve higher strength to weight ratios without impacting the safety of the occupants. One of the car components that received attention for achieving significant strength to weight ratio is the use of CFRP energy absorbers in the vehicle frame, as shown in Figure 1. The conventional approach involves the use of steel to manufacture the energy absorbing components such as crush cans behind the bumper, which absorbs energy through controlled collapse by folding and hinging, involving extensive local plastic deformation.

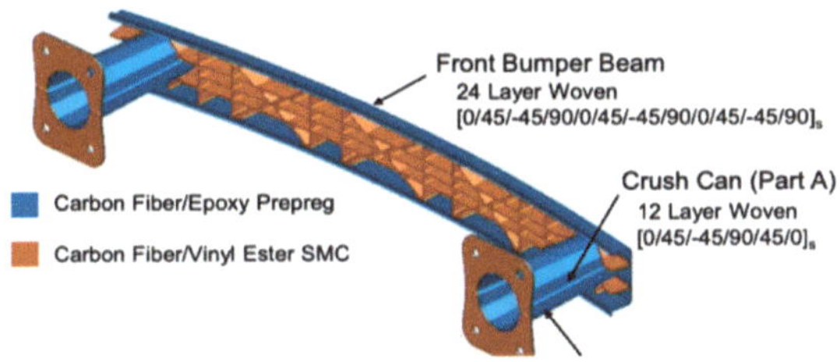

Figure 1. CFRP crush cans attached to composite bumpers [1].

As a result of new developments in composite materials industry, it was clear that the benefits of using composite materials could provide a perfect replacement for traditional materials like steel due to their high strength, stiffness, and low density. Due to more stringent safety legislation, the performance of composite materials during crushes became an important topic for researchers and manufacturers. It may be noted that the crush can be subjected to stone impact from the road. Any damage to the crush can lower its energy absorption capability. Whether they should be replaced due to stone impact creating perforations is the issue being addressed in this article.

The early research investigating the composite materials as crush energy absorption structures began in the mid-1970s. Three pioneering research groups are credited with major contributions establishing a solid understanding of how the composite materials behave and fail under crush loads and absorb energy: Thornton from Ford Motor Co., Hull from University of Cambridge, and Farley from the U.S. Army. Hull [2] was the first who identified three different crushing modes: Type I: a long single central interlaminar crack, whose length is greater than the thickness of the wall of the tube. Type II: a central interlaminar crack, whose length is less than the thickness of the tube wall. Type III: no central interlaminar crack with the tube wall bending outward from the center of the tube. Thronton provided great insight on the crushing mechanism through proper characterization and structural response of tubes under crush loads [3–5]. Farley [6–8] described Type I crushing mode as a lamina bending mode, while the Type II and III modes seem to be a combination of more fundamental crushing modes and geometry-related failure modes.

Thornton [4] reported that tubes made from unidirectional prepreg were found to produce higher energy-absorption capability than woven material. Other researchers like Hamada et al. [9–12] observed increasing the SEAs for a range of high temperature thermoplastic matrices and found a correlation with increasing matrix fracture toughness. Mamalis et al. [13,14] proposed a new method to perform the bending crushing test. Czaplicki et al. [15] reported that the energy absorbed by tulip triggered specimens was significantly higher than bevel triggered ones of the same geometry and material. Abdel Haq et al. studied crushing failure mechanisms in glass fiber tubular structures in detail, which showed similar failure modes as in CFRP tubes [16–18]. Farley [6–8] reported that the static and dynamic tests produced the same energy absorption, but it was reported that composite materials absorbed energy without catastrophic failure of the tubes. Farley also reported that higher strain at failure composite material system exhibited superior energy absorption capability. Farley [6] summarized the characteristics for tubes by four modes: transverse shearing, brittle fracturing, lamina bending, and local bucking, as shown in Figure 2.

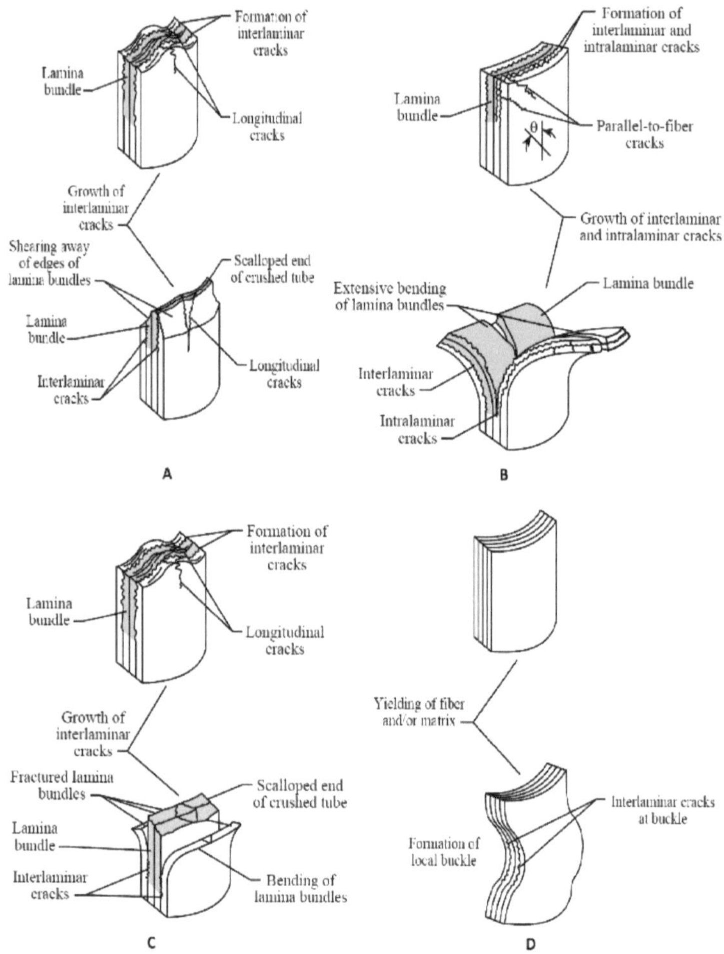

Figure 2. Typical failure modes observed in CFRP tubes undTheer axial crushing [6]. (**A–D**), represent typical key failure modes that are observed.

Kim et al. [19] reported that for tubes with hybrid unidirectional fibers, the crashing modes were determined by the types of the fibers in the axial direction. Chiu et al. [20] reported that strain rate will not affect the performance of composite structures response to crush loads. Ma et al. [21] concluded that for carbon/aramid CFRP capability of absorbing, energy increased after treatment while carbon/carbon CFRP did not show changes in energy absorption after treatment.

These key references point out that when increasing the fiber and matrix stiffness, the energy absorption capability increases. They also explain the effects of the fiber and matrix failure strain on the energy absorption capability.

The specific energy absorption (SEA) is a frequently used parameter to indicate the energy absorption capability, which is defined as the amount of energy, absorbed per unit mass crushed material, and it can be calculated by calculating the area under load-displacement curve, as shown in Figure 3.

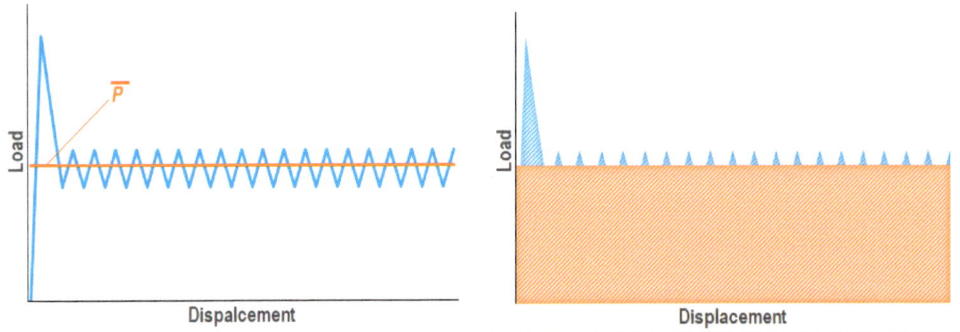

Figure 3. Typical load–displacement curve.

The $(\overline{P})$ is the mean crash load and is independent of the crash displacement. The *SEA* can be expressed in Equation (1) [10]:

$$SEA = \frac{\overline{P}}{A\,\rho} \quad (1)$$

where ρ is the material density, A is the cross-sectional area of the tube, and $\overline{P}$ is the average crush load. The load $\overline{P}$ times the entire displacement, say, d as in the orange rectangle (area under the average crushing load) represents total energy, $\overline{P} \times$ d. absorbed by the tube. Dividing the total mass of the crushed portion of the tube is M = A.d. ρ and then canceling d from numerator and denominator results in Equation (1).

The composite materials have anisotropic mechanical properties, which means that damages cause severe degradation, but it is difficult to predict the damage, because it is internal damage. Composites are known to be susceptible to damage caused by transverse loads even under low-velocity impacts. The composites can be damaged on the surface; importantly, they can be damaged below the surface by relatively light impacts causing barely visible damage. A hole or a cutout represents lack of load carrying capability in the axial direction in this local area of the tube akin to localized impact damage area and also serves as a stress concentration factor (SCF) for the composite. Although impact damage is not the same as holes or cutouts, the latter represents a more conservative estimate of the capability of the energy absorber, since in the impact damage case, some material in the damaged area similar to the size of the hole may still transmit load.

Many types of research investigated how circular holes and notches affect the flat composite plates; particularly, Saha et al. did a detailed study [22]. However, studies to find out how holes affect the composite cylinders and tubes are very limited. Initial studies on the effect of holes assumed that the curvature has no or little effect on stress concentration, but the thin shell theory required the shell to have critical length. By using bending and tension tests, it was concluded that a hole will add stresses to the shell if R/a is less than 4, where R is the inner diameter and a is the hole diameter. Van Tooren et al. [23] found that the hole's stress concentration is dependent on curvature. Quinn and Dulieu-Barton [24] confirmed that SCF can be approximated for a hole in the cylinder using flat plate data. SCF for composite analysis has been addressed by Wu and Mu [25]. Liang et al. [26] investigated the curvature effect for cylindrical shell with circular hole under pressure, where they reported that increasing the curved angle will lead to higher resistance to the external pressure load. In our investigation, we do not address strength characteristics due to the SCF in the tube. In this work, our goal is to assess the reduction in energy absorption capability of composite tubes due to macro-defects (holes) present in the tubes in CFRP to gain an understanding of how severely the energy absorption capability is compromised. Novelty for this work is in gaining new knowledge are two-fold: (a) how energy absorption is affected due to axial crushing of carbon fiber tubes in the presence of holes (macro-defects or discontinuities that may mimick stone impact),, where very little understanding is there in open literature, and (b) shedding light on failure mechanisms for such tubes subjected to axial crushing force and checking if they are similar to regular tubes without holes or present other modalities.

2. Experimental Aspects

2.1. Description of CFRP Tube Samples

The composite CFRP unidirectional tubes used in this experimental work were manufactured By Newport Inc. using Toray 700S-12K and Newport 301 epoxy, which is a 250° F to 300° F cure, semi-toughened, controlled flow epoxy resin system. Versatile processing, excellent mechanical properties, and long out-life make Newport 301 suitable for a variety of structural applications. The density of the CFRp material is 1.65 g/cc.

The mechanical properties of the composite are listed in Table 1. The tube geometry is shown in Figure 4; the lay-up is $[0_5/90_2/0_4]$. Each tube has a 45° chamfer end.

Table 1. Mechanical properties of the $[0_5/90_2/0_4]$ CFRP tube.

Tensile Modulus (GPa)	Tensile Strength (MPa)	Comp. Modulus (GPa)	Comp. Strength (MPa)	Flexural Modulus (GPa)	Flexural Strength (MPa)
127	2250	123	1247	98	1500

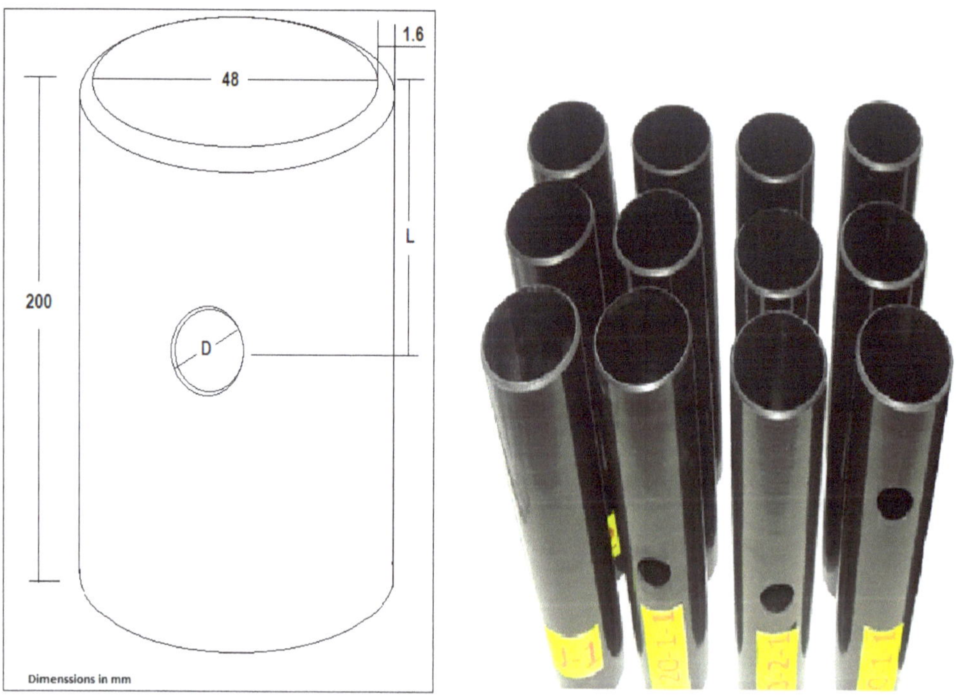

Figure 4. CFRP tube geometry and dimensions in mm.

The CFRP tubes used in these experiments were 200 mm long and 48 mm inner diameter with 1.6 mm in thickness. Tubes with holes were based on hole's location, hole's diameter, and number of holes; tube code 100-15-2 means the hole location is 100 mm from the chamfered end, hole diameter is 15 mm, and the number of holes is 2. Similarly, holes located 75 mm from top had the designation of 75-15-1, etc. Table 2 provides the average dimensions of the composite tubes. Figure 4 shows the tube geometry and dimensions.

Table 2. Tube sample types and dimensions.

Tube	Length (mm)	L (mm)	D (mm)	No. of Holes	No. of Samples
	200			0	5
100-15-1	200	100	15	1	5
100-15-2	200	100	15	2	5
100-20-1	200	100	20	1	5
100-20-2	200	100	20	2	3
75-15-1	200	75	15	1	5
75-15-2	200	75	15	2	5

2.2. Testing Setup

The composite tubes were tested using 810 MTS universal machine (245 kN) at room temperature. All specimens were crushed about 100 mm at a crosshead rate of 0.5 mm/s. Five repeated experiments were carried out for each set of tests to verify the repeatability of energy-absorption capability, except for the case where two holes of 20 mm dia were located at 100 mm from the chamfered end, which is discussed later. Repeatability of tests provided confidence in testing only three samples for this case only. All tests were run under the displacement-controlled condition.

3. Results and Discussion

The composite tube crushing process is progressive failure by nature, where the chamfered end will trigger the damaging processes when peak force is reached, then a slight decrease to stable crushing force is experienced. Damage spreads progressively throughout the tube. The advantage of this failure process is high energy absorption, because most of the tube is damaged over a large deflection range, while an average crushing force is maintained.

It was established by previous work [2–21] that composite tubes absorbed the energy via progressive damage in different modes. The damage initially starts at the top of the chamfered area where high hoop tensile strain exists, then propagates downwards along the tube wall. The damage propagates along the tube wall, causing the tube wall to split into fronds by matrix splits. In these tubes, the outer ply is 0° ply, which separates from the bulk of the tube and splays outwards due to delamination. With this splaying mode, debris starts falling and creates gaps. The inner-most plies exhibit similar behavior as the outer-most plies, but they splay inwards. In general, the inner plies show substantial damage due to internal constraints, which generate a large amount of debris. The fiber breakage is clearly observed for the 0° and 90° plies. The 0° plies break into smaller pieces compared to 90° plies, but the fragments in the 0° plies are smaller. The overall damage process is shown in Figure 5.

Out of all damage types, the brittle fracture (fiber rand matrix separation and fiber breaks in rapid succession) is the main damage. The axial load is carried mainly by the 0° fibers, but the 90° plies will also contribute to it, although axial stiffness for this layer is low. Due to axial loading, the 90° layers bend and break via shearing; these shear conditions are promoted due to relative axial displacement of 0° and 90° plies as a consequence of delamination between the layers. The test results show that crushing involves basic failure modes such as splaying, brittle fracture, and transverse shearing. The failure modes are shown in Figure 5.

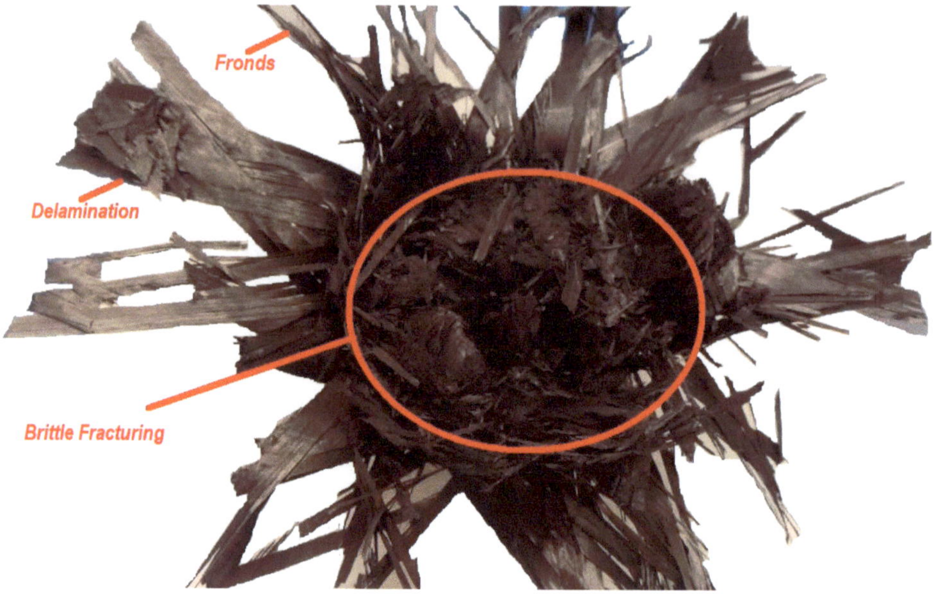

Figure 5. Representative collapse modes of 50 mm dia composite circular CFRP tubes, exhibiting many of the failure modes depicted by Farley [7] for regular CFRP tubes, as shown in Figure 2.

3.1. Response of Regular Tubes without Holes

For the regular tubes with no holes tested in this work, five samples were tested where the crushing behavior and force–displacement curves were quite similar. Using these results, an average curve was generated, as shown in Figure 6.

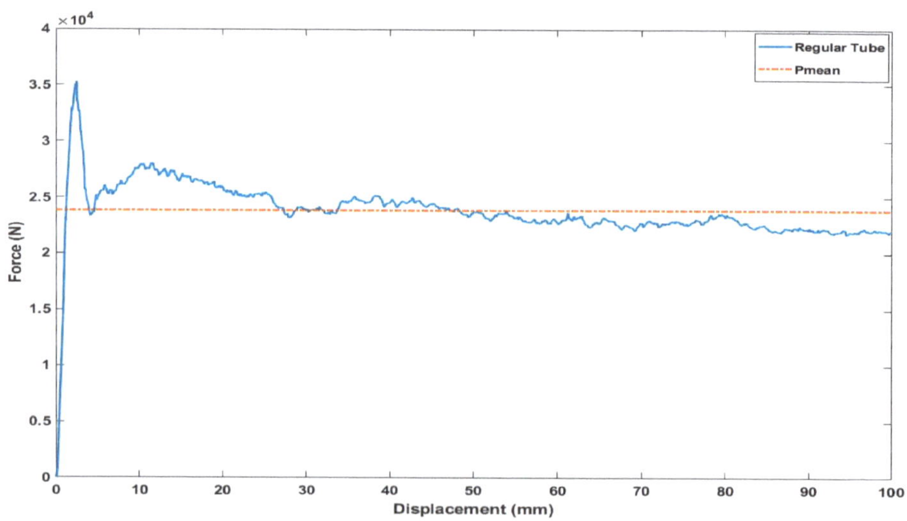

Figure 6. Average force–displacement experimental curve (blue) for the regular tube. The red line represents P_{mean} line for the entire load–deflection response (100 mm of deflection).

The average force–displacement represents the typical force–displacement curve for composite materials. The force will increase until it reaches the peak and then drop to a lower level and later increase to steady state level until the crash process is completed. The total energy absorption is equal to the area under the force–displacement curve, or by integrating the curve, the specific energy absorption (SEA) can be calculated by using Equation. 1, which is essential to compare the performance of tube material in relation to the weight of the system to show overall load carrying and energy absorption efficiency in relation to material mass. However, the SEA is dependent on the tube's material, geometry, ply orientation, and matrix. P_{max} is the peak force, and P_{mean} is the mean or average value of crushing force. This is estimated for the deflection level of 100 mm. Table 3 provides peak force, mean force, energy absorbed, and SEA for the five CFRP tubes tested.

Table 3. Peak force, mean force, and SEA.

Experimental			
P_{max} (kN)	P_{mean} (kN)	EA (kJ)	SEA (kJ/kg)
35.3 ± 3.25	23.9 ± 0.82	2.39 ± 0.08	57.5 ± 1.93

3.2. Response of Tube with Holes

It is expected that the mode of failure may be altered due to pre-existing holes according to the reported results by Corum et al. [27], who investigated the low energy impact effects on automotive structural composites, where they reported that the strength reduction using compression after impact (CAI) response of carbon fiber laminate due to central circular hole is similar to the strength reduction caused by impacted defects of approximately similar size, the hole providing a more conservative result. In this paper, the strength degradation due to holes was found to provide a reasonable lower bound to the strength degradation caused by impact damage. This result adds credence to the common assumption, often used in damage tolerance evaluations, that impact-induced defects introduce the same strength reduction as a hole of the same size

Each tube with an existing hole has a threshold size where there is no influence on the progressive crush response below this hole size. If the hole is located at 100 mm, a number of experiments confirmed that threshold diameter is approximately 12 mm; below this diameter size, no appreciable difference was found in P_{max} and P_{mean} values as long as the hole is located at least 100 mm from the chamfered end. The tubes with pre-existing holes are listed in Table 2. All the tubes are tested under a quasi-static test with 0.5 mm/sec load speed using 810 MTS machine (245 kN).

The samples experienced failure in various modes: the tubes with pre-existing holes started with splaying matching the regular tubes. After reaching the peak force, the tubes damaged via progressive failure. At a greater deflection level, cracks initiate at the hole, causing the tube to split and collapse. Hole initiated cracks propagate quickly and in sudden bursts. Once the platen contacts the dense damaged material pack, the load rises again due to debris compaction. This stage is not useful, as the tube has already collapsed and also does not contribute to energy absorption.

The presence of a hole makes the tube weaker at the point when the cracks initiate from the holes and cause load drops, and thus the absorbed energy will be less than the regular tube. This result matches what Karbhari et al. [28] reported, that post impact will cause a reduction in energy absorption by composite materials and the crush performance will change. It is observed that until the failure initiation from the crack, the peak p value (Figure 7), and the overall mean value, P_{mean} are not affected much. This clearly shows that if the location of the hole is removed away from the load application point, then original characteristics of energy absorption and P_{mean} may be preserved for the tube until precipitous load drop, when cracking at the hole dominates the tube response. Figure 7 is showing the damage as a function of the force–displacement curve at three points: A represents the P_{peak}, B represents the start of the steady state force, and C represents when the crack is initiated by the pre-existing hole. At Cm it is noted that the 15 mm circular hole is no longer circular due to the crack initiation.

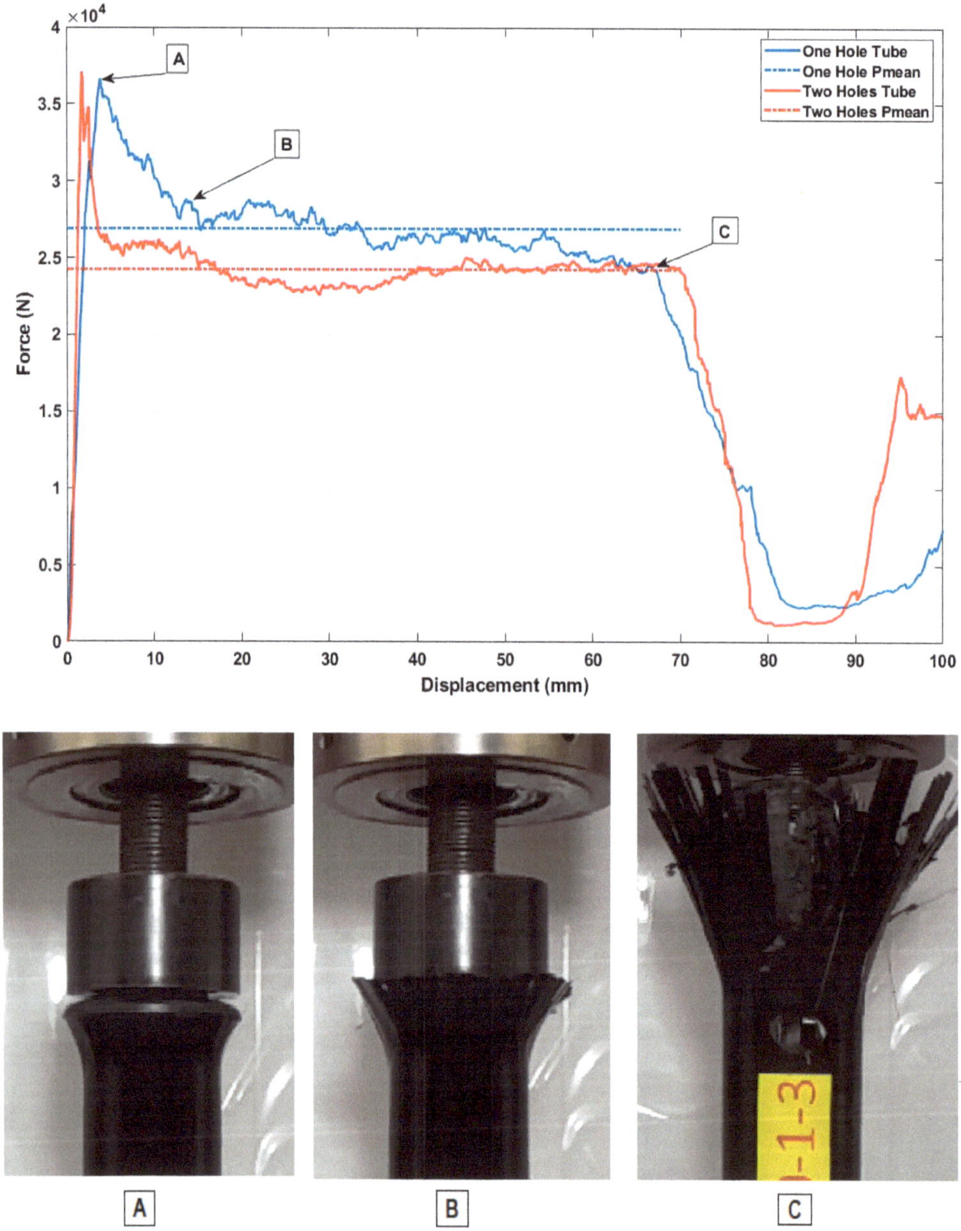

Figure 7. Force–displacement curves for the tubes with 15 mm hole, 100 mm from chamfered top (average).

The response of the tube for single hole vs. two holes is shown in Figure 7. The curves are average of five tests. One notes that the P_{mean} value for the two-hole tube is somewhat lower than the single-hole sample response. In addition, the precipitous drop in load carrying capability of the tube occurs around 70 mm vertical deflection. For the samples with 20 mm holes, the P_{mean} values (Figure 8) are lower than the 15 mm hole samples (Figure 7). Another important feature is that with larger hole size of 20 mm, the precipitous load drop occurs a little earlier at 66 mm deflection level compared to samples with 15 mm holes at 70 mm deflection level (compare Figures 7 and 8).

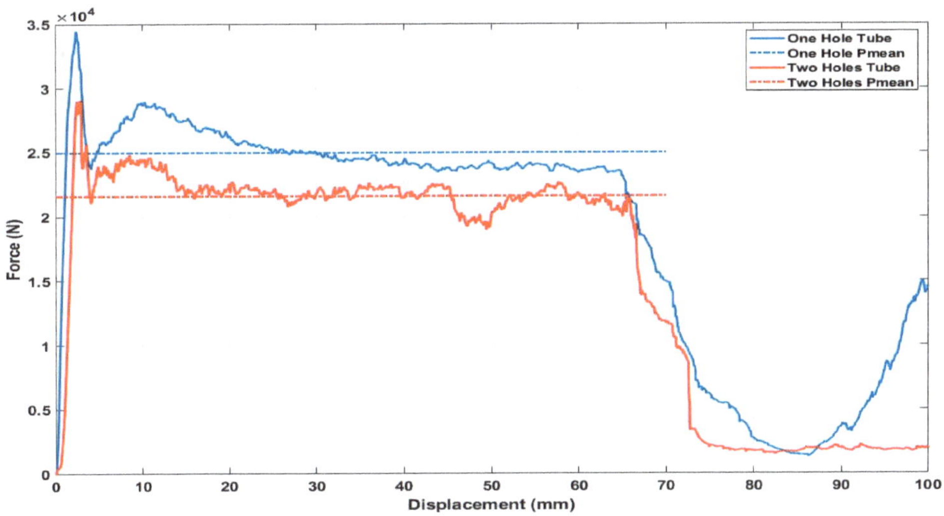

Figure 8. Force–displacement curves for the Tubes with 20 mm hole, 100 mm from chamfered top (average).

Figure 9 represents the load–displacement curves for the average results obtained from tubes with pre-existing 15 mm holes at 75 mm distance from the chamfered end. It is interesting to note that the location of the holes being closer from the top location of the tube have similar response and P_{mean} are very similar. However, the precipitous load drop occurs at about 50 mm of vertical deflection, much earlier than when the holes are 100 mm from the chamfered end. The force displacement curves for tubes with preexisting holes for the six tubes tested in this work showed same trend; the force will increase until it reaches the peak and drop to steady state level, and significantly drop when the crack is initiated by the hole and propagated around the circumference. P_{max}, P_{mean}, energy absorbed, and SEA are obtained from experimental data are listed in Table 4.

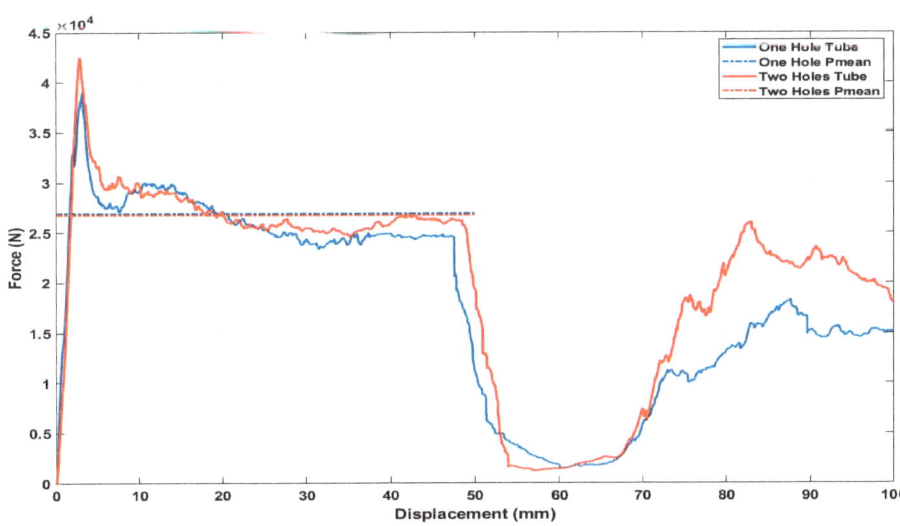

Figure 9. Force–displacement curves for the tubes with 15 mm hole, 75 mm from chamfered top (average).

Table 4. SEA and other data for tubes with holes.

Experimental					
Tube	P_{max} (kN)	P_{mean} (kN)	Energy (kJ)	SEA kJ/kg	% of Regular Tube SEA
100-15-1	36.6 ± 4.50	26.9 ± 1.85	1.82 ± 0.19	29.96 ± 2.7	52.1
100-15-2	37.1 ± 5.34	24.3 ± 1.17	1.71 ± 0.12	29.40 ± 2.3	51.1
100-20-1	34.4 ± 2.50	25.0 ± 1.34	1.63 ± 0.14	25.81 ± 2.0	44.9
100-20-2	29.1 ± 3.80	21.6 ± 0.95	1.43 ± 0.03	22.96 ± 1.5	39.9
75-15-1	39.0 ± 3.30	26.1 ± 0.50	1.24 ± 0.07	14.30 ± 0.6	24.9
75-15-2	42.5 ± 1.50	26.8 ± 0.96	1.30 ± 0.08	15.48 ± 1.2	26.9

The results listed in Table 4 show that the number of holes and the hole diameter did not play a significant role in the energy absorption capability of the tubes. For the tubes with one 20 mm hole located at 100 mm from the chamfered top, the drop in SEA is 7.2%, compared to a tube with one 15 mm hole located 100 mm from the chamfered top. Additionally, it is noted that the number of holes did not have dramatic effect on energy absorption. The tube with two 15 mm holes located 100 mm from the chamfered top has lower SEA by 0.26 kJ/kg compared to a tube with one 15 mm hole. The highest effect was caused by the hole location, where we can observe that the SEA for tubes with 15 mm hole located 75 mm from the chamfered top is approximately 50% of the SEA for tube with 15 mm hole located 100 mm from the chamfered top. In our review of even most recent papers in open literature, we find this important issue of presence of defects has not been addressed [29–32]. Our results provide some new information in this area that is relevant for application of these tubes for energy absorption in automotive applications.

4. Conclusions

In this study, the crashworthiness performance of CFRP composite tubes with and without pre-existing holes have been investigated experimentally. The work focused on the effect of holes on the energy absorption under quasi-static load in the tubes.

It was observed that tubes without holes crushed progressively with stable performance. The tube splayed, and long fronds were created, and small debris formation was observed. The tubes with pre-existing holes displayed similar failure modes as observed in tubes without holes until the damage progressed closer to the hole when the tube splayed and long fronds were created. When the load reached a critical point, a crack at the hole will be formed where the stress was the highest and then propagate around the circumference quickly. The crack leads to significant damage, which will cause the tube to split and collapse. It was found that specific energy absorption (SEA) reduces by 50% with one or two holes of 15 mm size, 100 mm distance from top of the tube. The SEA reduction is about 60% lower than the regular tube when the diameter of the hole is 20 mm located at 100 mm from top. The most severe reduction occurs if the location of single or double holes are 75 mm from top. In this case, a SEA reduction of 75% can be expected. Results indicate that holes can significantly alter the energy absorption capability of the tubes. It is also clear that in axial crushing of composite tubes, the location of the hole (100 to 75 mm) appears to create more pronounced effect than the size of the hole itself (15 vs. 20 mm) for the cases investigated. The failure modes for tubes with holes seem to preserve similar damage modes with delamination, frond creation, and brittle fracture, which are typically observed in regular composite tubes without holes under axial crushing load.

Author Contributions: O.A.: investigation, data curation and writing—original draft preparation; G.N.: writing—review and editing, supervision. All authors have read and agreed to the published version of the manuscript.

Funding: The research received no external funding.

Conflicts of Interest: The authors declare no conflict of interest.

References

1. Coppola, T.; Faruque, O.; Board, D. Validation of Material Models for Crash Simulation of Automotive carbon Fiber Composite Structures. In Proceedings of the Society of Plastics Engineers (SPE)—American Society for Composites (ACC) Meeting, East Lansing, MI, USA, 6 September 2017.
2. Hull, D. A Unified Approach to Progressive Crushing of Fiber Reinforced Composite Tubes. *Compos. Sci. Technol.* **1991**, *40*, 377–421. [CrossRef]
3. Thornton, P.H.; Jeryan, R.A. Crash Energy Management in Composite Automotive Structures. *Int. J. Impact Eng.* **1988**, *7*, 167–180. [CrossRef]
4. Thornton, P.H. Energy Absorption of Composite Structures. *J. Compos. Mater.* **1979**, *13*, 247–262. [CrossRef]
5. Thornton, P.H. The Crush Behavior of Pultruded Tubes at High Strain Rates. *J. Compos. Mater.* **1990**, *24*, 594–615. [CrossRef]
6. Farley, G.L. Energy Absorption of Composite Materials. *J. Compos. Mater.* **1983**, *17*, 267–279. [CrossRef]
7. Farley, G.L. Effect of Fiber and Matrix Maximum Strain on the Energy Absorption of Composite Materials. *J. Compos. Mater.* **1986**, *20*, 322–334. [CrossRef]
8. Farley, G.L. Effect of Specimen Geometry on the Energy Absorption Capability of Composite Materials. *J. Compos. Mater.* **1986**, *20*, 390–400. [CrossRef]
9. Hamada, H.; Ramakrishna, S. Effect of Fiber Material on the Energy Absorption Behavior of Thermoplastic Composite Tubes. *J. Thermoplast. Compos. Mater.* **1996**, *9*, 259–279. [CrossRef]
10. Ramakrishna, S.; Hamada, H.; Maekawa, Z.; Sato, H. Energy Absorption Behavior of Carbon-Fiber-Reinforced Thermoplastic Composite Tubes. *J. Thermoplast. Compos. Mater.* **1995**, *8*, 323–344. [CrossRef]
11. Hamada, H.; Ramakrishna, S.; Sato, H. Effect of Fiber Orientation on the Energy Absorption Capability of Carbon Fiber/PEEK Composite Tubes. *J. Compos. Mater.* **1996**, *30*, 947–963. [CrossRef]
12. Hamada, H.; Ramakrishna, S. Scaling Effects in the Energy Absorption of Carbon-Fiber/PEEK Composite Tubes. *Compos. Sci. Technol.* **1995**, *55*, 211–221. [CrossRef]
13. Mamalis, A.G.; Manolakos, D.E.; Viegelahn, G.L. Crashworthy Behavior of Thin-Walled Tubes of Fiberglass Composite Material Subjected to Axial Loading. *J. Compos. Mater.* **1990**, *24*, 72–91. [CrossRef]
14. Mamalis, A.G.; Manolakos, D.E.; Demosthenous, G.A.; Ioannidis, M.B. The Deformation Mechanism of Thin-Walled Non-Circular Composite Tubes Subjected to Bending. *Compos. Struct.* **1995**, *30*, 131–146. [CrossRef]
15. Czaplicki, M.J.; Robertson, R.E.; Thornton, P.H. Comparison of Bevel and Tulip Triggered Pultruded Tubes for Energy Absorption. *Compos. Sci. Technol.* **1991**, *40*, 31–46. [CrossRef]
16. Abdel-Haq, M.; Newaz, G.M. Role of Failure Modes on Energy Absorption in Unidirectional PMC Tubes. *J. Compos. Mater.* **2001**, *35*, 941–953. [CrossRef]
17. Abdel-Haq, M.; Newaz, G.M. Influence of Crush Mechanisms on Energy Absorption of PMC Square Tubes. *J. Thermoplast. Compos. Mater.* **2001**, *14*, 160–174. [CrossRef]
18. Abdel-Haq, M.; Broggiato, G.B.; Newaz, G.M. Constraint Effects on Energy Absorption in Unidirectional PMC Tubes. *J. Compos. Mater.* **1999**, *33*, 774–793. [CrossRef]
19. Kim, J.S.; Yoon, H.J.; Shin, K.B. A study on Crushing Behaviors of Composite Circular Tubes with Different Reinforced Fibers. *J. Impact Eng.* **2011**, *38*, 198–207. [CrossRef]
20. Chiu, L.N.S.; Falzon, B.G.; Ruan, D.; Xu, S.; Thomson, R.S.; Chen, B.; Yan, W. Crush Responses of Composite Cylinder under Quasi-static and Dynamic Loading. *Compos. Struct.* **2015**, *131*, 90–98. [CrossRef]
21. Ma, Y.; Sugahara, T.; Yang, Y.; Hamada, H. A study on the energy absorption properties of carbon/aramid fiber filament winding composite tube. *Compos. Struct.* **2015**, *123*, 301–311. [CrossRef]
22. Saha, M.; Prabhakaran, R.; Waters, W.A., Jr. Compressive Behavior of Pultruded Composite Plates with Circular Holes. *Compos. Struct.* **2004**, *65*, 29–36. [CrossRef]
23. Van Tooren, M.J.L.; Van Stijn, I.P.M.; Beukers, A. Curvature Effects on the Stress Distribution in Sandwich Cylinders with a Circular Cut-Out. *Compos. Part A* **2002**, *33*, 1557–1572. [CrossRef]
24. Quinn, S.; Dulieu-Barton, J.M. Determination of Stress Concentration Factors for Holes in Cylinders using Thermoelastic Stress Analysis. *Blackwell Sci. Strain* **2002**, *38*, 105–118. [CrossRef]
25. Wu, H.-C.; Mu, B. On Stress Concentrations for Isotropic/Orthotropic Plates and Cylinders with a Circular Hole. *Compos. Part B* **2003**, *34*, 127–134. [CrossRef]
26. Liang, C.C.; Hsu, C.Y.; Chen, W. Curvature Effect on Stress Concentrations around Circular Hole in Opened Shallow Cylindrical Shell under External Pressure. *Int. J. Press. Vessel. Pip.* **1998**, *75*, 749–763. [CrossRef]
27. Corum, J.M.; Battiste, R.L.; Ruggles-Wren, M.B. Low-Energy Impact Effects on Candidate Automotive Structural Composites. *Compos. Sci. Technol.* **2003**, *63*, 755–769. [CrossRef]
28. Karbhari, V.M.; Haller, J.E.; Falzon, P.K.; Herszberg, I. Post-impact crush of hybrid braided composite tubes. *Int. J. Impact Eng.* **1999**, *22*, 419–433. [CrossRef]
29. Chambe, J.E.; Bouvet, C.; Dorival, O.; Ferrero, J.-F. Energy absorption capacity of composite thin-wall circular tubes under axial crushing with different trigger initiations. *J. Compos. Mater.* **2020**, *54*, 1281–1304. [CrossRef]

30. Palanivelu, S.; Van Paepegem, W.; Degrieck, J.; Van Ackeren, J.; Kakogiannis, D.; Van Hemelrijck, D.; Wastiels, J.; Vantomme, J. Experimental study on the axial crushing behaviour of pultruded composite tubes. *Polym. Test.* **2010**, *29*, 224–234. [CrossRef]
31. Emadi, M.; Beheshti, H.; Heidari-Rarani, M. Multi-Objective Optimization of Hybrid Aluminum–Composite Tube Under Axial Crushing. *Int. J. Appl. Mech.* **2020**, *12*, 2050042. [CrossRef]
32. Hwang, S.-F.; Chang, Y.-C. Axial Crushing Behavior of Braided Carbon/Polyurethane Composite Tubes. *Appl. Compos. Mater.* **2019**, *26*, 1281–1297. [CrossRef]

C Journal of Carbon Research

MDPI

Article

Effects of the Stacking Faults on the Electrical Resistance of Highly Ordered Graphite Bulk Samples

Andreas Hentrich and Pablo D. Esquinazi *

Division of Superconductivity and Magnetism, Felix Bloch Institute for Solid State Physics, Linnéstrasse 5, University of Leipzig, 04103 Leipzig, Germany; andreas.hentrich@web.de
* Correspondence: esquin@physik.uni-leipzig.de

Received: 1 July 2020; Accepted: 23 July 2020; Published: 25 July 2020

Abstract: High anisotropy and the existence of two-dimensional highly conducting interfaces at stacking faults parallel to the graphene planes of the graphite structure influence, in a non-simple way, the transport properties of highly oriented graphite. We report two related effects on the electrical resistance of highly oriented pyrolytic as well as of natural graphite bulk samples, measured with the four points method in the temperature range 300 K $\leq T \leq$ 410 K. A qualitative and quantitative change in the temperature dependence of the resistance was obtained by simply enlarging the electrodes and contacting the edges of the internal interfaces on the same sample. Additionally, at temperatures $T \gtrsim$ 350 K the resistance can change with time. We show that this temperature-dependent annealing effect is related to the stacking faults and can irreversibly change the absolute value of the resistance and its temperature dependence. A partial recovery is obtained after leaving the sample at normal conditions for several days. The overall results stress the importance of the electrodes location on a bulk graphite sample, the contribution of the stacking faults in the interpretation of the measured transport properties and the need of systematic studies on the influence of high temperature annealing on the interfaces properties.

Keywords: graphite; stacking faults; interfaces; transport

1. Introduction

The electrical conductivity of carbon-based compounds, especially graphite, has been reported in a large number of studies in the last nearly 90 years, see in [1,2] and the references therein. Detailed studies of carbonaceous powders under different compaction pointed out the existence of different contributions to, and the complexity of, the electrical conductivity [3] as well as the effect of oxygen on it [4]. In highly oriented graphite samples with the narrowest rocking curve width, the typical size of the single crystalline regions, in the a, b direction parallel to the graphene planes, is of the order of micrometers [5]. Therefore, the electrical resistance of bulk samples of much larger size would have the intrinsic contribution of the graphene layers added to that of the grain boundaries. There is, however, a further effect that makes the estimate of the intrinsic resistivity difficult, namely, the large anisotropy of the electrical resistivity of graphite with its c-axis resistivity ρ_c being much larger than the in-plane one $\rho_{a,b}$. This anisotropy was already reported at the end of the 1930s [6]. Although its value is sample-dependent due to the influence of defective regions in the samples, there is consent among different publications that the ratio $\rho_c/\rho_{a,b} > 10^4$. This anisotropy implies that the input current has a relatively small penetration depth in the $c-$axis direction, when the electrodes are located, e.g., at the top surface of a well-ordered graphite sample. Would a graphite sample be homogeneous from the structural and electrical point of view, i.e., with only one of the two possible stacking orders and without twisted regions around the common $c-$axis, then we expect to have similar results for the normalized resistance obtained by placing the electrodes at the sample top surface or at the sample edges. As we report in this study, this does not occur in graphite samples of large enough thickness due to the contribution of highly conducting stacking faults to the total sample conductance.

The existence of stacking faults or two-dimensional (2D) interfaces in well-ordered graphite samples has been recently reported by transmission electron microscopy (TEM) [7,8]. These TEM images together with high-resolution X-ray diffraction (XRD) [9] as well as transport measurements [7,8] revealed that most, if not all, thick enough highly ordered graphite samples have 2D stacking faults located between twisted crystalline regions with the same or different stacking orders. These faults can be embedded within the multilayer graphene matrix and can be found even at the sample surface [7–12]. These "interfaces" show qualitative different transport properties than those of the graphene planes in any of the two graphite stacking orders, i.e., Bernal and rhombohedral. Taking into account the electronic band structure at the interfaces between twisted graphite regions [10,13–15] or between the two stacking orders of graphite [16–18] , such a difference in the transport properties is not surprising. For example, a recently published experiment was able to contact the edge of a relatively thin graphite sample, and thereby its embedded interfaces [19]. The obtained value of 10^2 for the (metalliclike) rest-resistance-ratio as well as the positive magnetoresistance of $\simeq 10^7$ at 21 T, at temperatures below 1K, are record-breaking values that emphasized the need to take into account these interfaces in the interpretation of the transport properties.

Previous transport studies revealed the influence of the 2D stacking faults on the transport properties below 300 K of mesoscopic, thin (thickness $\lesssim$ 100 nm and several μm^2 area) graphite samples by nanometer thickness reduction [20]. In this work, we are interested on the influence of the 2D interfaces on bulk samples, in particular in the temperature range above 300 K because of the expected superconducting [9,17,21–24] and/or magnetic transitions [25]. With a simple experimental approach and without the need of etching, lithography, or both processes, we show that the contribution of the interfaces can change qualitatively and quantitatively the resistance of a bulk graphite sample. Furthermore, at temperatures above ~350 K we found that the resistance shows a non-monotonic change with time. We show that this high temperature effects are directly related to changes at the interfaces.

2. Materials and Methods

The main characteristics and dimensions of the measured samples in this work are listed in Table 1. Figure 1 shows a scanning electron microscope image taken directly at the edge of a bulk HOPG sample. One can recognize openings between regions of different gray color. Each of these regions can have a thickness from several tens to hundreds of nm, mostly with the Bernal stacking order, as TEM images revealed [7,26,27]. Regions with the minority rhombohedral stacking order phase also exist but are in general thinner [9,28].

Table 1. Origin and dimensions of the measured samples. HOPG: Highly oriented pyrolytic graphite. Further characteristics of the internal structure of the HOPG sample, in particular TEM images, were reported in [7,8,26,29] and of the natural graphite sample in [9].

Sample	Origin	Dimensions (mm^3)
#1	HOPG-Advanced Ceramics	$4.63 \times 1.3 \times 0.47$
#2	Natural Graphite-Sri Lanka	$3.2 \times 0.5 \times 0.14$
#3	HOPG-Advanced Ceramics	$4.4 \times 1.28 \times 0.35$

The electrical resistance was measured using an AC LR700 resistance bridge (four points method) with a frequency of 16 Hz and input currents $\lesssim$ 10 μA. To measure the electrical resistance as a function of temperature we have used two devices. The graphite sample in the device called minicryo was in a vacuum of 10^{-7} mbar during the measurements and attached to a small and high efficient cryocooler, with small thermal mass and consumption; for more details see in [30]. The sample thermometer was fixed on the same Cu-block where the sample was attached. The heater for the temperature stabilization was fixed around the main cooling finger, where at its top the sample/thermometer holder was.

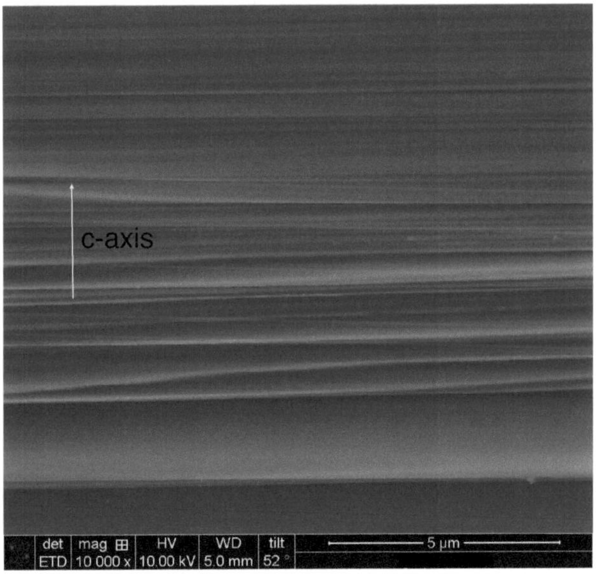

Figure 1. Scanning electron microscope image of the edge of a bulk HOPG sample where one can recognize interfaces. The $c-$axis of the graphite structure is normal to the graphene layers and the stacking faults or interfaces.

The second device used was the main sample temperature stabilization space of a He-flow cryostat from Quantum Design with a sample holder that allowed also four points measurements of the electrical resistance. The temperature was measured and set with the available thermometer and stabilization program. In this device, the sample is not in vacuum but in a ~0.01 mbar He atmosphere.

3. Results

3.1. Resistance before and after Contacting the Edge of a Graphite Sample

Several transport studies of different highly ordered graphite samples, from thin flakes of thickness $t \gtrsim 5$ nm to bulk samples, indicate that the intrinsic electronic transport in ideal graphite is semiconducting-like. This means that the resistance decreases with temperature following a typical exponential function having energy gaps between $\simeq 0.04$ eV for the Bernal stacking order and $\simeq$0.1 eV for the rhombohedral one [8,31]. The metallic-like part, which is measured in thick enough samples, is related to the contribution of 2D interfaces between regions of different stacking orders or twisted regions around a common $c-$axis [20]. This means that in a sample where the interfaces are located far away from the sample surface, i.e., at distances larger than the current penetration depth in the $c-$axis direction, we would expect that the resistance measured with the contacts at the top surface, as shown in the sketch of Figure 2a, shows a semiconducting-like behavior. Indeed, this is the case as we can see in Figure 2c. Note also that the irreversibility in the temperature loops is small and no time dependence is observed within experimental error, especially at high temperatures.

After we finished with the measurements shown in Figure 2c, we took out the sample from the cryostat and the electrical contacts were enlarged, contacting especially the side edge of the sample as sketched in Figure 2b. After waiting two days to dry the electrodes, the resistance was measured again during heating and cooling temperature cycles as shown in Figure 2d. Several remarkable changes were observed. First, the absolute value of the resistance at the beginning of the measurements was smaller by a factor 4. This decrease in the absolute resistance is not surprising because we are getting the contribution of several graphene layers in parallel contacting all their edges. However, the qualitative difference in the temperature dependence, a metallic-like dependence, can be interpreted as due to the contribution of certain stacking faults running in parallel to the graphene layers. It is also remarkable that the resistance decreases with time at temperatures above $\simeq$380 K, whereas, for this sample, the temperature dependence below 380 K does not change significantly. Evidently, this decrease at high enough temperatures is related to the contribution of certain interfaces and it is observed in all bulk samples, as we show below in the next sections. It is important to mention that the observed changes are not related to the contacts; these were already several days old and any drying process should not have any influence on the measured resistance. As we will see below, the changes produced by annealing are reversible after leaving the sample at normal conditions for several days, ruling out contact problems as well.

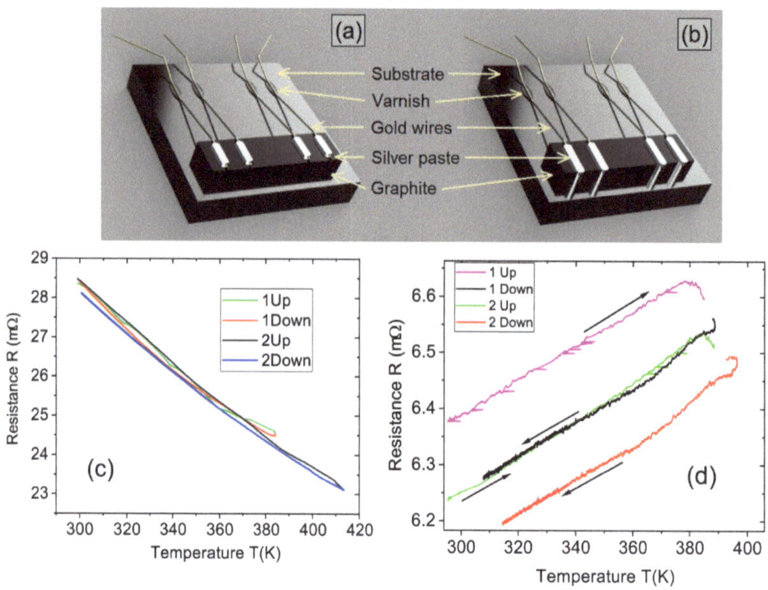

Figure 2. The upper pictures show sketches that describe the electrodes positions prepared on the same HOPG sample #1: (**a**) sample with top contacts and (**b**) top and side contacts. Panels (**c**,**d**) show the respective temperature dependence of the resistance obtained with the electrodes shown in the upper pictures. The labels indicate the sequential measurement step (1 or 2) and if it was measured by warming (Up) or cooling (Down). The measurements were done in the minicryo with a heating rate of 0.4 K/min and passive cooling with a maximum cooling rate of 0.4 K/min.

The observed changes in the absolute values of the resistance already emphasize the inaccuracy one has when one needs an estimate of the intrinsic resistivity of graphite. For example, assuming an electrically homogeneous sample without electrical anisotropy, from the results of Figure 2 at 300 K we estimate a resistivity $\rho(300\text{ K}) \sim 400\ \mu\Omega$cm (electrodes at the top, case (c)) and $\sim$100 $\mu\Omega$cm (electrodes at the top and side, case (d)). Those values are within those reported in the literature for highly ordered graphite bulk samples [1,2]. Much smaller values are obtained in case (c) if we assume a much smaller thickness due to the finite penetration depth ($\sim$20 nm at low temperatures [20]). The intrinsic resistivity value for ideal graphite was estimated in [20] to be $\rho(300\text{K}) \sim 580\ \mu\Omega$cm. The resistivity values in our sample are evidently smaller but not very different. The reasons for this relatively small difference are (1) the contribution of the 2D interfaces to the total conductance depends on temperature, getting weaker the higher the temperature [19,31], and (2) the electrical anisotropy also decreases with temperature.

Concluding this section, it should be clear that the temperature dependence of the resistance of bulk graphite samples is not unique but it depends on the location of the electrodes within the sample in a non-simple way, due to the high electrical anisotropy and the metallic-like contribution of certain interfaces embedded within the graphite structure. Upon the interface distribution within a sample, different and non-simple temperature behavior can be expected, in agreement with the results reported in [20], decreasing systematically the thickness of mesoscopic graphite samples.

3.2. High-Temperature Annealing Effects on the Resistance

Annealing effects in the diamagnetic susceptibility of bulk graphite at $T \gtrsim 400$ K have been already reported that showed an irreversibly decrease in the total diamagnetic signal [32]. We note that a significant part of the diamagnetic response of bulk graphite samples is not due to the graphene planes but to embedded interfaces at some stacking faults, as recently done studies reported [33]. In this section, we demonstrate and discuss the partially subtle effects of time and temperature on the resistance observed at $T \gtrsim 350$ K. For these measurements, we took a new natural graphite bulk sample (#2) with electrodes contacting its edges and fixed in the sample holder of the He-flow cryostat. Although in this cryostat there is no thermometer attached directly on the sample holder, it allows larger temperature changes in less amount of time. To check whether the sample reached the selected temperature, the measurement of its time dependence is necessary.

We programmed an increase in the thermometer temperature from 300 K to 390 K within 20 min, as shown by the red line in Figure 3a (right $y-$axis). After this temperature was reached at the thermometer, with a small overshoot recognized in Figure 3a, it remained constant within ± 0.1 K. As expected for this sample, the measured resistance (black curve in Figure 3a (left $y-$axis)) increases, from which we infer that the sample temperature also increases but with a certain delay. Note that the temperature overshoot at approximately $t \sim 20$ min is also observed in the resistance with a short delay, see inset in Figure 3a. After reaching a constant thermometer temperature (region of constant T labeled in the inset) the time-dependent slope of the resistance changes, and at $t \simeq 32$ min we observe a second change in the slope. From the data and the experience obtained from magnetization measurements with the same temperature inset, we estimate the delay between the thermometer and sample temperature to be of the order of $\lesssim 15$ min for this especial quick heating ramp. This means that at $t \gtrsim 32$ min, the sample temperature can be considered as constant. Nevertheless, the resistance increases following an exponential function $R(t > 32\text{min}) \simeq R(t \to \infty) - a\exp(-t/\tau)$ with a relaxation time of $\tau \simeq 23$ min, as shown by the dashed red line in the inset. At $t \gtrsim 140$ min, the resistance starts to decrease with a larger, temperature-dependent relaxation time, as we show in more detail in Figure 4 below. Note that in usual experiments even at these temperatures, one measures the resistance at smaller temperature steps taking a point after a waiting time of the order of 15 min. This means that this subtle increase in resistance is not always easily recognized. Annealing effects on the electrical resistance above room temperature are nowadays of importance because they can detrimentally influence the transition in the resistance observed at $T \sim 350$ K reported in [9,24].

Figure 3b shows the measured resistance vs. temperature from 300 K to 390 K with a warming rate of 5 K/min for sample #2. In the virgin, non-annealed state the sample resistance increases with T, while showing a change in its slope at ~350 K and at ~370 K. This behavior has been already reported and interpreted as a possible superconducting-like transition with a critical temperature of ~350 K [9,24]. We note that in those publications, a similar curve was obtained stabilizing the temperature for 15 min before taking a point. After the small temperature overshoot clearly observed in the resistance, see Figure 3b, we measured the resistance vs. time at 390 K. The sample temperature reached a constant value in the region described by the dashed line in Figure 3b. In the inset we show the time dependence of the resistance in a semilogarithmic scale in the time region where the sample temperature remained constant at $T = 390$ K, similar to that shown in the inset of Figure 3a.

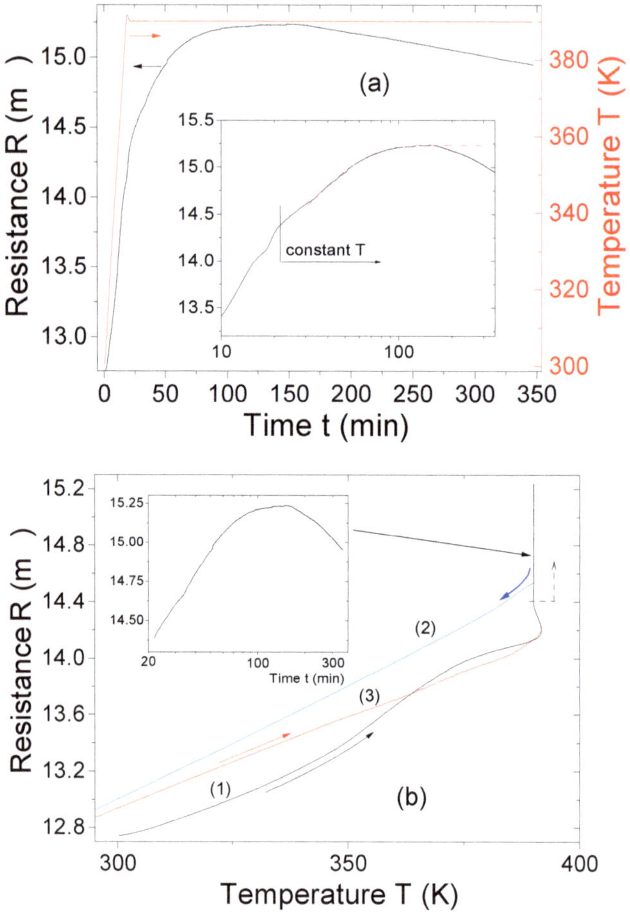

Figure 3. (**a**) Resistance of sample #2 as a function of the time (black curve, left $y-$axis) and the temperature measured by the thermometer in the He-flow cryostat (red curve, right $y-$axis) in double linear scale. The sample site edge was contacted. The inset shows the resistance vs. time in semilogarithmic scale. The time region depicted by the arrow indicates the period where the temperature of the sample remained constant. (**b**) Measured resistance of the same bulk sample in the virgin, non-annealed state, during heating (1), cooling afterwards (2), and a following second heating (3). Both heating and cooling rates were 5 K/min. The inset shows the change of the resistance with time measured in the region depicted by the dashed arrow in the main panel.

After $\simeq 5$ h at $T = 390$ K, we started to cool down the sample and measured its temperature dependence before the resistance decreases below a certain value, in order to check whether the increase in the resistance with time remains at lower temperatures. The blue curve (2) in Figure 3b shows that the resistance was larger than the one measured at the first warming, shown by curve (1). Note that the resistance in curve (2) does not show any change in its slope at $T \sim 350$ K, in contrast to curve (1). After cooling it down to 30 K the resistance was again measured to 390 K with similar rate. The result is shown by curve (3) in Figure 3b. We note that this curve does not show any change in its slope around 350 K, indicating that the anomaly at this temperature measured in the virgin state vanished after the temperature cycle.

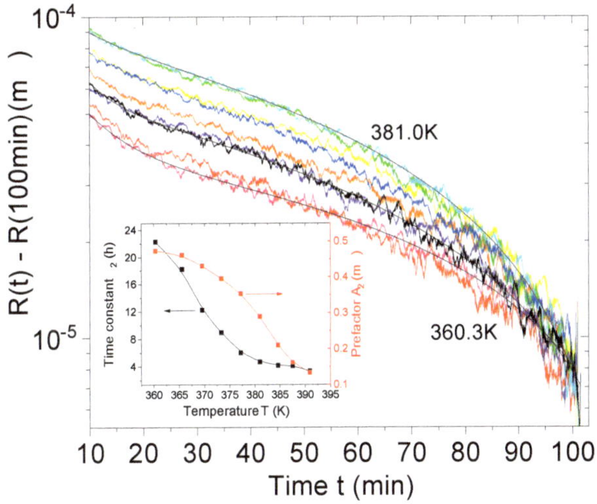

Figure 4. Change of the resistance with time at different constant temperatures of a natural graphite sample (#2) measured in the minicryo. Each of the experimental curves at a fixed temperature was obtained during 100 min. The sample temperatures were magenta (360.3 K), red (365.5 K), orange (369.5 K), yellow (373.3 K), green (377.2 K), light blue (381.0 K), dark blue (384.6 K), purple (387.5 K), and black (390.9 K). The three continuous lines are fits to two exponential time decay terms (see main text). The inset shows the respective temperature dependence of τ_2 (black points, left $y-$axis) and the prefactor A_2 (red points, right $y-$axis).

The results of this temperature cycle clearly indicate that the temperature annealing to 390 K for several hours was enough to wash out the transition at certain interfaces observed at $\sim$350 K in the virgin state. This effect of the temperature hysteresis cycle is in qualitative agreement with the partial vanishing of the diamagnetic signal after increasing the temperature above $\sim$400 K in a bulk HOPG sample [32]. All these results stress that the phase that is responsible for the step-like change in the resistance in the virgin curve is very sensitive to high-temperature annealing and care should be taken. This vanishing may provide an answer to the difficulties one has to observe and reproduce the transition in some specific samples.

We discuss now the characterization of the relaxation time of the decrease of the resistance at different temperatures at $T > 350$ K. For this characterization, we took sample #3 and measured the decrease of its resistance at different constant temperatures in a sequential manner. This means that after $\simeq$100 min, and after a certain decrease in the resistance was measured, we increased the temperature and started again with the time measurement. Note that the sample did not start at the different selected temperatures in its virgin state. The reason for the selection of this sequential procedure is the large relaxation time necessary for a full recovery of the sample, as we discuss in the next section.

Figure 4 shows the time dependence of the resistance at different constant temperatures. The three continuous curves through the experimental data are fits to the equation $R(t) = A_1 * e^{-t/\tau_1} + A_2 * e^{-t/\tau_2} + c$, where the prefactors, as well as the relaxation times, are parameters we obtain from the fits. The first relaxation term (subindex 1) represents a rather short relaxation term with a nearly temperature-independent relaxation time, $\tau_1 \simeq (7 \pm 2)$ min, with also a nearly constant prefactor, $A_1 \simeq (0.05 \pm 0.02)$ mΩ. This first relaxation is mostly related to a transient related to the temperature stabilization after a temperature step. The second relaxation process shows temperature-dependent relaxation times. The relaxation time as well as the prefactor of this process decrease with temperature, see the inset of Figure 4.

3.3. *Influence of Air Atmosphere: Recovery of the Resistance*

The observed time relaxation of the resistance at high enough temperatures when one contacts the edges of the samples, indicate that certain diffusion processes at the interfaces play a role. In this section, we demonstrate that there is a recovery of the resistance when one leaves the sample in air at normal atmosphere after the high-temperature annealing. The results indicate that at least part of the relatively large decrease of the resistance at high temperatures can be recovered. Figure 5 shows four temperature cycles (each cycle means $\simeq$ 300 K $\rightarrow$ 400 K $\rightarrow$ 300 K) measured after the sample preparation, i.e., starting with the sample in the virgin state. The results are similar to those shown in Figure 2d. We note, however, that the resistance of the sample decreased by a factor of two after the temperature cycles, in contrast to the few percent decrease observed when the maximum temperature of each cycle remains below 390 K; compare the results in Figure 2d with those in Figure 5.

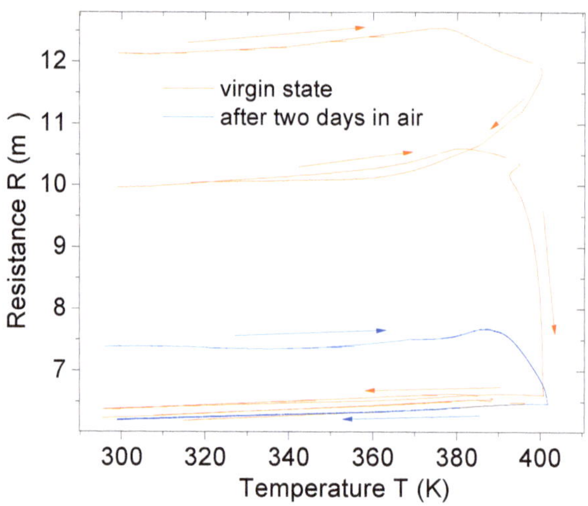

Figure 5. Two independent measurements of the same HOPG sample #3 in the minicryo. The orange curve shows the temperature-dependent cycles of the sample starting just after the sample preparation in the virgin state. At the starting point of the cycles, the resistance at 300 K of this sample was 12.1 mΩ decreasing to 6.2 mΩ at the end of cycles at the same temperature. Afterwards, the sample was left for two days in air at normal conditions. After this time, the resistance of the sample was measured again with one temperature cycle as show by the blue curve.

After we reached 290 K, at the end of the fourth temperature cycle starting with the sample in the virgin state, we took the sample out of the minicryo and left it at that temperature in air for two days. Afterwards, we measured its resistance and its temperature dependence within one temperature cycle. The result is the blue curve in Figure 5. We note that the initial resistance at 290 K was clearly higher than just after finishing the previous four cycles. The single temperature cycle measured for the sample, see blue curve in Figure 5, demonstrates the reversibility of the whole annealing process and suggests that the resistance might be fully recovered after a long period of time at normal conditions.

4. Discussion

Time-dependent effects in the resistance were reported in studies of graphene. For example, in [34], net charge density during air exposure leads to similar time dependence as described in this work. The interaction between O_2/H_2O redox couples and graphene has been proposed [35] as a possible origin, which leads to electrochemical p-doping [36].

Carbon nanotubes in humid atmosphere also showed time-dependent changes in the electrical resistance [37]. After the injection of water vapor onto a carbon nanotube sample measured in vacuum, the conductance decreases with small volumes of water. However, it increases again after injecting larger amount of water vapor. The authors assumed that water molecules can be absorbed on the surface and act as electron donors to the initially p-type semiconducting nanotube surface [37]. Thereby, until the compensation point is reached, the resistance will increase because the available free carriers will cancel each other out. Injecting even more water has the opposite effect: the additional free electrons do not get compensated anymore, and thereby the amount of free carriers increases. The time constant for this diffusion process is similar to those measured here in bulk graphite. Additionally, it was shown that the diffusion process of water into the sample can be inverted by heating the sample in vacuum, which is the situation in our experiments. The authors of [37,38] showed a variety of responses to NH_3, NO_2, CO_2, N_2, and air atmosphere in nanotubes. Because the effects of the atmosphere in graphene and carbon nanotubes are comparable, it suggests that water vapor alone can lead to n-doping, while water in combination with oxygen can lead to p-doping, thereby increasing the importance of the atmospheric humidity even more.

Another group of carbon-based materials, where related effects were measured, are carbonaceous powders. For example, in [4], the resistance of those materials was shown to be strongly dependent on the oxygen concentration at the surface. The closely related structure of graphite, carbon nanotubes, and graphene, and the similarity in the observed time dependence of the resistance, strongly suggest a similar origin and provides a possible explanation for the measured behavior in Figures 3–5. Therefore, we may assume that the origin of the observed changes in the resistance of bulk graphite is related to the diffusion of hydrogen, water, or oxygen molecules. Because no such annealing effects were measured when the contacts electrodes were located at the top of the selected bulk sample (see Figure 2c), the diffusion processes should happen mainly within the stacking faults, observed when the electrodes were located at the sample edge (see Figure 2d). Whatever the detailed origin for the observed large changes in the resistance at high enough temperatures, our results indicate that the superconducting properties of the interfaces can vanish after annealing.

5. Conclusions

Two effects related to the existence of stacking faults in bulk graphite samples were measured. The first effect is a further, clear proof of the importance of these interfaces in the temperature dependence of the resistance: By contacting the sample on the top sample surface, parallel to the a, b-plane and graphene layers, the influence of the interfaces is substantially diminished. This lead to a semiconducting-like behavior, which is the expected intrinsic one of ideal graphite. Afterwards, enlarging the same electrodes on the same sample to contact its edge leads to significant changes. Not only does the absolute resistance drop by a factor of four, but also the temperature dependence turned to metallic-like.

The second effect is related to the influence of temperature annealing and of the atmosphere on the transport properties of bulk graphite: When bulk graphite samples are prepared under standard atmosphere and measured in vacuum or low-pressure helium atmosphere, the sample undergoes significant non-monotonous changes, when it is heated to temperatures above 350 K over time spans of one to several hours. In a first stage, the resistance increases with time with a relaxation time of $\sim$20 min at 390 K. This first annealing effect already changes the temperature dependence of the sample resistance, in particular, the change of the slope of the resistance vs. temperature at $\sim$350 K and $\sim$370 K did vanish. Additionally, the resistance decreases exponentially with a temperature dependent time constant of several hours. This decrease in resistance is partially reversible when leaving the sample at normal conditions for several days.

By measuring and quantifying these two effects, the study done in this work emphasizes the importance of the atmosphere, the measuring time, and temperature when it is above 350 K, and the electrode positions on the resistance of bulk graphite samples with stacking faults. It appears compulsory for every researcher in this field to take care of the here observed changes. Disregarding the reported influence can result in unexpected large changes in resistance with a large impact on the sample transport parameters and can be a source of irreproducible results.

Author Contributions: Conceptualization, P.D.E.; Data curation, A.H.; Funding acquisition, P.D.E.; Investigation, A.H.; Supervision, P.D.E.; Validation, P.D.E.; Writing—original draft, P.D.E.; Writing—review & editing, P.D.E. All authors have read and agreed to the published version of the manuscript

Funding: Part of this research has been possible with the support of the Saechsischen Aufbaubank (SAB) and the European Regional Development Fund, Grant Nr.: 231301388.

Acknowledgments: We gratefully acknowledge the technical support of L. Botsch, J. Barzola-Quiquia, A. Setzer and W. Böhlmann.

Conflicts of Interest: The authors declare no conflict of interest.

Abbreviations

The following abbreviations are used in this manuscript.

XRD	X-ray diffraction
TEM	transmission electron microscopy
HOPG	highly oriented pyrolytic graphite
AC LR700	Alternating Current Linear Research 700
2D	two dimensional

References

1. Kelly, B.T. *Physics of Graphite*; Applied Science Publishers: London, UK, 1981.
2. Esquinazi, P. (Ed.) *Basic Physics of Functionalized Graphite*; Springer Series in Materials Science 244; Springer International Publishing AG: Cham, Switzerland, 2016.
3. Celzard, A.; Marêché, J.; Payot, F.; Furdin, G. Electrical conductivity of carbonaceous powders. *Carbon* **2002**, *40*, 2801–2815. [CrossRef]
4. Pérez-Rodríguez, S.; Torres, D.; Lázaro, M. Effect of oxygen and structural properties on the electrical conductivity of powders of nanostructured carbon materials. *Powder Technol.* **2018**, *340*, 380–388. [CrossRef]
5. González, J.C.; Muñoz, M.; García, N.; Barzola-Quiquia, J.; Spoddig, D.; Schindler, K.; Esquinazi, P. Sample-Size Effects in the Magnetoresistance of Graphite. *Phys. Rev. Lett.* **2007**, *99*, 216601. [CrossRef] [PubMed]
6. Krishnan, K.S.; Gangulin, N. Large Anisotropy of the Electrical Conductivity of Graphite. *Nature* **1939**, *144*, 667–667. [CrossRef]
7. Barzola-Quiquia, J.; Yao, J.L.; Rödiger, P.; Schindler, K.; Esquinazi, P. Sample size effects on the transport properties of mesoscopic graphite samples. *Phys. Status Sol. (A)* **2008**, *205*, 2924–2933. [CrossRef]
8. García, N.; Esquinazi, P.; Barzola-Quiquia, J.; Dusari, S. Evidence for semiconducting behavior with a narrow band gap of Bernal graphite. *New J. Phys.* **2012**, *14*, 053015. [CrossRef]
9. Precker, C.E.; Esquinazi, P.D.; Champi, A.; Barzola-Quiquia, J.; Zoraghi, M.; Muiños-Landin, S.; Setzer, A.; Böhlmann, W.; Spemann, D.; Meijer, J.; et al. Identification of a possible superconducting transition above room temperature in natural graphite crystals. *New J. Phys.* **2016**, *18*, 113041. [CrossRef]
10. Kuwabara, M.; Clarke, D.R.; Smith, A.A. Anomalous superperiodicity in scanning tunnelling microscope images in graphite. *Appl. Phys. Lett.* **1990**, *56*, 2396. [CrossRef]
11. Miller, D.L.; Kubista, K.D.; Rutter, G.M.; Ruan, M.; de Heer, W.A.; First, P.N.; Stroscio, J.A. Structural analysis of multilayer graphene via atomic moiré interferometry. *Phys. Rev. B* **2010**, *81*, 125427. [CrossRef]

12. Flores, M.; Cisternas, E.; Correa, J.; Vargas, P. Moiré patterns on STM images of graphite induced by rotations of surface and subsurface layer. *Chem. Phys.* **2013**, *423*, 49–54. [CrossRef]
13. Brihuega, I.; Mallet, P.; González-Herrero, H.; de Laissardière, G.T.; Ugeda, M.M.; Magaud, L.; Gómez-Rodríguez, J.M.; Ynduráin, F.; Veuillen, J.Y. Unraveling the Intrinsic and Robust Nature of van Hove Singularities in Twisted Bilayer Graphene by Scanning Tunneling Microscopy and Theoretical Analysis. *Phys. Rev. Lett.* **2012**, *109*, 196802. [CrossRef] [PubMed]
14. Esquinazi, P.; Heikkilä, T.T.; Lysogoskiy, Y.V.; Tayurskii, D.A.; Volovik, G.E. On the superconductivity of graphite interfaces. *JETP Lett.* **2014**, *100*, 336–339. [CrossRef]
15. Volovik, G.E. Graphite, Graphene, and the Flat Band Superconductivity. *JETP Lett.* **2018**, *107*, 516–517. [CrossRef]
16. Kopnin, N.B.; Heikkilä, T.T. *Carbon-Based Superconductors: Towards High-T_c Superconductivity*; CRC Press: Boca Raton, FL, USA; 2015; Chapter 9, pp. 231–263. [CrossRef]
17. Kopnin, N.B.; Ijäs, M.; Harju, A.; Heikkilä, T.T. High-temperature surface superconductivity in rhombohedral graphite. *Phys. Rev. B* **2013**, *87*, 140503. [CrossRef]
18. Muñoz, W.A.; Covaci, L.; Peeters, F. Tight-binding description of intrinsic superconducting correlations in multilayer graphene. *Phys. Rev. B* **2013**, *87*, 134509. [CrossRef]
19. Precker, C.E.; Barzola-Quiquia, J.; Esquinazi, P.D.; Stiller, M.; Chan, M.K.; Jaime, M.; Zhang, Z.; Grundmann, M. Record-Breaking Magnetoresistance at the Edge of a Microflake of Natural Graphite. *Adv. Eng. Mater.* **2019**, *21*, 1970039. [CrossRef]
20. Zoraghi, M.; Barzola-Quiquia, J.; Stiller, M.; Esquinazi, P.D.; Estrela-Lopis, I. Influence of interfaces on the transport properties of graphite revealed by nanometer thickness reduction. *Carbon* **2018**, *139*, 1074–1084. [CrossRef]
21. Kopnin, N.B.; Heikkilä, T.T.; Volovik, G.E. High-temperature surface superconductivity in topological flat-band systems. *Phys. Rev. B* **2011**, *83*, 220503. [CrossRef]
22. Ballestar, A.; Barzola-Quiquia, J.; Scheike, T.; Esquinazi, P. Evidence of Josephson-coupled superconducting regions at the interfaces of highly oriented pyrolytic graphite. *New J. Phys.* **2013**, *15*, 023024. [CrossRef]
23. Ballestar, A.; Esquinazi, P.; Böhlmann, W. Granular superconductivity below 5 K in SPI-II pyrolytic graphite. *Phys. Rev. B* **2015**, *91*, 014502. [CrossRef]
24. Stiller, M.; Esquinazi, P.D.; Barzola-Quiquia, J.; Precker, C.E. Local Magnetic Measurements of Trapped Flux through a Permanent Current Path in Graphite. *J. Low Temp. Phys.* **2018**, *191*, 105–121. [CrossRef]
25. Henck, H.; Avila, J.; Ben Aziza, Z.; Pierucci, D.; Baima, J.; Pamuk, B.; Chaste, J.; Utt, D.; Bartos, M.; Nogajewski, K.; et al. Flat electronic bands in long sequences of rhombohedral-stacked graphene. *Phys. Rev. B* **2018**, *97*, 245421. [CrossRef]
26. Esquinazi, P.D.; Lysogorskiy, Y. *Experimental Evidence for the Existence of Interfaces in Graphite and Their Relation to the Observed Metallic and Superconducting Behavior*; Springer Series in Materials Science 244; Esquinazi, P., Ed.; Springer International Publishing AG: Cham, Switzerland, 2016; Chapter 7, pp. 145–179._7. [CrossRef]
27. Esquinazi, P.D.; Precker, C.E.; Stiller, M.; Cordeiro, T.R.S.; Barzola-Quiquia, J.; Setzer, A.; Böhlmann, W. Evidence for room temperature superconductivity at graphite interfaces. *Quantum Stud. Math. Found.* **2017**. [CrossRef]
28. Lin, Q.; Li, T.; Liu, Z.; Song, Y.; He, L.; Hu, Z.; Guo, Q.; Ye, H. High-resolution TEM observations of isolated rhombohedral crystallites in graphite blocks. *Carbon* **2012**, *50*, 2369–2371. [CrossRef]
29. Spemann, D.; Esquinazi, P. *Evidence for Magnetic Order in Graphite from Magnetization and Transport Measurements*; Springer Series in Materials Science 244; Esquinazi, P., Ed.; Springer International Publishing AG: Cham, Switzerland, 2016; pp. 45–76._3. [CrossRef]
30. Steinmann, R. About Continuous Flow Cryostats. Available online: http://as-specialdevices.com/index.php/fr/news-and-announcements (accessed on 24 July 2020)
31. Zoraghi, M.; Barzola-Quiquia, J.; Stiller, M.; Setzer, A.; Esquinazi, P.; Kloess, G.; Muenster, T.; Lühmann, T.; Estrela-Lopis, I. Influence of rhombohedral stacking order in the electrical resistance of bulk and mesoscopic graphite. *Phys. Rev. B* **2017**, *95*, 045308. [CrossRef]
32. Scheike, T.; Esquinazi, P.; Setzer, A.; Böhlmann, W. Granular superconductivity at room temperature in bulk highly oriented pyrolytic graphite samples. *Carbon* **2013**, *59*, 140–149. [CrossRef]
33. Semenenko, B.; Esquinazi, P. Diamagnetism of Bulk Graphite Revised. *Magnetochemistry* **2018**, *4*, 52. [CrossRef]

34. Levesque, P.L.; Sabri, S.S.; Aguirre, C.M.; Guillemette, J.; Siaj, M.; Desjardins, P.; Szkopek, T.; Martel, R. Probing charge transfer at surfaces using graphene transistors. *Nano Lett.* **2011**. [CrossRef]
35. Xu, H.; Chen, Y.; Zhang, J.; Zhang, H. Investigating the mechanism of hysteresis effect in graphene electrical field device fabricated on SiO2 substrates using raman spectroscopy. *Small* **2012**. [CrossRef]
36. Pinto, H.; Jones, R.; Goss, J.P.; Briddon, P.R. Mechanisms of doping graphene. *Phys. Status Sol. (A)* **2010**. [CrossRef]
37. Zahab, A.; Spina, L.; Poncharal, P.; Marlière, C. Water-vapor effect on the electrical conductivity of a single-walled carbon nanotube mat. *Phys. Rev. B Condens. Matter Mater. Phys.* **2000**. [CrossRef]
38. Kong, J.; Franklin, N.R.; Zhou, C.; Chapline, M.G.; Peng, S.; Cho, K.; Dai, H. Nanotube molecular wires as chemical sensors. *Science* **2000**. [CrossRef] [PubMed]

C Journal of Carbon Research

Article

Expendable Graphite as an Efficient Flame-Retardant for Novel Partial Bio-Based Rigid Polyurethane Foams

Felipe M. de Souza [1], Jonghyun Choi [1], Sanket Bhoyate [2], Pawan K. Kahol [3] and Ram K. Gupta [1,4,*]

1 Department of Chemistry, Pittsburg State University, Pittsburg, KS 66762, USA; felipe.ms@ufu.br (F.M.d.S.); jonghyunchoi09@gmail.com (J.C.)

2 Department of Materials Science and Engineering, University of North Texas, Denton, TX 76203, USA; sanketbhoyate931@gmail.com

3 Department of Physics, Pittsburg State University, Pittsburg, KS 66762, USA; pkahol@pittstate.edu

4 Kansas Polymer Research Center, Pittsburg State University, Pittsburg, KS 66762, USA

* Correspondence: ramguptamsu@gmail.com or rgupta@pittstate.edu; Tel.: +1-620-235-4763; Fax: +1-620-235-4003

Received: 9 April 2020; Accepted: 28 April 2020; Published: 1 May 2020

Abstract: The rigid polyurethane foam (PU) is a versatile material, used especially for construction and household applications. The current situation demands a facile, cost-efficient, and greener approach for developing the polyurethanes from bio-derived materials. In this study, we present a novel bio-polyol synthesized using carvone, an extract from caraway, spearmint, or dill seeds via facile thiol-ene reaction. Our one-step reaction uses a UV irradiation to allow the room temperature conversion of the carvone to a high purity bio-polyol, as confirmed from the standard analytical characterizations. The hydroxyl number of 365 mg KOH/g close to its theoretical limit confirms the high conversion yield of the polyol for rigid PU synthesis. To overcome the flammability issues in PU, expandable graphite (EG) powder was used as an additive flame-retardant during the synthesis step. The resulting foams with EG maintained the uniform closed cell structure (>95%) with a high compression strength of 175 kPa. The addition of EG in PU results in the formation of a protective char layer during the flammability test and reduces the weight loss from 40.70% to 3.55% and burning time from 87 to 11 s. Our results confirm that the carvone-based polyol can be a novel alternative to the petroleum polyols for an industrial-scale application.

Keywords: flammability; flame-retardant polyurethane foams; bio-polyol; expandable graphite

1. Introduction

Polyurethane (PU) foams are largely produced in the industry due to their versatile applications in construction, automobiles, footwears, and coatings [1,2]. Polyurethane foams can be synthesized by the reaction between an oligomeric polyol and a di- or poly-isocyanate in the presence of blowing agent [3,4]. Increasing global concerns regarding petroleum resources have raised the interest for sustainable synthesis of PU foams using bio-derived alternates. Our group has presented several alternative routes for synthesizing bio-derived PU foams to promote sustainability and reduce the carbon footprints [2,5–9]. Unsaturated double bonds available in vegetable oils from soybean, orange, castor, sunflower, and cardanol can be readily converted into polyols by reacting with a thiol-containing polyfunctional compound. Such a compound can introduce functional groups, such as the hydroxyl group via thiol-ene reaction resulting in polyol with reactive hydroxyl groups [10,11]. Similarly, carvone is one of the widely available bio-oils extracted from caraway (*Carum carvi*), spearmint (*Mentha spicata*), and dill, and can be used as a raw material for the synthesis of polyol [12]. Use of carvone for polyol synthesis has not been previously explored. Hence, it would be advantageous to study the properties of carvone-based polyol for the synthesis of PU foams.

All the PU foams obtained either via petrochemicals or renewable resources are made up of hydrocarbon compounds and possess a highly porous structure [9,13]. A small fire accident to such a structure can, therefore, result in a major fire. Especially when the rigid PU foams are used as thermal insulators in buildings, such accidents can cause serious casualties and property damage [13]. For this reason, several efforts have been made to obtain flame-retardant (FR) PU foams using a blended FR compound [2,9]. They can be of several types such as halogenated or non-halogenated, additive or reactive, and solid phase or gas phase FR compounds [14]. Non-halogenated additive flame-retardants are widely used since they can be added directly in predetermined quantity to impart flame-retardant properties in foams. A few examples of additive flame-retardants are expandable graphite (EG), melamine, dimethyl phosphate, ammonium phosphate, and aluminum trihydroxide [15–20]. Specifically, EG is a low-cost non-halogenated additive flame-retardant and is widely studied in the literature for various applications [21–23]. Its flame retardancy occurs due to the formation of a protective char layer on the surface, inhibiting fire growth. It is also used as a filler within the foams to reduce the overall cost of the foams [18,24]. However, extensive quantities can result in a systematic decrease in mechanical properties of the foams. Hence, the convention is to obtain high flame retardancy with a low possible concentration of flame-retardant without deteriorating the properties of PU foam.

In this study, we have synthesized a novel carvone-based polyol using a facile thiol-ene reaction. The polyol was characterized by Fourier transformed infrared (FT-IR), gel permeation chromatography (GPC), hydroxyl number, and viscosity. The PU foams characterizations made were apparent density, closed-cell content, compressive strength, scanning electron microscopy, and flammability test.

2. Experimental

2.1. Materials

Carvone, expandable graphite (EG), and photoinitiator, 2-hydroxy-2-methyl propiophenone, were obtained from Sigma-Aldrich, St. Louis, MO, USA. 2-mercaptoethanol (2-ME) was obtained from Acros Organic, Pittsburgh, PA, USA. Both catalysts NIAX A-1 and DABCO T-12 were obtained at Air Products, Allentown, PA, USA. Tegostab B-8404, used as a surfactant, was acquired from Evonik, Parsippany, NJ, USA. Methylene diisocyanate (MDI) and Jeffol 522, which is a commercial polyol, were kindly donated from Huntsman, the Woodlands, TX, USA. DI water was used as a blowing agent obtained from the local Walmart (Pittsburg, KS, USA).

2.2. Synthesis of Carvone-Based Polyol

For the synthesis of carvone-based polyol, one mol of carvone was reacted with 2 mol of 2-mercaptoethanol in the presence of 2-hydroxy-2-methyl propiophenone as a photoinitiator. The reaction was exposed to UV light (365 nm) under continuous stirring for 8 h. Figure 1 shows the thiol-ene reaction for the synthesis of polyol.

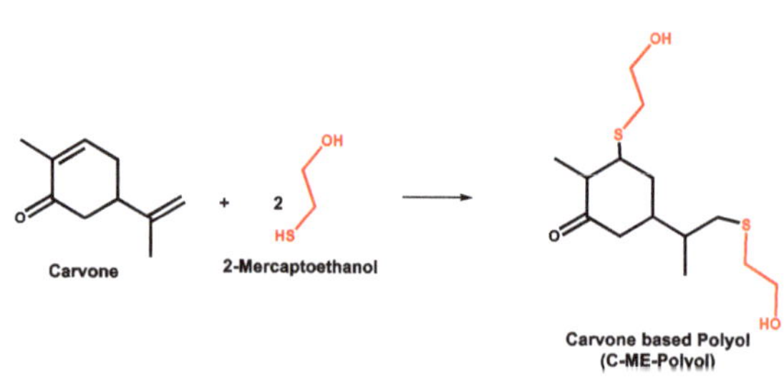

Figure 1. The chemical reaction for the synthesis of carvone-based polyol.

2.3. Polyol Characterizations

The characterizations of synthesized carvone-based polyol were based on standards from ISO (International Organization for Standardization) and ASTM (American Society for Testing and Materials). By using the Hanus method, the iodine number for carvone was determined. The hydroxyl number of the polyol was obtained according to the ASTM E1899-16 Method by reacting the polyol with *p*-toluenesulfonyl isocyanate (TSI), followed by potentiometric titration (888 Titrando, Tiamo Software system by Metrohm, AG, Switzerland). The viscosity of carvone and the polyol was measured using a TA Instruments rheometer from New Castle, Delaware, at room temperature by increasing the shear stress from 0.1 to 100 Pa. A cone plate with a 2° angle and 12.5 mm radius was used. The FT-IR spectra were obtained using a PerkinElmer Spectrum two from Waltham, Maryland. Gel Permeation Chromatography (GPC) was performed into the polyol and starting materials using a Waters system from Milford, Massachusetts. The system consists of a column with the dimensions of 300 × 7.8 mm with different pores from 50 and 102 to 104 Å. The standard eluent used was tetrahydrofuran, injecting 20 μL for 50 min per sample at 30 °C.

2.4. Synthesis of Flame-Retardant Rigid Polyurethane

FR-PU foams were synthesized by using different concentrations of EG, shown in Table 1. Firstly, the polyol and EG were manually mixed in a 500 mL plastic cup and then vigorously stirred (3000 rpm) for a few minutes to form a homogenous mixture. All the components but MDI were added and stirred for 30 s. Next, 33.39 g of MDI was added and vigorously stirred to obtain the PU foam. The samples were designated as EG 0, EG 0.5, EG 1.5, EG 3, EG 5, EG 8, and EG 10. The wt.% of the biocomponent (biopolyol) in EG 0, EG 0.5, EG 1.5, EG 3, EG 5, EG 8, and EG 10 was 18.3, 18.1, 17.8, 17.3, 16.7, 15.9, and 15.4, respectively. A foam (Jeffol) containing only Jeffol 522 was also prepared for comparison.

Table 1. The formulation in grams of foam's components.

Compound	Jeffol	EG 0	EG 0.5	EG 1.5	EG 3	EG 5	EG 8	EG 10
C 2 ME Polyol	0	10	10	10	10	10	10	10
Jeffol 522	20	10	10	10	10	10	10	10
A-1	0.14	0.14	0.14	0.14	0.14	0.14	0.14	0.14
Water	0.8	0.8	0.8	0.8	0.8	0.8	0.8	0.8
T-12	0.04	0.04	0.04	0.04	0.04	0.04	0.04	0.04
B8404	0.4	0.4	0.4	0.4	0.4	0.4	0.4	0.4
MDI	37.14	33.39	33.39	33.39	33.39	33.39	33.39	33.39
EG	0	0	0.5	1.5	3	5	8	10
Wt.% of EG in the foam	0	0	0.91	2.67	5.19	8.37	12.75	15.44

2.5. Characterizations of Polyurethane Foams

The foams were cut in a cylinder of 45 × 30 (mm) of diameter and height, respectively. These cylinders were used to measure apparent density and closed-cell content. Gravimetric density was obtained according to ASTM 1622-14. A Humi Pyc Volumetric and relative humidity (RH) Analyzer instrument was used to acquire the closed-cell content. The compressive strength of the PU foams was obtained by cutting them into 50 × 50 × 25 (mm^3) length, width, and height, respectively. The strain at 10% and yield at break were determined by using an Instron tensile instrument (Raleigh, NS, USA) operated by the Blue Hill system, adopting the ISO 844:2016. The compressive strength applied was at a strain rate of 3 cm/min, from the top of the sample. Thermogravimetric Analysis (TGA) was performed using a Q500 Discovery (New Castle, DE, USA) to analyze the foam's thermal decomposition. TGA was performed in nitrogen and air starting from 20 to 700 °C, at a 10 °C increase per min. Fire retardancy was determined by adopting the ASTM D 4986-18 standard test, which consisted of applying a direct flame into the sample with dimensions of 150 × 50 × 12.5 (mm^3) height, length, and thickness respectively, in a horizontal position for 10 s. After that, the time taken for the sample to quench the fire by itself along with the weight loss percentage was recorded. Cubicles of 0.5 cm^3 dimensions were cut and used to obtain images for the Scanning Electron Microscopy (SEM) (Phenom, Netherlands). A Magnetron Sputtering along with a monitor from Kurt J. Lesker Company (Jefferson Hills, PA, USA) was used to perform a gold sputtering on the surface of the samples to coat them with a thin conductive layer of gold (Au). The system was flushed several times with Ar gas. A vacuum of 50 mBar was applied, and the samples were put under gold plasma for 1 min.

3. Results and Discussion

The GPC chromatogram for carvone, 2-ME, and C-2-ME-Polyol are ascribed in Figure 2. The reduced retention time for C-2-ME-Polyol (36 min) as compared to carvone (42 min) suggests an increase in the molecular weight of polyol. No extra peaks were observed in GPC curves for unreacted compounds, which confirms that the reaction was complete. The carvone structure consists of two double bonds that have a difference in reactivity due to steric hindrance within the structure. Hence, the peak at 38 min could be linked to the external double bond of carvone reacted with 2-ME because of less steric hindrance, while the peak at 36 min could correspond to the occurrence of reaction on both the double bonds, as reported by others [25,26]. Further, the reaction proceeding for polyol synthesis can also be correlated to an increase in viscosity of C-2-ME-Polyol (2.702 Pa.s) compared to that of carvone (0.006173 Pa.s.). The viscosity of a commercial polyol (Jeffol 522) was 26.59 Pa.s. Jeffol 522 is a sucrose-based polyol with a hydroxyl number of 520 mg KOH/g and a hydroxyl functionality of 5. Although the viscosity value for polyol showed an obvious increase as compared to carvone, it was low in terms of materials processing and mixing, and therefore allowed for easy processing of PU foams [9]. The hydroxyl number for the synthesized polyol was 365 mg KOH/g, while the theoretical value was 367 mg KOH/g, which suggested a high conversion of essential oil into polyol structure. Since the observed hydroxyl number of the synthesized polyol matches well with the theoretical value calculated using a reaction of one mol of carvone with two moles of 2-mercaptoethanol, the possibilities of reacting with only one double bond could be minimum. It is likely to form oligomers of the polyol. Figure 3 shows the possible reaction of the formation of minor products. Similar behavior was reported in thiol-ene chemistry by other researchers also [25,26].

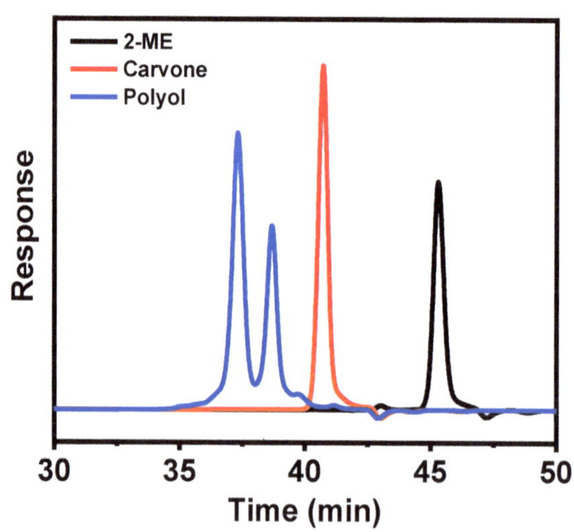

Figure 2. Gel Permeation Chromatogram for carvone, 2-mercaptoethanol (2-ME), and carvone-based polyol.

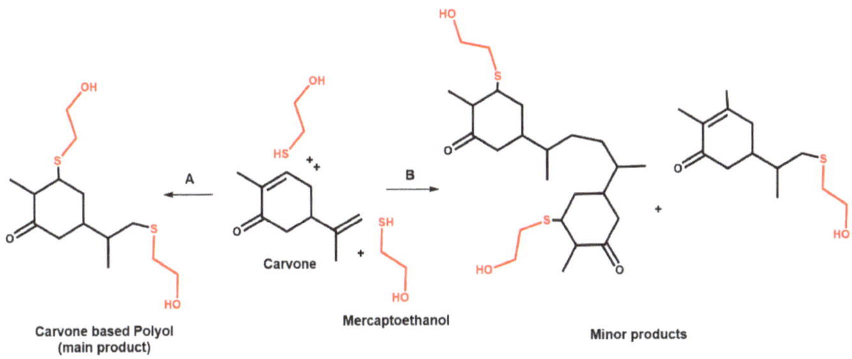

Figure 3. The chemical reaction for the synthesis of carvone-based polyol (**A**) along with expected byproducts (**B**).

The reaction proceedings for the thiol-ene reaction between carvone and 2-ME was analyzed using FT-IR spectra, shown in Figure 4. For carvone, a peak near 900 cm^{-1} correlates to an axial stretch for C=C [27]. While, for 2-ME, a peak near 2568 cm^{-1} represents an axial stretch for S-H bond and 3500 cm^{-1} for C–O–H bond [28]. After the thiol-ene reaction, the S–H peak near 2568 cm^{-1} and C=C peak near 900 cm^{-1} disappeared by forming C-S bond near 600–700 cm^{-1}, with a simultaneous appearance of C-O-H axial stretch at 3500 cm^{-1} for C-2-ME-Polyol [5,25,29]. Similar behavior was reported by Feng et al. for the synthesis of soybean oil and 2-ME-based polyol via thiol-ene reaction [28].

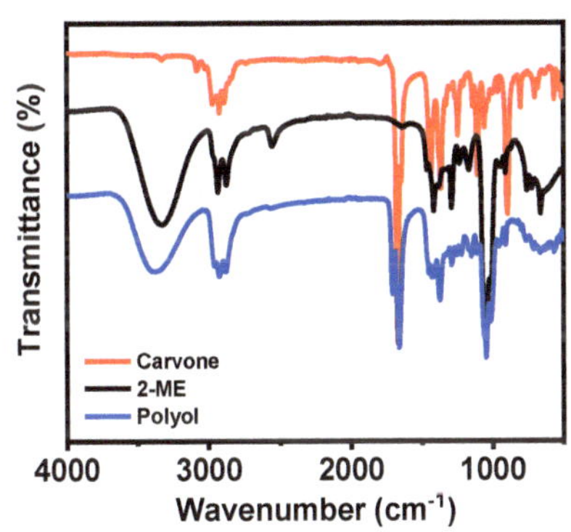

Figure 4. Fourier Transformed Infrared spectra for carvone, 2-mercaptoethanol, and carvone-based polyol.

The effect of EG concentration on an apparent density of PU foams is summarized in Figure 5. All the foams showed the apparent density in the range of 30–45 kg/m^3. The density of polyurethane foam prepared using commercial polyol (Jeffol 522) was 43 kg/m^3. As the amount of the EG exceeded 8.37 wt.% in the polyurethane foams, the density values for PU foams increased. This could be due to the presence of acidic groups on EG's surface that milds the effect of catalyst and slows down the rise time of the foams [21,30]. This caused a slight increase in the apparent density of the foams. The addition of a higher amount of catalyst could increase the rise time and maintain the density of foams [5]. However, all the densities were following industrial standards and can be used for rigid foam applications [26]. Hence, no further modifications were made to the concentration of the catalyst. The rise time for the foams EG 0, EG 0.5, EG 1.5, EG 3.0, EG 5.0, EG 8.0, and EG 10.0 was about 14, 14, 12, 12, 11, 9, and 8 s, respectively. The rise time of the foam prepared using Jeffol 522 was about 13 s. The closed-cell content of the prepared foams was measured using the ASTM D2856 method. Figure 6 shows the closed-cell content of all the foams. The closed-cell content of all the foams was observed in the range of 90%–95%. The closed-cell content of polyurethane foam prepared using Jeffol 522 was 93%. The FT-IR analysis of the prepared polyurethane foams was also performed (Figure 7). The peak around 3308 cm^{-1} is associated with the amide (–NH) group, while the peak around 1699 cm^{-1} is due to the carbonyl urethane group (–C=O). The peaks around 1598, 1515, and 1225 cm^{-1} are indicative of –C–N, –C=C, and –C–O–C, respectively.

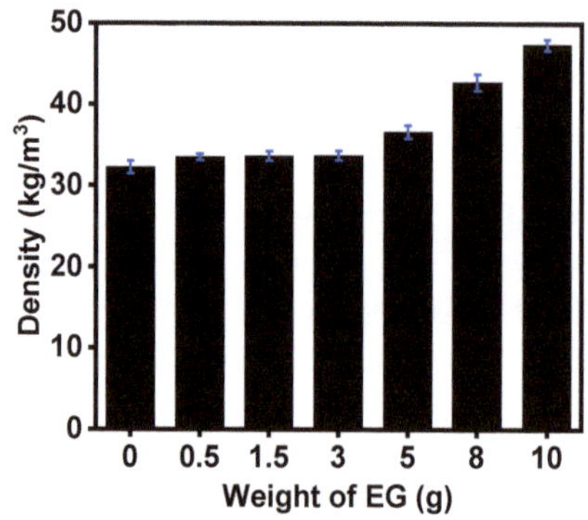

Figure 5. Density of polyurethane foams with carvone-based polyol with an increasing amount of EG.

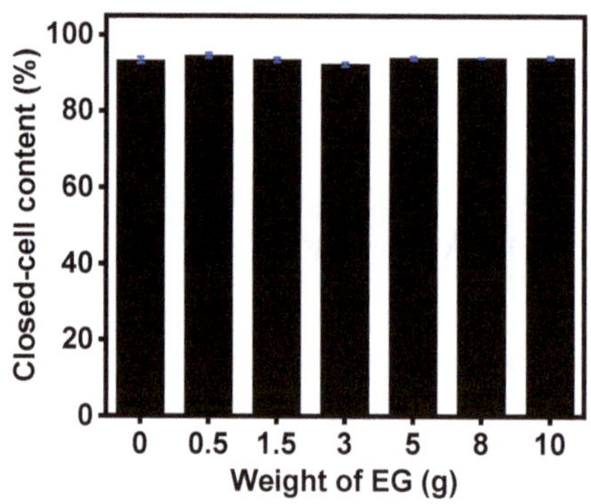

Figure 6. Closed-cell content of polyurethane foams with carvone-based polyol with an increasing amount of EG.

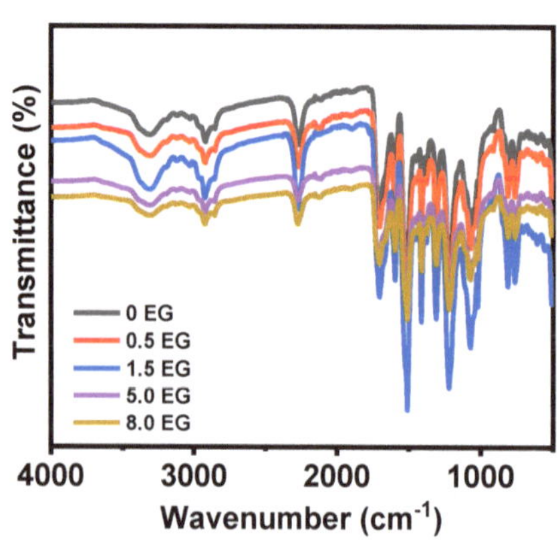

Figure 7. Fourier Transformed Infrared spectra for polyurethane foams.

The SEM images for all the synthesized PU foams with uniform cellular structure can be summarized from Figure 8. Increasing the concentration of EG from 0 (C-EG-0) to 15.44 wt.% (C-EG-10) decreases an average cell size from 160 to 120 micron. These results are in accordance with the density values. As the amount of EG exceeds 8.37 wt.% in the foam, the acid content within the reaction mixture of PU foam increases, which slows down the rise time of the foam and raises the apparent density of the foams. Unlike previous reports for PU foams containing EG, we observed a highly closed-cell structure > 95% for our EG containing PU foams, which could be due to compatibility of EG with carvone-based polyol structure [31,32]. A high and nearly constant value for closed-cell content with low variance in density can present rigid PU foams with good thermal insulation properties [2,33].

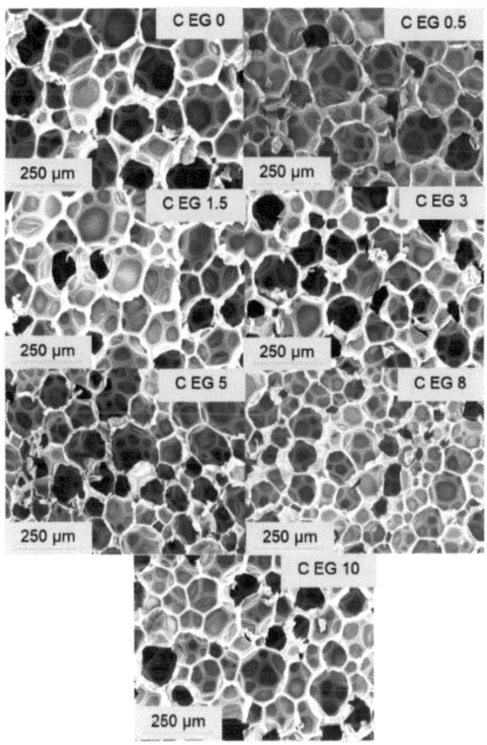

Figure 8. Scanning Electron Microscopy (SEM) of carvone-based PU with an increasing amount of EG.

A compression test was performed to analyze the effect of an increasing amount of EG on the mechanical properties of PU foams (Figure 9). EG is being reported as a cost-effective FR with the drawbacks of harder processing and a decrease of mechanical strength [24,34]. The ultimate compressive strength value was found to be 210 kPa and 209 kPa for neat PU foam prepared using synthesized biopolyol and Jeffol 522, respectively. As we increased the content of EG, the compression strength of PU foams slightly decreased. The 10 EG-PU foam showed a yield at the break of 175 kPa. Although EG acts as an effective FR when used in the form of flakes, its higher concentration in PU foams may cause defects into segments of the foam, hence it deteriorates some of its mechanical properties, causing the compressive strength to vary [21,24,35–38]. Even though there was a decrease in mechanical properties with increasing concentration of PU foams, all the EG-PU foams showed higher compression strength than foams originated from castor, cardanol, and corn oils that were 150, 130, and 75 kPa, respectively [5,39].

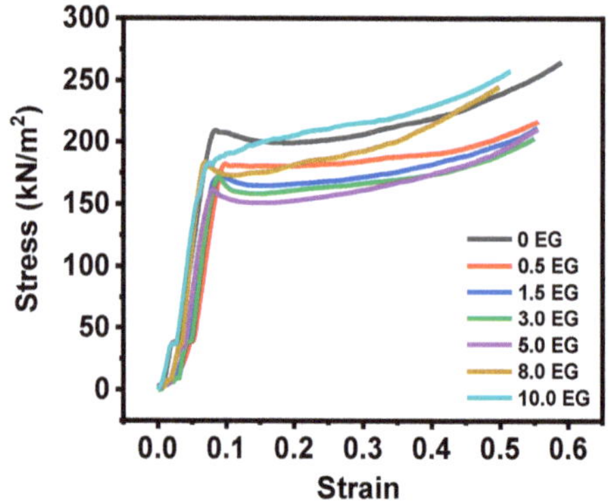

Figure 9. Compressive strength for carvone-based PU with an increasing amount of EG.

The flame-retardant properties of EG containing PU foams were analyzed using a horizontal burning test. In this study, the foams were held in the horizontal position and subjected to a fire for 10 s. The flame time and weight loss after the self-extinction of fire were recorded (Figures 10 and 11). The neat PU showed the longest flame time of 98 s and the highest weight loss of 46%. By adding 15.44 wt.% of EG (10 EG), the lowest flame time of 11 s and weight loss of 3.55% were achieved. A systematic reduction in the burnt area of samples with increasing EG content is summarized in Figure 12. However, with a further increase in EG concentration (>12.75 wt.%), there were no major changes observed in flame-retardant properties of rigid PU foams. The foam prepared using Jeffol 522 displayed the burning time of 95 s with a weight loss of 44%. The observed flame time was comparable to previously reported PU foams. For example, Ranaweera et al. prepared limonene-based FR-PU foams using DMMP as a flame-retardant [40]. The self-extinguish time for the best limonene-based FR-PU foam was reported to be 19 s. Similarly, Ramanujam et al. reported a functionalized corn-based FR-PU foam using DMMP and presented remarkable fire-retardant properties by reducing the weight loss from 38 wt.% from the neat PU to 5.5 wt.% by addition of 1.94 wt.% of PU. Unlike EG, DMMP can release toxic fumes containing PH_3 and it is almost five times more expensive than EG, which significantly increases the cost during the large-scale production. In contrast, EG also acts as a filler and can be used to reduce the overall cost of PU foams.

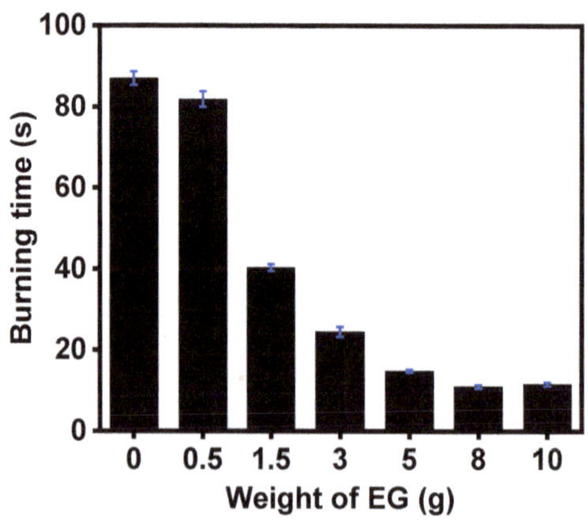

Figure 10. Burning time for the carvone-based PU foams with an increasing amount of EG.

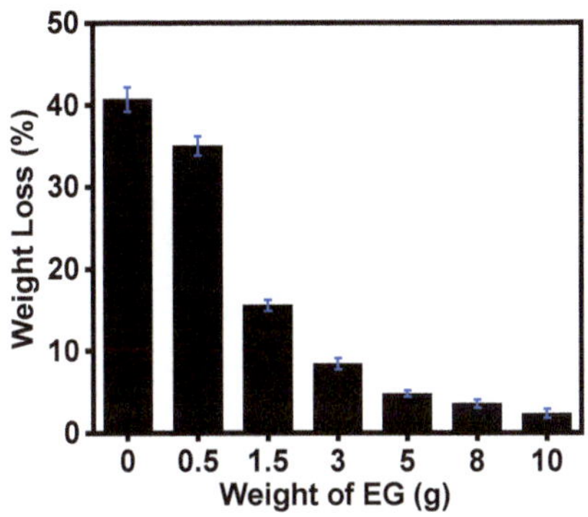

Figure 11. Weight loss for the PU foams with an increasing amount of EG after burning.

Figure 12. Photographs of the PU taken prior to and after the horizontal burning test.

The thermal analysis of FR-PU foams was performed using thermogravimetric analysis (TGA) and derivative thermogravimetric analysis (DTGA) in both a nitrogen and air atmosphere. Figure 13 shows the thermal properties of all the foams studied under a nitrogen atmosphere. The weight loss curve in TGA and DTGA shows two-step thermal decomposition behavior. At 250 °C, the hydrocarbon chains in PU foam start to decompose and EG activates and starts to expand, forming a protective char layer on the foam surface [21]. The first maximum decomposition occurs at 285 °C (Figure 13b). It can be noted that with an increasing amount of EG, there was a systematic decrease in the weight loss of the foams. This means that protective char formation increases with EG concentration and stabilizes the surface of the foam. The EG pyrolysis comprises of SO_2, CO_2, and water that gets entrapped while escaping near the edges of the graphitic structure, leading to the irreversible expansion of the graphene structure [21]. The expansion then creates a graphitic char layer on the surface of the foam that prevents the heat transfer through the solid phase and stops the fire [21,24,35]. This type of phenomenon is observed by the expansion of foam surface during the burning test, which may sometimes form blisters of charred graphene. A second degradation peak near 550 °C corresponds to the disintegration of the PU structure along with char, resulting in weight loss for all the foams [35,41]. As compared to neat PU foams, EG containing foams showed relatively lower weight loss, suggesting increased char stability with EG concentration. The thermal behavior of the prepared foams studied under an air atmosphere is shown in Figure 14. The foams showed similar behavior under nitrogen and air atmosphere; however, the maximum decomposition temperature for the foams was observed to shift towards a lower temperature (270 °C) in air. In addition, at the higher temperature, the foams under air seem to reach maximum decomposition around 600 °C. Some of the important thermal characteristics of the foams in nitrogen and air atmosphere are given in Table 2.

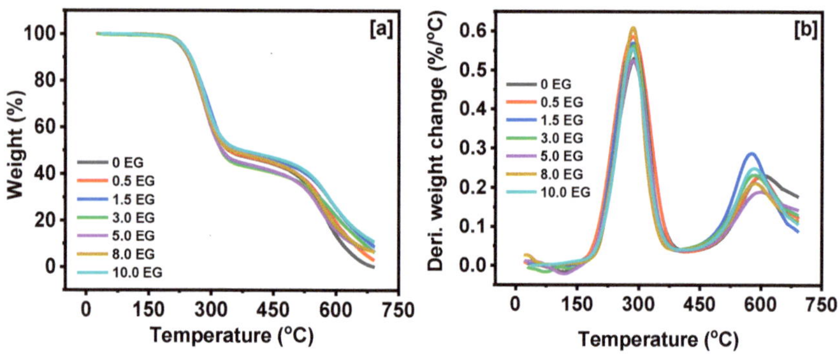

Figure 13. Thermogravimetric analysis (**a**) and derivative weight loss curves (**b**) of PU foams in a nitrogen environment.

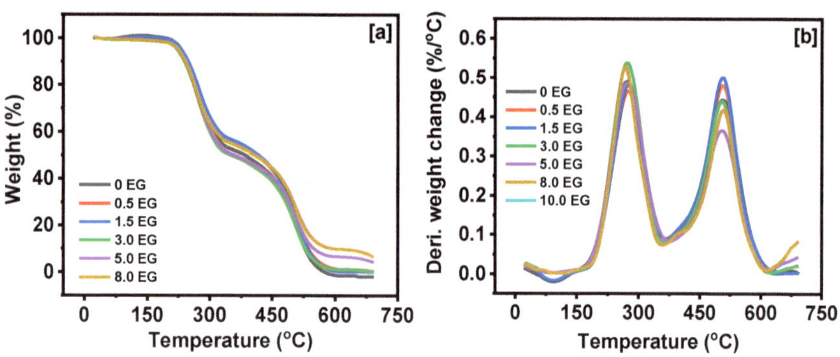

Figure 14. Thermogravimetric analysis (**a**) and derivative weight loss curves (**b**) of PU foams in an air environment.

Table 2. Thermal properties of the polyurethane foams.

	N_2	N_2	N_2	Air	Air	Air
Foam	$T_{d5\%}$	$T_{d10\%}$	T_{dmax}	$T_{d5\%}$	$T_{d10\%}$	T_{dmax}
0 EG	231.2	250.0	700	231.6	246.3	700
0.5 EG	233.0	251.2	700	226.9	245.5	700
1.5 EG	228.0	249.7	700	233.5	249.7	700
3.0 EG	227.8	246.4	700	224.0	240.9	700
5.0 EG	229.6	245.8	700	225.7	242.2	700
8.0 EG	231.4	247.9	700	225.9	243.1	700

4. Conclusions

The bio-based PU foams were successfully synthesized using a bio-oil named carvone and converted into a polyol using a facile and single-step thiol-ene reaction. The synthesized polyol was used as a sustainable alternative to petroleum-based polyol to make polyurethane foams. To enhance the flame retardancy in PU foams, we added different concentrations of EG during the synthesis step. All the synthesized PU foams showed an average apparent density of 45 kg/m^3 and a closed-cell content higher than 95%. SEM analysis confirmed that the addition of EG had no adverse effect on the cellular structure of the PU foams. Although an increase in EG content showed a slight decrease in the ultimate compressive strength of the PU foams, the obtained values were comparably higher than commercially available PU foams. Moreover, the horizontal burning test showed an effective drop-in self-quenching fire from 98 to 11 s, and respective weight loss from 40.7% to 2.36%, as the EG concentration increased from 0 to 15.44 wt.%. Both carvone and EG are widely available at relatively low-cost, therefore making them favorable to synthesize low-cost FR-PU foams for an industrial application.

Author Contributions: R.K.G. conceived the project and designed the experiments. F.M.d.S. performed the experiments. J.C. helped in flammability testing. S.B. took the SEM images and revised the manuscript. F.M.d.S. wrote the original draft. All authors reviewed the manuscript and provided their inputs. All authors have read and agreed to the published version of the manuscript.

Funding: This research received no external funding.

Acknowledgments: Ram K. Gupta expresses his sincere acknowledgment of the Polymer Chemistry Program and Kansas Polymer Research Center, Pittsburg State University, for providing financial and research support.

Conflicts of Interest: The authors declare no conflict of interest.

References

1. Zhang, C.; Yan, M.; Cochran, E.W.; Kessler, M.R. Biorenewable Polymers Based on Acrylated Epoxidized Soybean Oil and Methacrylated Vanillin. *Mater. Today* **2015**, *5*, 18–22. [CrossRef]
2. Zhang, C.; Bhoyate, S.; Ionescu, M.; Kahol, P.K.; Gupta, R.K. Highly Flame Retardant and Bio-Based Rigid Polyurethane Foams Derived from Orange Peel Oil. *Polym. Eng. Sci.* **2018**, *58*, 2078–2087. [CrossRef]
3. Zhang, C.; Madbouly, S.A.; Kessler, M.R. Biobased Polyurethanes Prepared from Different Vegetable Oils. *ACS Appl. Mater. Interfaces* **2015**, *7*, 1226–1233. [CrossRef] [PubMed]
4. Javni, I.; Zhang, W.; Petrović, Z.S. Effect of Different Isocyanates on the Properties of Soy-Based Polyurethanes. *J. Appl. Polym. Sci.* **2003**, *88*, 2912–2916. [CrossRef]
5. Ramanujam, S.; Zequine, C.; Bhoyate, S.; Neria, B.; Kahol, P.; Gupta, R. Novel Biobased Polyol Using Corn Oil for Highly Flame-Retardant Polyurethane Foams. *C—J. Carbon Res.* **2019**, *5*, 13. [CrossRef]
6. Bhoyate, S.; Ionescu, M.; Kahol, P.K.; Gupta, R.K. Castor-Oil Derived Nonhalogenated Reactive Flame-Retardant-Based Polyurethane Foams with Significant Reduced Heat Release Rate. *J. Appl. Polym. Sci* **2018**, 47276. [CrossRef]
7. Bhoyate, S.; Ionescu, M.; Kahol, P.K.; Gupta, R.K. Sustainable Flame-Retardant Polyurethanes Using Renewable Resources. *Ind. Crops Prod.* **2018**, *123*, 480. [CrossRef]
8. Bhoyate, S.; Ionescu, M.; Radojcic, D.; Kahol, P.K.; Chen, J.; Mishra, S.R.; Gupta, R.K. Highly Flame-Retardant Bio-Based Polyurethanes Using Novel Reactive Polyols. *J. Appl. Polym. Sci.* **2018**, *135*. [CrossRef]
9. Bhoyate, S.; Ionescu, M.; Kahol, P.K.; Chen, J.; Mishra, S.R.; Gupta, R.K. Highly Flame-Retardant Polyurethane Foam Based on Reactive Phosphorus Polyol and Limonene-Based Polyol. *J. Appl. Polym. Sci.* **2018**, *135*, 16–19. [CrossRef]
10. Ding, H.; Huang, K.; Li, S.; Xu, L.; Xia, J.; Li, M. Synthesis of a Novel Phosphorus and Nitrogen-Containing Bio-Based Polyol and Its Application in Flame Retardant Polyurethane Foam. *J. Anal. Appl. Pyrolysis* **2017**, *128*, 102–113. [CrossRef]
11. Montero de Espinosa, L.; Meier, M.A.R. Plant Oils: The Perfect Renewable Resource for Polymer Science?! *Eur. Polym. J.* **2011**, *47*, 837–852. [CrossRef]
12. De Carvalho, C.C.C.R.; da Fonseca, M.M.R. Carvone: Why and How Should One Bother to Produce This Terpene. *Food Chem.* **2006**, *95*, 413–422. [CrossRef]

13. Babrauskas, V.; Lucas, D.; Eisenberg, D.; Singla, V.; Dedeo, M.; Blum, A. Flame Retardants in Building Insulation: A Case for Re-Evaluating Building Codes. *Build. Res. Inf.* **2012**, *40*, 738–755. [CrossRef]
14. Levchik, S.V.; Weil, E.D. A Review of Recent Progress in Phosphorus-Based Flame Retardants. *J. Fire Sci.* **2006**, *24*, 345–364. [CrossRef]
15. Wang, L.; He, X.; Wilkie, C.A. The Utility of Nanocomposites in Fire Retardancy. *Materials (Basel)* **2010**, *3*, 4580–4606. [CrossRef] [PubMed]
16. Liu, F.; Ding, X.; Su, Y. Properties of Rigid Polyurethane Foams Produced by the Addition of Phosphorus Compounds. *Am. J. Mater. Res.* **2014**, *1*, 14–19.
17. Chen, Y.; Jia, Z.; Luo, Y.; Jia, D.; Li, B. Environmentally Friendly Flame-Retardant and Its Application in Rigid Polyurethane Foam. *Int. J. Polym. Sci.* **2014**, *2014*. [CrossRef]
18. Xing, W.; Yuan, H.; Zhang, P.; Yang, H.; Song, L.; Hu, Y. Functionalized Lignin for Halogen-Free Flame Retardant Rigid Polyurethane Foam: Preparation, Thermal Stability, Fire Performance and Mechanical Properties. *J. Polym. Res.* **2013**, *20*, 234. [CrossRef]
19. Konig, A.; Malek, A.; Fehrenbacher, U.; Brunklaus, G.; Wilhelm, M.; Hirth, T. Silane-Functionalized Flame-Retardant Aluminum Trihydroxide in Flexible Polyurethane Foam. *J. Cell. Plast.* **2010**, *46*, 395. [CrossRef]
20. Wu, D.; Zhao, P.; Liu, Y. Flame Retardant Property of Novel Intumescent Flame Retardant Rigid Polyurethane Foams. *Polym. Eng. Sci.* **2013**, *53*, 2478–2485. [CrossRef]
21. Camino, G.; Duquesne, S.; Delobel, R.; Eling, B.; Lindsay, C.; Roels, T. Mechanism of Expandable Graphite Fire Retardant Action in Polyurethanes. *ACS Symp. Ser.* **2001**, *797*, 90–109. [CrossRef]
22. Bhoyate, S.; Mensah-Darkwa, K.; Kahol, P.K.; Gupta, R.K. Recent Development on Nanocomposites of Graphene for Supercapacitor Applications. *Curr. Graphene Sci.* **2017**, *1*, 26. [CrossRef]
23. Bhoyate, S.; Mensah-Darkwa, K.; Kahol, P.K.; Gupta, R.K. Advancement in Light Energy Harvesting by Using Tailored Graphene in DSSC and BHJ Solar Cells. *Curr. Graphene Sci.* **2019**, *3*, 1. [CrossRef]
24. Gupta, R.; Ionescu, M.; Wan, X.; Radojcic, D.; Bilic, N. New Polyols with Isocynauric Structure by Thiol-Ene "Click" Chemistry Reactions. *J. Cell. Plast.* **2017**, *53*, 639–662. [CrossRef]
25. Wilbon, P.A.; Chu, F.; Tang, C. Progress in Renewable Polymers from Natural Terpenes, Terpenoids, and Rosin. *Macromol. Rapid Commun.* **2013**, *34*, 8–37. [CrossRef] [PubMed]
26. Elbers, N.; Ranaweera, C.K.; Ionescu, M.; Wan, X.; Kahol, P.K.; Gupta, R.K. Synthesis of Novel Biobased Polyol via Thiol-Ene Chemistry for Rigid Polyurethane Foams. *J. Renew. Mater.* **2017**, *5*, 74–83. [CrossRef]
27. Desroches, M.; Caillol, S.; Lapinte, V.; Auvergne, R.M.; Boutevin, B. Synthesis of Biobased Polyols by Thiol–Ene Coupling from Vegetable Oils. *Macromolecules* **2011**, *44*, 2489. [CrossRef]
28. Feng, Y.; Liang, H.; Yang, Z.; Yuan, T.; Luo, Y.; Li, P.; Yang, Z.; Zhang, C. A Solvent-Free and Scalable Method To Prepare Soybean-Oil-Based Polyols by Thiol–Ene Photo-Click Reaction and Biobased Polyurethanes Therefrom. *ACS Sustain. Chem. Eng.* **2017**, *5*, 7365–7373. [CrossRef]
29. Kolb, H.C.; Finn, M.G.; Sharpless, K.B. Click Chemistry: Diverse Chemical Function from a Few Good Reactions. *Angew. Chem. Int. Ed.* **2001**, *40*, 2004–2021. [CrossRef]
30. Liu, Y.; He, J.; Yang, R. Effects of Dimethyl Methylphosphonate, Aluminum Hydroxide, Ammonium Polyphosphate, and Expandable Graphite on the Flame Retardancy and Thermal Properties of Polyisocyanurate–Polyurethane Foams. *Ind. Eng. Chem. Res.* **2015**, *54*, 5876–5884. [CrossRef]
31. Wang, W.J.; He, K.; Dong, Q.X.; Fan, Y.; Zhu, N.; Xia, Y.B.; Li, H.F.; Wang, J.; Yuan, Z.; Wang, E.P.; et al. Influence of Aluminum Hydroxide and Expandable Graphite on the Flammability of Polyisocyanurate-Polyurethane Foams. *Appl. Mech. Mater.* **2013**, *368*, 741. [CrossRef]
32. Gharehbaghi, A.; Bashirzadeh, R.; Ahmadi, Z. Polyurethane Flexible Foam Fire Resisting by Melamine and Expandable Graphite: Industrial Approach. *J. Cell. Plast.* **2011**, *47*, 549. [CrossRef]
33. Hejna, A.; Kirpluks, M.; Kosmela, P.; Cabulis, U.; Haponiuk, J.; Piszczyk, Ł. The Influence of Crude Glycerol and Castor Oil-Based Polyol on the Structure and Performance of Rigid Polyurethane-Polyisocyanurate Foams. *Ind. Crops Prod.* **2017**, *95*, 113–125. [CrossRef]
34. Hu, X.-M.; Wang, D.-M. Enhanced Fire Behavior of Rigid Polyurethane Foam by Intumescent Flame Retardants. *J. Appl. Polym. Sci.* **2013**, *129*, 238–246. [CrossRef]
35. Wang, Y.; Wang, F.; Dong, Q.; Xie, M.; Liu, P.; Ding, Y.; Zhang, S.; Yang, M.; Zheng, G. Core-Shell Expandable Graphite @ Aluminum Hydroxide as a Flame-Retardant for Rigid Polyurethane Foams. *Polym. Degrad. Stab.* **2017**, *146*, 267–276. [CrossRef]

36. Shi, L.; Li, Z.-M.; Xie, B.-H.; Wang, J.-H.; Tian, C.-R.; Yang, M.-B. Flame Retardancy of Different-Sized Expandable Graphite Particles for High-Density Rigid Polyurethane Foams. *Polym. Int.* **2006**, *55*, 862–871. [CrossRef]
37. Tan, S.Q.; Abraham, T.; Ference, D.; Macosko, C.W. Rigid Polyurethane Foams from a Soybean Oil-Based Polyol. *Polymer (Guildf)* **2011**, *52*, 2840. [CrossRef]
38. Zhang, L.; Zhang, M.; Hu, L.; Zhou, Y. Synthesis of Rigid Polyurethane Foams with Castor Oil-Based Flame Retardant Polyols. *Ind. Crops Prod.* **2014**, *52*, 380–388. [CrossRef]
39. Liu, C.; Wang, C.; Hu, Y.; Zhang, F.; Shang, Q.; Lei, W.; Zhou, Y.; Cai, Z. Castor Oil-Based Polyfunctional Acrylate Monomers: Synthesis and Utilization in UV-Curable Materials. *Prog. Org. Coat.* **2018**, *121*, 236–246. [CrossRef]
40. Ranaweera, C.K.; Ionescu, M.; Bilic, N.; Wan, X.; Kahol, P.K.; Gupta, R.K. Biobased Polyols Using Thiol-Ene Chemistry for Rigid Polyurethane Foams with Enhanced Flame-Retardant Properties. *J. Renew. Mater.* **2017**, *5*, 1. [CrossRef]
41. Ye, L.; Meng, X.-Y.; Liu, X.-M.; Tang, J.-H.; Li, Z.-M. Flame-Retardant and Mechanical Properties of High-Density Rigid Polyurethane Foams Filled with Decabrominated Dipheny Ethane and Expandable Graphite. *J. Appl. Polym. Sci.* **2009**, *111*, 2372–2380. [CrossRef]

C Journal of Carbon Research

MDPI

Article

Power Cycling and Reliability Testing of Epoxy-Based Graphene Thermal Interface Materials

Jacob S. Lewis *, Timothy Perrier, Amirmahdi Mohammadzadeh, Fariborz Kargar and Alexander A. Balandin *

Phonon Optimized Engineered Materials (POEM) Center, Department of Electrical and Computer Engineering, Materials Science and Engineering Program, Bourns College of Engineering, University of California-Riverside, Riverside, CA 92521, USA; tperr003@ucr.edu (T.P.); amoha027@ucr.edu (A.M.); fkarg001@ucr.edu (F.K.)

* Correspondence: jlewi014@ucr.edu (J.S.L.); balandin@ece.ucr.edu (A.A.B.)

Received: 24 March 2020; Accepted: 21 April 2020; Published: 25 April 2020

Abstract: We report on the lifespan evolution of thermal diffusivity and thermal conductivity in curing epoxy-based thermal interface materials with graphene fillers. The performance and reliability of graphene composites have been investigated in up to 500 power cycling measurements. The tested composites were prepared with an epoxy resin base and randomly oriented fillers consisting of a mixture of few-layer and single-layer graphene. The power cycling treatment procedure was conducted with a custom-built setup, while the thermal characteristics were determined using the "laser flash" method. The thermal conductivity and thermal diffusivity of these composites do not degrade but instead improve with power cycling. Among all tested filled samples with different graphene loading fractions, an enhancement in the thermal conductivity values of 15% to 25% has been observed. The obtained results suggest that epoxy-based thermal interface materials with graphene fillers undergo an interesting and little-studied intrinsic performance enhancement, which can have important implications for the development of next-generation thermal interface materials.

Keywords: thermal interface materials; graphene; reliability; power cycling; temperature cycling; accelerated aging; thermal conductivity

1. Introduction

The consistent miniaturization of semiconductor-based circuit elements has, despite efficiency gains, resulted in the unintended increase in waste power density in such products. Semiconductor circuits designed specifically for signal amplification have seen improvements in power at higher operating frequencies at the cost of substantial and detrimental waste heat production. Decreasing the operating temperature of GaN transistors by 20 °C increases the device's mean time to failure by an order of magnitude [1]. Additionally, electronic products that directly interact with radiative energy, such as photovoltaic cells and light-emitting diodes, suffer either decreased performance or reduced lifespan at elevated operating temperatures [2,3]. Due to the comparative difficulty of directly reducing the production of waste heat in electronics, a thermal dissipative solution is the most popular, next-best option for control of device operating temperature [4]. Due to inevitable surface imperfections and the subsequent inadvertent production of air pockets between any solid-solid junction in such a solution, it is imperative to use a special material to replace the catastrophically thermally insulating air. Bridging the physical junctions between two imperfect surfaces in a thermal dissipative solution with a thin layer of material to replace these interstitial air pockets is of critical importance and serves to substantially improve thermal dissipation of the overall heat sink solution [5,6]. A material used in this manner is called a thermal interface material (TIM).

Though state-of-the-art TIMs can substantially reduce the overall thermal resistance of the dissipative solution, there remains considerable room for improvement of these materials across a wide range of parameters. One of the best classes of TIMs for reduction in junction thermal resistance are metal-based solder materials. They provide extremely low contact resistances at physical junctions and boast a typical metallic level of thermal conductivity. However, this type of TIM is difficult to process and suffers from poor reliability and a high risk of cracking failures. Investigations into these materials is often concerned with addressing lifespan reliability [7]. These drawbacks in metal solder-based TIMs are substantial enough to result in the dominance of polymeric TIMs in industry. Work to improve the performance of these polymeric TIMs often leads researchers to focus on developing more thermally conductive polymers to be used between junctions in a thermal dissipative solution and for the direct encapsulation of smaller chips in an integrated circuit [8]. In the former application, a viscous fluid capable of thermal expansion and contraction without structural damage is best suited, and in the latter, a solid, impermeable polymer to provide the chip's sensitive circuitry with unvarying protection from the environment is preferable [9,10]. Determination of reliability and end-of-life performance of polymeric TIMs is of great interest to industry because these are the two aspects of TIMs that provide polymeric TIMs with such a preferential usage over solder-based TIMs.

There remains considerable room for improvement in the performance of polymeric TIMs. A common tactic to enhance their performance is to load polymer matrices with filler materials of considerable thermal conductivity such as aluminum oxide, silver, carbon nanotubes, and graphite [11–16]. The most important considerations in terms of used fillers are their thermal conductivity, geometry, coefficient of thermal expansion, and concentration in the mixture. At high loading fractions, $f \approx 30$ vol%, the onset of a percolation threshold—the point at which any additional filler results in dramatic thermal conductivity improvement—has been observed [17,18]. At the microscopic level, the thermal percolation corresponds to the point when thermally conductive fillers form a continuous network. In this case, heat mostly propagates via the filler network rather than matrix material. Additionally, as has been recently shown, composites that are composed of multiple types of fillers with dissimilar geometries can exhibit synergistic improvement of thermal conductivity relative to composites of either single filler at identical total filler loading percentage [19–28].

Due to the exceptional thermal conductivity of graphene, found to be well above 2000 $Wm^{-1}K^{-1}$, and its potential to be a cheap, mass-producible material, it has seen extensive study as a potential filler material for next-generation polymeric TIMs [29–38]. Few-layer graphene retains very high thermal conductivity while offering a larger cross-sectional area, useful for both heat transfer and robustness upon contact with the matrix material. In the context of TIM research, the term "graphene fillers" is typically used to refer to a mixture of single-layer graphene and few-layer graphene flakes derived through chemical exfoliation or reduction in graphene oxide. Graphene fillers are expected to have lateral dimensions from hundreds of nm to a few μm in order to preserve the intrinsic heat conduction properties, principally influenced by the phonon mean free path. It has been demonstrated previously that TIMs with a high loading of graphene possess unprecedentedly high thermal conductivity, reaching as high as $\approx$11 $Wm^{-1}K^{-1}$ at a loading fraction of 45 vol% in cured and 7.1 $Wm^{-1}K^{-1}$ at 27 vol% in non-curing TIMs [17,39,40]. Additionally, graphene-filled polymers have been shown to be a very effective shielding material for electromagnetic interference, a property very useful for sensitive chip encapsulation [41]. One can expect that the thermal contact resistance of TIMs with graphene will be lower than that of TIMs with carbon nanotubes. Owing to its two-dimensional and flexible nature, graphene fillers couple better with both matrix materials and contacting surfaces. It has also been demonstrated by this research group that graphene can be used effectively in combination with boron nitride to achieve independent control of thermal and electrical conductivities [19,42].

During normal use, TIMs are subjected to a difficult environment. They are routinely operated at different temperatures, with varying junction surface morphology due to thermal expansion and warping, and must always maintain quality contact. A TIM suffers catastrophic failures over the course of its lifespan because, in case the of non-curing types, it has been pumped out of the junction, and in the case of curing types, it has cracked due to thermal stresses—both stemming from expansions and contractions [6,43]. However, the long-term, intrinsic performance of TIMs irrespective of the junctions in which they are applied has received far less consideration [44–52]. TIM intrinsic performance may alter due to numerous factors, including chemical and physical alterations of either the polymer matrix or the filler materials. Despite a tremendously large number of reports on the use of graphene in TIMs and other composites, we are not aware of any study concerned with the reliability or thermal cycling of graphene TIMs over the course of a realistic device lifespan. Knowledge of the long-term performance of graphene-filled TIMs is imperative for any practical application or adoption of these materials by industry.

In this paper, we report the salient thermal parameters of mono-filled graphene epoxy-based TIMs over the course of up to 500 power cycle treatments. An epoxy matrix was chosen because it is a standard material system used frequently in thermal composite investigations. It is also practically relevant, being one of the most commonly used base materials in curing TIMs. Unexpectedly, we found that instead of a degradation of the thermal conductivity of the cured graphene TIMs, it actually improved over the course of power cycling treatments. The obtained data suggest that epoxy-based thermal interface materials with graphene fillers undergo an interesting and little studied intrinsic performance enhancement. We argue that our results can have important implications for the development of next-generation TIMs with graphene and other fillers. The rest of this paper is organized as follows. In Section 2, we describe the sample preparation; in Section 3, we provide a brief background for the thermal cycling procedures and introduce the procedure used in this study; in Section 4, we present the thermal diffusivity and conductivity of graphene TIMs after a certain number of power cycle steps; in Section 5, we discuss the results of this study; in Section 6, we offer our conclusions of this study.

2. Methods

Sample preparation: The graphene and epoxy composite sample preparation procedure was in most ways identical to a previous study by this research group, save for the absence of any h-BN materials, the application of as little pressure as possible during curing, and the use of few-layer graphene (XG Sciences, Lansing, MI, USA) with a vendor-defined average lateral dimension of 25 μm [19]. The specific type of graphene product used was xGNP- H-25. According to vendor specifications, it had less than 1% of residual oxygen content, which most likely came from atmospheric contamination. This type of graphene powder is preferable to reduced graphene oxide (rGO), which has substantially larger oxygen content in the form of residual graphene oxide that degrades its thermal transport properties. First, the epoxy resin was weighed, placed in a rough vacuum environment for approximately two minutes to remove air bubbles, and then weighed once more. This final weight was then used to determine the necessary weights of all other components to achieve the desired constituent composition. Graphene powder was then added to the resin and mixed in a bladeless high-speed, planetary centrifuge (Flacktek Inc., Landrum, SC, USA) at rotational frequencies between 1500 and 2000 RPM for 90 s. At graphene loading fractions above 10 wt%, an additional mixing process of manually breaking agglomerations by needle was conducted as needed to ensure a homogeneous mixture. Then, the hardening agent was added immediately prior to a final planetary centrifuge cycle at 1500 RPM. For samples of graphene loading fractions over 15 wt%, a paper wadding-covered hand-pressed piston is applied into the mold to flatten the composite. The sample was then left to cure at room temperature for approximately 24 h before removing it from the mold. Finally, the samples were polished down to the desired dimensions of a diameter of 25.4 mm and a thickness between 1.5 and 1.8 mm. Supplemental Figures S2 and S3 show a SEM image of a fractured surface and the Raman spectrum, each of a 5.4 vol% sample without any cycling treatments. The G and 2D peaks in Supplemental Figure S3 indicate the presence of graphene and few-layer graphene in the composite. A noticeable disorder-induced D peak is explained by the defects and impurities present in graphene flakes and light scattering from the edges. It is known that lattice defects, disorder, and sample edges lead to the relaxation of the phonon scattering selection rules imposed by translation symmetry [52–57].

Density: The finalized sample densities were measured with Archimedes' Principle using an electronic scale (Mettler Toledo, Columbus, OH, USA) for use in determining their thermal parameters.

This technique relies upon measurements of mass both in and outside of water and determines density through the difference in measured mass of these two readings resultant from the object's buoyancy force. For better accuracy, the sample's weight in water was recorded after manually scraping off any adsorbed air bubbles and the recorded weight stabilized over time. For the purpose of this study, we converted the filler mass fraction ratio (φ) to the volume fraction (f) following the procedure described in detail to our previous published work [17].

Power cycling: The power cycling treatment procedure employed two programmable power supplies (Chroma 62000P, Foothill Ranch, CA, USA and Keithley 2400, Solon, OH, USA), a thermocouple-reading programmable benchtop multimeter (Keithley 2182A), a small electronics fan (Intel Corp., Santa Clara, CA, USA), and a Type-J thermocouple (OMEGA Engineering, Inc., Norwalk, CT, USA). A custom Nichrome wire loop heating element, with a typical diameter negligibly smaller than that of the sample, at 25.6 mm, sandwiched between two pieces of Kapton tape. A coil resistance of approximately 95 Ω was targeted in manufacture to achieve the desired power range, considering the output I-V curves of the particular power supply used. All equipment was controlled through a custom python script passing standard commands for programmable instruments (SCPI) to the appropriate equipment to allow for unattended 24 h operation. The power cycling procedure begins by finding the necessary power that power supply A needs to apply to the wire loop to reach the desired temperature of 120 °C and then holds for 8 min. It is important to note that this temperature is above the glass transition temperature of the polymer matrix, which has been reported in the literature at ≈100 °C [58,59]. Supplemental Figure S4 shows differential scanning calorimetry results, showing that the glass transition temperature does occur around this temperature. After the prescribed time is elapsed, power supply A removes all power to the wire loop and power supply B turns on the electronics fan pointed at the sample to speed the cooling phase. The temperature of the thermocouple is continuously polled and once its readings drop below 35 °C, the software pauses for another 8 min. After this time elapses, power supply A directly applies the power determined adequate in the prior step scaled to reach an equilibrium temperature of approximately 120 °C. The requisite power is re-calibrated every 50 runs should any unforeseen dramatic shift in the thermal properties of the sample come to pass. The end result of this procedure is temperatures fluctuating between 120 and ≈28 °C, depending on the local room temperature at that time. The equipment used in this procedure managed to heat and cool the sample at a rate of approximately 1 °C per second, with a much slower rate of change the closer to the temperature equilibrium point. Due to the fact that the TIM sample is subjected to a temperature gradient due to being heated from one side, most researchers refer to this procedure as power cycling instead of classic temperature cycling [6,60]. Supplemental Figure S5 shows the controlling software's own plot of the temperature at selected times throughout the cycling procedure.

Laser flash analysis: The thermal diffusivities were directly measured laser flash analysis (LFA 467, Netzsch GmbH, Selb, Germany) experiments [61]. In this technique, a light excitation source such as a Xenon lamp irradiates upon one surface of a sample. The absorption of this light results in a local increase in temperature and a subsequent temperature gradient between the opposing surfaces of a sample. This temperature gradient produces a heat wave moving to the colder, opposing surface, with the time at which it arrives being governed by the material's thermal diffusivity. As a black body, when the initially cold surface heats, it emits more infrared flux. The instrument records the voltage across a low-bandgap detector (InSb) over time to determine when the back surface begins to heat, determining the response time of the material to the applied temperature gradient. The thermal parameters extracted from LFA have been confirmed previously for other material systems with the "hot disk" and "3-omega" techniques [62–64]. The standard error in the diffusivity measurements was +/−3%. The overall uncertainty in thermal conductivity measurements was estimated to be less than 5% [65].

3. Thermal Cycling Treatment Protocols

There are multiple prevalent accelerated thermal aging schemas found in the literature: high-temperature storage (HTS), temperature cycling (TC), and power cycling (PC) [6]. HTS-accelerated aging treatments are procedures in which a sample is stored for an extended length of time at elevated temperatures and often at high humidity. The humidity in the environment has the ability to chemically harm the adhesion properties of some polymer matrix materials [60,66]. The reported results of studies based on this method can vary widely depending greatly on the type of TIM studied, showing both improvements and reductions in thermal performance over the course of treatment [44,45,66–68]. The wide variation in reported performance alterations is primarily due to the many slight variations in testing procedure and the different altering mechanisms acting on different TIMs, such as temperature-influenced wetness changes, chemical degradation, and physical form alterations [66,69]. Perhaps more representative of real-world TIM usage are thermal cycling-accelerated aging treatments. In TC treatments, the TIM is altered between uniform temperature environments in numerous iterations. As opposed to HTS treatments, this method better simulates the temperature alterations seen from a device powering on and off or simply a device under a varying workload. It is quite common for classic, non-curing TIMs to actually improve over the course of this treatment [69,70].

Even more realistic to real-world TIM usage still is the power cycling-accelerated aging treatment. This technique is similar to the TC method except that the PC method cycles the TIM from a single, localized heat source, as opposed to being in a uniform-temperature environment. This adds a temperature gradient and subsequent mechanical stress that is seen in real-world scenarios but not seen in TC treatments. Classic, non-curing TIMs often show substantial reduction in performance in this type of test, amounting to a 20%–60% increase in thermal resistance, verifying this method's relative value in predicting real-world long-term performance [6]. Due to the fact that heat is being applied to the TIM in a non-uniform manner, consideration of the rate of heat increase is even more important to consider in this method. Should one component heat up faster than another due to the local placement of the heater source, it would increase the effective thermal expansion mismatch. Indeed, in curing TIMs, it is theoretically possible that if the power were too great, that the TIM could crack alone by itself much like non-uniformly heated glassy materials. The susceptibility of a TIM to this type of failure would clearly be inhibited by a low coefficient of thermal expansion and a high thermal conductivity.

This study used a power cycling schema because it is best designed to simulate the real-world operating conditions of these materials. The procedure used in this study consisted of altering the temperature of the TIM between room temperature and an elevated temperature of 120 °C, a temperature range which encompasses the majority of applications of TIMs. This temperature change is achieved with a planar heat source. At set cycle counts, the samples' thermal parameters were measured with a laser flash analysis instrument (LFA 467, Netzsch GmbH). Figure 1 illustrates the power cycling treatment procedure used. More details of this procedure can be found in the Methods section. Any impact of environmental humidity should be largely neutralized in this study due to the samples being cured and not bound to any surface.

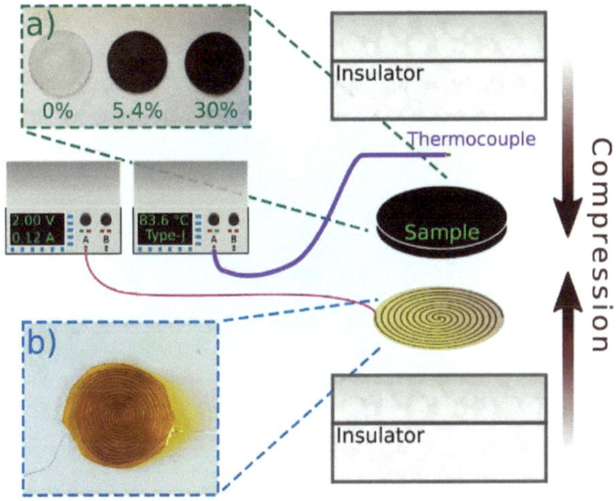

Figure 1. Simplified schematic of the temperature treatment experiment. A custom heating coil through which a current is passed is placed beneath the sample undergoing treatment. As a temperature feedback element, a Type-J thermocouple is used on the opposing surface of the sample to track its temperature. These are all then clamped under light compression between two thermal insulator materials. (**a**) Image of selected samples used in study. (**b**) Image of the custom heating element.

4. Thermal Diffusivity and Conductivity of the Graphene Composites

The ultimate metric for the effectiveness of a TIM is the reduction in the thermal resistance of a physical junction that it is placed between ($R_{TIM} < R_J$). R_{TIM} is expressed by the following relationship:

$$R_{TIM} = \frac{H_{BLT}}{\text{K}} + R_{C1} + R_{C2}$$

where H_{BLT} is the bond line thickness of the applied TIM, K is the TIM's thermal conductivity, and R_{C1} and R_{C2} are the contact resistances of the TIM at each of the junction's adjoining surfaces [71]. The thermal conductivity of a material expresses the ability of that material to pass an amount of heat energy over a given distance and temperature gradient. It will be the primary metric of TIM performance considered in this study. Maximizing the thermal conductivity and minimizing the bond line thickness and contact resistances achieves a lower thermal resistance.

Alternative to the more classic TIM view presented just prior, thermosetting plastics are frequently used to coat electronics and protect them from environmental contamination but inadvertently introduce a substantial thermal resistance hindering the radiative and convective removal of heat into adjoining heat sinks or directly into the device chassis. In this case, the R_{TIM} mathematical expression has only a single contact resistance component—between the chip and encapsulating TIM—and the thickness of the TIM is intentionally relatively thick, both factors often resulting in an even greater relative influence of thermal conductivity on the TIM's thermal dissipative performance. The reduction in overall thermal resistance in chip packaging would result in a smaller temperature difference between the heat-producing chip and the outside surface of the attached TIM, resulting in more radiative heat flow and convective heat flow through air. In this case, the color and morphology of the outside surface become parameters in heat flow due to their impacts on the encasing emissivity and surface area for convective exchange. Of course, thermal energy in these chips is still passed through the electrical connections. However, this channel of dissipation is mostly unaffected by the polymeric TIM encasing.

Thermal conductivity in this study was determined through the relationship $\text{K} = \alpha\rho C_p$, where α is the thermal diffusivity, ρ is the volumetric mass density, and C_p is the heat capacity of a material.

The heat capacity was calculated for each sample at each relevant temperature by the rule of mixtures using data from a differential scanning calorimetry (DSC 214 Polyma, Netzsch GmbH) experiment of a pure epoxy sample and heat capacities reported in the literature for graphite. It has been shown previously that the heat capacity of graphene and graphite only meaningfully deviate from one another below 100 K—well below temperatures considered at present—because of the existence of low-frequency ZA phonons in graphene, which contribute to the heat capacity of graphene [72,73]. The diffusivity was measured with a laser flash analysis instrument. In Figure 2, we present the evolution of the thermal diffusivity with power cycles.

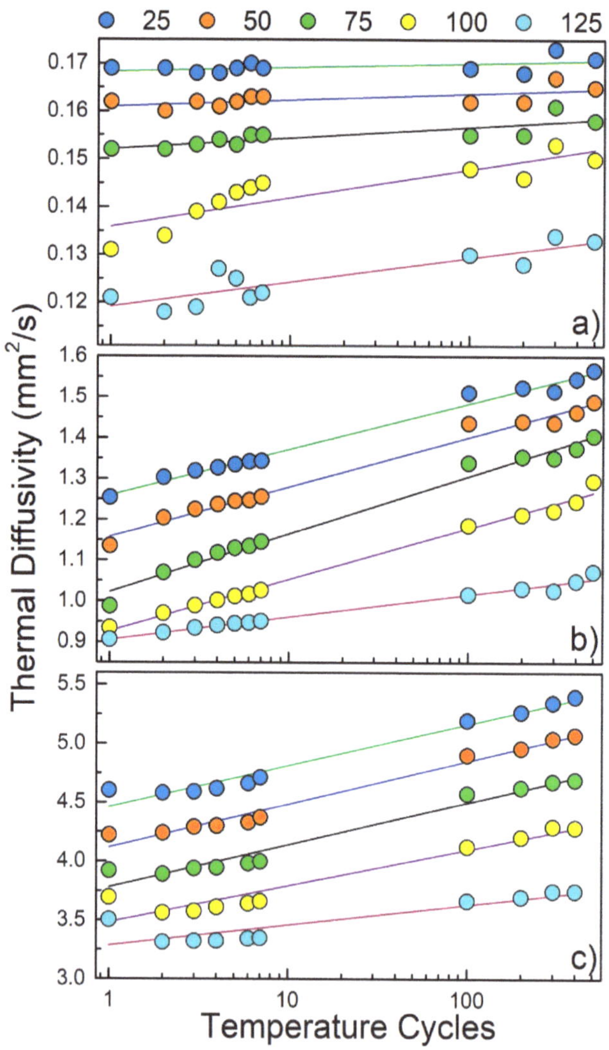

Figure 2. Thermal diffusivity for (**a**) pure epoxy matrix, (**b**) thermal interface materials (TIMs) with 5.4 vol% loading of graphene fillers, and (**c**) 30 vol% loading of graphene fillers. In each of the three samples, the measurements conducted at 25 °C resulted in the highest diffusivities. Additionally, increasing the number of heat treatment cycles either enhanced the thermal diffusivities or, in the case of the pure sample, had little to no effect depending on the temperature studied.

The thermal diffusivities were measured for samples with graphene concentrations of 0 vol%, 5.4 vol%, and 30 vol% and are shown in Figure 2. One should note here that higher loadings of graphene, up to 45 vol%, have been achieved via a multistage loading process [17]. Here, we limited the loading fraction to 30 vol% to be more representative of the commonly studied graphene-enhanced composites. The room temperature (RT) thermal diffusivity of the samples was 0.17, 1.25, and 4.6 mm^2s^{-1}, respectively. The enhancement in the thermal diffusivity as the graphene loading changes from 0 vol% to 30 vol% constitutes a factor of ~27. At the end of power cycling treatments, RT thermal diffusivities of 0.17, 1.57, and 5.40 mm^2/sec were recorded for 0 vol%, 5.4 vol%, and 30 vol%, respectively. This increase in diffusivity is a percentage increase of 0%, 25.6%, and 17.4% relative to their initial values. Supplemental Figure S1 provides a visualization for the improvement of the 30 vol% sample. After 400 cycles applied to the 30 vol% sample, the thermal diffusivity enhancement relative to 0 vol% would increase to ≈31.4 at RT. These results, in addition to the very modest changes in thermal diffusivity of the pure epoxy matrix over the full range of treatment cycles, suggest that the graphene fillers play an integral role in this observed behavior.

Figure 3 shows the thermal conductivities of the three tested samples. The mass densities of the pure epoxy composite with 5.4 vol% of graphene, and composite with 30 vol% loading of graphene were measured to be 1.18, 1.17, and 1.35 gcm^{-3}, respectively. The mass density of the samples did not show noticeable changes during the course of power cycling. Impressively, after power cycling treatments, the 30 vol% sample achieved a RT thermal conductivity of 9.3 $Wm^{-1}K^{-1}$. This value is above that in commercially available TIMs and at the level of the highest reported for graphene enhanced TIMs [17,41]. The 0 vol% TIM sample's enhancement in thermal conductivity at 100 °C was a modest 7.7%, from 0.39 to ≈0.42 $Wm^{-1}K^{-1}$ and did not display conclusive changes at other temperatures. In contrast, the 5.4 vol% and 30 vol% samples experienced strong enhancements in thermal conductivity over the course of treatment for all tested temperatures, achieving 24.9% and 17.3% improvements at room temperature, respectively. The thermal conductivity behavior with temperature is primarily influenced by the marked increase in composite heat capacity at increasing temperature. Detailed Raman and SEM inspection did not reveal any noticeable structural changes in the composites.

It is interesting to note that in each sample and at all temperature cycle counts, the thermal conductivities measured at 125 °C were lower than those measured at 100 °C, despite an otherwise uniform tendency for the thermal conductivity to increase with temperature. The failure of the increasing trend found in the lower-temperature regimen could be explained by experimenting past the glass transition temperature, at which point polymers are known to have degraded thermal transport properties, occurring in this epoxy around 100 °C [50,74]. It has been reported that the glass temperature of polymers can be raised through the inclusion inert materials, such as graphene [75,76]. In this instance, the elevation of the glass transition temperature would be substantially more modest than the 30 °C alteration seen previously in a poly-(methyl-methacrylate) matrix. Such an outcome would suggest that the graphene filler does not strongly inhibit the cross-linking of polymer chains such as Bisphenol-A [77,78]. The low dimensionality of graphene could allow for minimal reduction in cross-linking density.

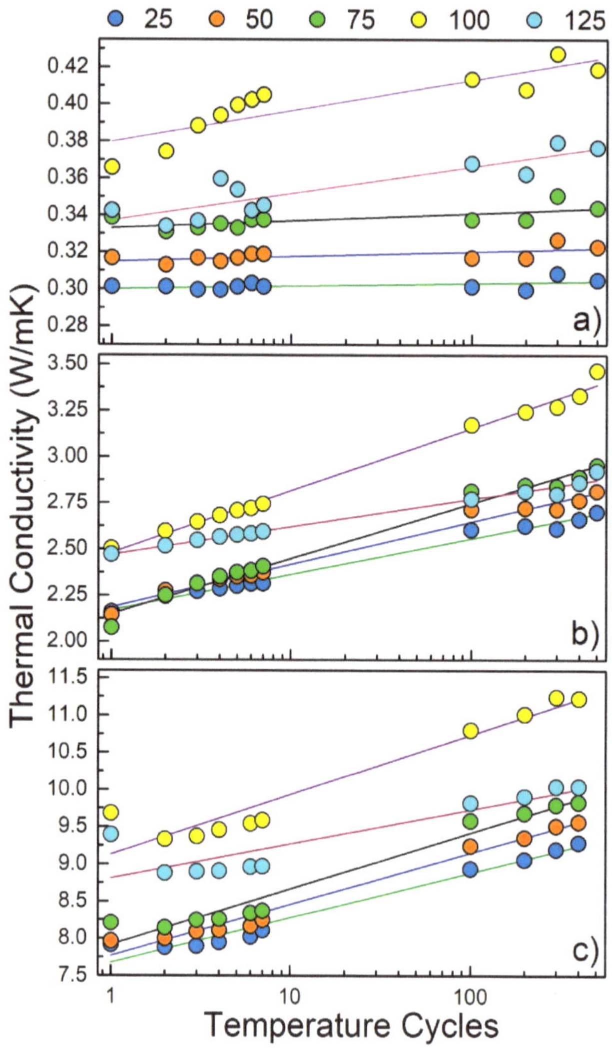

Figure 3. Thermal conductivity of (**a**) pure epoxy matrix, (**b**) TIMs with 5.4 vol% loading of graphene fillers, and (**c**) TIMs with 30 vol% loading of graphene fillers.

5. Discussion

The obtained results show a clear enhancement in the performance of cured graphene TIMs over a realistic device lifespan isolated from an overall heat sink solution. Furthermore, it is evident that the performance improvement is greater than that of the pure epoxy sample control, especially at lower device temperatures. In a previous study, changes in the thermal resistance over the course of over 2000 temperature cycles were observed for multiple silver-filled epoxies. A significant decline in thermal transport performance was found when composites were bonded between surfaces of disparate coefficients of thermal expansion. In such cases, the increase in the thermal resistance was approximately a factor of nine [79]. These results and others like them should not be considered contradictory to the present findings due to the removal of all interfacial effects in the present study.

In other studies, the behavior-altering mechanisms of thermal expansion mismatch and mechanical cracking are present but have been intentionally suppressed in this study by design. Due to the overwhelming contribution to the thermal resistance that TIM cracking and detachment can introduce, an intrinsic TIM behavior analysis of thermal conductivity is not possible or at least extremely challenging to determine in classical studies on the topic.

In another report, a pure epoxy adhesive TIM displayed thermal resistance reductions of 8% [44]. The observed slight reduction in the epoxy TIM thermal resistance was tentatively attributed to increased epoxy cross-linking. In our measurements, any improvement in the thermal conductivity of pure epoxy is inconclusive; there is only a small alteration over 100 °C and there is no noteworthy change below that temperature. The present results for pure epoxy are largely in agreement with the very modest reductions in thermal resistance of pure epoxy TIMs seen previously. However, upon the inclusion of graphene into the epoxy matrix, a clear increase in RT thermal conductivity over the course of power cycling treatments of 24.9% in 5.4 vol% and 17.3% in 30 vol% is observed. It is evident from these present results that graphene plays an integral role in the observed enhancement in power cycling treatments.

It is proposed that the mechanism of enhancement of thermal conductivity is indeed increased epoxy cross-linking at elevated temperatures, resulting in more tightly bound graphene to the more rigid polymer matrix. In this case, the length of the polymer molecules would increase, pressing increasingly upon the already dispersed graphene fillers. Likely aiding this process is the polymer matrix exhibiting thermal expansion owing to the treatment procedure, passing its glass transition temperature, pressing polymer molecules and graphene flakes closer together and cross-linking permanently in that position [76]. This would increase the vibrational coupling between the polymer matrix and the graphene fillers, allowing for better transfer of the lattice vibrations between graphene and the polymer matrix. In other words, the microscopic interfacial Kapitza [80] resistance of internal TIM components is decreased by time spent at elevated temperature. This proposed mechanism of enhancement would likely be highly dependent on the geometry of the filler material used.

There is little consistency in the literature that one could use to draw definite conclusions about performance of TIMs in accelerated aging studies. This can be attributed to different testing protocols and different materials examined. Previously, a universal tendency for the thermal resistance of all tested low-melting-temperature alloy TIMs comprising indium, gallium, bismuth, and tin between copper, nickel, and tungsten surfaces to modestly increase with isothermal elevated temperature treatment aging (HTS and TC methods) was shown, signifying at least a slight reduction in performance [49]. Some prior studies demonstrated substantial changes in the thermal resistance of phase change TIMs after power cycling over 10,000 times. Interestingly, the lack of any notable degradation of this TIM material over the examined cycles was attributed to the material's ability to re-conform to the thermal interface during melting and solidifying processes [67]. A similar reliability study on phase change material TIMs showed little change in thermal impedance over 2000 cycles [68]. Another study also examined the thermal interface resistance of a similar, graphene-filled epoxy cycled in a temperature range between 27 and 97 °C—a cycling process below glass transition of the matrix at all times—and found no noteworthy changes, even when tested over only 10 full cycles [50]. In light of the inconsistency in the reported data, briefly sampled just prior, even amongst studies of similar TIM materials, a more simplified approach in which the isolation of individual parameters of the TIM was used [6]. By negating any effect of interfacial resistance, we were able to directly analyze the thermal conductivity TIM parameter with confidence. Simplification of the power cycling-accelerated aging schema and resultant isolation of parameters of interest could be an effective method to increase the consistency among similar future studies. The isolation of the changes in the thermal conductivity from any potential changes in the contact resistances allows for a direct observation of the salient contributing factors to thermal resistance. The latter is unfeasible in more conventional lifespan TIM performance studies.

6. Conclusions

We reported the results of the first reliability study of epoxy matrix TIMs loaded with graphene fillers over the course of a realistic device lifespan. The systematic testing of the graphene-enhanced TIMs confirms their potential for practical applications in thermal management. Moreover, it was found, unexpectedly, that the thermal conductivity of graphene epoxy TIMs improves over the course of cycling. The obtained results clearly show an increase in the thermal diffusivity and conductivity of cured graphene and epoxy polymer composites over the course of up to 500 thermal cycle treatments, reaching an enhancement of up to ≈25%. In contrast, the reference sample-pure Bisphenol-A epoxy—only received a modest enhancement in thermal conductivity when at elevated temperatures (≥100 °C) of ≈7.7%. The graphene-loaded samples displayed enhanced thermal diffusivity and conductivity behaviors at all temperatures, importantly including room temperature. It is suggested that the mechanism for this marked enhancement in thermal conductivity of graphene-epoxy composites is related to increased cross-linking, resulting in graphene being better coupled to the polymer matrix.

Supplementary Materials: The following are available online at http://www.mdpi.com/2311-5629/6/2/26/s1, Figure S1: A three-dimensional visualization of the thermal diffusivity enhancement in the 30 vol% composite, Figure S2: A SEM micrograph of a fractured surface of a 5.4 vol% composite, Figure S3: Raman spectrum of a 5.4 vol% composite, Figure S4: Differential scanning calorimetry experimental results of our pure epoxy matrix, and Figure S5: Plot of temperature over time for the power cycling procedure.

Author Contributions: A.A.B. coordinated the project and contributed to the experimental and theoretical data analysis. J.S.L. designed the testing protocols, built the experimental setup, synthesized composites, and conducted thermal conductivity measurements and thermal cycling tests. T.P. assisted with sample preparation and thermal measurements and conducted thermal cycling tests. A.M. conducted material characterization and assisted with measurements. F.K. contributed to the testing protocol design and thermal data analysis. All authors contributed to the manuscript preparation. All authors have read and agreed to the published version of the manuscript.

Funding: The work at UCR was supported, in part, by the Office of Technology Partnerships (OTP), University of California via the Proof of Concept (POC) project "Graphene Thermal Interface Materials." A.A.B. also acknowledges the UC—National Laboratory Collaborative Research and Training Program—University of California Research Initiatives LFR-17-477237.

Conflicts of Interest: The authors declare no conflict of interest.

References

1. Trew, R.; Green, D.; Shealy, J. AlGaN/GaN HFET Reliability. *IEEE Microw. Mag.* **2009**, *10*, 116–127. [CrossRef]
2. Saadah, M.; Hernandez, E.; Balandin, A.A. Thermal Management of Concentrated Multi-Junction Solar Cells with Graphene-Enhanced Thermal Interface Materials. *Appl. Sci.* **2017**, *7*, 589. [CrossRef]
3. Biber, C. LED Light Emission as a Function of Thermal Conditions. In Proceedings of the Annual IEEE Semiconductor Thermal Measurement and Management Symposium, San Jose, CA, USA, 16–20 March 2008; pp. 180–184. [CrossRef]
4. Prasher, R. Thermal Interface Materials: Historical Perspective, Status, and Future Directions. *Proc. IEEE* **2006**, *94*, 1571–1586. [CrossRef]
5. Chung, D.D.L. Thermal Interface Materials. *J. Mater. Eng. Perform.* **2001**, *10*, 56–59. [CrossRef]
6. Due, J.; Robinson, A.J. Reliability of Thermal Interface Materials: A Review. *Appl. Therm. Eng.* **2013**, *50*, 455–463. [CrossRef]
7. Deppisch, C.; Fitzgerald, T.; Raman, A.; Hua, F.; Zhang, C.; Liu, P.; Miller, M. The Material Optimization and Reliability Characterization of an Indium-Solder Thermal Interface Material for CPU Packaging. *JOM* **2006**, *58*, 67–74. [CrossRef]
8. Narumanchi, S.; Mihalic, M.; Kelly, K.; Eesley, G. Thermal Interface Materials for Power Electronics Applications. In Proceedings of the 2008 11th Intersociety Conference on Thermal and Thermomechanical Phenomena in Electronic Systems, Orlando, FL, USA, 28–31 May 2008; pp. 395–404. [CrossRef]
9. Prasher, R.S. Surface Chemistry and Characteristics Based Model for the Thermal Contact Resistance of Fluidic Interstitial Thermal Interface Materials. *J. Heat Transfer* **2001**, *123*, 969–975. [CrossRef]
10. Prasher, R.S.; Matayabas, J.C. Thermal Contact Resistance of Cured Gel Polymeric Thermal Interface Material. *IEEE Trans. Compon. Packag. Technol.* **2004**, *27*, 702–709. [CrossRef]

11. Yu, A.; Ramesh, P.; Itkis, M.E.; Bekyarova, E.; Haddon, R.C. Graphite Nanoplatelet–Epoxy Composite Thermal Interface Materials. *J. Phys. Chem. C* **2007**, *111*. [CrossRef]
12. Xu, J.; Fisher, T.S. Enhancement of Thermal Interface Materials with Carbon Nanotube Arrays. *Int. J. Heat Mass Transf.* **2006**, *49*, 1658–1666. [CrossRef]
13. Tong, T.; Zhao, Y.; Delzeit, L.; Kashani, A.; Meyyappan, M.; Majumdar, A. Dense Vertically Aligned Multiwalled Carbon Nanotube Arrays as Thermal Interface Materials. *IEEE Trans. Compon. Packag. Technol.* **2007**, *30*, 92–100. [CrossRef]
14. Lin, C.; Chung, D.D.L. Graphite Nanoplatelet Pastes vs. Carbon Black Pastes as Thermal Interface Materials. *Carbon* **2009**, *47*, 295–305. [CrossRef]
15. Yu, H.; Li, L.; Kido, T.; Xi, G.; Xu, G.; Guo, F. Thermal and Insulating Properties of Epoxy/Aluminum Nitride Composites Used for Thermal Interface Material. *J. Appl. Polym. Sci.* **2012**, *124*, 669–677. [CrossRef]
16. Uetani, K.; Ata, S.; Tomonoh, S.; Yamada, T.; Yumura, M.; Hata, K. Elastomeric Thermal Interface Materials with High Through-Plane Thermal Conductivity from Carbon Fiber Fillers Vertically Aligned by Electrostatic Flocking. *Adv. Mater.* **2014**, *26*, 5857–5862. [CrossRef]
17. Kargar, F.; Barani, Z.; Salgado, R.; Debnath, B.; Lewis, J.S.; Aytan, E.; Lake, R.K.; Balandin, A.A. Thermal Percolation Threshold and Thermal Properties of Composites with High Loading of Graphene and Boron Nitride Fillers. *ACS Appl. Mater. Interfaces* **2018**, *10*, 37555–37565. [CrossRef]
18. Shtein, M.; Nadiv, R.; Buzaglo, M.; Regev, O. Graphene-Based Hybrid Composites for Efficient Thermal Management of Electronic Devices. *ACS Appl. Mater. Interfaces* **2015**, *7*, 23725–23730. [CrossRef]
19. Lewis, J.S.; Barani, Z.; Magana, A.S.; Kargar, F.; Balandin, A.A. Thermal and Electrical Conductivity Control in Hybrid Composites with Graphene and Boron Nitride Fillers. *Mater. Res. Express* **2019**, *6*, 085325. [CrossRef]
20. Chun, K.-Y.; Oh, Y.; Rho, J.; Ahn, J.-H.; Kim, Y.-J.; Choi, H.R.; Baik, S. Highly Conductive, Printable and Stretchable Composite Films of Carbon Nanotubes and Silver. *Nat. Nanotechnol.* **2010**, *5*, 853–857. [CrossRef]
21. Zhou, T.; Wang, X.; Liu, X.; Xiong, D. Improved Thermal Conductivity of Epoxy Composites Using a Hybrid Multi-Walled Carbon Nanotube/Micro-SiC Filler. *Carbon* **2010**, *48*, 1171–1176. [CrossRef]
22. Ma, P.-C.; Liu, M.-Y.; Zhang, H.; Wang, S.-Q.; Wang, R.; Wang, K.; Wong, Y.-K.; Tang, B.-Z.; Hong, S.-H.; Paik, K.-W.; et al. Enhanced Electrical Conductivity of Nanocomposites Containing Hybrid Fillers of Carbon Nanotubes and Carbon Black. *ACS Appl. Mater. Interfaces* **2009**, *1*, 1090–1096. [CrossRef]
23. Li, H.; Dai, S.; Miao, J.; Wu, X.; Chandrasekharan, N.; Qiu, H.; Yang, J. Enhanced Thermal Conductivity of Graphene/Polyimide Hybrid Film via a Novel "Molecular Welding" Strategy. *Carbon* **2018**, *126*, 319–327. [CrossRef]
24. Teng, C.-C.; Ma, C.-C.M.; Chiou, K.-C.; Lee, T.-M. Synergetic Effect of Thermal Conductive Properties of Epoxy Composites Containing Functionalized Multi-Walled Carbon Nanotubes and Aluminum Nitride. *Compos. Part B Eng.* **2012**, *43*, 265–271. [CrossRef]
25. Choi, S.; Im, H.; Kim, J. Flexible and High Thermal Conductivity Thin Films Based on Polymer: Aminated Multi-Walled Carbon Nanotubes/Micro-Aluminum Nitride Hybrid Composites. *Compos. Part A Appl. Sci. Manuf.* **2012**, *43*, 1860–1868. [CrossRef]
26. Choi, S.; Kim, J. Thermal Conductivity of Epoxy Composites with a Binary-Particle System of Aluminum Oxide and Aluminum Nitride Fillers. *Compos. Part B Eng.* **2013**, *51*, 140–147. [CrossRef]
27. Zhou, W.; Wang, C.; An, Q.; Ou, H. Thermal Properties of Heat Conductive Silicone Rubber Filled with Hybrid Fillers. *J. Compos. Mater.* **2008**, *42*, 173–187. [CrossRef]
28. Barani, Z.; Mohammadzadeh, A.; Geremew, A.; Huang, C.; Coleman, D.; Mangolini, L.; Kargar, F.; Balandin, A.A. Thermal Properties of the Binary-Filler Hybrid Composites with Graphene and Copper Nanoparticles. *Adv. Funct. Mater.* **2019**, 1904008. [CrossRef]
29. Balandin, A.A.; Ghosh, S.; Bao, W.; Calizo, I.; Teweldebrhan, D.; Miao, F.; Lau, C.N. Superior Thermal Conductivity of Single-Layer Graphene. *Nano Lett.* **2008**, *8*, 902–907. [CrossRef]
30. Balandin, A.A. Thermal Properties of Graphene and Nanostructured Carbon Materials. *Nat. Mater.* **2011**, *10*, 569–581. [CrossRef]
31. Ghosh, S.; Calizo, I.; Teweldebrhan, D.; Pokatilov, E.P.; Nika, D.L.; Balandin, A.A.; Bao, W.; Miao, F.; Lau, C.N. Extremely High Thermal Conductivity of Graphene: Prospects for Thermal Management Applications in Nanoelectronic Circuits. *Appl. Phys. Lett.* **2008**, *92*, 151911. [CrossRef]
32. Seol, J.H.; Jo, I.; Moore, A.L.; Lindsay, L.; Aitken, Z.H.; Pettes, M.T.; Li, X.; Yao, Z.; Huang, R.; Broido, D.; et al. Two-Dimensional Phonon Transport in Supported Graphene. *Science* **2010**, *328*, 213–216. [CrossRef]

33. Cai, W.; Moore, A.L.; Zhu, Y.; Li, X.; Chen, S.; Shi, L.; Ruoff, R.S. Thermal Transport in Suspended and Supported Monolayer Graphene Grown by Chemical Vapor Deposition. *Nano Lett.* **2010**, *10*, 1645–1651. [CrossRef] [PubMed]
34. Shahil, K.M.F.; Balandin, A.A. Graphene–Multilayer Graphene Nanocomposites as Highly Efficient Thermal Interface Materials. *Nano Lett.* **2012**, *12*, 861–867. [CrossRef] [PubMed]
35. Renteria, J.; Legedza, S.; Salgado, R.; Balandin, M.P.; Ramirez, S.; Saadah, M.; Kargar, F.; Balandin, A.A. Magnetically-Functionalized Self-Aligning Graphene Fillers for High-Efficiency Thermal Management Applications. *Mater. Des.* **2015**, *88*, 214–221. [CrossRef]
36. Dmitriev, A.S.; Valeev, A.R. Graphene Nanocomposites as Thermal Interface Materials for Cooling Energy Devices. *J. Phys. Conf. Ser.* **2017**, *891*, 012359. [CrossRef]
37. Tang, B.; Hu, G.; Gao, H.; Hai, L. Application of Graphene as Filler to Improve Thermal Transport Property of Epoxy Resin for Thermal Interface Materials. *Int. J. Heat Mass Transf.* **2015**, *85*, 420–429. [CrossRef]
38. Yan, Z.; Liu, G.; Khan, J.M.; Balandin, A.A. Graphene Quilts for Thermal Management of High-Power GaN Transistors. *Nat. Commun.* **2012**, *3*, 827. [CrossRef]
39. Naghibi, S.; Kargar, F.; Wright, D.; Huang, C.Y.T.; Mohammadzadeh, A.; Barani, Z.; Salgado, R.; Balandin, A.A. Noncuring Graphene Thermal Interface Materials for Advanced Electronics. *Adv. Electron. Mater.* **2020**, 1901303. [CrossRef]
40. Mahadevan, B.K.; Naghibi, S.; Kargar, F.; Balandin, A.A. Non-Curing Thermal Interface Materials with Graphene Fillers for Thermal Management of Concentrated Photovoltaic Solar Cells. *C J. Carbon Res.* **2019**, *6*, 2. [CrossRef]
41. Kargar, F.; Barani, Z.; Balinskiy, M.; Magana, A.S.; Lewis, J.S.; Balandin, A.A. Dual-Functional Graphene Composites for Electromagnetic Shielding and Thermal Management. *Adv. Electron. Mater.* **2018**, 1800558. [CrossRef]
42. Kargar, F.; Salgado, R.; Legedza, S.; Renteria, J.; Balandin, A.A. *A Comparative Study of the Thermal Interface Materials with Graphene and Boron Nitride Fillers*; Razeghi, M., Lee, Y.H., Ghazinejad, M., Eds.; International Society for Optics and Photonics: San Diego, CA, USA, 2014; Volume 9168, p. 91680S. [CrossRef]
43. Li, J.; Myllykoski, P.; Paulasto-Krockel, M. Study on Thermomechanical Reliability of Power Modules and Thermal Grease Pump-out Mechanism. In Proceedings of the 2015 16th International Conference on Thermal, Mechanical and Multi-Physics Simulation and Experiments in Microelectronics and Microsystems, EuroSimE 2015, Budapest, Hungary, 19–22 April 2015; Institute of Electrical and Electronics Engineers Inc.: New York, NY, USA, 2015. [CrossRef]
44. Khuu, V.; Osterman, M.; Bar-Cohen, A.; Pecht, M. Effects of Temperature Cycling and Elevated Temperature/Humidity on the Thermal Performance of Thermal Interface Materials. *IEEE Trans. Device Mater. Reliab.* **2009**, *9*, 379–391. [CrossRef]
45. Gowda, A.; Esler, D.; Paisner, S.N.; Tonapi, S.; Nagarkar, K.; Srihari, K. Reliability Testing of Silicone-Based Thermal Greases. In Proceedings of the Annual IEEE Semiconductor Thermal Measurement and Management Symposium, San Jose, CA, USA, 15–17 March 2005; pp. 64–71. [CrossRef]
46. Chen, C.I.; Ni, C.Y.; Pan, H.Y.; Chang, C.M.; Liu, D.S. Practical Evaluation For Long-term Stability of Thermal Interface Material. *Exp. Tech.* **2009**, *33*, 28–32. [CrossRef]
47. Chiu, C.P.; Chandran, B.; Mello, M.; Kelley, K. An Accelerated Reliability Test Method to Predict Thermal Grease Pump-out in Flip-Chip Applications. *Proc. Electron. Compon. Technol. Conf.* **2001**, 91–97. [CrossRef]
48. Wang, T.H.; Chen, H.Y.; Lee, C.C.; Lai, Y.S. High-Power-Used Thermal Gel Degradation Evaluation on Board-Level HFCBGA Subjected to Reliability Tests. In Proceedings of the 2009 4th International Microsystems, Packaging, Assembly and Circuits Technology Conference, Taipei, Taiwan, 21–23 October 2009; pp. 465–468. [CrossRef]
49. Roy, C.K.; Bhavnani, S.; Hamilton, M.C.; Johnson, R.W.; Knight, R.W.; Harris, D.K. Accelerated Aging and Thermal Cycling of Low Melting Temperature Alloys as Wet Thermal Interface Materials. *Microelectron. Reliab.* **2015**, *55*, 2698–2704. [CrossRef]
50. Park, W.; Guo, Y.; Li, X.; Hu, J.; Liu, L.; Ruan, X.; Chen, Y.P. High-Performance Thermal Interface Material Based on Few-Layer Graphene Composite. *J. Phys. Chem. C* **2015**, *119*, 26753–26759. [CrossRef]
51. Skuriat, R.; Li, J.F.; Agyakwa, P.A.; Mattey, N.; Evans, P.; Johnson, C.M. Degradation of Thermal Interface Materials for High-Temperature Power Electronics Applications. *Microelectron. Reliab.* **2013**, *53*, 1933–1942. [CrossRef]

52. Nylander, A.; Hansson, J.; Kabiri Samani, M.; Chandra Darmawan, C.; Borta Boyon, A.; Divay, L.; Ye, L.; Fu, Y.; Ziaei, A.; Liu, J. Reliability Investigation of a Carbon Nanotube Array Thermal Interface Material. *Energies* **2019**, *12*, 2080. [CrossRef]
53. Ferrari, A.C. Raman Spectroscopy of Graphene and Graphite: Disorder, Electron–Phonon Coupling, Doping and Nonadiabatic Effects. *Solid State Commun.* **2007**, *143*, 47–57. [CrossRef]
54. Ferrari, A.C.; Basko, D.M. Raman Spectroscopy as a Versatile Tool for Studying the Properties of Graphene. *Nat. Nanotechnol.* **2013**, *8*, 235–246. [CrossRef]
55. Calizo, I.; Bao, W.; Miao, F.; Lau, C.N.; Balandin, A.A. The effect of substrates on the Raman spectrum of graphene: Graphene-on-sapphire and graphene-on-glass. *Appl. Phys. Lett.* **2007**, *91*, 201904. [CrossRef]
56. Calizo, I.; Miao, F.; Bao, W.; Lau, C.N.; Balandin, A.A. Variable temperature Raman microscopy as a nanometrology tool for graphene layers and graphene-based devices. *Appl. Phys. Lett.* **2007**, *91*, 71913. [CrossRef]
57. Parvizi, F.; Teweldebrhan, D.; Ghosh, S.; Calizo, I.; Balandin, A.A.; Zhu, H.; Abbaschian, R. Properties of graphene produced by the high pressure–high temperature growth process. *Micro Nano Lett.* **2008**, *3*, 29. [CrossRef]
58. Plazek, D.J.; Choy, I.C. The Physical Properties of Bisphenol-a-Based Epoxy Resins during and after Curing. II. Creep Behavior above and below the Glass Transition Temperature. *J. Polym. Sci. Part B Polym. Phys.* **1989**, *27*, 307–324. [CrossRef]
59. DGEBA Epoxy Resin. Available online: https://polymerdatabase.com/polymers/bisphenol-adiglycidyletherepoxyresin.html (accessed on 5 January 2020).
60. Goel, N.; Anoop, T.K.; Bhattacharya, A.; Cervantes, J.A.; Mongia, R.K.; Machiroutu, S.V.; Lin, H.L.; Huang, Y.C.; Fan, K.C.; Denq, B.L.; et al. Technical Review of Characterization Methods for Thermal Interface Materials (TIM). In Proceedings of the 2008 11th IEEE Intersociety Conference on Thermal and Thermomechanical Phenomena in Electronic Systems, I-THERM, Orlando, FL, USA, 28–31 May 2008; pp. 248–258. [CrossRef]
61. Parker, W.J.; Jenkins, R.J.; Butler, C.P.; Abbott, G.L. Flash Method of Determining Thermal Diffusivity, Heat Capacity, and Thermal Conductivity. *J. Appl. Phys.* **1961**, *32*, 1679–1684. [CrossRef]
62. Liu, W.L.; Shamsa, M.; Calizo, I.; Balandin, A.A.; Ralchenko, V.; Popovich, A.; Saveliev, A. Thermal Conduction in Nanocrystalline Diamond Films: Effects of the Grain Boundary Scattering and Nitrogen Doping. *Appl. Phys. Lett.* **2006**, *89*, 171915. [CrossRef]
63. Balandin, A.A.; Shamsa, M.; Liu, W.L.; Casiraghi, C.; Ferrari, A.C. Thermal Conductivity of Ultrathin Tetrahedral Amorphous Carbon Films. *Appl. Phys. Lett.* **2008**, *93*, 043115. [CrossRef]
64. Ghosh, S.; Teweldebrhan, D.; Morales, J.R.; Garay, J.E.; Balandin, A.A. Thermal Properties of the Optically Transparent Pore-Free Nanostructured Yttria-Stabilized Zirconia. *J. Appl. Phys.* **2009**, *106*, 113507. [CrossRef]
65. Shamsa, M.; Liu, W.L.; Balandin, A.A.; Casiraghi, C.; Milne, W.I.; Ferrari, A.C. Thermal conductivity of diamond-like carbon films. *Appl. Phys. Lett.* **2006**, *89*, 161921. [CrossRef]
66. Dal, S.L.B. Degradation Mechanisms of Siloxane-Based Thermal Interface Materials under Reliability Stress Conditions. In Proceedings of the IEEE International Reliability Physics Symposium Proceedings, Phoenix, AZ, USA, 25–29 April 2004; Institute of Electrical and Electronics Engineers Inc.: New York, NY, USA, 2004; Volume 2004, pp. 537–542. [CrossRef]
67. Bharatham, L.; Wong, S.F.; Torresola, J.; Chen, C.K. Qualification of Phase Change Thermal Interface Material for Wave Solder Heat Sink on FCBGA Package. In Proceedings of the 7th Electronics Packaging Technology Conference, EPTC 2005, Singapore, 7–9 December 2005; Volume 2, pp. 537–542. [CrossRef]
68. Ramaswamy, C.; Shinde, S.; Pompeo, F.; Sablinski, W.; Bradley, S. Phase Change Materials as a Viable Thermal Interface Material for High-Power Electronic Applications. In Proceedings of the Thermomechanical Phenomena in Electronic Systems -Proceedings of the Intersociety Conference, Las Vegas, NV, USA, 1–4 June 2004; Volume 2, pp. 687–691. [CrossRef]
69. Paisner, S.N.; Touzelbaev, M.; Refai-Ahmed, G.; Yang, Y. New Developments for a No-Pump-out High-Performance Thermal Grease. In Proceedings of the 2010 12th IEEE Intersociety Conference on Thermal and Thermomechanical Phenomena in Electronic Systems, ITherm 2010, Las Vegas, NV, USA, 2–5 June 2010. [CrossRef]

70. Gowda, A.; Zhong, A.; Esler, D.; David, J.; Sandeep, T.; Srihari, K.; Schattenmann, F. Design of a High Reliability and Low Thermal Resistance Interface Material for Microelectronics. In Proceedings of the 5th Electronics Packaging Technology Conference, EPTC 2003, Singapore, 12 December 2003; Institute of Electrical and Electronics Engineers Inc.: New York, NY, USA, 2003; pp. 557–562. [CrossRef]
71. Prasher, R.S.; Shipley, J.; Prstic, S.; Koning, P.; Wang, J. Thermal Resistance of Particle Laden Polymeric Thermal Interface Materials. *J. Heat Transfer.* **2003**, *125*, 1170–1177. [CrossRef]
72. Nika, D.L.; Cocemasov, A.I.; Balandin, A.A. Specific Heat of Twisted Bilayer Graphene: Engineering Phonons by Atomic Plane Rotations. *Appl. Phys. Lett.* **2014**, *105*, 031904. [CrossRef]
73. Cocemasov, A.I.; Nika, D.L.; Balandin, A.A. Engineering of the Thermodynamic Properties of Bilayer Graphene by Atomic Plane Rotations: The Role of the out-of-Plane Phonons. *Nanoscale* **2015**, *7*, 12851–12859. [CrossRef]
74. Dos Santos, W.N.; De Sousa, J.A.; Gregorio, R. Thermal Conductivity Behaviour of Polymers around Glass Transition and Crystalline Melting Temperatures. *Polym. Test.* **2013**, *32*, 987–994. [CrossRef]
75. Ramanathan, T.; Abdala, A.A.; Stankovich, S.; Dikin, D.A.; Herrera-Alonso, M.; Piner, R.D.; Adamson, D.H.; Schniepp, H.C.; Chen, X.; Ruoff, R.S.; et al. Functionalized Graphene Sheets for Polymer Nanocomposites. *Nat. Nanotechnol.* **2008**, *3*, 327–331. [CrossRef] [PubMed]
76. Yasmin, A.; Daniel, I.M. Mechanical and Thermal Properties of Graphite Platelet/Epoxy Composites. *Polymer* **2004**, *45*, 8211–8219. [CrossRef]
77. Bansal, A.; Yang, H.; Li, C.; Cho, K.; Benicewicz, B.C.; Kumar, S.K.; Schadler, L.S. Quantitative Equivalence between Polymer Nanocomposites and Thin Polymer Films. *Nat. Mater.* **2005**, *4*, 693–698. [CrossRef]
78. Rittigstein, P.; Priestley, R.D.; Broadbelt, L.J.; Torkelson, J.M. Model Polymer Nanocomposites Provide an Understanding of Confinement Effects in Real Nanocomposites. *Nat. Mater.* **2007**, *6*, 278–282. [CrossRef]
79. Bjorneklett, A.; Tuhus, T.; Halbo, L.; Kristiansen, H. Thermal Resistance, Thermomechanical Stress and Thermal Cycling Endurance of Silicon Chips Bonded with Adhesives. In Proceedings of the Ninth Annual IEEE Semiconductor Thermal Measurement and Management Symposium, Austin, TX, USA, 2–4 February 1993; pp. 136–143. [CrossRef]
80. Kapitza, P.L. The Study of Heat Transfer in Helium II. *J. Phys.* **1941**, *4*, 181.

MDPI AG
Grosspeteranlage 5
4052 Basel
Switzerland
Tel.: +41 61 683 77 34

C Editorial Office
E-mail: carbon@mdpi.com
www.mdpi.com/journal/carbon

www.ingramcontent.com/pod-product-compliance
Lightning Source LLC
Chambersburg PA
CBHW040853230926
51850CB00014BA/3427
* 9 7 8 3 7 2 5 8 3 6 8 5 7 *